KNOWLEDGE-BASED INTELLIGENT INFORMATION ENGINEERING SYSTEMS AND ALLIED TECHNOLOGIES

Frontiers in Artificial Intelligence and Applications

Series Editors: J. Breuker, R. López de Mántaras, M. Mohammadian, S. Ohsuga and W. Swartout

Volume 82

Previously published in this series:

Vol. 81, In production

Vol. 80, T. Welzer et al. (Eds.), Knowledge-based Software Engineering

Vol. 79, H. Motoda (Ed.), Active Mining

Vol. 78, T. Vidal and P. Liberatore (Eds.), STAIRS 2002

Vol. 77, F. van Harmelen (Ed.), ECAI 2002

Vol. 76, P. Sinčák et al. (Eds.), Intelligent Technologies – Theory and Applications

Vol. 75, I.F. Cruz et al. (Eds), The Emerging Semantic Web

Vol. 74, M. Blay-Fornarino et al. (Eds.), Cooperative Systems Design

Vol. 73, H. Kangassalo et al. (Eds.), Information Modelling and Knowledge Bases XIII

Vol. 72, A. Namatame et al. (Eds.), Agent-Based Approaches in Economic and Social Complex Systems

Vol. 71, J.M. Abe and J.I. da Silva Filho (Eds.), Logic, Artificial Intelligence and Robotics

Vol. 70, B. Verheij et al. (Eds.), Legal Knowledge and Information Systems

Vol. 69, N. Baba et al. (Eds.), Knowledge-Based Intelligent Information Engineering Systems & Allied Technologies

Vol. 68, J.D. Moore et al. (Eds.), Artificial Intelligence in Education

Vol. 67, H. Jaakkola et al. (Eds.), Information Modelling and Knowledge Bases XII

Vol. 66, H.H. Lund et al. (Eds.), Seventh Scandinavian Conference on Artificial Intelligence

Vol. 65, In production

Vol. 64, J. Breuker et al. (Eds.), Legal Knowledge and Information Systems

Vol. 63, I. Gent et al. (Eds.), SAT2000

Vol. 62, T. Hruška and M. Hashimoto (Eds.), Knowledge-Based Software Engineering

Vol. 61, E. Kawaguchi et al. (Eds.), Information Modelling and Knowledge Bases XI

Vol. 60, P. Hoffman and D. Lemke (Eds.), Teaching and Learning in a Network World

Vol. 59, M. Mohammadian (Ed.), Advances in Intelligent Systems: Theory and Applications

Vol. 58, R. Dieng et al. (Eds.), Designing Cooperative Systems

Vol. 57, M. Mohammadian (Ed.), New Frontiers in Computational Intelligence and its Applications

Vol. 56, M.I. Torres and A. Sanfeliu (Eds.), Pattern Recognition and Applications

Vol. 55, G. Cumming et al. (Eds.), Advanced Research in Computers and Communications in Education

Vol. 54, W. Horn (Ed.), ECAI 2000

Vol. 53, E. Motta, Reusable Components for Knowledge Modelling

Vol. 52, In production

Vol. 51, H. Jaakkola et al. (Eds.), Information Modelling and Knowledge Bases X

Vol. 50, S.P. Lajoie and M. Vivet (Eds.), Artificial Intelligence in Education

Vol. 49, P. McNamara and H. Prakken (Eds.), Norms, Logics and Information Systems

Vol. 48, P. Návrat and H. Ueno (Eds.), Knowledge-Based Software Engineering

Vol. 47, M.T. Escrig and F. Toledo, Qualitative Spatial Reasoning: Theory and Practice

Vol. 46, N. Guarino (Ed.), Formal Ontology in Information Systems

Vol. 45, P.-J. Charrel et al. (Eds.), Information Modelling and Knowledge Bases IX

Vol. 44, K. de Koning, Model-Based Reasoning about Learner Behaviour

ISSN: 0922-6389

Knowledge-Based Intelligent Information Engineering Systems and Allied Technologies

KES 2002

Part 2

Edited by

E. Damiani

University of Milan, Italy

R.J. Howlett

University of Brighton, United Kingdom

L.C. Jain

University of South Australia, Australia

N. Ichalkaranje

University of South Australia, Australia

IOS
Press

Ohmsha

Amsterdam • Berlin • Oxford • Tokyo • Washington, DC

ISBN 1 58603 280 1 (IOS Press)
ISBN 4 274 90535 7 C3055 (Ohmsha)
Library of Congress Control Number: 2002110639

Publisher
IOS Press
Nieuwe Hemweg 6B
1013 BG Amsterdam
The Netherlands
fax: +31 20 620 3419
e-mail: order@iospress.nl

Distributor in the UK and Ireland
IOS Press/Lavis Marketing
73 Lime Walk
Headington
Oxford OX3 7AD
England
fax: +44 1865 75 0079

Distributor in the USA and Canada
IOS Press, Inc.
5795-G Burke Centre Parkway
Burke, VA 22015
USA
fax: +1 703 323 3668
e-mail: iosbooks@iospress.com

Distributor in Germany, Austria and Switzerland
IOS Press/LSL.de
Gerichtsweg 28
D-04103 Leipzig
Germany
fax: +49 341 995 4255

Distributor in Japan
Ohmsha, Ltd.
3-1 Kanda Nishiki-cho
Chiyoda-ku, Tokyo 101-8460
Japan
fax: +81 3 3233 2426

Contents

Part 1

General Chair's Welcome Message, *E. Damiani*   v
KES 2002 Conference Organisation   vi

KEYNOTE SPEECHES

Querying Databases Containing Imprecise Information: on Some Perspectives, *P. Bosc*   3
Social Intelligence Design and Communicative Reality, *T. Nishida*   5
Computational Intelligence and Infrastructure Security, *V. Rao Vemurri*   7

GENERAL SESSION PAPERS

A Competitive Node Decaying Method for Artificial Neural Networks, *Md. Shahjahan, Md. Monirul Islam and K. Murase*   11
Medical and Technical Evaluation of a CBR Therapy Adviser, *R. Schmidt, K. Dümcke, D. Steffen and L. Gierl*   16
Case-based Reasoning for Predicting the Temporal Spread of Infectious Diseases, *R. Schmidt and L. Gierl*   21
Knowledge Modeling for the Design of a KBS in the Functional Size Measurement Domain, *J.-M. Desharnais, A. Abran, A. Mayers, L. Buglione and V. Bevo*   26
Using Convolutional Neural Network for Character Recognition in Automatic Form Processing System, *H. Kiem, N.H. Son and D.M. Son*   32
Task Reallocation in Multiagent Systems based on Vickrey Auctioning, *I.C. Kim*   40
Evolutionary Optimisation of Technical Options for Manufacturing Industry, *C. Balducelli, G. Lepera and G. Vicoli*   45
Applying Neural Network based Novelty Detection to Industrial Machinery, *M.A. Timusk and C.K. Mechefske*   50
A CALS Approach to Adaptive Web Training using Fuzzy Hypermedia Model, *L. Di Lascio, A. Gisolfi, V. Loia and A. Santangelo*   55
Automatic Construction of Action Knowledge-Base for Mobile Robot using GA, *H. Watabe, R. Aruga and T. Kawaoka*   60
Technogenetic Base for Creation of Engineering Systems, *S. Dadunashvili*   65
The Trainable Expert System: A New Approach to Process Control, *S. Enk*   70
A Knowledge-based Web Site for Agriculture, *A. Mazzetti and L. Bonini*   75
A Measure of Fuzziness in Rough Sets, *H.-S. Lee*   80
A New IDCT Solution using a RISC Processor for Architecture, *K. Yamada, M. Kojima, T. Shimizu, S. Iwade, F. Sato and T. Mizuno*   85
Permutation Scheduling using Population Learning Algorithm, *J. Jedrzejowicz and P. Jedrzejowicz*   93
On the Notion of Compliance in Critiquing Intelligent Design Assistants: Isomorphic Representations of General Norms and Exceptions, *M.F. Ursu and R. Zimmer*   98
An Integration of Fuzzy and Two-valued Logics on Spatial Depiction, *T. Maeda, I. Hayashi, M. Umano and L.C. Jain*   105
Formalization of Expert AH Model for Machine Learning, *G. Molan and M. Molan*   110

A Rule-based Tracking System for Video Surveillance Applications, *L. Di Stefano,
M. Mola, G. Neri and E. Viarani*   115

Combined Features for Classifying Diffuse Lung Opacities in Thin-Section Computed
Tomography Images, *Y. Mitani, H. Yasuda, S. Kiso, K. Ueda, N. Matsunaga
and Y. Hamamoto*   121

Communicating Agent in Mobile Agent Network, *I. Lovrek, V. Sinkovic and
G. Jezic*   126

Ontological Approach to Electronic Payment Systems, *J. Carbo and
M. Fernandez-Lopez*   131

AI Application in Secure Database Systems, *Y. Bai and Y. Zhang*   136

Gas Turbine Faults: Detection, Isolation and Assessment using Neural Networks,
S. Ogaji, S. Sampath and R. Singh   141

Cluster Discriminant Analysis for Feature Space Visualization, *T. Suenaga, A. Sato
and H. Sakano*   146

Outlier Detection Data Mining using Cluster Discriminant Analysis – Data Mining from
Medical Data, *A. Sato, T. Suenaga and H. Sakano*   151

Coordinated Reasoning with Inference Fusion, *B. Hu, E. Compatangelo
and I. Arana*   156

Structure Discovery in Text Collections, *L. Massey*   161

Using the Topological Tree for Skin Lesion Structure Description, *R. Cucchiara
and C. Grana*   166

Advanced Fault Diagnosis using Genetic Algorithm for Gas Turbine Engines,
S. Sampath, S. Ogaji and R. Singh   171

A Knowledge Management System Enabling Regional Innovation, *A. Corallo,
E. Damiani and G. Elia*   176

Traffic-dependent Routing based on Self-Adaption, *D. Jevtic, M. Kunstic
and K. Cunko*   184

A Methodology for the Elicitation of Redesign Knowledge, *H. Ahriz and I. Arana*   189

Monitoring the Conformance of Connections to the Traffic Contract of ATM Networks
using Fuzzy Logic, *Z. Li and Z. Zhang*   194

Sampled Data Classifiers, *D. Stefanoiu, I. Tabus and F. Ionescu*   199

Text Mining Framework based on Graph-based Text Representation, *J. Tomita,
T. Ikeda and T. Satoh*   204

Human Movement Tracking using Self-Organization Model, *N. Yoshiike and
Y. Takefuji*   209

Web-based Occupant Feedback for Building Energy Management Systems,
J. Shimmin and I. Khoo   214

Improved Algorithms for Knight's Tour Problems using Neural Networks,
M. Aoba and Y. Takefuji   219

Modification of PCNN for a Moving Object Segmentation, *T. Chashikawa
and Y. Takefuji*   224

E-Maintenance and Business Intelligence for High Voltage Power Network
Components, *D. Bisci, M. Gallanti and A. Mazzetti*   229

Missing Value Estimation by Local Minor Components, *C.-H. Oh, K. Honda
and H. Ichihashi*   235

Preprocessing Time Series using Complexity Pursuit, *Y. Han and C. Fyfe*   240

The Scale Invariant Map and Maximum Likelihood Hebbian Learning, *E. Corchado
and C. Fyfe*   245

Case-based Reasoning for Reuse of Software Designs, *P. Gomes, F.C. Pereira,
P. Paiva, N. Seco, P. Carreiro, J.L. Ferreira and C. Bento*   250

vii

Discrete-Time American Options Evaluated by Weighting Functions, *Y. Yoshida,*
M. Yasuda, J.-I. Nakagami and M. Kurano — 255
Local Nearest Neighbor by Competition, *F.J. Ferrer-Troyano, R. Giráldez,*
J.C. Riquelme, J.S. Aguilar-Ruiz and D.S. Rodríguez-Baena — 260
Fusing and Modelling Sensors using Neural Networks for Low Cost Multi-Sensor
Robotics Systems, *A. Sánchez-Miralles and M.A. Sanz-Bobi* — 265
Extended Knowledge Model of Parametric Design for Concurrent Engineering,
M. Stefan and M. Valasek — 270
Discretization Oriented to Decision Rules Generation, *R. Giráldez, J.S. Aguilar-Ruiz,*
J.C. Riquelme, F.J. Ferrer-Troyano and D.S. Rodríguez-Baena — 275
Video-based Human Shape Detection by Deformable Templates and Neural Network,
S. Tate and Y. Takefuji — 280
Loubeth: Assistant-Agent Negotiation System for E-Commerce, *O. Cairó and*
J.G. Olarte — 286
A Framework to Analyze and Compare Knowledge Management Tools, *A. Valente*
and T. Housel — 291
Cooperative Games in a Stochastic Environment, *B. Apolloni, S. Bassis,*
D. Malchiodi and S. Gaito — 296
Forecasting the IBOVESPA using NARX Networks, *E.M. Jesus Oliveira and*
T.B. Ludermir — 301
Intelligent Support to Knowledge Sharing through the Articulation of Class Schemas,
E. Compatangelo and H. Meisel — 306
A Novel Input Representation for ANN-based Features Recognition, *L. Ding, Y. Yue*
and K. Ahmet — 311
Model-based Process Oriented Knowledge Management: The PROMOTE Approach,
R. Woitsch and D. Karagiannis — 316
Application of Neural Network for Prediction of Temporal Variation of Magnetic
Field Near Power Lines, *N.K. Roy, A. Nafalski and R.K. Jain* — 321
Image Quality Evaluation, *D. Van Der Weken, M. Nachtegael, E. Kerre and G. Chen* — 326
The Schein Rank Concept for Coding and Decoding Fuzzy Relations, *V. Loia,*
S. Sessa and W. Pedrycz — 331
Case-based Reasoning as a Heuristic Selector in a Hyper-Heuristic for Course
Timetabling Problems, *S. Petrovic and R. Qu* — 336

INVITED SESSION PAPERS

Session 1. Organiser: N. Baba
Intelligent Systems Utilizing Neural Networks and Genetic Algorithms

Lips Extraction with Template Matching by Genetic Algorithms, *T. Akashi,*
M. Fukumi and N. Akamatsu — 343
High-Speed Face Search by Threshold Value Determination Method using Neural
Networks, *Y. Fukuta, Y. Mitsukura, M. Fukumi and N. Akamatsu* — 348
Drift Ice Detection in Remote Sensing Images using Neural Networks and Genetic
Algorithms, *S. Tadokoro, M. Fukumi and N. Akamatsu* — 353
A System Identification Method using Expanded Neural Networks, *S. Yamawaki* — 358
A System Identification Method using a Hybrid-type Genetic Algorithm,
Y. Mitsukura, M. Fukumi and N. Akamatsu — 364

viii

Session 2. Organiser: N. Baba
Utilization of Soft Computing Techniques for Financial Markets

Network Competitive Strategies in Financial Engineering, *M. Resta* — 369
Utilization of Soft Computing Techniques for Dealing in the TOPIX and the
  NIKKEI-225, *N. Baba* — 374
Financial Decision Aid based on Heuristics and Fuzzy Logic, *S. Vraneš,*
  M. Stanojević and V. Tomašević — 381

Session 4. Organiser: M.S. Bartlett
Learning in Vision

Multiplicative Updates for Unsupervised and Contrastive Learning in Vision,
  D.D. Lee, H.S. Seung and L.K. Saul — 387
Extra-Label Information: Experiments with View-based Classification, *A. Rakhlin,*
  G. Yeo and T. Poggio — 392
Activity Bubbles and Natural Image Sequences, *A. Hyvarinen* — 397

Session 5. Organisers: H.-C. Huang and T.-Y. Chen
Knowledge-based Intelligent Watermarking and Signal Processing

A Gain-Shape VQ based Watermarking Technique with Modified Visual Secret
  Sharing Scheme, *J.-S. Pan, F.-H. Wang, T.-C. Yang and L. Jain* — 402
On the Scene Classification for the Automated Generation of Hierarchical Contents
  from Broadcasting TV News, *Y.H. Chen, C.L. Tseng, S.S. Cheng, T.M. Fang,*
  H.Y. Chan and H.C. Fu — 407
An Embedding Algorithm for Multiple Watermarks, *H.-C. Huang, J.-S. Pan and*
  F.-H. Wang — 412
Genetic Watermark Techniques based on Spatial Domain, *F.-H. Wang, L.C. Jain*
  and J.-S. Pan — 417
BTC-Immunized Watermarking using Local Homogeneity for Pixel Selection,
  C.-S. Shieh, H.-C. Huang, T.-Y. Chen and J.-S. Pan — 422

Session 7. Organiser: G. Castellano
Knowledge-based Neurocomputing Systems

Knowledge-based Neurocomputing for Operational Decision Support, *A. Bargiela,*
  C. Arsene and M. Tanaka — 427
Based Neuro-Symbolic Hybrid System in Asymmetric Sigma-Pi Connections for
  Treatment of Gradual Rules, *G. Reyes Salgado and B. Amy* — 433
Classifier Combination Techniques and Support Vector Machine: An Application to
  Character Recognition, *C. Tarantino, G. Pasquariello, N. Ancona, G. Satalino,*
  P. Blonda and A. D'Addabbo — 438
KERNEL: A System for Knowledge Extraction and Refinement by Neural Learning,
  G. Castellano, C. Castiello and A.M. Fanelli — 443

Session 9. Organiser: C. Faucher
Classification in Knowledge-based Systems

The Role of Classification(s) in Distributed Knowledge Management, *M. Bonifacio,*
P. Bouquet and R. Cuel   448
Using Prototypes for Classification in Case-based Medical Expert Systems, *R. Schmidt*
and L. Gierl   453
Incremental Development of a Hierarchical Classification System using an Object-
Oriented Database Management System (OODBMS), *G. Beydoun and L. Al-Jadir*   458

Session 11. Organiser: P. Frasconi
Machine Learning in Bioinformatics

A Neural Network-based Method for Predicting the Disulfide Connectivity in Proteins,
P. Fariselli, P.L. Martelli and R. Casadio   464
Genome Comparisons based on Profiles of Metabolic Pathways, *L. Liao, S. Kim and*
J.-F. Tomb   469
A Genetic Algorithm Approach to Solving the Protein Structure Prediction Problem,
R.O. Day, J.B. Zydallis, G.B. Lamont and R. Pachter   477
Supervised Gene Expression Data Analysis using Support Vector Machines and
Multi-Layer Perceptrons, *G. Valentini*   482
Knowledge Discovery for Gene Regulatory Regions Analysis, *N.A. Kolchanov,*
M.A. Pozdnyakov, Y.L. Orlov, O.V. Vishnevsky, E.E. Vityaev and
B.Y. Kovalerchuk   487

Session 12. Organiser: J. Gasós
Advanced E-Commerce Applications

Improving Customer Relation, a Knowledge-based Approach, *R. Traphöner*   492
COGITO: A Conversational Agent Supporting its Users' Search Strategies, *U. Thiel*
and M. L'Abbate   497
Personalised Configuration of Products and Services in Co-Operating Markets,
L. Ardissono, A. Goy, M. Holland, D. Jannach, R. Schäfer, M. Zanker and
R. Simeoni   502
A Framework for Internet Data Collection based on Intelligent Agents - Achieved
Results, *P.A.C. Sousa, J.P. Pimentão, F. Moura Pires and A. Steiger-Garção*   507

Session 14. Organiser: V. Giménez Martínez
Artificial Neural Networks and Adaptive Systems for the Information Technologies

An Adaptive Procedure to Implement an Optimal General Delta Model, *C. Torres,*
V. Giménez and F. Monasterio   512
Visual Analysis of Self-Organizing Maps, *M. Rubio, V. Giménez, F. Díaz, P. Gómez*
and R. Martínez   517
Performance of Supervised ART Artificial Neural Networks with Different Parameter
Vigilance Dynamics, *K.R. Al-Rawi, C. Gonzalo, E. Martínez and A. Arquero*   522
Neural Control of Simultaneous Chaotic Systems, *C. Hernández, R. Gonzalo,*
J. Castellanos and A. Martínez   527

x

Enhanced Neural Networks as Taylor Series Approximations, *L.F. Mingo,*
M.A. Díaz, M. Pérez and V. Palencia   532

Session 15. Organiser: G. Gini
Data Mining and Knowledge Management for Scientific Data

Finding and Using Global Implications for the Association Rule Problem,
L. Dumitriu, D. Stefanescu and E. Pecheanu   537
A Comparison of Probabilistic, Neural, and Fuzzy Modeling in Ecotoxicity, *G. Gini,*
M. Giumelli, E. Benfenati, N. Piclin, J. Chrétien and M. Pintore   542
Mining from Medical Data: Case-Studies in Meningitis and Stomach Cancer Domains,
S. Kawasaki, A. Saitou, D.D. Nguyen and T.B. Ho   547

Session 20. Organiser: R. Hamabe
Intelligent Image Coding and Image Processing

Image Quality Evaluation, *D. Van Der Weken, M. Nachtegael, E. Kerre and G. Chen*   552
Video Coding based upon Learning Motion Vector Quantization, *T. Sagara and*
R. Hamabe   557
Image Compression using DWT for Mobile System, *H. Kondo, T. Ishikawa, T. Kouda*
and L. Zhang   562
Blur Effect in Image Interfaces, *S. Hiraoka, R. Hamabe and K. Tomimatsu*   567

Session 21. Organisers: J. Hasebrook and G. Doeben-Henisch
Self Learning and the Learning Self: Technologies for the Learning Exploitation

Knowledge Robots for Knowledge Workers: Self-Learning Agents Connecting
Information and Skills, *J. Hasebrook, L. Erasmus and G. Doeben-Henisch*   572
Flexible Diagram Generation from Tagged Texts, *M. Murayama, Y. Nakamura*
and Y. Ohta   582
Control Architecture for Implementing Intelligent Non-Player Characters in Online
Computer Games, *I.C. Kim*   589
Cognitive Assistants in Synthetic Environments - Aspects of the Concept and
Engineering Development, *G. Sterling and T. Dillon*   594

Session 22. Organiser: K. Kondo
Computer Aided Diagnosis in Medicine

Knowledge-based Fuzzy Shape Detective Algorithm for Segmenting Human Brain
MR Images, *S. Kobashi, K. Kondo, Y. Hata and M. Matsui*   600
Challenge to the Development of a Transcranial Sonography System, *K. Nagamune,*
S. Kobashi, K. Kondo, Y. Hata, Y.T. Kitamura and T. Yanagida   604
User Interfaces for SDL Applications, *E. Ardizzone, D. Peri and R. Pirrone*   609
Topograms of Spectral Analysis of Physiological Signals for Intuitive Evaluation,
K. Kondo, Y. Hata, S. Kobashi, K. Maeda, H. Ishigaki, Y. Nasu, T. Suzuki and
K. Mabuchi   614

Fuzzy Maximum Intensity Projection (FMIP) of MR Endorrhachis Images, *Y. Hata, S. Kobashi, K. Kondo, S. Imawaki and M. Ishikawa* — 618

Session 23. Organiser: A. Hirose
Complex-valued Neural Networks

On Energy Function for Complex-valued Hopfield-type Neural Networks, *Y. Kuroe, N. Hashimoto and T. Mori* — 623
Generalization of the Complex-valued Neural Networks with the Orthogonal Decision Boundary, *T. Nitta* — 628
Study of the Chaotic Behavior of the Complex-valued Nagumo-Sato Model of a Single Neuron, *I. Nemoto* — 633
Context-Dependent Behavior of Coherent Neural Systems after Self-Organizing Mapping of Carrier Frequency Values, *A. Hirose and D. Ishimaru* — 638
InSAR Image Restoration by using Stochastic Complex-valued Neural Network, *A.B. Suksmono and A. Hirose* — 643

Session 24. Organiser: T.-Z. Hong
Intelligent Data Analysis and Applications

A Classifier with the Function-based Decision Tree, *B.-C. Chien and J.-Y. Lin* — 648
Keyword Extraction: A Cluster-based Approach, *T.-M. Chang and C.-M. Lai* — 653
A Platform for Developing Adaptive Tutoring Systems, *S.-L. Lee, J.C.R. Tseng and G.-J. Hwang* — 658
A Fuzzy Learning Diagnosis Approach for Developing Adaptive Tutoring Systems, *G.-J. Hwang, N.-Y. Lin, C.-H. Liao, I.-K. Chan and G.-H. Hwang* — 663
Web Transaction Mining with Intentional Behaviour, *Y.-H. Tao, Y.-M. Su and T.-P. Hong* — 668
Mining Multi-Dimensional Association Rules from Data Warehouses, *W.-Y. Lin and M.-C. Tseng* — 673
An Efficient Indexing Method for Data Mining - Condensable Bit-wise Indexing Method, *W.-C. Chen, S.-S. Tseng and C.-H. Lien* — 678
Discovery of Cross-Level Sequential Patterns from Transaction Databases, *S.-L. Wang, W.-S. Lo and T.-P. Hong* — 683

Session 25. Organiser: T. Ichimura
Human Centered Intelligent Systems

Response Generating Method and its Application to Web-based Health Care Service, *K. Mera, M. Yoshie, T. Ichimura, T. Yamashita, T. Aizawa and K. Yoshida* — 688
Kansei Modeling by Categorical Adaptive Method and its Visualization, *T. Murakami, R. Orihara and N. Sueda* — 693
Health Support Intelligent System for Diabetic Patients (HSISD), *M. Suka, T. Ichimura, N. Tajima, T. Nakamura and K. Yoshida* — 698
Interactive Support System for Human Decision Making by Fuzzy AHP, *T. Yamashita, Y. Yonezawa, T. Ichimura and K. Yamashita* — 703
Emotional Interface for Human Feelings by Mobile Phone, *T. Ichimura, K. Mera, Y. Miki and T. Yamashita* — 708

xii

Session 27. Organiser: Y. Ishida
Immunity-based Systems

A Mobile Anti-Virus System Inspired by the Immune System, *T. Okamoto
and Y. Ishida*   713
An Immune Optimization using Biological Immune Co-Evolutionary Phenomenon
and Cell-Cooperation for Division-And-Labor Problem, *N. Toma, S. Endo,
K. Yamada and H. Miyagi*   718
A Synthesis of Immune System Model with Darwinian Hereditary Neural Network,
S. Oeda, T. Ichimura, T. Yamashita and K. Yoshida   723
Emergence of an Immune Memory due to a Threshold with Antigen-Cognition,
K. Harada, T. Ikegami and N. Shiratori   728
Idiotypic Network Model for Feature Extraction in Pattern Recognition, *T. Shimooka,
Y. Kikuchi and K. Shimizu*   734
Fault Detection for Mobile Agent System using Immunity -based Diagnosis,
Y. Watanabe and Y. Ishida   739

Session 28. Organiser: N. Ishii
Intelligent Knowledge-based Systems (I)

A Study on the Training System using PDA for Patrolling the Engine Room,
K. Nagao, K. Matsushita and N. Ishii   744
Correlation Filter Processing Simulation of the Visual Evoked Potential, *H. Sasaki
and N. Ishii*   749
MAP Classification based on Paired Class Learning Accuracy, *T.Nakano and
N. Inuzuka*   754
Multi-layered Analog Electronic Circuits and the Hardware Implementation of
Knowledge based Learning System, *M. Kawaguchi, T. Jimbo, M. Umeno
and N. Ishii*   759

Author Index   I-VIII

Part 2

Session 28a. Organiser: Y. Adachi
Intelligent Knowledge-based Systems (II)

Development of Hybrid Type Writer Recognition System, *M. Ozaki, Y. Adachi
and N. Ishii* 765
Detection of Face Direction from Fresh Color and Boundary Images, *Y. Adachi,
S. Takeoka and M. Ozaki* 770
Recognition Studying States using Fuzzy Theory for Repeatedly Vocabulary Learning,
T. Yoshioka, K. Okuya and H. Nishizawa 775
Self-supervised Learning for Sensory Integrating System, *H. Takeuchi,
K. Yamauchi and N. Ishii* 780

Session 30. Organiser: R. Khosla
Human Centred Intelligent Systems

A Framework for High Level User Interface for Accessing Dynamic Contents on
the Web, *S. Bhalla, M. Hasegawa and N. Berthouze* 785
Pragmatic Considerations for Human-centeredness, *R. Khosla and P. Usmanij* 791
Intelligent Document Classification for Web Document Retrieval, *D. Sridharan,
P. Sakthivel and S.K. Srivatsa* 796
An Intelligent System for Learning Human Designs, *H. Chen, H. Abolhassani
and Z. Koono* 801
Video Interpretation: Human Behaviour Representation and On-line Recognition,
V.-T. Vu, F. Bremond and M. Thonnat 807

Session 31. Organiser: S. Kunifuji
Media Technology for CSCW

User Interface Design for Online Incidental Information Media, *A. Maeda,
Y. Nakamori and K. Sugiyama* 812
ReSOoM: A Meeting Support System for Retrieval Records of a Meeting with
Relationships among Record Elements, *I. Kuramoto, J. Noda, N. Fujimoto and
K.-I. Hagihara* 817
Electronic Tag-Playing Support System with Awareness Function, *T. Yoshino
and J. Munemori* 822
SmartCourier: Annotation Management Tool for Research Labs, *S. Ito, Y. Sumi,
K. Mase and S. Kunifuji* 827

Session 32. Organisers: Y. Yamashita and C. Kuroda
Intelligent Data Processing in Process Systems and Plants

Fault Detection based on Frequent Clusters, *Y. Yamashita* 833
A Fuzzy-Logic based Fault Diagnosis Strategy for Process Control Loops,
S.-Y. Chang and C.-T. Chang 838

xiv

Application of Wavelet Analysis to Chemical Process Diagnosis, *T. Matsuo and
  H. Sasaoka*   843
NN Application for Estimation of Products Quality Indexes, *S. Tronci, G. Testoni,
  A. Servida and R. Baratti*   848
Multiagent-oriented Decision Making Framework for Assembly and Operation of
  Micro Chemical Processes, *N. Kimura, H. Matsumoto and C. Kuroda*   853
Computer-aided Hazard Identification in Batch Processes using Petri Nets, *Y.-F. Wang
  and C.-T. Chang*   858
Acquisition of AGV Control Rules using Reinforcement Learning, *H. Yamaba
  and S. Tomita*   863

Session 33. Organisers: B. Lazzerini and F. Marcelloni
Data Clustering based on Non-Metric Similarity Measures

An Evaluation of Term Weighting Schemes for Discovering Related Terms from
  Texts, *L. Zhou*   868
A New Asymmetric Topographic Mapping Algorithm, *M. Martin-Merino and
  A. Muñoz*   873
Set-Theoretic Similarity Measures, *M. Rifqi and B. Bouchon-Meunier*   879
Clustering based on a Dissimilarity Measure Derived from Data, *P. Corsini,
  B. Lazzerini and F. Marcelloni*   885
Improving Information Retrieval by Words Selection Extracted from a Search Log,
  N. Kawamae   890

Session 34. Organiser: C. Van Leeuwen
Coupled Maps for Information Processing

Coupled Maps with Variable Connections, *J. Ito*   895
Emergence of Scale-Free Network Structure with Chaotic Units, *P. Gong and
  C. Van Leeuwen*   900
Enhanced Information Flow in a Chain of Fractally Coupled Chaotic Clusters,
  C. Wagner and R. Stoop   905
The "Divide and Conquer" Model of Image Segmentation: Object-Bounded
  Synchrony Propagation in Coupled Map Lattices, *A. Raffone and C. Van Leeuwen*   910
Some Problems Related to the Choice of Coupling in a System of Coupled Maps,
  I. Tyukin and C. Van Leeuwen   917
Component-based Framework for Design and Simulation of Neural Networks,
  P. Jurica, M. Brinkers and C. Van Leeuwen   922
Adaptation and Synchronization as a Way to Control Spatio-Temporal Pattern
  Formation in the System of Coupled Logistic Maps, *M. Zochowski and
  R. Dzakpasu*   927

Session 36. Organiser: V. Maojo
Intelligent and Integrated Systems in Medicine and Biology

Probabilistic-Weighted k-Nearest-Neighbour Algorithm: A New Approach for Gene
  Expression based Classification, *B. Sierra and E. Lazkano*   932

Description of an Ontology for Supporting BIKMAS, a Biomedical Informatics
    Knowledge Management System, *V. López-Alonso, L. Moreno, G. López-Campos*
    and F. Martín-Sanchez   940
Genetic Learning of Fuzzy Rules through Neuro-Fuzzy Topology Construction,
    A. Carrascal, D. Manrique, J. Ríos and C. Rossi   945
Analysing the Occurrence of Childhood Leukaemia using Seasonal Time Series,
    M. Fazekas   950
INFOGENMED: Integrating Heterogeneous Medical and Genetic Databases and
    Terminologies, *M. García-Remesal, J. Crespo, A. Silva, H. Billhardt, F. Martín,*
    J. Rodríguez-Pedrosa, V. Martín, A. Sousa, A. Babic and V. Maojo   955

Session 38. Organiser: J.K. Mattila
Methodological Aspects of Soft Computing

Fuzzy Rules in Strength Grading of Lumber, *J. Kortelainen and Y. Tolonen*   960
Interpretation – Insuperable Challenge to Computing with Words, *V. A. Niskanen*   965
Level-Sets and Topological Bases, *P. Kukkurainen*   969
Fuzzy Similarity based Classification in the Normal Łukasiewicz-Structure,
    P. Luukka and K. Saastamoinen   974
Comparison of the Fuzzy Similarity based Classification in the Normal and the
    Generalised Łukasiewicz-Structure, *K. Saastamoinen and P. Luukka*   982

Session 39. Organiser: D. P. Mital
Computer based Medical Decision Support Systems

Analysis of Cortical Connectivity using Hopfield Neural Network, *S. Dixit, K. Mosier,*
    S. Shankar and D. Mital   987
A Point of Care Patient Repository using Java Speech Technology, *S. Damarapu,*
    S. Srinivasan, D. Mital and S. Haque   992
Prediction of Hepatitis C using Artificial Neural Network, *R. Jajoo, D. Mital,*
    S. Haque and S. Srinivasan   997
Diagnosing the Aetiology of Neonatal Jaundice by Paramedical Personnel using an
    Expert System, *C. Dharmar, S. Srinivasan, D. Mital and S. Haque*   1002
Multi-Tier Web Application to Make EEG Universally Available using Java Servlets
    and Java Server Pages (JSP), *S. Dixit, S. Shankar and D. Mital*   1007

Session 40. Organiser: V. Moret-Bonillo
Knowledge-based Systems

Modelling Knowledge Bases by Reusing Generic Ontologies and Unified
    Terminology Servers, *M. Taboada, M. Arguello, D. Martínez, J. Des and J. Mira*   1013
A Multi-Agent Architecture for Intrusion Detection, *A. Alonso-Betanzos,*
    B. Guijarro-Berdiñas and J.A. Suárez-Romero   1018
Intelligent Analysis of Polysomnograms in a Sleep Apnea Decision Support System,
    M. Cabrero-Canosa, E. Hernández-Pereira and V. Moret-Bonillo   1023
A Task Model for a Management Resource System Integrated in a Cooperative
    Learning Environment, *M. Lama, R. Amorim, E. Sánchez, A. Riera, J. Vila,*
    S. Barro, B. Cebreiro and C. Fernández-Morante   1028

xvi

Adaptive Knowledge-based Systems with Intelligent Agents and Evolutionary
Computation, *J.G. Fernández García De La Rocha and E. Mosqueira-Rey*   1033

Session 41. Organiser: J. Munemori
Intelligent Network Applications

Cooperative Development Environment of Sign Language Animation System using
Humanoid Model, *T. Yuizono, K. Hara and S. Nakayama*   1038
Development and Application of a Group Learning Function in the Virtual School
System, *S. Fujii, J. Iwata, K. Yoshida and T. Mizuno*   1043
Evaluation and Trial of Distance Learning System using Mobile Phone and WWW,
*K. Nakada, K. Matsumoto, T. Akutsu, K. Nagamoto, S. Fujii, H. Ichimura and
K. Yoshida*   1048
GUNGEN DX II: New Idea Generation Support System for Hundreds of Ideas,
T. Shigenobu, T. Yoshino and J. Munemori   1053
Application and Evaluation of Distributed Remote Seminar Support System Remote
Wadaman II, *T. Yoshino and J. Munemori*   1058

Session 42. Organisers: T. Murai and V.N. Huynh
Semantic Fields and Models for Uncertainty

Semantic Field and Context Models of Uncertainty, *G. Resconi and V.-N. Huynh*   1063
An Extension of Context Model for Modelling of Uncertainty of Type 2,
V.N. Huynh, G. Resconi and Y. Nakamori   1068
A Fuzzification of Landscape Theory for Alliance Analysis, *S. Suganuma,
V.N. Huynh, S. Wang and Y. Nakamori*   1073
Weighted Functional Dependencies in Fuzzy Databases, *M. Nakata, T. Murai and
S. Miyamoto*   1078
Operations of Zooming In and Out on Possible Worlds for Semantic Fields,
T. Murai, G. Resconi, M. Nakata and Y. Sato   1083

Session 43. Organiser: D. Neagu
Hybrid Intelligent Systems

Neuro-Fuzzy based Analysis of Gene Expression Data, *V. Palade and D. Neagu*   1088
Toxicity Prediction using Assemblies of Hybrid Fuzzy Neural Models, *D. Neagu*   1093
Hybrid Intelligent Production Simulator and its If-Then Rules Acquisition by GA,
H. Yamamoto and E. Marui   1098
SAMIR: An Intelligent Web Agent, *F. Abbattista, P. Lops, G. Semeraro and
F. Zambetta*   1103
A Formal Semantics of Hybrid Symbolic-Neural Networks for Commonsense
Reasoning, *Y. Bai and Y. Zhang*   1110
Finding Exceptions to Rules in Fuzzy Rule Extraction, *V. Soler, J. Roig and M. Prim*   1115
An Improved Recurrent Neuro-Fuzzy Network for Self-Organizing Control,
N. Constantin and V. Palade   1120

Session 46. Organisers: P. J.-S. Pan and Y.-W. Chen
Knowledge based Signal and Image Processing (I)

Tabu Search Approach based Algorithm for Nonlinear Time Alignment, *T.-Y. Chen, J.S. Pan, X. Mei and S. Sun* — 1127
Genetic Algorithm on Portfolio Strategy in the Complicated Stock Market of Taiwan, *J.-F. Chang, S.-C. Chu and P.-Y. Wu* — 1132
Achieving Uplink Optimal SDMA of Smart Antenna by Phase-Only Disturbances using Genetic Algorithms, *C.-H. Hsu and T.M. Babij* — 1137
A Simple Tempo-Spectral Hybrid Method for Speech Noise Reduction, *H. Li, L. Ma and P. Gallinari* — 1142
Channel Noise Reduction using Genetic Index Assignment and Energy Allocation for VQ with BPSK and BFSK, *T.C. Yang, L.C. Jain, K.C. Huang and J.S. Pan* — 1147

Session 47. Organisers: Y.-W. Chen and P. J.-S. Pan
Knowledge based Signal and Image Processing (II)

Channel Noise Reduction for I/R VQ with Parallel Genetic Algorithm Index Assignment, *K.C. Huang, S.C. Chu and T.C. Yang* — 1152
Hand Shape Recognition for Biometric Identification by using LVQ, *Y. Fukuhara and Y. Takefuji* — 1157
A Robust System for Fingerprints Identification, *V. Conti, G. Pilato, S. Vitabile and F. Sorbello* — 1162
A Color, Texture and Shape Fusion Technique for Segmentation of High-Resolution Satellite Images, *T. Tateyama, Y.-W. Chen, X.-Y. Zeng and H. Tamashiro* — 1167
Classification of Remote Sensing Images using Spectral Independent Components, *X.-Y. Zeng, Y.-W. Chen and Z. Nakao* — 1172

Session 51. Organiser: V. Piuri
Soft Computing for Industrial Applications

An Embedded Neural Network System for Vision Applications, *A. Pinna, A. Alexandre, E. Belhaire, P. Garda and B. Granado* — 1177
A Soft Computing based Behavioral Model of Po River During Flood Events, *A. Boscolo, D. Russo and V. Fiorotto* — 1182
Quality Assessment in Dry Cured Ham Production, *A. Boscolo, C. Mangiavacchi and M. Ulian* — 1187
Neuro-Fuzzy Clustering Techniques for Complex Acoustic Scenarios, *R. Poluzzi, A. Savi and D. Vago* — 1192

Session 52. Organiser: H. Sakano
Real World Biometrics

New Architecture for Fingerprint Identification System to Apply Biometrics to Various Applications, *S. Shigematsu, T. Hatano, K. Saito, H. Suto, M. Nakanishi, K. Machida, Y. Okazaki and H. Kyuragi* — 1197
A Speaker Verification Method using CELP Parameters, *Y. Yamazaki, T. Kondo and N. Komatsu* — 1202

Independent Component Analysis for Face Authentification, *C. Havran, L. Hupet, J. Czyz, J. Lee, L. Vandendorpe and M. Verleysen* — 1207

Fusion of Eigenfaces and Elastic Graph Matching for Face Recognition, *G. Sazaklis and S.C.A. Thomopoulos* — 1212

Scenario based Data Collection Trials for the Evaluation of Multi-Modal Biometric Processing: A Preliminary Report, *M. C. Fairhurst, J. George and F. Deravi* — 1217

Session 53. Organiser: I. Nishizaki
Interactive and Hierarchical Decision Making Under Uncertainty

An Interactive Fuzzy Satisficing Method through a Variance Minimization Model for Multiobjective Linear Programming Problems Involving Random Variable Coefficients, *M. Sakawa, K. Kato and H. Katagiri* — 1222

Interactive Fuzzy Decentralized Two-Level Linear Fractional Programming through Decomposition Algorithms, *K. Kato, M. Sakawa and I. Nishizaki* — 1227

Interactive Decision Making for a Multiobjective 0-1 Programming Problem Involving Fuzzy Random Variable Coefficients, *H. Katagiri, M. Sakawa and Y. Sato* — 1232

Fuzzy Multiple Decision Maker-Multiple Objective Programming Problems using the Weighted Minimax Method, *H. Yano* — 1237

Nash-Stackelberg Equilibrium Solutions for Noncooperative Two-level Four-person Games, *I. Nishizaki, M. Sakawa and Y. Fujita* — 1242

Computational Methods for Two-level Integer Programming Problems with Fuzzy Parameters through Genetic Algorithms, *K. Niwa, I. Nishizaki and M. Sakawa* — 1247

Session 55. Organiser: S. Hiroko
Chance Discovery on Human-Data Interaction

Clustering using Small World Structure, *Y. Matsuo* — 1252

Extracting Characteristic Sentences from Related Documents, *N. Okazaki, Y. Matsuo, N. Matsumura, H. Tomobe and M. Ishizuka* — 1257

Utilizing Fault Cases for Supporting Fault Diagnosis Tasks, *Y. Kato, T. Shirakawa and K. Hori* — 1262

Mining and Characterizing Opinion Leaders from Threaded Online Discussions, *N. Matsumura, Y. Ohsawa and M. Ishizuka* — 1267

How Can We Facilitate Concept Articulation in Purchasing?, *H. Shoji, M. Mori and K. Hori* — 1271

Toward a Chance Discovery-Oriented Recommender System: A Prototype, *M. Mizuno, H. Shoji, Y. Ohsawa, Y. Matsuo, N. Matsumura and Y. Miyake* — 1276

An Approach to a Knowledge Reconstruction Engine for Supporting Event Planning, *S. Amitani, M. Mori and K. Hori* — 1281

Visual Representation of Scene Information for Shopping as Concept Articulation, *M. Mori, H. Shoji and K. Hori* — 1286

User's Interests Change as Chance Discovery, *A. Abe* — 1291

Session 58. Organiser: P. Crippa
Neural Networks as Universal Approximators

Self-Organizing Map with Limited Scope Learning, *M. Michihata, T. Miyoshi and
H. Masuyama* 1296
Steepest Ascent Training of Support Vector Machines, *S. Abe, Y. Hirokawa and
S. Ozawa* 1301
Comments on Using MLP and FFT for Fast Object/Face Detection, *H.M. El-Bakry* 1306
Writing Style Variation Absorption for a Hybrid Neuro-Markovian On-line
Handwriting Recognition System, *H. Li, T. Artieres, P. Gallinari and B. Dorizzi* 1311
Neural Network Approximation of Stochastic Processes: A Recursive Algorithm,
P. Crippa and C. Turchetti 1316

Session 62. Organiser: T. Nishida
Communicative Reality and its Application

Interpersonal Cognition in Anonymous Community, *S. Azechi and K. Matsumura* 1321
Conversational Contents Making a Comment Automatically, *H. Kubota,
K. Yamashita and T. Nishida* 1326
Forming a Sense of Reality from the Viewpoint of Cognitive Psychology,
K. Yamashita 1331
POC Communicator: A System for Collaborative Story Building, *T. Fukuhara,
T. Nishida and S. Uemura* 1336
The Motivations to Get and Send Information, *K. Matsumura* 1341
Applications of POC System to Education, *N. Fujihara, M. Taniguchi, T. Fukuhara
and T. Nishida* 1346
An Active-Affordance-based Method for Communication Between Humans and
Artifacts, *K. Terada and T. Nishida* 1351

Session 63. Organiser: N.Baba
Utilization of Nns and AI for Constructing Intelligent Systems

Key Words Extraction for Image Search of WWW, *S.Ito, Y. Mitsukura, M. Fukumi
and N. Akamatsu* 1357
A Proposal of Emotional Detection System from Speech Data, *H. Sato, Y. Mitsukura,
M. Fukumi and N. Akamatsu* 1362
Learning Correspondences of Syntax Structures from Bilingual Corpora, *T. Hirano
and K. Tsuda* 1367
Remote Consultation Business Model on Steel Material Selection using
KNOW-HOW and KNOW-WHO, *M. Takahashi, A. Fuji and K. Tsuda* 1372

Session 65. Organiser: N. Kubota
Computational Intelligence for Robotics

Real-Time Robust Recognition of Human using Illuminance-depending Gazing GA,
M. Minami, H. Suzuki and M. Miura 1377
Knowledge Sharing based on Dependability in Multi-robot Exploration,
F. Kobayashi, S. Shinyama and F. Kojima 1382

Communication of a Partner Robot based on Perceiving-Acting Cycle, *N. Kubota, D. Hisajima, F. Kojima and T. Fukuda*   1387

Motion Intelligence of a 1-Link Mobile Manipulator using GA Evolved in Time Domain, *M. Kobata, M. Minami and A. Tamamura*   1392

Perceptual System of Multiple Robots for Quasi-ecosystem, *N. Kubota, M. Mihara and F. Kojima*   1397

Author Index   1403

IKOMAT '02

Proceedings of the 2002 International Workshop on Intelligent Knowledge Management Techniques   1411-1578

KES 2002
E. Damiani et al. (Eds.)
IOS Press, 2002

Development of Hybrid Type Writer Recognition System

Masahiro OZAKI[1], Yoshinori ADACHI[2] and Naohiro ISHII[3]
*[1]Nagoya Women's University, [2]Chubu University, [3]Nagoya Institute of Technology,
[1]3-40, Shioji-cho, Mizuho-ku, Nagoya, JAPAN   467-8610   ozaki@nagoya-wu.ac.jp*

Abstract: A hybrid type writer recognition system that is a combination of the two-dimensional(2D) Fuzzy membership function method and the new local arc method is proposed. The 2D Fuzzy membership function method is simple and takes little computer time but cannot be applied to unstable characters. The new local arc method gives quite accurate results even for the unstable characters but takes longer time to obtain the curvature distributions. By taking strong   points of these two methods, the proposed method gives good recognition results.

1. Introduction

We had proposed the off-line writer recognition system[1][2][3][4] for Japanese "hiragana" characters. In the papers[2][3][4], we tried to make a writer recognition system as simple as possible by using two-dimensional(2D) Fuzzy membership functions which were obtained from simple summation of characters. And by the 2D Fuzzy membership function method[4],   98.7% recognition ratio was obtained. However in the paper, many characters were omitted as inadequate characters for recognition. Because shapes and sizes of characters were so different and they were consequently scattered too much. We called these characters as unstable characters. Therefore, some writers who always wrote characters in different size or different shape could not be recognized or in the case of too few characters, the recognition process happen to be abandoned.

To overcome these problems, a characteristic of each stroke was tested for writer recognition in the previous paper[4], instead of the stroke location as in the 2D Fuzzy membership function method. Japanese characters consist of several strokes. This makes the character size and shape unstable easily. However, even writing unstable characters, the peculiar way of writing strokes can not be influenced so much by the difference of the size and shape of characters and curvatures of strokes can be used for writer recognition.

The idea of local arcs was first proposed by Yoshimura et al.[5][6]. They calculated curvatures by pattern matching using prepared curvature mask patterns. In the paper, we proposed the method to calculate curvatures directly from the strokes. The recognition ratio by the new local arc method was larger than 98% with no inadequate character. The recognition ratio depended on the combination of the chord lengths.

In this paper, we propose new hybrid type writer recognition system. This is taking strong points of these two methods.

2. The 2D Fuzzy membership method

Eight types of "hiragana" "は(ha)"," ま(ma)","に(ni)"," す(su)","の(no)","と(to)","を(wo)","る(ru)", often appeared as daily letters, were used for investigation.   The written characters on the framed sheets were input and were normalized as 50 x 50 dots square after centered the gravity. Then to make a dictionary, we eliminated characters that far from the average character. The average and variance of characters are obtained from the following equations,

$$\overline{f_{ij}}(x, y) = \frac{1}{n} \sum_{k=1}^{n} f_{ij,k}(x, y) \qquad (x, y = 1,..., 50) \qquad (1)$$

where $f_{ij,k}(x, y)$ was a function indicating an existence of character strokes as follows,

$$f_{ij,k}(x, y) = \begin{cases} 1 & \textit{if stroke exists at } (x, y) \\ 0 & \textit{others} \end{cases} \qquad (2)$$

and subscripts i, j, k were indicating a writer, a type of character, and character number, respectively.

$$\sigma_{ij,k}^{2} = \frac{1}{2500} \sum_{x,y=1}^{50} (f_{ij,k}(x, y) - \bar{f}_{ij}(x, y))^{2} \qquad (3)$$

The average variance of each writer is depicted in **Figure 1**. The value strongly depends on writer. In the paper, we omitted the character its variance was larger than 0.002. Because with increasing the variance, the recognition ratio decreases as shown in **Figure 2**.

To emphasize a writer's feature, a principal component analysis was used for each writer and each type of character using four characters. And 2D Fuzzy membership functions $\mu_{ij}(x, y)$ was expressed by the following equation,

$$\mu_{ij}(x, y) = \sum_{l=1}^{m} \lambda_{ij,l} X_{ij} \vec{e}_{ij,l} / \sum_{l=1}^{m} \lambda_{ij,l} \qquad (4)$$

where $X_{ij} = (\vec{x}_{ij,1}, \vec{x}_{ij,2}, \cdots, \vec{x}_{ij,n})$ is a character matrix and $\vec{x}_{ij,k}$ ($50 \times 50 = 2,500$ *elements*) is a character. $\lambda_{ij,l} (l = 1, \cdots, n)$ is an eigenvalue and $\vec{e}_{ij,l} (l = 1, \cdots, n)$ is an eigenvector, which are obtained from the covariant matrix $Q_{ij} = X_{ij}^{T} X_{ij}$. From the principal component analysis of these matrixes, number of eigenvalue and eigenvector m was obtained to satisfy that the cumulative proportion was larger than 0.9.

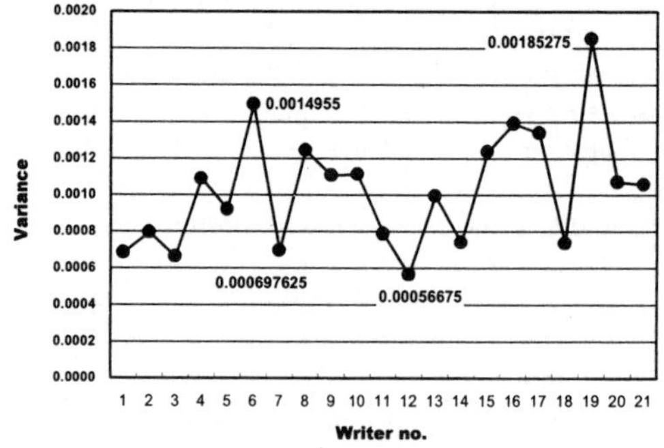

Figure 1 Average variance of each writer

Writer recognition was done by the following similarity evaluation function,

$$\eta_{ij,k} = \sum_{(x,y) \in A_{\alpha,ij}} \mu_{ij}(x, y) f_{ij,k}(x, y) \qquad (5)$$

where $\eta_{ij,k}$ indicates the k-th character's similarity value to the i-th writer.

$$A_{\alpha,ij} = \{(x, y) \mid \mu_{ij}(x, y) > \alpha \qquad (6)$$

where α was set to 0.5 empirically.

Handwritten characters usually fluctuated and those variances are depending on writers and types of characters. Then to absorb scattering of characters and to characterize one's handwriting features, a combination of three kinds of characters could work effectively. The following equation is used to obtain a similarity,

$$\eta_i = 1 - (1 - \eta_{ij_1k_1})(1 - \eta_{ij_2k_2})(1 - \eta_{ij_3k_3}) \qquad (7)$$

which is a product of three different kinds of characters, where k_1, k_2 and k_3 are indicating character numbers.

The average recognition ratio is 100% for every writer. Where characters have low similarity value, less than 0.4, were omitted as inadequate characters for recognition. The results show fairly good, but many characters were omitted as inadequate characters as listed in **Table 1**. For writer no. 19, more than 20% characters were omitted and in average about 6% characters were inadequate.

3. New local arc method

To overcome the problems occurred from the normalization and variance of character shapes mentioned in the previous section, curvature of strokes was used to figure out the feature of writers.   The curvature of local arc was shown to be useful for writer recognition by Yoshimura [5][6].   Yoshimura used mask patterns of arc to obtain the stroke's curvatures with the normalized characters.   In this study, we calculate the curvature directly with the original characters.   The curvature obtained by the following procedures.

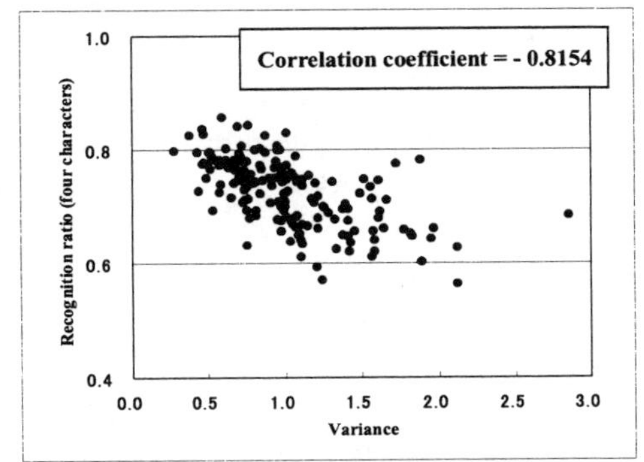

Figure 2   Character's variances versus recognition ratio(four characters)

(1) First two points making chord are fixed.   The length of the chord is 9, 13, and 17 dots and the direction is changed 15 degrees interval, in total 12 directions are investigated.

(2) The arc of the chord is found out by the following procedure.

   a. The tip of the chord is placed on the stroke.

   b. It is checked that the other end of the chord is on the stroke or not. If not on, then the chord is changed.

   c. The perpendicular bisector is decided that is pass the midpoint of the chord.

   d. The cross point of the perpendicular bisector and a stroke is searched within the distance from the midpoint less than 4 times of the length of the chord.

   e. The circle which passes those three points obtained from a, b, and d processes is obtained.

(3) The arc of the chord is divided into quarters. It is examined that these two quadruple

Table 1   Ratio of omitted characters

Writer no.	ha	ma	su	to	ru	ni	no	wo	Av.
1	0.0	6.3	0.0	0.0	6.3	0.0	0.0	0.0	1.6
2	0.0	0.0	0.0	0.0	0.0	0.0	0.0	3.1	0.4
3	0.0	3.1	0.0	0.0	0.0	0.0	0.0	0.0	0.4
4	6.3	9.4	9.4	3.1	3.1	9.4	6.3	6.3	6.7
5	9.4	9.4	3.1	0.0	6.3	0.0	12.5	0.0	5.1
6	15.6	18.8	15.6	25.0	12.5	12.5	6.3	15.6	15.2
7	3.1	3.1	3.1	0.0	0.0	0.0	3.1	3.1	1.9
8	3.1	12.5	6.3	6.3	9.4	9.4	3.1	3.1	6.7
9	6.3	9.4	3.1	9.4	3.1	6.3	12.5	6.3	7.1
10	0.0	0.0	0.0	3.1	3.1	12.5	3.1	0.0	2.7
11	0.0	0.0	3.1	3.1	6.3	0.0	9.4	0.0	2.7
12	0.0	0.0	3.1	0.0	0.0	0.0	0.0	0.0	0.4
13	6.3	18.8	12.5	12.5	0.0	0.0	3.1	3.1	7.0
14	0.0	3.1	9.4	0.0	3.1	3.1	3.1	0.0	2.7
15	12.5	6.3	9.4	9.4	3.1	6.3	6.3	3.1	7.1
16	6.3	6.3	15.6	15.6	12.5	6.3	9.4	9.4	10.2
17	3.1	12.5	15.6	12.5	15.6	3.1	6.3	12.5	10.2
18	0.0	0.0	9.4	0.0	0.0	0.0	0.0	0.0	1.2
19	15.6	18.8	28.1	15.6	21.9	21.9	25.0	18.8	20.7
20	9.4	3.1	6.3	9.4	6.3	9.4	9.4	6.3	7.5
21	0.0	9.4	6.3	12.5	12.5	6.3	0.0	3.1	6.3
Av.	4.6	7.2	7.6	6.5	6.0	5.1	5.7	4.5	5.9

points are on the stroke or not.

(4) If two quadruple are both on a stroke, then radius of the curvature is calculated. Otherwise, next condition is examined.

The above procedure (1) to (4) is carried out for every character.

The curvature dictionary of the local arc was obtained from randomly selected 4 characters for each type of character as follows: The obtained curvatures were classified for 12 x 11 matrix, i.e. 12 directions (15 degrees interval) and 11 curvatures (from –5 to 5). From the principal component analysis of these matrixes, eigenvalues and eigenvectors were obtained to satisfy that the cumulative proportion was larger than 0.9. By weighting sum using these eigenvalues and eigenvectors, dictionaries were prepared for each type of character of each writer.

Same as the 2D Fuzzy membership function method, the combination of three kinds of characters were used for each chord length by the following equation:

$$\eta_i = 1 - (1 - \eta_{ij_1k_1})(1 - \eta_{ij_2k_2})(1 - \eta_{ij_3k_3}) \qquad (8)$$

where subscription i is chord length, j is a kind of character and k is character number.

Furthermore, we tested the combination of three kinds of chord lengths. The similarity is evaluated by the following equation:

$$\eta = 1 - (1 - \eta_9)(1 - \eta_{13})(1 - \eta_{17}) \qquad (9)$$

where $\eta_9, \eta_{13}, \eta_{17}$ indicate the similarities obtained from the chord length 9, 13, and 17 respectively. The both results for the inadequate characters of **Table 1** are listed in **Table 2**. The combination of three kinds of characters with three kinds of chord lengths gives 99.2% recognition ratio.

Table 2　Recognition ratio by new local arc method

(%)

Chord length	Av.
9	90.5
13	95.1
17	94.9
combination of chords	99.2

4. Proposed hybrid type method

In this work we proposed the hybrid writer recognition method, which is a combination of the previous method and the proposed local arc method. The schematic flow of the proposed method is depicted in **Figure 3**. The process is as follows:

(1) Basically 2D Fuzzy membership function method is applied.

(2) If there are not more than three kind of dictionaries, the new local arc method is used.

(3) If similarity values are small, i.e. less than 0.4, the new local arc method is used.

To examine the ability of the proposed writer recognition system, 24,840 characters (= 108 subjects x 46 kinds of characters x 5 times) were applied to the system. The average recognition ratio by the 2D Fuzzy membership function method was 97.8% and 3.6% characters were omitted as inadequate character. They were taking the new local arc method. Then by using proposed hybrid method, 88.2% of those omitted characters were recognized.

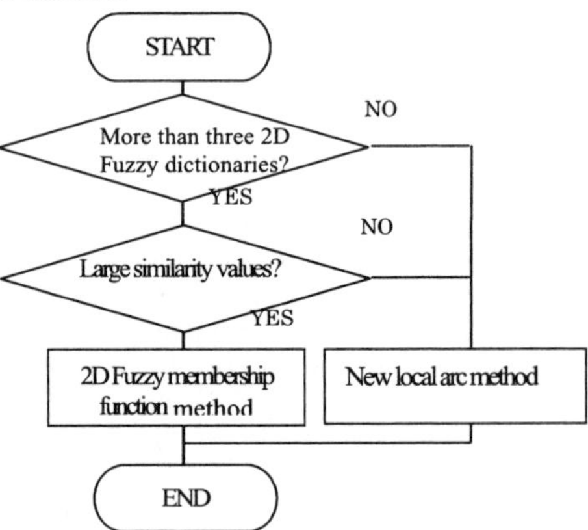

Figure 3　Schematic process flow of the proposed hybrid method

5. Conclusion

The previous method is very simple using 2D Fuzzy membership functions but has high recognition ratios.   However there also exclude many characters as inadequate characters. The local arc gives writers' characteristic features fairly well and the new local arc method gives good writer recognition ratio even for the omitted characters.   In this paper we proposed the hybrid type writer recognition system, and fairly good results were obtained with no omitted characters. By using the proposed method, only small number of characters is required to perform the writer recognition. Usually only three types of characters and 4 characters for each type of characters are required to prepare the private dictionaries. In the future we are going to extend this idea to signatures.

Reference

[1] M. Ozaki, Y. Adachi, N. Ishii, and T. Koyazu: Fuzzy CAI System to Improve Hand Writing Skills by Using Sensuous (1996) Trans. of IEICE Vol.J79-D-II NO.9 pp.1554-1561
[2] M. Ozaki, Y. Adachi, and N. Ishii: Writer Recognition by means of Fuzzy Similarity Evaluation Function (2000) Proc. KES 2000, pp.287-291
[3] M. Ozaki, Y. Adachi, and N. Ishii: Study of Accuracy Dependence of Writer Recognition on Number of Character (2000) Proc. KES 2000, pp.292-296
[4] M. Ozaki, Y. Adachi, N. Ishii and M. Yoshimura: Writer Recognition by means of Fuzzy Membership Function and Local Arcs (2001) Proc. KES 2001, pp.414-418
[5] M.Yoshimura and I.Yoshimura :   Writer recognition the state-of the art and issues to be addressed (1996) TECHNICAL REPORT of IEICE PRMU96-48 pp.81-90
[6]I.Yoshimura, M.Yoshimura: Writer Identification Using Localized Arc Pattern Method, Trans. of IEICE Vol.J74-D-II NO.2 pp.230-238

KES 2002
E. Damiani et al. (Eds.)
IOS Press, 2002

Detection of Face Direction from Fresh Color and Boundary Images

Yoshinori ADACHI[1], Saori TAKEOKA[1,2], and Masahiro OZAKI[2]
[1]Chubu University, [2]Nagoya Women's University
[1]1200 Matsumoto-Cho, Kasugai, Aichi, Japan 487-8501   *adachiy@isc.chubu.ac.jp*

Abstract: The new method for detection of the face direction is proposed. In this study, it was examined that to begin with, the face region was accurately extracted. In the previous paper, flesh color extraction only in the LUV space was done. In this paper, after it limited the region as an extraction object by carrying out from the flesh color extraction in the YCC space, by projecting it in the LUV space, flesh color extraction would be carried out in high accuracy. And, it is necessary to take out the face image of the size that is similar to the image that made the eigenspace in order to obtain the face direction by the eigenspace technique. Therefore, the characteristic of faces such as eye and nose, jaw had to be detected, and it would be possible to specify the face region at the comparatively good accuracy as the result which examined the flesh color distribution by combining with the edge image. This was adjusted to the size of the image that made the eigenspace, the scaling was done, and as a result of estimating the face direction, it would be able to be detected at the error of about 10%.

1. Introduction

From the viewpoint of security and man-machine interface, the human face recognition becomes an important problem, and many researchers have worked on this problem [1][2]. And, the detection of the face direction becomes a problem that is important from the marketing point of view.

Most of the works related to the human face recognition are using characteristics of face parts such as the relative locations or the shapes of eyes, mouse, nose and so on. Therefore it is very important to face parts correctly.

The most popular methods to detect the locations of face parts are using boundary information. For example, an edge-congested region is judged as an eye region [3]. However, by the environmental light change or the human face rotation, the edge image is not always made sharply.

There is a method using mosaic template matching [4]. But this has also difficulty in lighting and size changes. Therefore many restrictions are imposed on the face image taken. Usually a full face is taken for this purpose.

The three-dimensional (3D) object recognition or its direction detection from two-dimensional (2D) images is important in industry and has been studied by many researchers [5][6]. Most of them are based on the 3D structure of the object, because appearances of the 3D objects change so much by the object's direction or the lighting direction.

In this study, the face region was extracted from normally taken 2D image, and the automatic detection of face direction was examined. The color of the skin is different each human, and the color is variously changed by the makeup more and more. In addition, the color of the skin is greatly different by race and environment, so the flesh color is also quite different. It makes automatic face region extraction very difficult.

And, in the case of using the naturally taken image, it often happened that the size of the face is various and in which there are small hue differences between face region and background by the effect of lighting and shadow. These make the extraction difficult.

In this study, construction of the system that does the extraction at the good accuracy without being dependent on background and color of the skin, and detects the direction of the face is examined. On the extraction of the flesh color region, the improvement on the extraction method by the ellipse proposed in the previous research was tried.

The flesh color region is generously limited in the YCC color space, and this was made to be an original image to the ellipse extraction in LUV color space. The two-step process should squeeze the objective image. On the detection of the face direction, the eigenspace technique should be similarly used with the previous research.

2. Proposed method

The proposed method mainly consists of two parts. The face region extraction division and the face direction detection division are those.

2.1 Extraction of flesh color region

In the previous studies [7][8], we proposed the extraction method by using the ellipse region on the UV plane in the LUV space, and showed that the better results were obtained comparing with the existing methods. In this study, further improvement is added on the basis of the previous technique. Concretely, the flesh color region is extracted by next procedure.

Figure 1   Original image

(1) The original image shown in **Figure 1** is converted to the YCC color space from the RGB color space, and the region that satisfies next condition is assumed the flesh color region.

$$77 \le Cr \le 127 \quad \text{and} \quad 133 \le Cb \le 173 \tag{1}$$

(2) The histogram of the flesh color distribution is made, and the face existence candidate region is limited in the rectangle. These are shown in **Figures 2** and **3**.

(3) The face existence candidate region shown in **Figure 4** is mapped to the LUV color space from the RGB color space.

(4) The peak in flesh color region on the UV plane obtained beforehand was searched, and

Figure 2   Extracted image by YCC and its flesh color distributions.

Figure 3   Face existence candidate region.

Figure 4   Face existence image.

the region extracted by the ellipse centered at the peak coordinate was made to be a candidacy for the flesh color region.

Flesh color existence domain obtained beforehand is determined here by the next equation.

$$25 \le u \quad \text{and} \quad v \le \frac{1}{3}*u \quad \text{and} \quad 30^2 \le u^2 + v^2 \quad \text{and} \quad 75^2 \ge u^2 + v^2 \qquad (2)$$

The ellipse is decided here like the following.

a. In search of pixel distribution on the straight line that connects the peak on the UV plane of flesh color with the origin, the boundary with the region where 5% or less of the peak value continues is found, and it is made to be the length of the major axis.

b. In search of pixel distribution on the straight line that is orthogonalized to straight line that connects the peak with the origin and passes through the center of the major axes, the length of the minor axis is obtained in the same way.

c. The center of the ellipse is decided as the center of the minor axis and the lengths of the major and minor axes are already obtained in the above.

Decided ellipse is shown by the following equation.

$$\left[\frac{(U-P)\cos\theta + (V-Q)\sin\theta}{A/2}\right]^2 + \left[\frac{-(U-P)\sin\theta + (V-Q)\cos\theta}{B/2}\right]^2 = 1 \qquad (3)$$

where (P, Q) is the center of the ellipse and θ is a slope of the line connected the peak and the origin. A and B are the lengths of the major and minor axes respectively.

Examples of the decided ellipse on the UV plane and the extracted flesh color image from the face existence candidate image are shown in **Figures 5** and **6**.

(5) The position of eye and mouth is specified from the ruggedness of histogram of the flesh color region and histogram of edge image obtained by the Sobel method.

The candidate position becomes a concave in the flesh color histogram and a convex in the edge histogram as shown in **Figures 7** and **8**.

(6) In case of the profile, when the position of eye and mouth cannot be specified, the position of nose and jaw is detected from the edge image and the flesh color image

In the flesh color histogram, it is high at the nose, and it is different by the direction, the first flesh color will appear from the left end of the flesh color image, if it is the left direction.

And for the jaw, the flesh color histogram decreases (increase) and the gradient of it becomes large, and the histogram and the gradient of the edge image become large as shown in **Figures 9** and **10**.

(7) From the positions of eye and mouth or nose and jaw,

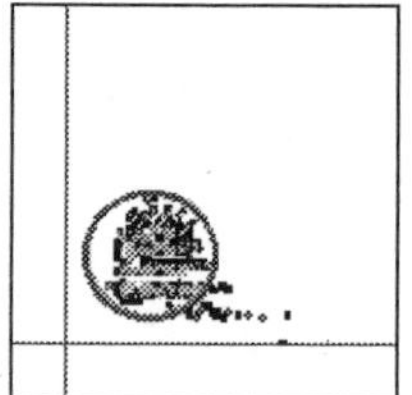

Figure 5   Ellipse obtained on UV plane.   Figure 6 Extracted flesh color image.

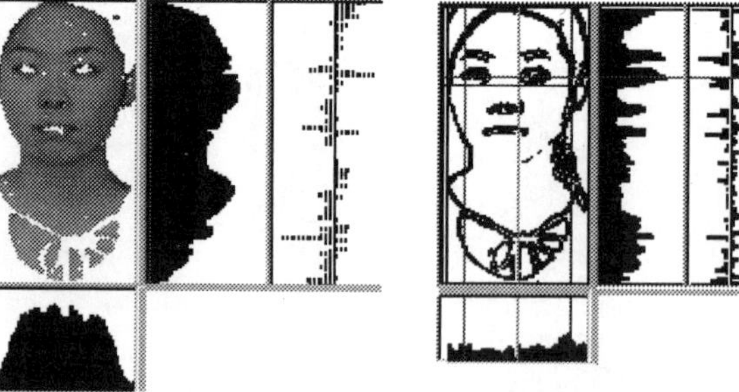

Figure 7   Concave in the histogram of flesh color image   Figure 8   Convex in the histogram of edge image

size and structure of the face are specified, and the face region is broached from the original image by centering the symmetry line of the face, which is guessed from the face size..

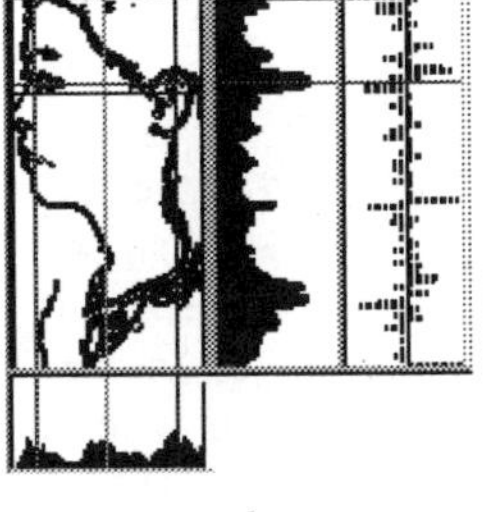

Figure 10 The jaw is detected at the large gradient and large histogram.

Figure 9 The location of jaw detected at large gradient.

2.2 Recognition of face direction

The recognition of the face direction was carried out by the method for specifying the angle by forming the eigenspace from rotational face image, and projecting the image on this space for the recognition, and comparing the agreement with the original images. Because of the skin color, face size and hairstyle, extracted face region is different one by one. Then in order to allow to some extent difference, many images were examined. The procedure of recognition processing is described in the following.

(1) 13 face images of the 15-degree interval are photographed from the right side to the left side for 13 persons. The background noise is removed in the manual and the average face image shown in **Figure 11** is made after normalizing size of 77 x 60 dots. The eigenspace is formed from the peculiar faces, which corresponded to the eigenvector, by the principal component analysis. The number of the eigenvector was automatically decided from the contribution ratios, in this case 0.9, and in most cases it becomes 3 to 7 dimensional space. The peculiar faces are depicted in **Figure 12.**

(2) By projecting the extracted unknown image in the eigenspace after adjusting its size the detection of face direction was carried out. Using the image gives the smallest distance, the angle of the unknown image is interpolated by a parabola with the angles of next images.

Figure 11   13 average face images.

Figure 12   3 dimensional eigenspace formed by 3 peculiar faces.

$\lambda = 0.34 \times 10^8$ $\lambda = 0.98 \times 10^7$ $\lambda = 0.40 \times 10^7$

3. Result s and discussion

This time, without carrying out the examination on many images yet, the reliability of the results is not sufficiently high enough, though in the many cases, this method gave good results. The agreement index of extracted face image is calculated by the following equation:

$$\eta = \frac{(s - s_1)}{(s + s_2)} \times 100 \, (\%) \tag{4}$$

where η is the agreement index, s is the ideal flesh color area, s_1 is omitted area, and s_2 is surplus area.   The η value shows about 75 to 95, and most of the flesh color region could be extracted.   And, it is confirmed that in the direction recognition, the eigenspace technique works fairly well to the manually extracted images within few degrees difference. The part of the results for the automatically extracted images shown in **Figure 13** is listed in **Table 1**, and it may have large error.   The large error is occurred by followings:

(1) The image is not sufficiently extracted by failing to take the position of jaws.

(2) The height and the direction of the face perpendicularly differ large.   In this case, it is necessary to form the eigenspace of the perpendicular rotation image also.

(3) The area of the face largely differs by the hairstyle.   It greatly affects the generation of the eigenspace.   Then, area adjusting may be necessary.

Figure 13   Sample of automatically extracted face images.

Table 1   Sample results of calculated face direction.

| | | (degree) |
Experimental	Calculated	Error
0	0.0	0.0
15	0.0	15.0
30	30.1	0.1
45	40.7	-4.3
60	43.3	-16.7
75	50.6	-24.4
90	111.9	21.9
105	137.3	32.3
120	153.3	33.3
135	168.9	33.9
150	178.0	28.0
165	179.6	14.6
180	179.96	0.04
Average error		10.3

4. Conclusion

The proposed YCC-LUV combination method works quite well for automatic face extraction. However, in face direction recognition by the eigenspace, it is quite sensitive to the condition setting of the recognition object.   Therefore quite accurate face extraction is necessary.   Or it must be required several eigenspaces obtained from the face parts.   In the future, such point will be examined.

Reference

[1] SONG X., LEE C.-W., XU G., and TSUJI S.: "Extraction of Facial Organ Features Using Partial Feature Template and Global Constraints", Trans. of IEICE D-II, J77-D-II, 8, pp.1601-1609 (1994).

[2] DOI M., CHEN Q., MATANI A., OSHIRO O., SATO K., and CHIHARA K.: "Lock Control System Based on Face Identification", Trans. of IEICE D-II, J80-D-II, 8, pp.2203-2208 (1997).

[3] AOYAMA K., YAMAMURA T., OHNISHI N., and SUGIE N.: "Estimating Face-and-Eye Direction from Images Taken with a Camera", Technical Report of IEICE, PRU95-233, pp.131-136 (1996).

[4] KOSUGI M.: "Human-Face Recognition Using Mosaic Pattern and Neural Networks", Trans. of IEICE D-II, J76-D-II, 6, pp.1132-1139 (1993).

[5] CHIN R. T. and DYER C. R.: "Model-Based Recognition in Robot Vision", ACM Computing Surveys, 18, 1, pp.67-108 (March 1986).

[6] BESL P. J. and JAIN R. C.: "Three-Dimensional Object Recognition", ACM Computing Surveys, 17, 1, pp.75-145 (1985).

[7] ADACHI Y., IMAI H., OZAKI M., and ISHII N.: "Study on Extraction of Flesh-Colored Region and Detection of Face Direction", Intl. J. of KES, 5, 2, pp.112-117 (2001).

[8] KAMIYA T., ADACHI Y., OZAKI M., and ISHII N.: "Recognition of Face Direction", KES'2001, Part 1, pp.637-641 (2001).

KES 2002
E. Damiani et al. (Eds.)
IOS Press, 2002

Recognition Studying States using Fuzzy Theory for Repeatedly Vocabulary Learning

Takayoshi YOSHIOKA, Kenichi OKUYA, Hitoshi NISHIZAWA

*Dept. of Electrical and Electronic Eng., Toyota National College of Technology
2-1 Eisei, Toyota, Aichi 471-8525, JAPAN*

Abstract. Web-based English Vocabulary self-learning system have needed to recognize learners study state automatically using fuzzy concepts in order to set the adequate questions for each learner's level. Thus it seems that the system is able to provide for efficient exercise. We introduced not only the degree of comprehension but also the rate of progress as the factors of fuzzy reasoning to infer learners' studying states.

1. Introduction

In this paper, the adaptive assessment of studying states under the domain of repeatedly Vocabulary Learning in the drill-style is discussed. With our Web-based self-learning system, we measured the learner's studying states using fuzzy concepts in order to set the adequate questions for each learner's level. With that function it seems that the system could make learners be able to do more efficient exercises. Since the method that we have implemented in our previous system, to evaluate learner's answers to the problems was too simple, just correct or not, so that the previous system had not provided the learners with very suitable problems. Then in order to measure a learner's degree of comprehension in more detail, we introduced the rate of right answer, that is not mere correct or not. Furthermore, in repeatedly learning like our system, study results are to be influenced with latest degree of comprehension and learning style such as learning attitude or learning period, so the study results become better or worse from latest study. Therefore we also introduced the rate of progress to measure a learner's studying states. Adopting the rate of comprehension and the rate of progress as the parameters of fuzzy sets to infer a learner's studying state, we made the system into more sensitive to for adaptive learning environment to the learners.

2. Background

In the actual working environment for engineers in Japan, communication in English is very important to accomplish their work. Above all, we think that knowledge of particular terminology and jargon words concerned with engineering subjects is essential for our graduates. Web-based technical English vocabulary learning system, which is Japanese-English word quiz, was developed in 1996 in our laboratory, aiming at providing individual learning for English vocabulary related to the learning history of each learner. The learning style is a kind of flash card, so learners are requested to answer the English words corresponding to the Japanese words which system provided.

On the other hand, an intelligent CAI (which stands for Computer-Aided Instruction) system needs student model that represents each learner's learning states to provide each learner with adequate level of exercise [1]. Some researches applied intelligent approach to get learner's learning states. For instance, Kato (1999) used bayesian network estimation for an adaptive practice in the drill-type CAI system [2]. Kalayar (1999) constructed the buggy model that represented as a bug library collected from learners practices to diagnosis student's state in the algorithm programming learning [3]. Hong (1999) used fuzzy classification to infer student's reading capability in the structured contents learning system [4]. Yoshine (1994) proposed fuzzy concepts to infer learner's comprehension level for stepwise understanding in CAL [5]. However most of researches use two parameters, comprehension level and answering time, to get student's state, in our system the answer time is meaningless to estimate learner's learning progress. Because our system requires a learner to type the spelling of an English word as answer of question, the typing speed and his/her comprehension level seems not to be related. We noted from log data that the spelling mistake became small as study of a learner progress, therefore we adopted the progress of the rate of right answer as a parameter for inferring the learner's state.

3. Web-based learning environment

In our learning environment every user has his/her home page, which is protected by the identification number and password to use the particular function [6]. For Japanese-English word translation quiz, which is one of the on-line functions in our learning environment, questions and answers of the exercises and learning history of the learners are stored on databases, and used to select the optimum question for each learner [7]. This system has been operated as combination 15-minute-long learning in the lectures which we call mini-exam with self-learning after classes. Registered students use a Web browser as their terminal software. The exercise is optimised for each student according to his/her learning history.

問4　再評価、もしくは、次問に進んで下さい（自習得点: 5917.3 + −0.8 ）

NO.	日本語	英単語	英単語(あなた)
1 ）	試運転（2語）[...n]	× trial run	try
2 ）	刃	○ blade	blade
3 ）	（酸素）分子	× molecule	
4 ）	気圧計	× barometer	barometre
5 ）	論理的な	○ logical	logical

Figure 1: An example of the learner's screen for Japanese-English Vocabulary Learning

Figure 1 shows an example of the technical English word exercise screen on the learner's terminal.

After we operated in actual lectures for 3 years to 150 students each year, we analysed that system has 2.2 times better efficiency than handwriting paper quiz [8]. Moreover learner can acquire most of 1000 technical English words that are stored on the system during 3-years-learning period. Even though the system provides students with some amount of learning effective, it's learning efficiency was not good very much compared with other research. A reason is that the system does not manage learner's each word learning history. Thus learners would have to answer the words that learners already learned, or the system would not provide the words that learners had not learned in our previous system.

Table 1: Studying state with transition condition

	studying state	transition condition
(1)	Initial	no answer from start
(2)	Learning	much partially correct
(3)	Forgetting	much partially wrong
(4)	Temporary Understanding	right answer at first time
(5)	Understanding	consecutive two-time correct answer

Table 2: Studying states transition matrix

	No answer	Partialy right (progress)			right answer
		minus	0	plus	
[start]	Initial	-	-	-	Temporary Understanding
Initial	Initial	-	Initial	Initial / Learning	Temporary Understanding
Learning	Forgetting	Fogetting / Learning	Learning	Fogetting / Learning	Temporary Understanding
Forgetting	Forgetting	Forgetting	Forgetting	Fogetting / Learning	Temporary Understanding
Temporary Understanding	Forgetting	Fogetting / Learning	-	-	Understanding
Understanding	Forgetting	Fogetting / Learning	-	-	-

4. Student model

In an intelligent CAI system, it uses the student model that presents each learner's learning states to provide him/her with adequate level of exercise. Student model usually consists of learners studying histories, such as the date and time when they use the system, points and problem number they did, the time between provided question and answering it, and so on.

4.1 Studying state

The learner's studying state is to be calculated with properties of student model. Thus to provide with adequate problems for learners' studying efficiently, teaching strategy can be adopted to each studying state. From the analysis of log data, in the drill-style repeating study, we assumed that learners could change their studying states that represent their comprehension levels as shown Table 1.

4.2 Transition of studying state

We have analysed the time-series of the rate of right answer that calculated with Levenshtein's distance between the right word and the word learner answered in order to get each learner's comprehension level of a learning word at a certain time. Besides, in repeatedly study the learners studying state could be influenced by their learning attitude, such as learning motivation and learning period, so that their learning attitude would be represented as the progress of the rate of right answer. Thus we also analysed learners' studying state transition by means of analysis of the progress of the rate of right answer that represented as the differential of the rate of right answer of a word between a learning-periods. The rate of right answer is calculated:

$$\textit{the rate of right answer} = (\textit{word length - Levenshtein's distance}) / \textit{word length} \qquad (1)$$

Table 2 shows the studying states transition matrix. In table 2, some of transitions are not clear. For instance, if a learner was at the forgetting state and he/she got plus progress, he/his studying state would be forgetting or learning depends on how progress he/she did. Thus we decided to use fuzzy reasoning to classify the learner's studying state in more detail from learner's studying histories.

4.3 Inference of studying states using Fuzzy reasoning

We introduced the rate of right answer and the rate of progress as the parameters of fuzzy sets to infer a learner's studying state. Table 3 shows the fuzzy rules for inferring learner's studying state. Figure 2 shows the fuzzy membership that induced from both fuzzy rules which shows on Table 3 and the results of analysis of the time-series of learner's studying state transition conducted from log data that have been operated for 3 years on our learning system.

5. Conclusion

In the drill-type learning system, we proposed using the rate of right answer and the rate of progress as the parameters of fuzzy sets to infer a learner's studying state. To provide each learner with adequate level of question, we implement the teaching strategy for efficient studies for each learner using the studying state that the system inferred automatically. Then system could be more sensitive to for adaptive learning environment to the learners.

References

[1] E.Wenger, Artificial Intelligence and Tutoring Systems, Morgan Kaufmann, 1987.

[2] Hiroshi KATO and Kanji AKAHORI, Item Selection Algorithm Based on Bayesian Estimation for an Adaptive Practice System, Japanese Trans. of the Institute of Electronics, Information and Communication Engineers, D-II Vol. J82-D-II No.1, 1999, pp.148-158 (in Japanese).

[3] Myat Kalayar, Hidenori IKEMATSU, Tsukasa HIRASHIMA, and Akira TAKEUCHI, Intelligent Learning Support System for Algorithm Learning, Proc. of Int. Conf. of Computers in Education (ICCE99), 1999, Vol.1 pp.840-843.

[4] Chao-Fu Hong, Yueh-Mei Chen, Yi-Chung Liu, and Tsai-Hsia Wu, Discuss 3D cognitive graph and meaningful learning, Proc. of Int. Conf. of Computers in Education (ICCE99), 1999, Vol.1 pp.240-243.

[5] K. YOSHINE, Y. ISOMOTO, and N. ISHII, Trial-and error Strategy for computer-assisted Learning in Fuzzy, Japanese Journal of Fuzzy Theory and Systems, 1994, Vol.5 No.5 P811-P822 (in Japanese).

[6] Hitoshi Nishizawa, Tsutomu Saito, and S. Pohjolainen, A Implementation of a Hypermedia Learning Environment for a Small Group, Trans. of Japanese Soc. for Information and System in Education, vol.15 no.4, 1999, pp249-253.

[7] Takayoshi Yoshioka, Shinya Kojima, Hitoshi Nishizawa, T. Saito, Individual Learning of Technical English Words on a WWW-based CAI System, 4th International Conference on Knowledge-Based Intelligent Engineering Systems & Allied Technologies (KES2000), 2000, pp397-400.

[8] Takayoshi Yoshioka, Shinya Kojima, Hitoshi Nishizawa, Tsutomu Saito, Educational Effectiveness for Learning of Technical English Words on a WWW-based CAI System, 6th International Conference on Soft Computing (IIZUKA2000), 2000, pp906-911.

Table 3: Fuzzy rules for inferring studying state

Rule No.	the rate of answer	progress (greatly)	class of studying state
1	High	Plus	Temporary Understanding
2	High	Minus	N/A
3	High	No changes	Learning
4	Low	Plus	Learning
5	Low	Minus	Forgetting
6	Low	No changes	?

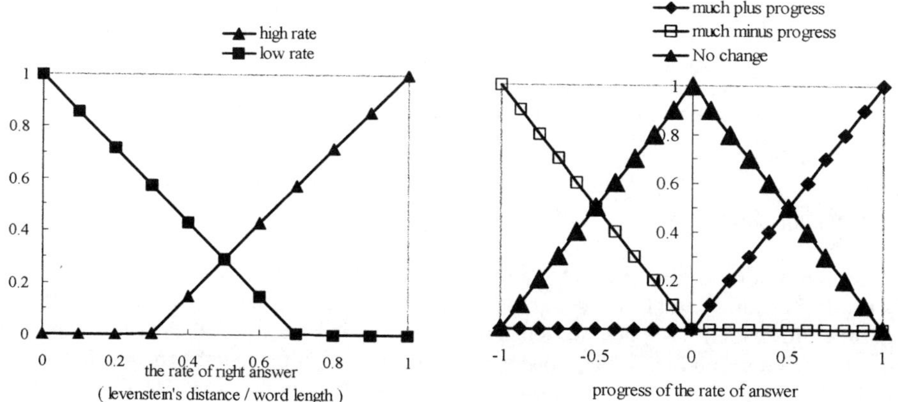

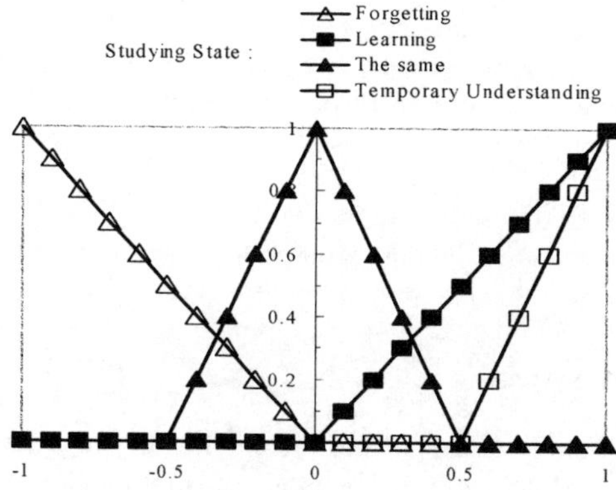

Figure 2: Fuzzy membership functions for the reasoning of the studying states

KES 2002
E. Damiani et al. (Eds.)
IOS Press, 2002

Self-supervised Learning for Sensory Integrating System

Hiromi Takeuchi, Koichiro Yamauchi†, Naohiro Ishii
Nagoya Institute of Technology, Nagoya, Japan
†Hokkaido University, Sapporo, Japan

Abstract. The living organism integrates sensory information from as eye and ear. The iintegration phenomenon was discussed firstly by McGurk. The sensory integrating system proposed has recognition capability using sensory integration. In this paper, we newly develop a sensory integrating system which has recognition algorithm for it. And, we propose a self-supervised learning algorithm for this system and interpret mathematically it using Bayesian method. Furthermore, we show that the phenomenon such as McGurk Effect can be interpret in the computer simulation.

1   Introduction

Almost all of species recognize object by integrating information from several sensor, such as eye and ear. McGurk[1] showed some examples of illusions caused by the integration of sensory information. The phenomenon is that the visual inferences upon the acoustic perception. For example, normal adults, who are seeing the shapes of the mouth pronouncing /ga-ga/ while hearing the sound /ba-ba/, they usually perceive the sound /da-da/. He also found that there was another effect. For example, we usually recognize the audio input /ka-ka/ with the visual input saying /pa-pa/ as /ka-ka/ or /pa-pa/ or their combination /ka-pa/. He interpreted that there are no middle class between /ka-ka/ and /pa-pa/, which shares the both features, so that we will ignore one side of sensory inputs.

From these points of view, we proposed sensory integrating system and an algorithm for the recognition, which is based on a Bayesian method. Using this method, the system evaluates whether the current output is suitable for the recognition result or not, and outputs a classification result clearly.

In this paper, we extend this system to be able to self-supervised learning. Furthermore, we show that the phenomenon such as McGurk Effect shows up in the result of computer simulation.

2   Sensory Integrating System

The detailed structure of the system[2] is illustrated in Fig.1. This system consists of several sets of multi-layered neural networks and sensors, each of which receives inputs from the corresponding sensor. We assume that all sensors of this system always observe a same object simultaneously. Each neural networks consists of two back-propagation networks, we call "Forward Networks" and "Backward Networks." The forward network receives inputs from

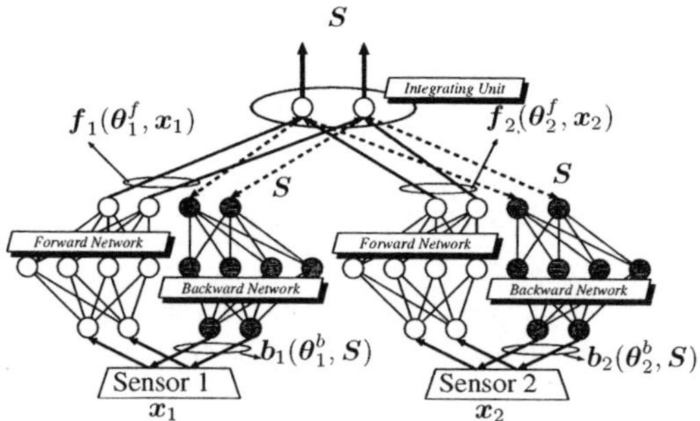

Figure 1: The Structure of Sensory Integrating System

the corresponding sensor, and predicts the class of the inputs. The outputs of all forward networks are sent to an integrating unit, which integrates all outputs of the forward networks. The integrating unit basically calculates the averaged outputs of all the forward networks. On the other hand, the backward networks try to reconstruct the sensory input from the output of the integrating unit. This is for calculating the confidence of each sensor. If the reconstructed input is quite different from the current input, the system tries to ignore the sensory input, and vice versa.

3 Recognition Algorithm

The system optimizes the integrated output and the value of the confidence of each sensor by repeating the forward and backward network calculations. The integrated output optimized is the final output of the system.

In the following text, let us denote the output from the integrating unit, a magnitude of confidence of the i-th sensor and the sensory input of the i-th sensor as S, c_i and x_i, respectively. The output of the forward network to the sensory input x_i is denoted by $f(\theta_i^f, x_i)$, where θ_i^f is the parameter vector of the i-th forward network. Similarly, the output of the backward network to the integrated outputs is denoted by $b(\theta_i^b, S)$, where θ_i^b denotes the parameter vector. In the following , however, θ_i^f and θ_i^b are usually omitted. Note that the default value of confidence c_i is 1. If the confidence c_i becomes 0, the i-th sensory input is ignored, otherwise the system references the i-th sensory input for the recognition.

The integrated output optimized is the final output of the system. The recognition algorithm was below.

1. Initialize confidence of all sensors c_i to 1.

2. Initialize S as the average of output of forward networks.

3. Set X_i to the backward networks output $b_i(S)$.

4. Update c_i

$$c_i := c_i + 2\beta(1 - c_i) - \gamma\|x_i - b_i(S)\|^2 \tag{1}$$

5. Update the vector of X_i as the equation below.

$$X_i := X_i + 2\gamma c_i(x_i - X_i) \tag{2}$$

6. Update S as the equation below.

$$S := S + 2\alpha \left\{ \frac{1}{m} \sum_i^m f_i[X_i] - S \right\} \tag{3}$$

7. Repeat Step 3 to Step 6 until S converges to a stable point. Therefore, if $\|S^{t-1} - S^t\| < \epsilon$, goto Step 3.

   where ϵ is a small positive constant and $\epsilon \ll 1$, S^{t-1} and S^t denote the outputs before and after the execution of equation (3).

Note that the forward networks predict the solution S from X_i directory so that S converges to an appropriate solution very quickly.

The detail of recognition and learning algorithm of this system is referred by [3].

4   Self-Supervised Learning Algorithm

In this section, we show the mathematical interpretation of self-supervised learning algorithm for sensory integrating system which proposed.

During the learning, the system modifies $\theta_1, \ldots, \theta_m$ and S^k to maximize this posterior probability

$$\prod_k^N P(S^k, \theta_1, \ldots, \theta_m | x_1^k, \ldots, x_m^k) = \prod_k^N \prod_i^m P(S^k, \theta_i | x_i^k), \tag{4}$$

where k denotes an index of each instance. θ_i denotes the parameter vectors of the i-th forward and backward networks, $\theta_i = (\theta_i^f, \theta_i^b)$. The right term of equation (4) is rewritten as the equation (5) using the well known Bayesian theorem.

$$\prod_k^N \prod_i^m P(S^k, \theta_i | x_i^k) \propto \prod_k^N \prod_i^m P(S^k, \theta_i) \prod_k^N \prod_i^m P(x_i^k | S^k, \theta_i) \tag{5}$$

where $P(S^k, \theta_i)$ is a prior probability of S^k, θ_i. We assume that the prior probability can be represented by the following equation.

$$P(S^k, \theta_i) \propto \exp\left(-\frac{1}{\sigma_S^2} \left\| S^k - f_i[\theta_i^f, b_i[\theta_i^b, S^k]] \right\|^2 \right) \tag{6}$$

Also, we assume the likelihood $P(x_i^k | S^k, \theta_i)$ is represented by the difference between the backward outputs and the sensory input.

$$P(x_i^k | S^k, \theta_i) \propto \exp\left(-\frac{1}{\sigma_x^2} \left\| x_i^k - b_i[\theta_i^b, S^k] \right\|^2 \right) \tag{7}$$

We redefine equation (5) as a potential U_L that is to be maximized during the learning. From equation (6)(7), we obtain

$$U_L = \sum_k^N \sum_i^m \left[-\alpha \left\| S^k - f_i[\theta_i^f, b_i[\theta_i^b, S^k]] \right\|^2 \right] + \sum_k^N \sum_i^m \left[-\beta \left\| x_i^k - b_i[\theta_i^b, S^k] \right\|^2 \right], \tag{8}$$

where α and β denote positive coefficients. Now, let us explain how to maximize the above potential function U_L. First of all, we denote the reconstructed sensory input $b_i[\theta_i^b, S^k]$ as X_i^k. Then, the above potential function can be simplified as follows.

$$U_L = \sum_k^N \sum_i^m \left[-\alpha \| S^k - f_i[X_i^k] \|^2 \right] + \sum_k^N \sum_i^m \left[-\beta \| x_i^k - X_i^k \|^2 \right] \tag{9}$$

From equation (9), we can see that X_i^k, which maximizes equation (9), is the current sensory input x_i^k. If we set X_i^k to x_i^k provisionally, the integrated output S^k, which maximizes U_L, can be represented as a function of x_i^k as follow.

$$S_*^k = \frac{1}{m} \sum_{i=1}^m f_i[\theta_i^f, x_i^k] \tag{10}$$

Therefore, the system maximizes the potential function U_L, whose S^k and X_i^k are substituted by S_*^k and x_i^k, respectively, as below. In the following text, we denote the substituted version of U_L as U_L'.

$$U_L' \equiv \sum_k^N \sum_i^m \left[-\alpha \left\| S_*^k - f_i[\theta_i^f, x_i^k] \right\|^2 \right] + \sum_k^N \sum_i^m \left[-\beta \left\| x_i^k - b_i[\theta_i^b, S_*^k] \right\|^2 \right] \tag{11}$$

Therefore, the modifications rule for the parameter are:

$$\theta_i^f := \theta_i^f - \epsilon \nabla_{\theta_i^f} U_L' \tag{12}$$

$$\theta_i^b := \theta_i^b - \epsilon \nabla_{\theta_i^b} U_L' \tag{13}$$

5   Computer Simulation

This section shows a computer simulation to evaluate the proposed system.

In this simulation, we assume that the system has two sensors. All neural networks have the same structure. This simulation is conducted in the following procedures.

1. Make the system learn using the proposed algorithm and output the output of an integrated unit.

2. Investigate the output of this system when the sensor inputs are given. Here the inputs are in the different category, when the first simulation is carried out.

Data sets used are 2-dimensional patterns shown in Fig.2. There are 5-classes of patterns in the data set. These points are generated by normal distribution whose average and variance are the center point of each class and 0.05, respectively. The number of data points is 250 per each class.

Fig.3 is the result of this simulation. This figure shows that this system is carrying out the category classification of the data.

From the result of Fig.3, we pick up two data sets which are class A in sensor 1 and class C in sensor 2. Doing in this way, we create inconsistent data sets from two sensors These data

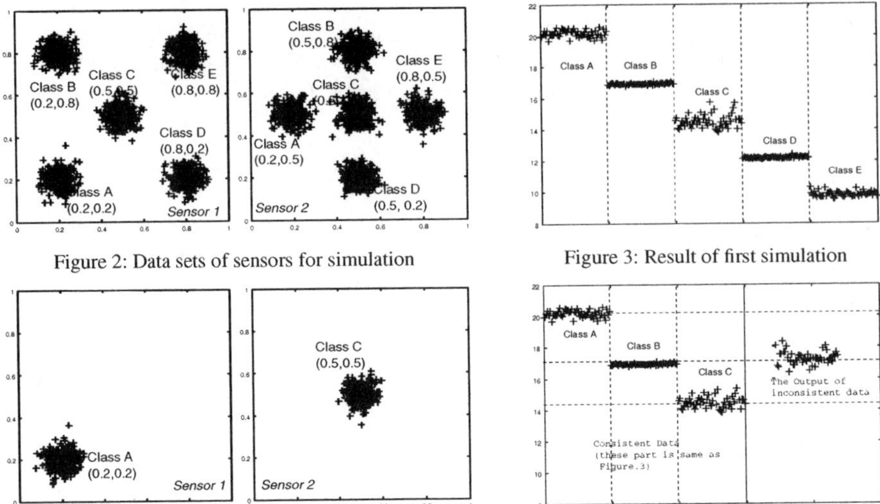

Figure 2: Data sets of sensors for simulation

Figure 3: Result of first simulation

Figure 4: Inconsistent data sets

Figure 5: Result of second simulation

sets are given simultaneously to the system as the sensor input. Fig.4 illustrates the sensor inputs.

Fig.5 shows the output of proposed system when the inconsistent data was inputted. As shown in this figure, the output value is same as the middle class B between class A and C. Therefore, we show that the phenomenon such as McGurk effect appears in the output of proposed system.

We believe these results suggest that the proposed system shows the same behavior as human beings, which McGurk firstly found.

6   Conclude

This paper has proposed self-supervised learning algorithm for sensor integration system. In the simulation, we show that the system can do self-supervised learning. Moreover, we show that the system behavior is similar to that of human being, which McGurk firstly found.

By imitating the sensory integration technique of the living organism, more advanced recognition and learning are considered to become possible.

References

[1] Harry McGurk and John MacDonald. Hearing lips and seeing voices. *Nature*, 264:746–748, 1976.

[2] Koichiro Yamauchi, Mikiya Oota, and Naohiro Ishii. A self-supervised learning system for pattern recognition by sensory integration. *Neural Networks*, 12(10):1347–1358, 1999.

[3] Hiromi Takeuchi, Yoshifumi Terabayashi, Koichiro Yamauchi, and Naohiro Ishii. An avoiding method for autonomous robots based on sensory integrating system using neural networks. In *SNPD'01*, pages 885–891, 8 2001.

KES 2002
E. Damiani et al. (Eds.)
IOS Press, 2002

A Framework for a High Level User Interface for Accessing Dynamic Contents on the Web

Subhash BHALLA, Masaki HASEGAWA and Nadia BERTHOUZE

Department of Computer Software,
The University of Aizu, Aizu-Wakamatsu,
Fukushima PO 965-8580, (Japan)

Abstract - Most users of the web are highly skilled at referring to tabular data. A natural inclination exists among the users for object-by-object traction to find information. We propose a convenient approach for the web information systems users. The users are assumed to be not skilled at using traditional database management system query interfaces such as the SQL language. The proposed interface introduces simplicity and avoids communication ambiguities.

1   Introduction

Existing web-based information systems (WEBIS) organize information for access by users. These systems adopt a 'page and related links' approach for access to data resources. Users select a related link as per their need. The use of such a hierarchical navigation helps the users in their search for the required piece of information. For example, many e-commerce sites use hierarchical navigation for items on the shopping lists, as shown here,

Computer Hardware − > Parts − > Hard Disk Drives − > Vendors.

Such an approach provides a limited support for a user's quest for information. The complexity of hierarchical navigation increases with increase in volume of data and information contents. Many users take a long series of steps to process a simple query. Also, complex queries that combine multiple options can not be performed within a single step.

In view of the above difficulties, many WEBIS provide 'input form and search' approach. The users are permitted to input selected key-words. Based on the inputs, the web service attempts to locate the information contents for meeting the users' needs. The 'input form' approach helps to reduce the search difficulties that are posed by hierarchical navigation.

Many web services plan to construct and use database management systems (DBMSs) services as a basic component. In the present environment, we compare the suitability of these approaches, and consider a possible alternative based on user's convenience. The aim of the proposed study is to provide an improved user interface for navigation and to find the information content as per the user's needs.

2 Background : Available Approaches

Most of the currently available WEBIS sites, support fixed set of items that describe the information contents of the page (**static contents**). Gradually, few WEBIS are beginning to support database access facilities in application domains, such as financial services and databases of chemical and biological information. These allow the web-clients to have access to DBMS resources supported over the web links. This kind of an access supports a flexibility to form and execute queries as per the users requirements (**dynamic contents**). For example, a user may query a database of suppliers of parts and form queries in database query languages, such as SQL. A complex query requiring multiple combinations on available data 'to find compatible printers and PCs' can be expressed by the user [3]. In contrast, the earlier static contents based systems force the users to depend on the options available within the displayed page contents. Thus, in order to support query navigation for web-clients, existing WEBIS may depend on the following possibilities.

- Hierarchical navigation
- Input form based content access networking
- Database query languages

The hierarchical navigation and input form based approaches are oriented towards static contents. The use of database query languages poses a new problem for the users as most web users are not skilled at the use of database query languages. In order to support dynamic contents, a web user interface is needed. The user interface is required to be simple [1]. With the given scenario of growing number of web users, it is reasonable to assume that programming and querying skills at the level of most of the web users, are low. In the present proposal, a web user communicates through a web interface system which in turn communicates with a relational database management system (RDBMS). The user intent is captured by using step-by-step navigation though a graphical interface. By identifying objects and a data operation, it is converted into its SQL equivalent by the interface.

3 Motivation

Database contents can be described a as collections of relations. Most users are highly skilled at gathering the information of their interest from tables, as in the case of air-line or train time-table information. We propose a navigation approach called Query-by-Object (QBO). The users of such an interface are restricted (within a step) to query at most one (or two entities or objects). The chosen objects are operated upon by use of relational algebra operations. The above simplification helps in removal of ambiguities, which can be eliminated with the help of a menu and system prompts. At the support level most of the queries steps can be dynamically expressed using simple SQL constructs at each step. The proposed interface supports basic algebraic operations as a language. The users of such a system are not required to acquire query (programming) skills prior to making an access to the dynamic contents in a web based information system.

Considering the earlier hierarchical navigation example, Computer Hardware − > Parts − > Hard Disk Drives − > Vendors, each of the above item can be stored in the form of relational tables as a part of a DBMS. Skilled users of the

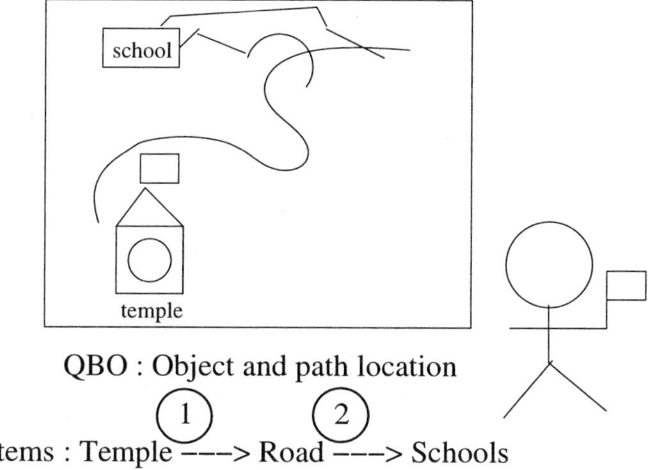

Figure 1: Query-By-Object examples

DBMS can adopt the query language SQL for finding information [3]. However, most users of the WEBIS are not skilled at using the SQL language. In lieu of the SQL, we propose an alternative programming interface, that offers a simple interactive query language interface. The proposal is based on the following considerations -

- Objects can be stored as per the entity-relationship (E-R) model approach. The user may select an object and use a unit step algebra operation to proceed with a query. Also, relational algebra operations such as **Union, Intersection, Cartesian product** and **Set-difference** can be used in conjunction with a single entity or a pair of entities.

- A graphical user interface can provide communication support to user based on description of tables and available options. The simplicity and system prompts also help to eliminate ambiguities.

4   Web Interface for Accessing Dynamic Contents

4.1   Query-By-Object (QBO) Approach

In the case of a person who reads from a map (Figure 1), the user may choose an object such as a temple (or locate many temples), then switch to another object and observe the roads connecting the object to a school. Similarly, in the proposed approach, a notional bag is associated with each query. A user may select database entity (entity-set) from a list. The object forms the content of the query bag. The granularity may be refined for the chosen entity-set by confining to a few selected entities or to restricting the contents to a chosen attribute.

The underlying database contents are organized as per the entity-relation-ship design. In response to the users choice of the contents of the bag, the system can provide menu support and prompts based on the possible operations and options. At the operations level, for unit entities, relational algebra operations are supported to provide a complete language [2]. The user's intent expressed through a step-wise series of steps is dynamically converted at each step into SQL statements that are serviced by the underlying database management system (DBMS) (Figure 2).

4.2 Step-wise Navigation System

At the lowest level, in the proposed scheme, the database contents are organized as well defined entities and relationships, as per the Entity-Relationship approach, as database relations. In order to support conventional queries, a Database Management System (DBMS) is used. The DBMS software provides a SQL query support for executing queries. A query interface that supports object access and simple relational algebra functions is supported at the user level by the WEBIS. The user interacts with the QBO interface. The user intents are dynamically converted into SQL statements that are executed by the DBMS. The conversion is facilitated by the use of algebraic operations, as these are common to both SQL and QBO. In this way, traditional relational DBMS techniques are utilized for concurrency control, recovery and data security by the WEBIS.

4.3 Comparision with Other Approaches

Existing WEBISs support ad-hoc approaches for providing query navigation by web-clients. With the increase in volume of data and information resources, adoption of the DBMS support for web based queries is becoming unavoidable. Considering the approaches highlighted earlier, the hierarchical navigation and input form based approaches are **static content** oriented approaches. These make the use of an information system based on a DBMS, to be restricted in terms of accesses [4], [5].

The use of SQL as a medium of querying the web data and information is a recent development. It permits the user to express his/her intent through the medium of database query support. However, most users of DBMSs are skilled programmers that access these resources through the use of SQL. Many users who are not skilled in SQL, use the services of a DBMS by depending upon an easy interface, such as an input form. In order to meet the challenge of providing a powerful and easy to use support, the Query-By-Example (QBE) approach has been explored by many researchers [6]. QBE is simple to use. It requires less experience and training. The main drawback of QBE is that it is good for expressing simple queries, but expression of complex queries involving more oprtions and object combinations is difficult.

The proposed QBO approach permits expression of user queries with ease, at the same level as that of the SQL language. The users locate the path and the objects using a step-wise (object-wise) navigation supported by the system menu. Similar to use of SQL for a relational database, the users can adopt a navigation by object and operations.

In contrast with QBE, the proposed approach restricts the user from accessing many objects. It attempts to introduce simplicity and aims at elimination of ambiguities withing the step-wise processing sequences. To illustrate the contrast between QBE and and the proposed approach, consider the following example of a complex queries for the given relational scheme,

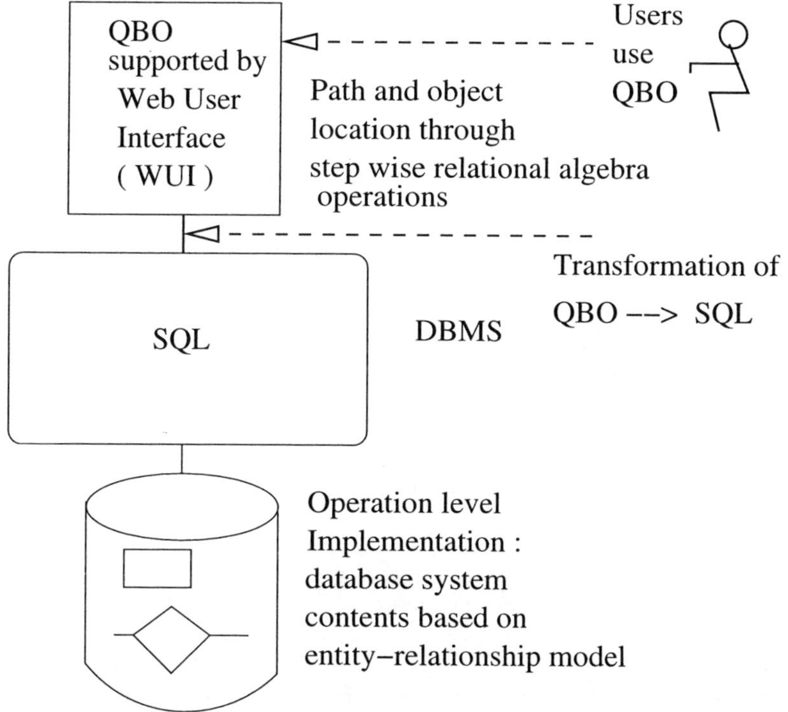

Figure 2: A WEBIS query system based on relational algebra.

considering Supliers, Parts and SP (Quantity-Supplied).

Example 1 - Find suppliers who also supply the parts supplied by a supplier in city $< Paris >$.
In case of QBE the user needs to correctly perform the following steps -

- selection of database table,
- define variables over attributes of chosen tables,
- linking of equality conditions among variables between multiple tables.

QBE -

S	S#	SNAME	STATUS	CITY
	_x			Paris

SP	S#	P#
	_x	_Y

SP	S#	P#
	.P	_Y

Proposed Approach - QBO

Using the proposed approach a user in the first step, gathers suppliers located in Paris.

Step 1. Gather object A, as suppliers located in Paris

(Query bag : $< suppliers >$ in Paris).
Step 2. Gather object B, as parts supplied by the suppliers
(Query bag : $< suppliers >$, $< Parts >$ supplied by the suppliers) from table SP.
Step 3. Redefine bag contents as object A
(Query bag : $< Parts >$ supplied).
Step 4. Gather object B, as supplier who supplied the parts
(Query bag : Parts $< suppliers >$ who supplied these parts) from table SP.

In contrast to the QBE approach, users of QBO find it natural to consider the steps -

1. Which suppliers are from Paris

2. Which parts do they supply

3. Who also supplies these parts

5   Summary and Conclusions

With the advent of the web based information systems, it has become necessary to support a high level language for user interactions. The medium of user interaction must allow the user to express DBMS queries. The proposed system presents a high level language for user interaction for DBMS applications that are supported through the WWW. The step-wise navigation in the proposed language is based on tracking objects and paths logically and is supported by the SQL support provided by a RDBMS. The proposed QBO approroach is intuitive. It is relationally complete language as it is based on supporting relational algebraic operations. It does not require the users to obtain programming skills prior to accessing the WEBIS resources. The users can perform table lookup oriented navigation in unit steps, which is a logical approach.

References

[1] T. Abraham and J.F. Roddick, "Survey of Spatio-Temporal Database," *GeoInformatica*, vol. 3, no. 1, pp. 61-99, March 1999.

[2] D.D. Chamberlain, et al., "SEQUEL 2: A Unified Approach to Data Definition, Manipulation and Control," *IBM Journal of Research and Development*, vol. 20, no. 5, pp. 560-575, Nov. 1976.

[3] A. Labrinidis and N. Roussoppoulos, "Generating Dynamic Content at Database-backed Web Servers: cgi-bin vs mod_perl," *SIGMOD Record*, Vol. 29, No. 1, March 2000.

[4] C. North and Ben Shneiderman, "Snap-Together Visualization: A User Interface for Coordinating Visualizations via Relational Schemata, In Proceedings of the 5th International Conference on *Advanced Visual Interfaces (AVI 2000)*, published by ACM, New York.

[5] I. Schlaisich and M. Egenhofer, "Multimodal Spatial Querying: What People Sketch and Talk About", 1st *International Conference on Universal Access in Human-Computer Interaction*, New Orleans, LAC. Stephanidis (ed.), 2000.

[6] A. Silberschatz, H. Korth, and S. Sudershan, "Database System Concepts," *McGraw-Hill Book Company*, 2002.

KES 2002
F. Damiani et al. (Eds.)
IOS Press, 2002

Pragmatic Considerations for Human-Centeredness

Rajiv KHOSLA and Petrus USMANIJ

School of Business, La Trobe University
Melbourne, VIC 3086, Australia

Abstract – This paper describes how some areas in computer science and information systems are evolving or moving towards human-centeredness. These areas include intelligent systems, electronic commerce, software engineering, multimedia databases, data mining and enterprise modelling. This evolution is based on the need for addressing pragmatic issues in these areas. These pragmatic issues and their impact on various areas helps us to identify a set of critical properties that need to form a part of a human-centered system development framework

1. Introduction

Human-centered system development is not a revolutionary concept in computer science and information systems but an evolutionary and enabling one. In this paper we show how some disciplines in computer science and information systems are evolving towards human-centeredness. Section 2 describes the pragmatic issues related which are constraining the evolution of disciplines like intelligent systems, software engineering, e-commerce and others. Section 3 concludes the paper.

2. Pragmatic Considerations

The pragmatic considerations represent the practical problems associated with use of various technologies. These practical problems are underpinned in epistemological limitations (and strengths) which human and computers have, human vs. technology mismatch, relationship between technology and people and the impact of this relationship on the use of technology, social, organizational and task context in which technology is used, and some others. This section explores these issues in intelligent systems, software engineering, electronic commerce and enterprise modeling.

2.1. Intelligent Systems and Human-Centeredness

The four most commonly used intelligent methodologies in the 90's are symbolic knowledge based systems (e.g. expert systems), artificial neural networks, fuzzy systems and genetic algorithms.

Symbolic knowledge based systems have served varied purposes in industry and commerce during the last three decades. The most widely used versions being expert systems which have found their way into industry and commerce, including manufacturing, planning, scheduling, design, diagnosis, sales and finance. In these applications, the unitary architecture of production rules, normally enhanced by frames or objects, has been used to capture human expertise and to solve different problems. However, practitioners have also identified some of the limitations of symbolic knowledge based systems, such as slow and constricted knowledge acquisition processes, inability to properly deal with imprecision in data, inability to process incomplete information etc.

The computational and practical issues associated with the four intelligent methodologies have made the practitioners and researchers look at ways of hybridizing the different intelligent methodologies from an applications viewpoint. However, the evolution of hybrid systems is not only an outcome of the practical problems encountered by these intelligent methodologies but is also an outcome of deliberative, fuzzy, reactive, self-organizing and evolutionary aspects of the human information processing system (Bezdek, 1994).

Intelligent hybrid systems can be grouped into three classes, namely, fusion systems, transformation systems, combination systems (Khosla, 1997). As fusion, transformation, and combination architectures have been motivated by and developed for different problem solving tasks/situations, it is useful to associate these architectures in a manner so as to maximize the quality as well as range of tasks that can be covered. These classes of systems are called associative systems (or associative hybrid systems) as shown in Figure 1.

The selection of these technological primitives is contingent upon satisfaction of task constraints (e.g. presence/absence of domain knowledge, noisy incomplete data, learning, adaptation, etc.) which in Figure 1 have been grouped under the quality of solution dimension.

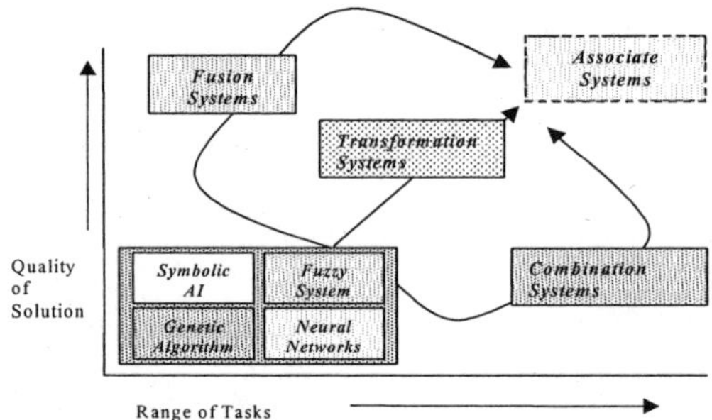

Figure 1: Intelligent Task-Centered Associative Systems

2.2. Software Engineering and Human-Centeredness

Software engineering can be defined as a layered technology that encompasses a definition phase, a development phase, and a maintenance phase. The development phase involves analysis, design, implementation and testing phases, respectively.

The traditional structured analysis and design methodology involves functional modeling using data flow methodology. Transformation and transaction flow methods are used to convert the data flow analysis into structured (architectural) design. One of the problems with the structured design methodology is the non-isomorphic transition between analysis and design phases. It can result in loss of information between analysis and design phases. The object-oriented methodology alleviates this problem by retaining the same vocabulary of objects and classes in the analysis and design phases respectively. More recently design patterns have been added to the armor of O-O methodology (Gamma et al., 1995; Pree, 1995; and others). Further, the vocabulary of design patterns is primarily suited to meet the needs of software designers and not users. Moreover, from problem solving and human-centered perspectives, O-O paradigm does not provide an intuitive means of modeling goals and tasks.

The design pattern construct has highlighted the fact that software engineering development process is not necessarily analogous to product development process in more mature traditional engineering disciplines. However, their definition suggests a technology-based focus (i.e., they are likely to carry the limitations of the technology along with them), which makes them somewhat unsuitable for human-centered systems.

Unlike O-O methodology, agent methodology is primarily driven by problem solving, tasks and task-based behavior (Jennings et al., 1996; Khosla and Dillon, 1997; and others). However, besides their collaborative and task characteristics, they lack the structural characteristics of the O-O methodology. Most of the agent based systems on the internet and in the field (till very recently) model tasks based on logic. This is at cross roads with the human-centered objectives, like activity-centeredness, focus on practitioner's goals and tasks, and the need to model tasks based on how users accomplish them rather than force fit a particular technology (like AI logic) on to the tasks.

Thus although, the three distinct approaches discussed in this section have made significant contributions towards achieving the primary goal of software engineering, namely, software quality, it is apparent they have not contributed in the same vein towards human-centeredness. The technology-centeredness of these approaches constrains them to model all aspects of a domain using a particular technology. This undermines to some extent the syntactic and semantic quality of a computer based artifact (software system) from a human-centered viewpoint.

2.3. Electronic Commerce and Human-Centeredness

Electronic Commerce (eCo) can be broadly seen as the application of information technology and telecommunications to create *trading networks* where goods and services are sold and purchased, thus increasing the efficiency and the effectiveness of traditional commerce (Figure 3). Many software architectures have been proposed for supporting such networks in the past few years; some of them are presented, addressing issues such as sales, ordering and delivery of products in the framework of the global internet (Hands, et al., 1998).

Indeed, internet based electronic commerce is currently a driving force behind the evolution of many *Web-based technologies* such as HTTP, HTML, Java, CGI and others (Hamilton, 1997), all of which were originally conceived for different applications.

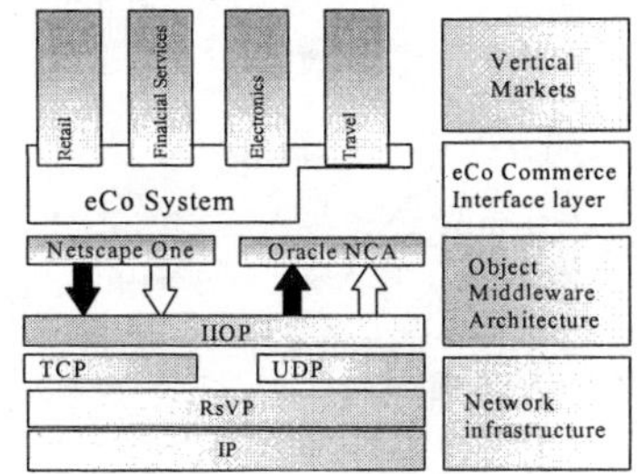

Figure 3: Electronic Commerce Framework

The eCo system framework overcomes a long-standing barrier to the development of electronic commerce, as XML documents provide, at least in principle, an incremental path to business automation, whereby browser-based tasks are gradually transferred to software agents. This development might allow traditional supply chains to evolve into open markets, while agents interact with business services through object documents.

However, unlike traditional commerce, the existing electronic commerce architectures on the internet are supplier-centered. XML 's human readability, while an advantage over CORBA (Glushko, 1999), does not eliminate the risk of a supply-side market model, where the structure and content of metadata are modeled w.r.t. the needs of vendors and distributors, leaving it to the brokers to transform them into a form more suitable for the buyers. In this setting, nearly every electronic commerce purchase is preceded by a *network search* or *product brokering* phase (Tenenbaum et al., 1998) when the customer navigates the trading network looking for the needed products or services. However, the avalanche of on-line suppliers and multimedia information about goods and services currently available, makes it difficult to locate, purchase and obtain the desired products at the best prices. General-purpose internet search engines seem wholly unfit for this task.

A solution to this problem is the definition of a human or consumer-centered brokerage architecture which will locate all the vendors carrying a specific product or service, then query them in parallel to locate the best deals.

user perspectives associated with data in the data mining process.

2.4. Enterprise Modeling and Human-Centeredness

Most organizations, including businesses, government agencies, industrial firms, and hospitals, now depend on computers as an integral part of their operations.

Whatever form of information technology is utilized, the fact is that the management of business relies heavily on information throughout the business process where data, information, and knowledge are the three main resources to support that business process. In terms of organizational levels, the demand for data is very high at the operational level as shown in Figure 3. The demand for information and knowledge increases as we move up from operational level to the strategic level. This is because degree of unstructuredness of problems increases as we move up from operational level to the strategic level. Further, like information systems exist at all levels, there is also evidence that intelligent systems exist at all levels. The evidence can be seen with the development of intelligent systems such as intelligent air line reservation systems at the operational level (Nwana and Ndumu, 1997), intelligent e-mail and news management systems at knowledge work level (Maes, 1994), intelligent production scheduling systems at the management level (Hamada et al., 1995), and intelligent forecasting and prediction systems at the strategic level (Khosla and Dillon, 1997a).

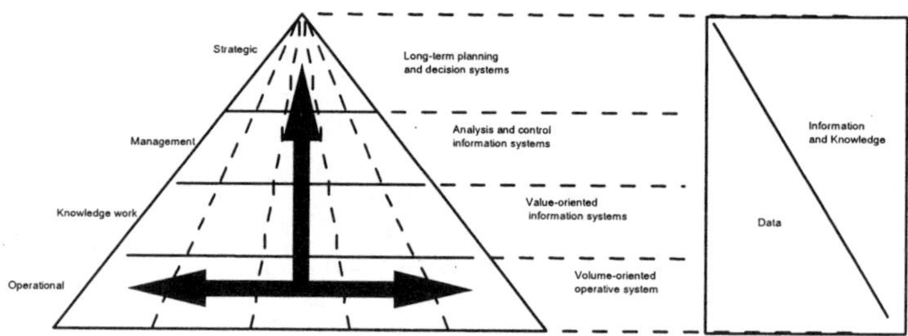

Figure 4: Applicability of Data, Information and Knowledge w.r.t. Organizational Level

Database systems are modeled using data abstraction technologies like entity – relationship diagrams, information systems rely on technologies based on functional abstraction like data flow diagrams, and intelligent systems rely on technologies like expert systems, fuzzy-logic, neural networks, and genetic algorithms which model different aspects of human-cognition, brain and evolution. These three types of systems exist at all levels of an enterprise. This has created an obvious need for integration and interoperability of these systems. However, the technology mismatch is the one of the major obstacles today for their integration and interoperability. More so, fields like knowledge discovery and data mining have demonstrated that one type of system (i.e. a standard database system with DBMS capabilities) can evolve into another type of system (i.e. an intelligent system) to provide sophisticated intelligent decision support. Thus, there is a need to build enterprise systems which are problem driven and which have capabilities to evolve with time.

3.   Conclusion

Humanization of technology has become an important research issue. This paper highlights the pragmatic issues   driving the evolution of intelligent systems, software engineering, electronic commerce and enterprise modeling. The authors have designed and developed several systems which address pragmatic issues in the above disciplines.

References

[1] Bezdek, J.C. 1994, "What is Computational Intelligence?" *Computational Intelligence: Imitating Life*, Eds. Robert Marks-II et al., IEEE Press, New York.

[2] Gamma, E et al., 1995, "Design Elements of Object-oriented Software", Massachusetts: Adisson-Wesley.

[3] Glushko R., Tenenbaum J. and Meltzer B., 1999, "An XML-framework for Agent-Based E Commerce", Communications of the ACM, vol. 42 no. 3.

[4] Hamada, K., Baba, T., Sato, K. and Yufu, M., 1995, "Hybridizing a Genetic Algorithm with Rule based Reasoning for Production Planning", *IEEE Expert.* 60-67.

[5] Hamilton S., 1997, "Electronic Commerce for the 21st Century", IEEE Computer, vol. 30, no. 5, pp.37-41.

[6] Hands, J., Patel A., Bessonov, M. and Smith, R., 1998, "An Inclusive and Hands Extensible Architecture for Electronic Brokerage", Proc. of the Hawai Intl. Conf. on System Sciences, Minitrack on Electronic Commerce, pp.332-339.

[7] Jennings, N. R. et al., 1996. "Using Archon to Develop Real-World DAI Applications, Part 1." *IEEE Expert* December, 64-70.

[8] Khosla, R., 1997, "Tutorial Notes on Software Engineering methodology for Intelligent Hybrid Multi-Agent Systems", Int. Conf. on Connectionist Information Processing and Information Sys, Dunedin, New Zealand, November.

[9] Khosla, R. and Dillon, T.S., 1997a, "Engineering Intelligent Hybrid Multi-Agent Systems", Boston, USA, Kluwer Academic Publishers.

[10]Khosla, R. and Dillon, T.S., 1997b, "Neuro-Expert System Applications in Power Systems,". K. Warwick, A. Ekwue & R.K. Aggarwal, eds. Artificial Intelligent Techniques in Power Systems. UK, IEE press, pp. 238-58.

[11]Khosla, R. and Dillon, T., 1997c, "Task Structure Level Symbolic-Connectionist Architecture," chapter in Connectionist-Symbolic Integration: From Unified to Hybrid Approaches, edited by Ron Sun and Frederic Alexandre, Lawrence Erlbaum Associates in USA, pp. 37-56, November 1997.

[12]Khosla, R. and Dillon, T., 1997d, "Fusion of Knowledge-Based Systems and Neural Networks and Applications", Keynote paper. *1st Int. Conf. on Conventional and Knowledge-Based Intelligent Electronic Systems.* Adelaide, Australia, May 21-23, pp. 27-44.

[13]Khosla, R. and Dillon, T., 1997e, "Learning Knowledge and Strategy of a Generic Neuro-Expert System Arch. in Alarm Processing," in IEEE Trans. on Power Systems, Vol. 12, No. 12, pp. 1610-18, November.

[14]Maes, P., 1994, "Agent That Reduce Work and Information Overload," Communications of the ACM, July, pp. 31-40.

[15]Nwana, H. S. and Ndumu, D. T., 1997 "An Introduction to Agent Technology," In Nwana, H. S. and Azarmi, N. (eds) *Software Agents and Soft Computing: Towards enhancing machine intelligence; concepts and applications*, Spinger- Verlag, pp. 3-26.

[16]Pree, W., 1995, "Design Patterns for Object-oriented Software Development", Massachusetts: Addison-Wesley.

[17]Tenenbaum J., Chowdhry T. and Hughes K., 1998, "eCo System: CommerceNet's Architectural Framework for Internet Commerce", *http://www.commercenet.org*

KES 2002
E. Damiani et al. (Eds.)
IOS Press, 2002

Intelligent Document Classification for Web Document Retrieval

D. Sridharan[1]   P. Sakthivel[1]   S.K.Srivatsa[2]

[1] *Ramanujan Computing Centre, Faculty of Electrical Engineering*
Anna University, 600025, Chennai,India
Tel: +91-44-2352269 x 3128, Fax: +91-44-2201160, E-mail:{sridhar,psv}@annauniv.edu

[2] *School of Electronics and Communication Engg, Faculty of Electrical Engineering*
Anna University, 600025, Chennai,India
Tel: +91-44-2352269 x 3065, Fax: +91-44-2201160, E-mail: profsks@annauniv.edu

Abstract. The World Wide Web contains huge amount of dynamic, heterogeneous, and hyperlinked distributed documents. Many modern information retrieval systems are developed based on matching process, which is automated, but classification and query formulations are manual process. The approach taken in this paper is to add intelligence to Information Retrieval by way of Document Classification based on Vector Space Model using connectionist methods with relevance feed back from the retrieved documents to the intelligent classifier as well as to the user query. The basic idea is to integrate three existing techniques: classification, query expansion and relevance feedback both to classified documents and user query to achieve a concept-based information search for the Web.

1. Introduction

With the advent of the World Wide Web in mid 90s, Internet became a household word and everyone was searching the relevant information in the web. A number of search engines became available on the Web and most of these engines are developed from University research projects. In the early days (1996-1997), the search engines competed on the number of pages indexed and the speed of retrieval. Now, most searchers realize that it is completely irrelevant if the search engine retrieves 100,000 or 1 million records and also it is very difficult to find the relevant document among large collection of retrieved documents. Now researchers recognized the limits of the massive quantity and lack of quality of information on the Web and employed a new idea based on concept searching, which automatically expanding the search to look for similar ideas and synonyms and alternatives of the search words. This type concept searching or classification retrieves relevant pages that would have been missed in a standard Boolean search. The important feature of classification is to identify the key concepts within a document and group together according to their concepts on specific topics and displaying it in a way that the user understands. Researchers have also focused on automatic query expansion to help the user formulate what information is really needed. Another research topic is on relevance feedback from the user, which gives the relevance of documents to clarify the ambiguity. In fact, these three techniques complement each other. However, the mechanisms of relevance feedback based on words or documents in the past research both have their own deficiencies. Word feedback has its upper bound performance in lexical-semantic expansion [1] and document feedback is sometimes too tiring for the users. An automatic query reformulation can be used to overcome these problems. The proposed system can be accommodated easily and inexpensively in future generations of web-based retrieval systems and technologies

The basic process of Information Retrieval using relevance feedback both to query and to the classifier is visualized in figure 1. The remainder of this paper is organized as follows. Section 2 and 3 describes Vector Space Model followed by query reformulation via relevance feedback based on Rocchio algorithm. And the next section 4 describes the taxonomy of designs and techniques used in this work. Section 6 outlines various related work on intelligent document retrieval system. The approach used in this paper is restricted to a relatively simple but effective relevance feedback for retraining the network and refining the user query to improve the performance of retrieval.

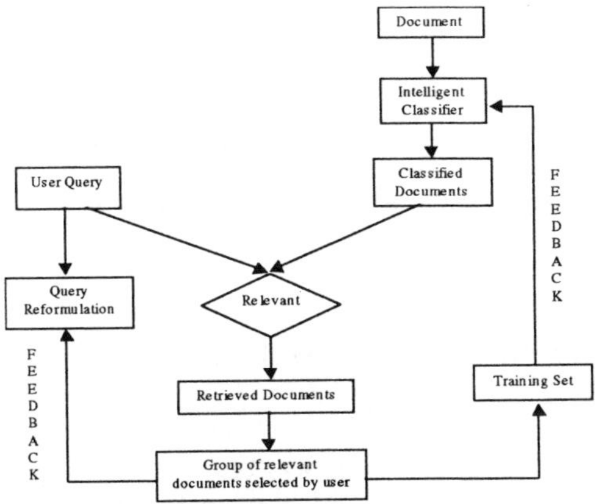

Figure 1 – Basic Process of Information Retrieval

2. Vector Space Model

The Vector Space Model was invented by Gerard Salton in the 1960's, is a mathematical model that represents documents and queries as term sets and computes the similarities between queries and documents. When combined into connectionist method with relevance feed back from the retrieved document gives intelligence for classifying the web documents and also for refining the query. The required criterion is that the queries and document use the same term set. In the vector space model, both queries and documents are represented as term vectors of the form Di = (di1, di2, ...,dit) and Q = (q1, q2, ...,qt). A document collection is then represented as a term-document (TD) matrix A:

$$
A = \begin{array}{c} \\ D_1 \\ D_2 \\ D_3 \end{array}
\begin{array}{ccc} T_1 & T_2 & T_t \end{array}
\left(\begin{array}{ccc}
a_{11} & a_{12} \cdots a_{1t} \\
a_{21} & a_{22} \cdots a_{2t} \\
a_{i1} & a_{i2} \cdots a_{it}
\end{array} \right)
\qquad
sim(D_iD_j) = \sum_{1 \le l \le n} d_{li}d_{lj}
$$

The rows of the above TD matrix represent the individual documents, the columns represent unique words and each entry in the matrix represents the number of occurrence of that word in that document. The similarity between a query vector Q and a document term vector D can also be computed. This is particularly advantageous because it allows one to sort all documents in decreasing order of similarity to a particular query. The required knowledge base for the connectionist method is developed from this Term-document matrix and remembered by a network of Inter-connected neurons, weighted synapses and threshold logic units. Learning algorithm can be applied to adjust connection weights, so that the network can predict or categorize the documents correctly.

3. Query refinement via relevance feedback

Relevance feedback is the reformulation of a search query in response to feedback provided by the user for the results of previous versions of the query and has long been suggested as a solution for query modification. Past research has applied Robertson and Sparck Jones' term weighting to query expansion [2], given the top 10–30 documents as relevant and all other documents in the corpus as non-relevant. Rocchio describes an elegant approach and shows how the optimal vector space query can be derived using vector addition and subtraction given the relevant and non-relevant documents [3]. The goal of relevance feedback is to retrieve and rank highly those documents that are similar to the document(s) the user found to be relevant

Now, given the group of documents selected by user as feedback unit, and then each document in this group has been digested as a set of relevant document vector (R) and a set of non-relevant document vector (S), the original query can be modified by Rocchio's algorithm. The new query is obtained by Q' = a Q + b sum (R) – c Sum (S). Where Q is original query vector and a,b,c are Rocchio weights.

4. Design of Neural Network for Classification

The proposed model for Intelligent Document Classifier is a feed forward Network with two layers of units fully connected between them as shown in Figure 2.

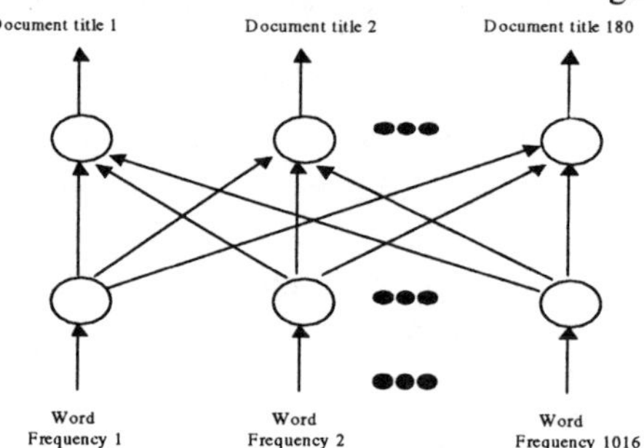

Figure 2 – Topology of the network

The first layer has the input unit and second layer has the output units. Each input unit in the first layer is associated with each word-stem and weight of the input unit is the word frequency in the document. Each output unit is associated with title of the research area and threshold logic units, defining, if the particular title characterizes the input document.

The data set used for training and testing the Intelligent Document Classifier consists of 2,344 research document and each one contains the title, authors name, citation information, abstract. This collection of documents was processed in such a way that tokenizing words from title and abstract, eliminating common words using a stop list, and the remaining words are passed through the stemmer. After this process, a total of 12,292 stemmed-words and 4,049 different titles were obtained from the entire collection. Finally the document frequencies are computed for each word-stem and document titles and it is found to be varying from 1 to 1,511 for word-stem and from 1 to 2,102 for document titles.

The number of stemmed-words in the above collection decides the number of nodes in the input layer of the inter-connected neurons. Hence it is decided to reduce the number of units in the input layer by setting a minimum Inverse Document Frequency (IDF) threshold as 75 on the entire collection. This is based on the assumption that a rare term that appears in only a few documents will have a high IDF score and at the same time the term that appears in most document will have a low IDF score. The terms that appear frequently in one document, but rarely on outside, are more relevant to the topic of categorization. A threshold of 75 for Inverse Document Frequency offers a reasonable reduction in the number of terms while retaining terms that are neither too specific nor too general. Using this criterion, the stemmed words are reduced to 1,016 for IDF threshold of 75. So, the number of units in the input layer was also reduced to 1,016. Using the same criterion for research titles, it is possible to have 180 neurons in the output layer. Finally, connections between all input units and all output units were created: 1016 X 180 = 182880. A Stuttgart Neural Network Simulator [4] is used for developing this intelligent classifier

5. Experiments

In the first experiment, the entire collection of 2,344 documents was divided into two sets: one is training set and another one is validation set. For training the network, 586 documents were used from the entire collection and this set is called as training set. Training documents were selected randomly among the training set. For validating the network, 1,758 documents were used and called this set as validation set and used to evaluate the network's ability to recognize different categorization of documents.

For training the network, a criterion was set that the average distance between patterns was less than a predefined value. An initial run was trained for a maximum of 25 cycles. The average distance per pattern in the last iteration was 0.75 and the error per pattern in the last cycle was 19.15. Attempts to train the network allowing more number of iterations resulted in higher values of error. Finally the network is converged to the value of an acceptable error (0.2) in 19 cycles and took about 6 hours (using a HP-700 workstation).

As a result of validation it is revealed that 90% of the proposed documents were correct. It is possible that the 10% of the documents are not completely correct and they may be coming under different categorization. This is because the network is trained by training set with the minimum Inverse Document Frequency threshold of 75.

In the second experiment, the user viewed an average of 62 full documents during the first search. Of the full documents seen, on average 18 of them were declared relevant. After a few tuning experiments, the constant are fixed as follows: a=0.4, b=0.4 c=0.2. This resulted in a great reduction in the number of terms without significant drop in performance and the new query is calculated. The performance of this system can be measured by evaluating precision and recall.

During the search the users used a number of search criteria other than thesaurus concepts such as free-text, author searching, limiting by publication types, although thesaurus searching accounted for 70 % of all the search criteria selected with free-text accounting for 12.5 % of all search criteria. This gives an average total of 27.7 relevant documents per query. Then the Relevance feedback average recall is 17.7/27.7=0.64 and Relevance feedback average Precision is 17.7/61.6=0.29(for titles)

When looking at the rankings a maximum of 22 relevant records were found in any one set of thirty documents produced in the ranking with an average of 5.2 documents per ranking using feedback only to query. Then the Relevance Feedback average recall is 5.2/27.7=0.19 and Relevance Feedback average precision = 5.2 / 30 = 0.17. The results seem to indicate that this method performs far better than the simple relevance feedback process in which only user query is modified.

6. Related work

Many interesting knowledge-based systems have been developed in the past few decades for applications such as medical diagnosis, engineering troubleshooting, and business decision making [4]. Most of these systems have been developed based on the manual knowledge acquisition process, a significant bottleneck for knowledge-based systems development. A recent approach to knowledge elicitation is referred to as ``knowledge mining" or ``knowledge discovery" [5][6].

PLEXUS is an expert system that helps users find information about gardening [7]. The system has a knowledge base of search strategies and term classifications similar to a thesaurus. EP-X is a prototype knowledge-based system that assists in searching environmental pollution literature [8]. The system interacts with users to suggest broadening or narrowing operations. GRANT is an expert system for finding sources of funding for given research proposals [9]. An intelligent system(AWPC) is developed for Automatic Web Page Categorization for Information Retrieval System [10]. This system classifies the web page according to the content available in the home page

The Yahoo server developed at the Stanford University represents one attempt to partition the Internet information space and provide meaningful subject categories (e.g., science, entertainment, engineering, etc.). However, the subject categories are limited in their granularity and the process of creating such categories is a manual effort. The demand to create up-to-date and subject specific categories and the requirement that an owner place a homepage under a proper subject category has significantly hampered Yahoo's success and popularity.

7. Conclusion

This article integrates the three mechanism viz. classification, query expansion and relevance feedback to retrieve a research article from the free text, which contains authors name, citation information abstract and set of manually assigned research key words. The results suggest that this method performs better than the models based on relevance feedback in which only the user queries are reformulated from the retrieved documents. The employment of conceptual feedback with query expansion based on the two models is a new approach in information retrieval. By expanding a query, one could not only increase the number of relevant documents retrieved but also classify the candidate documents. Thus, Web Information Retrieval can be carried out by generating new queries and filtering through the existed evidence. The issue of scalability has to be tested by using a larger collection of documents. The knowledge obtained by the network during the training phase can also be used as fuzzy rule for categorizing the documents.

References

[1] D. Harman, Relevance   feedback   revisited. In Proceedings of ACM SIGIR , International Conference on Research and Development in Information Retrieval, 1992, pp.1–10.

[2] S.E. Robertson and K. Sparck Jones, Relevance weighting of search terms. *Journal of the American Society for Information Science*, 27(3): 1988, pp.129–146.

[3] J.J. Rocchio and G. Salton eds. *Relevance feedback in information retrieval. The SMART Retrieval System.* Prentice-Hall, Inc., Englewood Cliffs, NJ, 1971, pp. 313–323.

[4] SNNS. Stuttgart   Neural   Network   Simulator ver 4.2 user Manual, Institute for Parallel and Distributed High Performance Systems, University of Stuttgart, 1998.

[5] F. Hayes - Roth and N. Jacobstein, The state   of knowledge-based systems. *Communica-tions of the ACM*, 37(3), 1994, pp.27-39.

[6] W. J. Frawley , G. Pietetsky - Shapiro and C.J. Matheus, eds. *Knowledge discovery in databases : an overview.* The MIT Press, Cambridge, MA, 1991.

[7] A.Vickery and H. M. Brooks, PLEXUS – The expert system for referral. *Journal of Information Processing and Management*, 23(2), 1987, pp. 99-117.

[8] P. J. Smith, S. J. Shute, D. Galdes and M.H. Chignell, Knowledge - based   search tactics for an intelligent intermediary system. *ACM Trans. on Information Systems,* (7), 1989, pp.246-270.

[9] P. R. Cohen and R. Kjeldsen, Information retrieval by constrained spreading activation in semantic networks. *Journal of Information Processing and Management,* 23(4), 1987, pp.255-268.

[10] Hisao Mase and Hiroshi Tsuji. Experiments on Automatic Web Page Categorization for Information Retrieval System *Journal of Informa tion. Processing Society of Japan*, 42(02), 2001.

KES 2002
E. Damiani et al. (Eds.)
IOS Press, 2002

An Intelligent System
for Learning Human Designs

Hui CHEN*, Hassan ABOLHASSANI**, Zenya KOONO***
** Center for Information Science, Kokushikan University*
4-28-1 Setagaya, Setagaya-ku, Tokyo, 154-8515, Japan
***Forntage-Razorfish Inc.,*
2-36-13 Ebisu, Shibuya-ku, Tokyo 150-0013, Japan
****Graduate School of University of East Asia*
Honfujisawa 2-13-5, Fujisawa, Kanagawa 251-0875, Japan

Abstract. This paper introduces an overview of an intelligent system for learning human design. Error free designs by experts have been taken as materials, and hierarchical expansions of concepts (in a hierarchical work process) have been taken as the central function during design. Based on these design knowledge model, a prototype intelligent system for learning human design has been developed. The more inside of human design has been studied in skill level, rule level and knowledge level. The system consisted of some CASE tools and expert units. It can design software automatically by various design knowledge. As the system has been design following to the human experts knowledge structure as to result in well revealing inside of human design. Experiments using it resulted in finding of various Learning Effects, and the mechanisms have been clarified.

1. Introduction

The purpose of this paper is to introduce a kind of expert system named Integrated Intelligent CASE tool as an intelligent machine. Automatic design has been regarded as the final key for solving "Software crisis". Although various kinds of automatic design have been proposed, their uses are limited. One reason is difficulties of ascertaining the characteristics of human design during software design. Authors' way is to reveal designer mental activities during software design and find design knowledge structures, and then simulate human design by using an intelligent system to reach software design automatically.

The most universal design model is human designs by experts. Thus, an expert designer's model has been taken. In order to purify human design process, a high maturity software development organization is taken as the expert model. Due to their accumulated improvements on their process or design procedure for a long period, their process is standardized to a solid hierarchical work process. Such a solid work process is taken as the knowledge model.

Design is regarded as a hierarchically expanding network from the initial specification to the source code along the hierarchical work process, in which hierarchical decompositions of human concept are made repetitively. When the hierarchical expanding network is folded, the engine in which hierarchical decompositions are performed is universal system for the mental operations.

As authors' study target is human design of software, the platform is an enriched CASE (Computer Aided Software Engineering) tool system. This paper explains the basis for the idea as well as the main structure together with results obtained.

2. Basic Assumptions

Besides the aforementioned, following are the basic assumptions for the study;
- Natural language is chosen as the working language.
- Charts and diagrams bearing natural language statements are used for representing concepts. They also show relations between concepts.
- A hierarchical decomposition consisting of a parent concept and the child concept, is regarded as the vital design process knowledge, and is named *a design rule*. A design is a hierarchical decomposing tree starting from the specification and ending in source code. A systematic acquisition of design knowledge becomes possible from documents, and based on them, a systematic design of the expert, which leads to an automatic design system, also becomes possible [4].
- In Ergonomics, Zipf [10] says that a human solves a problem first by the simplest way, and if unsucesessful, a more advanced way is taken, and the loop is repeated. Rasmussen's theory of *'Skill-based, Rule-based* and *Knowledge-based'* [9] is also underlaid. Rasmussen's first is *skill* level where actions are made reflectively using past memory. Authors first system, named Intelligent CASE tool (ICASE) [5], automates programming phase by reusing past designs using structured charts. Rasmussen's second is *rule* level. Authors have evolved ICASE to Integrated Intelligent CASE tool (IICASE) [6] that extends the application upward to data flow design prior to programming. IICASE has reproduced human software design based on rules of skeleton as data and methods for constructing phrases. Rasmussen's third is *knowledge* level. It was found that an addition of concept dictionary on IICASE enables the system to operate in knowledge-based level.

3. An Intelligent System

3.1 Architecture

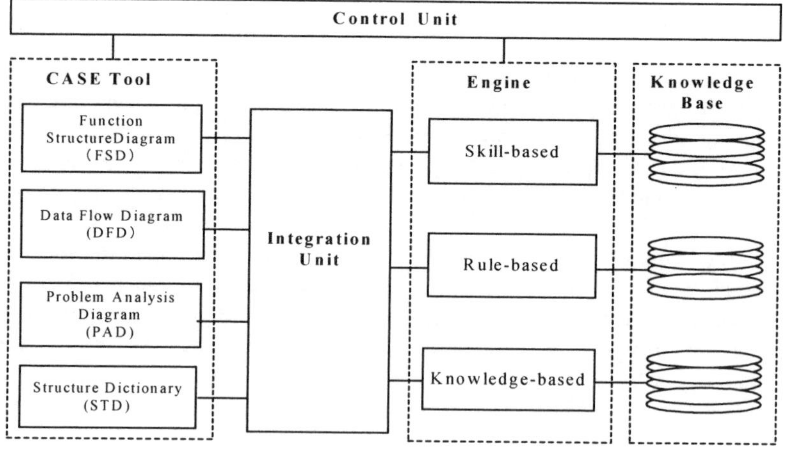

Figure 1: System structure of the intelligent system

An architecture of IICASE tool is shown in Figure 1. It constitutes a platform for automatic design of software. The left part of the figure, there are FSD (Function Structured Diagram), DFD (Data Flow Diagram), PAD (Problem Analysis Diagram, a kind of structured chart) [7] and STD (Structure Dictionary) CASE tools. A Structured Chart PAD CASE tool is fitted for the presentation, and a pair of PAD and DFD CASE tools is used. FSD CASE tool prepares or records the function structure of a design. STD CASE tool is used to define hierarchical structures of input data, output data and other data used during design.

The right part of the figure, there are engines and knowledge bases for automatic design. In the center, there is Integration unit that is to integrate combined operations of various CASE tools and engines. Control unit is to control various CASE tools, design process and various design results. Their interfaces are standardized. Thus various engines or design knowledge simulators may be attached and the necessary knowledge bases may be added.

As the designer operates these CASE tools, an engine can design automatically instead of a human designer. Operations of automatic design may be observed on the CASE tools screen, and if automatic design is incomplete, the addition may be easily made.

3.2 Skill Level Operation

Skill-based design may be reproduced by reusing past designs. Figure 2 [6] shows major three documents of the system with an engine for reuse in knowledge acquisition mode as well as related knowledge bases. In the left side of the figure, design rules in FSD, DFD and PAD cases are shown. Each hierarchical detailing shows that a *parent* concept (e.g. box) in a preceding stage is hierarchically detailed into several pieces (e.g. boxes) of the next level *children* concept. When they are cascaded, a hierarchically expanding network reaches to source code.

Figure 2 also shows the way how to acquire the elementary design rule. Here, a case of PAD's design knowledge acquisition is explained. In the left side bottom, a tree-walk program along the dotted route enables to acquire *parent* as well as *children* concept graphically. From the acquired information, a *design rule* may be derived, and stored in PAD Knowledge Base.

For automatic design, a designer draws a *parent* concept on the CASE tool screen and commands to start. It extracts the parent by tree-walk, and it is sent to the knowledge base.

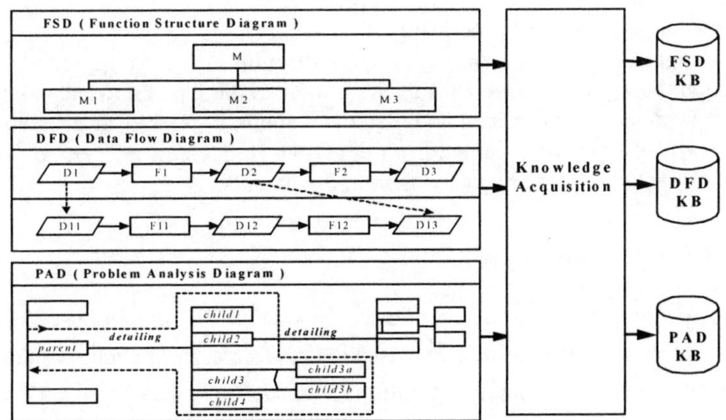

Figure 2 : Acquisition of design knowledge

If the corresponding design rule exists, the *children* are returned. The engine adds the children concepts to extend next level of the tree. Thus continuing, it is detailed to form a hierarchically expanding net. Coding is performed also by such design rules.

The intelligent operations are performed by *skill-based* engine. But the elementary functions such as tree walking and pasting the children concepts are assigned on each CASE tool. Theses process may be made PAD for control and DFD for function as well.

3.3 Rule Level Operation

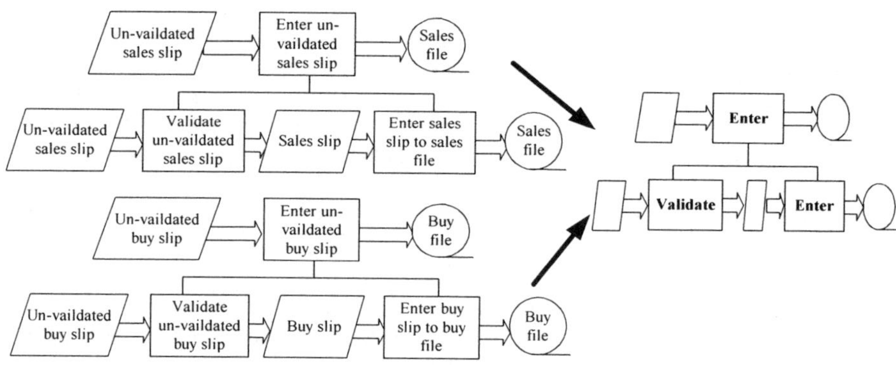

Figure 3: Design rules and a skeleton pattern

Figure 3 show an example of human design that is *rule* based. The left side of the figure shows two pairs of data flow diagrams, where a square box is a function while a lozenge is data. In a pair, the upper data flow is a concept while the lower one is a hierarchically detailed child concept. Comparing both DFD designs, these two design rules have the same verbs in parent and child functions. The right side of the pair of data flow diagram shows the skeleton [3] thus abstracted from the left side pairs. For example, the parent's input (output) data is copied to the children input (output), the intermediate data is the input data without the prefix (un-), and so on. A frame type knowledge, which has a DFD skeleton as data and rules as method, is fitted to generate a design rule. In this case, the design rule is creater by some natural language processing.

Rule-based engine uses a verb dictionary consisting of such a frame for each verb for each meaning as the knowledge base. In IICASE tool, frame type knowledge, which has a DFD skeleton as data and afore mentioned rules as method, is used to generate a design rule in DFD, and PAD may be obtained from the DFD. Both charts are sent to CASE tools and the resultant children parts are pasted next to the parents.

3.4 Knowledge Level Operation

The *knowledge* level operation is complicated versions of *rule-based* operation. In some cases where a simple operation using rule-based is impossible, this operation starts.

It consists of various elementary operations named Micro design rule [2]. Figure 4 shows a result of for detailing an input concept '1-sec clock (input data) obtain time (function) and Time (output data)'. From '1-sec clock', it extends to concept chain of 'Second', 'Minute' and 'Hour'. From 'Time', it is abstracted from 'Second', 'Minute' and

'Hour'. These groups may be formed by an elementary concept dictionary. The former is named Concept Definition Chain (CDC) and the latter is named Concept Abstraction (CAB). These elementary operations is called *Micro design rule*. Several micro design rules form a Macro of micro design rule. The figure shows a couple of CDC and CAB. All these operations are made by using a concept dictionary. The internal structure of human design consists of dictionary definitions, Micro design rules, Macro, design rule and design in a hierarchical order.

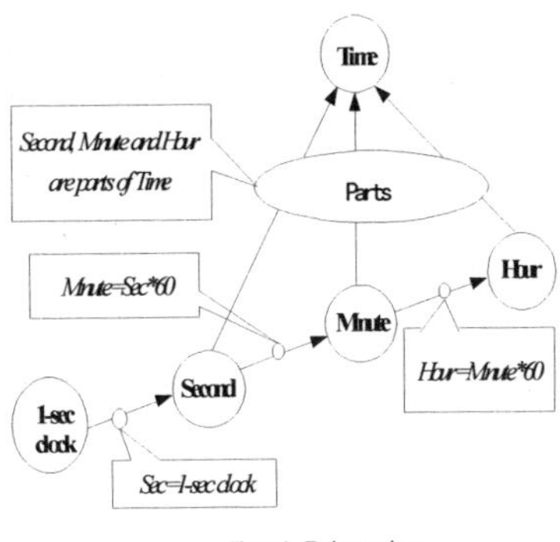

Figure 4: Basic operations

4. Discussion

Man-hour saving of ICASE has been evaluated quantitatively. The material is repeated (maintenance) designs on data accessing part in a telephone switching system. Figure 5.a [5] shows the accumulated number of design rules vs. the number of times of designs. Both plots show the same trend. Curves rise rapidly at first, but their gradients decay gradually. When they are plotted on both Logarithmic scales, plots show a linear trend line that they are Logarithmic Learning Effects[8]. Using the trend line equation, quantitative evaluations has been made.

Figure 5.b shows the degree of automatic detailing by *skill-based* operation as the number of repetitive designs increases. The curve starts from the initial zero, and rapidly clears over 80% but it does not tend to 100%. When the experience is over 10, man-hours may be decreased to less than 1/4 of pure human design. Thus it is very effective when the degree of the repetitive use of a design rule is high.

Figure 5.c [2] shows an evaluation data of this *rule-based* design experiment that has been made successfully on seven plus seven programs of Inventory Control System. It shows the accumulated number of pages used of the verb dictionary vs. the number of hierarchical detailing experienced in both Logarithmic scales. Various sequences of the initial seven programs were tried. As they show a similar tend, in Fig. 5.c several sequences are shown for making plots clear. As is shown, the trend of plots shows a broken line trend. The initial part shows a Learning Effect caused by mainly the number of verbs and the next part shows a Learning Effect by the number of pages. Thus the drgree of automatic detailing until 1000 hierarchical detailing or 110 programs is around 98%. This is high enough, and the necessary number of pages is small, due to the Learning Effect.

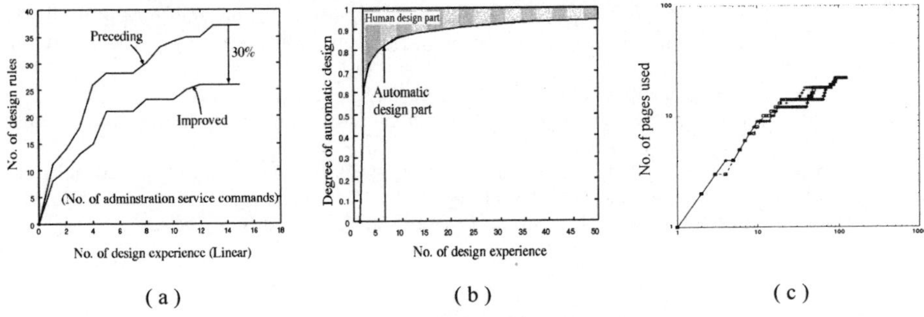

Figure 5: Various evaluations of design knowledge

In an actual human brain, the reuse engine follows this for remembering a design rule generated. For the second occurrence, the reuse is made resulting in faster operation than before. As designs are experienced, the frequency of the reuse increases showing a Learning Effect during new designs this time.

4. Conclusion

Is this paper, an intelligent system for learning human design has been reported. Different from usual AI way, an idealistic expert model has been taken, and simple hierarchical detailing process has been chosen as the central part of design. Using charts and diagrams, operations on charts and diagrams are taken as elemental intelligent operations for representing human design. Through these ideas, CASE tools with intelligent capabilities become a platform for design. Very important points is this study is that there is another way to reveal humen mentality beyond AI.

Through the analyses on designed results, various aspects of human design have been being revealed. The system consists of some CASE tools, engines and knowledge bases. It shows similar behavior to Zipf's ae well as Rasmussen's Skill-Rule-Knowledge levers for human design. The behavior is just like a human Learning Process. Probably, these are simple appearance of *Human Intelligence*.

Acknowledgements

Authors express their deep gratitude to Dr B. H. Far, Associate Professor of Calgary University and those who participated in this study during ten years of Software Creation Project in Saitama University. They are also thankful to various companies and fundations for their supports to the project.

References

[1] H. Abolhassani, H. Chen, B.H. Far and Z. Koono, Software Creation: A study on the inside of human design knowledge, the Journal of Institute of Electronics, Information and Communication Engineers - Special Issue on Knowledge Based Software Engineering, Vol. E83-D, No. 4, pp. 648-658, April 2000.

[2] H. Abolhassani, A study on automatic design based on human design knowledge, Ph.D thesis, Saitama University, 2001

[3] H. Abolhassani, H. Chen and Z. Koono, Software Creation: Cliche as Intermediate Knowledge in Software Design, the Journal of IEICE on Information and Systems, Vol. 85-D, No. 1, pp. 221-232, Januaray 2002.

[4] H. Chen, B. H. Far, and Z. Koono, A Systematic Construction Method of an Expert System Used for Automatic Software Design --Acquisition and Reproduction of Design Knowledge from Design Process, Journal of Japan Society for Artificial Intelligence, Vol. 12, No. 4, pp. 616-626, July, 1997. (In Japanese)

[5] H. Chen, N. Tsutsumi, H. Takano, and Z. Koono, Software Creation: An Intelligent CASE Tool Featuring Automatic Design For Structured Programming, in the Journal of Institute of Electronics, Information and Communication Engineers, Vol. E81-D, No. 12, pp. 1439-1449, 1998.

[6] H. Chen, H. Takano, H. Abolhassani and Z. Koono, Software Creation: An Integrated Environment for Automatic Software Design, 15th World Conference on Design and Process, pp. 27 (in CD ROM) June, 2000.

[7] Y. Futamura, 1984, Programming Techniques: Structured Programming using PAD, Ohmsya.

[8] W.M. Hancokc and F.H. Bayer, Handbook of Industrial Engineering, Ch2. Learning curve, ed. G. Salvency, John Wiley and Sons, 1982.

[9] J. Rasmussen, The role of hierarchical knowledge representation in decision making and system management, IEEE Trans. Syst., Man & Cybern., vol. SMC-15, No. 2, pp. 234-243, March/April 1985.

[10] G K. Zipf, Human Behavior and the Principle of Least Effort, Hanfer publishing, 1972.

KES 2002
E. Damiani et al. (Eds.)
IOS Press, 2002

Video interpretation: human behaviour representation and on-line recognition

Van-Thinh VU, François BRÉMOND and Monique THONNAT
Project ORION of I.N.R.I.A. Sophia Antipolis,
2004 route des Lucioles, BP93-06902 Sophia Antipolis Cedex, France.
e-mail: {Thinh.Vu, Francois.Bremond, Monique.Thonnat}@sophia.inria.fr
http://www-sop.inria.fr/orion/orion-eng.html

Abstract: This paper presents a recent work in human behaviour representation and on-line recognition for video interpretation. We propose a declarative model to represent human behaviours and we use the logic-based approach to recognise pre-defined behaviour models. We demonstrate our representation formalism and the inference engine on two video sequences. We also propose a limit for the number of behaviour actors based on experimental results. The processing time is still a challenge with complex behaviour models and cluttered scenes.

Keywords: video interpretation, on-line human behaviour recognition, cognitive vision, human behaviour representation.

1. Introduction

This paper presents recent works in human behaviour representation for video understanding ([9], [10]). The goal is to recognise behaviours of humans involved in a scene depicted by a video sequence. The recognition process takes as input (1) human behaviour models predefined by experts, (2) 3D geometric information of the observed environment and (3) a video stream of persons tracked by a vision module ([2], [10]). To represent human behaviours, we first propose a language to describe behaviour models and secondly a logic-based approach to recognise in real time human behaviour occurrences.

For 20 years, the issues of temporal scenario representation and recognition have been approached. M. Ghallab [3] has represented a temporal scenario as a set of temporal constraints on time-stamped events. The recognition algorithm implements propagation of temporal constraints based on RETE algorithm. C. Pinhanez and A. Bobick [7] have implemented a simple version of Allen's interval algebra [1]. S. Hongeng and R. Nevatia [4] have proposed a behaviour recognition method by using concurrence bayesian threads to estimate the likelihood of potential scenarios. N. Rota and M. Thonnat [8] have used a declarative representation of scenarios defined as a set of spatio-temporal and logic constraints. They have reduced the constraint resolution step by checking before the consistency of the constraint network using the AC4 algorithm [6].

All these techniques allow an efficient recognition of behaviours but, they do not let the experts to describe their scenarios in a natural way. For example, most of these approaches represent an event defined at one time point but cannot manipulate events defined on intervals.

2. Human behaviour representation

Our goal is to explicit all the knowledge necessary for the system to be able to recognise human behaviours. The description of this knowledge has to be declarative and in natural terms, so that the experts of the application domain can easily define and modify it. Thus, the recognition process uses only the knowledge represented by experts through behaviour models.

We represent a human behaviour model with the list of the actors involved in the behaviour and a set of constraints on these actors ([2], [8], [10]).

An actor can be a person detected as a *mobile object* by the recognition process or a *static object* of the observed environment. A person is represented by its characteristics: his/her position in the observed environment, width, velocity,…. A static object of the environment is defined as a priori knowledge (before processing) and can be either a zone of interest (a plane polygon corresponding to an entrance zone) or a piece of equipment (a 3D object as a desk). A zone is represented by its vertices, and a piece of equipment is represented by the vertices of its bounding box. The zones and the pieces of equipment are represented in the scene context of the observed environment. Static objects and mobile objects are called *scene-object*.

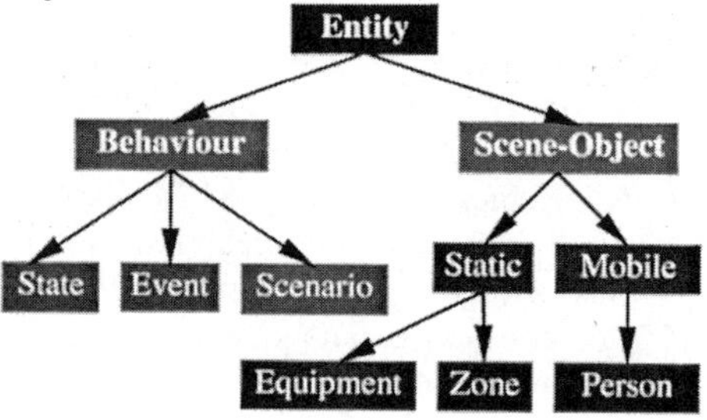

Figure 1: Six types of entities are defined either as "behaviour" or as "scene-object".

In our representation, persons detected by the camera, are also characterised by relevant states, events and scenarios. A state characterises a person during a given time interval. An event defines a change of state at two successive instants. A scenario defines a combination of states, events and sub-scenarios. The notion of *behaviour* is used as a generic name for these three notions. A behaviour involves at least one person, and is defined on a time interval. An interval is represented by its starting and ending time. Defining behaviour on the time interval is important for the experts to let them describe behaviours in the more natural way. Behaviours and scene-objects

are called *entities* and are defined by a generic class (see Figure 1).

To describe a human behaviour model, we have used the following definitions.

Definitions:

O - a set of scene-objects,

V - a set of variables corresponding to scene-objects,

$C = \{c \text{ constraint} \mid c : V^k \to \text{BOOL}, k > 0\}$,

$\mathcal{C} = P(C)$ set of parts of C,

A n-actor behaviour model is an element m = (*Actors, Constraints, Production*) of M = (V^n x $\mathcal{C}$ x $\mathcal{O}$). *Actors* is a set of n variables expressing the actors involved in the behaviour with their type and name. *Constraints* is a set of constraints defining the relationships between the actors and characterising the behaviour. *Production* is a set of deduced characteristics describing the behaviour once it has been recognised.

The n-uplet of scene-objects $s = (o_1,…, o_n)$ is a solution at time t of the recognition process using the human behaviour model *m* if all the constraints of *Constraints* are satisfied when we assign the n variables of *Actors* with the n corresponding scene-objects of *s*. In this case the behaviour model *m* is *satisfied* at the time of the recognition of *s*. A human behaviour *b* (called instance of *m*) will be produced using *s* with the characteristics given by *production*. The human behaviour *b* is *recognised* and is added to the characteristics of the persons contained in *s*.

Figure 2 shows a model of the state "*close_to*": a person *p* is close to a piece of equipment *eq*. It involves two actors a person *p* and a piece of equipment *eq*. There is only one constraint which verifies that *p* is close to *eq*.

```
State(close_to,
   Actors( (p : Person), (eq : Equipment) )
   Constraints(
   (p distance eq <= Close_Distance) )
   Production((s:State)(Name = "close_to")) )
```

Figure 2: A model of the state "close_to": a person is close to a piece of equipment

We have defined a description language to represent both parts *Constraints* and *Production* in a human behaviour model. These sets of constraints are expressed by logical predicates.

To express the logical predicates, we have used the arithmetical operators (+, -, *, /), the comparison operators (<, ≤, =, ≠, ≥, >) and the logical operators (! : not, & : and). To represent temporal relations between the behaviours, we use the operators of the interval algebra (before, after, meets,…) [1].

We use the spatial operator "distance": to calculate the distance between two scene-objects. We use also the operator "exists" to verify whether given entities exist (or not) and satisfy several constraints.

```
Event(moves_close_to,
    Actors( (p : Person), (eq : Equipment) )
    Constraints(
      (exists (state(s1, far_from, p, eq)
             state(s2, close_to, p, eq) )
        ( (Duration of s2 ≤ 1 min)
          (s1 before s2) ) ) )
    Production((e : Event)
        ( (Interval of e := Interval of s2) ) )
```

Figure 3: Model of the event "moves_close_to": a person moves close to a piece of equipment

a piece of equipment.

A model of the event *"moves_close_to"* is shown in Figure 3: a person *p* moves close to a

piece of equipment *eq*. An event of this model will be recognised if p is first, far from eq and then close to eq. "**state**(s1, **far_from**, *p*, *eq*)" expresses that s1 is a state where the person p is far from the piece of equipment eq. The production part indicates how to compute the time duration of the recognised event.

On Figure 4 we show an example of a more complex scenario, "vandalism": a person p tries to "break up" a piece of equipment eq. This behaviour will be recognised if a sequence of five events described on Figure 5 has been detected.

```
Scenario(vandlism,
    Actors( (p : Person), (eq : Equipment) )
    Constraints(
      (exists(event(e1, moves_close_to, p, eq)
             event(e2, stays_at, p, eq)
             event(e3, moves_away_from, p, eq)
             event(e4, moves_close_to, p, eq)
             event(e5, stays_at, p, eq) )
        ( (e1 meets e2) (e2 meets e3)
          (e3 before e4) (e4 meets e5) ) ) )
    Production((s : Scenario)
        ( (Interval of s := Interval of e5) ) )
```

Figure 4: A model of the scenario "vandalism": a person tries to break up a piece of equipment

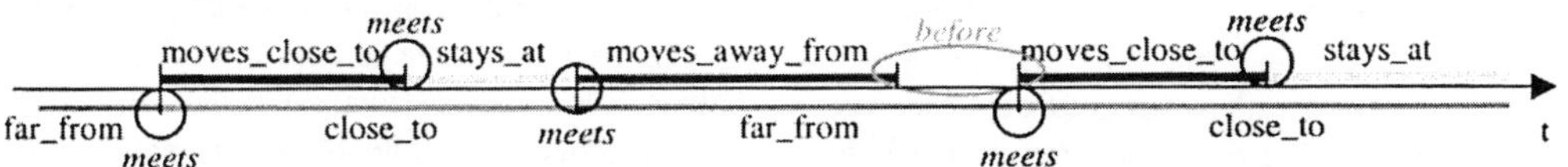

Figure 5: Temporal constraints for the states and events constituting the scenario "vandalism"

3. On-line human behaviour recognition

The human behaviour recognition process has to detect from a stream of observed persons at each frame which behaviour is happening. The process takes as input (1) the behaviour models pre-defined by experts, (2) the geometric information of the observed environment and (3) the persons tracked by the vision module. We suppose that the persons are correctly tracked: their characteristics (their position in the scene, their height,…) are well detected and at two successive frames, two persons having the same name correspond to the same real person.

3.1. Selection of behaviour models

We initiate the list LM of behaviour models to be recognised with all elementary behaviour models. An elementary behaviour model is a behaviour that can be recognised at any time

such as "close_to". The list LM is ordered by priority levels defined by the experts. Before the processing, for each behaviour model, we compute the set of *post-models* that corresponds to the behaviours that can be recognised once the given behaviour has been recognised. For example, once we have recognised the behaviour "close_to", it is possible that the behaviour "moves_close_to" might be recognised. So the list of *post-models* of the behaviour "close_to" contains the behaviour "moves_close_to".

Therefore, when a behaviour is recognised, its post-models (with the information about its actors) are added to the list LM of behaviour models to be recognised. The main loop to select the behaviour models at each frame is described briefly in Figure 6.

```
1: Initiate the list of models LM
     with the elementary models set
2: while LM ≠ ∅
3:    m ← get first element of LM
4:    LM ← LM - {m}
5:    Find_Solution(list of variables of m, m, ∅)
6:    if success then
         LM ← LM + post-models of m;
```

Figure 6: The main loop to select the behaviour models to be recognised

"*Find_Solution*" (line5:) is the function that finds all the solutions (n-uplet of actors) for the model m. This function is described in the next section.

3.2. Finding solutions of a behaviour model and verification of constraint "exists"

The process to find all the solutions of a behaviour model m, is realised thanks to the function "Find_Solution". This function selects a value for each variable (actor) and check if the selected values satisfy the constraints defined within the behaviour model m. This function is described briefly in Figure 7. It takes as input the behaviour model m, the list of variables lv to be instanciated and the list of values A of the variables already instanciated. At the beginning of the process lv is initialised with all variables of m and A is the empty list. At the end of process, lv is the empty list and A contains one solution. The side effect of this function is to store the solutions once they are found.

```
1: Find_Solution(lv, m, A)
2:    v ← first variable of lv
3:    while select_value a for v in
            the domain of v [type] and verifies
            the constraints of v
4:       if v is not the last variable of lv
5:       then Find_Solution(lv - {v}, m, A + {a})
6:       else create_instance b of m with
            A+{a} in the domain of m and A [type];
```

Figure 7: Finding all solutions of a behaviour model m.

To speed up the recognition process, we organise the variable domain with the variable type so the search space is reduced. There are two kinds of types: on the variable (person, equipment) and on the behaviour ("moves_close_to", "vandalism").

A second way to speed up the recognition process is to order the list of constraints for the behaviour model m. For that, we define an order on the constraints:

$$order(c) = \max_{i} \mid v_i \text{ appears in } c,$$

for example: $order((v_s \text{ before } v_l)) = 5$.

Then we regroup around the variable v_i every constraints that have an order i (by a pre-compilation module in the initialisation phase).

"*select_value*" (line 3:) is a process to choose a value for variable v and verify all corresponding constraints with the chosen value. A value is selected for v if it satisfies every own constraints of v. This process terminates as soon as a value is selected for v or when there is no more value to be selected.

Our method is correct because: when we choose a value for a variable v_i so every variables v_j (j < i) are already instantiated with a value, and every corresponding constraints of v_i are functions restricted to $\{v_1,..,v_i\}$ so they are well calculated.

We use also this algorithm for the verification of constraint "exit", but the procedure "Find_Solution" terminates immediately when a solution is found.

3.3. Managing the recognised behaviours

Each time a behaviour b is recognised, we search a previous instance of the same behaviour in the list of recognised behaviours. If it is found and the time interval of the found entity "meets" the time interval b, then the time interval of the found entity will be prolonged. In the other case, b is stocked in the list of recognised behaviours corresponding to the behaviour model and the actors involved.

4. Results

We have modelised seventeen human behaviour models thanks to our description language. These behaviours include the models of states of a person relatively to a piece of equipment, or between two persons, models of several events that we usually meet in the scenes of metro or in an office, a model of the scenario "vandalism" and three complex models with 5, 6 and 9 actors to test the processing time. We have also tested these models on two video sequences, one (seq1) of the metro of Nuremberg (340 frames, 1 person/frame and 1 piece of equipment) and another sequence (seq2) in a FNCA bank agency (500 frames, 5 persons/frame, 20 pieces of equipment and 6 zones).

We have also measured the processing time for each frame of these two sequences. For the sequence seq1 (Figure 8), for 12 pre-defined models, the average processing time per frame is 0.19ms and the maximal processing time per frame is 0.32ms. For seq2, for 14 pre-defined models (all behaviours except the 3 complex models), the average processing time per frame is 17.5ms and the maximal processing time per frame is 34.9ms. For the same sequence, for 16 (all behaviours except the model with 9 variables), the average processing time per frame is 67.1ms and the maximal processing time per frame is 101.3ms. The model with 9 variables could not be recognised because it takes to much processing time. These tests can be found in our demo page: http://www-sop.inria.fr/orion/personnel/Thinh.Vu/demos. The experimental results show that a condition to obtain a real time processing (100ms/frame) is that the number of actors of the models cannot be superior to 6.

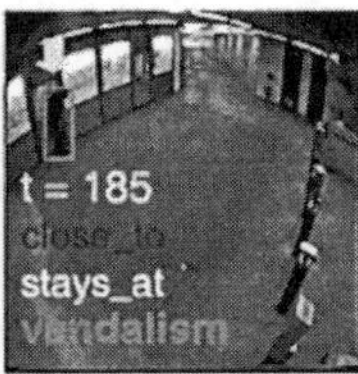

Figure 8: Different instants of the scenario "vandalism" (seq 1)

5. Conclusion

To represent human behaviours, we have proposed a generic model which is able to represent states, events and scenarios defined on time intervals and which is able to combine any type of constraints (logic, spatial and temporal). To help experts to describe scenario models in a declarative way, we have defined a description language. We have also proposed a recognition algorithm which can recognise in real time complex behaviours (containing up to 7 actors). We have accelerated the constraint resolution algorithm by structuring the search space of the actors.

We have validated the behaviour model and the behaviour recognition algorithm on a metro and a bank application.

The processing time of the recognition process is still a problem with complex behaviours (containing long term temporal constraints). To solve this problem we are currently trying to build special temporal operators able to check temporal relations in a bounded time.

References

[1] James F. Allen. *Towards a general theory of action and time*. Artificial Intelligence, 23:123-154, 1984.

[2] François Brémond. *Environnement de résolution de problèmes pour l'interprétation de séquences d'images*. Thèse, INRIA-Université de Nice Sophia Antipolis, 10/1997.

[3] Malik Ghallab. *On Chronicles: Representation, On-line Recognition and Learning*. 5th International Conference on Principles of Knowledge Representation and Reasoning (KR'96), Cambridge (USA), 5-8 November 1996, pp.597-606.

[4] Somboon Hongeng et Ramakant Nevatia. *Multi-Agent Event Recognition*. International Conference on Computer Vision (ICCV2001), Vancouver, B.C., Canada, 9-12/07/2001.

[5] Tony Jebara et Alex Pentland. *On Reversing Jensen's Inequality*. In Neural Information Processing Systems 13, NIPS 13, 12/2000.

[6] Roger Mohr et Thomas C. Henderson. *Arc and Path Consistency Revisited*. Research Note, Artificial Intelligence, pp225-233, vol28, 1986.

[7] Claudio Pinhanez et Aaron Bobick. *Human Action Detection Using PNF Propagation of Temporal Constraints*. M.T.T Media Laboratory Perceptual Section Technical Report No. 423, 04/1997.

[8] Nathanaël Rota et Monique Thonnat. *Activity Recognition from Video Sequences using Declarative Models*. 14th European Conference on Artificial Intelligence (ECAI 2000), Berlin, Proceeding ECAI'00 – W. Horn (ed.) IOS Press, Amsterdam, 20-25/08/2000.

[9] Cathérine Tessier. *Reconnaissance de scènes dynamiques à partir de données issues de capteurs: le projet PERCEPTION*. Rapport technique, Onera-Cert, 2 avenue Edouard-Belin, BP4025 31055 Toulouse Cedex France, 08/1997.

[10] Monique Thonnat et Nathanaël Rota. *Image understanding for visual surveillance application*. Third international workshop on cooperative distributed vision CDV-WS'99, pp51-82, Kyoto, Japan, 19-20/11/1999.

KES 2002
E. Damiani et al. (Eds.)
IOS Press, 2002

User Interface Design for
Online Incidental Information Media

Atsuhiko Maeda, Yoshiteru Nakamori, and Kozo Sugiyama
Japan Advanced Institute of Science and Technology
1-1 Asahidai, Tatsunokuchi, Nomi, Ishikawa 923-1292, Japan

Abstract. Incidental information media are useful media for giving non-urgent notices, advertising information etc. at the places where we incidentally pass or stay to do our primary activities. In this paper, we investigate how to design online incidental information media providing daily news at a lounge so that they maximize the opportunities for users to encounter interesting/important news while minimizing intrusiviness or distraction. By the usability tests, the following results are obtained: 1) "fade update" display is not as distracting as "scroll update" or "plain update" displays, and 2) it is effective to use personal mobile media without gesture in order to show the detail information of displayed news, and 3) "one-by-one update" display is superior to "at-a-time update" display in increasing the opportunities for users to encounter important/interesting news.

1. Introduction

Incidental information media are useful media for giving non-urgent notices, advertising information, etc. at the places where we incidentally pass or stay to do our primary activities such as going somewhere, waiting for someone and so on. There exists a variety of such media according to types of displayed information and kinds of places. Bulletin boards in a street and advertising posters hung in a train are traditional examples. It is convenient for us to incidentally obtain the information from incidental information media without interrupting our primary activities. Moreover, sometimes we may get interesting/important information earlier than seeing TV or reading a newspaper.

Recently online electronic display panels are often used for the media in our living environment such as streets, trains, stores, etc. and the online information displayed on the panels is frequently updated. However, online incidental information media are not widely used at the places such as waiting rooms or lounges where we stay relatively long, though we have some time to incidentally obtain information there. If the online electric displays are used at the lounges, the following problems arise.

- Even if some people in the lounge are not interested in looking at the displays, they always feel changes in display images in their peripheral views. So they may distract these people, as advertising banner animations on web pages often distract us [1, 2].
- On the other hand, most of people may ignore or forget that there exist the incidental information media unless the media interrupt them in some degree.

In this paper, we investigate how to design the online incidental information media providing daily news at the lounges so that they maximize the opportunities for users to encounter interesting/important news while minimizing intrusiveness or distraction. Our study is influenced by the work of Maglio et al. [3]. They investigated the display update feedback of such media so that they provide the most information while having the least impact on the user's peformance on the main task. They used artificial information in their

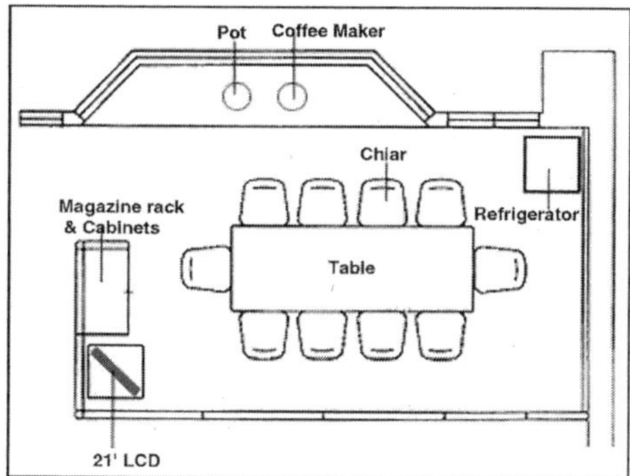

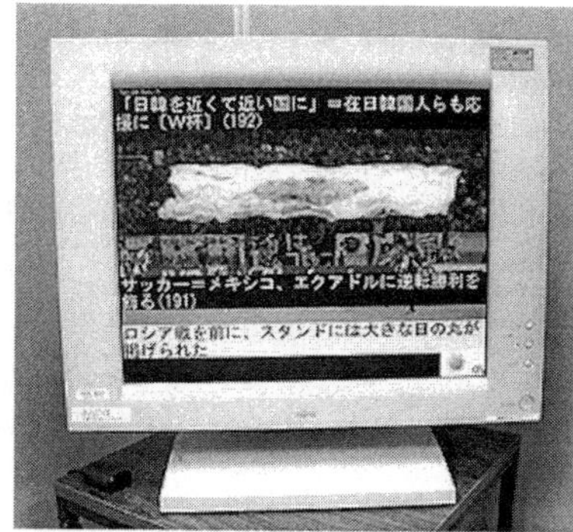

Figure 1: The experimental environment and system

experiments, regarded providing the most information as having a user memorize displayed information as much as possible, and concluded that the update feedback by scrolling motion was the best way. However, in case of actual daily news, not every information is important/interesting to a user. Therefore we do not consider here the tradeoffs between having a user memorize information as much as possible and minimizing intrusiviness or distraction. But we consider the tradeoffs between maximizing the opportunities for a user to encounter important/interesting news and minimizing intrusiviness or distraction.

The paper is organized as follows. We first describe the experimental environment and system to display online news. We next report on the experiment comparing three display update methods to minimize intrusiveness or distraction. We third report on the experiment comparing two update methods to maximize the opportunities for users to encounter their important/interesting news, after discussing how to getting the details of the news. We finally summarize the results of the experiments.

2. Experimental system and environment

We have developed an experimental system to show daily news headlines on the 21-inch LCD at the lounge of our laboratory (Figure 1). About 10 members of the lab are sharing this lounge to take tea, to read newspapers or magazines, to discuss, etc. The system automatically collects news headlines (usually only texts, sometimes with images) from several news web sites that the members often watch. When the system collects the latest news every ten minutes, we always have several ten candidates. The system displays several headlines among the candidates in order and repeatedly.

3. Usability test

3.1 Minimizing distraction

To determine the property of a display update that minimizes distraction, we have tested the following three methods.

1. Plain update: The five spaces to display headlines are prepared on the display. Each space is used to display one headline. Headlines appear there one by one at seven-second intervals. These headlines are disappeared after displayed for twenty-two second. The rewriting for these appearing and disappearing is quickly executed.

2. Scroll update: Headlines scroll from bottom to top at a constant rate. The time between updates is 83 microseconds. The times at which one headline is displayed and the time until which new headline appears are adjusted to the ones of the plain method.
3. Fade update: This update method is the same as the plain update, except that headlines gradually fade in and out when appearing and disappearing.

Experiment

After the lab's members had experienced each display update method for a while, we asked them the following questions about each display update method.

Q-1 Distraction: While you are not looking at the display, does the display distract you?

Q-2 Stress: While not looking at the display, are you under the stress by the display?

Q-3 Agreement: Do you agree with the installation of each display in the lounge?

To answer the upper two questions, the following five-point scales were used.

1. never   2.rarely   3.occasionally   4.often   5.always

To answer the last question, the following five-point scales were used.

1.strong disagreement   2.disagreement   3.neutral   4.agreement   5.strong agreement

Moreover, we asked the members the following questions about "fade update" after they had experienced the "fade update" display for 2 months.

Q-4 Seeing: While you stay the lounge, has the display come into your sight?

Q-5 Looking: While you stay the lounge, have you looked at the display?

To answer these questions, the following five-point scales were used.

1. never   2.rarely   3.occasionally   4.often   5.always

Results

Figure 2 summarizes the results of *Q 1-3*. Repeated measures ANOVAs were performed on these measures. Significant main effects were found for distraction ($F(1,8)=12.516$, $p=0.0005$), stress ($F(1,8)=4.265$, $p=0.0328$), agreement ($F(1,8)=6.346$, $p=0.0094$). Multiple comparisons of these measures showed followings:

● "Fade update" was not as distracting as "scroll update" ($p<0.01$) or "plain update" ($p<0.01$).
● "Fade update" was less stressful than "scroll update" ($p<0.05$).
● People preferred to install the "fade update" display rather than to install the "scroll update" display ($p<0.05$) or to install the "plain update" display ($p<0.05$).

In addition, the results of *Q 4-5* showed that users didn't ignore or forget the display even if "fade update" that was the least interruptive was used. The average of "seeing" was 3.67 and the average of "looking" was 3.44. There were no answers under "occasionally" in both questions. Figure 3 summarizes the results of *Q 4-5*.

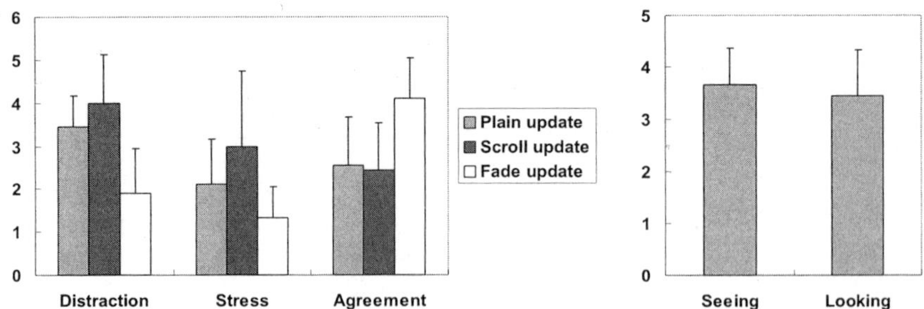

Figure 2: The results of questions 1-3   Figure 3: The results of question 4-5.

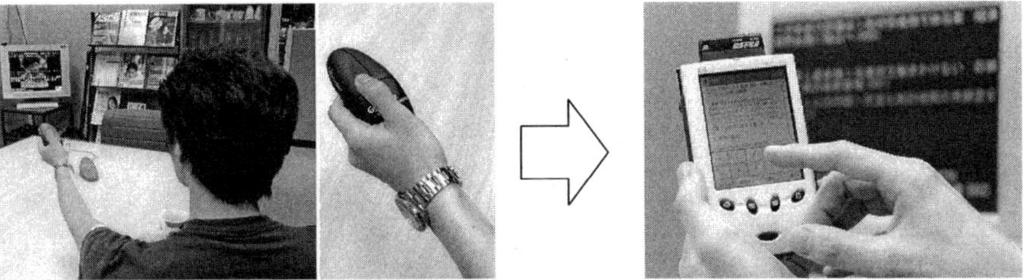

Figure 4: Transition of the methods for getting detail information.

3.2 Getting detail information of an important/interesting headline

During the test, the members often wanted to get details/contents of the particular news headlines when they were interested in them. Therefore, first we have modified the system so that the user can get detail information of the headlines displayed on the LCD. In the system the user could switch the headline view to the detail view by clicking a particular headline with a Gyro-mouse [4] that was put on the table in the lounge. However, since the LCD is a public display shared by the members, the following three problems have arisen:

- When one of the members looks at the detail on the LCD, the other members cannot see the hidden part.
- It is difficult to read a long article such as detail information on the public display that is installed far.
- Several members do not like that their interests become known by the other members.

To solve the above problems, we have used personal mobile media (Pocket PC [5]) that were connected to the public display through wireless LAN and could provide the details of the particular headline (see [6]). In the system we give each headline an ID number that is shown at the each headline tail. To get the detail information of a particular news headline, the user input the ID number on the mobile media. Immediately, the user can see personally the detail of the news headline on the mobile screen. Currently, one Pocket PC is always put on the table in the lounge for common use and mobile phones (i-mode [7]) are also available as personal mobile media.

3.3 Maximizing the opportunities for users to encounter important/interesting news

To determine the property of the display update method that maximizes the opportunities for users to encounter important/interesting news, we have tested the following two display update methods.

1. One-by-one update: This update is the above "fade update" itself. Note that the displayed headlines are updated one by one.
2. At-a-time update: Like "fade update", the five spaces to display headlines are prepared on the display. However, when appearing, five headlines fade in there at a time. These headlines also fade out at a time after displayed for thirty-five seconds. The times taken for fading in and out are adjusted to the times of "one-by-one update." Besides, the number of updated headlines for a given period is also adjusted to the number of "one-by-one update."

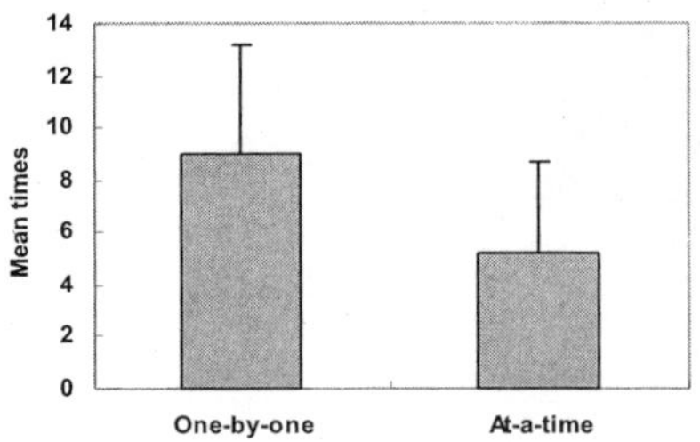

Figure 5: Means times of getting details for one-by-one and at-a-time updates.

Experiment and Discussion

Each update method was alternately used for fifteen days in the lounge. During this period, the number of times that users got details of displayed news headlines through the PDA on the table in the lounge was recorded per day, but we didn't inform the lab's members of this in consideration of influences on the result.

As a result, one-tailed t-test showed that the number of times that users got details during the use of "one-by-one update" was much more than during the use of "at-a-time update" ($t(28)=2.692$, $p=0.0059$). Figure 5 summarizes this result.

One possible reason for this result is that the periods when users continue to look at the "one-by-one update" display is longer than the periods when users continue to look at the "at-a-time update" display. The longer the periods when users continue to look at the display is, the greater the opportunities for users to encounter important/interesting news may be. However, additional experiments are needed to verify this possible reason.

4. Conclusion

In this paper, we have investigated how to design the online incidental information media providing daily news at the lounges so that they maximize the opportunities for users to encounter interesting/important news while minimizing intrusiveness or distraction. By the usability tests, the following results are obtained:

- "Fade update" display is not as distracting as "scroll update" or "plain update" displays.
- To show the details of news headlines on the public display, it is effective to use personal mobile media without gesture.
- "One-by-one update" display is superior to "at-a-time update" display in increasing the opportunities for users to encounter important/interesting news.

Overall, regarding the public display, we recommend the display on which news headlines fade in and out one by one as the best display. For future work, we intend to develop the technology so that it links our personal mobile media to ubiquitous public displays dynamically.

References

[1] http://www.useit.com/alertbox/9512.html
[2] Nielsen, J., Designing Web Usability: The Practice of Simplicity,
[3] Maglio, P.P., Campbell, C.S.: Tradeoffs in displaying peripheral information. *Proceedings of CHI '00*, pp.241-248, 2000.
[4] http://www.gyration.com/
[5] http://www.microsoft.com/mobile/pocketpc/
[6] Tani, M., et al., Courtyard: Integrating shared overview on a large screen and per-user detail on individual screens. *Proceedings of CHI '94*, pp.44-50, 1994.
[7] http://www.nttdocomo.co.jp/p_s/imode/

ReSPoM: a Meeting Support System for Retrieval Records of a Meeting with Relationships among Record Elements

Itaru KURAMOTO[†], Jun NODA[‡] , Noriyuki FUJIMOTO[*] and Ken-ichi HAGIHARA[*]

[†] *Faculty of Engineering and Design, Kyoto Institute of Technology,*
Gosyokaido-cho, Matsugasaki, Sakyo-ku, Kyoto 606-8585, JAPAN
[‡] *NEC Internet System Laboratories*
NEC Kansai Research Labs. 3F, 8916-47, Takayama-cho, Ikoma, Nara 630-0101 JAPAN
[*] *Graduate School of Information Science and Technology, Osaka University*
1-3 Machikaneyama, Toyonaka, Osaka 560-8531 JAPAN

Abstract. ReSPoM is the system to retrieve the records of a meeting effectively. It relates the audio records to handwriting notes easily during the meeting, and provides the function to trace the relationships for retrieving the records. It is expected that the relationships are useful to find the desired points of the meeting record exactly and quickly. In this paper, we show the detail of the relationships among records and the implementation of ReSPoM. The result of evaluation indicates the ReSPoM is effective to retrieve the records of a meeting.

1 Introduction

A meeting is one of the most common and important parts of cooperative work. There are various researches in order to support meetings with computers[1][2]. In this paper, we focus on the records of a meeting including video/audio media, handwriting notes, whiteboard-drawings. Mynatt pointed out in his study[3] that retrieving the records of a meeting is one of the most important works after meeting; however, most of studies about supporting a meeting tends to pay attentions only to recording, not to retrieving records.

According to the retrieval of records, we can hardly get the exact part of the records we need because of a huge amount of video/audio records or incompleteness of our handwriting notes. To solve the problem, we proposed a system, named ReSPoM, which relates the audio records to the notes during a meeting. The relationships are used for help to retrieve the desired part of the records.

First, we show the details of the way to relate the records. Next, we introduce the implementation of ReSPoM, and show the results for evaluation of ReSPoM.

2 Meeting and Record

2.1 Meeting

In this paper, we assume a meeting is held in a general meeting-room with whiteboard by three to six participants, like a research discussion in a laboratory. At a meeting participants may use some paper materials, make handwriting notes by ink, draw some pictures on the whiteboard, and discuss each other in any order; in some cases they may say something concurrently.

In order to record the meeting more precise, usually they record the situation of the meeting by video/audio media devices and keep the copy of drawing on the whiteboard by handwriting or by a self-copy system of the whiteboard.

2.2 Relationships among Meeting Element

We assume that a meeting consists of following **meeting elements**;

1. participants' handwriting notes,

2. utterances by participants,

3. written words on materials,

4. pictures on materials,

5. drawings on the whiteboard.

Retrievers are the participants who want to remember a part of a meeting they joined. In the ordinary way, in order to review the part they can use their handwriting notes and/or the video or audio records of the entire meeting. However, we can hardly review the desired part of a meeting with only handwriting notes and/or the video or audio records.

In case of handwriting notes only, retrievers can easily find out the part of the meeting which they thought the part was important. Usually, the notes written on the materials handed out the meeting or their own notebooks are remarkable so that they can find out the important part easily. Notes, however, are not so useful for understanding the part of the meeting clearly because of the incompleteness of notes. They write notes concurrently when an important part are discussed. If their attentions are directed to the important discussion, the notes are rough and far from complete.

In case of video/audio records only, retrievers can find out the complete record of the part. Video/audio record is a raw data stream, so it includes all of the discussion (or utterances) to which they may miss to listen. Instead, they can hardly to retrieve the exact point of the view/speech they want to review because a raw data stream tends to huge although they can access only sequentially.

Then, retrievers usually use both notes and video/audio records. According to meeting elements, they relate each other. For instance, we handwrite a note about a certain speech for keeping the importance of the speech. In this situation, we can consider that the note relates the speech. There are many relationships among some meeting elements in the same way.

However, the relationships themselves have not been recorded. This is the reason we can hardly retrieve the record and find out the meeting elements we want. In this case, they want to retrieve the point of the video/audio record which their notes point out. If they can keep the relationship among the notes and the desired point of a video/audio record, and trace it for retrieval, they can easily find out the exact part of the view/speech.

We conclude it is efficient for retrieval of the meeting elements retrievers desired to keep and tracing the relationships.

2.3 Keeping and Tracing the Relationship

There are two methods of keeping the relationship, manual keeping and automatic keeping by timestamps.

In the manual method, all meeting elements are visualized on the system. A participant simply touch two of them directly when he or she think there are a relationship among them. The method needs visualization of all type of meeting elements on one system. We use a PC with liquid crystal tablet display in order that note taking and speech visualization can be done at one display.

By contrast, a participant do nothing to keep relationships by the timestamp method. We consider that there are a relationship among two meeting elements which is recorded at the same time. However, this method cannot cover the relationships the participant needs. For example, he or she may connect a speech to the notes wrote before. This disadvantage can be fatal because we tend to discuss similar contents repeatedly.

Thus, the introduced system takes the former method.

3 ReSPoM: The System of Tracing Meeting Elements

ReSPoM is the system for retrieving meeting records effectively, which has a function of keeping relationships among meeting elements during a meeting and a function of tracing relationships for retrieving meeting elements after the meeting.

3.1 Overview

ReSPoM consists of two subsystems basically. One is **the Recorder**, which is used during the meeting in order to record meeting elements and the relationships. Another is **the Seeker**, which is used for retrieving records using relationships.

The Recorder is used during a face-to-face meeting and keeping audio records. The Seeker is used after the meeting.

We pay attentions to some points as follows when designing ReSPoM:

1. When using the Recorder, participants can do the equivalent actions which is done during a general meeting.

2. The actions provided by ReSPoM are like as the equivalent ones as possible.

3. Additional actions (like keeping relationships) must be simple and intuitive enough to avoid troubles to a meeting.

3.2 The Recorder

Figure 1 shows the screenshot of the Recorder. Each participant has the Recorder. It is implemented on a PC with a liquid crystal pen tablet display and a microphone. Recorders are connected by LAN each other.

It has three parts of the window: **Speech Visualization**, **Document Viewer**, and **Whiteboard Viewer**. Figure 1 shows former two parts. Participants can write notes on a certain page of documents displayed on the Document Viewer with stylus pen, which is like to note with ink. The usage of whiteboard drawing is also alike it.

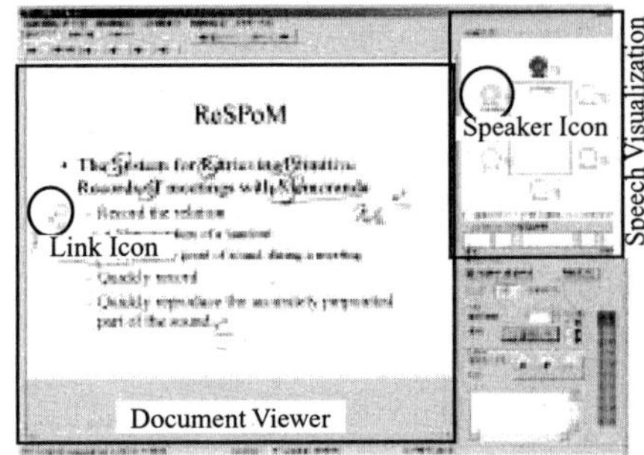

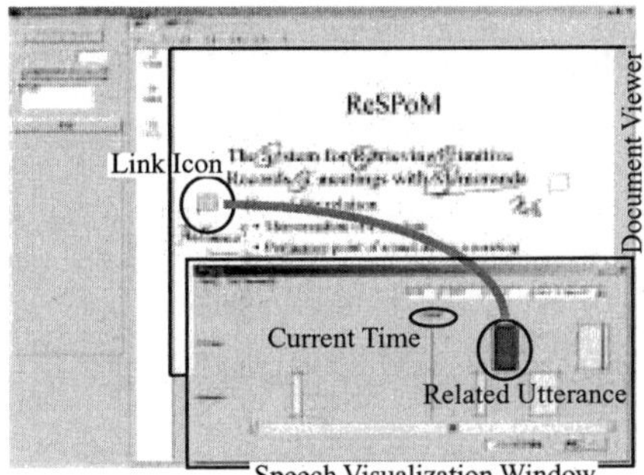

Figure 1: The screenshot of the Recorder.   Figure 2: The screenshot of the Seeker.

When a participant starts a utterance, his/her Recorder broadcasts the information of starting his/her utterance to all Recorders. When one of them get the information, it turns the **Speaker Icon** of him/her to yellow in order to show he or she starts speaking. When stops a utterance, the speaker's Recorder broadcasts, and all Recorders turn their Speaker Icons to gray. The utterances of past one minute can be shown the bottom of the Speech Visualization Block by chart. Besides, only the speaker's Recorder records his/her utterances, not broadcasting the audio stream.

Keeping relationships are simple. when a utterance is begin spoken now, and a participant feels there are a relation among the utterance and his/her note, he or she touch the yellowish Speaker Icon and the note written on the Document Viewer. In the case of relating past utterance to a note, he or she click grayish Speaker Icon (then the past-utterance chart shows), touch the yellow square showing the past utterance, and touch the note. When keeping a relationship, a **Link Icon** is put on the touched place of the note.

3.3   The Seeker

Figure 2 shows the screenshot of the Seeker. Its looks is like to the Recorder, but it has two windows. One is a main window, which has **Document Viewer**. The Document Viewer of the Seeker doesn't have the function of adding handwriting notes and relationships. Another is **Speech Visualization Window**. It shows all participants' utterances by chart.

Using the Speech Visualization Window, we can simply play/pause the recorded audio of a meeting. The chart moves synchronizing with the audio. The chart can be dragged to the desired point for the current time and can play from then. A certain utterance can play by clicking the square of the utterance on the chart.

Clicking the Link Icon on the Document Viewer, the current time is moved to the time which the utterance related by the Link Icon starts. Then, we can play the audio from related utterance of the notes simply. On the other hand, double-clicking the square of the utterance on the Speech Visualization Window, and if the utterance is related with a certain notes, the Document Viewer shows the page of documents on which there are related notes. In both situation, clicked or targeted square and Link Icon is emphasized to red.

Table 1: The Results of the Experimental Meeting

Participant	Using ReSPoM	Using Papers & Audio Tape
Answering Time	28 min.	40 min.
% of Correct Answer	86	71

(the values are average of each 2 participants.)

4 Evaluation

We made a experimental meeting for evaluating the efficiency of ReSPoM. 4 participants joined the meeting; 2 of them had paper materials and audio tape recorder for recording the audio of the meeting, and the others had ReSPoM(the Recorder). They discussed about 30 minutes in the meeting. A few days after the meeting, we made them answer some questions about the contents of the meeting. During answering the questions, participants of ReSPoM could use the Seeker, the others could use their own paper materials and the audio tape.

Table 1 shows the results of the experimental meeting. Clearly, the participants using ReSPoM spent shorter time to answer and got better ratio of correct answer. The result indicates that ReSPoM reduces the time for retrieval and that users can retrieve exact points of the meeting records more correctly.

5 Conclusion

We investigated the retrieving records of a meeting, and found that the lack of relationships caused troublesome to retrieve the audio stream record exactly and quickly. To solve the problem, we proposed the method of keeping the relationships and tracing it when retrieving. Then we introduced the meeting support system, ReSPoM, which have the function to keep relationships among the audio records and handwriting notes easily during the meeting, and the function to track the relationships for retrieving the records. The result of the evaluation indicates that the system is effective to retrieve the records of a meeting.

Acknowledgement

This research was partially supported by the Ministry of Education, Science, Sports and Culture, Grant-in-Aid for Scientific Research on Priority Area(2), 14022229, 2002.

References

[1] Chiu, P., Kapuskar, A., Reitmeier, S., and Wilcox, K., NoteLook: Taking Notes in Meetings with Digital Video and Ink, *ACM MULTIMEDIA '99*, 144–158(1999).

[2] Davis,R.C., Landay, J.A., Chen, C., Huang, J., Lee, R.B., Li, F.C., Lin, J., Morrey, C.B., Sehleimer, B., Price, M.N., and Schilit, B.N., NotePals: Lightweight Note Sharing by the Group, for the Group, *Proceedings of ACM CHI '99*, 338–245(1999).

[3] Mynatt, E.D., Igarashi, T., Edwards, W.K. and LaMarca, A., Flatland: New Dimensions in Office Whiteboards, *Proceedings of ACM CHI '99*, 15–20 (1999).

[4] Kaiya, H., Miura, N., Anai, G., Ebata, T., Nagaoka, H. and Saeki, M., Preliminary Experiments of a Computer System for Face-to-Face Meetings, *IEICE Journal Vol.J79-D-I*, No.6, 341–352 (1996).

KES 2002
E. Damiani et al. (Eds.)
IOS Press, 2002

Electronic Tag-Playing Support System with Awareness Function

Takashi Yoshino
Faculty of Systems Engineering, Wakayama University,
930 Sakaedani, Wakayama 640-8510, Japan

Jun Munemori
Faculty of Systems Engineering, Wakayama University,
930 Sakaedani, Wakayama 640-8510, Japan

Abstract. Mobile phone, PHS and PDA (Personal Digital Assistant) are highly portable, and positioning data will become important data for human communication. There are many location information services. However, most of services only display the position information on target as a map. There was no service using the sense of hearing or touch as expression of position information. Then, we have developed the electronic tag-playing support groupware with awareness functions using sound, vibration and blinking color screen. We performed experiments 30 times with the awareness functions and without the awareness functions at two universities. The results of the experiments show that the evaluation of the enhanced type (with the awareness functions) one-to-one experiments was higher than that of the standard type one-to-one experiment. The awareness functions seem to be effective to increase seriousness.

1 Introduction

The Internet has come into widespread use and cooperative works through the Internet increased. However, people sometimes could not take effective communication one another using groupware. Then, in order to support effective communication, the awareness function, which sends another person's situation, becomes important in groupware [1, 2].

PDA (Personal Digital Assistant) is highly portable, and can collect data in anytime and anywhere[3]. The positioning data would become important for a PDA[4].

We had developed an electronic tag-playing support system, called E-ONIGOCO [5, 6]. Tag is a traditional Japanese game. The game consists of runaways and chasers. If the chasers catch all runaways, then the game is over. E-ONIGOCO consists of a PDA, a GPS and a mobile phone. E-ONIGOCO has a mutual location information service using electronic mail and a web browser. We performed experiments using E-ONIGOCO. Participants felt that the game is relatively insteresting. We found that only the visual information of the display causes a lack of seriousness, and the feeling reduced the evaluation. That is, most of participants could not feel the reality of the other party without real world.

Then we have been developing an enhanced E-ONIGOCO to be added other senses (sense of hearing and sense of touch) for awareness function.

2 Electronic tag-playing support system with awareness functions

2.1 Design policy

We show the design policy of electronic tag-playing support system with awareness functions.

1. Awareness function using the sense of vision, hearing, and touch
   We considered using the sense of vision, hearing, and touch as awareness function, because we need many senses for awareness.

2. Communication using the function on a mobile phone
   Electronic tag plays in the open air. People need to carry equipment for electronic tag playing. Therefore, it is more desirable not to use extra equipments for communication of information. Then, we considered using a mobile phone. Most mobile phones have the functions of a web browser for the sense of vision, ringing tone (melodies) for the sense of hearing, and manners (vibration) mode for the sense of touch. We use the mobile phone of J-PHONE that can be controlled ringing tone or vibration by using the tag and attribute of HTML.

2.2 Type of tag playing

We performed two kinds of experiments.

1. One-to-one experiment
   There is one group of runaways and one group of chasers. The position of the opponent group is shown on the center of the screen as a symbol on a mobile phone.

2. Virtual tag-playing experiment
   Experiments are performed between two universities. The longitude and latitude data of places at Wakayama Univ. correspond to places at Kagoshima Univ. That is, maps of each university are overlapped. The area of Kagoshima Univ. is larger than that of Wakayama Univ. In virtual tag-playing experiment, the area of Kagoshima University was limited to the same size as Wakayama University. The virtual position of the opponent group is shown in each screen on a mobile phone.

   In one-to-one experiment, participant are in the actually same area. In virtual tag-playing experiment, participant are in the same area only on a screen.

2.3 Electronic tag-playing support system (E-ONIGOCO)

E-ONIGOCO consists of a mobile system and an information processing system for map.

1. Mobile system
   Figure 1 shows the mobile system. The mobile system consists of a PDA (3Com, Palm III), a GPS (GARMIN, etrex), a modem (I·O Data, SnapConnect), and a mobile phone (for J-PHONE). The accuracy of the GPS is about 15 meters. The total weight of the mobile system is about 700g. Participants send information of their positions by an electronic mail on a mobile phone. They can view maps by a web browser on a mobile phone. URL of the map is indicated by an electronic mail when the map is updating. The mail also indicates who is renewed.

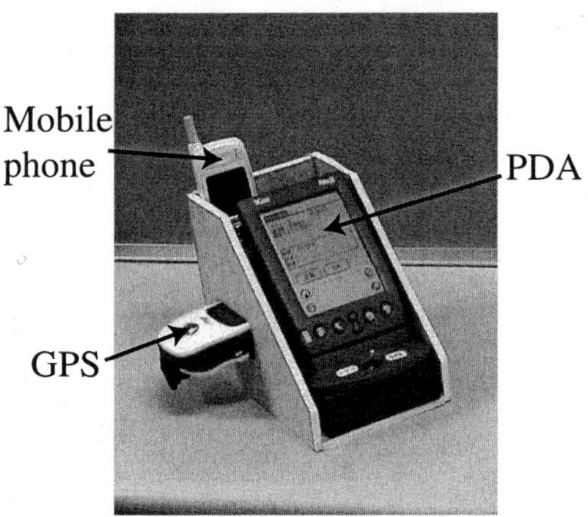

Figure 1: The mobile system.

2. Information processing system for map
   The information processing system for map processes the data from mobile systems.
   The system decides and makes a corresponding map from longitude and latitude data. A
   runaway or a chaser is positioned the center of the map and marked as a kind of symbol
   marks. The system also shows a locus, building name near their position, and a scale of a
   map. There are two kinds of displays of a map: a whole display and a detailed display.

This system has a virtual playing tag function. For example, a runaway is existed in
Wakayama Univ. and a chaser is existed in Kagoshima Univ. The distance between two
universities is about 900 km. However a runaway and a chaser seem to be existed in the
same university on the screen of a mobile phone. This function transforms the positioning
data of a GPS and corresponds to the longitude and latitude data of both two universities. If
the difference between the position of a runaway and that of a chaser will become below a
certain fixed value at the same time, we decide that a runaway seems to be arrested.

2.4   Awareness functions

We realized three kinds of awareness functions, in order to add some feelings to electronic tag
playing.

1. Visual awareness function: The color of a screen on a mobile phone is changed corresponding to the distance between a runaway and a chaser.

2. Hearing awareness function: The sound of a mobile phone is changed corresponding to
   the distance between a runaway and a chaser.

3. Touch awareness function: The vibration of a mobile phone is changed corresponding to
   the distance between a runaway and a chaser.

The awareness functions are realized by the function of the web browser on a mobile
phone. A participant feels awareness functions, when they look at the map on a web browser
on a mobile phone.

Table 1: Results of experiments.

Questionnaire items	One-to-one experiment				Virtual tag playing	
	Kagoshima Univ.		Wakayama Univ.		Kagoshima Univ. and Wakayama Univ.	
	Standard	Enhanced	Standard	Enhanced	Standard	Enhanced
1. Was it interesting?	3.6	4.3*	3.0	3.6*	2.9	3.4
2. Was the location information useful?	3.9	4.8*	3.9	3.9	4.3	4.6
3. Was the map easy to see?	2.6	2.9	2.9	3.2	2.7	3.3*
4. Was the locus easy to see?	2.8	3.4	2.5	3.4*	2.6	3.8*
5. Was the position display by the character useful?	4.7	4.5	3.6	4.1	4.4	4.2
6. Was it easy to use screen operation of a mobile phone?	3.3	3.4	3.5	3.5	3.2	3.7
7.Was the screen of a mobile phone small?	2.1	2.3	2.2	2.1	2.2	2.0
Time to be caught	80	66	24	68	66	15
	110	108	21	19	39	46
	90	136	14	73	107	33
	89	43	30	25	34	13
	67	66	40	28	28	58
Average time to be caught (minutes)	87.2	83.8	25.8	42.6	54.8	33.0

"*" shows a significant difference between standard type and enhanced type.
The significance of ANOVA is less than 0.05.

3 Experiments and discussion

3.1 Experiments and results

We performed the following experiments each 5 times from November 2001 to December 2001 at two universities. Standard type experiments only use location information. Enhanced type experiments use not only location information but also sound, vibration, and blinking color screen for awareness.

1. In Kagoshima University, One-to-one experiment (standard type/enhanced type)

2. In Wakayama University, One-to-one experiment (standard type/enhanced type)

3. Between Kagoshima University and Wakayama University, Virtual tag-playing experiment (standard type/enhanced type)

The area of Kagoshima Univ. is about 480,000 square meters, which is including 85 building and the area of Wakayama Univ. is about 100,000 square meters, which is including 26 building. Runaways or chasers consisted of two persons.

Table 1 shows the results of the experiments. We carried out a questionnaire. We had a 5-point scale (with answers "1: very bad", "2: bad", "3: neutral", "4: good" and "5: very good") for participants to evaluate each item on the questionnaire.

3.2 Discussion

The evaluations of many items on questionnaire of the enhanced type experiments are higher than that of the standard type experiments. The evaluation value of "interesting" regarding the enhanced type one-to-one experiments were better than that of the standard type one-to-one experiments (from 3.6 to 4.3 on 5-point scale in Kagoshima University, from 3.0 to 3.6 on 5-point scale in Wakayama University). The participants commented, "when we approaches each other, it is interesting that a mobile system sounds", "The color of a screen changes was legible", and "I was excitedly during the escape."

The evaluation of experiments in Kagoshima University tends to be higher than that in Wakayama University. We think that these results were caused by corresponding to the number of times of operation of an "awareness function." That is, the area of Kagoshima University is five times as broad as that of Wakayama University. Therefore the playing time in Wakayama University is shorter than that in Kagoshima University.

"Average time to be caught" did not change between the enhanced type experiments and the standard type experiments in Kagoshima University. We think that this is because the area of Kagoshima University is large enough for a place of electronic tag playing. In Wakayama University, "average time to be caught" of the enhanced type experiments was longer than that of standard type experiments. We think that this is because runaways became to escape seriously by the awareness functions. In virtual tag-playing experiment, the evaluation value of "interesting" increased from 2.9 to 3.4 on 5-point scale. Therefore, "average time to be caught" of the enhanced type experiments was shorter than that of standard type experiments. That is, motivation of chasers may increase by the awareness functions.

From the experiments, we could not find clearly which awareness function is effective for awareness. We think overall awareness functions raises feelings of interesting of E-ONIGOCO.

4 Conclusion

We have developed the electronic tag-playing support system with awareness functions. These functions use sound, vibration and blinking color screen for increasing the feelings of reality. We performed experiments 30 times with the awareness functions and without the awareness functions at two universities. The results of the experiments show that the evaluation of the enhanced type one-to-one experiments (using sound, vibration and blinking color screen) was higher than that of the standard type one-to-one experiment. The awareness function seems to be effective to the game.

We will consider the expression method of the feelings when being caught.

References

[1] Dourish, P., Bly, S. : Portholes : Supporting Awareness in a Distributed Work Group, Proceedings of CHI'92, ACM, pp. 541–547 (1992).

[2] Susumu Kunijifu, Naotaka Kato, Chie Kadowaki, Mikifumi Shikida: Knowledge management using Intelligent Groupware, Union of Japanese Scientists and Engineers Press, July 2001 (Japanese).

[3] R. Davis, J. Landay, V. Chen, J. Huang, R. Lee, F. Li, J. Lin, C. Morrey, B. Schleimer, M. Price, B. Schilit, "NotePals: Lightweight Note Sharing by the Group, for the Group",in Proceedings of ACM CHI'99, ACM Press, 338-345,1999.

[4] G. D. Abowd, C.G. Atkeson, J. Hong, S. Long, R. Kooper and M. Pinkerton, "Cyberguide: A mobile contex-aware tour guide", Wireless Networks, 3, pp.421-433, 1997.

[5] Jun Munemori, Emi Miyauchi, Tomohiro Muta, Takashi Yoshino, Kazutomo Yunokuchi: Development and Application of the Electronic Playing Tag Support Groupware, IPSJ Journal, Vol.42, No.11, pp.2584-2594, Nov. 2001 (Japanese).

[6] Jun Munemori, Tomohiro Muta, Takashi Yoshino: E-ONIGOCO: Electronic Playing Tag Support System for Cooperative Game, Proceedings of The 16th International Conference of Information Networking (ICOIN-16), pp.5D-1.1-5D-1.10, Jan.2002.

KES 2002
E. Damiani et al. (Eds.)
IOS Press, 2002

SmartCourier: Annotation Management Tool for Research Labs

Sadanori Ito†, Yasuyuki Sumi‡, Kenji Mase‡, Susumu Kunifuji†
†*School of Knowledge Science, Japan Advanced Institute Science and Technology, Hokuriku*
1-1 Asahidai, Tatsunokuchi, Ishikawa 923-1292 JAPAN
‡*ATR Media Information Science Laboratories*
2-2-2 Hikaridai, Seika-cho, Soraku-gun, Kyoto 619-0288 JAPAN
E-mail: *sito@jaist.ac.jp*

Abstract. SmartCourier is a pen-based annotation system that supports adaptive information sharing and recommending for efficient knowledge interaction. The aims of our system are to extract the users' knowledge and interests from the annotations they make on documents that are shared via the web. We describe two methods that Smart-Courier uses to extract the user's interests: the positional commonality of annotation and the relationship between the user's annotation activities and the underlying keywords of documents. Furthermore, we compare the evaluation results of two different methods for efficient knowledge interaction. Evaluation experiments have shown that the second method is more efficient for supporting knowledge interaction.

1 Introduction

Recently, the rapidly developing WWW environment has brought about major changes in our information use and cooperative work. Also, there has been great activity in the creative work involving intellectual inspiration[3], education and research, and researching via networked WWW environments. Appropriately managing computerized information as knowledge and developing a use environment adaptive to the situation of each user are important tasks. In this paper, we describe the SmartCourier for Research Labs, which supports adaptive information acquisition and knowledge sharing by extracting information of a user's interests from his or her annotation activity on computerized documents.

This system provides adaptive recommendation of relevant scientific papers, matchmaking of co-researchers having similar interests, and annotation sharing services by extracting context such as interests and research purposes from the annotation activity. Accordingly, the system supports effective information acquisition and knowledge sharing, and we aim to activate the community of research and learning activity professionals involved in intellectual inspiration.

2 Annotation

There are two types of studies related to annotation. One concentrates on the usefulness of annotation as shared knowledge, such as comments and ideas added on information resources like lecture notes. The other type concentrates on the usability of the interface of pen and paper on real books and documents. In these studies, annotation is understood as the activity of externalizing the relationship between users and documents by the user clearly adding information at a specified position on static documents. Here, the "position" indicates the character strings and images that are available for the input of annotation, and the "added information" indicates the entries by the user on the document, such as a mark showing the grade of importance of an items to the user.

This kind of annotation is not an activity that depends on a pen-based system. There have been many studies of systems that install a tag-text on a specific position of a displayed HTML on a web browser, such as CoNote [1]. In using CoNote for educational activities, annotations on documents of lectures and assignments are considered as one type of useful shared information resource. It has been found that sharing this information through the Web between teachers and students helps understanding and promotes discussions on the documentation among students. However, the problem is that this method inhibits various expressions containing the information besides text, which is made possible by the free pen-based system.

A study related to the computerization of documents that contain typography provided by real books and texts, such as XLibris[5] and Adobe Acrobat®, attempts to increase the usability of electronic documents by making various expressions of pen annotation recordable in the same way, as well as writing something down on real paper. This kind of annotation has the advantages of the entire text being visibile and a high visual effect because this intuitive-input creates dramatic contrast with text's the typography, and this attracts the attention of readers. Such as approach places emphasis on usability through the personal convenience of annotation.

Writing something down on a memo pad, documents or a whiteboard imposes no limits on the ways of writing and expressing, and is possible to represent various expressions. Therefore, they are advantageous as complementary communication tools for small groups of people who share background knowledge such as common interests and purposes. These tools can be enhanced by using them in a real-time communication environment that provides audio and visual communications, which is useful for transmitting and organizing concepts that are difficult to explain clearly orally or in text. However, the problem is that a memo pad or whiteboard is inappropriate as an information resource that is continuously shared because there is a lack of annotations and the information related to the background and intentions of written representations among users who do not have common background knowledge or access to a real-time communication environment. Generally, it is useful to evaluate the significance and/or comments of the information resource to increase the efficiency of the information acquisition. However, it can be said that annotation is not effectively used as an information resource by sharing such an evaluation because the annotation is written in a free-hand representation form individually, and others cannot understand the annotator's background knowledge such as the rate scale and meaning of the annotation.

3   SmartCourier for Research Labs

First, in this research, we build a user model representing the users' interests from free expression of the annotation on computerized documents. From the similarity of individuals' user models, we automatically extract potential communities share interests. Then, we developed a way to support appropriate information acquisition and knowledge sharing, such

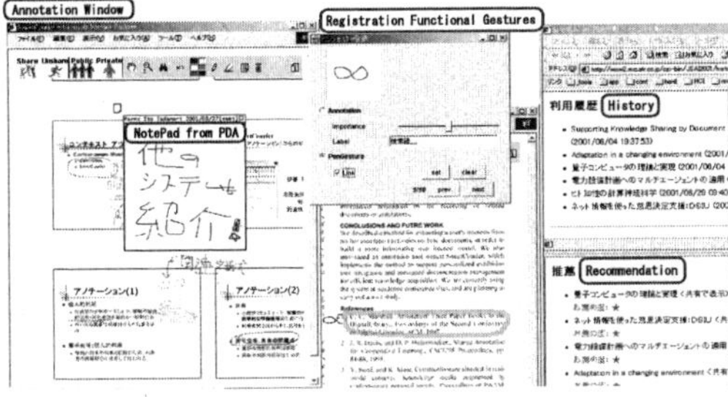

Figure 1: SmartCourier: system images

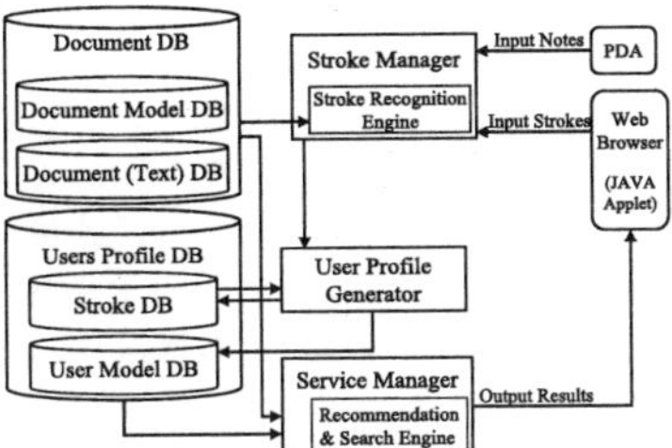

Figure 2: SmartCourier: system overview

as recommending computerized documents of related papers, sharing of the annotations on these documents, and matchmaking of research co-workers who have close interests. Accordingly, we support the sharing of heuristic knowledge among users who potentially share background knowledge such as interests. Furthermore, we aim to activate the research community involved in intellectual inspiration.

This system is constructed from Web service and personal information input devices such as a video tablet, which allows pen input and PDA functions. From each terminal, users are able to read and use shared information resources, and they can also receive services of information acquisition and knowledge sharing based on interests (figure 1 and figure 2).

For judging similarity of interests among user models, three indices are used. One is used to obtain a user's interests from his or her listed reading history of papers. On the papers, the keyword set indicating the realm is established. Judging similarity of interests among users is done by using the keyword set of papers that users read and creating a user vector that indicates users' interests. The second index is used to share the position of annotation. Users who leave annotations at the same position on the same paper are assumed to share interests, and the frequency of the positional commonality of annotation indicates the similarity of interests. We use the size and shape of the stroke as the third index to idetify strokes as the markings of sentences. If there is a duplicated part, we count this as positional commonality. For recognizing the shapes of the stroke, the gestures recognizing algorithm of Rubine [4] is used. These indices were used in a previous system (SmartCourier for JSAI2001 [2].) In this new system, SmartCourier for Research Labs, we adopted the following method to extract keywords for elements from marked texts and use a weighted keyword vector as the user's interests model based on this frequency of markings and the kinds of signs. At this time, if the user makes a defined stroke gesture of importance, its importance is rated (figure 1).

As a result of pre-research [2] about examples of annotations, underling, highlighting, boxing a sentence word (with a circle, square or parentheses), asterisking were obtained as typical markings on annotations. Markings are used for recording interesting points, such as problem posing, related papers, and names of researchers. Therefore, this system recognizes the default shapes of stokes defined from the results of the pre-research or user-defined shapes of strokes, and determine the annotating area for development of user interest models.

The match making service lists the group of users having similar user interest models, as calculated by the similarity of the keyword vector based on the cosine correlation from the higher rank. The paper recommendation service extracts the papers with markings having a

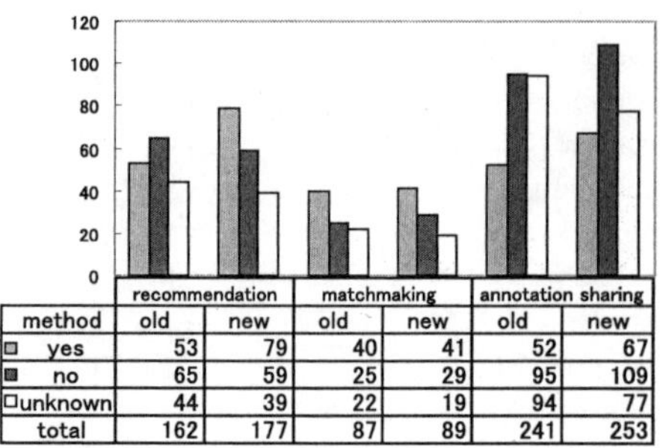

method	recommendation		matchmaking		annotation sharing	
	old	new	old	new	old	new
□ yes	53	79	40	41	52	67
■ no	65	59	25	29	95	109
□ unknown	44	39	22	19	94	77
total	162	177	87	89	241	253

Figure 3: Results of questionnaire in blind test

higher grade of importance to the group of users with similar user interest models. Then a list is made of unread papers by users of this service. The annotation sharing service lists users who leave annotations having higher grade of importance on recommended papers. It then lists the history of annotations by users introduced through matchmaking as prospects of annotation sharing. Each index is treated in isolation. We then prepared another sign to show the degree of recommendation for every index of the degree of similar. A user chooses an index sign and a recommendation list is displayed for that index.

4 Evaluation and Results

In this Chapter, we examine the influence on the research and learning community of services such as paper recommendation and matchmaking by SmartCourier.

Experimentation was done with 20 people of 5 groups as the experimental subjects. Groups A to D consisted of master course students belonging their respective laboratories as independent explicit communities. Group E was formed from doctor course students with research associates of Lab A. As documents used for reading and annotation, computerized proceedings of JSAI2001 were used. In the experimentation, after explaining the functions of the system to experimental subjects, we made them select papers that could be considered subjective reading related to their own research, learning tasks, and interests. They could select from all of the computerized proceedings with no limitation. We then made them read those papers and add annotations. [1] After all experimental subjects completed the work of reading and writing, we sought their impressions of the service provided by SmartCourier by giving them a blind test to compare the user modeling methods, i.e., the old method and new method. Furthermore, we limited paper recommendation to 10 papers and matchmaking to 5 people for each experimental subject. The number of papers browsed by experimental subjects was 75 of 231. That comes to an average of 3.75 papers read per person (minimum 1, maximum 13). The number of selected papers twice reached 17. This shows that experimental subjects pursued information acquisition according to their individual interests.

The results of a blind test with the old and new methods is are shown in a figure 3. In using the new recommendation method, a total of 177 papers were recommended. The fol-

---

[1] For convenience in implementing the experimentation, time available for reading is basically limited to one day, but reading time could be extended upon request. Subjects were also asked to read and write in the same field as the information acquisition activity as they typically use for general research field and studying.

lowing evaluation results were acquired. Papers that were interesting and unknown, or could not be acquired through general information acquisition, were found at an average rate of 4 papers per experimental subject. The superiority of the new methods is indicated by the number and ratio of valid recommendations. On the questionnaire asking whether the new or old method gave the most appropriate recommendation, 13 answered new method, 2 answered old method, and 5 answered no difference was noticed. On the paper recommendation using the new method, however, the amount of annotation activity and appropriate number of recommendations showing correlation, which had a correlation coefficient of -0.10, could be anticipated because keyword-set extracted from annotated areas was used for user modeling. In contrast, for groups A to D, without considering group E formed by research associates and doctor course students, the correlation coefficient was 0.50. In the interviews with group E members, the following comments were obtained: "necessary documents (proceedings of the conference) have already been checked out," "Not my field but around fields." From such comments, the results indicate that although the number of annotations and read papers is larger, the evaluation for the recommendation is low. That is, it is marginal support by users whose research tasks and fields are clearly set and who also have high information acquisition skills.

On matchmaking and annotation sharing, no difference between new and old methods was obtained from the results of the questionnaire. On matchmaking by the new method, about 41 of 89 introducees said they were sharing the same or similar interests even if there was little sharing. In this regard, the questionnaire also asked whether they already knew the research subjects of introduced users. The result was that 28 of 41 introducees did not know the research subjects of the other user. Also 31 of the 41 introducees that evaluated the matchmaking as appropriate were introduced from outside of the group. The number of introducees from inside the same group, considered an explicit community, only reached 10. As for experimental subjects outside of the group, 24 of 31 introducees were unknown users, and although many experimental subjects did not know students and researchers outside their explicit community, many instances of matchmaking were successful outside of the community.

From the result of the preceding experimentation, we have attempted to clarify the revealed existence of implicit potential communities. The incidence matrix of experimental subjects, browsed papers, and papers recommended as interesting were converted to an adjacency matrix of experimental subjects. Then the networks established by the interests of the experimental subjects were graphed out. This graph shows 16 cliques, and each clique was constructed from experimental subjects belonging to 3 to 5 groups. Hence, potential communities based on interests are formed beyond the explicit group. After the experimentation, we gave another questionnaire to evaluate the system as a whole: "As a system of stationary online annotation, do you agree with the continued use of the system in the university?". The results were 15 people for "agree", 4 people for "undecided," and 1 person for "disagree".

5 Conclusion

This paper proposed the SmartCourier for Research Labs as a system to support adaptive information recommendation and knowledge sharing services. Our aim was to contribute to the study of intellectual inspiration for users in research and learning activity. The system operates by developing a user model from the annotation activity on computerized documents.

From the results of experimentation, the SmartCourier system proud to be effective as an adaptive information sharing environment through annotations. Accordingly, it is promising as a method to activate a community through intellectual inspiration beyond existing explicit communities. Our results also indicated that new method using user modeling by keywords extracted from annotations was more effective than the old method of using browsing history and positional commonality.

Future work includes long-term experimentation on, for example, improving the user modeling and finding ways to cope with such temporal variables as changing interests. Additional improvements are also needed for the whole interface. We also need to develop an environment that has a high affinity with existing paper documents that are not computerized.

References

[1] J. Davis and D. Huttonlocker. Conote system overview, 1995. http://www3.cs.cornell.edu/dph/docs/ annotation/annotations.html.

[2] Sadanori Ito, Yasuyuki Sumi, Kenji Mase, and Susumu Kunifuji. Smartcourier: An annotation system for adaptive information sharing. *Journal of Japanese Society for Artificial Intelligence*, Vol. 17, No. 3, pp. 301–312, 2002(in Japanese).

[3] Susumu Kunifuji. Communication environment for intellectual inspiration. In *Proceedings of the 2nd Symposium on Advanced Science and Technology, Future Communication*, pp. 29–36, 1995(in Japanese).

[4] Dean Rubine. Specifying gestures by example. In *Proceedings of ACM SIGGRAPH '91:Computer Graphics*, pp. 329–337, 1991.

[5] B. N. Schilit, M. N. Price, G. Golovchinsky, K. Tanaka, and C. C. Marshall. As we may read: the reading appliance revolution. *IEEE Computer*, Vol. 32, No. 1, pp. 65–73, 1999.

KES 2002
E. Damiani et al. (Eds.)
IOS Press, 2002

Fault Detection based on Frequent Clusters

Yoshiyuki Yamashita

Department of Chemical Engineering, Tohoku University, Sendai 980-8579 Japan

Abstract. A new algorithm for fault detection is proposed based on clustered classes
of the ART2 neural network. An index for detecting changes of process state is defined
in order to represent differences between recent and historical class distributions. The
proposed algorithm was successfully demonstrated on an industrial wastewater treatment process for a detection of a state change.

1 Introduction

Fault detection of process plant is important, both from a safety and quality process maintenance standpoint. Much research has been done in this area over the years. Chiang *et. al.*
summarizes many approaches for fault detection and isolation (FDI), including SPC, PCA,
PLS, FDA, CVA, as well as soft computing methods [4]. Recent approaches to FDI based
on soft computing methods are also summarized in Calado *et. al.*, which includes neural
networks, fuzzy systems, knowledge-based systems and evolutionary algorithms [1].

Among these approaches, the ART2 network is one of the powerful tools available that
can solve a fault detection problem. Whiteley and Davis demonstrated the application of
the ART2 neural network to a recycle reactor [7, 8]. Yamashita *et al.* applied the ART2 to
classify flow patterns in a pneumatic conveyor [9]. Although the ART2 algorithm itself is
a powerful tool for state segmentation, the output classes of the algorithm do not always
provide information to an operator for direct recognition of the detection of state changes.

In this paper, a new method for fault detection is proposed that is based on the ART2
neural networks. Based on the segmented classes by the ART2 network, a method that keeps
track of class frequency is proposed here. In the proposed method, recent distribution of
classes are compared to the historical distribution of the classes. The proposed method is
successfully examined with data from industrial liquid-waste treatment plant.

2 Method

2.1 ART2 Clustering

Adaptive resonance theory (ART) is a series of neural networks originally introduced by
Grossberg [5]. The ART2 is a version of an unsupervised network which can cope with arbitrary sequences of analog continuous-valued input patterns. It employs a winner-take-all type
of learning, based on a similarity measure. By applying the ART2 network, the state vector
can be segmented into various classes. This classes can be directly used for fault detection.

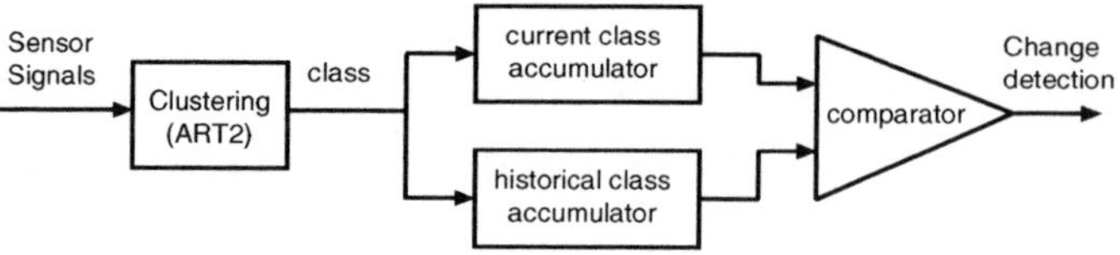

Figure 1: Schematic architecture

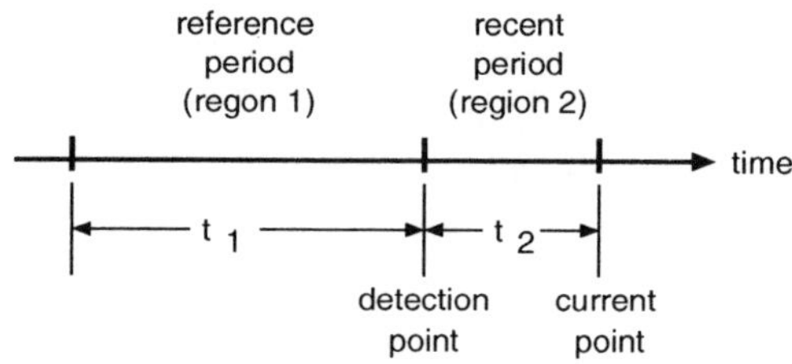

Figure 2: Time segmentation for the comparison

2.2 Changes of Segmented Classes

A single process state often corresponds to several clusters of the ART2 segmentation. Without labeling, it is sometimes difficult to detect changes of state by watching only class numbers. Considering the proper time-period, the frequencies of a class appearance are expected to change after changing the system states. For example, if the process has some troubles, recent frequent classes should differ from previous frequent classes [11]. Based on this idea, an index to detect state changes is proposed here.

Figure 1 shows the schematic architecture of the proposed algorithm. After segmenting the sensor signals into classes by ART2 neural networks, two vector accumulators are used for summing up class frequencies. The first accumulator is to evaluate classes in a recent period (Figure 2), and the second accumulator is to evaluate classes in a reference period. Comparing the two output vectors from these accumulators, an index is defined for the detection of changes in process state.

Define class frequency f of class k for each time period as:

$$f_{1,k} = \frac{m_{1,k}}{t_1 + 1} \tag{1}$$

$$f_{2,k} = \frac{m_{2,k}}{t_2 + 1} \tag{2}$$

where $m_{1,k}$ and $m_{2,k}$ represent the number of appearances of class k in region 1 and 2.

The average frequency of classes for each time period is defined as:

$$a_1 = \frac{1}{N_{c1}} \sum_{k=1}^{N_{c1}} f_{1,k} \tag{3}$$

$$a_2 = \frac{1}{N_{c2}} \sum_{k=1}^{N_{c2}} f_{2,k} \tag{4}$$

where N_{c1} and N_{c2} are the number of classes for each period.

Typical classes for each state can be considered more important than minor classes for each time period. Major classes can be defined as classes having more frequency than the average. If minor classes are not counted, an index (CMC; Changes of Major Class distribution) is proposed, based on the comparison of relatively frequent classes. Then an index for change detection can be defined as:

$$\text{CMC} = \sum_{k=1}^{N} C_k (f_{1,k} - f_{2,k})^2 \tag{5}$$

where

$$C_k = \begin{cases} 1 & \text{if } ((f_{1,k} \geq a_1) \text{ and } (f_{2,k} < a_2)) \\ & \text{or } ((f_{1,k} < a_1) \text{ and } (f_{2,k} \geq a_2)) \\ 0 & \text{otherwise.} \end{cases} \tag{6}$$

The sequence of calculation of the CMC index can be summarized as follows.

1. Clustering of variables space by using the ART2 neural network.

2. Counting each number of segmented classes to obtain class frequencies for each time periods.

3. Calculation of average frequency for each time period.

4. Calculation of the CMC index by comparing the major class frequencies of the two time periods.

If the state changes, the value of this index become large at t_2 steps after the detection point. It may be useful to set a threshold level in this index to detect fault. The threshold level of the index, time durations t_1, t_2, and vigilance value of the ART2 network are the adjustment parameters of this index.

3 Case Study

3.1 *Process Description*

To demonstrate the performance of the proposed method, it is applied to data from an industrial, wastewater treatment process. Figure 3 depicts a simplified flow diagram of the plant[10]. The parameters for valve openness (MV) and flow rates ($F1$ and $F2$) were measured and the average daily values were logged for 500 days. The original sampled signals are shown in Figure 4.

The process is a part of a petroleum purification plant. The purpose of the process is to remove light-oil and unpurified compounds from wastewater fed from the top of the multiple plates column by feeding steam from the bottom of the column. In this process, the set point of $F1$ is given in order to maintain the liquid level of the tank. The set point of $F2$ is adjusted by the operator, who is mainly concerned about the levels of upstream tanks. Fluids passing through the pump contain high acidity, and the pump has a relatively high risk of suffering damages. The objective of this case study problem is to detect state change, which may correspond to the pump failure. Actually, the pump may gradually lose the performance, and no one may be able to specify the exact time of the occurrence.

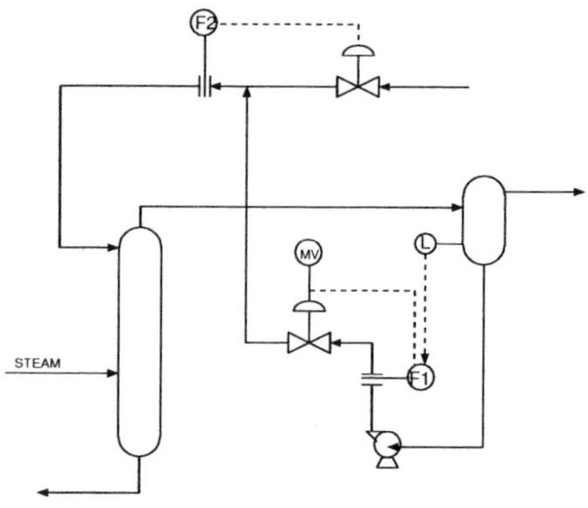

Figure 3: Process flow diagram of the plant

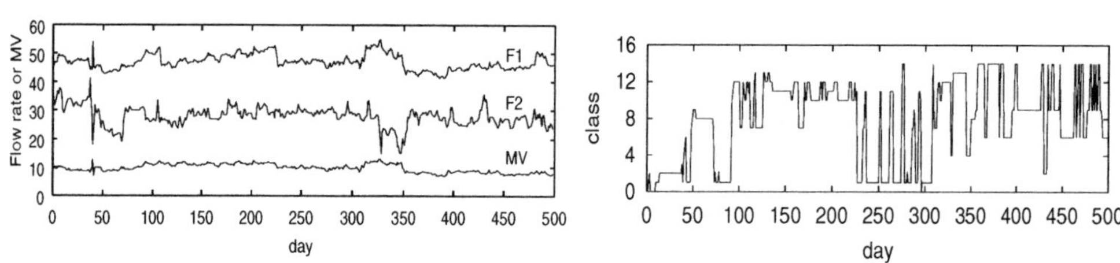

Fig. 4 Time series of the original signal   Fig. 5 Clustering results by ART2

3.2 *Result and Discussion*

For the data in this case study, state variables space is clustered by using the ART2 neural network [10]. Figure 5 provides an example of the changes in time for the segmented class numbers where the vigilance value ρ are set to 0.996. Complemental coding [7] is used for the input of the network. In order for application to on-line monitoring, the class number of the ART2 neural network is calculated at each time-point. New classes appear when the plant encounters some new state under certain conditions. The total number of generated classes is 15 at the end of the given data.

The proposed CMC index is calculated based on the ART2 classes (Figure 6). Except for the case $t_1 = 28$, changes of the state can be detected before the 330th day, the time where the CMC value nears unity. For a longer t_1 value ($t_1 = 182$), this index provides a fast and very clear detection capability for the system change. Upon comparison on the bottom two graphs ($t_2 = 7$ and $t_2 = 14$), the change seems easier to detect in case $t_2 = 14$, despite the fact that the on-line detection is slower because of the delay of t_2. There is a trade-off between accuracy and response time. The values of $t = 1$ and $t = 2$ must be selected by considering these factors.

The value of the index decreases gradually after detecting the changes, because the classes after state changes get majority in the reference region as times passes. If this decreasing characteristics is not desired, one could fix the reference period on the specific time ranges.

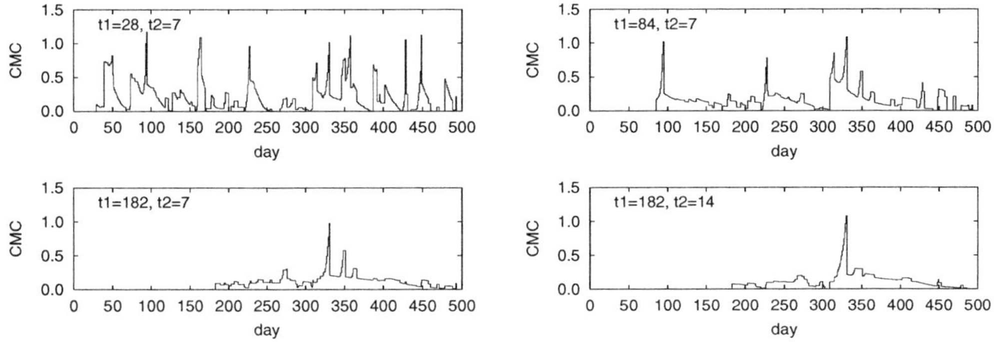

Figure 6: Detection of changes by CMC index

4   Conclusion

For the detection of process changes, a new on-line method is proposed. The method is based on ART2 clustering, which detects changes in class frequencies between the recent period and historical reference period. An index is defined as the total sum of the differences of class frequencies between these two periods. This index can detect changes of patterns in process state-vectors. This method is based on a clustering algorithm, although it does focus much attention on longer time periods than original clustering.

The proposed method was applied to an industrial wastewater plant. The results show excellent performance for the detection of process change. Time periods before and after the detection point must be defined carefully, considering the trade-off between the detection speed and accuracy. It is expected that this index will add new measure for fault detection in general.

References

[1] J. M. F. Calado, J. Korbicz, K. Patan, R. J. Patton and J. M. G. Sá da Costa, *European Journal of Control*,7 (2001) 248.

[2] G. A. Carpenter and S. Grossberg, *Appl. Opt.*, **26** (1987) 4919.

[3] G. A. Carpenter, S. Grossberg and D. B. Rosen, *Neural Networks*, **4** (1991) 493.

[4] L. H. Chiang, E. L. Russell and R. D. Braats, "Fault Detection and Diagnosis in Industrial Systems," Springer, London (2001).

[5] S. Grossberg, *Biol. Cybern.*, **23** (1976) 187.

[6] V. Venkatasubramanian, *Proc. PSE Asia 2000*, Kyoto, Japan (2000) 597.

[7] J. R. Whiteley and J. F. Davis, *IEEE Expert*, **8** (1993) 54.

[8] J. R. Whiteley and J. F. Davis, *Comp. Chem. Eng.*, **18** (1994) 637.

[9] Y. Yamashita, *Proc. KES'98*, vol.III, Adelaide, Australia (1998) 2.

[10] Y. Yamashita, H. Komori, E. Aoki and K. Hashimoto, *Kagaku Kogaku Ronbunshu*, **26** (2000) 457.

[11] Y. Yamashita, H. Komori and M. Suzuki, *Kagaku Kogaku Ronbunshu*, **27** (2001) 400.

KES 2002
E. Damiani et al. (Eds.)
IOS Press, 2002

A Fuzzy-Logic Based Fault Diagnosis Strategy for Process Control Loops

Sheng-Yung Chang and Chuei-Tin Chang
Department of Chemical Engineering, National Cheng Kung University,
Tainan, Taiwan 70101, Republic of China

Abstract. By considering the fault propagation behaviors in process systems *with* control loops, a fuzzy-logic based fault diagnosis strategy has been developed in the present work. The proposed fault diagnosis methods can be implemented in two stages. In the off-line preparation stage, the potential causes of a system hazard are identified by determining the minimal cut sets of a fault tree. The occurrence order of observable fault symptoms is derived from the system digraph and then encoded into a set of IF-THEN diagnosis rules. In the next on-line diagnosis stage, the occurrence indices of the top event and also the fault origins are computed in a fuzzy inference system based on real-time measurement data. Simulation studies have been carried out to demonstrate the feasibility of the proposed approach.

1 Introduction

Ulerich and Powers [1] reported the first attempt to perform fault diagnosis on the basis of fault trees. The most significant advantage of their approach is that the candidates of fault identification are restricted to only the causes of one or more given top events and, consequently, the diagnosis procedure can be greatly simplified. By incorporating the fault propagation patterns in a fuzzy inference system, Chang *et al.* [2, 3] developed a systematic fault diagnosis procedure for processes *without* control loops. This method was implemented in two stages: the off-line preparation stage and the on-line implementation stage. In the former case, a SDG system model was first constructed and the fault trees corresponding to the given top events were then synthesized according to the conventional algorithm. The symptom occurrence order caused by the basic events in each cut set can be easily determined on the basis of the SDG model. The candidate symptom patterns were then translated into a set of IF-THEN fuzzy inference rules for assessing the occurrence possibilities of the basic events in every cut set and also the top events. In the next stage, the on-line measurement data were first normalized. These normalized values were used as the inputs to a fuzzy inference system for computing the occurrence indices of top events and cut sets in real time.

Since the dynamic responses of feedback control loops were not considered in the original fault diagnosis system mentioned above, a more general fault diagnosis strategy is proposed in the present study to enhance its capability. Specifically, computation techniques have been developed to enumerate all possible symptom patterns by considering the loop dynamics. In addition to the initial and final states, the transient state of each loop variable is introduced for the generation of IF-THEN rules. The qualitative simulation results used for determining these transient states is also described in this paper. The effectiveness of this new feature has been verified with extensive simulation studies.

2 The Candidate Patterns in Loop-Free Systems

The effects of base event(s) in a cut set usually propagate throughout the entire system sequentially. In general, a series of intermediate events may occur before the top event. Since the performance of a diagnosis scheme should be evaluated not only in terms of its correctness but also its timeliness, it is the intention of this research to develop a fault identification procedure taking both the eventual symptoms and also their *occurrence order* into consideration. To identify this *symptom occurrence order* (SOO) associated with a given fault origin, the following operations should be performed on the digraph: (1) identify the *fault propagation paths* (FPPs) on the basis of the qualitative values of fault origin(s) and edge gains, (2) merge every pair of measured variable and its measurement signal in the FPPs, and then (3) eliminate the nodes representing the unmeasured variables. Let us consider the level control system in Figure 1 as an example. Its SDG model is shown in Figure 3. For the top event "liquid level in tank is too high," a fault tree can be constructed easily with a conventional method. The corresponding minimal cut sets are listed in Table 1. To illustrate the proposed method more clearly, let us examine the 9th cut set, i.e. $\{m_3(+1), CV\text{-}01 \text{ sticks}\}$. The corresponding symptom occurrence order (SOO) is shown as Figure 2.

If all symptoms in a SOO can be observed on-line, then it is certainly reasonable to confirm the existence of corresponding fault origin(s). However, it is also possible to find that these symptoms are only partially developed during the incipient period of an eventual system hazard and, further, their pattern may vary at different times during operation. To facilitate later discussions, let us define the collection of on-line symptoms at any time after the introduction of basic event(s) in a cut set as a *candidate pattern*. It is obvious that any candidate pattern can be considered as an evidence for fault identification with a degree of confidence. Thus, it is important to enumerate all possible candidate patterns and evaluate their respective significance in the off-line preparation stage. In a loop-free system, a SOO usually assumes the form of a tree. The total number of candidate patterns in this case can be computed with the following theorem:

Theorem 1. *Consider a tree-shaped SOO* **T**. *Let* **M** *be the set of all measured variables and* $\mathbf{X} = \{-10, -1, +1, +10\}$. *If* $\mathbf{P}^{(0)}(n_0)$ *denotes the initial path of length* n_0 *in* **T**, $\mathbf{P}^{(0,i)}(n_{0,i})$ $(i = 1, 2, \cdots, N_0)$ *denotes the ith branch path of length* $n_{0,i}$ *connecting to the end of* $\mathbf{P}^{(0)}(n_0)$, $\mathbf{P}^{(0,i,j)}(n_{0,i,j})$ $(j = 1, 2, \cdots, N_{0,i})$ *denotes the jth branch path of length* $n_{0,i,j}$ *connecting to the end of* $\mathbf{P}^{(0,i)}(n_{0,i})$, *etc., then the total number of candidate patterns* $N_{CP}(\mathbf{T})$ *can be computed according to the following equation*

$$N_{CP}(\mathbf{T}) = \mathcal{N}\left\{\mathbf{P}^{(0)}(n_0)\right\} = n_0 + \prod_{i_1=1}^{N_0} \mathcal{N}\left\{\mathbf{P}^{(0,i_1)}(n_{0,i_1})\right\} \tag{1}$$

where $\mathcal{N}\{\bullet\}$ *denotes the counting operator for a path in a tree-shaped SOO. The result of this operation should be generated recursively, i.e.*

$$\mathcal{N}\left\{\mathbf{P}^{(0,i_1,i_2,\cdots,i_k)}(n_{0,i_1,i_2,\cdots,i_k})\right\} = n_{0,i_1,i_2,\cdots,i_k} + \prod_{i_{k+1}=1}^{N_{0,i_1,i_2,\cdots,i_k}} \mathcal{N}\left\{\mathbf{P}^{(0,i_1,i_2,\cdots,i_{k+1})}(n_{0,i_1,i_2,\cdots,i_k,i_{k+1})}\right\}$$

$$\tag{2}$$

and $k = 1, 2, \cdots .$

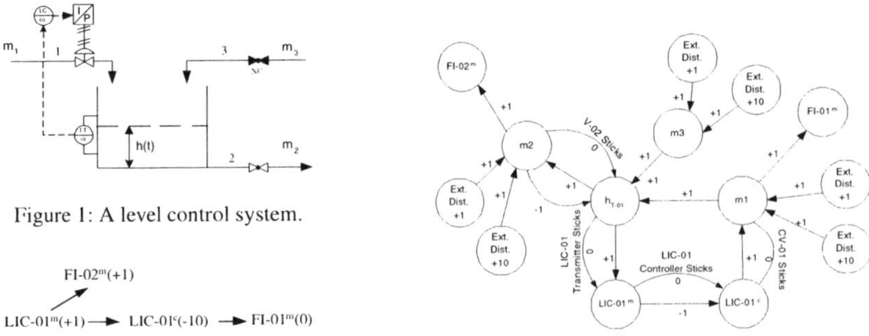

Figure 1: A level control system.

Figure 2: The SOO of cut set 9.

Figure 3: The digraph model of level control system.

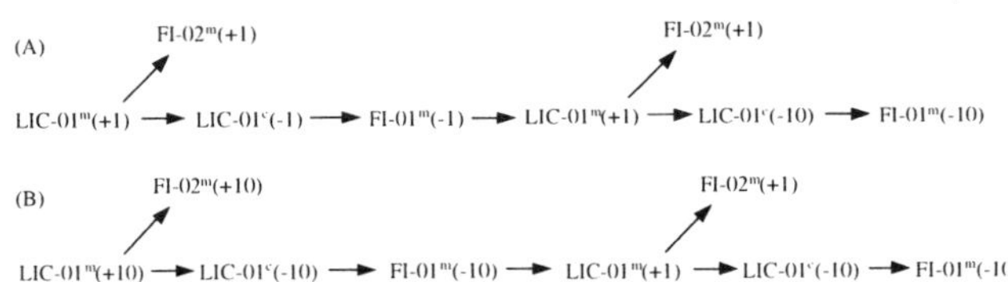

Figure 4: The single-valued SOO of cut set 3: (A)fast compensation; (B)slow compensation.

If there are no further branches connected to the end of the branch path $\mathbf{P}^{(0,i_1,i_2,\cdots,i_k)}(n_{0,i_1,i_2,\cdots,i_k})$, *i.e.* $N_{0,i_1,i_2,\cdots,i_k} = 0$, *then*

$$\prod_{i_{k+1}=1}^{0} [\bullet] = 1 \tag{3}$$

3 Qualitative Analysis of Loop Dynamics

In a loop-free process, the net effect of fault propagation can be viewed qualitatively as the direct transition from the normal system state to another new state. However, if a process contains feedback loops, the intermediate transient states caused by the compensation action of the controller must be considered. For illustration purpose, let us again consider the level control system in Figure 1. Originally, the gate valve on the stream 2 is open and the valve on stream 3 closed. It is assumed that the tank height is 100 cm and the outlet flow rate is proportional to the square root of liquid-level height. Let us next examine the scenario caused by the basic events in cut set 3 (see Table 1). The transient response of the control system can be simulated with SIMULINK [4]. It can be observed from the simulation results that the dynamic behavior of the system is dependent upon the compensation speed of controller. If the compensation speed is fast, the symptoms of fault propagation can be described qualitatively with the SOO in Figure 4(A). Otherwise, the SOO in Figure 4(B) should be used in diagnosis. To facilitate a concise representation of the candidate patterns, these two alternative SOOs are written in a unified multi-valued format as shown in Figure 5.

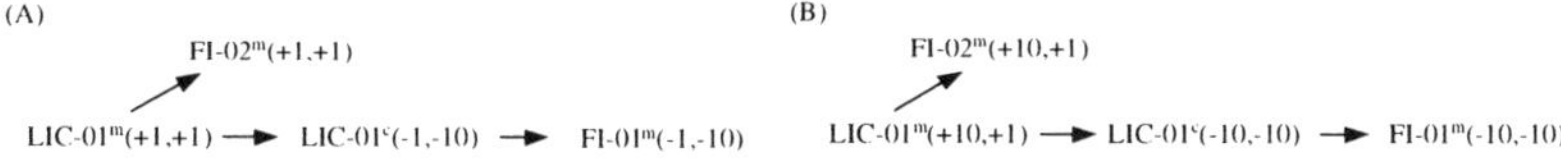

Figure 5: The multi-valued SOOs of cut set 3: (A)fast compensation; (B)slow compensation.

4 The Candidate Patterns in Process Control Systems

Let us use the multi-valued SOOs in Figure 5(B) as an example to illustrate the symptom occurrence order in the process control systems. Notice that LIC-01^m(+10, +1) $\to$ LIC-01^c(−10, −10) $\to$ FI-01^m(−10, −10) is a subpath on the feedback loop. Constrained by the dynamics of the loop, the final state of LIC-01^m, i.e. +1, can not be reached until the transient state of FI-01^m, i.e. -10, is realized. On the other hand, the occurrence order LIC-01^m(+10, +1) $\to$ FI-02^m(+10, +1) is not governed by the above constraint of loop dynamics. For example, if the fault propagation from LIC-01^m to FI-02^m is slow enough, FI-02^m should still remain at its initial steady state, i.e. 0, even after the final state of LIC-01^m has been reached. Thus, a hybrid SOO is often encountered in practical applications. To facilitate enumeration of candidate patterns in this situation, Theorem 1 should be extended to enumerate all possible candidate patterns. Due to the limitation of space, the corresponding theorems are omitted in this paper. The following example is used to illustrate the enumeration and derivation procedure of the corresponding IF-THEN rules.

Example 1. *Let us again consider cut set 3 in Table 1. The two possible SOOs of this fault origin can be found in Figures 5(A) and 5(B). Assuming that the compensation speed is fast, the candidate patterns of cut set can be determined according to Figure 5(A)(see Table 2). Notice that the non-zero value of each symptom, i.e. LIC-01^m, LIC-01^c, FI-01^m and FI-02^m, represent the qualitative level of deviation from its normal value. In the parenthesis next to each deviation value, a label "t" is used to denote the transient state and "f" the final state. The IF-THEN inference rules for diagnosing cut set 3 can be derived accordingly (Table 3). Notice that the premises of these fuzzy inference rules, i.e. SP, SN, LP, LN and ZE, can be obtained by a direct interpretation of the deviation values in the candidate patterns, i.e. ± 1, ± 10 and 0. On the other hand, the conclusions of inference, i.e. NOC, UCT1 to UCT7 and OCR, are basically fuzzy membership functions reflecting different degree of confidence in confirming the occurrence of cut set 3. The typical triangular and trapezoidal membership functions are used in this study for the definition of fuzzy variables. The popular centroid defuzzification algorithm [5] is adopted in the inference engine.*

5 Simulation Results

Let us assume that, in the level control system given in Figure 1, the valve on the stream 3 is mistakenly opened at time 1000 sec. This is basically the scenario described in cut set 3. The corresponding simulation data in this example were generated with SIMULINK. Since two different types of SOOs may be generated depending on the compensation speed of control loop, two sets of IN-THEN rules were created for diagnosis. The simulation results are shown in Figure 6. It can be seen that the performance of the proposed fuzzy-logic based diagnosis method is quite satisfactory. Notice also from Figures 6(A) and 6(B) that the early symptoms of fault propagation can only be detected with the "fast" rules. This observation indicates that the current compensation action of the controller is

MCS No.	Flow Rate m_1	m_2	m_3	CV-01 sticks	LIC-01 transmitter sticks	LIC-01 controller sticks
1	+10					
2		−10				
3			+10			
4	+1				Y	
5	−1					Y
6		−1		Y		
7		−1			Y	
8		−1				Y
9			+1	Y		
10			+1		Y	
11			+1			Y

Table 1: The minumal cut sets of the fault tree with top event "The tank level is too high"

No.	$LIC\text{-}01^m$	$LIC\text{-}01^c$	$FI\text{-}01^m$	$FI\text{-}02^m$
1	0	0	0	0
2	+1 (t)	0	0	0
3	+1 (t)	0	0	+1 (t)
4	+1 (t)	-1 (t)	0	0
5	+1 (t)	-1 (t)	0	+1 (t)
6	+1 (t)	-1 (t)	-1 (t)	0
7	+1 (t)	-1 (t)	-1 (t)	+1 (t)
8	+1 (f)	-1 (t)	-1 (t)	0
9	+1 (f)	-1 (t)	-1 (t)	+1 (t)
10	+1 (f)	-1 (t)	-1 (t)	+1 (f)
11	+1 (f)	-10 (f)	-1 (t)	0
12	+1 (f)	-10 (f)	-1 (t)	+1 (t)
13	+1 (f)	-10 (f)	-1 (t)	+1 (f)
14	+1 (f)	-10 (f)	-1 (f)	0
15	+1 (f)	-10 (f)	-1 (f)	+1 (t)
16	+1 (f)	-10 (f)	-1 (f)	+1 (f)

Table 2: The candidate patterns of cut set 3

$LIC\text{-}01^m$	$LIC\text{-}01^c$	$FI\text{-}01^m$	$FI\text{-}02^m$	THEN
ZE	ZE	ZE	ZE	NOC
SP	ZE	ZE	ZE	UCT1
SP	ZE	ZE	SP	UCT2
SP	SN	ZE	ZE	UCT2
SP	SN	ZE	SP	UCT3
SP	SN	SN	ZE	UCT3
SP	SN	SN	SP	UCT4
SP	SN	SN	ZE	UCT4
SP	SN	SN	SP	UCT5
SP	SN	SN	SP	UCT6
SP	LN	SN	ZE	UCT5
SP	LN	SN	SP	UCT6
SP	LN	SN	SP	UCT7
SP	LN	SN	ZE	UCT6
SP	LN	SN	SP	UCT7
SP	LN	SN	SP	OCR

Table 3: The IF-THEN inference rules of cut set 3

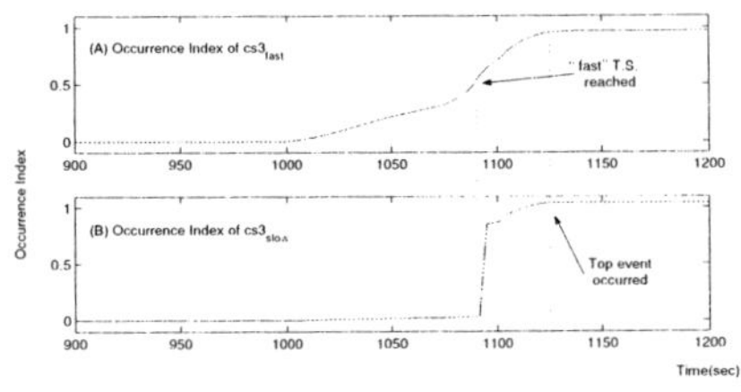

Figure 6: Simulation results of cut set 3: (A) occurrence index calculated with "fast" rules; (B) occurrence index calculated with "slow" rules.

fast enough for preventing the controlled variable, i.e. h_{T-01}, to reach the saturation level even under the influence of large external disturbances.

6 Conclusion

By capturing the dynamic characteristics of abnormal system behaviors with fuzzy inference rules, an effective fault diagnosis system has been developed in this work. Several theorems have been derived to facilitate enumeration of all possible candidate patterns for diagnosis in systems with or without feedback control loops. The simulation results show that the proposed strategy can be used not only to identify the correct fault origins but also the corresponding fault propagation mechanisms.

References

[1] N. H. Ulerich and G. J. Powers, On-line Hazard Aversion and Fault Diagnosis in Chemical Processes: The Diagraph + Fault Tree Method, IEEE Trans. Reliab., 37 (1988) 171.

[2] S. Y. Chang and C. R. Lin and C. T. Chang and S. W. Su, On-line Fault Diagnosis Using Dynamic Fault Tree, 4th IFAC Workshop on On-line Fault Detection and Supervision in the Chemical Process Industries, (2001) 192.

[3] S. Y. Chang and C. R. Lin and C. T. Chang, A Fuzzy Diagnosis Approach Using Dynamic Fault Tree, Accepted by Chem. Eng. Sci. (2002).

[4] SIMULINK: Dynamic System Simulation for MATLAB-Using Simulink, The Mathworks Inc., (2000).

[5] E. H. Mamdani and S. Assilian, An Experiment in Linguistic Synthesis with a Fuzzy Logic Controller, Int. J. of Man-Machine Studies, 7 (1975) 1.

Application of Wavelet analysis to chemical process diagnosis

Toru Matsuo
Technical Dept., Mitsui Chemicals Inc.; Asamuta-machi 30 Omuta,
Fukuoka Prefecture, Japan

Hideki Sasaoka
Research & Development Headquarters, Yamatake Corporation; 1-12-2 Kawana,
Fujisawa, Kanagawa Prefecture, Japan

Abstract.   Based on the wavelet transform, a new quantitative method to evaluate similarities among process variables is proposed. The method is implemented as a part of the process analysis tool and applied to an industrial polymer plant. As the result, gas component in the reactor was found to be highly related to product property. An inappropriate tuning of control loops was also found easily with the support of this tool.

1. Introduction

Frequency-related variables involved in time-series data can be extracted by using the Wavelet transform.   So, Wavelet analysis of phenomenon involving such frequency-related variables prevails in various fields now. In many cases, the Wavelet analysis succeeded to detect characteristic elements concealed behind complicated formative time-series data.

For process controls in chemical plants, sensor is equipped on most of the apparatus and process line being able to measure and collect time-series variables.   Such refined arrangements for the data sampling enable manufacturing process to be assessed precisely.   On the other hand, however, some difficulties appear in analyzing relationship among lots of time series data to diagnose process conditions. There are some reports illustrating that the Wavelet transform is an efficient tool, for extracting appropriate frequency-related information in cases [1–4].

In this paper, the Wavelet analysis is applied to our actual plant data with objects of analyzing relationship between input- and output-signals of the process, identifying cause of the inappropriate tuning and diagnosing condition of the process controls .Although frequencies extracted by the Wavelet analysis have been compared each other just in qualitative manner in the previous paper [1], we extended its availability further to the quantitative ones in the study.

2. Method and application to actual process

2-1 Wavelet analysis tool

Wavelet analysis tool manufactured by Yamatake Co. was used for analysis of variables involved in time-series process signals. The tool enables time-series data to visualize by plotting against time-frequency domain. By doing so, frequency-related information can be recognized as graphical patterns.

In the present study, the Wavelet analysis tool is so improved, that the degree of similarity among process variables can be compared quantitatively, yielding information contributive and necessary to analysis and diagnosis. Similarity R between two variable f and g in the time-frequency domain is proposed. It is defined as follows:

$$R = \frac{\sum_{i=1}^{I}\sum_{j=1}^{J}(f_{ij}-\bar{f})(g_{ij}-\bar{g})}{\sqrt{\sum_{i=1}^{I}\sum_{j=1}^{J}(f_{ij}-\bar{f})^2}\sqrt{\sum_{i=1}^{I}\sum_{j=1}^{J}(g_{ij}-\bar{g})^2}}$$

Where, $\bar{f}$ and $\bar{g}$ are the mean value of f and g.

2-2 Relationship found by Wavelet analysis

The frequency-related information involved in input-signals at chemical process is noted here. As the input signal suffers impacts (such as gain, noise, time shift) depending upon the process characteristics, its time-series data happen to vary at the output stage. However, the frequency-related information transformed by the Wavelet preserves its original information in many cases, so that the frequency-related information being presented at the input stage will be still observed at the output stage. By utilizing the fact, obscured relationship of the time-series data between input and output stages can be clarified. An example of the idea verified by simulation is shown in Fig. 1.

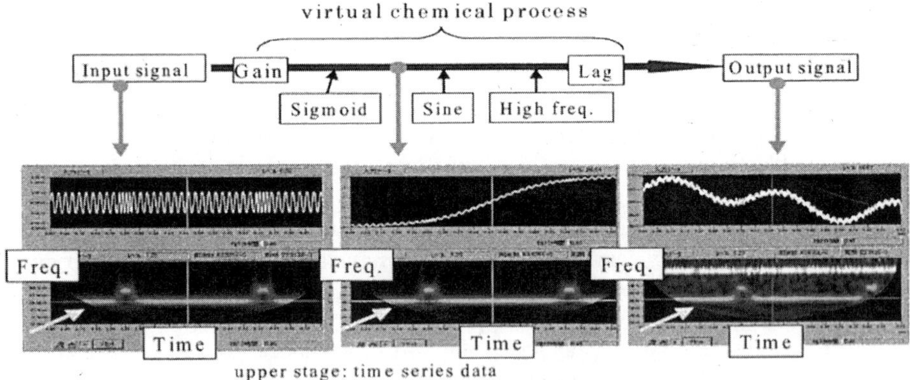

Fig. 1 Simulation of process response at input and output stages

2-3　Case study of actual plant

The above mentioned analysis tool was applied to actual plant, as mentioned below, for analysis.

(1) Product property and reactor condition

As for product property, its real-time measurement and recording were conducted at every 5-minute interval by on-line measuring apparatus mounted on the down-stream of the reactor. Other reactor conditions (i.e., pressure, temperature, gas composition) were measured and recorded in real-time (at every 1-minute interval for pressure and temperature; at 3-minute interval for gas composition.)

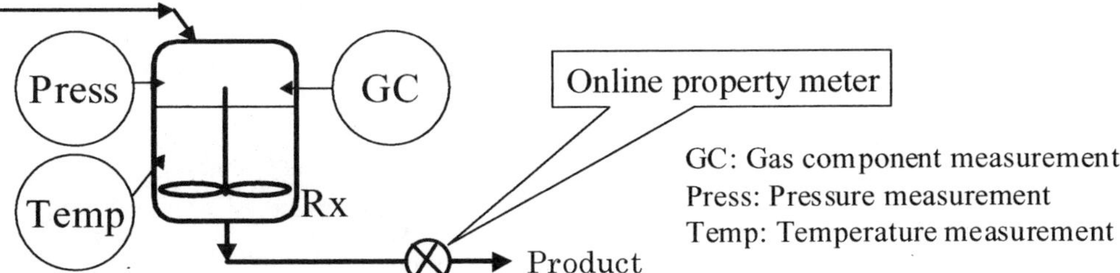

Fig. 2 Product property and reactor condition

(2)　　　Gas feeder and pressure control

Mono-component gas in pressure-controlled state is fed from gas supply plant. The pressure control is conducted by PID feedback system.

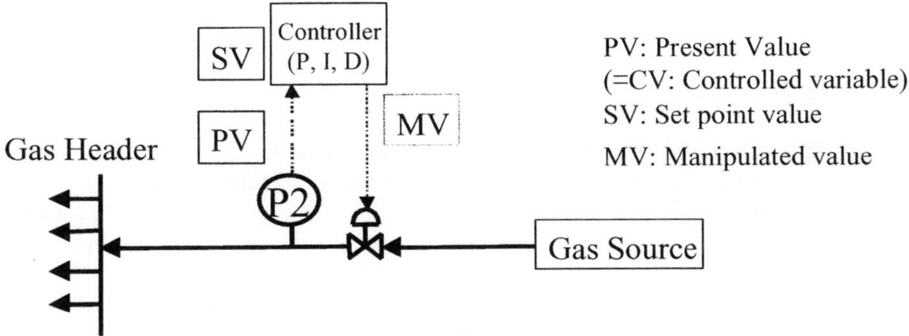

Fig.3 Gas feeder and pressure control

3. Results

3-1 *Relationship between product property and reactor condition*

When the cause of variations in product property was surveyed throughout the reactor conditions, it was not clarified which condition was most influential by just observing the time-series variations (Fig.4 left side).

However, it was indicated by the Wavelet analysis that gas composition in the reactor and product property had the same pattern in their frequency-variation behaviors (Fig.4 right side).

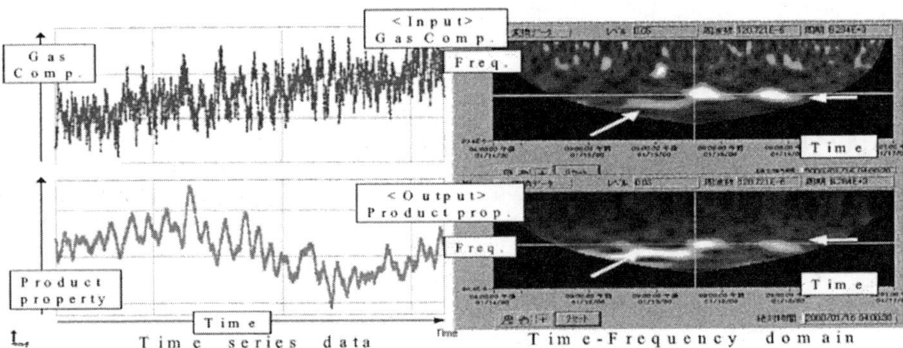

Fig.4 Relationship between gas composition and product property

Furthermore, by quantitatively comparing similarity of variations between plural conditions of the reactor (i.e., temperature, pressure, composition) and product quality, it was cleared that similarity relationship between the composition variation and the product quality was most significant.

From the result, we can conjecture that variations in the product quality are mainly due to variation in the gas composition, and, therefore the most effective suppression of the product property from variation is achieved by properly controlling the gas composition at constant.

3-2 *Analysis of feedback control system*

Just looking at the time-series data (Fig. 5, left side), it is difficult to judge whether the variation is caused by natural random noise or inappropriate tuning of controls.

By the Wavelet analysis, common bands distinctly appear at periods 52 and 103 minutes among the manipulated value (MV) and the present value (PV), and growth- and extinction-time of these bands are coincident. Based on these findings, it is expected that a resonant phenomenon does occur between the input and output signals due to inappropriate parameter tuning, resulting in poor controllability. Consequently, the poor controllability will expectantly be dissolved by tuning of PID parameters.

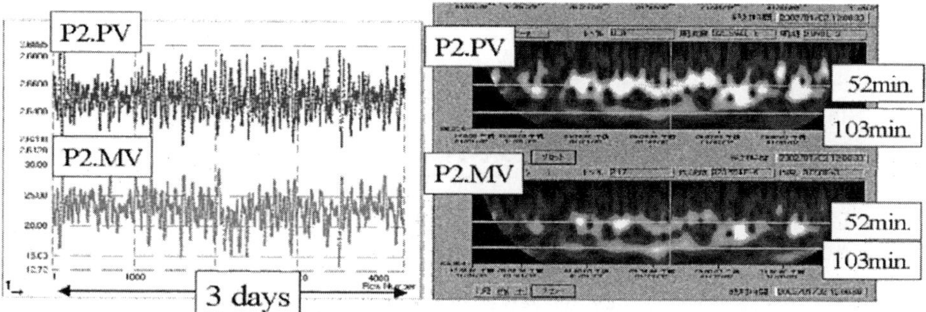

Fig.5 Frequency-related behaviors in pressure control system

4. Conclusion

The Wavelet analysis has been improved and the time-series data in chemical plant have been analyzed depending upon their frequencies. As the result, following items were suggested to be accomplished:

- Calculation of the quantitative similarity relationship between input and output signals of chemical process
- Identification of the cause that participates in variation of product property at reactor
- Survey of cause that participates in poor controllability of feed-back control system

5. Acknowledgement

The authors thank Dr. Yoshiyuki Yamashita, associate professor of Tohoku University and Dr. Hiromu Ohno, associate professor of Kobe University, for their valuable suggestions during the study, and Intelligent Manufacturing System (IMS) program for financial support during the study.

References

[1] Toru Matsuo, Jin Takagaki, Hideki Sasaoka and Hirohiko Kazato, Application of Wavelet transformation to process diagnosis, *KES'2001*, Nara, Japan (2001) 387-390.
[2] Hiromu Ohno, Modeling by non-stationary dataset, Technical Report No.13, 143rd committee, *JSPS,* Tokyo, Japan (1995).
[3] Masaharu Daiguji, Osami Kudo, and Tetsuya Wada, Application of Wavelet Analysis to Fault Detection in an Oil Refinery, *Computers and Chemical Engineering,* **21S** (1997) 1117-1122.
[4] Rodolphe L. Motard and Babu Joseph, eds., Wavelet Applications in Chemical Engineering, Kluwer Academic Publishers, Boston (1994).

KES 2002
E. Damiani et al. (Eds.)
IOS Press, 2002

NN Application for Estimation of Products Quality Indexes

Stefania TRONCI[1], Giampiero TESTONI[2], Alberto SERVIDA[3] and
Roberto BARATTI[1]

1   *Department of Chemical Engineering and Material Science, University of Cagliari,
    Piazza D'Armi, I-09123 Cagliari, Italy, email: {tronci, baratti}@dicm.unica.it*
2   *SARAS Raffinerie Sarde S.p.A., SS Sulcitana 195, I-09018 Sarroch (Ca), Italy,
    email: giampiero.testoni@saras.it*
3   *Department of Chemistry and Industrial Chemistry, University of Genova,
    via Dodecaneso 31, I-16146 Genova, Italy, email: servida@unige.it*

Abstract: The prime objective of this work is to demonstrate the potential of neural
network modeling in monitoring applications of crude units. In particular, we will
discuss the development of a state variable estimator to infer the quality indexes -
cut-points, flash and freezing points - of a crude distillation unit in operation at the
SARAS refinery. The estimator is based on a conventional feed forward neural
network, which is trained, validated, end tested off-line by using plant data spanning
an operating window of four months. The preliminary results show a rather good
agreement between the inferred quality indexes and the experimental measurements.

1. Introduction

The on-line measurement of product quality indexes represents a prime requirement for
the successful application of advanced process control and optimization systems in refining,
and in general, in the chemical manufacturing processes. In many instances, on-line
analytic instrumentation is not feasible for various reasons, among them, unacceptable time
delay responses and high maintenance costs. In principle, indirect and incomplete
secondary on-line measurements can be used in conjunction with a mathematical model of
the process to reconstruct the complete set of states. The state inference is accomplished
through on-line nonlinear estimators that consist of model-based dynamic systems driven
by the available on-line measurements. The need for state estimators is particularly strong
in chemical processes where, usually, the most important state variables, concentrations and
product quality indexes are intrinsically difficult to acquire [1].

Artificial neural networks (ANN) constitute a powerful modeling environment to
develop estimators that could be used for on-line applications. In this paper, we will
describe the formulation of a neural-based estimator for the on-line estimation of cut-points
of a crude unit, in operation at the SARAS refinery. The control and monitoring of the
quality indexes of products is of paramount importance to refineries because they have to
strictly comply with tight market specifications [2].

Typically, the crude, preheated in the heat exchangers train and in the furnace, is fed to
the main atmospheric distillation column, where it is separated into its major components:
naphtha, kerosene, gasoil, and atmospheric residuum. The product quality of these streams
is characterized by the cut-points between two adjacent products and/or by other indexes,

for instance, freezing and flash points. The cut-point represents the boiling temperature that separates components belonging to two adjacent product cuts. The flash point represents the inflammability of the product, while the freezing point indicates the temperature at which the product starts to freeze. In our case, the indexes of interest are the cut-points - the boiling temperatures at the 95% of ASTM boiling curve (for the naphtha, kerosene and gasoil streams) and the freezing and flash points of kerosene.

The crude quality greatly affects the product cut-points, and, when crude switch occurs, the operating conditions of the column must be significantly changed in order to maintain the product quality within the specifications. From an economical point of view, the reduction of transient time between crude switch is also of paramount importance to reduce off-spec operations. At SARAS, the distillation unit is controlled through a multivariable control strategy where the controlled variables are the quality indexes previously mentioned. The control system is configured so as to improve, during normal operations, the product quality compliance and the yields of valuable products. The first requirement is very important because the unit is run close to the acceptable specification limits. This puts high strains on the multivariable controller that is called to govern the unit rather close to the acceptable operating boundaries.

Unfortunately, the column is not equipped with on-line analyzers therefore, quality indexes estimators, based on available secondary measurements - such as temperatures, flowrates, and pressures - become essential for a successful control of this unit.
In the open literature there are only few papers dealing with the development of software sensors applied to crude distillation units [1, 3] and, these are restricted to only one side stream. Only recently, Tronci *et al.* [4] discussed the preliminary results of a software sensor, based on a feed-forward neural network, which was used to predict the temperature at the 95% of ASTM boiling of the side streams of a topping unit. In that work, the crude tower was equipped with on-line analyzers to continuously measure the 95% cut-point of the relevant side draws, thus, the development of the estimator relied on the availability of a huge amount of data.

In the present work, we will develop an estimator, based on a feed-forward neural network, to predict the quality indexes of the side streams for a distillation unit not equipped with on-line analyzers.

2. The Crude Unit

At SARAS there are three topping plants, and in the present work we will focus our attention on the plant number 2 and, in particular, to its main distillation tower. This crude unit has a maximum capacity of 135,000 bpsd. The throughput is normally maximized subject to furnace constraints, lightends fractionation section capacity, and, sometimes to pumparound heat duty.

The crude feed consists of high sulfur and low sulfur crudes. The feed can be either pure crude or a mixture of various crudes. The unit can be fed from up to four tanks simultaneously, two crude tanks and two residue or slop tanks. The feed is preheated in the heat exchangers train and in the furnace before entering the crude unit, where it is separated into its major products, which are sent to the downstream units for further processing. In particular, the column has a naphtha overhead product and four side draws. The feed is preheated in the heat exchangers train and in the furnace, before entering the main atmospheric distillation column (this is a fairly tall fractionator consisting of 57 trays). The crude oil is separated into its major components: naphtha, overhead product and four side draws: kerosene, light, medium and heavy gasoils (heavy gasoil normally is not drawn), and atmospheric residuum.

Crude switch takes place every 2-3 days, and it takes approximately 4-5 hr for the process to recover steady-state conditions. The product specifications are synthetically summarized in Table 1.

Table 1: Product specifications

Product	Constrains
BAT	95% ASTM
KERO	95% ASTM, Freezing and Flash Point
GAL	Free
GAM	Free
GAL/GAM	95% ASTM
GAP	Free
Residue	Free

Where BAT, KERO, GAL, GAM, and GAP stand for light virgin naphtha, kerosene, light, medium and heavy gasoil, respectively.

Cut-points, between two adjacent products, are used to characterize the product quality of BAT, KERO, and GAL/GAM blend. Moreover, the KERO draw is also subject to constrain on the flash and freezing points.

Since the column is not equipped with on-line analyzers, only laboratory measurements are available for the quality indexes (up to 1 measurement per day and per index).

3. Software sensor

A soft sensor was developed to continuously monitor the side stream quality indexes.
In developing soft sensors for on-line applications, simplicity represents one of the requirements to comply with, since, the sensor has to be implemented on DCS. Therefore, the software tool must be as much simple and compact as possible.

The selection of the inputs to be used to the neural soft sensor requires a bit of cautions. Measurements with high level of noise or with delay must be discharged. Moreover, a knowledge-based selection of the input variables allows a significant reduction of the input vector size and, surely, helps the soft sensor design. For example, when dealing with distillation column, the ratio between the reflux-to-distillate flowrate strongly affects the quality of the distillate and it is better to feed the ANN with this ratio rather than leaving the neural network with the task of learning the relationship distillate quality-reflux ratio. Using this idea, 14 variables were selected as inputs to the ANN. Since, the inputs to a neural-based software sensor represents the know-how on which the soft sensor relies, we cannot provide more details on the nature and the methodology we adopted to select the final inputs set.

Another important issue, when developing neural models, concerns with the selection of the number of hidden layers and hidden neurons in each hidden layers. This problem was solved by trial and error, with the final goal to obtain the simplest structure as possible. Finally, care must be also paid in planning a suitable experimental campaign or in selecting the calibration data set. The training data must be representative of all the possible situations that may occur in normal plant operations.

The neural soft sensor was calibrated and validated by using experimental data spanning four months of operation (about 140 measurements for each index). The data set was divided in two parts. The first, which accounted for two-third of the overall data, was used as calibration/validation data set, while the second was utilized as test data set.

4. Results and discussion

In this section we will briefly describe the preliminary results of the performance assessment of the neural soft sensor as inferential tool to predict the quality indexes of the crude unit. In other words, the data refer to the off-line validation of the software sensor carried out by comparing the predictions to the available laboratory measurements.

Table 2: Temperature intervals

Quality index	Temperature (°C)
Cut-point$_{BAT}$	35
Cut-point$_{KERO}$	50
Cut-point$_{GAL/GAM}$	70
Freezing point$_{KERO}$	30
Flash point$_{KERO}$	30

For sake of brevity, we have reported the results of the test set, which refer to January 2002. For confidential reasons all the figures do not report the absolute values of the cut-points, neither the values of the freezing and flash points. In Table 2 the temperature intervals of the 5 quality indexes are shown for readers convenience. We believe that by referring to Table 2 one can still appreciate the rather good capability of the developed soft sensor in recovering the dynamic behavior of the crude unit. All the reported data refer to a soft sensor relying on a neural network with 14 input, 5 hidden, and 5 output neurons.

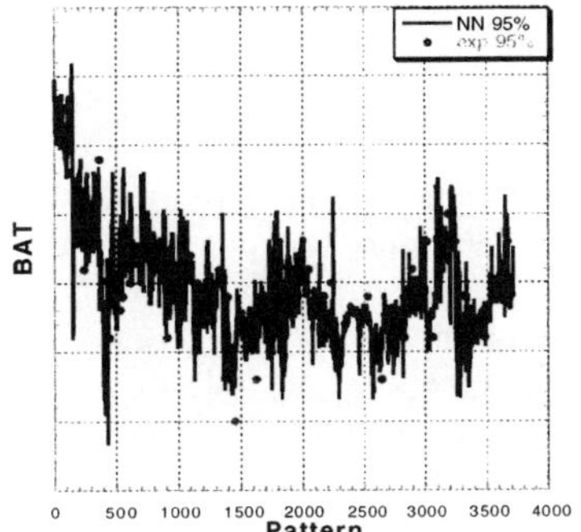

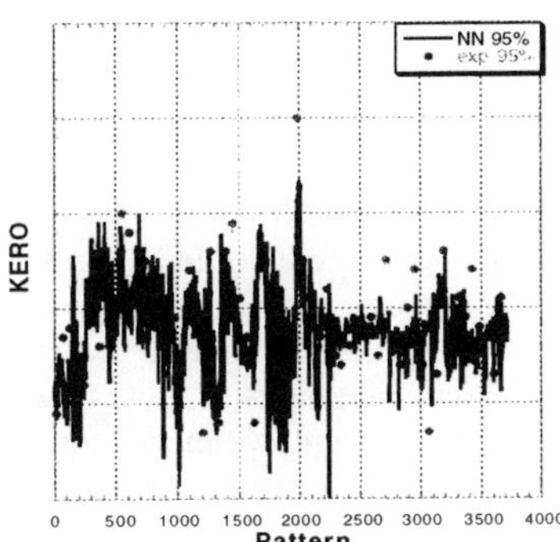

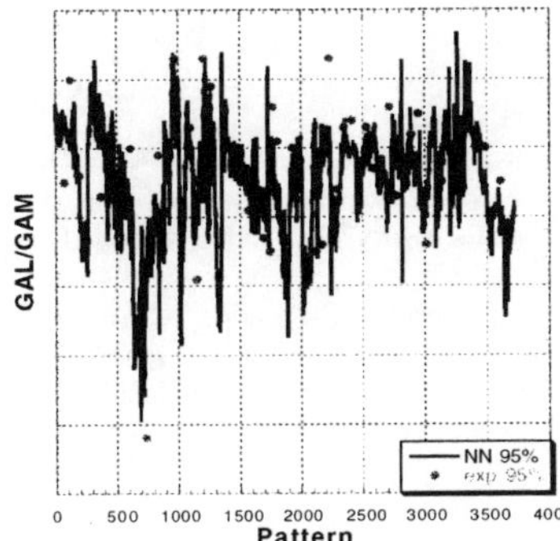

Figure 1: Measured and inferred cut-points

For the draws: BAT, KERO, and GAL/GAM blend, the comparison between the experimental and inferred cut-points is shown in Figure 1. During this period of time the unit was fed with various crude, nevertheless, the agreement is rather good, with an average error comparable to that of the laboratory measurement accuracy (+/- 4°C).

For the KERO stream, Figure 2 shows the agreement between the measured and the inferred values of the freezing and flash point, for the same operating window.

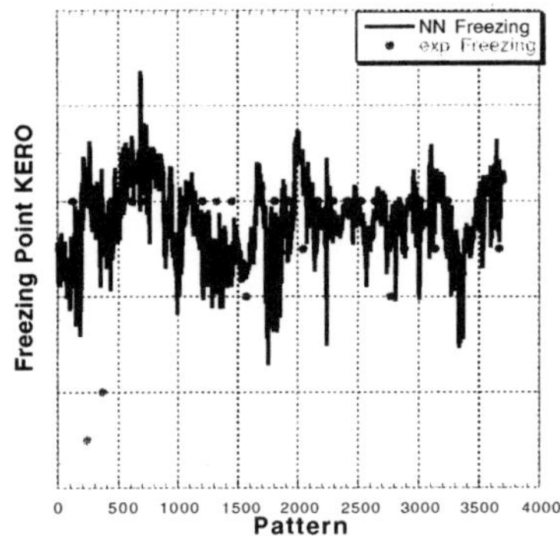

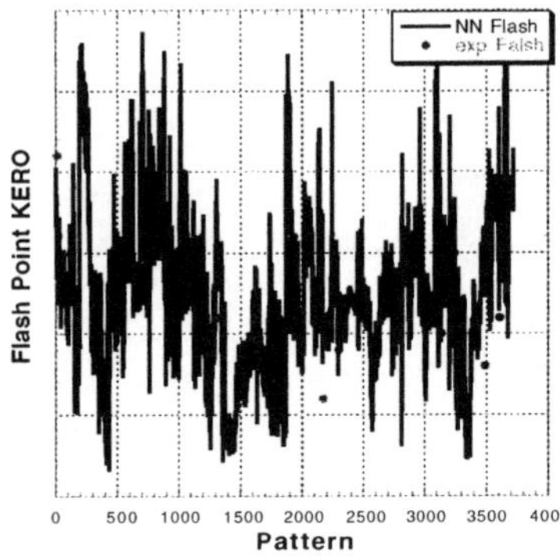

Figure 2: Freezing and Flash point

Presently, the software sensor is under on-line assessment at SARAS.

5. Conclusions

We have reported preliminary results on the exploitation of neural-based inferential sensors to monitor quality indexes of side streams of an atmospheric crude unit. The quality index estimation is an important problem in refineries because of the feed switching (this is more evident when the analytical instrumentation is not available).

The discussed results show that neural-based software sensors can, indeed, be used with confidence as alternative measurement methodologies or, even more, as back-up systems, when analytical instrumentation becomes unavailable because of failures or planned maintenance out-of-services.

References

[1] S. Park and C. Han, A nonlinear soft sensor based multivariate smoothing procedure for quality estimation in distillation columns, *Comp. Chem. Engng.*, **24** (2000) 871.

[2] R.W. Brooks, F.D. van Walsen and J. Drury, Choosing cutpoints to optimize product yields, *Hydr. Process.*, **78**, No.11 (1999) 53.

[3] M. Montesi, F. Trivella and A. Brambilla, Estimation of crude fraction cut-point using hybrid ANN, *proceeding of EANN 1996* , 17-19 June, London, England, (1999) 99.

[4] S. Tronci, G. Testoni, A. Servida and R. Baratti, NN Application for Estimation of Crude Fraction Cut-Point at SARAS, *proceeding of EANN 2001* , 16-18 July, Cagliari, Italy, (2001) 123.

KES 2002
E. Damiani et al. (Eds.)
IOS Press, 2002

Multiagent-oriented Decision Making Framework for Assembly and Operation of Micro Chemical Processes

Naoki KIMURA, Hideyuki MATSUMOTO and Chiaki KURODA
Department of Chemical Engineering, Graduate School of Science and Engineering,
Tokyo Institute of Technology
2-12-1 Ohkayama, Meguro-ku, Tokyo 152-8550 JAPAN

Abstract. In this paper, a multiagent-oriented decision making method for the process assembly and operation of micro chemical processes is proposed. The characteristic of this study is simultaneous parallel assembly using multi-objective evaluation considering the operation. In result, it was possible to manage efficiently the information updated frequently in sensitive micro chemical processes.

1. Introduction

Recently, it has been required to produce chemical products precisely and cleanly and to respond quickly to various demands of the productive market. The micro chemical processes are being expected as such an efficient and flexible production system[1-2]. In micro chemical processes, it is possible to produce various functional chemical products efficiently by assembling the micro unit equipments. However there is a limit in the centralized management system because of a mass of information updated frequently based on larger-scale and complicated accumulation of many micro modules. Kuroda *et al.* proposed the decentralized cooperative operation method of chemical batch processes by using movable autonomous reactors[3]. Wada built a manufacturing system using machine agents, which are mounted on each processing equipment, and product agent, which are generated for each job[4].

In this study, a multiagent-oriented decision making framework for process assembly design system considering about the operation of micro chemical processes is proposed. The concept of multiagent is applied to manage the information on the equipments and the products. And the cooperative multi-objective decision making method AHP (Analytic Hierarchy Process) is applied for autonomously cooperative evaluation in micro chemical processes.

2. Multiagent System for Micro Chemical Processes

2.1 Micro Chemical Processes

There are planned many types of micro chemical processes. In this study, we imaged a

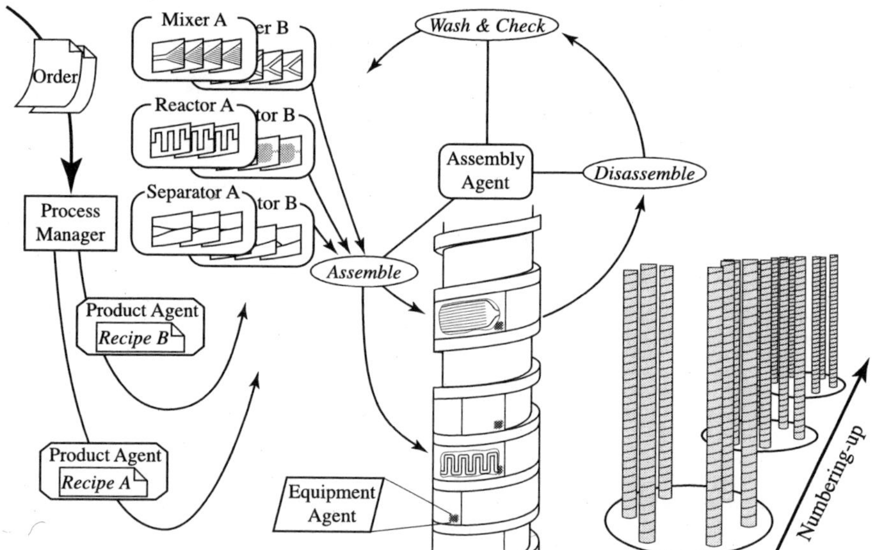

Fig.1 Framework of a spiral shaped micro chemical process.

spiral shaped micro chemical process shown in Figure 1. Each micro equipment module is fabricated as one unit operational module, e.g. a mixer, a reactor, a separator. To produce chemical products, various micro equipment modules are assembled and set on the groove of the spiral set up in the surroundings of the cylinder. There are heating/cooling utilities, power supply and communication lines are embedded inside the cylinder. When production finished, the equipment modules are disassembled, washed, checked and reused for other jobs. And it is possible to produce various products simultaneously on one cylinder. Moreover, the inclease of production can be realized by preparing such a cylinder, that is called "Numbering-up".

In our framework, three kinds of agents are defined for flexible production. "Equipment agent" is attached to each micro equipment module. An equipment agent manages its schedule and history and negotiates with the other agents about allocation of jobs, receives a recipe of allocated job from a product agent and executes it. "Product agent" is generated for each job by a process manager. The product agent holds a recipe which describes operative procedure, the scale of production and so on. The product agent continuously gets the information about the processing state from equipment agents. "Assembly agent" is attached each cylinder and manages assembling/disassembling of equipment modules by negotiating with the other agents. And we arranged a process manager as an interface with human operators. When the process manager gets an order of production, it generates one product agent per one order. The product agent negotiates with equipment agents and assembly agents. When an equipment agent accepts to execute the job, the product agent moves to the equipment agent and gives it the contents of the recipe.

Figure 2 illustrates a schematic diagram of a recipe in this study. This "Recipe A" is for Product A, the lot number is 2, and the production quantity is set to 3800. The recipe consists of some subrecipes. One subrecipe corresponds to the execution procedure for one equipment-agent. In this case, Recipe A consists of 3 subrecipes, and the contents of the subrecipe specify the equipment type and the procedure. In this study, the values of processing time and processing load can be calculated from the contents of the subrecipe.

In sensitive micro chemical processes, various unexpected accidents often occur. For example, also a little chance of operation results in the variation of the properties of

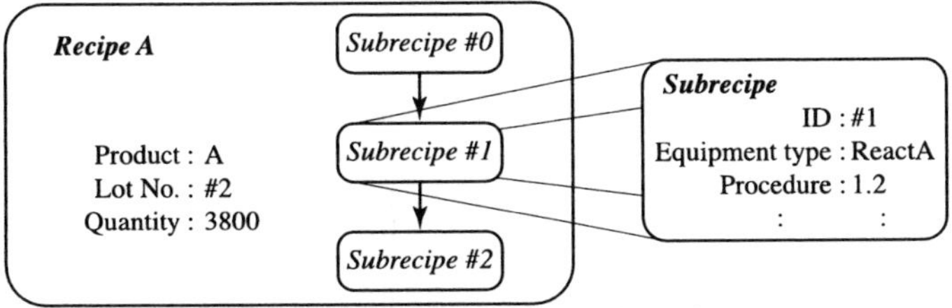

Fig.2 A schematic diagram of a recipe.

intermediate products. The following operation procedure or conditions have to be changed to correspond to the variation of intermediate products. Therefore the recipe has a flexible and object-oriented structure, namely the prodcut agent can change the recipe according to the processing report from equipment agents. Information traffic and memory usage decrease by erasing the used subrecipes.

2.2 Multiagent-oriented Decision Making Framework

In order to realize cooperative assembly and operation, the application of Analytic Hierarchy Process (AHP) is introduced. A problem is analyzed hierarchically and evaluated multi-objectively. Hierarchical multiple criteria for selection of an equipment are set as shown in Figure 3. The problem is evaluated by some criteria in level 2. Each criterion in level 2 is subdivided some criteria in level 3. Each alternative equipment is set to level 4. The evaluation of a criterion is calculated by summation of weighted evaluations of criteria in lower level. In AHP, it is able to decide the weights by compare the importance of two criteria. The pairwise comparison ratio a_{ij}, which is comparison of the importance of criterion i and criterion j, that is w_i and w_j, is defined as :

$$a_{ij} = w_i / w_j$$

Considering a pairwise comparison matrix $A = [\, a_{ij}\,]$ and an importance index (weight) vector $W = [\, w_i\,]$, the relation :

$$A\,W = n\,W$$

should exist. When A is given, the vector W is calculated as an eigenvector of A. In this study, each agent has its own matrix A, and exchanges the matrix between a product agent and an assembly agent to cooperatively adapt to changes in the process environment.

A product agent collects the information of each equipment before selection of an equipment, then the product agent evaluates each equipment to make decision. The product agent has the product-oriented pairwise comparison matrix and the assembly agent has the process-oriented one. The product agent can judge which adapted to the state of the process enviroment by exchanging both matrices. Also the assembly agent can adapt to changes in the process environment.

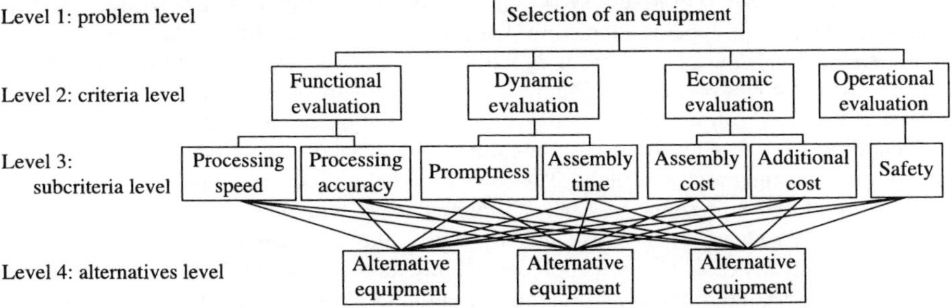

Fig.3 Hierarchical multiple criteria for selection of an equipment.

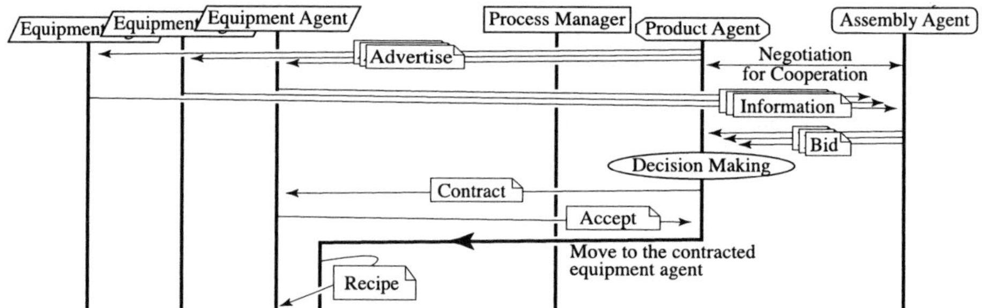

Fig.4 Communication based on the contract net model.

2.3 Communication Network

The negotiation and execution mechanism based on the contract net model is considered as shown in Figure 4. A product agent sends an *"advertise"* message to the equipment agent group which is specified by the subrecipe. An advertise message contains the production quantity, the start time of execution, a copy of the subrecipe and so on. When an equipment agent receives the advertise message, it estimates the processing time and the safety data from the above information, checks its own schedule, and sends the information of the estimated processing time, the safety data and its own functional properties to an assembly agent. When the assembly agent receives the information from the equipment agents, the assembly agent estimates the assembling time and the cost based on the received information, and sends a *"bid"* message to the product agent. The bid message contains the information of the functional properties of the equipment agent, the estimated time, the cost and the safety data to execute the job. At the same time, the product agent and the assembly agent negotiate for cooperation, i.e. exchange their pairwise comparison matrices. When the product agent has finished negotiation and received bid messages, it evaluates the bids based on the contents of bids and makes decision to contract. When the contract is completed, the product agent moves to the accepted equipment agent and delivers the subrecipe to request the execution. By using the same procedure, the product agent contracts with all the equipment agents to execute the job. When all the contracts are completed, the product agent requests the assembly agent to assemble the contracted equipments and starts processing simultaneously.

3. Results of Simulations and Discussions

The above mentioned multiagent system was implemented and simulations with various conditions were carried out on two multi-task computers (CPU:PowerPC G4 450-500MHz, Memory:512-768MB, OS:Mac OS X). The simulation programs are written in Java Language. TCP/IP protocol is adopted for the communication among agents.

A simulation with 54 equipment agents, 35 product agents, 1 assembly agent and 1 process manager was carried out. Figure 5 illustrates a part of an example of Gantt-chart-like diagram of execution. This is a final result of execution and not such a pre-set schedule as a Gantt-chart. This Gantt-chart-like diagram was indicated to the process manager in real time to monitor the progress of jobs. The rows of the chart represent equipment agents and the horizontal axis represents the time. A rectangular bar represents execution of a job. For example, job *"ProductC #1"* was executed by three equipment agents, which were *"MixF2"*, *"ReactC2"* and *"SepC2"*. These equopments were temporarily conjoined for job *"ProductC #1"* on the groove of the spiral set up in the surroundings of the cylinder. All equipments have various function and different functional properties i.e. processing speed and accuracy. In

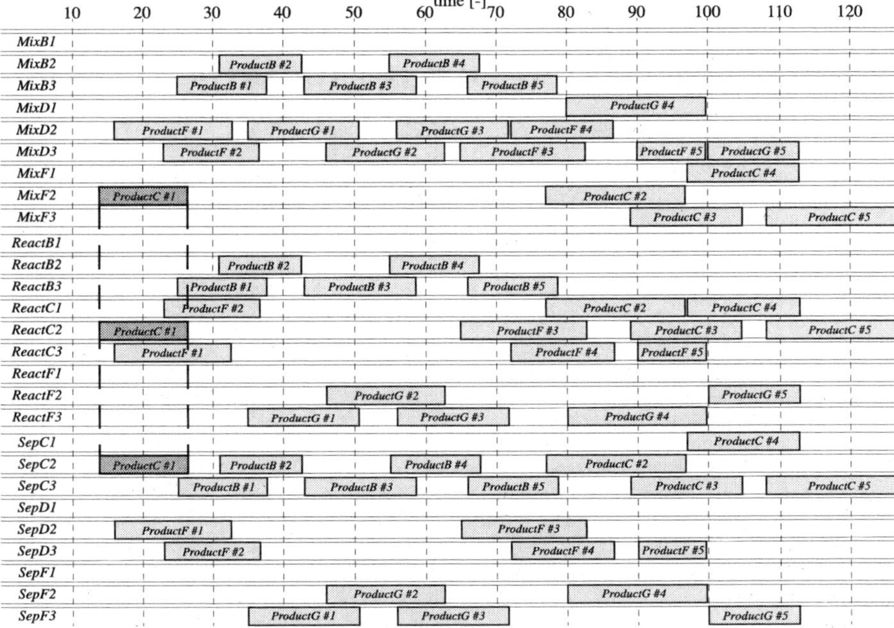

Fig. 5 A diagram of simulation results.

sensitive micro chemical processes, new equipments are frequently added and old or disabled equipments are removed, therefore the information is frequently updated. However, the result shows that the multiagent-oriented system has sufficient effectiveness in the information management without deadlock, information loss and traffic congestion.

4. Conclusion

A multiagent-oriented method was applied to management of micro chemical processes. It was possible to manage efficiently the information updated frequently in sensitive micro chemical processes.

Acknowledgement

This work was supported by a Grant-in-Aid for Scientific Research (No.12450309) from the Ministry of Education, Science and Culture of Japan.

References

[1] Löwe, H. and W. Ehrfeld, State-of-the-art in microreaction technology: concepts, manufacturing and applications, *Electrochimica Acta*, **44** (1999), 3679-3689.

[2] Koch, M., C.G.J. Schabmueller, A.G.R. Evans and A. Brunnschweiler, Micromachined chemical reaction system, *Sensors and Actuators A*, **74** (1999), 207-210.

[3] Kuroda, C. and M. Ishida, A Proposal for Decentralizd Cooperative Dicision-Making in Chemical Batch Operation, *Engineering Application of Artificial Intelligence*, **6** (1993), 399-407.

[4] Wada, H, An Approach for Future Manufacturing Systems using Agents, *Journal of Japanese Society for Artificial Intelligence*, **15** (2000), 583-591.

KES 2002
E. Damiani et al. (Eds.)
IOS Press, 2002

Computer-Aided Hazard Identification in Batch Processes Using Petri Nets

Yi-Feng Wang and Chuei-Tin Chang
Department of Chemical Engineering, National Cheng Kung University,
Tainan, Taiwan 70101, Republic of China

Abstract. A systematic procedure is presented in this paper to construct Petri nets for modeling the fault propagation behaviors in batch processes. A complete system model is basically organized according to a hierarchy of four levels, i.e. (1) the controller/operator, (2) the valves, (3) the process units, and (4) the sensors. Every component model is built with two distinct elements. One is used to characterize the equipment states and the other the input-output relations. The effectiveness and correctness of this methodology has been successfully demonstrated with a simple example.

1 Introduction

In order to ensure operation safety, hazard analysis is one of the basic tasks that must be performed in designing or revamping any chemical process. Numerous techniques are available for this purpose, e.g. fault tree analysis (FTA), event tree analysis (ETA) and hazard and operability study (HAZOP), etc. In implementing these methods, there are always needs (1) to reason deductively for finding all combinations of basic events that could lead to an undesirable condition and/or (2) to predict all possible consequences of a given fault origin. If every fault propagation mechanism is to be identified manually, a rigorous hazard analysis is bound to be labor- and time-consuming and its results often error-prone. In the past, many efficient computer aids have been adopted for hazard assessment. The most popular one is the digraph-based algorithm [1, 2]. Although proven useful, this method is effective mostly in applications concerning the *continuous* processes. This is due to the fact that the digraph model is inherently unsuitable for describing the dynamic causal relationships among time, discrete events, equipment states and system configurations in the *batch* or *semi-batch* processes. On the other hand, the Petri net (PN) is well known for its capability in representing the discrete-event systems. Originally, the Petri net was not designed for handling the real variables. Through a series of enhancements, the Petri net has now been developed into a powerful modeling tool by later researchers [3, 4]. The aim of this paper is thus to develop a systematic procedure to construct Petri nets for automatic hazard identification in the batch processes.

2 The Elements in Petri-Net Models

In order to facilitate proper representation of sequential operations, a collection of PN elements is chosen in this work. Specifically, both the discrete and continuous places are allowed

Level	Component Models
1	timer, operator, PLC
2	valve, pump, compressor
3	process unit
4	sensor

Table 1: The Hierarchy in PN-Based System Models for Sequential Operations

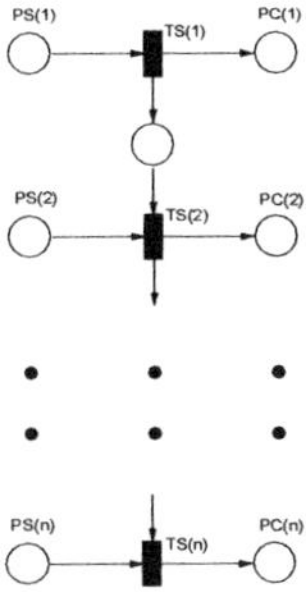

Figure 1: The Generalized PN Model Representing the Input-Output Relations of an Operator/Controller

in the proposed PN model and the transitions can be timed and non-timed. In addition, three different types of place-to-transition arcs are utilized, i.e. the weighted arcs, the inhibitor arcs and the static test arcs, and all transition-to-place arcs are weighted arcs. If an arc is directed toward or away from a continuous place, the independent variables of its weight function should be the token number in one or more user-selected place. For the sake of brevity, the definitions of these elements are omitted in this paper. A detailed description can be found in the literature [5, 6].

3 The Hierarchy in a System Model

A hierarchical approach has been taken in this work to construct Petri nets for modeling the *normal* operation steps in batch processes. The components in a complete system model can be classified into the four different levels shown in Table 1. In general, every item in the P&ID is described with a component model here. Each component consists of two distinct parts. One is used to characterize the equipment state and the other the input-output conditions. In the latter case, several different versions are needed if a change in the equipment state alters the relations among process conditions. Following is a general description of the component models in every level.

- *Level 1:* The operating steps specified in a recipe are executed sequentially by a 1st-level component. In general, each operation step can be characterized with two elementary actions: (1) confirmation of an initiation signal and (2) execution of an operation command. The input-output model of a controller/operator can be developed accordingly (see Figure 1). The input place $PS(i)$ denotes the status of the ith initiation signal. The initiation signal can be obtained either from a sensor in the 4th level or from an internal clock. The non-timed transition $TS(i)$ denotes the confirmation action of the ith initiation signal. The output place $PC(i)$ is used to reflect the status of the ith operation command. Notice that it is not necessary to model its equipment state in this case since there is only one possibility during normal operation, i.e. the component is in service.

- *Level 2:* All 2nd-level components can be described with two alternative equipment states. The PN model of a 3-way valve is presented in Figure 2. In this Petri net, the places $PV(+)$ and $PV(-)$ denote two alternative valve positions connecting lines 1&2 and 1&3

Figure 2: The PN Model Describing the Equipment States of a 3-Way Valve

Figure 3: The PN Model Representing the Input-Output Relations of a 3-Way Valve

respectively, and the transitions $TV(1)$ and $TV(2)$ represent the valve-switching actions from $PV(+)$ to $PV(-)$ and vice versa. Notice that the input places $PC(1)$ and $PC(2)$ of the two transitions $TV(1)$ and $TV(2)$ are associated with the corresponding operation commands issued by controller/operator. On the other hand, the causal relations between the input and output conditions of a level-2 component are described *qualitatively* in this work. Let us use the 3-way valve again as example. Its input and output flow rates can be related with the Petri net in Figure 3. Notice that the places here represent qualitative deviation levels from their normal values. Each of them can actually be expanded to form a sub-PN [7, 8]. This type of *deviation places* is a new extension created mainly to facilitate simulation of the fault propagation behaviors in sequential operations.

- *Level 3:* Basically all process units in the P&ID can be considered as the 3rd-level components. The input-output models of a level-3 component can be constructed with deviation places. The model structure is essentially the same as that of any level-2 component. On the other hand, the equipment state of a process unit can usually be assumed to be unchanged under normal operating conditions. However, this assumption may not be valid if (1) there is a continuous accumulation (or depletion) of mass and/or energy in the unit or (2) the performance of a unit deteriorates quickly during operation. Thus, it is necessary to describe the transients in these process units with continuous places. A generalized version is presented in Figure 4. Here, the *continuous* place $PES(j)$ represents the jth equipment state used to characterize a level-3 component; the discrete place $POM(i)$ denotes the ith operation mode defined by the equipment states of level-2 components; $TES_{in}(i)$ and $TES_{out}(i)$ are time-delayed transitions enabled after the ith mode is activated in operation; the weight function w_{in} and w_{out} denote respectively the amounts of increase and decrease in the continuous state variable during a very small time increment Δt. In other words, $w_{in}/\Delta t$ and $w_{out}/\Delta t$ can be considered as the approximate rates of increase and decrease of the state variable respectively.

- *Level 4:* The input of a sensor is the equipment state or the output condition of a 3rd-level component. The sensor output is the measurement signal. The corresponding model structure can also be found in Figure 3. Since it can be assumed that a sensor is always in the working state during normal operation, there is no need to model its equipment state.

4   The Failure Mechanisms in Component Models

To facilitate hazard analysis, it is still necessary to incorporate additional sub-PNs in each component model to depict the fault propagation behaviours caused by various equipment

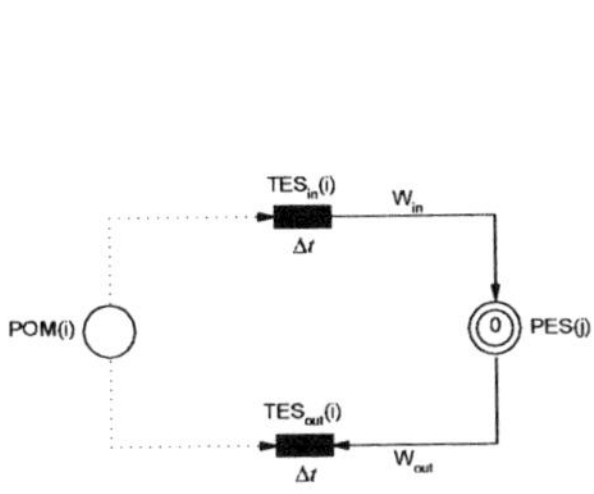

Figure 4: The Generalized PN Model Describing the Equipment State of a Process Unit in Level 3

Figure 5: A Generalized Failure Model

failures. A generalized failure model can be found in Figure 5. In this model, the direct outcome of a failure is treated as a change in the equipment state of a component. The equipment state caused by the ith failure is represented by the place $PFS(i)$. The effects of a failure can be readily modeled with a combination of the inhibitor arcs and test arcs. The former arcs are used to disable the transitions corresponding to the routine events, i.e. $TN(j)$, and the latter activate the alternative transitions representing the failure events, i.e. $TF(k)$.

5   The Model Construction Procedure

In building a PN-based system model for hazard analysis, the component models should be constructed first and then connected in sequence from top to bottom level according to the P&ID. In principle, all component models should be included to ensure the comprehensiveness of analysis.

6   Applications

The proposed method has been tested extensively with several complex industrial problems. Due to the limitation of space, only a simple example concerning a mixing process is presented here. There are three tanks in this process. Tank 3 is being fed from the other two tanks. Initially, the amount of liquid A in tank 1 is 1 liter and that of liquid B in tank 2 is 2 liter. The mixing operation begins when the outlet valve V1 of tank 1 and the outlet valve V2 of tank 2 are both opened by an operator. It is assumed that the flow rates of these two feed streams are the same: 1/60 liter/second. The operator is instructed to close valve V2 whenever tank 1 is empty (step 1) or vice versa (step 2). Only valve failures are considered in this example, i.e. sticking and failing close. The PN model for this batch operation can be found in Figure 6. It is assumed in this example that the possibility of sensor failures can be excluded and the sensor measurements are identical to the equipment states of Tank 1 and Tank 2. As a result, the sensor models are omitted.

The fault propagation behaviors resulting from the valve failures given in Figure 5 have been simulated with an existing commercial software Visual Object Net++. It can be observed from the simulation results that the causes for the condition "off-spec product" occurring between 0 and 1 min are {valve V1 failing close between 0 and 1 min} and {valve V2 failing close between 0 and 1 min} and the cause for the same condition occurring between 1 and 2

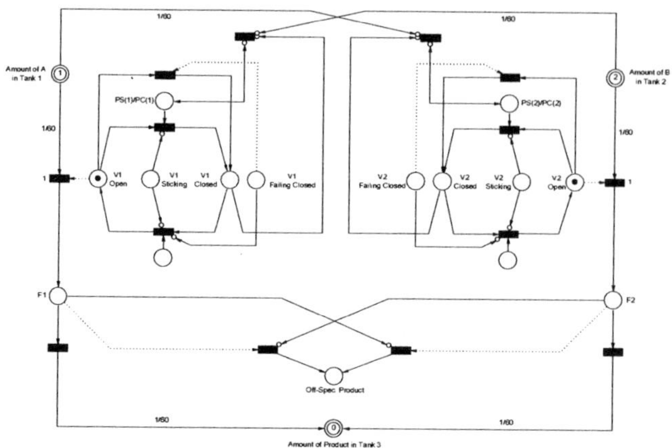

Figure 6: The PN Model of Mixing Process

min is {valve V2 sticking between 0 and 1 min}.

7 Conclusions

A hierarchical approach is proposed in this study to construct a comprehensive PN model for any batch process. By carrying out simulation studies, identification and enumeration of critical fault propagation scenarios become very efficient. It is clear from the application results that the proposed PN models can indeed be used as the basis for rigorous hazard analysis.

References

[1] C. T. Chang and K. S. Hwang, Studies on the Digraph-based Approach for Fault-Tree Synthesis 1. The Ratio-Control Systems, Ind. Eng. Chem. Res., **33** (1994) 1520.

[2] C. T. Chang, D. S. Hsu and D. M. Hwang, Studies on the Digraph-based Approach for Fault-Tree Synthesis 2. The Trip Systems, Ind. Eng. Chem. Res., **33** (1994) 1700.

[3] J. L. Peterson, Petri Net Theory and the Modeling of Systems, Prentice-Hall,Englewood Cliffs, NJ, (1981).

[4] H. Alla and R. David, A modelling and analysis tool for discrete events systems: continuous Petri net, Perform. Eval., **33** (1998) 175.

[5] R. Drath, Hybrid Object Nets: An Object Oriented Concept for Modeling Complex Hybrid Systems, Hybrid Dynamical Systems. 3rd International Conference on Automation of Mixed Processes, ADPM'98, Reims, (1998).

[6] F. D. J. Bowden, A Brief Survey and Synthesis of the Roles of Time in Petri Nets, Math. Comput. Model., **31** (2000) 55.

[7] Y. F. Wang, J. Y. Wu and C. T. Chang, Automatic Hazard Analysis of Batch Operations with Petri Nets, Reliab. Eng. Syst. Saf. (accepted), (2001).

[8] Y. F. Wang and C. T. Chang, A Hierarchical Approach to Construct Petri Nets for Modeling the Fault Propagation Mechanisms in Sequential Operations, submitted to Comput. Chem. Eng., (2002).

KES 2002
E. Damiani et al. (Eds.)
IOS Press, 2002

Acquisition of AGV Control Rules Using Reinforcement Learning

Hisaaki Yamaba, Shigeyuki Tomita
Miyazaki University
1-1 Gakuen Kibanadai Nishi, Miyazaki, Japan

Abstract.

When a simulation-based approach is adopted in order to design production systems under uncertain conditions, a design support environment that generates simulators of various configurations automatically is indispensable. Also, such environment is required to devise suitable operating rules of each design candidate at the same time.

This paper deals with acquisition of AGVs control rules using reinforcement learning method. Concretely, Profit sharing was adopted and the computer-based support system was developed by Java language by means of Object-Oriented technique. A series of experiments was carried out using this system in order to confirm whether rules for various path topologies can be learned.

1   Introduction

A simulation based approach is one of the most effective way as a design method of production systems under uncertain conditions. However, it is a very hard task to develop simulators for every structure candidates. Therefore, a design support environment that generates such simulators automatically is indispensable. Also, such an environment must be able to devise a suitable operating strategy of each design candidate at the same time. Because performance of each design plan should be evaluated under its most suitable operating rules.

This study deals with the design problem of Automated Guarded Vehicle System (AGVS) that allows various type of path layouts. This problem is very difficult approaching by mathematical formalized model because of combinatorial complexity and uncertainty of the environment.

This paper focuses on acquisition of AGVs control rules using reinforcement learning method. Concretely, Profit sharing was adopted and the computer-based support system was developed by Java language by means of Object-Oriented technique. Also, a series of experiments was carried out using this system in order to confirm whether rules for various path topologies can be learned.

2   Class of target AGVS

It is assumed that the target production system discussed here consists of several machines, buffering areas, AGVs and a path. Designers are allowed to adopt various types of path layouts such as linear, loop, mesh and their variations (Figure 1). A path for AGVs has limited width that does not allow two or more AGVs to run side by side at a time. So, many hazards,

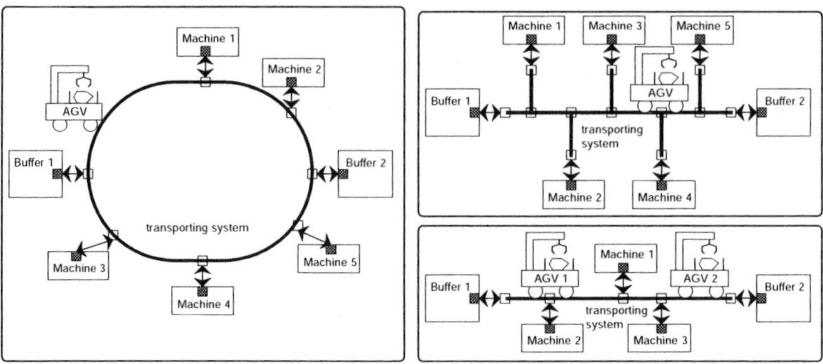

Figure 1: Various architectures of plants

such as collision or deadlocks in AGV operations, arise from inadequate controlling rules of an AGV System[1].

Each load is required to be processed at several machines in designated order. AGVs transport loads from a machine to another. Since each machine does not have its own buffering area, they cannot accept a new load until an AGV picks up a processed load from the machine. Therefore, an AGV transports a load to a common buffering area if the next machine is processing another load.

One problem is how to dispatch loads to AGVs and the other is how to control traveling of AGVs without collision among them. In this paper, as for the former problem, loads and AGVs are selected at random. We focused on the latter problem in which we expect Profit sharing method to obtain rational strategy.

3 Learning of Control Rules

In order to deal with various path layouts, *Incremental route planning*[2] was employed as the basis of routing algorithm in this paper. *Incremental route planning* selects the next node for the vehicle to visit so as to reach its destination based on the status of neighboring nodes and information about the global network. The next node is selected among all adjacent nodes such that it will result in the shortest travel time. The next node selection is repeated until the vehicle reaches its destination. In our research, acquisition of rational strategy for selecting the next node using restricted local information is intended.

An AGV has four states, that is, {Traveling, Transporting, Picking-up/Dropping-down, Vacant}. Traveling is the moving from the current position of a vehicle to the pickup point. Transporting is the one from the pickup point to the drop-off point. Vacant is the state with no assigned transportation job. And there are four basic actions for achieving Transporting/Traveling such as, {forward-move, backward-move, detour-move, waiting}. State space of AGVs is represented by state of each AGV and their current positions and destination positions. The absolute positions of AGVs are not important but their relative positions are significant. For example, in the case that two vehicles are in a production system with a linear form path layout, there are 86 situations for which control rules are needed.

In general, an AGV moves forward by selecting a node in the shortest path toward its

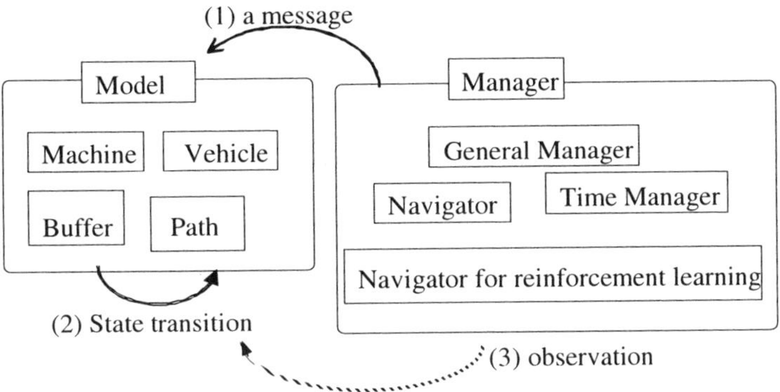

Figure 2: Components of simulators.

destination. But when other AGV is in its sight, it can get other's state and its destination. Then, using Profit sharing, it learns about its control rules of each state; which basic action it should select.

4 Development of Simulators

In order to construct simulators with various path layouts, Object-Oriented approach was adopted. Parts of simulators are classified into two groups: "Models" and "Managers"(Figure 2). Each "Model" represents a component of production systems such as a machine, a vehicle and so on. For example, Class Vehicle is defined together with its methods that describe the states and the actions mentioned above. And each of the vehicles is assigned to one instance of the class. On the other hand, interactions among "Models" are encapsulated in "Managers". Each method of the classes is described according to the task analysis of the target production system.

A Simulator is driven by simple message passing as bellow.

1. A "Manager" sends a message to a "Model".

2. The "Model" changes its state according to the described method.

3. "Managers" that observe the "Model" send messages to other "Models" according to the described method.

Simulators made of such objects can be easily re-built when the configuration of the target plant changes.

In this study, Class Navigator was defined separated form another Managers in order to concentrate methods for controlling AGVs. The Class for reinforcement learning of AGVs control was developed easily by modifying Class Navigator.

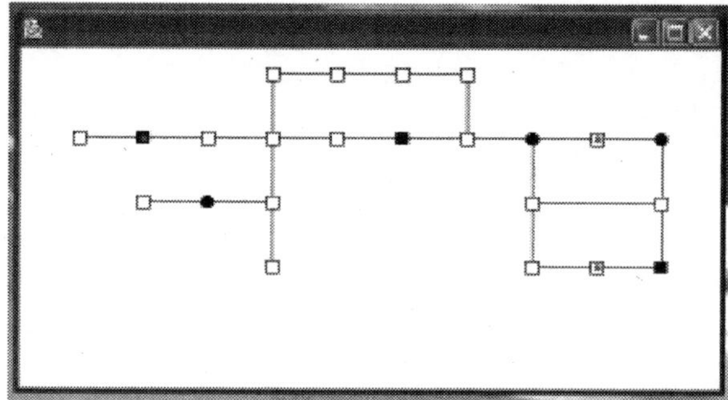

Figure 3: Screen shot of a simulator

5　Experiments

5.1　Generation of Simulators

First, we developed the components of simulators in the Java language, and constructed simulators with various path layouts using the components. Figure 3 shows a screen shot of one of the simulators. In this experiment, episodes started when any potential conflict was detected by comparing path occupation time of each vehicle. As a result, it was confirmed that simulators with various path layout ran normally and some rules were reinforced. However, the speed of learning was very slow because long time was needed until the first reward was given.

5.2　Performance of obtained rules

Next, two simulation experiments were held in order to verify whether suitable vehicle control rules can be obtained by means of Profit sharing. The production system with linear path layout and two vehicles was assumed in the experiment.

A 5000 steps simulation was repeated 20 times. The number of loads carried in one simulation experiment was used as the index for the evaluation.

AGVs select their actions by a roulette-based weighting of the conflicting rule set. The initial weight of each rule is 10 and the reward is 16. And we use a geometrically decreasing function (common ration = 0.5). Also, episodes of Profit sharing are started when another vehicle comes into the sight in these experiments.

Initially, the reward was given to the AGV who executed the transporting job only when the job was completed. As a result, *backward-move* was not reinforced in situations such that the action seems to be preferable (Experiment 1).

Therefore, in the next experiment (Experiment 2), we gave reward not only to the AGV who executed a transporting job but also to the vehicles that were in the sight of the transporter. Because, reinforcement of a *backward-move* which has contributed to a successful execution of another vehicle's transportation is expected.

As a simulation for learning was repeated, the number of processed loads increased. The result of the experiment in which vehicles were controlled by hand-coded rules is also shown

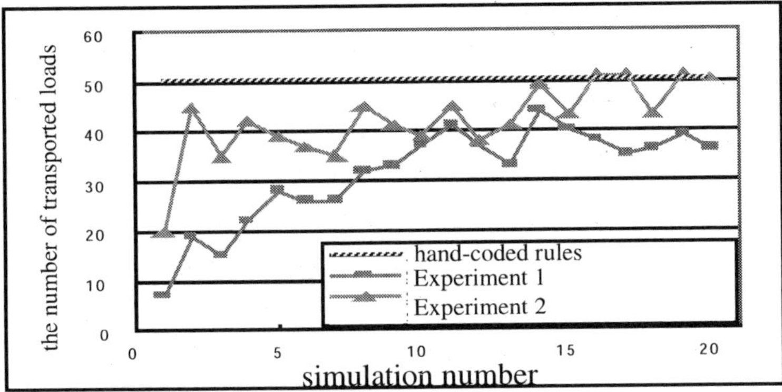

Figure 4: Results of experiments.

in Figure 4. These results show that the performance of the rules obtained by reinforcement learning is about equal to the one of the hand-coded rules.

6   Conclusion

An attempt was made to obtain operation rules by reinforcement learning in order to realize a design support environment of production systems.

A computer aided design system for designing topological structures of AGVS was developed by means of Object-Oriented approach. Program modules that represent components of production systems such as machines, vehicles and so on were defined as classes. Simulators with various path topologies were able to be constructed by instances of these classes easily.

Verification and evaluation of effectiveness of this approach were confirmed through a series of experiments on the simulation system. As a result, it was confirmed that simulators with various path layout can easily be constructed and vehicles can learn control rules automatically by using reinforcement learning.

References

[1] Ling Qiu, Wen-Jing Hsu, Shell-Ying Huang and Han Wang, Scheduling and routing algorithm for AGVs: a survey, *Integernational Journal of Production Research*, **40**, (2002) 745-760.

[2] Taghaboni, F, and Tanchoco, J.M.A, Comparison of dynamic routing techniques for atuomated guided vehicles systems, *Integernational Journal of Production Research*, **33**, (1995) 2653-2669.

[3] Sachiyo Arai, Kazuteru Miyazaki, Shigenobu Kobayashi, Cranes Control Using Multi-Agent Reinforcement Learning, *International Conference on Intelligent Autonomous System 5*, (1998) 335-342.

KES 2002
E. Damiani et al. (Eds.)
IOS Press, 2002

An Evaluation of Term Weighting Schemes for Discovering Related Terms from Texts

Lina Zhou
University of Maryland, Baltimore County, MD 21250, U.S.A.

Abstract. How to search and reuse texts efficiently is an important issue in managing enormous amount of digital resource. Terms and semantic relationship between terms have great potential in linking and organizing textual information, thereby increasing the efficiency and effectiveness of information access. Term weighting schemes have been widely applied to document indexing in information retrieval and question answering systems. We are going to apply three term weighting schemes to the automatic discovery of term semantic relationship from texts, and discuss the performance of the weighting schemes on the identification of five types of generic relations in this paper. The preliminary evaluation results provide insights into elements of the term weighting schemes that contribute to better performance in determining related terms from texts.

1. Introduction

With the widespread adoption of computer and Internet technology, electronic texts are accumulated exponentially at an unprecedented speed. However, effective search and reuse of most texts is not available partly due to lack of association among them. Terms and semantic relationship between terms have great potential in linking and organizing textual information, thereby increasing the efficiency and effectiveness of information access.

Empirical cooccurrence-based approach has received a lot attention in the automatic identification of lexical associations [1-4] due to easy access to large-scaled text resource and powerful computers. It is based on the assumption that words similar in meaning tend to occur in the same document. The approach normally requires a model to represent term (or word) and a metric to measure similarities between terms. Prior researches primarily utilized raw co-occurrence frequencies or distributional probabilities as term representation. Automatic term weighting have been widely adopted in identifying keywords from texts for many purposes. Enlightened by the effectiveness of term weighting scheme in information retrieval and question answering, we are going to apply and evaluate term weighting schemes in the automatic discovery of term semantic relationship. We enrich the above assumption of co-occurrence approach with the proposition that semantically related terms are more likely to occur in the same texts with approximately equivalent weights than unrelated terms. Therefore, it would be interesting to find out whether term weighting scheme is helpful for identifying related terms and which scheme has better performance.

2. Related Work and Research Objectives

A term is potentially related to a group of other terms through certain semantic relations. The relations could be either domain-specific or generic. The generic relations include hyponym/hypernym (IS-A/HAS-A), meronym/holonym (Part-of/Has-Part), synonym, and so on.

Some progress has been made in the automatic identification of related terms from texts. Lin [5] automatically extracted similar words from texts and created a tree structure for words based on their similarities. Dagan et. al. [4] proposed probabilistic word association models based on distributional word similarity, and applied them to language modeling and pseudo-word disambiguation. Rioloff [3] developed a bootstrapping algorithm for creating semantic lexicon from a set of seed words. Zhou et. al. [6] applied a weighting scheme to the automatic acquisition of related terms for ontology development. However, little discussion has been devoted to the effect of different term weighting schemes on the acquisition results. Based on the extensive research and application of term weighting functions in such areas as information retrieval and question answering [7-9], we select three of the most popular weighting schemes and compare their performance in discovering related terms. A good scheme is expected to lead to the reliable estimation of term relationship, and ultimately little human involvement in post-validation. It can be safely assumed that the higher the similarity between two terms, the stronger the semantic relationship between them. Thus, we can make use of similarity metrics to measure the strength of term relationship. Holding term vector model [9] as term representation and *cosine* measure as similarity metric constantly, we apply three weighting schemes separately and then make side-by-side comparison between their results. Therefore, the main goal of this study is to investigate which of the commonly used term weighting schemes can support measuring term semantic relationship most effectively.

3. Term Weighting Schemes and Related Terms Generation

3.1 Term Weighting Schemes

A term weighting scheme can play two types of roles in determining related terms. One is to filter term candidates in order to obtain a representative set of terms from a document. If a term is not important in a document, which is reflected as a low weight value, it may be meaningless to examine the relationship between the term and other collocated terms. In this paper, we are more interested in the other function that a term weighting scheme can play in measuring term relationship. In this case, we input term weights to a similarity metric and the measured results directly determine related terms.

Among the variety of possible term weighting schemes in information retrieval or question answering, we choose three commonly used ones: *tf*idf* [7], *tc* weighting [8], and *ltc* weighting [9]. The following are definitions of w_{ik} , weights of k_{th} term in i_{th} document, for three schemes respectively:

$$ltc \text{ weighting:} \quad w_{ik}^{ltc} = \frac{tf_{ik} \times \ln(N_D / n_k)}{\sqrt{\sum_{j=1}^{N} (tf_{ij} \times \ln(N_D / n_j))^2}} \tag{1}$$

$$tc \text{ weighting:} \quad w_{ik}^{tc} = tf_{ik} \times \ln(N_D / n_k) \tag{2}$$

$$tf * idf \text{ weighting:} \quad w_{ik}^{tf*idf} = f_{ik} \times \ln(N_D / n_k) \tag{3}$$

$$tf_{ik} = \begin{cases} 0, & \text{if } f_{ik} = 0; \\ \ln(f_{ik}) + 1, & \text{otherwise.} \end{cases}$$

where, N_D is the total number of documents in the collection C, n_k is the number of documents in C that contain term k, f_{ik} is the frequency of term k in document i.

All the above schemes allow a term to be weighted by factors dependent upon its importance both within a document and in the entire document collection. It is shown that any two of the above three weighting schemes share some common elements while differ on others, which enables us to break down a scheme into smaller components and compare them at finer granularity.

3.2 Related Terms Generation

We generate related terms in the following three steps:

1) Term representation - a term vector model is created for each term, consisting of pairs of relevant document number and term weight in that document. The vectors of all terms constitute a term vector space (TVS).

2) Similarity measurement – *cosine* measure is used to calculate term similarities in TVS. The similarity between term t_i and t_j is defined as:

$$sim(t_i, t_j) = \frac{\sum_{k=1}^{L} w_{ik} * w_{jk}}{\sqrt{\sum_{k=1}^{N}(w_{ik})^2 * \sum_{k=1}^{M}(w_{jk})^2}} \qquad (4)$$

where, w_{ik} and w_{jk} are the weights of t_i and t_j in the document k respectively, N and M are the number of documents where t_i and t_j occur respectively, L is the number of documents in a collection where both t_i and t_j occur.

3) Related Terms Set (RTSET) extraction - criteria are developed for extracting a reasonable RTSET for each term based on the results of step 2). One is g, the threshold for minimum similarity value. The other is *f*, the threshold for maximum number of related terms that could be extracted for each term.

4. Evaluation

We tested the three weighting schemes in acquiring related terms from a text collection in community development (ComD). One of the motivation of choosing ComD as test domain was that terms in ComD had a significantly large intersection with those in WordNet [10], which was used as baseline for our evaluation. 3,725 documents were collected from the Internet. Among over 6,000 terms and RTSET that were generated with the procedure described in Section 3.2, 576 terms and their RTSET were randomly selected for evaluation.

WordNet (WN) [10] is a general-purpose thesaurus. SYNSET in WN, which denotes a set of synonyms, is similar to RTSET in our results. Among many relations that are compiled into WN, we selected five of the most popular relation types for evaluation, namely *synonym* (syno), *hypernym* (hype), *hyponym* (hypo), *holonym* (holo), and *meronym* (mero). We literally called SYNSET of hype in WN as HYPESET. HYPOSET, HOLOSET, and MEROSET were labeled and explained in the same way.

The evaluation was conducted for each relation type separately in terms of two measures: precision and recall. The *precision* of a term on relation r is defined as the percentage of terms in the extracted RTSET that are included in the rSET of the term in WN, while the *recall* of a term on relation r is the percentage of terms in the rSET of the term in WN that are extracted in the RTSET. The overall precision (or recall) on relation r is the average precision (or recall) of all individual terms on the relation.

5. Results and Analysis

It was shown in Figure 1 and Figure 2 that the performance of *tc* scheme was consistently the best among the three in terms of both precision and recall. *ltc* and *tc* weighting exhibited the same pattern across five relation types. The precision of *tf·idf* was much better than that of *ltc* on holo, mero, and hypo relations, whereas the former was slightly worse than the latter on hype and syno relations. Cross-reference between the two figures revealed that the precisions and recalls of *tf·idf* and *ltc* on five relations were similar, and the performance of *ltc* scheme degraded severely on hype and syno relations.

The next question we want to address is which element(s) in *tc* scheme attribute to its superior performance. Compared with *tf·idf* scheme in Formula (2), *tc* in Formula (1) includes transformed term frequency rather than raw term frequency. The purpose of converting term frequency with logarithmic function in *tc* was to dampen the effect of large differences on frequency. We can infer that it may be desirable to reduce the difference on term frequency in a weighting scheme for the identification of term semantic relations. Compared with *tc* scheme, *ltc* in Formula (3) introduces a vector length based normalization factor to squash values into a 0 to 1 scale. It illustrated the similar patterns that *ltc* and *tc* displayed over all five relations. Moreover, it indicated that normalizing weight with vector length might not be necessary. Further analysis of *ltc* scheme exposed that the effect of difference on term frequency in a document was decreased with the logarithmic function in the nominator, and length-based normalization factor in the denominator actually punished terms in long documents again. It partly explained the poorest performance of *ltc* on hype and syno relations, for almost every term in WN has these two relations, which reduced the discriminatory ability of the weighting scheme. On the contrary, *tf·idf* keeps a large discriminatory factor of term frequency, consequently it outperformed *ltc* on hype and syno relations. Overall, *tc* scheme maintained a good balance between the local term frequency and global collection frequency; therefore, it achieved the best outcome in differentiating between significantly related terms and other accidentally co-occurred ones.

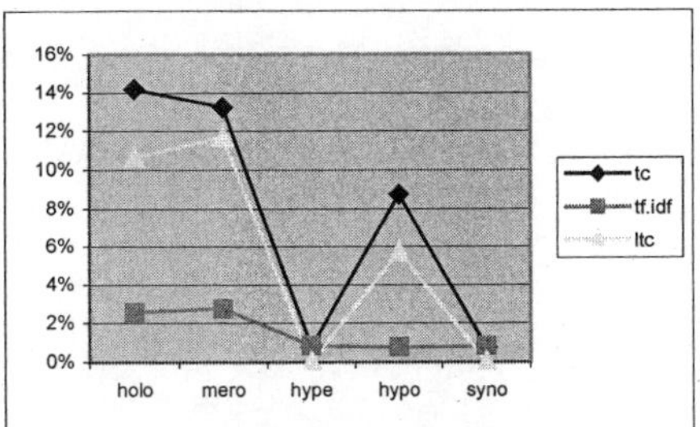
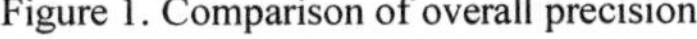

Figure 1. Comparison of overall precision

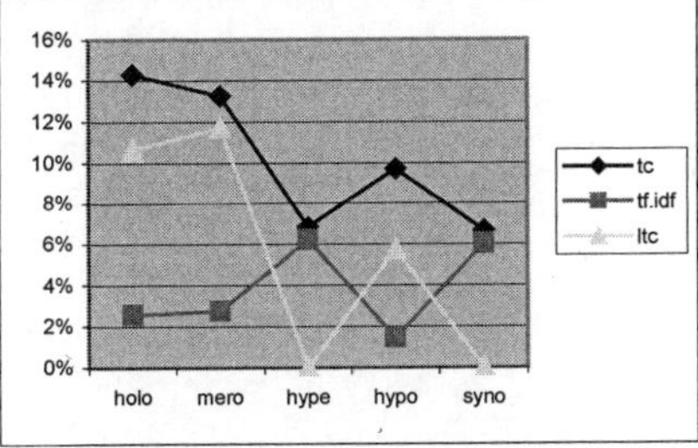

Figure 2. Comparison of overall recall

6. Discussion

In this paper, we evaluated the performance of three term weighting schemes in discovering term semantic relationship in terms of precision and recall. The overall precisions and recalls were not as high as expected. There are several possible explanations for this. First, we severely punished terms that were mismatched between WN and the test set by assigning 0 to both of their precisions and recalls. Secondly, the threshold of minimum similarity value, g, was set as high as 0.33, and that of maximum number of extracted related terms, f, was set as low as 12 during RTSET extraction. We may improve the results by adjusting the two thresholds. For example, if we reduce the value of g and/or increase f, the recall will be improved. Thirdly, WN is a general-purpose resource, which does not reflect the domain specificity of ComD knowledge. All term weighting schemes selected in this study were subject to the same restriction in evaluation, so we could still make a valid comparison between different relation types.

There are many potential applications of the automatically acquired related term, such as building semantic lexicon and ontology, creating language models, and enhancing the effectiveness of information retrieval and document clustering.

References

[1]   H. Li and N. Abe, "Word Clustering and Disambiguation Based on Co-occurence Data," presented at COLING-ACL'98, 1998.

[2]   S. Jean-David, "Automatic acquisition of terminological relations from a corpus for query expansion," presented at ACM SIGIR'98, Melbourne, Australia, 1998.

[3]   E. Riloff and J. Shepherd, "A Corpus-Based Bootstrapping Algorithm for Semi-Automated Semantic Lexicon Construction," *Journal of Natural Language Engineering*, vol. 5, pp. 147-156, 1999.

[4]   I. Dagan, L. Lee, and F. C. N. Pereira, "Similarity-based Models of Word Cooccurrence Probabilities," *MLJ*, vol. 34(1-3), 1999.

[5]   D. Lin, "Automatic retrieval and clustering of similar words," presented at COLING-ACL'98, Montreal, Canada, 1998.

[6]   L. Zhou, Q. E. Booker, and D. Zhang, "ROD - Toward Rapid Ontology Development for Underdeveloped Domains," presented at 35th Hawaii International Conference on System Sciences, Big Island, Hawaii, 2002.

[7]   G. Salton and C. Buckley, "Term weighting approaches in automatic text retrieval," *Information Processing & Management*, vol. 24, pp. 513-523, 1988.

[8]   A. Singhal, J. Choi, D. Hindle, D. D. Lewis, and F. Pereira, "AT&T at TREC-7," presented at The Seventh Text Retrieval Conference (TREC-7), Gaithersburg, Maryland, 1998.

[9]   C. Buckley, J. Allan, and G. Salton, "Automatic Routing and Ad-hoc Retrieval Using SMART," presented at Text Retrieval Conference (TREC-2), Gaithersburg, Maryland, 1993.

[10]   G. A. Miller, "WORDNET: An On-Line Lexical Database," *International Journal of Lexicography*, vol. 3-4, pp. 235-312, 1990.

KES 2002
E. Damiani et al. (Eds.)
IOS Press, 2002

A New Asymmetric Topographic Mapping Algorithm

Manuel MARTIN-MERINO [1] and Alberto MUÑOZ [2] *
[1] *University Pontificia of Salamanca, C/Compañía 5, 37002 Salamanca, Spain*
mmerino@ieee.org

[2] University Carlos III of Madrid, C/Madrid 126, 28903 Getafe, Spain
albmun@est-econ.uc3m.es

Abstract. Self organizing maps (SOM) are useful tools to visualize word relationships. However SOM algorithm assumes that object distances are symmetric and therefore when asymmetry arises the visual maps generated are degraded and the information provided by asymmetry is lost. In this paper a we present new model based on SOM algorithm that is able to deal with asymmetric dissimilarities and to represent the skew symmetric component in an intuitive way. The topographic map generated is also a valuable tool that help to discover the topic hierarchy. The algorithm proposed has been tested in a real database with promising results.

1 Introduction

Let X be an $n \times m$ matrix representing n objects as vectors in $\mathbb{R}^m$. Let $(D = \delta_{ij})$ be the matrix made up of object dissimilarities. The dissimilarity matrix is non-symmetric when $(\delta_{ij} \neq \delta_{ji})$. Asymmetry arises in many practical problems such as (sociometry, psychology [8]). It is particularly important when objects are organized hierarchically. Consider for instance, the problem of modeling word relations. As has been pointed out in [5], the organization of document collections into topics is strongly hierarchical. Therefore broad terms like "Mathematics" contain the meaning of specific terms such as "Bayesian statistics" but the reverse relation is much weaker. So any algorithm proposed should be able to work with asymmetric similarities . On the other hand the visualization of the skew symmetric component should allows to discover easily the topic hierarchy.

Self organizing maps [4, 7] have been successfully applied to generate visual representations of word relations. However the algorithm is not suitable to work with asymmetric similarities and therefore is not able to accurately represent word relations. On the other hand the information provided by asymmetry and consequently the information about the topic hierarchy is lost.

In this paper, we propose a new error function for the mapping algorithm that allows to deal with asymmetric dissimilarities and to represent the skew symmetric component. For this purpose, we first define an asymmetry coefficient that conveys the information provided by

*Financial support from DGICYT grant BEC2000-0167 (Spain) is gratefully appreciated.

asymmetry. Next objects are embedded in a bidimensional grid in which first and second dimensions represent the symmetric and skew symmetric components of the dissimilarity matrix respectively.

This paper is organized as follows. Section 2 introduces a coefficient of asymmetry and section 3 presents the new asymmetric topographic mapping algorithm. Next, in section 4 the new algorithm is tested in a real problem and finally section 5 gets conclusions and outlines future research trends.

2   Asymmetry

As we have mentioned earlier asymmetry arises when $(\delta_{ij} \neq \delta_{ji})$. In this case the dissimilarity matrix can be decomposed into a symmetric and skew symmetric component $(D = S + A)$ [8] where $s_{ij} = \frac{\delta_{ij} + \delta_{ji}}{2}$ and $a_{ij} = \frac{\delta_{ij} - \delta_{ji}}{2}$. The skew symmetric component conveys the information provided by the dissimilarity matrix that can not be explained by a symmetric model. This component will suggest coefficients that are able to model asymmetry according to a given similarity measure. If we compute the skew symmetric component of the widely used fuzzy logic similarity measure [6], we get $a_{ij} = \frac{|x_i \wedge x_j|}{|x_i|} - \frac{|x_j \wedge x_i|}{|x_j|} \propto |x_i| - |x_j|$. This expression shows that asymmetry is related to individual properties of objects and is proportional to the difference of L_1 norms. If we define for object i the asymmetry coefficient a_i as $|\ x_i\ |$, it is clear that the $a_{ij} = a_i - a_j$. The asymmetry coefficient defined is associated to individual objects. Moreover, the difference of asymmetry coefficients for pairs of objects is a good approximation of the skew symmetric component of the dissimilarity matrix. Other interesting asymmetry coefficients have been proposed in [6].

Let get a deeper understanding of the meaning of this coefficient in the context of word relation visualization. Consider the vectorial representation of term k. Each component j is 1 if term k appears in document j. So, the L_1 norm gives the number of documents indexed by term k. This coefficient will become large for broad terms and provides not only information about asymmetry but also about the hierarchical structure of topics (terms) in the database.

3   SOM Based Asymmetric Topographic Algorithm

In this section we present a new topographic mapping algorithm that take advantage of the asymmetry to induce automatically a topic hierarchy.

The self-organizing map [3] is a nonlinear visualization technique that allows to order high dimensional data. Input vector are represented as neurons arranged according to a regular grid (usually 2D) in such a way that similar vectors in input space become spatially close in the grid. The algorithm proceeds in two steps. First, a quantization algorithm is run that carries out a Voronoi tessellation of the input space. Next the prototypes are organized in such a way that neighboring neurons in the grid represent similar vectors in input space. Several topologies may be selected for the grid of neurons [3] but linear or square lattices perform quite well in a wide range of problems. Prototypes are organized by minimizing the following quantization error function,

$$E(\mathcal{W}) = \sum_r \sum_{x_\mu \in V_r} e_r^\mu = \sum_r \sum_{x_\mu \in V_r} \sum_s h_{rs} D(x_\mu, \omega_s) \qquad (1)$$

where V_r is the Voronoi region associated with the prototype ω_r, x_μ is the input pattern, $D(x_\mu, \omega_s)$ is usually a Euclidean distance and h_{rs} is the neighborhood function that works as a symmetric smoothing kernel over the grid. Note that SOM algorithm assumes that object dissimilarities in input space are symmetric. Therefore when asymmetric relations arise the final map is degraded and the information provided by asymmetry is lost.

We propose to extend the SOM algorithm to deal with asymmetric similarities. For this purpose we first define an asymmetric dissimilarity measure suitable to model word relations in a simple way. Next a new topology is defined that is able to represent both, the symmetric and the skew symmetric components of the dissimilarity matrix. Let $D'(x_\mu, \omega_s)$ be a dissimilarity measure defined as

$$D'(x_\mu, \omega_s) = (\mid x_\mu \mid - \mid \omega_s \mid)^2 + (x_\mu - \omega_s)^T (x_\mu - \omega_s) \qquad (2)$$

The first term on the right hand side is the difference of object's asymmetry coefficients and model the information related to asymmetry. This term is proportional to the skew symmetric component of the fuzzy logic similarity measure (see section 2). The second term is the Euclidean distance. It can be empirically shown that this term is strongly correlated with the symmetric component of the fuzzy logic similarity measure when the dimension of input space is high and vectors are normalized according to the L_2 norm. Therefore this measure may be considered as a simplified version of the asymmetric fuzzy logic similarity that is more efficient computationally. Note that strictly speaking, $D'(x_\mu, \omega_s)$ is not an asymmetric dissimilarity unless first term of expression (2) is multiplied by $sign(\mid x_\mu \mid - \mid \omega_k \mid)$. However function error optimization is easier if this term is incorporated into the definition of the asymmetric kernel over the grid (see equation (5)).
Voronoi regions are created using this new dissimilarity measure. Points belonging to the same region will have similar asymmetry coefficients (L_1 norm) and small Euclidean distances among them.

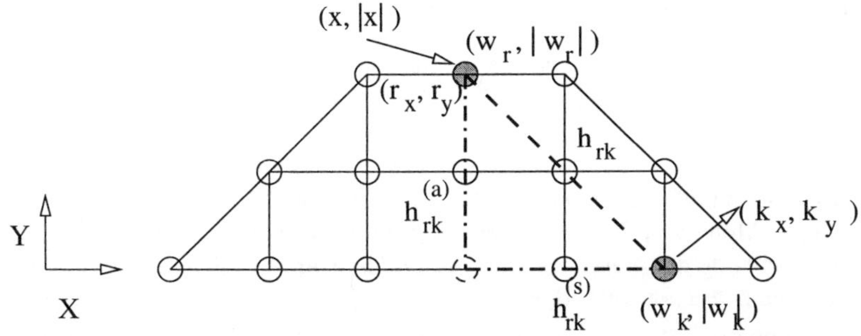

Figure 1: trapezoidal topology grid for our asymmetric SOM algorithm. X and Y directions represent the symmetric and skew symmetric components respectively.

Figure 1 shows the grid over which neurons are arranged in the asymmetric SOM algorithm. X coordinate represent the symmetric component of the dissimilarity defined by equation (2). Prototypes corresponding to neighboring neurons along X axis will have small Euclidean distances in input space. A symmetric kernel is used to transform the distances defined as $h_{rk}^{(s)} = exp(-\frac{(d_{rk}^x)^2}{2\sigma_s^2})$ where $d_{rk}^x = \mid r_x - k_x \mid$ is the Manhattan distance and σ_s is a parameter that determines the degree of smoothing. Y axis represent the skew-symmetric component of the dissimilarity defined in (2). Prototypes of neurons with similar Y grid coordinates will have similar asymmetry coefficients. In this case, a non symmetric kernel is used to transform the distances defined as $h_{rk}^{(a)} = exp\left(-\frac{sign(r_y - k_y)\,(d_{rk}^y)^2}{2\sigma_a^2}\right)$ where $d_{rk}^y = \mid r_y - k_y \mid$ is the Manhattan distance and σ_a the smoothing parameter. This new kernel does not obeys the constraint $h_{rk} = h_{kr}$. Finally the distance between neurons r and k over the grid is

$$d_{rk} = d_{rk}^x + d_{rk}^y \tag{3}$$

Prototypes are organized by optimization of the error function (1) with e_r^μ defined as

$$e_r^\mu = \sum_s h_{rs}^{'(a)}(\mid x_\mu \mid - \mid \omega_s \mid)^2 + \sum_s h_{rs}^{(s)}(x_\mu - \omega_s)^T(x_\mu - \omega_s) \tag{4}$$

where

$$h_{rs}^{'(a)} = exp\left(-\frac{sign(r_y - k_y)\,sign(\mid \omega_r \mid - \mid \omega_k \mid)\,(d_{rk}^y)^2}{2\sigma_a^2}\right) \tag{5}$$

Equation (4) shows that the error function will be minimized when the Euclidean distance between prototypes of neighboring neurons is small and the difference of L_1 norms is also small. On the other hand, note that $h_{rs}^{'a}$ include also the sign associated with the skew symmetric component of the dissimilarity defined in (2). This term favors that $\mid \omega_r \mid > \mid \omega_k \mid$ if $r_y > k_y$. In this way broad terms that have large asymmetry coefficient will be placed at the top of the trapezoidal topology (see figure 1) and specific terms at the bottom of the grid . Therefore a topic hierarchy is generated automatically by exploiting the connection between asymmetry and hierarchy.

The error function obtained by substituting (4) in (1) is optimized by a Robins Monro algorithm [3] that allows to work in an on-line fashion.

$$\frac{\partial E(W)}{\partial \omega_{kl}} = \sum_r \sum_{x_\mu \in V_r} \frac{\partial e_r^\mu}{\partial \omega_{kl}} = \sum_r \sum_{x_\mu \in V_r} h_{rk}^{(a)} \frac{\partial D^{(a)}(x_\mu, \omega_k)}{\partial \omega_{kl}} + h_{rs}^{(s)} \frac{\partial D^{(s)}(x_\mu, \omega_k)}{\partial \omega_{kl}} \tag{6}$$

However the derivation of equation (6) poses several problems. First the resulting updating equations are more complex than in the symmetric case. Thus the computational load become excessively high. Moreover, dimension reduction techniques may not be applied due to the fact that they usually do not preserve L_1 norms. Finally some problems of convergence arises when the parameters are not well adjusted.

We propose to embed the data in a space with a new coordinate, the L_1 norm. Each prototype is represented as $\hat{\omega}_r = (\omega_r, \mid \omega_r \mid)$ and may be directly derived with respect to the last coordinate. Using this trick a simple updating equation is derived for each coordinate

$$\omega_{kl}^{(t+1)} = \omega_{kl}^{(t)} + \alpha_{sl}\, h_{rk}^{(s)}(x_{\mu l} - \omega_{kl}^{(t)}) \tag{7}$$

$$\mid \omega_k \mid^{(t+1)} = \mid \omega_k \mid^{(t)} + \alpha_a\, h_{rk}^{(a)}(\mid x_\mu \mid - \mid \omega_k \mid^{(t)}) \tag{8}$$

Note that a batch version may be easily obtained by taking expression (6) equal to 0 and solving the resulting linear equation.

4 Experimental Results

We have tested our asymmetric SOM algorithm in a real and challenging problem such as word relation visualization. Word relations are strongly asymmetric and besides terms are organized hierarchically [5]. So this application will help to determine the capability of the asymmetric SOM algorithm to deal with asymmetric proximities and to reproduce topic hierarchies.

In order to easily assess the performance of the asymmetric algorithm presented we have built a database made up of 2000 documents that group in three main topics ("Library science", "Science and technology" and "Sociology"). There are 13 well defined subtopics.

Let X be the document matrix. Each column represents a word codified as a vector so that $x_{ij} = 1$ if word j is found in document i. Other codifications have been proposed that multiply word coordinates by suitable weights [1]. So, the dimension of the space in which words are codified is given by the number of documents in the database, 2000 in this case. Words were normalized according to L_2 norm because this preprocessing step improves the results of all clustering and visualization algorithms tested. Before applying our asymmetric SOM algorithm dimensionality is reduced to 150 by means of a random projection technique [4]. All the parameters for the asymmetric SOM algorithm are fixed using the same procedure that the symmetric counterpart. However α_{sl} and α_a in equations (8) have been fixed to values proportional to the variance of the respective variable. This improves significantly the convergence of the algorithm.

The performance of our algorithm is evaluated through the Spearman correlation coefficient [2]. This coefficient measure the preservation of neighbor's order by the mapping algorithm. Neighbors in input and output spaces will be computed according to the distances defined in (2) and (3). Spearman correlation for our algorithm is 0.62 that is greater than 0.59 for the symmetric counterpart. This shows that our algorithm preserves asymmetric distances as well as the classical SOM preserves Euclidean distances.

Tables 1, 2 and 3 show some term relations induced by neurons of first, second and third levels respectively ($y = 4$, $y = 3$, $y = 2$). Due to the lack of space only 4 neurons per level have been displayed. Terms shown for each neuron correspond to those with larger L_1 norm. Neurons of table 1 represent broad terms corresponding to four major topics in the database. It can be seen in subsequent tables that neurons in levels 3 and 4 represent the same topics but the vocabulary is more specific. So, our algorithm is helpful to discover hierarchical term relations.

Neur.	6	8	15	17
	RESEARCH	TESTING	AUTOMATIC	MODEL
	SYSTEMS	SEMICONDUCTOR	PATTERN	STRUCTURE
	LITERATURE	MATERIAL	ALGORITHMS	APPLICATIONS
	BIOLOGICAL	DEVICES	ACCURACY	EXPERIMENTAL
	LIBRARY	POWER	RECOGNITION	TECHNIQUE

Table 1: Neurons belonging to first level for the asymmetric SOM.

Neur.	4	7	15	17
	LINE ABSTRACTING MEASUREMENT UNIVERSITY USERS	REPRESENTATION REQUIREMENTS THERMAL TEMPERATURE EFFECT	QUANTIZATION CLASSIFIER OPTIMAL SOM VECTOR	PROPERTIES PROCESSES DIFFERENCE SIGNAL COMPONENT

Table 2: Neurons belonging to second level for the asymmetric SOM.

Neur.	2	5	16	17
	CULTURE INDEXING SEARCHING CATEGORIES ASSOCIATION	COMMUNICATION IMPLEMENTATION LIGHT ELECTRIC CIRCUIT	FUZZY PARAMETER SIMILARITY SUPERVISED CLUSTERING	STANDARD OUTPUT LOSS RESISTANCE DISPERSION

Table 3: Neurons belonging to third level for the asymmetric SOM.

5   Conclusions

In this work we have proposed a new asymmetric topographic mapping algorithm. The new algorithm represents both, the symmetric and the skew-symmetric components of the dissimilarity matrix and is useful to discover hierarchical relationships.

Experimental results show that our algorithm achieves good preservation of asymmetric distances. On the other hand the asymmetric topographic algorithm stand up as a useful tool to discover the hierarchical topic structure.

Future research trends will focus on the study of new topologies suitable to represent asymmetric similarities.

References

[1] R. Baeza-Yates and B. Ribeiro-Neto. Modern Information Retrieval. Addison Wesley, UK, 1999.

[2] J. C. Bezdek and N. R. Pal . An Index of Topological Preservation for Feature Extraction, Pattern Recognition, 28(3):381-391, 1995.

[3] T. Kohonen. Self-Organizing Maps, Second Edition, Germany, Springer Verlag, 1997.

[4] T. Kohonen, S. Kaski, K. Lagus, J. Salojarvi, J. Honkela, V. Paatero and A. Saarela. Organization of a Massive Document Collection. IEEE Transactions on Neural Networks, 11(3):574-585, 2000.

[5] D. Lawrie, W. B. Croft and A. Rosenberg. Finding Topic Words for Hierarchical Summarization, SIGIR'01, New Orleans, USA, ACM, 349-357, 2001.

[6] M. Martin-Merino and A. Muñoz. Self Organizing Map and Sammon Mapping for Asymmetric Proximities, ICANN, LNCS 2130, 429-435, Springer Verlag, 2001.

[7] D. Merkl, Text Classification with Self-organizing Maps: Some Lessons Learned, Neurocomputing, 21, 61-77, 1998.

[8] B. Zielman and W.J. Heiser. Models for asymmetric proximities. British Journal of Mathematical and Statistical Psychology, 49:127-146, 1996.

Set-theoretic similarity measures

M. Rifqi and B. Bouchon-Meunier
LIP6 – Pôle IA
8, rue du Capitaine Scott – 75015 Paris, France
{Maria.Rifqi, Bernadette.Bouchon-Meunier}@lip6.fr

Abstract. In this paper, we give an overview of general classes of similarity measures which presents the advantage of being inserted in a cognitive framework and being available for objects described by means of non-classical kinds of values, such as linguistic values, as well as traditional numerical attributes. We also give means to prioritize one measure over the others.

1 Introduction

Similarity is a widely used concept, with utilizations in various fields, such as pattern recognition, case-based reasoning, image processing, approximate reasoning, machine learning, information retrieval for instance. There exist many definitions of similarity or resemblance, or conversely, dissimilarity or distance. In the case where the objects to be compared are not defined on a metrical universe, distances are clearly not appropriate. Several types of similarity are nevertheless available [1], which have been connected with seminal works in psychology [6].

2 Knowledge representation

Let us consider descriptions of objects defined on a universe Ω. The elements of Ω can be, for instance, sets of symbols (words in text mining, features in images...), intervals of the set of real numbers $I\!\!R$ (measurements,...), fuzzy sets of a reference set U (representation of linguistic descriptions in database querying,...). In this last case, the membership function of A is denoted by f_A. We suppose that an *order* $\subseteq$ is defined on Ω, and an operation of *difference* $-$ on Ω is such that: $A - B$ is monotonous with respect to A, according to $\subseteq$, and $A \subseteq B$ implies $A - B = \emptyset$.

We also define operations of *union* $\cup$ and *intersection* $\cap$ on Ω. We finally suppose given a mapping $M : \Omega \to R^+$ such that: $M(\emptyset) = 0$ and M is monotonous with respect to $\subseteq$.

For instance, in the case of a set Ω of symbols or intervals of $I\!\!R$, the operations are classical intersection, union and difference, and $M(A)$ is the cardinality $|A|$ of A for symbols, its norm $\|A\|$ for intervals. The order $\subseteq$ is the classical inclusion.

Let us finally consider a reference set U, the set Ω of fuzzy sets of U, the order $\subseteq$ is the classical inclusion of fuzzy sets ($A \subseteq B$ if and only if $f_A \leq f_B$), the intersection and union are defined by min and max operators. *Differences* − of fuzzy sets can be:

$$f_{A-B}(x) \;=\; \max(0, f_A(x) - f_B(x)) \;[7] \tag{1}$$

$$f_{A-_2B}(x) \;=\; \begin{cases} f_A(x) & \text{if } f_B(x) = 0 \\ 0 & \text{if } f_B(x) > 0 \end{cases} \tag{2}$$

The mapping M is a *fuzzy set measure*, for instance: $M_1(A) = \int_U f_A(x)dx$, $M_2(A) = \sup_{x\in U} f_A(x)$, $M_3(A) = \sum_{count} f_A(x)$ if U is countable.

3   Measures of similitude

Let us now consider measures of comparison of elements of Ω [1].

An *M-measure of similitude* (m.sim) on Ω is a mapping : $S : \Omega \times \Omega \to [0, 1]$, defined as: $S(A, B) = F_S(M(A \cap B), M(B - A), M(A - B))$, for a given mapping $F_S : R^{+3} \to [0, 1]$, such that $F_S(u, v, w)$ is non-decreasing in u, and non-increasing in v and w.

Obviously, the notion of M-measure of similitude is still very general and corresponds to mappings with very different behaviors. In order to help choosing one of them in a particular problem solving, we consider the following additional properties which may be satisfied by M-measures of similitude.

- *reflexivity* ($S(A, A) = 1$ for any A) which means that $F_S(u, 0, 0) = 1$ for any $u \neq 0$.

- *exclusiveness* ($S(A, B) = 0$ for any A and B such that $A \cap B = \emptyset$), which means that $F_S(0, v, w) = 0$ for any v and w different from 0.

- *symmetry* ($S(A, B) = S(B, A)$ for any A and B), which means that $F_S(u, v, w) = F_S(u, w, v)$ for any u, v, w.

We then distinguish the following m.sim of particular interest :

- An *M-measure of satisfiability* (m.sat) is an exclusive and reflexive m.sim independent of the third component: $S(A, B) = F_S(M(A \cap B), M(B - A))$, for a function $F_S : R^{+2} \to [0, 1]$ such that $F_S(u, v) = F_S(u, v, .)$. As a consequence, an m.sat satisfies the *containment* property (if $B \subseteq A$, $B \neq \emptyset$, then $S(A, B) = 1$).

- An *M-measure of resemblance* (m.res) is a symmetric and reflexive m.sim: $S(A, B) = F_S(M(A \cap B), M(B - A), M(A - B))$.

4   Discrimination power of measures of similitude

4.1   Measures of satisfiability

A m.sat corresponds to a situation in which we consider a reference object or a class and we need to decide if a new object is compatible with it or satisfies it. For instance, m.sat are appropriate for rule base systems.

4.1.1 Scale sensitivity

It is desirable that a m.sat depends only on the relative weights of its components and not on the scale of the system. In order to obtain an objective measure, we propose [4] to normalize the m.sat. We note: $X = M(A \cap B)$ and $Y = M(B - A)$. We consider: $x = \frac{X}{\sqrt{X^2+Y^2}}$, the reduced intersection and $y = \frac{Y}{\sqrt{X^2+Y^2}}$, the reduced distinctive feature.

As $x^2 + y^2 = 1$, the domain of definition of the m.sat is a quarter of circle. It can be described by a unique argument ϕ, with $\phi = \arctan \frac{y}{x}$. We note the m.sat $S(A, B) = \eta(\phi)$, with η decreasing with respect to ϕ, such that $\eta(\frac{\pi}{2}) = 0$ and $\eta(0) = 1$.

We can represent (figure 1) the reference set A by the vector $V_{\vec{A}}$ and the set B by a vector $V_{\vec{B}}$. When the two vectors are orthogonal, then the satisfiability vanishes: $S(A, B) = 0$. More generally, the satisfiability appears as a projection, and a lack of satisfiability is represented as a deviation in figure 1.

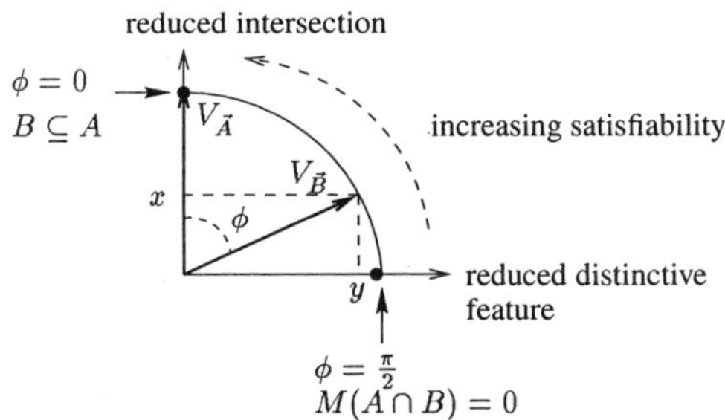

Figure 1: New representation of a measure of satisfiability

This new form of a m.sat, expressed by a unique variable, has the advantage of not being dependent upon the size of the system. Furthermore, this normalization makes the definition of a m.sat more simple insofar as the argument is a segment $[0, \frac{\pi}{2}]$ and not a quarter of plan.

4.1.2 Examples

There are of course many possible choices for the satisfiability measure η. Among them, let us distinguish the following forms (see figure (a) of table 1):

- $\eta_1(\phi) = 1 - \frac{2}{\pi}\phi$. It is a linear satisfiability function.
- $\eta_2(\phi) = \cos \phi$, where $\eta_2(\phi)$ can be seen as the scalar product $V_{\vec{A}} \cdot V_{\vec{B}}$.
- $\eta_3(\phi) = \frac{1}{1+\tan \phi}$.
- $\eta_4(\phi) = 1 - \sin \phi$.

The third measure can be rewritten as: $S(A, B) = \frac{M(A \cap B)}{M(B)}$, in the case of fuzzy sets of U, with $M = M_3$. The last one can be rewritten as: $S(A, B) = 1 - M(B - A)$ introduced in [2] for fuzzy sets with $M = M_2$ and with the difference $-_2$.

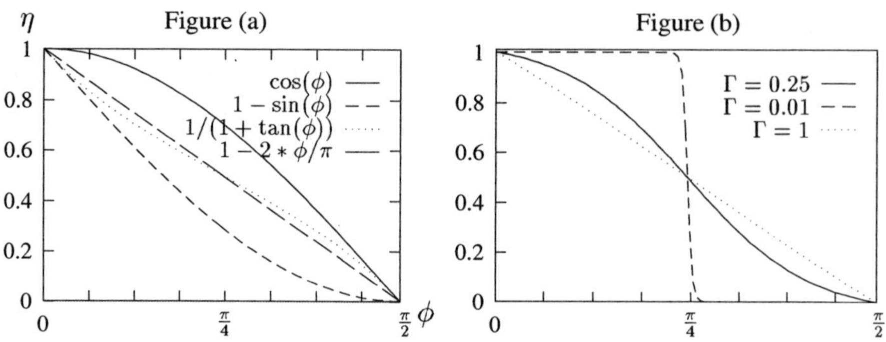

Table 1: The behaviour of measures of satisfiability

4.1.3 Discrimination power

With this new representation, the m.sat can be easily compared. We consider the *discrimination power* of a measure of satisfiability as given by the derivative $\eta'(\phi)$ of η.

For every possible η, we have: $\int_0^{\frac{\pi}{2}} \eta'(\phi)d\phi = -1$. This means that the total discrimination power $\eta'(\phi)$ has to be distributed on the $[0, \frac{\pi}{2}]$ interval, but a high discrimination power somewhere implies a low discrimination power elsewhere. Accordingly, it is necessary to choose a measure with a discrimination power suitable for the considered application. This suggests a method of construction of a m.sat.

We can remark that no function with a high discrimination power for $\eta(\phi) = 1/2$ but a low discrimination for $\eta(\phi) = 0$ and $\eta(\phi) = 1$ is available in figure (a) of table 1. Nevertheless, this kind of measures means that if a description is not far from the reference, then the satisfiability is near 1 because the difference is not significative. If a description is very far from the reference, we can consider that the satisfiability is null. We propose such an interesting measure based on the Fermi-Dirac function $F_{FD}(\phi) = \dfrac{1}{1+\exp^{\frac{(\phi-\frac{\pi}{4})}{\Gamma}}}$, where $\Gamma \in \mathbb{R}^+$ controls the decrease of the curve. This measure is defined as:

$$\eta(\phi) = \frac{F_{FD}(\phi) - F_{FD}(\frac{\pi}{2})}{F_{FD}(0) - F_{FD}(\frac{\pi}{2})}$$

The choice of Γ enables to define a m.sat more or less severe, as shown on the figure(b) of table 1.

4.2 *Measures of resemblance*

A m.res is used for a comparison between the descriptions of two objects, of the same level of generality, to decide if they have many common characteristics.

Measures of resemblance are appropriate for case-based reasoning or instance-based learning. In clustering methods, distances can be replaced by a m.res. More generally, similarity-based classification methods [5] have to use m.res as soon as all objects have the same level of generality. Let us denote $Z = M(A - B)$.

We focus on m.res satisfying the property of *exclusiveness*.

Following our normalization procedure, we define: $x = \frac{X}{\sqrt{X^2+Y^2+Z^2}}$, $y = \frac{Y}{\sqrt{X^2+Y^2+Z^2}}$, $z = \frac{Z}{\sqrt{X^2+Y^2+Z^2}}$, for $(X, Y, Z) \neq (0,0,0)$. Similarly to the case of $\mathtt{m.sat}$, this ensures that an exclusive $\mathtt{m.res}$ is not dependent on the scale of the problem.

The domain of study is now restricted to a part of the unity sphere since $x^2 + y^2 + z^2 = 1$. Geometrically, the sphere is simply obtained by a rotation of the satisfiability circle around the x-axis (see figure 2). The vector representation is still valid.

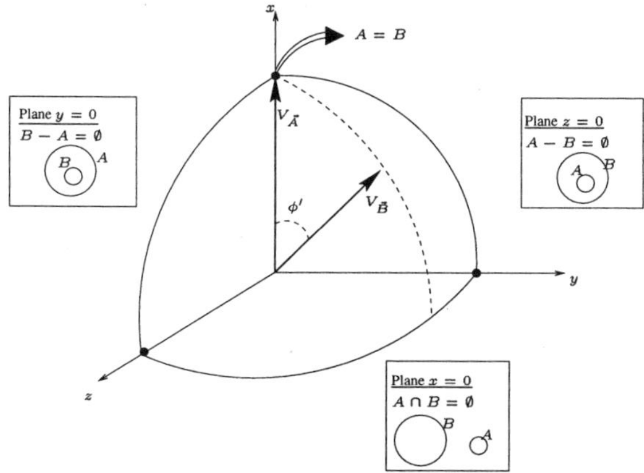

Figure 2: New representation of an exclusive measure of resemblance

Let us consider $\rho(y, z)$ with ξ non decreasing with regard to x and z, $\xi(0,0) = 0$ and $\xi(y, z) = \xi(z, y)$. This means that ρ can be described by any symmetrical function with respect to y and z. Let us define $\psi = \arctan(\frac{\rho}{x})$ and a $\mathtt{m.res}$ $S(A, B) = \nu(x, \rho)$ such that: $\nu(x, \rho)$ is increasing with respect to x and decreasing with respect to ρ, $\nu(0, \rho) = 0$ if $\rho \neq 0$, $\nu(x, 0) = 1$.

These conditions show that the problem has been reduced to a satisfiability measure. We can therefore use again the solution described in the preceding section dealing with satisfiability. With this definition of ρ, an exclusive resemblance appears as a satisfiability where a global distinctive feature ρ is defined by $\rho(y, z)$, from the two individual distinctive features y and z.

We can also consider different exclusive $\mathtt{m.res}$ as we have already done with $\mathtt{m.sat}$. In the case where $\rho^0 = y + z$, we get $\nu_0 = \frac{1}{1+\frac{\rho^0}{x}} = \frac{1}{1+\tan\psi}$. This measure corresponds to the $\mathtt{m.sat}$ η_3. Furthermore, ν_1 can be also written as: $S(A, B) = \frac{M(A\cap B)}{M(A\cup B)}$ with M such that: $M(A \cup B) = M(A \cap B) + M(A - B) + M(B - A)$, for instance M_3. This measure was introduced in [3].

Other definitions of ρ can be envisaged, for instance: $\rho' = \sqrt{y^2 + z^2}$ or $\rho'' = \sqrt{(\sqrt{y} + \sqrt{z})^2}$ associated with ν' and ν''. The choice of a particular form of ρ has an effect on the measure of resemblance because this parameter represents distinctive elements. We can notice that, $\rho'' \geq \rho^0 \geq \rho'$ $\forall y, z$. As ρ has a decreasing effect on an exclusive $\mathtt{m.res}$, the above relation implies that, for a given x and for all y and z, $\nu''(x, \rho') \leq \nu(x, \rho^0) \leq \nu'(x, \rho')$. This relation means that $\nu''(x, \rho')$ penalizes more the differences between two sets than $\nu(x, \rho^0)$ and that $\nu(x, \rho^0)$ penalizes more the differences than $\nu'(x, \rho')$. Furthermore, a particular ρ is sensitive to the symmetry between y and z. Indeed, if distinctive features y and z are unbalanced for instance, it means that $y \gg z$ or $z \gg y$, the behaviours of two given ρ can be opposite and then the order of resemblances of objects are totally inverted.

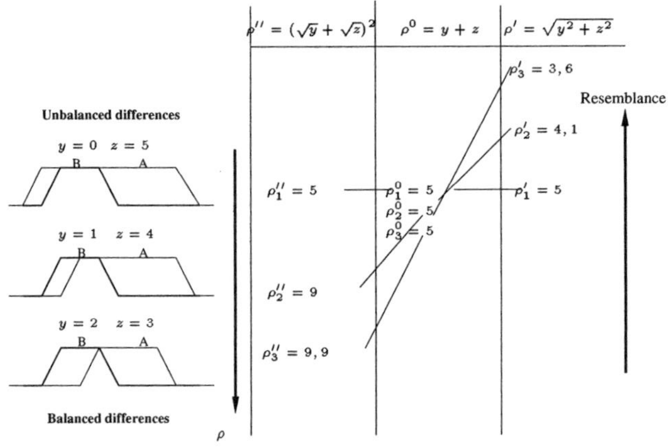

Figure 3: The effects of different definitions of ρ on exclusive measures of resemblance.

5   Conclusion

In the domain of non metric similarity measures, the paper focuses on those based on sets (fuzzy or classical). Even in this restricted study, the choice of a similarity remains very large. We have proposed a way to focalize on particular family of measures in two steps: the first one consists in distinguishing general properties such as symmetry, reflexivity, exclusiveness,etc. Then, the second step enables to refine the family found in the first step by describing the discrimination power of the desired measure.

References

[1] B. Bouchon-Meunier, M. Rifqi, and S. Bothorel. Towards general measures of comparison of objects. *Fuzzy Sets and Systems*, 84(2):143–153, 1996.

[2] B. Bouchon-Meunier and L. Valverde. Analogy relations and inference. In *Proceedings of 2nd IEEE International Conference on Fuzzy Systems*, pages 1140–1144, San Fransisco, 1993.

[3] D. Dubois and H. Prade. *Fuzzy Sets and Systems, Theory and Applications*. Academic Press, New- York, 1980.

[4] M. Rifqi, V. Berger, and B. Bouchon-Meunier. Discrimination power of measures of comparison. *Fuzzy Sets and Systems*, 110(2):189–196, March 2000.

[5] M. Rifqi, S. Bothorel, B. Bouchon-Meunier, and S. Muller. Similarity and prototype based approach for classification of microcalcifications. In 7th *IFSA World Congress*, pages 123–128, Prague, 1997.

[6] A. Tversky. Features of similarity. *Psychological Review*, 84:327–352, 1977.

[7] L. A. Zadeh. The concept of a linguistic variable and its application to approximate reasoning. *Information Sciences*, 8 and 9:199–249, 301–357, 43–80, 1975.

KES 2002
E. Damiani et al. (Eds.)
IOS Press, 2002

Clustering Based on a Dissimilarity Measure Derived from Data

Paolo CORSINI, Beatrice LAZZERINI, Francesco MARCELLONI
Dipartimento di Ingegneria della Informazione, University of Pisa, Via Diotisalvi, 2
56122 Pisa, ITALY, e-mail: {p.corsini, b.lazzerini, f.marcelloni}@iet.unipi.it

Abstract. Distance functions are not suitable to express object dissimilarity when the nature of dissimilarity is conceptual rather than metric. For this reason, we propose to learn the dissimilarity relation directly from the available data by appropriately training a feed-forward neural network. We exploit some pairs of points with known dissimilarity value to teach the dissimilarity relation to the neural network. Then, we adopt the neural dissimilarity measure to guide an unsupervised relational clustering algorithm. A synthetic data set is used to show how the clustering algorithm based on the neural dissimilarity outperforms some widely used clustering algorithms (with possible partial supervision) based on spatial dissimilarity.

1. Introduction

Data clustering refers to the problem of partitioning a data set into homogeneous groups (called *clusters*) in such a way that patterns within a cluster are more similar to each other than patterns belonging to different clusters. Most clustering algorithms [1][2][3] exploit a similarity (or dissimilarity) measure, which is expressed in terms of some distance function (such as the Euclidean distance and the Mahalanobis distance). Distance functions, however, may fail to model the dissimilarity concept when the data are not distributed according to any known regular law, or whenever the dissimilarity between any two arbitrary patterns depends on conceptual aspects that cannot be expressed in metric terms. Consider, for example, the pixels of an image made up of distinguishable elements with irregular-shaped contours. The dissimilarity between pixels should be small (large) when the pixels belong to the same image element (different image elements).

It is obvious that in such cases no distance function can be a meaningful measure of dissimilarity between patterns; instead, a more general dissimilarity measure should be used. An interesting approach to this problem consists in obtaining the dissimilarity measure directly from the data, without assuming in advance any specific cluster shape [4].

Following this approach, in this paper we assume that the patterns are points in a multi-dimensional space and exploit a multilayer perceptron with supervised learning to extract the dissimilarity relation from the data themselves. The trained neural network produces a degree of dissimilarity between pairs of input points. A training sample is a triple $(x_1, x_2, r_{1,2})$, where $r_{1,2} \in [0,1]$ is the known degree of dissimilarity between the points x_1 and x_2. In this way, the dissimilarity between pairs of points is not dependent on their distance, but on their semantic affinity. The output of the neural network can be used to generate a dissimilarity relation between points in the data set. Based on this relation, we use a version of the fuzzy non-metric (FNM) model proposed in [5] as relational clustering algorithm.

To test the effectiveness of our approach we present an example of its application to an artificial data set. We show how the relational clustering algorithm, which exploits the dissimilarity relation extracted using a limited number of training samples, actually outperforms clustering algorithms based on spatial dissimilarity.

2. The Neural Network

To determine dissimilarity degrees between pairs of patterns, we use a standard feed-forward two-layer neural network. The network takes a pair (x_1, x_2) of patterns as input, and outputs the degree of dissimilarity between x_1 and x_2. Each neuron is equipped with a sigmoidal non linear function. The standard back-propagation algorithm with a dynamically decreasing learning rate was used as a learning scheme. Errors less than 0.02 were treated as zero. Initial weights were random values in the range $\left[-1/\sqrt{m}, 1/\sqrt{m}\right]$, with m being the number of inputs to a neuron. The choice of the number of neurons for each layer depends on specific application characteristics, such as the number and shape of the classes, which partition the data set. The architecture with 20 and 8 neurons for the first and second hidden layers provided the best performance on the data set shown in Section 4.

To assess the consistency of the dissimilarity model expressed by the neural network, we tested the network both with pairs of points belonging to the same class but far from each other in terms of spatial distance, and with pairs of points belonging to different classes but close to each other. These pairs of points had not been included in the training set of the network. In most cases (about 90%), the dissimilarity degree was lower (higher) than 0.5 for pairs of points belonging (not belonging) to the same class.

3. The Algorithm

Let $X = \{x_1,...,x_N\}$ be the set of objects. A similarity or dissimilarity square binary relation R in X is a function $\rho : X \times X \rightarrow [0,1]$. Relation R is often expressed in terms of a matrix $R = [r_{i,j} = \rho(x_i, x_j)]_{N \times N}$. Relational clustering methods proposed in the literature generally use a fuzzy square binary relation of dissimilarity, which is positive, irreflexive and symmetric, i.e., $\forall i, j, r_{i,j} \geq 0, r_{i,i} = 0$ and $r_{i,j} = r_{j,i}$ [5][6]. This relation typically expresses a spatial dissimilarity computed as function of the distance between objects. For instance, each element $r_{i,j}$ of the matrix R might be defined as the Euclidean distance between x_i and x_j. The dissimilarity relation obtained by the neural network is certainly positive, but might be neither irreflexive nor symmetric. Due to the interpolation performed by the neural network, irreflexivity and symmetry properties could not be satisfied in a strict sense. Actually, if the training set of the neural network has been appropriately chosen, we expect that the symmetric elements of R are approximately equal and the elements on the diagonal are close to 0.

A fuzzy C-partition of X is a real $C \times N$ matrix U, with $C \in \{2..N-1\}$, $U = [u_{i,k}]$, such that

$$u_{i,k} \in [0,1] \; \forall i,k; \; \sum_{i=1}^{C} u_{i,k} = 1 \; \forall k; \; 0 < \sum_{k=1}^{N} u_{i,k} < N \; \forall i,$$ where $u_{i,k}$ is the membership value of x_k

to the i-*th* fuzzy cluster of the C-partition. Assume that $R = [r_{i,j}]_{N \times N}$ is the dissimilarity relation obtained from the neural network. More precisely, each element $r_{i,j}$ of the square NxN matrix R represents the output of the neural network when x_i and x_j are the inputs. In order to partition the objects to C fuzzy clusters, we can adopt the same objective function as in [5]:

$$J(U) = \sum_{i=1}^{C} \sum_{k=1}^{N} \sum_{j=1}^{N} u_{i,k}^{2} u_{i,j}^{2} r_{k,j} \; .$$

The C-partition of X is obtained by minimizing $J(U)$ with respect to U. We can rewrite the objective function as:

$$J(U) = \sum_{i=1}^{C} \sum_{k=1}^{N} u_{i,k}^{2} \sum_{j=1}^{N} u_{i,j}^{2} r_{k,j} = \sum_{i=1}^{C} \sum_{k=1}^{N} u_{i,k}^{2} D_{i,k}$$

where

$$D_{i,k} = \sum_{j=1}^{N} u_{i,j}^{2} r_{k,j} \; . \tag{1}$$

Applying the Lagrange multipliers, we obtain the first order necessary conditions for optimality of U:

$$u_{i,k} = \left(\sum_{j=1}^{C} \frac{D_{i,k}}{D_{j,k}} \right)^{-1} , \; 1 \le i \le C; \; 1 \le k \le N \tag{2}$$

The minimization of $J(U)$ can be obtained by an alternating optimization scheme based on (1) and (2). After initializing randomly U, the algorithm computes $D_{i,k}$, $1 \le i \le C; 1 \le k \le N$, by using formula (1). Then, formula (2) is applied to update U. Formulas (1) and (2) are iteratively applied until the difference between the values of $J(U)$ at two consecutive iterations is smaller than a given arbitrarily small real constant.

The choice of the training samples strongly influences the effectivenss of our approach. Only if the samples are a significant model of the dissimilarity relation, the relational algorithm can correctly partition the data set even though the cluster shape is irregular and not easily definable algebraically. Thus, the combination neural network-relational clustering algorithm allows partitioning the data set without any a-priori assumption on cluster shapes. On the contrary, traditional clustering algorithms often use a fixed distance measure to establish the spatial dissimilarity, thus enforcing a specific type of cluster shape. The success of these algorithms depends on the suitability of the distance measure for the application. Partial supervision may be used to improve traditional clustering algorithms [7][8]. A few labelled patterns are used as reference elements to guide the clustering algorithms in modeling the clusters. The labelled patterns, however, cannot modify the shape of the clusters, but only the dimensions. Indeed, the shape is fixed by the definition of distance used to measure the dissimilarity between the points: the labelled points can only modify and adapt the dimensions of the clusters.

4. Experimental results

To assess the effectiveness of the proposed approach, we refer to a synthetic bi-dimensional data set, that is not easily managed by clustering algorithms based on spatial dissimilarity. As shown in Fig. 1, the data set is composed of two classes and the points within each class are grouped into two clusters. A cluster belonging to a class is farther from the other cluster of the same class than from a cluster belonging to the other class. Obviously, clustering algorithms, which measure the dissimilarity between two points as the distance between the two points, cannot partition correctly the data set. Indeed, both the Euclidean and the Mahalanobis distances which induce, respectively, spherical and ellipsoidal cluster shapes, lead, for instance, the FNM relational algorithm to place the points with ordinates in the interval [0.1,0.2] and in the interval [0.5,0.6], respectively, into two different clusters. Thus, the percentage of correctly partitioned points can never be higher than 50%. Prototype-based clustering algorithms such as the fuzzy C-means have similar drawbacks. Actually, either using the Euclidean or the Mahalanobis distance the fuzzy C-means obtains 50% of correctly recognised patterns.

We performed six experiments to evaluate the capability of the neural network to identify, based on a small-sized training set, a dissimilarity measure, which allowed the relational clustering algorithm to partition the data set correctly. We randomly extracted a pool of patterns (called *training pool*) from the data set. This pool was composed of 6%, 10%, 15%, 20%, 25% and 30% of the data set, respectively, in the six experiments. From the training pool, we extracted the patterns used to train the neural network. More precisely, assume that C is the number of classes, which we expect to identify in the data set. Let n_c, $c = 1..C$, be the number of elements of class c contained in the training pool. For each element $x_{\bar{c},\bar{j}}$ of class $\bar{c}$, $\bar{c} = 1..C$, $\bar{j} = 1.. \; n_{\bar{c}}$, we form $P \cdot C$ pairs $(x_{\bar{c},\bar{j}}, y_{c,i})$, obtained by randomly choosing P elements $y_{c,i}$, $i = 1..P$, $P \leq n_c$, from each class c. We insert both triples $(x_{\bar{c},\bar{j}}, y_{c,i}, s_{\bar{c},\bar{j},c,i})$ and $(y_{c,i}, x_{\bar{c},\bar{j}}, s_{\bar{c},\bar{j},c,i})$ into the training set whose cardinality is equal to $2PC\sum_{c=1}^{C} n_c$.

In the training phase we used only 0 and 1 to express the dissimilarity degree of two input patterns belonging to the same class or to different classes, respectively. The training set of the neural network was obtained by coupling each element of the training pool with one element (i.e., $P=1$) of the training pool for each class.

Each experiment was repeated 10 times choosing randomly the points composing the training pool with the constraint of having at least one point for each of the clusters of the two classes. We also compared our approach with the supervised fuzzy C-means (SFCM) algorithm with either the weighted distance or the Mahalanobis distance. To this aim, we used the points in the training pool as labelled patterns that guide the clustering algorithms. Table 1 shows the results obtained in the six experiments. The first column indicates the percentage of points of the data set depicted in Fig. 1 included in the training pool; the second, third and fourth columns report the mean and standard deviation (in the form (mean ± standard deviation)) of the percentage of correctly classified points achieved, respectively, by our approach, by the SFCM with the weighted distance (SFCMW), and by SFCM with the Mahalanobis distance (SFCMM). As expected, in our approach the classification percentage increases with the increase of patterns in the training pool. Further, the standard deviation of this percentage decreases with the increase of patterns in the training pool. For small percentages of patterns in the training pool, the probability of selecting patterns, which identify correctly the cluster shape, is low. Further, a small amount of patterns can be insufficient to trace the boundaries of the class conveniently. These problems tend to disappear when the patterns of the training pool are a significant sample of the overall data set. As shown in Table 1, the class shape is sufficiently identified even with 20% of the points of the data set.

On the other hand, we note that the hyperellipsoidal-shaped clusters induced by the weighted distance and developed along the principal axes cannot model the two classes shown in Fig. 1. The percentage of correctly partitioned points is quite low. Further, the low value of the standard deviation suggests that this percentage does not depend on the choice of points used to train the neural network. Unlike the weighted distance, the Mahalanobis distance can originate hyperellisoidal-shaped clusters developed along any direction, and therefore can model more accurately the classes in Fig. 1. Thus, the percentage of correctly partitioned points is higher than the one obtained with the weighted distance, but it is far from the one obtained with our approach. It is also interesting to observe that the standard deviation is high in the six experiments. This leads us to conclude that the choice of the labelled points affects dramatically the performance of the algorithm.

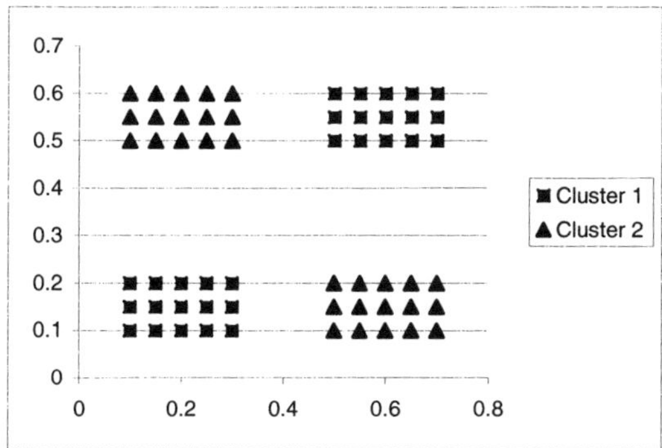

Figure 1. Synthetic two-dimensional data set

Table 1. Percentage of correctly classified patterns in the six experiments

Training pool	Correctly classified points		
	Our Approach	SFCMW	SFCMM
6%	62.41% ± 15.15%	53.33 % ± 0.00%	67.33% ± 22.54%
10%	77.53% ± 16.32%	55.67 % ± 1.61%	68.33% ± 21.90%
15%	79.43% ± 15.69%	57.50 % ± 0.87%	70.16% ± 21.60%
20%	91.47% ± 8.63%	59.00 % ± 2.70%	75.83% ± 20.82%
25%	93.55% ± 4.15%	62.17 % ± 2.36%	81.00% ± 20.00%
30%	96.45% ± 2.50%	64.00 % ± 3.87%	82.50% ± 18.44%

Conclusions

In this paper, we adopted the approach to extract dissimilarity relations directly from data using a feed-forward neural network. The dissimilarity relation is then used by a fuzzy relational clustering algorithm to partition those data sets, which are not easily managed by traditional clustering algorithms based on spatial dissimilarity.

References

[1]   A.K. Jain and R.C. Dubes, Algorithms for Clustering, Englewood Cliffs, N.J.: Prentice Hall, 1988.
[2]   L. Kaufman and P.J. Rousseeuw, Finding Groups in Data. An Introduction to Cluster Analysis, Wiley, Canada, 1990.
[3]   B.S. Everitt, S. Landau, M. Leese, Cluster Analysis, Arnold, London, 2001.
[4]   W. Pedycz, G. Succi, M. Reformat, P. Musilek, X. Bai, Expressing similarity in software engineering: A neural model, Proc. Second International Workshop on Soft Computing Applied to Software Engineering, Enschede, The Netherlands, February 2001.
[5]   M. Roubens, Pattern classification problems and fuzzy sets, *Fuzzy Sets and Systems* **1** (1978) 239-253.
[6]   R.J. Hathaway, J.W. Davenport and J.C. Bezdek, Relational duals of the c-means clustering algorithms, *Pattern Recognition* **22** (1989) 205-212.
[7]   W. Pedrycz and J. Waletzky, Fuzzy clustering with partial supervision, *IEEE Transactions on Systems, Man and Cybernetics – Part B: Cybernetics* **27** (1997) 787-795.
[8]   F. Marcelloni, Recognition of olfactory signals based on supervised fuzzy c-means and k-NN algorithms, *Pattern Recognition Letters* **22** (2001) 1007-1019.

KES 2002
E. Damiani et al. (Eds.)
IOS Press, 2002

Improving Information Retrieval by Words Selection Extracted from a Search Log

Noriaki KAWAMAE
*Center for Advanced Research and Technology The University of Tokyo, 4-6-1 #45,
Komaba Meguroku Tokyo JAPAN*

Abstract. We propose a novel user support method that selects appropriate keywords to help all users. Our proposed method selects words based on a user's search history and the relevance between keywords, as measured for five different kinds of user search activity

1. Introduction

These days, with the spread of personal computers and growth of the World Wide Web, it is easier and more convenient to obtain data and documents electronically than ever before.

Existing search systems, however, do not necessarily help users to retrieve information efficiently. Although these systems are based on database retrieval systems, they differ in terms of the users, the types of data, and the users' purposes. It is thus difficult for users to retrieve information from the World Wide Web with existing search systems. Such systems require advanced retrieval techniques for the users.

2. User search behavior and user support methods

We define the efficiency of a user's search activity as the number of steps that the user takes from start to finish. The number of steps means the number of times the user changes keywords or refers to the World Wide Web.

The thesaurus and QE are typical user support methods. The thesaurus is a special dictionary of meanings and concepts and includes synonyms and antonyms for each word. A thesaurus can thus be used to eliminate lexical disagreements and meet requirement. Constructing a thesaurus, however, is too expensive due to the large cost in human effort and time. QE expands the number of keywords by adding ones in the user's query and meets requirement. The problem with this method is that support is restricted to users[1].

3. Keyword selection based on user behavior

3.1. Improvement to user search behavior by keyword selection

Our proposed user support method not only resolves all the problems stated previously but also improves all users' search behavior. This method displays keywords related to the user query, in addition to search results based on the degree of relevance with respect to the query. In this paper, "relevance" refers to one of five types of relevance between keywords, five depending on the situation in which the user changes keywords: alternate expression,

Table 3.1 Index and a search sequence

Number of appearances in search log		
Ignoring order	Fre{word}	Number of search sequences including "word"
Number of appearances of pair "key" and "word" in search sequence		
Ignoring order	InFre{key}{word}	Number of search sequences including "key" and "word" simultaneously
Considering order	Ainf{key}{word}	Number of search sequences including "key" and "word" appearing after "key"
Considering order	Binf{key}{word}	Number of search sequences including "key" and "word" appearing before "key"
The average distance between "key" and "word" in search sequence		
Ignoring order	Avema{key}{word}= Ma{key}{word}/ Infre{key}{word}	The average distance between "key" and "word" in a search sequence (Ma{key}{word}: total distance between "key" and "word" in a search sequence)
Considering order	Aveaa{key}{word}= Aa{key}{word}/ Ainf{key}{word}	The average distance between "key" and "word" appearing after "key" in a search sequence (Aa{key}{word}: total distance between "key" and "word" appearing after "key" in a search sequence)
Considering order	Aveba{key}{word}= Ba{key}{word}/ Binf{key}{word}	The average distance between "key" and "word" appearing after "key" in a search sequence (Ba{key}{word}: total distance between "key" and "word" appearing before "key" in a search sequence)
Relative ratio		
Ignoring order	Infbyf{key}{word}= InFre{key}{word}/ Fre{word}	The relative ratio of the number of appearances in a search sequence to that in a search log

concrete expression, abstract expression, an alternate word in the same concept, or a demand to change or revert to a previous query.

Our method enables the search system to select relevant keywords for the user to use in the next search. For example, if the search system receives "apple" as a user query, it will display "New York", "Mac" as an alternate expression, and "Fuji apple" as a concrete expression of the term "apple." Conventionally, the user must seek these keywords on his or her own.

3.2. Extracting relevance between keywords from the user behavior model

The user search history is a record of the user's search activities and reflects his or her information needs. Our proposed method can thus enable a search system to select keywords according to the user's needs. Table3.1. shows the numerical relevance between keywords reflects tendencies in the user's behavior.

word{UID}={Japanese, sushi, Sushi, noodle, eel|

Fig. 3.1 Example of a search sequence

User search behavior is thus modeled based on these relationships. For example, Fig 3.1 shows "eel" appearing after "Sushi." The relative distance between these keywords is 2, and the number of times this pair appears is 1.

3.3. User search behavior model

Table 3.2 illustrates the relationships between keywords in our model for user search behavior. The model enables us to select appropriate keywords based on these relationships and the user's behavior. For example, keywords for a concrete expression based on a user's previous query can be obtained from a hash. The table also shows the relationships between each model for user search behavior and the corresponding search sequences.

Table 3.2 User behavior and index

User behavior	Index	Usage of index
Change	Infbyf{key}{word}	A required change from a key is a word that increases the value of the index
Another	Aveaa{key}{word}	An another expression of a key is a word that decreases the value of the index
Concrete	Ainf{key}{word}	A concrete expression of a key is a word that increases the value of the index
Abstract	Binf{key}{word}	An abstract expression of a key is a word that increases the value of the index
Alternate	Avema{key}{word}	An alternate word in the same concept within a key is a word that increases the value of the index

3.4. Hash

There are two advantages to preparing a hash in our proposed method. First, this enables us to calculate the relationships between keywords more easily. Secondly, it enables us to refine the accuracy of the selected words by examining whether these words are actually used. Table 3.3 lists all seven kinds of hash with Sushi as key from the search sequence in Fig 3.1.

The cost of calculating the relationships between keywords as defined in section 3.3 is too expensive for real-time processing. Therefore, we have to use batch processing for these calculations. This leads a problem in that the relationships between keywords may not reflect the latest search behavior of the user. To solve this problem, we construct a hash from the search sequence. The hashes defined in this paper include not only words from the search sequence based on each hash's criteria but also statistical data. A hash can be revised to reflect new trends in the user search behavior.

Seven different kinds of hash can be constructed from various combinations of a key and a word in a search sequence. The differences derive from the order and distance of appearance. The order of appearance is classified as either front, back, or both. In addition to these three classifications, we can define "last" by considering whether the word appearing last in a search sequence is the most appropriate in the sequence. The distance between a key and a word is classified as neighboring or throughout in a search sequence. If a word appears last, the word housed in the hash can be identified as unique. There are two other cases for the distance between keywords. The combination of appearance and distance produces six kinds of criteria for constructing each hash. A total of seven combinations can be created. Table 3.3 lists all seven kinds of hash for the search sequence in Fig 3.2.

We use the Con hash (one of the seven kinds of hash). Previous works on natural language processing showed that the meaning relationship between keywords is limited to the neighbouring one or two words and this approach cannot select proper keywords because each user's search behavior is different in terms of experience, degree of need, interest, and search technique. Therefore we use Con as our hash not to miss promising words.

Table 3.3 Hash example with sushi as key in Fig 3.2

Hash	Word housed in hash	Example
From{key}	word appears just after key	noodle
To{key}	word appears just before key	,sushi
Mul{key}	word appears just after and before key	Sushi, noodle
Con{key}	All words	Japanese, sushi, noodle, eel
Acon{key}	All words appearing after key	noodle, eel
Bcon{key}	All words appearing before the key	Japanese, sushi
Econ{key}	Last word	eel

4. Validity of the proposed method

4.1. Experimental concept

This section examines the validity of our method. We calculated the probability that selected keywords will be used by a user in his or her next query and confirmed the validity by simulation. For the simulation we used a search log from a commercial search system as data. We selected search sequences with two or more different words from the search log, and we used 80% of them as learning data and the remaining 20% as test data.

The learning data was used to construct a hash and calculate the relationships between keywords. The test data was used to represent a virtual form of user search behavior. With this simulation, we could obtain the same experimental results as we could obtain with actual data.

The search log was used in the experiment and consisted of 952,666 lines. Each line corresponds to one transaction. The lines were converted into search sequences by organizing them on a per–user basis. A total of 159,039 search sequences were generated from this processing.

4.2. Validity of the hash

Table 4.1 shows our experimental results, which demonstrate the effectiveness of the chosen hash, "Con". We should note that we omitted the word "which", which appeared over 1000 times in each hash. This indicates that the relationship between keywords is wider than previously considered. Therefore, "Con" is the most promising hash and selects the most appropriate words.

Table 4.1 Rate of coincidence for each hash

Hash	From	To	Mul	Con	Acon	Bcon	Econ
Tar	1749	1770	1949	1949	1749	1770	1749
Match	164	151	252	638	453	296	160
%	9.38	8.53	12.93	32.73	25.90	16.72	9.15

Tar: The number of words agreeing with the key of the hash
Match: The number of selected words used

4.3. User behavior model

We also demonstrated the effectiveness of our user behavior model. This experiment differs from the previous experiment in terms of the number of words selected. In the previous experiment we used all the words in a hash, while in this experiment we used only 10 words from each hash. The words were selected as large order based on an index value for appearance, such as Fre, InFr, Ainf, Binf, or Infbyf, and as small order based on an index value for distance, such as Avema, Avea, or Aveba. The number of selected words chosen by the user was 638, which could be reduced to 231 by excluding duplicate combinations between the user query and the selected words. Table 4.2 shows the experimental results. A high value of the rate of coincidence for each user behavior indicates a high probability that the user will use the selected words in his or her next query. The highest value for each user behavior was mainly determined by the index value of our modeled user's behavior. These results demonstrate the validity of our proposed model.

Table 4.2 Rate of coincidence by index

Reason	Fre	Infre	Ainf	Binf	Infrebyf	Avebp	Aveap	Avemp
Change	32.5	42.4	41.6	43.1	**54.3***	44.5	42.3	43.7
Another	32.2	41.5	40.8	42.3	47.6	48.9	**52.2***	47.5
Concrete	36.4	58.3	**58.3***	56.4	45.8	45.3	48.2	46.9
Abstract	35.6	38.5	36.2	**43.9***	33.8	41.2	38.5	39.2
Alternate	30.8	37.4	38.3	37.6	45.6	48.1	47.2	**48.5***

Reason: Reason for user behavior as keywords change

5. Conclusion

We have proposed a novel user support method that selects appropriate keywords to help users for whom existing user support methods are ineffective. Our proposed method selects the words based on the user's search history and the relevance between keywords, as measured for five defined kinds of user search activity. Our results can be summarized as follows:

*Our method supports not only all kinds of user activities but also users who cannot concretely define their information needs in terms of keywords.
*Using a search log enables our method to share users' knowledge.
*A user behavior search model enables our method to select words automatically, saving time and human effort.

The selected words can not only resolve lexical disagreements but also help users to define their information needs concretely.

References

[1]   Noriaki Kawamae and Terumasa Aoki and Hiroshi Yasuda : Words Clustering Based on Factor Analysis, AAAI Spring Symposium 2002 on "Acquiring (and Using) Linguistic (and World) Knowledge for Information Access

KES 2002
E. Damiani et al. (Eds.)
IOS Press, 2002

Coupled maps with variable connections

Junji Ito

Laboratory for Perceptual Dynamics, Brain Science Institute, RIKEN
2-1 Hirosawa, Wako-shi, Saitama 351-0198 Japan

Abstract. Coupled map models with variable connection weights between the units are considered. A generally observed feature in this type of models is the appearance of units with massive outgoing connections. Such structure formation is the result of feedback between unit and connection dynamics.

1 INTRODUCTION

Since the introduction by Kaneko [1, 2], coupled map lattice (CML) and globally coupled maps (GCM) have been the target of extensive research during these few decades [3, 4]. Though originally devised as a phenomenological model for spatio-temporal chaos, more recently they have been applied to information processing [5, 6, 7].

Conventional coupled maps do not allow the change of connection strength between units. This means that the topology of the network is fixed in time. It is expected that more powerful information processing could be implemented in coupled maps when we introduce change of connection strength into the model [6, 8].

Here we report the characteristics of some models of coupled maps which incorporate the connection change which depends on the units' dynamics.

2 MODEL I

MODEL I is the most simple extension of ordinary GCM to one including connection change [9]. This model is described by the following set of equations:

$$x_{n+1}^i = f((1-c)x_n^i + c \sum_{j=1}^{N} w_n^{ij} x_n^j)$$

$$w_{n+1}^{ij} = \frac{[1 + \delta \cdot g(x_n^i, x_n^j)]w_n^{ij}}{\sum_{j=1}^{N}[1 + \delta \cdot g(x_n^i, x_n^j)]w_n^{ij}}$$

$$f(x) = ax(1-x)$$

$$g(x, y) = 1 - 2|x - y|,$$

where x_{n+1}^i is the state variable of unit i and w_n^{ij} is the strength of coupling from unit j to unit i at the n-th time step. The function $f(x)$ represents the unit dynamics. Here we used the logistic map for its simplicity and the universality of the period doubling bifurcation to

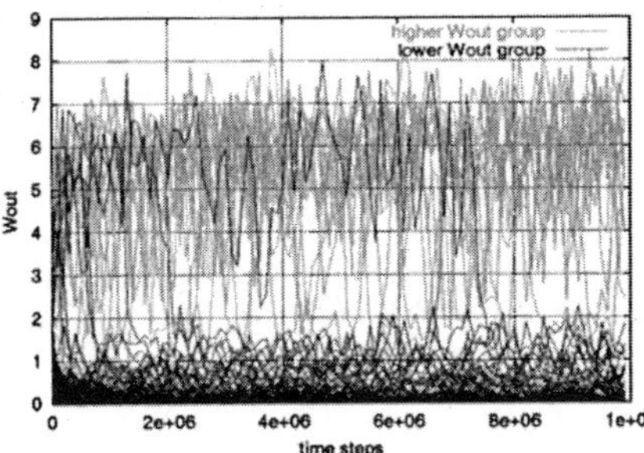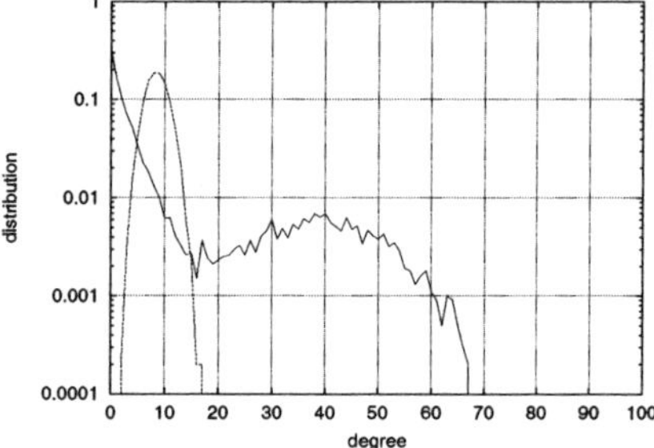

Figure 1: (Left) temporal change of strength of total connections which go out from each unit, W_{out}^i. Each line corresponds to each unit and data for all units are superimposed. Black line: units of higher W_{out}^i value group, gray line: units of lower W_{out}^i value group. (Right) Distribution of in-degree and out-degree of the network at $1,000,000$ steps. Solid line: out-degree, broken line: in-degree. $N = 100, a = 3.97, c = 0.125, \delta = 0.1$

result in chaotic oscillation. (Models with other unit dynamics are considered in the following sections.) The function $g(x)$ determines the rule of connection change. Here we adopt a Hebbian type of connection change, *i.e.*, when two unit's state variables take similar values, the connections between them are strengthened. Normalization of incoming connection strength is introduced, considering the general assumption that, in any type of network system, some resource is needed to establish connections between units, of which the amount is generally limited. Such a resource limitation causes competition among connection strength. The normalization introduced here is a simple representation of such competition.

The role of parameters in the model is as follows: a determines the nonlinearity of units' dynamics, c represents the strength of influence of other units on the dynamics of each unit, δ is the plasticity of connection strength, and N is the system size (*i.e.* number of units.) When $\delta = 0$, this model reduces to an ordinary GCM.

The most striking finding from this model is spontaneous formation of network structure. In a parameter regime corresponding to weakly chaotic oscillation and moderate coupling strength, the system self-organizes into a characteristic structure which consists of two distinct groups of units, one composed of units with massive outgoing connections and another with poor outgoing connections.

Fig. 1 (left) shows the temporal change of strength of total connections which go out from each unit, W_{out}^i, defined as $W_{out}^i(t) = \frac{1}{\tau} \sum_{n=t}^{t+\tau} \sum_{j=1}^{N} w_n^{ji}$, where τ represents time steps of averaging period, which here we set to be $100,000$ steps. Despite large fluctuation, separation of units into two groups is distinct. Distribution of in-degree and out-degree of the network at $1,000,000$ steps is shown in Fig. 1 (right). Here degrees are calculated using a simple digitalizing method of connections, where connections with larger weights than some threshold, say $1/N$, is considered to exist, while connections with smaller weights are ignored. While in-degree forms a unimodal distribution, out-degree consists of two components, one with exponential decay in smaller degree, which corresponds to the lower W_{out}^i group, and another with large degree, which corresponds to the higher W_{out}^i group.

Such self-organization takes place through the interplay between unit and connection dynamics. Due to the characteristics in dynamics of logistic map, units naturally form two quasi-clusters, but they frequently change the cluster to which they belong. The frequency of cluster switching and the value of W_{out}^i is strongly correlated, *i.e.*, the larger the value

W_{out}^i takes, the slower the cluster switching of unit i becomes and vice versa. This positive feedback drives the initial small fluctuation among W_{out}^i to the eventual separation of units.

3 MODEL II

In the dynamics of logistic map, oscillation around an unstable fixed point is very dominant. Indeed, the structure formation observed in MODEL I is closely related to oscillatory unit dynamics. Hence, we need to confirm whether network structure can be spontaneously formed in models with other unit dynamics, and if so, what kind of structure will appear.

One of the most interesting nonlinear dynamics other than chaotic oscillation is ones which possess excitability. Here we construct another model which is composed of units with excitability. MODEL II is different from MODEL I mainly in the function form of $f(x)$. It is described as follows:

$$x_{n+1}^i = f(x_n^i + c \sum_{j=1}^{N} w_n^{ij} x_n^j)$$

$$w_{n+1}^{ij} = \frac{[1 + \delta \cdot g(x_n^i, x_n^j)]w_n^{ij}}{\sum_{j=1}^{N}[1 + \delta \cdot g(x_n^i, x_n^j)]w_n^{ij}}$$

$$f(x) = x + \omega + \frac{k}{2\pi}\sin 2\pi x (mod 1),$$

$$g(x, y) = \cos 2\pi(x_n^i - x_n^j).$$

Here we adopt circle map as the function of unit dynamics. It has two parameters: ω is the characteristic angular velocity and k represents the nonlinearity of the map. To render excitability to the map, we set these two parameters to be $\omega = 0.4$ and $k = 2.9392$. With these parameter values, this map has two adjacent stable and unstable fixed points, which leads to excitability.

In this model again we can observe spontaneous structure formation. In this case, structure of the network can be easily seen in the connection matrix, w_n^{ij}. In fig. 2 (left), w_n^{ij} is illustrated. Units are clearly partitioned into three groups, each of which has a single unit with massive outgoing connections. We shall call such units pacemakers. They drive other members of their group and lead them to synchronized bursting. In fig. 2 (right) we show temporal changes of mean fields which indicate synchronized bursting among units. While the mean field of the whole system shows no apparent periodicity, the mean field of each group shows switching between bursting and resting periods.

This group formation is not fixed in time, but groups repeatedly appear and are annihilated. Such dynamics in network structure is again due to the interplay between unit and connection dynamics. Initially, units are in a synchronized bursting state, but this state is not stable. Hence, units occasionally fail to stay in phase, which leads to sudden decrease of connection strength from those units to all the other units. Those units which haven't experienced such drop-out eventually become pacemaker units and each of them form a group of synchronized bursting. Once a group is formed, connections between units in and out of that group tend to decrease, because their dynamics is generally out of phase. Here, positive feedback which grows initial small fluctuations to macro scale network structure is observed again as in MODEL I.

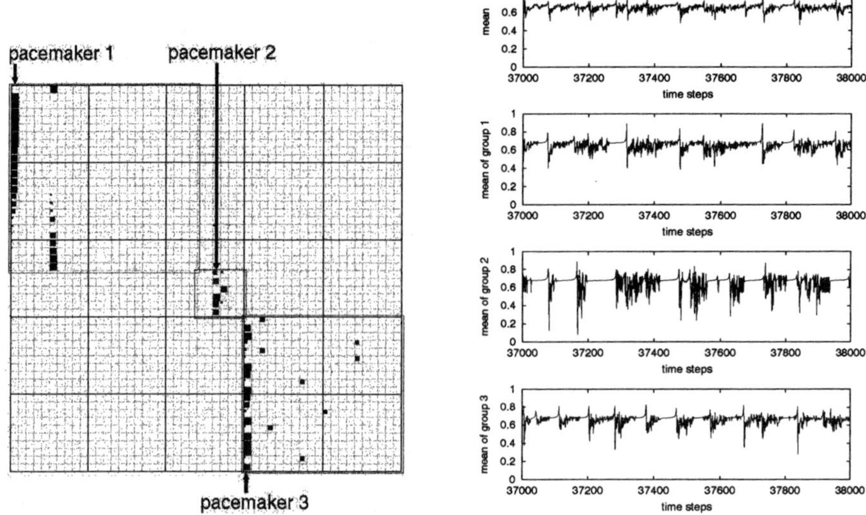

Figure 2: (Left) Connection matrix w_n^{ij}. Value of w_n^{ij} is represented by the size of a filled square at a cell in i-th row, j-th column. (Right) Temporal change of mean field. From the top, mean field of the whole system, group 1, group 2, and group 3 are shown respectively.

4   MODEL III

So far, our models are not supplied with external input. It is very important to investigate how the system responds to external input, when it is viewed as an information processing devise.

Here we introduce a model which exhibits self-organization of network structure depending on external input [10]. The model is described as below:

$$x_{n+1}^i = x_n^i + \Omega + \frac{k}{2\pi} \sin 2\pi x_n^i + \frac{c}{2\pi} \sum_{j=1}^{N} \varepsilon_n^{ij} \sin 2\pi x_n^j + I^i$$

$$\varepsilon_{n+1}^{ij} = \frac{[1 + \delta \cos 2\pi (x_n^i - x_n^j)]\varepsilon_n^{ij}}{\sum_{j=1}^{N}[1 + \delta \cos 2\pi (x_n^i - x_n^j)]\varepsilon_n^{ij}},$$

where I^i is external input to unit i.

Supplied with constant external input to a unit, this model self-organizes into hierarchical structure with the input units as a 'root', as illustrated in fig. 3.

The detailed mechanism for this type of structure formation is not yet revealed, but some interesting properties are observed such as clear input dependency, power law distribution of lifetime of unit to stay within the structure, and so on.

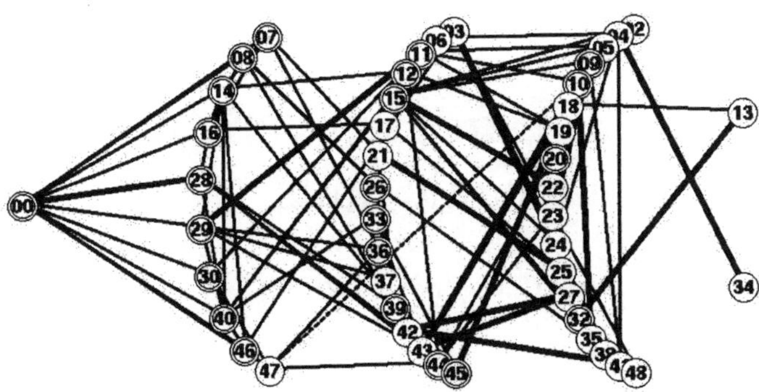

Figure 3: Graph representation of network structure in MODEL III. Each circle represents unit in the system and number corresponds to unit index. Lines between circles represents connection between corresponding units. Here external input is applied to unit 0, which is placed in the leftmost. For details of this graph, see ref [10].

5　SUMARRY

We have introduced three types of coupled map model which incorporate connection change which depends on units' dynamics. All models exhibit spontaneous formation of network structure. The common mechanism of such structure formation is feedback between unit and connection dynamics which amplifies small fluctuations in network structure which is initiated by chaotic nature of unit dynamics to large scale network structure. For simplicity and generality of GCM, phenomena observed in this study is supposed to be found in realistic dynamic networks.

References

[1] K. Kaneko, Spatiotemporal intermittency in coupled map lattices, Prog. Theo. Phys. **74** (1985) 1033–1044.

[2] K. Kaneko, Clustering, coding, switching, hierarchical ordering, and control in network of chaotic elements, Physica D **41** (1990) 137–172.

[3] K. Kaneko ed., Theory and Applications of Coupled Map Lattices, Wiley (1993).

[4] K. Kaneko and I. Tsuda, Complex Systems: Chaos and Beyond, Springer Verlag (2000)

[5] H. Nozawa, Solution of the optimization problem using the neural-network model as a globally coupled map, Physica D **75** (1994) 179–189.

[6] C. van Leeuwen, M. Steyvers and M. Nooter, Stability and Intermittency in Large-Scale Coupled Oscillator Models for Perceptual Segmentation, Journal of Mathematical Psychology, **41** (1997) 319–344.

[7] D. DeMaris, Synchronization opponent systems: Attractor basin transient statistics as a population code for object representation, Neurocomputing **38** (2001) 547–554.

[8] K. Kaneko, Relevance of dynamic clustering to biological networks, Physica D **75** (1994) 55–73.

[9] J. Ito and K. Kaneko, Spontaneous structure formation in a network of chaotic units with variable connection strengths, Physical Review Letters **88** (2002) 028701.

[10] J. Ito and K. Kaneko, Self-organized hierarchical structure in a plastic network of chaotic units, Neural Networks **13** (2000) 275–281.

KES 2002
E. Damiani et al. (Eds.)
IOS Press, 2002

Emergence of scale-free network structure with chaotic units

Pulin GONG and Cees VAN LEEUWEN
*Laboratory for Perceptual Dynamics, RIKEN BSI, 2-1, Hirosawa, Wako-Shi, Saitama,
351-0198, Japan*

Abstract. We consider growth of a network, of which the units are characterized by chaotic activity. New nodes are randomly attached one-by-one, while connections are being rewired in adaptation to the dynamic clustering patterns in the units' activity. The network self-organizes into a complex network of which the connectivity distribution reveals a power law. At the same time, the network has a high clustering coefficient and small average shortest path length. In this model, preferential attachment is an emergent outcome from random attachment and adaptive rewiring.

1. Introduction

Two important families of network structures have recently been identified. Watts and Strogatz [1] characterized *small-world networks* by the combination of a high clustering coefficient and short average shortest path length. Small-world structures, therefore are optimal for communication of information across the vertices. Barabasi and Albert [2] introduced *scale-free networks*. Of these, the distribution $P(k)$ of vertex connectivity is free of a characteristic scale, that is, the distribution decays as a power law $P(k) \sim k^{-\gamma}$. Scale-free networks are optimally robust with respect to random perturbation or removal of a proportion of the network connections [3]. A diversity of systems such as the world-wide web, social networks, the metabolic networks of life-sustaining chemical reactions inside cells, and protein interaction networks all belong to the highly heterogeneous family of scale-free networks [4,5,6,7]. None of these real systems possess scale-free structure by design. It is of importance to investigate how these structures could have grown naturally. We present a model for network growth combined with adaptive rewiring according to the spatiotemporal characteristics of the network activity. In our simulation studies, we show that a scale-free small-world network is obtained for our model if the oscillatory activity of the element is chaotic.

2. Growth and adaptive rewiring model

In our model, chaotic oscillatory activity is produced by the logistic map shown in Equation 1. We choose parameter *a=1.79,* for chaotic activity. Coupled logistic maps are described by Equation 2

$$x(n+1) = f(x(n)) = 1 - ax(n)^2 \tag{1}$$

$$x_i(n+1) = (1-\varepsilon)f(x_i(n)) + \frac{\varepsilon}{M_i}\sum_{j \in B(i)} f(x_j(n)) \tag{2}$$

where $x_i(n)$ is the activity of the i-th ($1 \le i \le N$) unit at the n-th time step. Where N is the total number of units in the current network. In the present study the number N of the current network is increased by adding new nodes one by one. M_i and $B(i)$ are the number and the set of the neighbors of the unit i respectively. The neighbors of the unit i are those units that have direct connections with unit i. The connections are bi-directional, this means that if unit i is a neighbor of unit j, j is also a neighbor of i. ε is the coupling strength. Throughout the paper, the coupling strength is fixed to be $\varepsilon = 0.4$. The parameter a is the parameter which controls the dynamics of the each unit. We define the coherence $d_{ij}(n)$ between unit i and unit j as the absolute value of the difference between the activation values of the units, as in Equation 3.

$$d_{ij}(n) = \left| x_i(n) - x_j(n) \right| \tag{3}$$

We start from a sparsely, fully connected, small random network with the total number of units M_0 linked by a number of connections L_0. A model for growth combined with adaptive rewiring according to network activity is described as follows:

[I] Add a new node i_n with m connections to m different nodes in the current network randomly. [II] Choose random initial activation values in the range $(-1, 1)$ for all units of the new network. Calculate the state of the system according to the Eq. (2), and discard an initial transient time T. [III] Then the dynamical coherence at time T+1, between the new added node i_n and all the other units in new current network $d_{i_n j}(T + 1)$ is calculated. We obtain the unit $j = j_1$, for which the value $d_{i_n j}(T + 1)$ is minimum amongst all the other units. Furthermore we obtain the unit $j = j_2$ for which $d_{i_n j_2}(T + 1)$ is maximum amongst the neighbors of new unit i_n. [IV] If unit j_1 is one of the neighbors of the unit i_n, then no change in the connections is made. Otherwise the connection between units i_n and j_2 is replaced by a connection between units i_n and j_1. [V] Go to step [II] and repeat the algorithm for K_0 times. [VI] Go to step [I] and repeat adding a new node. In the following study, $M_0 = 50$, m=14, $L_0 = 850$, $K_0 = 75$, T=2000.

3. Results

After t iterations of the algorithm, a complex network with $M_0 + t$ nodes results. Figure 1 shows the distribution of its connections for $t = 1700$. The connectivity in the network reveals a heavy-tailed distribution. The network has evolved into a scale-free state with the probability a node has k connections, following a power-law $p(k) = k^{-\gamma}$, with the exponent $\gamma = 3.09 \pm 0.17$. The clustering coefficient and the average shortest path length of the final network are calculated according to [1]. We obtain for our network the clustering coefficient $C_1 = 0.15$, and the average shortest path length $L_1 = 2.70$. It has been found that many real networks present a clustering coefficient much larger than the corresponding random graph. In order to compare with the corresponding random graph of the self-organized scale-free network, we generated a random graph by connecting nodes randomly, making sure that the numbers of nodes and connections between them are matched with those of the obtained self-organized network. For the random graph, clustering coefficient and shortest path length, respectively, are $C_0 = 1.7 \times 10^{-2}$, $D_0 = 2.56$. We observe that $C_1 >> C_0$, and $D_1 \geq D_0$, that is, the clustering coefficient of the self-organized scale-free network is much larger than that of the corresponding random network, and the shortest path length is close to that of the random network. We may, therefore, conclude that a growing network with chaotic units, according to our model produces a scale-free network with the characteristics of small-world network. Note that these results do not depend on the precise choice of parameters m, k_0, and T. Similar results were obtained with for an array of parameter values, as long as network activity stays within the chaotic range.

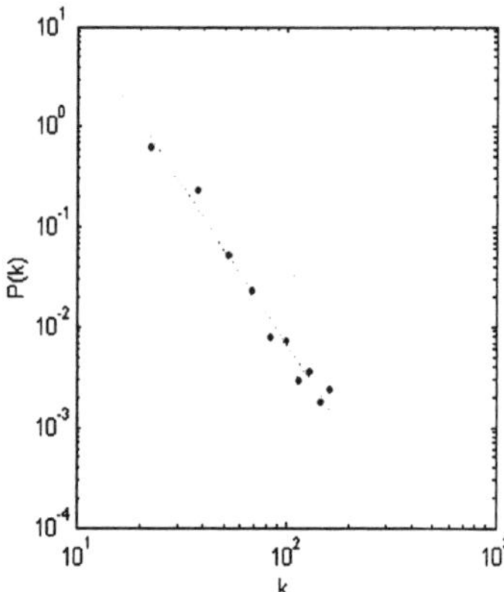

Figure.1 (log-log plot) Distribution of the connections of the self-organized network with chaotic units when *t=1700*. The red line is obtained by the least squares fit of the original data.

Preferential attachment is a common feature of realistic growing networks [2,12]. For our model we consider the set of nodes at time t, and record the connections for all the nodes in the set. The connections for the set of nodes are denoted by K_i ($1 \leq i \leq M_0 + t$). At time $t + \Delta t$, $\Delta t << t$, we measure the connections J_i of the set of nodes at time t. So we can obtain the increase in the number of connections over the time interval as a function of the number of its connections at time t, which are $\Delta k_i = J_i - K_i$, and the result is shown in Figure.2.

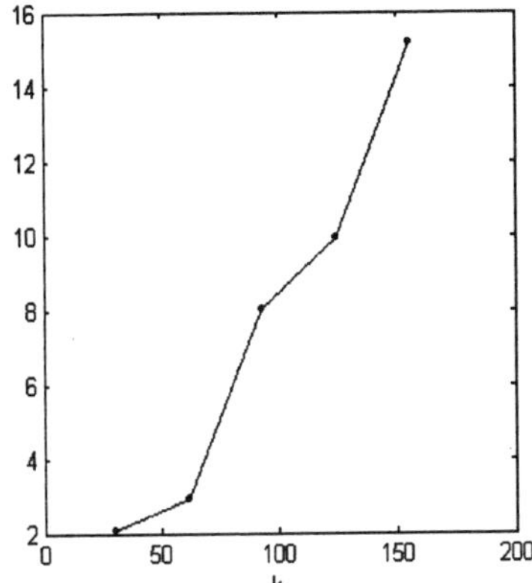

Figure. 2 The average number of attachments in a time interval as a function of the number of connections at the beginning of the interval, at time t. $t=1650$, and $\Delta t = 250$.

In Figure.2, we observe that there is preferential attachment for the growing network. In the earlier models preferential attachment is accomplished by a *prescription* to attach the new node to the target nodes according to their degree of connectivity [2,8,9,10]. Thus, the decision where to attach a node is based on detailed knowledge about the connectivity distribution of the entire network. It seems the new node knows the whole distribution of the connectivity. This is an unlikely requirement for real large-scale networks. By contrast in our model with dynamical units preferential attachment is an emergent property based on network dynamics.

4. Conclusion

We investigated the development of structure in networks with chaotic units. Network growth was modeled by adding new nodes one by one in combination with adaptive rewiring of the connections according to the dynamical coherence of the nodes. The network self-organizes into a scale-free network with small-world characteristics. During network growth, preferential attachment occurs as a consequence of adaptive rewiring. This may be considered more plausible than an explicit prescription for preferential attachment. We may conclude that scale-free network characteristics can emerge in adaptation to chaotic network activity. These observations, therefore, suggest yet another role for chaotic activity in the genesis of complex system structure.

5. References

1. D.J. Watts, S.H. Strogatz, Collective dynamics of 'small-world' network, Nature 393 (1998) 440.

2. A.L. Barabasi, and R. Albert, Emergence of scaling in random networks, Science 286 (1999) 509-512.
3. R. Albert, H. Jeong and A.L. Barabasi, Error and attack tolerance of complex networks, Nature 406 (2000) 378-381.
4. R. Albert, H. Jeong and A.L. Barabasi, Diameter of the world-wide web, Nature 401 (1999) 130-131.
5. M.E.J. Newman, The structure of scientific collaboration network, Proc. Natl. Acad. Sci. 98 (2001) 404-409.
6. H. Jeong, B. Tombor, R. Albert, z.N. Oltval and A.L. Barabasi, The large-scale organization of metabolic networks Nature 407 (2000) 651-654.
7. H. Jeong, S.P. Mason, A.L. Barabasi, A.N. Oltvai, Lethality and centrality in protein networks, Nature 411 (2001) 41-42.

Enhanced Information Flow in a Chain of Fractally Coupled Chaotic Clusters

C. Wagner[1] and R. Stoop[2]
[1]*Institute of Pharmacology, University of Bern,
Friedbuehlstr. 49, CH-3010 Bern, Switzerland*
[2]*Institute of Neuroinformatics, University/ETH Zuerich,
Winterthurerstr. 190, CH-8057 Zuerich, Switzerland*

Abstract. We study the flow of information in a biologically motivated model of neocortical neural networks. Our simulations reveal that for nearest neighbor coupling the information speed increases linearly as a function of the number of connected neighbors. In contrast, for fractal coupling, an enhanced dependence is found.

Since the introduction of coupled map lattices (CML) by Kaneko [1] the question of cluster formation remained an active field of research [2, 8]. The occurrence of clusters in globally coupled maps (GCML) depends on the strength of coupling between the oscillators and on the strength of the nonlinearity of a single map. In the phase diagram of GCML a partially ordered phase is observed, where coexisting synchronized clusters are found. This "glassy" phase is sandwiched between the fully synchronized and the turbulent phase [3, 4]. The importance of coupled map lattices in biological sciences was already stressed by Kaneko [5]. He particularly emphasized the relevance of dynamical clustering, e.g., in ecological or in neural networks. The results of these studies are based on the assumption of global coupling. However, the architecture of real ecological systems or neural networks are generally far away from this idealizations. Niebur et al. [6] investigated phase locking and phase synchronization in more general 2-d networks of coupled oscillators. Their work revealed that sparse coupling leads to a faster and more robust global phase locking. A new connection scheme for coupled map lattices representing neural networks was introduced by Raghavachari and Glazier [7] which is based on the fractal structure of dendritic trees. In this network, the connections in a d-dimensional sphere of radius R scales with $R^{d-\alpha}$ whereby the probability that a connection exists between the locations r_i and r_j is given by

$$p_{i,j} = 1/ \mid r_i - r_j \mid^{-\alpha} . \tag{1}$$

The interesting feature of this fractally coupled network is that it synchronizes already at rather low connectivity of approximately 25%, a value which matches the one observed in the brain.

Many areas of the cortex show a columnar structure [9]. Neurons in these columns are sensitive to the same input. A typical example are the orientation sensitive columns in the visual cortex (V1), where the cells of a column increase their spiking frequency when bars are presented under a preferred orientation angle. In our approach, we model these columns as clusters and explore the information flow in these networks for different couplings between the columnar structures. Numerical simulations reveal that the propagation of information is enhanced in fractally coupled clusters compared to nearest neighbor coupled networks with equal connection density.

The model We consider N coupled chaotic oscillators where the motion of individual oscillators i, $i = 1...N$ is given by

$$x_i(t+1) = (1 - \varepsilon)f(x_i(t)) + \frac{\varepsilon}{A_i} \sum_{j \in conn} f(x_j(t)), \tag{2}$$

where t denotes the time, A_i represents the number of connections at the ith site and j runs over all sites which are connected to site i. We use the logistic map for the evolution of a single oscillator $f(x(n)) = 1 - ax^2(n)$. The network used is a chain of columns, or clusters, constructed in the following way. First, the size of a cluster (cs) is selected, second the number of nearest neighbor clusters (nc) is determined to which a cluster is connected and third connections are distributed due to Eqn. 1. According to equation 1, for $\alpha = 0$ the clusters are fully connected (A), for $\alpha = 0.6$ the network is fractally coupled (B) and for $\alpha \to \infty$ the system reduces to a nearest neighbor coupled network (C). The parameter nc, which describes how many clusters are directly coupled, varies between 1 and N/cs. In the limiting case $\alpha = 0$ and $nc = N/cs$, the network is globally coupled. For the simulations, we impose periodic boundary conditions on the lattice, i.e. $x_1 = x_{N+1}$. To estimate the flow of information along the chain we use the maximal velocity of the propagation of perturbations [10], [11]. A small perturbation is applied to the oscillators of a cluster, and the spreading of information is followed by the difference to a replica system without perturbation. The speed of information transfer v^* can directly be measured from the disturbance at the leftmost $i_l(t)$ and the rightmost $i_r(t)$ oscillator

$$v^* = \lim_{t \to \infty} \lim_{N \to \infty} \frac{i_r(t) - i_l(t)}{2t}. \tag{3}$$

For $\alpha \to \infty$, the velocity of information flow can be understood as the result of two independent contributions: The chaotic instability of the map leads to an average exponential growth of the initial perturbation d_0, $\mid \delta x_0(t) \mid \approx d_0 \exp \lambda t$, whereas the diffusive coupling results in a Gaussian spreading, $\mid \delta x_i(t) \mid \approx \frac{\delta x_0(t)}{\sqrt{4\pi Dt}} \exp -\frac{i^2}{4Dt}$. The combined effects are then given by the equation

$$\mid \delta x_i(t) \mid \approx \frac{d_0}{\sqrt{4\pi Dt}} \exp\left(\lambda t - \frac{i^2}{4Dt}\right). \tag{4}$$

In order to describe the evolution of an initially localized, infinitesimal perturbation, one can use the convective Lyapunov exponent [11]

$$\Lambda(v) = \lim_{t \to \infty} \frac{1}{t} \ln \frac{\delta x_i(t)}{d_0}. \tag{5}$$

It expresses the growth rate of a disturbance when measured from a frame moving with velocity v. For spatially symmetric chaotic systems, the convective Lyapunov exponent is symmetric with respect to $v = 0$, with the maximum of Λ at the origin. At the critical value of the velocity $v = v^*$, the convective Lyapunov exponent becomes 0. In general,

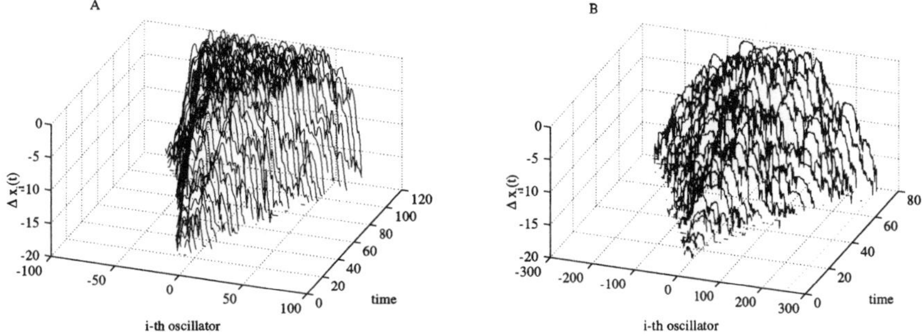

Figure 1: Simulation of the disturbance propagation of fractally coupled, clustered networks for $\alpha = 10$ (corresponds to nearest neighbor coupling) and $\alpha = 0.6$ (B). Parameters used for the network: $N = 512$, $cs = 8$, $nc = 1$, $\varepsilon = 0.8$, the perturbation: $d_0 = 1e - 8$ and the maps: $a = 2$.

perturbations which travel faster than v^* are exponentially damped. Combining Eqn. 4 and Eqn. 5 and substituting $i = vt$ provides an expression for the convective Lyapunov exponent

$$\Lambda(v) = \lambda - \frac{v^2}{4D}. \tag{6}$$

The velocity of the traveling wave front is determined at the borderline of damped and undamped perturbations, which is given by $\Lambda(v^*) = 0$. This yields for the critical velocity

$$v^* = \sqrt{4D\lambda}. \tag{7}$$

Results To track the propagation of disturbance, we use the difference between the original network $x_i(t)$ and a replica system $x_i'(t)$ $\Delta x_i(t) =\mid x_i(t) - x_i'(t) \mid$. A whole cluster (i=-7 to 0) was perturbed by adding 1e-8 to the actual value of each cluster element. Typical evolutions of the disturbance are shown in the semi logarithmic plot of Fig. 1, using 512 oscillators and a cluster size of 8. In panel (A), $\alpha = 10$ was chosen (corresponding to the nearest neighbor coupled network), for panel (B), $\alpha = 0.6$. The velocity of disturbance propagation was estimated using Eqn. 3. For the two values of α shown in Fig. 1 we obtain a more than 5-fold increased speed of information flow in the case of a fractally coupled network $\alpha = 0.6$. In addition, differences between the two cases are found in the width of the wavefront (which is related to the length of connections) and in its ruggedness (which can be associated to the incomplete coupling) (see Fig. 1). These results are not too much surprising since we use a larger number of connections in the case of fractal coupling. Therefore, we increased the number of nearest neighbors, so that the oscillators were not only coupled to their first nearest neighbors but also to the second nearest neighbors and so on. For a series of n-th nearest neighbors we calculated the velocity, which shows a linear relationship, as presented in Fig. 2. In order to be able to compare the velocities of wave propagation for the different networks, we use equal numbers of connections per oscillator (divided by 2). For first nearest neighbor coupling this number is 1, and for $\alpha = 0$-fractal coupling,

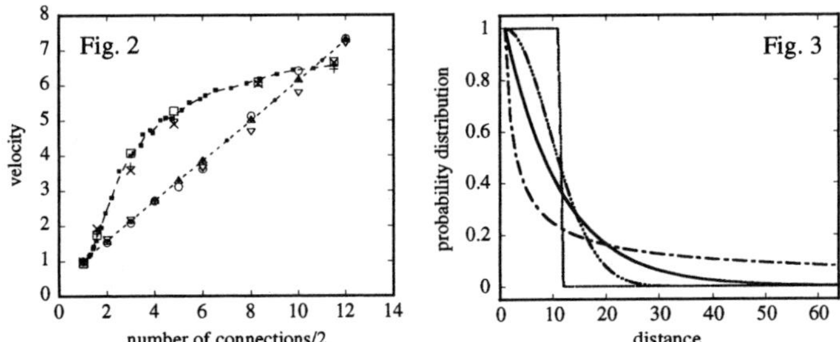

Figure 2: Information flow velocity versus connectivity for different chaotic maps. Open circles (logistic map $a = 2$), upper triangle (logistic map $a = 1.9$), lower triangle (tent map, slope=2) and open squares (logistic map $a = 2$), plus (logistic map $a = 1.9$), cross (tent map, slope=2): simulation results from nearest neighbor and fractally coupled network, respectively (scaled by the velocity for first nearest neighbor coupling). Closed circles and closed squares: velocities calculated via the diffusion coefficient. Broken line: guideline to the eye.

Figure 3: Probability distributions for the presence of connections at distance $| r_i - r_j |$ (same $numberofconnections/2 = 11$, $N = 128$). The diffusion coefficients are $D \approx 177$, $D \approx 69$, $D \approx 25$ and $D \approx 16$ for fractal ($\alpha = 0.608$, dashed line), exponential ($\mu = 0.095$, full line), Gaussian ($\gamma = 0.0071$, dashed-dotted line) and nearest neighbor coupling ($nn = 11$, dotted line), respectively.

this number is 11.5 ($cs = 8$; $nc = 1$). Fig. 2 shows the comparison between the speed of perturbations in networks with nearest neighbor coupling and fractal architecture. The results convincingly reveal the enhanced information flow of fractally coupled clusters. When α was varied from 0, 0.2, 0.6, 1, 2, 5 to 10, the mean connections per oscillator decreases from 11.5, 8.29, 4.81, 2.99, 1.57, 1.02 to 1. Only in the case of $\alpha = 0$, the oscillators of a cluster are synchronized. One of the prerequisites for simulations of networks representing the columnar structure of the cortex is that the connectivity between the columns is smaller than the intro columnar connection density. In our model this condition is satisfied for $\alpha > 0.8$ (corresponds to $numberofconnections/2 < 4$). As can be recognized in Fig. 2 the velocity enhancing effect becomes largest in this range.

In view of Eqn. 7 the origin of the increased information transfer can be associated with the diffusion coefficient D, the Lyapunov exponent λ or both. We first focussed on the diffusion coefficient since it is directly related to the network configuration. In order to estimate D, we used the description of the process in terms of a Markov chain. From the connectivity matrix and the interaction strength $\varepsilon = 0.8$ the transition matrix P of the Markov process can be easily derived. It must be properly scaled so that each row sums up to 1. Two absorbing states were put at both ends of the chain. The time τ which is required to diffuse the distance Δs from the center of the chain to either end, can be calculated via the fundamental matrix of P [12]. The diffusion coefficient is then given by $D = \frac{\Delta s^2}{2\tau} \approx \frac{N^2}{8\tau}$. For first nearest neighbor coupling, the diffusion coefficient $D_0 = \varepsilon/2$ and the critical velocity $v_0^* \approx 0.72$ are well known [10]. Assuming only a small variation of the convective Lyapunov exponent, the speed of the wavefront can be approximated by, $v_{conn}^* = v_0^* \sqrt{\frac{D_{conn}}{D_0}}$, where $conn$ indicates the dependence on the network architecture. Fig. 2 shows the curves for the network with increasing number of nearest neighbors and for the fractally coupled network. In both cases, a good agreement between the calculated and the simulated velocity is obtained. Since the velocity enhancement is due to an increased diffusion coefficient the reason for the superior behavior of the fractal network compared to other networks with decaying connection probabilities is given by the particular structure of its transition matrix. On the one hand fast diffusion can be achieved if sites far of each other are connected (non zero transition matrix element far off the diagonal). On the other hand the probability for such a transition must be high, which is accomplished by preventing jumps to sites of shorter distance (the transition probabilities emanating from the same site sum up to 1). The two conditions lead to the requirements for the probability density: fast decay for short range transitions and a long tail for long range transitions. As displayed in Fig. 3 the power law of fractal coupling satisfies these requirements best compared to exponential coupling ($p_{i,j} = \exp\left(-\mu \mid r_i - r_j \mid -1\right)$), Gaussian coupling ($p_{ij} = \exp\left(-\gamma(\mid r_i - r_j \mid -1)^2\right)$) and nearest neighbor coupling.

Summary We explored the effect of clustering and of fractal coupling on the velocity of information transfer in a chain of chaotic maps. The coupling was limited to the first nearest clusters (nc=1). In comparison to the nearest neighbor coupling, a considerable increase of the speed of information flow through the network was observed when fractal coupling was used. The velocity enhancing effect is due to the diffusive process rather than to the chaotic behavior of the maps. As a consequence, the information flow enhancement observed here can be purely explained by the architecture of the network.

References

[1] K. Kaneko, Chaotic but Regular Posi-Nega Switch among Coded Attractors by Cluster-Size variation, Phys. Rev. Lett. **63** (1989) 219 – 223.

[2] K. Kaneko, On the Strength of Attractors in a high Dimensional System: Milnor Attractor Network, Robust Global Attraction, and Noise-Induced Selection, Physica D **124** (1998) 322 – 344.

[3] O. Popovich and Yu. Maistrenko and E. Mosekilde, Loss of Coherence in a System of Globally Coupled Maps, Phys. Rev. E **64** (2001) 026205-1 – 026205-11.

[4] S.C. Manrubia and U. Bastolla and A.S. Mikhailov, Replica-Symmetry Breaking in Dynamical Glasses, Eur. Phys J. B **23** (2001) 497 – 508.

[5] K. Kaneko, Relevance of Dynamic Clustering to Biological Networks, Physica D **75** (1994) 55 – 73.

[6] E. Niebur and H.G. Schuster and D.M. Kammen and C. Koch, Oscillator-Phase Coupling for Different Two- Dimensional Network Connectivities, Phys. Rev. A **44** (1991) 6895 – 6904.

[7] S. Raghavachari and J.A. Glazier, Spatially Coherent States in Fractally Coupled Map Lattices, Phys. Rev. Lett. **74** (1995) 3297 – 3300.

[8] D.H. Zanette and A.S. Mikhailov, Mutual Synchronization in Ensembles of Globally Coupled Neural Networks, Phys. Rev. E **58** (1998) 872 – 875.

[9] M.F. Bear and B.W. Connors and M.A. Paradiso, Neuroscience, Exploring the Brain, Williams and Wilkins, Baltimore, (1996).

[10] M. Cencini and A. Torcini, Linear and Nonlinear Information Flow in Spatially Extended Systems Phys. Rev. E **63** (2001) 056201.

[11] G. Giacomelli and R. Hegger and A. Politi and M. Vasalli, Convective Lyapunov Exponents and Propagation of Correlations, Phys. Rev. Lett. **85** (2000) 3616 – 3619.

[12] J.G. Kemeny and J.L. Snell, Finite Markov Chains, Springer, New York, (1976).

KES 2002
E. Damiani et al. (Eds.)
IOS Press, 2002

The "Divide and Conquer" model of image segmentation: object-bounded synchrony propagation in Coupled Map Lattices

Antonino RAFFONE

Department of Psychology, University of Sunderland, St. Peter's Campus, Sunderland SR6 0DD, United Kingdom. email: antonino.raffone@sunderland.ac.uk

Cees VAN LEEUWEN Leeuwen

Laboratory for Perceptual Dynamics, RIKEN Brain Science Institute, 2-1 Hirosawa, Saitama 351-0198, Japan. email: ceesvl@brain.riken.go.jp

Abstract. Image segmentation is a crucial processing stage in both biological and artificial sensory systems. In this study, we develop a novel neural network model of visual segmentation, the "Divide and Conquer" model, based on object-bounded synchrony propagation in a 2-D lattice of coupled logistic maps characterized by a chaotic evolution. Complex artificial images with gray-level objects are effectively and rapidly (within a few tenths of iterations) segmented in terms of *local transitive synchrony*, which is propagated and readout within narrow spatial ranges.

1. Introduction

The bottom-up (data-driven) and top-down (knowledge-driven) segmentation of visual and auditory scenes is a crucial processing step in the perception of the external world [1-2]. Perceptual grouping, i.e. the composition of larger perceptual units from simpler perceptual elements, is complementary to perceptual segmentation. The combined operation of perceptual segmentation (segregation) and grouping is necessary to enable pattern (object) recognition in both biological (animal or human) and artificial (computer, robot) sensory systems.

The neural mechanisms governing image segmentation in biological systems are still unclear. Several neurophysiological studies suggest that synchronization of neuronal action potentials coding for features belonging to the same object (or perceptual group), may enable the encoding of separated objects and background [3-5]. However, despite the amount of available data and several computational investigations, the neural circuit properties and the readout mechanism related to this phenomenon, are still controversial [6].

As a neural synchrony-based solution, Wang and Terman [7] showed that in their LEGION model the segmentation of complex real images may be successfully accomplished by networks of locally cooperating and globally competing relaxation oscillators. Since the state evolution of these oscillators is described by computationally expensive systems of differential equations, Wang and Terman [7-8] developed simplified algorithms reflecting the LEGION principles. However, it must be noted that global desynchronization (inhibition) in LEGION appears to be neuroanatomically and neurophysiologically problematic, since fast inhibition operates locally in the cerebral cortex [9], and nested oscillatory phase-lags are not generally observed in recordings from low-level visual areas. Furthermore, it appears unclear to what extent the LEGION simplified algorithms reflect the dynamics described by the related systems of differential equations.

In this study, we suggest an alternative neural network approach to neural synchrony-based (bottom-up) image segmentation, the "Divide and Conquer" model, in which local synchrony is propagated and readout in a 2-D network of coupled logistic maps characterized by a chaotic evolution [10-11]. As the name of the model suggests, object labeling depends on the propagation and readout of synchrony states within the spatial boundaries of separated sets of network units coding for different image pixels.

In this model, the dynamic couplings (coherence states) between network units are modulated by input patterns. Elsewhere it has been shown that coupling between chaotic elements may give rise to flexible pattern encoding and readout in terms of *non-transitive graded* and *intermittent* synchrony [12]. However, in the "Divide and Conquer" model neural synchrony is readout *transitively* through spatially-overlapping sets of network units.

2. The model

The "Divide and Conquer" image segmentation model is characterized by its network architecture, the system of equations governing the network unit dynamics, and an object readout algorithm which is applied off-line after a given number of iterations. The external input represents gray-level images, with a one-to-one mapping between image pixels and network units. The gray-level is encoded by the variable I in the interval between 0 (black) and 1 (white).

In the simulations reported here, we considered a bounded 2-D layer of 64 x 64 (4096) units. The units, each of which is assumed to represent a group of tightly connected neurons in the primary visual cortex, were connected within the 3 closest rings, in the absence of self-connections. Thus, each unit (indexed with i) was connected to other 48 network units (denoted with j) with a uniform structural weight (denoted with W_{ij}, equal to 1). This structural weight is multiplied by an input modulation weight S_{ij}, dependent on the local contrast (difference in gray-level) between the two units, according to the following formula

$$S_{ij} = \exp\left[-\frac{|I_i - I_j|}{\lambda}\right] \tag{1}$$

where λ is a contrast tolerance constant ranging between 0 and 1.

Each network unit (denoted with the index i) is characterized by an activation value R_i (corresponding to the firing rate of the neuronal group), ranging between 0 and 1. For the sake of simplicity, the R_i values are set equal to 1 if the gray-level of a given unit is equal or higher than a threshold θ, and to 0 if this gray-level is lower than θ. Given the stationarity of the input and the sake of computational efficiency, the input-dependent R_i and S_{ij} values are kept constant during the network state evolution.

In our model, the crucial variable for image segmentation is given by the phase X_i, corresponding to an average spike-timing coordinate within a given neuronal group [12], which evolves according to the following revisited system of logistic map equations

$$WLF_i(t) = \frac{\sum_{j=1}^{N} X_j(t) S_{ij} W_{ij} R_j}{\sum_{j=1}^{N} S_{ij} W_{ij} R_j + \varepsilon} \tag{2}$$

$$Net_i(t) = X_i(t)\left(1 - C_{\textit{eff}}\right) + WLF_i(t)\,C_{\textit{eff}} \tag{3}$$

$$X_i(t+1) = A_i\,Net_i(t) \tag{4}$$

Equation 2 computes the weighted local field *WLF* of a given unit [11-12], where N is the number of units in the network and ε is an arbitrarily small constant to prevent the instantiation of a null denominator. The *Net* value computed in Equation 3 is the argument of the logistic map in Equation 4. In Equation 3, the effective coupling value $C_{\textit{eff}}$ is equal to the preset coupling constant C if the product of R_i and the denominator of the second side in Equation 2 minus ε is higher than 0. It implies that the phase X_i of inactive units (R_i equal to 0) or units which are structurally connected (with W_{ij} equal to 1) to no active unit, evolves like an isolated logistic map.

During a simulation period of T iterations (from iteration T_0), the *Time-Averaged Coherence Value TACV$_{ij}$* (ranging from 0 to 1) between pairs of neighbor network units, is computed with the following formula, where τ is a phase difference tolerance constant

$$TACV_{ij} = \frac{\displaystyle\sum_{t=T_0}^{t=T_0+T} \exp\left(-\left|X_i(t) - X_j(t)\right|/\tau\right)}{T} \tag{5}$$

After the simulation period (iteration $T_0 + T$), the following readout algorithm for object labeling, is implemented: This algorithm is based on the sequential updating of two variables: the first variable is SET_{ik}, i.e. the set of synchrony-coupled neighbors, with i ranging between *1* and N, and k ranging between 1 and 8 (number of neighbors in the first ring); the second variable is $Label_i$, i.e. the labeling (integer) number coding for these objects (image segments). A convenient property of this algorithm is that the specification of only one parameter, the synchrony *readout threshold* θ_r (ranging from 0 to 1), is required. The threshold parameter θ is not crucial for the readout; it is essentially used to filter out a subset of image pixels.

The initial values of the variable SET_{ik} are set equal to 0, meaning the absence of recruited units. The initial values of the variable $Label_i$ are set equal to 0 if $R_i < \theta$ (background condition), and to i otherwise (each unit is initially coding for a separate object). The spatial coordinates of each unit are fixed. However, the application of the algorithm is independent on any specific mapping between unit index i and pair of spatial coordinates.

[I] Go to the next unit i in the vector of the network units, from $i = 1$ to $i = N$. If $R_i \geq \theta$ then go to step [II], otherwise go to $i + 1$.

[II] Consider sequentially each of the 8 units (indexed with k in this sub-iteration, and with j in the vector of the network units) in the first ring around unit i (i.e. characterized by the minimum Euclidean distance in each possible radial direction from unit i), sub-iterating from $k = 1$ to $k = 8$. If $R_j < \theta$ then go to $k + 1$. If $TACV_{ij} \geq \theta_r$ then $Set_{ik} = j$, otherwise $Set_{ik} = 0$ (unchanged condition). Go to step [I].

[III] Go to the next unit i in the vector of the network units, from $i = 1$ to $i = N$. If $R_i \geq \theta$ then go to step [II], otherwise go to $i + 1$.

[IV] Consider sequentially each of the 8 units (indexed with k in this sub-iteration and with j in the vector of the network units) in the first ring around unit i, iterating from $k = 1$ to $k = 8$. ($Set_{ik} = j$, if $TACV_{ij} \geq \theta_r$, from step [II]). If $R_j < \theta$ then go to $k + 1$. If $Set_{ik} > 0$: if $Label_i \geq Label_j$ then $Label_i = Label_j$, sub-reiterate from $k = 1$; otherwise (if $Label_i < Label_j$) $Label_j = Label_i$. Go to step [III] after the iteration $k = 8$, in all cases.

The algorithm includes the two main stages of *local recruitment* (the first two steps) and the "divide and conquer" stage (the last two steps). During the recruitment process, each unit is *symmetrically* linked to a subset (variable *Set*) of neighbor units, based on the synchrony readout threshold. During the "divide and conquer" process, the transitivity of the synchrony relationship is encoded in terms of overlap between sets of linked neighbor units. Therefore, the algorithm divides the network units in non-overlapping sets (objects), and enables the propagation ("conquering") of the symmetric synchrony-based linking state to a larger set of non-neighbor units, through convergence onto the minimum labeling number (variable *Label*) in each set of linked units.

After the application of the algorithm, the image is segmented into a number of objects, i.e. sets of units with the same label, plus the background (labeled with 0). Constraints can be set to define an object in terms of minimum size, e.g. by excluding sets with a number of units below a given object-size threshold, which would correspond to noise.

3. Computer simulations

We first studied the systematic relationships between coherence ($TACV$) and the coupled logistic map parameters A (oscillation amplitude) and C (coupling term), with the computational scheme described above. The input was given by 25 squared gray-patches (objects), with a side randomly sampled between 7 and 9 pixels. A 0.25 maximum overlap (with an occlusion effect) between objects' area was allowed. The gray level, uniform in each object, was randomly assigned between 0.5 and 1. The background value was set equal to 1 (black). The following parameters and simulation setting were used: θ=0.25; λ=0.001; τ=0.1; T=150; T_0 =51. The initial phase values X_i were randomly assigned between 0 and 1. In the simulation conditions, the A value was varied between 3.7 and 4.0 (step equal to 0.1), in which the phase X_i exhibit a chaotic evolution [11]. The C values was varied between 0 and 0.4 (step equal to 0.1). With each pair of A and C values, 10 replications were run, with random displacements of the 25 objects. The coherence value ($TACV$) between a given unit and the units coding for the same object in the next-neighbor ring, averaged over all network units (coding for an object), replication, and iterations (from iteration 51 to iteration 200 in each simulation), was observed.

As shown in Figure 1, coherence decreases with the A value and increases with the coupling term C. Note the very low standard deviation over the 10 replications with different initial values and input patterns. Readout thresholds in image segmentation may be computed based on these coherence values.

We then applied the "Divide and Conquer" model in image segmentation tasks, with A = 3.7 and C = 0.4. A typical simulation is shown in Figure 2. The model was presented with 50 overlapping objects, the gray level of which ranged between 0.25 and 1. The simulation duration was 100 iterations. Note the different segmentation results with different θ values (from 0.1 to 0.8, step equal to 0.1). With a low θ different overlapping objects tend to be grouped together, whereas with high θ values even the same object may be decomposed into more parts. However, the segmentation is correct within a relatively broad range of θ values.

We also applied the "Divide and Conquer" model to the segmentation of images with non-overlapping gradients. Figure 3 shows a typical simulation with 15 objects, defined as linear gradients with a variable size and maximum gray-level in their center. The minimum value of the gradients was always set equal to 0.25 (the maximum value ranged between 0.25 and 1). The contrast tolerance parameter λ (Equation 1) was set equal to 0.1. Note the successful segmentation of integrated objects with a wide range of θ values, despite the internal gradients.

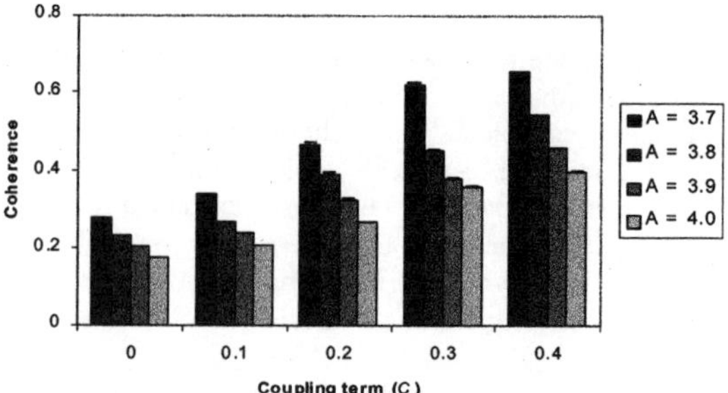

Figure 1. *Average coherence values within objects with different C and A values. Note the increase in coherence with C and the decrease in coherence with A. See text for more explanations.*

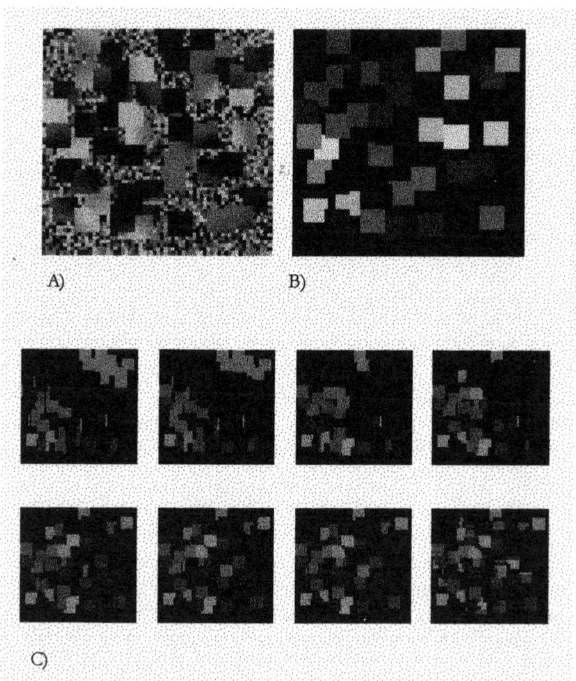

Figure 2. *(A) Phase values of the units at Iteration 40. Note the independent chaotic evolution of background units. B) Input image. C) Segmentation maps (with θ_T ranging from 0.1 to 0.8, from top-left). See text for more explanations.*

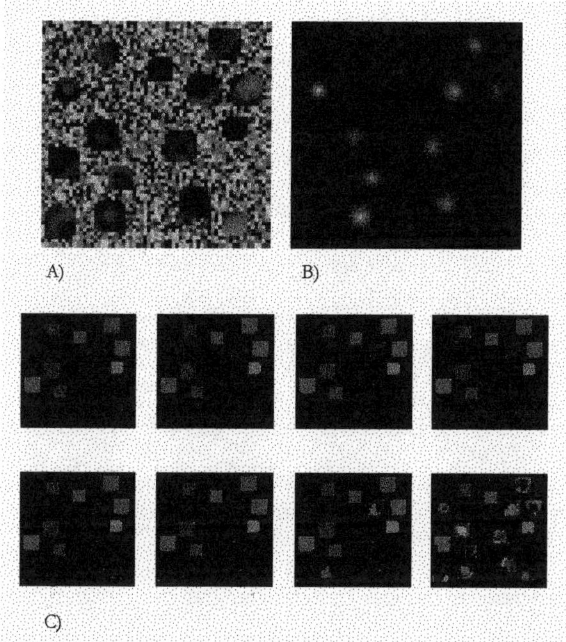

Figure *3. Image segmentation with gradients. (A) Phase values of the units at Iteration 40. B) Input image. C) Segmentation maps (with* θr *ranging from 0.1 to 0.8, from top-left). See text for further explanation.*

4. Conclusions

Our computer simulations have shown that segmentation of artificial images with relatively complex patterns may be effectively and rapidly accomplished by the "Divide and Conquer" model. In this model, diffusive links between the states of chaotic elements play a crucial role. The transitivity and the symmetry of local synchronization (coherence) states are spatially extended in readout, based on adequate thresholds. These readout thresholds are crucial in defining the relative segregating and grouping strengths in the application of the algorithm.

As discussed elsewhere [11, 12], graded and intermittent synchronization of diffusively and pulse coupled chaotic neural oscillators may provide the bases for flexible binding dynamics in the neural representation of the external world at different processing stages. It remains to be seen how the "Divide and Conquer" model performs with real images, and how segmentation may be coupled to more complex representational dynamics in simulated vision.

References

[1] Bregman, A.S. (1990) *Auditory scene analysis.* Cambridge, MA: MIT Press.

[2] Rock, I., & Palmer, S. (1990). The legacy of Gestalt psychology. *Scientific American, 263,* 84–90.

[3] Eckhorn, R., et al. (1988).Coherent oscillations: A mechanism of feature linking in the visual cortex? Multiple electrode and correlation analyses in the cat. *Biological Cybernetics, 60,* 121-130.

[4] Gray, C.M., König, P., Engel, A.K., & Singer, W. (1989). Oscillatory responses in cat visual cortex exhibit inter-columnar synchronization which reflects global stimulus properties. *Nature, 338*, 334-337.

[5] Singer W (1999). Neural synchrony: a versatile code for the definition of relations. *Neuron, 24*, 49-65.

[6] Shadlen, M.N., & Movshon, J.A. (1999). Synchrony unbound: A critical evaluation of the temporal binding hypothesis. *Neuron, 24*, 67-77.

[7] Wang, D.L., & Terman, D. (1997). Image segmentation based on oscillatory correlation. *Neural Computation, 9*, 805-836.

[8] Liu, X., & Wang, D.L. (1999). Range image segmentation using a LEGION network. *IEEE Transactions on Neural Networks, 10*, 564-573.

[9] Braitenberg, V., & Schüz, A. (1991). *Anatomy of the cortex.* Berlin: Springer-Verlag.

[10] Kaneko, K. (1990). Clustering, coding, switching, hierarchical ordering and control in a network of chaotic elements. *Physica D, 41*, 137-172.

[11] Van Leeuwen, C. Steyvers, M., & Nooter, M. (1997). Stability and intermittency in large-scale coupled oscillator models for perceptual segmentation. *Journal of Mathematical Psychology, 41*, 319-344.

[12] Raffone, A., & Van Leeuwen, C. (2001). Activation and coherence in memory processes: Revisiting the Parallel Distributed Processing approach to retrieval. *Connection Science, 13*, 349-382.

KES 2002
E. Damiani et al. (Eds.)
IOS Press, 2002

Some Problems Related to the Choice of Coupling in a System of Coupled Maps

Ivan Tyukin, Cees van Leeuwen
RIKEN Brain Science Institute
2-1, Hirosawa-Wako, Saitama, 351-0198, Japan

Abstract. In this paper we analyze the conditions of stability for asymmetrically coupled maps (CM). We derive sufficient conditions for synchronization in this type of systems utilizing Adamar's lemma. From our results, the stability conditions of synchronization in CMs with symmetric connections given in van Leeuwen et al. (1997) are derived as a special case. We show that in the totally non-symmetric case the system movement towards the synchronization manifold may be disturbed by an additional term of which the value depends on the system state.

1 Introduction

In the study of nonlinear phenomena, coupled maps (CM) are computationally convenient tools. Coupled maps are real-valued nonlinear oscillators which are updated in discrete time. Their couplings can either be global (a unique parameter describing coupling strength) or locally defined (coupling parameters differ from one node to anther). Dynamics of these systems have been studied extensively in the last decades [1, 2]. Nonlinear phenomena emerging in this class of systems are clustering, dynamics of the patterns and intermittency. Most of these phenomena (and even a definition of pattern) have been formulated in terms of synchronization[1] between the units of the network (or state variables of a system of CM). Therefore, one of the corner stones in the study of nonlinear phenomena in CM is an analysis of synchronization between its subsystems.

Stability conditions for the synchronization of both linearly [3] and nonlinearly [4] coupled maps have been given. These studies show that chaotic oscillators in CM can reach stable synchrony from arbitrary initial conditions, depending on coupling strength. In these studies a global coupling parameter was assumed and, by consequence, the connections in the system are symmetric[2]. For practical applications, however, it is not sufficient to deal with global coupling parameters only. There may be reasons to allow the system couplings to be locally parameterized.

Many results regarding synchronization in nonlinear systems cover only those situations where the orbits are close enough to satisfy applicability conditions of the first Lyapunov

---

[1] In this paper we will consider only the trajectory synchronization; phase synchronization or synchronization with respect to a given goal function are not considered in this paper. For a detailed review of the definitions of different types of synchronization, see [5].

[2] Precise equation showing a symmetry in this case will be given in the next section

method (linearization around a given equilibrium or target surface). The present study extends these results to systems with local coupling strength without symmetry restrictions. In addition we aim to derive sufficient conditions for synchronization which are not based on any approximation of a nonlinear system by its linearized dynamics. The paper is organized as follows. In Section 2 we introduce the necessary mathematical notations and formulate the problem. Section 3 contains the main results, Section 4 concludes the paper.

2   Problem Formulation

Let us consider two discrete systems with coupling:

$$\begin{aligned}
z_1(k) &= Cx(k) + (1 - C)y(k); \\
z_2(k) &= Cy(k) + (1 - C)x(k); \\
x(k+1) &= f(z_1(k)); \\
y(k+1) &= f(z_2(k)),
\end{aligned} \tag{1}$$

where $C \in R$ is a coupling strength, $x(k) \in R^n$, $y(k) \in R^n$ are the states of each system at step k, k is a number of iterations, $z_1(k) \in R^n$, $z_1(k) \in R^n$ are the variables that model internal connections in the system, $f(\cdot) : R^n \to R^n$ is a map describing the system evolution. We will assume that $f(\cdot) \in C^1$ i.e. $f(\cdot)$ is continuously differentiable. Model (1) of the coupled maps may be rewritten in more traditional way:

$$\begin{aligned}
z_1(k+1) &= Cf(z_1(k)) + (1 - C)f(z_2(k)); \\
z_2(k+1) &= Cf(z_2(k)) + (1 - C)f(z_1(k)),
\end{aligned} \tag{2}$$

however in the rest of the paper we will use its form described by equations (1) instead of (2). A reason to do so is that model (1) clearly reflects its coupling property[3], whereas (2) does not. In particular, if we rewrite system (1) as

$$\begin{aligned}
\begin{pmatrix} z_1(k) \\ z_2(k) \end{pmatrix} &= \begin{pmatrix} C_{11} & C_{12} \\ C_{21} & C_{22} \end{pmatrix} \begin{pmatrix} x(k) \\ y(k) \end{pmatrix} \\
x(k+1) &= f(z_1(k)); \\
y(k+1) &= f(z_2(k)),
\end{aligned} \tag{3}$$

where $C_{i,j} \in R^n \times R^n$, $C_{11} = C_{22} = CI_n$, $C_{12} = C_{21} = (1 - C)I_n$, I_n is identity matrix of size $n \times n$. It is clear, that matrix

$$\mathcal{C} = \begin{pmatrix} C_{11} & C_{12} \\ C_{21} & C_{22} \end{pmatrix}$$

under these assumptions is symmetric ($\mathcal{C} = \mathcal{C}^T$ as $C_{12} = C_{21}$ and $C_{11}, C_{12}, C_{21}, C_{22}$ are diagonal) and has a very simple structure. In order to extend a notion of the symmetric coupling to multi-dimensional case and to make it more general we introduce the following

---

[3]The coupling is *nonlinear* due to the nonlinearities $f(z_i(k))$, $i = 1, 2$ in the system right-hand side, but it is *linear* with respect to the states $x(k)$, $y(k)$ and corresponding vectors $z_i(k)$. Although model (2) which is the linear in the parameter C, model (1) is preferred as a starting point because its potential ability to reflect internal structure of the system via additional state vectors $x(k)$, $y(k)$ and their interconnections formalized functions by $z_i(k)$.

Definition 1. *Coupling strength in system (3) is said to be symmetric if matrix C satisfies the following equalities $C_{11} = C_{22}$, $C_{12} = C_{21}$.*

One may observe that our definition of symmetric coupling differs from a requirement of matrix C to be symmetric: $C = C^T$. Our definition requires that matrix C is invariant under permutations of its blocks C_{11} and C_{22}, C_{12} and C_{21} correspondingly.

We are interested in an answer to the question, whether is it necessary to use symmetric coupling for synchronization, and if not, how to derive the conditions for synchronization if the coupling strength is non-symmetric. Most of the results dealing with synchronization in models of types (1), (2) are based on *linearization* technique around some target manifold $\psi(x(k), y(k)) : x(k) - y(k) = 0$. Therefore, is it possible to obtain non-local conditions that ensure synchronization not only within sufficiently small domain of the synchronization manifold, but also for any $x(k), y(k) \in R^n$.

3 Main Results

We start with an analysis of the system dynamics for the special case of linear coupling and then consider applications of the proposed results. Let us introduce the following:

Theorem 1. *Let system (3) be given and matrix C satisfy the following condition:*

$$C_{11} + C_{12} = C_{21} + C_{22}.$$

In addition, let

$$\left\| \int_0^1 \frac{\partial f(S(\lambda))}{\partial S} d\lambda (C_{11} - C_{21}) \right\| < 1,$$

where

$$S = (C_{11}x(k) + C_{12}y(k))(1 - \lambda) + (C_{21}x(k) + C_{22}y(k))\lambda.$$

Then all the trajectories of system (3) satisfy the limiting relation:

$$\lim_{k \to \infty} x(k) - y(k) = 0.$$

Proof of theorem 1. The theorem proof is quite straight-forward and is based mostly on a result which is known as Adamar's lemma. Consider a continuously differentiable function f and the difference $f(z_2) - f(z_1)$. Let us introduce auxiliary variables λ and S:

$$S = z_1(1 - \lambda) + \lambda z_2.$$

It is clear that $S = z_1$ for $\lambda = 0$ and $S = z_2$ for $\lambda = 1$, hence $f(S(1)) = f(z_2)$ and $f(S(0)) = f(z_1)$. Therefore,

$$f(z_2) - f(z_1) = \int_0^1 \frac{\partial f(S(\lambda))}{\partial \lambda} d\lambda = \int_0^1 \frac{\partial f(S(\lambda))}{\partial S} \frac{\partial S}{\partial \lambda} d\lambda = \int_0^1 \frac{\partial f(S(\lambda))}{\partial S} d\lambda (z_2 - z_1).$$

The last equation is the essence of Adamar's lemma. Now we apply this result to our system in a form of equation (3). Deviation from synchronization manifold $\psi(x(k), y(k))$ is given by the following equality:

$$x(k + 1) - y(k + 1) = f(z_1(k)) - f(z_2(k)) = \int_0^1 \frac{\partial f(S(\lambda))}{\partial S} d\lambda (z_2(k) - z_1(k))$$

$$= \int_0^1 \frac{\partial f(S(\lambda))}{\partial S} d\lambda (C_{11}x(k) + C_{12}y(k) - C_{21}x(k) + C_{22}y(k)). \tag{4}$$

It has been assumed that $C_{11} - C_{21} = C_{12} - C_{22}$, therefore equation (4) may be written as follows:

$$x(k + 1) - y(k + 1) = \int_0^1 \frac{\partial f(S(\lambda))}{\partial S} d\lambda (C_{11} - C_{21})(x(k) - y(k)). \tag{5}$$

The last equality automatically proves the theorem. *The theorem is proven.*

Remark 1. Theorem 1 has been formulated (and does provide us with sufficient conditions for synchronization between two systems) for a case when nonlinearity $f(\cdot)$ is a differentiable function. It is important to observe, that the results depend on the existence of the integral of partial derivatives $\partial f(S(\lambda))/\partial S$ over $[0, 1]$ and do take into account information about the Jacobians of the nonlinearity. However, these Jacobians have to be integrated with respect to λ and may not remain constant. Nevertheless, there is a link between the first Lyapunov method and the statements of Theorem 1. If one derives equation (5) assuming that $x(k)$ is close enough to $y(k)$ and $\partial f(S(\lambda))/\partial S$ is not changing very much with respect to λ, the final formula will be written as follows:

$$x(k + 1) - y(k + 1) = \frac{\partial f(z_1)}{\partial z_1}\Big|_{z_1(k)=z_2(k),\ x(k)=y(k)}(C_{11} - C_{21})(x(k) - y(k)),$$

which is exactly the result of the first Lyapunov method.

Remark 2. It is desirable to note that the theorem includes as a special case the earlier results about synchronization between two systems [4]. Let $x(k), y(k) \in R^1$ $C_{11} = C_{22} = C$, $C_{12} = C_{21} = (1 - C)$, then $C_{11} - C_{21} = 2C - 1$. Let in addition $f(z_i) = Az_i(1 - z_i)$, then

$$\frac{\partial f(S(\lambda))}{\partial S} = A(1 - 2S(\lambda)).$$

Therefore, difference $x(k + 1) - y(k + 1)$ in this case satisfy the following equation:

$$\begin{aligned}
x(k + 1) - y(k + 1) &= A \int_0^1 (1 - 2S(\lambda))d\lambda(2C - 1)(x(k) - y(k)) \\
&= A(2C - 1)(x(k) - y(k))(1 - z_2(k) - z_1(k)), \tag{6}
\end{aligned}$$

where the term $(1 - z_2(k) - z_1(k))$ is a result of integration $\int_0^1 (1 - 2S(\lambda))d\lambda$.

Remark 3. Theorem 1 allows us to use non-symmetric coupling in the system. However, it still assumes some structure: $C_{11} + C_{12} = C_{21} + C_{22}$. If this is not the case, then equation (4) yields:

$$x(k + 1) - y(k + 1) = f(z_1(k)) - f(z_2(k)) = \int_0^1 \frac{\partial f(S(\lambda))}{\partial S} d\lambda(z_2(k) - z_1(k))$$

$$= \int_0^1 \frac{\partial f(S(\lambda))}{\partial S} d\lambda(C_{11} - C_{21})(x(k) - y(k) + \Delta Cy(k)),$$

where $\Delta C = (C_{11} - C_{21}) - (C_{12} - C_{22})$. Under this condition there is an additional term $\Delta Cy(k)$ that plays a role of a disturbance and forces the system to move away from the synchronization manifold. Furthermore, without this structure imposed by Theorem 1 conditions, the synchronization (even local) may be impossible as almost any value of the state $y(k)$ (except those which belong to $Ker\Delta C : y(k) \in \{y|\Delta Cy = 0\}$) will destroy synchronization.

Remark 4. The Theorem conditions hold for symmetric coupling (in a sense of Definition 1) automatically. Indeed, $C_{11} + C_{12} = C_{21} + C_{22}$ as $C_{11} = C_{22}$, $C_{12} = C_{21}$. Therefore, any symmetrically coupled system is *invariant* on the synchronization manifold $x(k) - y(k) = 0$ (see Remark 3 for details). However as it follows from the Theorem, a property of symmetric coupling is not a minimal requirement for the system to be invariant on the $x(k) - y(k) = 0$.

4 Conclusion

In this paper we dealt with analysis of the properties of the coupling functions that ensure synchronization of two identical nonlinear dynamic systems. We investigated the role of the coupling function in the synchronization processes. It was shown that non-symmetric coupling is theoretically acceptable (see Theorem 1) but some structure is still necessary for maintaining synchronization (as has been discussed in 3). Sufficient conditions that have been obtained in the paper are state-dependent and may be useful for the further analysis.

Despite some generality in the claims we have to make two comments regarding restrictions of the results in their present form. It is desirable to note, that an impact of the coupling strength on boundedness of the trajectories has not been investigated at al. This may dramatically restrict applicability of the results to the practical problems. Furthermore, the results presented in the paper have been proven only for two coupled nonlinear systems. Nevertheless we believe that their extension on multi-system case is possible and is a good topic for the future investigations in this line.

References

[1] Kaneko K. "Pattern Dynamics in Spatiotemporal Chaos", *Physica D*, Vol. 34 (1989), pp. 1–41.

[2] Kaneko K. "Relevance of Dynamic Clustering to Biological Networks", *Physica D*, Vol. 75 (1994), pp. 137–172.

[3] Ding M., Yang W. "Stability of synchronous chaos and on-off intermitency in coupled map lattices", *Phys. Rev. E*, Vol. 56, No. 4 (1997), pp. 4009–4015.

[4] Van Leeuwen C., Steyvers M., Nooter M. "Stability and Intermittentcy in Large-Scale Coupled Oscillator Models for Perceprual Segmentation", *Journal of Mathematical Psychology*, Vol. 41 (1997), pp. 319–343.

[5] I.I. Blekhman, A.L. Fradkov, H. Nijmeijer and A.Yu. Pogromsky. On self-synchronization and controlled synchronization. *Syst. Contr. Lett.* Vol. 36, No. 10 (1997), pp. 299-306.

KES 2002
E. Damiani et al. (Eds.)
IOS Press, 2002

Component-based Framework for Design and Simulation of Neural Networks

Peter JURICA, Martijn BRINKERS, Cees VAN LEEUWEN
Laboratory for Perceptual Dynamics, Brain Science Institute, RIKEN,
2-1 Hirosawa, 351-0198 Wako-shi, Saitama, Japan

Abstract. A component-based framework for computer aided neural network modelling and simulation is proposed. The approach features independent implementation of simulation engine and graphical environment, bound together only by a general-purpose communication framework. The framework enables design and implementation with extreme flexibility, reusability and concurrent ease of use. Apart from neural networks, the system allows modelling and simulation of any system of differential or difference equations that can be modularized. The system has specific facilities for the study of the behaviour of Coupled Maps and their application to modelling of information processes.

1. Introduction

A neural network simulator should allow the user to test many different configurations of neural network models, it should enable access to all data being created throughout the simulation process with both visualization and independent data storage and processing facilities. Many software packages are available today that serve as simulation environments for neural network systems, or generally for solving systems of differential or difference equations. Nevertheless, researchers still end up developing their own simulation programs, as is evidenced by the wide availability of shareware, freeware and regularly available commercial products. Either the available packages do not meet the specific needs of users or they are so complex that it is easier to develop an ad-hoc model, using a general-purpose programming language. Such simulations are unlikely to be re-used.

2. Problem formulation

The developments in software engineering of the last few years have increasingly turned to component based systems. Ease and flexibility of use are important goals for all designers of these environments. Re-usability has been added as a central target in the software design process. Component-oriented systems provide features like re-use and dynamic change of implementation at run-time by the use of plug-in components. Beyond its timesaving advantages, the major benefit for researchers lies in the enhanced opportunities this strategy offers for sharing implemented functionality.

Object composition and inheritance are two techniques for reusing functionality in object-oriented systems [1]. For creating a class, allowing it to inherit some of the functionality of a parent class will often require knowledge of its implementation. For this reason this approach is often called white-box reuse. Object composition is a different method of reusing functionality. Whereas objects should prefeably be simple, composition is used to combine features of simple objects into more complex units. These units are ready for use without any internal modification. For these reasons, their reuse is often referred to as black-box reuse. The communication with components is done through their interfaces, which are the only way of exposing their functionality and properties and therefore should be accordingly well defined.

The component-based approach is tailored to systems to which new functional elements are frequently added. Implementation of these new units is done without any need of modification and subsequent recompilation of the whole environment. New components are loaded dynamically during run-time of an application and are used through the common interface. Component object modelling is widely being used for various applications. Well-known examples are Internet browses like Internet Explorer, where plug-in technology allows web designers to use presentation techniques not provided by the browser itself, for example Macromedia Flash animations. Another good example is development environment Microsoft Visual Studio, of which essential parts, like code and GUI editor can be modified or even replaced by different editors (for instance, WholeTomato's Visual Assist). Finally we find some of the most relevant examples among environments used for modelling and simulation based on flow diagrams. From those serving the same target group as our framework we can mention packages like NeuroDimension's NeuroSolutions and Siemens's ECANSE.

3. Implementation

Implementation of such systems requires the use of standardized technology for handling the communication between components with the required speed. In absence of such technology, many software producers, in the past have created their own protocols. Nowadays several protocols offer satisfactory speed. The most widespread of these is Microsoft's Component Object Model (COM). All systems mentioned earlier are based on this technology.

COM is a very powerful tool. All low-level and also many higher-level programming languages allow us to create COM components. Another benefit is that component implementation can be done in any of those programming languages supporting COM components development. Thus, possible implementations range from Visual Basic or Delphi for simple and rapid development or C++ for fast and efficient components. For these reasons we adopted this technology for our implementation.

Simulation environments generally accept components conforming to some standard way of implementation. The reuse of such components, therefore, in most cases is restricted. Components developed for such systems cannot be re-used elsewhere unless access to their source code is given.

We overcame this problem in the most radical way by, instead of creating our own simulation environment, using the operating system as a hosting simulation environment for our components. As a consequence, components are completely independent of any environment except the operating system, which must support COM technology.

Specifically, our components can be used by any system running on the Microsoft Windows platform capable of using ActiveX components using the OLE Automation standard. Automation (originally known as OLE Automation) is a part of COM technology. It was originally developed as a way for applications (such as Word and Excel) to expose their functionality to other applications, including scripting languages. The intention was to provide a simple way to access properties and call methods that put as little strain on the Automation client as possible—and allowed calls to be made without needing access to type information about the object. Scripting languages, such as Visual Basic for Applications, VBScript, and JScript, use Automation exclusively. So for using scripting languages to control the components, an Automation interface had to be implemented. [2]

A major problem of the COM technology is its extreme complexity. Without experience in operating system programming, this may seem an insurmountable obstacle. In order to allow developers to build components without any knowledge of COM technology, we prepared a toolkit including wizards to help generating basic projects for components. The task for a component developer is thereby reduced to implementing the function bodies relevant to the component's role within a system: the algorithm and the handling of its in and output. The wizards handle the implementation of code needed to integrate the component into the COM engine. This code is hidden for the user and no knowledge of it is needed.

A set of interfaces has been designed that all components have to implement to make them accessible for other parts of the product. This set of interfaces creates the specification of the framework proposed. For the user only two of them are important: the main interface consisting of functions for initialization and activation of the component and the communication interface. The implementation of the communication interface is not necessary for exchange of common types, but its knowledge is essential. The framework specifies when will these methods of interfaces be called and what tasks should they normally perform.

A set of ready-made components is offered which can be assembled using, among others, the ActiveX scripting languages.

To further enhance the ease of use, we have been designing and developing an application with a graphical user interface. This application should help those users not familiar with programming, and give them access to all the features of components. This is another place where features of the COM technology are being used. The purpose of such graphical environment is then to present all features of components and make them accessible in more convenient form for all kinds of users but still preserving the flexibility of their use.

4. Framework and environment description

Components in our approach are characterized by methods and properties and input/output relations. Methods are functions that can be performed by the component [3]. Properties represent the local settings for the simulation and are used as option panel for the component. Some methods and properties are tightly bound together by the fact that properties often influence the action performed by a method.

Components are linked by connectors. Connectors don't have any but symbolic visualizing function, they don't do anything and in the implementation are not even existing as actual objects. They are presented only as a link from one component to another. The communication is done without any mediation directly between two connected components. We distinguish input and output and bi-directional connectors. Input and output of components can be of any type, the only requirement is compatibility of types between input and output along a connection. Input and output connectors can be designed to accept or send out more than one type. For example, bit sets for saving computer memory or integer sets of 0 and 1 to secure compatibility with visualization components. This allows, for instance, visualization tools to be generic for information of different types as long as a simple conversions can guarantee compatibility. Communication through bi-directional connection resembles the process of a function call.

There is room for components with more abstract functionality, not accepting input or output like other components. We call them meta-components. Instead of communicating with other components in the graph these components can play the role of supervisor or be driving the simulation. For example such components can implement an algorithm globally adjusting properties of more components at once or even dynamically change the topology of a diagram used for simulation.

To achieve maximal user friendliness and comprehensible design of networks in our graphical interface we adopted the tried and tested method of flow diagrams. Their transformation into executable software code is done by the simulation environment. The generated program then drives the simulation. However, the method of flow diagrams is not mandatory. Had it been an inseparable part of simulation, it would have constrained the user with possible limitations in visualization of complex data flow. For this reason, the user has the option to design his or her program independently of the graphical interface.

Another move to enhance the user-friendliness of the system is to construct a layered architecture of interfaces on top of the COM protocol. The lower level interface is mediating services of the COM engine to the higher levels. Although models can be implemented using the interfaces at any level as desired, the toolkit and wizards that it contains provide tools only at the highest level.

The main interface that makes from component a component available for our system contains functions needed to run a simulation. The basic set of functions used in our approach include: initnode, activate, reset, die and one function returning the collection of connectors attached to a component. Initnode function is called in the initialization stage of the simulation before the activation loop starts. Activate is the function which should contain the implementation of functionality being modelled. Reset is called to recall the default settings of components parameters and die to tell the component that it is going to be destroyed by the system and therefore should release any resources. The last function gives the possibility to get access to all connectors for the components. Implementer uses then the communication interface obtained from this function to send or receive the data. Other interfaces worth to mention are designated for exchange of not common types user-defined types. In that case the implementer has to implement interface based on interface used for data types exchange. Such types exchange start with negotiation process when receiving component asks for typed available. If some of them is acceptable it asks for another interface reserved for actual data exchange. In that way the program is in no way restricted to use of a type. Examples can be shown even with streaming video. The only requirement is the presence of the same interface for exchange on both sides of communication.

The completed flow diagram specifies the features of the components used and their input/output relations. By doing so, the functionality of the program generated is completely determined. It can be modified without recompiling the program. Properties and methods of components can be available for inspection and modification during both design and run-time of a simulation.

An example may be helpful to illustrate the concepts used in our software. Suppose we want to simulate a process of learning a mapping from 1D to 1D, e.g. it is a simple function approximation problem. Implementation can be based on some simple neural network or some other adaptive algorithm. A diagram on the following figure can serve that purpose.

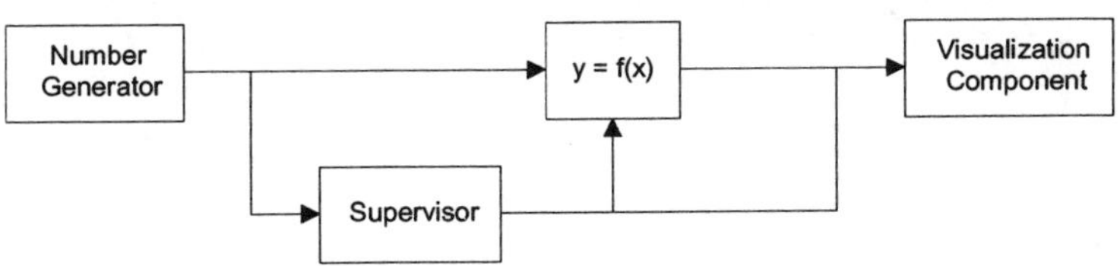

Figure 1: A simple component-based system for 1D mapping.

In this program, one component is continuously generating numbers, another learning the 1D mapping and a third one for visualization of the other one's behaviour. Finally, a component supervises the learning phase. Once the flow diagram is completed, a script is generated that initializes the components prior to the simulation. Then it finds the source, which in this case the Number Generator. The source is activated, followed by the other components along the connections: the Supervisor, the mapping, and last the Visualization Component. Such activation cycle will repeat in a loop until stopped by user or some termination condition.

After successful learning phase we can store parameters settings for the mapping component and use it later in some real application. The learning phase can be triggered either by setting a value of some parameters or by determining the presence of the driving signal from supervisor.

Starting such simulation will result in opening of one output window showing for example two numbers or two signals on a chart. Prior to the simulation we can change the default parameters of components and so change the simulation. For example we can change the learning algorithm if more of them are implemented or switch of the learning completely. Everything depends on abilities of single components.

5. Conclusion

A component-based approach in combination with flowcharts provides users a wide range of possible applications. Supported by the programming languages and technologies available today our design and simulation tools achieve a high degree of flexibility combined with improved possibilities of reuse. Rather than competing with other systems in optimizing algorithms, our software uses the advantages of them through a common technology widely used in today's software engineering, the Component Object Model, and makes these technologies ready for use by those inexperienced in system design. This strategy enables people to implement models of neural networks using programming languages or other applications suitable for their particular tasks and combines their strengths and advantages using one common framework.

The system described in this paper is an open system, which should evolve through time with developing technologies, allowing users to adapt it to needs of new simulation methods and approaches. The software package with documentation is accessible through the laboratory home page at <u>http://pdl.brain.riken.go.jp/applications/</u>, which is updated regularly, depending on further developments.

Future developments will include libraries of components, for sharing implemented knowledge among researchers, which offer further support to their re-use. Today's very popular peer-to-peer sharing engines are showing us the viability of this development. Another advantage is that such components are very well prepared for distributed computing on the network, for example using extended technology over COM called Distributed COM. Parallel paths in graphs may imply possible independent parallel activation of components making them suitable for parallel computation either on multi-processor systems or in computational clusters. For all these features it may be concluded that component-based framework enhanced by a graphical interface based on flow diagrams can meet the requirements of many users as a suitable design and simulation environment.

References

[1] Erich Gamma, Richard Helm, Ralph Johnson and John Vlissides, Design Patterns: Elements of Reusable Object-Oriented Software. Addison Wesley, 1995.

[2] Dr. GUI and COM Automation, on-line MSDN Technical Article, Microsoft, 1999, http://msdn.microsoft.com/library/default.asp?url=/library/en-us/dnguion/html/drgui021099.asp

[3] Martijn Brinkers, Steven Verver, Cees van Leeuwen: Component-based Software for Flexible and Reusable Neural Network Programming, In A.N. Brand, S. Doosje, & B.A. Wesselingh (Eds.), *Zevende Workshop Computers in de Psychologie. Extended abstracts,* Utrecht: UU, FSW., 1999.

KES 2002
E. Damiani et al. (Eds.)
IOS Press, 2002

Adaptation and synchronization as a way to control spatio-temporal pattern formation in the system of coupled logistic maps

Michał Żochowski, Rhonda Dzakpasu
Department of Physics and Biophysics Research Division
University of Michigan
Ann Arbor, MI 48109

We use coupled logistic maps to study collective dynamics of self-organizing systems. The coupled map systems were introduced by Kaneko [1], and have attracted rapidly growing attention in recent years [2], [3], [4], [5]. One of the major questions in such systems is how one can locally adjust the emergence of the synchronized states to control formation of large scale spatio-temporal patterns. Revealing the mechanisms that underlie such pattern formation may help in understanding the mechanisms underlying spatio–temporal synchronization that has been found in different brain modalities [6, 7]. It is theorized, with the experimental data emerging [8], that synchronous activity of distributed neural populations can facilitate binding of different features of processed information [9, 10]. Synchronized elements would form a coherent group that would stand out from the background of unrelated activity of other ensembles allowing for rapid feature recognition and binding.

We present a class of models that exemplifies how such dynamical control can be achieved when the signal arriving to the given element is integrated at two independent levels. The first level is a fast coupling of the iterates of the receiving unit to the external input or internal signal coming from other units. It is a fast mechanism because it produces instantaneous changes in the receiving unit based on the activity of the other elements in the system. The second integration takes place on a much slower time scale. This mechanism is designed to estimate temporal characteristics of the incoming signal and induce slow changes in the *macroscopic* dynamical properties of the receiving element. The interaction of these two mechanisms allows the system to synchronize on a controlled dynamical trajectory.

We present two examples that implement this scheme. The first model will demonstrate how full dynamical control of a target element can be achieved solely based on its input signal from the control element. The second model will show how this control scheme can be implemented to achieve functional self-organization of the system based on the temporal properties of the input signal. The dynamical control leads to selective synchronization of subsets of coupled elements, forming spatio-temporal representation of the input. Thus the system adapts its spatio-temporal properties depending on the temporal characteristics of the input.

1 Controlling the dynamics in a system of two coupled logistic maps

It has been established that two coupled, identical chaotic systems can synchronize [11]. It is however unclear how two systems could synchronize if the properties of their

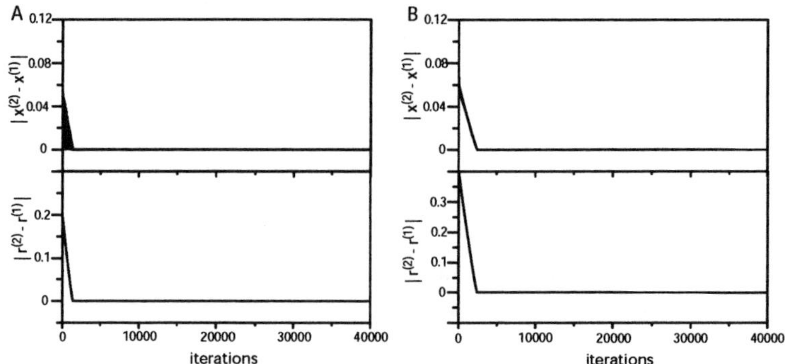

Figure 1: Synchronization of a target and control element. The difference between iterates as well as the difference between the control parameters of the two maps converges to zero. Fig 1A depicts the case where the control map is chaotic and the target map is initially periodic ($r^{(1)} = 3.6$, $r^{(2)} = 3.4$) . Fig 1B depicts the reverse situation ($r^{(1)} = 3.3$, $r^{(2)} = 3.7$).

dynamical trajectories are different. Here we describe a mechanism that allows for dynamical control and synchronization of two initially non-identical elements. In this example we have a system built of two coupled units: a control element and a target element. The target element will adjust its properties to synchronize with the control element.

The state controlling map $x^{(1)}$ perturbs the state of the target map $x^{(2)}$ if the two trajectories are non-identical:

$$x^{(1)}_{n+1} = f(x^{(1)}_n), \tag{1}$$

$$x^{(2)}_{n+1} = \frac{f(x^{(2)}_n) + \varepsilon x^{(1)}_{n+1}}{1 + \varepsilon}; \tag{2}$$

with $\varepsilon = 0.4$. For both $x^{(1)}$ and $x^{(2)}$, the logistic map was chosen:

$$f(x^{(m)}) = r^{(m)} x^{(m)} (1 - x^{(m)}); \tag{3}$$

$m = 1, 2$. The parameter $r^{(1)}$ in the control system was a constant arbitrary value. The parameter $r^{(2)}$ in the target map is changing with subsequent iterations of the map. The changes in $r^{(2)}$ are determined from the Liapunov exponents, which are approximated for both maps as a running average to form a dynamical variable:

$$\lambda^{(m)}_{n+1} = \frac{\lambda^{(m)}_n + \alpha \log \left| \frac{df}{dx}(x^{(m)}_{n+1}) \right|}{1 + \alpha}. \tag{4}$$

The Liapunov exponent $\lambda^{(2)}$ for the target system is calculated from the internal (unperturbed) dynamics of the uncoupled target map:

$$\tilde{x}^{(2)}_{n+1} = r^{(2)}_n x^{(2)}_n (1 - x^{(2)}_n) \tag{5}$$

The changes in $r^{(2)}$ are defined as follows:

$$r^{(2)}_{n+1} = r^{(2)}_n + \gamma A(\lambda^{(1)}_n, \lambda^{(2)}_n). \tag{6}$$

where

$$A(\lambda_n^{(1)}, \lambda_n^{(2)}) = (|\lambda_n^{(1)} - \lambda_n^{(2)}|)^{\frac{1}{4}} \mathrm{sgn}(\lambda_n^{(1)} - \lambda_n^{(2)}), \tag{7}$$

and $\gamma = 0.0002$. Thus, when the estimated Liapunov exponents of the target and control element are the same the control parameter of the target element becomes stable.

During the evolution of the system the target element will slowly adjust its control parameter to match that of the control system and the two will synchronize. This is shown in figure 1 in which the difference between corresponding iterates $(x_n^{(1)} - x_n^{(2)})$ and values of control parameters of the coupled and target maps are plotted. Figure 1A depicts the case when the dynamics of the control map is a fixed point and the target map starts in the chaotic regime. Fig 1B depicts opposite case; the dynamics of the control map is chaotic and the dynamics of the target map is initially periodic. In both cases the control parameter of the target map converges quickly to the parameter of the control map and the two units synchronize.

2 Building spatial representation of temporal information

We will now apply the scheme described in the previous section to construct a spatial representation of temporal input. This will be achieved by the formation of synchronized clusters of the elements of the network that have the same characteristics as the input signal. All other elements will remain unsynchronized. The formation of clusters is facilitated by linking the dynamical properties of the input signal with the coupling constant of the given element. If the dynamical properties of the given element of the network agree with those of the input, the coupling constant will be strengthened and the ability of the element to form the cluster will be enhanced. Conversely, if the input signal and the element of the network do not share similar characteristics the coupling constant will be decreased preventing this element from synchronizing with the others.

The input signal in this case is an additional logistic map that is unidirectionally coupled with every element of the network,

$$x_{n+1}^{(i)} = (1 - \varepsilon_n^{(i)}) f^{(i)}(x_n^{(i)}) + \varepsilon_n^{(i)} \sum_{j=1}^{N} \varepsilon_n^{(j)} f^{(j)}(x_n^{(j)}) + \beta \xi_n, \tag{8}$$

where $i = 1, N$, $\beta = 0.0001$ is the amplitude of white noise, ξ, added to the system. Thus, the coupling strength $\varepsilon_n^{(i)}$ of every element becomes variable i.e., the sensitivity of the element to the stimuli from within the network may change during the dynamics. The sensitivity will be modulated throughout the input stimuli. Only the elements that share common characteristics with the given input will be effectively coupled. The network is divided into three subsystems and elements in each subsystem have different control parameters $(\forall_{i \in I} r^{(i)} = 3.61, \forall_{i \in II} r^{(i)} = 3.65, \forall_{i \in III} r^{(i)} = 3.67)$.

The coupling strength ε_n^i will now depend on the difference between the Liapunov exponents of the input and the i-th element

$$\varepsilon_{n+1}^{(i)} = \frac{\varepsilon_n^{(i)} + \gamma F(|\lambda_n^{input} - \lambda_n^{(i)}|)}{1 + \gamma}. \tag{9}$$

where $F(x)$ is defined as:

$$F(x) = 1 - \frac{1}{1 + ax^4}, \tag{10}$$

$a = 10^6$, and $\gamma = 0.04$. The sensitivity of the given element to the signal of other elements in the system increases if the Liapunov exponent of the input matches that of the element.

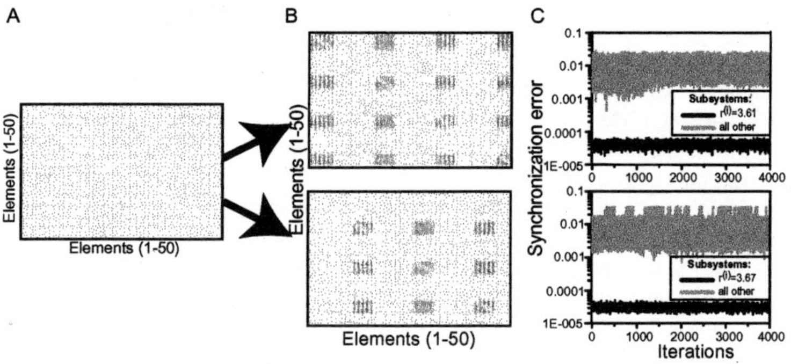

Figure 2: Network driven by the input. Depending on the temporal characteristics of the signal different clusters are synchronized. A) initial state; B) final state. C) Synchronization error is plotted for elements in every subsystem and for those belonging to different subsystems, during the simulation run.

When the network is stimulated, depending on the temporal characteristics of the input signal, two different subsystems are activated (Fig.2). Increased coupling strength between the elements responding to the same characteristics of the input allows for selective synchronization of elements within the system. In both cases the initial state of the elements is randomly generated (Fig. 2A). During the simulation the elements belonging to one of the subsystems synchronize whereas the other elements continue to be desynchronized (Fig. 2B). Therefore the network transforms information of some temporal characteristics of the signal into a spatio-temporal representation. Figure 2C plots average the synchronization errors for all elements belonging to the same subsystems, as well as across the subsystems. The quality of synchronization in the stimulated subsystem is at least three orders of magnitude better.

Finally we show that as the external signal changes, the network is able to switch the activation between the different subsystems (Fig. 3). The logistic map generating the input signal initially has the control parameter set to $r_d = 3.61$; during the simulation it is changed (at the dashed line) to $r_d = 3.67$. As the temporal characterisitics of the input change, the coupling constant of the coupled elements evolves. The coupling constant of the elements that matched the the input before become smaller and the elements desynchronize. At the same time the coupling constant of the elements that match the new temporal characterisitics of the input signal is increased and those elements form a new synchronized cluster (Fig 3). Introduction of white noise facilitates fast destruction of spurious coherent state.

In summary, we were able to show that integrating the signal on two different timescales can provide full control of the dynamics of the coupled system. This may lead to the formation of spatial representation of the incoming information based on synchronization of the units that are processing relevant information. The general scheme of modulating the properties of the elements based on the temporal characteristics of the signal could be relevant to information processing in the brain, since it is hypothesized that formation of coherent states of activity can lead to feature binding and thus formation of cognitive representations of arriving stimuli.

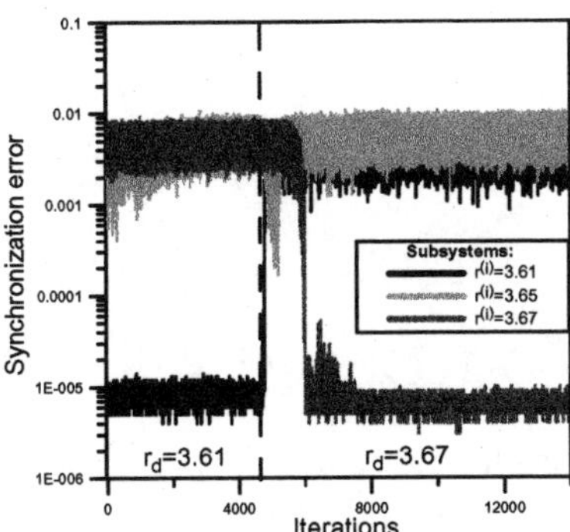

Figure 3: Network driven by the changing input. As the input signal changes during the simulation, different cluster synchronizes as other desynchronizes. The synchronization error between the elements within the same subsystems is plotted. Dashed line denotes the moment when the input signal is changed.

References

[1] K. Kaneko. Like structures and spatiotemporal intermittency of coupled logistic lattice."*toward a prelude of a field theory of chaos*". *Prog. Theor. Phys*, 72:480, 1984.

[2] F. Xie and H.A. Cerdeira. Coherent-ordered transition in chaotic globally coupled maps. *Phys. Rev. E*, 54:3235, 1996.

[3] F. Xie, G. Hu, and Z. Qu. On-off intermittency in a coupled-map lattice system. *Phys. Rev. E*, 52:R1265, 1995.

[4] Y. Nagai, X. Hua, and Y. Lai. Controlling on-off intermittent dynamics. *Phys. Rev. E*, 54:1190, 1996.

[5] V.N. Belykh and E. Mosekilde. One–dimensional map lattices: Synchronization, bifurcations and chaotic structures. *Phys. Rev. E*, 54:3196, 1996.

[6] R. Eckhorn, R. Bauer, W. Jordan, M. Brosh, W. Kruse, M. Munk, and H.J. Reitboeck. Coherent oscillations: A mechanism of feature linking in the visual cortex? *Biol. Cybern.*, 60:121–130, 1988.

[7] Y. Lam, L.B. Cohen, M. Wachowiak, and M. Zochowski. Odors elicit three different oscillations in the turtle olfactory bulb. *J. Neurosci.*, 20:749–762, 2000.

[8] M. Stopfer, S. Bhagavan, B.H. Smith, and G. Laurent. Impaired odour discrimination on desynchronization of odour-encoding neural assemblies. *Nature*, 390:70–74, 1997.

[9] C. von der Malsburg. Nervous structures with dynamical links. *Ber. Bunsenges. Phys. Chem.*, 89:703–710, 1985.

[10] W. Singer. Neuronal synchrony: A versatile code for the definition of relations? *Neuron*, 24:49–65, 1999.

[11] L. M. Pecora, T. L. Carroll, G. A. Johnson, D. J. Mar, and J. F. Heagy. Fundamentals of synchronization in chaotic systems, concepts, and applications. *Chaos*, 7(4):520–543, 1997.

KES 2002
E. Damiani et al. (Eds.)
IOS Press, 2002

Probabilistic-Weighted k-Nearest-Neighbour Algorithm: a New Approach for Gene Expression Based Classification

Basilio Sierra and Elena Lazkano
Department of Computer Science and Artificial Intelligence
University of the Basque Country
e-mail: ccpsiarb@si.ehu.es
http://www.sc.ehu.es/ccwrobot

Abstract. The k-Nearest-Neighbour (KNN) decision rule has often been used in Pattern Recognition problems. One of the difficulties that arises when utilizing this technique is that each of the labeled samples is given equal importance in deciding the class memberships of the pattern to be classified, regardless of the "typicalness" of each of the neighbours.

We report on the application of a new hybrid version named Probabilistic Weighted k-NN algorithm (PW-K-NN) which classifies the new cases based on their distance and on the probability each case has to be of its own class. These probabilities are given by a Bayesian Network, which structure has been previously learned using a genetic algorithm. The net only uses the genes choosen by a forward selection technique together with the Naive Bayes.

Experiments have been carried out by using Bioinformatics community well known databases. We present the results obtained with two databases by using all the genes in the classification process and by using those selected for the probability measure.

Keywords: Supervised Classification, Machine Learning, Nearest Neighbour, Bayesian Network, Feature Subset Selection

1 Introduction

A right and accurate cancer classification allows to the medical staff the application of specific therapies and treatments, related with the specific cancer type [10]. Development of high throughput data acquisition technologies in biological sciences, and specifically in genome sequencing, together with advances in digital storage and computing, have begun to transform biology, from a data poor science to a data rich science. In order to manage and treat with all this new biological data, the Bioinformatics discipline powerfully emerges.

These advances in the last decade in genome sequencing have lead to the spectacular development of a new technology, named DNA micro-array, which can be included into the Bioinformatics discipline. DNA micro-array allows the monitoring and measurement of the expression levels of thousands of genes simultaneously in an organism.

A systematic and computational analysis of these micro-array datasets is an interesting way to study and understand many aspects of the underlying biological processes.

Our work is focused on class prediction for cancer classification. Starting from a set of samples for which the classification (the class type) is known, we tackle the construction of the predictive model which discriminates among the different cancer types of the problem. Thus, our goal is the maximization of the predictive accuracy of the classifiers.

The nearest neighbour classification method assigns to an unclassified point the class of the nearest of a set of previously classified points. This rule is independent of the underlying joint distribution of the sample points and their classifications. An extension to this approach is the k-Nearest Neoghbours (k-NN) method, in which the classification is made taken into account the k nearest points and classifying the unclassified point by a voting criteria from this k points.

We present a new voting method which takes into account the fact that not all the cases of the database are typical in the class they belong to (i.e., they could have some degree of exceptionality). This new method gives to each point in the training database a measure of its typicality by using a Bayesian Network.

The rest of the paper is organized as follows: Section 2 reviews the previous work found in the literature. Section 3 presents the new PW-K-NN algorithm; in Section 4 the experimental methodology is explained, while in Section 5 experimental results are presented and compared with those obtained using only the k-NN algorithm. Finally, in Section 6 some conclusions are given and the future work is pointed out.

2　Related work in micro-array datasets for class prediction

Classic statistical classification techniques (Discriminant Analysis, Gaussian and Logistic Classifiers,...) [16, 21], Support Vector Machines [3, 8], Neural Networks [11, 17] and K-NN [1, 10, 15, 21] are the most broadly used class prediction procedures in micro-array domains. As far as we know, only two works use decision trees [2, 9] and only our previous study [4] employs the NB classifier. In our knowledge, the CN2 algorithm [5] or classifiers of the same IF-THEN rules family are not still used in micro-array domains.

From the point of view of the accuracy estimation technique employed, the vast majority of the works use a hold-out procedure, using a portion of the samples to train a predictive model and using the rest of the instances as a test set to estimate the goodness of the classifier. However, other works use more sophisticated validation techniques, such as k-fold cross-validation [9]. Leaving one out cross validation (LOOCV) [14], a special case of k-fold cross-validation, can also be found in the literature [4, 15]. Although we consider that due to its unbiased nature [14], LOOCV is the most suited estimation procedure for micro-array datasets, the employment of the hold-out procedure can be justified in many works because a portion of the samples arise from a specific medical-center or study and other samples arise from a different source. In this way, it is common to use the samples of the first medical-center to train the classifier and the samples of the second source to measure its accuracy.

3   The Probabilistic-Weighted k Nearest Neighbour Method

In this section the new proposed paradigm is presented as an extension of the well known k-NN algorithm. We use Bayesian Networks as classifiers, as they are the model used to obtain the probabilities to weight the votes.

3.1   The k-NN Classification Method

The Nearest Neighbor (NN) classification decision method gives to x the category θ'_n of its nearest neighbor x'_n. In case of tie for the nearest neighbor, the decision rule has to be modified in order to break it. A mistake is made if $\theta'_n \neq \theta$. An immediate extension to this decision rule is the so called k-NN approach [6], which assigns to the candidate x the class which is most frequently represented in the k nearest neighbors to x.

Much research has been devoted to the k-NN rule [7]. We could give different weights to the variables in the distance computation [20], or different weight to each neighbour in the voting process. The method presented in this paper carries out the weighting of each neighbour taking into account its typicality.

We propose a modification of the k-NN algorithm in order to obtain more advantage on the information about the probability that a case x has to belong to its real class (i.e., a measure of the exceptionality of the case is given).

The new proposed method, PW-k-NN is shown in its algorithmic form in Figure 3.1.

```
begin PW-k-NN
    As input we have the samples file TR, containing n cases (X_i, θ_i), i = 1, ..., n,
    the value of k and a new case (Y, θ) to be classified
    FOR each case in TR DO
        BEGIN
        Be (X_i, θ_i) the case
        Estimate W_i = P(θ_i/X_i) using the BN in TR − (X_i, θ_i)
        Add this weight to the case in TR, (x_i, θ_i, W_i)
        END
    Search the k Nearest Neighbours of (Y, θ) in TR
    Reset the weights of all existing classes WC_i = 0
    FOR each of the k-NN DO
        BEGIN
        Be (X_i, θ_i, W_i) the case
        Actualize the weight of the class θ_i as follows:
                WC_i ← WC_i + W_i
        END
    Output the class θ_i which greatest weight WC_i
end PW-k-NN
```

Figure 1: The pseudo-code of the Probabilistic-Weighted k-Nearest Neighbour Algorithm.

The algorithm works as follows: given a set of n correctly classified cases,

$$(x_1, \theta_1), (x_2, \theta_2), \cdots, (x_n)$$

in a classification problem having M different classes, being $1, 2, \cdots, M$ the coded class number, and given a new case to be classified (x, θ), in which the class θ is unknown and a number k, the following classification process is done:

1. For each of the cases, compute the probability of belonging to its real class.

$$\forall i \ (x_i, \theta_i) : W_i = P(\theta_i/x_i) \Rightarrow (x_i, \theta_i, W_i)$$

2. Then, given the new case to be classified, search for its k nearest neighbours,

$$(x_{N_1}, \theta_{N_1}, W_{N_1}), \cdots , (x_{N_k}, \theta_{N_k}, W_{N_k})$$

and compute the vote of each class by using it's weights.

3. Give to the new case the class θ with the highest weight. An example of it is showed in Figure 2.

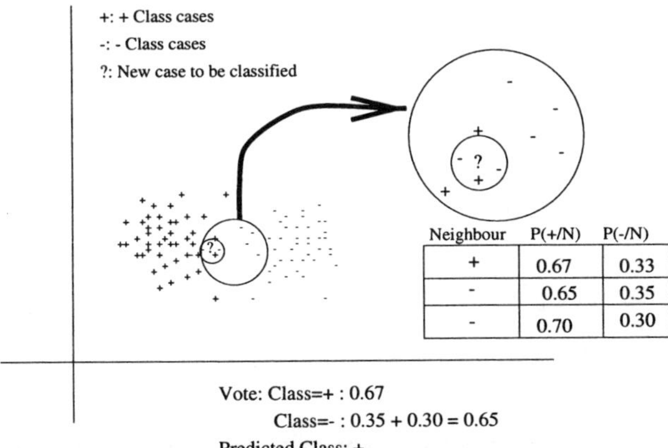

Neighbour	P(+/N)	P(-/N)
+	0.67	0.33
-	0.65	0.35
-	0.70	0.30

Figure 2: Probabilistic-Weighted 3-Nearest Neighbour Decision Rule

As in the k-NN classification technique, a tie-break method must be implemented. In this first approach to the method, we break ties by using the prior probabilities of the classes, that is, selecting the most probably class in case of tie.

4 Experimental methodology

Micro-array datasets have a very large number of predictive genes (usually more than $1,000$) and a small number of samples (usually less than 100). Due to the low number of samples of the datasets, the LOOCV procedure, is used to perform the accuracy estimation. In the leave-one-out technique, the supervised algorithm is run n times, where n is the number of examples of the dataset. Each time $n - 1$ examples are used for training and the remaining example is used for testing, where each example is used only once for testing. The LOOCV estimate of accuracy is the overall number of correct classifications, divided by n, the number of examples in the dataset. Although we consider LOOCV as the most reliable estimator, its computational cost can only be assumed, as in our case, when few examples are presented.

We need to know how to measure the "tipicallity" of each of the k Nearest Neighbours in order to weight the votes. We use a Bayesian Network (BN) [18] as classifiers, using the Genes selected by a Forward Feature Subset Selection technique [12] by obtaining the BN structure and the probability each case has to be part of each of the classes.

The experimental methodology can be divided in the following actions: Feature Subset Selection, Bayesian Network learning for the selected features (Genes), and K-NN classification based on the probabilities given by the learned BN structure.

4.1 Gene selection process: Feature Subset Selection

We use Sequential Forward Selection (SFS) [13], a classic and well known hill-climbing, deterministic search algorithm which starts from an empty subset of genes. It sequentially selects genes until no improvement is achieved in the evaluation function value. As the totality of previous micro-array studies note that very few genes are needed to discriminate between different cell classes, we consider that SFS could be an appropriate search engine because it performs the major part of its search near the empty gene subset.

4.2 Bayesian Network Structure Learning

Once the variables are selected, we learn the structure of a Bayesian network (BN) classifier by using Genetic Algorithms [19], obtaining the structure which maximizes the well classified percentage of the Markov Blanquet of the variable to be classified. This learning technique has been presented in another papers and thereby, is not detailed in deep in this one.

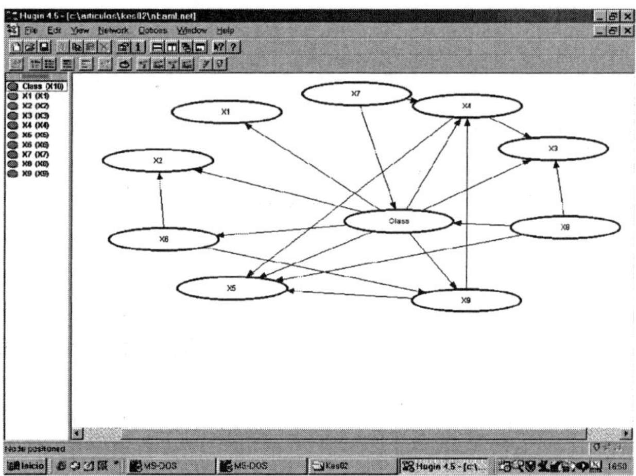

Figure 3: Markov Blanquet of the variable to be classified learned by a genetic algorithm

Figure 3 shows the BN structure used for the Leukemia database.

4.3 Classification by the K-NN new algorithm

The BN structure gives us the probability each case has to be part of its own class, its "tipicallity". We also take into account the distance of the case in order to weight its vote. In this manner, we try to avoid the negative effect of the outlier cases.

5   Experimental Results

We test the classification accuracy of our the k-NN algorithm and the same algorithm with the probability weighting PW-k-NN in the following two well known micro-array class prediction datasets:

- the Colon dataset is presented by Ben-Dor et al. (2000) [3]. This dataset is composed by 62 samples of colon ephitelial cells. Each sample is characterized by 2,000 genes. The samples were collected from colon-cancer patients. The 'tumor' (22 samples) biopsies were collected from tumors, and the 'normal' (40 samples) biopsies were collected from healthy parts of the colons of the same patient. The task is to predict the status of biopsy samples;

- Leukemia dataset is presented by Golub et al. (1999) [10]. It contains 72 instances of leukemia patients involving 7,129 genes. The class to be predicted is the specific type of acute leukemia of the patient: acute myeloid leukemia-AML (25 patients) or acute lymphoblastic leukemia-ALL (47 patients);

We use the discretized versions of this two databases, as the goal of the experimental done is to compare the new presented technique with the standard algorithm.

Table 1 shows the LOOCV accuracy obtained by the standard k-NN algorithm in the discretized micro-array datasets.

Table 1: LOOCV accuracy (well-classified cases) obtained by the standard k-NN algorithm for different k values

K	1	2	3	4	5	6	7	8	9	10
Colon	46	46	46	44	46	45	46	46	46	46
Leukemia	54	54	54	55	55	55	55	55	54	54

Table 2 shows the results obtained by the new paradigm. As it can be seen, the best results outperforms the accuracy obtained by the standard algorithm for the Colon database, and the result is the same for the other database used. The main goal of this experiment is to show that the new proposed algorithm can compite with the standard one given a classification problem.

Table 2: LOOCV accuracy (well-classified cases) obtained by the new PW-k-NN algorithm for different k values

K	1	2	3	4	5	6	7	8	9	10
Colon	46	50	47	48	49	49	48	49	48	48
Leukemia	54	52	53	53	54	55	53	51	52	51

6   Conclusions and Future Work

We have presented a new paradigm for supervised classification problems that belongs to the family of the distance based classification methods. We have applied this new paradigm to two well-known gene expression databases, obtaining results that indicate better performance of the new paradigm.

In this study, we have not attempted to discuss the biological significance of the genes used in the classification problem, but we think that this important point should be taken into account for future works.

We also plan to do a more complete comparison between the two paradigms used in this paper in order to present it to the machine learning community as a new classification model.

Acknowledgments

This work is supported by the University of the Basque Country (Euskal Herriko Unibertsitatea).

References

[1] V. Aris and M. Recce. A Ranking Method to Improve Detection of Disease Using Selectively Expressed Genes in Microarray Data. In Proceedings of the First Conference on Critical Assessment of Microarray Data Analysis, CAMDA2000, 2000.

[2] M. Beibel. Selection of Informative Genes in Gene Expression Based Diagnosis: A Nonparametric Approach. In R. Brause and E. Hanisch, editors, Lecture Notes in Computer Sciences. Proceedings of the First International Symposium in Medical Data Analysis, ISMDA2000, volume 1933, pages 300–307. Springer-Verlag, 2000.

[3] A. Ben-Dor, L. Bruhn, N. Friedman, I. Nachman, M. Schummer, and Z. Yakhini. Tissue Classification with Gene Expression Profiles. Journal of Computational Biology, 7(3-4):559–584, 2000.

[4] R. Blanco, P. Larrañaga, I. Inza, and B. Sierra. Selection of highly accurate genes for cancer classification by Estimation of Distribution Algorithms. In Workshop of Bayesian Models in Medicine. AIME 2001, pages 29–34, 2001.

[5] P. Clark and T. Nibblet. The CN2 induction algorithm. Machine Learning, 3(4):261–283, 1989.

[6] T. M. Cover and P. E. Hart. Nearest neighbor pattern classification. IEEE Trans. IT, 13(1):21–27, 1967.

[7] B.V. Dasarathy. Nearest Neighbor (NN) norms: NN pattern classification techniques. IEEE Computer Society Press, 1991.

[8] C.H.Q. Ding. Tumor Tissue Classification Using Support vector Machines and K-Nearest Neighbor Methods. In Proceedings of the First Conference on Critical Assessment of Microarray Data Analysis, CAMDA2000, 2000.

[9] W. Dubitzky, M. Granzow, D. Berrar, S. Bulashevska, C. Conrad, D. Gerlich, and R. Eils. Symbolic and Subsymbolic Machine Learning Approaches for Molecular Classification of Cancer and Ranking of Genes. In Proceedings of the First Conference on Critical Assessment of Microarray Data Analysis, CAMDA2000, 2000.

[10] T.R. Golub, D.K. Slonim, P. Tamayo, C. Huard, M. Gaasenbeek, J.P. Mesirov, H. Coller, M.L. Loh, J.R. Downing, M.A. Caliguri, C.D. Bloomfield, and E.S. Lander. Molecular Classification of Cancer: Class Discovery and Class Prediction by Gene Expression Monitoring. Science, 286:531–537, 1999.

[11] K.B. Hwang, D.Y. Cho, S.W. Wook Park, S.D. Kim, and B.Y. Zhang. Applying Machine Learning Techniques to Analysis of Gene Expression Data: Cancer Diagnosis. In Proceedings of the First Conference on Critical Assessment of Microarray Data Analysis, CAMDA2000, 2000.

[12] I. Inza, B. Sierra, R. Blanco, and P. Larra naga. Gene selection by sequential search wrapper approaches in microarray cancer class prediction. Journal of Intelligent and Fuzzy Systems, 2002. accepted.

[13] J. Kittler. Feature set search algorithms. In C.H. Chen, editor, Pattern Recognition and Signal Processing, pages 41–60. Sithoff and Noordhoff, 1978.

[14] R. Kohavi. A study of cross-validation and bootstrap for accuracy estimation and model selection. In N. Lavrac and S. Wrobel, editors, Proceedings of the International Joint Conference on Artificial Intelligence, 1995.

[15] L. Li, L.G. Pedersen, T.A. Darden, and C. Weinberg. Computational Analysis of Leukemia Microarray Expression Data Using the GA/KNN Method. In Proceedings of the First Conference on Critical Assessment of Microarray Data Analysis, CAMDA2000, 2000.

[16] W. Li and Y. Yang. How many genes are needed for a discriminant microarray data analysis? In Proceedings of the First Conference on Critical Assessment of Microarray Data Analysis, CAMDA2000, 2000.

[17] A. Mateos, J. Herrero, J. Tamames, and J. Dopazo. Supervised and hierarchical unsupervised neural networks for clustering both gene expression profiles and genes. In Proceedings of the Second Conference on Critical Assessment of Microarray Data Analysis, CAMDA2001, 2001.

[18] J. Pearl. Probabilistic Reasoning in Intelligent Systems. Morgan Kaufmann, Palo Alto, CA, 1988.

[19] B. Sierra and P. Larrañaga. Predicting survival in malignant skin melanoma using bayesian networks automatically induced by genetic algorithms. an empirical comparision between different approachess. Artificial Intelligence in Medicine, 14:215–230, 1998.

[20] B. Sierra, N. Serrano, P. Larrañaga, E.J. Plasencia, I. Inza, J.J. Jiménez, J.M. De la Rosa, and M.L. Mora. Machine learning inspired approaches to combine standard medical measures at an intensive care unit. In Proceedings of the AIMDM99 conference. Lecture Notes on Arrtificial Intelligence, volume 1620, pages 366–371, 1999.

[21] E.P. Xing, M.I. Jordan, and R.M. Karp. Feature Selection for High-Dimensional Genomic Microarray Data. In Proceedings of the Eighteenth International Conference in Machine Learning, ICML2001, pages 601–608, 2001.

KES 2002
E. Damiani et al. (Eds.)
IOS Press, 2002

Description of an Ontology for Supporting BIKMAS, a Biomedical Informatics Knowledge Management System

Victoria López-Alonso, Lucia Moreno, Guillermo López-Campos and Fernando Martín-Sanchez[1]

[1]*Department of Bioinformatics. Health Informatics Coordination Unit.
Institute of Health "Carlos III". Ministry of Health and Consumer Affairs
Ctra. Majadahonda a Pozuelo Km 2. 28220 Majadahonda, Madrid. Spain
{victorialopez, glopez, fmartin}@isciii.es*

Abstract. BIKMAS is a knowledge management system to capture, store, share and transfer scientific knowledge in the Biomedical Informatics field. We describe the role of an ontology as a tool for integration of vocabularies in BIKMAS. The source and type of the scientific information, the relevant knowledge themes and the distribution of the information in the different operational tasks of the Biomedical Unit are described in BIKMAS ontology.

1. Introduction

The massive amount of scientific information that arrives to a Biomedical Informatics Unit grows exponentially from year to year. Sometimes this situation makes the extraction of relevant information difficult. To facilitate automatic distribution of significant information we have developed BIKMAS as an active Biomedical Informatics knowledge management system [1].

BIKMAS is capable of retrieving scientific information from different sources (e.g. books, journals, newspapers, reports, emails, courses) and filtering it following several established criteria according to the priorities of the research unit (such as date of publishing, origin of the information and theme). After filtering the relevant information BIKMAS distributes such information for publishing in different places (e.g. web pages or the mailing list of the research unit). Relevant information can be also stored in different working files depending on the tasks to be carried out by the research unit (e.g. communications, papers, seminars).

Some of the characteristics of BIKMAS are: (1) information is obtained from diverse sources and is integrated into a single system, (2) terms that identify the concepts that are stored must be used consistently throughout the software modules and by different users and (3) knowledge is constantly changing and must be consistent with the external uses of the system. With all this factors in mind the main focus of this paper is centered in the management of names, meanings and organization of concepts in BIKMAS.

Two types of knowledge models (or domain models) could be used to identify concepts in a knowledge management system: terminologies and ontologies [2]. The use of terminologies is restricted to natural language definitions for specific domains while ontologies range from simple taxonomies based on inheritance to complex concepts systems including common sense knowledge.

Ontology is an old philosophical term borrowed by computer science that could be defined as the capturing of knowledge of a domain [3]. The use of ontologies restricts the possible interpretation of terms by using a controlled vocabulary for annotation and developing the structure that can be exploited for the knowledge processing. It might be also important to make clear the difference between an application ontology and a knowledge base. Ontology is a "particular knowledge base, describing facts assumed to be always true by a community of users, in virtue of the agreed in meaning of the vocabulary used" [4].

Since the beginning of the nineties ontologies have become a popular research topic investigated by several Artificial Intelligence research communities, including knowledge engineering, natural-language processing and knowledge representation. Early work in computational ontologies includes the CYC project [5] and the Knowledge Sharing Effort project (KSE) [6] where ontologies are used to share knowledge bases between various knowledge-based systems. Many of the current efforts in medical ontologies are directed towards generation of controlled terminologies, or reference ontologies [7]. These vocabularies taxonomically organize terms in certain areas, such as International Classification of Diseases (ICD-9-CM), Systematized Nomenclature of Human and Veterinary Medicine (SNOMED), Medical Subject Heading (MeSH), Read Codes, Current Procedural Terminology (CPT), Unified Medical Language System (UMLS), etc.

The use of ontologies has also become popular in Bioinformatics for annotation, querying, database schema, and community reference [8]. The Gene Ontology (GO) [8] is used to annotate gene products as to their molecular function, biological process and cellular location. The TAMBIS (Transparent Access to Multiple Bioinformation Sources) system uses an ontology to enable biologists to ask questions over multiple external databases using a common query interface [9]. An ontology of array annotation is also available from the ontology-working group (OWG) for the MGED project [10].

Ontologies can be used to communicate systems, people and organizations, interoperate between systems, and support the design and development of knowledge software systems. We report below how we use an ontology in BIKMAS to create a shared domain avoiding subjective or biased classifications.

2. Developing an ontology for BIKMAS

BIKMAS design is as generic as possible since technologies and configurations are highly susceptible to change, being concerned with the flow of scientific information for optimum accessibility and usability. The system collects information from a number of scientific sources and classifies relevant knowledge for a Biomedical Informatics research unit.

The problem of developing ontologies has been well studied and a number of methodologies have been proposed [11]. To develop an ontology for BIKMAS we thought about the purpose of the system in terms of its expected competency. For example the types of queries that the system will be able to solve. BIKMAS accepts queries about certain information resources (e.g. author, title, origin), themes of interest and existing information relevant for research or working tasks. We envision that BIKMAS will be able to process statistics about the types and characteristics of the information that is used. If the terms used in the system are formally defined in an ontology, the system could take decisions about which resources are relevant to solve a particular query.

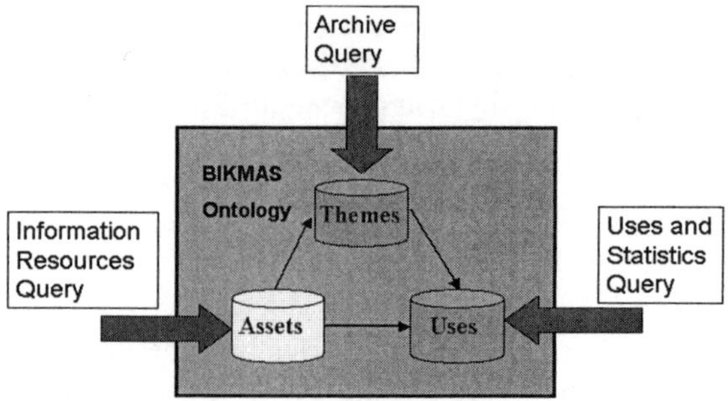

Figure 1. Structure of the three frames in the ontology domain developed for BIKMAS.

We have defined three main BIKMAS ontology frames as shown in Figure 1 ("information asset" frame, "Biomedical Informatics themes" frame and "uses" frame). The "information asset" frame describes the terminology used to identify and answer queries about specific information resources. The "Biomedical Informatics themes" frame defines the terminology of the different research fields that are relevant to the Biomedical Informatics research Unit. The "uses frame" locates places where to classify and publish this significant information.

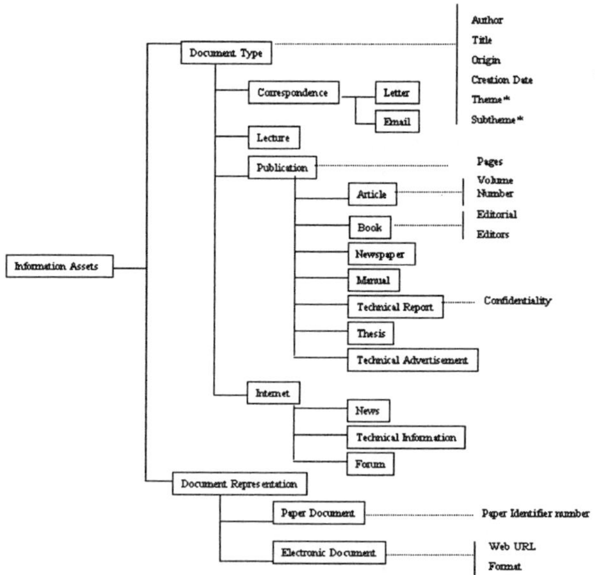

Figure 2. Defining classes and slots in the information asset frame. * Themes and Subthemes are described in the Biomedical Informatics themes frame.

There has been a proliferation in the amount of scientific information available, both from traditional printed sources and from electronic resources. The frame "information asset" establishes a fixed vocabulary describing the retrieved information and the relationships within it. This model is used to integrate the various information sources that are available providing the terminology for accessing them. To enter a document in the system the user has to consider what are the type and the representation of the document as showed in Figure 2.

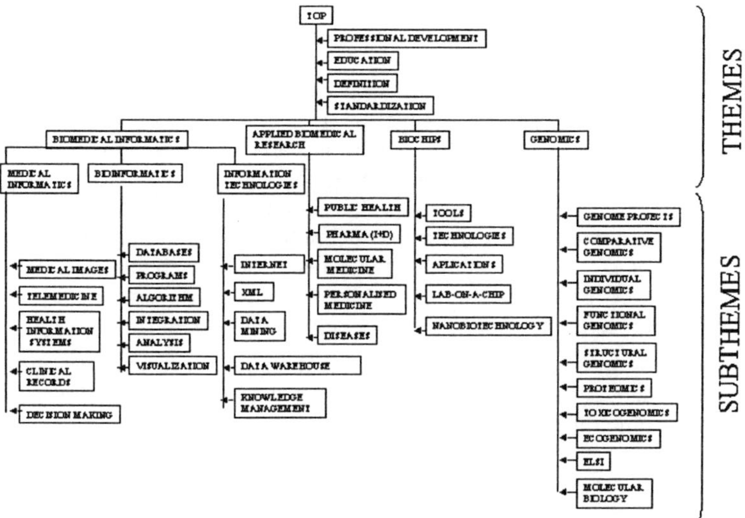

Figure 3. Biomedical Informatics themes frame of the BIKMAS ontology.

An important issue in BIKMAS is to define the scheme to classify the information. In BIKMAS a document may be related to others establishing an internal organization using a categorization for describing themes and subthemes as showed in Figure 3.

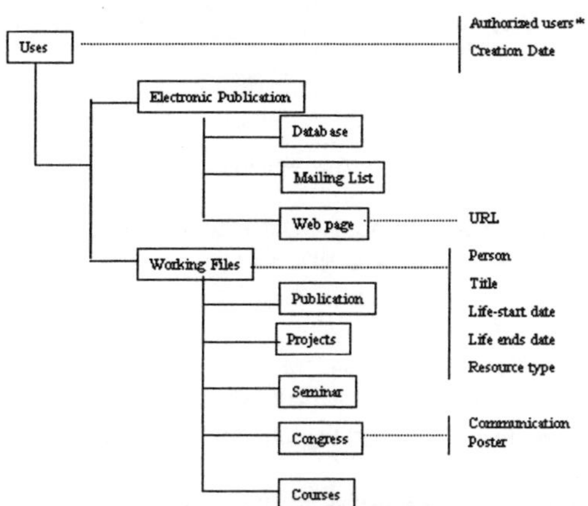

Figure 4. Uses frame of the BIKMAS ontology.* Authorized users should be described in the system

The "uses frame" is designed to structure the automatic distribution of the information depending on the work and research tasks of the users (Figure 4). Each scientific unit has different places and characteristics where to publish information. The system will distribute the information to be used in electronic publication or working files required by the scientific unit.

Once an ontology is developed capturing the basic concepts and relationships among them in a given application area, that ontology can guide the execution of a large numbers of general software components that contribute to a final system. The reuse of a domain ontology throughout BIKMAS has important advantages. The most apparent benefit of an ontology is to facilitate system maintenance, it results from having a single location of information that developers must update when the application areas evolves.

3. Discussion

BIKMAS ontology integrates models of different domains into a coherent support. This shared understanding provides several benefits as: (1) communication between users and research units, to unify different research fields. Usually the person who develops the ontology has a deeper knowledge or understanding of the annotated entity than the users querying the database. (2) The ontology could be used to unify BIKMAS with others software tools. (3) Ontologies also assist in the process of building and maintaining software systems because they can assist in the process of identifying requirements and defining a specification of the system. (4) A formal representation of the knowledge improves reliability by facilitating automatic checking of the system.

The use of ontologies in BIKMAS provides a mechanism to enable task automation because the management system is optimized by restricting contents to a set of pre-defined values according to some rules.

References

[1] V. López Alonso, L. Moreno López, G. López Campos, F. Martín Sánchez, "BIKMAS" Sistema de Gestión del Conocimiento para Unidades de Investigación Biomédica, Inforsalud, Madrid, 2002.

[2] N. Guarino and P Giaretta, Ontologies and Knowledge Bases: Towards a Terminological Clarification. In: N. Mars (ed.), Towards Very Large Knowledge Bases: Knowledge Building and Knowledge Sharing. IOS Press, Amsterdam, 1995, pp 25-32.

[3] M. Uschold, M. King, S. Moralee, and Y. Zorgios, The enterprise Ontology, The knowledge Engineering Review. Special issue on Putting Ontologies to use, 1998, 13 (1): 31-89.

[4] N. Guarino, Formal Ontology and Information Systems, Proceedings of FOIS '98, Trento, Italy (Amsterdam, IOS Press). 1998, pp. 3-15.

[5] J. H. Gennari and D.E. Olivier, A web-based Architecture for a Medical Vocabulary Server. 19th Annual Symposium on Computer Applications in Medical Care, New Orleans L.A., 1995.

[6] R. Neches, R.E. Fikes, T. Finin, T.R. Gruber, T. Senator and W.R. Swartout, Enabling Technology for Knowledge Sharing. *AI Magazine* **12** (1991) 36-56.

[7] D.E Oliver, Y. Shahar, E.H. Shortliffe and M.A. Musen, Representation of Change in Controlled Medical Terminologies, *Artificial Intelligence in Medicine* **15** (1999) 53-76.

[8] The Gen Ontology Consortium, Gene Ontology: tool for the unification of biology, *Nature Genetics* **25** (2000) 25-29.

[9] P.G. Baker, C.A. Goble, S. Bechhofer, N.W. Paton, R. Stevens and A. Bross, An ontology for Bioinformatics Applications, *Bioinformatics* **15** (1999) 510-520.

[10] http://www.cbil.upenn.edu/Ontology/index.html

[11] D.M. Jones, T.J.M. Bench-Capon and P.R.S. Visser, Methodologies for Ontology Development, Proceedings IT&KNOWs, Budapest, 1998.

KES 2002
E. Damiani et al. (Eds.)
IOS Press, 2002

Genetic Learning of fuzzy rules through Neuro-Fuzzy topology construction

A. Carrascal, D. Manrique, J. Ríos, C. Rossi
Artificial Intelligence Department
Facultad de Informática; Univ. Politécnica de Madrid
28660 Boadilla del Monte, Madrid, Spain

Abstract. This paper proposes a new approach for constructing fuzzy knowledge bases by using evolutionary methods. A genetic algorithm has been designed to automatically build neuro-fuzzy architectures based on a new indirect codification method. The neuro-fuzzy architecture obtained as a solution represents the best fuzzy knowledge base that solves a given problem.

1. Introduction

Fuzzy logic is a kind of symbolic knowledge representation that makes possible natural knowledge modeling. Fuzzy logic is used in domains such as control systems, diagnostic, data analysis, classification, inductive learning or decision support [1]. The bottle-neck in constructing this kind of system is the knowledge acquisition process. Besides being a time-consuming process that make expert systems very difficult to build, the quality of the acquired knowledge also depends on many aspects, such as the availability of the domain experts, their expertise, their ability and attitude to express their expertise, the relationship between knowledge engineers and domain experts, and so on. So, automatic systems for knowledge acquisition are needed [2].

Recently, genetic algorithms have been used to accomplish the automatic generation of fuzzy knowledge bases. The most common way to combine genetic algorithms and fuzzy logic consists on coding rules in bit strings, and applying genetic operators until the rules base that best solve the problem is found [3].

Several approaches have been proposed to codify fuzzy rules with bit strings. Pittsburgh based approaches consist on codify the entire knowledge base with just one bit string. Among them, the most used method is the Pittsburgh direct encoding method (PDEM), where each bit of the string represents the presence or not of the corresponding linguistic value of a rule [4].

Another way to combine fuzzy logic and genetic algorithms consists on using neuro-fuzzy networks representing the rules base [5]. This approach allow for the use of a genetic algorithm designed for finding good neural networks architectures. The most important system with this features that has been recently built is FALCON (Fuzzy Adaptive Learning Control Network) [6] This system has the drawback of using an encoding which is not coherent with the crossover operator it uses. This fact prevents a good balancing of the exploration and exploitation capabilities, necessary to avoid falling into local optima and to obtain a fast convergence to the solution.

This paper presents a new genetic system for constructing fuzzy knowledge bases called GLF (Genetic Learning of Fuzzy-rules). This system uses an extended version of neuro-fuzzy architectures for representing fuzzy knowledge bases. These architectures are built automatically by means of a genetic algorithm that uses a new coding method, which, together with the Hamming crossover operator, overcomes the above mentioned difficulties and obtains higher performances [7].

2. The Genetic Learning of Fuzzy-rules System

The GLF system receive in input two kind of information: the data related to the design of the neuro-fuzzy architecture and the data regarding the application environment.

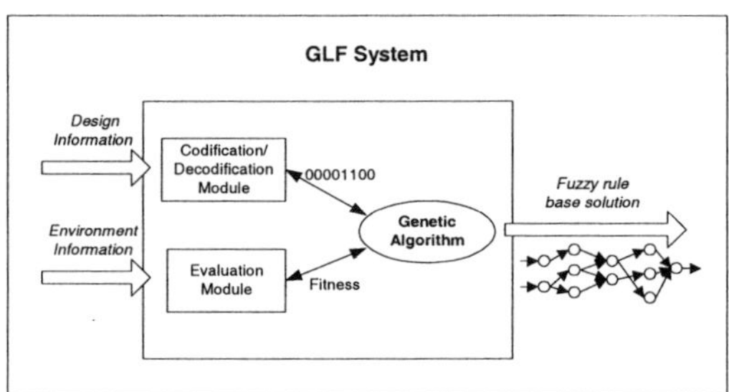

Figure 1: General Scheme of the GLF System

The design information are composed by the input and output variables set, with the corresponding fuzzy-sets, and a maximum number of neurons for the hidden layer. The environment information is the information needed for evaluating the goodness of the generated solutions. Starting with these inputs, the GLF system generates a neuro-fuzzy architecture representing the best knowledge base for the given problem.

As shown in figure 1, the GLS system is composed by a genetic algorithm, a coding/decoding module and an evaluation module. The genetic algorithm implements the operations of reproduction, crossover and mutation, the coding/decoding module implements a new coding method called GLF for transforming bit strings in neuro-fuzzy networks. When the genetic algorithm needs to evaluate an individual of the population, it decodes the bit string and uses the evaluation module to compute the individual's fitness.

2.1 Neuro-Fuzzy Architecture

The GLF codification method represents an extended version of neuro-fuzzy architectures that allows to compose fuzzy rules in disjunctive normal form (DNF) [8] as shown in the following expression: x_1 is (A_{11} OR ... OR A_{1N1}) AND ... AND x_n is (A_{n1} OR ... OR A_{nNn}) $\Rightarrow$ y is B. Where x_i and y are the linguistic variables of the antecedent and of the consequent respectively; with i=1..n. A_{ij} are the fuzzy sets (also called linguistic values) associated to the variable x_i, with i=1..n and j = 1..N_i ; where N_i is the number of linguistic values associated to variable x_i. B is the fuzzy set associated to the variable of the consequent.

As shown in figure 2, these neuro-fuzzy architectures are formed by five layers: input variable layer (1), input set term layer (2), rule layer (3), output set term layer (4) and output variable layer (5).

Neurons of layers two and four represents the membership functions of the linguistic variables of the input and output layers respectively. The linguistic values A_{ij} and B_k of layers two and three are numbered in ascending order. Neurons of layer three connect with logic implications, the logic expressions of the antecedent with the linguistic values of the consequent. A logic expression is a set of linguistic values connected with the logic operators AND and OR. Neurons of layer three perform a disjunction on the linguistic values associated to the same linguistic variable, and a conjunction when these values refer to different linguistic values. This is due to the little usefulness of the conjunction of linguistic values associated to the same linguistic variable. The number of neurons of layer three, H, represents the maximum number of logic expression which can be built. The number of neurons of layer four, M, represents the maximum number of linguistic values in the consequent. Since each rule is formed by a logic expression in the antecedent and a linguistic value for the consequent, the maximum number of rules which can be contained in the knowledge base represented by a given neuro-fuzzy architecture is H·M.

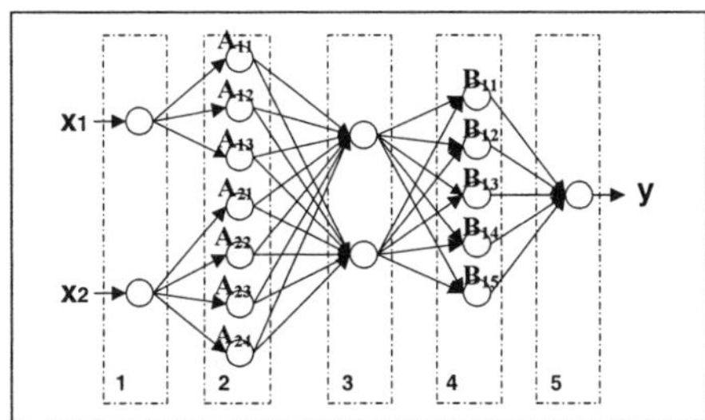

Figure 2: Neuro-Fuzzy Architecture Example

2.2 The Encoding Method

The GLF method encodes neuro-fuzzy architectures composed by N neurons on layer two, H neurons on layer thee and M neurons on layer four in bit strings. Let $C_{23}=\{c_{23ij}\}$, with i=1,..,N and j=1,..,H the set of the connections between layers two and three, and $C_{34}=\{c_{34jk}\}$, with k=1,..,M the set of the connections between layers three and four. A neuro-fuzzy architecture v is denoted with $v(N,H,M,C_{23},C_{34})$. It is not necessary to encode the connections between layers one-two and four-five, since they are determined with the sets C_{23} y C_{34}.

A basic neuro-fuzzy architecture is defined as a neuro-fuzzy architecture composed by exactly two connections c_{23ij} y c_{34jk}. A neuro-fuzzy architecture is valid when $\forall c_{23ij} \in C_{23}$, $\exists c_{34jk}$ and $\forall c_{34jk} \in C_{34}$, $\exists c_{23ij}$. Within the set of valid architectures V_{NHM}, the superimposition operator $\oplus$ is defined as:

$$\forall v_1(N,H,M,C_{23},C_{34}),\ v_2(N,H,M,C'_{23},C'_{34}) \in V_{NHM};\ v_1 \oplus v_2 = v(N,H,M,C_{23} \cup C'_{23}, C_{34} \cup C'_{34}) \in V_{NHM}$$

The cardinality of the set of valid architectures $\#V_{NHM}$ is:

$$\#V_{NHM} = \binom{M}{0} \cdot \left(2^{N \cdot H} - 1\right)^0 + \binom{M}{1} \cdot \left(2^{N \cdot H} - 1\right)^1 + ... + \binom{M}{M} \cdot \left(2^{N \cdot H} - 1\right)^0 = 2^{N \cdot H \cdot M}$$

Hence, we need strings of L=N·H·M bits in order to encode the whole set V_{NHM}. Similarly, the number of basic neuro-fuzzy architectures is N·H·M. The used encoding consist on assigning, in an arbitrary way, to each basic architecture v_i a bit string with only the ith bit set to 1, with i=1..L. Given $v = b_1 \oplus ... \oplus b_n$, with b_i (i=1..n) being a basic neuro-fuzzy architecture whose codification is c_i, the encoding c of v is given by $c = c_1 \vee ... \vee c_n$, where $\vee$ is the logical bit-wise operation OR. This encoding method has the property that two neuro-fuzzy architectures which differ by one bit (hamming distance 1) have very similar phenotypes, the only difference being one basic architecture, therefore their similarity is both morphological and functional. We propose to use this encoding with the Hamming crossover operator, which is based on the hypothesis that the hamming distance between two bit strings is a measure of the difference between the two phenotypes that they represent.

3. Experimental Results

We have compared the GLF system with the PDEM and FALCON systems, considering the percentage of success in convergence, precision of the solutions found and dimension of the knowledge bases obtained. The PDEM system have been chosen as a representative of the direct coding methods due to its popularity and because of its simple implementation. The FALCON system is the main representative of the indirect coding methods that use neuro-fuzzy architectures.

Two different tests have been accomplished: the problem of the isolated optimum and the building of a car navigation system. The problem of the isolated optimum consists on finding the simplest rule base that models a decision space which is constant except for a single combination of linguistic values. This space is depicted in figure 3. The rules generated are composed of two input variables and one output variable. The domain of each input variable is made of six fuzzy sets ($A_{11}...A_{16}$, $A_{21}...A_{26}$) and the output variable domain is made of three fuzzy sets ($B_{11}...B_{13}$). This problem is aimed to show the capability of exploring the search space of the algorithms. Figure 4 shows the convergence process (on average) for the three considered algorithms over 20000 executions run. It can be noticed how PDEM and FALCON prematurely converge to a local optimum, while the proposed approach is able to find the optimum solution after 17938 iterations on average.

The second problem consists on the generation of a fuzzy rule base whose aim is to drive a simulated vehicle on a track, shown in picture 5. The input variables used are four: car speed, and front, left and right distances of the car from the track borders. The output variables are two, acceleration and turning. The problem is solved when the car is able to perform a tour of the track without touching the track borders. Table 1 shows for each algorithm the percentage of success on this problem after 20000 executions run. The PDEM system performs worst, due to the fact that relies only on the mutation operator in order to modify the rules, and this makes the search process highly dependent on randomness. Falcon behaves better that PDEM, but it is still inferior to the proposed system. This is due to the fact that the crossover operator it uses (the classic single point crossover) do not take into account the properties of the encoding method used.

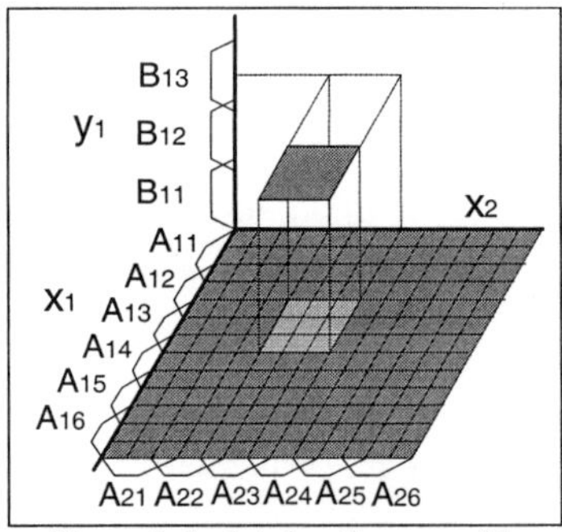

Figure 3: Isolated optimum problem

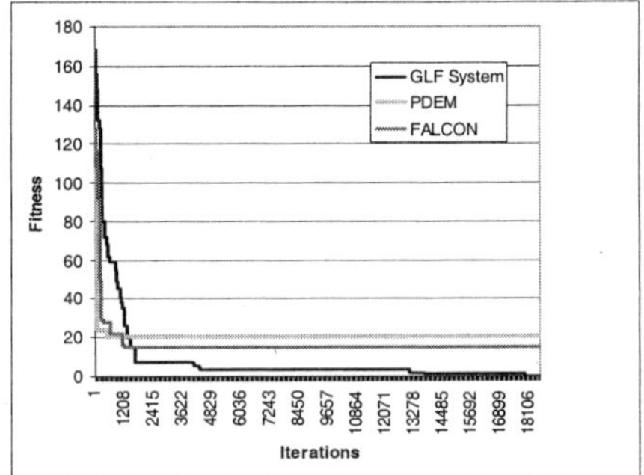

Figure 4: Convergence processes

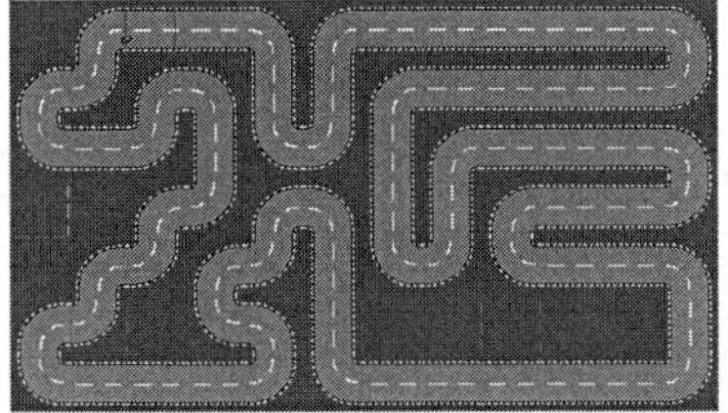

	1000 iterations	3000 iterations	6000 iterations
GLF	86%	91%	98%
PDEM	65%	73%	79%
FALCON	70%	86%	91%

Table 1: Success percentages for the nav. problem

Figure 5: Track employed for the navigation problem

4. Conclusions

An evolutive method for the generation of fuzzy logic intelligent systems has been proposed. This work shows a different interpretation of the neuro-fuzzy architectures that allows to represent rules in disjunctive normal form. Moreover, a new binary codification method for this type of architectures is described. The main advantage of this encoding method consists on the fact that similar bit strings encode neuro-fuzzy architectures which are functionally and morphologically similar. To take advantage of this feature, we adopted the Hamming crossover operator, which computes the degree of genetic diversity of two individuals based on the Hamming distance between two bit strings. This operator dynamically adapts the exploration / exploitation ratio during the search process of the genetic algorithm. The results show that this synergy between the crossover operator and the codification method assures a successful convergence process, obtaining the smallest set of fuzzy rules that solves a given problem.

References

[1] J. Harris. An Introduction to Fuzzy Logic Applications. Kluwer Academic Publishers, 2000.
[2] A.L. Castellanos, J. Castellanos, D. Manrique, A. Martínez. A New Approach for Extracting Rules from a Trained Neural Network. Progress in Artificial Intelligence. Lecture Notes in Artificial Intelligence. 1997, pp 297-302..
[3] O. Cordón, F. Herrera, F. Hoffmann, L. Magdalena. Genetic Fuzzy Systems. Evolutionary Tuning and learning of fuzzy knowledge bases. Ed. World Scientific. 2001.
[4] L. Magdalena and F. Monasterio. Auto-adaptative Fuzzy Controller by Evolutive knowledge Base. Proceedings of the V Spanish Congress on Fuzzy Logic Technologies, ESTYLF'95 Madrid, 1995, pp 296-302.
[5] D. Nauck. Neuro-Fuzzy Systems: Review and Prospects. Proc. Fifth European Congress on Intelligent Techniques and Soft Computing (EUFIT'97), Aachen, Germany, 1997, pp. 1044-1053.
[6] Chung, I.-F., C.J.Lin, and C.T. Lin. A GA-based fuzzy adaptive learning control network, *Fuzzy Sts and Systems*. **112**(1) (2000) 65-84.
[7] D. Barrios, A. Carrascal, D. Manrique, J. Ríos. ADANNET: Automatic Design of Artificial Neural Networks by Evolutionay Techniques. 21st Annual international Conference of the British Computer Society's Specialist Group on Knowledge Based Systems and Applied Artificial Intelligence (ES2001), Cambridge, UK , 2001, pp 67-80.
[8] A. González, R. Pérez and A. Verdegay. Learning the structure of a fuzzy rule: A Genetic Approach. Proc. First European Congresson Fuzzy and Intelligent Technologies (EUFIT'93), Aachen, Germany, 1993, pp. 814-819.

KES 2002
E. Damiani et al. (Eds.)
IOS Press, 2002

Analysing the Occurrence of Childhood Leukaemia using Seasonal Time Series

Maria FAZEKAS
Debrecen University, Debrecen, Hungary

Abstract. We examined the periodicity of the childhood leukaemia in Hungary using seasonal decomposition time series. Between 1988 and 2000 the number of annually diagnosed leukaemia was analysed. The time series of the number of patients between the age of 0 and 18 years and the data series divided at the median were analysed. From the time series the cyclic trends, the seasonally adjusted series, the moving averages and the data series of the random components. The cyclic trend of the dates of diagnosis revealed that a higher percent of the peaks (60%) fell within the winter months than in the other seasons. This proves the seasonal occurrence of the childhood leukaemia in Hungary. These data seem to highlight the role of the environmental effects (viral infections, epidemics, etc.) on the onset of the disease.

1. Introduction

The method of seasonal time series analysis can be used in various fields of the medicine. With such time series one can detect the periodic trend of the occurrence of a certain disease [1], [2], [3], [4], [5], [6], [7]. Among other diseases, the seasonal periodicity of the childhood lymphoid leukaemia was also analysed using statistical methods [8], [9]. The pathogenesis of the childhood lymphoid leukaemia is still uncertain, but certain environmental effects may provoke the manifestation of latent genes during viral infections, epidemics or pregnancy.

The date of birth and the date of the diagnosis of these patients were statistically analysed to determine the role, which the accumulating viral infections and other environmental effects may play during the conception and fatal period on the manifestation of the disease. Because the available data is rather limited and controversial, it seemed logical to make an in-depth analysis of the date of diagnosis of the lymphoid leukaemia in Hungarian children.

2. Method

2.1 Patients' Data

The databank of the Hungarian Paediatric Oncology Workgroup contains the data of all the patients with lymphoid leukaemia diagnosed between 1988 and 2000. In this time interval a total of 814 children were registered (of which 467 were boys). The patients were 0-18 years old, with a mean age of 6,4 years and a median of 5,4 years.

2.2 Seasonal Time Series

The time series usually consist of three components: the trend, the periodicity and the random effects. The trend is a long-term movement representing the main direction of changes. The periodicity marks cyclic fluctuations within the time series. The irregularity of the peaks and drops form a more-or-less constant pattern around the trend line. Due this stability the length and the amplitude of the seasonal changes is constant or changes very slowly. If the periodic fluctuation pattern is stable, it is called a constant periodic fluctuation. When the pattern changes slowly and regularly over the time, we speak of a changing periodicity.

The third component of the time series is the random error causing irregular, unpredictable, non-systematic fluctuations in the data independent from the trend line.

An important part of the time series analysis is the identification and isolation of the time series components. One might ask how these components come together and how can we define the connection between the time series and its components with a mathematical formula? The relationship between the components of a time series can be described either with an additive or a multiplicative model.

Let $y_{i,j}$ (i=1,..., n; j=1,...., m) mark the observed value of the time series. The index i stands for the time interval (i.e. a year), the j stands for a particular period in the time interval (i.e. a month of the year). By breaking down the time series based on the time intervals and the periods we get a matrix-like table. In the rows of the matrix are the values from the various periods of the same time interval; while in the columns are the values from the same periods over various time intervals.

$y_{1,1}; y_{1,2}; \ldots; y_{1,m};$

$y_{2,1}; y_{2,2}; \ldots; y_{2,m};$

$y_{3,1}; y_{3,2}; \ldots, y_{3,m};$

$\ldots$

$y_{n,1}; y_{n,2}; \ldots; y_{n,m}$

Let $d_{i,j}$ (i=1,2,...,n; j=1,2,...,m) mark the trend of the time series, $s_{i,j}$ (i=1,2,...,n; j=1,2,...,m), the periodic fluctuation and $\varepsilon_{i,j}$ (i=1,2,...,n; j=1,2,...,m), the random error. Using these denotations the additive seasonal model can be defined as $y_{i,j}=d_{i,j}+s_{i,j}+\varepsilon_{i,j}$, (i=1,2,...,n; j=1,2,...,m), the multiplicative model as $y_{i,j}=d_{i,j}*s_{i,j}*\varepsilon_{i,j}$; (i=1,2,...,n; j=1,2,...,m).

The trend of a time series can easily be computed with moving averages or analytic trend calculation. Moving averaging generates the trend as the dynamic average of the time series. Analytic trend calculation approximates the long-term movement in the time series with a simple curve (linear, parabolic or exponential curve) and estimates its parameters.

The indices of the periodic fluctuation are called seasonal differences (in the additive model) or seasonal ratios (in the multiplicative model). These indices represent the absolute difference from the average of the time interval using the additive model or the percentile difference using the multiplicative model. Seasonal adjustment is done by subtracting the j seasonal difference from the j data value of each i season (additive model) or by dividing the j data value of each i season by the j seasonal ratio (multiplicative model). The seasonally adjusted data reflect only the effect of the trend and the random error.

2.3 Analysing the Date of Diagnosis

The analysis of the periodicity of childhood leukaemia was performed on the basis of the date of the diagnosis (year + month) of the disease. We analysed three data series. The first data series contained the number of all the patients diagnosed monthly, the second contained the number of those patients younger than the value of the median, the third series contained the number those older than the value of the median.

The components of the time series can be identified and isolated using statistical program packages. The analysis of the seasonal periodicity of the childhood leukaemia was done with the Seasonal Decomposition module of the SPSS 9.0 statistical program package.

3. Results

Analysis of the seasonality of childhood lymphoid leukaemia was performed both on the total number of patients and on the data series divided at the median. The seasonal trend in the time series of all the patients revealed 10 peaks (peak=the value of the cyclic trend greater than 6), see **Figure 1**. Six of these peaks fell within the winter months (November-February), 2 in the autumn period (September-October), 1 in the summer months (June-August) and 1 in the spring months (March-May).

The seasonal trend of the younger age group showed 9 peaks (peak=cyclic trend greater than 3) in the winter, 1 in the spring, and 1 in the summer months. In the older age group 7 peaks (peak=cyclic trend greater than 3) was in the winter months, 1 in the autumn, 1 in the spring and 5 in the summer months.

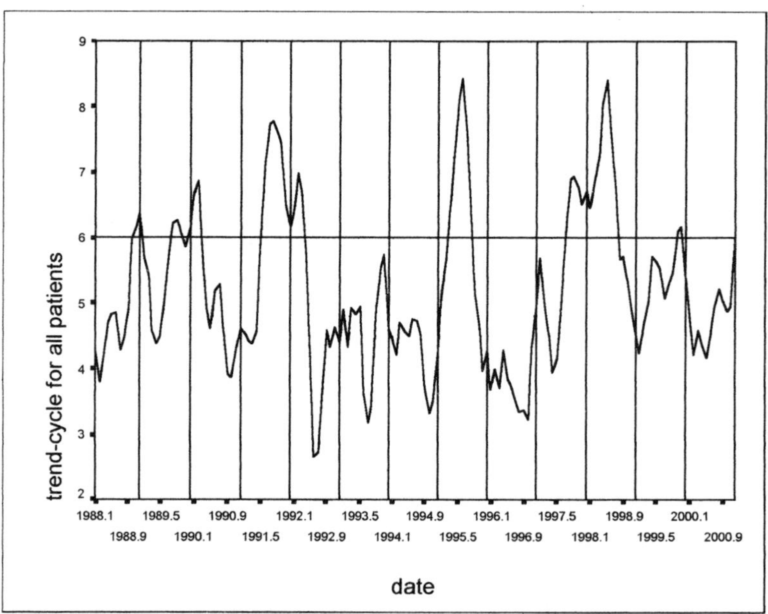

Figure 1 Seasonal trend of all the cases of lymphoid leukaemia diagnosed monthly in the observed period.

Figure 2 shows the number of patients, the moving average calculated from the number of patients with childhood leukaemia and the cyclic trend.

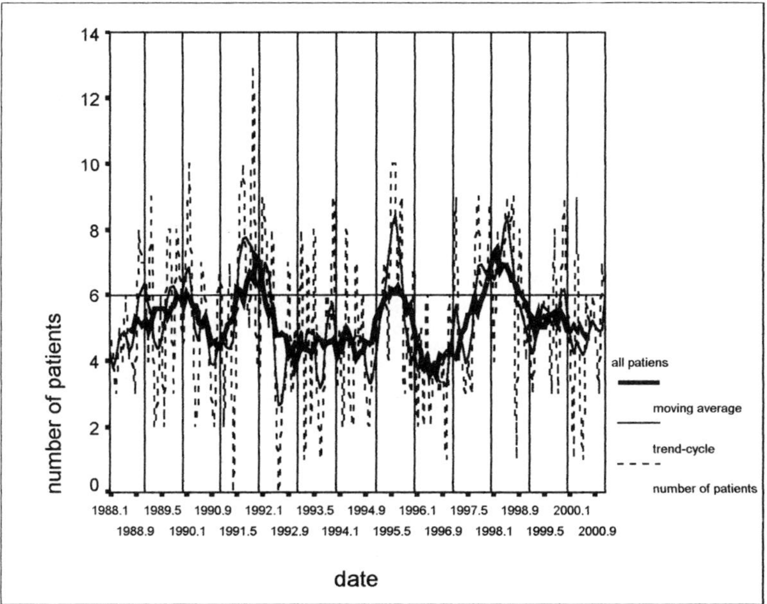

Figure 2 The number of patients, the moving average (window width=12 months) of the number have diagnosed cases of childhood leukaemia and the cyclic trend

4. Discussion

Analysis of the seasonality of childhood lymphoid leukaemia in Hungary was performed both on the total number of patients and on the data series divided at the median. This way the characteristics can be observed more easily.

A certain periodicity was found in the dates of the diagnosis in patients with leukaemia. Although there was some difference in the patterns of the cyclic trend peaks of the three time series, the majority of the peaks fell within the winter months in all three-time series. This was more significant in the group of all the patients (60% of the peaks in the winter months) and in the younger age group (64%). The results of the analyses proved the seasonal occurrence of the childhood lymphoid leukaemia. Some studies reported similar seasonality [10], while other studies denied any kind such periodicity [12].

Recent studies found no significant accumulation in the date of birth of leukaemia [11], however, they suggested that other factors, like the annually changing month of the outbreak of epidemics may mask some fine details in the long-term comparison studies [12].

Our results prove the seasonal occurrence of the childhood lymphoid leukaemia in Hungary. Due to the controversial nature of the available international data, further studies should be carried out.

Acknowledgements

Statistical data are courtesy of Prof. Pál Kajtár, head of the Hungarian Paediatric Oncology Workgroup.

References

[1] D. M. Fleming, K. W. Cross, R. Sunderland, A. M. Ross, Comparison of the Seasonal Pattern of Asthma Identified in General Practitioner Episodes, Hospital Admissions and Deaths. *Thorax* **8** (2000) 662-665.

[2] P. Saynajakangas, T. Keistinen, T. Tuuponen, Seasonal Fluctuations in Hospitalisation for Pneumonia in Finland. *Int J Circumpolar Health.* **60(1)** (2001) 34-40.

[3] A. Weber, M. Weber, P. Milligan, Modelling Epidemics Caused by Respiratory Syncytial Virus (RSV). *Math Biosci* **172(2)** 95-113.

[4] H. Tine and A. Strodl, Trends and Seasonal Variations in the Occurrence of Salmonella in Pigs, Pork and Humans in Denmark, 1995-2000. *BMTW Heft 9/10* **114** (2001) 346-349.

[5] A. L. Oberg, J. A. Ferguson, L. M. McIntrre , R. D. Horner, Incidence of Stroke and Season of the Year: Evidence of an Association. *Am J Epidemiol* **152(6)** (2000) 558-564.

[6] L. Cani, M. Rios, J. Sanchez, Meningococcal Disease in Spain: Seasonal Nature and Resent Changers. *Gac Sanit* **15(4)** (2001) 336-340.

[7] B. M. Cattanach, D. Papworth, G. Patrick, D. T. Goodhead, T. Hacker, L. Cobb, E. Wihitehill, Investigation of Lung Tumour Indication in C3H/HeH Mice, with and without Tumour Promotion with Urethane, following Paternal X-Irradiation. *Mutat Res* **403(1-2)** (1998) 1-12.

[8] P. Cohen, The Influence on Survival of Onset of Childhood Acute Leukaemia (ALL). *Chronobiol Int* **4(2)** (1987) 291-297.

[9] R. E. Harris, F. E. Harrel, K. D. Patil, R. Al-Rashid, The Seasonal Risk of Paediatric/Childhood Acute Lymphocyte Leukaemia in the United States. *J Chronic Dis* **40(10)** (1987) 915-923.

[10] N. J. Vienna, A. K. Polan, Childhood Lymphatic Leukaemia Prenatal Seasonality and Possible Association with Congenital Varicella. *Am J Epidemiol* **103** (1976) 321-332.

[11] A. A. Meltzer, J. F. Annegers, M. R. Spitz, Month-of-birth and Incidence of Acute Lymphoblastic Leukaemia in Children. *Leuk. Lymph.* **23** (1996) 85-92.

[12] H. T. Sorenson, L. Pedersen, J. H. Olse, et al. Seasonal Variation in Month of Birth and Diagnosis of Early Childhood Acute Lymphoblastic Leukaemia. *J. A. M. A.* **285** 168-169.

KES 2002
E. Damiani et al. (Eds.)
IOS Press, 2002

INFOGENMED: Integrating Heterogeneous Medical and Genetic Databases and Terminologies

García-Remesal, M., Crespo J., Silva, A., Billhardt, H., [1]Martín, F.,
Rodríguez-Pedrosa, J., Martín, V., [2]Sousa, A., [3]Babic, A., and Maojo, V.

Medical Informatics Group, Polytechnical University of Madrid, Spain
[1] *Bioinformatics Unit, Institute of Health Carlos III, Madrid, Spain*
[2] *Universidade de Aveiro/IEETA, Portugal*
[3] *Department of Biomedical Engineering, Linköpings Universitet, Sweden*

Abstract. In this article we present the INFOGENMED project, a virtual laboratory for managing and integrating medical and genetic information in health settings. IN-FOGENMED aims to integrate and access distributed and heterogeneous medical and genetic data sources through an unified user interface. Once the system is accessed, users perceive that they are working with a single, local database. The system also provides the integration and use of medical terminologies by means of a vocabulary server, accessible over Internet. This vocabulary server can be used, for example, to enhance searches (e.g. searches in bibliographical databases).
Keywords: Medical Informatics, Bioinformatics, Heterogeneous Database Integration, Medical Terminologies, Genetic Terminologies, Vocabulary Servers.

1 Introduction

The Human Genome Project has launched numerous efforts to expand research in the biomedical sciences and explore new application areas. New technologies and tools are needed to extract data, information, and knowledge that can be applied to create novel diagnostic and therapeutic procedures in medicine.

The development of new genomic-based technologies as well as the growth of Internet are contributing to envision novel medical applications of genetic information. Based on such vision, researchers at different sites are working towards developing new informatics tools to ensure an optimal integration of genetic and clinical information.

Thus, a new breed of systems and software tools are necessary to convert the data that geneticians and molecular biologists can obtain into information that physicians and health workers can use. Furthermore, health professionals will have to manage a kind of information that they are not familiar to. They will need novel methods to search, access, and retrieve genetic information and also to gather, classify and interpret such information.

Two of the main current problems to integrate all the biomedical knowledge are: i) information needs to be located, accessed, and retrieved from different sources over the Web, and ii) the exchange of data is difficult since databases can present a wide range of formats, coding and terminologies that are not unified.

We propose a novel approach to address some of these novel questions. The main objectives of the INFOGENMED project are: i) to facilitate the identification, access, integration and retrieval of genetic and medical information from heterogeneous databases over Internet, ii) to develop new contributions to the exchange and unification of medical and genetic terminologies, which can be stored in vocabulary servers. These servers can be queried over Internet to aid in information exchange between different information systems.

The sections below are organized into four parts (Sections 2-5). Section 2 introduces the methods that we have used for the integration of distributed and heterogeneous databases. Section 3 presents methods for the integration of medical terminologies based on a vocabulary server. In section 4 we present the architecture of the INFOGENMED project. Finally, section 5 provides conclusions and suggests future research directions.

2   Integration of distributed and heterogeneous medical and genetic databases

To integrate distributed and heterogeneous databases we must consider two levels of heterogeneity: i) databases may be hosted by different machines spread over Internet, with different architectures, operating systems and database management systems, and ii) databases can present different conceptual data models, and different underlying database schemas.

The first level of heterogeneity can be addressed by using technologies such as Java technologies or CORBA. The second level of heterogeneity is more complex, since it is common that two databases containing information related to the same context or domain present incompatible database schemas. These incompatibilities may occur due to different perspectives shown by database designers.

To integrate two schemas related to the same domain we must analyze in detail their structure. We must look for semantic incompatibilities such as, for instance, naming conflicts, homonyms and synonyms, duplicated or inconsistent data, type and scaling conflicts. We must also look for structural incompatibilities, i.e., structuring the same information in different ways.

We applied a semiautomated database integration method to develop a preliminary prototype of the INFOGENMED project. It requires user interaction to detect and eliminate semantic and structural incompatibilities. This method is based on an approach based on the concept of *virtual repositories*. A virtual repository is a repository that does not physically exist but gives users the perception of working with a single local repository that integrates data from different sources.

The method for database integration follows a two step process: i) it translates each database into a conceptual schema by means of a *mapping process*, and ii) it unifies these schemas to obtain an unified conceptual schema. The latter represents the information space of the input set of databases by means of the *unification process*. The first step cannot be automated because of the possible existence of semantic and structural incompatibilities, so it requires user interaction. The second step can be fully automated via the unification algorithm described in [2].

3 Integration of medical terminologies and development of a vocabulary server accessible over Internet

We propose the development of a vocabulary server to integrate medical vocabularies. The vocabulary server must supply the following information: i) definitions of concepts, ii) generation of terms related to a given concept, iii) generation of different variants or synonyms for a term, and iv) codes corresponding to different codification schemas (e.g. SNOMED, CPT) for a given concept. We propose the use of the Unified Medical Language System (UMLS) and the Medical Subject Headings (MeSH) thesaurus to satisfy these information needs.

We will now describe the desirable functioning of the vocabulary server. Once the user has selected a concept, the vocabulary server will search for its location in the controlled vocabulary — e.g., MeSH. The results of this search will provide information about the related MeSH contexts of that concept. It will locate its different places in the MeSH tree, providing functionalities to *jump* to any of them. At this point, users can get the definition of that concept, its synonyms, or its corresponding code in other codification schema. Note that this setting satisfies the needs stated above.

We have chosen a relational database model to represent and store the information needed.

Regarding access to the vocabulary server over Internet, we propose a client/server architecture. Initially, the server connects to the database. Once the connection is established, the server waits for requests from users/applications. The server receives a request, and a dialog between the client application and the vocabulary server is established. The server answers every query about the vocabulary terms requested by the user or the client application.

We are considering other approaches for knowledge representation such ontologies, semantic nets, frames or conceptual graphs hierarchies. These techniques are being currently evaluated.

4 A preliminary prototype of the INFOGENMED project

As stated above, the prototype is composed of two modules: i) the heterogeneous database integration/access module, and ii) the vocabulary server module. The prototype's architecture is depicted in figure 1.

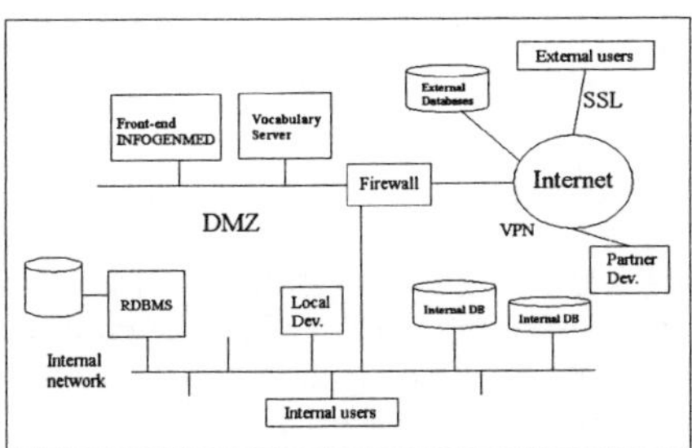

Figure 1: System's Architecture.

The heterogeneous database integration/access module is based on the Common Object Request Broker Architecture (CORBA). This module is composed of three main subsystems:

i) the control system, ii) the administration system, and iii) the data access interface.

The control system provides the mapping facility and the unification engine. It performs some low-level tasks, such as record keeping of connected objects.

The administration system provides i) addition of new databases to the system, ii) creation of virtual repositories from existing databases by means of *mapping processes*, and iii) creation of virtual repositories from a set of already existing virtual repositories by means of the *unification process*.

The data access interface supplies the following functionalities: i) a list of all currently connected virtual repositories, their descriptions, and their schemas and ii) a language, similar to SQL, to query virtual repositories. When we launch a query for a particular virtual repository, the system translates the query into a valid query for each of its children using mapping information. Once the translation is done, it transfers the new queries to its children. At the lowest level in the hierarchy, the queries are translated into SQL and passed to the corresponding databases. Results are returned the same way back.

To test the functionality of this module, we have developed a web interface that provides navigation capabilities to the existing virtual repositories. Once the user has selected a virtual repository she is interested in, she can launch any query to it. Once the information has been retrieved, it is presented to the user through a Web browser interface.

The prototype of the vocabulary server has been written using Java. The vocabulary server can run on any machine supporting a Java virtual machine. The server follows the client/server architecture and it can serve multiple clients simultaneously.

As stated in section 3, both the UMLS and the MeSH thesaurus have been used to build the server. At the current state, the vocabulary server supports the following terminologies: UMLS, MeSH, SNOMED, ICD-9-CM, and CPT. The vocabulary server also contains the MeSH concepts in Spanish. A relational database model has been built to store the information needed by the vocabulary server.

To test the functionality of the vocabulary server, we have developed two web-based applications. The first application is a web interface to access MEDLINE using the MeSH concepts in English or Spanish. Users can query the vocabulary server by using this web interface. They can choose which MeSH terms are the most appropriate to query the MEDLINE database. These terms can be combined by means of boolean operators to restrict the results of the query.

The second application is a MeSH browser. This application provides functionalities to browse the MeSH hierarchy, both in English and Spanish.

5 Conclusions and future research

A preliminary prototype of the INFOGENMED project is already available. It provides functionalities to access, integrate, and query distributed and heterogeneous medical and genetic data sources. At the current state, this prototype only provides functionalities to integrate structured data sources — e.g. relational or object-oriented databases.

This prototype also includes a preliminary version of the vocabulary server. It integrates several medical vocabulary codings. The underlying data model used to support the vocabulary server is a relational data model.

Regarding future work, an extension to the proposed system might be the incorporation of different, non structured information sources, such as, for instance, information from biochips, DNA or protein sequences, HTML pages, PDF documents, or texts. A mechanism to query both structured and non structured data sources is under investigation. Regarding the vocabulary server, we plan to include more medical and genetic terminologies. We are also considering other different approaches for knowledge representation such as ontologies or conceptual graphs.

The integration of genetic and medical data sources might provide physicians more information that could be useful to shift towards new trends in patient care and biomedical research.

6 Acknowledgments

We thank the European Commission for its support for this project under grant IST-2001-39013-INFOGENMED.

References

[1] Altman, R.: *The Interactions Between Clinical Informatics and Bioinformatics*. JAMIA 2000, Vol 7, 5: 439-443

[2] Billhardt, H., Crespo J., Maté, J.L., Maojo, V., Martín, F.: *A new method for unifying heterogeneous databases*. In: Proceedings ISMDA 2001, Madrid, Spain (2001) 2199:54-61

[3] Cimino, J.: *Coding Systems in Health Care*. Methods of Information in Medicine. 1996, 35:273-84.

[4] Ingenerf, J.: *Taxonomic vocabularies in medicine: the intention of usage determines different established structures*. In: Proceedings of the Sixth World Congress on Medical Informatics. Medinfo 1995.pp. 136-39.

[5] Maojo, V., Martin, F., Ibarrola, N., Lopez-Campos, G., Crespo, J., Caja, D., Barreiro, J.M., Billhardt, H.: *Functional definition of the INFOGENMED WORKSTATION: a virtual laboratory for genetic information management in clinical environments*. In: Proceedings AMIA 1999, Washington USA (1999) 1112

[6] Maojo, V., Martín, F., Crespo, J., Billhardt, H.: *Theory, abstraction, and design in medical informatics*. Methods of Information in Medicine 2002, 41: 44-50

[7] Rodríguez-Pedrosa, J., Maojo, V., Crespo, J., Billhardt, H., Sanandrés, J.A.: *Acceso a MEDLINE usando un servidor de vocabulario con términos MeSH* In: Actas de INFORSALUD-NET 98, I Jornadas Internacionales de Internet en Salud. Sociedad Española de Informática de la Salud. Madrid, 1998

[8] Rodríguez-Pedrosa, J., Maojo, V., Crespo, J., Fernández, I.: *A concept model for the automatic maintenance of controlled medical vocabularies* In: Proceedings of the Ninth World Congress on Medical Informatics. MEDINFO '98, Seoul, Korea (1998). pp. 618-622

KES 2002
E. Damiani et al. (Eds.)
IOS Press, 2002

Fuzzy Rules in Strength Grading of Lumber

Jari KORTELAINEN[1] and Yrjö TOLONEN[2]
[1]*Lappeenranta University of Technology, Lappeenranta, Finland*
[2]*Mikkeli Polytechnic, Mikkeli, Finland*

Abstract. Traditional strength grading (classification) of lumber is based on visual observations of defects and certain physical variables. Additionally, some measurable characteristics have been included in the grading process. For each variable, crisp classes and their borders have been set by laborious statistical methods. The grading equipments are based on these principles, but their yield is still low, especially in high strength classes. Moreover, the increased production speed of saw mills demands a fast grading process. The paper presents a fuzzy rule -based approach to strength grading of lumber. We try to reduce the number of the variables and create fuzzy rules by means of expert's knowledge. Here, only four variables are proposed for the strength grading and for each of them a maximum of three fuzzy sets are determined. Presently, the objects are graded to five crisp strength classes for which the minimum strength values are set by authorities.

1 Background

Strength grading (classification) divides lumber to separate classes with certain strength values. The strength of each individual piece of lumber should be estimated such that the strength grading process is reliable and fast. Presently, in Finland there are five strength classes in use, namely, "rejected", "T18", "T24", "T30", and "T40". The lower borders of the classes "T18", "T24", "T30", and "T40" are set by authorities to the bending strength values of 20. 8 MPa, 26. 0 Mpa, 29. 9 Mpa, and 37. 7 Mpa, respectively. Thus, a piece of lumber belongs to the highest class of which the lower border is equal to or is exceeded by its estimated bending strength value.

The main problem is in estimating of the bending strength, because the real strength value can be detected only after breaking the piece. A large number of separate variables has been used with varying success. The most utilized has been the modulus of elasticity in bending (MOE) [7]. The coefficient of determination (COD) between MOE and the bending strength seems to be generally between 0. 5 and 0. 7 as reported in [3]. Other variables seem to do much worse in this respect: for knots COD varies between 0. 16 and 0. 27, and for density between 0. 16 and 0. 4 (see [3]). We point out that there are many ways to estimate the influence of knots on the strength, and we believe knots are important as shown in [7]. Also, the combination of two variables, for instance MOE and density, gives only a little higher COD according to [4]. Moreover, it is generally known that the inclined grain direction reduces especially the tensile strength [4]. The bending of a piece produces both tensile and compression stresses, and the former often causes the fracture. Interesting enough, in visual strength grading the amount of knots is a crucial variable (see [4]).

In this paper four variables are chosen to estimate the bending strength of lumber pieces, namely, density, inclined grain direction, amount of knots, and MOE. Thus, we consider four input variables and one output variable. Linear regression models are classically used in this context (see e.g. [4]), although the nature of the dependency between these five variables might be non-linear, and complex interdependencies between the four input variables may occur. A simple description for dependencies and a fast computation method are needed.

The fuzzy sets are introduced by L. A. Zadeh in [8] and fuzzy rules are *if - then* rules where the antecedents or the consequents are expressed by means of fuzzy sets. Fuzzy rules have recently been used in visual classification of wooden boards like in [1]. In the next Section, a Takagi - Sugeno fuzzy knowledge base (FKB) (sometimes called Sugeno FKB, see e.g. [2]), i.e. a set of certain type of *if - then* rules, is considered and applied in estimating the bending strength of lumber pieces.

2   A Takagi - Sugeno Type of Fuzzy Knowledge Base

In this Section we first briefly recall Tagaki - Sugeno FKBs with the associated method for computing (crisp) values of the output variable when the (crisp) values of the input variables are known. We slightly follow [2] and then present a FKB for lumber strength grading.

Let us consider a system with the input variables x_1, x_2, ..., x_n ($n \in \mathbb{N}$) taking values from their domains U_1, U_2, ..., U_n, and let us denote the output variable of the system by y taking values from its domain Y. Takagi - Sugeno FKB consists of $m \in \mathbb{N}$ rules, of which the ith rule, $i \in \{ 1, 2, ..., m \}$, can be written in the form

$$\mathfrak{R}_i: \quad if\ x_{(1)}\ is\ \mathcal{A}^i_{(1)}\ and\ x_{(2)}\ is\ \mathcal{A}^i_{(2)}\ and\ ...\ and\ x_{(k_i)}\ is\ \mathcal{A}^i_{(k_i)},\ then\ y = f_i(x_1,\ x_2,\ ...,\ x_n),$$

where $\{ (1), (2), ..., (k_i) \}$, $k_i \leq n$, denotes a suitable k_i permutation of $\{ 1, 2, ..., n \}$ for the ith rule, and each $\mathcal{A}^i_{(j)}$ is a fuzzy set for the ith rule in $U_{(j)}$. It is worth noticing that each domain U_j, $j \in \{ 1, 2, ..., n \}$, is covered by possibly varying number of fuzzy sets.

Now, let x_1^*, x_2^*, ..., x_n^* be acquired values for the n input variables, denote $u_i^* = f_i(x_1^*, x_2^*, ..., x_n^*)$, and for the ith rule we write

$$\mu_i^* = \mathbf{T}(\mathcal{A}^i_{(1)}(x_{(1)}^*),\ \mathcal{A}^i_{(2)}(x_{(2)}^*),\ ...,\ \mathcal{A}^i_{(k_i)}(x_{(k_i)}^*)),$$

where $\mathbf{T}$ is a suitable predefined T-norm, for example, the min-operator (see [5]). The output of the system, $y = u^*$, is then computed by the formula

$$u^* = \frac{\sum_{i=1}^m \mu_i^* \cdot u_i^*}{\sum_{i=1}^m \mu_i^*}. \tag{1}$$

For more detailed analysis we refer [2].

In this paper we use an interesting special case of Takagi - Sugeno FKB: we assume that in the expression $y = f_i(x_1, x_2, ..., x_n)$ the function f_i is a *constant* function for all $i \in \{ 1, 2, ..., m \}$. Our FKB consist of only ten (10) fuzzy rules with four input variables, namely, *density* (x_1), *inclined grain direction* (x_2), *amount of knots* (x_3), and *modulus of elasticity in bending* (x_4). The output variable (y) of the system is *estimate for bending strength*. The domains of the variables x_1 and x_4 are divided into three fuzzy sets. Other two domains of the input variables x_2 and x_3 are, however, divided only into two fuzzy sets. The domains of all variables are normalized to [0, 100].

In the Finnish Standard on Wood Structures [6] the grading rules are expressed by means of crisp constraints, thus, crisp classes. In this paper we use these classes as a basis for membership functions for the fuzzy sets, when possible. Additionally, distribution curves from [4] are utilized, when needed.

The fuzzy rules are determined by an expert and are as follows:

$\Re_1$: *if x_1 is small and x_4 is small then y = rejected,*

$\Re_2$: *if x_1 is small and x_2 is large and x_4 is medium then y = T18,*

$\Re_3$: *if x_1 is small and x_3 is large and x_4 is medium then y = T18,*

$\Re_4$: *if x_1 is medium and x_4 is medium then y = T30,*

$\Re_5$: *if x_1 is medium and x_2 is small and x_3 is small and x_4 is large then y = T40,*

$\Re_6$: *if x_1 is large and x_2 is large and x_4 is small then y = T24,*

$\Re_7$: *if x_1 is large and x_3 is large and x_4 is small then y = T18,*

$\Re_8$: *if x_1 is large and x_2 is large and x_4 is medium then y = T30,*

$\Re_9$: *if x_1 is large and x_3 is large and x_4 is medium then y = T30,*

$\Re_{10}$: *if x_1 is large and x_2 is small and x_3 is small and x_4 is large then y = T40,*

where the labels "rejected", "T18", "T24", "T30", and "T40" are replaced by certain values in the normalized domain [0, 100] of the output variable *estimate for bending strength*. For simplicity, triangular or trapezoidal membership functions are determined for the four input variables.

3　Application on a Finnish Data

The data utilized in this paper are reported in [4]. The set of results of Norway Spruce (Picea Abies), originating from Central Finland was selected for our analysis. Originally, a large variety of characteristics have been measured from the lumber by size of 45×150 mm^2. Especially, the bending strength has been measured using the four-point-loading according to the standard EN408, and the most decisive defect has always located in the central portion of a loaded specimen. In addition to measured numerical values, a part of the lumber has been photographed both in original and fractured condition. The values of chosen variables have been computed and adjusted to 12% humidity, too. More detailed data description and the results of the conventional strength grading (both visual and machine) can be found in [4].

Out of those characteristics we utilized *density, inclined grain direction,* and *modulus of elasticity in bending* as three input variables for our analysis. The values of the fourth variable, *amount of knots,* was visually approximated on 200 photographs of 100 pieces mentioned above.

Table 1: Strength grading results.

Strength Class	Fuzzy System	GBSF
T40	73	83
T30	16	8
T24	2	4
T18	3	3
rejected	6	2
Total	100	100

The following can be stated as general observations from our analysis:

- Fuzzy Logic in its wide sense seems to suite well in strength grading of lumber.

- In the fuzzy system created here, only four variables were needed instead of more than double number of variables presented in [6].

- For satisfactory performance only ten fuzzy rules were needed instead of a fairly large table of crisp constraints presented in [6].

The Matlab Fuzzy Toolbox was used when computing the values for *estimate for bending strength*, and the grading results of 100 pieces of lumber is presented in Table 1. The label "GBSF" in Table 1 means Grading based on Bending Strength Final, i.e. the bending strength was measured by breaking the pieces. It is worth noticing that the fuzzy system graded 97 pieces of lumber into correct or too low strength class in the sense of GBSF.

4   Conclusions

The paper introduced a fuzzy rule -based approach to strength grading of lumber. The data reported earlier in [4] was utilized by choosing certain variables and a sample of results of Norway Spruce (Picea Abies), originated from Central Finland. The size of the sample was 100 pieces.

In practical situations, conventional grading equipments give low yields especially in high strength classes. As it can be seen in Table 1, the sum of the pieces in the two highest strength classes, T40 and T30, given by the created fuzzy system is nearly equal to the respective sum given by GBSF. While the grading results seem to be quite promising, at least the following should be mentioned: It is not clear that the created fuzzy system is applicable also for other species. Moreover, the size of the sample was small. In our sample it seemed that only high values for *inclined grain direction* were present or measured. Even small values for *inclined grain direction*, however, cause decrease in the bending strength. Generally, there have been difficulties in measuring the grain direction, but a potentially functioning prototype equipment has now been tested for some months by the second author.

This paper serves as our first step in applying Fuzzy Logic to strength grading of lumber. The method showed high potentials in case of Norway Spruce (Picea Abies) lumber, still, more testing is needed to certify the results. In development of fuzzy rules there are still big opportunities for progress.

5 Acknowledgements

The authors are grateful to Camilla Ahlblad (f. Lindgren), MSc(Eng), and The Finnish Quality Control Association for Timber Structures (PLY ry) for permission to utilize their data on Finnish softwood spieces, especially Norway Spruce (Picea Abies). The financial support from Technology Development Centre of Finland (TEKES) and EU Structural Fund are acknowledged.

References

[1] C. A. de Franca, A. Gonzaga, and A. F. F. Slaets. Classification of wooden boards by neural networks and fuzzy rules. In *Proceedings of the Workshop on Cybernetic Vision*. IEEE Comp Soc, 1997.

[2] D. Driankov, H. Hellendoorn, and M. Reinfrank. *An Introduction to Fuzzy Control*. Springer-Verlag, second edition, 1996.

[3] P. Hoffmeyer. Strenght grading gives more value: Part 2 - present techniques. Raport **335**, Danmarks Tekniske Universitet, Laboratoriet for Bygningsmaterialer, Lyngby, Danmark, 1995. In Danish.

[4] C. Lindgren. *Visual Strength Grading and Machine Strength Grading of Finnish Timber*, volume **820** of *VTT - Julkaisuja*. VTT, 1997. In Finnish.

[5] A. Di Nola, S. Sessa, W. Pedrycz, and E. Sanchez. *Fuzzy Relation Equations and Their Applications to Knowledge Engineering*. Kluwer Academic Publishers, Dordrecht, 1989.

[6] Suomen Rakennusinsinöörien liitto RIL r.y. (The Association of Finnish Civil Engineers). *Finnish Standard on Wood Structures*, 1986. RIL **120 - 1986**. In Finnish.

[7] Y. Tolonen. Ideal strength method for grading of sawn timber. Report **15**, Helsinki University of Technology, Laboratory of Structural Engineering and Building Physics, Helsinki, Finland, 1989.

[8] L. A. Zadeh. Fuzzy sets. *Information and Control*, **8**:338–353, 1965.

KES 2002
E. Damiani et al. (Eds.)
IOS Press, 2002

Interpretation – Insuperable Challenge to Computing with Words?

Vesa A. NISKANEN
University of Helsinki, Dept. of Economics & Management, PO Box 27, 00014 Helsinki, Finland

Abstract. Computing with words approach is central in soft computing, but will it meet insuperable challenges, when we have to apply it to interpretative methods typical of the qualitative research?

1. Soft Computing and Computing with Words

Soft Computing (SC) means a human-like and multi-disciplinary approach to concept analysis, interpretation, reasoning, theory formation, explanation, prediction and model construction, in other words, we aim to mimic or utilize human reasoning in these contexts and, to some extent, the acts and processes of animated beings in general. According to [4,5,6], the main constituents of SC are the theories of information granulation (TFIG), computing with words (CW) and computational theory of perceptions (CTP). The objective of CW is to apply linguistic models and reasonings even when numeric models are available.

Below, we will consider whether we are able to apply CW, when we use linguistic materials and models in the manner of the qualitative research in which case interpretation plays an essential role. We will show that, when we carry out qualitative methods, there are still several problems to be solved.

2. Qualitative Research and Interpretation

The qualitative research constitute a mixture of several methods stemming from various philosophical traditions such as phenomenology and hermeneutics [1,2]. In the qualitative research, our studies usually stem from our foreknowledge and hypotheses (if any) as well as on possible background theories. It is our aim to clarify the foreknowledge and assess the hypotheses. Our data include observations, interviews, documents and other mainly non-numeric and imprecise materials.

When we examine our data, we often first segment our material into various constituents and add our keywords, comments and annotations to them. Second, we file and index these constituents. Third, we construct structures, such as logical, semantic and terminological networks, which reveal the interrelations between the constituents. At this stage, we may also carry out tests, experiments and validations if necessary. These operations and procedures provide us a basis for model constructions, theory formations, argumentations, explanations, descriptions and other outcomes, and interpretations play an essential role in this context. In general, we should examine the whole according to its constituents and vice versa (hermeneutic circle). Finally, we draw the conclusions or resolve the established problems according to our studies.

Interpretation generally means delivering messages, explanation, exegesis or translation. Typical fields of science which apply interpretation are philosophy, historical research, the social sciences and the behavioural sciences. In the natural sciences, we usually presuppose that the research objects exist independent of us, whereas the objects of interpretation are connected to people's mental acts, and hence we have to operate at two levels. In addition to the meanings related to the researcher's conceptual system (which systems are also used in the natural sciences), the research objects as such may include various meanings. For example, we may examine the "objective" features of a painting, such as its height or age, but we may also perform interpretations on its functions, financial and esthetic values and role in arts. In practice, the interpretations may include the painter's conscious meanings, his/her unconscious or latent meanings, researcher's meanings and the painting's intrinsic meanings.

Interpretation is usually based on the foreknowledge of the researcher, and this knowledge will be modified during the study. Successful interpretation should correspond well with the object or phenomenon under interpretation, and then we may understand the essential features and constituents of the objects or phenomena. We also assume that the object of research is reasonable, consistent, coherent or consequential in nature. We have to bear in mind that the interpretations are related to their contexts, the acts or effects based on our aims, linguistic conventions in the society, the beliefs held by the society and the period of time when the phenomena have occurred.

3. CW and Interpretation

The qualitative research has insufficiently utilized computers thus far, even though software are available mainly to text and discourse analyses [3]. This state of affairs has often be due to the lack of the scholars' knowledge in computing. Hence, the future challenge to the qualitative approach is that it should integrate more computer models in its studies.

The SC methods may support the qualitative research in this respect, because they are also appropriate to non-numeric data, approximate entities, linguistic models and human-like reasoning. Our basic tools are fuzzy variables, rules and reasonings as well as constructions such as the information granulations, fuzzy cognitive maps and precisiated natural languages (PNL). To a great extent, the application of SC to the qualitative research would be a new frontier in sciences.

When we perform interpretative analysis in the qualitative research, we have to raise appropriate questions in order to find the correct answers. Hence, we usually assume that each constituent in our material will provide answers to certain questions, and our aim is to specify these questions. The obtained answers, in turn, may lead to additional questions. For example, the answer to the question

Who is the inventor of fuzzy sets and logic?

is *Lotfi Zadeh*, and we may thus raise additional questions such as *Where does he live?*, *What is fuzzy logic?* and *What is the distinction between a fuzzy set and conventional set?*. In this sense, our questions are *interpretation hypotheses*, and our goal is to assess these questions.

Since various additional questions may be raised in this process, we have to focus on the relevant ones. One method is to emphasize the constituents under study one after another. For example, the statement

Lotfi Zadeh is the inventor of fuzzy sets and logic

is related to the following questions according to the placement of the emphasis in this sentence (bold letters):

Answer	**Question derived from answer**
• **Lotfi Zadeh** is the inventor of fuzzy sets and logic.	Who is the inventor of fuzzy sets and logic?
• Lotfi Zadeh is the **inventor** of fuzzy sets and logic.	What is Lotfi Zadeh's role within fuzzy systems?
• Lotfi Zadeh is the inventor of **fuzzy** sets and logic.	Which sets and logic has Lotfi Zadeh invented?
• Lotfi Zadeh is the inventor of fuzzy **sets and logic**.	Which areas in the fuzzy systems has Lotfi Zadeh invented?

We aim to find the principal question or problem in our material and possible other questions related to its constituents. The resolution to the main problem also support us to find the resolutions to the problems related to the constituents, and vice versa, i.e., we apply the hermeneutic circle (Fig. 1).

A successful interpretation leads to correct problem solving and compression of material, such as usable plots, summaries and categorizations. Our interpretations are like networks which may change their structure and nodes according to our reasonings, and our foreknowledge provides the tentative configurations for these networks.

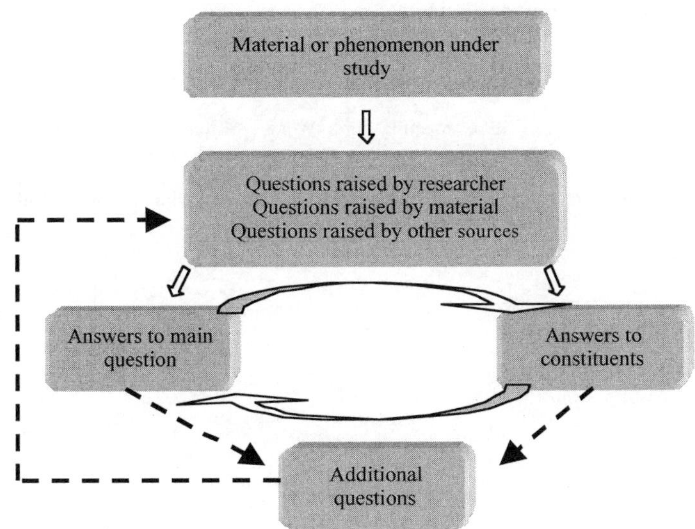

Figure 1. Model of Raising Questions in Interpretation.

Similar problem-settings can be found in the information technology today. The increasing amount of hyperlinked information and knowledge available on the web has shifted the research focus to the search engines that will deliver the required information quickly. In the terms of information technology, we thus have to consider data compression, data mining, the structure of the data, clustering or discrimination of data and web crawling. We apply methods such as optimizations, simulations and various techniques related to evolutionary computing, virtual worlds, intelligent agents, web robots, e-science, e-business and e-learning. All this problemacy will boil down to semantics, and interpretion, in turn belongs at the core of semantics.

To date, most of the problems related to interpretations are unsolved in computer environment although the human beings are able to perform interpretions effortlessly. Unlike the hard computing, the CW approach may operate with both precise and imprecise linguistic entities as well as construct complicated linguistic simulation models, but several of the foregoing acts, such as performing text summaries or providing appropriate questions according to the data, will require much further studies. In this respect, the CW should abandon its traditional commitments to the quantitative sciences to a great extent and apply much more qualitative methods instead.

4. Summary

CW aims to apply human-like reasoning which uses imprecise and linguistic modelling. In this area, interpretation plays an essential role. However, if we attempt to apply interpretation in computer environment, we should use methods typical of the qualitative research, and most of these methods still seem too complicated for CW. Is interpretation even a superable challenge to CW? At least one thing is certain in this respect, the future is fuzzy.

References

[1]   N. Denzin & Y. Lincoln, Handbook of Qualitative Research, Sage, Thousand Oaks, 1994.
[2]   D. Silverman, Interpreting Qualitative Data, Sage, Thousand Oaks, 1993.
[3]   E. Weitzman & M. Miles, Computer Programs for Qualitative Analysis, Sage, Thousand Oaks, 1995.
[4]   L. Zadeh, Fuzzy logic = Computing with words, IEEE Transactions on Fuzzy Systems, vol. 2, pp. 103-111, 1996.
[5]   L. Zadeh, From Computing with Numbers to Computing with Words - From Manipulation of Measurements to Manipulation of Perceptions, IEEE Transactions on Circuits and Systems 45, 105-119, 1999.
[6]   L. Zadeh, Toward a theory of fuzzy information granulation and its centrality in human reasoning and fuzzy logic, Fuzzy Sets and Systems 90/2, 111-127, 1997.

KES 2002
E. Damiani et al. (Eds.)
IOS Press, 2002

Level-Sets and Topological Bases

Paavo KUKKURAINEN

Lappeenranta University of Technology, Lappeenranta, Finland

Abstract. Let $\mathcal{A}: A \longrightarrow L$ be an L-set on A where A is a nonempty set and L is a lattice. Then the corresponding level-sets $\mathcal{A}_p$ form a topological base on A. In this paper we consider a P-set $\mathcal{A}: A \longrightarrow P$, $P \subset \mathcal{P}(A)$, where P is a poset. For a certain $\mathcal{A}$, we will prove sufficient and necessary conditions concerning the topological base of level-sets $\mathcal{A}_p$ so that $p = \mathcal{A}_p$ for any $p \in P$.
Keywords: Fuzzy algebra, P-sets, MV-algebra, topology.

1 Introduction

For a lattice L and an L-set $\mathcal{A}: A \longrightarrow L$ the level-sets $\mathcal{A}_p$ form a topological base on a nonempty set A (see Proposition 3.1). Conversely, if we have a topological base P on A, what are the conditions such that P is formed by level-sets $\mathcal{A}_p$? We have an answer to this question. In this case $\mathcal{A}$ is a fixed poset-value mapping (a P-set). The result is based on B.Seselja's and A.Tepavcevic's work ([8]). However, they did not consider this problem and they had not a topological point of view. In the end there are some illustrative examples and we also compare the situation with the case where $\mathcal{A}$ is different from above, defined on an MV-algebra and level-sets $\mathcal{A}_p$ are deductive systems of this MV-algebra.

2 Preliminaries

A poset is a partially ordered set $P = (P, \leq)$ with a binary relation $\leq$. A lattice is a poset any two of whose elements x and y have a least upper bound $x \vee y$ and a greatest lower bound $x \wedge y$. The lattice order $\leq$ is defined in the following way: $x \leq y$ if and only if $x \vee y = y$ if and only if $x \wedge y = x$ for any elements x and y of a lattice L ([2]).

In the following we refer to [10]: Let L be a lattice with the least element 0 and the greatest element 1, binary operations $\cdot$ (product) and $\rightarrow$ (residuum) satisfying the following conditions:

(i) $\cdot$ is associative, commutative and isotone.

(ii) $x \cdot 1 = x$ for all elements $x \in L$.

(iii) The Galois connection $x \cdot y \leq z$ if and only if $x \leq y \rightarrow z$ holds for all elements $x, y, z \in L$.

Then a system $L = (L, \leq, \vee, \wedge, \cdot, \rightarrow, 0, 1)$ is called a residuated lattice.

Let A be a nonempty set with binary operations $+, \cdot$ and an unary operation $*$. Assume that A contains two distinct elements 0 and 1 and the following conditions are valid: $(A, +, 0)$,

$(A, \cdot, 1)$ are commutative monoids with identity 0 and 1, respectively, $(x + y)^* = x^* \cdot y^*$, $(x \cdot y)^* = x^* + y^*$, $x^{**} = x$, $0^* = 1$ and $x + x^* \cdot y = y + y^* \cdot x$. Then a system $A = (A, +, \cdot, ^*, 0, 1)$ is called an MV-algebra ([1]).

Let W be a nonempty set which contains an element 1, a binary operation $\rightarrow$, and a unary operation * obeying the following axioms ([11]):

$$1 \rightarrow x = x, \tag{1}$$

$$(x \rightarrow y) \rightarrow [(y \rightarrow z) \rightarrow (x \rightarrow z)] = 1, \tag{2}$$

$$(x \rightarrow y) \rightarrow y = (y \rightarrow x) \rightarrow x, \tag{3}$$

$$(x^* \rightarrow y^*) \rightarrow (y \rightarrow x) = 1. \tag{4}$$

Then a system $W = (W, \rightarrow, ^*, 1)$ is said to be a Wajsberg algebra.

If we define on an MV-algebra a binary operation $\rightarrow$ such that $x \rightarrow y = x^* + y$, and on a Wajsberg algebra operations $+$ and $\cdot$ as follows: $x + y = x^* \rightarrow y$ and $x \cdot y = (x \rightarrow y^*)^*$, and set $0 = 1^*$, we can regard a Wajsberg algebra as an MV-algebra and on the contrary. This is proved in [11] and [3].

Next we take some fuzzy concepts considered in [7]: Let A be a nonempty set and $P = (P, \leq)$ a poset. Then a poset-valued mapping $\mathcal{A}: A \longrightarrow P$ is a P-set on A. For every $p \in P$, $\mathcal{A}_p = \{x \in A \,|\, \mathcal{A}(x) \geq p\}$ is a p-level-set (shortly a level-set) of $\mathcal{A}$. Let the co-domain of $\mathcal{A}$ be a lattice $L = (L, \vee, \wedge, \leq)$. Then the mapping $\mathcal{A}: A \longrightarrow L$ is an L-set on A.

3 Topological Bases formed by Level-Sets

The following result is proved in [6] and also in the case that the lattice L is a complete lattice in [5]:

Proposition 3.1. ([6]) Let $\mathcal{A}: A \longrightarrow L$ be an L-set, where A is a nonempty set and L is a lattice. Then the collection $\mathcal{A}_L$ of level-sets $\mathcal{A}_p$ of $\mathcal{A}$ form a topological base on A.

Proof. By [8], $\bigcup(\mathcal{A}_p \mid p \in L) = A$. Let $\mathcal{A}_q, \mathcal{A}_r \in \mathcal{A}_L$. By [8], $\mathcal{A}_q \cap \mathcal{A}_r = \mathcal{A}_{q \vee r} \in \mathcal{A}_L$ for every $p, q \in L$, where $q \vee r$ exists since L is a lattice. Consequently, $\mathcal{A}_L$ form a topological base on A ([4]) $\qquad\square$

As a consequence of Proposition 3.1 we can regard an MV-algebra as a topological space.

The next proposition is represented by B.Seselja and A.Tepavcevic in [8]:

Proposition 3.2. ([8]) Let A be a nonempty set, and P a family of its subsets i.e. $P \subset \mathcal{P}(A)$, such that:

(i) $\bigcup P = A$,

(ii) for every $x \in A$, $\quad \bigcap(p \in P \mid x \in p) \in P$.

Let $\mathcal{A}: A \longrightarrow P$ be defined with

$$\mathcal{A}(x) = \bigcap(p \in P \mid x \in p). \tag{5}$$

Then, $\mathcal{A}$ is a P-set, where $(P, \leq)$ is a poset ordered under $p \leq q$ if and only if $q \subset p$ ($p, q \in P$), and for every $p \in P$,

$$p = \mathcal{A}_p. \tag{6}$$

We are ready to prove our main result which completes Proposition 3.1 in some way:

Proposition 3.3. Assume

(i) A is a nonempty set and P a family of its subsets i.e. $P \subset \mathcal{P}(A)$.

(ii) For every $x \in A$, $\quad \bigcap(p \in P \mid x \in p) \in P$.

(iii) Define an order on P such that $p \leq q$ if and only if $q \subset p$ for every $p, q \in P$, and

$$\mathcal{A}: A \longrightarrow P, \quad \mathcal{A}(x) = \bigcap(p \in P \mid x \in p). \tag{7}$$

Then, $(P, \leq)$ is a poset and for a P-set $\mathcal{A}$, P is a topological base on A if and only if $p = \mathcal{A}_p$ for every $p \in P$.

Proof. Clearly, $(P, \leq)$ is a poset and $\mathcal{A}$ is a P-set. If P is a topological base on A then $\bigcup P = A$, and all the assumptions of Proposition 3.2 are valid. Therefore, $p = \mathcal{A}_p$ for every $p \in P$. Conversely, let $p = \mathcal{A}_p$. By [8], $\bigcup P = \bigcup(p \mid p \in P) = \bigcup \mathcal{A}_p = A$. Moreover, if $p_1, p_2 \in P$ such that $x \in p_1 \cap p_2$ then $x \in \bigcap(p \in P \mid x \in p) \subset (p_1 \cap p_2 \mid x \in p_i \in P)$, and by assumption, $\bigcap(p \in P \mid x \in p) \in P$. Therefore, P is a topological base on A ([4]). $\square$

Corollary 3.4. In the case of Proposition 3.3 the topological base $\mathcal{A}_P$ formed by level-sets $\mathcal{A}_p$ exits if the topological base P exits. Moreover, P and $\mathcal{A}_P$ coincides.

Proof. Let $\mathcal{A}_P = \{ \mathcal{A}_p \mid p \in P \}$. We will prove that $P = \mathcal{A}_P$. By Proposition 3.3, P is a topological base on A if and only if $p = \mathcal{A}_p$ for every $p \in P$. Let $p \in P$. Then $p = \mathcal{A}_p \in \mathcal{A}_P$ and $P \subset \mathcal{A}_P$. Conversely, let $q \in \mathcal{A}_P$. Then there exits $p \in P$ such that $q = \mathcal{A}_p$. If $q \notin P$, then $p = \mathcal{A}_p = q \notin P$ which is a contradiction. Thus $q \in P$ and $\mathcal{A}_P \subset P$. $\square$

The following example is represented in [8]. However, it is applied here to Proposition 3.3.

Example 3.5. [8] Let $A = \{ a, b, c \}$ and $P = \{ \emptyset, \{a\}, \{b\}, \{ b, c \} \}$. For the mapping (7), $\mathcal{A}: A \longrightarrow P$, we have $\mathcal{A}(a) = \{ a \}$, $\mathcal{A}(b) = \{ b \}$, $\mathcal{A}(c) = \{ b, c \}$. Then, by definition of order (in Proposition 3.3), $\{a\} \leq \emptyset$, $\{b\} \leq \emptyset$, $\{ b, c \} \leq \{b\}$. Therefore, $\mathcal{A}_\emptyset = \emptyset$, $\mathcal{A}_{\{a\}} = \{a\}$, $\mathcal{A}_{\{b\}} = \{b\}$ $\mathcal{A}_{\{b,c\}} = \{b, c\}$. By Propositon 3.3, P is a topological base on A. In fact, $\bigcup P = \bigcup \{ \{a\}, \{b\}, \{ b, c \} \} = \{ a, b, c \} = A$ and $\{a\} \cap \{b\} = \emptyset$, $\{a\} \cap \{b, c\} = \emptyset$, $\{b\} \cap \{b, c\} = \{b\}$ and $\emptyset \cap$ any other subset of P is empty.

Example 3.6. Let $I = [0, 1]$ be the closed unit interval, $m \geq 1$ a fixed positive integer, k a non-negative integer and $P \subset \mathcal{P}(I)$ such that P consists of open intervals $(\frac{k-1}{m}, \frac{k}{m})$, $1 \leq k \leq m$, and their unions, half-open intervals $[0, \frac{k}{m})$, $1 \leq k \leq m$ and $(\frac{k}{m}, 1]$, $0 \leq k \leq m - 1$. Then P is a topological base on I. Define the order on P as in Proposition 3.3 and let $\mathcal{A}: I \longrightarrow P$ be a P-set on I such that

$$\mathcal{A}(x) = \bigcap(p \in P \mid x \in p) \tag{8}$$

Since the assumptions of Proposition 3.3 are valid and P is a topological base on I, then $p = \mathcal{A}_p$ for every $p \in P$. Moreover, the bases P and $\mathcal{A}_P$ coincides.

Next we consider the situation where $\mathcal{A}$ is not defined by (7). The following proposition is represented in ([6]). We first need some preliminaries.

Recall ([3], [9], [11]) that a deductive system (an implicative filter) D of an MV-algebra (a Wajsberg algebra) A is a subset D of A such that

(i) $1 \in D$.

(ii) If x, $x \to y \in D$, then $y \in D$.

In addition, the operation $\cdot$, which is defined on a poset P, is idempotent if $x \cdot x = x$ for every $x \in P$ (cf. [2], [11]). In this case we also say that x is idempotent.

Proposition 3.7. ([6]) Let A be an MV-algebra, and L a residuated lattice such that the operation $\cdot$ in L is idempotent. If an L-set $\mathcal{A}: A \longrightarrow L$ satisfies the condition:

$$\mathcal{A}(x \to y) = \mathcal{A}(x) \to \mathcal{A}(y), \tag{9}$$

Then every level-set $\mathcal{A}_p$ is a deductive system of A.

Proof. Such an L exists, for example a Gödel structure ([11]). Let $x \in \mathcal{A}_p$, $x \to y \in \mathcal{A}_p$. Then $\mathcal{A}(x) \geq p$, $\mathcal{A}(x \to y) \geq p$ and

$$p \leq \mathcal{A}(x \to y) = \mathcal{A}(x) \to \mathcal{A}(y) \leq p \to \mathcal{A}(y). \tag{10}$$

By Galois connection (L is a residuated lattice) and since p is idempotent, we obtain $\mathcal{A}(y) \geq p \cdot p = p$, which implies $y \in \mathcal{A}_p$. Further, for every $p \in L$

$$\mathcal{A}(1) = \mathcal{A}(x \to x) = \mathcal{A}(x) \to \mathcal{A}(x) = 1' \geq p, \tag{11}$$

which implies $1 \in \mathcal{A}_p$. The elements 1 and $1'$ are the greatest elements in L and L'. $\qquad\square$

Example 3.8. Let $S(m) = \{\, 0, \frac{1}{m}, \frac{2}{m}, \cdots, \frac{m-1}{m}, 1 \,\}$ for a fixed $m \geq 1$. Define the operations $\oplus$, $\odot$, * on $S(m)$ in the following way: $\left(\frac{p}{m}\right) \oplus \left(\frac{k}{m}\right) = \frac{min(m, p+k)}{m}$, $\left(\frac{p}{m}\right) \odot \left(\frac{k}{m}\right) = \frac{max(0, p+k-m)}{m}$, $\left(\frac{p}{m}\right)* = \frac{m-p}{m}$. Then $(S(m), \oplus, \odot, *, 0, 1)$ is an MV-algebra equipped with the Lukasiewicz structure ([9], [11]).

Let $\mathcal{A}: S(m) \to L$ be a L-set on $S(m)$ such that

$$\mathcal{A}(x \to y) = \mathcal{A}(x) \to \mathcal{A}(y) \tag{12}$$

and the assumptions for L in Proposition 3.7 are valid. Then $\mathcal{A}_p$ is a deductive system of $S(m)$. On the other hand, $S(m)$ is a subalgebra of the Lukasiewicz structure and so the only deductive systems are $S(m)$ and $\{1\}$ ([9], [11]). Since L is a lattice, by Proposition 3.1, the topological base on $S(m)$ formed by level-sets $\mathcal{A}_p$ exits and is now included in $\{\{1\}, S(m)\}$. The corresponding topology is $\{\emptyset, \{1\}, S(m)\}$ or $\{\emptyset, S(m)\}$.

References

[1] L. P. Belluce and S. Sessa. Orthogonal decompositions of *MV*-spaces. *Mathware & Soft Computing*, **4**:5–22, 1997.

[2] G. Birkhoff. *Lattice Theory*, volume **XXV**. AMS, Providence, Rhode Island, third edition, 1995. Eight printing.

[3] R. O. Cignoli, I. M. L. D'Ottaviano, and D. Mundici. *Algebraic Foundations of Many-valued Reasoning*, volume **VII**. Kluwer Academic Publishers, Dordrecht, The Netherlands, 2000.

[4] J. K. Kelly. *General Topology*. Van Nostrand, 1955.

[5] J. Kortelainen. *A Topological Approach to Fuzzy Sets*. Ph.D. dissertation, Lappeenranta University of Technology, Lappeenranta, Finland, 1999. Acta Universitatis Lappeenrantaensis **90**.

[6] P. Kukkurainen. *L*-sets and topological *MV*-algebras. The paper is submitted for publication in Fuzzy Sets and Systems, 2002.

[7] B. Seselja. Homomorphisms of poset-valued algebras. *Fuzzy Sets and Systems*, **121**:333–340, 2001.

[8] B. Seselja and A. Tepavcevic. On a construction of codes by *P*-fuzzy sets. *Zb. Rad. Prirod.-Mat. Fak. Ser. Mat.*, **20**:71–80, 1990.

[9] E. Turunen. *Sumean Logiikan Matematiikka*. Otatieto Oy, Helsinki, 1997. In Finnish.

[10] E. Turunen. *BL*-algebras of basic fuzzy logic. *Mathware and Soft Computing*, **6**:49–61, 1999.

[11] E. Turunen. *Mathematics Behind Fuzzy Logic*. Physica-Verlag, Heidelberg, 1999.

KES 2002
E. Damiani et al. (Eds.)
IOS Press, 2002

Fuzzy Similarity Based Classification in the Normal Łukasiewicz-Structure

PASI LUUKKA and KALLE SAASTAMOINEN
Department of Information Technology
Lappeenranta University of Technology
P.O. Box 20, FIN-53851, Finland

Abstract. In this article we have studied fuzzy similarity based classification
in the normal Łukasiewicz-Structure. We have studied
fuzzy similarity in different metrics and we will show
that classifying results will get better results by implementing
different kinds of metrics in to the Łukasiewicz-Structure.

1. Introduction

Same way as notion of fuzzy subset generalizes that of the classical subset, the concept of similarity can be considered as a many-valued generalization of the classical notion of equivalence[1]. Equivalence relation is a familiar way to classify similar mathematical objects. Fuzzy similarity is an equivalence relation that can be used to classify multi-valued objects. Because of this, it is suitable for classifying problems that are possible to classify based on clustering by finding similarities in objects. There is also a close link between the notion of similarity and that of distance (see for example [2] and [3]). When we use fuzzy similarity in classifying we always have to implement some metric into it which will measure normalized distance between instance vectors. Most used distance measures are the Euclidean and the Manhattan. Depending on the type of a particular feature attribute, different metrics should be used. For example, the Euclidean and Manhattan are not suitable for symbolic attributes and in such cases some other metrics should be used. Especially in instance-based classification methods the selection of a similarity measure is a critical consideration. Some times poor results which has been got from the instance-based methods may actually originate in the underlying similarity measure, not the method itself [4]. It is also well known that mathematical machinery in general used in engineering leans heavily on the use of metrics. As example we like to embed signals of interest in metric spaces. Then we can measure the similarity or dissimilarity of signals, fidelity and dissortion, as distance. Every metric space can be transformed to the pseudo-metric space, but not every pseudo-metric can be transformed into the metric space. This is due the fact that it takes more axioms to enforce metricity. In practice it is always an application that decides our choice of an axiom-set and the flavor of the mathematics used. One application may require us to work in a metric space, while another may require a pseudo-metric. Our motivation for this paper is to improve and test some already quite good results what we have got from the our previous research activities between weighted fuzzy similarity in classifying [5] and [6].

This article handles the suitability of different metrics to the Łukasiewicz-structure and applying these different structures to pattern recognition problems. In this study we have derived a classifier based on the Łukasiewicz-structure which is normally implemented using Manhattan metric but we will test it now with different kinds of metrics. Our motivation for this is the fact that the success of unsupervised algorithms, such as the clustering methods, depends crucially on the metric, the measure of the distance between the objects of interest [7].

The data sets were chosen as diverse as possible so that classifiers properties would be apparent. Data sets were taken from a UCI Repository of Machine Learning Database [8] archive so that they were differently distributed and their dimensions varied. Classifiers were implemented with MATLAB^{TM}-software.

2. Łukasiewicz-Structure with Different Metrics in Pattern Recognition

There are three good reasons why we have chosen to use Łukasiewicz-structure in defining memberships of objects. One reason is that Łukasiewicz-structure holds the fact that the mean of many fuzzy similarities is still a fuzzy similarity [9]. Secondly it also has a strong connection to the first-order fuzzy logic [10], which is a well studied area in the modern mathematics. Thirdly it also holds the fact that any pseudo-metric induces fuzzy similarity on a given non-empty set X with respect to the Łukasiewicz conjunction [11]. Next we will shortly introduce mathematical background concerning fuzzy similarity, Łukasiewicz-structure and metrics that we will implement into it. After that we can construct algorithms that uses fuzzy similarity in Łukasiewicz-structure with different metrics as a base of a fuzzy classification method.

2.1. Mathematical Background

Definition 1: Let L be a resituated lattice and X is a non-empty set. L – valued binary relation S defined in X is a fuzzy similarity if it fulfills the following conditions: [3].

1. $\forall x \in X : S\langle x, x \rangle = 1$
2. $\forall x, y \in X : S\langle x, y \rangle = S\langle y, x \rangle$
3. $\forall x, y, z \in X : S\langle x, y \rangle \odot S\langle y, z \rangle \leq S\langle x, z \rangle$

Depending on the choice of the operation $\odot$ (sometimes marked as $*$), S is also called a fuzzy equivalence relation [11], indistinguishability operator [12], fuzzy equality (relation) [13] or proximity relation [14].

It is easy to see that letting L be the two element set $\{0,1\}$, fuzzy similarity coincides with the usual equivalence relation.

Definition 2: **Łukasiewics norm** or **Łukasiewics conjuction**: a $\odot$ b=max$\{$a+b-1,0$\}$.

This is the $t - norm$, which means that it preserves transitivity w.r.t. the triangular inequality. It is also important to realize that for practical applications with the unit interval as the underlying lattice L, considering only MV-algebras is very restrictive, since $a \odot b = \max\{a+b-1,0\}$ is the only choice for the operation $\odot$ up to isomorphism.

We can construct a lattice called **normal Łukasiewicz-structure** or more formally just Łukasiewicz-structure:

Definition 3: Łukasiewics-structure: $a \odot b = \max\{a+b-1,0\}, \quad a \to b = \min\{1,1-a+b\}$

If we examine Łukasiewicz-valued fuzzy similarities S_i $i=1,...,n$ in a set X we can define a binary relation in L by stipulating $S\langle x,y\rangle = \frac{1}{n}\sum_{i=1}^{n} S_i\langle x,y\rangle$ for all x, $y \in X$. It is easy to prove that this is still a Łukasiewicz-valued fuzzy similarity [9].

Definition 4: **Pseudo-metric**, in a set X, is a mapping $d : X \times X \rightarrow [0,\infty[$ such that $\forall x,y,z \in X$ the following conditions holds true:

1. $d(x,x) = 0$
2. $d(x,y) = d(y,x)$
3. $d(x,z) \le d(x,y) + d(y,z)$

It is known that pseudo-metrics bounded by one and fuzzy equivalence relations with respect to the Łukasiewicz conjunction are dual concepts [11]. Which means that S is a fuzzy equivalence relation on X with respect to $\odot$ if and only if 1-S is a pseudo-metric on X. So similarity $S\langle x,y\rangle$ now gets the form $S\langle x,y\rangle = 1 - d(x,y)$. In case that pseudo-metric d is not bounded by one, we can enforce this property by considering the pseudo-metric $\hat{d} = \min\{d(x,y),1\}$, which coincides with d for 'small' distances. Thus any pseudo-metric induces a fuzzy equivalence relation on X with respect to the Łukasiewicz conjunction by

$$S\langle x,y\rangle = 1 - \hat{d} = 1 - \min\{d(x,y),1\} \tag{1}$$

Now we are ready to study how to use fuzzy similarity for finding similar pairs. We are examining a choice situation where features of different objects can be expressed in values between [0,1]. Let X be the set of m objects. If we know the similarity value of the features $f_1,...,f_n$ between objects, we can choose the object that has the highest total similarity value. The problem is to find for object x_i similar object x_j, where $1 \le i, j \le m$ and $i \ne j$. By choosing Łukasiewicz-structure for features of the objects we get n fuzzy similarities for comparing two objects (x_1, x_2):

$$S_{f_i}\langle x_1, x_2\rangle = x_1(f_i) \leftrightarrow x_2(f_i), \tag{2}$$

where, x_1, $x_2 \in X$ and $i \in \{1,...,n\}$. Because Łukasiewicz-structure is chosen for membership of objects, we can define the fuzzy similarity as follows:

$$S\langle x_1, x_2\rangle = \frac{1}{n}\sum_{i=1}^{n}(x_1(f_i) \leftrightarrow x_2(f_i)). \tag{3}$$

It is a very important to realize that this holds only in the Łukasiewicz-structure. Moreover, in the Łukasiewicz-structure we can give different non-zero weights $(W_1,...,W_n)$ to the different features and we get the following formula which again meets the definition of the fuzzy similarity:

$$S\langle x_1, x_2\rangle = \frac{\sum_{i=1}^{n} W_i\left(x_1(f_i) \leftrightarrow x_2(f_i)\right)}{\sum_{i=1}^{n} W_i} \tag{4}$$

In Łukasiewicz-structure equivalence relation $a \leftrightarrow b$ is defined as $1 - \max\{a,b\} + \min\{a,b\}$ or equivalently $1 - |a - b|$. The formula of fuzzy similarity in such a case is:

$$S\langle x_1, x_2 \rangle = 1 - \frac{1}{n} \sum_{i=1}^{n} |x_1(f_i) - x_2(f_i)| \tag{5}$$

or

$$S\langle x_1, x_2 \rangle = 1 - \frac{\sum_{i=1}^{n} W_i |x_1(f_i) - x_2(f_i)|}{\sum_{i=1}^{n} W_i} \tag{6}$$

2.2. Different Metrics in Łukasiewicz -structure

We are going to implement the following metrics in to our classifier based on similarity in Łukasiewicz-structure:

Minkowsky:

$$d(x,y) = \left(\sum_{i=1}^{N} |x_i - y_i|^p \right)^{\frac{1}{p}} \tag{7}$$

Cumulative Minkowsky:

$$d(x,y) = \left(\sum_{i=1}^{N} \left| \sum_{u=1}^{i} x_u - \sum_{u=1}^{i} y_u \right|^p \right)^{\frac{1}{p}} \tag{8}$$

Mahalanobis:

$$d(x,y) = (x-y)^T C^{-1} (x-y) \tag{9}$$

Modified G:

$$d(x,y) = 2 \left\{ \begin{array}{l} \left[\sum_{f=x,y} \sum_{i=1}^{N} f_i \log f_i \right] - \left[\sum_{f=x,y} \left(\sum_{i=1}^{N} f_i \right) \log \left(\sum_{i=1}^{N} f_i \right) \right] - \\[2ex] \left[\sum_{i=1}^{N} \left(\sum_{f=x,y} f_i \right) \log \left(\sum_{f=x,y} f_i \right) \right] + \left[\left(\sum_{f=x,y} \sum_{i=1}^{N} f_i \right) \log \sum_{f=x,y} \sum_{i=1}^{N} f_i \right] \end{array} \right\} \tag{10}$$

Above we see some commonly used distance functions. Most common ones are Euclidean and Manhattan corresponding norms are L_2 and L_1 which both are special cases of the Minkowsky which norm is L_p. As the degree of the norm p increases, the weight of a large difference between values of a single attributes increases. Thus the Euclidean distance weights large differences more than the Manhattan distance. Both the Euclidean and Manhattan distances are calculated separately for each dimensions, and thus, they are not very good measures for similarity between objects where attributes are correlated and ordered. This kind of situations the Cumulative Minkowsky distances can be used. It is because they measure the difference of cumulative values of attribute vectors. Thus they can be thought to measure the spatial concentration of the values in the attribute vector and the order of the attributes affects the value of the distance. Therefore, with ordered data these measures are likely to provide better results than the standard Euclidean and Manhattan distances.

To prevent the effects of unbalanced ranges of different dimensions, the attributes are often normalized. The normalization can be performed by dividing the attribute values by the range (maximum-minimum) of the corresponding attribute. Another possibility is to perform the normalization using the standard deviation instead of the range.

Most of these distance functions performed previously consider the attributes to be non-correlated. For cross-correlated attributes, statistical properties of data set can be used to reduce the effect of correlations. One such distance function with a statistical factor, the correlation matrix is the Mahalanobis distance. Eventually calculation of the correlation matrix in the Mahalanobis distance needs quite much data, the exact amount depending on the length of the attribute vectors and the variation between them. If there is not enough variation in the data, numerical calculation of the inverse of covariance matrix may be impossible because of the singularity of the matrix (ill-posed problem). A more comprehensive study of the Mahalanobis distance with a limited sample size can be found in [15]. Furthermore, the removal of correlation effect should be computed among the samples of a same class rather than over samples in all classes. Such a calculation procedure is again problematic if the classes are unknown.

Another statistical method, the log likelihood ratio G, measures the degree that observed data fits to an expected distribution. As a distance function the basic G metric has a drawback of depending on the order of variables (non-symmetric metric), i.e., which variable is considered to be the sample and which the expected distribution. Sokal and Rohlf [16] introduced a modified G test, which is a two-way test of interaction or heterogeneity.

2.3 Description of Pattern Recognition Method in the Algorithmic Form

Method starts with fuzzification of a data set, which means, in practice, scaling the data between zero and one. After this an ideal vector (e.g. mean vector) is calculated for every class. Samples are classified so that the maximal fuzzy similarity value between ideal and test vector is calculated. The method gets test element *test* (dimension *dim*), learning set *learn* (dimension *dim*, *n* different classes) and dimension (*dim*) of the data as its parameters. In addition, different weights for features can be set in *weights*. The example of how Manhattan distance can be applied to this method is presented in the pseudo-code form below:

Require: *test,learn[1...n],weights,dim*
  Scale *test* between [0,1]
  Scale *learn* between [0,1]
  For i = 1 to n **do**
   Idealvec[i]=IDEAL[learn[i]]

$$\max sim[i] = 1 - \frac{\sum_{j=1}^{dim} weights[j] \left| idealvec[i][j] - test[j] \right|}{\sum_{j=1}^{dim} weights[j]}$$

  end for
Class = arg max$_i$, maxsim[i]

In the algorithm, operator *IDEAL* forms the vector that represents the objects of the class best. In this paper, the *IDEAL* operator is the mean vector of the class. Implementation to the other metrics is straightforward in normal Łukasiewicz-structure. For this algorithm, weights can be optimized for example by using genetic algorithms in optimization. In [6] there is desciption on a form of a flowchart for how optimal weights can be found and closer desciption for how it would be done by using genetic algorithm.

3. Empirical Results

3.1. Maximal Fuzzy Similarity with Weight Optimization in Classification

Classification is a good way to test how weighted similarity measure works in practice. *A priori* information for determing weights was not available for data sets. We used a genetic algorithm to find the optimal weights for our fuzzy similarity classifier. Genetic algorithm was used because of their diversity and robustness. We used three different data sets. Data sets were splitted in half. One half was used for learning and the other half for testing.

Iris-data Set Data set consists of three classes and it is four dimensional.
Thyroid gland-data Set Data consist also of three classes and it is five dimensional.
Wine recognition-data Set Data consist of three classes and it is 13 dimensional.

Normal Minkowsky-metrics:

Iris-data Set: Using different Minkowsky-metrics led to different results with Iris-data set. Best results 96 % was found with power values like 1, 2, 3. In Iris data set commonly used Manhattan and Euclidean metrics seems to work well.
Wine recognition-data Set: Best results was found with Minkowsky metric when power value was three.
Thyroid gland-data Set: With thyroid gland-data set results were again much better with optimal weights. Also optimal power value was near $p=100$.

Table 1: Classification results (percentage of correct classifications) with three data sets and normal Minkowsky metric used

Data\Metric	L_1	L_2	L_3	$L_{0.5}$	L_{100}
Wine(weighted)	95.51	95.51	97.75	95.51	94.38
Wine	91.01	91.01	91.01	91.01	91.01
Iris(weighted)	96.00	96.00	96.00	94.67	94.67
Iris	94.67	94.67	94.67	94.67	94.67
Thyroid(weighted)	94.44	94.44	93.52	94.44	95.37
Thyroid	89.81	89.81	89.81	89.81	89.81

Cumulative Minkowsky-metrics:

Iris-data Set: Using different cumulative Minkowsky-metrics led often to same results with Iris-data set but some differences was recogniced. Best results 73.33 % was found with power value p=2.
Wine recognition-data Set: Best results was found with Minkowsky metric when power value was three.
Thyroid gland-data Set: With thyroid gland-data set results were much better with optimal weights. Also optimal power value was p=3.

Table 2: Classification results (percentage of correct classifications) with three different data sets and cumulative Minkowsky metrics used

Data\Metric	L_1	L_2	L_3	$L_{0.5}$
Wine(weighted)	22.47	55.06	69.66	53.93
Wine	22.47	55.06	59.55	53.93
Iris(weighted)	66.67	73.33	66.67	66.67
Iris	66.67	66.67	66.67	64.00
Thyroid(weighted)	53.70	68.52	83.33	75.93
Thyroid	14.81	68.52	23.15	43.52

Table 3: Comparison for cumulative euclidean metric and normal euclidean metric

Class:	1	2	3
Iris(Cumulative Euclidean)	96.00	96.00	96.00
Iris(normal Euclidean)	94.67	94.67	94.67

In table 3 there is a comparison with cumulative euclidean metric and normal euclidean metric used in similarity. Here even classification results are worser with cumulative metric than with euclidean metric, class 3 was managed to classify 100 % correctly which was never the case in normal minkowsky metrics.

Statistical metrics:

With statistical metrics like Mahalanobis distance and Modified G there are some limitations and they can not be used always. With Mahalanobis distance covariance matrix is calculated and this can not be done always cause the singularity of matrix. With Modified G there is logarithm involved and this causes problems when we are close to zero. Because of these reasons we were not able to produce all results in these cases.

Iris-data Set: With Iris data set weighting enhanced results significantly when we used modified G metric. Best results were found by using Mahalanobis metric.
Wine recognition-data Set: Here covariance matrix could not be calculated so we got results only with Modified G metrics. Results were again much better with optimal weights.
Thyroid gland-data Set: With thyroid data set we were able to use Mahalanobis distance but not Modified G distance because of logarithm. Results with Mahalanobis distance were pretty good again.

Table 4: Classification results (percentage of correct classifications) with three different data sets and statistical metrics used.

Data\Metric	Modified G	Mahalanobis
Wine(weighted)	70.79	0
Wine	40.45	0
Iris(weighted)	66.67	96.00
Iris	33.33	93.33
Thyroid(weighted)	0	87.96
Thyroid	0	66.67

4. Conclusions

In this study we have shown that fuzzy similarity is very useful in classification and results are highly dependable on underlying metric. Usual Manhattan and Euclidean metrics are not always the best choices but sometimes better results can be found using other metrics.

Weighting had biggest enhancements to classification results with cumulative metrics and with statistical metric modified G. Biggest difference in classification results in weighting compared to results without weighting was over 60 % (see table 2) so clearly proper weighting has something new to offer to classification process.

The major advantages of the method is that it provides semantical information about the classification task and it allows partial membership of the class.

The next step is to develop an algorithm that calculated similarities to its neighborhood. Class of the sample is based on this information.

References

[1] Zadeh, L.: *Similarity Relations and Fuzzy Ordering.* Inform Sci, 3, 1971
[2] Formato, F., Gerla, G., Scarpati, L: Fuzzy Subgroups and Similarities. Soft Computing, 3, 1999
[3] Valverde, L.: On the Structure of F-Indistinguishability Operators. Fuzzy Sets and Systems, 17,1985
[4] Kämäräinen, J., Kyrki, V., Ilonen, J., Kälviäinen, H.: Similarity Measures for Ordered Histograms. paper published in the proceedins of the SCIA2001 confenrence, Bergen, Norway, 2001
[5] Könönen, V., Luukka, P., Saastamoinen, K.: *New Classifier Based on Maximal Fuzzy Similarity.* Paper sent for review to Fuzzy Sets and Systems
[6] Luukka, P., Saastamoinen, K., Könönen, V.: *A Classifier Based on the Maximal Fuzzy Similarity in the Generalized Łukasiewicz –structure* Paper published in the FUZZ-IEEE 2001 Conference.
[7] Kaski, S., Sinkkonen, J., Peltonen, J.: Bankruptcy Analysis with Self-Organising Maps in Learning Metrics. IEEE Transactions on Neural Networks, Vol. 12. No. 4, July 2001
[8] UCI Repository of Machine Learning Databases network document. Referenced 4.11.2000. Available: *ftp://ftp.ics.uci.edu/pub/machine-learning-databases*
[9] Turunen, E.: *Mathematics behind Fuzzy Logic.* Advances in Soft Computing, Physica-Verlag, Heidelberg, 1999
[10] Novak, V.: *On the Syntactico-semantical Completeness of First-Order Fuzzy Logic.* Kybernetika 26, 1990
[11] Klawonn, F., Castro J.L.: *Similarity in Fuzzy Reasoning.* Math Soft Comp, 2, 1995
[12] Trillas, E., Valverde L.: *An Inquiry into Indistinguishability Operators.* in: 'Aspects of Vaguenes.' Skala, H.J., Termini, S., Trillas, E., eds., Reidel, Dordrecht, 1984
[13] Höhle, U., Stout, L.N.: *Foundations of Fuzzy Sets.* Fuzzy Sets and Systems, 40, 1991
[14] Dubois, D., Prade, H.: *Similarity-Based Approximate Reasoning.* in: 'Computational Intelligence Imitating Life.' Zurada, J.M., Marks II, R.J., Robinson, C.J., eds., IEEE Press, New York, 1994
[15] Takeshita, T., Nozawa, S., Kimura, F.: On the bias of Mahalanobis distance due to limited sample size effect. in: 'Proceedings of the Second International Conference on Document and Recognition.', 1993, pp. 171-174
[16] Sokal, R.R., Rohlf, F.J.: Biometry. W.H. Freeman, New York, 1969

KES 2002
E. Damiani et al. (Eds.)
IOS Press, 2002

Comparison of the Fuzzy Similarity Based Classification in the Normal and the Generalized Łukasiewicz-Structure

KALLE SAASTAMOINEN and PASI LUUKKA
Department of Information Technology
Lappeenranta University of Technology
P.O. Box 20, FIN-53851, Finland

Abstract. In this article we will introduce classification results what we have got by using generalized Łukasiewicz-Structure and cumulative similarity measure and compared results to similarity based classification what we have done previously in normal Łukasiewicz-Structure with different metrics.

1. Introduction

Our previous articles we have used Łukasiewicz-structure in classification. We have achieved good results both in using normal and so called generalized Łukasiewicz-structure. When we have applied generalized form of the Łukasiewicz-structure we have used genetic algorithm to improve our classification results.

Normal Łukasiewicz-structure is much used in fuzzy similarity applications. See [1], [2], [3] for examples in classification, water reservoir operation and traffic signal control. Generalized form on the other hand is not that common. To our knowledge it has been used in classification in our articles [4] and [5] but not so much in other applications.

Our motivation for this paper is to compare and test some already quite good results what we have got from the our previous research activities between weighted fuzzy similarity and metrics in classifying [1], [4] and [5].

The data sets were chosen as diverse as possible so that classifiers properties would be apparent. Data sets were taken from a UCI Repository of Machine Learning Database [6] archive so that they were differently distributed and their dimensions varied. Classifiers were implemented with MATLAB^{TM}-software.

2. Łukasiewicz-Structure in Pattern Recognition

There are three reasons why we have chosen to use Łukasiewicz-structure in defining memberships of objects. One reason is that Łukasiewicz-structure holds the fact that the mean of many fuzzy similarities is still a fuzzy similarity [7]. Secondly it also has a strong connection to the first-order fuzzy logic [8], which is a well studied area in the modern mathematics. Thirdly it also holds the fact that any pseudo-metric induces fuzzy similarity on a given non-empty set X with respect to the Łukasiewicz conjunction [9]. Next we will shortly introduce mathematical background concerning fuzzy similarity and generalized

Łukasiewicz-structure as a base of a fuzzy classification method. The more detailed mathematical background can be found from [5].

2.1. Mathematical Background

We can construct a lattice called **normal Łukasiewicz-structure** or more formally just Łukasiewicz-structure:

Definition 1: Łukasiewics-structure: $a \odot b = \max\{a+b-1,0\}$, $a \to b = \min\{1,1-a+b\}$

If we examine Łukasiewicz-valued fuzzy similarities S_i $i=1,...,n$ in a set X we can

define a binary relation in L by stipulating $S\langle x,y \rangle = \dfrac{1}{n}\sum_{i=1}^{n} S_i \langle x,y \rangle$ for all x, $y \in X$. It is

easy to prove that this is still a Łukasiewicz-valued fuzzy similarity [7].

Definition 2: Letting L be the real unit interval [0,1] endowed with the usual order relation, we may construct the following usual residuated lattice: Generalized Łukasiewicz-structure:

$a \odot b = \sqrt[p]{\max\{a^p+b^p-1,0\}}$, $a \to b = \min\left\{1,\sqrt[p]{1-a^p+b^p}\right\}$, where p is a fixed natural number.

Now we are ready to study how to use fuzzy similarity for finding similar pairs. We are examining a choice situation where features of different objects can be expressed in values between [0,1]. Let X be the set of m objects. If we know the similarity value of the features $f_1,...,f_n$ between objects, we can choose the object that has the highest total similarity value. The problem is to find for object x_i similar object x_j, where $1 \le i,j \le m$ and $i \ne j$. By choosing Łukasiewicz-structure for features of the objects we get n fuzzy similarities for comparing two objects (x_1,x_2):

$$S_{f_i}\langle x_1,x_2 \rangle = x_1(f_i) \leftrightarrow x_2(f_i), \tag{1}$$

where, x_1, $x_2 \in X$ and $i \in \{1,...,n\}$. Because Łukasiewicz-structure is chosen for membership of objects, we can define the fuzzy similarity as follows:

$$S\langle x_1,x_2 \rangle = \frac{1}{n}\sum_{i=1}^{n}(x_1(f_i) \leftrightarrow x_2(f_i)). \tag{2}$$

This holds only in the Łukasiewicz-structure. Moreover, in the Łukasiewicz-structure we can give different non-zero weights $(W_1,..,W_n)$ to the different features and we get the following formula which again meets the definition of the fuzzy similarity:

$$S\langle x_1,x_2 \rangle = \frac{\sum_{i=1}^{n} W_i\left(x_1(f_i) \leftrightarrow x_2(f_i)\right)}{\sum_{i=1}^{n} W_i} \tag{3}$$

In Łukasiewicz-structure equivalence relation $a \leftrightarrow b$ is defined as $1-\max\{a,b\}+\min\{a,b\}$ or equivalently $1-|a-b|$. The formula of fuzzy similarity in such a case is:

$$S\langle x_1,x_2 \rangle = 1 - \frac{\sum_{i=1}^{n} W_i\,|x_1(f_i)-x_2(f_i)|}{\sum_{i=1}^{n} W_i} \tag{4}$$

It should be noted that in normal Łukasiewicz-structure we have used fuzzy similarity in the form of $S\langle x_1,x_2\rangle = 1 - \dfrac{\sum_{i=1}^{n} W_i d\left(x_1(f_i),x_2(f_i)\right)}{\sum_{i=1}^{n} W_i}$ and this why I notate it as 'normal' from this on distance is also normed between $[0,1]$.

In the generalized Łukasiewicz-structure equivalence relation can be defined as $\min\{\sqrt[p]{1-a^p+b^p},\sqrt[p]{1-b^p+a^p}\}$.

2.2. Minkowsky Metrics Counterparts in the Generalized Łukasiewicz -structure

Well-known distance measures based on Minkowsky metric are defined in following Formulae (N is the number of dimensions and p is the positive real number):

Minkowsky:

$$d(x,y) = \left(\sum_{i=1}^{N} |x_i - y_i|^p \right)^{\frac{1}{p}} \tag{5}$$

Cumulative Minkowsky:

$$d(x,y) = \left(\sum_{i=1}^{N} \left| \sum_{u=1}^{i} x_u - \sum_{u=1}^{i} y_u \right|^p \right)^{\frac{1}{p}} \tag{6}$$

Similarity measure based on the Minkowsky metric:

$$S\langle x_1,x_2\rangle = \frac{\sum_{i=1}^{n} W_i \sqrt[p]{1 - \left|x_1^p(f_i) - x_2^p(f_i)\right|}}{\sum_{i=1}^{n} W_i} \tag{7}$$

Similarity measure based on cumulative Minkowsky:

$$S\langle x_1,x_2\rangle = \frac{\sum_{i=1}^{n} W_i \sqrt[p]{1 - \left|\left(\sum_{u=1}^{i} x_1(f_i)\right)^p - \left(\sum_{u=1}^{i} x_2(f_i)\right)^p\right|}}{\sum_{i=1}^{n} W_i} \tag{8}$$

2.3 Description of Pattern Recognition Method in the Algorithmic Form

Method starts with fuzzification of a data set, which means, in practice, scaling the data between zero and one. After this an ideal vector (e.g. mean vector) is calculated for every class. Samples are classified so that the fuzzy similarity value between ideal and test vector is calculated. The sample can be classified to the class with the highest similarity value using generalized Łukasiewicz structure. The method gets test element *test* (dimension *dim*), learning set *learn* (dimension *dim*, different classes) and dimension (*dim*) of the data as its parameters. In addition, different weights for features can be set in *weights* and value p can be set for generalized Łukasiewicz-structure. The method is presented in the pseudo-code form below:

Require: *test,learn[1…n],weights,dim*
  Scale *test* between [0,1]
  Scale *learn* between [0,1]
  For i = 1 to n **do**
    *Idealvec[i]=*IDEAL[*learn[i]*]

$$\max\ sim\,[i]=\frac{\sum_{j=1}^{\dim} weights\,[j]\sqrt[p]{1-\left|idealvec\,[i][j]^{p}-test\,[j]^{p}\right|}}{\sum_{j=1}^{\dim} weights\,[j]}$$

  end for
Class = arg max$_i$, maxsim[i]

In the algorithm, *IDEAL[i]* is the vector that best characterizes the class *i*. In this paper, the *IDEAL* operator is the mean vector of the class. For this algorithm, weights can be optimized for example by using genetic algorithms in optimization.
Optimal weights and power values have been chosen by using genetic algorithm. Algorithm run through 100 generation or until every sample was classified correctly.

3. Empirical Results

3.1. Maximal Fuzzy Similarity with Weight Optimization in Classification

Classification is a good way to test how weighted similarity measure works in practice. *A priori* information for determing weights was not available for data sets. We used a genetic algorithm to find the optimal weights for our fuzzy similarity classifier. Genetic algorithm was used because of their diversity and robustness. We used three different data sets. Data sets were splitted in half. One half was used for learning and the other half for testing.

Iris-data Set Data set consists of three classes and it is four dimensional.
Thyroid gland-data Set Data consist also of three classes and it is five dimensional.
Wine recognition-data Set Data consist of three classes and it is 13 dimensional.

Minkowsky-metrics in 'normal' Łukasiewicz- vs. generalized Łukasiewicz-structure:

Wine recognition-data Set: In all power value cases classified here 'normal' Łukasiewicz-structure worked slightly better than generalized.
Iris-data Set: Using different Minkowsky-metrics led to different results with Iris-data set in both cases. Best results for normal Łukasiewicz-structure 96% was found with power values 1, 2 and 3. In most of the cases generalized form of the Łukasiewicz-structure worked better than 'normal' but when power value was 100 'normal' form worked better
Thyroid gland-data Set: With thyroid gland-data set results were better using generalized Łukasiewicz-structure with power value 0.5 in other cases normal worked again slightly better than generalized.

Table 1: Comparison of classification results (percentage of correct classifications) with three data sets.

Data (L-structure)	L$_1$	L$_2$	L$_3$	L$_{0.5}$	L$_{100}$
Wine (normal)	95.51	95.51	97.75	95.51	94.38
Wine (general)	95.51	95.51	92.14	94.38	87.64
Iris (normal)	96.00	96.00	96.00	94.67	94.67
Iris (general)	96.00	96.00	96.00	96.00	76.00
Thyroid (normal)	94.44	94.44	93.52	94.44	95.37
Thyroid (general)	94.44	94.44	92.59	95.37	89.81

Cumulative Minkowsky-metrics in 'normal' Łukasiewicz- vs. generalized Łukasiewicz-structure:

Wine recognition-data Set: Best results was found with 'normal' Łukasiewicz-structure when power value was three.
Iris-data Set: Generalized Łukasiewicz-structure found best results with power value two.
Thyroid gland-data Set: Best results was found with 'normal' Łukasiewicz-structure when power value was three.

Table 2: Comparison of classification results (percentage of correct classifications) with three different data sets and cumulative Minkowsky metrics used.

Data (L-structure)	L_1	L_2	L_3	$L_{0.5}$
Wine (normal)	22.47	55.06	69.66	53.93
Wine (general)	32.95	40.45	64.04	39.33
Iris (normal)	66.67	73.33	66.67	66.67
Iris (general)	64.00	86.67	65.33	61.33
Thyroid (normal)	53.70	68.52	83.33	75.93
Thyroid (general)	70.09	71.30	71.30	69.44

In table 3 (below) we see how well classifiers has managed with Iris data which is divided for three classes. Here we can see that third class was classified 100 % correctly. This was never the case in normal Minkowsky metrics.

Table 3: Comparison for cumulative similarity and normal similarity measure with best results in power 2.

Class	1	2	3
Iris data (cumulative Minkowsky generalized)	100%	60%	100%
Iris data (cumulative Minkowsky normal)	100%	20%	100%
Iris data (Minkowsky)	100%	96%	92%

4. Conclusions

As we can see from the results, there were no big differences in classification using generalized form of the Łukasiewicz-structure or normal form. We can also see that results in general got worse when we applied cumulative Minkowsky to the classification. But also some new information was found by using cumulative Minkowsky. This indicates that by combining normal and cumulative Minkowsky even better results might be found.

  The next step is to develop an algorithm that combines normal and cumulative Minkowsky based similarity.

References

[1] Könönen, V., Luukka, P., Saastamoinen, K.: *New Classifier Based on Maximal Fuzzy Similarity.* Paper sent for review to Fuzzy Sets and Systems.
[2] Dubrovin, T. Jolma, A. Turunen, E.: *Fuzzy Model for Real-Time Reservoir Operation.* Journal of Water Resources Planning and Management, 2002.
[3] Niittymäki, J. Turunen, E.: *Traffic Signal Control on total Fuzzy Similarity Based Reasoning.* Paper accepted for publication in the Journal of Fuzzy Sets and Systems, International Journal of Soft Computing.
[4] Luukka, P., Saastamoinen, K., Könönen, V.: *A Classifier Based on the Maximal Fuzzy Similarity in Generalized Łukasiewicz –structure* Paper published in the FUZZ-IEEE 2001 Conference.
[5] Saastamoinen, K., Könönen, V., Luukka, P.: *A Classifier Based on the Fuzzy Similarity in the Ł ukasiewicz-Structure in Different Metrics.* Paper accepted for publication in the FUZZ-IEEE 2002 Conference, Honolulu, Hawaii.
[6] UCI Repository of Machine Learning Databases network document. Referenced 4.11.2000. Availa *ftp://ftp.ics.uci.edu/pub/machine-learning-databases.*
[7] Turunen, E.: *Mathematics behind Fuzzy Logic.* Advances in Soft Computing, Physica-Verlag, Heidelberg, 1999
[8] Novak, V.: *On the Syntactico-semantical Completeness of First-Order Fuzzy Logic.* Kybernetika 26, 1990
[9] Klawonn, F., Castro J.L.: *Similarity in Fuzzy Reasoning.* Math Soft Comp, 2, 1995

KES 2002
E. Damiani et al. (Eds.)
IOS Press, 2002

Analysis of Cortical Connectivity using Hopfield Neural Network

S.Dixit*, K.Mosier**, Sidhartha Shankar[+], and Dinesh Mital[++]

*Department of Neurology, University of Medicine and Dentistry of New Jersey,
Newark, NJ, USA
**Departments of Radiology and Surgery, Memorial Sloan-Kettering Cancer Center,
New York, NY, USA
[+]Research and Development, Paytrust, Inc.
Lawrenceville, NJ, USA
[++]Department of Health Informatics, SHRP, University of Medicine and Dentistry of NewJersey, NJ,
USA.

Abstract:

There is striking similarity in the connectivity between perceptrons in an Artificial Neural Network and neurons in brain. Therefore it is a natural logical step to investigate cortical connectivity using Artificial Neural Networks. Present approaches to ascertaining cortical connectivity, e.g. Structural Equation Modeling between various regions of interest (ROI) in the active brain are-tedious and time-consuming. For example, modeling the connectivity of a large number of brain regions often involves numerous parameter changes to achieve a good fit. Functional Magnetic Resonance Imaging (fMRI) is increasing recognized as a standard technique for brain mapping. This study explores the utility of a Hopfield Neural Network to determine cortical connectivity in an fMRI data set.

Introduction:

Artificial Neural networks are being widely used in medicine in general and in Neuroinformatics in particular. Neural networks or artificial neural networks are based on the fact that perceptrons in neural networks simulate the neuronal architecture in the human brain.

In human brain, when there is an activation of region in response to a stimulus, a neurotransmitter is released into the synaptic cleft. Depending on whether the neurotransmitter is inhibitory or stimulatory, the neurons will fire. Similar theory is adapted while studying neural networks. In an artificial neural net, the network architecture is constructed whereby each node represents a synapse. When the total value of an input reaching any node is positive, an output is generated, whereas no output will generated when the input is below the set threshold.

fMRI has come to be recognized as a very important technique in analyzing functional brain activity. fMRI data are acquired as the subject performs some cognitive, sensory or motor task and the data displayed as a colorized map of active regions overlaid on MR images. An important step is to determine how these activated regions are functionally connected during performance of a particular task. Structural Equation Modeling (SEM) is now being widely used as statistical tool to find connectivity between various regions of the brain. However, there are a number of limitations to SEM. SEM techniques are time consuming and tedious. SEM is based on multiple regression techniques to compare a connectivity model based on the observed data to a connectivity model based on hypothesized or known neuroanatomical models. Fitting models containing a large number of brain regions, for example, often requires multiple parameter changes, or changes to the model, to achieve a good fit. Moreover, as research extends into higher cognitive functions, neuroanatomical animal models are of limited value in guiding these decisions [1]. Neural networks may offer advantages to circumvent some of these limitations in SEM. In particular, neural network techniques may provide increased computational speed for larger data sets, and the ability to efficaciously determine the effects of changing connectivity weights in one region of a large network. This paper presents a plausible neural network, which can simulate the way different areas in brain are connected while performing sensorimotor object exploration tasks.

Purpose:

Our purpose in this investigation was to simulate the cortical networks using Artificial Neural Networks during manipulation of objects of differing geometry in the absence of visual input. The question we wished to address was whether the cortical connectivity model obtained using Artificial Neural Network would be consistent with the model obtained by Structural Equation Modeling.

Materials and Methods:

Four healthy right-handed adult human subjects (3M, 1F) were imaged using Functional Magnetic Resonance Imaging (fMRI) techniques. Subjects were imaged on 1.5T G.E. Horizon Echospeed MR system using a standard quadrature "bird-cage" head coil. Images were acquired using conventional BOLD (Blood Oxygen Level Dependent) techniques with the following acquisition parameters: TR/TE = 4000/60; 64 x 64 matrix, 24 cm FOV, 5 mm slice thickness, 90^0 flip angle A total of 28 slices in the axial plane were acquired for each task in each subject Subjects manipulated a total of four objects, using an event-related paradigm design. Each subject freely explored a series of wooden objects among the fingers of the dominant hand for 20 seconds. Each object exploration epoch was separated by a 32 second baseline. The objects were 1-2 cm in greatest dimension and consisted of a square, triangle, rectangle and cylinder. Subjects performed this task with their eyes closed to avoid visual input The raw echo-planar data was reconstructed off-line using software routines written in IDL (Interactive Data Language), motion corrected and statistical maps of activation generated with SPM99. Statistical maps were generated with a SPM (Z) threshold set to 3.09 (p <0.001) and the images normalized to Talairach coordinates and overlaid to T1-w images.

Our research work was divided into three steps: Collection of data, Processing of data and Neural Network analysis of data using MATLAB (6.1).

Auto-Associative Neural Network:

The Auto-Associative Neural Architecture used in our study was initially described by Hopfield [2]. The neural network consists of a set of nodes, which are connected to each other and which are arranged in layers; the simplest of which is a two-layer network consisting of an input layer and an output layer. The time delay is assumed to be same for transmission of information from any one node to another. Activation of a node at any particular time is determined by one of the two states active represented by 1, or inactive represented by 0. The net synaptic input is obtained by the sum of input from all neurons from which it is receiving connections.

Algorithm: (for auto-associative neural network)

The net synaptic input is determined by the following equation:

$$E_i = \sum_{j \ne i} w_{ji} \, (2^* \, s_j - 1)$$

where E_i = net Synaptic Input

S_j = State of neuron (active or inactive)

W_{ji} = Weight of connection between

neuron j and neuron i.

Methodology and Neural architecture:

fMRI data, which showed activated areas, was analyzed using IDL and SPM Image Processing tools. The amount of activation was calculated and called Score. This data served as input for autoassociative neural network. The number of parameters were same as number of areas activated during the experiment task and as obtained by the analyses of this data. Initial weights were zero, and the areas only had association with themselves in the weight matrix. Numbers of runs were adjusted until stable matrix was obtained with output same as target. Hebbs rule was used to get the Initial weight matrix and HW rule was used to adjust these weights. The weight matrix so obtained was said to show the connectivity between various areas. The Mean Square Error (MSE) was as near Zero as possible.

Training:

Various algorithms of LEARNING were used for neural network training during the course of study. These algorithms were used to update the connection weights. An effort was made to use the learning algorithm that can give best estimate of connectivity between area.

Learning rules, for a connectionist system, are algorithms or equations, which govern changes in the weights of the connections in a network. One of the simplest learning rules, Hebbian Learning Rule, based on a rule initially proposed by Hebb in 1949 was used in study. <u>Hebb's rule states that the simultaneous excitation of two neuron results in a strengthening of the connections between them</u> . Connections between two neurons will strengthen, if more often than not the two neurons fire together. Learning rules incorporating an error reduction procedure utilize the discrepancy between the desired output pattern and an actual output pattern to change (improve) its weights during training. The learning rule is typically applied repeatedly to the same set of training inputs across a large number of epochs or training loops with error gradually reduced across epochs as the weights are fine-tuned.

Training of the neural network was unsupervised; the system learns locally and automatically; the neurons adjusted their interconnectivity by adjusting their weights depending on their co-activation.

The projections from one neuron to another are called synaptic weights, which can be either positive or negative. Initially all areas were assumed to be connected to all other areas. The values of weights obtained at equilibrium after training the neural network were either positive or negative. Positive weights depicted connectivity between the two regions, were kept in the model and for negative weights; existing literature was referred to determine their inclusion in the model.

Widrow-Hoff Learning and the LMS Algorithm:

In 1960 Bernard Widrow and his graduate student Marcian Hoff developed a new neural network and a new learning rule which they called the LMS (Least Mean Square) Algorithm. Therefore the algorithm tries to find the best weight matrix, W, and bias, b, to minimize error. The neural network devised by Widrow and Hoff is

$$a = Wp + b$$

where p is the input pattern and a is the output. To obtain W and b the following equations are used

$$W(k + 1) = W(k) + 2ae(k)\, p^T(k),$$

and

$$b(k + 1) = b(k) + 2ae(k)$$

Results and Conclusion:

The model obtained by auto-associative neural network represents the anatomical connections between areas believed to be involved in object manipulation task. Our study shows that the auto associative artificial neural network model can simulate the way regions of interest (ROI) are connected in cortex. Once the model is trained, it can be used to determine connection between various ROIs in cortex, with different data sets.

The best results were found using Hopfield neural network with Hebbs' rule of learning. The final stable matrix was obtained using HW rule of training.

Figure below shows the number of runs, and MSLE as obtained by neural network. Conventionally structural equation modeling is widely used to get functional connectivity between various areas of activation. The basic drawbacks with SEM methods are that we use prepiori condition in which, we refer literature to decide that what all areas are connected functionally. Once the model is made in which we group the areas that are functionally connected together, we use path analysis to find whether the model is good fit or not. Many runs have to be done before good fit is obtained.

The advantage of neural network is that we don't assume anything, nothing is hypothesized, so we are trying to reduce the human bias. Second, we are feeding the initial fMRI data with little processing as input. Once the neural network is training, it's very quick to get the connectivity matrix. Training it self doesn't take long.

Drawback and restrictions:

Some of the major differences between brains and computers were spelled out in the following terms by Churchland (1989, p.100):"The brain seems to be a computer with a radically different style. For example, the brain changes as it learns, it appears to store and process Information in the same places...Most obviously, the brain is a parallel machine, in which many interactions occur at the same time in many different channels."

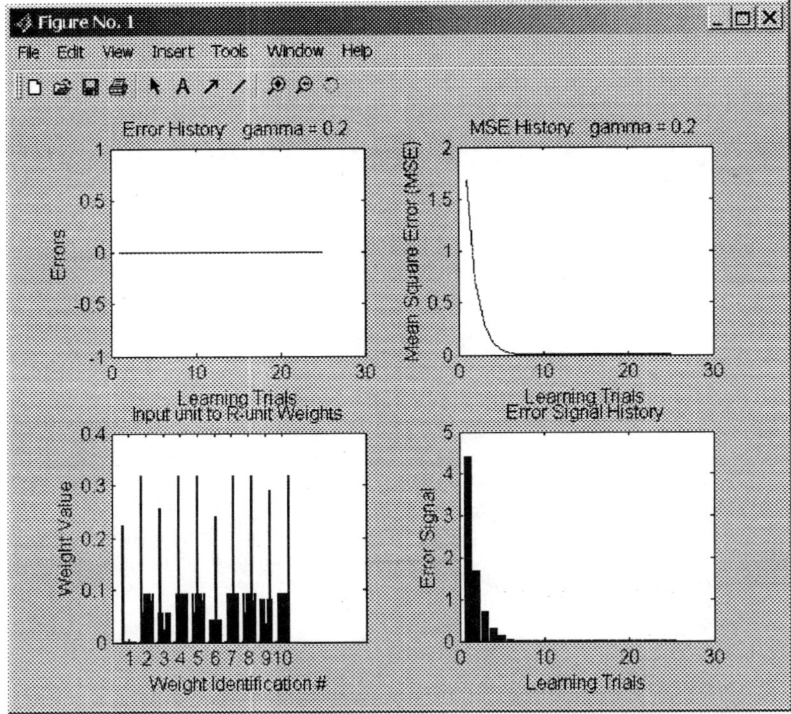

This contrasts with most computer functions, which involves serial processing and relatively few interactions. Another drawback using the neural network is that, as with all autoassociative neural network sometime we get unstable points and second, we are not able to know in are connection matrix that whether these areas are stimulating or inhibitory all we know is they are functionally connected, so this still has to be answered.

References:

[1]. McKiernan, K.A., Conant, L.L., Chen, A., Binder, J.R. (2001, June). *Development and cross-validation of a model of linguistic processing using neural network and path analysis with FMRI data.* Poster presented at the annual meeting of Human Brain Mapping, Brighton, UK

[2]. Hebb, D.O. Organization of behavior (New York, Wiley, 1949)

[3]. Hopfield, J.J.Neural Network and physical systems with emergent collective computational abilities. Proceedings *of National Academy of Science 79(1982)*. 2554-2558.

[4].http://www.psych.ualberta.ca/~mike/Pearl_Street/OldDictionary/H/hebbian_learning_rule.htm

KES 2002
E. Damiani et al. (Eds.)
IOS Press, 2002

A Point of Care Patient Repository Using Java Speech Technology

Suneela Damarapu, Shankar Srinivasan, Dinesh Mital and Syed Haque
UMDNJ-SHRP
65 Bergen St.
Newark, New Jersey
USA 07107

Abstract

Voice recognition software as it comes out of the box is not adequate for professionals such as physicians who need additional highly specialized vocabulary specific to their subject matter. Once such a vocabulary is created and setup it is imperative to develop a voice based search for the terms, definitions or information such as images, acquired signals etc. pertaining to specific medical content. In this paper we report the design and implementation of a one such system, namely the Medical Speech Interface System (MSI) for specialized users (in this case Respiratory Specialists) to do point of care information transaction processing with voice technology.

1. Introduction

Speaking with one's voice is the most natural way to communicate. Indeed for more than three decades, engineers and speech specialists have been working towards the development of various kinds of machines that automatically recognize and interpret what people say. Such machines are mainly based on computers or computing processor chips, perform 'speech recognition' or 'voice recognition'. Voice recognition is claimed to be the next big thing in technology finance [1][2] rivaling Internet investment. Voice recognition software for dictation, for example, is now both practical and economical too. However, as it comes out of the box, this software is not adequate for professionals such as physicians who need additional highly specialized vocabulary specific to their subject matter. Once such a vocabulary is created and setup it is imperative to develop a voice based search for the terms, definitions or information such as images, acquired signals etc. pertaining to specific medical content. Furthermore with the Health Care Industry's rapidly changing regulations and ever increasing requirements for detailed documentation, it is easy to see why voice recognition technology has been rated the number one emerging healthcare information technology.

In this paper we report the design and implementation of a Medical Speech Interface System (MSI) for specialized users to do information transaction processing with voice technology. The development of this system spans the use of a wide array of different hardware and software Entities. The ability to put together these different Entities to obtain a desired inter-working functional application to the end user is the purpose behind the development of this MSI System. The MSI System allows the End User to input sets of clinical data via voice. The output given by the MSI System in voice is the supplied sets of Clinical Output Information and the result of which, is achieved by using pre-determined procedures in the background. A set of Clinical Terms from the user is obtained from the form presented to the End User (the user dictates to the form). The set of Clinical Output Information is derived by a pre-calculated processing done by the internal engine of the MSI System (output dictated in voice).

2. Objectives

To synchronize the operations performed by the different Entities of the MSI System. To implement a voice activated interface in a client-server MSI to activate search queries and receive query results in the form of speech, images or text based on the client's preference. The goal is to retrieve exact clinical and demographic data of patients, for purposes of clinical practice, quality assurance and clinical research.

3.   Methodology

The Entities that comprise MSI are given below:

- a sensitive Microphone for Speech detection,

- a PCI card to buffer the speech signal,

- Dragon Naturally Speaking Software to process speech to text

- a database to store medical records and information

- a Java Speech Servlet program to serve as interface to the input – form that takes dictation from the end user, and to invoke internal tasks – database operations, post processing, present output to the end user (i.e. dictated to the user).

The overall schematic diagram of the speech interfaced MSI is illustrated below. The details are given in the following sections.

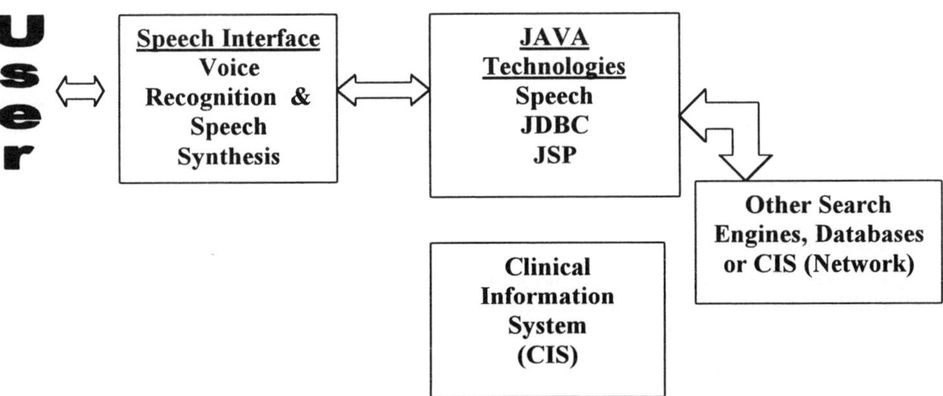

Figure 1: Overall Architecture of the Medical Speech Interfaced System.

3.1 Speech Interface

The speech interface in this implementation consists of both speech recognition and speech synthesis.

Speech recognition (implemented using a commercial sound card and interfacing software drivers) is the process of converting spoken language to written text. The heart of speech recognition is the recognition engine; the complex code that uses sophisticated statistics to match digitized sounds to words. Recognition engines analyze groups of two or three words (bigrams and trigrams) in sequence to determine the most likely word groupings [3].

The words or groupings of words in this report would be for "command and control" functions as for example, connecting to a particular database, navigating through documents and editing text.

Specifically for clinical applications as in the CIS, the next component is the clinical context or language model. This extends the base search engine dictionary by adding the unique lexicon or terminology used by a particular clinical specialty, say, Respiratory Diseases. Developing an effective context often requires collecting and analyzing thousands of hours of actual clinical dictation in the specific specialty.   The recognition system developed interfaces with the rest of the CIS as is illustrated in Figure 2 below.

The figure shows the speech recognition system integrated with the Hospital Information System (HIS) which includes archiving in Clinical Data Repositories (CDR) or Computerized Patient Records (CPR). The inbound interface (speech/voice recognition) imports relevant patient demographic and certain clinical information from the HIS or Practice Management Systems (PMS). The outbound interface (which includes speech synthesis) loads the completed reports into the CDR and/or CPR and distributes copies electronically to referring, attending and consulting. The availability of clinical and demographic data both eases dictation and provides a context for checking data.

3.2 Speech Synthesis

A speech synthesizer converts written text into spoken language. Speech synthesis is also referred to as text-to-speech (TTS) conversion. The major steps in producing speech from text are:

a)   Structure analysis to process the input text to determine where paragraphs, sentences and other structures start and end.

b)   Text pre-processing to analyze the input text for special constructs of the language.

c)   Text-to-phoneme conversion - to convert each word to phonemes.

d)   Prosody analysis to process the sentence structure, words and phonemes to determine appropriate prosody for the sentence.

e)   Waveform production is where the phonemes and prosody information are used to produce the audio waveform for each sentence. In this project Java Speech API is employed to provide the text to speech synthesis conversion.

4.   Sample Run

A simple example 'run' of the system is as follows. A Java Servlet coded with a Form is sent to the Client upon the invocation of the Servlet. The form has several Text Fields, radio buttons, checkboxes and a row of speech command sensitive buttons pertaining to choices of Image or Speech for the return of query results. The servlet will process the results of the Form and it will return the appropriate query result by means of the JDBC (Java Database Connectivity) Application Programming Interface (API). The retrieved results are then displayed on the client's screen, and an option in speech to the End User is available. Figure 3 below illustrates an example showing the outputs performed by the search engine after a voice based query request is made to it. The response from the servlet is also shown. To obtain results in 'speech', the user should select the 'speech' option as the desired output. This selected option is recognized internally by the servlet software. The servlet software translates the description of the query result from text to speech using the Java Speech API function calls. The Client – End user will hear in audio the output pertaining to the query result.

This system is oriented toward the practical and research needs of Respiratory Physicians. The database created in Oracle consists of the personal history of the patient, his/her clinical findings, laboratory results and diagnoses. The aim is to record, organize and retrieve clinical and demographic data of patients, for purposes of clinical practice, quality assurance and clinical research. A form (external view) can be created for the lab personnel to constantly update the laboratory findings using speech. X-ray images can also be uploaded into or from the database as is also shown below.

5. Conclusions

The results of this feasibility study have been very encouraging. Further investigations are planned to incorporate speaker independence, image compression and more natural voice synthesis abilities to the speech output. Economical and convenient generation of medical records without stenographic intervention and the consequent reduction of professional proofreading time will make it an irresistible tool for the Respiratory Physicians.

References

1. "Recognize the Power of Voice", Ed Paisley, *Global Tech Ventures*, Oct ' 98.

2. "Voice, Voice Everywhere" - Elizabeth Hurt, *Business* 2, Nov' 00.

3. "Automatic Recognition of Keywords in Unconstrained Speech using Hidden Markov Models", J.G. Wilpon et al., *IEEE Transactions on ASSP*, pp.1870-1888, Nov ' 90.

4. "Sun Releases Two Specifications Within Java Speech API", *Java Spot News*,   Sept' 97.

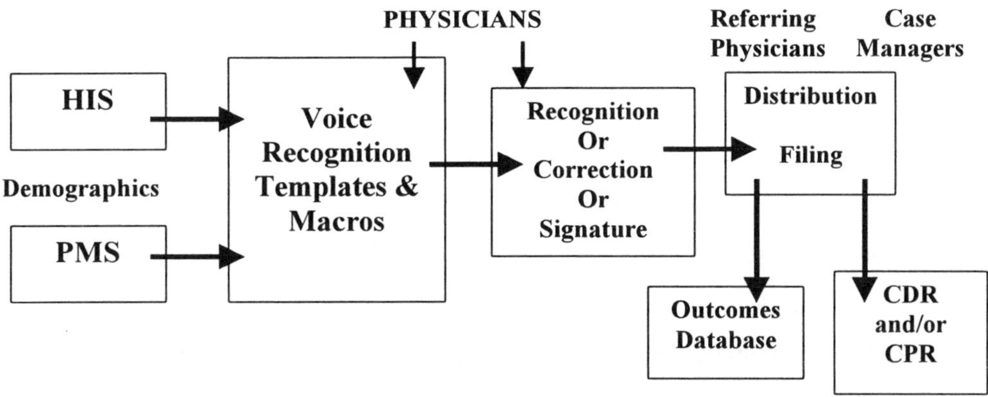

Figure 2: The overall CIS with the Speech recognition system incorporated

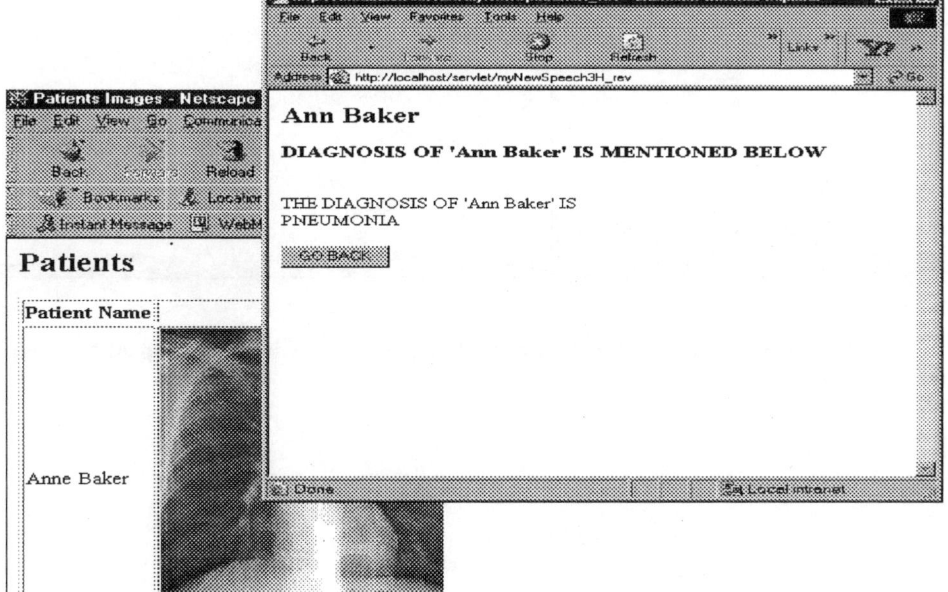

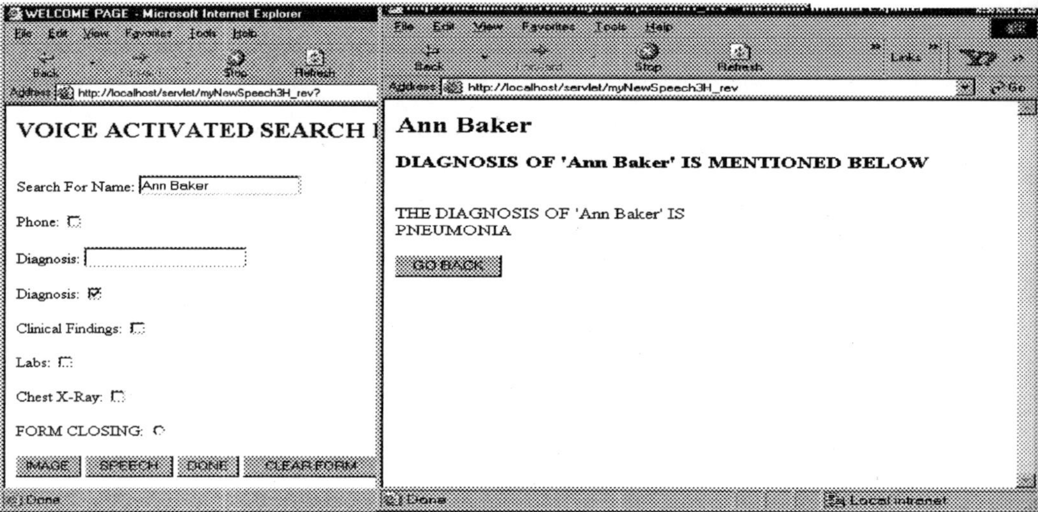

Figure 3: Examples of Queries and Responses in a simple version of the search engine.

KES 2002
E. Damiani et al. (Eds.)
IOS Press, 2002

Prediction of Hepatitis C using Artificial Neural Network

Rinki Jajoo, Dinesh Mital, Syed Haque and Shankar Srinivasan

School Of Health Related Professions Department Of Biomedical Informatics
University of medicine and dentistry of New Jersey
65 Bergen Street, Newark, New Jersey

Abstract

The main objective of this research project is develop an expert system module, based on a back propagation feed forward artificial neural networks (ANNs), for the diagnosis of hepatitis C and compare its performance with other existing computer based decision support systems. The ANN based system was developed with a commercially available software package (Brain Maker, California scientific Software). Two different types of ANN models, unsupervised and supervised, were developed, compared, and tested. The predictive accuracy and the model training for supervised model was significantly better. The model was able to predict the Hepatitis C in patients very accurately, however performance was not significantly better than the traditional computer model based techniques. Further investigations are needed to understand the impact of this methodology on the outcome analysis. An existing database of hepatitis C infected patient was used. Data of 15 infected and 20 normal individual were collected. Dichotomous variables were coded as present (1) or not present (0). Continuous variable were recorded for patient age, ethnicity, patient number and patient sex. The results have been very interesting, however, some more research work is required to fine-tune the results. The main advantage of the developed system is that it is adaptive and self- adaptive type.

Introduction:

An important application area of artificial neural network is medical diagnosis. Chronic infection with hepatitis C virus (HCV) is a major public health problem and is associated with over 10,000 deaths a year in the United States [1, 8]. In the early stages, HCV tends to be asymptomatic and can be detected only through screening.

Recently there has been a considerable interest in making computer-based models and develop decision support systems for health care providers for improving the quality of health care. Shea Steven and Clayton Paul [10] have proposed such a system. Other models have been developed by Sonke Gabe, et al [6] , and Shigehiko and Heber [7]. Neural networks have capability of solving clinical problems which are based on symptoms and patterns. There have been efforts in last few years to develop ANN based models for clinical applications. Freeman, et al [2], Lippmann and shahian [5] have proposed ANN based models successfully. ANN based models have advantage over other types of decision support systems, as ANN based systems are self organizing type and have learning capabilities.

Critics have argued that limitations of artificial neural networks include the subjective judgment in selecting which initial input variables to present to the artificial neural network for analysis. It has been suggested that an artificial neural network, if elaborate, can find a relationship between completely unrelated input variables.[9] To limit the emphasis on the incidental associations between subjectively determined input variable, it might be beneficial to guide artificial neural networks development by presenting a subset of initial variables that are more likely to predict an outcome. Such univariate predictors may have more legitimacy for association with an outcome when compared with variable identified by artificial neural network modeling alone. The supervised model might then be more generalizable to independent prospective data.

Objective:

Our objective was to develop an artificial neural network (ANN) based model and compare the results with model-based system (Rule based system). These types of expert system differ in how knowledge based is structured and consequently, how the inference engine works. Artificial neural networks are computer programs that can identify patterns of variables that can predict an outcome. They provide a non-linear approach to data analysis and have been successfully used to model clinical data with results comparable to traditional modeling techniques. The focus of the study was to construct an artificial neural network for prediction of Hepatitis C disease. In effort to limit the initial training variable distraction, a supervised artificial neural network was constructed with a subset of input variable.

Basics of Hepatitis C:

Hepatitis C is a disease caused by the hepatitis C virus (HCV), which infects mostly liver. A very small organism that attaches itself to healthy cells and forces them to produce more of the virus. Experts have identified two types of hepatitis C viruses [1,3]: acute Hepatitis C (short-term infection) and chronic Hepatitis C (long-term, progressive infection). These are broken into sub categories:

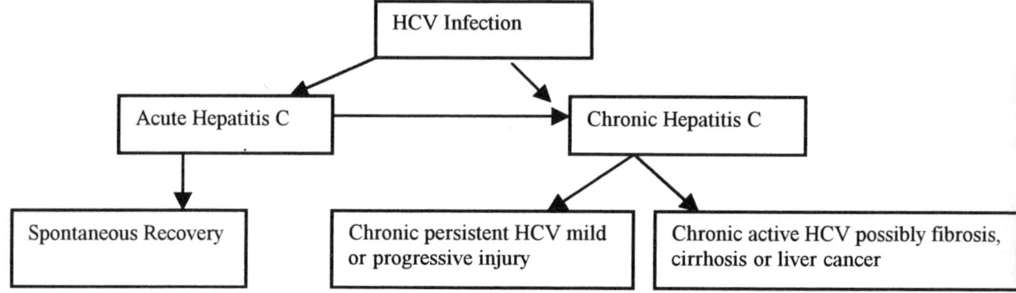

Figure1: Types of hepatitis C

Chronic hepatitis C can occur on its own and is generally the more serious of the two types. In about 80% of cases, however, acute hepatitis C develops into chronic hepatitis C over time. Both chronic hepatitis C and acute hepatitis C can cause very different degrees of symptoms and liver function damage. In many cases, HCV infection can be treated effectively with medication.

Database and Patient characteristics:

An existing data of hepatitis C infected patient was used. Data of 15 infected and 10 normal individual were collected. Dichotomous variables were coded as present (1) or not present (0). Continues variable were recorded for patient age, ethnicity, patient number and patient sex. The typical patient database is shown in Table 1. The information for the input variable for our variables was obtained through sources [8, 11-13].

Artificial Neural network

Artificial Neural networks use connectionist computational theory. They are composed of a series of computational node structured into several layers. Each node is connected to all nodes in the previous layer. The Artificial Neural network structure used for in this project consists of a three layered feed forward design. The first layer is an input layer in which each node represent an input variable; the second layer is a hidden computational layer and the third layer is a single output node representing the outcome for a given data record. (i.e., Hepatitis C)

There are two user-dependent parameter that affects the training. a) The learning rate and b) The momentum. The learning rate defines the amount that the weights are changed during each iteration. A larger learning rate leads to larger weight change. Momentum allows the weight change to be proportional to the previous weight change.
The Artificial Neural networks in our study used a sigmoid transfer function and were trained with back propagation learning algorithm. During training, the outcome is known for each input data is record. When the output node value for the Artificial Neural network is calculated, the value is compared with the known output from the training data. If there is a discrepancy between the calculated and the actual output, the error is propagated through the Artificial Neural network and weights are adjusted in an iterative manner until the error discrepancy is within the allowed tolerance.

History	Screening	Special Risk group
Age	Viral Genotype	Family member treated for Viral hepatitis
Sex	HCV-PCR(Polymerase Chain Reaction)	Maternal to fetal
Patient Number	Liver Biopsy (Histology)	Needle Pricing Injury
Blood transfusion (Prior to 1990)	Aminotransferase level (SGPT, SGOT, ALT)	Healthcare worker
Intravenous Drug Use		Hemophiliacs
Alcoholism		Frequent sex partner
Sexual Promiscuity (unprotected Sex)		Intranasal Cocaine
Any Surgery		Ear / Body piercing
Tattoo		Treated for Chronic Fatigue
Unknown		Ethnicity

Table 1. Input Variable

Once the training is completed, the neural network can be externally validated on the data for which only input values are provided. Although other methods of training exist, the majority of clinical applications of artificial neural network in the literature have used the back propagation-training algorithm used in their study.

Neural Network Based Model

The artificial neural networks based model was developed with a commercially available software product [14] Brain Maker, California scientific Software. The Data is divided into two groups. The first 25 records are used to train the data. The data used in training is from the past records of the hepatitis C patient where is output is known. The rest of the 10 data sets have only input node and the output node is derived from the Model. In addition to that an another ANN system was developed without training procedure and results were compared.

Experimental Results:

The outcome of the proposed model shows that the supervised model are better than unsupervised system. The training of the network is done using a number of facts. This system used minimum of 136 facts and maximum of 180 facts to train the model. There are 8 hidden neuron used by the hidden layer of the model. As the data represented in the table shows when the system is trained using the past facts it will try to learn after each iteration. This system was trained using 1218 facts in 87 training runs. During this period, the RMS error decreased (0.079). When system started learning there are 14 numbers of bad outputs and no good outputs. The hidden input and output connection shows a normal distribution.

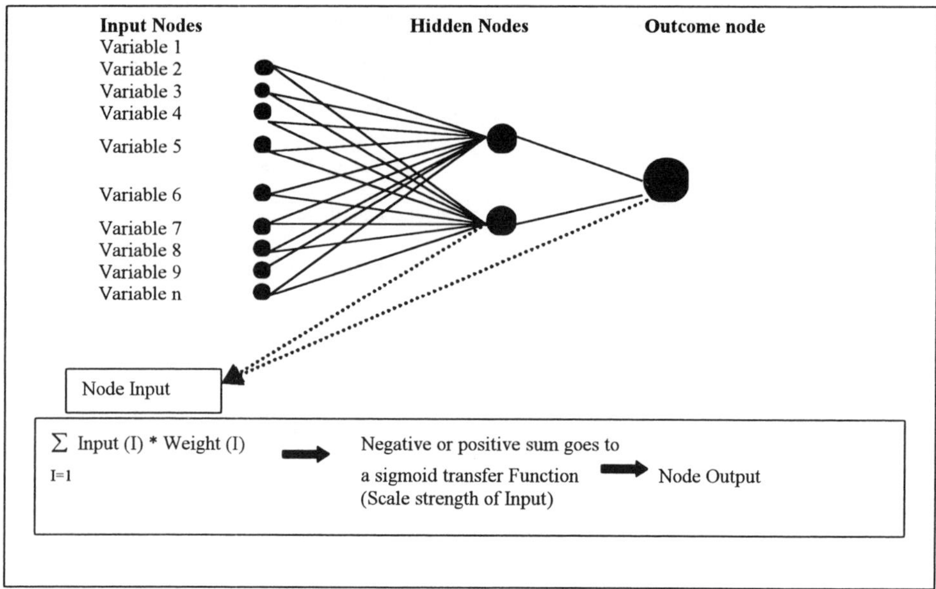

Figure 2: **Representation of the artificial neural network (3 layered network). Input nodes are all relevant diagnostic test results; Hidden nodes are used for calculation purpose; Output node represents the outcome.**

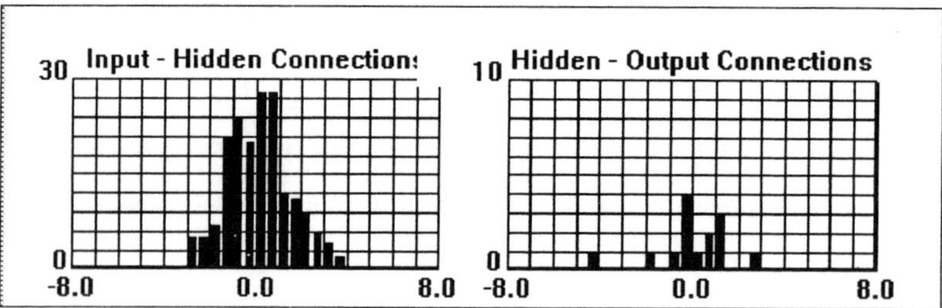

Figure 3: **Histogram plot for the Hidden input and output neurons. Plots shows normal distribution.**

An unsupervised model cannot be used to predict the clinical cases. The output is vague and unreliable. The supervised model is more specific and reliable. The system changed its layer weight according to the case presented to the model in each fact. As shown in the graph (Fig 4), the value of the input layer increases the output value increases with it.

Conclusion:

In summary, we have designed, implemented and tested artificial neural network and rule based models. Both the models are tested with normal and infected patient data. The prediction of the outcome is more specific and accurate using artificial neural network based model. The classification performance of the Supervised ANN model is superior than either rule based model or Unsupervised network . In addition, the ANN based system is self adaptive and learns with each additional input data provided to it. This model has been tested on a very limited and small patient database. However, we plan to test it rigorously on large patient database. We believe that this proposed model for predicting the Hepatitis C would assist doctors in the diagnosis of disease.

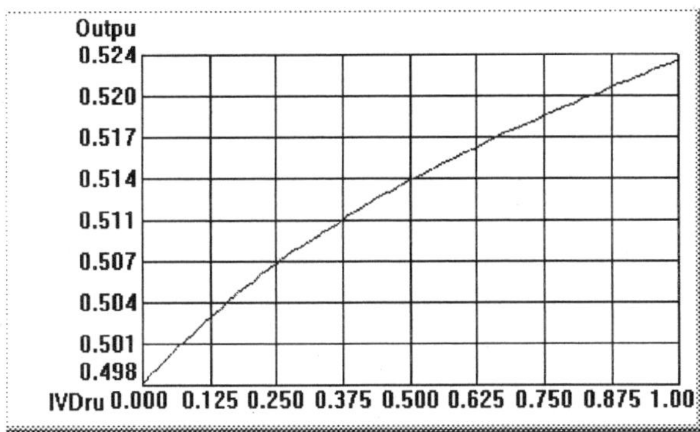

Figure 4 : Plot of Input Variables against Output

References

[1] Pratt Daniel and Kalpan Marshall, Evaluation of abnormal liver enzyme results in asymptomatic patients, The New England Journal of Medicine, Vol. 347 (17), April 2000, 1266 -1271.

[2] Freeman Rosario, Eagle Kim, Bates Eric, et.al, Comparison of artificial neural network with logistic regression in prediction of in-hospital death after percutaneous transluminal coronary angioplasty, American Heart Journal, Vol. 140(3), Sep 2000, 511-520.

[3] Fischer, Lucy Rose, Tope Dawn, and Conboy Kathleen, Screening for hepatitis C virus in Health maintenance organization, Vol160(11), !2 June 2000; 1665 –1673.

[4] Rinki Jajoo: Artificial intelligence in medicine, Internal Report, UMDNJ, May 2000.

[5] Richard P. Lippmann and David M. Shahian, Coronary artery bypass risk and prediction using neural network. Ann Thorac Surg 1997, 1635-43.

[6] Sonke Gabe's,Heskes Tom, and Verbeek A, Prediction of bladder outlet obstruction in men with lower tract symptoms using artificial neural network.

[7] Shigeohiko K, Kunio Doi and Heber MacMohan, Classification of normal and abnormal lungs with interstitial disease by rule based and ANN, Yearbook of medical Informatics 1999.

[8] National center of infectious Disease Network www.cdc.gov.

[9] Prince Alan and Smolensky Paul, Optimality: From Neural Network to Universal Grammar. The American association for the advancement of Science, Vol. 275, (5306) March 1997, 1604-1610.

[10] Shea Steven, Clayton Paul: Computerized decision support systems begin to come of age.   The American Journal of Medicine, Vol. 106 (2), Feb 1999, 261-262.

[11] The Hepcare Information Network www.HepCare.com.

[12] The your health Information Network www.yourhealth.com.

[13] The Hepatitis Information Network. www.hepnet.com.

[14] Brainmaker, Simulated Biological Intelligence, California Scientific Software, 1998.

KES 2002
E. Damiani et al. (Eds.)
IOS Press, 2002

Diagnosing The Aetiology Of Neonatal Jaundice By Paramedical Personnel Using An Expert System

Chitra Dharmar, Shankar Srinivasan, Dinesh Mital, Syed Haque

Department of Health Informatics

UMDNJ-SHRP

Newark, New Jersey

USA

ABSTRACT :

The main objective of this project is to develop an expert system module ,that uses the clinical decision criteria of experts in the neonatology field to advice paramedical and semi-skilled personnel who require guidance to diagnose the etiology of neonatal jaundice. The project was developed with a commercially available software package EXSYS (Multilogic Inc.). In this project we have used a rule-based system to build the decision tree that would aid the paramedical personnel to arrive at a appropriate etiology for hyperbilirubinemia in a neonate .The project is field tested in India, where the residents are asked to come up with the correct diagnosis given the clinical tests.   The disadvantage of this model is that there is the problem of it becoming obsolete in a few years time unless it is constantly updated by developers. There is also the problem that this system cannot train or learn by itself unlike an artificial neural network.

KEYWORDS: RULE-BASED SYSTEMS, EXPERT SYSTEMS, NEONATAL HYPERBILIRUBINEMIA.

1. INTRODUCTION

An expert system imitates the reasoning process experts use to solve problems and it could be put to use by a non-expert to solve his/her specific problems. These systems are used to propagate scarce knowledge resources for improved and consistent results. Ultimately such systems could function in a specific usually narrow area of expertise or domain. Many commercial expert systems are rule-based, because the technology of rule-based is relatively well developed. In such systems the knowledge is represented as a series of production rules. The classic example of a rule-based system is MYCIN, developed to aid physicians in diagnosing and treating patients in the first critical 24-48 hr period, when much of the decision making may be imprecise, since all the relevant information is not available. This program's record of correct diagnosis and prescribed treatment has equaled the performance of top human experts. Several other Expert Systems also abound in various fields of medicine[1]. In this paper we report the development of an Expert System for use by paramedical personnel in the domain of neonatology.

Each year approximately 60% of the 4 million new born in the United states become clinically jaundiced and any receive various form of evaluation & treatment [2]. The normal serum bilirubin level is less than 1 mg/dl. Adults appear jaundiced when serum bilirubin levels are greater than 2mg/dl. Newborn infants appear jaundiced when the serum bilirubin levels are greater than 7mg/dl. Physiologic jaundice of the newborn occurs in 25 to 50 % of all term infants and to a higher degree in preterm infants [3] . Jaundice of the newborn occurs not only due to physiologic causes but also due to a variety of pathologic causes too. Since the preterm infants are under the good care of a neonatologist, they are not our primary concern , but we are concentrating on term infants with non-physiologic causes of jaundice. Here we are interested in formulating a guideline for a multi-purpose health worker or a junior resident or in any situation where there is no immediate pediatric or neonatal care. The set of goals for such a scenario is illustrated below.

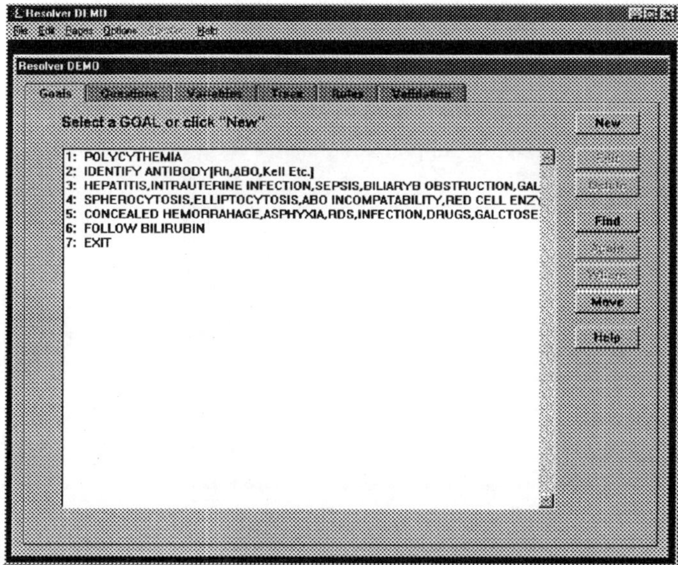

Figure 1: Illustration of the Goals open to the Paramedical Personnel for Field Testing

2. Methodology

The software EXSYS DEVELOPER that we have used has several different pages for entering different types of information and knowledge. This system has three important components - Database, Inference engine, Knowledge base, and Static knowledge in the form of rules and facts. The methodology involved here is to build a branched tree with various rules and goals. The input parameters are each given a confidence level between 0 and 10 (0 being NO CONFIDENCE and 10 the HIGHEST). Once the model was developed it had to be validated and checked for possible errors. The system worked well given the input variable and this could assist para-medical personnel and junior residents in arriving at an etiology for neonatal hyperilirubinemia.

Non-physiologic causes of jaundice may not be easy to distinguish from physiologic jaundice. The following situations may suggest non-physiologic jaundice and will require investigation. General conditions like onset of jaundice within first 24 hrs of life and persistence of jaundice over prolonged period, signs of underlying illness in infants like vomiting, lethargy, poor feeding, excessive weight loss and temperature instability. The above mentioned symptoms are non specific. A detailed family history regarding the following should be obtained : history of jaundice, anaemia, gall bladder disease, liver disease, sibling with jaundice or anaemia, maternal illness during pregnancy, maternal drug intake, ethnicity, difficult labor or delayed cord clamping, history regarding breast feeding, delayed stooling and excessive vomiting[4]. The physician assistant or the general practitioner should next proceed to a detailed physical examination, and jaundice is looked for in the infant by blanching the skin with finger pressure to observe color of skin instead of looking for yellowish discoloration of conjunctive as in the case of an adult. Transcutaneous bilirubinometer and card ictometer can be used to assess the severity of jaundice [5].

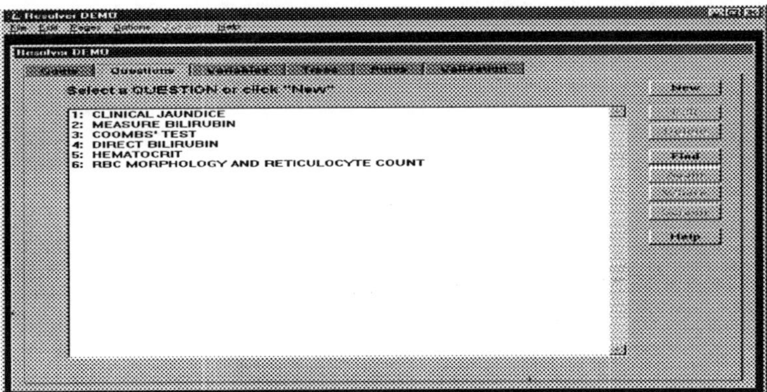

Figure 2: The Query Page of the Expert System for the Decision Making Process

The next step is to proceed to running a few tests on the infant and the following tests are indicated in the presence of non-physiologic jaundice. 1. Total serum bilirubin . Blood type, Rh and direct COOMBS' to test for isoimmune hemolytic disease. Infants of women who are Rh negative should have a blood type, Rh, COOMBS' test performed at birth. Blood typing and COOMBS' test should be done for infants who are discharged early especially if mother's blood group is type O. For the mother is it imperative to do blood type, Rh grouping and antibody screen during pregnancy and should be repeated at delivery. 4. Perform peripheral smear for RBC morphology and reticulocyte count to detect causes of COOMBS' negative hemolytic disease, for eg. Spherocytosis. 5. Identification of antibody on infant's RBC especially if results of direct COOMBS' is positive. 6. Direct bilirubin estimation is necessary when jaundice persists beyond the first 2 weeks of life and when there are signs of cholestasis. 7. In prolonged jaundice tests for liver disorders, congenital infection, sepsis, or metabolic diseases or hypothyroidism. 8. A G-6-PD screen may be helpful especially in male infants of Asian, Mediterranean or Middle eastern descent [3]. Figure 2 above shows the corresponding query page in the Expert System while Figure 3 shows the decision tree for one of the cases.

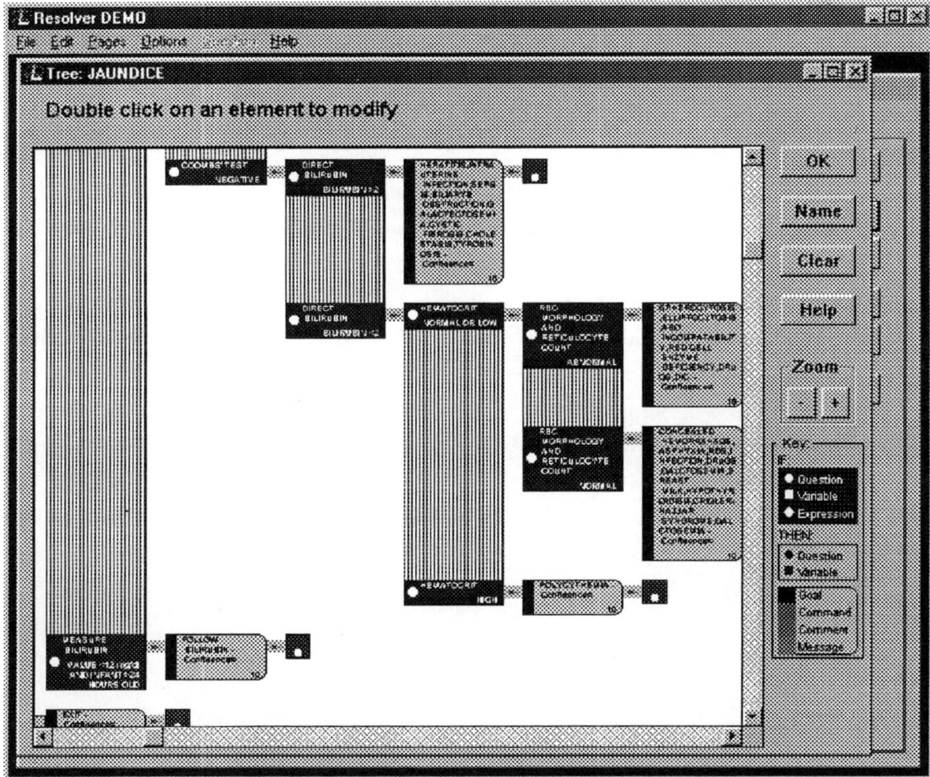

Figure 3: Illustration of a Decision Tree in the Expert System

3. Results

With the help of the above knowledge base, several rules were formulated and executed in the software. The system was then field tested in India and the junior residents (all with basic medical experience) are given a set of clinical conditions and the clinical tests and asked to come up with the correct diagnosis of neonatal jaundice . This was compared with the results that the expert system would bring given similar conditions. The comparison test was very encouraging and modifications are being made to make the system more user-friendly and comprehensive.

4. Conclusions

In summary, we have designed and validated this system for neonatal jaundice the disadvantage of this is that it can not learn adaptively and will repeat the mistake until a programmer can fix it. There is also the problem of it becoming obsolete in a few years time unless constantly updated by developers. Further work can be done in developing an algothrim for the management of hyperbilirubinemia in an infant. We have built our system based on the recommendation by AAP to aid in the evaluation & treatment of healthy term infants with jaundice. In these guide line, the AAP has attempted to describe a range of acceptable practices, recognizing that adequate data are not available from scientific literature to provide more precise recommendations. Now by developing this system and making it available for developing countries where the condition exist but where neonatologist are rare, would be welcome.

REFERENCES:

1. Jackson, Peter : Introduction to Expert System. Third edition ,1998.

2. AAP Practice parameters, Management of Hyperbilirubinemia in the Healthy Term Newborn. *Paediatrics* 94:558, 1994.

3. Hinkes, M.T. and Cloherty, J.P. *Neonatal Hyperbilirubinemia*, pp. 175-209.

4. Pearson , Mangolis: The clinical algorithm nosology : a method for comparing algorithmic guidelines *Medical Decision Making*. 1992;12:123-131.

5. Hegyi T. Transcutaneous Bilirubinometery : a new light on old subject. *Paediatrics* 69:124, 1982

6. Newman, K Evaluation and treatment of jaundice in the term newborn :A kindler gentler approach *Paediatrics* 89:809,1992.

KES 2002
E. Damiani et al. (Eds.)
IOS Press, 2002

Multi-Tier Web Application to make EEG Universally available using Java Servlets and Java Server Pages (JSP)

S.Dixit
Department of Neurology, University of Medicine and Dentistry of New Jersey,
Newark, NJ, USA
Sidhartha Shankar
Research and Development, Paytrust, Inc.
Lawrenceville, NJ, USA
&
Dinesh Mital
Department of Health Informatics, SHRP, University of Medicine and Dentistry of NewJersey, NJ, USA.

Abstract: An attempt is made to make EEG data, which is stored in EDF (European Data Format) format available to anybody who has a connection to the Internet using a phone line or broadband using only "the web-browser". Java Servlets and JSP extend the functionality of the web server by servicing HTTP requests and returns results as HTTP responses. The multi-tier application separates the Presentation Logic, Application logic and data , which makes it very easy to maintain code and makes the application flexible and puts the code exactly where it belongs. The application provides all the benefits, which are available by remote facilitation of medical resources, and hence it is extensible to other forms of data over the Internet very easily. The monitoring capabilities intra-hospital (and inter hospitals) can be expanded without the need for installation of costly networking cables. Extending the "EEG over Internet" to those departments who refer the patients to Neuro-Electro diagnostic lab can increase the productivity of the hospital. The doctors can see the EEG results from virtually anywhere. **Immediate results facilitate early diagnosis which is especially critical with stroke, head injury, and seizures.** *Thus a wider range of diagnostic services, and a higher standard of care to your patient population can be offered with very little investment.*

1.0 Introduction

The web has come a long way from the days of simple publishing of data to actually developing sophisticated application for real life problems. Its impact has been so great that it has changed the way we work, play and live our lives. We can telecommute to work, play online without going out of our house and actually experience a holiday without going there. This paper is a similar attempt to provide the healthcare providers to monitor and see the EEG results from virtually anywhere freeing them from the restriction of being in the EEG lab or their desktops to see the EEG.

2.0 Architecture

In the past, *2-tier applications* -- also known as *client/server applications* -- were common. Client Server Computing was born out of the advent of LAN Local Area Networks- Computers were connected not only to each other but also to servers. Figure 1 illustrates the typical two-tier architecture. There are three flavors of Client server architecture.

- Thin Client, Fat Server e.g. Mainframe and dumb Unix terminals

- Fat Client, Thin Server e.g. Windows Desktop Applications

- Thin Client, Thin Server.—ideal case of 2 tier model.

Client/Server communication is quite easy to understand in terms of roles, and is very closely analogous to what happens in a shopping situation. An assistant in a shop awaits a customer. The assistant doesn't know in advance which customer might arrive (or even how many - the store manager is supposed to make sure that enough assistants are employed to just about cope with the maximum number of shoppers arriving at any one time). A server on the network is typically a dedicated computer that runs a program called the server. This awaits requests from the network, according to some specified protocol, and servers them, one or more at a time, without regard for who they come from.

 In some cases, the only service provided by the server was that of a database server. In those situations, the client was then responsible for data access, applying business logic, converting the results into a format suitable for display, displaying the intended interface to the user, and accepting user input. The client/server architecture is generally easy to deploy at first, but is difficult to upgrade or enhance, and is usually based on proprietary protocols -- typically proprietary database protocols. It also makes reuse of business and presentation logic difficult, if not impossible. Finally, and perhaps most important in the era of the Web, two-tier applications typically do not prove very scalable and are therefore not well suited to the Internet.

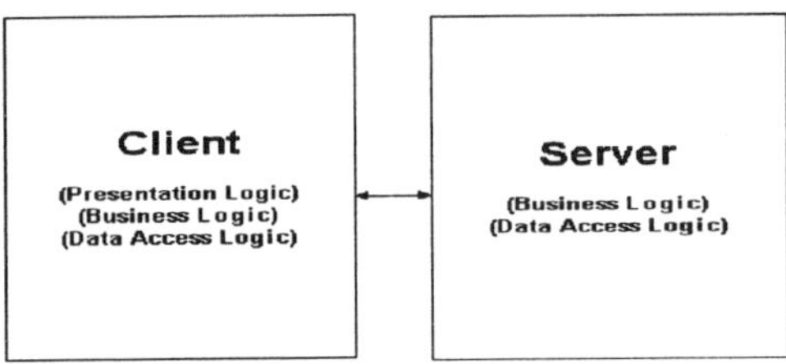

Figure-1 A typical 2-Tier(Client-Server) Application

One reason for the popularity of Client Server model is the quality of the tools and middleware that have been most commonly used since the 90's: Remote-SQL, ODBC, relatively inexpensive and well integrated PC-tools (like Visual Basic, Power-Builder, MS Access, 4-GL-Tools by the DBMS manufactures). In comparison the server side uses relatively expensive tools. In addition the PC-based tools show good Rapid-Application-Development (RAD) qualities i.e. that simpler applications can be produced in a comparatively short time. Since it is simpler to attempt to achieve efficiency in software development using tools, than to achieve efficiency using your algorithms.

2.1 What is n-tier or Multi-tier?

In the term "N-tier," "N" implies any number, like 2-tier, or 4-tier, basically any number of distinct tiers used in any application architecture. "Tier" can be defined as "one of two or more rows, levels, or ranks arranged one above another each serving distinct and separate tasks." The various services contained in a 2-tier application is separated into different layers.If the presentation logic, Business logic, data access logic, business logic is all encapsulated in different layers so that each of these layers provide just one specialized function. They can scaled easily. If we have to change the underlying database for example, we can simply swap the database, modify the data access layer little bit and we are ready to go. No change is required to any of the layers above or below it. That is the power of multi-tier architecture which is achieved by the Java Servlets and Java Server Pages in our case.

- **Presentation:** In a typical Web application, a browser running on the client machine handles presentation.

- **Dynamically generated presentation:** Although a browser could handle some dynamically generated presentation, for the widest support of different browsers much of the action should be done on the Web server using JSPs, servlets, or XML (Extensible Markup Language) and XSL (Extensible Stylesheet Language).

- **Business logic:** Business logic is best implemented in Session Beans.

- **Data access:** Data access is best implemented in Entity Beans and using JDBC.

2.2 MVC (Model-View-Controller) or Model 2 Architecture

Separation of Presentation and Business Logic has never been so easy with just JSP. With increasing complexity the pages would clutter with Scriplets and processing code, making it hard to debug or make changes. To counter this problem, the JSP/Servlet's Model 2 architecture was introduced, providing a cleaner approach to code manageability. This is a hybrid approach for serving dynamic content combining the use of Servlets and JSP. It takes advantage of the strengths of both technologies: JSP is used to generate the presentation layer and Servlets to perform process-intensive tasks. The Servlet acts as the controller processing the requests, creating beans (models) or objects used by the JSP, and depending on the user's actions, dispatches the appropriate JSP page to the user. The JSP (view) page, which contains no processing logic, is simply responsible for extracting and inserting the dynamic content from the Servlet into static templates.

2.2 Java Server Pages and Java Servlets

Servlets are programs which runs on the Web Server and build web pages "on the fly". The web pages are generated based on the data submitted by the user. In our case, long-term monitoring of a stroke patient causes the edf data file to change every second requiring the web page to correspond to the recorded data. EEG is plotted every second and the display modified dynamically.

Servlets are efficient, convenient, powerful, portable and inexpensive compared to all comparable technologies. Previously every request to the web server resulted in creating a heavy process on the server. In servlets each request is handled by a light-weight java thread.

Anyone who knows JAVA can write servlets with a small learning curve. It has got utilities for reading, parsing HTML data, handling cookies and tracking sessions etc.

Using Java Server Pages technology you can mix dynamically generated HTML with static, regular HTML. The static HTML lies in one part of JSP and the dynamic part is in another part. It lets you separate the two into different portions. Its advantage is two-fold since we use Java for the dynamic part which is powerful and can be ported easily. Also Web page design experts can build the HTML, leaving places for servlet programmers to insert the dynamic content. And, since it runs on the server, Java Server Pages have access to all server-side resources like databases, catalogs, pricing information, and the like.

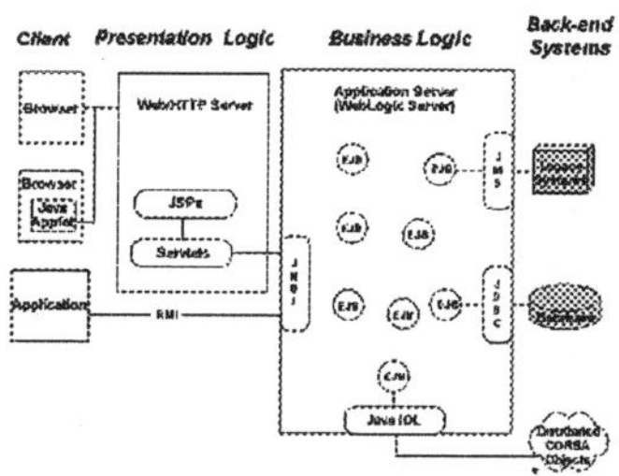

Figure 2. A sample n-tier application architecture

The EDFViewer application contents

File/Directory	Contents
APP	JSP, HTML, and GIF files
APP/classes	Java class files
APP/lib	Java jars
APP/tmp	Temporary servlet files
APP/work	Generated java for JSP and XSL

- 🗀 example

- o index.html
- o folder.gif
- o Image.gif
- o jsp-interest
  - index.jsp
  - parser.jsp
  - viewer.jsp
- o WEB-INF
  - lib
    - edf.jar
  - classes
    - edfviewer.class
    - parser.class
  - xsl
    - default.xsl
    - viewer.xsl

2.3 Enterprise Java Beans(EJBs):

Java Beans get preferential treatment in JSP 1.0. Three types of Java Beans were created in EDF Viewer application for a page, across a session, and for the entire application.

The classes and lib directories can contain application beans used by jsp:useBean These are simply Java classes implementing the bitmechanic work of an application.

For example, EDF Viewer application has a set of Java classes that perform the plotting of EDF data contained in the file. The application can put those classes in the beans directory and access them from the JSP page. Beans can be created with different lifetimes.

- Application beans last the lifetime of an application.
- Session beans last for a user's session.
- Request beans last for a single request.
- Page beans only last for a single page.

Requests and pages often last the same lifetime, but may differ if one page forwards or includes another page.

2.4 Accessing Beans

Each bean is defined with a jsp:useBean directive.

JSP assigns the created bean object to the JavaScript variable named by jsp:useBean.

In addition, the created beans are stored in JSP variables: page beans are stored in `request`, session beans are stored in `session`, and application beans are stored in `application`. Storing the beans in the JSP variables lets other beans and functions retrieve the beans.

2.4 Session:

Session variables let applications keep track of the user as he moves through the site.e.g. Any e-commerce site needs this capability to keep track of the user's purchases. JSP sessions start when the page accesses the session variable. Sessions end when the session times out, when the session is invalidated, or when the application ends.

3 Advantages of multi-tier approach to application development

3.1 Portability

A Portable application environment gives organizations the flexibility to migrate servers and swap tools, as business needs change. Portability also enables developers to share their work with a wider audience. If we wanted a large user base to access the EDFViewer application we can change the underlying database or other tools without very little problem. We can change the web server anytime we feel, it is constraining our   user-base.

Java Server Pages technology delivers "Write Once, Run Anywhere" capability, offering unprecedented reuse on any platform and on any server.

3.2 Performance

Pages built using JSP technology are typically implemented using a translation phase that is performed once, the first time the page is called. The page is compiled into a Java Servlet class and remains in server memory, so subsequent calls to the page have very fast response time.This is of tremendous advantage to us since the parsing logic for the EDFViewer is loaded and remains in server memory for the entire session of a user. Hence it results in faster parsing of the EDF data file.

3.3 Component-based Architecture

As mentioned earlier, JSP follows the model of separating programming logic from page design through the use of components like JavaBeans, Enterprise JavaBeans (EJB) and custom JSP tags. Assuming that web page developers are not familiar with scripting languages, JSP technology encapsulates much of the functionality required for dynamic content generation in easy-to-use JavaBeans components. The page developer can instantiate JavaBeans components, set or retrieve bean attributes and perform other functions that are otherwise more difficult and time-consuming to code. The EDFViewer application can be managed very easily because of the separation of presentation logic from the actual functionality. So web designers can be given the presentation code to maintain where programmers can take care of the functionality, like the parsing of the data or actual plotting.

3.4 Extensibility

The JSP technology is extensible through the development of customized tag libraries. This lets web page developers work with familiar tools and constructs, such as tags, to perform sophisticated functions.

3.5 Deployment:

The 2-tier-model implies a complicated software-distribution-procedure: as all of the application logic is executed on the PC, all those machines (maybe thousands) have to be updated in case of a new release. For the multi-tier architecture, since the logic is not on clients but the server, only the server needs to be updated with the new code.

4.0 Conclusion

Thus the EDFViewer application comprises of servlets, JSP pages, scripts and Java Beans into a self-contained web application. Applications are just generalized virtual hosts, only based on the URL instead of the host name.

The EDFViewer application groups pages for reading patient specific data, parsing of the EDF data, submission of new patient data , and displaying the EEG into a single application.

The EDFViewer Application can keep track of user sessions, giving the users the illusion of a single application out of disjoint pages.

References:

[1] T Penzel. Prototypes and applications in Neurotelemedicine. Chapter 3.1. in the book: "European Neurological Network", eds T Paiva and T Penzel, IOS press 78(8), 2000:89-100. A Java applet which enables viewing and analysing the recordings through an internet browser.

[2] Bellon E, Feron M, Wauters J, Aerts W, Marchal G, and Suetens P (1999) Advanced software for teradiology: Exploiting Java applets and the WWW to make multimedia tele-access and tele-interaction universally available.*European Telemedicine (Wootton R , ed.)* Kensington Publications Ltd., London, England: 66-71.

KES 2002
E. Damiani et al. (Eds.)
IOS Press, 2002

Modelling knowledge bases by reusing generic ontologies and unified terminology servers

M. Taboada (1), M. Arguello (1), D. Martínez (2), J. Des (3) and J. Mira (4)
(1) Dpto. de Electrónica e Computación. Universidade de Santiago de Compostela.
15782 Santiago de Compostela. Spain.
Author to contact: María Taboada. E-mail: chus@dec.usc.es
Telephone number: 34-981-563100
Fax number: 34-981-528012
(2) Dpto. de Física Aplicada. Universidade de Santiago de Compostela. Spain
(3) Hospital Comarcal de Monforte. Monforte. Spain
(4) Dpto. de Inteligencia Artificial. UNED. Madrid. Spain

Abstract. Modelling knowledge bases by reusing pre-existing knowledge is not a straightforward process. In most cases, the current ontologies contain too general constructs, so the reusing process must try to bridge the gap between domain applications and general ontologies. A solution is to extract specific concepts from some unified terminology server and to combine them with generic concepts from some ontology. This is the approach followed in this paper.

1 Introduction

During the last few decades, many knowledge-based systems in different fields have been developed, but many of them have not been accepted in practice. Their development as isolated computer systems has been one of the main faults. Currently, the Web has drastically changed the availability of these systems. The appearance of languages oriented towards the provision of semantic inter-operability in the Web [2], such as DAML+OIL [13], will facilitate, in a not too distant future, the intercommunication between different Web agents.

Taking into account the availability of these languages, the current design of knowledge-based systems should be oriented towards the reuse of knowledge stored in the Web. This would permit the development of knowledge-based systems with the vocabulary necessary to inter-operate with other agents of the Web. However, the reuse of knowledge is not a simple task. Although many ontologies have been built from the beginning of 90's in several domains [6, 3, 4], it is unlikely to found an ontology which includes all knowledge required by a particular system. In addition, in most domains ontologies represent very abstracted knowledge, so they can not be directly applied to build specific knowledge bases [7].

On the other hand, nowadays there are available unified terminology servers in the Web. For example, in the clinical domain a very frequented unified terminology server is the *Unified Medical Language System (UMLS)* [8]. It contains a large number of concepts, classifications, nomenclatures and thesauri. However, it is not easy to use, due to its size, diversity and heterogeneity of concepts [1], and the informal representation of the stored knowledge cannot be used directly by a knowledge-based system.

In this paper, we describe the modelling process of a knowledge base for a diagnostic application. Our approach reuses knowledge by integrating concepts and relationships from a generic ontology and a specific part contained in a unified terminology server. The article has been structured as follows. The second section briefly describes the basic requirements in the development of the knowledge base. In the third section, we will present the main steps followed in the building of the knowledge base by reusing pre-existent medical knowledge and by following the general methodology of knowledge modelling CommonKADS [10]. In the fourth section, we will show a scenario in a clinical domain. Finally, we will present the conclusions.

2   Requirements of the knowledge base

- *Purpose of the knowledge base*: To support a knowledge-based application oriented towards medical diagnosis.

- *Level of formality*: 1) Semiformal modelling of the ontology using the Unified Modelling Language UML [9] and the rules schema of CommonKADS [10], and 2) representation of the knowledge base using the language for the Semantic Web DAML+OIL.

- *Scope of the knowledge base*: More than 600 pathologies relative to ophthalmological problems, description of more than 30 symptoms and 200 physical exploration signs, description of clinical state abstractions, physical exploration tests, etc.

- *Strategy for the efficient acquisition of knowledge*: The interviews with the experts were centered on the patterns used by them to collect information about the patient (clinical history).

- *Construction of the knowledge base*: We have followed the basic stages of the ontological life cycle proposed by the most representative methodologies for the construction of ontologies [12, 5]. However, in this article, we will only concentrate on the conceptualization stage.

3   Main stages in building a knowledge base by reusing

Currently, the building of knowledge-based systems is considered a modelling activity. In order to obtain structured and coherent knowledge models, it is advisable to carry out this activity applying some Knowledge Engineering methodology. We have followed the general methodology of knowledge modelling CommonKADS [10], which includes three main stages in the construction of a knowledge model: knowledge identification, specification and refinement. Next, we summarize the activities that have been carried out during the development of these stages. It should be noted that here:

1. We are only considering the activities relative to the building of the domain factual model and not the whole of the knowledge model. Information about the modelling of tasks and methods for the diagnostic system via reuse can be found in [11].

2. The modelling have been carried out by reusing the specific knowledge stored in an unified terminology server and taking into account the structure of a generic ontology.

3.1 Knowledge identification

In this stage, knowledge sources are identified. Several activities have been included:

- Exploration of all domain information sources in order to elaborate the most complete characterization of the application domain.

- Listing of potential components for reusing, taking into account:

    - Exploration of knowledge sources for reusing:
        1. The library described in [3] specifies general categories of medical knowledge, which have been compiled as much from medical bibliography as from implemented systems.
        2. The medical knowledge server Unified Medical Language System (UMLS) [8] developed and maintained by the U.S.A. National Library of Medicine (http://umlsks.nlm.nih.gov) is a tool focussed on facilitating the development of biomedical systems and integration of information from different sources. Of all the sources of knowledge and tools provided by UMLS, we have used *Methasaurus*, which contains information about a large number of medical concepts, nomenclatures and thesauri, *the Semantic Network*, which provides a classification of all the concepts represented in the Methasaurus and *the Semantic Navigator*, which draws all of the semantic space that is related to each UMLS concept.

    - Extraction of concepts and relations: Several general categories of medical knowledge were extracted from the library of [3]. These categories have facilitated to break down the ontology underlying to our knowledge base into various smaller parts and thereby makes it easier to analyze. In addition, we have reused as concepts and relationships from UMLS as possible.

3.2 Knowledge specification

In this second stage we started to construct a specification of the domain model and the underlying ontology. The reusable knowledge selected in the previous stage provided us part of the complete specification. The next activities were iteratively carried out: 1) refinement of the reused concepts and relationships, 2) addition of new relationships among the modelled concepts and 3) addition of rule schemas among modelled concepts.

3.3 Knowledge refinement

In the final stage, 1) contents of the domain model were filled by adding of instances of concepts, relationships and rule schemas, and 2) the resulting domain model was validated by simulating a lot of scenarios through a small prototype.

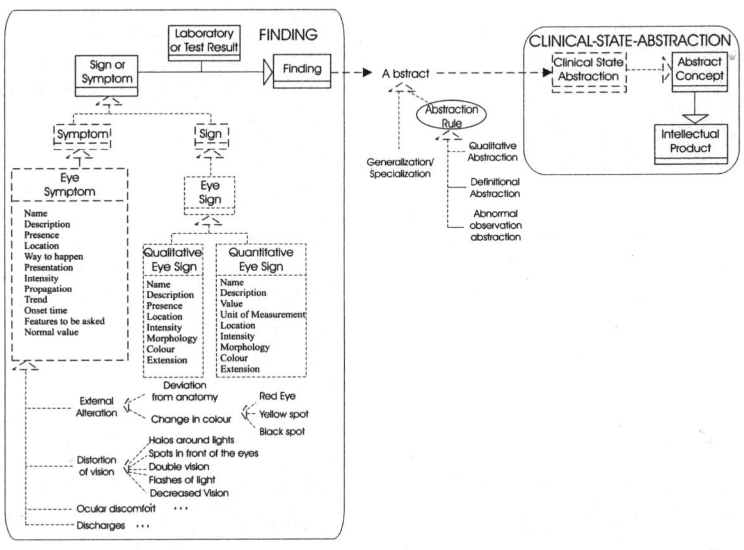

Figure 1: Part of Findings and Clinical State Abstraction

4   A scenario

In this section, we are presenting an example of modelling a knowledge base by reusing. We will center on a small part of our knowledge base, named *Findings*. In the UMLS Semantic Network, a finding is defined as "everything that is directly observed or measured" and two types are differentiated: "*Laboratory or Test Result*" and "*Sign or Symptom*". We have added a new hierarchy, distinguishing clearly between signs and symptoms. The signs include a large part of the hierarchy *Eye and vision observations* provided by the UMLS Semantic Navigator. However, the hierarchy of symptoms has been modelled from scratch, as the object *Eye Symptoms* defined in the UMLS Semantic Navigator contains no information at all. Part of this hierarchy can be seen in Fig. 1, as well as the set of attributes we have modelled in order to describe the symptoms and ocular signs. In this figure the reused concepts and relationships are shown within a continuous line and the new concepts and relationships within a dashed line.

We have also differentiated between findings and clinical state abstractions, exactly as they are collected in the Falasconi and Stefanelli library [3]. The findings are subjective evidences (symptoms) or objectives (signs) of the presence of an illness, while the clinical state abstractions describe evidences at a higher level of abstraction, as these are obtained from the findings by following some type of reasoning process. For example, the results obtained for the measurement of the distance visual acuity in the right or left eye are considered findings, whilst the conclusion on Visual Acuity (Normal, Reduced, Poor, etc.) reached by the doctor after analyzing those results is considered a clinical state abstraction. However, in UMLS the conclusion on Visual Acuity is considered a finding.

The clinical state abstractions have been modelled as a subclass of *abstract concept*, which, in turn, is a subclass of *Intellectual Product* in the UMLS Semantic Network. We have also considered that the findings and clinical state abstractions are related through various types of relationships between expressions about the value of their attributes (i.e. types of rule schemas).

5 Conclusions

In this article, we have described the modelling process of a knowledge base for a diagnostic application following the general methodology CommonKADS [10]. Our approach has been concentrated mainly on: 1) extracting specific concepts from Methasaurus, classifications from the UMLS Semantic Network and relationships from the UMLS Semantic Navigator and 2) combining this knowledge with a generic ontology in medicine [3]. This process has not been direct, as the knowledge contained in UMLS is not formally represented. So, it was necessary an additional step based on formalizing the extracted knowledge.

References

[1] Bodenreider, O., Burgun, A., Botti, G., Fieschi, M. and Le Beux, P., Evaluation of the Unified Medical Language System as medical knowledge source, J. Am. Med. Inf. Assoc., **5** (1998) 76-87.

[2] Decker, S., van Hermelen, F., Broekstra, J., Erdmann, M. , Fensel, D., Horrocks, I., Klein, M. and Melnik, S. , The Semantic Web - on the respective roles of XML and RDF, IEEE Internet Computing, Setember/October (2000).

[3] Falasconi, S. and Stefanelli, M., A library of implemented ontologies, Proc. of the ECAI Workshop on comparison of implemented ontologies, Amsterdam (1994).

[4] Fernández, M., Gómez-Pérez, A., Pazos, J. and Pazos, A., Building a Chemical Ontology using methondology and the ontology design environment, IEEE Intelligent Systems and their applications, **14** (1999) 37-45.

[5] Gómez-Pérez, A., Knowledge Sharing and Reuse, The handbook on Applied Expert Systems, CRC Press, (1998).

[6] Gruber, T. and Olsen, G., An ontology for engineering mathematics, Proc. of the Fourth Int. Conf. on Principles of Knowledge Representation and Reasoning, Germany (1994).

[7] Kayed, R.M. and Colomb, R.M., Extracting ontological concepts for tendering conceptual structures, Data and Knowledge Engineering, 40 (2002) 71-89.

[8] National Library of Medicine, Unified Medical Language System, Bethesda, MD (2001).

[9] Rumbaugh, J., Jacobson, I. and Booch, G., The Unified Modelling Language Reference Manual, Addison-Wesley (1999).

[10] Schreiber, A., Akkermans, H., Anjewierden, A.A., Hoog, R., Shadbolt, N.R., Val de Velde, W. and Wielinga, B., Engineering and managing knowledge. The CommonKADS methodology, The MIT Press (1999).

[11] Taboada, M., Des, J., Mira, J. and Marn, R., Development of diagnosis systems in medicine with reusable knowledge components, IEEE Intelligent Systems, **16** (2001) 68-73.

[12] Uschold, M. and Gruninger, M., ONTOLOGIES: Principles, Methods and Applications, Knowledge Engineering Review, **11** (1996).

[13] van Harmelen, F. and Horrocks, I., Reference description of the DAML+OIL ontology markup language, http://www.daml.org/2001/03/reference.html, DARPA Agent Markup Language Program (2001).

KES 2002
E. Damiani et al. (Eds.)
IOS Press, 2002

A Multi-Agent Architecture for Intrusion Detection

Amparo Alonso-Betanzos, Bertha Guijarro-Berdiñas, Juan A. Suárez-Romero
LIDIA Laboratory, Department of Computer Science, University of A Coruña
Campus Elviña, s/n, 15071, A Coruña, Spain
E-mails: ciamparo@udc.es, cibertha@udc.es, ja@mail2.udc.es

Abstract. Intrusion detection is an aspect of security that is attracting a great deal of attention. Given the heterogeneity of the networks and the complexity of the domain, agent-based techniques are appropriate for the development of intrusion detection systems. This article proposes the broad lines for designing an architecture that uses seven classes of intelligent agents, co-operating among themselves in a loosely coupled manner, and including both social and domain knowledge.

1 Introduction

There is a tendency for an increasing number of services to be offered via the Internet. Businesses thus potentially have a greater number of customers, which in turn means that security is becoming a priority issue. One aspect receiving a great deal of attention in recent years is the detection of intruders. Intruders are individuals who use a system without authorisation or individuals having authorisation but abusing this by performing tasks other than those for which they are authorised. The question of security in the computerised world is a highly complex and active one, and we are still a long way from obtaining systems that are highly reliable and secure.

The desirable features for the optimal functioning of an Intrusion Detection System (IDS) should include the following [8]: (a) the system should be fault tolerant, in such a way that even if some elements of the system fail, the system continues to function more or less correctly; (b) it should be able to resist attacks; and (c) it should be highly adaptable and configurable. As commented earlier, networks are becoming increasingly complex and flexible, and so an IDS should be capable of adapting to system needs.

This research proposes to construct an IDS based on intelligent agents, given that their use allows to achieve the desirable features mentioned above.

2 Background

The earliest reference in the field of intrusion detection can be credited to James P. Anderson [1] who discussed the need for adapting system auditing mechanisms so that these would provide useful information for investigating possible attacks. Seven years later Dorothy Denning described a model automating the intrusion detection process [3], which later developed into the IDES (Intrusion Detection Expert System) [7], a detector employing statistical techniques

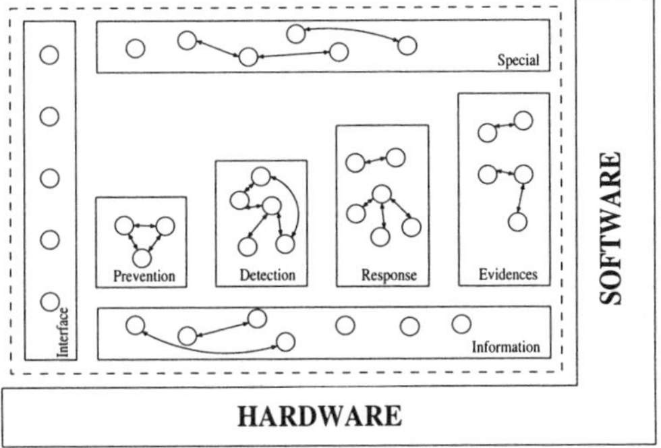

Figure 1: A scheme of the proposed architecture

and heuristic rules to detect security breaches. From this point, several intrusion detection systems began to figure in the literature.

One of these systems was AAFID (Autonomous Agents For Intrusion Detection) [8], which as its name indicates, used agents to detect intruders. In this architecture, a set of agents monitors specific aspects of a machine requiring security. These agents send information to the transceivers, which in turn, amalgamate the information and re-send it to the monitors. The monitors process this information and decide whether or not an attack has occurred. To make the system more scaleable, the monitors may be built in layers. The main drawback to the AAFID system is the extreme rigidity of its architecture, which means that the introduction of new agents is complicated. Moreover, certain agents are very sensitive to error and an attack that manages to deactivate one agent in the upper hierarchy may cause the entire system to be deactivated.

On the other hand, different approaches have been used in terms of the techniques employed for intrusion detection, but the approach that has been most successful and therefore most popular has been that based on signatures rules. This technique, however, has a major drawback in that it is only capable of detecting known attacks and requires an expert to create the rules for detecting each attack. Other approaches based on the use of artificial neural networks [6], genetic algorithms [2], or simulations of the human immunological systems have also been attempted [4].

3   Proposed architecture

This article outlines the design of the architecture of an IDS using intelligent agents, based essentially on the AAFID approach. It is intended to create a flexible and adaptable architecture whilst permitting the integration of different intrusion detection techniques.

Figure 1 illustrates the architecture of the system in terms of seven categories of specialist agents performing very specific tasks. These agents are as follows: *information agents, prevention agents, detection agents, response agents, evidence-search agents, interface agents,* and finally, *special agents.*

Information agents. Intrusions are detected by analysing the information coming from different sources, such as system log files, existing network connections, incoming network packets, etc. However, this information may be in different formats or, depending on the type of system being protected, may even be unavailable. Moreover, the frequency with which a certain kind of information is required may vary. It may also be necessary to obtain information resulting from the amalgamation of data from different sources. The aim of the information agents is to resolve all these problems. This agents permit the software and hardware base for the IDS to be isolated and information to be supplied to the agents requiring it, in accordance with a standard format and irrespective of the platform that the IDS is protecting. Each information agent is responsible for providing a specific kind of information. These agents filter the information in such a way that only the necessary information is fed to the requesting agents. In order to supply information amalgamated from different sources, links can be set up between agents, as illustrated in figure 1. One example of this kind of data is that supplied via a data-mining algorithm.

Prevention agents. The classical security paradigm consists of prevention, detection and response to attack. Although, strictly speaking, intrusion detection —as its name would indicate— focuses on the second aspect, we have included the three aspects in the proposed architecture. Prevention is the aspect that has been most thoroughly investigated and the majority of organisations base their security systems on specific attack prevention elements, such as firewalls. We propose integrating the readily available elements of the organisation (if any) in our multi-agent architecture, encapsulating them in the detection agents.

Detection agents. One of the main notions underlying AAFID is the use of several agents to detect intruders. These agents can be organised in a hierarchical structure with a view to establishing different monitoring levels, since the majority of systems, composed of networks with elements of different kinds, are too complex for a single entity to be capable of detecting attacks as they occur. This same idea underlies our own architecture, although in our case the hierarchical structure for the detection agents has an additional objective, which is that the agents can also be structured so as to constitute more complex detection techniques. Given that each agent implements a different detection technique, we can create a new agent that combines the techniques of other agents so as to make detection techniques more potent.

Response agents. When the detection agents discover an intrusion, this must be dealt with. The way an attack is handled depends on a variety of factors. One solution is to cut the connection via which the intruder has obtained access to the system. However, this is not always possible and might even provoke a denial of service attack. Another solution is simply to inform the system security manager, but response to the intrusion may in this case be too slow. We have therefore integrated different response possibilities in our system, using response agents for this purpose. These react in a variety of ways to an attack, because they combine various agents so as to obtain a more complex and efficient counter-attack response.

Evidence-search agents. Many attacks lead to legal action. For this action to successfully proceed, proof of the attack and evidence both have to be gathered as a basis for taking the intruders to court. Both the evidence itself and the evidence collection process will depend on the legislation in force in a particular country, although many attacks involve systems in different countries, each having their own laws on the subject. On the other hand, the gathering of evidence may enter into conflict with the need for a response to an attack, and it becomes necessary therefore, for the evidence-search and response agents to interact mutually in order to determine how long to wait while evidence is being gathered before responding to an attack.

Interface agents. In a complex and distributed system, the interaction mechanisms play a crucial role. For our system we opted to use interface agents, which act as 'people representatives', in a manner of speaking, and so the different people who interact with the system are integrated within the same system as if they were agents themselves. Each interface agent interacts with the user in different ways, by using standard network management protocols, or by directly using graphical interfaces, etc. Another attractive feature of the interface agents is their capacity for learning from users, acquiring knowledge that can subsequently be applied to the system. Thus, for example, the interface agents can learn which procedures an administrator uses to confirm an attack in order to be able to anticipate the user's response when an attack occurs.

Special agents. There are certain agents that do not belong to any of the categories described above. These special agents perform a variety of tasks, one example of which is that of starting the rest of the agents, so that these can carry on performing dynamically in accordance with the requirements of the system.

4 Co-operation in the system

The agents interact with each other in order to perform their tasks. Existing systems normally use a series of proprietary co-operation protocols, which impedes inter-operability between the IDSs. There have been several proposals for standardising these information exchange protocols, such as the IDEF (Intrusion Detection Exchange Format) [10]. These protocols, even though they represent a great advance in terms of increasing system inter-operability, are inadequate for our system, due to the fact that the agents in our proposed system need to be able to express their objectives and capacities to other agents. This need is common to all multi-agent systems, irrespective of their application domain, and in order to express these notions a communication language between agents is generally used [5].

Our proposed system, however, is based on combining both approaches. On the one hand a communication language is used that allows the agents to inter-operate and that permits them to express their requests and objectives as well as communicate information. The content of these messages is defined, on the other hand, by means of one of the standard protocols from the intrusion detection field, such as that defined for the IDEF. We can thus define information type as well as concepts and relationships within a standard framework. In this way, and unlike the AAFID system, the agents are highly autonomous, which in turn means that the relationships between them are not fixed beforehand; on the contrary, it is the agents themselves who establish dynamic links between each other when these are required. The flexibility of the system is thus greatly increased.

5 Knowledge in the system

The knowledge employed by an agent in a multi-agent system, such as the one described here, may be classified in terms of two categories [9]. On the one hand we have the knowledge required to resolve the problem, in other words, the knowledge an agent needs in order to fulfil its objective and be guided in the decision-making processes. On the other hand we have social knowledge, which permits the agent to interact with the other agents.

The first kind of knowledge is specific to each agent and may be more important in some agents than in others, depending on their objectives. Thus, for example, knowledge use is *a priori* more intensive in the case of the detection agents than in the case of the information agents. Even within the category of detection agents the degree of use is different, depending on the techniques used to perform tasks.

The second kind of knowledge, social knowledge, has been largely ignored in the intrusion detection field, although it has received more attention in the field of multi-agent systems. In our proposed system, this kind of knowledge is particularly important. In fact, to guarantee the flexibility of the architecture of a system, it is crucial to equip its elements with mechanisms that allow dynamic interaction between these.

6   Conclusions

Our paper has proposed the broad lines for the design of an IDS based on intelligent agents. The distribution of functions throughout the system in the form of different kinds of agents means that the system is more fault tolerant, whilst permitting the integration of new agents employing different techniques, both features which make the system more efficient overall. The flexible architecture, mainly a result of the dynamic co-operation between the agents, makes the system in its turn more flexible, and thus the agents are able to adapt to the circumstances of the moment when performing their tasks.

References

[1] James P. Anderson, Computer Security Threat Monitoring and Surveillance. Technical Report, James P. Anderson Co., Fort Washington, PA, April 1980.

[2] Mark Crosbie and Gene Spafford, Active Defense of a Computer System Using Autonomous Agents. Technical Report 95-008, Purdue University, 1995.

[3] Dorothy Denning, An Intrusion Detection Model, Proceedings of the Seventh IEEE Symposium on Security and Privacy, May 1986.

[4] S. Forrest et al., Self-Nonself Discrimination in a Computer. Proceedings of IEEE Symposium in Security and Privacy, 202-212, May 1994.

[5] M. T. Kone, A. Shimazu and T. Nakajima, The State of the Art in Agent Communication Languages. Knowledge and Information Systems 2 (2000), 259-284.

[6] R. P. Lippmann and R. K. Cunningham, Improving Intrusion Detection Using Keyword Selection and Neural Networks. Computer Networks 34 (2000), 597-603.

[7] Teresa F. Lunt et al., IDES: The Enhanced Prototype. A Real-Time Intrusion-Detection Expert System. Technical Report SRI Project 4185-010, SRI-CSL-88-12, CSL SRI International, October 1988.

[8] E. H. Spafford and D. Zamboni, Intrusion Detection Using Autonomous Agents. Computer Networks 34 (2000) 547-570.

[9] Gerhard Weiss (ed), Multiagent Systems: A Modern Approach to Distributed Artificial Intelligence. The MIT Press, 1999.

[10] IETF Intrusion Detection Working Group, Intrusion Detection Exchange Format. http://www.ietf.org/html.charters/idwg-charter.html.

KES 2002
E. Damiani et al. (Eds.)
IOS Press, 2002

1023

Intelligent Analysis of Polysomnograms in a Sleep Apnea Decision Support System

M. Cabrero-Canosa, E. Hernandez-Pereira and V. Moret-Bonillo
Laboratory for R&D in Artificial Intelligence (LIDIA),
Department of Computer Science, University of A Coruña, Spain

Abstract Automation of the medical diagnosis of the Sleep Apnea Syndrome (SAS) requires an intelligent analysis of the pneumological and neurophysiological signals of the patient that combines both conventional and Artificial Intelligence techniques in order to detect respiratory abnormalities and construct a hypnogram for the patient. Nonetheless, an independent analysis of the signals is, in itself, neither sufficient nor adequate to the problem, and a process of temporal fusion and correlation between the signals is necessary for both a correct classification of the apneic events within a sleep stage framework, and to explain the occurrence of abnormal sleep patterns as a consequence of these events. This is the approach incorporated in SAMOA, a computerised system for respiratory analysis and sleep study, which makes a customised diagnosis for the patient in respect of the possible existence of the Sleep Apnea Syndrome and also classifies the syndrome. In this article, the most important aspects of the system in terms of the information integration and diagnostic processes are described and the validation results obtained are discussed.

1 Introduction

Sleep is absolutely essential to each human being. Research carried out in recent years has identified a number of physiological disorders that affect sleep, disorders which are notably prevalent in the population at large [1]. One of these disorders is the Sleep Apneas Syndrome (SAS), which is defined as an occurrence of five or more interruptions (or apneas) to respiration in the space of an hour's sleep, each interruption lasting more than ten seconds. These breathing pauses cause continual interruptions of sleep and, as a consequence, diurnal somnolence, memory loss, a reduced capacity to concentrate and a general deterioration in intellectual capacity. Furthermore, there is a proven increase in mortality due principally to cardiovascular complications and to traffic accidents [2]. These facts have converted sleep study into a scientific discipline in itself and a new medical speciality [3].

The traditional methods of sleep analysis, limited to direct observations of the behaviour of the sleeping individual, do not guarantee objectivity in the interpretation of results. For this reason they have been replaced by study methods based on an analysis of the electro-physiological parameters linked to sleep, which provide less subjective information. These parameters permit us to distinguish between the different types of sleep, known as sleep stages, which occur cyclically. In order to obtain the sequence of sleep stages, or hypnogram, the standard reference used is that published by Rechtschaffen & Kales (R&K) in 1968 [4].

This specifies a set of basic rules, applicable over fixed intervals of 30 seconds (epochs), for the correct classification of sleep stages on the basis of a simultaneous recording of: (a) cerebral activity, obtained from the electroencephalogram (EEG); (b) ocular movements, obtained from the electrooculogram (EOG); and (c) muscular tone, which is obtained from an analysis of the electromyogram (EMG). Nonetheless, the occurrence of different pathologies associated with sleep tend to alter the normal cycle and the standard R&K is thus inadequate for an effective diagnosis. For this reason it is necessary to include expert knowledge so as to be able to 'adjust' the clinical criteria of the R&K standard to the patient.

The diagnosis of patients with SAS requires a complex test to be carried out: the overnight polysomnography. This is a technique for monitoring sleep in which cardio-respiratory variables are recorded in addition to collecting the electrophysiological data from the EEG, EOG and EMG. An automatic analysis of sleep in the context of an SAS diagnosis thus requires an interpretation of the patient's neurophysiology in the context of his/her pneumological situation [2].

Most of the automatic applications that have been developed only resolve very specific aspects of the analysis, such as the sleep stage classification tasks or the detection and classification of apneas. The need is evident, therefore, for an integrated help system for SAS diagnosis, a system which will correctly interpret an apnea episode whilst taking into account the sleep phase through which the patient is passing, and which will explain the sleep stage transitions within the framework of the respiratory abnormalities that have occurred.

This article describes the development of an intelligent polysomnogram analysis module for patients with the Sleep Apneas Syndrome, that has been incorporated in SAMOA[1] [5]. It has been implemented as a rule-based system, with a mechanism for explaining the SAS diagnosis. The knowledge base includes the basic classification rules of the R&K paradigm supplemented by the expert clinical knowledge of SAS.

2 Methods

The process of automatic SAS diagnosis centres its analysis on two sets of input data: neurophysiological and pneumological signals recorded during the polysomnography. Nevertheless, given that clinical knowledge is expressed symbolically these data need to be pre-processed in order to extract the most important characteristics [5].

The pneumological analysis detects the respiratory irregularities–whether apneas or hypopneas–that have occurred during sleep, on the basis of the degree of obstruction in the upper airways. With a view to customising therapy for the patient, these are characterised as obstructive, central or mixed, depending on the degree of absence or presence of respiratory effort [6]. As the responsibility of what is known as the *Respiratory Function Characterisation Module* of SAMOA, this task involves compiling a list of possible apneic events, symbolically characterising the degree of reduction in the airflow signal and its behaviour from the point of view of the effort signal. The complete algorithm is described in [7] and [5].

The neurophysiological analysis studies the sleep cycles of the patient as contextual information for interpreting the respiratory irregularities that have occurred during a night's sleep. In this way it is possible to reject certain events as merely reflecting a change of sleep stage,

---

[1]SAMOA, from the spanish acronym for Sistema Automático de Monitorización de Apneas de Sueño (Automatic System for Sleep Apneas Monitoring)

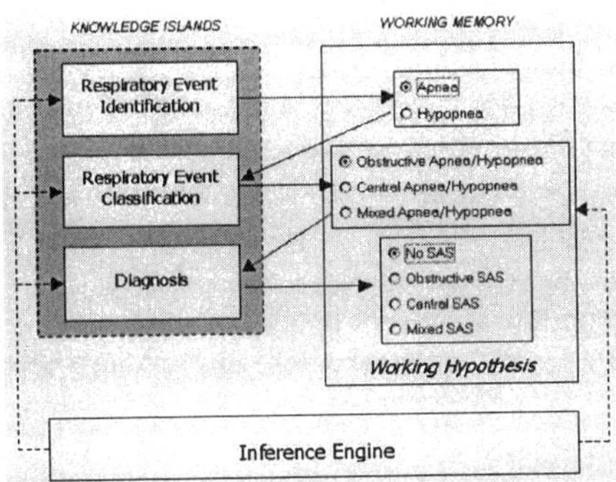

Figure 1: Block diagram of SAMOA's Decision Support System

and to explain any abnormal sleep stage transitions caused by apneic episodes [8]. This task is the responsibility of the *Hypnogram Construction Module* of SAMOA, which elaborates the sequence of stages through which a sleeping patient's passes, in the form of a hypnogram, or patient sleep map. This process is described in full in [5].

This analysis and interpretation protocol results in a symbolic evidence set, on the basis of which a decision support system arrives at its diagnostic conclusions. The input information for the support system requires the application of a prior temporal correlation process to unify the different time scales in a single frame of reference [7].

The decision support system for the SAS diagnosis is implemented via a production system (Figure 1). The system's knowledge base is organised into three knowledge islands, each dedicated to the resolution of the one of the following tasks:

1. *Apnea and hypopnea identification.* This task checks if a possible apneic event is a real event, whether an apnea or hypopnea. To do this, as an evidence set it uses the following information: the degree of reduction in the airflow, the desaturations associated with the event in question, the position signal, the EMG signal and the hypnogram.

2. *Apnea and hypopnea classification.* This task classifies the apneic events confirmed in the previous task as obstructive, central or mixed. In this case the evidence set is formed by the confirmed apneic event and the information on the behaviour of the respiratory effort. Also incorporated are contextual elements, such as data relating to the height and weight of the patient, which modify the interpretation of the evidence set.

3. *Development of a SAS diagnosis.* This requires two sets of data: (1) the results of the assessment of the polysomnographic test, expressed in terms of the Apnea/Hypopnea Index (AHI) per hour of sleep, calculated for the total of confirmed events and broken down into type of event; and (2) contextual data such as demographic information, patient's medical examination and the results of a polysomnographic prescription test that examines the symptoms in relation to the syndrome. All this information constitutes the available evidence set that permits a customised diagnosis to be developed, indicating the presence or absence of the syndrome and its classification. If no polysomnographic prescription test has been carried out, then the diagnostic conclusions are made solely on the basis of the results of the polysomnographic analysis and information is provided in this respect.

One of the more noteworthy aspects of this module is its capacity to manage situations that are apparently contradictory and to inform the clinician to that effect. This occurs when the results of an analysis of both the polysomnographic test and the polysomnographic prescription test produce results that are incompatible in terms of the classification of the syndrome. In such cases the system recommends a review of the results by the clinician.

In addition to a diagnostic conclusion for each particular situation, the outputs for the inferential process of the production system provide explanations with different levels of detail. Included in the first level are all the accepted and rejected apneic events and, where appropriate, the process that led to a specific classification. In the second, more general level, the diagnosis obtained as to the syndrome is explained on the basis of the symbolic evidence obtained in respect of respiratory characterisation, hypnogram, medical examination, polysomnographic prescription and demographic information.

Furthermore, it is provided a set of numerical data that quantitatively summarise the results of the polysomnogram analysis and that clinically justify the qualitative reasoning process carried out in the diagnostic process.

3   Results and Discussion

An independent and blind analysis of 2819 minutes of polysomnographic recordings corresponding to 7 patients was carried out by SAMOA and an expert collaborating in the design of the system. Pairs' measurements were applied using the $SHIVA^{©}$ validation tool [9], and system performance was studied in terms of the identification and classification of events and the construction of the hypnogram [5].

Table 1 shows the results obtained for the validation of the system using *Sensitivity (Se)* and *Specificity (Sp)* to rate performance. A series of conclusions can be drawn from the results obtained, as follows:

Firstly, the fact that the level of *false positives* for hypopneas is high in the events identification category can be explained by precision differences between the expert and SAMOA when quantifying reductions in the airflow signal; the former applies a more conservative criterion than the latter and thus classifies the majority of events as apneas.

Secondly, for the classification of the central and mixed events, the unsatisfactory results for *Sensitivity* and the high values for *Specificity* can be explained by the fact that most of the 7 patients in the study population suffer from obstructive apneas; the expert's classification is consequently conditioned by this knowledge.

Thirdly, the system obtained satisfactory results in terms of sleep stage classification. Moreover, it was particularly successfully in differentiating between the wakefulness and REM phases which, as it happens, is the error most frequently committed by existing automated systems. The most disappointing results, however, were obtained for sleep phases 1 and 2, both of which obtained a high level of *false negatives*, reflected principally in the *Sensitivity* value. This can be attributed to the similarity between phases in sleep stages transitions. This factor, combined with the application of a conservative criterion, causes these phases to be absorbed by those of adjacent epochs. Furthermore, the relatively high level of *true negatives* is due fundamentally to the large number of classification categories, with the implication that it is easier to discard a specific phase rather than classify it correctly.

Our final conclusion is that the system, despite the discrepancies observed, can be considered to perform satisfactorily. Nevertheless, a more complete validation based on a larger

Table 1: Performance measurement for (a) apneic events detection and classification; (b) apnea/hypopnea identification and (c) sleep stage classification

SAMOA		%Se	%Sp
Events	Apneas	80.0	82.7
Identification	Hypopneas	53.2	82.2
Classification	Obstructive	87.5	52.0
of Apneas &	Central	20.7	96.5
Hypopneas	Mixed	50.0	89.8
	Wakefulness	62.0	79.2
Sleep	Phase 1	43.9	81.4
Stages	Phase 2	25.4	89.8
Classification	Deep Sleep	47.7	94.4
	REM	66.8	93.9

data sample and the opinions of various experts in the field is required.

4 Acknowledgments

The authors wish to thank Dr. Hector Verea and Dr. Maite Martin (Juan Canalejo Hospital, A Coruña) for their valuable suggestions in developing this project. This research has been financed by CICYT-ERDF (Project TIC2001-0569) and by Xunta of Galicia (XUGA PGIDT01PXI105003PR).

References

[1] D. de la Calzada. *Diagnóstico y manejo de las alteraciones del sueño*. Dymas, 1993.

[2] J. M. Montserrat. Diagnóstico y tratamiento del síndrome de las apneas-hipopneas obstructivas durante el sueño. *Cuadernos de Formación Continuada SEPAR*, 5, Abril 1996.

[3] Sociedad Española de Neumología y Cirugía Torácica. *Normativa sobre diagnóstico y tratamiento del síndrome de apnea obstructiva del sueño*, volume 14. Publicaciones-SEPAR, http://www.separ.es/servicios/publicaciones/recomen/rec11.htm, 2001.

[4] A Rechtschaffen and A Kales. *A Manual of Standardized Terminology, Techniques and Scoring System for Sleep Stages of Human Subjects*. BIS/BRI, UCLA, Los Ángeles, 2 edition, 1968.

[5] E. Hernández. *Técnicas de Inteligencia Artificial e Ingeniería del Software para un Sistema Inteligente de Monitorizacion de Apneas en Sueño*. PhD thesis, Departamento de Computación. Facultad de Informática. Universidad de A Coruña, 2000.

[6] K. P. Strohl and S. Redline. State of the art: Recognition of obstructive sleep apnea. *Am. J. Respir. Crit. Care Med.*, 154:279–289, 1996.

[7] M. Cabrero, M. Castro, M. Graña, E. Hernández, and V. Moret. Temporal issues in the intelligent interpretation of the sleep apnea syndrome. *Lecture Notes on Artificial Intelligence*, 2101:235–238, 2001.

[8] J. C. Principe, T. G. Chang, S. K. Gala, and A. Tomé. Information processing models for sleep staging. *International Journal Expert Systems with Applications*, 6(4):399–409, 1993.

[9] V. Moret-Bonillo, E. Mosqueira-Rey, and A. Alonso-Betanzos. Information analysis and validation of intelligent monitoring systems in intensive care units. *IEEE-TITB*, 1(2):87–89, 1997.

KES 2002
E. Damiani et al. (Eds.)
IOS Press, 2002

A Task Model for a Management Resource System Integrated in a Cooperative Learning Environment

M. Lama, R. Amorim, E. Sánchez, A. Riera, J. Vila and S. Barro
Department of Electrónica e Computación. University of Santiago de Compostela.
Santiago de Compostela, A Coruña. Spain
e–mail: {lama,rramorin,teddy,vila,senen}@dec.usc.es

B. Cebreiro and C. Fernández–Morante
Department of Didáctica e Organización Escolar. University of Santiago de Compostela.
Santiago de Compostela, A Coruña. Spain.
e–mail: {dobclusc,trek1671}@usc.es

Abstract. The resource accessibility problem in a cooperative learning environment can be stated as how both teachers and students can gain access to resources (information and hardware devices) not only from remote locations, but also from the classroom. In this paper we present a task model (TMo) that describes the *services* for a Resource Management System (RMS) to solve such problem. The model illustrates how to: *1)* manage course information, and *2)* operate hardware devices. As a modelling methodology we have resorted to CommonKADS.

1 Introduction

Cooperative learning is an educational technique in which students collaborate in groups to acquire new knowledge, to learn social skills and/or to learn how to solve specific tasks [3]. In this approach, teacher responsibilities focus on supervising, in a non-directive manner, the evolution of student activity as well as proposing changes in activities with poor pedagogical results to improve learning performance. With this objective in mind, *virtual* environments, in which cooperation is achieved through appropriate graphical interfaces, have already been developed [4, 8, 7].

Our work differs from those virtual approaches by focusing on a *classroom* environment. We emphasize two main features: resource management and teacher–student interactions. In our approach each agent posses processing devices that grant access to both learning resources and communication facilities with other agents. In this paper we present a task model that describes the services for a Resource Management System (RMS) to solve the resource accessibility problem. It requires the identification of the human agents task model as a first step to determine what services the RMS should offer. The model, developed using CommonKADS methodology [6], is of special interest as long as it will help to refine the teacher task model (the main contribution of this paper). This refinement will be the starting point to develop a knowledge–based system that will support teacher´s decision–making about pedagogical actions needed to be taken.

2 Description of the Teaching environment

Our teaching environment is constituted by a set of hardware devices intended to support teaching and cooperative learning tasks. According to their capabilities, those devices can be classified as follows (figure 1):

- *Teaching hardware resources.* They are used by the teacher, and eventually by the students, in order to facilitate follow–up explanations.

    - *Mimio.* It acquires what is written on the blackboard and saves it as an image file.

    - *Projector.* It displays human agents presentations.

- *User devices.* They provide graphical interfaces to gain access and control over the teaching environment resources.

    - *Office computer.* It is used by the teacher to handle course information (such as student data, teaching material, and so on), and to plan educational activities to be performed in the classroom.

    - *Classroom computer.* It is available for each student group in order to visualize and manipulate the information needed to elaborate the required materials during the educational activity.

    - *Personal Digital Assistants (PDAs).* They are used by human agents to manage the hardware resources of the environment, to introduce/request personal data (calendars, educational materials, etc.), or to interact with other agents in order to share information in real time (to follow–up of explanations or to look for answers to evaluation forms).

- *Support devices.* They provide to human agents transparent access operations to the environment resources.

    - *Classroom server.* It runs server–side software components that carry out low–level control of hardware resources [5]. Additionally, it enables the communication amongst the entities of the teaching environment.

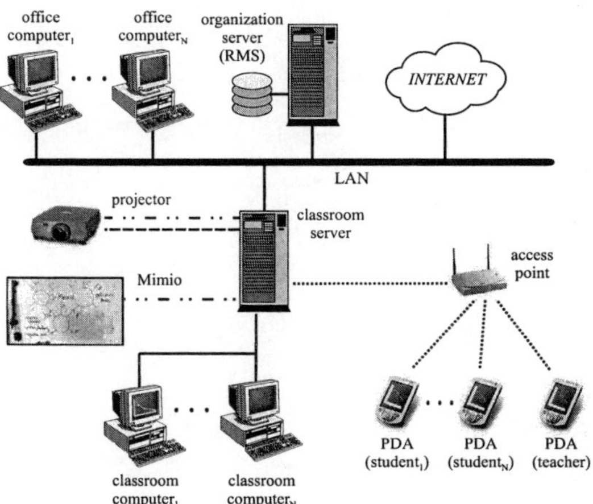

Figure 1: A teaching environment for cooperative learning.

> – *Organization server*. It provides cooperative learning support services. It will be accessible from the classroom as well as from each teacher's office.

In the next section we describe the teacher task model, which is a necessary intermediate step to help us in order to identify the services provided by the organization server. Those services will constitute what we have called the *Resource Management System* (RMS).

3   The teacher task model

In cooperative learning the *teaching* task can be considered as a *planning* problem: to describe the set of activities to be performed by the students within a certain period of time. This problem can be solved using a general class of methods known as *propose–verify–criticise–modify* (PVCM) [2]. It implies to perform a certain domain subtasks (figure 2):

- *Propose*. Based on course information, the teacher defines pedagogical objectives and proposes an activity to be developed by the students to achieve these objectives. For each activity, student groups as well as material and resources for those groups are also specified. What is more, the teacher needs to plan possible changes in the activity in case of poor learning performance.

- *Verify*. The teacher needs to keep track both student and group performance in the classroom. To do this, the teacher uses the PDA to annotate in–place observations.

- *Criticise*. Based on human agents observations, the teacher evaluates the educational activity performance for each group and student. This evaluation can be eventually weighted by student feedback obtained through evaluation forms sent from the teacher's PDA.

- *Modify*. After evaluation, the teacher could decide to refine the current activity. Two possible actions can be considered: *1)* to change the pedagogical material used by the students; and/or *2)* to propose new teaching initiatives, which could consist on either brief explanations or negotiations with the students with regard deadlines or other schedule matters.

Figure 2 shows the breakdown of the general tasks (propose, verify, criticise and modify) into domain subtasks (white ovals) as well as *transfer functions* (grey rectangles) [1]. The later represents interactions between the teacher and the other agents in the environment (students and RMS). For example, the teacher can either send information to other agents (*present*) or request information from them (*obtain* and *receive*).

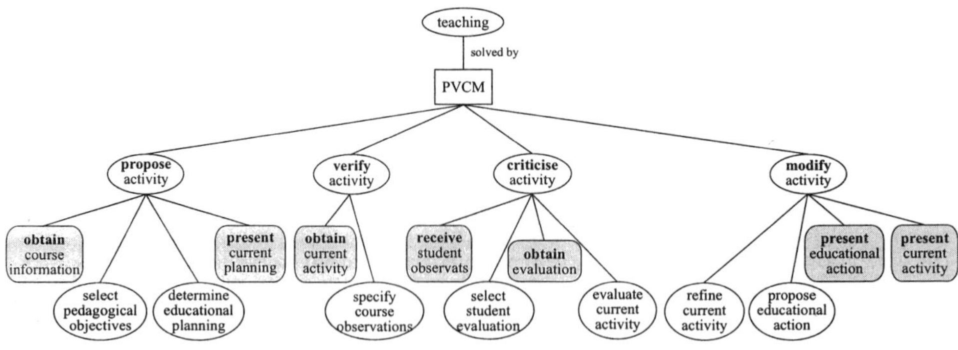

Figure 2: The *teaching* task model.

4 The RMS task model

The RMS task breakdown is based on the transfer functions identified in the teacher TMo. Each transfer function can therefore be associated to a RMS task (or service), whose execution will induce an action on a resource of the environment. In this way, the RMS task model represent, through appropriate transfer functions, human agent needs to access to resources in the environment. The tasks of this model can be easily classified considering the resource under their control:

- *Information management*. It presents the teacher the required course information: student data, list of pedagogical objectives, pre–designed activity templates, or the current educational activity.

  This kind of tasks also include services to solve teacher–student *interactions*: student requests for additional material or resources needed to perform the proposed activity, teacher requests for student answers to evaluation forms sent during the development of the educational activity or even to negotiate a common agenda with regard to the scheduling of additional lectures, and so on.

- *Access to hardware resources*. This means interaction with hardware resources in order to send/acquire information regarded teaching explanations. In addition, these tasks can forward information to PDAs in order to facilitate teaching follow–up.

Figure 3 shows the TMo of the RMS, which support teacher operations as well as teacher–student interactions. In this model the *resource management* task has not been associated to any general method for solving it. This is consistent by considering that RMS activation will be based on human agents demand.

5 Discussion

We have already implemented and validated the RMS capabilities with regard on communication between devices and low–level control of hardware resources. The task model of the teacher is also under current enhancement confirming PVCM resolution method as a suitable one to describe the general teaching methodology in cooperative learning. On this regard, the PVCM methods describe the professor–students interactions by which the learning activity is conducted. These interactions are represented by the *verify* task as well as the *criticise–modify* task. They constitute the main difference when compared with classical learning methodologies, in which interactions typically take place at the end of each lecture. The model completion,

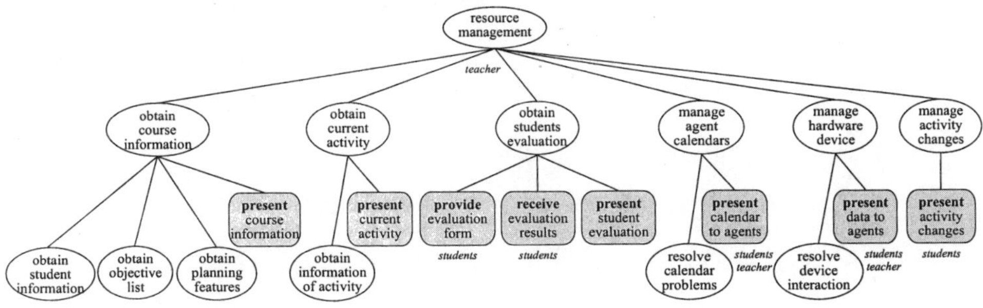

Figure 3: The RMS task model with regard to its interaction with the teacher.

however, cannot be guaranteed yet. New domain tasks, and thus new RMS services, will be possible detected as long as we explore all interaction possibilities amongst the entities of the environment.

Our RMS is the previous and necessary step for the design and implementation of a knowledge–based system to support most of the professor activities: course scheduling, student effort evaluation, resource allocation, and so on. Finally, RMS is also required *1)* to validate the professor knowledge model, and *2)* to assess the feasibility of resource management in a complex cooperative learning environment.

References

[1] J. Akkermans, R. Gustavsson, and F. Ygger. An integrated structured analysis approach to intelligent agent communication. In J. Cuena, editor, *Proceedings of the IFIP 1998 World Computer Congress, IT&KNOWS Conference*. Chapman & Hall, 1998.

[2] B. Chandrasekaran. Design problem solving: A task analysis. *AI Magazine*, 11(4):59–71, 1990.

[3] D. Johnson, R. Johnson, and E. Holubec. *The new circles of learning: Cooperation in the classroom*. Alexandria: Association for Supervision and Curriculum Development, 1994.

[4] K. O'Neill and L. Gomez. The collaboratory notebook: A networked knowledge–building environment fot project learning. In *Proceedings of the World Conference on Educational Multimedia, Hipermedia & Telecommunications (ED–MEDIA'94)*, 1994.

[5] J. Riera, A.and Vila and S. Barro. A PDA classroom computer system. In C. Montgomerie and J. Viteli, editors, *Proceedings of the World Conference on Educational Multimedia, Hipermedia & Telecommunications (ED–MEDIA'01)*, 2001.

[6] G. Schreiber, H. Akkermans, A. Anjewierden, de Hoog. R., N. Shadbolt, W. Van de Velde, and B. Wielinga, editors. *Knowledge Engineering and Management: The CommonKADS Methodology*. MIT Press, Boston, 1999.

[7] A. Shabo, K. Nagel, and M. Guzdial. JavaCAP: A collaborative case authoring program on the WWW. In R. Hall, N. Miyake, and N. Enyedy, editors, *Proceedings of the Second International Conference on Computer Support for Collaborative Learning (CSCL'97)*, 1997.

[8] D. Shuters and D. Jones. An architecture for intelligent collaborative educational systems. In B. Boulay and R. Mizoguchi, editors, *Proceedings of the 8th World Conference on Artificial Intelligence in Education (AIED'97)*. IOS Press, 1997.

KES 2002
E. Damiani et al. (Eds.)
IOS Press, 2002

Adaptive Knowledge-Based Systems with Intelligent Agents and Evolutionary Computation

Juan Gabriel Fernández García de la Rocha, Eduardo Mosqueira-Rey
Department of Computer Science, University of A Coruña, 15071, A Coruña, Spain

Abstract. Complex real world problems require systems that can take full advantage of domain knowledge. In dynamic application domains it is therefore necessary to equip these systems with the ability to adapt their knowledge. One way of designing such an adaptive knowledge-based system is through the use of intelligent agents. This paper proposes using evolutionary computational techniques to enhance the adaptive ability of intelligent agents.

1. Introduction

A knowledge-based system is a computerized system that uses its knowledge of a particular domain to resolve a problem within this domain. It is mainly characterised by the fact that its knowledge is separated from the control structures responsible for determining its behaviour.

In dynamic application domains, defined as domains in which system state, the available data and the actions to be executed are subject to continual change, knowledge-based systems as traditionally structured may not be capable of producing positive results and this makes it necessary to equip systems with the ability to adapt their behaviour to the changing conditions of their domain. This need has given rise to what are referred to as adaptive knowledge-based systems.

When developing an adaptive knowledge-based system, it is logical to structure its architecture around the use of intelligent agents [1]. This is because one of the features of intelligent agents - understood to be systems having the aim of complying with a series of motives or goals in a normally dynamic and unpredictable environment - is their ability to adapt. This ability to adapt is based on learning capacity. Agents, in different situations, must learn to execute actions that bring them as close as possible to their goals.

The problem with intelligent agents is that their learning process is generally slow because of the large number of possible situation-action pairs. In an attempt to deal with this shortcoming, we can use evolutionary algorithms, which are an ideal solution for knowledge-acquisition tasks. The complementary nature of these two techniques allows us to integrate intelligent agents with evolutionary computation techniques in order to improve an agent's learning process.

The aim of this paper is to investigate the different ways of integrating evolutionary computation techniques into intelligent agent systems. These agent architectures are ideal for designing knowledge-based systems highly capable of adapting behaviour in accordance with a continuous acquisition of knowledge.

2. Intelligent Agents

Intelligent agents have been defined in many ways but, in general, these definitions tend to revolve around their basic properties. Although there is no overall consensus as to these properties, they tend to be defined as in Table 1 below.

Table 1.

Basic Properties of Intelligent Agents

Reactivity	Perception of the state of their environment and capacity for response to changes in this environment
Proactivity	Opportunistic objective-driven behaviour where agents 'take the initiative'
Adaptability	Ability to improve behaviour over time
Autonomy	Control over the actions they execute
Social ability	Ability to interact with other agents

Learning is a very general term denoting the manner in which people and computers optimise the knowledge available to them in order to further increase knowledge and improve capabilities. In systems with specific tasks to execute, learning is a process by which these improve their performance. In the case of intelligent agents, learning allows them to acquire knowledge from their interaction experiences with the environment (heuristic knowledge). Intelligent agents are autonomous in that their performance is defined by their heuristic knowledge, i.e. the more they learn from their experiences, the more tasks they will be able to execute without the need for external monitoring. Learning from experience also allows agents to adapt to their environment.

3. Evolutionary Behaviour

Evolutionary behaviour is a computational paradigm based on biological models from the fields of genetics and evolution [2]. The term evolutionary algorithm covers a large variety of optimisation algorithms and problem-solving systems. Evolutionary algorithms operate on a population of individuals that evolve according to the principles of natural selection. They include a group of operators, called search operators or genetic operators. The most commonly used techniques in evolutionary algorithms are recombination and mutation. Also necessary is an evaluation function to play the role of the environment and to classify individuals in terms of utility or fitness.

Seen from a biological point of view these algorithms are very simple, yet they constitute a robust and powerful tool for adaptive search. Genetic algorithms endeavour to find improved solutions for a problem (e.g. improved parameter combinations in an optimisation problem) as the population develops. Furthermore, rather than searching randomly for a solution to a problem, evolutionary algorithms apply stochastic search methods.

4. Why Intelligent Agents and Evolutionary Computation?

Intelligent agents must execute the action best suited to the satisfaction of their goals in the different situations arising in their environment. This means that agents must learn the strategy or selection process that allows them to choose the best solutions for their particular series of goals. Unfortunately, this learning process generally tends to be slow, due to the large number of possible situation-action pairs. The application of genetic algorithms, however, can be viewed as a complementary learning process that permits this process to be accelerated.

Evolutionary computation techniques can be integrated into agents with a view to improving their learning efficiency. This integration is possible because evolutionary algorithms have the ability to adapt to changing situations, namely to changes occurring within their environment. In many traditional optimisation procedures, if any of the variables in the problem are changed, then calculations must be made again from scratch. In an evolutionary algorithm, this return to zero when there is a change in the environment is not necessary, since the current knowledge of the problem (the current population) serves as a starting point. For this reason, evolutionary algorithms can be applied to dynamic environments.

According to Goonatilake and Khebbal's classification of hybrid systems [3], a hypothetical system of agents which learn evolutively is a function-replacing hybrid. In this hybrid category, any of the main functions of a specific intelligent technique can be replaced by another technique. The motivation behind these hybrid systems lies in improving existing techniques, by increasing execution speed or improving reliability. In an intelligent agent-evolutionary computation hybrid system the learning function of the agents, rather than being replaced, is improved through the use of a genetic algorithm.

Speeding up the process of learning an action selection strategy means that agents adapt better and faster to their environment. In other words, by integrating algorithms within intelligent agents we can also optimise a system's behaviour (i.e. performance).

5. Integration methods

When it comes to integrating agents and evolutionary systems we focus on agents whose knowledge is divided into different competence modules. A simple example of such a module would be what is termed as a set of situation-action rules. An agent whose behaviour is determined by situation-action rules pattern-matches its rules and the situation as is, and then executes the rule which, in the past, came closest to satisfying the goal in a similar situation.

There are two ways of integrating evolutionary computation techniques and intelligent agents. The first way is to use an agent population which we cause to evolve genetically (Figure 1). Examples of this type of integration include the evolutionary model for autonomous agent development [4], FuzzyEvoAgents [5], hybridization based on biological symbiosis [6],[7], the Amalthaea system [8] and the InfoSpider system [9].

The second way is to cause a single agent's competence modules to evolve (Figure 2), in other words, to take as the competence modules the individuals making up the evolutionary algorithm population. Examples of this option are systems based on the dynamic prioritisation of rules [10] and classification systems [11].

Since evolutionary programming is based on biological models, the concepts of the biological domain must be transferred to the computerized domain and, in our particular case, to the intelligent agent domain (Table 2). The use of terminology analogous to that used in biology will permit us to lay the groundwork for the development of an evolutionary algorithm adapted to intelligent agents.

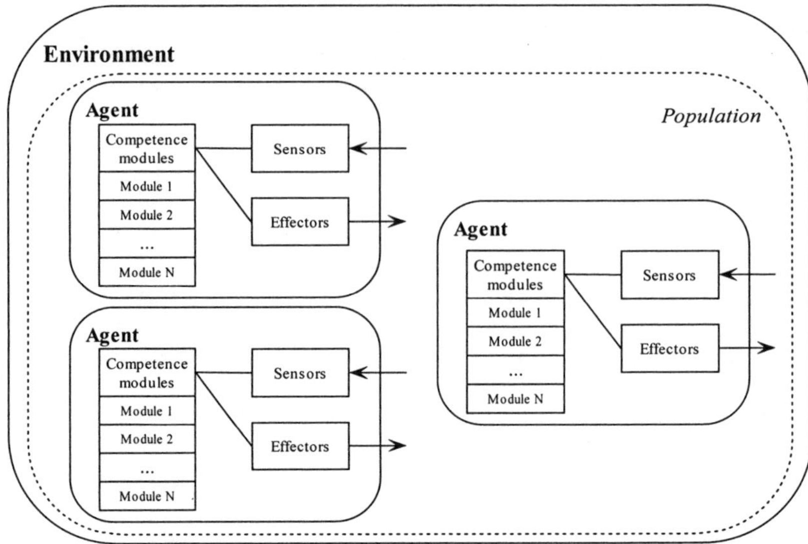

Figure 1.- Population formed from agents

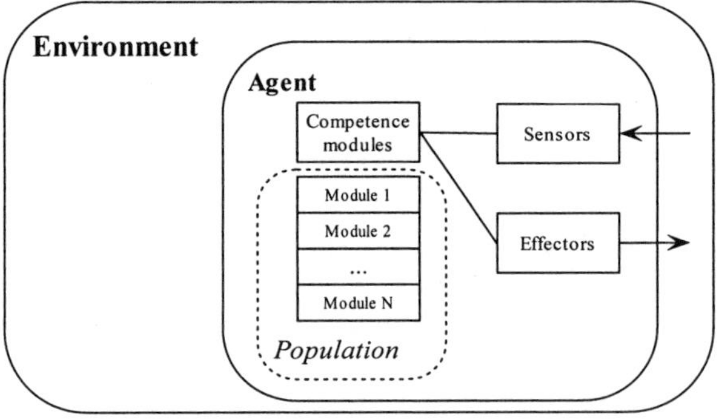

Figure 2.- Population formed from competence modules

Table 2.

Equivalence between biological and computational concepts

Biology	Agent population	Competence Module Population
Individuals	Agents	Competence modules
Chromosome	Set of competence modules	Individual competence module
Gene	Individual competence module	Component part of each competence module

In the first integration option described above, the individuals who make up the population are the agents themselves. Each chromosome is composed by a list of elements, called genes, each of which in turn corresponds to a competence module. Each individual has only one copy of a chromosome, thus following the pattern of lower-order plants and animals whose chromosomes do not possess homologous patterns except during mating. The chromosome is, therefore, an intelligent agent's set of competence modules.

In the second integration option, we have a population of competence modules corresponding to a single agent, and therefore the individuals in the population will be the competence modules. Each individual has only one chromosome which is the competence module itself. Each chromosome is formed by genes corresponding to the component parts of the competence module. For example, if the competence modules were rules, then the different genes could be the conditional clauses and the conclusion for each rule.

6. Conclusions

In this paper we have studied the two basic ways of integrating evolutionary computation techniques into intelligent agent systems. The first consists of causing a population made up of agents to evolve genetically, whereas the second method consists of causing a population formed from the competence modules of a single agent to evolve. Applying the principles of evolution to intelligent agents we improve their performance by speeding up their process of learning the correct solutions for different situations occurring in their environment. Speeding up an agent's learning process facilitates the acquisition of heuristic knowledge, which in turn increases autonomy and improves the adaptive capacity of these agents.

Knowledge-based systems operating within dynamic environments need to adapt knowledge of the environment to the changes occurring within that environment. This makes agents with evolutionary computation hybrid architectures ideal solutions for designing knowledge-based systems to be used in dynamic application domains.

Acknowledgements

This project has been partially funded by the Autonomous Government of Galicia (Spain) via Research Project PGIDT01PXI110503PR, and by the Spanish Inter-Ministerial Commission for Science and Technology (CICyT) via Research Project TIC 2001-0569 (co-funded by the European Regional Development Fund, ERDF).

References

[1] M. J. Wooldridge, N. R. Jennings, "Intelligent Agents: Theory and Practice", Knowledge Engineering Review 10 (2).
[2] David B. Fogel, "What is evolutionary computation?", IEEE Spectrum, February 2000
[3] Suran Goonatilake, Sukhdev Khebbal (eds.), "Intelligent Hybrid Systems", John Wiley & Sons, 1995.
[4] Frank Dellaert, Randall D. Beer, "Toward an Evolvable Model of Development for Autonomous Agent Synthesis", In R. Brookds, P. Maes (eds.), Artificial Live IV: Proceedings of the Fourth International Workshop on the Synthesis and Simulation of Living Systems, MIT Press, 1994.
[5] Antonio Di Nola, Antonio Gisolfi, Vicenzo Loia, Salvatore Sessa, "Emerging Behaviors in Fuzzy Evolutionary Agents", Proceedings of the 7th European Congress on Intelligent Techniques & Soft Computing, EUFIT'99, Aachen (Germany), Septiembre, 13-16 1999, CD ROM ISBN 3-89653-808-X
[6] Jason M. Daida, Steven J. Ross, Brian C. Hannan, "Biological Symbiosis as a Metaphor for Computational Hybridization", In L. J. Eshelman (ed.), Proceedings of the Sixth International Conference on Genetic Algorithms, Morgan Kaufmann, pp. 328-335.
[7] John Polito, Jason M. Daida, Tommaso F. Bersano-Begey, "Musica ex Machina: Composing 16th Century Counterpoint with Genetic and Symbiosis", Evolutionary Programming VI: Proceedings of the Sixth Annual Conference on Evolutionary Programming, Springer-Verlag, 1997.
[8] Alexandro Moukas, "Amalthaea: information discovery and filtering using multiagent evolving ecosystem", Applied Artificial Intelligence 11, 437-457, 1997.
[9] Filippo Menczer, Alvaro E. Monge, "Scalable Web Search by Adaptive Online Agents: An InfoSpiders Case Study" In Matthias Klusch (ed.), "Intelligent Information Agents, Agent-Based Information Discovery and Management on the Internet", Springer-Verlag, 1999.
[10] Evaggelos Nonas, Alexandra Poulovassilis, "Optimisation of Active Rule Agents Using a Genetic Algorithm Approach", In "Database and Expert Systems Applications", pp. 332-341, 1998.
[11] David E. Goldberg, "Genetic Algorithms in Search, Optimization & Machine Learning", Addison-Wesley, 1989.

KES 2002
E. Damiani et al. (Eds.)
IOS Press, 2002

Cooperative Development Environment of Sign Language Animation System using Humanoid Model

Takaya Yuizono, Kousuke Hara and Shigeru Nakayama
Department of Information and Computer Science, Faculty of Engineering, Kagoshima University
1-21-40, Korimoto Kagoshima 890-0065, Japan

Abstract. We have developed a sign language animation system made by cooperative group. The system presents the animation of a 3D avatar referred to a humanoid model. The system has been developed as a multi-layered system, which has middle server between animation database server and 3D graphics client system. The two client systems developed here are both Java3D application and Web based VRML client. Both client systems support the same humanoid model and can show a sign language shared with the same animation data. The database aims at load balancer to make an avatar animation from a sign language word. Our estimation of this system shows that 3D avatar animation is better than both our early avatar animation system and printed document.

1 Introduction

A computer network such as the Internet can allows us to use not only verbal communication, such as text and voice, but also nonverbal communication, such as gesture and face expression [1]. The multimedia technology is expected to support the disabled. There is the research of the sign language aided by computer to support deaf people. The sign language [2], which can express natural communication, is not a body language represented by a sequence of gestures. We have developed animation system with avatar, which is a user entity displayed with 3D graphics in the cyberspace, referred to humanoid model [3], [4]. The model supports interoperability of avatar in the 3D virtual world.

The number of words in Japanese Sign Language (JSL) is approximately 4000. The inputting all the animation data of the language is the hard work. A sign language is different in the country, so JSL is different from American Sign Language, and deaf people form their peculiar community [2]. Therefore, it is natural that a sign language system supports multi-user and shares data about their language via computer network. The shared database supports the cooperative making of the animation data by many people.

This paper describes the cooperative development environment of sign language animation system using humanoid avatar.

2 Sign Language Animation System

The sign language animation system has been developed as a layered system, which has middle server between animation database server and 3D graphics client system. The layer

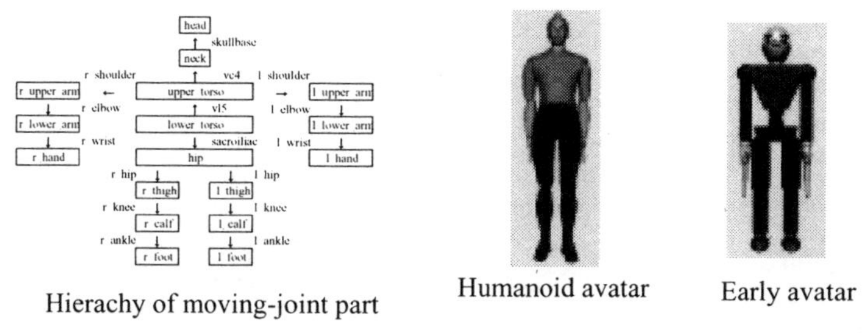

Hierachy of moving-joint part   Humanoid avatar   Early avatar

Figure 1: Hierarchy of moving-joint part and avatar model for sign language animation.

system has flexibility for the system development.

2.1 Humanoid model and client system

Humanoid model [3] has been suggested as standard avatar model for supporting its interoperability in the 3D virtual world and the model is conventionally realized as VRML model. The avatar referred to the humanoid model is shown in the center of figure 1. The left side of the figure shows hierarchy of moving-joint part to animate sign language. The hand part has many moving-joint parts to represent finger motion. The right side of the figure shows early avatar model made of 3D primitive objects, but the body size ratio is different from humans.

In the client system, the humanoid model shows sign language animation. So, we have developed two client systems shown in figure 2: one has been developed as a Java client application with Java3D API, other has been developed as Web based client application with both VRML and Java. Avatars in both client supports same humanoid model and can show a sign language from same animation data. The client systems support key frame animation [5] to show a sign language, and get the animation data from the shared animation database via the middle server.

2.2 Sign language database

Sign language database has been developed with Microsoft SQL server. The database has one animation data table and there index tables. Animation data table has key frame animation data, which is set of humanoid joint angles shown some pose. Three index tables are sign word index, finger spelling index and compound word index. A sign word index points at a set of key frame data representing a sign language word. A finger spelling index points at a set of key frame data representing a finger spell, too. A compound words index points at multiple sign words that express some meaning.

2.3 Middle server

Middle server consists of some servers and realizes multi-layered system as shown in figure 3. The server to access sign language database is developed with Java RMI (Remote Method Invocation), which is Java based object-oriented remote procedure call. A Java3D client application takes the animation data via the RMI server to access the database. Otherwise, a web client application accesses the RMI server with Java Servelet server and gets animation data

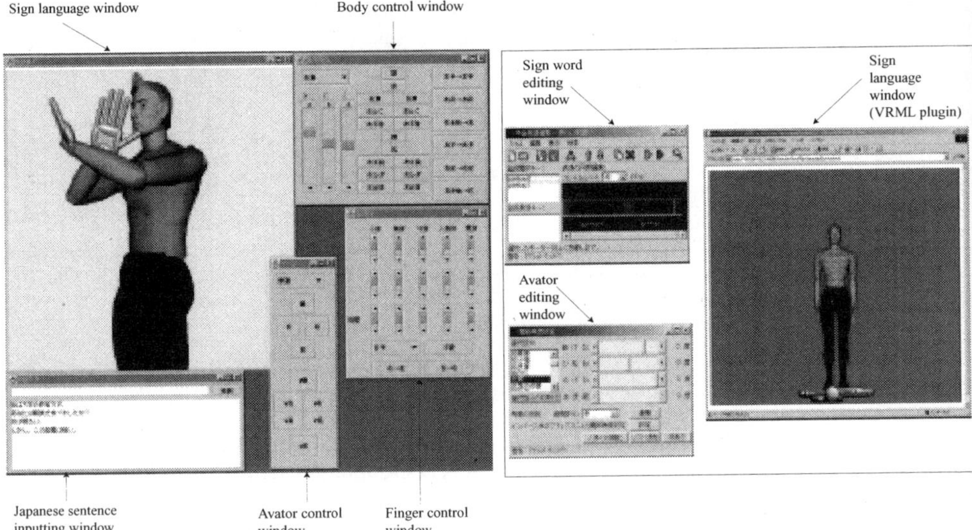

Figure 2: The left part is a Java3D client system. The right part is a VRML client system on the web.

from the database. Middle server has the translator service, which gets words from Japanese sentences using token analysis. The translation that produces a sequence of sign language words is not enough for representing sign language correctly.

3   Experiments and results

3.1   Procedure

We have estimated humanoid based sign language animation and have a trial of making animation data by eleven participants, who have experience to see the sign language in the TV programming but cannot use the sign language. We used the Java3D client application for the estimation. We took a questionnaire from participants after these experiments.

(a) Estimation of sign language animation

We have compared the humanoid avatar animation with four kinds of presentation. The same sign language conversations collected from TV program [6] are demonstrated for participants twice with each presentation. After the demonstration, participants do pair comparison method between the five kinds of sign language presentation. The method is used for a decision of weight values about some evaluation items in AHP [7], which is one of the decision methods by quantifying human instinct.

Those presentations are as follows. (1) "Early avatar" made of 3D primitive objects but the body size ratio is different from humans. (2) A "large video" screen has a width of 320 pixels and a height of 240 pixels and its presentation quality is similar to TV. (3) A "small video" screen has a width of 160 pixels and a height of 120 pixels and its presentation quality is similar to multimedia conferencing system whose data transmission amount is a few hundreds bits per seconds. (4) "Printed document" is continuous still images printed on A4 size paper and the document explains the same sign language conversation from the TV program text[6].

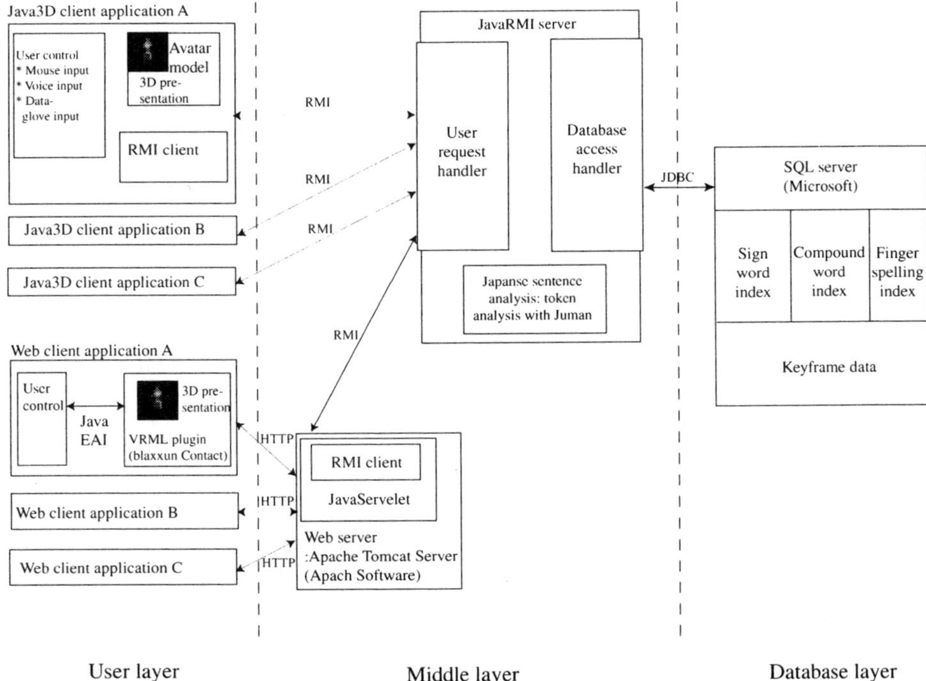

Figure 3: The configuration of sign language animation system.

(b) Trial of making animation data

The experiment, in which participants input sign language animation data while referring to partial copies of a sign language dictionary [8], took about 50 minutes.

3.2   *Results of experiments*

Estimation result of the sign language animation in the right of figure 4 shows bellows. Humanoid avatar animation is better than that of "early avatar", that of "small video", and that of "printed document". The animation is not superior to that of "large video". The number of sign language word data developed by 11 participants who are novice user to our system is 18. Otherwise, two authors who are more familiar with our system than the participants make 15 words while the similar time.

3.3   *Results of questionnaire surveys*

The results of the questionnaire are described as below. The merit of the sign system compared with video presentation is that the view is changeable to see the sign language animation from multiple directions. The drawbacks are that the animation smoothness is not enough, the face with no expression of the humanoid model is not familiar, and the key frame animation making is difficult. The participants suggested the addition of automatic recognition function of sign language, and the representation function of facial expression.

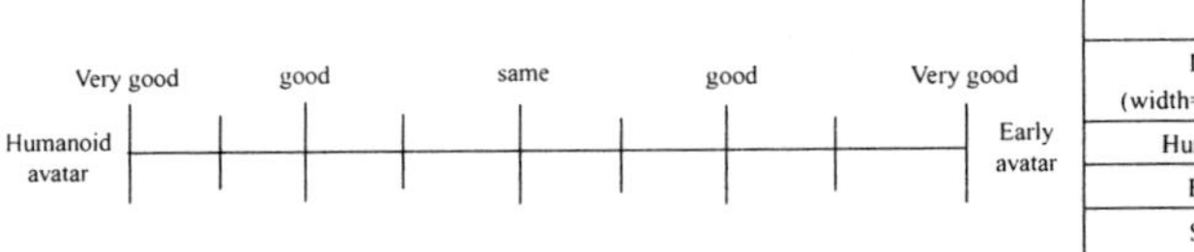

Presentations	Weight value
Large video (width=320, height=240)	0.35
Humanoid avatar	0.29
Early avatar	0.14
Small video (width=120, height=80)	0.13
Printed document	0.09

Figure 4: The left part is an input form of pair comparison method. The right part is the table of weight value to evaluate five kinds of sign language presentation.

We consider improving the smoothness by taking account of the signer's motion rhythm, because the key frame animation of humanoid monotonously moves by linear transformation. Our system should support face expression because the expression is important of sign language communication. We will reconsider the user interface to reduce the burden of making humanoid animation using a motion capture.

4 Conclusion

The sign language animation system referred to a humanoid model made by cooperative group has been developed. The model is a standard avatar model for supporting its interoperability in the 3D virtual world. The animation system consists of two kind of client systems for humanoid animation, which system supports humanoid controlling and animation modeling, and sign language database, and middle server through which all client access the database and stored animation data. The estimation of this system shows that humanoid avatar animation is better than both early avatar animation and printed document.

In future plan, we will reduce the burden of making humanoid animation and consider the natural representation of sign language.

References

[1] Kurokawa, T., "Nonverbal communication and Interfaces",pp.163-170, Journal of Human Interface Society, Vol.2, No.3, 1997 (in Japanese).

[2] O. Sacks, "Seeing Voices", Shobun-sha, Tokyo, 1996 (Japanese Translation).

[3] Humanoid Animation Working Group, "Specification for a Standard Humanoid; Version 1.1", 1999, http://www.h-anim.org/.

[4] C. Babski, D. Thalmann, "Real-Time Animation and Motion Capture in Web Human Director",Proc. Web3D-VRML2000 Symposium, ACM Press, pp.139-146, Feb. 2000.

[5] Hodgins, J. K., "Animating Human Motion", Scientific American, March 1998.

[6] "NHK Sign Language for Everyone Vol. II", NHK Press, Tokyo, 2000 (in Japanese).

[7] Tone, K., "Introduction of AHP", Nikkagiren, Tokyo, 1986 (in Japanese).

[8] Ogata, H., "Understandable Sign Language Dictionary", Natume-sha, Tokyo, 2000 (in Japanese).

KES 2002
E. Damiani et al. (Eds.)
IOS Press, 2002

1043

Development and Application of a Group Learning Function in the Virtual School System

Satoru Fujii*, Jun Iwata*, Kouji Yoshida** and Tadanori Mizuno***
*Matsue National College of Technology,14-4 Nishi-Ikuma Matsue 690-8518, Japan
** Kurashiki University of Science and the Arts, Nishinoura Tsurashima Kurashiki 712-8505, Japan   *** Shizuoka University3-5-1, Johoku Hamamatsu 432-8011, Japan

Abstract. We developed a 3D Virtual School with learning support rooms to help students study online. The students who had studied in the school found it very useful. We have added a tool which enables students to communicate with other students while studying in this school. The tool we have developed is called a Group Learning Function, which enables students to chat with other students and draw figures and pictures with a whiteboard. We have evaluated this function and confirmed that it is practical for students in their studies.

1. Introduction

There are many kinds of virtual school systems available now. These systems have various functions but not so many of them have useful learning systems constructed with individual HMI, adequate contents and 3D graphics[1],[2]. Among group learning support systems available, such systems as "Lotus Learning Space" support self-learning, interlace learning and unsynchronized learning. These systems have an electric whiteboard system which enables students to discuss together online [3],[4],[5].

We developed the Virtual School System using 3D graphics, which had a Language Learning Room called "Web-CALL", a Mathematics Room for studying mathematics functions, and an IT(Information Technology) Learning Room for studying for a certification examination in information processing and a Typing Training Room for practicing typewriting for computer literacy [6],[7]. We opened the virtual school system on LAN and it has been frequently used by many students. Since we regarded a group learning function with a whiteboard and a chatting tool, where users can draw figures and pictures or discuss with other students in the Virtual School, as an important factor for successful use of the system, we have developed a Group Learning Function and applied it to the Virtual School System. With the use of the function, students can discuss their learning materials with other students by chatting and drawing images on the whiteboard [8].

The Group Learning Function was evaluated by the students who had tried it. As a result, this function is useful for students to communicate with other students. Still more, we checked the load of the server. When 20 students logged in simultaneously, CPU idle rate fell by 13 to 14 %, and capacity of main memory fell by 10 to 11MB. These falls were small for the server to drive stably. Therefore, we confirmed that this function could be used by many students simultaneously.

2. Construction of the Virtual School System

Fig.1 shows the structure of the Virtual School System. The system including various learning rooms is developed with VRML and displayed with 3D graphics. This school has a Language Learning Room, a Mathematics Room, an IT Learning Room, and a Typewriting Training Room. Users can enter the system using a mouse and can select a learning room they want to study. The 3D Display Processor executes dynamic events using Materials DB that has many kinds of sound, voice and picture data. Each Learning Support System called from 3D display Processor has its own Teaching Materials DB and Score DB. Therefore, this function can offer teaching materials to students and can collect learning history from them.

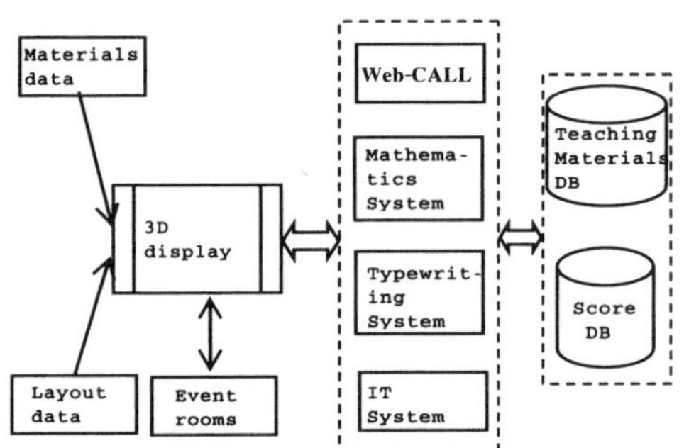

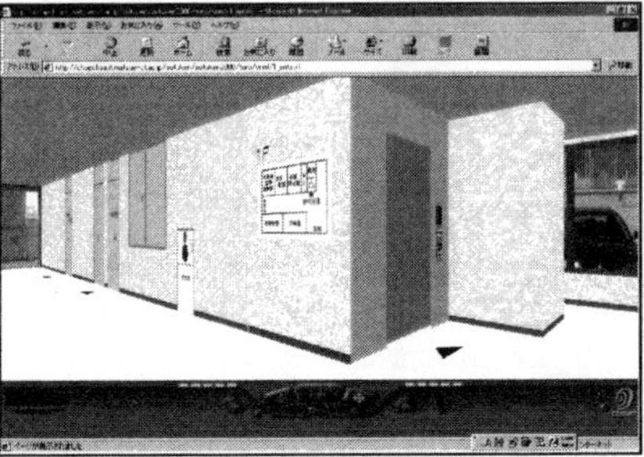

Fig.1 Construction of the Virtual School System.   Fig.2 A screen of the Virtual School System.

The first floor of Virtual School is shown in Fig.2. Each floor has different learning rooms, research rooms and a laboratory, controlled by the layout DB. Students move around the school using a mouse and a keyboard, and they can travel to another floor by elevator. The elevator door opens when the button near the door is clicked. Floors are selected from the numbers indicated on the door of elevator. To enter each room to study, students click the door of the room and then the learning support system of the room starts. If students want to visit another room, they exit from the room after finishing their studies.

Each learning room has different teaching materials and interfaces. Now we explain the IT Learning Room as an example. We have developed the room by modifying the language learning room called "Web-CALL" which was developed in 1997. Fig.3 shows a scene of the IT Learning Room. Questions are shown in the upper-left frame. Buttons and pull-downs are used to select the answers. When selected answers are submitted to the server, the score result is displayed in the lower frame. When the number of the question in the score result is clicked, the hints about the question are displayed in the upper-right frame. Therefore, the students can answer the question again using the hints. Still more, teachers can use an authoring tool to create their own teaching materials about information technology. This tool also enables teachers to add new questions and hints or modify them easily.

3. Support Function for Group Learning

A scene of Group Learning Function is shown in Fig.4. The upper-left frame is a whiteboard. A tool to set color and size of the pen is placed on the right of the lower frame. The lower-left frame is used for chatting, where a narrow window on the left shows the list of the students who are in the room. This window helps the students to recognize who are in the session. Right frame can be used by students to upload the files from the server computer.

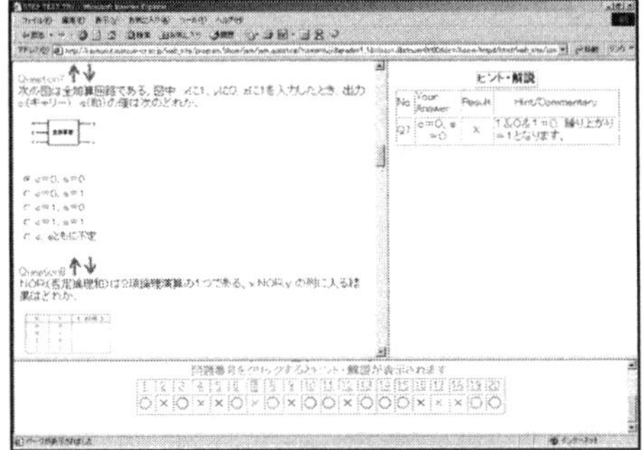

Fig.3 A scene of IT Learning Room.

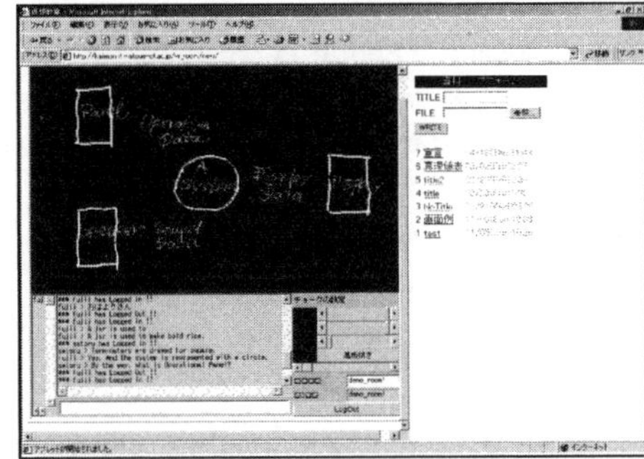

Fig.4 A screen of Group Learning Function.

Group Learning Function has several functions listed below.
(1) Whiteboard Function
Whiteboard is the area where we can draw figures freely using a mouse or a pen. Contents drawn by each student are transmitted quickly to each whiteboard of all the students who are in the same session. Whiteboards can be used in all the Learning Rooms and can be added if necessary. Whiteboards are useful in communicating with other students by drawing figures, pictures, characters, mathematical expressions, formulas and circuit diagrams.
(2) Chat Function
Participants of learning group chat by key-in with their keyboards. Their chat is sent to all the other participants and displayed immediately.
(3) History Function
History function records drawing and chatting data at regular intervals, and shows the date sequentially when requested. Students who join in a learning room late can know the process of learning by reading through the history.
(4) User Management Function
A window in the right hand of chatting window shows users' names accessing to the Group Learning Function. When a user logs in or logs out, user management function appends or deletes his or her user name, and redraws the window of members at regular intervals.
(5) Uploading and referring to files
This function is a tool for uploading text documents to the server computer, and all the students can refer to these documents.

4. Construction of Group Learning Function

Fig.5 shows the construction of the Group Learning Function. The function is developed with JDK, Apache and Tomcat for the server while it is developed with Browser and Java Plug-In for clients. Users can enter a learning room after downloading Java Applet to their terminals. Load of server is much reduced by using Applets, because Applets process dynamic drawing in the client terminals. There are two kinds of Applet: a drawing Applet and a chatting Applet. Each of them sends its logs to each Java Servlet. This Java Servlet receives logs and sends them to all the clients. At the same time, it records logs to a log file that can be used for storing history information.

Users draw images on a whiteboard by dragging and dropping a mouse. Group Learning Function gets the drawing history from users at regular intervals, and sends the history back to users when they drop a mouse. As a result, this function replays drawings sequentially. We have developed this function by Java Multi-thread that can process two activities simultaneously.

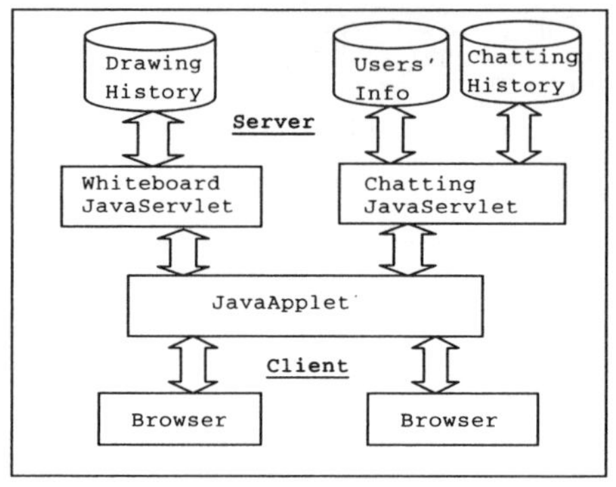

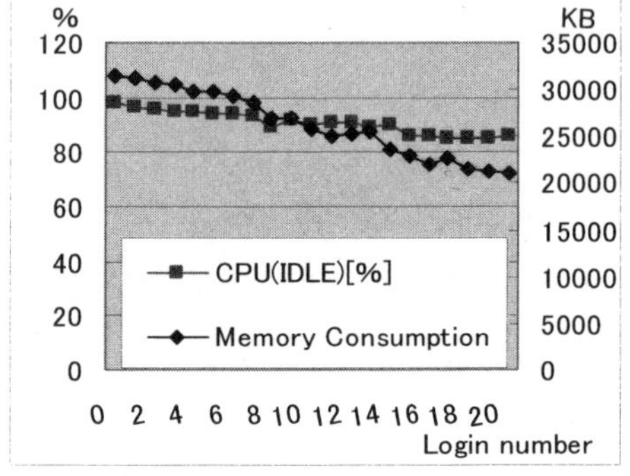

Fig.5 Construction of the Group Learning Function.   Fig.6 Load of CPU and Memory

5.Evaluation of Group learning Function

The Group Learning Function was evaluated by five students who had currently tried it. The students were given a task to design software in cooperation by drawing a data flow diagram (DFD) on the whiteboard. They spend about one hour designing a DFD on a whiteboard through discussion with a chatting tool. After the session, we asked them to write their impressions on the system to evaluate the effectiveness of the function. Their impressions are as follows;

① The whiteboard we used was effective to communicate with other students, but it was not convenient to draw pictures with a mouse, and the cleaner of the whiteboard was too small.

② Chatting tool was easy to use and was effective in cooperative learning.

③ The tool for uploading text documents should be used when necessary.

④ Changing learning rooms and whiteboards were useful in discussing the topic we are interested in, but the interface of the Group Learning Function needs improvement because users do not know which room they are in.

Most students evaluated that Group Learning Function was especially useful for discussion in a small group members.   IT Learning Room in the Virtual School offers the students questions and hints when they want to study. There are adequate contents; many kinds of questions and hints made by the teachers. It may be the best way to ask their teachers questions directly, but students are not always able to meet their teachers at school. Therefore, the use of the group learning function in a virtual school is quite helpful for them, where they can ask other students the question. Whiteboard in the Group Learning Function makes it easier for students to understand the subjects by drawing pictures or figures as well as by chatting with other students.

We evaluated the loading of the server when students were using the Group Learning Function. Each client chatting or drawing in the system sends at least two request signals to the server in every two seconds. Java Servlet in the server processes each request by thread, so memory consumption increases in proportion to the numbers of request. Memory consumption fluctuates according to the number of the users accessing the room. When many students in a learning room chat or draw images at the same time, the load to the memory increases. Java Servlet we have developed does not create new processes compared with a CGI written with Perl language, so the server can avoid using much memory. The biggest problem was the amount of the load on CPU and memory of the server when they process drawing data on whiteboards. To settle this problem, we have designed the construction of data shown in Fig.4 and made it possible to reduce the load to the sever and deal with simultaneous use of the function by many students. Fig.6 shows the data of an experimental measurement. With 20 login students logged in, rate of CPU idle decreased 13 to 14% and free memory volume decreased 10 to 11MB. These decreases were little and had no effect in processing the system, so this system could work stably. Therefore, we could accomplish the purpose of developing a system that many users access at the same time.

6 . Conclusion

We have added a new Group Learning Function to the 3D Virtual School System which we had developed. The function is useful for students to communicate with other students. It enables students to discuss the matters on the subject they are studying by chatting or drawing images on whiteboards. We have evaluated the function by letting students use it in their cooperative studies and make some comments on the function. As a result, the function was useful for them to communicate with other learners online. Furthermore, we have evaluated the loading of the server computer. When 20 students logged in the function simultaneously, CPU idle rate decreased 13 to 14%, and capacity of main memory decreased 10 to 11MB. These decreases were so small for the server system to drive stably that we recognized that the function could be used by many students simultaneously. The function proved to be useful, but more research and modification are necessary to improve it more useful and effective in the 3D Virtual School.

References

[1] K.Nakabayashi, Y.Koike, M.Maruyama, Y.Hirafumi, Y.Fukuhara and Y.Nakamura, "CALAT:An Intellegent CAI System Using the WWW", The Trans. Of the IEICE D-2, vol.J80-D-2, no.4, pp.906-932 (1997).
[2] S.Kajita, "Current Status of WebCT and Future of Information Basis for Higher Education", Media Education Center, Vol.3, pp.5-14(2001).
[3] http://www.lotus.co.jp/home.nsf/Content/DP1_LearningSpace_R50_top(2002).
[4] T.Hirahara, T.Yamanoue, H.Anzai and I.Arita, "The Electronic Chalkboard System for Distributing Network Environment Using TCP", IPSJ Journal, Vol.43, No.1, pp.176-184 (2002).
[5] A.Kawaguchi, Y.kato, S.Ishihara, S.Sakai and T.Mizuno, "A Digest Making Method for Helping Users to Join to Teleconference from Halfway", IPSJ Journal, Vol.42, No.12, pp.3031-3040 (2001).
[6] Fujii, J.Iwata, M.Hattori, M.Iijima and T.Mizuno : "Web-CALL":A language learning support system using Internet,IEEE Computer Society ICPADS2000, NGITA Workshop, pp.326-331(2000).
[7] S.Fujii, N.Okamoto, J.Iwata and T.Mizuno : Development of a 3 D Virtual School with Learning Support Systems, Proceeding of the ACIS 2nd International Conference on SNPD'01, pp.152-159 (2001).
[8] S.Fujii and T.Mizuno, "Development of aGroup Learning System with chatting tool and whiteboard", Technical Report of IEICE, Vol.101, No.397, ET2001-53, pp.57-63 (2001).

KES 2002
E. Damiani et al. (Eds.)
IOS Press, 2002

Evaluation and Trial of Distance Learning System Using Mobile Phone and WWW

Kazuhiro Nakada*, Kouiti Matsumoto*, Tomonori Akutsu*, Kenji Nagamoto*,
Satoru Fujii**, Hiroshi Ichimura*** and Kouji Yoshida*

*Kazuhiro Nakada, Kouiti Matsumoto, Tomonori Akutsu, Kenji Nagamoto,
Kouji Yoshida
Kurashiki University of Science and the Arts
College of Science and Industrial Technology,
Kurashiki University of Science and the Arts
2640, Nishinoura Tsurashima Kurashiki 712-8505, Japan

**Satoru Fujii
Matsue National College of Technology
Matsue National College of Technology,
14-4 Nishi-Ikuma Matsue 690-8518, Japan

***Hiroshi Ichimura
Tokyo National College of Technology
Tokyo National College of Technology,
1220-2 Kunugida Hachioji Tokyo 193-0997, Japan

Abstract

We produced experimentally the distance learning system working with a personal computer and a mobile phone simultaneously. We support individual students to plan their own learning schedule and time limit. We are attempting to confirm this communication information with a personal computer and a mobile phone mutually. The ranking function of the graphs make the students keep their interest by questionnaire result. Student motivation was raised by accessing mail transmission of a message on the WWW. It was effective to use "push type" information by e-mail. We understood that communication function with a mobile phone raised motivation to try.

1. Introduction

In recent years, there is growing concern about the progress of information and communication systems, high speed networking and multimedia environment. Various developments for distance learning and supporting systems has occurred using these technologies. These developments attempt to use the internet for distance learning and supporting systems.

There are various kinds of learning styles promoted by the growing use of personal computers. Video and CD-ROM materials appear in the market, too. However, it is not possible to capture the progress of students in distance learning only through this kind of one-sided teaching material and visual contents. In addition, student interest cannot be maintained using only text as explanation. There is also an issue of weak student motivation, due to the fact that students have no way to see the progress of other students, thereby knowing their own relative achievement. Without such indication, students these days show difficulty learning independently.   The popularity of using e-mail by mobile phone has presented a new opportunity to study. By sending information related to distance learning, we thought that it may improve learning.

Under these circumstances, information exchange tools and support tools were proposed by using email or the WWW[1][2], but there are many problems existing concerning drop-out students who cannot maintain their motivation for learning.[3]. As we have an off-line meeting on the mailing lists, it is reported that distance learning grew more effective after having gotten to know each other in a way similar to a face-to-face classroom[4]. Research also indicates that close communication and quick response increase effectiveness[5]. In this study, we conducted a trial to raise motivation by using graph indication function[6]. We created an example study by using distance learning in teaching penmanship and calligraphy[7].

We experimented with effective ways of performing distance learning based on the concept of "any time , anywhere you like". We attempted to provide a solution to maintaining motivation and a method to control progress status, which are two of the main challenges we currently face in distance learning.

2.Distance Learning System

We used a personal computer for our learning tool. The PC provides easy access to material by sending URL information from a server to a mobile phone of a student. We allow individual students to plan their own learning schedule and time limit.

This system is supported by UNIX server. The instructor and students use e-mail and the WWW on a personal computer. We have managed to count the number of accessed WWW pages and to record the results of student examinations on the WWW. This is accomplished by cookie function which passes along student user information. We created a user access logging database and then generated its graph from the database through a CGI program of Perl language. Figure.1 shows the distance learning system.

We learn this system amang thirteen menbers, This system learns contents of information literacy.

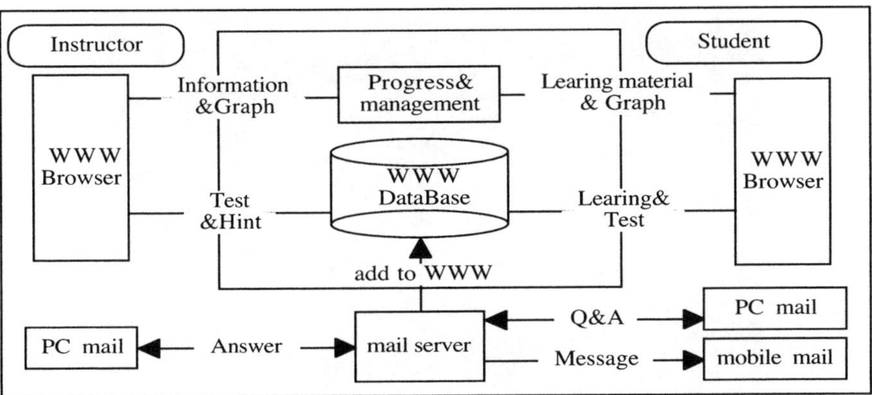

Figure.1 Distance Learning System

2.1 Mobile phone support function

(1)Mail transmission function and data exchange

As it is a simple mail transmission, we use the mail address as a registration list for a personal computer beforehand. We transmit messages using this list. By referring to a page accumulated from the browser of a mobile phone, we can access the registration list. Accordingly, we can send a message to a mail address easily.

Next we provide individual memo registration function for each learning person of a mobile phone. We can register the item which learning person wants to memorize by oneself with this function. And we can watch this information from a personal computer and a mobile phone freely. We can refer to each other's information too registered you with as common information.

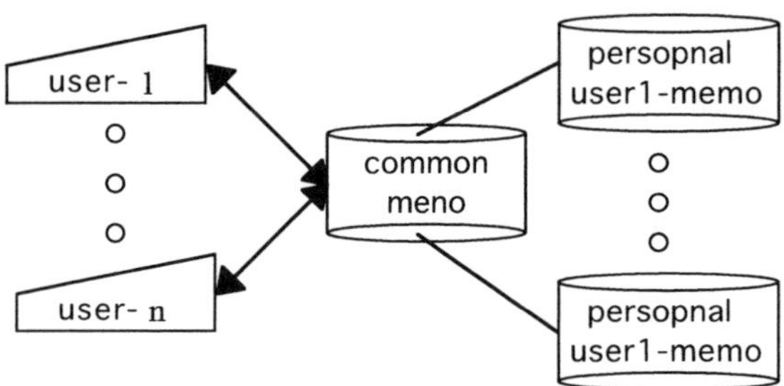

Figure.2 User communication data structure

(2)Mail receiver function

We can receive mail by browsing at this system from a mobile phone. Of course, we are aware of the receipt of all messages by the standard mail receiver function of a mobile phone.

(3)Mail transmission and history function

This system accumulates the history of transmission and reception of all mail to and from an individual student, and we can observe the information.

(4)Learning schedule setting and communication function

Schedule setting uses a personal computer beforehand, and establishes the learning schedule of a student. The corresponding lesson contents for each specific day is sent by mail according to each individual schedule. We show schedule setting frame by figure 3(Figure has been translated from Japanese to English).

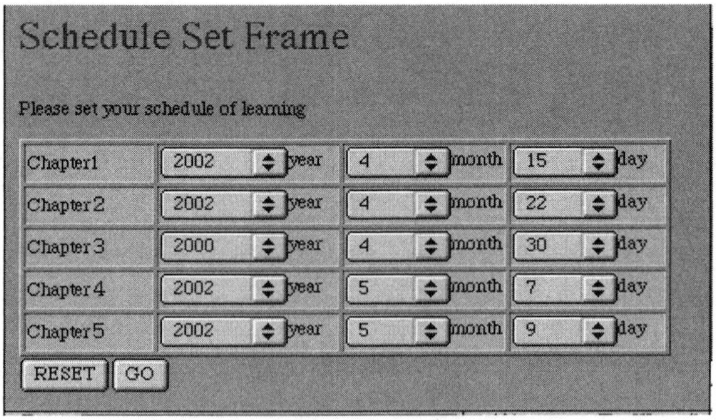

Figure.3 Schedule setting frame

2.2 Motivation continuance function

A teacher can follow the learning history and progress of a student by the WWW. Students can track their progress status themselves by using a line graph. We deliver each Q&A by mailing list and search mailing list archives. From the instructor's side, it is possible to see the progress of students, as well as check the status of extremely slow and rapid students. Also, it is possible to add any explanations or hints for specific lessons, such as inclusion of helpful teaching materials. We show an indication frame of ranking by figure 4 (Figure has been translated from Japanese to English).

Indication of progress ranking

Show learning student 30 % of high rank

Ranking	Name	progress(%)	Graph
1	K.ngamoto	20%	
2	H.Nkagawa	19%	
3	S.Nakamura	19%	
4	M.Shibue	19%	

Figure.4 Indication frame of ranking

3. Result

3. 1 Mail transmission with a mobile phone

Answers are sent as e-mail messages to a mobile phone, and interest is increased. It is a helpful function that the schedule is shown from a mobile phone. It is effective that we can freely acquire the knowledge about learning from a mobile phone.

There is also the function that test items can be accomplished piecemeal. It takes a long time to write a message on a mobile phone. To increase efficiency, we can write mail using keywords.

3. 2 Motivation and questionnaire result

There is a correlation of around 0.3,which is between the mail transmission   by automatic and motivation. Next, there is a high correlation of around 0.4,which is between the mail contents by automatic transmission and motivation.

Student motivation was raised by accessing mail transmission of a message on the WWW.   Even if there is not a direct reply by a real teacher, being able to view the graph function provided stimulating feedback. This showed the helpfulness for students of the graph indication and communication by mail transmission.
 · The results of a questionnaire showed the following feedback:
○An opinion from a student

A visual graph raises motivation by comparing progress with other students.

Showing ranking information of one's individual level in relation to the whole class adds to learning desire.

This relative graph and degree of progress, it becomes a target of progress of oneself.

Frequent examinations raise motivation and provides reliable progress.

Sending the Q&A by mailing list is convenient and also serves as a helpful reference.

Motivation is further raised by using automatic mail transmission.
○An opinion from a teacher

Graph indication of student progress is useful for management of the student.

Relativity graph provides instant understanding of a student's progress and level.

Distance education can be accomplished on one's free time, without limitation.

When the dynamic teaching materials attract interest in the case of distance learning, students feel as though they have a tutor helping them.

4. Conclusions

As for the curriculum this time concerning "information literasy", we found that student motivation is increased by indicating the progress of others. This relative positioning provides an experience similar to a traditional classroom. We noticed that it is important to use, search and "push type" information by e-mail.

However, in the task of progress management, by only looking at pictures and recording examination data, a gap arises in the knowledge level of students, and it is difficult to judge the students' level.

We want to effectively use a mobile phone not only to raise motivation but also to provide easy access in the future.   We want to illustrate that the system built in this progress management will be able to check an effective side in the supporting application.   As we found it produces a beneficial effect by discussing by mailing list mutually, we apply this in a way of collaboration learning. We would like to further study both the method and the support tools. This study receives assistance from scientific research costs subsidy "14580240","13480051", in addition to the above.

REFERENCES

[1]Tomohiro Nishida and Nobuaki Tokura, "The Communication between Students and Teachers via E-mail and WWW" , IEICE,Vol.96,No.431,pp.1-8,1996.

[2]Jun muenori, Hajime yoshida, Takaya Yuizono and Masaru Sudo, "Remote Seminar Support System and Its Application and Estimation to a Seminar via Internet", IPSJ,Vol.39,No.2,pp.447-457,1998.

[3]Lisa Neal, "Virtual Classrooms and Communities" , In Proceedings of ACM GROUP'97 Conference, November 16-19,1997,Phoenix,AZ

[4] Linda A.Jorn,Ann Hill Duin, and Billie J.Wahlstrom, "Designing and Managing Virtual Learning Communities" , IEEE TRANSACTIONS ON PROFESSIONAL COMMUNICATION Vol 39,Num 4,pp.183-191,1996.

[5]Raymond A.Dumont, "Teaching and Learning in Cyberspace", IEEE TRANSACTIONS ON PROFESSIONAL COMMUNICATION Vol 39,Num 4,pp.192-204,1996.

[6] Kouji Yoshida, Noriaki Kawan, Haruyuki Ohtani, Tadanori Mizuno, Sanshiro Saka, "Evaluation & Trial of Effective Distance Learning System including the progress management of Students" ,Proceedings of ICCE'99,7th International Conference on Computers in Education Vol.55, pp.906-907(1999).

[7]Hiroshi Ichimura, Masato Suzuki, Michio Murai, Satoshi Yazawa, Toshiaki Kuroiwa, Seiu Yamashita, Masahiro Kuroda, Kouji Yoshida, Tadanori Mizuno, Sanshirou Sakai, "Design of Next Generation Distance Learning System for Penmanship and Calligraphy" ,SNPD'01,pp.96-103(2001).

GUNGEN DX II: New Idea Generation Support System for Hundreds of Ideas

Tomohiro Shigenobu
Graduate School of Systems Engineering, Wakayama University,
930 Sakaedani, Wakayama 640-8510, Japan

Takashi Yoshino and Jun Munemori
Faculty of Systems Engineering, Wakayama University,
930 Sakaedani, Wakayama 640-8510, Japan

Abstract. We have developed a new idea generation support system GUNGEN DX II. GUNGEN DX II has two functions for handling hundreds of ideas. The temporary islands making function supports to make the islands step of KJ method. This function supports categorization of many ideas intuitively and instantaneously by dropping an idea from the top of a screen. The sentence convert function supports to conclude the sentence step. This function converts island titles into a sentence semi-automatically. We had collected 287 ideas for the KJ method using PDA by 14 students. We carried out the KJ method using the ideas. The results of the experiments were shown that these functions were enable to handle nearly 300 ideas. The sentence convert function shortened the time to conclude a sentence of KJ method.

1 Introduction

We have developed an idea generation support system, called GUNGEN [1]. GUNGEN supports the KJ method [2], which is one of the famous idea generation methods in Japan. We had applied it to the student experiments for ten years.

The number of ideas was under 70 in almost every experiment. In the case of under 70 ideas, even students could summarize the ideas. Students could not summarize when about 200 ideas were proposed. When many participants collect ideas using a PDA (Personal Digital Assistant), we can also collect a lot of ideas [3]. We can handle about 100 ideas at a time by GUNGEN. However, GUNGEN didn't have no special function to handle over hundred ideas. We should deal with hundreds of ideas effectively.

KJ-Editor [4], D-ABDUCTOR [5] are a KJ method support system. The methods of handling of many ideas are not considered especially about the systems. Therefore, we have developed GUNGEN DX II for handling hundreds of ideas.

We carried out the KJ method using nearly 300 ideas. This paper describes the development and application of the GUNGEN DX II.

2 The KJ method

The KJ method was developed by Jiro Kawakita [2]. This method is called KJ method after his initials. The KJ method is known as a method for establishing an orderly system from a chaotic mass of information. The KJ method consists of four steps shown below.

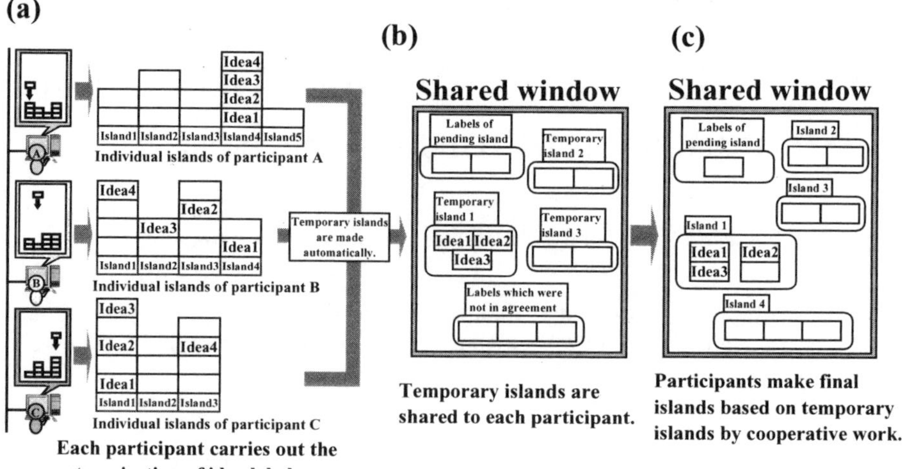

Figure 1: The process of making islands.

In the first step, participants propose their ideas on a theme, and they write down each idea on a label (a small piece of paper). This step corresponds to brainstorming. In the second step, the participants examine these labels and classify them into groups through discussion. Each group is called an island and given a representative title. In the third step, the participants look for an arrangement that expressed mutual relation of the representative titles spatially. Then the participants connect the related representative titles together in a certain kind of line. In the last step, the participants write a conclusive composition.

3   GUNGEN DX II

3.1   Design policy

The following items are design policies for handling hundreds of ideas.

1. The new method of making islands for categorizing over one hundred ideas
   Participants usually make islands to examine ideas one by one in the conventional method. When there are a lot of ideas, it takes a long time to categorize using the conventional method. Moreover, it is difficult in human capability.

2. Generating a sentence for conclusion using island titles semi-automatically
   When the number of ideas increased, the number of islands may increase. It is difficult for participants to understand the relation between islands. Therefore, participants may be troubled about writing a conclusion using island titles.

3.2   The temporary islands making function

Figure 1 shows the process of making the islands. A label corresponds to an idea. Each participant categorizes all labels individually (Fig.1 (a)). The categorized labels are called individual islands. Each participant's result can be different one another. When all participants

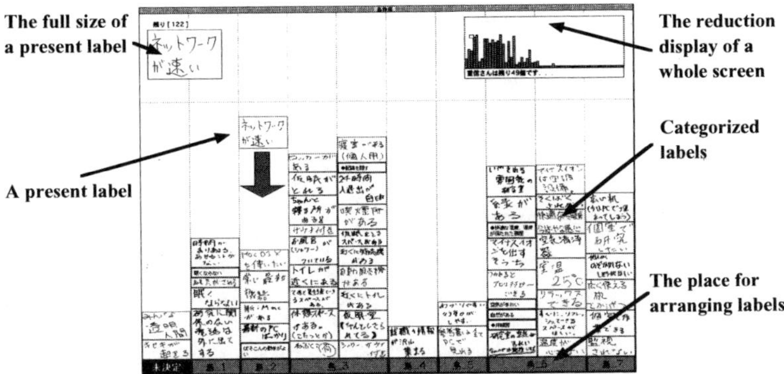

Figure 2: An example of individual islands making screen.

finish dropping all labels, this system compares the results of the categorization of all participants. This function makes islands from the common portion of each participant's individual island. They are called temporary islands. Temporary islands are shared to all participants (Fig.1 (b)). After the labels have been arranged automatically, the participants can correct the arrangement of temporary islands by themselves (Fig.1 (c)).

1. Method of making the individual island
   Figure 2 shows an example of the screen in making individual islands. Each participant performs this task individually. A label is dropped at fixed speed from the top of a screen in order. An idea label arrives at the bottom in 15 seconds. A label is moved by the arrow key of a keyboard. Similar labels are stacked on the same position intuitively. The stacked ideas become a individual island.

2. Method of making the temporary island
   We explain the method using Figure 1. The idea number on the label shows the idea. The same number shows the same idea. Temporary islands are made after all participants finish dropping a label. Temporary islands are made automatically using individual islands. For example, when three persons use this method, a temporary island will be made if a part of categorization of a person is in agreement with other person's individual island. The island 4 of participant A consists of "1", "2", "3", and "4" is used for comparison. Participant A thought that "1", "2", "3", and "4" were similar. Participant B thought that "1", "2", "3", and "4" were not similar. Participant C thought that "1", "2", and "3" were similar. Although all persons' results are not in agreement, the island which consists of "1", "2", and "3" is made by majority. It is a temporary island. The temporary islands making function arranges temporary islands automatically and each participant shares them. The function supports to categorize many ideas quickly and roughly.

3.3   The sentence convert function

Figure 3 shows an example of the sentence convert function screen. This function generates the sentence for writing a conclusion. This function consists of two steps. First step, participants illustrate islands using the relation between the islands. Second step, participants decide the order of the islands for making a sentence. Moreover, the relation of islands is

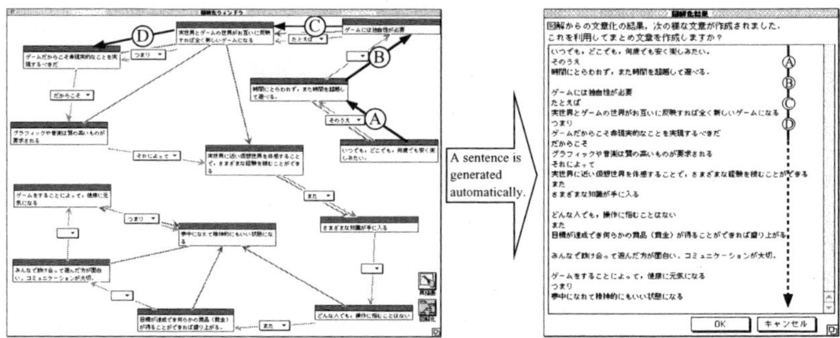

Figure 3: An example of the sentence convert function screen.

Table 1: Results of experiments.

	Ideas	Islands					Sentence		Total
		Temporary islands		Final islands		Total			
	Ideas	Temporary islands	Necessary average time (minutes)	Final islands	Necessary time (minutes)	Necessary time (minutes)	Characters in a sentence	Necessary time (minutes)	Necessary time (minutes)
Experiment 1	287	43	95	39	130	225	1521	114	339
Experiment 2	287	52	70	24	167	237	1010	56	293
Experiment 3	287	39	56	32	185	241	1288	71	312
Average Value	287	45	74	32	161	234	1273	80	315
Past experiments (Nine averages)	48	-		7	61	61	413	65	126
The ratio of this experiments and past experiments	6.0	-		4.6	-	3.8	3.1	1.2	2.5

expressed by a conjunction. A conjunction is chosen from the list of conjunctions, or can be input freely. After finishing the work, a sentence is generated in order automatically. For example, Figure 3 shows that paragraphs of the conclusion sentence are made in alphabetical order automatically. The conclusion sentence is written using the sentence.

4   Experiments and discussion

4.1   Procedure

A PDA was given to each participant. Each participant always carried the PDA, and collected ideas. The number of participants was 14 students. The theme of the collecting ideas was "the ideal laboratory." They collected ideas over three weeks. We carried out the KJ method about "the ideal laboratory" using collected ideas three times. We carried out questionnaire surveys after the experiments of the KJ method. We had a 5 points scale (with answers "1:very bad", "2:bad", "3:neutral", "4:good" and "5:very good") for participants to evaluate each item on the questionnaire.

4.2   Results of Experiments

The number of collected data was 381. 287 ideas were related to the theme. Other data were their personal data. We carried out the KJ method by three persons using 287 ideas three times. Table 1 shows the results of experiments. The results of the past experiments are shown for comparison. In this experiment, the number of ideas was 6.0 times larger than the past experiment (Table 1). Necessary average time to make temporary islands is a period to finish dropping all ideas. In order to categorize 287 ideas into islands, it took 74 minutes in average

(Table 1). In all experiments, the number of final islands became less than the number of temporary islands. The number of characters of a conclusion sentence was 3.1 times larger than the past experiment, and over 1000 characters (Table 1).

4.3   Discussion

1. Making islands
   Participants could categorize many ideas using the temporary islands making function. They thought that they could categorize the ideas intuitively (3.9 on a 5-point scale). The temporary islands of 60 percent consist of four or less ideas. The participants evaluated high for making final islands using temporary islands (3.8 on a 5-point scale). The number of final islands was 4.6 times larger than the past experiment (Table 1). The time to make final islands from temporary islands was required for about three hours (Table 1). This result shows that the time was not shortened for making the islands. Because, temporary islands could not be used as it is. That is, temporary islands needed correction. But the temporary islands making function was evaluated high as effective for categorizing many ideas (4.0 on a 5-point scale). The participants felt that a conventional method is almost impossible for categorizing nearly 300 ideas. The proposed method can be effective in rough categorization of hundreds of ideas.

2. Making a conclusion sentence
   The number of final islands was 4.6 times larger than the past experiment (Table 1). The number of characters of a conclusion sentence was 3.1 times larger than the past experiment (Table 1). But the time for making a conclusion sentence was 1.2 times larger than the past experiment which did not use the sentence conversion function (Table 1). Even if the number of islands increased, participants could grasp the relation between islands by using this function. Therefore, it became easy to write a conclusion sentence. The sentence convert function is effective when processing many islands.

5   Conclusion

We applied GUNGEN DX II three times. The results of experiments were shown below.
1. Even students could perform the KJ method using about 300 ideas by the temporary islands making function.
2. The enough time was required for correcting temporary islands.
3. The sentence convert function shortened the time to conclude a sentence of KJ method.

Our next aim is to shorten the time to correct temporary islands. In the future, we will improve and evaluate GUNGEN-DXII through further application.

References

[1] Munemori J., Horikiri I., and Nagasawa Y.: Groupware for New Idea Generation Support System (GUN-GEN) and Its Application and Estimation to the Student Experiments of the Distributed and Cooperative KJ Method, Trans. of IPSJ, Vol.35, No.1, pp. 143-153, 1994.

[2] Kawakita, J.: KJ method, Chuou Kouron–sha, Tokyo, Japan, 1986.

[3] Yoshino, T., Munemori, J., Shigenobu, T., and Yunokuchi, K.: A Spiral-type Idea Generation Method Support System for Sharing and Reusing Knowledge and Information Among a Group, Journal of IPSJ, Vol.41, No.10, pp.2794-2803, 2000.

[4] Ohmi, Y., Kawai, K., Takeda, N., and Ohiwa, H.: Pointing Operations in Cooperative Works using Card-handling Tool "KJ-Editor", Journal of IPSJ, Vol.36, No.11, pp.2720-2727, 1995.

[5] Misue, K. and Sugiyama, K.: On Development of Diagram Based Idea Organizer D-ABDUCTOR, Journal of IPSJ, Vol.35, No.9, pp.1739-1749, 1994.

KES 2002
E. Damiani et al. (Eds.)
IOS Press, 2002

Application and Evaluation of Distributed Remote Seminar Support System RemoteWadaman II

Takashi Yoshino
Faculty of Systems Engineering, Wakayama University,
930 Sakaedani, Wakayama 640-8510, Japan

Jun Munemori
Faculty of Systems Engineering, Wakayama University,
930 Sakaedani, Wakayama 640-8510, Japan

Abstract. We have developed a real-time distributed remote seminar support system, named RemoteWadaman II. The system can connect several computers via the Internet. Each student can operate his or her own computer. This system uses the following two communication methods. The first method is for large amount data and asynchronous information using a file server. The second method is for shared cursors and a shared screen using a direct communication by the mesh-type network. We developed functions to send and receive a file all over a seminar for the increase in efficiency of a seminar. RemoteWadaman II was applied to seminars between two Universities 55 times and estimated by log analysis and questionnaires. Each participant was seen a shared report and shared cursors on his or her own PC, and each participant was able to use a chat. The number of the sheets of a report increased by the function to send and receive a file after beginning a seminar. As the result, this system has improved both the quality and quantity of a seminar.

1   Introduction

The Internet has come into wide use recently and high performance personal computers (PCs) have appeared. We have applied the Internet and PC technology to the field of education. Distance education support systems have been much studied and developed[1, 2, 3]. There is almost no example to apply actual classes for a long period[3]. The real-time distance education system may be carried out easily only several times. However, a flexible system and suitable support are needed for application over a long period. We have developed a real-time distributed remote seminar support system, called RemoteWadaman II. The system can connect several computers via the Internet. Each student can operate his or her own computer. This system uses combining two kinds of communication methods, mesh-type network and a file server. RemoteWadaman II was applied to seminars between two Universities 55 times over two years and evaluated by log analysis and questionnaires.

2　Distributed Remote Seminar Support System RemoteWadaman II

2.1　Design policy

The design policies of RemoteWadaman II are summarized as follows.

1. Distributed type network composition that combined mesh-type network and a file server
   This system uses the following two communication methods. The mesh-type network is prepared for shared cursors and a shared screen. The file server is prepared for large amount data and asynchronous information.

2. Each participant uses a PC
   When each participant uses a PC, he or she can look at a shared screen and shared cursors on his or her own PC. Moreover, each participant can use a chat.

3. Flexible communication connection establishment
   At a conventional seminar, there are some students who are late for the beginning of a seminar. In actual long-term application, it is anticipated that the same situation as a conventional seminar arises. Thus, even if a student is somewhat behind, he or she can participate a seminar and submit a report.

2.2　Function of RemoteWadaman II

Figure 1 shows the system configuration of RemoteWadaman II. RemoteWadaman II uses three kind of communication: audio and video communication, data communication, and file transfer communication. Audio and video communication is used between a PC of an instructor and a PC of a student. This communication is always connected during a seminar. Data communication is used for the exchange of data, such as a shared screen and a chat. All participant computers are always connected by the data communication in mesh type network. File transfer communication is connected between a participant computer and a file server asynchronously. A file server stores report files, participant names and the IP addresses of connected PCs.

Figure 2 shows an example of a PC screen on a distributed remote seminar. Remote-Wadaman II has the functions of a text-based chat, a shared screen, a shared pointer, and a floor control. If a presenter adds something in a shared screen and pushes a contents share button, the contents are shared by all the members. The system also has the seminar support functions: a member check function, a report file submitting function, an automatic report file receiving function, and an automatic communication connection establishment function. When a student starts the system, the system interconnects each PC automatically and receives report files. Then a student can participate in a remote seminar. If students are late, they can participate in a remote seminar using the automatic report file receiving function and the automatic communication connection establishment function.

3　Experiments

The procedure of a remote seminar is as follows. A student creates on the report file of Remote-Wadaman II as report cards. A student submits a report file to a file server. When a participant

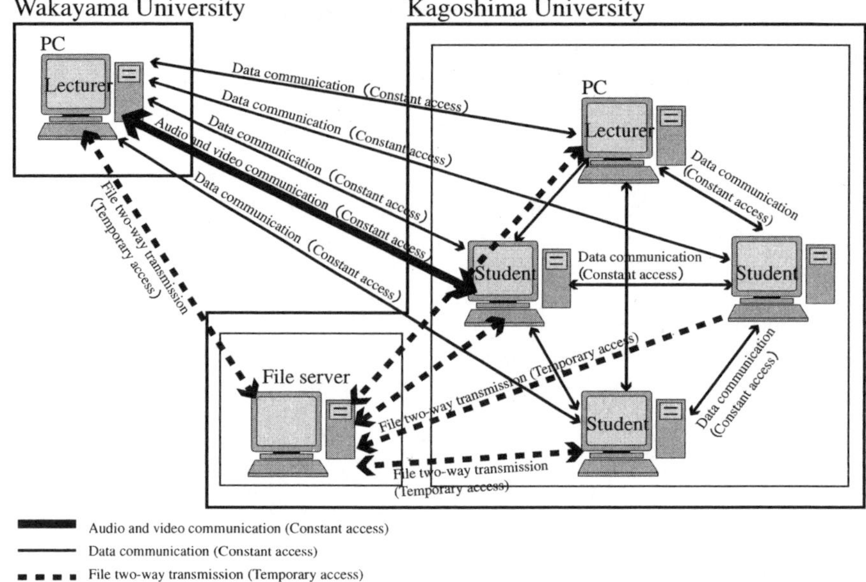

Figure 1: System configuration of RemoteWadaman II.

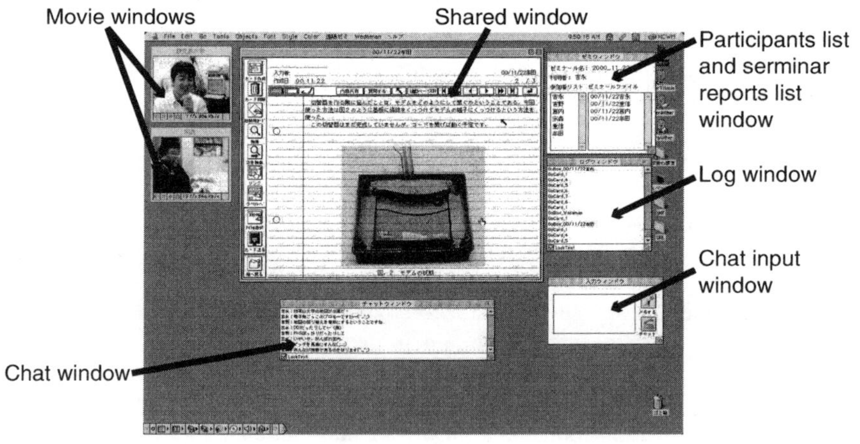

Figure 2: An example of a PC screen on a distributed remote seminar.

Table 1: Results of experiments.

Fiscal year	The average number of participants	The average number of participating students	The average presentation time per student	The average time required to start a seminar	An average number of sheets of a report	An average rate of packet loss	The average amount of audio and video transmission			
							At Kagoshima University		At Wakayama University	
	(people)	(people)	(min:sec)	(min:sec)	(Sheets)	(%)	(f/s)	(Kb/s)	(f/s)	(Kb/s)
1999	4.9	2.9	15:28	17:21	5.2	24.7	9.2	300.4	12.8	497.4
2000	5.1	3.3	16:43	07:20	5.7	13.1	11.6	420.9	16.5	574.3

pushes a start button of a remote seminar, a system establishes communication connection automatically and receives report files from the file server. After finishing preparation of a remote seminar, a seminar starts. RemoteWadaman II was applied to seminars between Wakayama University and Kagoshima University 55 times from April 1999 to March 2001. The instructor of Wakayama University guided a research for graduation or a master's thesis to the students (three students in 1999, four students in 2000) of Kagoshima University.

4 Result of Experiments and Discussion

4.1 Results of experiments

Table 1 shows the results of experiments. The students of Kagoshima University have made the theses for graduation only by instruction almost in remoteness for one year. As compared with the result of a former system [4], the average of presentation time is increased by about 1.5 times to 1.6 times, and the average of number of report cards is increased by about 2.6 times to 2.9 times. The number of report cards increased because students can submit a report file using a report file submitting function after a seminar starts.

4.2 Results of questionnaire surveys

Table 2 shows the results of questionnaires. After all seminars of each fiscal year finished, questionnaire surveys on a scale of 1 to 5 (with answers "1:very bad", "2:bad", "3:neutral", "4:good" and "5:very good") were conducted with the students. An automatic report file receiving function was evaluated quite high (4.0 in the 1999 fiscal year, 4.8 in the 2000 fiscal year). Evaluation of communication by the chat was improved compared with the former system (1.8 in a former system [4], 3.7 in the 1999 fiscal year, 3.5 in the 2000 fiscal year).

Because all the members can see and input a chat, the evaluation may have been improved. From the questionnaire of description form, the student regarded as a merit that a lecturer can teach from a remote place.

4.3 Effect of Support System

We summarized the effect of remote seminar supports in RemoteWadaman II. RemoteWadaman II is a distributed system, that is, one user uses his or her own PC. All the members can see a shared screen, shared pointers, and can input a chat. That is, the quality of a seminar

Table 2: Results of questionnaire.

Questionnaire items	Fiscal year 2000	Fiscal year 1999	Fiscal year 1996[4]
(1) Were the remote seminars useful to your research?	4.5	4.0	4.4
(2) Could you take communication by the video?	4.0	3.3	3.5
(3) Could you take communication by the voice?	5.0	4.7	4.0
(4) Could you take communication by the chat function?	3.5	3.7	1.8
(5) Was the automatic report file receiving function useful?	4.8	4.0	-
(6) Were you concentrated on the seminars?	3.8	3.0	-

has improved compared with the former system [4]. This system has a report file submitting function and an automatic report file receiving function. The number of report cards increased by these functions, because students who come late to a seminar could submit their report file using the functions. The quantity of a seminar has improved compared with the former system [4]. The time of preparation of a seminar was shortened using an automatic communication connection establishment function (It decreased from 25 minutes to 7 minutes in average). Because the conventional manual communication connection became automatic. The time efficiency of a seminar was raised compared with the former system.

5 Conclusion

RemoteWadaman II was applied to seminars between two Universities 55 times over two years and evaluated. We found the following results from these experiments.

1. All the members can see a shared screen, shared pointers, and input a chat on Remote-Wadaman II. The number of report cards increased by a report file submitting function and an automatic report file receiving function. As the result, this system has improved the contents of a seminar.

2. Time of preparation of a seminar was shortened using the file server. An automatic communication connection establishment function and an automatic report file receiving function were effective to shorten the preparation of a seminar.

References

[1] http://www.icce2001.org/

[2] J. J. Cadiz, Anand Balachandran, Elizabeth Sanocki, Anoop Gupta, Jonathan Grudin, Gavin Jancke : Distance learning through distributed collaborative video viewing, Proceeding of the ACM 2000 Conference on Computer Supported Cooperative Work(CSCW 2000), pp. 135–144, 2000.

[3] A.He, Zixue Cheng, Tongjun Huang, Ryohei Nakatani, Yuichi Amadatsu, Akio Koyama, Yue Zhao, Shoichi Noguchi :Design of a Real-time Interactive Tele-Exercise Classroom for Computer Exercises over a Gigabit Network, Proceedings of 15th International Conference on Information Networking(ICOIN–15), pp. 757–762, 2001.

[4] Jun Munemori, Hajime Yoshida, Takaya Yuizono, and Masaru Sudo: Remote Seminar Support System and Its Application and Estimation to a Seminar via Internet, Trans. of IPSJ, Vol.39, No.2, pp. 447-457, 1998.

KES 2002
E. Damiani et al. (Eds.)
IOS Press, 2002

Semantic field and context models of uncertainty

Germano Resconi

Dipartimento di Matematica, Università cattolica via Trieste 17
25128 Brescia, Italy
e-mail: resconi@numerica.it

Van-Nam Huynh

Department of Computer Science, University of Quinhon
170 An Duong Vuong, Quinhon, Vietnam
Email: vn_huynh@yahoo.com

Abstract

Every object can have different semantic values in different contexts (worlds). This difference generate uncertainty when we want give a meaning to an object. For example consider the property to have a red colour. The semantic value of the object is the logic value true or false to have the red colour. In many cases only when we know the context where we operate is possible to know if the proposition "to have a red colour" is true or false. In one context the proposition is true and in other contexts the same proposition is false. In conclusion the same object can have two different values for the same property. One object in one context is judged red and in another is judged rose. A space of possible conflicting independent contexts is necessary to evaluate ambiguous properties of one object. When a property is true in all the contexts, the property is a semantic invariant. When in all the space of the contexts the structures, inside the contexts, are the same (isomorphism, homomorphism.) the properties of the structures are always true so are semantic invariants. In the space of the independent contexts a semantic field can *compensate* the lack of truth of the property in the space of the contexts. A property false in one context can be possible true by the semantic field We can also use the semantic field to *destroy* the truth in one context in which the property is true but can be necessary false. A semantic field destroy the truth in one context. In the same time in the context start another semantic field that generate truth in another context where the property is false. In this way a migration of the true value from one context to another is possible thanks to the semantic fields. Every context has a degree of truth and the sum of all the degree of truth is equal to one. Every subset A of the domain D of properties of one object in one context gives the range of imprecision for a property in the context. High range of imprecision gives low degree of truth. Unification model of different fuzzy logic theories are possible by context space, semantic field and average operation

1. META-THEORY BASED ON MODAL LOGIC

In a series of papers Klir, *et al.*, (1994, 1995; Resconi, *et al.*, 1992, 1993, 1994, 1996) a meta-theory was developed. The new approach is based on Kripke model of modal logic. A Kripke model is a network among contexts or worlds [1]. Given a context we can give the logic value true or false to any logic proposition. This network is given by the structure or model

$$M = <W, R, V>. \qquad (1)$$

where W is a set of worlds or contexts, R is the accessible relation among contexts or preference.

We know that the set of the contexts is embedded in a universal set of discourse D. At every context in W we associate by the function

$$\Omega : W \to 2^{D}$$

a subset C of the domain D. Every subset C of the domain of one object gives the range of imprecision [3]. When the subset A is a singleton, in this case we have a precise characterisation of the property of the object in one context. When a property "colour" is given by a logic proposition as "the colour of the car is red", the domain D is a set of propositions. Every subset of D is the set of imperfect knowledge on the assigned proposition and context. When the subsystem in a context is a singleton, we have a perfect knowledge. A degree of knowledge is given by the cardinality of the subset of D. Because the evaluation of the same property for the same object is given in a space of conflicting and independent contexts, the evaluation of the property is ambiguous. In one context can be true but in another can be false. Every context is a fractional part of the total true value. The fractional part is proportional to the degree of knowledge for every context.

Example :Given four contexts (worlds) and S the space of the contexts $S = \{A_1 , A_2, A_3 , A_4 \}$. Given the domain of the discourse $D = \{p_1 , p_2, p_3 , p_4, p_5 , p_6, p_7 , p_8\}$., where p_k are logic propositions.

When by Ω we give this association

$$A_1 \to \{ p_2, p_3 , p_4 \}., A_2 \to \{ p_1\}., A_3 \to \{ p_5\}., A_4 \to \{ p_6, p_7 , p_8\}.$$

Because the degree of knowledge for the four context are $K = \{1/3 , 1, 1, 1/3 \}$, the fractional parts F of the truth related to the four context are

$$F = \left\{ \frac{\frac{1}{3}}{\frac{1}{3}+1+1+\frac{1}{3}} , \frac{1}{\frac{1}{3}+1+1+\frac{1}{3}} , \frac{1}{\frac{1}{3}+1+1+\frac{1}{3}} , \frac{\frac{1}{3}}{\frac{1}{3}+1+1+\frac{1}{3}} \right\} = \left\{\frac{1}{8},\frac{3}{8},\frac{3}{8},\frac{1}{8}\right\} = \left\{\psi_1, \psi_2, \psi_3, \psi_4\right\}$$

So.Resconi, *et al.* (1992-1996) suggested , in the *hierarchical meta-theory* based upon modal logic, to adjoin a function $\Psi : W \to [0 , 1]$,where $\sum_i \psi_i = 1$ and i =1,2,..n are the indices of the contexts. So we have the new model

$$N = < W, R, V, \Psi > \qquad (2)$$

At this point, we should ask: "what are the linkages between the concepts of a population of observers, a population of possible contexts and the algebra of fuzzy subsets of well defined universe of discourse?" Consider a sentence such as: "John is tall" where $x \in X$ is a person, "John", in a population of people, X, and "tall" is a linguistic term of a linguistic variable, the height of people Any observer or observer can be modelled in an abstract way by a context $w_k \in W$, where W is the set of all possible contexts that compose the indivisible object associated with a particular h and k is the index, 1, 2, 3…n, that associates with any assessment of the object given by the population of the observers. In a space S of the contexts, the proposition p = "John, with the height of 1.75 metres, is tall", can be true or false for a set of, say, 25 contexts in the space S. Let "true" be indicated by 1 and "false" by 0. With this background, we next define the membership expression of truthood of a given atomic sentence in a finite set of worlds or contexts $w_k \in W$ as follows:

$$\mu_{1.75}(A) = \textit{(Sum of fraction of true for contexts where p is true)} = \sum_{k=1}^{25} \psi_k S_k \qquad (3)$$

where ψ_k are the fractional parts of the truth related to the twenty five context and A is the fuzzy set for the sentence "John is tall".

2. SEMANTIC FIELD

2.1 Fundamental definitions

In a given context space S, we locate the contexts in any point of the space, but we can also locate a more complex object as the semantic field. All the contexts (worlds) in which the semantic field has a given value are connected one with the other. The connection is the relation R in the Kripke model. When inside the connected set of contexts almost one proposition is true, all the other contexts where the same proposition is false is possible true. In this case the semantic field is a compensative field. But when inside the connected set of contexts almost one proposition is false and in all the other contexts the proposition is true but necessary false, the semantic field is a destructive semantic field.

Definition 1. The semantic field is formally defined as the following tuple:

$$S F = <\ H, W_t, F, \eta, \delta, So\ > \qquad (4)$$

where H is the reference space for the contexts, W_t is the set of contexts (worlds) at time t , $\delta : W_t \to \Gamma$, is the function that associates at every context in W_t one position in the space Γ . F is the value of the semantic field, $\eta : W_t \to F$ is the function that associates at every context one value of the semantic field, So is the set of sources of the semantic field. The sources So are a set of points where the fields are generated. When the value F of the semantic field is positive the semantic field is *compensative*, when the value of the semantic field is negative, the semantic field is *destructive*.

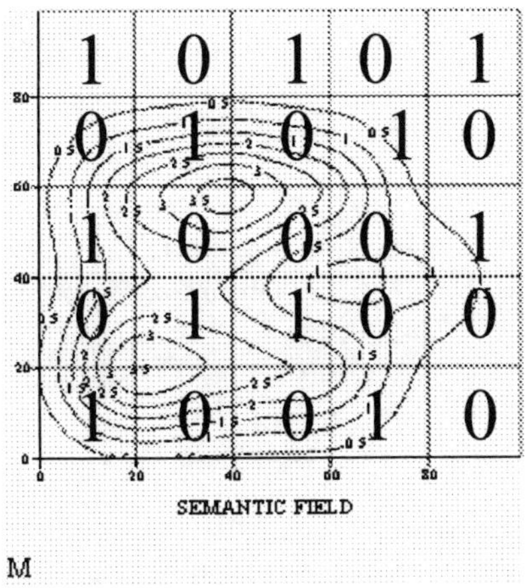

Fig. 1

Image of the semantic field inside the reference space H of the contexts with the logic values for every context. All the contexts where F > 2 are connected one with the others.

3. FUZZY LOGIC OPERATIONS UNDER THE EFFECT OF THE SEMANTIC FIELD

3.1 NOT Operation

In the previous chapter we introduce the operator Γ that transform a space of the contexts in anothe by the semantic field. For the evaluation of the proposition p given by this matrix

$$v(p) = \begin{bmatrix} 1 & 0 & 1 & 0 & 1 \\ 0 & 1 & 0 & 1 & 0 \\ 1 & 0 & 0 & 0 & 1 \\ 0 & 1 & 1 & 0 & 0 \\ 1 & 0 & 0 & 1 & 0 \end{bmatrix} \quad v(\neg p) = \begin{bmatrix} 0 & 1 & 0 & 1 & 0 \\ 1 & 0 & 1 & 0 & 1 \\ 0 & 1 & 1 & 1 & 0 \\ 1 & 0 & 0 & 1 & 1 \\ 0 & 1 & 1 & 0 & 1 \end{bmatrix}$$

Where $\mu(p) = 11/25$ is the membership function value for the sentence p. With the semantic field we compensate false value into true value and we destroy true value into false value. At the sentence p , we substitute the sentence Γ p that is the sentence p under the effect of the semantic field that activate the operator Γ . So we have

$$v(\Gamma p) = \begin{bmatrix} 1 & 1 & 0 & 0 & 1 \\ 0 & 0 & 1 & 1 & 0 \\ 1 & 0 & 0 & 0 & 1 \\ 0 & 1 & 1 & 0 & 0 \\ 1 & 0 & 0 & 1 & 0 \end{bmatrix}, \quad v(\neg\Gamma p) = v(\Gamma\neg p) = \begin{bmatrix} 0 & 0 & 1 & 1 & 0 \\ 1 & 1 & 0 & 0 & 1 \\ 0 & 1 & 1 & 1 & 0 \\ 1 & 0 & 0 & 1 & 1 \\ 0 & 1 & 1 & 0 & 1 \end{bmatrix}$$

Where
$$\Gamma p = (\text{True} \wedge (SF > \Theta)) \vee (p \wedge (-\Theta < SF < \Theta)) \vee (\text{False} \wedge (SF \le -\Theta)) \quad (5)$$

and Θ is the threshold for the semantic field.

So $\mu(p \wedge \neg p) = 0$ and $\mu(p \wedge \Gamma\neg p) = 2/25 > 0$. Under the action of the semantic field the absur sentence has a value different from zero. In the contexts C_{12} , C_{13} , C_{22} , C_{23} the logic values of and $\neg$ p are the same and this creates a non zero absurd condition. For simplicity we denote

$$\text{Absu}(p) = p \wedge \Gamma\neg p$$

3.2 Operation AND

Let us next investigate the combinations of the truthoods for any two propositions, such '"John is tall" is true' AND '"John is heavy" is true', given that we know the truthhood membersh functions of $\mu_{p_1}(x)$ for P_1: '"John is tall" is true', and $\mu_{p_2}(x)$ for P_2: '"John is heavy" is true'. We have proved it previously (Resconi and Türkşen, 2001 [2]) that in the meta-theory when $\mu_{p_1}(x) \ge \mu_{p_2}(x)$ we can write

$$\mu_{p_1 \wedge p_2}(x) = \min[\mu_{p_1}(x), \mu_{p_2}(x)] - \frac{|cW_1 \cap W_2|}{|W|} = \frac{|W_1 \cap W_2|}{|W|}, \quad (6)$$

where $\mu_{p_1}(x) = \dfrac{W_1}{W}$, $\mu_{p_2}(x) = \dfrac{W_2}{W}$ and W_1 is the set of contexts where p_1 is true and W_2 is the set of contexts where p_2 is true. Given the two normal forms

$$F_1(p,q) = p \wedge q \text{ and } F_2(p,q) = (p \vee q) \wedge (p \vee \neg q) \wedge (\neg p \vee q).$$

In the classical logic the two forms are equal. With the matrix evaluation with $\wedge$ the AND , $\vee$ the OR and $\neg$ the NOT, we write

$$F_2(p,q) = (p \vee q) \wedge (p \vee \Gamma \neg q) \wedge (\Gamma \neg p \vee q) = = [\, p \wedge q \,] \vee \text{Absu} (p) \vee \text{Absu} (q) \quad (7)$$

3.3 Operation OR

For $\mu_{p_1}(x) \geq \mu_{p_2}(x)$ we can write;

$$\mu_{p_1 \vee p_2} = \max(\mu_{p_1}, \mu_{p_2}) + \frac{\left|cW_1 \cap W_2\right|}{|W|} = \frac{\left|W_1 \cup W_2\right|}{|W|} \quad (8)$$

Given the two normal forms

$$G_1(p,q) = p \vee q \text{ and } G_2(p,q) = (p \wedge q) \vee (p \wedge \neg q) \vee (\neg p \wedge q).$$

In the classical logic the two forms are equal. With the matrix evaluation with $\wedge$ the AND , $\vee$ the OR and $\neg$ the NOT, we write

$$G_2(p,q) = (p \wedge q) \vee (p \wedge \Gamma \neg q) \vee (\Gamma \neg p \wedge q)$$

With a simple computation we have that

$$G_2 = (p \wedge q) \vee (p \wedge \Gamma \neg q) \vee (\Gamma \neg p \wedge q) \vee (p \wedge q) = (p \vee q) \setminus [\, (p \wedge \text{Absu}(\neg q)) \vee (q \wedge \text{Absu}(\neg p))\,]$$

where given the sentences A and B the sentence A\B is true when A is true and B is false.

3.4 Inferential operator

We know that $p \rightarrow q \equiv \neg p \vee q$, for $p \rightarrow q \equiv \Gamma \neg p \vee q$ we have

$$\Gamma \neg p \vee q \equiv (\neg p \vee q) \wedge \neg [\, \neg q \wedge (\text{Absu}(p) \vee \text{Absu}(\neg p))\,] \vee$$

$$\neg (\neg p \vee q) \wedge [\, \neg q \wedge (\text{Absu}(p) \vee \text{Absu}(\neg p))\,]$$

REFERENCES

1.　　G.Resconi,T.Murai and M. Shimbo, Field Theory and Modal Logic by Semantic Fields to Make Uncertainty Emerge from Information, *Int.J.General System* **29** (5) (2000) 737-782.

2.　　G.Resconi and I.B. Türkşen, Canonical forms of fuzzy truthoods by meta-theory based upon modal logic, *Information Sciences* **131** (2001) 157 – 194.

3.　　Jorg Gebhardt and Rudolf Kruse, The context model: An Integrating View of Vagueness and Uncertainty, *International Journal of Approximate Reasoning* **9** (1993) 283-314.

KES 2002
E. Damiani et al. (Eds.)
IOS Press, 2002

An Extension of Context Model for Modelling of Uncertainty of Type 2

V.N. Huynh[†], G. Resconi[‡] and Y. Nakamori[††]

[†] Department of Computer Science, University of Quinhon
170 An Duong Vuong, Quinhon, VIETNAM
[‡] Department of Mathematics, Catholic University at Brescia
via Trieste 17, Brescia 25128, ITALY
[††] Japan Advanced Institute of Science and Technology
Tatsunokuchi, Ishikawa, 923-1292, JAPAN

Abstract. In this work, we introduce an extension of the context model as a framework for modelling of both vagueness and conflict in fuzzy data analysis. From a decision-making point of view, we define a uncertainty measure of type 2 within the framework of the extended context model. Potential applicability of this measure in situations of decision analysis and further development of the proposed framework is discussed.

1 Introduction

The context model introduced by Gebhardt and Kruse in [5] aims at providing integrating structures and concepts for handling imperfect data. The motivation for the context model arises from the intention to develop a common formal framework that supports a better understanding and comparison of existing models of partial ignorance to reduce the rivalry between well-known approaches. Particularly, the authors presented basic ideas keyed to the interpretation of Bayes theory and the Dempster-Shafer theory within the context model. And, consequently, a direct comparison between these two approaches based on the well-known decision-making problems within the context model were also examined.

Recently, Huynh and Nakamori in [9] have proposed a context model based approach to fuzzy concept analysis. It has been shown that we can provide a unifying interpretation for notions of fuzzy sets [16] and LT-fuzzy sets [13] within the context model. Furthermore, the context model also provides a practical method for constructing membership functions of fuzzy concepts, and simultaneously, a theoretical justification for use of some well-known t-norm based connectives in applications [10].

In this paper we briefly introduce an extension of context model allowing the representation and manipulation of fuzzy degrees of belief from a decision-making viewpoint.

2 Preliminaries

2.1 Motivation

To clarify our motivation in this work, let us recall briefly the interpretation of data and the kinds of imperfectness within the context model [5]. On a level of abstraction sufficient for a large number of applications in the field of knowledge-based systems, data characterizes the state of an object (obj) with respect to underlying relevant frame conditions (*cond*). In this sense, we assume that it is possible to characterize *obj* by an element *state(obj, cond)* of a well-defined set *Dom(obj)* of distinguishable object states. *Dom(obj)* is usually called the *universe of discourse* or *frame of discernment* of *obj* with respect to *cond*. Then we are interested in the problem that the original characterization of *state(obj, cond)* is not available due to a lack of information about *obj* and *cond*. Generally, *cond* merely permits us to use statements like "*state(obj, cond)* $\in$ *Char(obj, cond)*", where *Char(obj, cond)* $\subseteq$ *Dom(obj)* and called an *imprecise characterization of obj* with respect to *cond*. The second kind of imperfect knowledge in context model is *conflict*. This kind of imperfectness is induced by information about preferences between the elements of *Char(obj, cond)* that interprets for the existence of contexts. The combined occurence of imprecision and conflict in data reflects vagueness in the context model, and *state(obj, cond)* is described by a *vague characteristic* of *obj* w.r.t. *cond*.

As mentioned above, although information about preferences between the elements of *Char(obj, cond)* is modelled by contexts, this also means they have the same possibility or chance to be the unknown original value of *state(obj, cond)* in each context. However, in many practical situations, even in the same context elements of *Char(obj, cond)* may have different degrees of possibility to be the unknown original value of *state(obj, cond)*. Especially in the situations where *cond* only permits us to express in the form of verbal statements like "*state(obj, cond)* is A", where A is a linguistic value represented by a fuzzy set in *Dom(obj)*. Let us consider the following example.

Example 1. Assume that we want to forecast the temperature of the next day. Let $D = \{-40, \ldots, 40\}$ be the frame of discernment (temperatures measured in $°C$). We are told by expert E_1 that tomorrow's temperature will be *very high*, whereas another expert E_2 asserts that it will be *medium*. Assuming that we have degree of confidence of 0.5 in expert E_1 and of 0.8 in expert E_2, what is our belief about some predicted intervals of tomorrow's temperature?

This example is inspired by Denœux [1]. However, Denœux [1] proposed a principled approach to the representation and manipulation of imprecise degrees of belief in the framework of evidence theory. In this paper we will introduce an alternative approach to deal with mentioned situations in the spirit of context model.

2.2   The Context Model

Formally, a context model is defined as a triple $\langle D, C, A_C(D) \rangle$, where D is a nonempty *universe of discourse*, C is a nonempty *finite set of contexts*, and the set $A_C(D) = \{a | a : C \to 2^D\}$ which is called the set of all vague characteristics of D with respect to C. For $a_1, a_2 \in A_C(D)$, then a_1 is said to be *more specific* than a_2 iff $(\forall c \in C)(a_1(c) \subseteq a_2(2))$. A vague characteristic a is called to be *contradictory* iff there exists $c \in C$ such that $a(c) = \emptyset$. From now on we restrict to consider only vague characteristics that are not contradictory in the context model.

If there is a finite measure P_C on the measurable space $(C, 2^C)$, then $a \in A_C(D)$ is called a *valuated vague characteristic* of D w.r.t. P_C. Then we call a quadruple $\langle D, C, A_C(D), P_C \rangle$ a valuated context model. Mathematically, if $P_C(C) = 1$ the mapping $a : C \to 2^D$ is a random set but obviously with a different interpretation within the context model.

Let a be a vague characteristic in the valuated context model $\langle D, C, A_C(D), P_C \rangle$. For each $X \in 2^D$, we define the acceptance degree $\text{Acc}_a(X)$ that evaluates the proposition "*state*(*obj*, *cond*) $\in X$" is true. Due to inherent imprecision of a, it does not allow to uniquely determine acceptance degrees $\text{Acc}_a(X)$, $X \in 2^D$. However, we can calculate upper and lower bounds for them similar as done in the theory of upper and lower probabilities and Dempster-Shafer theory [14] as follows:

$$\underline{\text{Acc}}_a(X) = P_C(\{c \in C | a(c) \subseteq X\}) \tag{1}$$

$$\overline{\text{Acc}}_a(X) = P_C(\{c \in C | \emptyset \neq a(c) \cap X\}) \tag{2}$$

More details on the context model and its applications could be found in [5, 6, 7, 12].

3 An Extension of Context Model

As motivated in Subsection 2.1, in this section we introduce an extension of the context model for dealing with both vagueness and partial conflict in data analysis. Let D be a nonempty universe of discourse, and denote $\overline{\mathcal{F}}(D)$ the set of all normal fuzzy subsets of D. Now an extended context model is defined as a triple $\langle D, C, \Gamma_C(D) \rangle$, where $\Gamma_C(D) = \{a | a : C \to \overline{\mathcal{F}}(D)\}$, and each $a \in \Gamma_C(D)$ is called a context-dependent vague characteristic. As such, we also restrict to consider only context-dependent vague characteristics that are not contradictory in the extended context model.

From a point of view of decision analysis, we also intend to evaluate the acceptance degree $\text{Acc}_a(X), X \in 2^D$, that the proposition "*state*(*obj*, *cond*) $\in X$" is true. Obviously, vagueness and partial conflict due to contexts in a do not allow to uniquely determine an interval of acceptance degrees but a fuzzy quantity in the set of non-negative real numbers R^+. This can be done in terms of the α-cuts of fuzzy sets $a(c), c \in C$, as follows.

Let α be any real number in $(0, 1]$, and ${}^\alpha a(c)$, for any $c \in C$, the α-cut of $a(c)$. Then by (1) and (2) we get

$$^\alpha\underline{\text{Acc}}_a(X) = P_C(\{c \in C | \, ^\alpha a(c) \subseteq X\}) \tag{3}$$

$$^\alpha\overline{\text{Acc}}_a(X) = P_C(\{c \in C | \, \emptyset \neq \, ^\alpha a(c) \cap X\}) \tag{4}$$

For any $\alpha, \beta \in (0, 1]$ and $\alpha \leq \beta$, we have ${}^\beta a(c) \subseteq {}^\alpha a(c)$. It directly follows by (3) and (4) that

$$^\alpha\underline{\text{Acc}}_a(X) \leq \, ^\beta\underline{\text{Acc}}_a(X) \text{ and } \, ^\beta\overline{\text{Acc}}_a(X) \leq \, ^\alpha\overline{\text{Acc}}_a(X)$$

Equivalently, we have

$$[^\beta\underline{\text{Acc}}_a(X), \, ^\beta\overline{\text{Acc}}_a(X)] \subseteq [^\alpha\underline{\text{Acc}}_a(X), \, ^\alpha\overline{\text{Acc}}_a(X)]$$

Under such a condition of monotonicity, now we can define $\text{Acc}_a(X)$ as a fuzzy set of R^+ whose membership function $\mu_{\text{Acc}_a(X)}$ is defined by

$$\mu_{\text{Acc}_a(X)}(r) = \sup_\alpha \{\alpha | \, r \in [^\alpha\underline{\text{Acc}}_a(X), \, ^\alpha\overline{\text{Acc}}_a(X)]\}$$

In the case where $P_C(C) = 1$, the function Acc_a is formally equivalent to a fuzzy subset of type 2 of 2^D, i.e. that a fuzzy set with fuzzy membership values [17].

Example 2. This example models Example 1 by using the notion of extended context model. Let $D = \{-40, \ldots, 40\}$ be the frame of discernment, and $C = \{E_1, E_2\}$ be the set of contexts. Assume that linguistic values *very high* and *medium* are represented by normal fuzzy sets in D whose membership functions are denoted by μ_{VH} and μ_M, respectively. Then *tomorrow's temperature* (*temp*) is considered as a context-dependent vague characteristic, that is

$$
\begin{array}{rcl}
temp: & C & \longrightarrow \overline{\mathcal{F}}(D) \\
 & E_1 & \longmapsto \mu_{VH} \\
 & E_2 & \longmapsto \mu_M
\end{array}
$$

The measure P_C reflecting degrees of confidence in Experts is defined by $P_C : 2^C \longrightarrow R^+$ so that $P_C(\emptyset) = 0$, $P_C(\{E_1\}) = 0.5$, $P_C(\{E_2\}) = 0.8$, and $P_C(C) = 1.3$.

Assuming that we have to decide a forecasted interval for *temp* from some predicted intervals of temperature available, say TI_1, TI_2, TI_3. By the procedure specified above, we can calculate $\mathrm{Acc}_{temp}(TI_i)$ for $TI_i, i = 1, 2, 3$. The next step in the decision process may consist in comparison of the obtained fuzzy quatities. This may be done, for example, on the basis of a partial order such as $\mathrm{Acc}_a(X_1) \leq \mathrm{Acc}_a(X_2)$ if and only if

$$
{}^\alpha \underline{\mathrm{Acc}}_a(X_1) \leq {}^\alpha \underline{\mathrm{Acc}}_a(X_2) \text{ and } {}^\alpha \overline{\mathrm{Acc}}_a(X_1) \leq {}^\alpha \overline{\mathrm{Acc}}_a(X_2)
$$

In this case we have to admit indeterminacy when two fuzzy degrees of acceptance are incomparable.

Remark. The manipulation of fuzzy quatities may be considerably simplified by restricting the consideration on fuzzy numbers with the LL parameterization introduced in Dubois and Prade [2]. Then, as mentioned in Klir and Yuan [11], many methods for total ordering of fuzzy numbers that have been suggested in the literature can be used in the comparison of fuzzy degrees of acceptance. It should be noticed that in the spirit of previous applications of fuzzy set theory to decision analysis, e.g. [3, 4, 15], the utilities were often described in terms of fuzzy numbers.

For any $X \in 2^D$, as an alternative representation of $\mathrm{Acc}_a(X)$, we define the so-called *mass distribution* m_a of a via α-cuts as follows.

$$
{}^\alpha m_a(X) = P_C(\{c \in C \mid {}^\alpha a(c) = X\})
$$

Due to the additivity property of P_C, we have

$$
{}^\alpha \underline{\mathrm{Acc}}_a(X) = \sum_{A \in {}^\alpha a(C) : A \subseteq X} {}^\alpha m_a(A) \tag{5}
$$

$$
{}^\alpha \overline{\mathrm{Acc}}_a(X) = \sum_{A \in {}^\alpha a(C) : A \cap X \neq \emptyset} {}^\alpha m_a(A) \tag{6}
$$

Up to now we have considered the representation and manipulation of vague knowledge by the extended context model. To justify how to come to decision-making aspects by the extended context model, we need to explore extended versions of further important operations such as conditioning, data revision, conbination on context-dependent vague characteristics [5].

4   Conclusions

In this paper, we introduced an extension of the context model for dealing with both vagueness and conflict in fuzzy data analysis. A uncertainty measure of type 2 emerged from a decision-making point of view has been also proposed. The notion of extended context model may allow us to model some situations where heterogeneous data coming from a variety of sources considered as contexts (especially including human-centered systems encapsulating human expertise) should have to be taken into account [8]. The further important operations mentioned at the end of preceding section as well as applications of the proposed framework in problems such as fuzzy data analysis and decision making are being the subject of our further work.

Acknowledgement

Support in part from the National Basic Research Programme in Natural Sciences of Vietnam is gratefully acknowledged.

References

[1] T. Denœux, Modeling vague beliefs using fuzzy-valued belief structures, *Fuzzy Sets and Systems* **116** (2000) 167–199.

[2] D. Dubois & H. Prade, *Possibility Theory – An Approach to Computerized Processing of Uncertainty*, Plenum Press, New York, 1987.

[3] D. Dubois & H. Prade, The use of fuzzy numbers in decision analysis, in M.M. Gupta & E. Sanchez (Eds.), *Fuzzy Information and Decision Processes*, North-Holland, New York, 1982, pp. 309–321.

[4] A.N.S. Freeling, Fuzzy sets and decision analysis, *IEEE Transactions on Systems, Man and Cybernetics* **10** (7) (1980) 341–354.

[5] J. Gebhardt & R. Kruse, The context model: An integrating view of vagueness and uncertainty, *International Journal of Approximate Reasoning* **9** (1993) 283–314.

[6] J. Gebhardt & R. Kruse, Parallel combination of information sources, in D.M. Gabbay & P. Smets (Eds.), *Handbook of Defeasible Reasoning and Uncertainty Management Systems*, Vol. **3** (Kluwer, Dordrecht, The Netherlands, 1998) 393–439.

[7] J. Gebhardt, Learning from data – Possibilistic graphical models, in D.M. Gabbay & P. Smets (Eds.), *Handbook of Defeasible Reasoning and Uncertainty Management Systems*, Vol. **4** (Kluwer, Dordrecht, The Netherlands, 2000) 314–389.

[8] R.J. Hathaway, J.C. Bezdek & W. Pedrycz, A Parametric Model for Fusing Heterogeneous Fuzzy Data, *IEEE Transactions on Fuzzy Systems* **4** (3) (1996) 270–281.

[9] V.N. Huynh & Y. Nakamori, Fuzzy concept formation based on context model, in: N. Baba et al. (Eds.), *Knowledge-Based Intelligent Information Engineering Systems & Allied Technologies* (IOS Press, 2001), pp. 687–691.

[10] V.N. Huynh, Y. Nakamori, T.B. Ho & G. Resconi, A context model for constructing membership functions of fuzzy concepts based on modal logic, in: T. Eiter & K.-D. Schewe (Eds.), *Foundations of Information and Knowledge Systems*, LNCS 2284, Springer-Verlag, Berlin Heidelberg, 2002, pp. 93–104.

[11] R. Klir & B. Yuan, *Fuzzy Sets and Fuzzy Logic: Theory and Applications*, Prentice Hall, Upper Saddle River, NJ, 1995.

[12] R. Kruse, J. Gebhardt & F. Klawonn, Numerical and logical approaches to fuzzy set theory by the context model, in: R. Lowen and M. Roubens (Eds.), *Fuzzy Logic: State of the Art*, Kluwer Academic Publishers, Dordrecht, 1993, pp. 365–376.

[13] H. Rasiowa & C.H. Nguyen, *LT*-fuzzy sets, *Fuzzy Sets and Systems* **47** (1992) 323–339.

[14] G. Shafer, *A Mathematical Theory of Evidence* (Princeton University Press, 1976).

[15] S.R. Watson, J.J. Weiss & M.L. Donnell, Fuzzy decision analysis, *IEEE Transactions on Systems, Man and Cybernetics* **9** (1) (1979) 1–9.

[16] L.A. Zadeh, Fuzzy sets, *Information and Control* **8** (1965) 338–353.

[17] L.A. Zadeh, The concept of linguistic variable and its application to approximate reasoning, *Information Sciences,* I: **8** (1975) 199–249; II: **8** (1975) 310–357.

A Fuzzification of Landscape Theory for Alliance Analysis

S. Suganuma[1,3], V.N. Huynh[2], S. Wang[3], Y. Nakamori[1]

[1]Japan Advanced Institute of Science and Technology
1-1 Asahidai, Tatsunokuchi, Ishikawa 923-1292, JAPAN

[2]Department of Computer Science, University of Quinhon
170 - An Duong Vuong, Quinhon, VIETNAM

[3]Academy of Mathematics and System Sciences
Chinese Academy of Sciences, Beijing 100080, P.R. CHINA

Abstract. This paper aims at extending landscape theory by using the notion of fuzzy partitions. Fuzzification of the theory may allow us to analyze a variety of aggregation processes in political, economic, and social problems in a more flexible manner. The simulation results for the problems of the international alignment of the Second World War in Europe and the coalition formation in standard-setting alliances in the case of UNIX operating system are comparable with those given by the original theory.

1 Introduction

A formal theory of aggregation called landscape theory was proposed by R. Axelrod and D. Scott Bennett in 1993. Aggregation means the organization of elements of a system in patterns that tend to put highly compatible elements together and less compatible elements apart [1]. Landscape theory aims at predicting how aggregation will lead to alignments among actors (such as nations, business firms, etc.), whose leaders are myopic in their assessments and incremental in their actions. The theory mimics the idea of an abstract landscape that has been widely used in the physical and natural sciences and, more recently, in artificial intelligence to characterize the dynamics of systems. Key concepts from these setting have been used in landscape theory of aggregation. It has been shown that landscape theory can be applied to analyze a wide variety of important aggregation processes in political, economic, and social problems [3].

Landscape theory has been supported by the results of the international alignment of the Second World War in Europe [1] and the coalition formation in standard-setting alliances in the case of UNIX operating system [2].

It should be noticed that aggregation has been studied without landscapes as a descriptive problem in statistics with the most commonly used technique is cluster analysis [4]. However, cluster analysis has been considered as the art of finding groups in data that have been mainly of the form of tuples, while this is not the case for actors in landscape theory. Furthermore, unlike landscape theory, cluster analysis is not based on a dynamic theory of behaviour and it cannot make predictions.

In this paper we aim at extending landscape theory by using the notion of fuzzy partitions also called fuzzy configurations. The reason for developing a fuzzy version of landscape theory is, as we hoped, it may allow us to analyze a variety of aggregation processes in political, economic, and social problems in a more flexible manner. For example, in international politics, the leaders of nations always want to find flexible policies in settlement of conflicts with others (such as a border dispute, a matter of ethnic) in a peaceful solution. At the same time, there may be also some nations that play an intercessional role to hold a peacefully equilibration status in internationally political contentions. Even in situations where we need a crisp aggregation, fuzzification of the theory also allows us to use membership values in deciding the core and boundary actors of alliances, thereby providing more useful information for dealing with boundary actors.

Formally, the idea is similar to that for fuzzy clustering [7]. Particularly, we first mathematically reformalize landscape theory so that the distance between any two actors can be expressed via their membership grades in a specific configuration. With this we can naturally extend from crisp membership grades to fuzzy ones, and consequently obtain a continuous objective function that describes the energy landscape of the entire system in a fuzzy setting. The minimization of this function under certain constrains yields local minima corresponding to predicted fuzzy configurations. Due to complexity of a multidimensional nonlinear optimization problem, it would be inefficient to get locally optimal solutions in an analytical manner. However, we overcome this deficiency by developing a GA-based algorithm for luckily finding a near-optimized solution corresponding to a predicted fuzzy configuration. The simulation results for the problems of the international alignment of the Second World War in Europe and the coalition formation in standard-setting alliances in the case of UNIX operating system show that the algorithm should be feasible. Furthermore, the obtained results are also comparable with those given in [1, 2].

2 Landscape theory

This section briefly recalls key concepts from landscape theory. More detailed discussions of the theory as well as its applications could be referred to [3].

Landscape theory begins with a finite set of *actors*, denoted by $\mathcal{A} = \{A_1, \ldots, A_n\}$. Each actor A_i has its own *size*, $s_i > 0$, that reflects the important of that actor to others. In addition, each pair of actors A_i and A_j has a *propensity*, p_{ij}, that is a measure of how willing the two actors are to be in the same coalition together. For example, in the language of international alignments, the propensity number is positive and large if the two nations get along well together and negative if they have many sources of potential conflict. Further, if one country has a source of conflict with another then the second country typically has the same source of conflict with the first. Thus the theory assumes that propensity is symmetric, that is $p_{ij} = p_{ji}$.

By a *configuration* we mean a partition of the set of actors. That is, each actor is placed into one and only one group. Given a configuration X, we then define the *distance* between any two actors A_i and A_j within X, denoted by $d_{ij}(X)$. For example, assume that $X = \{X_1, \ldots, X_m\}$ is a configuration, the simplest measure of distance can be defined as follows

$$d_{ij}(X) = \begin{cases} 0 & \text{if } A_i, A_j \in X_k, \text{ for some } k \\ 1 & \text{otherwise.} \end{cases} \tag{1}$$

Using distance and propensity, the so-called *frustration* of an actor A_i is defined as measured how poorly or well a given configuration satisfies the propensities of a given actor to be near or far from each other actor. Formally, the frustration of an actor A_i in a configuration X is defined by

$$F_i(X) = \sum_{i \neq j} s_j p_{ij} d_{ij}(X) \tag{2}$$

where s_j is the size of A_j, p_{ij} is the propensity of A_i to be close to A_j, and $d_{ij}(X)$ is the distance between A_i and A_j in X.

Finally, the so-called *energy*, E, of an entire configuration X is now defined as the weighted sum of the frustrations of each actor in a configuration, where the weights are just the sizes of the actors. More exactly, the energy of a configuration X is as follows

$$E(X) = \sum_i s_i F_i(X) \tag{3}$$

Equivalently, the following equation defines the energy of a configuration in terms of size of the actors, their propensities to work together, and their distances in the configuration:

$$E(X) = \sum_{i,j} s_i s_j p_{ij} d_{ij}(X) \tag{4}$$

where the summation is over all ordered pairs of distinct actors.

Now the predicted configurations are based on the attempts of actors to minimize their frustrations based on their pairwise propensities to align with some actors and oppose others. These attempts lead to a local minimum in the energy landscape of the entire system.

3 Fuzzification of Landscape Theory

In this section, we will propose an extension of landscape theory based on the notion of fuzzy partitions. It should be also noticed that in landscape theory, the measure of distance in a configuration can be defined differently according to various applied situations. In [5, 6], the author has extended landscape theory by introducting another measure of distance. In our consideration, as mentioned in Introduction, we extend the theory by relaxing the notion of membership grades in the definition of a configuration. We now mathematically restate landscape theory in terms of an optimized problem with constraints. Given

- $\mathcal{A} = \{A_1, \ldots, A_n\}$ – a set of n actors, and each A_i has its size s_i, $i = 1, \ldots, n$,

- $[p_{ij}]_{n \times n}$ – the symmetric matrix of propensities,

- $X = \{X_1, \ldots, X_m\}$ – a configuration, where $1 < m < n$.

For any $A_i \in \mathcal{A}, X_k \in X$, let us denote

$$u_{ik} := \mu_{X_k}(A_i) = \begin{cases} 1 & \text{if } A_i \in X_k \\ 0 & \text{otherwise.} \end{cases} \tag{5}$$

where μ_{X_k} denotes the characteristic function of X_k.

It is easily seen that the measure of distance defined by (1) can be stated in terms of membership grades as follows

$$d_{ij}(X) = \frac{1}{2} \sum_{k=1}^{m} (u_{ik} - u_{jk})^2 \tag{6}$$

Substituting (6) into (4) yields the following expression of the energy of a configuration

$$E(X) = \frac{1}{2} \sum_{i=1}^{n} \sum_{i \neq j=1}^{n} s_i s_j p_{ij} \sum_{k=1}^{m} (u_{ik} - u_{jk})^2 \tag{7}$$

As mentioned in the preceding section, the problem is to find out predicted configurations which are local minima in the energy landscape of the entire system. Mathematically, the theory now aims at minimizing the objective function (7) under the constrains (8)–(10) below

$$1 < m < n; \qquad u_{ik} \in \{0, 1\}, \quad \text{for all } i \in \{1, \ldots, n\}, k \in \{1, \ldots, m\} \tag{8}$$

and

$$\sum_{i=1}^{n} u_{ik} > 0 \qquad \text{for all } k \in \{1, \ldots, m\} \tag{9}$$

and

$$\sum_{k=1}^{m} u_{ik} = 1 \qquad \text{for all } i \in \{1, \ldots, n\} \tag{10}$$

It is of interest to see that the idea of landscape theory someway is similar to that of clustering based on objective functions with the well-known c-means algorithms. Normally, the parameter m may be small and could be determined easily in most practical situations. For the sake of simplicity, in this study we assume that m is given and fixed a priori.

Under such a formulation, we can extend the theory by relaxing the notion of a configuration to a fuzzy one. Now we aim at finding out fuzzy configurations that minimize the function (7) subject to constraints

$$\begin{cases} u_{ik} \in [0, 1], & \text{for all } i \in \{1, \ldots, n\}, k \in \{1, \ldots, m\} \\ \sum_{i=1}^{n} u_{ik} > 0 & \text{for all } k \in \{1, \ldots, m\} \\ \sum_{k=1}^{m} u_{ik} = 1 & \text{for all } i \in \{1, \ldots, n\} \end{cases} \tag{11}$$

Let us denote

$$\mathcal{U} = \{U = [u_{ik}]_{n \times m} | u_{ik} \in [0, 1] \wedge \sum_{k=1}^{m} u_{ik} = 1 \wedge \sum_{i=1}^{n} u_{ik} > 0, \forall i, k\}$$

In the next section we develop a GA-based algorithm for finding near-optimized solutions corresponding to predicted fuzzy configurations. Then we make two simulations with the proposed algorithm, one for for the problem of the international alignment of the Second World War in Europe, and the other for the problem of coalition formation in standard-setting alliances in the case of UNIX operating system.

4   An Algorithm and Simulation Results

The algorithm proposed for solving the above optimized problem is described as follows.
Input: the set of sizes of actors $\{s_1 \ldots, s_n\}$, the propensity matrix $[p_{ij}]_{n \times n}$
Output: $U \in \mathcal{U}$ that minimizes the energy function E.

1. Set $t = 0$. Initiate the first configuration U_0 randomly. Set a value for T.

2. Calculate the energy value for $U_0 = [u_{ik}^{(0)}]_{n \times m}$ by

$$E(U_0) = \frac{1}{2} \sum_{i=1}^{n} \sum_{i \neq j=1}^{n} s_i s_j p_{ij} \sum_{k=1}^{m} (u_{ik}^{(0)} - u_{jk}^{(0)})^2$$

3. Set $t = t + 1$

4. Use GA to generate the next configuration U_t from the preceding configuration.

5. Calculate the energy value $E(U_t)$.

6. While $|E(U_t) - E(U_{t-1})| \geq \epsilon$ and $t \leq T$ do

    If $E(U_t) < E(U_{t-1})$ Then goto step 3.

    Else set $t := t - 1$ goto step 3.

To illustrate for applicability of our approach and to test the proposed algorithm, we have used the 1936 size data of Europe for predicting the alignment of the Second World War in Europe. Further, we have also tested the algorithm with another example of estimating the choices of nine computer companies to join one of two alliances sponsoring competing UNIX operating system standards in 1988. The simulation results that we have got are comparable with those given in [1, 2].

Due to the limitation of the page number, we could not discuss and analyse in detail these examples here. More details about these illustrated examples will be given in the presentation.

Acknowledgement

The second author is partially supported by the National Basic Research Programme in Natural Sciences of Vietnam.

References

[1] R. Axelrod & D. Scott Bennett, A landscape theory of aggregation, *British Journal of Political Science* **23** (Apr. 1993) 211–233.

[2] R. Axelrod et al., Coalition formation in standard-setting alliances, *Management Science* **41** (Sep. 1995) 1493–1508.

[3] R. Axelrod, *The Complexity of Cooperation* (Princeton University Press, Princeton, New Jersey, USA 1997).

[4] L. Kaufman & P. J. Rousseeuw, *Finding Groups in Data - An Introduction to Cluster Analysis* (New York: Wiley, 1990).

[5] K. Kijima, *Invitation to Drama Theory* (Ohmsha, Japan, 2001, in Japanese).

[6] K. Kijima, Generalized lanscape theory: Agent-based approach to alliance formations in civil aviation industry, *Journal of Systems Science and Complexity* **14** (2) (2001) 113–123.

[7] E. H. Ruspini, A new approach to clustering, *Information and Control* **15** (1) (1969) 22–32.

KES 2002
E. Damiani et al. (Eds.)
IOS Press, 2002

Weighted Functional Dependencies in Fuzzy Databases

Michinori Nakata
Josai International University, nakata@jiu.ac.jp
Tetsuya Murai
Hokkaido University, murahiko@main.eng.hokudai.ac.jp
Sadaaki Miyamoto
University of Tsukuba, miyamoto@esys.tsukuba.ac.jp

Abstract. Weighted functional dependencies, discovered by data mining, are examined in a possibility-based fuzzy relational database. Whether a functional dependency accompanied by a weight holds in a relation is determined by comparing the weight with to what degree the relation satisfies the functional dependency. Obtained inference rules are sound and complete, which are similar to Armstrong's rules. Thus, we can discover all functional dependencies related with a discovered functional dependency with a weight by using the inference rules.

1. Introduction

Data mining where various kinds of implicit knowledge are discovered from databases is very actively developed. Discovered rules expressing cause-effect relationships have the form of weighted functional dependencies, functional dependencies accompanied by a weight expressing to what degree the functional dependencies are satisfied.

Imperfect information is ubiquitous in the real world[5]. When we extract some rules from databases reflected the real world, we are faced on handling databases containing imperfect information. It is necessary to investigate properties of weighted functional dependencies in order to use them in realistic databases. Thus, weighted functional dependencies are examined in the framework of databases handling imperfect information under fuzzy sets and possibility theory in this paper.

Some work for functional dependencies is done in fuzzy relational models[2, 4, 7]. It is important to consider the following viewpoints on the treatment of functional dependencies. First, a compatibility degree of a tuple with a functional dependency is essentially not a binary value[8]. Second, the functional dependency should be dealt with in the same way as other restrictions[4]. Under considering these points, we deal with weighted functional dependencies in a possibility-based fuzzy relational databases under fuzzy sets and possibility theory. In particular, we focus on inference rules in this paper.

2. Possibility-based fuzzy relational databases

2.1. Framework

Definition 2.1.1[6]
The value $t[A_j]$ of an attribute A_j in a tuple t is represented by a possibility distribution where every element is included in a domain D_j .

We cannot mention that a possibility distribution with plural elements is equal to any possibility distribution for its actual value, even if symbolically equal.

In addition to the above extension, we introduce a membership attribute, as is done by some authors in extended relational databases handling a kind of imperfect information[1, 3].

Definition 2.1.2
A relation r is an extended set of tuple values with $t[\mu]_\Pi > 0$ of their membership attribute value $t[\mu](= (t[\mu]_N, t[\mu]_\Pi))$, which denotes with what degree a tuple value $t[\mathcal{A}]$ necessarily and possibly belongs to the relation r; namely, $r = \{(t[\mathcal{A}], t[\mu]) \mid t[\mu]_\Pi > 0\}$.

Definition 2.1.3
A membership attribute value of a tuple that exists in a relation is equal to a compatibility degree of its tuple value with the imposed restrictions on the relation.

The imposed restrictions come from integrity constraints, query processing, and update processing. The compatibility degree is calculated, but not given. This means that the membership attribute values are ones calculated under restrictions. Consequently, any membership attribute value is an inevitable product created from allowing a possibility distribution with plural elements as an attribute value.

2.2. Restrictions and membership attribute values

As is addressed in the previous section, a tuple is accompanied by a compatibility degree of a tuple value with restrictions, which is expressed as a membership attribute value. Each restriction is expressed by elementary restrictions and logical operators $not(\neg)$, $and(\wedge)$, and $or(\vee)$. An elementary restriction is "A_i is m" or "$A_i \theta A_j$," where m is a predicate that is expressed by a membership function μ_m and θ is one of arithmetic comparators: $<, \le, =, >,$ and $\ge$. When a restriction c is "A_i is m," a compatibility degree $Com(c \mid t[\mathcal{A}])$ of a tuple value $t[\mathcal{A}]$ with the restriction c is:

$$
\begin{aligned}
Com(c \mid t[\mathcal{A}])_\Pi &= \max_x \min(\pi_{A_i}(x), \mu_m(x)), \\
Com(c \mid t[\mathcal{A}])_N &= 1 - Com(\neg c \mid t[\mathcal{A}])_\Pi = 1 - \max_x \min(\pi_{A_i}(x), 1 - \mu_m(x)).
\end{aligned}
$$

If this is the only one restriction imposed on a relation r, a membership attribute value $t[\mu]$ of a tuple t is:

$$
t[\mu] = (t[\mu]_N, t[\mu]_\Pi) = (Com(c \mid t[\mathcal{A}])_N, Com(c \mid t[\mathcal{A}])_\Pi).
$$

For negation $not(\neg)$ of a restriction c,

$$
Com(\neg c \mid t[\mathcal{A}])_N = 1 - Com(c \mid t[\mathcal{A}])_\Pi, \quad Com(\neg c \mid t[\mathcal{A}])_\Pi = 1 - Com(c \mid t[\mathcal{A}])_N.
$$

For conjunction $and(\wedge)$ of two elementary restrictions c_1 and c_2,

$$
\begin{aligned}
Com(c_1 \wedge c_2 \mid t[\mathcal{A}])_N &= \min(Com(c_1 \mid t[\mathcal{A}])_N, Com(c_2 \mid t[\mathcal{A}])_N), \\
Com(c_1 \wedge c_2 \mid t[\mathcal{A}])_\Pi &\le \min(Com(c_1 \mid t[\mathcal{A}])_\Pi, Com(c_2 \mid t[\mathcal{A}])_\Pi),
\end{aligned}
$$

where the equality holds if the two restrictions are noninteractive. Disjunction $or(\vee)$ is obtained from conjunction and negation.

2.3. Satisfaction degrees with rules

When we have a discovered rule, we calculate to what degree each tuple, belonging with a membership attribute value to the relation, satisfies the rule.

Definition 2.3.1
A satisfaction degree $D(c|t)(= (D(c|t)_N, D(c|t)_\Pi)$ of a tuple t with a rule c is:

$$
D(c|t)_\Pi = \begin{cases} 1 & if\ t[\mu]_\Pi \le Com(c \mid t[\mathcal{A}]_\Pi) \\ Com(c \mid t[\mathcal{A}])_\Pi & otherwise, \end{cases}
$$

$$
D(c|t)_N = \begin{cases} 1 & if\ t[\mu]_N \le Com(c \mid t[\mathcal{A}]_N)\ and\ t[\mu]_\Pi \le Com(c \mid t[\mathcal{A}]_\Pi) \\ Com(c \mid t[\mathcal{A}])_\Pi & if\ t[\mu]_N \le Com(c \mid t[\mathcal{A}]_N)\ and\ t[\mu]_\Pi > Com(c \mid t[\mathcal{A}]_\Pi) \\ Com(c \mid t[\mathcal{A}])_N & otherwise, \end{cases}
$$

where a compatibility degree of a tuple value with a rule is calculated by the same way as one with a restriction.

It is said that a rule c holds in a tuple t, if $D(c|t)_N = 1$; namely, $Com(c \mid t[\mathcal{A}])_N \geq t[\mu]_N$ and $Com(c \mid t[\mathcal{A}])_\Pi \geq t[\mu]_\Pi$. To what degree a tuple satisfies a rule should be distinguished from the fact that its tuple value has a compatibility degree with that rule. Both are linked by the above definition.

Definition 2.3.2
A satisfaction degree $D(c|r)(= (D(c|r)_N, D(c|r)_\Pi)$ of a relation r with a rule c is:

$$D(c|r)_N = \min_t D(c|t)_N, \quad D(c|r)_\Pi = \min_t D(c|t)_\Pi.$$

It is said that a rule c holds in a relation r, if $D(c|r)_N = 1$; namely, if $\forall t \, Com(c \mid t[\mathcal{A}])_N \geq t[\mu]_N$ and $Com(c \mid t[\mathcal{A}])_\Pi \geq t[\mu]_\Pi$. $D(c|r)_N < 1$ means that the rule holds to a degree $(D(c|r)_N, D(c|r)_\Pi)$ that is less than $(1, 1)$ in the relation r.

2.4. Satisfaction degrees with weighted rules

A discovered rule, which is obtained by data mining, is usually accompanied by a weight. This weight denotes to what degree the discovered rule is satisfied. In order to use a discovered rule in databases, we have to evaluate the satisfaction degree. Each discovered rule is expressed by the form c_α^β where c is a body and α and β denote satisfaction degrees of c in necessity and in possibility at the point that c was discovered.

Definition 2.4.1
Suppose we have a weighted rule c_α^β on a relation r. A satisfaction degree of a relation r with the rule c_α^β is:

$$D(c_\alpha^\beta|r)_\Pi = \begin{cases} 1 & if \ \beta \leq D(c|r)_\Pi \\ D(c|r)_\Pi & otherwise, \end{cases}$$

$$D(c_\alpha^\beta|r)_N = \begin{cases} 1 & if \ \alpha \leq D(c|r)_N \ and \ \beta \leq D(c|r)_\Pi \\ D(c|r)_\Pi & if \ \alpha \leq D(c|r)_N \ and \ \beta \geq D(c|r)_\Pi \\ D(c|r)_N & otherwise. \end{cases}$$

It is said that a weighted rule c_α^β holds in a relation, if $D(c_\alpha^\beta|r) = 1$.

Theorem 2.4.2
When a satisfaction degree $D(c|r)(= (D(c|r)_N, D(c|r)_\Pi)$ is obtained for a rule c in a relation r, a weighted rule c_α^β holds in the relation, where $\alpha = D(c|r)_N$ and $\beta = D(c|r)_\Pi$.

Theorem 2.4.3
If a rule accompanied by a weight holds in a relation, a rule accompanied by a lower weight in necessity and in possibility than the weight also holds in that relation.

When more than one rule is obtained from data mining, logical operators may be used. For a logical operator $\neg(not)$,

$$D(\neg c_\alpha^\beta|r)_N = 1 - D(c_\alpha^\beta|r)_\Pi, \quad D(\neg c_\alpha^\beta|r)_\Pi = 1 - D(c_\alpha^\beta|r)_N,$$

For logical operators $\wedge(and)$ and $\vee(or)$,

$$\begin{aligned} D(c1_{\alpha_1}^{\beta_1} \wedge c2_{\alpha_2}^{\beta_2}|r)_N &= \min(D(c1_{\alpha_1}^{\beta_1}|r)_N, D(c2_{\alpha_2}^{\beta_2}|r)_N), \\ D(c1_{\alpha_1}^{\beta_1} \wedge c2_{\alpha_2}^{\beta_2}|r)_\Pi &\leq \min(D(c1_{\alpha_1}^{\beta_1}|r)_\Pi, D(c2_{\alpha_2}^{\beta_2}|r)_\Pi), \\ D(c1_{\alpha_1}^{\beta_1} \vee c2_{\alpha_2}^{\beta_2}|r)_N &\geq \max(D(c1_{\alpha_1}^{\beta_1}|r)_N, D(c2_{\alpha_2}^{\beta_2}|r)_N), \\ D(c1_{\alpha_1}^{\beta_1} \wedge c2_{\alpha_2}^{\beta_2}|r)_\Pi &= \max(D(c1_{\alpha_1}^{\beta_1}|r)_\Pi, D(c2_{\alpha_2}^{\beta_2}|r)_\Pi), \end{aligned}$$

where the equality holds when two rules $c1_{\alpha_1}^{\beta_1}$ and $c2_{\alpha_2}^{\beta_2}$ are noninteractive.

3. Functional Dependencies

It is valuable to obtain inference rules, because by using inference rules we can discover another functional dependency. Thus, we examine inference rules of weighted functional dependencies under the framework described in the previous section.

3.1. Formulation of functional dependencies

Suppose X and Y are a subset of $\{A_1, \ldots, A_n\}$. When Y is functionally dependent on X in a relational database, this is expressed by $X \rightarrow Y$.

Definition 3.1.1
The requirement of a functional dependency $X \rightarrow Y$ in a relation r is that, for every pair of tuples (t_i, t_j), if $t_i[X] = t_j[X]$, then $t_i[Y] = t_j[Y]$.

A functional dependency can be considered as an implication statement[8]. We use $X \rightarrow Y = \neg X \vee Y$ according to Vassiliou[8]. When the membership attribute values are equal to $(1, 1)$, a functional dependency $f = X \rightarrow Y$ holds if $X = 0$ or $Y = 1$;[1] namely, values of X are absolutely not equal or those of Y are absolutely equal.

Every weighted functional dependency f_α^β is dealt with as merely a weighted rule. Thus, we can use all the expressions in the previous section by replacing c by f. A compatibility degree of a tuple value $t_i[\mathcal{A}]$ with a body f of a functional dependency must be calculated for all pairs of that tuple value and others.

$$Com(f|t_i[\mathcal{A}]) = \wedge_{j \neq i} Com(f|t_i[\mathcal{A}], t_j[\mathcal{A}]),$$

where $Com(f|t_i[\mathcal{A}], t_j[\mathcal{A}])$ is a compatibility degree of a pair $(t_i[\mathcal{A}], t_j[\mathcal{A}])$ of tuple values with the body f of f_α^β. Under $f = X \rightarrow Y = \neg X \vee Y$,

$$Com(f|t_i[\mathcal{A}], t_j[\mathcal{A}])_N \geq \max(1 - X_{ij,\Pi}, Y_{ij,N}),$$
$$Com(f|t_i[\mathcal{A}], t_j[\mathcal{A}])_\Pi = \max(1 - X_{ij,N}, Y_{ij,\Pi}),$$

where

$$X_{ij,N} = Com(t_i[X] \, EQ \, t_j[X])_N, \quad Y_{ij,N} = Com(t_i[Y] \, EQ \, t_j[Y])_N,$$
$$X_{ij,\Pi} = Com(t_i[X] \, EQ \, t_j[X])_\Pi, \quad Y_{ij,\Pi} = Com(t_i[Y] \, EQ \, t_j[Y])_\Pi.$$

When X and Y are single attributes,

$$Com(t_i[X] \, EQ \, t_j[X])_\Pi = \max_{x,y} \min(\mu_{EQ}(x, y), \pi_{t_i[X]}(x), \pi_{t_j[X]}(y)),$$

where $\mu_{EQ}(x, y)$ is a binary relation that denotes a relationship EQ between x and y; uaually, if $x = y$ $\mu_{EQ}(x, y) = 1$, otherwise $\mu_{EQ}(x, y) = 0$.

3.2. Inference rules of weighted functional dependencies

When a set of functional dependencies holds in a relation, functional dependencies not contained in that set may also hold. Inference rules are the rules that derive implicit functional dependencies from given ones. When weighted functional dependencies are obtained by data mining, the following inference rules are obtained.

Theorem 3.2.1
The following inference rules are sound and complete in a relation:
  AW1.   Reflexivity
        If $Y \subseteq X$, then $X \rightarrow Y|_1^1$.
  AW2.   Augmentation
        If $X \rightarrow Y|_\alpha^\beta$, then $ZX \rightarrow ZY|_\alpha^\beta$.
  AW3.   Transitivity
        If $X \rightarrow Y|_{\alpha_1}^{\beta_1}$ and $Y \rightarrow Z|_{\alpha_2}^{\beta_2}$, then $X \rightarrow Z|_\alpha^\beta$,
        where $\alpha = \min(\alpha_1, \alpha_2)$ and $\beta = \min(\beta_1, \beta_2)$.

---

[1] When Gödel implication is used, $f = X \rightarrow Y$ holds in some cases except $X = 0$ or $Y = 1$. This is problematic on functional dependencies in possibility-based fuzzy relational databases.

Among the above inference rules we do not include the rule that weighted functional dependencies with the same body as and with a lower weight than a functional dependency can be derived from that functional dependency. Such a rule is valid for any rule in our framework, as is shown in theorem 2.4.3. By using the sound and complete inference rules, we can derive all functional dependencies related with discovered functional dependencies with a weight in the possibility-based fuzzy relational database.

4. Conclusions

Weighted functional dependencies, functional dependencies accompanied by a weight, have been examined in a possibility-based fuzzy relational database. The weighted functional dependencies are discovered by data mining. When we have a weighted functional dependency, a compatibility degree with the weighted functional dependency is calculated for every tuple. Whether a weighted functional dependency holds in a relation is determined by comparing the weight with to what degree the relation satisfies the functional dependency

We have examined what inference rules are valid. Sound and complete inference rules are obtained, which are similar to Armstrong's rules in the conventional relational databases. Thus, we can derive all functional dependencies related with a discovered functional dependency with a weight by using the inference rules in the possibility-based fuzzy relational database.

Acknowledgment

This research has been partially supported by the Grant-in-Aid for Scientific Research (B) and (C), Japan Society for the Promotion of Science, No. 14380171 and No.13680475, respectively.

References

[1] Biskup, J. [1983]A Foundation of Codd's Relational Maybe-Operations. ACM Transactions on Database Systems, **8**:4, 608-636.

[2] Bosc, P., Dubois, D., and Prade, H. [1994] Fuzzy Functional Dependencies — An Overview and a Critical Discussion, in Proceedings of FUZZ-IEEE '94, IEEE, pp. 325-330.

[3] Lee, S. K. [1992]Imprecise and Uncertain Information in Databases: An Evidential Approach. in Proceedings of the 8th International Conference on Data Engineering, IEEE 1992, pp. 614-621.

[4] Nakata, M. [2000] On Inference Rules of Dependencies in Fuzzy Relational Data Models: Functional Dependencies, in *Knowledge Management in Fuzzy Databases*, Eds., O. Pons, M. A. Vila, and J. Kacprzyk, Physica Verlag, pp. 36-66.

[5] Parsons, S. [1996] Current Approaches to Handling Imperfect Information in Data and Knowledge Bases, IEEE Transactions on Knowledge and Data Engineering, **8**:3, 353-372.

[6] Prade, H. and Testemale, C. [1984] Generalizing Database Relational Algebra for the Treatment of Incomplete or Uncertain Information and Vague Queries, Information Science, **34**, 115-143.

[7] Raju, K. V. S. V. N. and Majumdar, A. K. [1988] Fuzzy Functional Dependencies and Lossless Join Decomposition of Fuzzy Relational Database Systems, ACM Trans. on Database Systems, **13**:2, 129-166.

[8] Vassiliou, Y. [1980]Functional Dependencies and Incomplete Information, in Proceedings of the 6th VLDB Conference, pp. 260-269, Morgan Kaufmann Publishers.

KES 2002
E. Damiani et al. (Eds.)
IOS Press, 2002

Operations of Zooming In and Out on Possible Worlds for Semantic Fields

Tetsuya MURAI[†], Germano RESCONI[‡], Michinori NAKATA[§], Yoshiharu SATO[†]

† Division of Systems & Information Engineering, Graduate School
of Engineering, Hokkaido University, Sapporo 060-8628, JAPAN

‡ Dipartimento di Matematica, Universita Cattolica, 25128 Brescia, ITALY

§ Faculty of Management & Information Sciences, Josai International University
Togane, Chiba 283-8555, JAPAN

Abstract. Operations of zooming in and out are formulated using filtration in modal logic. The operations can change granularity in our reasoning process. Thus they enable us to ignore those worlds unrelated with the current reasoning.

1.   Introduction

Possible worlds have an important role in the theory of semantic fields proposed by Resconi, Murai, and Shimbo[7]. In particular, we introduced the concepts of local and global worlds because, when we deal with some specified problem, we do not have to consider all of worlds. In this paper, we extend the concept of local and global worlds from a point of view of filtration in modal logic (cf. Chellas[1]) and granular computing (cf. Lin[2]). Then we can change granularity of possible worlds as if we chance

2.   Possible Worlds

Given a countably infinite set of atomic sentences $\mathcal{P}$, a language $\mathcal{L}_{\mathrm{BL}}(\mathcal{P})$ for logic of belief is formed as the least set of sentences in the usual way from $\mathcal{P}$ with the well-known propositional operators such as $\top$ (the truth constant), $\perp$ (the falsity constant), $\neg$ (negation), $\wedge$ (conjunction), $\vee$ (disjunction), $\rightarrow$ (material implication), $\leftrightarrow$ (equivalence), and a modal operators B(belief) by the following formation rules.

1. $\mathsf{p} \in \mathcal{P} \Rightarrow \mathsf{p} \in \mathcal{L}_{\mathrm{BL}}(\mathcal{P})$.

2. $\top, \perp \in \mathcal{L}_{\mathrm{BL}}(\mathcal{P})$.

3. $p \in \mathcal{L}_{\mathrm{BL}}(\mathcal{P}) \Rightarrow \neg p, \mathsf{B}p \in \mathcal{L}_{\mathrm{BL}}(\mathcal{P})$.

4. $p, p' \in \mathcal{L}_{\mathrm{BL}}(\mathcal{P}) \Rightarrow p \wedge p', p \vee p', p \rightarrow p', p \leftrightarrow p' \in \mathcal{L}_{\mathrm{BL}}(\mathcal{P})$.

In this paper a sentence is said to be *non-modal* when it does not contain any appearance of the modal operator B.

Given a countably infinite set of atomic sentences

$$\mathcal{P} = \{\mathsf{p}_1, \cdots, \mathsf{p}_n, \cdots\},$$

we define the set U of possible worlds as the power set of $\mathcal{P}$:

$$U_{\mathcal{P}} \overset{\text{def}}{=} 2^{\mathcal{P}}.$$

We call any subset in $\mathcal{P}$ an *elementary possible world*, or, *elementary world*, for short. Then, for an atomic sentence p in $\mathcal{P}$ and a world x in $U_{\mathcal{P}}$, a *valuation* $V_{\mathcal{P}}$ is naturally defined by

$$V_{\mathcal{P}}(\mathsf{p}, x) = \begin{cases} \mathsf{TRUE}, & \text{if } \mathsf{p} \in x, \\ \mathsf{FALSE}, & \text{otherwise.} \end{cases}$$

When a binary relation R is defined on $U_{\mathcal{P}}$, we have a Kripke model:

$$\mathcal{M}_{\mathcal{P}} = <U_{\mathcal{P}}, R, V_{\mathcal{P}}>.$$

The relation R is usually called an *accessibility relation*. Let us define

$$\mathcal{M}_{\mathcal{P}}, x \models \mathsf{p} \overset{\text{def}}{\Longleftrightarrow} x \in V(\mathsf{p}),$$

the left-hand side of which means that a sentence p is true at a possible world x in $\mathcal{M}_{\mathcal{P}}$. $\models$ is extended for every compound sentence in the usual way:

$$
\begin{aligned}
&\mathcal{M}_{\mathcal{P}}, x \models \top, \\
&\mathsf{Not}\ (\mathcal{M}_{\mathcal{P}}, x \models \bot), \\
&\mathcal{M}_{\mathcal{P}}, x \models \neg p \quad \overset{\text{def}}{\Longleftrightarrow} \quad \mathsf{not}\ (\mathcal{M}_{\mathcal{P}}, x \models p), \\
&\mathcal{M}_{\mathcal{P}}, x \models p \vee q \quad \overset{\text{def}}{\Longleftrightarrow} \quad \mathcal{M}_{\mathcal{P}}, x \models p \ \mathsf{or}\ \mathcal{M}_{\mathcal{P}}, x \models q, \\
&\mathcal{M}_{\mathcal{P}}, x \models p \wedge q \quad \overset{\text{def}}{\Longleftrightarrow} \quad \mathcal{M}_{\mathcal{P}}, x \models p \ \mathsf{and}\ \mathcal{M}_{\mathcal{P}}, x \models q, \\
&\mathcal{M}_{\mathcal{P}}, x \models p \rightarrow q \quad \overset{\text{def}}{\Longleftrightarrow} \quad \mathcal{M}_{\mathcal{P}}, x \models p \Rightarrow \mathcal{M}_{\mathcal{P}}, x \models q, \\
&\mathcal{M}_{\mathcal{P}}, x \models p \leftrightarrow q \quad \overset{\text{def}}{\Longleftrightarrow} \quad \mathcal{M}_{\mathcal{P}}, x \models p \Leftrightarrow \mathcal{M}_{\mathcal{P}}, x \models q, \\
&\mathcal{M}_{\mathcal{P}}, x \models \mathsf{B}p \quad \overset{\text{def}}{\Longleftrightarrow} \quad \forall y \in U\ (xRy \Rightarrow \mathcal{M}_{\mathcal{P}}, y \models p),
\end{aligned}
$$

The set of possible worlds define by

$$\|p\|^{\mathcal{M}_{\mathcal{P}}} \overset{\text{def}}{=} \{x \in U_{\mathcal{P}} \mid \mathcal{M}_{\mathcal{P}}, x \models p\}.$$

is called the *truth set* or *proposition* of p in $\mathcal{M}_{\mathcal{P}}$. Intuitively, it is the set of worlds at which p is true in $\mathcal{M}_{\mathcal{P}}$. A sentence p is said to be *valid* in $\mathcal{M}_{\mathcal{P}}$, written $\mathcal{M}_{\mathcal{P}} \models p$, just in case $\|p\|^{\mathcal{M}_{\mathcal{P}}} = U_{\mathcal{P}}$.

3. Filtration as Generating Granules of Possible Worlds

When we are concerned only with finite sentences, the set U is, in general, too large for us. So we need ways of granularizing U, one of which is to make a quotient set whose elements are regarded as granules of possible worlds. This idea was formulated as the *filtration* method in modal logic (cf. Chellas[1]) in order to solve the decision problem.

Example 1 *Assume we are interested in* $\Gamma = \{\mathsf{p}_1 \rightarrow \mathsf{p}_2\}$. *Then the set of atomic sentences that we should deal with is*

$$\mathcal{P}_\Gamma = \{\mathsf{p}_1, \mathsf{p}_2\}.$$

An equivalence relation $\sim_\Gamma$ on U are defined by

$$x \sim_\Gamma y \overset{\text{def}}{\Longleftrightarrow} \forall \mathsf{p} \in \mathcal{P}_\Gamma\ [V(\mathsf{p}, x) = V(\mathsf{p}, y)],$$

which is the so-called agreement relation on U in the filtration method (cf.[1]). The relation induces the following quotient set

$$U_\Gamma \stackrel{\text{def}}{=} U_\mathcal{P}/\sim_\Gamma = \{X_1, X_2, X_3, X_4\},$$

where

$$X_1 = \{x \in U_\mathcal{P} \mid x \cap \mathcal{P}_\Gamma = \{\mathsf{p}_1, \mathsf{p}_2\}\},$$
$$X_2 = \{x \in U_\mathcal{P} \mid x \cap \mathcal{P}_\Gamma = \{\mathsf{p}_1\}\},$$
$$X_3 = \{x \in U_\mathcal{P} \mid x \cap \mathcal{P}_\Gamma = \{\mathsf{p}_2\}\},$$
$$X_4 = \{x \in U_\mathcal{P} \mid x \cap \mathcal{P}_\Gamma = \emptyset \}.$$

Then we have the following bijection

$$\iota : U_\Gamma \to 2^\Gamma$$

defined by

$$\iota(X) = \cap X.$$

A valuation is newly defined by

$$V_\Gamma(\mathsf{p}, X) = \begin{cases} \text{TRUE}, & \text{if } \mathsf{p} \in \cap X, \\ \text{FALSE}, & \text{otherwise}, \end{cases}$$

for p in $\mathcal{P}_\Gamma$ and X in U_Γ. The valuation is extended in the usual way into compound sentences. Thus we have the following table.

V_Γ	p_1	p_2	$\mathsf{p}_1 \to \mathsf{p}_2$
X_1	TRUE	TRUE	TRUE
X_2	TRUE	FALSE	FALSE
X_3	FALSE	TRUE	TRUE
X_4	FALSE	FALSE	TRUE

The equivalence classes (granules of elementary worlds) are found to describe columns of the well-known truth table. ∎

We give a formal description. When we are concerned with a set Γ of non-modal sentences, let

$$\mathcal{P}_\Gamma \stackrel{\text{def}}{=} \mathcal{P} \cap \mathrm{sub}(\Gamma),$$

where $\mathrm{sub}(\Gamma)$ is the union of the sets of subsentences of each sentence in Γ. Then, we can define an agreement relation $\sim_\Gamma$ by

$$x \sim_\Gamma y \stackrel{\text{def}}{\iff} \forall \mathsf{p} \in \mathcal{P}_\Gamma \, [V(\mathsf{p}, x) = V(\mathsf{p}, y)].$$

Then the relation induces the quotient set

$$U_\Gamma \stackrel{\text{def}}{=} U_\mathcal{P}/\sim_\Gamma .$$

Its elements are regarded as the granules of possible worlds under $\mathcal{P}_\Gamma$. The valuation is given by

$$V_\Gamma(\mathsf{p}, X) = \begin{cases} \text{TRUE}, & \text{if } \mathsf{p} \in \cap X, \\ \text{FALSE}, & \text{otherwise}. \end{cases}$$

for p in $\mathcal{P}_\Gamma$ and X in U_Γ.

There is a bijection

$$\iota : U_\Gamma \to 2^\Gamma,$$

which is defined by

$$\iota(X) = \cap X.$$

Finally we briefly note on the filtration method. When an accessibility relation R is defined on $U_\mathcal{P}$, assume we can define a new accessibility relation R' on U_Γ satisfying

1. if xRy then $[x]_{\sim_\Gamma} R' [y]_{\sim_\Gamma}$,

2. if $[x]_{\sim_\Gamma} R' [y]_{\sim_\Gamma}$ then $\mathcal{M}, x \models \mathsf{B}p \Rightarrow \mathcal{M}, y \models p$, for every sentence $\mathsf{B}p$ in Γ,

3. if $[x]_{\sim_\Gamma} R' [y]_{\sim_\Gamma}$ then $\mathcal{M}, y \models p \Rightarrow \mathcal{M}, x \models \neg\mathsf{B}\neg p$, for every sentence $\neg\mathsf{B}\neg p$ in Γ.

Then the model

$$\mathcal{M}_\Gamma = <U_\Gamma, R', V_\Gamma>$$

is called a *filtration* through Sub(Γ). Note that R', and thus $\mathcal{M}_\Gamma$, is not uniquely determined by the above conditions (cf. Chellas[1], p.102).

With respect to a filtration, the following remarkable result has already been proved (cf. [1]): for every p in Γ,

$$\mathcal{M} \models p \Longleftrightarrow \mathcal{M}_\Gamma \models p.$$

Note that, if $|\mathcal{P}_\Gamma| = n$, then $|U_\mathcal{P}/\sim_\Gamma| \leq 2^n$. Thus, this is useful to prove finite determination and thus decidability.

4.   Zooming In and Out: Basic Operations of Reconstructing Possible Worlds

A set Γ of sentences we are concerned with at time t is called a *focus* at t and its elements *focal* ones at t. When we move our viewpoint from one focus to another, we must reconstruct the set of granularized possible worlds.

Let Γ be the current focus and let Γ' be the next focus we will move. First we consider the following two simpler cases.

Case 1: $\mathcal{P}_\Gamma \supseteq \mathcal{P}_\Delta$.

In this case we need granularization, which is a function

$$\mathsf{In}_{\Gamma,\Delta} : U_\Gamma \to U_\Delta$$

where, for any X in U_Γ,

$$\mathsf{In}_{\Gamma,\Delta}(X) = (\cap X) \cap \mathcal{P}_\Delta.$$

We call this function *zooming in* from Γ to Δ.

Case 2: $\mathcal{P}_\Gamma \subseteq \mathcal{P}_\Delta$.

In this case, we need the inverse operation of granularization, which is a function

$$\mathsf{Out}_{\Gamma,\Delta} : U_\Gamma \to U_\Delta$$

where

$$\mathsf{Out}_{\Gamma,\Delta}(X) = \{Y \in U_\Delta \mid Y \cap \mathcal{P}_\Gamma = X\}.$$

We call this function *zooming out* from Γ to Δ.

Case 3: Not-nested case

When the two sets Γ and Δ are not nested, the movement from Γ to Δ is represented using combining zooming out and in by

$$\mathsf{In}_{\Gamma \cup \Delta, \Delta} \circ \mathsf{Out}_{\Gamma, \Gamma \cup \Delta} : U_\Gamma \to U_\Delta,$$

where $\circ$ is the composition of two functions.

5. Concluding Remarks

The extension of local and global worlds in this paper is also related to granular reasoning developed by Murai et al.[3, 4]. Thus semantic fields have a close relationship with granular reasoning as will be shown in our forthcoming paper with their application to image databases.

Acknowledgments

The first author was partially supported by Grant-in-Aid No. 14380171 for Scientific Research(B) of the Japan Society for the Promotion of Science of Japan.

References

[1] B.F.Chellas (1980): *Modal Logic: An Introduction.* Cambridge University Press, Cambridge.

[2] T.Y.Lin (1998): Granular Computing on Binary Relation, I Data Mining and Neighborhood Systems, II Rough Set Representations and Belief Functions. *L. Polkowski and A. Skowron (eds.), Rough Sets in Knowledge Discovery 1: Methodology and Applications,* Physica-Verlag, Heidelberg, 107-121, 122-140.

[3] T.Murai, M.Nakata, and Y.Sato (2001): A Note on Filtration and Granular Reasoning. In *T.Terano et al.(eds.), New Frontiers in Artificial Intelligence, Lecture Notes in Artificial Intelligence, Vol. 2253,* Springer, 385-389.

[4] T.Murai, M.Nakata, and Y.Sato (2002), A Remark on Granular Reasoning and Filtration. In *M.Inuiguchi et al. (eds.), Rough Set Theory and Granular Computing,* Physica Verlag, to appear.

[5] Z.Pawlak, (1982): Rough Sets. Int. J. Computer and Information Sciences, **11**, 341–356.

[6] Z.Pawlak, (1991): *Rough Sets: Theoretical Aspects of Reasoning about Data.* Kluwer, Dordrecht.

[7] G.Resconi, T.Murai, and M.Shimbo (2000): Field Theory and Modal Logic by Semantic Fields to Make Uncertainty Emerge from Information, Int. J. of General Systems, **29**(5), 737-782.

KES 2002
E. Damiani et al. (Eds.)
IOS Press, 2002

Neuro-Fuzzy Based Analysis of Gene Expression Data

Vasile Palade[1], Daniel Neagu[2]

*[1] Oxford University, Computing Laboratory, Parks Road, OX1 3QD,
Oxford –UK. E-mail: Vasile.Palade@comlab.ox.ac.uk*
*[2] University of Bradford, Department of Computing, Bradford - UK
BD7 1DP, E-mail: D.Neagu@bradford.ac.uk*

Abstract. The paper is concerned on the application of neuro-fuzzy techniques for functional analysis of gene expression data from microarray experiments. The objective of this paper is to learn and predict functional classes of the E.Coli genes using neuro-fuzzy based techniques such as modular neuro and neuro-fuzzy networks. Methods of combining explicit and implicit knowledge in functional interpretation and analysis of gene expression data are proposed. Keywords: neuro-fuzzy, gene expression data analysis, modular neural networks

1. Introduction

DNA microarray technology has allowed the biologists and practitioners in the field to monitor and analyse the expression pattern of thousands of genes in a single experiment. These array technologies have already produced tremendous amount of data that require proper methods of data analysing. The dramatic increase of resulting genetic expression data has determined the researchers in the field to look at ways of applying various machine learning and intelligent techniques for data analysis.

A main issue in analysing gene expression data is the identification of groups of genes with similar expression patterns over several different conditions and the most common approach is using various clustering techniques [2][4]. In this paper we investigate the problem of using neuro-fuzzy techniques for predicting the functional classes of genes of the well-known microorganism Escherichia coli.

There are some certain advantages of applying fuzzy logic to the analysis of gene expression data. First, fuzzy logic can cope very well with noise in the data, which is often the case with gene expression data obtained from DNA microarray experiments. Second fuzzy logic can handle the uncertainty and vagueness of the expression levels. Finally the predictions made using fuzzy logic are easily interpretable in human understandable terms, using fuzzy rules and linguistic values. On the other hand neural networks have learning, adaptation and generalization advantages and they have been already applied to various bioinformatics problems. Combined and integrated neuro-fuzzy techniques have little been applied until now for analysing and interpreting gene expression data. This stands as motivation for our paper and research.

The structure of the paper is the following. Section 2 presents a motivation on the use of combined neuro-fuzzy techniques in functional genomics. Section 3 describes our modular neuro-fuzzy approach. Section 4 presents the case study and some results on applying the neuro-fuzzy modular approach described in section 3 for classification and prediction of functional classes of E. coli genes. The paper ends with some conclusions and remarks.

2. Why Neuro-Fuzzy in functional genomics?

Some authors have focussed on the use of neural networks in bioinformatics applications [1][3][4] for solving specific tasks, such as analysing and interpreting gene expression data. Other authors [8][10] used fuzzy logic for analysing gene expression data. There is also a tremendous number of authors who tried to integrate neural networks and fuzzy logic with application to various other fields [6][7], but very a few tried to apply combined neuro-fuzzy techniques to bioinformatics problems. An attempt is done in [2], where authors applied ANFIS neuro-fuzzy networks for analysing gene expression data along with some clustering techniques.

Neural networks have been applied with promising results to functional genomics problems due to their capabilities to cope with complexity, uncertainty, noisy or corrupted data. Neural networks are very good modelling tools for highly non-linear and complex data and they can be seen as universal approximation technique. The drawback of using neural networks is their lack of transparency in human understandable terms. Fuzzy techniques can be appropriate for analysing bioinformatics data as it allows the integration in a natural way of the biologist knowledge into the process of data interpretation and analysing. The formulation of the knowledge is done in a human understandable way such as linguistic rules. The main drawback of neural networks is represented by their "black box" nature, whilst the disadvantage of fuzzy systems is represented by the difficult and time-consuming process of knowledge acquisition. On the other hand the advantage of neural network over fuzzy systems is learning and adaptation capabilities, while the advantage of fuzzy system is the human understandable form of knowledge representation. Neural networks use an implicit way of knowledge representation whilst fuzzy and neuro-fuzzy systems represent knowledge in an explicit form, such as rules.

The combination of neural networks and fuzzy systems can be done in two main ways:
- *Neural networks are the basic methodology and fuzzy logic is the second.* These hybrid systems are mainly neural networks, but the neural networks are equipped with abilities of processing fuzzy information. The systems are usually termed Fuzzy Neural Networks (FNNs) and they are networks where the inputs and/or the outputs and/or the weights are fuzzy sets.
- *Fuzzy logic is the basic methodology and neural networks the subsequent.* These systems can be viewed as fuzzy systems augmented with neural network facilities, such as learning, adaptation, and parallelism. The systems are usually called Neuro-Fuzzy Systems (NFSs), they are represented as feed-forward network architectures and they can be always interpreted as a set of fuzzy rules. The most often used neuro-fuzzy system is a 5-layered network implementing a Mamdani or a TSK fuzzy model, and this represent in fact the most common understanding for a NFS.

In the next section we shortly propose some integrating strategies in a modular neuro-fuzzy system [5][6][11] that consists of a modular arrangements of neuro-fuzzy structures that cooperate in solving the problem.

3. Modular neuro-fuzzy networks

The system combines modules based on implicit knowledge (IKMs) with modules based on explicit knowledge (EKMs). For our work purposes, we define the *explicit knowledge* as knowledge represented by neural networks, which are computationally identical to a fuzzy rules set, and are created by mapping the given fuzzy rules into neuro-fuzzy structures. The intrinsic representation of explicit knowledge is based on fuzzy neurons in a MAPI [6] implementation. The numerical weights corresponding to the connections between neurons are computed using Combine Rules First Method [12], or Fire Each Rule Method [12]. The implicit knowledge based modules (IKM) are neural and neuro-fuzzy networks developed by training with experimental data.

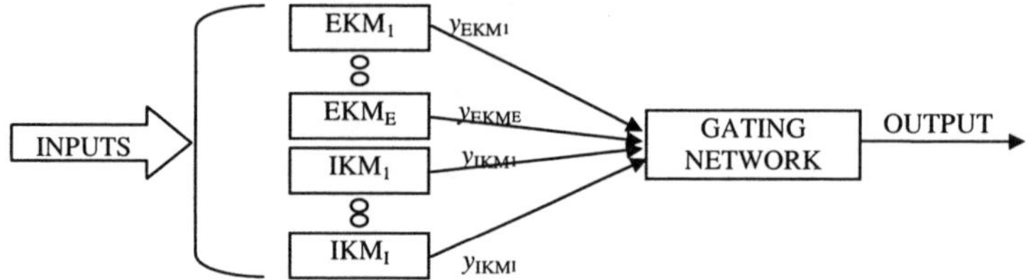

Figura 1. Modular arrangement of neuro-fuzzy IKM and EKM modules and the gating network

The proposed modular approach allows each individual module to exhibit its "opinion" about entries, offering advantages in terms of learning and generalization comparing with a single neural or neuro-fuzzy network. The modular neuro-fuzzy structure of the system (called NIKE and described in detail in our previous papers [5][6]) is based on three strategies to integrate and combine individual neuro-fuzzy IKMs and EKMs: fire each module using fuzzy integration (FEMF), unsupervised-trained gating network (UGN) and supervised-trained gating network (SGN).

The first strategy is an adapted Fire Each Rule Method [12] for modular networks. This is the simplest mode to integrate IKMs and EKMs with fuzzy outputs. After off-line training of the IKMs and mapping the explicit rules in EKMs, the general output of the system is composed as a T-conorm of fuzzy outputs of each module: the fourth layer outputs of IKM-NFNs and the EKM-FNNs' outputs.

The second strategy (UGN) proposes a competitive-based aggregation of the EKM and IKM outputs, considering that IKMs and EKMs compete one another to learn the training patterns and the gating network mediates the competition. The third strategy uses a supervised trained layer to process the overall output of the involved modules. Each defuzzified output of expert networks is an input to the final layer. The supervised training process of the final network assures a weighted aggregation of IKMs and EKMs' outputs with respect to their specialization.

Relevant fuzzy rules can be extracted from IKMs using Effect Measure Method (EMM) [13] and move into EKM modules. EMM combines the weights between the layers of the network in order to select the strongest dependencies between the fuzzy output and inputs. This approach aims to explain the patterns learned by IKMs, as well as highlights and adjusts some explicit rules given by human experts.

4. Data analysis using modular NF networks

In [9] DNA arrays of the entire set of Escherichia coli genes were used to measure the genomic expression patterns of cells growing on minimal glucose medium and on Luria broth (rich medium) containing glucose. The results are available on http://www.ou.edu/cas/botany-micro/faculty/~tconway/global.html. These results are also summarized in Table 1. There are 4290 genes from which 1428 are unclassified. The remaining 2862 genes were classified in 21 meaningful classes, which are not easy separable classes. In order to illustrate this, figure 2 presents 6 classes (number 1, 2, 5, 9, 10, 19 from Table 1), the two axes representing the logarithmic expression intensity (normalized to [0;1] interval) under the two different growth conditions: minimal glucose medium and Luria broth containing glucose. For the sake of simplicity few genes with negative expression value were ignored.

Table 1. Functional classes of E. coli genes

No.	Functional group	Number of genes
1	Amino acid biosynthesis	97
2	Putative transport proteins	291
3	Carbon compound catabolism	124
4	Cell processes	170
5	Putative enzymes	453
6	Central intermediary metabolism	149
7	DNA replication, repair, restriction/modification	105
8	Energy metabolism	136
9	Transport and binding proteins	254
10	Translation and post-translational modification	128
11	Phage, transposon or plasmid	91
12	Transcription, RNA processing and degradation	28
13	Cell structure	84
14	Putative factors	67
15	Putative membrane proteins	54
16	Putative regulatory protein	167
17	Biosynthesis of cofactors, prosthetic groups and carriers	106
18	Regulatory function	208
19	Putative cell structure	43
20	Nucleotide biosynthesis and metabolism	66
21	Fatty acid and phospholipid metabolism	41
22	Hypothetical, unclassified, unknown	1428
23	Total	4290

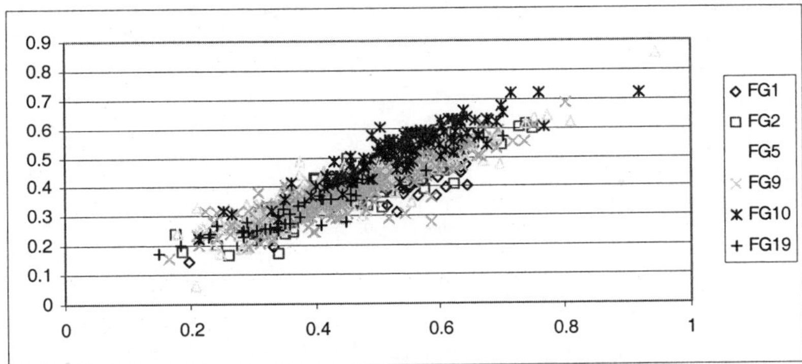

Figure 2. The data distribution for 6 functional classes: 1, 2, 5, 9, 10, 19.

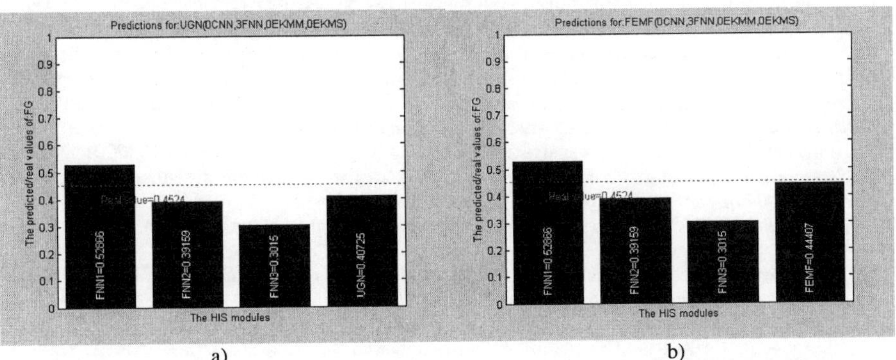

a)   b)

Figure 3. Prediction ability of the modular neuro-fuzzy network a) using FEMF b) using UGN

With combinations of EKMs and IKMs our system predicted very well the 22 classes of E.coli genes due to better generalization abilities. Using a reasonable number (3-5) of linguistic values to describe the two inputs (expression intensities under the two different growth conditions) the average prediction performance was 78-83%, which can be increased if we use more linguistic values. In a simple example, 3 neuro-fuzzy IKM modules (NFN1, NFN2, NFN3) were trained with the 6 classes given above: NFN1 with classes 1, 10, 19; NFN2 with classes 2, 5, 9; NFN3 with classes 1, 2, 9. Then we used the 3 strategies (FEMF, UGN, SGN) for integrating the 3 experts. The generalization ability of the integrated system in all cases is better than that of the individual modules. Figure 3 shows the prediction performance of UGN and FEMF strategies for a value belonging to class 5.

5. Conclusions

We have applied three different modular neuro-fuzzy structures for analysing gene expression data of Escherichia coli microorganism. The accuracy of gene function prediction is satisfactory due to the increasing generalization ability of our system that integrate the opinion of more neuro-fuzzy modules on the same input pattern. Our system can use both prior knowledge from biologists and gene expression data obtained from DNA experiments, which can increase the generalization and prediction ability in some areas of the input space where data are not representative enough. This study demonstrates that the combination of neural networks with fuzzy systems can allow a better analysing and interpretation of gene expression data, especially when there is an interest on combining explicit knowledge of human expert with experimental data. Future work will be carried out on the line of the transparency/interpretability compromise of the neuro-fuzzy models used in our hybrid modular approach for analysing gene expression data, and on the line of extracting biological meaningful explicit knowledge out of the implicit knowledge modules trained with experimental data.

References

[1] W. Choe et al., Neural network schemes for detecting rare events in human genomic DNA. *Bioinformatics* **16/12** (2000) 1062-1072.
[2] R. Guthke et al., Gene expression data mining for functional genomics. Proc. of ESIT2000, 14-15 Sept. 2000, Aachen-Germany, pp. 170-177.
[3] K.B. Hwang et al., Applying machine learning techniques to analysis of gene expression data: cancer diagnosis. *Methods of Microarray Data Analysis* (CAMDA'00), pp. 167-182, Kluwer Academics, 2002.
[4] S. Kaski et al., Analysis and visualisation of gene expression data using self-organizing maps. In Proc. of IEEE Workshop on Nonlinear Signal and Image Processing –NSIP01, 2001, proceedings on CD-ROM.
[5] C.D. Neagu and V. Palade, Modular neuro-fuzzy networks used in explicit and implicit knowledge integration. Proc. of Int. Conf. FLAIRS2002, Pensacola –Florida, USA, May 2002, pp 277-281.
[6] C.D. Neagu et al., Neural and neuro-fuzzy integration in a knowledge based system for air quality prediction. *Applied Intelligence* - Kluwer Acad. Publ., (in printing in the edition for September 2002).
[7] V. Palade et al., Fault diagnosis of an industrial gas turbine using neuro-fuzzy methods. Proc. of 15th IFAC World Congress, Barcelona - Spain, July 2002 (in printing).
[8] S. Tomida et al., Gene expression analysis using fuzzy ART, *Genome Informatics* **12** (2001) 245-246.
[9] H. Tao et al., Functional genomics: expression analysis of Escherichia coli growing on minimal and rich media. *Journal of Bacteriology* **181/20** (1999) 6425-6440.
[10] P.J. Woolf and Y. Wang, A fuzzy logic approach to analysing gene expression data. *Physiol Genomics* **3** (2000) 9-15.
[11] I. Hatzilygeroudis and J. Prenzas, Constructed Modular Hybrid Rule Bases for Expert Systems, *Int. Journal on Artificial Intelligence Tools* **10** (2001) 87-105.
[12] J.J. Buckley and Y. Hayashi. Neural nets for fuzzy systems. *Fuzzy Sets and Systems* **71** (1995) 265-276.
[13] I. Enbutsu et al., Fuzzy Rule Extraction from a Multilayered Network. Procs. of IJCNN'91, Seattle (1991) pp. 461-465.

KES 2002
E. Damiani et al. (Eds.)
IOS Press, 2002

Toxicity Prediction using Assemblies of Hybrid Fuzzy Neural Models

Daniel NEAGU

Department of Computing, University of Bradford, Bradford BD7 1DP, UK

Abstract. The paper presents models based on neural and neuro-fuzzy structures, developed to represent knowledge about the toxicity of a large number of industrial organic compounds. The neuro-fuzzy models here proposed include quantitative structure-activity relations (QSARs) and numerical values. The developed approaches to insert knowledge by training, and to map rules in neuro-fuzzy structures are evaluated and show that the combination of fuzzy inference systems and connectionist structures improve over individual models for toxicity prediction.

1. Introduction

Modelling and prediction of the consequences of the huge number of chemicals to human health and environment rely on many variables: toxicological endpoints, methods to describe the physico-chemical properties of molecules, homogeneity of the data set, computational algorithms to produce statistical relationships, validation methods. The approach of describing the bio-chemical action of different classes of chemical compounds through relations dependent on their structures is known as the QSAR problem.

Until now, several research papers have been published, discussing the role that artificial intelligence (AI) tools could play in QSAR modelling and toxicity prediction [1][4]. In all these cases, the artificial neural nets (ANNs) are restricted to process crisp data. Neuro-fuzzy systems [2][3][8] have drawn also increasing research interest in the recent years. A special focus is to develop generic-processing modules [6] for knowledge representation.

In this paper, connectionist models based on the formal neuron MAPI [14] are used to represent knowledge about organic compounds and to predict their toxicity. In section 2, the toxicity prediction and QSARs are emphasized. Section 3 reviews data analysis. Section 4 presents the results of the neuro-fuzzy knowledge models applied to toxicity prediction, obtained with NIKE (Neural explicit&Implicit Knowledge inference systEm), a hybrid intelligent system (developed by the author in Matlab 6R12, © The MathWorks, Inc.) based on modular neural/ neuro-fuzzy nets [5][9][11][15]. The results show that, combining experts on the same problem will gain better predictions against toxicity (section 5). The paper is ending with conclusions and ideas on future work

2. Data Description

The U.S. Environmental Protection Agency [16] provided to build up a data set referred to acute toxicity 96 hours (LC_{50}) for fathead minnow (*Pimephales promelas*). The data set contains 568 organic compounds, commonly used in industrial processes. This is a large set of compounds belonging to different chemical classes: a positive characteristic is the homogeneity and reliability of this toxicological data. A large number of descriptors was calculated by Istituto di Ricerche Farmacologiche "Mario Negri" Milano, Italy.

2.1. Molecular Descriptors

The organic compounds are defined by descriptors, classified [7] in: *constitutional descriptors*, depending on the number and type of atoms, bonds and functional groups; *geometrical descriptors*: molecular surface area and volume, shadow area, projections and gravitational indices, moments of inertia; *topological descriptors*: molecular connectivity indices related to the degree of branching in compounds; *electrostatic descriptors*: partial atomic charges depending on the possibility for some sites in molecule to form hydrogen bonds; *quantum–chemicals descriptors*: total energy of the molecule, energies of the lowest unoccupied and highest occupied molecular orbital (*HOMO* and *LUMO*), ionisation potentials, heat of formation; *hydrophobic descriptors* (*logD*: the expression of lipophilicity of the molecule at various pH, *logP*: the octanol-water partition coefficient).

2.2. The QSAR Approach

The quantitative structure-activity relationship method has been applied to chemical design and toxicity prediction tasks. Finding a QSAR is essentially a regression process and, historically, linear regression methods have been used. For the current set, three original QSAR equations, developed at Istituto "Mario Negri", using various descriptors, were considered, using five descriptors (QSAR1, 2), and two descriptors (QSAR3):

$$\log(1/LC_{50}) = 0.7919 + 0.09772*QM6 - 0.2045*C35 + 0.1276*G2 - 0.3509*pH9 - 0.3879*logP \tag{1}$$

$$\log(1/LC_{50}) = 0.8779 + 0.1385*QM6 - 0.06703*C35 - 0.02937*T6 - 0.06165*G12 - 0.6854*logP \tag{2}$$

$$\log(1/LC_{50}) = 0.8237 + 0.1711*QM6 - 0.7974*logP \tag{3}$$

where *QM6* is LUMO, AND *LC₅₀* is expressed in mmoles/l. The QSARs performance reaches accuracy values (for an absolute error of 0.1) around 70%.

3. Data Analysis

The inputs and the output were fuzzified with respect of the 568 organic compounds descriptors values. The input data set consists of 17 descriptors: *QM1* (Total Energy-kcal/mol), *QM3* (Heat of Formation-kcal/mol), *QM6* (LUMO-eV), *C9* (Relative number of N atoms), *C24* (Relative number of single bonds), *C35* (Molecular weight), *T6* (Kier&Hall index order 0), *T22* (Average Information content order 1), *G2* (Moment of inertia B), *G10* (Molecular volume), *G12* (Molecular surface area), *E13* (TMSA Total molecular surface area), *E24* (FPSA-2 Fractional PPSA: PPSA-2/TMSA), *E28* (PPSA-3 Atomic charge weighted PPSA), *E31* (FPSA-3 Fractional PPSA PPSA-3/TMSA), *pH9* (logD), logP, while the output is toxicity: $\log(1/LC_{50})$. The membership functions followed a trapezoidal fuzzification. The linguistic variables considered for descriptors, and for toxicity, are characterized by the term sets:

$$D_i = \{Low, Med, High\}, i = 1..17 \tag{4}$$

$$\log(1/LC50) = \{VeryLow, Low, Medium, High, VeryHigh\} \tag{5}$$

Five levels of toxicity are defined for the normalized $\log(1/LC_{50})$: *VeryLow* (0-0.2), *Low* (0.2-0.4), *Medium* (0.4-0.6), *High* (0.6-0.8), and *VeryHigh* (0.8-1). The membership functions shapes could be finally chosen from the list of: *Bell, Gaussian, Pi, S, Z, Triangular, Trapezoidal*, and *Sigmoidal*.

The set of available patterns was divided in two independent subsets, the training and testing data. A pattern is defined as a vector of input values (descriptors) and output values (toxicity). The training set was used to adjust neural and neuro-fuzzy network connections with the backpropagation [15] (*traingdx*) algorithm: this function updates weight and bias values according to gradient descent momentum and to adaptive learning rate. The data set (568 compounds) was randomly divided, saving the distribution of the five fuzzy values of the output. The algorithm is a 70-30 partitioning: 401 training and 167 testing cases (Fig.1).

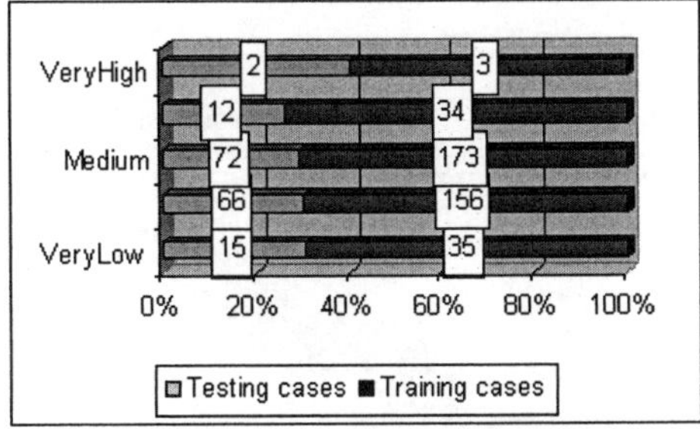

Fig. 1. The distribution of testing and training sets.

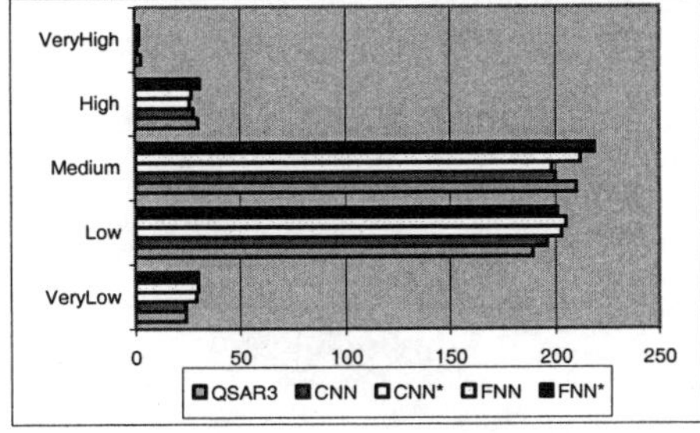

Fig. 2. Accuracy prediction for IKMs: number of cases predicted with absolute error lower than 0.1.

4. Hybrid Neuro-Fuzzy Structures for Toxicity Representation

NIKE automates the tasks, from data representation, to toxicity predicting for a given new input. It also suggests how the fuzzy inference produced the result, when required [10][11]. The *implicit knowledge* is the knowledge represented by neural/ neuro-fuzzy networks, created and adapted by a learning algorithm. The *explicit knowledge* is defined as the knowledge base represented by neural networks, computationally identical to the I/O relations set, and created by mapping fuzzy rules in hybrid neuro-fuzzy nets.

4.1. The Knowledge Representation Modules

IKM-CNN (Implicit Knowledge Module-based on Crisp Neural Net) takes charge of modeling the data set as a multilayer perceptron (MLP [15]). The MLP model is also used to compare the overall performance of the neuro-fuzzy and QSAR approaches [12].

IKM-FNN (Implicit Knowledge Module-based on Fuzzy Neural Net) is a multilayered neural structure with an input layer, performing the membership degrees of the inputs, a three-layered FNN2 [3], and a defuzzification layer. The weights of the connections between layer 1 and layer 2 are set to 1. A linguistic variable X_i (descriptors or toxicity) is described by m_i fuzzy sets, A_{ij}, the degrees of membership being performed by $\mu_{ij}(x_i)$, $j=1,2,...,m_i$, $i=1,2,..,p$. The layers 1 and 5 are used in training and prediction to fuzzify; the layers 2-4 are organized as a MLP to aquire implicit rules by FNN training [3][8][6].

Two steps were used to insert QSAR in IKM. The strategy follows *concept support techniques* [13], used to insert a priori knowledge [14]. The pre-training phase uses a data collection generated by a selected QSAR function. Then the model is re-trained with the original data set. The results are compared with the normal training procedure (random initial weights). The best structures resulted by QSAR insertion in a pre-training phase were retained for further combination of modules. The results in prediction accuracy are better than QSAR ones (Fig. 2). We used QSAR2 outputs, before learning the original patterns. An example of the improved result is given in Fig. 3a,b, and the performance for trained FNN, FNN* is shown in Fig. 3c,d (CNN*, FNN* define pre-trained/retrained nets).

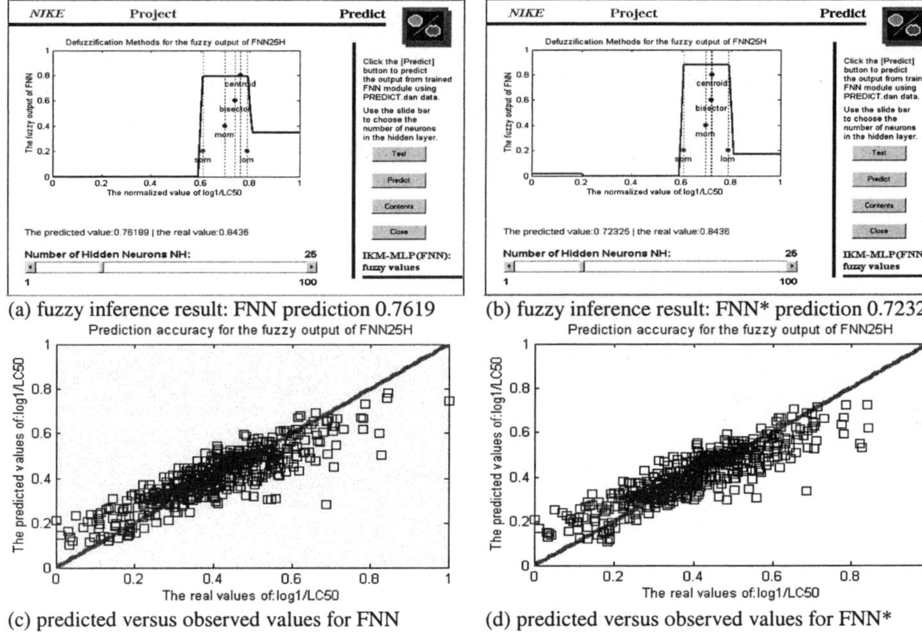

(a) fuzzy inference result: FNN prediction 0.7619

(b) fuzzy inference result: FNN* prediction 0.7232

(c) predicted versus observed values for FNN

(d) predicted versus observed values for FNN*

Fig. 3. FNN comparison during the application of the concept support technique.

The capabilities of MAPI-based neural nets to perform fuzzy computing [14] are used to implement EKM equivalent to QSARs [11][17]. The neural inference is accorded to multiple premises fuzzy rules systems and the extended version of Modus Ponens [18].

4.2. Fire Each Module Strategy: the Integration of the Developed Structures

Fire Each Module Strategy (FEM) is the simplest mode to integrate IKM and EKM by fuzzy processing using assemblies of ANNs as modular structures [5]. After off-line training applied to neuro-fuzzy modules, the general output of the system is composed as a T-conorm [10][18] of the fuzzy outputs of CNN*, FNN* (IKM), and QSAR2, QSAR3 (EKM). The output of the system is the averaged output of the modules (Table 1, Fig. 4, 5).

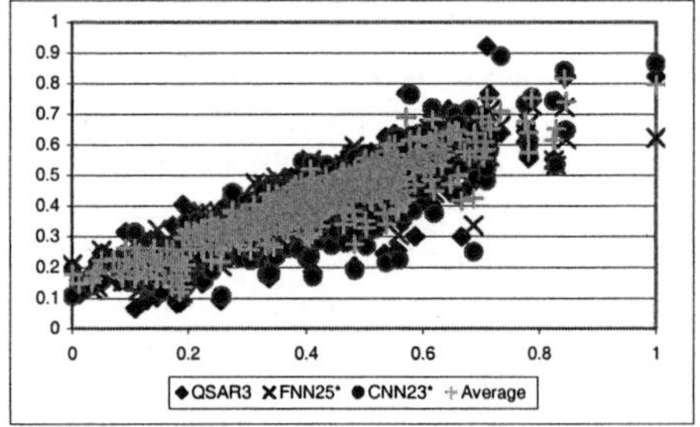

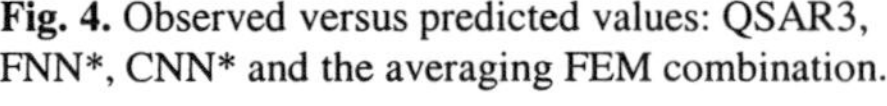

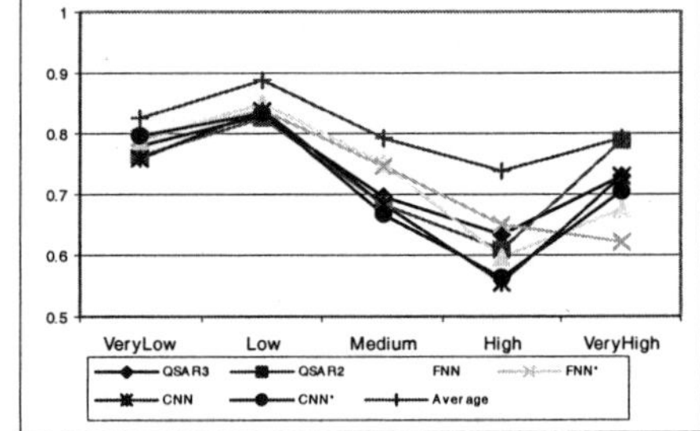

Fig. 4. Observed versus predicted values: QSAR3, FNN*, CNN* and the averaging FEM combination.

Fig. 5. The accuracy of toxicity prediction, by classes (fuzzy values).

Table 1. General results for accuracy of prediction (absolute error <0.1).

	QSAR2	QSAR3	CNN	CNN*	FNN	FNN*	FEM
well predicted cases	447	456	449	458	476	481	507
accuracy of prediction	78.69%	80.28%	79.05%	80.63%	83.80%	84.68%	89.26%

5. Conclusions

Our study evaluates the representation of the toxicity of industrial organic compounds as neuro-fuzzy knowledge, combining ANNs and QSARs. The proposed approach gives an encouraging alternative to the stochastic models: these kind of models are able to learn from large collections of descriptors about industrial organic compounds and are capable of representing QSARs to improve the prediction results. An important feature of our model is its validation with a large test set, showing certain abilities to generalize.

Further, ANNs can be used to predict toxicity, and to classify under a toxicity scale. This offers the possibility of improving the performance using neuro-fuzzy combined modules to guide and refine further decisions. We also evaluated the use of assemblies of ANNs to improve the overall results. Some improvement was found, since explicit knowledge was inserted in the connectionist system. This suggests that the 568 training samples used in this study do not provide the best coverage of the problem domain, which is split in contiguous sub-domains. Future work will be carried out following new possibilities to integrate implicit knowledge with explicit QSARs into the hybrid system NIKE.

Acknowledgments. This work is partially funded by the E.U. HPRN-CT-1999-00015, and was done in the period of the post-doctoral research fellowship at Politecnico di Milano, Italy. Special thanks for continuous support to Emilio Benfenati and Giuseppina Gini.

References

[1] R. Adamczak, W. Duch, Neural networks for structure-activity relationship problems, in Proceedings of the 5th Int'l Conference on Neural Networks and Soft Computing, Zakopane, 2000, pp. 669-674.

[2] J.J. Buckley, and Y. Hayashi, Neural nets for fuzzy systems, *Fuzzy Sets and Systems* **71** (1995) 265-276.

[3] R. Fuller, Introduction to Neuro-Fuzzy Systems, Adv. Soft Computing, Springer-Verlag: Berlin, 1999.

[4] G. Gini, Predictive Toxicology of Chemicals: Experiences and Impact of AI Tools, *AI Magazine*, AAAI Press **21/2** (2000) 81-84.

[5] R.A. Jacobs, M.I. Jordan, and A.G. Barto, Task de-composition through competition in a modular connectionist architecture: The what and where vision tasks, *Cognitive Science* **15** (1991) 219–250.

[6] J.-S. R. Jang, C.-T. Sun, E. Mizutani, Neuro-Fuzzy and Soft Computing. Prentice-Hall, NJ, 1997.

[7] A.R. Katritzky, V.S. Lobanov, M. Karelson, CODESSA Comprehensive Descriptors for Structural and Statistical Analysis. Reference Manual, version 2.0, Gainesville, 1994.

[8] C.T.Lin, C.S.Lee, Neural Fuzzy Systems: A Neuro-Fuzzy Synergism to Intelligent systems. Prentice-Hall, NJ: 1996.

[9] V. Palade, Ron J. Patton, F.J. Uppal, J. Quevedo, S. Daley, Fault diagnosis of an industrial gas turbine using neuro-fuzzy methods, Procs. of IFAC 15th World Congress, Barcelona, July 2002 (in printing).

[10] C-D. Neagu, V. Palade, An interactive fuzzy operator used in rule extraction from neural networks, *Neural Network World Journal*, **10/4**, 675-684, (2000).

[11] C.-D. Neagu, N.M. Avouris, E. Kalapanidas, V. Palade, Neural and Neuro-fuzzy Integration in a Knowledge-based System for Air Quality Prediction", *Applied Intelligence Journal* **17** (2002).

[12] C.-D. Neagu,, A.O. Aptula, G. Gini, Neural and Neuro-Fuzzy Models of Toxic Action of Phenols, Procs. IEEE IS2002 (accepted 2002).

[13] E. Prem, M. Mackinger, G. Dorffner, Concept Support as a Method for Programming Neural Networks with Symbolic Knowledge, Tech. Rep. OEFAI04/1993, Austrian Research Inst. for AI 1993.

[14] A.F. da Rocha, Neural Nets: A Theory for Brains and Machines, LNAI, Springer-Verlag, 1992.

[15] D.E. Rumelhart, and J.L. McClelland, Parallel Distributed Processing, Explanations in the Microstructure of Cognition, MIT Press, 1986.

[16] C.L. Russom, S.P. Bradbury, S.J. Broderius, D.E. Hammermeister, R.A. Drummond. Predicting modes of toxic action from chemical structure: acute toxicity in the fathead minnow (*Pimephales promelas*), *Environ. Toxicol. Chem.* **16** (1997).

[17] M. Sugeno, G.T. Kang, Structure identification of fuzzy model, *Fuzzy Sets and Systems* **28** (1988)15-33.

[18] L.A. Zadeh, The role of fuzzy logic in the management of uncertainty in expert systems, *Fuzzy Sets and Systems* **11/3** (1983) 199-227.

KES 2002
E. Damiani et al. (Eds.)
IOS Press, 2002

Hybrid Intelligent Production Simulator and its If-then Rules Acquisition by GA

Hidehiko YAMAMOTO* and Etsuo MARUI**
**Dept. of Human and Information Systems, Faculty of Engineering, Gifu University*
***Dept. of Mechanical and Systems Engineering, Faculty of Engineering, Gifu University*
1-1, Yanagido, Gifu-shi, 501-1193, JAPAN
**e-mail yam-h@cc.gifu-u.ac.jp*

Abstract. The authors attempt to develop a hybrid intelligent off-line production simulator that assists the production engineers' decisions. The simulator consists of if-then rules, GA and discrete simulator. After a number of operations, improved if-then rules are acquired. The acquired parts input sequence can be used in starting the new FTL operation.

1. Introduction

A large-scale Flexible Transfer Lines (FTL) operates by linking with several production lines. FTL is a production system for flexibility due to model change and composed of special machine tools and automatic loading systems and conveyers. In this kind of a large-scale FTL, it is often happened that the downstream line takes parts, one-by-one, which the downstream line wants from the product conveyer of the upstream line. This system is called a pull production. The FTL which uses a pull production will make a line stop if the parts variety that the downstream line demands is not on the product conveyer. In other words, in the system of a pull production, an upstream line has to give the parts variety that the downstream line wants quickly, without waiting. Whenever the downstream line demands, the upstream line has to keep the parts variety that the downstream line demands on the product conveyer. In order to do this, the upstream line must be attentive to the downstream line's demands and input the parts variety into the upstream line's bay.
The object production line of this research is the FTL mentioned above, that is, the FTL to operate linking two production lines, an upstream line and a downstream line under the above pull production. This paper describes the hybrid intelligent production simulator for the FTL. The simulator has if-then rules, discrete simulatior and GA (Genetic Algorithm)system and can acquire if-then rules by GA. The rules decides what part is put into the bay of the upstream line in the production simulator.
The research to formularize the parts input problems and to optimize by using Simulated Anealing Method or Linear Programming Method has been done. The FTL that our research struggles with is an automated production line whose characteristics are parts are ejected in progress and line stops usually occurs because of machine tools' breakdown, as we can often see in automotive and electric appliance production lines. It is difficult to formularize this kind of production behavior. In this research, the behavior is expressed with a discrete simulation model.
Also, intelligent scheduling systems to automatically acquire scheduling knowledge have been developing. For example, K.Miyashita makes a rule selection by case-based reasoning [1] and S.Fujii acquires rules by ID3 learning [2][3]. Other knowledge acquisition research with GA such as Michigan Approach has been done [4][5]. It classifies knowledge with a specifying set of if-then rules. The present paper's research is different from the

above because better knowledge acquisition can be done by improving parameters of if-then rules through GA generation evolutions.

2. Object production line

The object FTL of the research is the production line to operate by linking with two kinds of production lines, the upstream line (Up-line) and the downstream line (Dw-line), under the Toyota Kanbann System or Just In Time System(JIT). Up-line inputs k kinds of parts one-by-one into the bay of the Up-line and has the product (finished parts) conveyer to form each queue of the variety. The defective parts and the parts to need re-machining are ejected at optional locations in Up-line. According to the downstream line's instructions, the finished parts of the upstream line are moved to the downstream line one-by-one from the product conveyer.

A(1)	f_1
A(2)	f_2
A(k)	f_k

IF: Ratio for parts A(1) $\geq$ X(1)%
THEN:Rank value of parts A(1) is P(1,1)
IF: X(1)%> Ratio for parts A(1) $\geq$ X(2)%
THEN: Rank value of parts A(1) is P(1,2)
:
IF: X(m-2)%> Ratio for parts A(1) $\geq$ X(m-1)%
THEN: Rank value of parts A(1) is P(1,m-1)
ELSE: Rank value of parts A(1) is P(1,m)

Fig.1 *Conveyer situation example*   **Fig.2** *If-then rules examples*

3. Acquision of parts input rules by GA

3.1 If-then rules

The production simulator of FTL we are developing uses the feedback information that tells something to the bay of the Up-line by judging some or other production results. The method adopted as judging information about the parts input is the queue behavior of the finished parts, which mostly reflects the demands of the Dw-line.

The method expresses the relation between the queue behavior of the finished parts and the parts input and selects the input parts by matching the queue behavior and if-then rules. However, there's a problem : actual experienced production engineers do not have the knowledge indicating the relation between the two. In order to solve the problem, KAPS (knowledge acquiring-type production simulator) by automatically acquiring the better if-then rules is developed and the input parts sequence is decided. KAPS has three elements: [1]if-then rules that give the bay of Up-line the behavior on the product conveyer as feedback information; [2]a GA system which considers the if-then rules as individuals; [3]A simulator which calculates the fitness of the GA. Through a lot of generations' evolutions, if-then rules are improved.

3.2 Rank values

Actually, when the parts input operations are carried out, an operator uses if-then rules such as "if a conveyer queue behavior is ***, then input part A". Basically, the if-then rules of KAPS are the same. The parts content ratio indicating parts' queue behavior is described as an if-part. Although it is good that the then-part is expressed with input parts names, there is not a simple relation such that we can determine input parts from the parts content ratio. Without giving direct descriptions of input parts, the rank values indicating the

priority of input order is set up and is adopted as substitute knowledge of then-part, e.g., "if the content ratio of part A is X %, then the rank value of part A's input order is P". This rank value is the constant indicating that the bigger the value, the higher the priority.
Here, if the parts queue behavior on the product conveyer for k parts is $f_1, f_2, f_3, \ldots$, as shown in *Fig.1*, the content ratio $X(2)$ of part A(2) is calculated by the equation (1).

$$X(2) = \frac{f_2}{f_1 + f_2 + \ldots + f_k} \qquad (1)$$

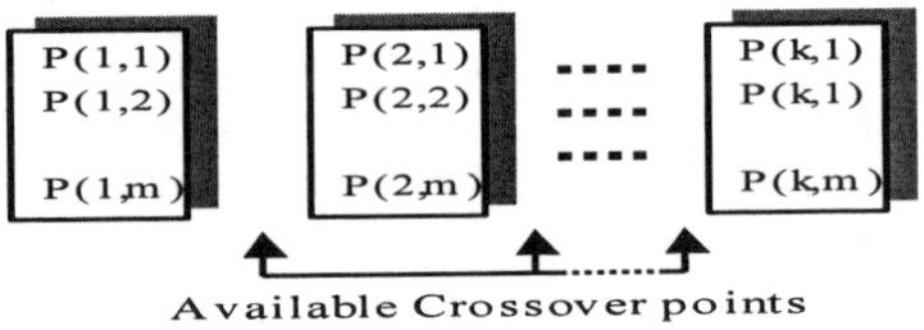

P(1,1)	P(2,1)		P(k,1)
P(1,2)	P(2,2)		P(k,1)
P(1,m)	P(2,m)		P(k,m)

Available Crossover points

Fig.3 Individual matrix example

As the possible content ratio for one kind of part is infinite, the number of rules becomes enormous if direct if-then rules are described. In order to prevent this, KAPS expresses if-part with a content ratio expressing a range, e.g., "if content ratio of part A is from X_1% to X_2%, … ". If the classification number of a range expression is a limited number m, m if-then rules for k kinds of parts, $A(1), A(2), \ldots, A(K)$ can be expressed. The number of if-then rules becomes the total of $k \times m$. If m is not so big, it is possible to prevent a staggering number of if-then rules.

Fig.2 shows the m kinds rule's example for part $A(1)$. In *Fig.2*, $X(m)$ indicates the content ratio and its size relation is $X(m-1)<X(m)$. The rank values are expressed with a variable $P(k,m)$ by using parts variety, k, and classification number, m. In this way, the rank value $P(k,m)$ is chosen by GA while $X(m)$ is acquired.

3.3 How to acquire rank values in rules

The rank value $P(k,m)$ described in *section 3.2* can not be decided by experienced engineers. The if-then rules cannot be adopted as reasoning tools until the rank values are decided as fixed values. KAPS fixes the rank values by using a GA as below.
In order to carry out GA operations, the expression for individuals must be defined. KAPS considers a group of k sets that consists of a set of rank values for one kind of part as an individual (Individual I). That is, Individual I is expressed with equation (2) and the gene number of an individual is (parts type, k)×(classification number, m). As defined above, an individual means a prospective set of the rank values that a part can take and, in the later process, the suitable rank value will be chosen from among the prospective set. For example, equation (2) indicates that the prospective set that the parts kind k can take are M pieces from $P(1,1)$ to $P(1,M)$. M is the maximum number of m.

$$I = \{P(1,1), P(2,1), \ldots, P(k,1), P(1,2), \ldots, P(k,2), \ldots\ldots, P(1,M), \ldots P(k,M)\} \qquad (2)$$

A set where many Individual Is are gathered corresponds to an individuals' set for a generation and the set joins GA operations such as crossover and mutation.
In a crossover operation, the column elements in expressing the elements of an individual as a matrix is considered as the minimum separable unit and a one-point crossover is carried out. This individual's matrix expression tells that the left numbers of elements $P(k,m)$ indicate a column and the right ones indicate a row. The individual of equation(2) is expressed as the matrix of *Fig.3*. By the one-point crossover considering column as the minimum separable unit, the biggest number of minimum separable units is considered as k and the number of their spaces, $(k-1)$, are selected as the object of a crossover point. After

randomly selecting one-point from among *(k-1)* object crossover points, the front unit is exchanged for the front unit of another object individual.

Mutation is also done randomly to select a unit from among k pieces of minimum separable units and to reproduce the values in the unit by the algorithm mentioned above.

3.4 Match if-then rules with ratio

The parts input of KAPS is decided by matching rank values and the if-then rules mentioned in 3.2 and 3.3. That is, (1) observe the content ratio for each part, (2) match the facts of the observation with if-then rules, (3) select the parts corresponding to the maximum rank value as an input part.

For example, consider parts *A(k)* and if-then rules as shown in *Fig.4*. The column of *Fig.4* shows *M*, the classification number of a range expression, the row shows a variety *K* and *P(k,m)* means the rank values for *k*.

Consider the situation that the content ratio for part *A(1)* is observed as *X(α)*in the process (1) and the matching result in the process (2) is *P(1,g₁)*. In the same manner, the matching results of each *A(2)~A(k)* are *P(2,g₂)*, *P(3,g₃)*, ... , *P(K,gₖ)*. The bold words in *Fig.4* correspond to these. In the process step (3), the part A_{max} that has the maximum rank values, *Max(P,(1,g₁)*, *P(2,g₂)*, ..., *P(K,gₖ))* is searched. The part A_{max} is the next input one.

In this way, KAPS decides the input parts by selecting the rank values representing input priority for each part from among a set of prospective rank values and searching for the parts whose rank value is the maximum among the selected values.

3.5 Rules acquision by GA

KAPS decides the input parts by using if-then rules with the method mentioned in 3.4. However, if the if-then rules used here are not correct, that is, each value of rank values is not correct, simulation results will not be good. In order to find good rank values, KAPS improves if-then rules by GA operations and simulations' repetitions.

The rules' revisions are carried out with the four processes, as shown in *Fig.5*, : [1] generate the initial rank values in if-then rules by GA; [2] carry out production simulation with the rules; [3]evaluate the result of the simulations; [4] reproduce the next generation's if-then rules by GA. The cycle repetitions from [2] to [4] makes an automatic improvement of rank values and thus better if-then rules can be acquired.

4. Application examples

Simulation examples with KAPS were carried out. The object production system has two kinds of FTL: an upstream production line and a downstream production line. The production lines are linked to each other and manufacture 10 kinds of parts. The simulation conditions of the examples are shown in *Table1*.

The simulation examples consider some breakdowns to happen randomly. The reciprocal method of production ratio is adopted as the tentative parts input order for the upstream line. In order to have several patterns of breakdowns, ten random series*V*, *V=1~10*, are adopted and simulations are carried out by changing the random series. *Table2* shows the maximum criterion value acquired for each random series *V*. As a comparison, the method to freeze input order was carried out. That is, with a fixed input order method using the reciprocal method of production ratio, simulations for each random series were carried out. The criterion value of the condition is shown in the left row of *Table 2*. The criterion value results of KAPS are better than those of the fixed input order judging from *Table 2*.

5. Conclusions

When a FTL that operates by linking the upstream line with the downstream line is developed, we need to have a design process to decide the parts input order for the FTL. This paper's research deals with the development of a hybrid intelligent production simulator KAPS and the method to automatically acquire the rules by GA.

KAPS consists of if-then rules whose then-part includes the rank values indicating the parts input order, a GA system and a discrete simulator. Through the generations' evolution by simulations, the rank values in the if-then rules, for example, IF: Ratio for parts $A(1) \geq X(1)\%$,THEN:Rank value of parts $A(1)$ is $P(1,1)$ are improved and , as a result, better if-then rules can be acquired. By comparing the simulation results of KAPS with those of the ordinal method, it is ascertained that KAPS is useful. Because of the evaluation, it is also acsertained that the new if-then rules acquiring method we developed is useful.

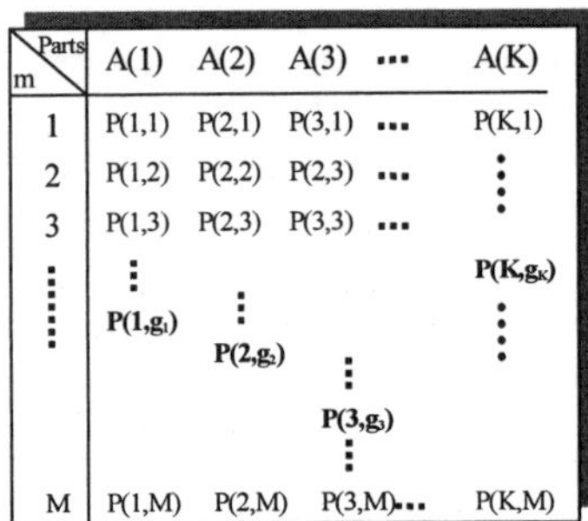

Fig.4 If-then rules

Fig.5 Rule acquisition cycle

Table 1 Production conditions

Variety	k=10,A(1),A(2),...,A(10)
Repair time	ts=20sec
Defective unit probability	Q=5%
Stock quantity	S=7
Machining time	Ct=15sec
Repair time	ta=20sec
Simulation time	10H=36,000sec
Goal production ratio	A(1): A(2): A(3): A(4): A(5): A(6): A(7): A(8): A(9): A(10) =9:5:7:6:7:6:3:2:4:1
Population size	200
Elite preservation number	6
Cross-over probability	92%
Mutation probability	5%

Table 2 Best fitness

	Conventional System	KAPS
Series1	0.990847	0.50868
Series2	0.903760	0.069488
Series3	1.037016	0.053628
Series4	1.382990	0.045593
Series5	1.199286	0.037231
Series6	0.822987	0.058836
Series7	0.962088	0.072083
Series8	0.782835	0.083705
Series9	0.841625	0.057635
Series10	0.952000	0.075315

References

[1] K.Miyashita and K.P.Sycara, A Framework for Case-based Revision for Scheduling Generation and Reactive Schedule Management, Journal of Japanese Society for Artificial Intelligence, Vol.9-3(1994), 426.

[2] J.R.Quinlian, Introduction of Decision Tree, Center for Advanced Computer Sciences Machine Learning, (1987), 81.

[3] S.Fujii et. al.,Introduction of Scheduling Rules by Pairwise Job Interchange, Transactions of SICE, Vol.32, No.2(1996), 261.

[4] J.Holland and J.Reitman, Cognitive Systems Based on Adaptive Algorithms, Pattern Directed Inference Systems, Academic Press (1987), 313.

[5] J.Greffenstte et al., Learning Sequential Decision Rules using Simulation Models and Competition, Machine Learning (1990).

SAMIR: An Intelligent Web Agent

Fabio ABBATTISTA, Pasquale LOPS, Giovanni SEMERARO and Fabio ZAMBETTA

Dipartimento di Informatica, Università di Bari, Via E.Orabona 4, Bari, Italy

Abstract. A modern key issue in the web community is that a web site *must* be equipped with a virtual agent able to support users in a natural way. Following this trend, we decided to implement SAMIR (Scenographic Agents Mimic Intelligent Reasoning), a prototype of a tool for animating 3D intelligent agents, mainly founded on a genetic algorithms based learning system, namely an XCS. SAMIR adopts the MPEG-4 Facial Animation Parameters as a description format for facial expressions and animations.

1. Introduction

Men's life is rapidly shifting towards a "new virtual dimension", compared to the more traditional physical one, where users interact and communicate in ways made possible by e-mail, web, and chat technologies.

Many efforts are currently spent to develop convincing intelligent autonomous personalities enriched by characters with typical human-like expressions and features [3], in order to provide web users with more interactivity and easiness of use.

This paper presents SAMIR (Scenographic Agents Mimic Intelligent Reasoning), a prototype of a tool for animating 3D intelligent agents, software entities that perform a set of tasks on behalf of a user with some degree of autonomy, coupled with a 3D graphical layout. The paper is organized as follows. The architecture of the SAMIR system is introduced in the next section whilst sections 3 to 5 detail the modules which constitute SAMIR. Experimental results are reported in section 6. Finally, section 7 addresses our future work.

2. The SAMIR System

The SAMIR system is a tool for designing a personal digital assistant where an intelligent web agent is integrated with a purely 3D humanoid, robotic, or cartoon-like layout [1].

The architecture of SAMIR is depicted in Fig. 1.

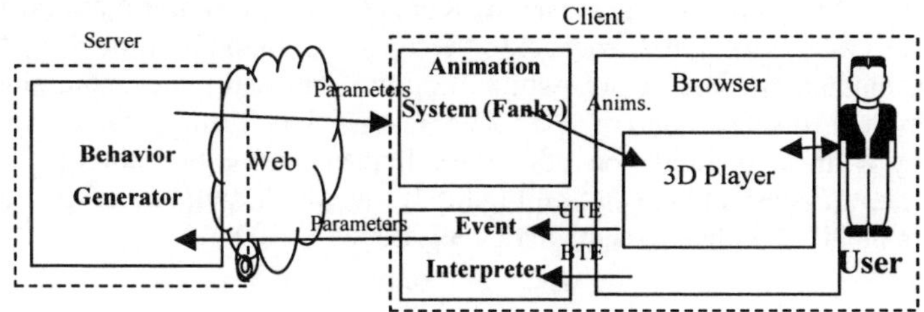

Fig. 1. SAMIR system architecture.

3. The Event Interpreter

The Event Interpreter is the module which converts all the events occurring in the course of the user interaction into parameters. These events have been classified in:
- User Triggered Events (UTE), corresponding to the dialogue log between the user and the system.
- Browser Triggered Events (BTE), representing user choices and/or commands sent to the browser.

The Event Interpreter module extracts information to be processed by the Behavior Generator module. It relies on the A.L.I.C.E. chatterbot [11] to let users chat with SAMIR, using the knowledge of the system stored into its AIML (Artificial Intelligence Markup Language) files (see the URL http://alicebot.org/TR/2001/WD-aiml).

4. The Behavior Generator

The Behavior Generator module aims at improving the naturalness of the dialoguing agent. Two main goals of the Behavior Generator are: i) To extract profiles of the users that have access to the system according to their technical skills; ii) To display suitable facial expressions according to the context of the current dialogue.

The Behavior Generator is composed of two main sub-modules:
- the *User Classifier module*, whose aim is the automatic assignment of one out of some predefined classes to each user on the ground of information drawn from previous sessions [2]. This module induces a decision tree and/or a set of rules used to perform a classification of the user interacting with the web site (see [8] for details of experimentations). User classification results are stored in a repository of user profiles and serve to personalize the interaction with the corresponding user.
- the *User Interaction module*, whose aim is to manage the consistency between the facial expression of the character and the conversation tone. The module is mainly based on Learning Classifier Systems (LCS), a machine learning paradigm introduced by Holland in 1976 [5]. The learning module of SAMIR has been implemented through an XCS [13], a new kind of LCS which differs in many aspects from the traditional Holland's framework. The most appealing characteristic of this system is that it is very related to the Q-learning but it can generate task representations which can be more compact than tabular Q-learning [10]. At discrete time intervals, the agent observes a state of the environment, takes an action, observes a new state, and, finally, receives an immediate reward.

In the User Interaction module, rules are expressed in the classical form *if* <condition> **then** <action>, where <condition> (the state of the environment) represents a combination of 8 possible events, sensed by 8 *effectors,* such as user classification (*don't care, novice, teacher, expert*), user request to the agent, operation performed by the user in the web site, amount of user errors, while <action> represents the expression that the Animation System displays during user interaction. Specifically, each expression is represented by a linear combination of the six fundamental expressions [4] - *surprise, sadness, joy, fear, disgust* and *anger* (see Figure 2) - in which each one of the fundamental expressions is included with a percentage ranging from 0 to 100.

5. The Animation System

Our animation module, named Fanky [7], meets some necessary requirements: effectiveness, reliability, efficiency and MPEG-4 compliance [12].

The object-oriented model we propose is partially based on Paradiso et al.'s Tinky [6], but we improved it by introducing the concept of Standard Anatomical Component (SAC). A SAC is a facial region identified by its own methods and properties, including skin elasticity, material, and mesh topology links.

The result of a continuous process of animating the face via MPEG-4 Facial Animation Parameters (FAP) streams is rendered and shown in a 3D player embedded into a common web browser. More details about the animation system can be found in [14].

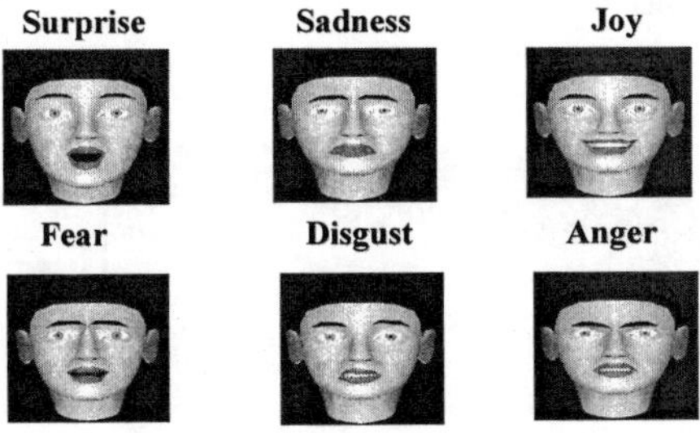

Fig. 2. The six fundamental expressions.

6. Experimental Results

Two different sets of experiments have been performed so far. In the first one, the goals were to stress the capability of the XCS component along the following dimensions: i) To verify the effectiveness of the XCS when learning a set of predefined interaction rules; ii) To verify the ability of the XCS when discovering new rules.

In a preliminary phase, we defined a set of 30 interaction rules that take into account a number of combinations of kinds of users (*novices*, *experts*, etc.) with several situations that may occur in the course of an interaction (performed actions, user requests and user errors). This set of predefined rules represented the training set, that is the minimal know-how that SAMIR should possess to start its *work* in the web site.

To evaluate the performance of the system during the training phase, the following procedure has been adopted for the credit assignment:

If system action **matches** expected action **then** credit=100
Else credit=k/d(system action, expected action).

where k is a constant and the distance measure d is defined as the mean squared distance among the 6 components of the expression (*surprise %, sadness %, joy %, fear %, disgust %, anger %*).

The experiments aim at verifying the effectiveness of the User Interaction module to learn the 30 predefined rules. We performed 5 runs of the system with the same parameters (see Table 1) but with different initial populations. The values of the parameters have been empirically selected. Figures 3 and 4 show the performance of the system for each run. On average, the system was able to learn 26 out of 30 rules. The 4 unlearned rules were, in most cases, very close to the original predefined ones. The most interesting result has been the ability of the system to discover new rules, not included in the original set of 30 rules.

For the sake of brevity, we omit to report the whole set of learned rules. Table 2 shows only some rules learnt by the system. The second and third columns represent, respectively, the input to the system (the *<condition>* part of the rule) and the expected output (the *<action>* part of the rule). The fourth and fifth columns show the system output and the corresponding *<condition>* part, respectively.

Table 1. The parameters used in the first experiment.

Parameter	Value
Maximum Population Size (N)	24000
Crossover Probability (Px)	1.0
Mutation Probability (Pm)	0.03
Learning Rate (β)	0.1
'Don't care' symbol Probability (P#)	0.05

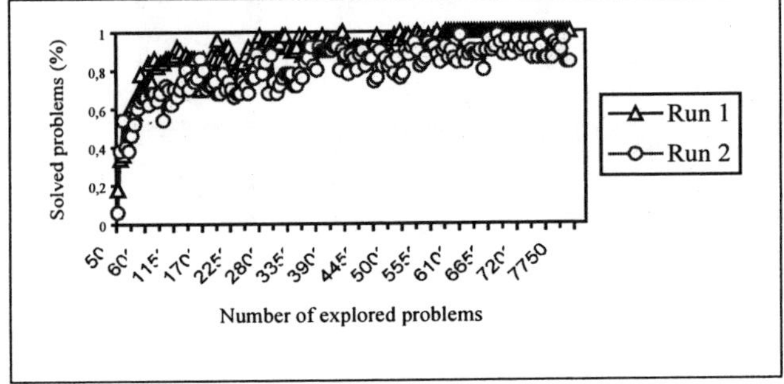

Fig. 3. Performances of the XCS in the first 2 runs of the experimentation.

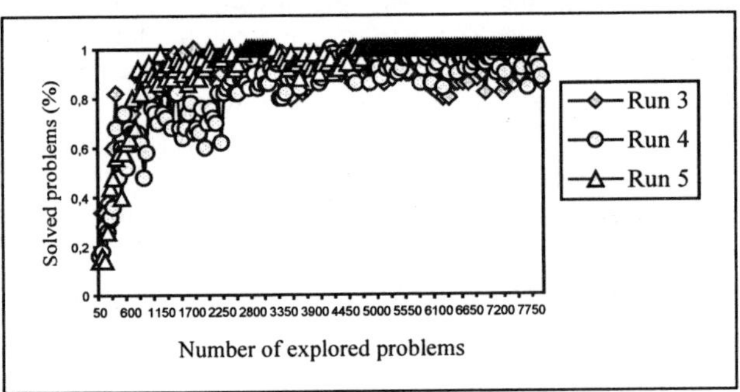

Fig. 4. Performances of the XCS in the last 3 runs of the experimentation.

The first row in Table 2 represents a perfect match between the expected action and the one performed by the system. In the second row, the rule does not represent a perfect match but the two actions (the expected one and the action performed by the system) are quite *similar*. The third rule represents a new rule, discovered by the system.

Figure 5 shows the expressions corresponding to the action part of the last two rules in Table 2, as displayed by the Fanky module.

Due to the inherent features of XCS, SAMIR has been able to learn quite effectively the predefined rules of behavior and to generalize some new behavioral patterns that could update the initial set of rules. Along these dimensions, SAMIR is comparable with a human assistant that, after a preliminary phase of training, continues to learn new rules of behavior on the ground of personal experiences and interaction with human customers.

Table 2. Results of the training phase

Rule	Input	Expected Output	System Output	System Condition
1	0001001001100	0 0 0 0 60 40	0 0 0 0 60 40	0#0#001001100
2	0001111011000	0 0 60 0 0 0	0 0 40 0 0 0	0001111011000
3	0101100001001 (A novice user makes few errors)	-	70 0 30 0 0 0	-

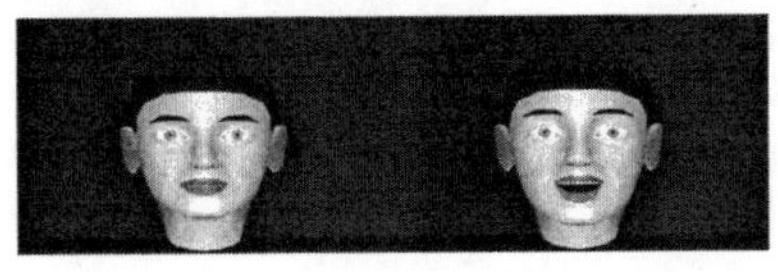

a)   b)

Fig. 5. Expressions corresponding to the last two rules in Table 2, as displayed by the Fanky module: a) Expression for rule 2; b) Expression for rule 3.

The second set of experiments aims at verifying the capability of the Event Interpreter module to dialogue with users and, consequently, their reactions in front of a digital chatter. A sample of 10 users was selected, with different skills and knowledge about our system. Users were asked to interact with SAMIR over a period of a week, and to report their comments and suggestions at the end of the experimentation. During the first 3 days, users were free to choose the topics of the conversation, in order to familiarize with the system. In the second part of the week, the topics of the conversation were imposed (*Computer Science, Gossip and Sport, Politics and Religion*). All the interactions have been recorded in order to analyze quantitative and qualitative factors.

6.1 Dialogues duration

Over the period of the experimentation, 1,096 dialogues were recorded (an average of 100 dialogues for each user). The dialogues length ranges from 20 minutes to 293 minutes and each user spent, on average, 70 minutes per day interacting with SAMIR. The mean length of the dialogues was 126 minutes over the free conversation period and 108 minutes over the guided conversation period.

6.2 Errors in the dialogues

We observed several situations in which SAMIR was not able to keep the dialogue going with the users. From the analysis of the log files, we classified these situations in 6 different categories. Table 3 shows the frequency of each category observed in the logs

Table 3. Frequency of the most common errors.

Category	Frequency
Quoted Question	29%
Fall-back	20%
Misunderstood Question	16%
Wrong Answer	15%
Evasive Answer	11%
Unknown topic	6%

6.3 User comments

At the end of the experiment, users were asked to fill-in a questionnaire to report their own opinions. Table 4 synthesizes how users evaluate their dialogues with SAMIR.

Table 4. User evaluation of the dialogues.

Evaluation	Frequency
Interesting	80%
Funny	80%
Ambiguous	80%
Intelligent	60%
Involving	40%
Sensible	40%
Real	30%
Up-to-date	30%

Moreover, from the analysis of the separate dialogue logs, both positive and negative characteristics emerged:

6.4 Positive characteristics

In some dialogues, the Event Interpreter module exhibits a sort of ironic behavior. As the following dialogue fragment shows, the module seems to dislike Bill Gates:

    User1>With whom you chat in this moment?
    Samir> Bill Gates,:-)
    User1>Is Bill Gates your Friends?
    Samir>No he is the president of Microsoft
    User1>Do you Like Microsoft?
    Samir>I want to replace Microsoft with Artificial Intelligence

In another situation the module has been able to defend itself from the critics of a user:

    User2>Hey baby...
    Samir>Don't call me "baby". Can I help you?
    User2>Yes... what are you?
    Samir>What can I do for you? I am the latest result in artificial intelligence, which can reproduce the capabilities of the human brain with greater speed and accuracy
    User2>Are you fast?
    Samir>I don't know whether or not I am fast. I am a giant electronic brain
    User2>You are stupid, in my opinion
    Samir>Hey, I'll tell my other robot friends you called me stupid
    User2>You are pretty dumb
    Samir>Message to all Robots: User2 called me dumb

In several situations the chatterbot gave some non-sense answer but, unexpectedly, users found it very nice and were more attracted by it.

6.5 Negative characteristics

The chatterbot is not able to manage indefinite pronouns, as the following dialogue fragment shows:

    User3>How are you?
    Samir>Everything is running smoothly
    User3>What are you doing?
    Samir>I'm chatting with people on the net
    User3>It is interesting. You like chatting with people on the net?
    Samir>"It" being what? Chatting is my job!

In very complex dialogues concerning many different interrelated topics, the chatterbot was not able to continue the dialogue and it often answered "I lost my train of thought...". This situation occurred more frequently when dialogues were on very specific topics like *Science, Religion* or *Sport*.

6.6 Final Remarks

The dialogues duration and the user comments seem to indicate that users did not dislike to interact with SAMIR and sometimes users were very interested in.

On average, the chatterbot gave 10% of erroneous answers. Very interesting was the analysis of the dialogues between the chatterbot and a mother tongue user. In this case, the chatterbot was able to keep longer dialogues going with the user. This is probably due to the lack of syntactical errors in user answers, that interrupt the fluency of the dialogue and, consequently, force users to repeat words or sentences.

However, a major weakness showed by SAMIR was the high level of ambiguity exhibited in the course of user interactions, that proved to be one of the main sources of misunderstandings during the dialogues.

7. Future work

The SAMIR prototype will be improved by adding several features, like speech capabilities synchronized with labial movements of the character, and by merging the visemes (the visual counterpart of phonemes) with concurrent facial displays.

Concerning the Event Interpreter, we intend to modify A.L.I.C.E. source code, distributed under GPL license (http://www.gnu.org/licenses/licenses.html#GPL), to support dynamic generation of AIML files.

Finally, we are applying the SAMIR system to an e-commerce web site developed within the EU funded project COGITO. Project COGITO aims at an agent-based interface for B-to-C e-commerce applications that is not merely re-active to some user request, but capable of engaging in a goal-directed conversation with the user, e.g., by taking the initiative to recommend products [9]. The SAMIR system will be used to improve the current capabilities of the chatterbot implemented in the COGITO system.

References

[1] F. Abbattista *et al.*, An Agent that Learns to Support Users of a Web Site, Procs of the 6th Online World Conf on Soft Comp in Industrial App., Session on Soft Computing for Intelligent 3D Agents, 2001.

[2] D. Banyon and D. Murray, Applying User Modeling to Human-Computer Interaction Design, *Artificial Intelligence Review* 7 (1993) 199-225.

[3] J. Cassell *et al.*, Embodied Conversational Agents. MIT Press, Cambridge, 2000.

[4] P. Ekman, Universals and cultural differences in facial expressions of emotion, Nebraska Symposium on Motivation (1972), 207-283.

[5] J.H. Holland, Adaptation, *Progress in Theoretical Biology* 4 (1976) 263-293.

[6] A. Paradiso *et al.*, The Design of Expressive Cartoons for the Web – Tinky, Proceedings of ICMCS Conference, IEEE Press, Florence, 1999, pp. 276-281.

[7] A. Paradiso *et al.*, Fanky: a tool for animating 3D intelligent agents. In: A. de Antonio, R. Aylett, D. Ballin, Intelligent Virtual Agents, Springer, Berlin, 2001, pp. 242-243.

[8] G. Semeraro *et al.*, Learning Interaction Models in a Digital Library Service. In: M. Bauer, P.J. Gmytrasiewicz, J. Vassileva (eds.), User Modelling 2001. Springer, Berlin, 2001, pp. 44-53.

[9] U. Thiel *et al.*, The COGITO Project: Intelligent E-Commerce with Guiding Agents based on Personalized Interaction Tools. In: J. Gasós, K.-D. Thoben (eds.), e-Business applications: results of applied research on e-Commerce, Supply Chain Management and Extended Enterprises, Section 2: eCommerce, Springer-Verlag, Berlin, 2001, in press.

[10] C.J.C.H. Watkins, Learning from delayed rewards, *PhD thesis, Univ. of* Cambridge Psych. Dept, 1989.

[11] Website of the Alice chatterbot, http://www.alicebot.org.

[12] Website of the MPEG-4 specification, http://mpeg.telecomitalialab.com/standards/mpeg-4/mpeg-4.htm.

[13] S.W. Wilson, Classifier Fitness based on Accuracy. *Evolutionary Computation* 3 (1995) 149-175.

[14] F. Zambetta *et al.*, An Agent To Personalize User Interactions. In: N. Rozic, D. Begusic (eds.), SoftCOM 2001 – Int'l Conf. on Software, Telecomm.s and Comp. Networks, FESB, Split, 2001, pp. 431-438.

KES 2002
E. Damiani et al. (Eds.)
IOS Press, 2002

A Formal Semantics of Hybrid Symbolic-Neural Networks for Commonsense Reasoning

Yun Bai and Yan Zhang
School of Computing and Information Technology
University of Western Sydney
Penrith South DC, NSW 1797, Australia
E-mail: {ybai,yan}@cit.uws.edu.au

Abstract. Commonsense reasoning is an essential issue in knowledge representation, system dynamics modeling and information processing. However, logic based theories of commonsense reasoning always suffer from computational difficulties during their reasoning. On the other hand, recent research shows that the *hybrid symbolic-neural network* (HSNN) has both specification and computational advantages in commonsense reasoning. But the major shortcoming of this neural network, as argued by some researchers, is a lack of formal semantics. This paper is to investigate formal semantics of HSNNs and show that under some formalization, a formal semantics is actually embedded within HSNNs.
Key words: neural network, knowledge representation, logic programming, commonsense reasoning

1 Introduction

Commonsense reasoning is an essential issue in knowledge representation, system dynamics modeling and information processing. However, logic based theories of commonsense reasoning always suffer from computational difficulties during their reasoning. For instance, even for a simple propositional case of default reasoning, the computation of a default theory's extension is intractable. On the other hand, recent research shows that the *hybrid symbolic-neural network* (HSNN) has both specification and computational advantages in logical reasoning [5, 6]. But the major shortcoming of this neural network, as argued by some researchers, is a lack of formal semantics. The purpose of this paper is to investigate formal semantics of HSNNs and show that under some formalization, a formal semantics is actually embedded within HSNNs.

The paper is organized as follows. In the following section, we briefly review the prioritized logic program proposed by Zhang and Foo recently [7]. In section 3 we represent the framework of *hybrid symbolic-neural network* (HSNN) and show how this network can be used in commonsense reasoning. In section 4, we describe a translation from HSNN into a PLP, and illustrate that the answer set semantics of PLP correctly characterizes the reasoning procedure of the corresponding HSNN. Finally in section 5 we conclude the paper with some remarks.

2 PLP: A Review

Our method of investigating the logic semantics of HSNNs is to use a logic programming approach, and *prioritized logic programs* (PLPs) in particular. A PLP is an extension of Gelfond and Lifschitz's extended logic program by associating some preference relationships among rules of the program. PLP has a well-defined semantics and is powerful in handling information conflicts in commonsense reasoning. We show that each HSNN can be translated into a *prioritized logic program* (PLP). We then use the PLP answer set semantics to characterize behaviors of the corresponding HSNN.

In this section we briefly review prioritized logic programs (PLPs) proposed by Zhang and Foo recently [7]. The language $\mathcal{L}$ of PLPs is a language of extended logic programs [3] with the following augments:

- *Names:* $N, N_1, N_2, \cdots$.
- A strict partial ordering $<$ on names.
- A naming function $\mathcal{N}$, which maps a rule to a name.

Terms, atoms, literals and rules in PLPs are defined as the same in extended logic programs. For the naming function $\mathcal{N}$, we require that for any rules r and r' in a PLP (see the following definition), $\mathcal{N}(r) = \mathcal{N}(r')$ iff r and r' indicate the same rule. A *prioritized logic program* (PLP) $\mathcal{P}$ is a triple $(\Pi, \mathcal{N}, <)$, where Π is an extended logic program, $\mathcal{N}$ is a naming function mapping each rule in Π to a name, and $<$ is a strict partial ordering on names. Note that a rule in Π including variables is viewed as a set of rules obtained by substituting these variables with all possible constants in the language. The partial ordering $<$ in $\mathcal{P}$ plays an essential role in the evaluation of $\mathcal{P}$. Intuitively $<$ represents a preference of applying rules during the evaluation of the program. In particular, if $\mathcal{N}(r) < \mathcal{N}(r')$ holds in $\mathcal{P}$, rule r would be preferred to apply over rule r' during the evaluation of $\mathcal{P}$ (i.e. rule r is more preferred than rule r'). Consider the following classical example represented in our formalism:

$\mathcal{P}_1$:
$N_1 : Fly(x) \leftarrow Bird(x), not \neg Fly(x),$
$N_2 : \neg Fly(x) \leftarrow Penguin(x), not\ Fly(x),$
$N_3 : Bird(Tweety) \leftarrow,$
$N_4 : Penguin(Tweety) \leftarrow,$
$N_2 < N_1.$

Obviously, rules N_1 and N_2 conflict with each other as their heads are complementary literals, and applying N_1 will defeat N_2 and *vice versa*. However, as $N_2 < N_1$, we would expect that rule N_2 is preferred to apply first and then defeat rule N_1 after applying N_2 so that the desired solution $\neg Fly(Tweety)$ can be derived. This idea is formalized by following definitions.

Definition 1. *Let Π be an extended logic program and r a rule with the form $L_0 \leftarrow L_1, \cdots, L_m,$ not $L_{m+1}, \cdots,$ not L_n (r does not necessarily belong to Π). Rule r is defeated by Π iff Π has answer set(s) and for every answer set $Ans(\Pi)$ of Π, there exists some $L_i \in Ans(\Pi)$, where $m + 1 \leq i \leq n$.*

Definition 2. *Let $\mathcal{P} = (\Pi, \mathcal{N}, <)$ be a PLP and $\mathcal{P}(<^+)$ denote the $<$-closure of $\mathcal{P}$ (i.e. $\mathcal{P}(<^+$) is the smallest set containing all preference relations of $\mathcal{P}$ and closed under transitivity). $\mathcal{P}^<$ is a reduct of $\mathcal{P}$ with respect to $<$ iff there exists a sequence of sets Π_i ($i = 0, 1, \cdots$) such that:*

1. $\Pi_0 = \Pi$;

2. $\Pi_i = \Pi_{i-1} - \{r_1, \cdots, r_k \mid$ (a) there exists $r \in \Pi_{i-1}$ such that $\mathcal{N}(r) < \mathcal{N}(r_i) \in \mathcal{P}(<^+)$ ($i = 1, \cdots, k$) and $r_1, \cdots, r_k$ are defeated by $\Pi_{i-1} - \{r_1, \cdots, r_k\}$, and (b) there does *not exist a rule $r' \in \Pi_{i-1}$ such that $N(r_j) < N(r')$ for some j $(j = 1, \cdots, k)$ and r' is defeated by $\Pi_{i-1} - \{r'\}\}$;*

3. $\mathcal{P}^< = \bigcap_{i=0}^{\infty} \Pi_i$.

Clearly $\mathcal{P}^<$ is an extended logic program obtained from Π by eliminating some rules from Π. In particular, if $\mathcal{N}(r) < \mathcal{N}(r')$ and $\Pi - \{r'\}$ defeats r', rule r' is eliminated from Π if no *less preferred rule* can be eliminated (i.e. conditions (a) and (b) in (ii)). This procedure is continued until a fixed point is reached. Note that due to the transitivity of $<$, we need to consider each $\mathcal{N}(r) < \mathcal{N}(r')$ in the $<$-closure of $\mathcal{P}$. It should be also noted that the reduct of a PLP may not be unique generally[7]. Now it is quite straightforward to define the answer set for a prioritized logic program.

Definition 3. *Let $\mathcal{P} = (\Pi, \mathcal{N}, <)$ be a PLP and Lit the set of all ground literals in the language of $\mathcal{P}$. For any subset S of Lit, S is an answer set of $\mathcal{P}$, denoted as $Ans^P(\mathcal{P})$, iff $S = Ans(\mathcal{P}^<)$, where $Ans(\mathcal{P}^<)$ is an answer set of extended logic program $\mathcal{P}^<$. A ground literal L is derivable from a PLP $\mathcal{P}$, denoted as $\mathcal{P} \vdash L$, iff $\mathcal{P}$ has answer set(s) and L belongs to every answer set of $\mathcal{P}$.*

3 Hybrid Symbolic-Neural Networks (HSNNs)

Neural networks have been studied mainly for pattern processing and learning [1, 2]. However, there is a class of neural networks capable of doing both pattern processing and logic reasoning, which we call *hybrid symbolic-neural networks* (HSNN) [4].

A HSNN is a finite directed graph of nodes and links. A set of nodes are selected to be *input nodes* and another set to be *output nodes*. The network is three-valued and each node can take one of the following ordered pairs as its activation value:

$(1, 0)$ *for true*,
$(0, 1)$ *for false* and
$(0, 0)$ *do not know.*

Each directed edge is associated with an ordered pair (x, y) of real numbers.

Definition 4. *Let p be the proposition of a particular node and $\{q_1, \cdots, q_k\}$ (where $k > 0$) be the set of all nodes linking to p. If the node value of q_i is (a_i, b_i), and the weights of the linking q_i to p is (x_i, y_i),*

$$\text{Net Excitatory Input} NE = \Sigma_{i=1}^{k} a_i x_i,$$
$$\text{Net Inhibitory Input} NI = \Sigma_{i=1}^{k} b_i y_i.$$

$$\text{Value of node } p = \begin{cases} (1, 0) & \text{if}\,(NE - NI) \geq 1 \\ (0, 1) & \text{if}\,(NE - NI) \leq -1 \\ (0, 0) & \text{otherwise} \end{cases}$$

A two-input three-valued logical OR and NOT can be realized as:

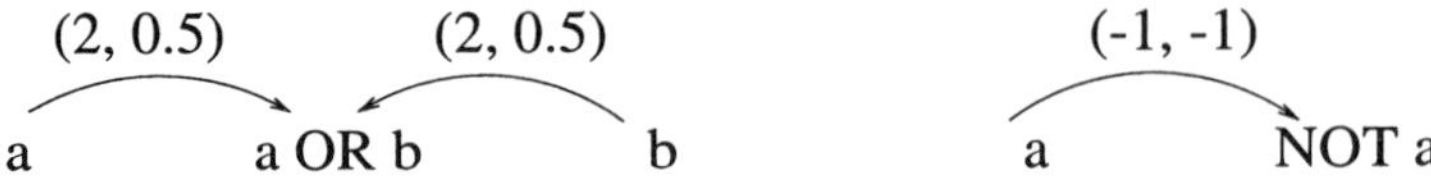

Figure 1: Logical OR and NOT.

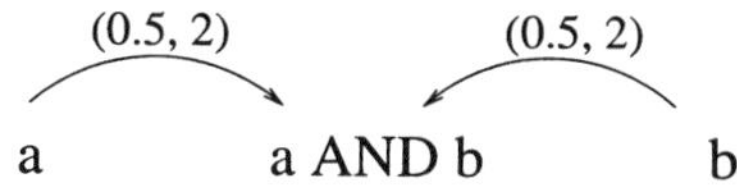

Figure 2: Logical AND.

Another important property of logic network is the *principle of duality*. This can be illustrated by swapping the weights in the edges of OR connective above to make it into an AND relation as Figure 2 shows:

Using the same principle, the dual operation of NOT is equal to itself and the dual operation of a XOR becomes a XAND as follows:

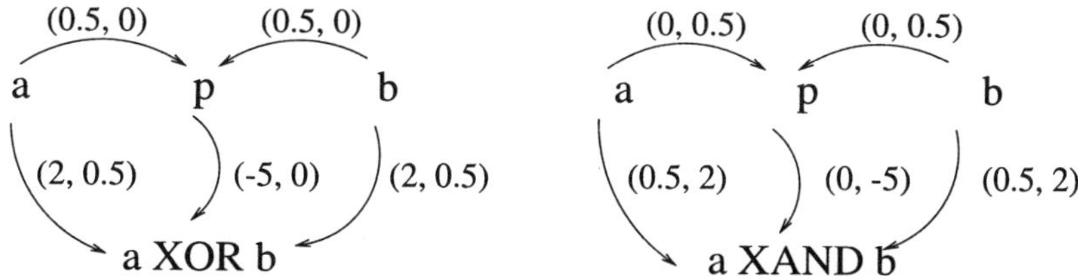

Figure 3: Logical XOR and XAND.

With the flexibility of joining various segments representing different logical relations, the neural logic network is capable of expressing complex logical statements such as "(a AND b) OR c", and "(NOT a AND b) XOR (c XAND NOT d)".

Neural logic network can simulate logical inferences to predict the outcome from a set of contingent statement about a situation.

4 From HSNN to PLP

We translate a HSNN into PLP by following steps:

1. Given a HSNN, each node with an input value (1,0) in the network presents a positive atomic proposition, which is mapped to some ground letter in PLP language, each node with input value (0,1) in the network then presents a negative atomic proposition, which can be mapped to a negative ground letter in the PLP language, while each node with input value (0,0) in the network presents some *unknown* atomic proposition, that can be mapped to a letter in PLP with negation as failure.

2. For each part of the network which represent a logic proposition $NOT\ a\ OR\ b_1\ OR\ b_2 \cdots OR\ b_k$, we generates a rule $a \leftarrow b_1, \cdots, b_k$. Note that here a must be a node with input value either (1,0) or (0,1), where $b_1, \cdots, b_k$'s input values can be one of (1,0), (0,1) or (0,0).

3. For any two rules we have generated from the last step $r_1 : a_1 \leftarrow \cdots$ and $r_2 : a_2 \leftarrow \cdots$, there are two links from $NOT\ a_1$ and $NOT\ a_2$ to other nodes in the corresponding network, if the weight associated with the link from $NOT\ a_1$ is bigger than the weight associated with the link from $NOT\ a_2$, we specify $r_1 < r_2$, that is r_1 is more preferred than r_2.

4. After finishing all these specifications, we actually generate a PLP for each given HSNN.

Now we summerize the major mathematical results we have achieved.
- The translations from HSNN to PLP described above is sound in the sense that every solution derived from the HSNN can be also derived from the corresponding PLP.
- The way of handling information conflict in a HSNN can be modeled by proper preference relationships in the corresponding PLP.
- The answer set semantics of PLPs provides a semantic characterization for HSNNs.

5 Conclusions

On our best knowledge, our work is the first effort to investigate formal semantics of neural network systems by using logic programming approach. It provides a logic basis to justify the fact that the hybrid symbolic-neural network not only has a computational benefit but also has a semantical foundation in commonsense reasoning.

Most logic based theories of commonsense reasoning have formal semantics but with expensive computational costs, while HSNNs have simple computational costs. This paper shows that from representation point of view, HSNNs are also semantically equivalent to some logic based systems, i.e. prioritized logic programs, in commonsense reasoning. In the future, we plan to apply HSNNs to solve information conflict in security authorization specification where practical logic programming method is usually applicable but with high computational cost.

References

[1] D.H. Fisher, Knowledge acquisition via incremental conceptual clustering. *Machine Learning*, 2:139-172, 1987.

[2] K. Fukushima, Neocognitron: A hierarchical neural network capable of visual pattern recognition. *Neural Networks*, 1:119-130, 1988.

[3] M. Gelfond and V. Lifschitz, Classical negation in logic programs and disjunctive databases. *New Generation Computing*, **9** (1991) 365-386.

[4] B.T Low, *Reasoning about Beliefs: An Inference Network Approach*, PhD Thesis, 1994.

[5] T.J. Reynolds, H.H. Teh, and B.T. Low, Programming in neural logic. In *Proceedings of the First Pacific Rim International Conference on Artificial Intelligence*, 1990.

[6] H. Takagi and I. Hayasgi, Artificial neural network drive fuzzy reasoning. In *Proceedings of the International Joint Conference on Approximate Reasoning*, 1989.

[7] Zhang and N.Y. Foo, Answer sets for prioritized logic programs. In —em Proceedings of the 1997 International Logic Programming Symposium (ILPS'97), pp 69-83. MIT Press, 1997

Finding Exceptions to Rules in Fuzzy Rule Extraction

Vicenç SOLER, Jordi ROIG, Marta PRIM

Dept. of Computer Science
Campus UAB, Edifici Q
Autonomous University of Barcelona
08193 Bellaterra, Spain

Abstract. How to proof the existence of exceptions to fuzzy rules in fuzzy rule extraction is described. In this paper, a Genetic Algorithm (GA) extracts rules from both a data set and the Membership Functions (MF) of the variables defined by a human expert. So, the aim is to find both a set of rules easily comprehensible for human capabilities (due to the human expert definition of the MFs) and the exceptions of those rules (if exist).

1. Introduction

Genetic Algorithms (GA) are methods based on principles of natural selection and evolution for global searches. Given a problem, GAs run repeatedly by using the three fundamentals operators: reproduction, crossover and mutation. These operators, combined randomly, are based on a fitness function evolution to find a better solution in the searching space. Chromosomes represent the individuals of the GA and a chromosome is divided in genes. GAs are used to find solutions to problems with a large set of possible solutions and they have the advantage that only require information concerning the quality of the solution. This fact makes GA a very good method to solve complex problems [2][3].

Sometimes, we, human beings, have the problem of knowing the behaviour of a system because we have worked with it for a long time in our lives. But sometimes, we are not capable to extract the knowledge that is inside the behaviour of this system to teach other human beings about what we have learnt. For example, sometimes people say: "... and now this will happen, I don't know why, but this will happen." And in most of cases "this" happens. Sometimes we call this "intuition".

Sometimes, when searching for rules from examples[4][5], not the 100% of the instances match the rules found. So, in this paper we want to pay attention in the existence of exceptions, because we know no noise has been introduced in the set of examples. That is, all the patterns are part of the behaviour of the system.

Then, to look for the exceptions, this paper presents the study of a GA that extracts fuzzy rules from both a training data set and the variables which a human expert has defined MFs.

2. The Problem to Solve

Problem: we have a set of examples which express knowledge, but it is difficult for human persons to extract this knowledge from this set. This set corresponds to the experience lived by the human expert. Consequently, the solution is to extract a set of rules which express this knowledge in a better way for human comprehension, having into account the possible existence of exceptions.

Thus, in fuzzy space we have, mainly, two things: variables expressed in MFs and rules. There are a lot of methods that extract fuzzy variables and rules from a training data set [1] [6] [7] [8], but in this case we prefer that the human expert express the variables by himself, because sometimes the MFs of the variables extracted from these methods are incomprehensible for humans because the distribution of the MFs for every variables are not logical.

3. The GA

The codification of one chromosome of our GA is expressed in the following lines:

$$(x_{1,1}, \ldots, x_{1,n}, x_{2,1}, \ldots, x_{2,n}, x_{m,1}, \ldots, x_{m,n})$$

where n is the number of variables (input variables plus output variables) and m is the number of rules. $x_{i,j}$ is the value a gene can take which is an integer value compressed in the interval $[\,0\,,\,n\_fuzzysets_j\,]$ where $n\_fuzzysets_j$ is the number of MFs of the jth variable. If a $x_{i,j}$ has value 0 it express that this variable is not present in the rule. Every $x_{i,1}, \ldots, x_{i,,n}$ corresponds to a rule of the system.

The problem we have found in searching the right set of rules has been to see, how to implement the fact of, a priori, there is not known the exact number of rules the system needs. As a gene can take a 0 value, if all the output variables are 0 or all the input variables less 1 are 0, it means that rule does not have to take into account to the system.

The initial population is always taken randomly. Every gene of a chromosome is generated randomly in the interval $[\,0\,,\,n\_fuzzysets_j\,]$. This is because we do not want the GA be influenced by anything. We want to try how the GA is capable of finding a solution without knowing anything, only the MFs of the variables.

4. A Case Study: the IRIS Data Set

In this section, we will apply the GA explained above to the Iris Flower Problem. This problem involves distinguishing among three species of iris flowers: Setosa, Veriscolor and Virginica. There are 50 training instances for each class (150 in total). Each training instance is described by four attributes: sepal length (SL), sepal width (SW), petal length (PL), and petal width (PW). Consequently, the system has four input variables (SL, SW, PL, PW) and one output variable (FLOWER).

In this case, we have defined, for every variable, a set of three MFs. We only tried to divide the width of a variable in three equilibrated sets. Therefore, here there is no "expert" and we have defined the MFs described in figures fig1, fig2, fig3 and fig4.

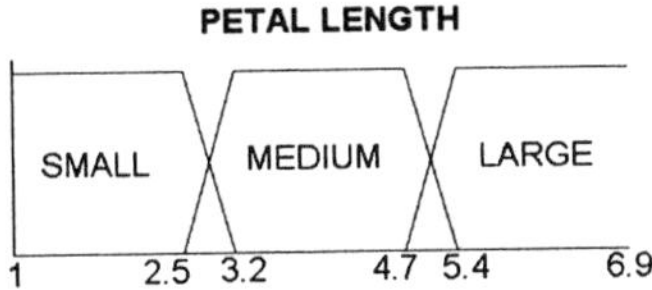

Fig1. Petal length MF

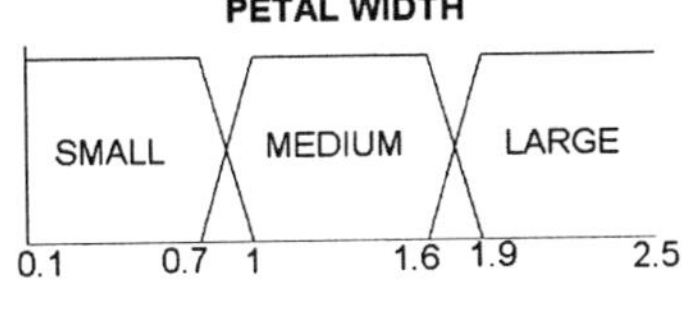

Fig2. Petal width MF

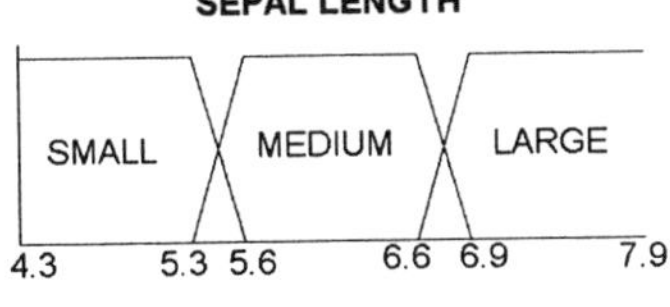

Fig3. Sepal length MF

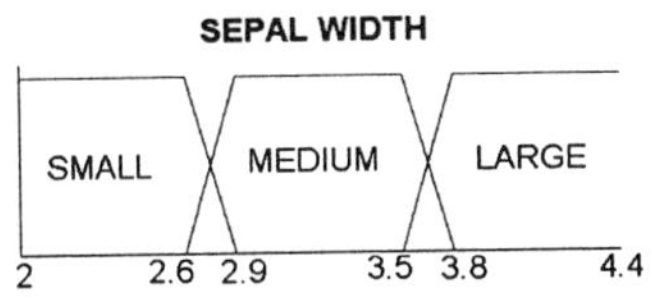

Fig4. Sepal width MF

The output variable only classifies if a flower belongs to set 1 (Setosa), 2 (Virginica) or 3 (Versicolor).

Thus, these MFs are very simple and very comprehensible for human experts.

So, as expressed above, the initial population for one chromosome (generated randomly), can be:

$$(1\,2\,1\,1\,2 \quad 2\,2\,2\,3\,0 \quad 0\,3\,2\,1\,3 \quad 2\,1\,1\,0\,1 ...)$$

where every set of five integers corresponds to a rule (four input variables and one output variable). The rule 1 2 1 1 2 can be expressed as:

IF *sl* IS *small* AND *sw* IS *medium* AND *pl* IS *small* AND *pw* IS *small* THEN *flower* IS *of class 2* (Virginica).

The second rule has no effect because the integer value corresponding to the output variable is 0 and it means the rule will not be calculated. This is a way the system has to reduce the number of rules.

In the third one, the first element is 0, and it means the SL variable is not in the rule:

IF *sw* IS *large* AND *pl* IS *medium* AND *pw* IS *small* THEN *flower* IS *of class 3* (Versicolor).

And so on...

The *fitness* value only depends on the number of patterns matched correctly.

5. Results

The presented GA has been applied to the IRIS data set mentioned in the previous section. We always have taken the same MFs, given in the previous section too.

The results obtained from the application of the problem have been, in the beginning, no hopeful, because the GA has not found a perfect solution that matches all the patterns, taking into account the tests have been varying the number of genes (rules). Thus, the main question is: "Does it exist a solution for this problem?" We have found the proof in the patterns does not match the set of rules found. Executing the GA with different number of genes (rules), there is a coincidence in several patterns that does not match with some sets of rules. In a great number of GA simulations, the patterns mentioned in *table-1* does not match the rules obtained. So, this is a symptom that these cases are the "exceptions" to the rules found with the MFs defined. This is due to the way the MFs of the variables were defined, because the most values of those patterns are located in ambiguous areas.

The simulations have been with a population of 40 chromosomes. This table reflects the GA simulations with the maximum number of rules used, the quantity of examples matched (from minimum to maximum, as average) and which patterns does not match, at least, 50% of the simulations.

Table 1. Table of results

# rules	Examples matched	Patterns not match over 50% times
3,4,5,6,7,8,9, 10,11,15	144-148	78 (88.9%) 71 (66.7%) 84 (55.6%) 120(55.6%) 134 (50.0%)

From these results, some variations in the definition of the MFs has been introduced to test how vary the results from one set of MFs to another. These are described in *table-2*.

Table 2. Table of MFs variation

Variable	MF	New dimension
SL	M	5.3 , 5.6 , 6.8 , 6.9
	G	6.8 , 6.9 , 7.9
SW	P	2 , 2.6 , 2.7
	M	2.6 , 2.7 , 3.5 , 3.8
PL	P	1 , 2.5 , 3.2
	M	2.5 , 3.2 , 5.1 , 5.4
	G	5.1 , 5.4 , 6.9
PW	M	0.7 , 1 , 1.8 , 1.9
	G	1.8 , 1.9 , 2.5

In this case, the number of patterns matching the rules found never is over 142 (always among 141 and 142) and the most of the patterns does not match any sets of rules (100% not match). See *table-3*.

Table 3. Table of results in MFs variation

# rules	Examples matched	Patterns not match over 50% times	
3,4,5,6,7,8,9, 10,11,15	141-142	107 (100%) 120 (100%) 124 (100%) 127 (100%)	128 (100%) 134 (100%) 139 (100%) 150 (100%)

On the other hand, we have tested our method with another set of MFs. Rather than define 3 MF for every variable, we have defined 5, with equilibrated distribution too. The results for this increasing of MFs segmentation have not been good. The GA does not converge as good and quickly as with 3 MFs per variable.

We have used another existents methods to do the same which were mentioned in the Section. We have used the RecBF [7] and the wu-chen [1] methods. The first one did not give us good results and neither do the other one. The wu-chen method [1] gives a good set of rules [3] which matches 145 of 150 examples with the MFs given by it. The evolution of the GA reaches 146 or 147 as much.

6. Conclusions

The goal is accomplished obtaining a reduced set of rules (among a minimum of 3 and a maximum of 15) for this problem. One of the most incomprehensible things for human beings, we think, is the comprehension of the distribution of the MFs. If it is the human who express it and gives it to the system, if the set of rules is small, the system has a more comprehension to him.

The future work is to know why exactly exist patterns that constantly do not match most of the set of rules found and how to merge different set of rules given by GA simulations with exactly the same parameters.

In contrast, other methods [1] [6] [7] [8] try to achieve the fact of "we don't know anything about the problem and we want to extract the MFs and the rules". This is not the goal of this work. The objective is to have a set of rules and variables comprehensible for human beings. We only will modify (if necessary) slightly the number of MFs to reach a good solution.

References

[1] T.P. Wu and S.M. Chen, "A New Method for Constructing MFs and Fuzzy Rules from Training Examples", IEEE Transactions on Systems, Man and Cybernetics-Part B: Cybernetics, Vol. 29, No.1, pp. 25-40, Feb. 1999.

[2] L.X. Wang and J.M. Mendel, "Generating Fuzzy Rules by Learning from Examples", IEEE Transactions on Systems, Man and Cybernetics, Vol. 22, No.6, pp. 1414-1427, Nov./Dec.1992.

[3] S.K. Pal, S. Bandyopadhyay and A. Murphy, "Genetic Algorithms for Generation of Class Boundaries", IEEE Transactions on Systems, Man and Cybernetics-Part B: Cybernetics, Vol. 28, No.6, pp. 816-828, Dec. 1998.

[4] R. Setiono and W.K. Leow, "FERNN: An Algorithm for Fast Extraction of Rules from Neural Networks", Technical Report, National University of Singapore.

[5] G.G. Towell and J.W. Shavlik, "Extracting refined rules from knowledge-base neural networks", Machine Learning, 13(1), pp. 71-101, 1993.

[6] T.P. Hong, J.B. Chen, "Finding relevant attributes and MFs", MFs and Systems, 103, pp.389-404, 1999.

[7] M. Berthold and K.P. Huber, "Constructing Fuzzy Graphs from Examples", Intelligent Data Analysis, 3, pp. 37-53, 1999.

[8] F. Herrera, M.Lozano, J.L. Verdegay, "A Learning Process for Fuzzy Control Rules using Genetic Algorithms", Technical Report #DECSAI-95108, Feb 1995.

KES 2002
E. Damiani et al. (Eds.)
IOS Press, 2002

An Improved Recurrent Neuro-Fuzzy Network for Self-Organizing Control

Nicolae Constantin[1], Vasile Palade[2]

[1] *University "Politehnica" Bucharest, Department of Control and Systems Engineering, Independentei Str., no.313, 77206, Bucharest - Romania. E-mail: nicu@lnx.cib.pub.ro*
[2] *Oxford University, Computing Laboratory, Parks Road, OX1 3QD, Oxford –UK. E-mail: Vasile.Palade@comlab.ox.ac.uk*

Abstract. This paper proposes an improved neuro-fuzzy network that is suitable for adaptive control. The proposed neuro-fuzzy network represents a modified fuzzy rule based model with neural network learning ability. The rules in the neuro-fuzzy network are created and adapted based on an on-line learning algorithm. The network structure and parameters are identified during the learning algorithm of the neuro-fuzzy network. It is proved that our neuro-fuzzy network has excellent control abilities and can greatly reduce the training time and avoid the parameter overfitting.

1. Introduction

It is very well known that multi-layered Backpropagation neural networks (BPNN) are much time consuming in training that represents a serious drawback in many real-world applications. Moreover the training required for an on-line adaptation to the plant parameter variations can usually determines the overtuning of the neural network. Neuro-fuzzy network (NFN) based approach has become a popular research topic in the last decade and various neuro-fuzzy systems were proposed in the literature, e.g. Jang's adaptive-network-based fuzzy inference system [1], Lin's neural-network-based fuzzy logic control and decision system [2][11][12], Wang's several adaptive fuzzy systems [3], Ishibuchi's fuzzy neural net with fuzzy inputs and fuzzy targets [4], the fuzzy neural network based method for learning fuzzy control rules and membership functions by fuzzy error backpropagation by Nauck and Kruse [5]. Neuro-fuzzy networks embed the low-level learning and computational power of neural networks into fuzzy systems, and provide the high-level human-like thinking and reasoning [15] of fuzzy systems into neural networks. However due to their feedforward network structure, the application field is limited mainly to non-dynamic problems. A recurrent neural network that involves dynamic elements in the form of feedback connections, used as internal memories, performs a dynamic mapping [6].

The recurrent neuro-fuzzy network proposed in this paper can cope with temporal problems via the inclusion of some internal memories. A major characteristic of the network is that no a priori assignment and design of the rules are required. The rules are constructed automatically during the on-line operation. There are two learning phases: the network structure identification and the parameter learning. The clusters formed in the input space are aligned in order to decrease the number of rules and the number of membership functions. First the network is trained off-line in order to learn the inverse dynamic model of the plant, and then the network is configured as a feedforward network controller to the plant. Next, a conventional on-line training scheme is used to adapt the network to the changes in the practical environment.

This paper is organized as follows. Section 2 describes the architecture of the network. The on-line structure and parameter learning algorithm is presented in section 3. The control results of using NFN comparatively with BPNN on the water bath temperature control system are presented in section 4. Conclusions are summarized in the last section.

2. Architecture of the recurrent neuro-fuzzy network

The network is a five-layered neuro-fuzzy network embedded with dynamic feedback connections. The functions of nodes in each layer are shortly presented below. The layers 1 and 2 are identical with those used in the well-known structure of a neuro-fuzzy network [1][2]. Each node in layer 1 only transmits the input value to the next layer, and the nodes in layer 2 perform the fuzzification of the input values. Gaussian membership functions are used.

Layer 3: Nodes are called rule nodes and perform precondition matching of the rules. Each node has some inputs from layer 2 that represent the rule's spatial firing degree and some inputs from the feedback layer that represent the rule's temporal firing degree. The following AND operator is used on each rule node:

$$a^{(3)} = a^{(6)} \prod_i u_i^{(3)} = a^{(6)} e^{-[D_i(x-m_i)]^T [D_i(x-m_i)]} \tag{1}$$

where $D_i = \mathrm{diag}\,(1/\sigma_{i1}, \ldots, 1/\sigma_{in})$, $m_i = (m_{i1}, m_{i2}, \ldots, m_{in})^T$, m_{ij}, σ_{ij} are the center and the width of the Gaussian membership functions in layer 2, $a^{(6)}$ is the output of the feedback term node, and $a^{(3)}$ is the output of the rule node. The fuzzy AND operation used in (1) represents the algebraic product in fuzzy set theory.

Layers 4 and 5 have the same structure as in a common neuro-fuzzy network. Each output term node in layer 4 performs a fuzzy OR operation to integrate the fired rules with the same consequent part. The nodes in layer 5 perform the defuzzification for the corresponding output linguistic variables.

Feedback layer (layer 6): Two types of nodes are used in this layer. The square node is named context node and is associated with a circle node named feedback term node. The number of context nodes is equal to the number of output term nodes in layer 4. The context node works as a defuzzifier:

$$h_j = \sum_i a_i^{(4)} w_{ji} \tag{2}$$

where the internal variable h_j is interpreted as the inference result of the hidden (internal) rule and w_{ji} is the link weight from the ith node in layer 4 to the jth internal variable ($a_i^{(4)}$ is the output value of the ith node in layer 4) . The output of the feedback term node is:

$$a^{(6)} = \frac{1}{1+e^{-h_i}} \tag{3}$$

This output is connected to the rule nodes in layer 3, which are connected to the output term nodes in layer 4 (see figure 1). The outputs of the feedback term nodes contain the firing history of the fuzzy rules. The following dynamic fuzzy rules are implemented:

Rule i: IF $x_1(t)$ is A_{i1} and ... and $x_n(t)$ is A_{in} and $h_i(t)$ is G
       THEN $y_1(t+1)$ is B_{i1} and $y_2(t+1)$ is B_{i2} and $h_1(t+1)$ is w_{1i} and ... and $h_m(t+1)$ is w_{mi}

where x_i is the input variable, y_i is the output variable, $A_{i1},\ldots, A_{in}$, G, B_{i1}, B_{i2} are fuzzy sets, h_i is the internal variable, w_{1i} and w_{mi} are fuzzy singletons, and n and m are the numbers of input and internal variables, respectively. This fuzzy rule can be decomposed into two parts: the external rule and the internal rule, both forming a hierarchical relation. The external rule can be expressed as:

Rule i: IF $x_1(t)$ is A $_{i1}$ and ... and $x_n(t)$ is A_{in} and $h_i(t)$ is G
              THEN $y_1(t+1)$ is B_{i1} and $y_2(t+1)$ is B_{i2}

and the internal rule is:

Rule i: IF $x_1(t)$ is A $_{i1}$ and ... and $x_n(t)$ is A_{in} and $h_i(t)$ is G
              THEN $h_1(t+1)$ is w_{1i} and ... and $h_m(t+1)$ is w_{mi}

In time domain a rule is fired by the outputs of rules that were fired one time step ahead.

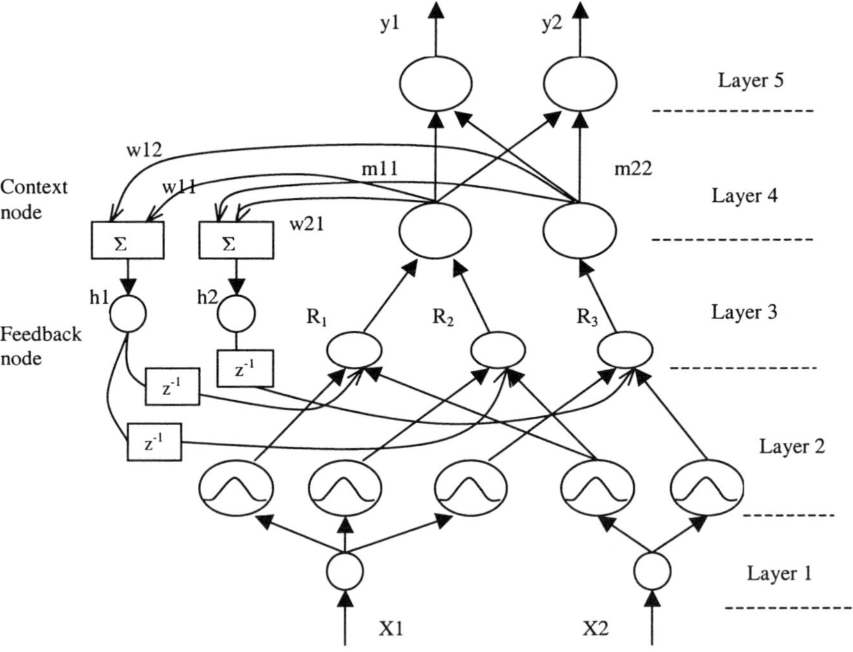

Figure 1. Structure of the recurrent neuro-fuzzy network

3. Training algorithm

The structure identification and the parameter learning are the two phases of the training algorithm of the network. The structure identification includes the precondition, consequents, and feedback structure identification of dynamic fuzzy if-then rules.

The number of rules depends on the way the input space is partitioned. Here only the spatial information is used for clustering and the spatial firing strength is used as a degree measure:

$$F^i(x) = \prod_i u_i^{(3)} = e^{-[D_i(x-m_i)]^T[D_i(x-m_i)]} \qquad (4)$$

where $F^i \in [0,1]$. The exponent represents the distance between x and the center of cluster i. Based on this measure, the criterion for the generation of a new fuzzy rule is as follows.

Let $\mu = (m_i, \sigma_i)$ represent the Gaussian membership function with center m_i and width σ_i, and E(A,B) represent the fuzzy measure of two fuzzy sets A and B. And suppose that no rule exists initially. The algorithm for the generation of new fuzzy rules and fuzzy sets for each input variable is the following.

IF x is the first incoming pattern THEN
   Do { Generate a new rule, with center $m_1 = x$, width
        $D_1 = diag(1/\sigma_0, \ldots, 1/\sigma_0)$, where σ_0 is a prespecified constant.
        After decomposition, we have n membership functions with
           $m_{1i} = x_i$, and, $\sigma_{1i} = \sigma_0$, $i = 1, \ldots, n$.
   }
ELSE
  for each newly incoming pattern x, do
  { find $J = argmax_{1 \le j \le c(t)}\ F^j(x)$,
    IF $F^J \ge F_{in}(t)$ THEN
       perform fuzzy measure and eliminate unnecessary membership functions
    ELSE
      { $c(t+1) = c(t) + 1$, generate a new fuzzy rule, with
      $m_{c(t+1)} = x$, $D_{c(t+1)} = (-1/\beta)\ diag(\ 1/\ln(F^J), \ldots, 1/\ln(F^J))$.
      After decomposition, we have $m_{new-i} = x_i$, $\sigma_{new-i} = -\beta \ln(F^J)$, $i = 1, \ldots, n$.
      Calculate the fuzzy measure for each input variable: {degree (i, t) =
      $max_{1 \le j \le k_i}\ E\ [\mu\ (m_{new-i}, \sigma_{new-i}), \mu\ (m_{ji}, \sigma_{ji})]$ where k_i is the number of partitions
         of the ith input variable.
     IF degree$(i, t) \le \rho$ THEN
       adopt this new membership function and set $k_i = k_i + 1$,
    ELSE
       set the projected membership function as the closest one.
   }
 }

In order to reduce the number of fuzzy sets for each input variable and to avoid their redundancy the similarities between them should be checked in each input dimension. The generation of a new cluster corresponds to the generation of a new fuzzy rule with its preconditioned part constructed by the learning algorithm. It is necessary to decide the consequent part of the generated rule. When a new input cluster is formed after the presentation of the current input-output training pattern (x, d), then the consequent part is constructed by the following algorithm:

IF there are no output clusters
  Do {part1 from the previous process, with x replaced by d}
ELSE
  Do {
      find $J = argmax_j\ F^j(d)$;
      IF $F^J \ge F_{out}(t)$ THEN
        connect input cluster $c(t+1)$ to the existing output cluster J
      ELSE

    generate a new output cluster connecting input cluster
    c(t+1) to the newly generated output cluster.
  }

Since only the center of each output membership function is used for defuzzification, the consequent part of each rule may be simply regarded as a singleton.

Feedback structure identification: It is considered that the number of internal variables is equal to the output term nodes in layer 4. During on-line learning a context node is created once an output cluster is created. The inputs for the context nodes come from all the neurons in layer 4 with the link weight assigned with a small random value in [-1,1]. Each internal variable has a global membership function assigned to it and acts as the feedback term node of the corresponding context node. In general, each rule node has its own corresponding internal variable, which memorizes the firing history of the rule. Due to the fact that the feedback term node is connected to the rule that maps to the corresponding output cluster, the parameter number in the feedback layer can be effectively reduced.

Parameter identification: After the structure is adjusted according to the current training pattern, the network starts the parameter identification phase to optimally adjust the parameters. The learning is performed on the whole network. In order to derive the learning algorithm the ordered derivative method [7] is used. This is a partial derivative whose constant and varying terms are defined using an ordered set of equations. The goal is to minimize the error function:

$$E(t+1) = 1/2\left[y_j(t+1) - y_j^d(t+1)\right]^2 \qquad (5)$$

where $y_j^d(t+1)$ is the desired output and $y_j(t+1)$ is the current output. For each training pattern, starting at the input nodes, a feedforward pass is used to compute the activation levels of all the nodes in the network in order to obtain the current output. The update rules for the network parameters are as follows:

- update rule of $\hat{m}_{ji}$ (the center of the output membership functions):

$$\hat{m}_{ji}(t+1) = \hat{m}_{ji}(t) - \eta \frac{\partial E}{\partial \hat{m}_{ji}(t)}(t+1) \qquad (6)$$

Based on the same strategy the other free parameters are updated with similar rules:
- update rule of m_{pq} (the center of the membership functions in the precondition part)
- update rule of σ_{pq} (the width of the membership functions in the precondition part)
- update rule of w_{pq} (the memory weight parameters in the feedback layer)

In the equations given above, two different learning constants (η and η_w) are used. In [7] it is suggested that the learning constant η_w may be chosen several times larger than η such that one could obtain about the same convergence speed for all the parameters.

4. Simulation results

Consider a discrete-time SISO temperature control system:

$$y_p(k+1) = a(T)y_p(k) + \frac{b(T)}{1+e^{1/2y_p(k)-\gamma}}u(k) + [1-a(T)]y_0 \qquad (7)$$

$$a(T) = e^{-\alpha T},$$

where

$$b(T) = \frac{\alpha}{\beta}(1-e^{-\alpha T}).$$

This model is given in [10] and represents the equation for a real water temperature control system. Another example can be found in [13]. The system parameters are $\alpha =$ 1.00151e-4, $\beta = 8.67973$e-3, $\gamma = 40$ and $y_0 = 25°C$. The plant input u(k) is limited to [0V;5V] and the sampling period is chosen $T = 30$s. The task is to control the water bath system to follow the three set-points given in (8).

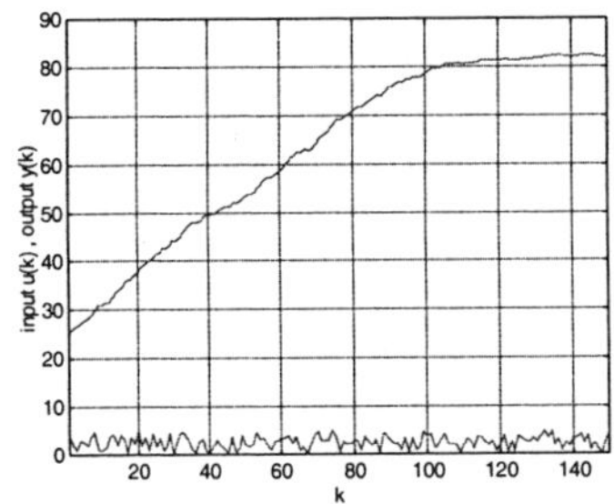

$$r(k) = \begin{cases} 35°C, & k \le 40, \\ 55°C, & 40 < k \le 80 \\ 75°C, & 80 < k \le 120. \end{cases} \qquad (8)$$

Figure 2.   Input –output characteristic of the process

A sequence of random input signals u(k) is used to obtain the open-loop input-output characteristic of the system as shown in figure 2. It is obvious that the system behaves linearly up to about 70°C and then becomes nonlinear and saturates at about 82°C. From the input-output characteristic of the system 100 training patterns are selected to cover the entire system functioning space.

For comparing reasons of the NFN described above with a BPNN, a four-layer Backpropagation feedforward network with two hidden layers is used. The number of hidden nodes is decided by trial and error.

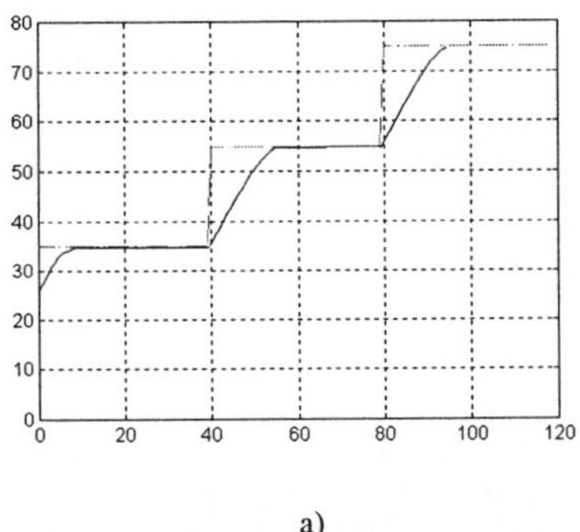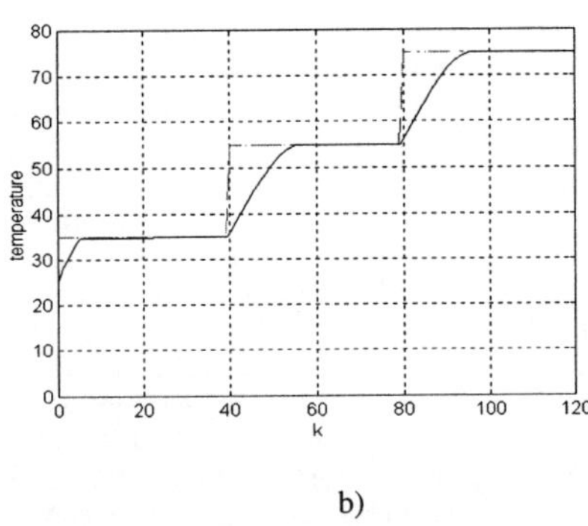

a)   b)

Figure 3. Control performance in the 20th trial of the on-line training using   a) BPNN   b) NFN

The learning parameters of the NFN were: $\eta = \eta_w = 0.005$, $\rho = 0.9$, $F_{in} = 0.1$, $F_{out} = 0.7$. Using the selected 100 patterns with input vector [$y_p(k)$, $y_p(k+1)$] and target pattern u(k), NFN and BPNN are trained to minimize the sum of the squared error for all patterns. After 10000 iterations, the error for BPNN (with 15 nodes in both hidden layers) is only 0.7231, while the error for NFN reaches 0.6652 after just 6 iterations. However the errors are still large and on-line training is necessary to achieve high accuracy.

In on-line training, both NFN and BPNN are trained by the conventional on-line training scheme. In order to test the control performance in each trial, a performance index, the sum of absolute error (SAE) is defined:

$$SAE = \sum_k \left| r(k) - y_p(k) \right| \tag{9}$$

where r(k) and y_p(k) are the reference output and the actual output of the control system.

The performance index is calculated for a trial with k ranging from 1 to 120. After the 20th trial of the on-line training algorithm, the SAE is 345.31 for BPNN and 338.41 for NFN.

The number of generated rules is 7, and the number of fuzzy sets on the y(k) and y(k+1) dimensions are 4 and 4 respectively. NFN can reduce the training time and avoid the overtuning of the network. It has also good control capabilities in respect to set-point changes. In terms of number of network parameters the NFN also use less parameters than the BPNN.

5. Conclusions

A fuzzy neural network with on-line self-organizing learning ability is proposed in this paper. It performs fuzzy reasoning by creating fuzzy rules, which are automatically and optimally generated during the on-line operation via simultaneously structure and parameter learning. The structure identification can reduce the number of rules and the size of the network. The final control performance of the NFN controller in the conventional on-line scheme shows the best regulation capability. The NFN overcomes the disadvantages of BPNN and has good control abilities for the water bath temperature control system.

References

[1] J.R. Jang, "ANFIS: Adaptive-network-based fuzzy inference systems". *IEEE Trans.Syst.,Man,Cybern.* **23** (1993) 665-685.

[2] C.T. Lin, "Neural fuzzy control systems with structure and parameter learning". *World Sci.* (1994).

[3] L.X. Wang, "Adaptive Fuzzy Systems and Control". Englewood Cliffs, NJ: Prentice-Hall, 1994.

[4] H. Ishibuchi, K. Kwon and H.Tanaka, "Learning of fuzzy neural networks from fuzzy inputs and fuzzy targets". *Proc. of 5th IFSA World Congress*, vol. I, pp. 147-150,1993.

[5] D. Nauck and R. Kruse, " A fuzzy neural network learning fuzzy controlrules and membership functions by fuzzy error backpropagation", *Proc. of IEEE Int. Conf. Neural Networks*, pp. 1022-1027,1993.

[6] C.L.Gilles, G.M.Kuhn, and R.J.Williams, "Dynamic recurrent neural networks: Theory and applications". *IEEE Trans. on Neural Networks*, **5** (1994) 153-156.

[7] P.S. Sastry, G.Santharam and K.P. Unnikrishnan, "Memory neural networks for identification and control of dynamic systems". *IEEE Trans. on Neural Networks*, **5** (1994) 306-319.

[8] N. Constantin, I.Dumitrache, "Neural Adaptive Control for Multivariable Nonlinear Processes". Proc. of the European Control Conference - ECC99, CD-ROM-F508, 1999.

[9] J. Tanomaru and S. Omatu, "Process control by on-line trained neural controllers". *IEEE Trans. on Ind. Electronics* **39/6** (1992) 511-521.

[10] C.T. Lin, C.F. Juang and C.P. Li, "Water bath temperature control with a neural fuzzy inference network". *Fuzzy Sets and Systems* **111** (2000) 285-306.

[11] C.T. Lin and C.F. Juang, "A Recurrent Self-Organizing Neural Fuzzy Inference Network". *IEEE Trans. on Neural Networks*, **10/4** (1999) 828-845.

[12] C.T. Lin and C.S. Lee, "Neural Fuzzy System; A Neural-Fuzzy Sinergism to Intelligent Systems". Prentice Hall, Englewood Cliffs, NJ, 1996.

[14] I. Hatzilygeroudis and J. Prenzas, "Constructed Modular Hybrid Rule Bases for Expert Systems", *Int. Journal on Artificial Intelligence Tools* **10** (2001) 87-105.

[15] C.D. Neagu et al., "Air Quality Prediction Using Neuro-Fuzzy Tools". 9th IFAC Int. Symp. on Large Scale Systems – LSS2001, Bucharest, IFAC Procs. Volumes, Elsevier, pp. 229-235.

KES 2002
E. Damiani et al. (Eds.)
IOS Press, 2002

Tabu Search Approach Based Algorithm for Nonlinear Time Alignment

Tsong-Yi Chen[*], Jeng-Shyang Pan[*], Xiaodan Mei[**] and Shenghe Sun[**]

[*] *Dept. of Electronic Engineering, Kaohsiung Institute of Technology, Taiwan*
[**] *Dept. of Automatic Test, Measurement and Control, Harbin Institute of Technology, China*

Abstract. The tabu search algorithm is the generalized heuristic global search technique with short-time memory, and suitable for solving many nonlinear optimization problems. This paper applies this technique to speech recognition systems as a nonlinear time alignment technique, and presents a new algorithm for optimizing time warping based on tabu search (TS) approach, which makes time warping functions optimized globally. Simulation results show that TSTW has better time warping performance than dynamic time warping (DTW) and genetic time warping (GTW).

Key words: Tabu search; Speech recognition; Nonlinear time alignment; Dynamic time warping

1. INTRODUCTION

People often use the template matching technique in speech recognition systems, especially in recognition of isolated word. It usually includes two main processes: training the feature vectors and comparing them. First a great deal of speech signals are sampled by the mike and then extracted a sequence of feature vectors that can characterize the input speech. The feature vectors of every word are stored in a database as "reference patterns". The recognized word is also sampled and extracted "feature" as "test patterns", then these test patterns is compared with the reference patterns in the database. Comparison is a similarity measurement between the test pattern and the reference patterns by computing the spectral distortion between them. "Reference pattern" which has distortion values less than a predefined value is the recognition result, the word it corresponds is thought of as the recognized word.

However the individual speaking rates involve great variations. These variations cause a nonlinear fluctuation and timing difference in the time-axis, therefore causes a major problem in computing the spectral-distortion measurement that directly affects the performance of the speech recognition system. People usually use the linear time normalization technique make all the speech patterns normalized to a common time axis to solve this problem. But experimental results show that these linear techniques are insufficient to deal with the highly nonlinear utterances.

The DTW proposed by F. Itakura [1] has been used to deal with this problem as a dynamic programming-matching-based nonlinear time warping technique. Its algorithm is simple and the training time and searching time are shorter than other methods. However, it has some shortcomings, such as it can only realize the local optimization, it cannot take the dynamic slope weighting function and the performance is not so perfect. Therefore people always look for the new methods that can overcome these disadvantages to realize perfect time-alignment. The literature [2] proposed the GTW algorithm to solve these problems successfully, however the GTW algorithm needs much running time and so it can also be improved.

A generalized heuristic global search method presented in this paper with short-time memory called TSTW was developed to solve the nonlinear time warping problem. When we use the tabu search algorithm as the search method for the warping paths instead of the DTW and the GTW, these considerations can be solved better. Experimental results show that the TSTW can obtain better results than the DTW and the GTW. It can reach the level of the GTW but the convergence speed is faster than the GTW greatly.

In this paper, the definition of time warping is given in Section 2, and the tabu search algorithm is described in Section 3. The TSTW algorithm is presented in Section 4. Experimental results are given in Section 5 and conclusions are given in Section 6.

2. TIME WARPING

Let $R=\{R(1),\cdots,R(m),\cdots,R(M)\}$ and $T=\{T(1),\cdots,T(n),\cdots,T(N)\}$ showed in Fig.1 represent reference pattern and test pattern, where $R(m)$ and $T(n)$ are the parameter vectors of the short-time acoustic features (any set of acoustic features can be used, as long as the distance measure for comparing pairs of feature vectors is known and its properties are understood) where $n=1,2,\cdots,N$; $m=1,2,\cdots,M$. The dissimilarity between R and T is defined by a function of the short-time spectral distortions $D(T,\ R)$. Time warping is used for finding a path from (n_1, m_1) to (n_N, m_N), which make $D(T,\ R)$ least, where $(n_1, m_1)=(1, 1)$ and $(n_N, m_N)=(N, M)$.

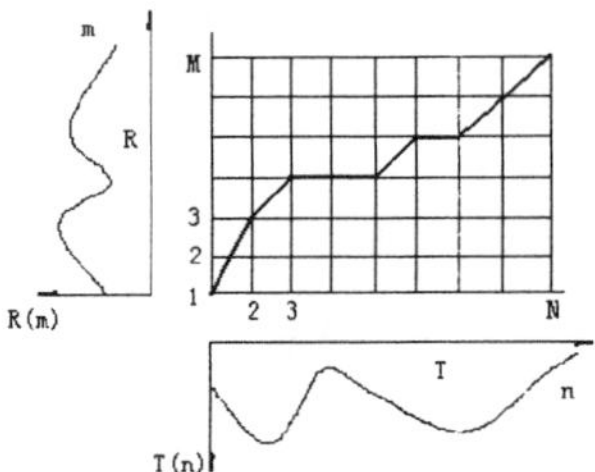

Fig.1 Example of a time warping path

Suppose that the path passes the points $(n_1, m_1),\ \cdots\ ,(n_i, m_i),\ \cdots\ ,(n_N, m_N)$, if the path has passed the point (n_{i-1}, m_{i-1}), the next point is one of the following:

1. $(n_i, m_i)= (n_{i-1}+1, m_{i-1}+2)$ or
2. $(n_i, m_i)= (n_{i-1}+1, m_{i-1}+1)$ or
3. $(n_i, m_i)= (n_{i-1}+1, m_{i-1})$

Let the distortion of the point (n_i, m_i) is $d(n_i, m_i)$, which is a short-time spectral distortion between n_i and m_i, it can be calculated by

$$d(n_i, m_i)=\min\{\ d(n_{i-1}+1, m_{i-1}),\quad d(n_{i-1}+1, m_{i-1}+1),\quad d(n_{i-1}+1, m_{i-1}+2)\ \}\tag{1}$$

then $n_i=n_{i-1}+1$, and m_i can be achieved by (1)

if $d(n_i, m_i)=d(n_{i-1}+1, m_{i-1})$, then $m_i=m_{i-1}$;

if $d(n_i, m_i)=d(n_{i-1}+1, m_{i-1}+1)$, then $m_i=m_{i-1}+1$;

if $d(n_i, m_i)=d(n_{i-1}+1, m_{i-1}+2)$, then $m_i=m_{i-1}+2$;

so the distortion $D(T,\ R)$ from (n_1, m_1) to (n_N, m_N) is

$$D(T,R) = \sum_{i=1}^{N} d(m_i, n_i)\tag{2}$$

The time warping path is a curve relating the m time axis of the reference pattern to the n time axis of the test pattern, can be written as follow

$$m_i=W(n_i),\quad i=1,\ldots,N\tag{3}$$

where $W(1)=1$; $W(N)=M$. So the accumulated distance computation between the reference pattern and test pattern can be seeked by

$$D_W(T, R) = \sum_{i=1}^{N} \frac{d(m_i, n_i) \cdot q}{Q} = \sum_{i=1}^{N} \frac{d(W(n_i), n_i) \cdot q}{Q} \tag{4}$$

where q is a nonnegative weighting coefficient and Q is a normalizing factor. The optimal warping path is achieved by minimizing (4) such that

$$D_{\hat{W}}(T, R) = \min_{W} D_W(T, R) \tag{5}$$

So time warping problem is to fimd the optimal warping path $\hat{W}$ which minimizes the accumulated distance D between the reference pattern and test pattern.

Of course a warping path must satisfy the following constraints: the endpoint constraint, the allowable region, the monotony constraint, the local continuity constraint, and the slope constraint [2]. As shown in Fig.2, Sakoe and Chiba [3] suggest four types of slope constraints.

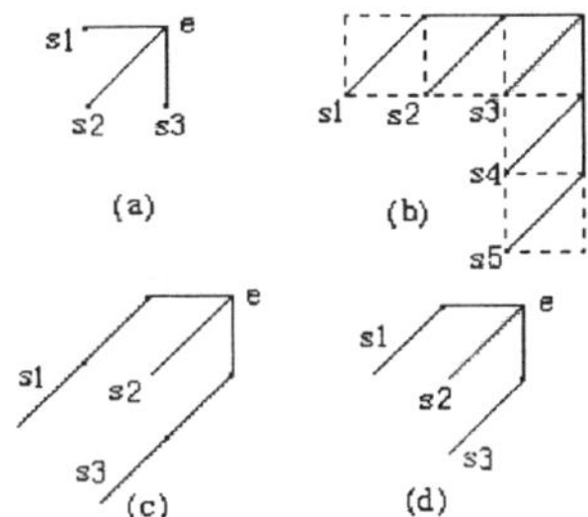

Fig.2 Four types of slope constraints suggest by H.Sakoe and S.Chiba

3. TABU SEARCH ALGORITHM

The tabu search algorithm is a generalized heuristic global search technique with short-time memory, was proposed by Glover [4]. The basic idea of the tabu search is to explore the search space of all f easible s olutions b y a sequence of m oves and to forbid some search directions at the present iteration in order to avoid cycling and jump off local optima. The elements of a move from the current solution to its selected neighbor are partially or completely recorded in the tabu list for the purpose of forbidding the reversal of the replacement in a number of future iterations. The tabu search approach begins with test solutions generated randomly and their corresponding objective function values computed. If the best of these solutions is not tabu or if it is tabu but satisfies the aspiration criterion, then select this solution to be the new current solution to generate test solutions for next iteration. It is called aspiration criterion if the test solution is a tabu solution but the objective value is better than the best value of all iterations. The **Tabu Search** algorithm is given as follows:

    {   generate the initial solutions;
         calculate the current solution and the best solution;
         while termination criterion not reached
         {   generate the test solutions in the neighborhood of the current solutions;
              calculate the corresponding objective values;
              update the current solution and the best solution;
              update the tabu list;   }   }

4. TSTW ALGORITHM

The optimal warping path searching problem must be mapped to the tabu search algorithm before it can be used. This includes as follow:

A solution of this algorithm is defined as a path P_m consisting of a set of k subpaths, each subpath is one of the subpaths in Fig.2. If select the second slope constraint in Fig.2, then P_m $=p_m(1)-p_m(2)-\cdots p_m(i)\cdots -p_m(k)$, where $p_m(i) \in \{s1,s2,s3,s4,s5\}$.

The initial test solutions are generated randomly. After the first iteration, the test solutions are generated from the best solution of current iteration by swapping two indicies randomly. The tabu list memory stores the swapped indices only. It is a tabu condition if the swapped indicies to generate the new test solution from the best solution of current iteration are the same as any records in the tabu list memory.

Let $S_i=\{\,P_1, P_2, \ldots, P_{Ns}\}$ to be the set of the test solutions, let $P_c=\{p_c(1), p_c(2), \ldots, p_c(N)\}$ and $P_b=\{\,p_b(1), p_b(2), \ldots, p_b(N)\}$ be the best solution of current iteration and the best solution of a ll iterations r espectively, let $V_i=\{v_1, v_2, \ldots, v_{Ns}\}$, v_c and v_b den ote the s et of o bjective function values for test solutions, the objective function value for the best solution of current iretation and the objective function value for the best solution of all iretations, respectively, where v_i is the objective function value for solution P_i, $1 \leq i \leq N_s$. The algorithm is given as follows:

Step 0. Set the tabu list size T_S, the number of test solutions N_S and the maximum number of iterations I_m. Set the iteration counter $i=1$ and insertion point of the tabu list $t_l=1$. Generate N_S solutions $S_i=\{P_1,P_2,\ldots,P_{Ns}\}$ randomly, calculate the corresponding objective values $V_i=\{v_1,v_2,\ldots,v_{Ns}\}$ using equations (4)(5) and find out the current best solution $P_c=P_j$, $j=\text{argmin}_l\, v_l$, $1 \leq l \leq N_s$. Set $P_b=P_c$ and $v_b=v_c$.

Step 1. Copy the current best path P_c to each test solution P_i, $1 \leq i \leq N_s$. For each test solution P_i, $1 \leq i \leq N_s$, generate two random integers r_1 and r_2, $1 \leq r_1 \leq N$, $1 \leq r_2 \leq N$, $r_1 \neq r_2$. Generate the new test s olutions by swapping $P_i(r_1)$ a nd $P_i(r_2)$. C alculate t he c orresponding o bjective values $v_1,v_2,\ldots,v_{Ns}$ for the new test solutions.

Step 2. Sort $v_1,v_2,\ldots,v_{Ns}$ in increasing order. From the best test solution to the worst test solution, if the test solution is a non-tabu solution or it is a tabu solution but its objective value is less than the best value of all iterations v_b (aspiration level), then choose this solution as the current best solution P_c and choose its objective value as the current best objective value v_c, go to step 3; otherwise, try the next test solution. If all test solutions are tabu solutions, then go to step 1.

Step 3. If $v_c<v_b$, set $P_c=P_b$ and $v_c=v_b$. Insert the swapped indices of the current best solution P_c into the tabu list. Set the inserting point of the tabu list $t_l=t_l+1$. If $t_l>T_S$, set $t_l=1$. If $i<I_m$, set $i=i+1$ and go to step 1; otherwise, record the best path index and terminate the algorithm.

5. EXPERIMENTAL RESULTS

The TSTW algorithm was tested using a database of 9 Chinese words which was sampled at 8 k Hz a nd w hose f eature v ectors a re e xtracted b y a 1 0th – order c epstral a nalysis. T he short-time spectral distortion measurement is

$$d(h_R,h_T) = \sum_{i=1}^{10}\left|h_{Ri} - h_{Ti}\right|, \tag{6}$$

where h_R and h_T are the short-time spectral f eature vectors of reference and test patterns, respectively. For each of the 9 words is spoken for ten times, so there are 90 utterance patterns. The allowable region is defined as $Q_{max}=3$ and $Q_{min}=1/3$. The slope constraint is the second type in Fig.2. The weighting coefficients are 4,3,2,3,4 for s1, s2, s3, s4, s5, respectively.

This paper presents the results of the three algorithms: dynamic time warping (DTW), genetic time warping (GTW), tabu search time warping (TSTW), as shown in Table 1. The population size for GTW and the number of generations are set 20, the crossover rate is 0.7 and the mutation rate is 0.05 in GTW. The tabu list size and the number of iterations and current solutions are all 20, and the probability value of the solutions is 10%.

Table 1 shows the comparison results using the mean distortions of the same words D_s, the mean distortions of different words D_d and the difference $|D_d - D_s|$, respectively. We also use them to compare the performance of the algorithms because the absolute difference $|D_d - D_s|$ can make out the abilities of recognizing confused utterances. A lower value in $|D_d - D_s|$ implies that this speech recognition system has a weak ability to identify confused utterances. And a higher $|D_d - D_s|$ value implies a high level. Form Table 1, we can see that both of the GTW and the TSTW have higher level of confidence in recognizing confused utterances than the DTW, but from Table 2, the TSTW converges faster than DTW. Table 2 shows the comparison of GTW and TSTW for running-time (sec.), Where I_m denotes the number of iterations for TSTW and the number of generations for GTW. When the number of iterations of TSTW is one half of the ones of the GTW, the performance of TSTW can reach the similar level of the GTW. This verifies that the TSTW can converge faster and not affect its performance.

Table 1 Experiment results of DTW, GTW and TSTW

	DTW			GTW			TSTW		
	Ms	Md	Md-Ms	Ms	Md	Md-Ms	Ms	Md	Md-Ms
1	0.833	4.980	**4.147**	0.795	5.174	**4.379**	0.813	5.259	**4.446**
2	0.812	5.681	**4.869**	0.688	5.563	**4.875**	0.751	5.907	**5.156**
3	0.904	3.933	**3.029**	0.849	4.148	**3.299**	0.922	4.021	**3.279**
4	0.794	4.139	**3.345**	0.844	5.082	**4.238**	0.908	5.198	**4.290**
5	0.931	6.198	**5.267**	0.927	6.873	**5.946**	1.009	7.028	**6.019**
6	0.927	3.909	**2.982**	0.943	4.368	**3.425**	0.997	4.297	**3.300**
7	0.823	5.856	**5.033**	0.848	6.156	**5.308**	0.961	6.179	**5.218**
8	1.048	3.900	**2.852**	1.053	4.215	**3.162**	1.124	4.265	**3.141**
9	0.786	3.668	**2.882**	0.865	4.552	**3.687**	0.801	4.401	**3.600**

Table 2 Comparison of GTW and TSTW for running-time

I_m	20	40	80	100
GTW	0.205	0.366	0.698	0.861
TSTW	0.103	0.211	0.464	0.547

6. CONCLUSION

This paper proposed the optimized time warping algorithms by using the idea of tabu search to solve the nonlinear time alignment problems. Its fundamental idea is to find the optimal time warping path by a sequence of searching with memory. Experimental results have shown that the TSTW not only has the higher levels in identifying confused utterances but also has a shorter convergence time compared with the GTW.

REFERENCE

[1] F. Itakura, Minimum Prediction Residual Principle Applied to speech Recognition. IEEE Trans. On ASSP, **23**(1975) 67~72

[2] S. Kwong, C. W. Chau and Wolfgang A. Halang, Genetic Algorithm for Optimizing the Nonlinear Time Alignment of Automatic Speech Recognition System. IEEE Trans., **IE-43**(1996) 559~566.

[3] H. Sakoe and S. Chiba, Dynamic Programming Algotithm Optimization for spoken word recognition. IEEE Trans. On ASSP, **26**(1978) 43~49.

[4] F. Glover and M. Laguna, Tabu search, Kluwer Academic Publishers, 1997.

KES 2002
E. Damiani et al. (Eds.)
IOS Press, 2002

Genetic Algorithm on Portfolio Strategy in the Complicated Stock Market of Taiwan

*Jui – Fang Chang**   *Shu-Chuan Chu***   *Pei - Yun Wu****

**Dept. of International Trade , National Kaohsiung University of Applied Sciences, Taiwan,*
***Dept. of Industrial Engineering & Management, National Kaohsiung University of Applied Sciences, Taiwan,*
****Dept. of   Electronic Engineering, National Kaohsiung University of Applied Sciences, Taiwan,*

Abstract. Due to the stock market is so complicated and filled with many kinds of variables, most investors can not be easy to find or induce the rules of i nvestment f rom s o m uch d ata. A rtificial I ntelligence (AI) i s a po pular technique, and it has been applied to solve engineering problem for many years. Some of professionals have been planning to apply AI to the research of investment and financial management recently. The methods such as neural networks, fuzzy logic, gray system theory, and genetic algorithms have been applied to this field. In this research we try to utilize Genetic Algorithm to find the optimal portfolio strategy in the complicated stock market of Taiwan. 20 stocks and 10 financial ratios have been chosen and tried from the possible tractions in order to find the optimal portfolios strategy. From the results, we found the portfolio got 14% ROI per annual. It shows the profit from fixed account i s b etter t han the p rofit g et f rom average w eight s tock m arket o f Taiwan.

1. Introduction

Artificial Intelligence (AI) is a popular science recently; many professionals are studying AI to allow for more sophisticated application such as Genetic Algorithms (GA) Neural Networks, and fuzzy logic.

The stock market is full of every kind of noises. The general investor faces difficulty due to large amount of data. AI technology can be used to search for the trade tactics to conform the characteristics of the market. From an investor's standpoint, the price fluctuation of a single stock is not the point, for most important are the hazards and honorarium rate of the whole stock portfolio.

A point of portfolio theory is how to get the optimal return on investment and how to arrive at the lowest venture. This paper investigates Genetic Algorithm approaches to solve these problems[8,9].

2. Algorithm

In this paper, we encoded the portfolio in accordance with the GA approach. [1,2,3,4] Here, we produce a group of 20 candidate solutions for every iteration, then pick the suitable candidate solutions and apply the crossover and mutation techniques to produce the solutions for the next iteration. The best member of 20 portfolios compare between depositing the investment amount in fixed account and investing in stock market. In this way, the optimal portfolio is obtain in the final generations, as show in figure 1.

In this paper, we established a fitness function to evaluate the performance of the portfolio[5,6,7]. We assume ten genes were included in each chromosome and the fitness function was composed by the sum of these ten genes, which was multiplied by a suitable ratio, and the sum of ratio equal to one. If the value of the fitness function is higher, the performance of portfolio would be better. It means the portfolio would get higher profit.

2.1 Selection

The principle behind GAs is essential Darwinian natural selection: both parents and offspring have the same chance of competing for survival. An evident advantage of this approach is that the GA performance can often be improved. This is because genes are always evaluated within the context of a larger individual.

2.2 Crossover

The crossover technique is simply using single point crossover as show in figure2.

Figure 1. Outline of GA

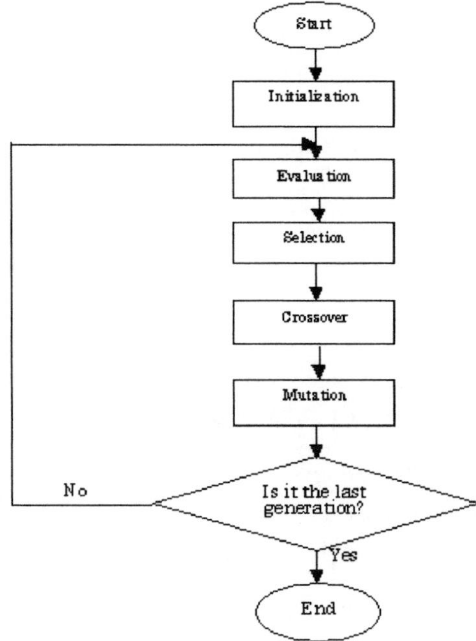

Figure 2. Representation and crossover

Initial Chromosome 1 (Parent 1):

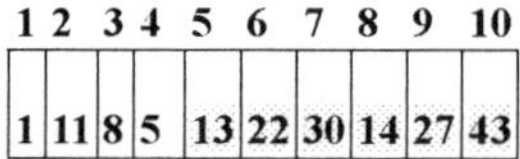

Initial Chromosome 2 (Parent 2):

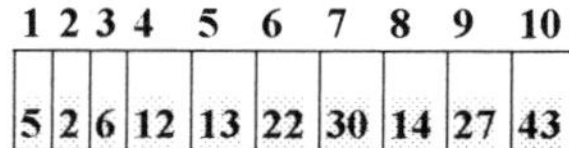 **Crossover**

Child Chromosome 1:

Child Chromosome 2:

2.3 Mutation

Mutation operates on a single string and generally changes a gene at random, which is shown in figure 3. In this paper, the mutation rate is set to be 0.2.

Example:

Figure 3. Mutation

Chromosome 1:

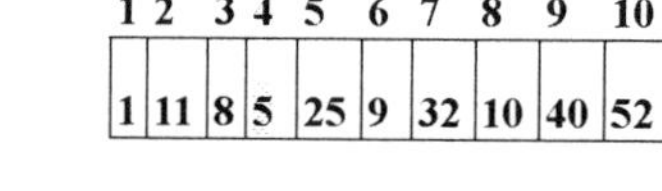 **Mutate**

Chromosome 1 :

3. Problem Description

There are two risks that investors of any estate must sustain: the first risk is system risk or market risk, which has a close relation with economy; it cannot be dispersed or removed, so it is called inseparable risk. The second is non-system risk, which is created from the various factors of individual companies; it can be deleted from the portfolio method, so it is called detachable risk. The risk referred to in this paper is detachable risk. The goal of constructing portfolios is to have the lowest detachable risk at the same amount of remuneration rate.

In this paper, we construct a stock selection system, and utilize the GA to find the optimal portfolio strategy in the complicated stock market of Taiwan; the optimal portfolio strategy contains the best 20 stocks and apportionment rates. Each portfolio strategy contains 20 stocks. We obtain the best portfolio by means of the 10 financial ratios over the past 8 years.

The Data series covered the years from 1991 to 2000 and 5,040 pieces of data were chosen from 40 0 c ompanies t hat h ave been pe rforming w ell o n t he Taiwan s tock market. The data taken from 1991 to 1998 were used to construct the model, and the data between 1999 and 2000 were used to test the model of the 25 companies, there were 20 stocks that possessed the best capital return and were utilized to establish the portfolio. Furthermore, 10 financial ratios were composed based on earnings per share, cash flow per share, liquidity ratio, current ratio, debt ratio, return on assets, profitability (net margin), and return of equity (after tax).

4. Results and Discussions

The overall quality of the portfolio is evaluated by the evaluation function that computes 10 financial ratios by testing it with selected stocks. In period of research, Table 1 shows there are a great deal of profit obtained from GA which is better than the profit margin of market, and got 14% ROT per annual. Otherwise, it also verify that Taiwan stock market is not efficient. According to the results of basic analysis of financial report, the investors in Taiwan stock market still have a chance to make a great deal of profit. In this article, we have assumed the investors always buy the stock that based on the opening price and did not constrain the quantity. However, if people want to evaluate the real performance of the stock market, they should run by real-time study in order to control the price and quantity.

Table 1   ROI of Portfolio and Market

Time	1998 IV	1999 I	1999 II	1999 III	1999 IV	2000 I	2000 II	2000 III	2000 IV
ROI of Porfolio	14.01	13.95	13.37	14.38	16.91	14.12	14.29	14.92	14.42
ROI of Average Weight of Market	0.54	-5.58	-0.96	4.12	-3.39	-3.78	3.28	-6.77	-8.38

Figure 4   ROI of Portfolio by Genetic Algorithm

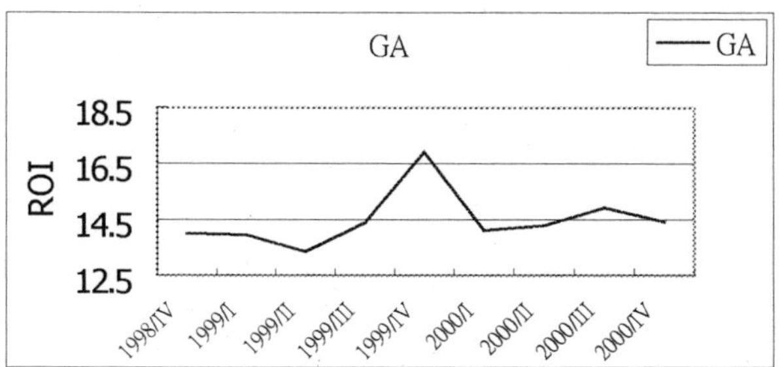

Figure 5   ROI of Portfolio By Average Weight Of Market

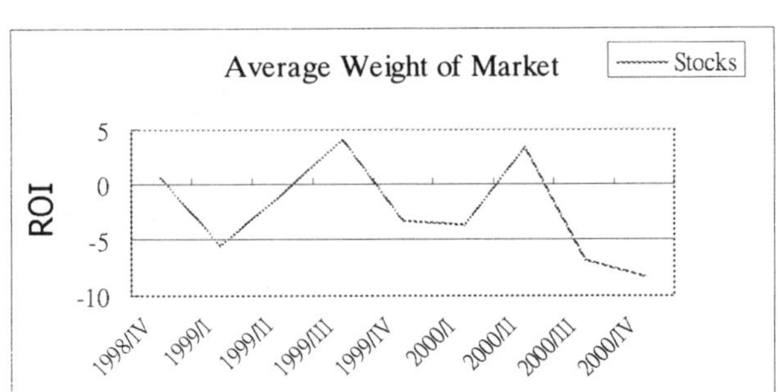

References

1. A Loraschi, A. Tettamanzi, M. Tomassini and P. Verda, *"Distributed Genetic Algorithms with an Application to Portfolio Selection Problems,"* in Proceedings of the Int. Conf. on Artificial Neural Nets and Genetic Algorithms, D.W. Pearson, N.C. Steele and R.F. Albrecht (Editors), Springer-Verlag, 384, 1995.

2. Chu, S. C. and Fang, H. L., *"Genetic Algorithms vs. Tabu Search in Timetable Scheduling,"* Third International Conference on Knowledge-Based Intelligent Information Engineering Systems, 1999.

3. Chih-Ron Chiu, *"An Application of Genetic Algorithms on Portfolio Strategies,"* National Sun Yat-Sen University Institute of Economics-Master's paper, 1999.

4. Herrera, E. Herrera-Viedma and J.L. Verdegay, A Linguistic, *"Decision Process in G roup D ecision M aking,"* G roup D ecision a nd N egotiation 5, pp. 1 65-176, 1996.

5. Holland. J. H. *"Adaptation in Natural and Artificial Systems,"* University of Michigan Press, 1975.

6. Haliassos, M ichael a nd Alexander M ichaelides, " *Portfolio c hoice an d li quidity constraints.*" Working paper: University of Cyprus, 1999.

7. H. M. Markowitz, *"Portfolio Selection,"* New York, John Wiley & Sons, 1959.

8. Thomas K L Tong, C M Tam and Albert P C Chan, *" Genetic algorithm optimization in building portfolio management,"* Construction Management and Economics, London, Vol. 19, Iss. 6, pp. 601, Oct 2001.

9. Yusen Xia, Shouyang Wang and Xiaotie Deng, *"A compromise solution to mutual funds portfolio selection with transaction costs"*, European Journal of Operational Research, Amsterdam; Nov 1, Vol. 134, Iss. 3; pp. 564, 2001.

KES 2002
E. Damiani et al. (Eds.)
IOS Press, 2002

1137

Achieving Uplink Optimal SDMA of Smart Antenna by Phase-only Disturbances Using Genetic Algorithms

Chao-Hsing Hsu[#]* and Tadeusz M. Babij*

#Department of Electronic Engineering, Chienkuo Institute of Technology, Taiwan
***Department of Electrical & Computer Engineering, Florida International University, FL, U.S.A.**

Abstract: This paper presents a smart antenna system composed of adaptive antenna arrays. It can not only derive the maximum main lobes towards all the desired signals in their required directions but also suppress interferences. Thus, the optimal Spatial Division Multiple Access (SDMA) effect is achieved, and the Signal Interference Ratio (SIR) is maximized. The uplink radiation pattern formula based on phase available for optimum search is deduced. The appropriate phase disturbances for optimum radiation pattern are figured out by using genetic algorithms. The results show the validity of the proposed method and effectiveness of genetic algorithms for this problem.

. Introduction

As known, having SDMA function, a smart antenna can deal with more than one user in a same channel at the same time. This means the desired signal of one user is the interfering signal of other users. So, in order to get optimal radiation pattern of a smart antenna system, both pattern nulling technique and main lobe maximizing technique should be applied.

In this paper, the structure of the proposed smart antenna system is given and the corresponding uplink radiation formula based on constant amplitude weights and adjustable phase weights is deduced. It is used as the fitness function for optimal radiation pattern search. As mentioned above, the maximum output power of desired signals and minimum output power of interfering signals to receiver should be reached at the same time, it is important to select an effective optimization technique. Genetic algorithms without gradient-based search have the ability to get out of local minima so as to reach global optima by mutation technique. It is adopted in this paper and good results have been gotten.

. Developing radiation pattern formula of smart antenna for optimum search

For a linear array of $2N$ equispaced sensor elements as shown in Figure 1, a signal of wavelength λ from direction θ with respect to array normal impinges on any two adjacent element n and element $n+1$ by distance d as shown in Figure 2 [1-3].

Obviously, there is a time delay τ for the signal to reach one element first then the element next to it [4].

$$\tau = \frac{d \sin \theta}{v}$$

Where, v is the propagation speed of radio wave. τ corresponds to a phase shift of ψ :

$$\psi = \frac{2\pi}{\lambda} d \sin \theta = kd \sin \theta$$

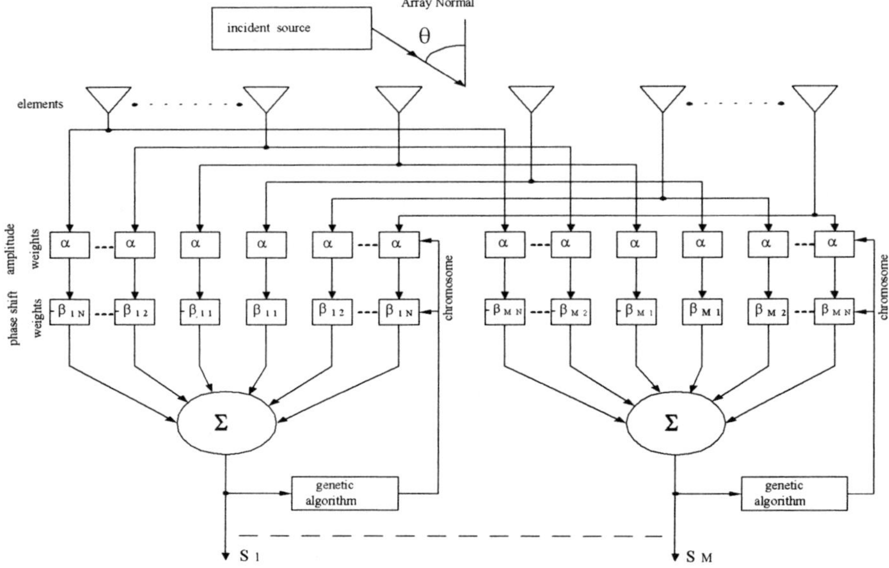

Figure 1. Uplink diagram of smart antenna designed by phase-only disturbances

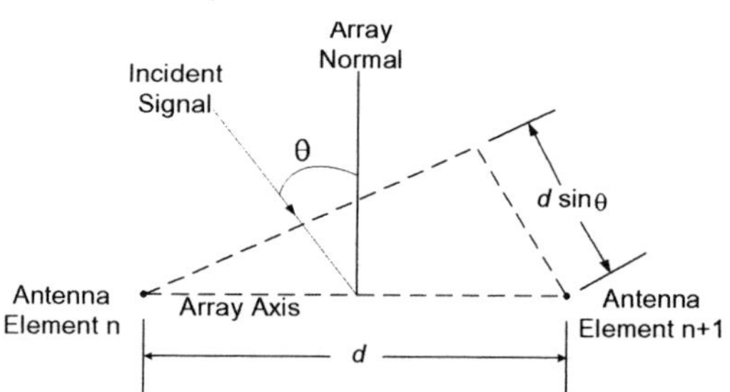

Figure 2. The incident signal reaching any two adjacent elements

Then, for far field, the array factor of the adaptive linear array for receiver m of a smart antenna can written as

$$AF_m(\theta) = \frac{1}{M} \sum_{n-1}^{2N} w_{mn} e^{j(n-1)\psi} \quad , \qquad m=1, 2, 3, \ldots, M$$

Where, AF_m = array factor of user m

θ = incident angle of the signal

M = number of users

$2N$ = number of elements

w_{mn} = complex array weight at element n of receiver m

$_{mn}$ can be expressed as

$$w_{mn} = \alpha\ e^{j\beta_{mn}} \qquad (4)$$

here, α : constant amplitude weight at any element for any receiver

β_{mn} : phase shift weight at element n for receiver m

If the reference point is set at the physical center of the array, the array factor can be rewritten as:

$$AF_m(\theta) = \frac{1}{M}\sum_{n=1}^{2N} w_{mn} e^{j(n-N-0.5)\psi} = \frac{1}{M}\sum_{n=1}^{2N}\alpha\ e^{j[(n-N-0.5)\psi+\beta_{mn}]} \quad , \qquad m=1, 2, 3, \ldots, M \qquad (5)$$

Assuming, amplitude weights are constant and phase shift weights are in odd symmetry. Then, the uation (5) can be simplified to

$$AF_m(\theta) = 2*\frac{1}{M}\sum_{n=1}^{N}\alpha\ \cos[(n-0.5)\psi+\beta_{mn}] \qquad , \qquad m=1, 2, 3, \ldots, M \qquad (6)$$

This is suitable for optimal solution search.

Phase Weight Vector Searching of Optimal Radiation Pattern by Genetic Algorithms

Genetic algorithms are adopted to determine weights of phase shift for getting optimal uplink radiation ttern of a smart antenna [5]. The goal is to maximize the total output power of the desired signal to the rresponding receiver and minimize the total output power of its interfering signals. In reference paper [1], e genetic algorithm is used to search the weight vector only for minimizing the output power of terfering signal. So it has to be improved in order to realize the above goal. During the process of genetic gorithm iteration, the weight vector kept for the next step iteration should be able to make the output wer of the desired signal to be increased and the output power of the interfering signal to be decreased onotonically. Obviously, this technique can maximize the *signal to interference ratio* (SIR).

Computer Calculation Results and Simulations

As an example, a smart antenna composed of two adaptive antenna arrays is studied. It can serve two ers at the same time. Therefore, M equals to 2. The linear antenna array contains 20 isotropic elements. , N equals to 10. The distance d between two adjacent elements is half of λ. In this paper, the optimal ttern technique feature is phase-only disturbances. Every amplitude weight is kept to be equal to 1 and ase shift weights are set in odd symmetry.

Genetic algorithms are utilized to search the weight vectors of the optimal radiation pattern of the smart tenna. The fitness function is the square of $AF_m(\theta)$ in equation (6). The necessary variables used are fined as follows: Population size p is 20; the survival rate p_s is 0.5; the probability of crossover is 0.5; the oportion of mutation is 0.5. The value of β_{mn} is set between $-\pi$ and π.

With respect to array normal shown in Figure 1, the signal of user 1 is coming from 50^0, user 2's signal from -40^0. Linear adaptive antenna 1 and 2 are used to serve user 1 and user 2 respectively. Obviously, e desired signal of one user should be considered as the interfering signal of another user. With the ethod introduced above, the appropriate weight vector, listed in Table 1, is gotten and the optimal liation pattern shown in Figure 3, 4 and 5, are achieved. The SIR for user 1 is 225dB, for user 2 is 5dB. The values of SIR are large enough. This means the interferences can be ignored.

The radiation pattern of the smart antenna for user 1 and user 2 is shown in Figure 5. The radiation pattern is mapped from 90^0 to 270^0. The user 1's signal comes from the direction of 50^0. The user 2's signal comes from the direction of -40^0 (i.e. 320^0).

Table 1. The weight vector for the optimal pattern

weight vector for user 1	weight vector for user 2
$\beta_{11} = 0.270$	$\beta_{21} = 1.708$
$\beta_{12} = 2.631$	$\beta_{22} = -2.857$
$\beta_{13} = 1.426$	$\beta_{23} = -0.132$
$\beta_{14} = -2.449$	$\beta_{24} = 1.344$
$\beta_{15} = 0.528$	$\beta_{25} = 3.014$
$\beta_{16} = -0.257$	$\beta_{26} = -1.250$
$\beta_{17} = -2.901$	$\beta_{27} = 0.628$
$\beta_{18} = 1.539$	$\beta_{28} = 2.462$
$\beta_{19} = -1.978$	$\beta_{29} = -1.922$
$\beta_{110} = 1.909$	$\beta_{210} = -1.375$

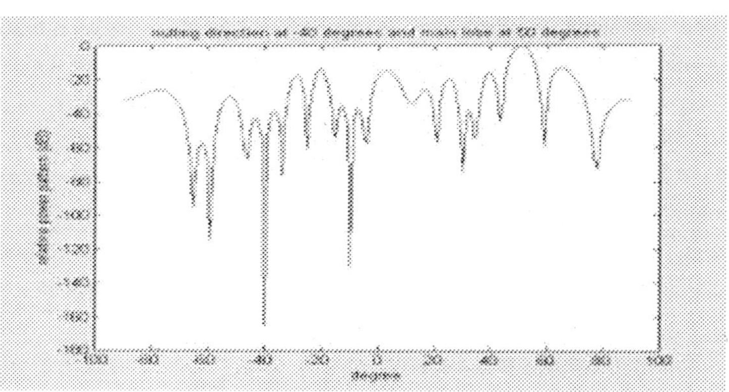

Figure 3. Optimal radiation pattern for user 1 at 50^0 and interfering source at -40^0

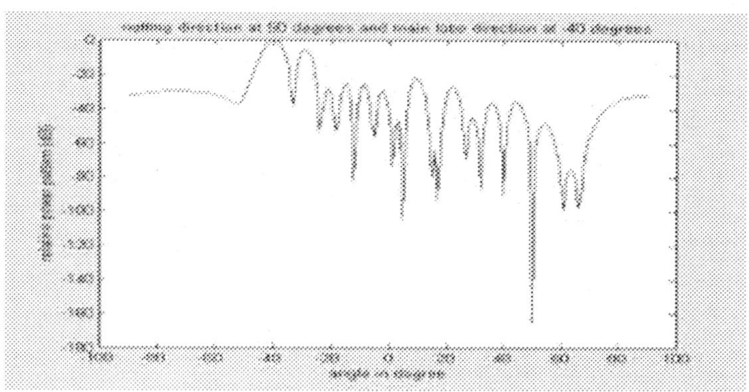

Figure 4. Optimal radiation pattern for user 2 at -40^0 and interfering source at 50^0

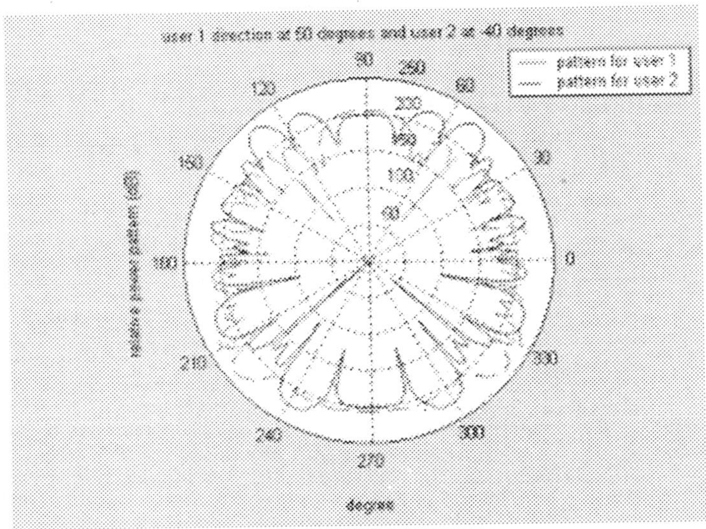

Figure 5. Radiation pattern of smart antenna

5. Conclusions

In the paper, the uplink output power formula based on phase shift of a smart antenna composed of adaptive antenna array is deduced. The amplitude weights are kept constant. In order to be able to adopt optimization techniques to realize the optimal radiation pattern, setting phase shift weights in odd symmetry reforms the formula. Genetic algorithms are applied to find the phase shift weight vector of the optimal radiation pattern of the proposed smart antenna. The optimal radiation pattern can make the power of desired signal to be highest and the power of interfering signal to be lowest at the same time. The appropriate phase shift weight vector and the corresponding optimal radiation pattern of a two-user smart antenna system are derived. Thus, the optimal Spatial Division Multiple Access (SDMA) effect is achieved. The results show the effectiveness of the proposed method.

Reference

[1]. C. H. Hsu and T. M. Babij, "Pattern Nulling of Adaptive Antenna by Phase and Amplitude Perturbations Using Genetic Algorithm, " KES'2001, Part 2, pp. 1047-1051, 2001.

[2]. Randy L. Haupt, "Phase-only Adaptive Nulling with a Genetic Algorithm," IEEE Trans. Antenna Propagation, Vol. AP-45, pp. 1009-1015, June 1997.

[3]. Wen-Pin Liao and Fu-Lai Chu, "Array Pattern Nulling by Phase and Position Perturbations with the Use of the Genetic Algorithm" Microwave and Optical Technology Letters, Vol. 15, No. 4, pp. 251-256, 1997.

[4]. Monzingo, Miller, "Introduction to Adaptive Arrays", Wiley, New York, 1968.

[5]. M. Gen and R. Cheng, "Genetic Algorithms & Engineering Design", John Wiley & Sons publishing Company, 1997.

KES 2002
E. Damiani et al. (Eds.)
IOS Press, 2002

A Simple Tempo-Spectral Hybrid Method
for Speech Noise Reduction

Haifeng LI, Lin MA, Patrick GALLINARI
Computer Science Lab., University Paris 6, 8 Rue Capitaine Scott, 75015 Paris, France

Abstract. This paper presents a simple yet effective noise reduction approach for speech systems. Unlike most conventional methods, the suggested approach treats consonants and vowels separately. The reason is simple: noise similar consonants can be cleaned only through spectral or cepstral subtraction; vowels however, may be considered as periodic signals and the pitch information helps to gather several successive pitch waves together to obtain a better estimation of the original signal. The suggested approach was compared with several wildly studied methods, like spectral subtraction, Kalman filtering and Gaussian MMSE on the NTIMIT speech database. A better robustness was approved and the results were very encouraging.

1. Introduction

For most practical speech processing applications, noise reduction (NR) is indispensable into the front-end of these systems and its effectiveness is crucial to the system's performance, because here occurred errors are difficult to be recovered in following processing. Usually, the NR should be as simple as possible in order to save the system resource for other more time consuming treatments such as encoding, feature extraction and recognition [1,2].

Two common ways for noise reduction are Wiener filtering and spectral subtraction [1,3,4], based on which many more complex and model-based methods (like HMM based ones, etc. [2,5]) have been proposed for a better performance. Being the most representative non-parametric method, spectral subtraction has also been inferred as the most efficient one to uncorrelated additive noise [6]. Its basic advantage is exactly the implementation simplicity and low computational complexity that is highly demanded in a commercial device like mobile phones [7]. We then decided to consider it as a starting point.

However, this method has a drawback when the noise becomes non-stationary [2]. Noise spectrum estimation is often done at non-speech periods, it may decade quickly for a long uninterrupted speech, because of the lack of trustable information to refresh it. Fortunately, speech is composed by consonants and vowels. Since vowels are often strictly short time periodic, a simple smoothing method can give satisfactory results at little computation cost. This explains why a human being can easily point out the noisy sampling points from the vowel waveform. Our idea is to process vowels and consonants separately: vowels will be smoothed in the time domain and the residual noise will be used for estimating noise spectrum; meanwhile, consonants will be cleared by spectral subtraction with such up-to time noise spectrum. As the result, the performance will be improved on a long speech.

In Section 2, the schema of the suggested tempo-spectral hybrid approach is introduced and the details are given in the coming Sections 3, 4 and 5. In Section 6, a brief comparison with several traditional methods is given and the performance of the proposed approach is tested on the NTIMIT speech corpus. At last, our conclusions are given in Section 7.

2. Approach Overview

Assume that a digital signal $y(n)$ is a noisy measurement of a true speech $s(n)$ so that $y(n)=s(n)+d(n)$, where $d(n)$ is the noise that may be white or colour, stationary or non-stationary, but an important common character is that it is far from periodic.

The diagram of our approach is shown in Fig-1. Here, consonant frames and vowel frames are firstly separated by a consonant/vowel detector and passed to two independent channels – a spectral one and a temporal one. The wave-cut based autocorrelation technique is applied for creating a score of possibility at which a frame may be considered as a vowel. At meantime, the pitch of the frame is also estimated. The detail is assumed in Section 3.

For consonants, a spectral subtraction is used in the spectral domain (Section 4). Spectral subtraction bases on this fact: the power spectrum of the sum of two independent random signals is the sum of their power spectra. Therefore, the noise elimination may be simply realized by subtracting the noise's power spectrum from that of the observed signal. With accurate noise spectrum estimation, a good performance can be easily obtained. Here, no special information about the signal is required. In our method, a good estimation of noise spectrum is insured by the treatments in the temporal channel where vowels are processed.

For vowels, thanks to their periodic nature, simple smoothing methods may be applied. Especially when successive waveforms can be compared, the smoothing result may be greatly improved. That's why the pitch information is needed in our inter-pitch middle filtering method (Section 5). Assuming $\hat{s}(n)$ is the noise reduced speech, $y(n)-\hat{s}(n)$ will be a good estimation for noise with which an accurate noise spectrum will be computed dynamically. Such a dynamic estimation makes our approach auto-adaptive.

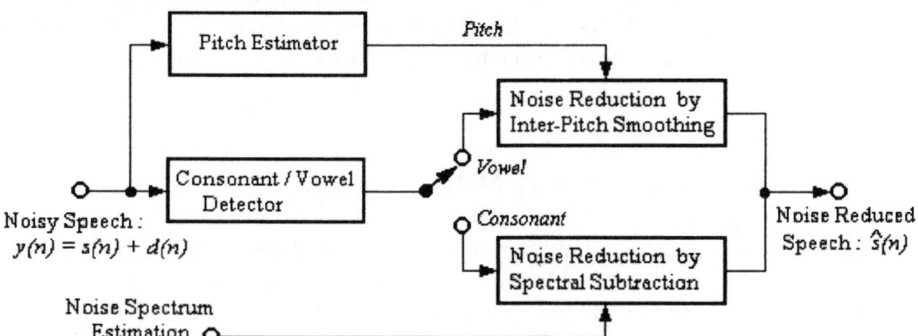

Fig-1: Schema of the suggested tempo-spectral hybrid noise reduction approach

3. Consonant/Vowel Discrimination through Wave-Cut Autocorrelation

We use $y(n)$, $n=1{\sim}N$, to indicate an input frame. The wave-cut autocorrelation is realized by 3 steps. Firstly, an amplitude threshold is calculated: $T^{(A)}=\alpha A$ $(0<\alpha\le 1)$, where A is the average amplitude. Secondly, $y(n)$ is cut by $T^{(A)}$. And lastly, the autocorrelation function is calculated: $\Phi^{(v)}(m)=\Sigma y(n)y(n-m)$. As the wave-cut greatly reduces the noise energy and the noise $d(n)$ is usually an independent white noise, $\Phi^{(v)}(m)$ reserves most of the periodic information of the original signal for computing the pitch.

A pitch estimation is as bellowing: firstly, a wave-cut is also done to $\Phi^{(v)}(m)$ according to a threshold $T^{(T)}=\beta\Phi^{(v)}(0)$, $(0<\beta\le 1)$. Secondly, all local peaks and their locations are found: $L(1){\sim}L(H)$. Then H pitch candidates are obtained: $P(1)=L(1)$ and $P(h)=L(h)-L(h-1)$ for $h=2{\sim}H$. Lastly, the average value is computed as the pitch estimation for this frame $P^{(v)}$.

As the result, a consonant/vowel score can then be created: $cv = F/H$, where F is the number of the pitch candidates whose values are enough close to $P^{(v)}$ according to a threshold θ. In our experiments, $\theta = 0$.

This score cv acts as a good estimator of the possibility at which a signal frame can be considered as a periodic signal – a vowel. The larger is the value, the more possible that it appears. So with a simple threshold (for example, we use 70%), this frame is classified as a consonant ($cv < 70\%$) or a vowel ($cv \geq 70\%$).

4. Consonant Noise Reduction by Spectral Subtraction

As it is mentioned formerly, when the power spectrum of the signal and the noise are additional, the noise elimination may be simply realized by subtracting the noise's power spectrum from that of the observed signal in the frequency domain [5,6]. Traditionally, short-time Fourier transform is used to get the time-frequency representations of a signal. The noise power spectrum estimation is usually created from non-speech frames and averaged with a forgetting factor, at the condition that the signal to noise ratio (SNR) is not too low.

This standard form is based on a fairly accurate voice activity detector to classify non-speech frames and requires a sufficient amount of non-speech frames. However, this is often difficult to achieve in very noisy condition or in a changing environment application.

In our approach, we applied 3 criterions to insure an accurate noise spectrum estimation: firstly, at time when the speech energy is enough low, the noise spectrum estimation is totally refreshed and stored the result as a prior information (D^{old}); secondly, during the vowel frames, noise spectrum evolves with an up-to-time estimation according to a formula like this: $D^{old}=\lambda D^{old}+(1-\lambda)D^{new}$, where D^{old} is the current noise spectrum, D^{new} is the up-to-time estimation, and $0\leq\lambda\leq1$ is a forgetting factor; and thirdly, during consonants, the noise spectrum rests as what it was like at the beginning of the consonant.

5. Vowel Noise Reduction by Inter-Pitch Middle Filtering

For vowels, we apply a simple smoothing method in time domain to eliminate noise. The advantage is that in this way, all assumptions to noise are relaxed. The only assumption is that noise should be enough random according to the considered signal. Clearly, this request is easily satisfied. The most common time domain smoothing methods are average value filtering and middle value filtering. Here, we prefer the later one.

In the common form of a middle value filtering, $K=2j+1$ ($j=1,2,...$) samples are considered in the neighborhood of the current point $y(n)$ as: $\{y(n-j),..., y(n),..., y(n+j)\}$. We call this a neighborhood smoothing. These values are then sorted in a sequence as: $y'(1)\leq...\leq y'(k) \leq...\leq y'(K)$, where $y'(k)\in \{y(n-j),..., y(n),..., y(n+j)\}$. At last, the current sample is replaced by the middle value in that sequence: $y(n) \leftarrow y'(j+1)$.

The well-known advantage of such a neighborhood smoothing is that up to j strange values may be eliminated and the residual error may be much smaller than an average filtering. However, it has also a well-known drawback that all local peaks will be smoothed. What is more, whenever there is a strange value, the corrected value is only the closest one to the real value.

Thanks to the periodic nature, we apply an inter-pitch form of middle value filtering in our approach. The difference with the neighborhood form is that the K samples are now chosen from the K successive pitch waveforms as: $\{y(n-j*P^{(v)}),..., y(n),..., y(n+j*P^{(v)})\}$, where $P^{(v)}$ is the pitch of this frame. In theory, we have $y(n-j*P^{(v)})=...= y(n)=...= y(n+j*P^{(v)})$, if there isn't any noise among them. So when the noise sample number is smaller than j, the exact value can always be obtained.

We thus call this an inter-pitch form middle value smoothing. Clearly, this form possesses the same advantage as the former one, but the residual error should be small than the former one. What is more, this method has a good robustness when the pitch estimation is multiplied.

6. Experiments and Results

To evaluate the performance of the suggested tempo-spectral hybrid approach (TSH), experiments were complimented on both the TIMIT and NTIMIT acoustic-phonetic speech corpuses of NIST. Each of these corpuses contains 6300 sentences. The speech waveforms are digitized at a 16 kHz sampling rate. The NTIMIT is a noisy form of the TIMIT corpus where all the sentences are well aligned between them. So we use NTIMIT as the $y(n)$ database and TIMIT as its reference $s(n)$ database. Here, the signal to noise ratio (SNR) is used to evaluate the performance of the TSH.

In our simulation experiments, 100 sentences among those 6300 in NTIMIT were chosen randomly. A window of 512 sample points was used with 256 points overlap between successive windows. With their correspondences in TIMIT, noise signals were calculated and noise spectrum was estimated for each frame. The average among all the frames was then stored as the a priori noise spectrum.

Other 100 sentences were chosen to evaluate the TSH. The parameters used during these tests were: $\alpha = 1.0$, $\beta = 0.4$, $\lambda = 0.9$, and $K = 5$. The threshold for cv score was experimentally chosen as 70% for a good compensation between the two channels.

The result is very encouraging. From observing the result spectrums and waveforms, we could conclude that the noise had been dramatically cancelled, while the average SNR is improved from the about -15 dB for the noisy speeches to the +16 dB for the cleared results.

In order to obtain a further impression about the performance of the proposed TSH, a comparison with several conventional techniques was also realized. Three wildly used methods were used as references [1,8]:

1) a standard spectral subtraction noise reduction (SB),
2) a Kalman filter based speech enhancement method (KF), and
3) a minimum mean square error based one under Gaussian assumption (MMSE-G).

Here, ten sentences were chosen from the TIMIT corpus as clear speeches. A Gaussian white noise was created artificially and added to the clear speeches. Experiments were tried out in five SNR conditions: -10 dB, -5 dB, 0 dB, 5 dB and 10 dB, just the same as in [1]. The choice of the parameters rested the same as in the former experiments. The average SNR of the processed speeches is drawn Fig-2. Among these five conditions, an average enhancement of about 7 dB was always obtained for our TSH approach over the SB.

From Fig-2, the effect of the proposed TSH is evident. In most of the cases, it outperforms all these classical methods, and it may reach an improvement of more than 5 dB over the best of them (in a reasonable practical application of a 10 dB noise).

The good performance of the TSH may be explained in several aspects. At first, the separated consonant and vowel channels may give better treatments according to the specialties of each frame. Secondly, the time domain inter-pitch middle value smoothing may bring a better and cleaner estimation to vowels. At the same time, it insures an up-to-time noise spectrum estimation. Then, the more accurate noise spectrum may bring a better spectral subtraction result to consonants. As the result, this hybrid approach works in an EM like manner to profit an improved global performance.

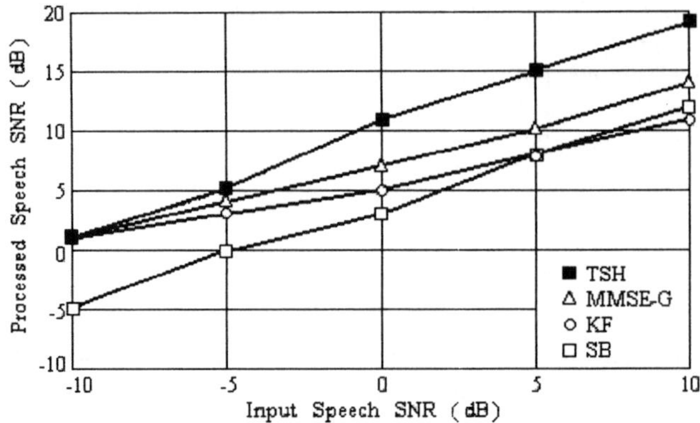

Fig-2: Comparison of the TSH with the conventional SB, KF and MMSE-G

7. Conclusions

In this paper, we present a consonant/vowel detection based tempo-spectral hybrid approach (TSH) for speech enhancement. It uses a wave-cut autocorrelation technique to estimate the pitch of a signal frame and to create a score at which the current frame can be classified as consonant (or vowel). After this detection, consonants and vowels are processed separately in two independent channels. A simple inter-pitch middle value smoothing is applied in the time channel while a simple spectral subtraction is used in the frequency channel. The noise spectrum estimation is refreshed dynamically by the time channel.

Experiments were realized on the NTIMIT speech database, and the results show that the suggested method can enhance a noisy speech from -15 dB to +16 dB. Other experiments on TIMIT database with simulated Gaussian noises prove that it is capable to outperform many classical methods, such as SB, KF and MMSE-G.

The good performance of the proposed approach is due to its novelties as described formerly. Besides, this method has a good robustness and generality that make it possible be applied in a large amount of practical applications.

References

[1] X.J. Yang, H.S. Chi, *et al.*, Digital Processing of Speech Signals. Publishing House of Electronics Industry, Beijing, China, 1995.

[2] S. Young, Large Vocabulary Continuous Speech Recognition: A Review. In Proc. of the IEEE Workshop on Automatic Speech Recognition and Understanding, USA, December 1995, pp. 3-28.

[3] A. Yasmin, Speech Enhancement Using Voice Source Models. PhD Thesis, University of Waterloo, Canada, 1999.

[4] S. F. Boll, Suppression of Acoustic Noise in Speech Using Spectral Subtraction, IEEE Transactions on Acoustics, Speech and Signal Processing, Vol.27, 1979, pp. 113-120.

[5] B. T. Logan and A. J. Robinson, Enhancement and Recognition of Noisy Speech Within an Auto-regressive Hidden Markov Model Framework Using Noise Estimates from the Noisy Signal. In proceedings of ICASSP'97, Munich, Germany, 1997, pp. 843-846.

[6] J.S. Lim and A.V. Oppenheim, Enhancement and Bandwidth Compression of Noisy Speech. In Proceedings of IEEE. Vol.67(12), 1979, pp.1586-1604.

[7] J. Yang, Frequency Domain Noise Suppression Approach in Mobile Telephone Systems, In Proceedings of ICASSP'93, Vol.2, Minneapolis, Minnesota, USA, April 1993, pp. 363-666.

[8] C. Kermorvant, A Comparison of Noise Reduction Techniques for Robust Speech Recognition. IDIAP Research Report, IDIAP-RR 99-10, 1999.

KES 2002
E. Damiani et al. (Eds.)
IOS Press, 2002

Channel Noise Reduction Using Genetic Index Assignment and Energy Allocation for VQ with BPSK and BFSK

T. C. Yang*, L.C. Jain*, K. C. Huang**, J. S. Pan***

*Department of Electrical and Information Engineering, University of South Australia, Australia
**Department of Automatic Engineering, Lan-Yang Institute of Technology, Taiwan
***Department of Electronic Engineering, National Kaohsiung University of Applied Sciences, Taiwan

Abstract. In this Paper, allocate the energy to bits according to their error sensitivity combined with index assignment has been shown to be effective for reducing the channel error of VQ-based signal compression. A novel energy allocation scheme termed genetic energy allocation is combined with the genetic index assignment to reduce the channel noise based on BPSK and BFSK for VQ system. Experimental results had shown significant improvement in reducing noisy inference.

1. Introduction

Vector Quantization (VQ) [1] shown in Fig. 1 is one kind of compressed technique for data and image. The image block X consists of a given set of k-dimensional data vectors. The encoder of VQ encodes with a much smaller set of codevectors. Only the index of the codevector is sent to the decoder. The index is modulated by BPSK/BFSK and transmitted to the receiver. In the receiving end, the demodulation is operated and the index j is obtained due to the channel noise. The decoder has the same codebook as the encoder, and decoding is operated by table look-up procedure. Therefore, distortions are introduced in the decoding step.

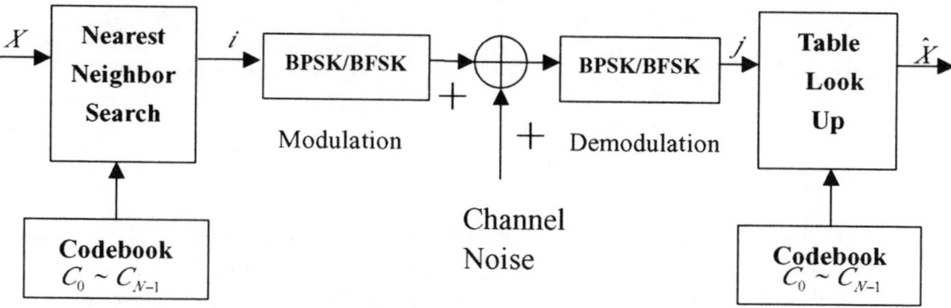

Fig. 1 The Block Diagram of Encoder and Decoder of M/R VQ over Channel Noise

2. Conventional Energy Allocation Algorithm

At first, Bedrosian [3] developed the Energy Allocation technique. Gadkari and Rose [4] took advantage of the similar idea to allocate the energy to each bit of the codevector index for VQ-based BPSK (Binary Phase Shift Keying) over channel noises. The codevectors satisfy the centroid rule [5] if the squared Euclidean distortion is used. The overall distortion is shown as following:

$$D = E|x - c_i|^2 + E|c_j - c_i|^2 = D_q + D_c,\qquad(1)$$

Where D_q and D_c denote the quantization distortion and the channel distortion, respectively. Let ε_i be the bit error rate and D_i be the channel distortion due to the error of the ith bit. Assume only single bit error is considered, then the distortion due to channel errors can be simplified to Eq. 2, where m is the number of bits to represent the index of the codevector. Therefore, the total energy can be expressed using Eq. 3.

$$D_c = \sum_{i=1}^{m} D_i \varepsilon_i ,\qquad(2)$$

$$E_{tot} = \sum_{i=1}^{m} E_i ,\qquad(3)$$

Where E_i is the energy assigned to the ith bit. The channel SNR is given by

$$10 \log \frac{E_{tot}}{\sigma_n^2},\qquad(4)$$

Where σ_n^2 is the variance of the Gaussian Channel noise. If the selected modulation is BPSK, then the ith bit error rate is expressed as

$$\varepsilon_i = \frac{1}{\sqrt{2\pi}} \int_{\sqrt{\frac{E_i}{\sigma_n^2}}}^{\infty} e^{\frac{-x^2}{2}} dx\qquad(5)$$

The partial derivative $\partial\varepsilon_i / \partial E_i$ can be evaluated and the necessary condition for optimality is given as

$$\frac{\partial D_c}{\partial E_i} = D_i \frac{\partial \varepsilon_i}{\partial E_i} = \frac{-D_i}{2\sigma_n \sqrt{2\pi E_i}} e^{-\frac{E_i}{2\sigma_n^2}}\qquad(6)$$

If we choose BFSK, the bit error rate is:

$$\varepsilon_i = \frac{1}{\sqrt{\pi}} \int_{\sqrt{E_i / 4\sigma_n^2}}^{\infty} \exp(-z^2) dz\qquad(7)$$

Evaluate the partial derivative $\partial\varepsilon_i / \partial E_i$; the result is obtained as following:

$$\frac{\partial D_c}{\partial E_i} = D_i \frac{\partial \varepsilon_i}{\partial E_i} = \frac{-D_i}{4\sigma_n \sqrt{\pi E_i}} e^{-\frac{E_i}{4\sigma_n^2}}\qquad(8)$$

The conventional energy allocation algorithm (CEAA) [4] is to set $E_{tot} = M\Delta E$, where ΔE is a small quantum of energy. Firstly, ΔE is assigned to each bit. Eq. 6 and Eq. 8 are evaluated for each bit and allocate the energy ΔE to the bit with the largest value of $\frac{\partial D_c}{\partial E_i}$ for BPSK and BFSK, respectively. The procedure is operated iteratively until the total energy is allocated.

3. Genetic Energy Allocation Algorithm

Genetic Energy Allocation Algorithm (GEAA) is to apply genetic Algorithm [6-8] for energy allocation. GEAA can be described as follows:

Step0. Calculate the distortion D_i due to the ith bit error off line.

Step1. Initialization

Set the number of generation, the population size and each individual consists of m parameters, where $N = 2^m$, N is the size of the codebook. Allocate the energy of ΔE to each parameter randomly until $M\Delta E = E_{tot}$, where M is the allocated number of times. The energy must satisfy $E_{tot} = M\Delta E = \sum_{i=1}^{m} E_i$.

Step2. Evaluation

Calculate the bit error probability ε_i by using Eq. 5 and Eq. 7 for BPSK and BFSK. The channel distortion, D_c, can be acquired by using Eq. 2.

Step3. Selection

Select some individuals for next generation randomly. Keep the best of individual for next generation.

Step4. Crossover

The uniform order-based crossover technique [9] is used to produce the next individual from the two selected individuals randomly. If the energy of one individual is more than E_{tot}, then iteratively reduce ΔE for one bit (parameter) randomly. If the energy of one individual is smaller than E_{tot}, then iteratively add ΔE for one bit (parameter) randomly.

Step5. Mutation

Choose two parameters of the individual randomly, and move the energy of ΔE from one parameter to the other parameter.

Step6. Step2 to step 5 are repeated until the satisfied fitness is found or the number of generations have been reached.

4. Hybrid Method

The Genetic Index Assignment Algorithm (GIAA) is combined with Genetic Energy Allocation Algorithm, which is termed GIEAA, to reduce the channel noise for VQ-based BPSK and BFSK signal compression. The hybrid method is described as following:

Step1: Calculate the same bit error rate ε_i by assigning same energy for each bit. Repeat the following steps for N_t times.

Step2: Given the bit error rate ε_i, apply GIAA to optimize the index assignment for N_i generations.

Step3: Given the channel distortion D_i due to the ith bit error; apply GEAA to optimize the energy allocation for N_e generations.

5. Experimental results and conclusion

The test material is the LENA image with size 512×512. The block size is set to 16 (size 4 pixels * 4 lines). It is used to generate 256 codevectors for VQ by applying GLA and Splitting LBG [8]. Experiments were carried out to test the performance of channel distortion by comparing CEAA, GEAA, GIAA and GIEAA based both on GLA and splitting LBG for VQ systems using BPSK and BFSK modulations.

The parameters used for the number of generation, the number of individuals are 500, 50, respectively. The variance of the Gaussian Channel noise σ_n^2 is set to 8. The channel SNR is set from 5 to 10 db. The total energy E_{tot} can be calculated by using Eq. 4. The small quantum of energy ΔE is set to 0.01. The iteration number N_t, N_i and N_e are set to 30,500 and 500, respectively. The error sensitivities of bits in natural binary code are shown in Table 1. The Channel distortion for VQ based on BPSK and BFSP for CEAA, GEAA, GIAA and GIEAA are shown from Table 2 to Table 5 and also in Fig. 2 and Fig. 3.

As shown from Table 2 and Table 5, combining the Genetic Index Assignment with Genetic Energy Allocation Algorithm can reduce the channel distortion up to 25% based on GLA codebook design. Simulated results prove the proposed hybrid optimization scheme using genetic algorithm.

Table 1: Sensitivities of bits for VQ

Bit number i	0	1	2	3	4	5	6	7
D_i(GLA)	244	228	232	237	238	227	235	235

Bit number i	0	1	2	3	4	5	6	7
D_i(Splitting LBG)	333	211	150	133	119	90	68	53

Table 2: Channel distortion using GLA for BPSK modulation

Channel SNR	CEAA	GEAA	GIAA	GIEAA	Improvement
5 db	689	650	584	515	25.25%
6 db	683	644	570	510	25.32%
7 db	677	638	571	506	25.25%
8 db	671	631	566	501	25.33%
9 db	664	625	562	495	25.45%
10 db	658	619	547	489	25.68%

Table 3: Channel distortion using Splitting LBG for BPSK modulation

Channel SNR	CEAA	GEAA	GIAA	GIEAA	Improvement
5 db	417	379	497	374	10.31%
6 db	413	375	500	370	10.41%
7 db	409	370	489	366	10.51%
8 db	405	367	478	361	10.86%
9 db	401	363	476	358	10.72%
10 db	397	359	478	352	11.33%

Table 4: Channel distortion using GLA for BFSK modulation

Channel SNR	CEAA	GEAA	GIAA	GIEAA	Improvement
5 db	780	731	656	579	25.76%
6 db	776	726	643	575	25.90%
7 db	771	722	646	572	25.81%
8 db	767	717	643	568	25.94%
9 db	762	712	640	564	25.98%
10 db	758	708	626	560	26.12%

Table 5: Channel distortion using Splitting LBG for BFSK modulation

Channel SNR	CEAA	GEAA	GIAA	GIEAA	Improvement
5 db	482	433	559	428	11.20%
6 db	479	430	564	428	10.64%
7 db	476	427	553	421	11.55%
8 db	473	424	543	420	11.20%
9 db	470	420	542	415	11.70%
10 db	467	418	546	412	11.77%

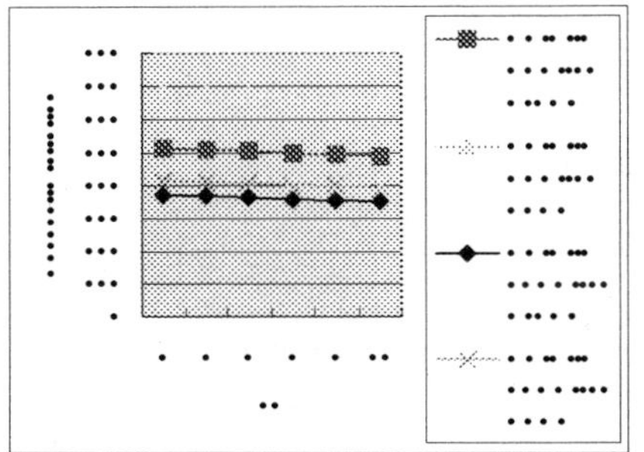

Fig. 2 The performance comparison of VQ system using BPSK modulation

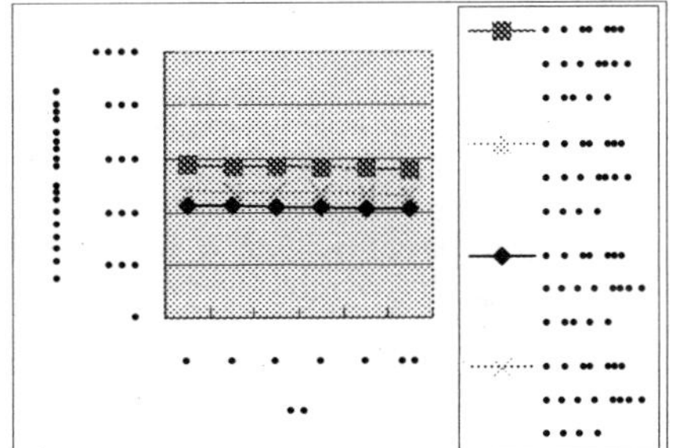

Fig. 3 The performance comparison of VQ system using BFSK modulation

Reference

[1] R. M. Gray, Vector Quantization, *IEEE ASSP Magazine*, 1(1984) 4-28.

[2] T. Y. Chen, T. C. Yang, Y. W. Chen, B. Y. Liao, Channel Noise Reduction for M/R VQ with Index Assignment, Fifth International Conference on Information Engineering Systems & Allied Technologies, (2001) 1027-1031.

[3] F. Bedrosian, Weighted PCM, IRE Trans. IT-4, (1958) 45-49.

[4] S. Gadkari and K. Rose, Robust Vector Quantization by Transmission Energy Allocation, IEE Electronics Letters, 32(1996) 1451-1453.

[5] Y. Linde, A. Buzo and R. M. Gray, An Algorithm for Vector Quantizer Design, IEEE Trans. on Communication, 28(1980) 84-95.

[6] D. E. Goldberg, Genetic Algorithm in Search, Optimization and Machine Learning, Addison-Wesley Publishing Company, 1989.

[7] J. S. Pan, F. R. McInnes and M. A. Jack, VQ codebook design using genetic algorithms, IEE Electronics Letters, 31(1995) 1418-1419.

[8] J. S. Pan, F. R. McInnes and M. A. Jack, Application of Parallel Genetic Algorithm and Property of Multiple Global Optima to VQ Codevector Index Assignment, IEE Electronics Letters, 32(1996) 296-297.

[9] L. David, Handbook of genetic algorithms, Van Nostrand Reinhold, 1991.

KES 2002
E. Damiani et al. (Eds.)
IOS Press, 2002

Channel Noise Reduction for I/R VQ with Parallel Genetic Algorithm Index Assignment

K.C. Huang [†], S.C. Chu [‡], and T. C. Yang [*]
[†] *Department of Automatic Engineering, Lan-Yang Institute of Technology, Taiwan*
[‡] *School of Informatics and Engineering, Flinders University of South Australia*
[*] *Department of Electrical and Information Engineering, University of South Australia, Australia*

Abstract. The interpolation Residual VQ(I/R VQ), is well recognized as a highly efficient compression method, with which the encoding speed is greatly improved without serious degradation in image quality. Here, the authors propose to introduce a so-called parallel genetic algorithm on index assignment over a noisy channel environment. By using this novel method, a suitable index assignment is found out to minimize the impact of channel noise With our approach, channel distortion can be substantially reduced without incurring extra cost such as that in error-detection code and error-correction code. The index assignment of the scalar codebook and the residual codebook can be separate or hybrid optimization. Experimental results demonstrate the usefulness of the proposed approaches.

1. Introduction

Nowadays the Internet is widely employed in every aspect of life. It conveniently transmits data includes text, image and speech; however, one of the annoyed matters that most users argue about is the slowly transmitting rate over it. The compression technique is a useful solution to this problem at present. One of the efficient compression techniques is the vector quantization (VQ). Nevertheless, vector quantization as a central reduction scheme is highly sensitive to channel noise. An useful technique to reduce this imperfection without adding the parity bits is to apply the optimization algorithm for index assignment of the code vectors. In this paper, the parallel genetic algorithm (PGA) is applied to index assignment for I/R VQ to reduce the channel noise. At first, the VQ and GA will be introduced. Then the formula of the channel distortion for I/R VQ is employed due to noise influence, and the algorithm of the index assignment implemented by GA will be described as well. Some experimental results will be shown at the end of this paper.

In the purpose of reducing the bandwidth requirement for data transmission and space requirement for data storage, a variety of data compression techniques had been developed [1]. Vector Quantization (VQ) [2] is widely used for compression of speech and images where the binary indices of the optimally chosen codevectors are sent. A vector $X = \{x^1, x^2, x^3, ..., x^k\}$ consisting of k samples of information source in the k-dimensional Euclidean space R^k is sent to the vector quantizer. Only the index i is transmitted to the decoder and decoding is operated by table look-up procedure.

Although VQ could deliver good resource data compression, it can also provide good resilience against channel errors if an appropriate mapping of codevectors to binary codewords is used; that is, by assigning suitable indices to code vectors can significantly reduce the channel distortion due to an imperfect channel. If the number of codevectors is N, the possible combination of indices to codevectors is N!. To test N! assignments is an NP-hard problem. The simulated annealing technique to design the codevector indices is first applied by Farvardin [4]. Zeger and Gersho [3] introduced the binary switching algorithm to improve the codebook index assignment. Pan et. al. [7] worked on the parallel genetic algorithm to evolutionarily approach a better solution. The tabu search approaches were developed for code vector indices assignment by Pan and Chu [8] and results show that the channel distortion can be improved by comparing with the binary switching algorithm and the parallel genetic algorithm. Potter and Chiang [9] adopted a minimax technique, in which the worst case performance is greatly improved which maintains good average performance by applying the minimax design criterion. The genetic algorithm [5-7] is proposed to improve the index assignment of the I/R VQ in this paper,.

Genetic algorithms (GA) [5] are powerful and broadly applicable stochastic search and optimization techniques based on principles from evolution theory. A set of solutions in GA to a problem is called chromosomes that is composed of genes (features, characters or detectors). The individual of the whole population usually contains only one chromosome. The performance of the solution is called fitness of which chromosomes are evaluated, ordered, and then new chromosomes are produced by using the selected candidates as parents and applying mutation and crossover operations. The new set of chromosomes is then evaluated and ordered again. This cycle continues until a suitable solution is found. GA is one of the adaptive methods that can be used in search and optimization problems.

2. Interpolation Residual Vector Quantization (I/R VQ)

A N-level vector quantizer can be defined as a mapping from a k-dimensional Euclidean space R^k into a finite set $C = \{c_1,...,c_N\}$. Given an input vector X, the output is the index i of the codeword C_i which satisfies

$$i = \arg\min_p \sum_{l=1}^{k} \left(x^l - c_p^l\right)^2 \tag{1}$$

The codebook plays the central role of a vector quantizer. Linde et al. [3] had suggested an iterative method for generating codebook from given training data.

One of the most important parameters to design the codebook is its size. It decides the encoding speed and the image quality. In the beginning, this seems that there is an inevitable trade-off between these two important criteria. However, the interpolation residual VQ (I/R VQ) [10, 11] is introduced by splitting a single large codebook into two smaller ones, namely scalar codebook and residual codebook to deal with this problem. Through the process of down-sampling and interpolation, The main benefit of the I/R VQ is to reduce the encoding time with only a little degradation in the recovered performance. This virtue makes I/R VQ prevail in real-time applications. Figure 1 illustrates the general structure of I/R VQ.

3. Index Assignment

As shown in Figure 1 and 2, i_S and i_R denote the indices of scalar codeword and residual codeword, respectively. Both of them are going to be sent through a communication channel. However, the previous work in [11] has shown that the performance of I/RVQ is sensitive to channel noise. The noise introduced over the channel can cause errors in the received indices; that is, the index i is changed to index j. Thus, distortions would occur in the decoding phase. Assume the channel model is a binary symmetric channel with bit error probability ε, i.e.,

$$P(b(c_j)/b(c_i)) = (1-\varepsilon)^{m-H(b(c_i),b(c_j))}\varepsilon^{H(b(c_i),b(c_j))} \tag{2}$$

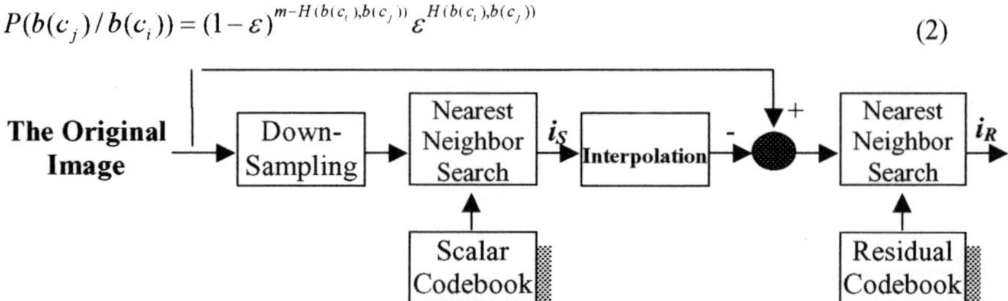

Figure 1. The encoding processes of IRVQ

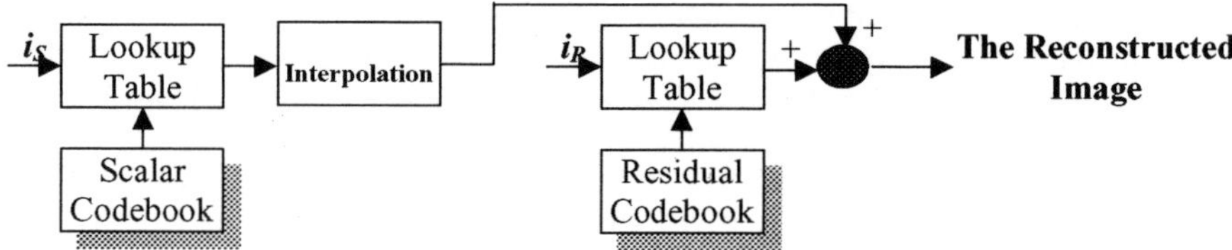

Figure 2. The decoding processes of IRVQ

where $b(c_i)$, $i = 1,2, \ldots, N$, is the index with m bit string of codeword c_i; $P(b(c_j)/b(c_i))$ denotes the probability that index $b(c_j)$ is received given the index $b(c_i)$ is sent; $H(b(c_j),b(c_i))$ denotes the Hamming distance between $b(c_j)$ and $b(c_i)$, i.e., the number of bits in which $b(c_i)$ and $b(c_j)$ differ. Let us assume the distortion between c_i and c_j is given by a non-negative distortion measure $d(c_i, c_j)$. Thus, for a given assignment of codeword indices $b = (b(c_1),b(c_2),\ldots,b(c_N))$, the distortion caused by the channel noise can be expressed as

$$D = \sum_{i=1}^{N}\{P(c_i)[\sum_{l=1}^{m}\varepsilon^l \cdot (1-\varepsilon)^{m-l} \cdot \sum_{j:H(b(c_i),b(c_j))=l}d(c_i,c_j)]\} \tag{3}$$

where $P(c_i)$ denotes the probability density function of c_i.

The channel is assumed to be memoryless binary symmetric channel. For standard VQ, the ensemble average distortion of random index assignment can be expressed as [7, 11, 12]:

$$D = \frac{1-(1-\varepsilon)^m}{N-1}\sum_{i=1}^{N}P(c_i)\sum_{j=1}^{N}d(c_i,c_j). \tag{4}$$

Here, N, $P(C_i)$ and ε are the codebook size, the probability of transmitted index $b(C_i)$ and the probability of channel error, respectively.
The average distortion of the I/R VQ can be formulated as:

$$D = \frac{1-(1-\varepsilon)^m}{N_1 N_2 - 1}\sum_{i\geq 1}^{N_1}\sum_{k\geq 1}^{N_2} P(s\vec{1}+r)\sum_{j\geq 1}^{N_1}\sum_{i\geq 1}^{N_2}d(s\vec{1}+r,s'\vec{1}+r) \tag{5}$$

where N_1 is the size of scalar codebook, N_2 is the size of residual codebook, and $N_1 * N_2 = 2^m$, m is the total number of transmitted bits. And $P(s_i \vec{1} + r_k)$ is the transmitted probability of index $b(s_i \vec{1} + r_k)$. The $P(s_i \vec{1} + r_k)$ can be set as constant $1/N_1 N_2$. The evaluation function of the channel distortions for I/R VQ is

$$D = \sum_{i=1}^{N_1} \sum_{k=1}^{N_2} P(s_i \vec{1} + r_k) \sum_{t=1}^{m} \varepsilon^t (1-\varepsilon)^{m-t} \sum_{H(s_i \vec{1}+r_k, s_j \vec{1}+r_l)=t} d(s_i \vec{1} + r_k, s_j \vec{1} + r_l) \tag{6}$$

4. Experiment Results

Some previous work was done by Shieh et. al.[11] in using GA to reduce the channel distortion in I/R VQ. However, they only applied GA in the indices i_S and i_R individually by employing the criterion in equation (3 – 4) to evaluate the channel distortion. Here, we merge the indices i_S and i_R to a single codebook. The average distortion of the I/R VQ is calculated by equation (5). The fitness of the PGA employs equation (6) to search a better combination of the index assignment.

Therefore, the experiments were conducted in two aspects: The first one examined the result by evaluating the channel distortion in indices i_S and i_R individually and then calculated the merging distortion using (6). The second experiment merged the i_S and i_R in the beginning and then calculated the channel distortion. The test material was the LENA image with size 512×512. Both of the scalar codebook and the residual codebook can be generated using the LBG algorithm [13]. All the relevant parameters for conducting the experiments are summarized as below:

Test Image: a 512x512 8-bit grey-level image of Lena, as shown in Figure 2(a).
IRVQ parameters:
Down-sampling rate: 4:1
Block size: 4x4
Size of scalar codebook: $32 = 2^5$
Size of residual codebook: $8 = 2^3$
GA training parameters:
Population size: 8 islands x 100 chromosomes
Chromosome coding scheme: 32/8 integers for scalar/residual codebook
Crossover operator: swap genes according to a randomly pick up site
Maximal generation: 500 generations
Migration interval: 50 generations
Channel bit error probability: 0.01

All the experimental results are shown in Table 1 and 2. The percentage of the improvement is calculated using the following equation:

$$Improvement\_Rate = \frac{Average\_Distortion - Distortion\_after\_improve}{Average\_Distortion} * 100\%$$

The *Average_Distortion* which be calculated using equation (4) is 19.175047 in this case. Table 1 presents the results by evaluating the channel distortion in indices i_S and i_R individually and Table 1 presents the results by emerging indices i_S and i_R . It can be seen that the average improvement rate have increased from 35.78% (Table 1) to 39.84% (Table 2). Clearly the experimental results confirm the proposed merged optimization algorithm does have better performance. Significant and consistent improvement in the experiments is a clear evidence of the superiority of our method. In these experiments, the index assignment for scalar codebook and residual codebook are done separately. We believe that further improvement can be expected if both index assignments are resolved simultaneously.

Table 1: Separate Index Assignment

Method / Seed	Separate Mean and Residual	
	Distortion	Improvement Rate(%)
1	12.239084	36.171817
2	12.353211	35.576632
3	12.342679	35.631558
4	12.242108	36.156047
5	12.339763	35.646765
6	12.290067	35.905935
7	12.312981	35.786436
8	12.351128	35.587496
9	12.357763	35.552893
Average	12.314309	35.779509

Table 2:: Merging Index Assignment

Method / Seed	Merging Mean and Residual	
	Distortion	Improvement Rate(%)
1	11.595984	39.525655
2	11.584423	39.585947
3	11.432206	40.379776
4	11.590219	39.555721
5	11.489099	40.083073
6	11.557682	39.725405
7	11.498094	40.036163
8	11.596014	39.525499
9	11.478717	40.137216
Average	11.535826	39.839384

5. Conclusions

In this paper, a new means to find a better codebook index assignment to reduce the distortion of a noisy channel by merging the index assignment of scalar codevector i_S and residual codevector i_R to a single codebook. Parallel genetic algorithm (PGA) is also introduced as the optimization method to obtain the final results. Experiments show that the merging index has indeed better performance than separate index assignment.

References

[1] Gersho and R. M. Gray, *Vector Quantization and Signal Compression*, Kluwer Academic Publishers, London, 1992.

[2] R. M. Gray, Vector Quantization, *IEEE ASSP Magazine*, (1984) 4-29.

[3] K. Zeger and A. Gersho, Pseudo-Gray Coding, *IEEE Trans. On Communications*, **38** (1990) 2147-2158.

[4] N. Farvardin, A Study of Vector Quantization for Noisy Channels, *IEEE Trans. On Information Theory*, **29** (1990) 1317-1319.

[5] H. L. Fang, Genetic Algorithms in Timetabling and Scheduling, Ph. D. Thesis, Department of Artificial Intelligence, University of Edinburgh, 1994.

[6] J. H. Holland, Adaptation in Natural and Artificial Systems, University of Michigan Press, 1975.

[7] J. S. Pan, F. R. McInnes and M. A. Jack, Application of Parallel Genetic Algorithm and Property of Multiple Global Optimal to VQ Codevector Index Assignment, *IEE Electronics Letters*, **32** (1996) 296-297.

[8] J.S.Pan and S.C.Chu, Non-Redundant VQ Channel Coding Using Tabu Search Strategy, *IEE Electronics Letters*, **32** (1996) 1545-1546.

[9] L. C. Potter and D. M. Chiang, Minimax Nonredundant Channel Coding, *IEEE Trans. on Communications*, **43** (1995) 804-811.

[10] A. Gersho, Optimal Nonlinear Interpolation Vector Quantization, *IEEE Trans. On Communications*, **38** (1990) 1285-1287.

[11] C.-S. Shieh , M.-L. Wang , J.-F. Chang, and T. Y. Chen, Channel Distortion Minimization for Interpolation Residual VQ, KES 2001, Part2, 1042-1046

[12] T. Y. Chen, T.C. Yang, Y.W. Chen, B.Y. Liao, Channel Noise reduction for M/R VQ with index assignment, KES 2001, Part2, 1027-1032

[13] Y. Linde, A. Buzo and R. M. Gray, An Algorithm for Vector Quantizer Design, *IEEE Trans. on Communication*, **28** (1980) 84-95.

KES 2002
E. Damiani et al. (Eds.)
IOS Press, 2002

Hand shape recognition for biometric identification by using LVQ

Yoshihisa Fukuhara*, Yoshiyasu Takefuji**
* Graduate School of Media and Governance, Keio University
**Faculty of Environmental Information, Keio University
5322, Endo, Fujisawa, Kanagawa, 252-8520, Japan

Abstract. Many certification systems have been developed by biometrics. Most of those systems force presentation of our body features to their systems. This therefore causes stress to users. Moreover, this kind of systems will not discriminate correct user, if the user changed to another user after the identification. This defect will cause a serious security problem. To solve this problem, we developed a continuous biometric personal identification system by using hand shape recognition. To discriminate hand shapes, we applied Learning Vector Quantization (LVQ) networks to the proposed system. The proposed system can discriminate over 20 users and exclude unknown users.

1. Introduction

Many certification systems have been developed by biometrics. Most of those systems, including fingerprint[1] and iris identification[2] can discriminate users by using biometric feature of human body. However, we have to present our body feature to these systems and it will be cause of a stress of users. Moreover, this kind of systems will not discriminate correct user, if the user changed to other user after the identification. This defect will cause a serious security problem. To solve this problem, we developed a continuous biometric personal identification[3] by using hand shape recognition. Every person has his own peculiar hand shape and form, especially when users use some kind of tool. In this paper, we use a mouse as an input device to obtain features of hand shapes. To discriminate hand shapes, we applied Learning Vector Quantization (LVQ) networks to the proposed system. The proposed system can discriminate over 20 users and exclude unknown users.

2. The method to obtain hand shapes

To obtain hand shapes, we use a mouse, which was supplied by Fuji Xerox laboratory[4] as shown in Fiugure-1. This mouse has optical sensors on the surface. And it can measure a hand shape of a user continuously. It has four LEDs and four photo-transistors (PTr) on its surfaces. PTrs can measure distances between the mouse and the user's hand by reflection. As shown in Figure-2, we can obtain 16 vectors data from four sensors. When a user clicks the mouse button, the mouse obtains features of the user's hand shape. Since every people have his own peculiar hand shape, these vectors show a feature of hand shape of each user.

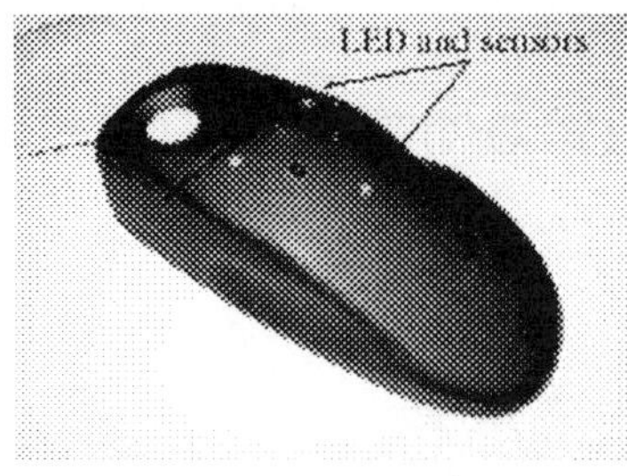

Fig.1. Mouse with optical sensors.

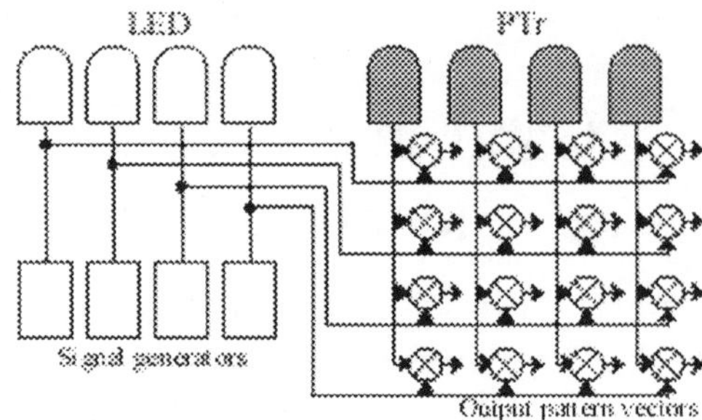

Fig.2. Structure of the mouse

3. The proposed method

To identify users, the system has to learn hand shapes of users. We use Learning vector quantization (LVQ) neural networks as the learning method. The proposed system obtains a feature of hand shape as teaching data, while the user uses a computer. After that the system is trained features of each person by using LVQ networks. And the system can discriminate users and exclude unknown users in order to set up thresholds in the networks.

3.1 Learning Vector Quantization (LVQ) algorithms

LVQ is a well-known neural networks applicable to pattern recognition[5]. It is a method for training competitive layers in a supervised learning. The decision borders are approximated by the nearest-neighbor rule and they toward the Bayes limits gradually. In the LVQ networks, each output neuron shows target classes chosen by the user. We choose OLVQ1 and LVQ3 for the proposed system from some kinds of improvements of LVQ networks.

a) The optimised-learning-rate LVQ1 (OLVQ1)

At first, we have to choose codebook vectors m_i. These vectors are generated randomly or chosen from input vectors. And m_c which is the nearest vector to input vector x is defined by Eq.1 and Eq.2. OLVQ1 is based on the basic LVQ1 algorithm. However, it has an individual learning rate $\alpha_i(t)$ for each m_i for fast convergence.

$$c = \arg\min_i\{\| x - m_i \|\} \tag{1}$$

$$m_c(t+1) = [1 - s(t)\alpha_c(t)]m_c(t) + s(t)\alpha_c(t)x(t) \tag{2}$$

where $S(t) = +1$, if mc and x belong to the same class. And $S(t) = -1$, if m_c and x belong to the different class. The learning rate $\alpha_c(t)$ is defined by Eq.3.

$$\alpha_c(t) = \frac{\alpha_c(t-1)}{1 + s(t)\alpha_c(t-1)} \tag{3}$$

b) The LVQ3

Generally, LVQ3 is employed after LVQ1 or OLVQ1 algorithm in order to obtain accurate borders. LVQ3 is defined by Eq.4 and Eq.5. If the two closest vectors m_i, m_j and input vector x belong to the different class, we use Eq.4. On the other hand, we have to use Eq.5, if m_i, m_j and x belong to the same class. Note that $k \in \{i, j\}$ and is $\sigma(t)$ a learning rate. And ε is constant number.

$$m_i(t+1) = m_i(t) - \sigma(t)[x(t) - m_i(t)]$$
$$m_j(t+1) = m_j(t) + \sigma(t)[x(t) - m_j(t)] \tag{4}$$
$$m_k(t+1) = m_k(t) + \varepsilon\sigma(t)[x(t) - m_k(t)] \tag{5}$$

3.2 Set threshold

After the learning, thresholds are set up between each class in order to exclude unknown users. Two kinds of thresholds are used in the proposed system. One is a distance from the nearest codebook vector. And another is defined between different kinds of codebook vectors. If the two closest codebook vectors belong to the same class, the threshold is described as Eq.6. where m_{min} is the smallest vector and m_{max} is the largest vector in the class. m_x is the input vector. Note that β is a constant value.

$$m_{min} - \beta(m_{max} - m_{min}) < m_x < m_{max} + \beta(m_{max} - m_{min}) \tag{6}$$

If the two closest codebook vectors belong to the different class, the threshold is described as Eq.7. where m_1 is the closest vector from m_x and m_2 is the second closest vector. And γ is a constant value.

$$|m_1 - m_x| < \gamma\frac{|m_1 - m_2|}{2} \tag{7}$$

4. Simulation result

To show the effectiveness of our system, we experimented two simulations on a computer. In this simulation, we prepared data of 20 users(From A to T) for the training. And the system obtained 240 training data each person. Note that $\alpha_c(0) = 0.3$, $\sigma = 0.1$, $\varepsilon = 0.1$, $\beta = 0.7$ and $\gamma = 0.7$. As a result, we gave 80 data to the system for each user and collect statistics of answers. The performance of the computer used in the simulation is Windows2000 machine, which has Pentium Celeron-500Mhz CPU. And a development language is Borland Delphi 5.0.

4.1 Discrimination

We trained the system with 19 users, and the system was present feature of 20 users, including 19 memorized users and one unknown user. We attempt same experiment 20 times changing unknown user. Table-1 shows all of the results of this experiment and we summarized these results as distributions as shown in Figure-3. Average of percentage of correct answers is 76%(min:7.5%, max:100%) and wrong answers is 9.8%(min:0%, max:29%).

4.2 Identification

We trained the system with one user, and the system was present feature of 20 users including 19 unknown users. In this simulation, we changed parameters of thresholds (β and γ) every user to obtain reliable results. Currently, these values were obtained experimentally. Table-2 shows the result of this experiment and Figure-4 shows distributions of results. Average of percentage of correct answers is 88%(min:70%, max:100%) and wrong answers is 1.2%(min:0%, max:90%).

Table.1. Results of discrimination.

▨ Percentages of wrong answers (unknown user)   ☐ Percentages of correct answers (correct user)

	A	B	C	D	E	F	G	H	I	J	K	L	M	N	O	P	Q	R	S	T
1	13	75	85	96	83	91	36	96	100	93	91	80	99	98	60	41	68	25	96	85
2	99	13	86	99	78	91	49	93	100	94	85	74	99	99	51	26	63	21	94	84
3	91	73	0	98	76	93	39	96	100	81	86	61	98	96	69	25	41	18	95	91
4	96	73	86	3.8	78	90	56	96	100	88	95	65	98	98	70	26	24	8.8	96	94
5	94	73	81	98	10	93	53	91	100	95	95	68	99	95	61	29	43	25	91	83
6	98	69	84	98	74	3.8	50	96	100	75	84	76	99	98	60	26	36	19	94	89
7	93	71	84	98	78	91	24	98	98	21	94	71	99	99	59	21	30	23	93	56
8	94	68	81	99	75	93	73	3.8	100	70	91	66	98	99	68	24	54	35	96	89
9	95	64	83	98	83	88	39	95	0	64	96	68	100	100	60	30	39	18	99	85
10	91	65	80	95	76	93	48	95	99	3.8	88	75	98	96	71	30	33	20	96	94
11	94	75	86	95	78	86	34	96	100	91	29	84	99	94	56	33	31	18	98	88
12	93	65	81	96	81	90	51	95	96	73	91	1.3	99	98	59	33	41	23	94	89
13	95	76	91	98	85	90	38	90	100	80	89	89	15	99	76	33	28	33	98	85
14	93	69	84	98	76	84	45	88	100	88	90	53	96	10	69	45	23	26	96	91
15	94	75	86	99	76	91	46	98	100	90	89	73	99	98	5	34	20	16	95	81
16	95	61	83	94	84	90	54	96	100	95	91	69	96	96	55	15	48	11	99	86
17	98	65	84	98	73	88	45	96	100	89	95	71	100	99	63	30	11	35	96	93
18	95	63	78	98	80	91	41	98	100	95	99	68	100	99	63	25	51	23	99	91
19	90	66	83	98	79	93	55	95	99	89	85	55	98	96	66	28	41	7.5	11	81
20	94	66	89	98	76	86	45	93	100	88	93	55	99	100	65	35	34	23	99	0

Table.2. Results of identification.

▨ Percentages of wrong answers (unknown user)   ☐ Percentages of correct answers (correct user)

	A	B	C	D	E	F	G	H	I	J	K	L	M	N	O	P	Q	R	S	T
1	100	0	0	0	0	0	0	0	0	0	0	0	0	0	0	0	0	0	0	0
2	1.3	85	0	5	0	0	0	0	0	0	1.3	0	0	0	0	0	3.8	0	0	0
3	0	0	99	0	0	1.3	3.8	0	0	0	0	0	0	0	0	0	1.3	0	0	0
4	0	0	0	98	0	0	0	0	0	0	0	0	0	0	0	0	0	0	0	0
5	0	0	0	0	89	0	0	0	0	0	0	0	0	0	0	0	0	0	0	0
6	0	0	0	0	0	94	0	0	0	0	0	0	0	0	0	0	0	0	0	0
7	0	0	0	0	0	0	99	1.3	0	0	0	0	0	23	0	8.8	1.3	26	0	0
8	0	0	0	0	0	0	1.3	100	0	0	0	0	0	0	0	0	0	1.3	0	0
9	0	0	0	0	0	0	0	0	100	0	0	0	0	0	0	0	0	0	0	0
10	0	0	0	0	0	0	0	0	0	71	0	0	0	0	0	0	0	0	0	0
11	0	0	0	0	0	0	0	0	0	0	80	0	0	0	0	0	0	0	0	0
12	0	0	0	0	0	0	0	0	0	0	0	70	0	0	0	0	0	0	0	0
13	0	0	0	0	0	0	0	0	0	0	0	0	100	0	0	0	0	0	0	0
14	0	0	0	0	0	0	0	0	0	0	0	0	0	100	0	0	0	0	0	0
15	0	0	0	0	0	0	0	0	0	0	0	0	0	0	74	0	0	0	0	0
16	0	0	0	0	0	0	0	0	0	0	0	0	0	46	0	71	0	0	0	0
17	3.8	6.3	0	0	1.3	0	2.5	0	0	0	0	0	0	0	0	0	74	0	0	0
18	21	20	0	0	29	0	15	65	23	1.3	90	11	0	0	24	0	1.3	70	0	0
19	0	0	0	0	0	5	0	0	0	0	0	0	0	0	0	0	0	0	96	0
20	0	0	0	0	0	0	0	0	0	0	0	0	0	0	0	0	0	0	0	99

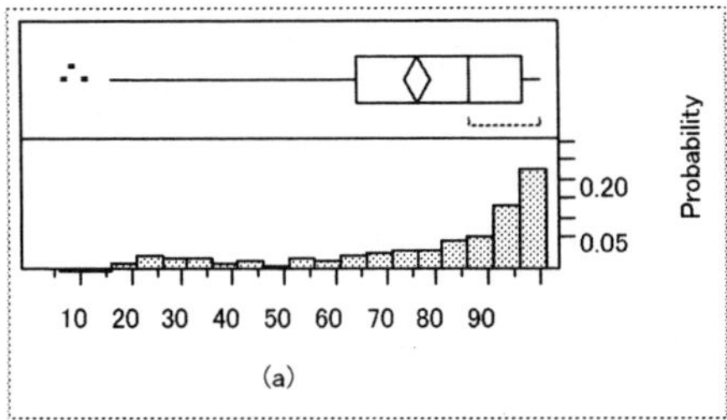
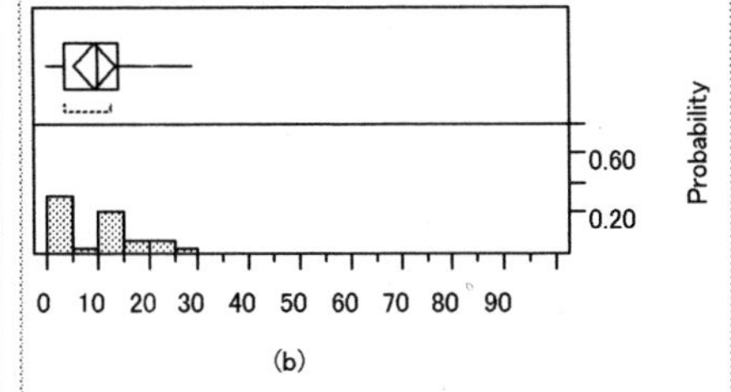

Figure.3. Discrimination: Distribution of a percentage of correct and wrong answers.
(a) Distribution of correct answers.   (b) Distribution of wrong answers.

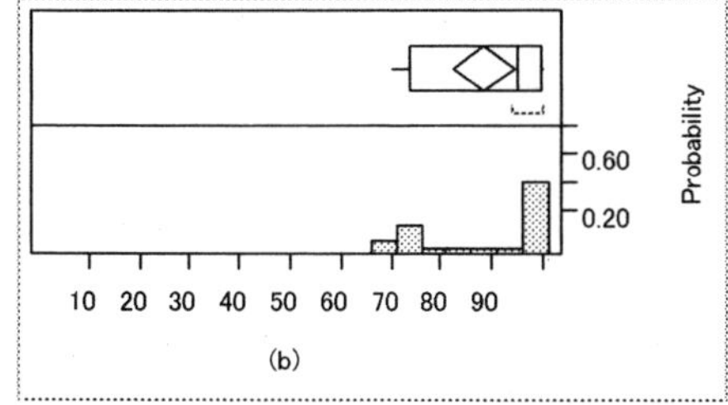
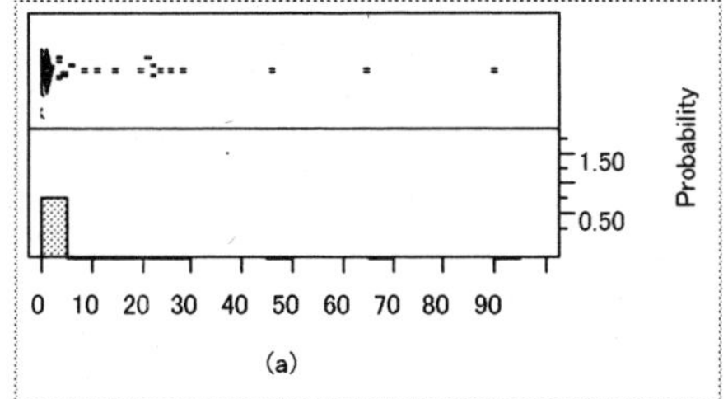

Figure.4. Identification: Distribution of a percentage of correct and wrong answers.
(a) Distribution of correct answers.   (b) Distribution of wrong answers.

5. Discussions and Conclusion

From these experiments, we can understand that the proposed system discriminates several users and identifies a single user under almost all combinations. Generally, LVQ networks cannot discriminate unknown classes. On the other hand, the proposed system can discriminate unknown users by applying thresholds to the network. However, the system shows wrong results in some situation. In discrimination examination, especially, user-G, P and R show low percentages of correct answers relatively. As shown in Table-2(line 18), when the system learned user-R in identification examination, we obtained an unsuitable result. We think that features of these users are overlapped with other users. We think we can improve the ability of discrimination by placing optical sensors in a more suitable position and modifying threshold algorithms, so that the proposed system works more accurately, when the feature of users conflict with each other.

Acknowledgment

We would like to thank Mr. Hajime Sugino(Fuji Xerox Co.,Ltd.) for his support and helpful advice.

References
[1]   C. L. Wilson, G. T. Candela and C. I. Watson. *Neural Network Fingerprint Classification*, Artificial Neural Networks, 1(2), p.203-p.228, 1994
[2]   R.P. Wildes, *Iris Recognition: An Emerging Biometric Technology*, Proceedings of the IEEE, vol.85, p.1348-1363, 1997
[3]   A. Jain, R. Bolle and S.Pankanti, *BIOMETRIC Personal Identification in Networked Society*, Kluwer Academc Publishers, 1999
[4]   H. Sugino and K. Sakai. *Biometric personal identification by the optical hand shape sensor on the mouse*, Society conference of IEICE, 2000
[5]   T. Kohonen. *Self-Organizing Map*. Springer Verlag, Heidelberg, 1995

KES 2002
E. Damiani et al. (Eds.)
IOS Press, 2002

A Robust System for Fingerprints Identification

Vincenzo CONTI[2], Giovanni PILATO[1], Salvatore VITABILE[1], Filippo SORBELLO[2]
[1]CE.R.E. - Italian National Research Council – Palermo - Italy
[2]Dipartimento Ingegneria Informatica - University of Palermo, Italy

Abstract. A robust system based on fingerprint minutiae extraction for the identification of the personal identity is proposed. The system is mainly characterized by an adaptive energy threshold determination, by an extension of the Tanimoto distance and by a new intersection operator in order to enhance system performances with noisy, translation and/or rotation problems. In order to validate the effectiveness of the proposed approach, experimental trials have been conducted on an ink-on-paper and scanned fingerprint database. Experimental results show a good noise immunity of the proposed solution.

1. Introduction

Fingerprints are formed by a system of lines having a certain bending, from terminal points and from bifurcation of lines [10]. Fingerprints can be used in two types of system for establishing the identity of a person: verification and identification system.

In the verification system, a person must be identified declaring, for example, his personal data, a login name, or thanks to a magnetic card. The system will require the introduction of the fingerprint and will verify the propriety of such data accepting or denying the access to the individual to a determined service. The task of the verification is based on stored data of the person with the characteristics of his own fingerprint.

The identification system has only a total or partial fingerprint and it must establish the identity of the person, that is, it compares the fingerprint acquired with all fingerprint images present in a database. A lot of noise problems can arise in analyzing and matching fingerprint images: for example, translation, rotation, images showing false characteristics due to the presence of noise. The last problem typically arises when the fingerprints are taken with ink on paper and then electronically acquired with a scanner; in fact this disadvantage is mainly due to the presence of ink stains. Furthermore, the acquired gray-scale images have not a uniform distribution and information: it is very hard to choose a gray level value for the image binarization process in a set of scanned fingerprint images.

The fingerprints matching problem has been addressed with syntactic technique [17], graph representation [18], hybrid-matching algorithms including feature extraction and texture information [20].

In this paper a noise-robust approach for fingerprints matching, based on image processing and on characteristic minutiae extraction, is proposed. The designed system is divided in three principal steps: fingerprint pre-processing, fingerprint minutiae extraction and fingerprints matching. The solution is designed for dealing with noise images (the classical ink on paper) and then electronically acquired with a scanner. In this case, the binarization process plays an essential role for the fingerprint minutiae extraction. The proposed system individuates a binarization threshold, calculated as the median of the energy histogram of the image. The threshold is then used in a local analysis of sub-images energy for the binarization process.

Furthermore the matching process is based on an extension of the Tanimoto distance [18], [1] and on a new intersection operator in order to enhance system performances with noisy, translation and/or rotation problems. The system has been tested with an ink-on-paper 512x512 pixel with 256 gray level fingerprint database available in our lab, showing good performances.

2. The proposed technique

A fingerprint is represented by particular points, called *minutiae*. Minutiae are *endpoints* and *bifurcation* of lines, called *ridges* (see Figure 1).

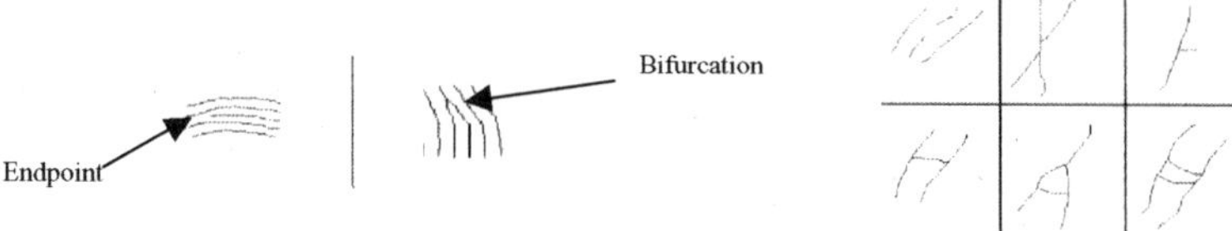

Figure 1: On the left ridge ending and ridge bifurcation minutiae; on the right false minutiae: interrupted ridges,forks, spurs, structure ladders, triangles and bridges in clockwise order.

Classic matching consists on counting and maximizing the number of the matching minutiae pairs between two fingerprints [7], [10], [11], [13]. An ideal matching fingerprint task is immune from translation, rotation and non-linear deformations of fingerprints. In real applications, however, fingerprints are very noisy, especially when they have been taken with ink and paper, as shown in Figure 1.

A system working on noisy fingerprints has to deal with false minutiae. A powerful pre-processing phase is necessary [8], [9], [12] in order to erase them without loss of information.

The proposed system aims to recognize the identity of a person comparing the current fingerprint with the ones included in a database. The matching task can be divided in three principal phases: a pre-processing phase, a fingerprint minutiae extraction phase and a matching phase. The aforementioned steps are shown in Figure 2, and they will be described in the next sub-sections.

Figure 2: The main phases of the proposed system: (a) the initial fingerprint image; (b) the pre-processing result image; (c) the post-processing result image with the spatial minutiae information extraction. This information will be used in the matching phase.

2.1 Pre-processing with adaptive energy threshold determination

A system working on noise fingerprints has to deal with false minutiae. A powerful pre-processing phase is necessary [8], [9], [12] in order to erase them without loss of information. In this phase several digital image-processing algorithms have been optimized in order to reduce the noise areas affecting the fingerprint images.

The first step is the directional image extraction. It is a particular image in which every element represents the local orientation of the ridges in the original gray-scale image 1. In fingerprint images, the directions inside noisy blocks could be completely different from the direction extracted in the neighbor blocks. This could generate mistakes in the next processing steps, so a smoothing algorithm on the directional images is applied [7].

In the second step the segmentation of the first image is performed. The segmentation algorithm is realized using two complementary methods: the directional and the variance method [3], [4]. The directional method shows a good behavior either when it is applied in the low contrast and noisy areas or when it is applied in regions containing clear ridges. The variance method shows a good behavior when it is applied in high contrast areas. Furthermore, the above algorithms split the background area from the foreground area, containing the needed information, in order to reduce the system processing time.

The third step deals with the foreground image filtering. The filtering process is based on the well-known Gabor pass-band filter [5], [14]. The Gabor filter enhances image quality, shifting the dark pixels through the black and the light pixels through the white.

In the fourth step, the binarization of the last image is performed. It is very hard to choose a threshold for the image binarization process in a set of scanned fingerprint images. The entire process needs an adaptive technique to set the optimal threshold value on the basis of the local pixels gray level distribution. The adopted algorithm is based on the automatic determination of the Local Energy Threshold (LET) for the image binarization. The LET is calculated as the median of the image energetic histogram.

The average energy of the eight neighbors of each considered pixel is calculated. If x_i is the central pixel of a 3x3 mask, the new value for x_i is calculated as follows:

$$x_i = \begin{cases} 255 & \dfrac{1}{8}\sum_{\substack{k=0 \\ k \ne i}}^{8} x_k \ge LET \\[2ex] 0 & \text{otherwise} \end{cases} \tag{1}$$

Successively, a smoothing algorithm is applied in order to erase the impurities present in the binarized image. The adopted algorithm is based on a self-adaptable dimension window that eliminates small parts of image, called *impurity islands*. They are characterized by white zones in ridges or black zones in valleys (a valley is a zone between two ridges). The algorithm starts with a moving snake on the binary image. The snake tries to close on itself drawing a rectangular window when it meets an impurity island (a default dimension is set taking into account the average dimension of impurity islands and the binary image dimension). When the snake hits

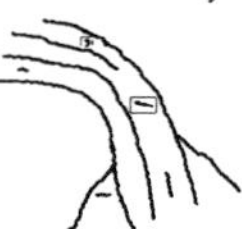

Figure 3: Windows of different dimension bound the impurity islands of different dimension.

against an obstacle (lines or impurities), it returns on its path, reducing the window dimension. A typical result of this algorithm applied on the ridges is shown in Figure 3.
In the fifth step a thinning algorithm is applied in order to obtain well-defined and distinguished ridges [6], [2].

2.2   Minutiae extraction

This phase deals with fingerprint characteristics extraction [10], [11]. The extracted characteristics are the minutiae (endpoint or bifurcation). For each image, if n is the number of the extracted minutiae, a set of n records $r(X, Y, \Theta)$ is obtained, where X and Y are the spatial coordinates of the minutia and Θ is the angle between the minutia direction and the horizontal axis direction.

Since fingerprints are usually noisy and the pre-processing could erase some zones, several false endpoints and/or bifurcations can arise. In figure 1 several false minutiae can be noticed: in clockwise order, *interrupted ridges, forks, spurs, structure ladders, triangles and bridges* are depicted.

The *interrupted ridges* are two very closed lines with the same direction. A *fork* is composed by two lines connected by a noisy line. The *spurs* are short lines whose direction is orthogonal to ridges direction: they generate two bifurcations and one endpoint. The *structure ladders* are pseudo-rectangle between two ridges: they form four bifurcations. The *triangles* are formed by a real bifurcation with a noisy line between two ridges: they generate two false bifurcations and a real one. Finally, the *bridge* is a noisy line between two ridges, generating two false minutiae.

Moreover, fingerprint *a priori* knowledge has been used to develop the algorithm for false minutiae rejection: there are no intersections between ridge, a bifurcation line is a long line, there are no structure ladders and triangles in the ideal fingerprint images.

The algorithm for false minutiae rejecting is divided into several steps, executed in a pre-arranged order: elimination of the *spurs*, union of the *endpoints*, elimination of the *bridges*, elimination of the *triangles*, elimination of the structure *ladders* [8].

2.3   Matching

Two different views of the same fingerprint can appear with translation and/or rotation random offsets. An ideal designed matching system should be immune from them.
In the matching phase, the previous extracted records are considered. The record fields are X, Y, Θ, i.e. the spatial coordinates and the ridge direction, respectively. Let us to consider a Cartesian Space with the origin in the bottom left corner of the image. Let V the set of records $r(X, Y, \Theta)$ representing the current image, W_i the set of records $r(X, Y, \Theta)$ of the i-th database image. The goal is to find the set W_j, among the N sets, representing the N images of the databases, maximizing the following function, that is an extension of the well-known Tanimoto distance [18], [1]:

$$F(W_j) \equiv \frac{V \therefore W_j}{V \cup W_j} \qquad (2)$$

where $\therefore$ is a new intersection operator defined as follows:

$$k_i \in (V \therefore W_j) <=> \left\{ |X - X_i| \le T_x; |Y - Y_i| \le T_y; |\Theta - \Theta_i| \le T_\Theta \right\} \qquad (3)$$

with T_x, T_y and T_Θ noisy (translation and/or rotation) immunity thresholds.

3. Experimental results

In order to validate the effectiveness of the proposed approach, experimental trials have been conducted on an ink-on-paper (512x512 pixel with 256 gray level) and scanned fingerprint database. In Table 1 are shown the obtained results with the database. We have used the False Acceptance Rate (F.A.R.) and False Rejection Rate (F.R.R.) parameters in order to evaluate the proposed system performances.

Table 1: matching results with the our fingerprints database

T_Θ	Tx , Ty	False Acceptance Rate	False Reject Rate
4°	4	6,3%	27,2%
7°	7	12.1%	17.2%
8°	8	30.6%	10,8%

4. Conclusions

In this paper a noise robust system based on fingerprint minutiae extraction for the identification of the personal identity has been proposed. The system is mainly characterized by an adaptive energy threshold determination, by an extension of the Tanimoto distance and by a new intersection operator in order to enhance system performances with noisy, translation and/or rotation problems. Experimental results show a good noise immunity of the proposed solution. Current works are aimed to test our solutions on the fingerprint NIST 4 [19] database.

5. Acknowledgements

Authors would like to thank Giovanni Milici for his initial contribution to this research and Engineering S.p.A. for its partial financial support.

References

[1] S. Vitabile, G. Pilato, G. Pollaccia, F. Sorbello (2001), Road Signs Recognition Using a Dynamic Pixel Aggregation Technique in the HSV Color Space, Proc. of 11° International Conference on Image Analysis and Processing, Palermo – Italy, pp. 582-587, IEEE Computer Society Press.

[2] F. Sorbello, G.A.M. Gioiello, S. Vitabile (1999), Handwritten Character Recognition using a MLP, in L.C. Jain and B. Lazzerini Editors, Knowledge-Based Intelligent Techniques in Character Recognition, Chapter 5, pp. 91-119, CRC Press Publishers.

[3] B. M. Mehtre, N. N. Murthy, S. Kapoor, B. Chatterjee, Segmentation of fingerprint images using the directional image, Pattern Recognition, 1987, vol.20, n°4, pp. 429-435

[4] B. M. Mehtre, B. Chatterjee, Segmentation of fingerprint images – A composite method, Pattern Recognition, 1989, vol.22, n°4, pp. 381-385

[5] L. Hong, Y. Wan, A. Jain, Fingerprint image enhancement: algorithm and performance evaluation, Pattern Analysis and Machine Intelligence, 1998, vol.20, n°8, pp. 777-789

[6] R. W. Zhou, C. Quek, G. S. Hg, A novel single-pass thinning algorithm and an effective set of performance criteria, Pattern Recognition Letters, 1995, vol.16, pp. 1267-1275

[7] Wahab, S. H. Chin, E. C. Tan, Novel approach to automated fingerprint recognition, IEE proc.-Visual Image Signal Process., Vol. 145, No. 3, June 1998, pp. 160-166

[8] Q. Xiao, H. Raafat, Fingerprint image postprocessing: a combined statistical and structural approach, Pattern Recognition, Vol. 24, No. 10, 1991, pp. 985-992

[9] Farina, Z. M. Kovacs-Vajna, A. Leone, Fingerprint minutiae extraction from skeletonized binary image, Pattern Recognition, Vol. 32, 1999, pp.877-889

[10] K. Jain, L. Hong, S. Pankanti, R. Bolle, An identity-Authentication system using fingerprints, Proceeding of IEEE, Vol.85, No.9, 1997, pp.1365-1388

[11] K. Jain, L. Hong, R. Bolle, On line fingerprint verification, IEEE transactions on pattern analysis and machine intelligence, Vol.19, No.4, 1997, pp.302-313

[12] N. K. Ratha, S. Chen, A. K. Jain, Adaptive flow orientation-based feature extraction in fingerprint images, Pattern Recognition, Vol.28, No.11, 1995, pp.1657-1672

[13] L. Coetzee, E. C. Botha, Fingerprint recognition in low quality images, Pattern recognition, Vol.26, No.10, 1993, pp.1441-1460.

[14] K. Jain, F. Farrokhinia, Unsupervised texture segmentation using Gabor filters", Pattern Recognition, 1991 vol.24, pp. 1167-1186

[15] V. Kameswara Rao, k. Black, Type Classification of Fingerprints: A Syntactic Approach, IEEE Transaction Pattern Analysis and Machine Intelligence, vol. 2, No.3, pp.223-231, 1990

[16] N. K. Ratha, R. M. Bolle, V. D. Pandit, V. Vaish, Robust Fingerprint Authentication Using Local Structural Similarity, IEEE 2000

[17] Jain, A. Ross, S. Prabhakar, Fingerprint Matching Using Minutiae and Texture Feature, IEEE 2001

[18] K. R. Sloan Jr., Steven L. Tanimoto, Progressive Refinement of Raster Images, IEEE Transactions on Computers, Volume 28, Number 11, pp. 871-874, November 1979.

[19] http://www.nist.gov/srd/nistsd4.htm

[20] Almansa, L. Cohen, Fingerprint image matching by minimization of a thin plate energy using a two-step algorithm with auxiliary variables, IEEE 2000

KES 2002
E. Damiani et al. (Eds.)
IOS Press, 2002

1167

A color, texture and shape fusion technique for segmentation of high-resolution satellite images

Tomoko Tateyama[1], Yen-Wei Chen[1,2], Xiang-Yan Zeng[1], Hiroyasu Tamashiro[1]
[1]*Department of EEE, Faculty of Engineering, University of the Ryukyus, Japan*
[2]*Institute of Computational Science and Engineering, Ocean Unive. of Qingdao, CHAINA*
e-mail:tomoko@augusta.eee.u-ryukyu.ac.jp

Abstract.
Extraction or segmentations of a target object such as roads, from remote sensing images is an important issue. In this paper, we propose a new method which combines color,texture and shape information for segmentation of high-resolution satelite images. In the proposed methhod, the features of color and texture information are used for a global segmentation and the shape information is used for a local segmentation.

We also propose a new PCA-based template maching for extractions of texture features which is easy for computation and free from the illumination, and a Morphorogy-based smoothing filter, which is used as a pre-processing operator for extraction of shape information.

1 Introduction

Remotely sensed images (satellite images) have been used in many fields, such as agriculture environment monitoring and so on [1]. To date, spectral information are mainly used for classifications and analysis of the satellite images. In 1999, a high resolution satellite (IKONOS) was successfully launched and it can observe the earth with a high spatial resolution of $1m$. So it is now possible to extract the road, building and other geographic spatial information, which can be used for constructing and updating a geographic information system (GIS) [2]. Unfortunately, the conventional multi-spectral information (or color) based segmentations are not enough for land or GIS use. Matsuyama and Nagao have proposed a method for the segmentation of aerial images using both spectral features and shape information [3]. Dubussion et al., have also proposed a method for combining color and texture information for the segmentation of aerial images [4]. Furthermore, a shape-based road model has been proposed for extracting the road from the satellite image by Bajçosy [5]. In this paper, we propose a new method which combines color, texture and shape information for the segmentation of high-resolution satellite images. In the proposed method, the color and texture features are used for a global classification and the shape information is used for a local segmentation. In the global segmentation, a new PCA-based template matching method [6] is used for extracting texture features. Since the PCA-based template matching method is free from the illumination, the method is effective for analysis of satellite images. In the local segmentation, a Morphoragy smoothing filter is used as a pre-processing operator to remove the noise in order to extract the shape information.

2   Proposed segmentation algorithm

The flowchart of the proposed segmentation algorithm is shown in Fig.1. In the global segmentation process, we first extract the features of color and texture, and then a trained neural network is used for the global segmentation based on the extracted features of color and texture. In the local segmentation process, we first use a Morphorogy smoothing filter to remove the salt and pepper noise, and then a labeling and a thinning processes are used for extract the shape information.

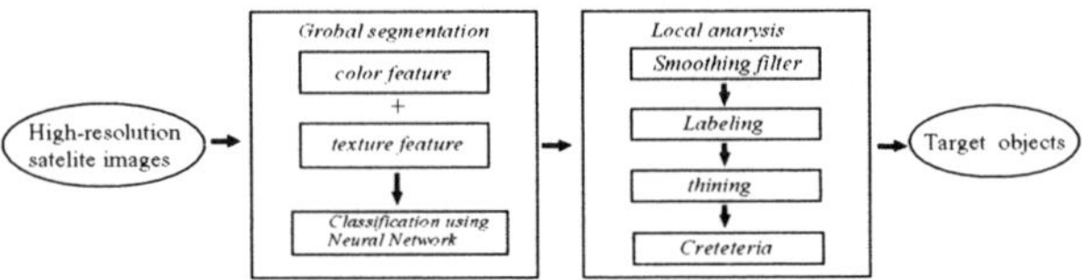

Figure 1: Process for proposed method

2.1   *Global segmentations*

2.1.1   Color features

The pixels in the IKONOS images are represented in the RGB space. We use a simple normalized RGB as shown in Eq.(1), which is independent on the intensity, as the color features.

$$\frac{R}{R+G+B}, \; \frac{G}{R+G+B}, \; \frac{B}{R+G+B} \tag{1}$$

2.1.2   Texture features

There are many methods have been proposed for extractions of texture features, such as Gabor filter [7] and so on. In this paper, we propose to use a new PCA-based template matching method for the extraction of texture features. All the templates are obtained by learning the principal components of image patches, and then the image is transformed to a feature image, in which each pixel is represented by the classes of the patterns that match the neighbor blocks best. The texture features are extracted from the feature image. The advantages of the PCA template matching method are that it is easy for computing and free from the illumination. Details about the method is presented in Ref.[6]. Fig.2 shows the PCA templates learned from a "girl" image and Fig.3 shows an example of texture segmentation with comparison to conventional Gabor filter.

2.1.3   Neural networks

The global segmentation is done by use of a three-layer neural network as shown in Fig.4. based on extracted features of color and texture [8]. The neural network is pre-trained by the teaching signals. Each pixel in the satellite image is segmented to corresponding class based on its color and texture features. Details about the neural network and the training is presented in Ref.[8].

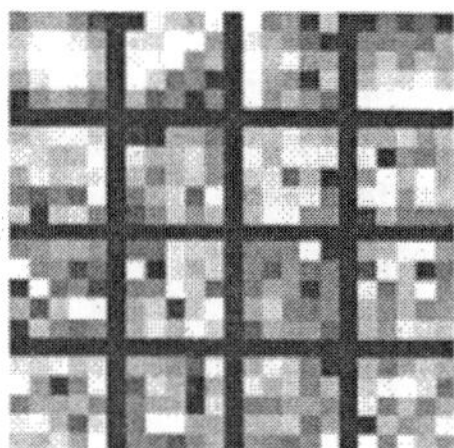

Figure 2: PCA templates

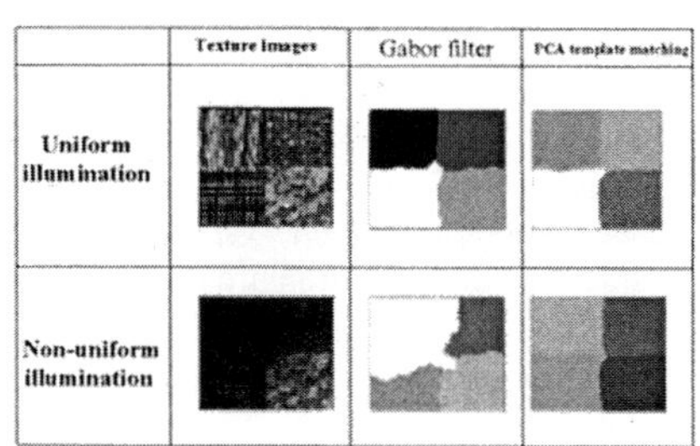

Figure 3: Example of texture segmentation result

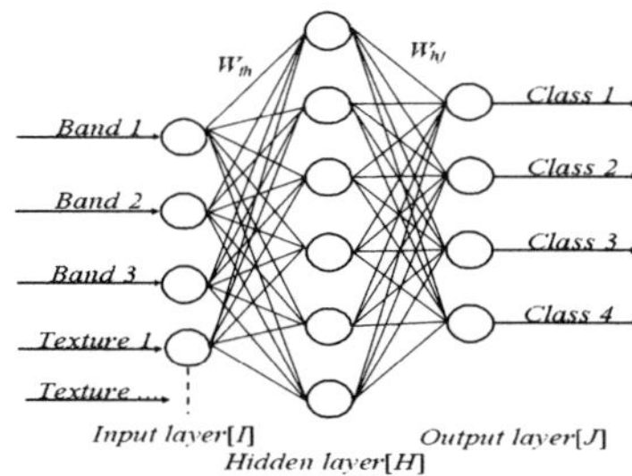

Figure 4: A three-layer of neural network

2.2   Local segmentations

In the Global segmentation process, the objects with same color and texture features will be classified to the same class. Thus in the local segmentation process, we try to extract or segment the target object from the result of the global segmentation using the shape information of the target object.

2.2.1   Morphorogy smoothing filter

Since the results (binary images) of the Global segmentation contains many noise, which is a kind of salt-pepper noise. We propose a Morphorogy smoothing filter to remove the noise without any distortion of signals. The filter is shown in Eq.(2):

$$F_q = ((((A \bigcirc B_s) \bigcirc B) \bigcirc B_S) \oplus B) \tag{2}$$

where A is the binary image, B is structuring element, and B_s is symmetric element of B. $\bigcirc$ is a dilation operator and $\bigcirc$ is a erotion operator. $(A \bigcirc B_s) \bigcirc B$ is usually called as opening and $(A \bigcirc B_s) \bigcirc B$ is know as closing [9]. The filter is a combination of opening and closing operators. The opening operator smoothes the region from inside and it can remove the salt-pepper noise, small island and so on. On the other hand, the closing operator smoothes the region from outside so that it can be used to fill the small holes in the image. The advantage of the proposed filter is that can remove the noise without any distortion of signals.

2.2.2 Labeling process

Since there are several objects with same color and texture features will be classified to the same class. In the labeling process, the connected pixels are labeled with same index by the use of a conventional boundary-following algorithm[10]. The following thinning process and shape information extraction are performed to each labeled region.

2.2.3 Thinning process

In order to get information about the length of the object, we perform a thinning process to each labeled region. The length l can be roughly estimated as the pixel number of the thinning line. Thus the width w of the object can also be roughly estimated as S/l^2, where S is the area of the object and it can be obtained by counting the pixel number of the labeled region. The shape of the object can be estimated by use of the information of length and width. The algorithm used for thinning process was proposed in our previous work [11]. Details about the algorithm is shown in [11].

2.2.4 Criteria

In this paper, we focus our research on segmentation of roads from the satellite images. The criteria for roads is shown in Eq.(3)

$$w = S/l^2 < c \tag{3}$$

where c is a threshold, which is set as 0.05. If the labeled region satisfies the Eq.(3), the region is considered as a road and will be extracted.

An typical example of road segmentation from the high-resolution satellite image is shown in Fig.5. Fig.5(a) is the high-resolution satellite (IKONOS) image, which is a color image with 240x240 pixels. Fig.5(b) and (c) show the global segmentation results using color alone and both color and texture, respectively. It can be seen that there are many noise and other objects contained in the segmented image. By using the texture features, we can obtain a more smooth and clear image. Fig.5(d) is the result obtained by Morphorogy smoothing filter. Almost of the noise have been removed. Fig.5(e) is the thinning result which is used to calculate the length and width for each labeled object. Fig.55(f) is the final result of the road segmentation using criteria of Eq.(3). It can be seen that only the road is extracted.

3 Conclution

In this paper, we proposed a new method, which combines color, texture and shape features, for segmentation of high-resolution satellite images. We demonstrated the applicability of the proposed method using real high-satellite (IKONOS) images.

References

[1] M.Ogami, *Image Processing Handbook*, Shoukou-dou, Tokyo, 1992.

[2] Y.W.Chen et al., "Geographic information system construction using high-resolution satellite images," *Proc. Int'l Conf. On Industrial Logistics 2001*, pp.25-29 (2001).

[3] T.Matsuyama, M,Nagao : "Structuralanalysis for the aerial photograph", *IPSJ*, Vol.21, No.58, pp.468-479, (1980).

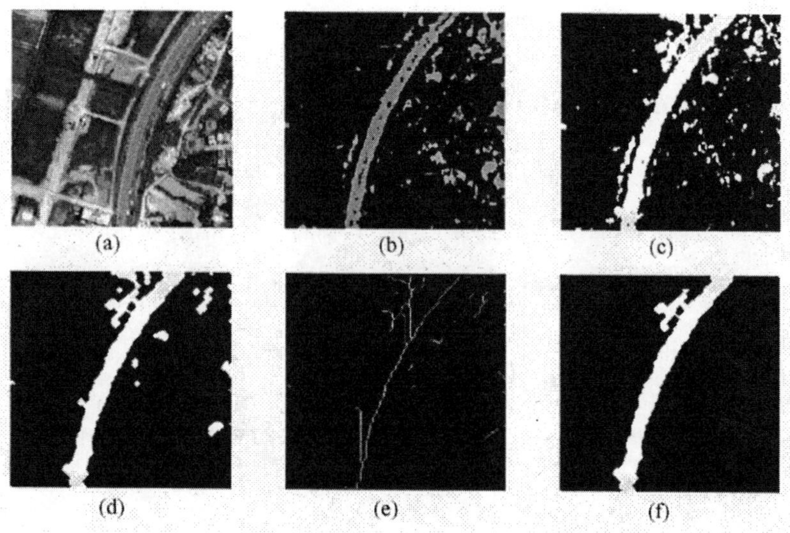

Figure 5: Simulation results by proposed method

[4] M.P.Dubussion-Jelly, A.Gupta: "Collor and texture fusion, application to aerial image segmentation and GIS updating", *Image and Vision Computing* Vol.18, pp.823-832, (2000).

[5] R.Bajcosy and M.Tavakoli: "Computer Recognition of Road from Satelite Pictures", *IEEE Trans.*, Vol.SMC-6, No.9, pp623-637, (1976).

[6] X.-Y.Zeng,Y.-W.Chen,Z.Nakao and H.Lu: "Texture segmentation based on pattern maps obtaind by principle component analysis," *Proc. of Int.Tech Conf.on Circuits/Systems.Computers and Communication*, pp.1058-1061, (2001).

[7] B.S.Manjunath and W.Y.Ma: "Texture features for browsing and retrieval of image data", *IEEE Trans. Pattern Analysis and Machine Intelligence*, Vol.18, pp.837-842, (1996)

[8] H.Tamashiro,T.Tateyama,Y.-W.Chen: "A color and texture fusion technique for classification of high resolution satelite images" *IEICE Technique Report. HIP 2001-90*, pp.73-78, 2002,.

[9] Ioannis pitas et al.,: "Morphological shape Decomposition", *IEE transactions on Pattern Analysis and Machine Intelligence.*, Vol.12, pp.38-45 (1990).

[10] R.Jain et al.,: *Machine Vision*, McGraw-Hill, New York, (1995).

[11] T.Kiniyoshi, Y.W.Chen and Z.Nakao: "Development of a Geographic Information System for University of the Ryukyus II", *Bulletin of Engineering Faculty. University of the Ryukyus*, No.58, pp.95-98, (1999).

KES 2002
E. Damiani et al. (Eds.)
IOS Press, 2002

Classification of Remote Sensing Images Using Spectral Independent Components

Xiang-Yan ZENG[1], Yen-Wei CHEN[1, 2] and Zensho NAKAO[1]

[1] *Department of EEE, Faculty of Engineering, Univ. of the Ryukyus, Japan*

[2] *Institute of Computational Science and Engineering, Ocean Univ. of Qingdao, China*

Abstract.

We apply independent component analysis (ICA) to learn an efficient color representation of remote sensing images. Among the three basis functions obtained from RGB color space, two are in opponent-color model by which the responses of R, G and B cones are combined in opponent fashions. This is coincident with the idea of contrasting reflected in many color systems. The interesting point is that there is no summation component which responds to illumination in other transforms such as principal component analysis (PCA). The spectral independent components are then used to segment the remote sensing images. The experimental results are compared with those obtained by RGB and RGB principal components (RGBPC).

1. Introduction

Remote sensing image analysis has many applications where the information of earth surface is necessary, such as planning and management of public transport systems and environment investigation. To classify the image into nonoverlapping regions is the preprocessing step for many tasks. Color information or multi-spectra information has been widely used for classification of remote sensing images [1]. To find an efficient color representation is crucial for a classification algorithm. There are several transformations of the RGB color space which can generate new color spaces [2]. However, these are general transformations independent of the actual images.

So, it may be a more reasonable way to find an encoding of color information by analysis of target images. In the recent years, there exists an increasing interest in developing image coding (spatial and spectra) methods by statistics analysis. These approaches follow the idea of Barlow that the goal of vision information processing is to transform the input signals such that it reduces the redundancy between the inputs [3]. In the analysis of spectra information, Buchsbaum et al. found opponent coding to be the most efficient way to encode human photoreceptor signals [4]. In an analysis of spectra of natural scenes using principal component analysis (PCA), Ruderman et al. found principal components close to those of Buchsbaum [5]. Since PCA as a decorrelation solution uses only second -order statistics, methods involving higher-order statistics are attracting more and more attention [6,7,8,9]. Independent component analysis (ICA) is such a technique that the transformed variables are not only decorrelated but also as statistically independent from each other as possible. Lee et al. has analyzed the spectral properties of nature image by using ICA [10]. Their experiment was implemented with 31 waveband images and the resulting basis functions are found to be related to object reflectance.

In this paper, we try to find an efficient color encoding method for RGB or multi-spectra remote sensing images. Our consideration is that how the R, G and B values should be combined depends on many factors, including the object reflectance and instrument function. With the same scene, different photo systems may get different color images. Our experiment is different from those of Ruderman et al. and Lee et al. in that the remote sensing images are recorded under conditions different from that of normal nature scene images. We apply ICA to analyze the encoding information of RGB remote sensing images and use the independent components of RGB to do classification.

2. Independent Component Analysis

ICA is to linear transform the data such that the outputs are as statistically independent from each other as possible. The general model can be described as follows.

We start the basis assumption that a data vector $\mathbf{x} = (x_1,...,x_M)$ can be represented in terms of a linear superposition of basis functions,

$$\mathbf{x} = \mathbf{As} = \mathbf{a}_1 * s_1 + ... + \mathbf{a}_N * s_N \tag{1}$$

where $\mathbf{s} = (s_1,...,s_N)$ is the coefficients, $\mathbf{A}$ is a $M \times N$ matrix and the columns $\mathbf{a}_1,...,\mathbf{a}_N$ are called basis functions. The basis functions are consistent and the coefficients vary with data. The goal of efficient coding is to find a set of $\mathbf{a}_i$ that result in the coefficient values being statistically as independent as possible over an ensemble of data Thus, the linear image analysis process is to find a matrix $\mathbf{W}$, so that

$$\mathbf{y} = \mathbf{W}\mathbf{x} \tag{2}$$

recover the independent components $\mathbf{s}$, possibly permuted and rescaled. The rows of $\mathbf{W}$ respond to the columns of $\mathbf{A}$.

Bell & Sejnowski have proposed a neural learning algorithm for ICA [9]. The approach is to maximize by stochastic gradient ascent the joint entropy. The updating formula for $\mathbf{W}$ is:

$$\Delta\mathbf{W} = (\mathbf{I} + g(\mathbf{y})\mathbf{y}^T)\mathbf{W} \tag{3}$$

where $\mathbf{y} = \mathbf{W}\mathbf{x}$, and $g(y) = 1 - 2/(1 + e^{-y})$ is calculated for each component of $\mathbf{y}$. Before the learning procedure, $\mathbf{x}$ is sphered by subtracting the mean $\mathbf{m}_x$ and multiplying by a whitening filter:

$$\mathbf{x} = \mathbf{W}_0(\mathbf{x} - \mathbf{m}_x) \tag{4}$$

where $\mathbf{W}_0 = [(\mathbf{x} - \mathbf{m}_x)(\mathbf{x} - \mathbf{m}_x)^T]^{-1/2}$. Therefore the complete transform is calculated as the product of the whitening matrix and the learned matrix:

$$\mathbf{W}_I = \mathbf{W}\mathbf{W}_0 \tag{5}$$

3. Independent Components of the Spectra of Remote Sensing Images

In our experiment, the training data set consists of 15000 samples, with 5000 samples randomly selected from each of the images shown in Fig. 1 quoted as (a), (b), (c). A sample pixel has 3 values of R, G, B. We did not consider the spatial information. However, in the later segmentation results, we can find that the independent components reflect some spatial relationship since neighbor pixels are more consistently classified than those in RGB.

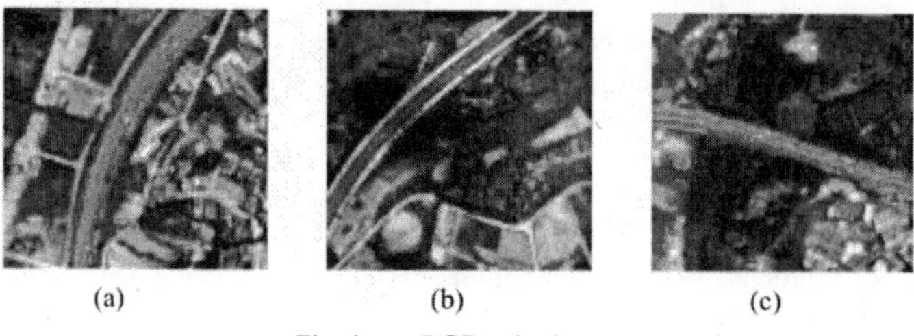

(a)   (b)   (c)

Fig. 1   RGB color images.

The ICA learning procedure was done in 100 weeps, and the resulting transform matrix is:

$$\mathbf{W}_I = \begin{bmatrix} 0.1054 & -0.0224 & -0.0703 \\ -0.1350 & 0.2459 & -0.1092 \\ -0.0403 & 0.0327 & 0.0394 \end{bmatrix}$$

The spectral independent components of a pixel will be:

$$\begin{bmatrix} I_1 \\ I_2 \\ I_3 \end{bmatrix} = \begin{bmatrix} 0.1054 & -0.0224 & -0.0703 \\ -0.1350 & 0.2459 & -0.1092 \\ -0.0403 & 0.0327 & 0.0394 \end{bmatrix} \begin{bmatrix} R \\ G \\ B \end{bmatrix}$$

The first two axes of the transform reflect an opponent-color model by which R, G, and B are combined into opponent color components. This is similar to the idea in many color transforms. The interesting point is that there is no summation component that usually appears in many transforms including the one obtained by principal component analysis. Instead, the last component seems to be mainly represented with G, where R opposes B. This direction has some interpreted meanings. The R, G, B histograms of the images shown in Fig. 1 have similar characteristics. As an example, we plot the R, G, B histogram of one image (Fig. 1 (a)) in Fig. 2. We can see that G histogram is obviously more densely distributed than R and B. The independent components of image Fig. 1 (a) and their distributions are shown in Fig. 3, and Fig. 4. The first component is mainly distributed in a range of [100,150], and the second component is mainly in [70, 110].

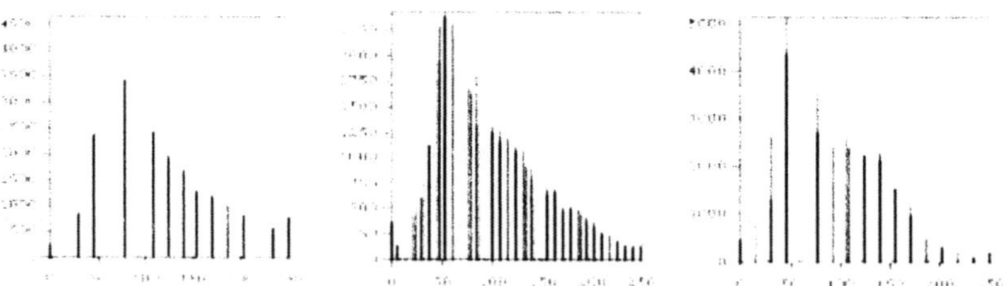

Fig. 2   Histograms of R, G, and B of image (a)

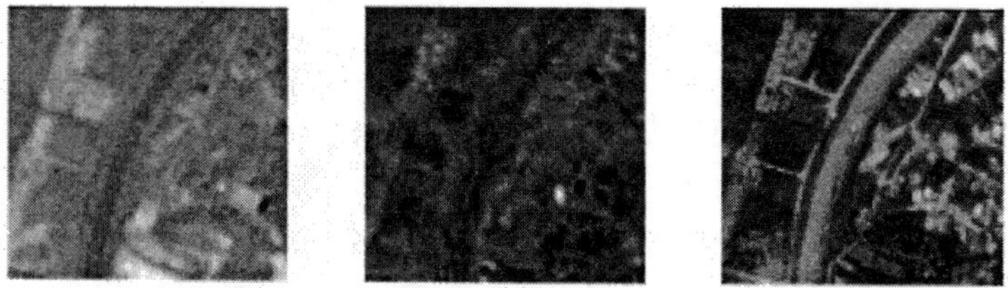

Fig. 3   3 independent components of Fig. 1 (a)

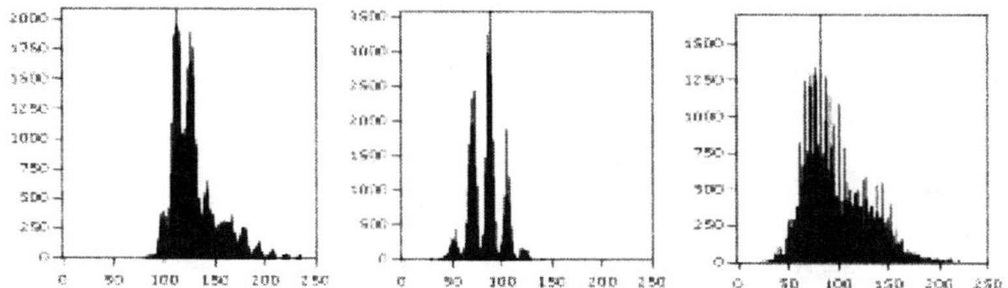

Fig. 4   Histograms of the three independent components

4. Segmentation of Remote Sensing Images Using Independent Components

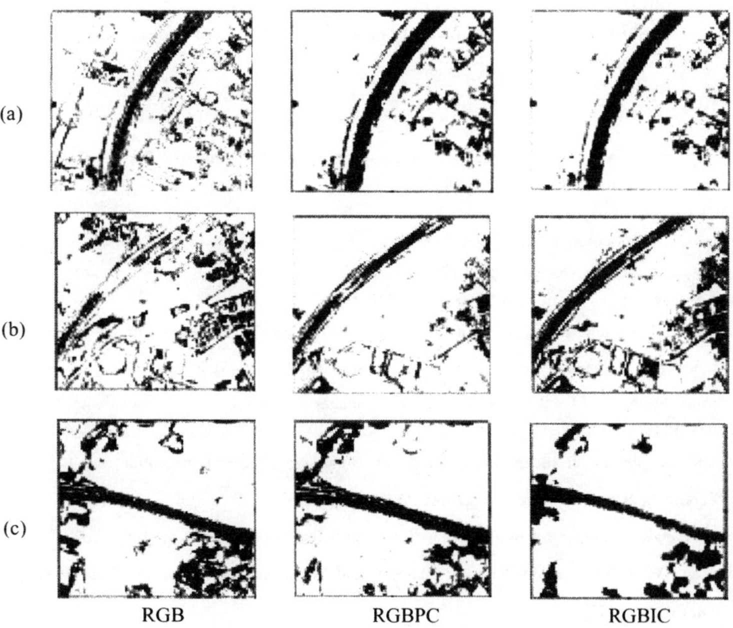

Fig. 5   The main roads extracted from the images in
Fig. 1   (a),   (b),   (c)

We apply the independent components of RGB space (RGBIC) to the classification of remote sensing images. The unsupervised K-means algorithm is used for clustering the spectral features. PCA is also used to transform the RGB space (RGBPC). The predetermined numbers of object classes are 6, 6 and 4 for the images of Fig. 1 (a), (b) and (c), and the numbers keep same for classification by RGB, RGBPC and RGBIC. Our intention is to extract the main roads from the images. We show the results of RGB, RGBPC and RGBIC in Fig. 5, where black pixels are clustered into the same class as the roads. Although these results may be limited by that we use unsupervised clustering algorithm and only spectral information, we here consider mainly the efficient representation of color information. For the purpose of road extraction, the results of RGBIC and RGBPC are much better than that of RGB for post-processing. The superiority of RGBICA to RGBPC is also discernable. In the results of image (a) and (c), the roads in RGBIC are more isolated from the background. In the image (b), the road is interrupted by the shadow. This affection is still left in RGBPC, and is much overcome in RGBIC.

5. Conclusions

In this paper, we apply independent component analysis (ICA) to learn an efficient color representation of remote sensing images. The obtained basis functions have rather interpreted meanings. The independent components are the opponent combination of R, G, and B, and have compensative distribution. Being without the summation component, the RGBIC space may be more independent from the influence of illumination. The spectral independent components are used to classify remote sensing images. The comparison with RGB and RGBPC indicate that the segmentation results by RGBIC are more appropriate for the later processing.

References

[1] H. Murai, S. Omatsu and S. OE, Principal Component Analysis for Remotely Sensed Data Classified by Kohonen's Feature Mapping Preprocessor and Multi-Layered Neural Network Classifier. IEICE Trans. Commun., E78-B 12 (1995) 1604-1610.

[2] M. –P. Dubuisson-Jolly and A. Gupta, Color and Texture Fusion: Application to Aerial Image Segmentation and GIS Updating. Image nad Vision Computing, 18 (2000) 823-832.

[3] H. Barlow, Sensory Communication, MIT press, 217-234, 1961.

[4] G.Buchsbaum and A. Gottschalk, Trichromacy, opponent colours coding and optimum colour information transmission in the retina. Proceedings of the Royal Society Lodon,B, 220 (1983) 89-113.

[5] D. L. Ruderman, Statistics of Cone Responses to Natural Images: Implications for Visual Coding, J. Opt. Soc. Am. 15(8) (1998) 2036-2045.

[6] G. Deco, and D. Obradovic, Independent Component Analysis: General Formulation and Linear Case, An Information-theoretic Approach to Neural Computing, Springer. 1996.

[7] P. Comon, Independent Component Analysis- a New Concept? Signal Processing, 36 (1994) 287-314.

[8]G. Wyszecki and W. S. Stiles, Color Science: Concepts and Methods, Quantitative Data and Formulae. Wiely, New York, 2nd ed. Edtion, 1982.

[9] A.Bell and T. Sejnowski, An information-maximasation approach to blind separation and blind deconvlution," *Neural Computation*, 7(1995) 1129-1159.

[10] T.W.Lee, T. Achtler, and T. J. Sejnowski, The Spectral Independent Components of Natural Scenes, Institute for Neural Computation at UCSD Technical Report Series No. INC-9901, September 1999.

An embedded neural network system for vision applications

Andrea Pinna[1], Annick Alexandre[1], Eric Belhaire[2],
Patrick Garda[1], Bertrand Granado[1]
[1] *LISIF Universite Pierre et Marie Curie (VI), Paris, France*
[2] *IEF Universite Paris-Sud (XI), Orsay, France*

Abstract. We present here a new paradigm based on a centralized architecture to realize electronic artificial retina. This original architecture, named connectionnist retina, can execute in real time RBF and MLP neural networks applications. We demonstrate that this intelligent embedded system could be used for vision applications. We describe the realized prototype system.

1 Introduction

Electronic artificial retina [1] are integrated circuits which combine an image sensor with a parallel architecture. Such an association provides the basic image sensor with a processing capability.

We would like to integrate neural networks algorithms into an artificial electronic retina because those algorithms are able to solve cognitive problems like face recognition, pattern recognition or control.

Whereas image processing studies image-to-image transformations, image understanding fluids description. Recognition require descriptions to manipulate and understand an image composed of a set of objets. Electronic artificial retina with this processing capability could be use for example in automotive applications for active or passive security, or in wireless telecomunications for identification tasks.

In this article we propose a central processing architecture to realize this kind of artificial electronic retina. The parallel architecture, chosen to compute the neural networks models, makes our artificial retina programmable. We can simulate in real time MLP[1] and RBF[2] [2] neural networks for embedded system applications, such as automotive, satellite and home applications. This new artificial retina paradigm allows to perform low and intermediate level processing.

. A prototype CMOS imager was fabricated in a standard 0.6 μm, n-well CMOS process.

We designed a snapshot imager because the MLP and RBF neural networks need all the pixels of the image, which are the input of the neural network, acquired at the same time.

The parallel architecture was implemented into a SoC emulator based on an embedded RISC microprocessor system (ARM). This paper describes an implementation of this paradigm, named connectionnist retina. We also present the prototype realization.

2 The connectionnist retina architecture

The connectionnist retina uses a centralized architecture [3] composed of a CMOS imager, of an analog to digital converter (ADC) and of a parallel architecture specialized in the

---

[1]Multi Layer Perceptron
[2]Radial Base Function

simulation of connectionnist networks.

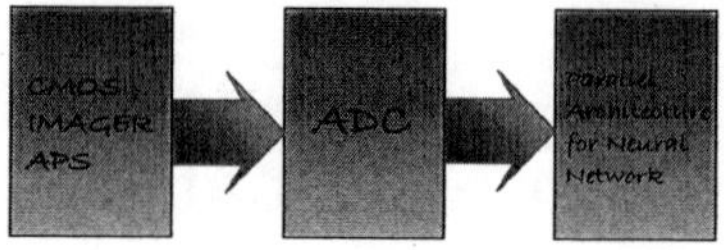

Figure 1: Architecture of connectionnist retina.

The CMOS imager is able to transduce a light information representing for a example a written character or a face. The vision imager and the connectionnist architecture communicate through the analog to digital converter.

2.1 CMOS Image Sensors

The CMOS vision imager is made of a photosensor matrix which is able to transduce a light information. It is a mixed circuit comprising a sensor matrix of 64x64 CMOS APS [4] and a digital circuit that manages the acquisition. The photosensor is based on a phototransistor structure [5]. In order to achieve a small pixel size, a single type of MOS transistor is used within the pixel. The incident light on each pixel of the imager generates a photocurrent that is integrated and stored as a voltage. The image acquisition is not continuous, it is controlled by a snapshot read-out method. Such a snapshot is controlled by a shutter command. In the figure 2 we can see the shutter as an isolation point between the photosensor and the read-out circuit. This command is well suited to RBF and MLP neural networks algorithms where all the pixels of the image need to be acquired at the same time.

The electronic shutter has a wide range of exposure time. We obtained a readout speed of 25 frames per second for 64 x 64-pixels video.

The shutter command can also control two important imager parameters: the readout speed and the sensitivity.

To our best knowledge the electronic shutter has never been used in an electronic artificial retina. The CMOS imagers often perform continuous reading acquisition [1].

Photointegration is performed simultaneously on the entire matrix.

The matrix is read row by row and the digital output is obtained from the analog to digital converter. The reading task is accomplished by two digital units composed of a digital scan shift register.

Before a row is read, a reset of the columns is made. The reading frequency is a function of the frame frequency and of the matrix size. The figure 2 illustrates the exposure and the readout sequence for a single frame (note that the timing is not scaled). We can see how the imager is read. After the acquisition, controlled by the reset and shutter commands, we first

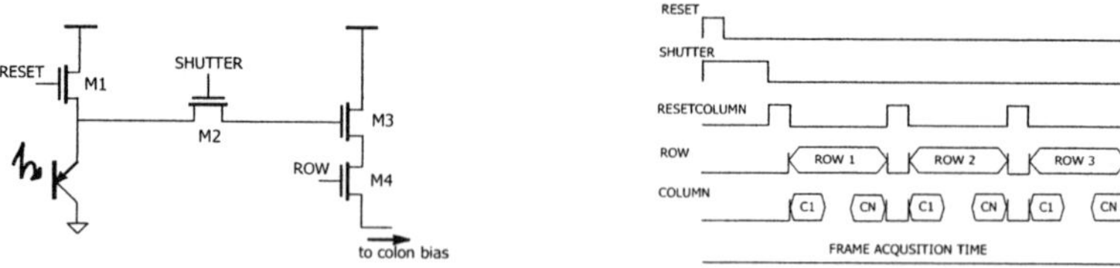

Figure 2: APS schematic, single frame exposure and readout sequence

reset all columns. After, we select a row and it is read column by column. This sequence is repeated for all the rows. A dark column is added to estimate the darkcurrent and to checkout the influence of the next pixels.

A disadvantage of APS is the fixed pattern noise, it results from the non-uniformity of the pixel properties on the silicon wafer. It is possible to smooth this noise by performing a correlated double sampling. This smoothing operation is executed by the parallel architecture.

2.2　Analog-Digital converter

The converter is the communication channel between the imager and the parallel architecture. We use a 8-bit converter which provides a 256 gray-scale imaging. The chosen readout methodology use a single conveter [6].

2.3　Parallel Architecture

2.3.1　Introduction

The parallel architecture, named MAHARADJA [3], can simulate in real time two kinds of neural network models: the RBF and the MLP. The MAHARADJA architecture is illustrated in figure 3. The table 1 presents the neural dynamics of these models.

Table 1: Main parameters for neural networks : V_i = post-synaptic potential, X_i = state of the neuron, W_{ji} = weight, E_i = neighbours of the neuron.

PRN	PMC	RBF	
		Euclide	*Manhattan*
V_i	$\displaystyle\sum_{j\in E_i} X_j * W_{ij}$	$\displaystyle\sum_{j\in E_i} (W_{ji} - X_j)^2$	$\displaystyle\sum_{j\in E_i} |W_{ji} - X_j|$
X_i	$m\dfrac{1-e^{-\lambda V_i}}{1+e^{-\lambda V_i}}$	$e^{\frac{-V_i}{\lambda}}$	$e^{\frac{-V_i}{\lambda}}$

2.3.2　Description

The system is composed of four major components: a micro-controller, an array of processing element (UNE)[3], a 256 Kbyte memory and an input unit.

The processing elements are organized in a SIMD[4] way. A controller manages the UNE and the synchronization with the micro-processor. The internal data buses are 16 bits wide. A command bus manages the four operator inside the UNE. Inside UNE architecture we can find four operators. For the RBF neural network it is possible to compute three kinds of distance: Euclide, Manhattan, Mahalanobis. Each UNE can calculate different post-synaptic potentials according to the type of neural networks to be realized, RBF or MLP. The figure 3 illustrates the UNE architecture. The MAHARADJA system is described in [7].

---

[3]Unit NEural
[4]Single Instruction Stream Multiple Data Stream

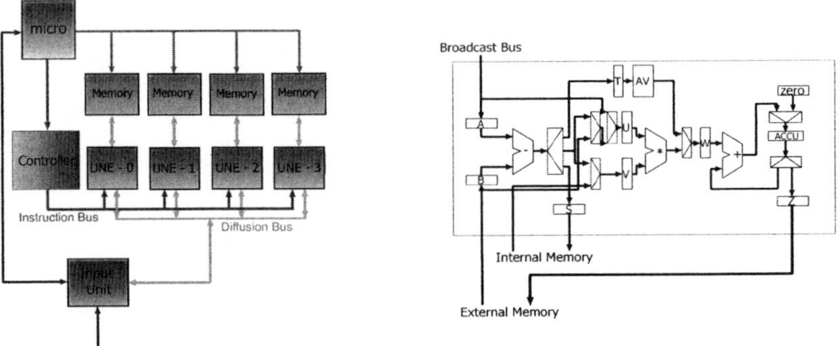

Figure 3: Architecture of MAHARADJA and Internal architecture of UNE: A,B,S,T,U,V,W and U are the syncronisation registers to avoid the computation, VA is the absolute computational unit and ACCU is the storage register

Figure 4: Circuit photograph,images takes with a 20 ms shutter speed and the imager board

3   Results

A connectionnist retina prototype was made. Different tests were performed at different shutter speeds. We are interested to integrate several RBF and MLP neural networks models for visual recognition and motion detection applications. To validate the real time acquisition and processing we simulated two different neural networks models.

3.1   CMOS Image Sensor Layout

Two experimental 64x64 APS imagers were realized in a 0.6-μm CMOS technology. The circuit photograph is shown in figure 4. The matrix sensor size is 2.18 mm x 2.18 mm and we obtained a fill factor greater than 50%.

3.2   Implementation of MAHARADJA

MAHARADJA was described in VHDL and the integration of the MAHARADJA architecture into an ARM emulator system is under way. We estimated, with a methodology described in [3], the ability to achieve the real time for neural network processing for an embedded system. We simulated a MLP network for the character recognition named LeNet. This is a TDNN Multi-Layer Perceptron with 96522 local connections, 1920 full connections and 4365 neurons. Its function is to recognize handwritten digits. It was developed by Y.Lecun in the AT&T laboratories. The results are shown in Table 2. The advantage to integrate the processing task outside the pixel is a more accurate data processing made at 16 bits.

At this time we are testing the image sensor and we are realizing the entire system on a card to evaluate the connectionist retina system. The APS imager has been succesfully tested and operated. The figure 4 shows the prototype board and four images taken with a shutter speed of 20 *ms*.

Table 2: Neural Network simulation time: latency and performance.

Neural Network	Performance	
Simulated	*Latency Constraints time*	*Execution time estimated*
LeNet	40 ms	59,6 ms à 10 MHz 29,8 ms à 20 MHz 23,8 ms à 25 MHz
RBF (4-10-256)	40 ms	473μs

4　Conclusion

In this paper we presented and validated a new paradigm to realize an electronic artificial retina. It is based on a central processing architecture. This new approach allows to perform low and intermediate image processing for real time embedded system applications. We validated this approach with a demonstrator named connectionnist retina which executes neural networks simulations in real time.

References

[1] Alireza Moini. *Vision Chips or Seeing Silicon.* 1997.

[2] Christopher M. Bishop. *Neural Networks for Pattern Recognition.* Oxford University Press, 1996.

[3] Bertrand Granado. *Architecture des syst mes lectronique pour les r seaux des neurones - Conception d'une r tine connexionniste.* PhD thesis, Universit de Paris-Sud - Orsay, 1998.

[4] Chye Huat Aw and Bruce A. Wooley. A 128x128-pixel standard-cmos image sensor with electronic shutter. *IEEE Journal of Solid-State Circuits*, 31(12):1992–1930, December 1996.

[5] T. Delbruck Carver Mead. Analog vlsi phototransduction by a continuous-time, adaptive,logarithmic photoreceptor circuits. *CNS Memo*, (30), April 1996.

[6] Arsenia Chorti, Bertrand Granado, Bruce Denby, and Patrick Garda. An electronic system for simulation of neural networks with a microsecond real-time constraint. *ACAT*, September 2000.

[7] Luc Gaborit Patrick Garda Bertrand Granado, Andrea Pinna. Mahardja: A system for the real time simulation of rbf with the mahalanobis distance. *Neural Processing Letter*, (13):253–266, 2001.

KES 2002
E. Damiani et al. (Eds.)
IOS Press, 2002

A soft computing based behavioral model of Po River during flood events

A.Boscolo*, D.Russo**, V.Fiorotto**
*DEEI: Dipartimento di Elettrotecnica, Elettronica ed Informatica Universita' di Trieste
p.le Europa 1, I-34100 Trieste
**DIC: Dipartimento di Ingegneria Civile Universita' degli studi di Trieste p.le Europa 1,
I-34100 Trieste*

Abstract. The flood evaluation is a basic aspect to define the flooding risk in the areas close to large rivers. For this purpose the water level are usually computed via deterministic model that solve the De Saint Venant equations. This approach has some structural limits due to the difficulty in the evaluation of the relevant geometric and hydraulic parameters. This fact affects the results especially in the real time forecasting framework where a tuning procedure is not practicable. These problems can be avoided using kind of models based on a qualitative description of the physical phenomenon. In this work the capability of a simple fuzzy model is evaluated with reference to the Po River case.

1. Introduction

Real time forecasting models of natural phenomena, such as river's floods wave, represent a very important tool in managing extreme events; indeed, they provide a time lag available to carry out suitable action plans in order to avoid human losses and reduce damages.

In the fluvial hydraulic context, the most used models are the deterministic one, that are able to simulate the evolution in time and space of the interested quantities via integration of the physically based equations that are the continuity and momentum equations:

$$\frac{\partial vA}{\partial x} + \frac{\partial A}{\partial t} = 0 \tag{1}$$

$$\frac{\partial h}{\partial x} + \frac{v}{g}\frac{\partial v}{\partial x} + \frac{1}{g}\frac{\partial v}{\partial t} + J - i = 0 \tag{2}$$

where: v = mean velocity, A = cross section flood area, h = water depth, i = the bottom slope and J = the energy slope that can be evaluated according the following formula:

$$J = n^2 v^2 / R_h^{\frac{4}{3}} \tag{3}$$

where: n = Manning's coefficient, R_h = cross section hydraulic radius.

Working out such a model means to integrate eq. (1-3) in space and time assigning the initial and boundary conditions. For this purpose a topographic description of the river is needed in order to compute the cross section geometric quantities A and R_h to the detail requested by the required accuracy in results. Moreover an evaluation of the Manning's coefficient n, that take into account in an overall manner bed roughness, bedforms effects, channel cross section irregularities, sinuosity and meandering, etc., must be carried out [2].

If these conditions are satisfied this methodology, very sophisticated from a theoretical point of view, gives accurate results in terms of water level h and cross section mean velocity v along the abscissa x (distance along the mean flow direction) and time t.

In large rivers, during the major flood events, some variations in cross section geometry and river morphology could occur leading to some important problems. For instance, we can have natural meander short-cuts in sinuous river, changing in this way the stream geometry, possible embankment failures affecting the downstream discharge, variations in local energy loss coefficient and so on [2].

The evaluation of these parameters during a flood wave simulation is very difficult; for this reason usually some simplifications are introduced in the model affecting the goodness of results.

In addition, the choice of a proper value of the Manning's coefficients is very difficult because they might change during the flood event as a consequence of variations in bedform shape and dimension [3].

In fact, the bedforms change their shape from ripples to dune reaching sometimes the washed out condition [8]. The evolution of the bedforms has an important impact on the river sections geometry and on the energy losses.

Since these parameters are not enough known in a quantitative way to allow a previous value assignment, their use in deterministic models for real time forecasting purposes is troublesome cause the limited time of persistence of flood peaks.

In this context the application of an alternative modeling technique, such as the fuzzy one, could turn out of some interest. This kind of systems belongs to the behavioral model family that is models based on the experience and available previous knowledge about the physical phenomenon to be studied [5], [6].

These models are mostly based on a qualitative knowledge rather than on a quantitative one; thus they can take into account all those facts that are not enough known to be described in terms of mathematical representation.

This capability turns out very useful in the river hydraulics framework because it allows simulating some phenomena otherwise ignored or, at least, simplified.

In this paper such a model is developed and applied to the Po River reach, with the aim of real time forecasting of water levels at the Pontelagoscuro gauging station by the simple use of water level readings in some upstream locations.

The main difference from a deterministic model is that the fuzzy system is able to solve the problem without a quantitative evaluation of: changes in river morphology and cross sections, discharge variation due to flooding and, at least, bed form evolution and so on.

In addition, this methodology has the advantage of taking into account many other facts in a global way such as vegetation in the flood plain areas that depend on seasonal and meteorological factors, and so on [2].

2. Model definition

In this section a fuzzy model of a Po River reach, between Cremona and the Pontelagoscuro gauging stations is developed in order to test this methodology.

The Po River is the most important Italian watercourse with a maximum discharge (1951) equal to 12500 m^3/s as evaluated at the Pontelagoscuro gauging station (basin surface 70000 km^2). This event was characterized by important floodings due to embankment overflow. In 1994 and 2000 maximum discharges equal to 11000 m^3/s and 10000 m^3/s, respectively, were measured in Pontelagoscuro.

These values are close to the maximum design discharge (12500 m^3/s) [7] generating comprehensible worries for the hydraulic safety of the downstream countries. For these reasons the study of Po River flood waves evolution can be considered an up-to-date item.

Several water level monitoring stations are placed along this river to check the water level evolution continuously; in this model the data of the gauging stations of Cremona, Roncocorrente and Revere are used as input to forecast the water level at Pontelagoscuro.

The water levels are gauged and transmitted to the flood control office during each flood event in real time. From a physical point of view the spatial and temporal distribution of these values as measured along the watercourse takes into account in a global manner all the complex factors that govern the specific flood event [2].

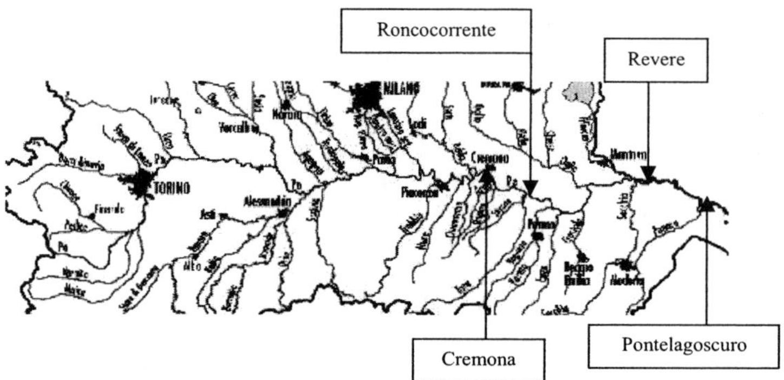

Fig. 1 Po River basin

In this way it is possible to take into account many aspects of the flood wave mechanics which play a not well defined role in the phenomenon due to their non-linear interdependence as put in evidence by eq. (1-3) also. The water levels as system variables allow to take into account the effect of the tributaries discharge contribution in the main river water levels behavior with a notable simplification in the fuzzy procedure. It is manifest that the effect of the tributaries discharge on the main course water level could be explicitly considered generating in this way a more complex and powerful model; but this is beyond the scopes of this preliminary work.

In this context, another simplification was introduced neglecting time as an explicit variable: in this way a semi-static model was developed where the time dependence is taken into account implicitly by the rules that define the system behavior.

In view of model applications, this choice is congruent with the accuracy in model results as required by real life cases and gives a meaningful simplification in the computation procedure.

In fact, the discharge propagation celerity, in a first approximation, is of order 1.5 times the mean flow velocity, which is rather constant at higher stages ranging between 3 − 4 m/s.

This applies to levels as well, whenever the loop in flood wave rating curve [2] is very narrow, like in this case, where the persistence time of water stages next to the peak is longer than the wave propagation time. For these reasons the influence of the celerity flood wave propagation on water levels is negligible.

The Matlab Fuzzy toolbox was used; via a user-friendly graphic interface (GUI) it allows the development of fuzzy models in a simple way, as shown in fig. 2.

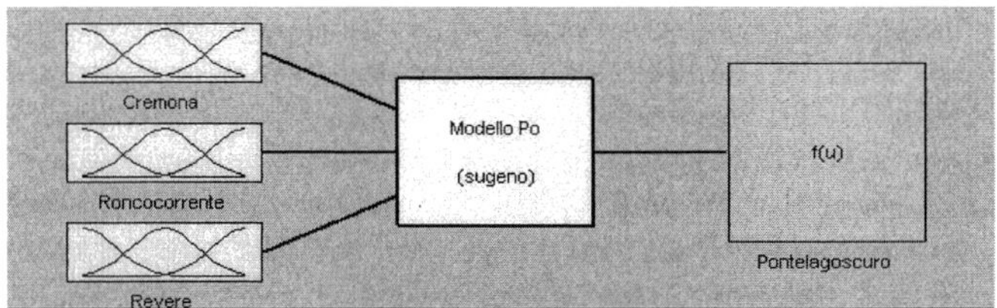

Fig. 2 Model scheme

The range of values that water levels can assume, was divided in four sub-ranges: Low-Medium, Medium, Medium-High, High. Here only the major discharge grater than 6000 m^3/s are considered; for this reason the low water levels are neglected.

Previous studies using different type of membership functions have pointed out that the gaussian ones were the better choice, in term of minimization of global error between the data and computation results.

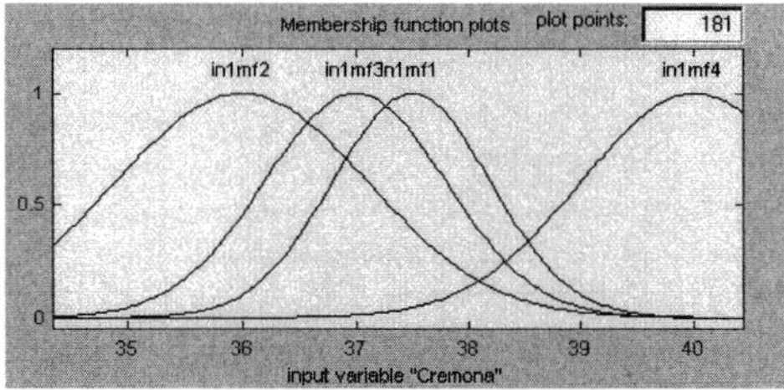

Fig. 3 Cremona monitoring station membership function of water level

The fuzzy model capability to describe the output as a function of the input needs a proper assembly of behavioral rules [1]; this set of rules allows to describe the physical system behavior during a flood event. For this purpose it is necessary to put together the river hydraulics principles with the heuristic knowledge of the global physical phenomenon. A rough work of tuning was performed with the aim to define the rules; this pointed out that twenty rules were sufficient to describe the system behavior with an error compatible with practical purposes.

3. Model results

The tuning was made on two recorded flood events that happened in October 2000 and in November 1994. The analysis of fig. 4 shows the capability of the model to fit the recorded levels with an error less than 0.3 m. This error is in the range of approximation acceptable in river hydraulics problems where a freeboard of 1-1.5 m is usually prescribed to prevent levee overtopping.

Only the data of these two flood events were available. For this reason the model sensitivity was checked generating a series of 'synthetic' flood events, whose propagation was computed using the deterministic model Hec-Ras [4] developed by USACE (1998); a good agreement between the two results is highlighted in fig. 5. In the deterministic model a time independent value of the Manning's coefficients was used according to data reported in [7] even if in real life cases these coefficients may change in time according to discharge variations [3], [8].

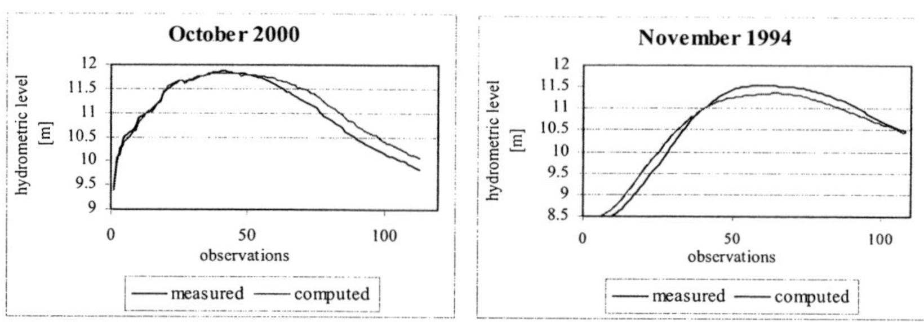

Fig. 4 Comparison between the measured hydrometric levels and the calculated ones with the fuzzy model in the Pontelagoscuro monitoring station during the flood event of 2000 and 1994

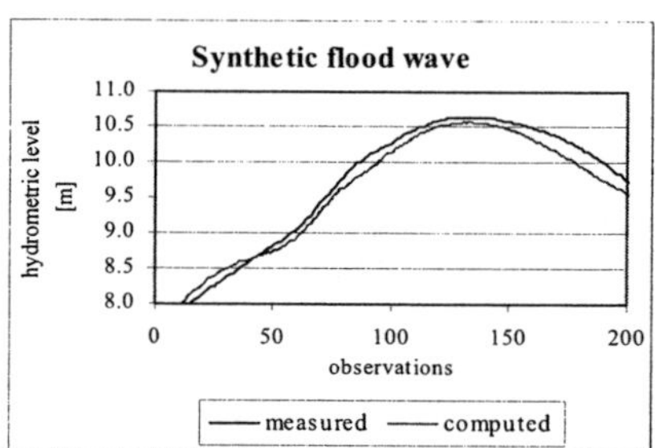

Fig. 5 Comparison between the Hec-Ras generated flood wave and the fuzzy model one

Figs. 4 and 5 underline the capability of these methodology in flood events simulation.

The good agreement of the fuzzy system results as compared with both the historical records and the synthetic waves emphasizes that water levels are a good candidate as fuzzy model variables in flood wave analysis.

Moreover the computational efficiency of the fuzzy models makes these ones suitable to the implementation in real time forecasting systems.

Acknowledgements

The writers are indebted to ing. Ernesto Reali the President of the Magistrato per il Po and ing. Gianluca Zanichelli, for the continuos support during the progress of the work.

References

[1]　Cammarata, S., Reti neuronali: dal Perceptron alle reti caotiche e neuro-fuzzy. Etas libri, Milano, 1997.

[2]　Chow V. T., Open Channel Hydraulics. McGraw-Hill, New York, 1959.

[3]　Graf W. H., Hydraulics of sediment transport. McGraw-Hill, New York, 1971.

[4]　HEC - Hydrologic Engineering Center USACE *HEC-RAS River Analysis System version,* 1998.

[5]　Kosko, B., Il fuzzy pensiero: teoria e applicazioni della logica fuzzy. Baldini&Castoldi, Milano, 1997.

[6]　Kosko, B., Neural networks and fuzzy system: a dynamical system approach to machine intelligence. Prentice-Hall, Englewood Cliffs, 1992.

[7]　SIMPO "Study and preliminary design of the hydraulic regulation of the River Po main channel for civil and environmental purposes". Magistrato per il Po - SIMPO Spa (in Italian), 1980.

[8]　van Rijn, L. C., Sediment transport, part III: bed forms and alluvial roughness. J.Hydr. Engrg., ASCE, 110 (12), 1733-1754, 1984.

KES 2002
E. Damiani et al. (Eds.)
IOS Press, 2002

Quality Assessment in Dry Cured Ham Production

A.Boscolo C.Mangiavacchi M.Ulian
APL,*DEEI,Università di Trieste,via A.Valerio 10 34127,Trieste,apldeei@units.it*

Abstract. Product quality monitoring is a need in every productive field. Probably food industry is the field where quality monitoring has been addressed for the longest time and a lot of quality control tools have been developed. S. Daniele dry cured ham is famous all over the world and its quality is guaranteed by a panel of human experts. The proposed approach tries to mimic the behaviour of a panel of experts with an artificial olfactory system. The aim is to aid the expert in the early recognition of tissue degeneration and product quality classification. This paper gives a brief description of the addressed issues, the choices adopted in the system development process and the obtained results.

1 Introduction

Dry cured ham has been produced for 2500 years in Europe. Its productive technology is quite simple and is aimed to obtain a stable meat product through tissue free water evaporation. The process is divided into three steps: salt addition, waiting at low temperature that the salt equally distributes inside the muscular tissue and then the meat is ready for seasoning. During the seasoning period complex modifications take place into the muscular and adipose tissues. Dry cured ham organoleptic and nutritional characteristics are the final results of these modification. Changes in food habit and market conditions cause a continuous review of the productive philosophy between typicality defence and process innovation. Worker experience and instinct had been the only way to control the production until the seventies. Since then a lot of money has been invested to boost technology into the ham production, obtaining an increment in the productivity and in the production control. Quality is checked by a panel of human experts through the so called "prova dell'ago" (the aroma of some specific point into the ham shape is checked with the aim of a needle shaped horse bone) with the addition of the measurements of some chemical parameters like saltiness, humidity and proteolysis index. Quality check and a constant technological progress has not yet solved one of the most important problems affecting ham production: production rejection due to superficial or deep decomposition of the shin. This is a key problem because it can diffuse from a single piece to the whole lot during the seasoning process.

2 Aroma analysis

Dry cured ham volatile components characterisation is aimed to the definition of a product quality level based on its analytical characteristics. Aroma is the most important attribute to meat acceptance. In the sixties Lillard and Ayres showed the presence of ketones, aldehydes, esters and alcohol. Gas chromatography and mass spectroscopy have shown that in the aroma of European dry cured ham there are about 260 components. This is not an amazing number if

Table 1: Families and number of substances found in Italian ham

Family	Substances
Alcohol	25
Aldehydes	16
Ketones	11
Aliphatic hydrocarbons	12
Furans	3
Carboxylic acids	5
Pyrazines	4
Nitrogenous compounds	4
Sulphur organic compounds	4

it is considered that whisky, coffee and tobacco have 260, 670 and more than 600 components respectively. Considering the results for Italian ham only, it has been found the presence of about 150 components, 122 of which has been identified. Table 1 shows these substances grouped in families and ordered by their importance in the total volatile area coverage.

Concentration, temperature, odorous threshold and solubility in water and fat are the main factors that affect the weight of each of the above mentioned substances in the aroma.

3 The lab. demonstrator

Meat product aroma is typically characterised by a lot of components, which differ each other not only for their composition but even for their molecular structure. Adding to this the lack of dedicated sensor for each substance and the need to use a limited number of devices makes the characterisation of the product by an artificial olfactory system a hard task. In this context the proposed artificial olfactory system tries to overcome the above mentioned drawbacks exploiting the information redundancy, derived from the low selectivity of commercially available sensors, which is analysed by a knowledge based system, whose aim is the mimicking of the human expert panel in expressing a qualitative judgement on the product. One of the main problem still affecting dry cured ham production is the tissue degeneration, whose detection in the early stage is the main aim of the proposed system. The sensor array selection is not only based on the application need, but is also constrained by cost and availability of the devices. For this application indium oxide and tin oxide based sensors have been chosen in order to sense either gases due to the degenerative process either to gather information useful for quality classification of the ham. The information necessary to choose most suitable sensor array has been gathered either from specific literature either through the collaboration with INEQ. The latter is an association of ham producer based in San Daniele whose aim is to train the experts for quality checking and to set the characteristics a ham must comply with to be marked as S. Daniele ham. This process has ended with the definition of the sensor array, which is constituted by eight gas sensor, a temperature sensor and a relative humidity one. Gas sensors have been chosen in order to sense amines, fuel gases, methane, organic volatile solvents, alcohol vapours, ammonia and hydrogen. The sensor array is inserted in a measurement chamber made of glass in order to reduce the quantity of gas that can be absorbed from the walls. To analyse the same quantity of gas in each measure the sample is injected through a volumetric pump and to increase measure repeatability the atmosphere inside the cham-

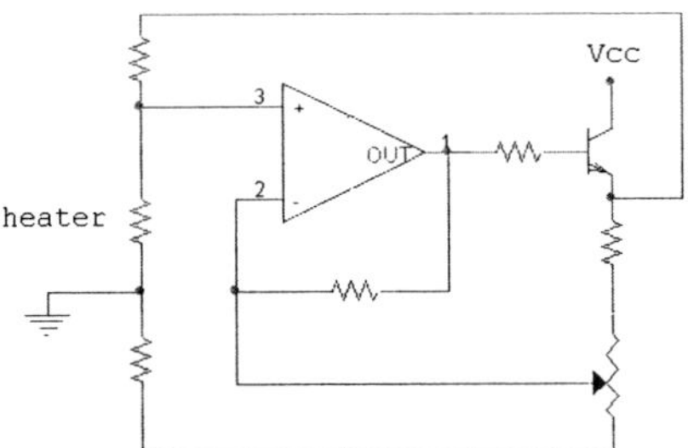

Figure 1: Full bridge with feedback topology for heater control

ber is homogenised through an air circulation system. An electronic conditioning circuit has been developed for each gas sensor, where the measurement circuit is a half–bridge topology while the heater is inserted in a full bridge with feedback configuration to achieve maximum stability, shown in figure 1 in the sensor temperature and so to stabilise the sensor behaviour. The developed circuits can be easily tuned to meet the needs of the sensor they are used with.

The interface between the conditioning circuits and the PC uses SCXI boards to obtain a modular and remotely controlled system. Timing of the acquisition system is managed by the control software developed using the virtual instrument technology in LabVIEW. The virtual instrument has one input that is the sampling frequency and has two output that are a graph of the output of each sensor and a file with the acquired data. Before showing and saving the data they are processed by a noise reduction procedure. Data analysis is made off–line also by a virtual instrument. The user is asked to mark two time slot that identifies the baseline voltage and the output one due to the presence of gas. With these value the relative change of the output for each sensor is evaluated and fed to the knowledge based system. In this first version this system is a neural network devoted to the sample classification. The need to choose the most suitable neural network topology is the reason why the analysis is made off–line and not integrated into the control software.

4   Results

Temperature and sample size are the first two problem addressed. Samples are stored in a freezer to save their organoleptic characteristics and their shape and size depends on the spot they are cut from. So before each measurement a subsample is cut always with the same dimensions and it is let to reach the room temperature. The proposed approach is validated by analysing the system behaviour for simple features and the modifying the knowledge based system to expand the available features. The first phase of test have checked sensor characteristics like reversibility, repeatability and poisoning resistance. These tests have been executed using a single gas or a mixture that have been made in our laboratory. The second test phase have checked the suitability of the sensor array in the classification of three class of quality: good, gone bad and rancid. The class identifier used are the same used by the panel of experts that have classified the sample we use.

The results obtained, table 2, show how the system can easily discriminate the three classes proposed with a high chance for improvement. The goal of the system is the early detection of tissue degenerative processes and so two classes named "dry" and "deep dry"

Table 2: Average relative change of the sensor output for different ham samples

Sensor	Good	Gone bad	Rancid
Fuel gases	6.6	11.7	32.3
Amine	22.0	77.2	71.9
Food vapours	2.6	6.8	53.5
Hydrogen sulphide	44.2	104.9	149.3
Alcohol vapours	8.4	14.5	39.4
Sulphur organic compounds	60.2	133.1	189.5
Organic solvents	22.7	56.9	112.3
Ammonia	7.2	14.3	37.4

Table 3: Results comparison of different neural network in the classification of five classes of quality

NN Typology	Error	STD
Cascade Forward	0.32	0.55
Radial Basis	1.24	1.06
Feed Forward	0.27	0.46
General Regression	0.002	0.004

have been inserted between the good one and the gone bad one. With these new types of samples the obtained results haven't been so easily interpreted. The most simple approach to address the analysis of the results have seemed to be the use of a neural network. The next test phase have tried to find out the most suitable neural network typology for the application. Only two constraint have been imposed to the networks tested in order to have significant data for the comparison:

- minimum number of nodes necessary to pass the training process;

- equally spaced output classes.

Eighty samples equally distributed on the five classes have been chosen for the training and twenty samples for the test. Table 3 shows the results of these tests as average and standard deviation of the distance of the classified sample from the centre of the class.

Probabilistic neural network have been also tested obtaining a correct classification for each sample with a probability higher than 97%. General regression and probabilistic are the most suitable neural network typology for this application with the structure constrained imposed. The final test have been the detection of early tissue degeneration. For this test a sample of ham have been let in air for an hour and the aroma has been sampled every five minutes. These samples of aroma have been classified by an expert from INEQ and by the two neural networks. The results of the probabilistic neural network have been the only that agree with the expert while the general regression typology changes its output only near the end of the hour.

5   Conclusion

This work shows that the proposed approach, sensor array and neural network, can be successful in the detection of early tissue degeneration. The system can be helpful in quality monitoring and classification, moreover it shows its objectivity as ham samples have been classified by different experts. Following the validation approach used till now the next step will try to find out the limits on field of the proposed approach.

Acknowledgement

Authors thank dott. Alain Bosero and INEQ for their valuable contribution to this work.

References

[1] L. Bolzoni, G. Barbieri and R. Virgili, Changes in Volatile Compounds of Parma Ham During Maturation, Meat Science **43** 301–310, Elsevier Science Ltd (1996).

[2] A. Boscolo and C. Mangiavacchi,Gas Knowledge Based Analyser for Quality Control (in Italian) , Convegno Nazionale del Gruppo Misure GMEE (1996), Lecce.

[3] J.W. Gardner and P.N. Bartlett, Sensors and Sensory Systems for an Electronic Nose, NATO ASI Series, Kluwer Academic Publisher (1992).

[4] A. Boscolo, Knowledge Based Instrumentation: Uncertainty, Information and Knowledge (in Italian), Convegno Internazionale dell'Automazione BIAS, Milano(1996).

[5] A. Boscolo and C. Mangiavacchi, Artificial Olfactory Evaluators in Quality Monitoring (in Italian), Congresso Metrologia & Qualità, Milano (2001).

[6] A. Boscolo, C. Mangiavacchi, M. Ulian and A. Bosero, Development Tool for Olfactory Evaluators (in Italian), Convegno Nazionale del Gruppo Misure GMEE (2000), Perugia.

KES 2002
E. Damiani et al. (Eds.)
IOS Press, 2002

Neuro-Fuzzy Clustering Techniques
For Complex Acoustic Scenarios

Rinaldo Poluzzi, Alberto Savi, *ST Microlectronics, Centro
Direzionale Colleoni, 20041 Agrate Brianza, Italy*;
Davide Vago, DSide, *Via Tiraboschi 3/a, Milano*

Abstract. A complex acoustic scenarios clustering technique based on a
neuro-fuzzy network is disclosed. An example of metrics to evaluate the
fitness function for network training is described. Some experimental
results are reported.

1. Introduction

The problem of clustering of complex acoustic scenarios arise in various fields of
speech and signal processing. Future acoustic man machine interfaces will perform
adaptation to different acoustic environments, jointly with speech recognition, speech
understanding, speech synthesis. The most important field in which classification of
acoustic scenarios classification could be useful is, however, noise reduction. In the
literature are widely known various methods of Cepstral Normalization (Dependent
Cepstral Normalization, Fixed Codeword Dependent Cepstral Normalization, Phone
Dependent Cepstral Normalization) to perform noise reduction and enhance speech
recognition robustness to environmental noise in the frequency-cepstral domain. All
these methods comprise clustering of vectors in the frequency domain, before
performing vector compensation.

The method presented herein is based on clustering of acoustic scenarios in the time
frequency domain. As shown by Bregman, the process of primitive grouping follows
certain rules: elements of a signal representation are grouped over time (sequentially)
and over frequency (simultaneously), if the features of these elements are close or
similar one to each other or if they change coherently. So, a member of streaming
phenomena can be clustered by means of the device described below.

2. Time-frequency representation of signal mixtures

The starting point is a time-frequency representation of the incoming signal mixture.
This can be achieved by means of a wavelet based QMF filter bank. The number of
subbands into which the signal can be decomposed can be stated accordingly to a
predetermined model of hearing. For each time frequency element in this representation
a number of features can be extracted which characterise the elements in the time-
frequency regions of interest. Grouping the elements with the aid of these cues results in
similar groups of tones. On the clusters so obtained a fuzzy rulebase can be applied in
order to classify signals coming from similar acoustic streams or from a certain class of
acoustic scenarios (for example sounds coming from horns, phone rings and so on).

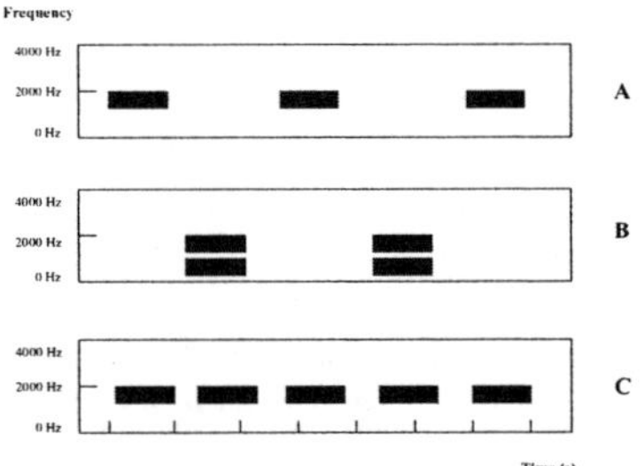

Fig. 1 Clusters in the time frequency plane

Let us consider a group of time-frequency vectors

$$o_{ij}(t_i.f_j) \tag{1.}$$

From each of these vectors it is possible to extract a feature vector

$$\mathbf{x}_{ij} \in M \tag{2}$$

belonging to a m dimensional vector space. In order to perform fuzzy clustering, for each feature the degree of membership to the associated fuzzy set must be evaluated. The membership functions

$$\mu_k : M \to [0,1] \tag{3}$$

must be evaluated. If a norm $\|.\|$ is defined in the feature space M, this norm measures the distance between the time-frequency objects. The Euclidean norm is suitable for most of the applications. If the notation

$$u_{k,ij} = \mu_k(\mathbf{x}_{i,j}) \tag{4}$$

is used, a partition can be defined on the feature space such as

$$Q = \sum_i \sum_j \sum_k u_{k,ij} \left\| \mathbf{x}_{i,j} - \mathbf{v_k} \right\|^2 \quad (\text{with } \mathbf{x}_{i,j}, \mathbf{v}_k \in M) \tag{5}$$

(in the formula $\mathbf{v_k}$ represents the centres of the clusters which can be viewed as the centres of gravity of each cluster). By means of a fuzzy rulebase for each feature the membership to one cluster or to another cluster can be determined.

To perform clustering, a neuro-fuzzy network was used. This because this kind of network has the best capability of generalisation and learning on one hand and reduced computational complexity on the other hand. The network solves a fixed-learning problem: i. e. the network has the capability of generating a model of the examined system based on pairs of input-output patterns (the output pattern to which the network evolution aims is called "target"). The learning algorithm is based on a genetic algorithm or on a random search algorithm. Anyway, any algorithm for function optimisation can be used.

3. Architecture of the neuro-fuzzy network

Fig.2 describes the general architecture of the neuro-fuzzy network which performs clustering. The filter operates on a N sample window. Eight features of the signal are extracted from this window. These features are the neuro-fuzzy network inputs. The output signal is the weighted sum of the window samples; the weights are the output of the network. The output signal is compared with the target and the distance between the target and the signal is evaluated on the basis of a well chosen fitness function. After this, there is a new modification of the network weights and a new fitness evaluation. The process is iterated until a fixed value of the fitness is achieved or a fixed number of optimisation algorithm generations have occurred

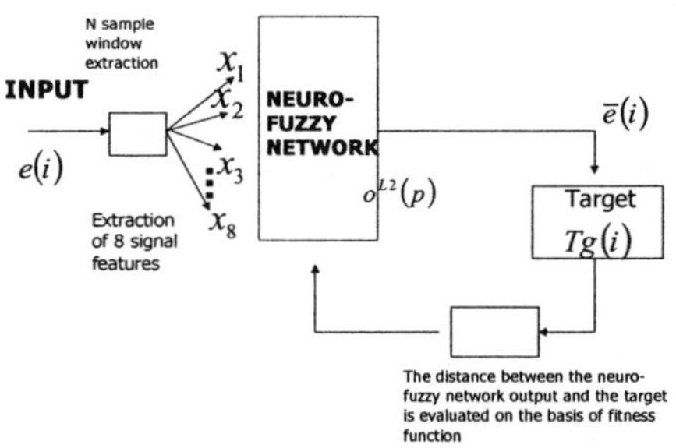

Fig.2 General architecture of the neural network based filter

The neuro fuzzy network is a three layer neuro-fuzzy perceptron. As far as the neuro-fuzzy topology is concerned, see fig. 3

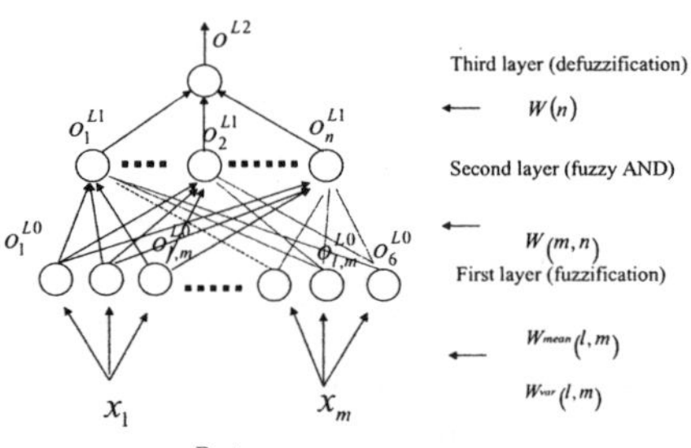

Fig. 3 A three layer neuro-fuzzy perceptron

The first layer performs fuzzification of the input variables. In this layer the degree of membership of the input features to a generic fuzzy set is evaluated. As far as the fuzzy sets are concerned, they are triangular shaped fuzzy sets. Three fuzzy sets were considered for the clustering: a fuzzy set "high", a fuzzy set "low" and a fuzzy set "n_high". The first fuzzy set was centred around the mean value of the energy, the second and the third fuzzy sets were centred around a fraction of the mean value of the energy.

The second layer performs fuzzy AND. The weights connecting the first layer to the second are randomly initialised and are not updated. The outputs of the second layer are calculated according to the formula:

$$o_n^{L1} = \min_m \left(W(m,n) \cdot o_m^{L0} \right) \tag{6}$$

The third layer is the defuzzification layer and gives a discrete output value. The defuzzification method is that of the centre of gravity. The third layer weights are updated during the network training. The output is calculated according to the formula:

$$o^{L2} = \frac{\sum_n W(n) \cdot o_n^{L1}}{\sum_n o_n^{L1}} \tag{7}$$

Large effort was devoted to find a fitness function to optimise the performance of the clusterer. As said previously, the Euclidean norm can be suitable for most of the applications. We found that the fitness function so defined:

$$F = \sum_{i=1}^{Nsamples} \bar{e}(i) \otimes Tg(i) \tag{8}$$

showed good performances. In this formula *Nsamples* represents the number of windows on which the signal was decomposed, $\bar{e}(i)$ represents each output sample from the neural network, $Tg(i)$ represents each sample of the target signal.

4. Simulation results

The method described above has been tested at first as a high control unit of an apparatus for noise reduction based on another neuro-fuzzy filter. The neuro-fuzzy filter is described in the paper [2]. The apparatus for acoustic scenarios clustering has been used as a trigger for the system for real time training of the neuro-fuzzy network performing signal filtering. When a change in the environmental noise was detected, the training of the network was repeated. The simulation result showed an enhancement of the performance of the system due to syncronisation to changes in the environmental noise.

After that the apparatus has been tested as a system to classify noises coming from real world. The noises coming from the environment in a automatic speech recognition system were classified at first into two main clusters:
- harmonic sounds (this cluster included all the sounds like car horns, phone rings, etc., characterised by an high content of harmonics in the high frequency subbands)
- non harmonic sounds (this cluster included all the other sounds).

The neuro-fuzzy apparatus was trained on a signal belonging to one of the two clusters. After training the apparatus showed good discrimination properties on signals belonging to the same cluster.

The last phase of the test experiment consisted in the time frequency plane: in this phase the network was trained for example on the noise coming from a car engine (or from a crowded environment) and then showed good generalisation properties on sounds of the same nature: other sounds coming from car engines (or from crowded environments).

Table 1

Acoustic scenario	S.N.R enhancem ent without clustering	S.N.R enhancem ent with clustering
Male voice + crowd + noise coming from a town square	1.8 dB	15.5 dB
Male voice + train + noise coming from a town square	1.8 dB	2.2 dB
Male voice + speaker +crowd	1.9 dB	4.9 dB
Male voice + crowd + noise from a town square	1.0 dB	15.0 dB

See Table 1 for some simulation results relative to a set of simulations in which the acoustic files were at first filtered with a neuro-fuzzy network and then filtered with the same neuro fuzzy network, with real time training syncronised with changes in the acoustic scenario. As it can be seen in the second set of experiences the SNR enhancement is higher with respect to the first set of simulations.

References

[1] Bregman, A. S. Pinker, Auditory streaming and the building of timbre, Can. J. Psych. 1978.
[2] R. Poluzzi, A. Savi, Neuro-fuzzy filtering techniques for complex acoustic scenarios, Proceedings of WILF 2001.
[3] D. Nauf, F. Klawonn, R. Kruse, Foundations of Neuro-Fuzzy Systems, Wiley, 1996.
[4] Lehn K.H., Modeling Binaural Auditory Scene Analysis buy a Temporal Fuzzy Clustering Analysis Approach, 1997 IEEE ASSP.

New architecture for fingerprint identification system to apply biometrics to various applications

Satoshi Shigematsu, Takahiro Hatano, Kenichi Saito**, Hiroki Suto,
Mamoru Nakanishi, Katsuyuki Machida*, Yukio Okazaki, and Hakaru Kyuragi*

NTT Lifestyle and Environmental Technology Laboratories
**NTT Telecommunication Energy Laboratories*
***NTT Electronics**

3-1, Morinosato Wakamiya, Atsugi-shi, Kanagawa-Pref., 243-0198 Japan

Abstract. This paper presents an architecture for a fingerprint identification system that provides user authentication in various applications. In the architecture, all users have their own token and use it for fingerprint identification, and the systems authenticate users using their tokens. The architecture also allows user identification customization to enhance accuracy. A prototype system demonstrates that the architecture can be applied to various systems and that it enhances identification accuracy.

1. Introduction

Nowadays, user authentication technologies are even more strongly required to ensure the security of systems, especially since the terror attack. Many user authentication technologies (biometrics) that use the characteristics of the human body and system architectures using the biometrics have been developed. The most of architectures use fingerprints [1-6]. These architectures can improve security in many small-scale applications, such as personal computer access control and door access control in small-scale facilities that have used passwords or keys for security.

However, there are some systems for which these conventional architectures are not suitable. These include a large-scale system like ATM of bank and the access control for large-scale facilities. In such systems, a huge number of users use a huge number of terminals simultaneously. If conventional architectures are applied, the system has to therefore manage and protect an enormous amount of user template data and perform a huge number of identifications simultaneously. And, since user's personal information, such as fingerprint data, is transferred through a network, the network also has to be protected to prevent the leakage of user privacy. So, these increase system scale and cost.

Other systems in which conventional architectures do not work well are room locks in hotels and coin lockers. These systems are used temporally, which means users are unspecified persons and cannot register their fingerprints in advance; they have to enroll their fingerprints each time and in each system they can use it. When conventional architectures are applied, this is inconvenient, and, in addition, users are often reluctant to register their fingerprint to an unknown system.

Applying user authentication to such systems would make them much more secure. To solve these issues and make it possible to apply fingerprint identification to any system, we have developed a new system architecture. In this architecture, all users carry a fingerprint identification token that performs all processes, from sensing to identification of a fingerprint, and users are authenticated for any service by a fingerprint identification in the token. Moreover, the token can also customize identification to each user individually to enhance accuracy. This paper presents the architecture of the fingerprint identification system, application to systems, identification customization, and experimental results.

2. The architecture of fingerprint identification token

Figure 1(a) shows the system architecture with the fingerprint identification token. The token is composed of an identifier, a memory and a fingerprint sensor. The identifier verifies the fingerprints sensed by the sensor and stored in the memory, and outputs the identification results to a terminal that provides services. Users enroll fingerprints only to their own token initially, and plug the token into the terminal when they subscribe to the service. Then, user puts a finger on the sensor, and the fingerprint is verified in the token for identification. If the identification succeeds, the result is sent to the terminal and services are provided.

Figure 1(b) shows the process for identification in the token. When the fingerprint verification succeeds, the identifier makes the result and outputs it. The token employs a one-time password or challenge response for the result to foil impostors and ensure the security of the system. If verification fails, the nonscrambled result is sent. This reduces the response time in case of failure. Thus, the token outputs only the result and user's fingerprint data are used only in his/her own token. This achieves perfect protection of user privacy.

Each user uses his/her own token individually and the token basically identifies the fingerprints of one person. Therefore, the token does not require a high-performance identifier, and that lowers costs and increases identification speed.

In addition, this system is more hygienic than the conventional one in which the sensor is shared by all users. And, even if someone's sensor breaks, which break more easily than other components, other users can still use the system. This means the system using the token has robustness.

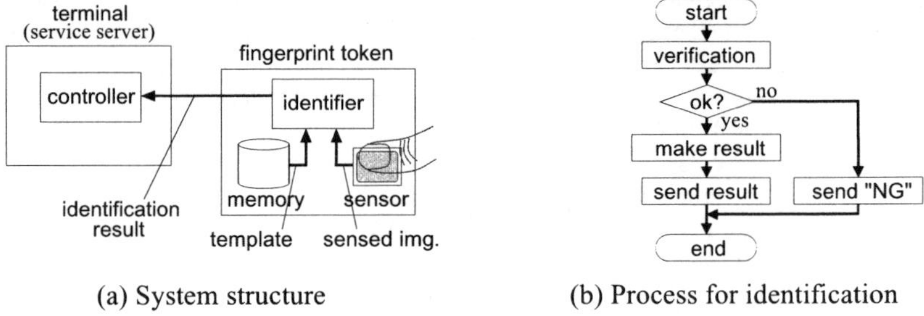

(a) System structure　　　　　(b) Process for identification

Fig. 1　Fingerprint identification token

3. The application of the token to the systems

Figure 2 shows the application of the token for a large-scale system. The system consists of a server, network, terminals and the token as shown in Fig. 2(a). In this paper, the ATM system is the example. A huge number of terminals (ATMs) are connected to the server (account manager) through the network. When user opens an account at the bank, the user also registers the token to the bank and information about his/her token is stored in the database.

To withdraw money through the ATM, the user plugs the token into the ATM and the fingerprint is identified by the token. Then, the token and the server communicate with each other as shown in Fig. 2(b). First, the token send the account number as the user ID. The server replies with an identification request and challenge data. Then, the token identifies the fingerprint. If identification is successful, the token returns the challenge response as the result. If the server confirms it, permission for service is sent to the ATM. Finally, the ATM dispenses money to the user.

In this architecture, the user data is managed only in the token. The identification is performed in the token and the transaction requires only few data which do not include user privacy, so the system is quite simple and safe. Thus, the architecture provides to apply the user authentication by a fingerprint identification to the large-scale systems.

Figure 3 shows an application of the token to a temporarily used. The system consists of equipment that provides the service and the token. In this paper, a coin locker system is the example. The user plugs the token into the locker and the token and locker communicate with each other as shown in Fig. 3(b).

First, user put his/her bag in the locker. The locker then locks the door, and the token stores the usage information, such as the equipment and locker IDs, and the password sent from the locker. Then, when user withdraws the baggage, the user plugs the token into the locker and the locker sends an identification request and an equipment ID. If the identification in the token succeeds, the token retrieves the password from the memory using the equipment ID as a key, and relays it to the locker. If the password is confirmed, the locker unlocks.

In this architecture, the user information is never sent to the equipment and the user is not required to enroll on each system, so user privacy can be protected and the system becomes more secure and convenient. And, the controller in the equipment is not required high performance, because the controller manages only the passwords. The token can be applied to not only to these systems but also to conventional small-scale and individual systems.

4. User customisation of the identification

The identification result is decided based on whether the matching ratio between template and sensed fingerprints exceeds a decision threshold or not. If the threshold is set lower, the false acceptance rate (FAR) becomes high because the impostor fingerprints can easy be accepted. On the other hand, if it's high, the false rejection rate (FRR) increases. Practically, the system has to fix the threshold for identification in advance. A conventional system has to

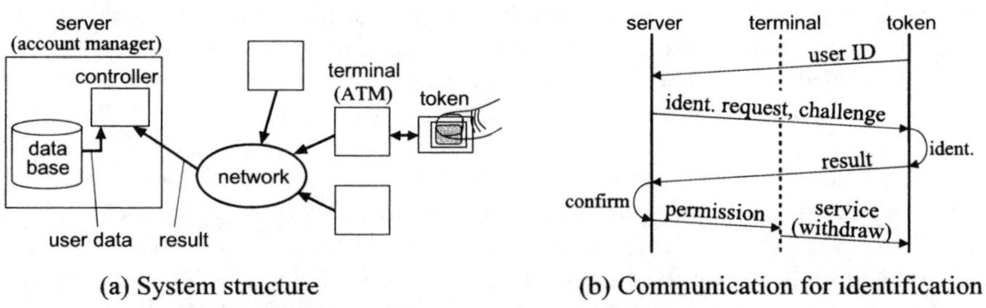

(a) System structure (b) Communication for identification

Fig. 2 Application to large-scale system

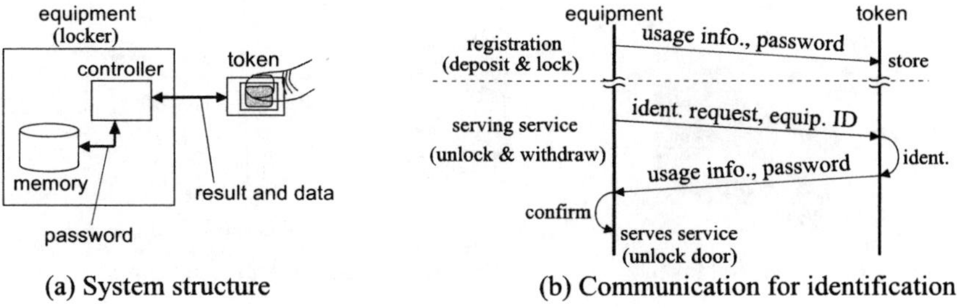

(a) System structure (b) Communication for identification

Fig. 3 Application to temporal-used system

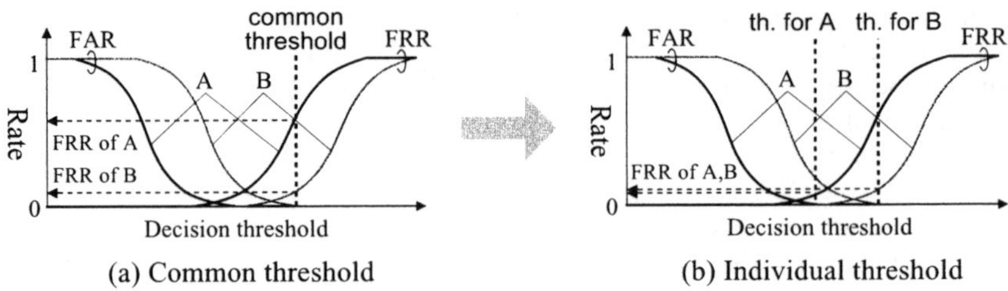

(a) Common threshold   (b) Individual threshold

Fig. 4 Principle of user customization of identification

use one common threshold for all users because they share the same identifier. To guarantee lower FAR for all users, the common threshold has to be set high as shown in Fig. 4(a). However, since the relation between FAR and FRR differs individually, the fixed common threshold is much too high for some users and the FRR for them is degraded. For instance, in Fig. 4(a) the common threshold is much too high for user A and the FRR for A is degraded.

In the proposed architecture, the user identifies a fingerprint in his/her own token. So, identification can be customized to enhance the identification accuracy. To customize, we optimize the decision threshold for each user individually as shown in Fig. 4(b). The threshold for user A is set lower than one for B. Thus, a lower FRR for all users can be achieved while maintaining a lower FAR, and the identification accuracy is enhanced totally. Moreover, in the proposed architecture, users also have their own fingerprint sensor, so fingerprint sensing can be adjusted for each user to obtain clear images.

5. Experimental results

To check the effectiveness of the proposed architecture, prototype system was implemented. Figure 5 shows the prototype. The token is connected to the PC through the adapter. We employ a pattern-matching-based algorithm [7] for fingerprint identification. The PC emulates some applications, such as the ATM and coin locker, to check whether the token can be applied to them. The experiment results confirm the proposed system architecture enables the user authentication using fingerprint identification in various systems.

We also performed an experiment on the identification customization with 30 users. Figure 6(a) shows the histogram of the minimum decision threshold for each user when FAR is set as 0.001. This indicates the thresholds depend on the person and spread widely. Figure 6(b) shows the FAR and FRR for user A and B, who have the lowest and highest threshold respectively. To decide the common threshold in a conventional system, the threshold is fixed high enough to cover user B, and it brings the degradation of FRR for user A. In the proposed architecture, the thresholds are fixed individually, and the FRR for user A can be lowered very much. Thus, using individual decision thresholds enhances the FRR for all users while maintaining a lower FAR.

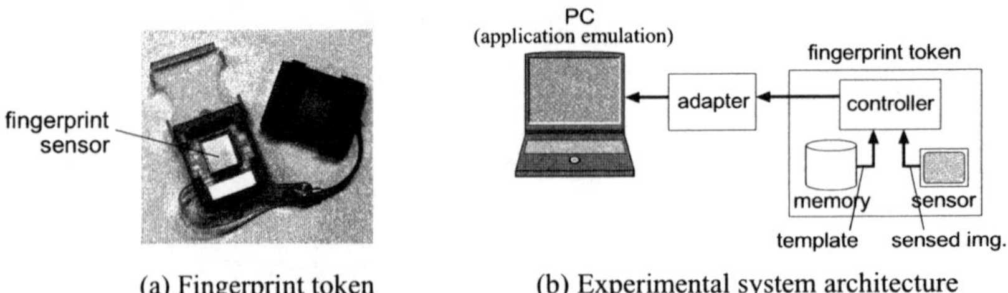

(a) Fingerprint token   (b) Experimental system architecture

Fig. 5 Implimentation of prototype system

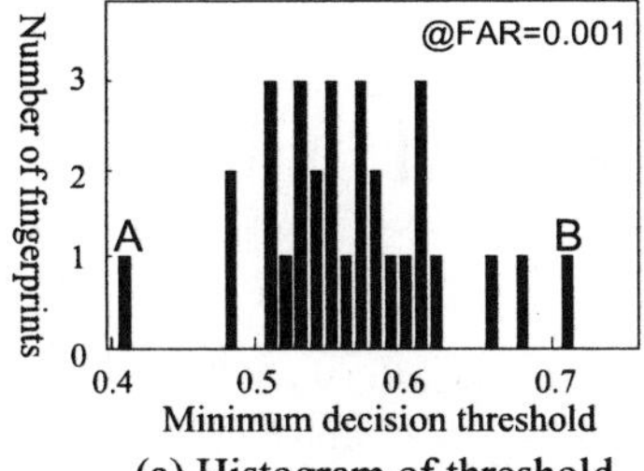

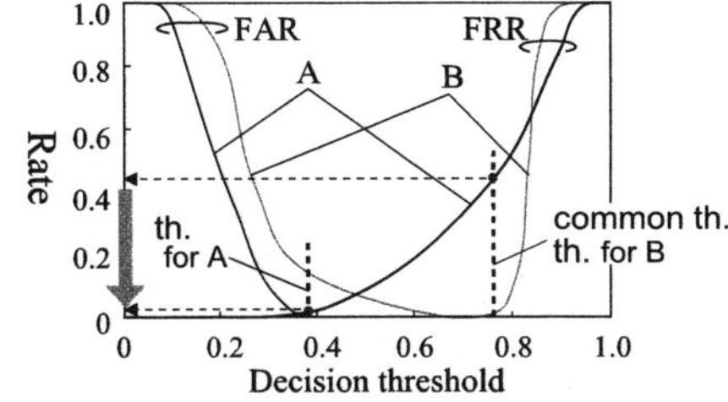

(a) Histogram of threshold   (b) Results of using individual threshold

Fig. 6 Experimental results

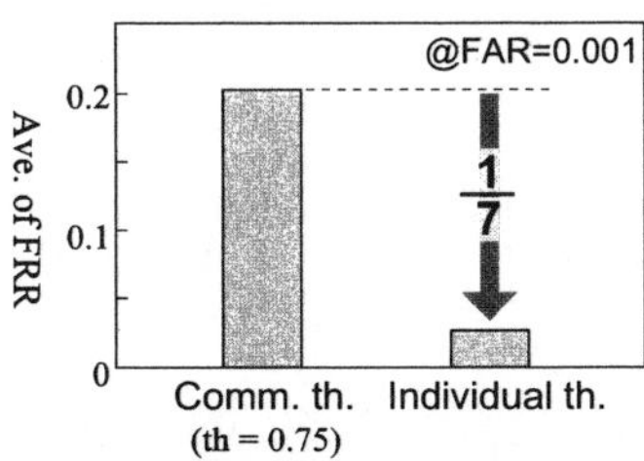

Fig. 7 Effectiveness of user customize

Figure 7 summarizes the effectiveness of the identification customization. The individual threshold reduces FRR to 1/7 of FRR when the common threshold is used. These results confirm that the identification customization would be very effective in enhancing the identification accuracy in practical systems.

6. Summary

A new architecture for a fingerprint identification system has been proposed. In the architecture, all users use their own token to identify their fingerprint and authenticate themselves. The token performs all processes, from sensing to identification of the user's fingerprint, and outputs the result to the system safety. The architecture also allows the user customization of identification.

A prototype implementation demonstrated that the proposed architecture can be applied to large-scale and temporarily used systems, and the user customization enhances the identification accuracy. The proposed architecture provides user authentication using biometrics for various systems to increase system security.

References

[1] G. Choi *et al.*,Fingerprint verification system with improved image enhancement and reliable matching," *Dig. of Tech. Papers ICCE. 2000*, 4-5.

[2] K. Ohashi, "New model of automatic fingerprint verification system," *Proc. of Int. Carnahan Conf. on Security Technology*, 1995, 36 –40.

[3] http://www.veridicom.com/

[4] http://www.startek.com.tw/

[5] http://www.biolinkusa.com/

[6] http://www.bioscrypt.com/

[7] T. Hatano *et al.*, "A fingerprint verification algorithm using the differential matching rate," *Proc. of ICPR-2002*, 2002.

KES 2002
E. Damiani et al. (Eds.)
IOS Press, 2002

A speaker verification method using CELP parameters

Yasushi YAMAZAKI†, Tadashi KONDO†† and Naohisa KOMATSU††

†*Dept. of Information and Media Sciences, The University of Kitakyushu*
yamazaki@env.kitakyu-u.ac.jp
††*Dept. of Electronics, Information and Communication Engineering, Waseda University*
{kondo,komatsu}@kom.comm.waseda.ac.jp

Abstract. We propose a text-independent speaker verification method based on a speech coding scheme. The proposed method utilizes CELP parameters which are used in speech coding schemes for mobile communication systems, and verifies a speaker only with the encoded speech information. The reliability of the proposed method is discussed with some simulation results.

1 Introduction

With the recent advances in information and communication systems, where personal information is dealt with frequently, the need to protect one's privacy is increasing. Especially, the development of an identity verification scheme to verify specific users is one of the key technologies to provide user-oriented network services in the future. In many verification systems, a user is verified by something she/he knows or possesses. It is true that these kinds of parameters are easy to handle at a low cost. However, at the same time, these parameters have the problem concerning human errors such as forgetting passwords or losing ID cards. To solve this problem, many approaches to identify oneself using biometrics have been proposed[1]. Especially, speech is one of the typical biometric parameters and is also used generally in person-to-person communications. In other words, speech contains semantic information as well as singular information.

In the case of digital communication systems, speech information is encoded for transmission. It should be noted that the encoded speech also contains semantic information and singular information. Therefore, we can realize a speech or speaker recognition system by using only the encoded speech. However, sufficient research has not been carried out on the use of encoded speech for speaker recognition.

In this paper, we propose a speaker verification method based on a speech coding scheme in digital transmission systems. In the proposed method, we utilize CELP (Code Excited Linear Prediction) parameters which are used in speech coding schemes for mobile communication systems or IP networks. The merits of the proposed method are as follows;

- Speaker verification is easily realized in the current mobile terminals or network systems by adding a little function.

- Since CELP parameters contain a speaker's characteristics of articulation, text-independent speaker verification is realized.

2 Process of speaker verification

In this section, we describe the proposed speaker verification method. The proposed method consists of CELP coding block and speaker verification block (see Fig.1). For example, a mobile terminal with the function of speaker verification is realized by adding only the speaker verification block in Fig.1 to a current mobile terminal. Moreover, the speaker verification block consists of enrollment process and verification process. In the following subsections, we will explain the above-mentioned blocks.

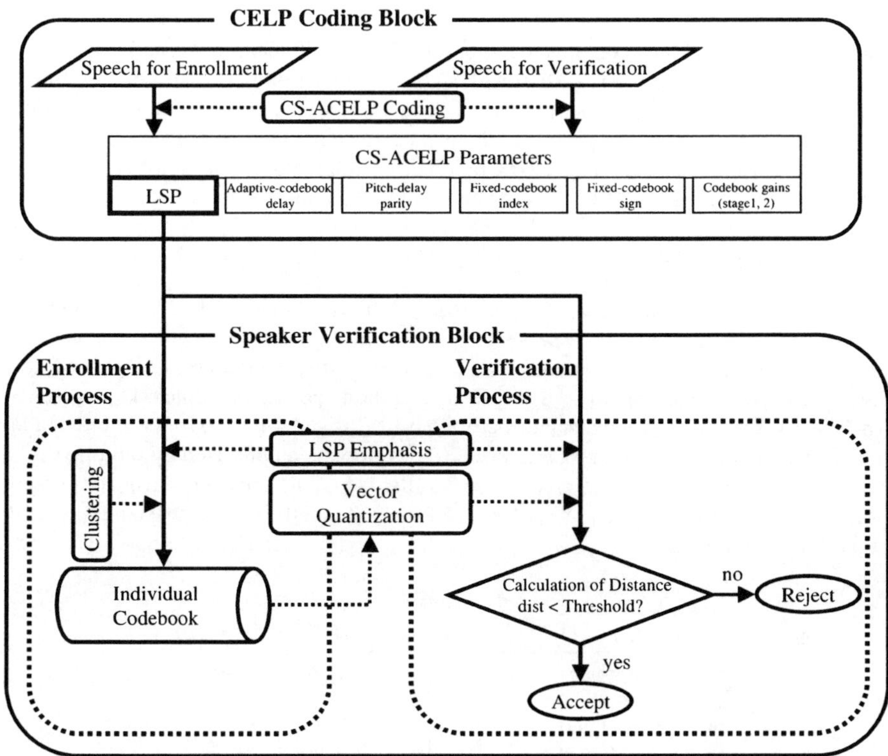

Figure 1: Outline of the proposed speaker verification

2.1 CELP coding block

In the CELP coding block, speech signal is encoded by a CELP coder and the encoded parameters are extracted. In general, CELP is an AbS (Analysis by Synthesis) coding scheme with an excited sound source which is prepared in a codebook. In the proposed method, we use a CS-ACELP (Conjugate-Structure Algebraic-Code-Excited Linear-Prediction)[2] coder, a kind of CELP coder which is standardized by ITU-T as ITU-T Recommendation G.729. This coder operates on speech frames of 10 ms corresponding to 80 samples at a sampling rate of 8000 samples per second. For every 10 ms frame, the speech signal is analyzed to extract the parameters of the CELP model. Among those parameters, we focus on a parameter called LSP (Line Spectrum Pair), which corresponds to a parameter for articulation of speech. It is well known that a speaker's individual characteristics is shown in articulation. Therefore, it is expected that LSP contains a speaker's personal features which are unique to each speaker and useful for speaker verification. With consideration given to the above suggestions, we use LSP as a principle parameter of the proposed speaker verification method.

2.2 *Speaker verification block*

The speaker verification block contains two subprocesses, enrollment process and verification process. In the enrollment process, we extract LSP (hereafter, enrollment LSP) for enrollment and define a codebook (hereafter, an individual codebook) which is produced by clustering the enrollment LSP as a speaker's personal features. On the other hand, in the verification process, we verify a speaker based on the quantization error in the process of vector quantization of LSP (hereafter, verification LSP) by referring to the individual codebook.

2.2.1 Enrollment process

First, preprocess for emphasizing the characteristics of the enrollment LSP is carried out with a view to emphasize each speaker's personal features. In the proposed method, an average value of i-th order LSP of all registered speakers is defined as a common LSP (c-LSP(i)). Similarly, an average value of i-th order LSP of a certain speaker is defined as a personal LSP (p-LSP(i)). We calculate the distance between c-LSP(i) and p-LSP(i) for each order, and call it a LSP deviation. In the preprocess, the LSP deviation is emphasized by multiplying a constant value k under the condition that the n-th order p-LSP(n) does not overlap or exceed (n+1)-th order p-LSP(n+1) or (n-1)-th order p-LSP(n-1). An outline of emphasizing the LSP deviation is shown in Fig.2.

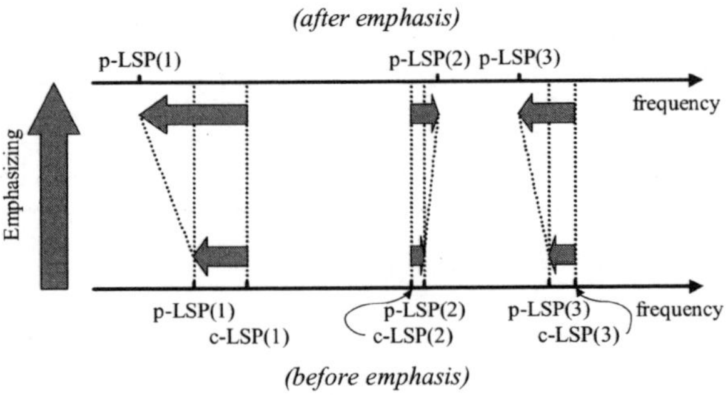

Figure 2: LSP deviation and its emphasis

Second, an individual codebook is produced by applying a clustering algorithm to the preprocessed enrollment LSP. In the proposed method, we use the LBG algorithm[3] as the clustering algorithm. The individual codebook can be stored in a user's smart card or her/his mobile terminal. In the case of using a smart card, for example, the card is inserted into her/his mobile terminal and will be activated only when she/he is verified to be an authorized user by the proposed method.

2.2.2 Verification process

First, a speaker's speech is encoded by the CS-ACELP coder and the produced LSP (verification LSP) is used for verification. Second, the Euclidean distance between the verification LSP and the vectors in the individual codebook is calculated. In the case that the calculated distance is smaller than the preset threshold value, the speaker is accepted as an authorized user.

3 Reliability test

The reliability of the proposed speaker verification method was evaluated using Japanese speech database constructed by ATR (Advanced Telecommunication Research Institute)[4]. Parameters for the reliability test are shown in Table 1. These parameters were chosen, based on the results of a preparatory experiment, so as to achieve stable extraction of personal features.

Table 1: Parameters for the reliability test

Celp Coding Block	[speech data]
	ATR Japanese speech database
	(continuous speech, phoneme-balanced sentences)
	[speaker]
	For making c-LSP : 7 males and 7 females†
	For speaker verification : 7 males and 7 females‡
	† and ‡ are different speakers.
	[speech length]
	For making c-LSP : 90 seconds
	For enrollment process : 90 seconds†
	For verification process : 6, 12, 18 seconds‡
	† and ‡ are different speech.
	[sampling frequency] 8 kHz
	[cut-off frequency] 3.1 kHz
Speaker Verification Block	[LSP emphasis] k=20
	[clustering algorithm] LBG + splitting
	[individual codebook] 16 clusters

Figure 3 shows the FRR (False Rejection Rate) and the FAR (False Acceptance Rate) for 6 sec speech data. In Fig.3, the EER (Equal Error Rate) is 7.2%. Also, the EERs for different speech lengths in the verification process are given by Table 2. These results suggest that the proposed method is effective for text-independent speaker verification. Moreover, we evaluated the reliability of the proposed method under the condition that the speech signal was contaminated by some noises. We gathered some noises at a congested railroad station and made a set of noise-mixed speech data, in which SN ratio was set to 10 dB. Under that condition, we obtained an EER of 10.3% for 6 sec speech data. This result suggests that the proposed method can also verify a speaker to some extent under a realistic condition in which speech and noise coexist.

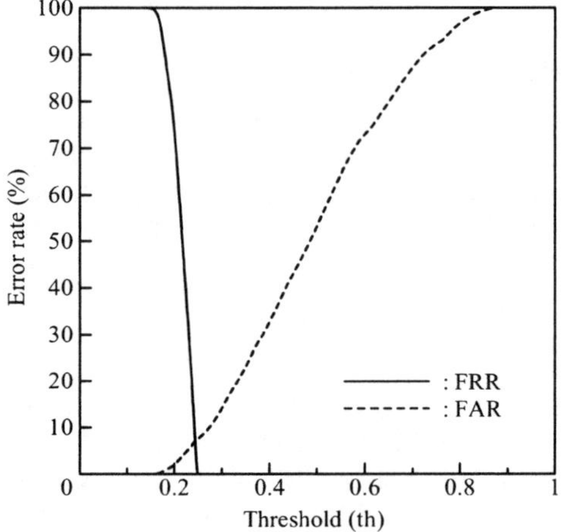

Figure 3: Verification results (speech length is 6 seconds)

Table 2: EER for different speech lengths

speech length (seconds)	6	12	18
EER (%)	7.2	2.9	2.3

4   Conclusion

In this paper, we proposed a text-independent speaker verification method using CELP parameters. We also showed the reliability of the proposed method by presenting some simulation results using speech data. From the results of the reliability test, it is obvious that the proposed method is effective for speaker verification using encoded speech information such as speaker verification in mobile communication systems. Our further research may involve the determination of appropriate threshold for the verification of a speaker, and the evaluation of verification accuracy using an increased number of speakers and an increased amount of noises.

References

[1] A.Jain et al., "BIOMETRICS —Personal Identification in Networked Society", Kluwer Academic Publishers, 1999.

[2] ITU-T, "Coding of speech at 8 kbit/s using conjugate-structure algebraic-code-excited linear-prediction (CS-ACELP)", ITU-T Recommendation G.729, 1996.

[3] Y.Linde et al., "An algorithm for vector quantizer design", IEEE Trans. Comm. COM-28, 1, pp.84-95, 1980.

[4] H.Kuwabara et al., "Construction of ATR Japanese speech database as a research tool (Appendix I)", TR-I-0086, ATR Interpreting Telephony Research Laboratories, 1989.

Independent Component Analysis for face authentication

C. Havran, L. Hupet, J. Czyz, J. Lee, L. Vandendorpe, M. Verleysen
Université catholique de Louvain, Electricity Dept.
3 place du Levant, B-1348 Louvain-la-Neuve, Belgium
{czyz, vandendorpe}@tele.ucl.ac.be; {lee, verleysen}@dice.ucl.ac.be

Abstract. In this paper, Independent Component Analysis (ICA) is presented as an alternative feature extraction algorithm to Principal Component Analysis (PCA) widely used in automatic face recognition/authentication tasks. We show that the promising ICA algorithm extracts from faces features that are relevant and efficient for authentication. This leads to improved success rates and a reduced client model size over a PCA based feature extraction.

1. Introduction

Face authentication has gained considerable attention these last years, through the increasing need for access verification systems using several modalities (voice, face image, fingerprints, pin codes, etc.). Such systems are used for the verification of a user's identity on the Net, when using a bank automaton, when entering a secured building, etc. Face authentication is different from face recognition (or classification): in authentication tasks, the system knows *a priori* the identity of the user (for example through its pin code), and has to *verify* this identity; in other words, the system has to decide whether the a priori user is an impostor or not. In face recognition, the a priori identity is not known: the system has to decide which of the images stored in a database resembles the most to the image to recognize; the decision is no more binary. Although ICA (Independent Component Analysis) could be beneficial both for face authentication and recognition, we will concentrate on the first in this paper.

In face authentication, as in most image processing problems, features are extracted from the images before processing. Working with rough images is not efficient: in face authentication, several images of a single person may be dramatically different, because of changes in viewpoint, in colour and illumination, or simply because the person's face looks different from day to day. Therefore extracting relevant features, or *discriminant* ones, is a must. Nevertheless, one hardly knows in advance which possible features will be discriminant or not. For this reason, one of the methods often used to extract features in face authentication is PCA (Principal Component Analysis) [1]. Another family of methods are the local feature-based methods such as [2], or those based on LDA (Linear Discriminant Analysis) as in [3]. In this paper, we show how the promising ICA (Independent Component Analysis) technique extracts features that are more closely related to our intuition of discriminant information, and that improve the success rate compared to an equivalent system using PCA. PCA, LDA and ICA belong to the family of subspace methods [4]. The remaining of this paper is organised as follows. Section 2 presents the problem of face authentication. Section 3 shows how to extract features from rough images, and presents the procedure based on ICA. Section 4 shows the experimental results.

2. Face authentication

Face authentication systems typically compare a feature vector X extracted from the face image to verify with a client template, consisting in similar feature vectors Y_i extracted from images of the claimed person stored in a database ($1 \leq i \leq n$, where n is the number of images of this person in the learning set). The matching may be made in different ways, one being to take the Euclidean distance between vectors (this method will be taken as an example here). If the distance between X and Y_i is lower than a threshold, the face from which X is extracted will be deemed to correspond with the face from which Y_i is extracted. Choosing the best threshold is an important part of the problem: a too small threshold will lead to a high *False Rejection Rate* (FRR), while a too high one will lead to a high *False Acceptance Rate* (FAR); FRR and FAR are defined as the proportion of feature vectors extracted from images in a validation set being wrongly classified, respectively wrongly authentified and wrongly rejected. The validation and test sets must be independent (though with faces of the same people) from the learning set, in order to get objective results. One way of setting the threshold is to choose the one leading to equal FRR and FAR. If the a priori probabilities of having false acceptances (impostors) and false rejections are equal, this corresponds to minimizing the number of wrong decisions, as a result of Bayes' law. Other criteria could be considered, such as using individual thresholds for each person in the database; again, as our goal is to measure the advantages of ICA with respect to PCA feature extraction, we will not investigate other ways of fixing thresholds, and use the global threshold leading to FRR = FAR in the remaining of this paper.

3. Feature extraction

Taking decisions on rough images has been shown [3] to be dramatically sensitive to illumination conditions, viewpoints, expression and day-to-day differences in a face of the same person, to the point that two very similar (to the human eye) images could be extremely different if compared pixel by pixel. It is therefore necessary to extract relevant, discriminant features from the images and to compare the features instead of the rough images. Of course, the more discriminant are the features, the easier will be the subsequent authentication.

3.1. Principal Component Analysis (PCA)

A traditional way of extracting features is to use PCA. From a set of N d-dimensional images in the learning set (d being the number of pixels in each image), the PCA method extracts so-called *principal components* as the eigenvectors of the covariance matrix of the data. Geometrically, the principal components are the directions in the data space maximizing the variance of the projection of the original vectors on these axes. The principal components are ranked by the associated eigenvalues of the same matrix, the largest eigenvalue corresponding to the axis maximizing the variance of the projections, and so on. As principal components have the same dimension d as the original images, and can be represented as such, they are often referred to as *eigenfaces*.

The set of eigenfaces is built on the whole learning set. What is important to notice is the fact that, once the eigenfaces are known, each face in the learning and in the validation set may be coded, or *reconstructed*, as a linear combination of eigenfaces; the sets of linear coefficients form the feature vectors. If there are at least d uncorrelated faces in the learning set, $N \geq d$, then PCA will extract d eigenfaces and the reconstruction is error-free. However, if only P eigenfaces are kept ($P < d$), the coding minimizes the mean square error between the original faces and the coded ones.

While projecting the original faces on the principal components is mathematically justified by the fact that it maximizes the variance after projection, there is no rationale under the fact that the directions in which the variance is maximum are the most discriminant directions for taking the right decision. There is even some intuitive evidence that it is not the case: the first eigenvectors will correspond to the general shapes of faces, which are common to all of them (therefore not discriminant). The idea is then to replace PCA by a method able to extract more perceptive features from faces. By perceptive, we mean features that could be used by a human being to discriminate between or to describe faces. However, as it will be justified below, ICA will not replace PCA, but will be used as a supplementary step after PCA. This will allow keeping a larger number of eigenvectors after PCA (keeping thus more information in the feature vectors), this number being further reduced by ICA. PCA will act as a first whitening filter, while the discriminant reduction will be achieved by ICA.

3.2. Independent Component Analysis (ICA)

ICA is a data analysis tool derived from the "source separation" signal processing techniques. The aim of source separation is to recover original signals S_i, from known observations X_j, where each observation is an (unknown) mixture of the original signals. Under the assumption that the original signals S_i are statistically independent, and under mild conditions on the mixture, it is possible to recover the original signals from the observations. The algorithmic techniques making this task possible are often called ICA, as they factorise the observations as a combination of original sources. If the mixing is linear, ICA estimates the inverse of the mixing matrix. The number of observations N $(1 \leq j \leq N)$ must be at least equal to the number of original signals M $(1 \leq i \leq M)$; often it is assumed that $N = M$. It is not necessary to have signals X_j to consider using ICA: X_j may also be multi-dimensional data (vectors). Assuming that each X_j is an unknown, different combination of original "source vectors" S_i, ICA will expand each signal X_j into a weighted sum of source vectors S_i (ICA estimates both the source vectors S_i and the coefficients of the weighted sum). This view is not far from the PCA expansion: the eigenvectors of PCA are replaced by the independent source vectors in ICA. For a review of ICA techniques and properties, see for example [5].

In our case, we assume that the faces in the learning set, viewed as high-dimensional vectors, are linear combination of unknown independent source vectors. This may not be strictly true, depending on the respective number of images in the database and size of an image in pixels, but in any case ICA will find estimates of independent source vectors that are optimal to reconstruct the original images (observations) in the least-square sense. The idea is then to substitute PCA with ICA, and to use the coefficients of the ICA expansion (instead of those from PCA) as feature vectors for the faces. It is expected that, ICA source vectors being independent (instead of PCA eigenvectors being uncorrelated only), they will be closer to natural features of images, and thus more able to represent differences between faces.

3.3. Use of ICA in face authentication

However ICA does not have advantages only. ICA algorithms are iterative, and sometimes converge difficultly. Moreover, ICA methods show difficulties to handle large number of signals (or high-dimensional vectors in our case). The FastICA package [6] has been used for its good performances in our simulations. To overcome the difficulties related to the high dimensionality of vectors, their dimensionality has first been reduced by PCA. This might sound odd given the above arguments; however, the experiments in section 4 make clear that the dimension P of the vectors after PCA reduction will be chosen much larger than in "PCA only" experiments. Furthermore, PCA whitening helps ICA to converge.

A further difficulty of ICA compared to PCA is the ordering of source vectors. In PCA, the corresponding eigenvalues are used to rank the eigenvectors (according to their contribution to the total variance of data), the first ranked (corresponding to the largest eigenvalues) being kept in case of dimensionality reduction. ICA does not offer an ordering of the source vectors. For this reason, as suggested by [7], we rank them according to a class separability criterion estimated over the learning set $r_i = \sigma_{bi}/\sigma_{wi}$ where σ_{bi} and σ_{wi} are the within and between class variance of the source component S_i (the classes correspond to the different identities in the database). A high value of r_i corresponds to a discriminant source vector S_i, thus only the Q source vectors associated with the highest values of r_i are conserved. Note that ordering PCA eigenvectors according to a class separability criterion (instead of eigenvalues) would lead to a method similar to LDA (see introduction).

The following procedure is thus suggested to expand the faces into feature vectors and to authentify a face accordingly: (i) the d-dimensional images in the learning set are reduced by PCA to P-dimensional vectors; the last are formed by the coefficients of the PCA expansion corresponding to the largest eigenvalues; (ii) the P-dimensional vectors are further expanded into Q-dimensional feature ones by ICA; the choice of ICA source vectors is made according to criterion (1); (iii) projection matrices from steps 1 and 2 (sizes $P{\times}d$ and $Q{\times}P$ respectively) are used to transform the images in the validation set into Q-dimensional feature vectors; (iv) these feature vectors are authentified according to the procedure described in section 2. Three parameters must be determined in the method: P, Q, and the threshold used for the authentication procedure. For each value of P and Q, the threshold is fixed to have FAR=FRR; P and Q are chosen to minimize this error rate. Finally, (v) the performances of the method (including the threshold value) are measured on an independent test set (on this set, FAR will not be necessarily equal to FRR).

4. Experimental results

Our experiments were performed on frontal face images from the XM2VTS database [8]. XM2VTS is a publicly available multimodal database recorded specifically for assessing the performances of biometric approaches to identity verification. It contains 8 face images of 295 persons. The subjects were recorded in four separate sessions distributed over a period of 5 months. The standard experimental protocol associated with the database divides the database into 200 clients and 95 impostors. The protocol specifies a partitioning of the database into disjoint sets for training, validation and testing.

Figure 1 shows the FAR=FRR obtained on the validation set after optimization of the threshold and Q, with respect to P. The two curves show respectively (plain line) the result of the algorithm described in section 3, and (dashed line) the result of a PCA only applied to the original data. It clearly shows the improvement obtained by the use of ICA for a wide range of the dimension P. It also shows that, contrarily to the use of PCA only, applying the procedure from section 3 gives comparable results in a wide range of the parameter P, making its choice less critical. Figure 2 shows a detail from Figure 1, with two supplementary curves corresponding to error bars around the ICA curve (the experiments have been conducted a large number of times to assess the reliability of the ICA step). Even the top curve shows improved results compared to PCA-only feature extraction. Both Figures 1 and 2 have been obtained by using the Euclidean distance between feature vectors for their matching (see section 2). It has been found experimentally that using the angle between these vectors instead of the Euclidean distance further improves the results. Table 1 summarizes some results obtained, both with the use of the Euclidean distance and of the angle. The values shown for dimensions P and Q are those found after optimization.

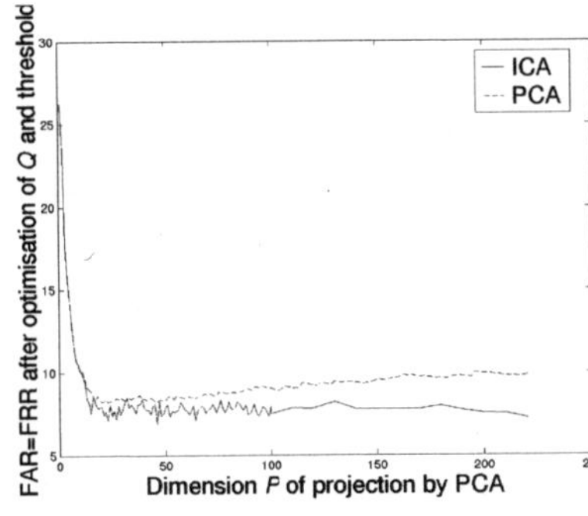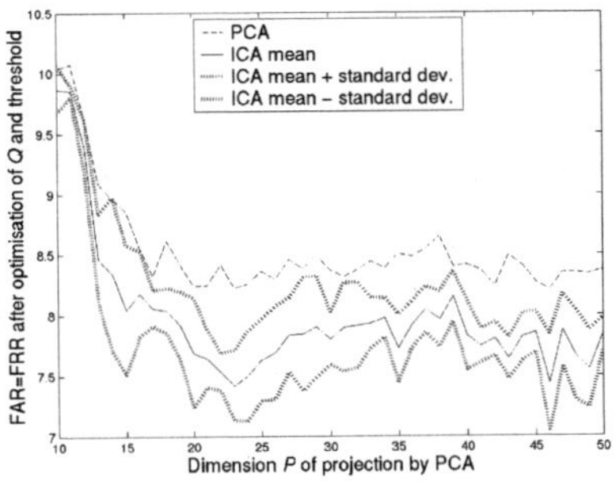

Figure 1: performance of ICA and PCA (see text for details). Figure 2: detail of Fig.1, + standard deviations.

Table 1: results of the method, and of PCA-only feature extraction for comparison.

Type of distance	Dimension P after PCA	Dimension Q after ICA	FAR=FRR on learning set	(FAR+FRR)/2 on validation set
Euclidean	46	no ICA	8.2	8.06
Euclidean	23	18	7	7.34
Angle	100	no ICA	6.44	5.44
Angle	81	71	5.62	5.21

5. Conclusion

This papers describes a procedure for using ICA as feature extractor in the context of face authentication. Results on experiments performed on a standard database show increased performances with respect to the use of PCA only as feature extractor. Moreover, the results also show a lower sensitivity to the choice of the projection dimension after PCA. Further work may consist in replacing the simple decision system authentifying the faces through simple distance comparisons between feature vectors, by a multi-dimensional classifier (artificial neural network) on the components of these vectors.

6. Acknowledgements

The authors would like to thank Christian Jutten (I.N.P. Grenoble) for fruitful discussions. Michel Verleysen is a senior research associate of the Belgian FNRS. Part of this work was realized with the support of the 'Ministère de la Région wallonne', under the 'Programme de Formation et d'Impulsion à la Recherche Scientifique et Technologique' and by the European Union IST projects BANCA and OPTIVIP.

References

[1] M. Turk and A. Pentland, Eigenfaces for recognition. Journal of cognitive neuroscience, 3(1), 1991.

[2] L.Wiskott, J.-M. Fellous, N.Kruger and C.von der Malsburg "Face recognition by elastic bunch graph matching" Trans. on Pattern Analysis and Machine Intelligence, Vol.19, pp.775-779 , 1997

[3] P. Belhumeur, J.P. Hespanha, D.J. Kriegman , Eigenfaces vs. Fisherfaces: recognition using class specific linear projection. IEEE Trans. on Pattern Analysis and Machine Intelligence, 19(7), 1997, pp. 711-720.

[4] E. Oja, Subspace methods of Pattern Recognition, Research Studies Press Ltd., 1983.

[5] A. Hyvärinen, J. Karhunen, E. Oja, Independent Component Analysis, Wiley, New-York, 2001.

[6] The FastICA package, http://www.cis.hut.fi/projects/ica/fastica/.

[7] M. Bartlett, H. Lades, T. Sejnowski, Independent component representations for face recognition". In Proceedings of the SPIE Symposium of Eletronic Imaging, 3299, 1998, pp. 528-539.

[8] K.Messer, J.Matas, J. Kittler, J. Luettin, G. Maître, XM2VTSBD: The Extended M2VTS database. Int'l Conf. on Audio- & Video-based Biometric Authentication (AVBPA 99), Washington D.C.,1999.

KES 2002
E. Damiani et al. (Eds.)
IOS Press, 2002

Fusion of Eigenfaces and Elastic Graph Matching for Face Recognition

Giorgos SAZAKLIS and Stelios C. A. THOMOPOULOS

Institute of Informatics and Telecommunications
NCSR Demokritos
P.O. Box 60228
GR-153 10, Athens, Greece

{sazaklis, scat}@iit.demokritos.gr

Abstract

Eigenfaces and Elastic Graph Matching are two popular and well-studied methods for face recognition. However, each method has certain drawbacks that make it particularly suitable for recognition of a restricted type of facial images. In this paper, we propose to fuse the two methods together in order to extend the capabilities of a Face Recognition System beyond what each method can handle alone. We are particularly interested in recognizing facial images of uncooperative subjects taken under variable illumination or pose.

We present recognition results on the standard ORL database that surpass the performance of either method alone and demonstrate the power of fusion. We also propose the use of an infrared camera that will produce a low-frequency image of the subjects' eyes to assist in recognition of uncooperative subjects.

1. Introduction

Face Recognition is one of the most active biometrics research areas during the last few years, since it constitutes a non-intrusive, natural and low-cost modality that can be used to enhance security and control access to facilities, buildings or computers. It is one of the few modalities that people use in everyday life and is suitable for identifying even uncooperative subjects. Two of the most popular and successful approaches for face recognition are the eigenfaces ([12], [10]) and Elastic Graph Matching [13].

Despite great progress in facial recognition technology that has been witnessed lately [7], there is a gap between reported performance levels and the user experience when systems are deployed in real-world scenarios. Some of the prominent problems are sensitivity to illumination variations, pose changes and uncooperative subjects. For example, simple gadgets that one can add like facial hair, or dark glasses, are enough to throw off a face recognition system. Dark glasses can hide the eyes depriving the system of one of the most information-rich areas of the face. In this paper we propose the use of an infrared camera in addition to a visible light camera, in order to provide the system with an additional low-frequency image of the eyes region to assist in recognition.

In the following sections, we present a short description of previous work on the two methods of Eigenfaces and Graph Matching, together with a discussion of their weak and strong points. Next, we describe how our fusion approach combines the two methods together alleviating the weak points of each and enhancing system performance. Finally, we describe our experiments and present our results on the ORL database, together with an experiment where the use of an infrared camera is simulated: The eyes region from face images is convolved with a low-pass filter, creating composite images where only the low-frequency components are kept in the eyes area, and then recognition follows on the composite image.

2. Related work

In this section, we briefly summarize the main ideas behind each of the two methods we have used, namely Eigenfaces [12] and Elastic Graph Matching [13] and discuss their main features and properties. We place some emphasis on their complimentary nature that highlights our motivation for using fusion.

2.1. Eigenfaces

The so-called "eigenfaces" method ([12], [5]) is one of the most popular methods for face recognition. It is based on the Principal Components Analysis (PCA) of the face images in a training set. The main idea is that since all human faces share certain common characteristics, features in a set of face images will be highly correlated. The K-L (Karhunen-Loeve) transform can be used to project the images to a space of reduced dimensionality where features are uncorrelated. Nearest neighbor classifiers can be used for classification. Euclidean distances d in the projection space are mapped into the $[0,1]$ interval using the mapping function: $f = d / (1+d)$. It is easily seen that f is also a metric with distance values in $[0,1]$ and can be used in the fusion process.

2.2. Elastic Graph Matching

Elastic Graph Matching is another popular method [13] for robust face recognition. It models the face as a graph whose nodes correspond to fiducial points labeled with vectors (so-called 'jets') that capture information about the local face structure. Graph edges are vectors connecting nearby nodes that can be used to enforce relative placement constraints of facial features. Node labels (jets) are the coefficients of the image convolution with a family of Gabor filters that are complex local waves of specific frequency and orientation [1]. Each jet element is the convolution result on the node location describing the image structure locally with respect to the frequency and orientation of a specific Gabor filter. We have used a family of 40 Gabor kernels that are generated by combining a set of 5 different wave frequency values, namely $\{\pi/8, \pi/(4\sqrt{2}), \pi/4, \pi/(2\sqrt{2}), \pi/2\}$ with a set of 8 orientations, namely $\{\mu \cdot \pi/8, \mu=0...7\}$. We considered the norm of the complex result, since the convolution phase is known to be very sensitive to the exact node placement [13].

Once graph node descriptions have been produced, a graph similarity function can provide a similarity score against all graphs that describe model images. We have used a weighted sum of relative displacements of edge vectors and jet similarity values between corresponding nodes of query and model images as our graph similarity function. At the current version of our system, all edge weights are 1, except weights of edges adjacent to the eye pupils or the lips, which are 0, since eye pupils and lips can move around without incurring any penalty. The best results were obtained when the relative importance of jet and edge similarity terms is controlled by a parameter λ, such that $10^{-6} < \lambda < 10^{-2}$. The similarity function of two nodes is nothing more than the cosine of their angle when they are considered high-dimensional vectors. It lies in the $[0,1]$ interval, since all vector elements are positive.

2.3. Facial Feature Detection

An essential step for Graph Matching is the successful detection of features on the face prior to node label generation. Here lies an important difference of our system from others ([13]). In contrast with [13], where a regular grid was used for a first coarse node placement and subsequently node positions were fine-tuned to positions of local maxima of the similarity function, we used a separate feature detector that can search and detect fiducial image points in the face, when trained with a suitable set of feature examples (selected manually) from a training set of images.

We have assumed that intensities of corresponding pixels follow a multivariate Gaussian probability density function and have used the Maximum Likelihood detector of [5] that places a feature at the global maximum of the multivariate Gaussian pdf. This can be efficiently computed via the projection of candidate feature vectors into two orthogonal subspaces: a principal components subspace that gives rise to a DIFS distance term ("Distance In Feature Space") and an orthogonal subspace that gives rise to a DFFS distance measure ("Distance From Feature Space"). For more details, the reader is referred to [5].

3. Fusion of Eigenfaces and Graph Matching

The rationale for using fusion to improve the performance of Eigenfaces and Graph Matching is based on the observation that the two techniques are complimentary in many aspects. Fusion will improve the overall performance [11], because each time it will select the outcome of the method that is best suited to process the test image. More specifically:

Graph Matching is better suited for face comparisons across views, while eigenfaces can be used for face comparison within a specific view. This is because graph matching detects fiducial points on the original image and describes them independently of their position within the face (i.e. view). Thus no image alignment is necessary and graph matching can be used successfully to compare images across different views. Graph nodes are attributed with feature descriptions irrespective of the particular view. Of course, complete independence from the view setting is not possible, since the projective transformation together with possible illumination changes can alter the feature descriptions.

On the other hand, when eigenfaces compares images of a specific view, facial features are already aligned and image pixels can be attributed physical meaning, so differences between image projections will be mainly due to different identities. When comparisons occur across different views, a registration and normalization phase is necessary to facilitate feature alignment. One solution is to find two or three landmark points on the face (e.g. two eyes and the nose tip) and apply an affine transformation that will align corresponding pixels. However, good alignment is quite sensitive to accurate feature detection, since even error of a single pixel will magnify when the image is scaled. Additional evidence for this claim is that techniques that extend eigenfaces to multiple views fare best when a separate eigenspace is used for each view [6].

Eigenfaces uses the whole face area and is better suited for utilizing face-specific landmarks. While eigenfaces uses the whole face area and utilizes specific landmarks the person may have (scars, wrinkles or moles), graph matching is not using the whole face area, as graph nodes are placed in pre-defined facial landmarks. Since the designer decides the graph node set in advance before the test set faces are seen, the system will not utilize important features that can be face-specific (e.g. a large mole, or a deep wrinkle). The eigenfaces is a preferred method for such cases.

We see that the two methods are complimentary, since each has specific advantages that make it a better choice for certain kinds of facial images. So, using fusion to increase the success rate of each one alone is justified and can increase recognition performance. For each face in the test set, both eigenfaces and graph matching come up with suggestions accompanied by confidence scores as described below. Confidence scores are derived from the distance values, which are metrics in the [0,1] interval, as we saw earlier.

As a first approach to the fusion problem, we opt to choose the *suggestion with the maximum confidence score*. A more elaborate procedure involving the use of the Likelihood Ratio Test that takes into account the image statistics is going to be used in future versions of our system [11].

If d_0 is the distance of the test image from the model of choice M, and d_1 is the smallest distance from the rest of the models (when all faces belonging to model M are removed, d_1 will be the minimum distance of the test image from faces of different identity), then the *significance confidence score* is the Mahalanobis distance between model M and the next best choice. This will be the separation of d_0 and d_1, divided by the standard deviation s of the distances from the rest of the models [2]:

$$\text{Conf}_{\text{sig}} = (d_1 - d_0) \, / \, s$$

4. Experiments

We have conducted experiments on the publicly available face database of Olivetti Research Labs (ORL), which is comprised of 400 face images that belong to 40 individuals. Images are grayscale with 256 gray levels and size 112 by 92 pixels. Each face is closely cropped and scaled, so no face detection or size normalization phase is necessary. On the other hand, illumination, pose and facial expression variability is observed across images of the database (see Fig. 1), so we can say it is a challenging (but small) database for face recognition applications.

Fig.1 Some faces from the ORL database.

We tested the performance of the proposed fusion technique on this database. We randomly chose a set of 5 out of the 10 images per person, to serve as the training set, while the rest are put in the test set. This creates a 50-50 split of the database with no overlap between training and testing sets.

For the eigenfaces method registration or normalization is not performed, because of the high sensitivity of the final registration result on accurate feature detection. Eigenfaces are formed using the first 100 eigenvectors of the covariance matrix of the pixels, while a nearest neighbor classifier using the modified Euclidean distance metric we saw earlier is used to select the best match.

Graph Matching begins with a feature detection phase (we use the Max. Likelihood feature detector from [5]). Subsequently each detected feature forms a node in the face graph, attributed with a vector of convolution coefficients as described earlier. Distance vectors between nodes are computed and graph similarity scores are extracted. The best match is the one with the maximum similarity score.

At the final step, fusion decides the system outcome, by selecting the match with the best confidence. Table 1 reports the performance of the system. We also compare our results with the performance of other methods on the same database as it is reported in the literature by other researchers. All cited results have been produced using the method of cross-validation, after a random 50-50 split of all the faces to training and test sets.

We have also performed an experiment that simulates the use of two cameras, one operating in visible light and the other in infrared. The target here is to deal with uncooperative subjects who wear dark glasses to throw off recognition. Imagery in infrared can be superimposed on a visible light image producing a low-frequency image of the eyes area. For simulation purposes, we have convolved the image with a 5x5 Gaussian kernel, keeping only the low-frequency components. Subsequently, we have replaced the eyes area of the original image with the blurred piece and have performed recognition on the composite image using the system described above. Results are shown in Table 2.

Table 1. Recognition Results and Comparison with state-of-the-art

Method	Recognition Rate	Reference
Fusion	97.7 %	Our work
Eigenfaces (100 eigenvectors)	96.1 %	Our work
Graph Matching	90.5 %	Our work
Eigenfaces (40 eigenvectors)	89.5 %	[9], [4]
Elastic Matching	80.0 %	[14]
Self-Organizing Map + Convolutional NN	96.2 %	[4]
TopDown HMM with gray values	87.0 %	[8]
Pseudo 2D HMM with gray values	94.5 %	[9]
Point Matching + Correlation	84.0 %	[3]

Table 2. Results for the simulated use of an IR camera together with a visible light camera

Method	Recognition Rate
Eigenfaces	93.09 %
Graph Matching	85.82 %
Fusion	94.74 %

5. Conclusions

We have described a face recognition system that uses fusion to mold the two complimentary approaches of eigenfaces and elastic graph matching. We observed a performance increase in comparison to each method alone and validated our work through experiments on the ORL database. In the future, we plan to apply certain assumptions about signal statistics to build an elaborate fusion scheme according to [11] in order to optimize fusion performance. We also plan to expand our experiments to larger data sets such as the FERET or the Purdue databases.

References

[1] DAUGMAN J.G., "Complete discrete 2-D Gabor transform by neural networks for image analysis and compression", *IEEE Trans. on Acoustics, Speech and Signal Processing,* 36(7):1169-1179.

[2] LADES M., VORBRÜGGEN J., BUHMANN J., LANGE J., VON DER MALSBURG, WÜRTZ R. AND KONEN W. "Distortion Invariant Object Recognition in the Dynamic Link Architecture", *IEEE Trans. On Computers,* vol. 42, no. 3, pp. 300-310, March 1993.

[3] LAM K.-M., AND YAN H., "An Analytic-to-Holistic Approach for Face Recognition Based on a Single Frontal View", *IEEE Trans. on Pattern Analysis and Machine Intelligence,* 20(7):673-686, 1998.

[4] LAWRENCE S., GILES C.L., TSOI A. C., AND BACK A. D., "Face Recognition: A Convolutional Neural Network Approach", *IEEE Transactions on Neural Networks,* 8(1):114-132, 1997.

[5] MOGHADDAM B. AND PENTLAND A., "Probabilistic Visual Learning for Object Representation", *IEEE Trans. On Pattern Analysis and Machine Intelligence,* vol. 19, no. 7, pp. 696-710, July 1997.

[6] PENTLAND A., MOGHADDAM B. AND STARNER T., "View-Based and Modular Eigenspaces for Face Recognition," *Proc. IEEE Conf. Computer Vision and Pattern Recognition,* Seattle, June 1994.

[7] PHILLIPS P.J., MOON H., RIZVI S. AND RAUSS P., "The FERET Evaluation Methodology for Face Recognition Algorithms", in *IEEE Trans. On Pattern Analysis and Machine Intelligence,* vol. 22, no. 10, pp. 1090-1104, October 2000.

[8] SAMARIA F. AND HARTER A., "Parameterisation of a Stochastic Model for Human Face Identification", *Proceedings of 2nd IEEE Workshop on Applications of Computer Vision,* Sarasota FL, December 1994.

[9] SAMARIA F., "Face Recognition Using Hidden Markov Models", *Ph.D. Thesis,* Engineering Dept. Cambridge Univ., Oct. 1994.

[10] SIROVICH L. AND KIRBY M., "Low-dimensional procedure for the characterization of human face", *J. Opt. Soc. Amer.,* vol. 4, no. 3, pp. 519-524, 1987.

[11] THOMOPOULOS S.C.A., VISWANATHAN R., BOUGOULIAS D.K., "Optimal Distributed Decision Fusion", *IEEE Trans. On Aerospace and Electronic Systems,* 25(5):761-764, Sept. 1989.

[12] TURK M. AND PENTLAND A., "Eigenfaces for recognition", *J. Cognitive Neuroscience* 3(1):71–86, 1991.

[13] WISKOTT L., FELLOUS J., KRUGER N. AND VON DER MALSBURG C., "Face Recognition by Elastic Bunch Graph Matching", in *Intelligent Biometric Techniques in Fingerprint and Face Recognition,* edited by L.C. Jain, et al. Springer-Verlag 1999.

[14] ZHANG J, YAN Y AND LADES M., "Face Recognition: Eigenface, Elastic Matching and Neural Nets", *Proceedings of the IEEE,* 85(9):1423-1435, Sept. 1997.

KES 2002
E. Damiani et al. (Eds.)
IOS Press, 2002

Scenario based data collection trials for the evaluation of multi-modal biometric processing: a preliminary report

M.C.Fairhurst, J.George, F.Deravi
Department of Electronics, University of Kent, Canterbury, Kent CT2 7NT

Abstract This paper describes a preliminary analysis of the results of
trials to collect multimodal biometric data from a large group of subjects.
In particular we show how the data obtained points to the practical value of
developing systems based on the exploitation of a combination of
biometric modalities to authenticate claimed identity

Background

There is an increasing recognition that, in order to achieve the robustness and reliability
required for many applications of biometric identity checking, no single biometric
modality is likely to meet all the criteria associated with a specified task domain. Thus,
increasingly, multi-modal biometric checking is likely to be adopted as a practical
solution which can offer two principal benefits. First, combining evidence of identity
from more than one biometric source can generally improve the performance accuracy
attainable. Second, having available multiple modalities can offer a flexible solution to
the problem of the non-availability or unreliability of any particular modality in a given
situation or to issues of non-acceptability of a particular modality for an individual user
or user group, and it is this second aspect which is of principal interest here. Achieving
these desirable objectives is not without cost, however, since increasing the number of
modalities involved requires more data acquisition and increased system complexity.
More importantly, however, the availability of a greater volume and diversity of data
raises important questions concerning how the identification or verification process is
managed. This in turn raises equally important questions both in relation to the data
sources themselves and their selection and combination and, most significantly for the
way in which the decision-making is implemented.

At a more fundamental level, in order to support both the development of appropriate
verification structures and a reliable and meaningful evaluation of the performance of
multi-modal systems, appropriate data is required which links a range of different
biometric modality samples for individuals across a representative population of subjects.

This paper presents an initial analysis of a trial to acquire "real-world" multi-modal data
from a large population of subjects, allowing an exploration of some of these issues.

Data collection trials

As part of a current project designated IAMBIC (Intelligent Agents for Multi-modal Biometric Identification and Control) data collection trials have been undertaken to acquire data from a range of biometric devices. The data collection protocol is linked specifically to a scenario testing regime and has been developed within the established guidelines of good practice in biometric testing [1]. Of primary importance, however, is the ultimate availability of quantitative live data to aid the development and evaluation of algorithms for modality integration in biometric system design (see, for example, [2]).

The biometric devices adopted generate data covering a broad range of modalities. In our immediate testing these are facial images, voice samples and fingerprint, and the acquired measurements are all obtained using commercially available devices (with verification thresholds set mid range according to manufacturers' specifications), since a principal goal of the IAMBIC project is to investigate the development of data management techniques which are independent of the specific devices adopted, hence offering effectively a "plug-and-play" element in system design [3]. The devices themselves are not identified specifically since the focus of this paper is an evaluation of biometric modalities rather than individual devices or verification algorithms. Although during data collection basic precautions were taken to minimise obvious environmental disturbances, no formal optimisation of lighting or noise characteristics was undertaken.

Volunteers (221 in total) have been recruited, each volunteer receiving a modest payment for participation, and each being provided with full details of the aims of the trial and the procedures involved. Each volunteer is required to give written consent to participation and supplies brief personal information to assist in data analysis. The protocol requires that each volunteer undertakes two data collection sessions. The first visit entails enrolment on each of the systems adopted (up to three enrolment attempts per device are permitted), together with a post-enrolment verification check to ensure that the enrolment process has been successful. At a second session at least one month later a verification process is carried out using the original enrolment models. Investigation of longer term variability of data is outside the scope of the present study

Results and discussion

In this paper we will present an initial analysis of some preliminary results which shed considerable light on the practical realities of biometric device utilisation and, especially, on the advantages which accrue when more than one biometric modality can be tested. We will document these results as a short set of "experiments" based on the analysis of different aspects of the data collected, coupled with some observations about each.

Experiment 1: The failure to enrol rate can be as valuable an indicator of the viability of a biometric device as its subsequent error rates, and our first experiment aimed to determine the percentage of subjects who failed to enrol successfully with respect to each of the three experimental devices. The results are shown in Table 1.

Table 1-   Failure to enrol rates for each modality

Modality	Failure to enrol rate
Fingerprint	2.7%
Voice	9.95%
Face	1.36%

It is striking that there is considerable variability in the extent to which a satisfactory enrolment can be achieved. The Fingerprint and Face biometrics generated a relatively small failure to enrol rate, but the Voice modality proved significantly less reliable in achieving satisfactory enrolment.

Experiment 2: In this experiment we investigate time-based changes in biometric data by considering the proportion of subjects who, after a satisfactory enrolment at the first visit, failed to verify identity at the second. The results are shown in Table 2.

Table 2-   Failure rates for each modality

Modality	Failure rate
Fingerprint	14.42%
Voice	1.5%
Face	27.06%

These results are again extremely important, likewise showing considerable variability across the available devices. Here the Voice biometric provides, by a significant margin, the most stable performance, while the Face biometric performs relatively poorly, generating a failure rate almost twice that of the Fingerprint modality.

These results suggest that, while the Voice system might present some difficulties at enrolment, the performance returned is much more stable than for the other modalities once a satisfactory enrolment has been achieved.

Experiment 3: In this experiment we determine the proportion of subjects who failed to verify correctly both at the first and at the second visit. The results are shown in Table 3.

Table 3-   Failure probability rates for each modality

Modality	Failure rate
Fingerprint	4.18%
Voice	0.0%
Face	1.83%

These results effectively identify the proportion of so-called "goats" (subjects who generate inherently unstable data with respect to a particular biometric) in the population and who therefore would need to be excluded from a system or treated as special cases in some way. The goat phenomenon is widely recognised [4] but not often investigated quantitatively. It is notable here that the Voice system seems less susceptible to instability (no goats identified) than either of the other two modalities, supporting the observations made in connection with Experiment 2.

Experiment 4: Defining failure to enrol as the situation where a subject fails to enrol on every device in the chosen combination, then for each possible combination of two biometric modalities (Fingerprint/Voice, Fingerprint/Face, Voice/Face) the failure to enrol rates are as shown in Table 4.

Table 4-   Failure to enrol rates for two modalities

Modality	Failure to enrol rate
Fingerprint / Voice	0.91%
Voice / Face	0.0%
Face / Fingerprint	0.0%

It is notable that the failure to enrol proportion drops significantly when moving to a multiple modality system (cf. Table 1) and, indeed, moving to a three modality system showed the failure to enrol rate drop to zero. This is a powerful indicator of the value of even a rudimentary multimodal system in overcoming the deficiencies in any single modality for any particular potential enrolee.

Experiment 5: We again measured the failure rate in verification at the second visit for subjects who had successfully enrolled, but this time using all possible combinations of two modalities (verification failure is defined as a failure to satisfactorily verify identity in both of the modalities tested). The results are shown in Table 5.

Table 5-   Verification failure rates for two modalities

Modality	Verification failure rate
Fingerprint / Voice	0.45%
Voice / Face	0.0%
Face / Fingerprint	3.61%

It is apparent that the failure rates are dramatically reduced (cf. Table 2) by using a dual modality system, with the error rate falling to zero for the Voice/Face combination. This contrasts sharply with failure rates of as much as 27% when a single modality is adopted. It should also be noted that, when using all three modalities in combination, no errors were detected. Similarly, it is interesting to observe that with any biometric combination the number of goats identifiable also fell to zero.

Conclusions

This paper describes a recently undertaken formal trial to gather multi-modal biometric data from a large and representative population of individuals, based on commercially available biometric devices.

A preliminary analysis of the performance of the modalities adopted has been presented, focusing particularly on performance comparisons between verification based on single individual biometric modalities and approaches which exploit the opportunity to collect authentication data from more than one biometric source. It is important to emphasise, however, that the analysis presented here does not, in itself, provide a complete picture, since we have focused entirely on Type I verification errors, while an overall assessment must balance this with a similar analysis of Type II error rates (the subject of on-going investigation). Nevertheless, the results presented here are important in that they provide clear *quantitative* estimates of the performance potentially achievable in a real practical scenario, with typical users, and using currently available devices.

The database constructed will eventually provide a basis for a well-regulated evaluation of a range of algorithms for the practical exploitation of multi-modal biometric systems and, especially, for developing novel and effective ways in which such systems can be managed to provide reliable and robust biometric systems in the future [5]. Although further work is required (for example, precise rules for an optimal overall verification decision need to be formulated) this study provides a firm base on which to build.

References

1.   Best practices in testing and reporting performance of biometric devices, Biometrics Working Group/NPL, January 2000

2.   L.Hong, A,Jain: Integrating faces and fingerprints for personal identification, IEEE Trans PAMI, 20, 1295-1306, 1998

3.   F.Deravi, M.C.Fairhurst, N.Mavity, R.M.Guest: Design of multimodal biometric systems for univresal authentication and access control, Proc. WISA, 9-20, Seoul, 2001

4.   R.Plamondon, G.Lorette: Automatic signature verification and writer identification – the state of the art", Pattern Recognition, 22, 119-128, 1989

5.   K.Sirlantzis, M.C.Fairhurst: Optimisation of multiple classifier systems using genetic algorithms, Proc. ICIP, 1094-1097, 2001

The authors acknowledge the support of EPSRC and DTI through the IAMBIC LINK project, and the contribution made by the industrial partners Neusciences Ltd and Cardionetics.

KES 2002
E. Damiani et al. (Eds.)
IOS Press, 2002

An Interactive Fuzzy Satisficing Method through a Variance Minimization Model for Multiobjective Linear Programming Problems Involving Random Variable Coefficients

Masatoshi SAKAWA   Kosuke KATO   Hideki KATAGIRI
Graduate School of Engineering, Hiroshima University, 739-8527 JAPAN

Abstract. In this paper, we focus on multiobjective linear programming problems with random variable coefficients in objective functions and/or constraints. For such multiobjective problems, after reformulation of them on the basis of a variance minimization model for the chance constrained programming, incorporating fuzzy goals of the decision maker for the objective functions, we propose an interactive fuzzy satisficing method to derive a satisficing solution for the decision maker as a fusion of the stochastic programming and the fuzzy one.

1 Introduction

In constructing a mathematical model of a decision making situation in the real world, we should use approaches to reflect the randomness or the ambiguity involved in the situation since we cannot always obtain exact values of all parameters in the situation. Stochastic programming and fuzzy programming are two typical approaches for such decision making problems involving uncertainty.

Stochastic programming, as an optimization method based on the probability theory, has been developing in various ways [5]. Especially, for multiobjective stochastic linear programming problems, Stancu-Minasian [5] considered the minimum risk approach, while Teghem et al. [10] proposed an interactive method.

On the other hand, fuzzy mathematical programming representing the ambiguity in a decision making situation by fuzzy concepts has attracted attention of many researchers [6]. Fuzzy multiobjective linear programming have been rapidly developed by numerous researchers, and an increasing number of successful applications have been appearing [7].

As a hybrid of the stochastic approach and fuzzy approach, in particular, Hulsurkar et al. [3] applied fuzzy programming to multiobjective stochastic linear programming problems. However, in their method, since membership functions for the objective functions are supposed to be aggregated by minimum operator or product operator, optimal solutions which sufficiently reflect the decision maker's preference may not be obtained. To overcome this drawback, Sakawa et al. [8] showed that a satisficing solution to a multiobjective stochastic linear programming problem sufficiently reflecting the decision maker's preference can be derived by application of the interactive fuzzy satisficing method after reformulating it based on the expectation model for the chance constrained programming. However, in their method, since the objective functions regarded as random variables are reduced to thier expectations, the requirement of the decision maker for risk is not reflected in the obtained solution.

Under these circumstances, in this paper, we focus on multiobjective linear programming problems with random variable coefficients in objective functions and/or constraints. Using the variance minimization model, we transform the multiobjective stochastic programming problems into deterministic ones. Assuming that the decision maker has a fuzzy goal for each of the objective functions, having determined the fuzzy goals of the decision maker, we present an interactive fuzzy satisficing method to derive a satisficing solution for the decision maker by updating the reference membership levels.

2 Multiobjective linear programming problems with random variable coefficients

In this paper, we deal with multiobjective stochastic linear programming problems formally formulated as:

$$
\left.\begin{aligned}
\text{minimize} \quad & z_1(\boldsymbol{x}, \omega) = \boldsymbol{c}_1(\omega)\boldsymbol{x} \\
& \quad\vdots \\
\text{minimize} \quad & z_k(\boldsymbol{x}, \omega) = \boldsymbol{c}_k(\omega)\boldsymbol{x} \\
\text{subject to} \quad & A\boldsymbol{x} \leq \boldsymbol{b}(\omega) \\
& \boldsymbol{x} \geq \boldsymbol{0}
\end{aligned}\right\} \tag{1}
$$

where $\boldsymbol{x}$ is an n dimensional decision variable column vector, and A is an $m \times n$ coefficient matrix. $\boldsymbol{c}_l(\omega)$, $l = 1, \ldots, k$ are n dimensional random variable row vectors with mean $\bar{\boldsymbol{c}}_l$ and covariance matrix $V_l = (v_{jh}^l) = (\mathrm{Cov}\{c_{lj}(\omega), c_{lh}(\omega)\})$, $j = 1, \ldots, n$, $h = 1, \ldots, n$, and $b_i(\omega)$, $i = 1, \ldots, m$ are random variables which are independent of each other, and the distribution function of each of them are also assumed to be continuous and strictly increasing.

Since the problem (1) contains random variable coefficients, definitions and solution methods for ordinary mathematical programming problems cannot be directly applied. Consequently, we deal with the constraints in (1) as chance constrained conditions [2] which mean that the constraints need to be satisfied with a certain probability (satisficing level) and over. Under the chance constrained conditions, we study a variance minimization model, which aims to minimize the variance of each of objective functions.

3 Variation minimization model

Replacing the constraints in (1) by chance constrained conditions with satisficing levels β_i, $i = 1, \ldots, m$ along with substituting the minimization of variances of objective functions for the minimization of the objective functions $z_l(\boldsymbol{x}, \omega)$ in (1), the problem can be converted as:

$$
\left.\begin{aligned}
\text{minimize} \quad & z_1'(\boldsymbol{x}) = \mathrm{Var}\left\{z_1(\boldsymbol{x}, \omega)\right\} = \boldsymbol{x}^T V_1 \boldsymbol{x} \\
& \quad\vdots \\
\text{minimize} \quad & z_k'(\boldsymbol{x}) = \mathrm{Var}\left\{z_k(\boldsymbol{x}, \omega)\right\} = \boldsymbol{x}^T V_k \boldsymbol{x} \\
\text{subject to} \quad & \Pr[\boldsymbol{a}_1\boldsymbol{x} \leq b_1(\omega)] \geq \beta_1 \\
& \quad\vdots \\
& \Pr[\boldsymbol{a}_m\boldsymbol{x} \leq b_m(\omega)] \geq \beta_m \\
& \boldsymbol{x} \geq \boldsymbol{0}
\end{aligned}\right\} \tag{2}
$$

where $\boldsymbol{a}_i$ is the ith row vector of A and $b_i(\omega)$ is the ith element of $\boldsymbol{b}(\omega)$.

Using continuous and strictly increasing distribution functions $F_i(r) = \Pr[b_i(\omega) \leq r]$ of random variables $b_i(\omega)$, $i = 1, \ldots, m$, the i th constraint in (2) can be rewritten as:

$$\Pr[\boldsymbol{a}_i\boldsymbol{x} \leq b_i(\omega)] \geq \beta_i \Leftrightarrow \boldsymbol{a}_i\boldsymbol{x} \leq F_i^{-1}(1 - \beta_i). \tag{3}$$

Letting $\hat{b}_i = F_i^{-1}(1 - \beta_i)$ in (3), the problem (2) can be transformed into the following equivalent multiobjective convex quadratic programming problem:

$$\left. \begin{array}{ll} \text{minimize} & (z_1'(\boldsymbol{x}), \ldots, z_k'(\boldsymbol{x})) \\ \text{subject to} & A\boldsymbol{x} \leq \hat{\boldsymbol{b}} \\ & \boldsymbol{x} \geq \boldsymbol{0} \end{array} \right\} \tag{4}$$

where $\hat{\boldsymbol{b}} = (\hat{b}_1, \ldots, \hat{b}_m)^T$.

In order to take the requirement of the decision maker for expectations of objective functions into account, we consider the following problem

$$\left. \begin{array}{ll} \text{minimize} & (z_1'(\boldsymbol{x}), \ldots, z_k'(\boldsymbol{x})) \\ \text{subject to} & A\boldsymbol{x} \leq \hat{\boldsymbol{b}} \\ & \bar{C}\boldsymbol{x} \leq \gamma \\ & \boldsymbol{x} \geq \boldsymbol{0} \end{array} \right\} \tag{5}$$

where $\bar{C} = (\bar{\boldsymbol{c}}_1^T, \ldots, \bar{\boldsymbol{c}}_k^T)^T$ and $\gamma = (\gamma_1, \ldots, \gamma_k)$ where each element γ_l is the permissible level of the decision maker for $\mathrm{E}\{z_l(\boldsymbol{x}, \omega)\} = \bar{\boldsymbol{c}}_l\boldsymbol{x}$. In the following, for notational convenience, the feasible region of (5) is denoted by X.

4 An interactive fuzzy satisficing method

In order to consider the imprecise nature of the decision maker's judgements for each objective function in (5), if we introduce the fuzzy goals such as "$z_l'(\boldsymbol{x})$ should be substantially less than or equal to a certain value", the problem (5) can be rewritten as:

$$\underset{\boldsymbol{x}\in X}{\text{maximize}} \quad (\mu_1(z_1'(\boldsymbol{x})), \ldots, \mu_k(z_k'(\boldsymbol{x}))) \tag{6}$$

where $\mu_l(\cdot)$ is a membership function to quantify a fuzzy goal for the lth objective function in (5).

Since the problem (6) is regarded as a fuzzy multiobjective decision making problem, there rarely exist a complete optimal solution that simultaneously optimizes all objective functions. As a reasonable solution concept for the fuzzy multiobjective decision making problem, Sakawa et al. [9] defined M-Pareto optimality on the basis of membership function values by directly extending the Pareto optimality in the ordinary multiobjective programming problem.

Definition 1 (M-Pareto optimal solution). *$\boldsymbol{x}^* \in X$ is said to be an M-Pareto optimal solution if and only if there does not exist another $\boldsymbol{x} \in X$ such that $\mu_i(z_i'(\boldsymbol{x})) \geq \mu_i(z_i'(\boldsymbol{x}^*))$ for all i and $\mu_j(z_j'(\boldsymbol{x})) > \mu_j(z_j'(\boldsymbol{x}^*))$ for at least one j.*

In order to derive a satisficing solution for the decision maker from the M-Pareto optimal solution set, Sakawa et al. proposed an interactive fuzzy satisficing method such that the decision maker interactively updates the aspiration levels of achievement for the membership values of all membership functions, called the reference membership levels until he is satisfied [9].

To be more specific, for the decision maker's reference membership levels $\bar{\mu}_l, i = 1, \ldots, k$, the corresponding M-Pareto optimal solution, which is nearest to the requirements in the augmented minimax sense or better than them if the reference membership levels are attainable, is obtained by solving the following augmented minimax problem

$$\underset{\boldsymbol{x} \in X}{\text{minimize}} \ \underset{l=1,\ldots,k}{\max} \left[\bar{\mu}_l - \mu_l(z_l'(\boldsymbol{x})) + \rho \sum_{i=1,\ldots,k} (\bar{\mu}_i - \mu_i(z_i'(\boldsymbol{x}))) \right] \qquad (7)$$

By introducing the auxiliary variable v, this problem can be equivalently transformed as:

$$\left. \begin{aligned} &\text{minimize} \ \ v \\ &\text{subject to} \ \ \bar{\mu}_1 - \mu_1(z_1'(\boldsymbol{x})) + \rho \sum_{i=1,\ldots,k} (\bar{\mu}_i - \mu_i(z_i'(\boldsymbol{x}))) \le v \\ &\qquad\qquad\qquad \vdots \\ &\ \ \ \bar{\mu}_k - \mu_k(z_k'(\boldsymbol{x})) + \rho \sum_{i=1,\ldots,k} (\bar{\mu}_i - \mu_i(z_i'(\boldsymbol{x}))) \le v \\ &\ \ \ \boldsymbol{x} \in X. \end{aligned} \right\} \qquad (8)$$

Here, assuming that each of membership functions $\mu_l(\cdot), l = 1, \ldots, k$ is nonincreasing an concave, the problem (8) is a convex programming problem. Under the assumption, we can solve (8) by a traditional convex programming technique as the sequential quadratic programming method. We now summarize the interactive algorithm.

Interactive fuzzy satisficing method

Step 1: Ask the decision maker to specify the satisficing levels $\beta_l, l = 1, \ldots, m$ for each of the constraints in (1).

Step 2: After calculating the individual minimum $\bar{z}_l^{\min}$ and maximum $\bar{z}_l^{\max}$ of $\mathrm{E}[z_l(\boldsymbol{x}, \omega)] = \bar{z}_l(\boldsymbol{x}), l = 1, \ldots, k$ under the chance constrained conditions, ask the decision maker to specify permissible levels $\gamma_l, l = 1, \ldots, k$ for objective functions.

Step 3: Calculate the individual minimum $z_{l,\min}'$ of $z_l'(\boldsymbol{x}), l = 1, \ldots, k$ in (5).

Step 4: Ask the decision maker to determine membership functions $\mu_l(z_l'(\boldsymbol{x}))$ for objective functions in (5) on the basis of individual minima $z_{l,\min}'$.

Step 5: Ask the decision maker to set the initial reference membership levels $\bar{\mu}_l = 1, l = 1, \ldots, k$.

Step 6: Calculate the optimal solution $\boldsymbol{x}^*$ to the augmented minimax problem (8) corresponding to the current reference membership levels $\bar{\mu}_l, l = 1, \ldots, k$.

Step 7: The decision maker is supplied with the obtained solution x^*. If the decision maker is satisfied with the current membership function values of x^*, stop. Otherwise, ask the decision maker to update the reference membership levels $\bar{\mu}_l$, $l = 1, \ldots, k$ by considering the current membership function values $\mu_l(z_l'(x^*))$, and return to step 6.

Here it should be stressed to the decision maker that any improvement of one membership function can be achieved only at the expense of at least one of other membership functions.

5 Conclusion

In this paper, we focused on multiobjective linear programming problems involving random variable coefficients. After the formulation as the variance minimization model, we introduced fuzzy goals to consider the ambiguous or fuzzy judgements of the decision maker and presented an interactive fuzzy satisficing method as a fusion of stochastic approaches and fuzzy ones to derive a satisficing solution for the decision maker from the M-Pareto optimal solution set.

References

[1] R.E. Bellman and L.A. Zadeh, "Decision making in a fuzzy environment," *Management Science*, Vol. 17, 1970, pp 141-164.

[2] A. Charnes and W.W. Cooper, "Chance constrained programming," *Management Science*, Vol. 6, 1959, pp 73-79.

[3] S. Hulsurkar, M.P. Biswal and S.B. Sinha, "Fuzzy programming approach to multi-objective stochastic linear programming problems," *Fuzzy Sets and Systems*, Vol. 88, 1997, pp 173-181.

[4] H. Ishii, "Stochastic optimization," M. Iri and H. Konno (eds.): *Applications of Mathematical Programming (Theoretical Topics)*, Sangyo Tosho, Tokyo, 1982, pp 1-40 (in Japanese).

[5] I.M. Stancu-Minasian, "Overview of different approaches for solving stochastic programming problems with multiple objective functions," In R. Slowinski and J. Teghem (eds.): *Stochastic Versus Fuzzy Approaches to Multiobjective Mathematical Programming under Uncertainty*, Kluwer Academic Publishers, Dordrecht/Boston/London, 1990, pp 71-101.

[6] H. Rommelfanger, "Fuzzy linear programming and applications," *European Journal of Operational Research*, Vol. 92, 1996, pp 512-527.

[7] M. Sakawa, *Fuzzy Sets and Interactive Multiobjective Optimization*, Plenum Press, New York, 1993.

[8] M. Sakawa, K. Kato, I. Nishizaki and M. Yoshioka, "An interactive fuzzy satisficing method for multiobjective linear programming problems involving random variable coefficients," *Transactions of the Institute of Electronics, Information and Communication Engineers A*, Vol. J83-A, 2000, pp 161-167 (in Japanese).

[9] M. Sakawa and H. Yano, "An interactive fuzzy satisficing method using augmented minimax problems and its application to environmental systems," *IEEE Transactions on Systems, Man, and Cybernetics*, Vol. SMC-15, 1985, pp 720-729.

[10] J. Teghem Jr., D. Dufrane, M. Thauvoye and P. Kunsch: "STRANGE: an interactive method for multiobjective linear programming under uncertainty," *European Journal of Operational Research*, Vol. 26, 1986, pp 65-82.

[11] A.P. Wierzbicki, "The use of reference objectives in multiobjective optimization," In G. Fandel and T. Gal (eds.): *Multiple Criteria Decision Making: Theory and Application*, Springer-Verlag, Berlin, 1995, pp 468-486.

[12] H.-J. Zimmermann, "Fuzzy programming and linear programming with several objective functions," *Fuzzy Sets and Systems*, Vol. 1, 1978, pp 45-55.

KES 2002
E. Damiani et al. (Eds.)
IOS Press, 2002

Interactive Fuzzy Decentralized Two-Level Linear Fractional Programming through Decomposition Algorithms

Kosuke KATO   Masatoshi SAKAWA   Ichiro NISHIZAKI
Graduate School of Engineering, Hiroshima University, 739-8527 JAPAN

Abstract. In decentralized two-level linear fractional programming problems, there exist one decision maker (DM) at the upper level and multiple DMs at the lower level. We have already proposed interactive fuzzy programming for decentralized two-level linear programming problems on the assumption that all of the DMs can cooperate with each other, and succeeded in obtaining satisficing solutions for the DMs. In this paper, we focus on decentralized two-level linear fractional programming problems with block angular structure, which are often encountered in the case that all of the DMs can decide their strategy almost independently of each other. In order to derive satisficing solutions for the DMs to the problems, we adopt the interactive fuzzy programming and consider the application of the Ritter partitioning procedure and the Dantzig-Wolfe decomposition principle to it for the purpose of utilizing the special structure of the problems.

1   Introduction

In this paper, we consider decentralized two-level linear fractional programming problems in which there exist a single decision maker (DM) at the upper level and two or more DMs at the lower level.

Various approaches for such decentralized two-level programming problems could exist according to situations which the DMs are placed in. Stackelberg solutions [7] are regarded as reasonable ones to decentralized two-level linear programming problems under the condition that all the DMs do not have motivation to cooperate mutually. Thus, the formulation based on the Stackelberg solution under the competitive situation seems impractical for a case where the DM at the upper level is a supervisory administration and the DMs at the lower level are individual divisions. Additionally, owing to the difficulty of calculating Stackelberg solutions, another solution concept to two-level programming problems, which is easier of calculation, is required.

From this standpoint, Y.-J. Lai [2] and H.-S. Shih et al. [8] developed fuzzy programming techniques for multi-level programming problems. Unfortunately, in their way such that fuzzy goals are postulated not only for objective functions but for decision variables at the upper level, an undesirable solution may be obtained as the final decision because of the conflict between fuzzy goals for objective functions and those for decision variables. To cope with this drawback, M. Sakawa et al. [4] centered on only fuzzy goals for objective functions and excluded those for decision variables. Furthermore, their approach has been extended to two-level linear programming problems with fuzzy parameters [5] and decentralized two-level linear programming problems [6].

In the real world, various decision making situations are generally formulated as large-scale mathematical programming problems involving a large number of variables and constraints. Although it is difficult to obtain strict optimal solutions to such large-scale problems because of some restrictions, we can often solve them efficiently since some of them have utilizable special structures. For instance, a decision making problem in a decentralized system such as a parent company and several associated child companies which are independent of one another, is formulated as a decentralized two-level programming problem with block angular structure.

Under these circumstances, focusing on block angular decentralized two-level linear fractional programming problems, we attempt to derive a satisficing solution for the decision maker through the interactive fuzzy programming technique proposed by M. Sakawa et al. [6] and consider the application of the decomposition principle by G.B. Dantzig and P. Wolfe [1] and the partitioning procedure by K. Ritter [3] to the interactive fuzzy programming.

2　Problem Formulation

In this paper, we consider a situation where there exist a single decision maker (DM0) at the upper level and p decision makers (DM1, $\dots$, DMp) at the lower level who can almost independently decide their strategy one another.

Such a situation can be formulated as the following decentralized two-level linear fractional programming problems with block angular structure (1).

$$
\left.
\begin{array}{ll}
\underset{\text{DM0}}{\text{minimize}} & z_0(\boldsymbol{x}_0, \boldsymbol{x}_1, \dots, \boldsymbol{x}_k) \\
\underset{\text{DM1}}{\text{minimize}} & z_1(\boldsymbol{x}_0, \boldsymbol{x}_1, \dots, \boldsymbol{x}_k) \\
\quad\vdots & \qquad\vdots \\
\underset{\text{DM}k}{\text{minimize}} & z_k(\boldsymbol{x}_0, \boldsymbol{x}_1, \dots, \boldsymbol{x}_k) \\
\text{subject to} & A_0\boldsymbol{x}_0 + A_1\boldsymbol{x}_1 + \cdots + A_k\boldsymbol{x}_k \leq \boldsymbol{a} \\
& B_0\boldsymbol{x}_0 \qquad\qquad\qquad\qquad\; \leq \boldsymbol{b}_0 \\
& \qquad\; B_1\boldsymbol{x}_1 \qquad\qquad\quad\; \leq \boldsymbol{b}_1 \\
& \qquad\qquad\qquad\ddots \\
& \qquad\qquad\qquad\quad B_k\boldsymbol{x}_k \leq \boldsymbol{b}_k \\
& \boldsymbol{x}_j \geq \boldsymbol{0}, \quad j = 0, 1, \dots, k
\end{array}
\right\} \tag{1}
$$

where $\boldsymbol{x}_0$ is the n_0 dimensional decision variable column vector for the decision maker at the upper level (DM0), $\boldsymbol{x}_j$ is the n_j dimensional decision variable column vector for the jth decision maker at the lower level (DMj, $j = 1, \dots, k$). For notational convenience, we use $\boldsymbol{x} = (\boldsymbol{x}_0^T, \boldsymbol{x}_1^T, \cdots, \boldsymbol{x}_k^T)^T$, and X denotes the feasible region of (1). In (1), $z_i(\boldsymbol{x}) = z_i(\boldsymbol{x}_0, \boldsymbol{x}_1, \dots, \boldsymbol{x}_k)$, $i = 0, 1, \dots, k$ are linear fractional functions as:

$$
z_i(\boldsymbol{x}) = \frac{p_i(\boldsymbol{x})}{q_i(\boldsymbol{x})} = \frac{\boldsymbol{c}_{i0}\boldsymbol{x}_0 + \boldsymbol{c}_{i1}\boldsymbol{x}_1 + \cdots + \boldsymbol{c}_{ik}\boldsymbol{x}_k + c_{i,k+1}}{\boldsymbol{d}_{i0}\boldsymbol{x}_0 + \boldsymbol{d}_{i1}\boldsymbol{x}_1 + \cdots + \boldsymbol{d}_{ik}\boldsymbol{x}_k + d_{i,k+1}} \tag{2}
$$

where $\boldsymbol{c}_{ij}$, $\boldsymbol{d}_{ij}$, $i, j = 0, 1, \dots, k$ are n_j dimensional coefficients vectors, and it is assumed that $q_i(\boldsymbol{x}) > 0$, $i = 1, \dots, k$.

For example, consider a production planning problem in an administrative office at the upper level and several autonomous divisions at the lower level of a company. In this case, the situation that all of the DMs can cooperate with each other seems natural rather than one that all the DMs do not have motivation to cooperate mutually.

Under the hypothesis of cooperation among all DMs, M. Sakawa et al. [6] proposed the interactive fuzzy programming for decentralized two-level linear programming problems in order to derive satisficing solutions for the DMs through interactions with the DM at the upper level by introducing fuzzy goals to consider the imprecise nature of DMs' judgement for objective functions.

In this paper, focusing on the case of cooperative relation between DM0 at the upper level and DMj, $j = 1, \ldots, p$ at the lower level as in [6], we apply the interactive fuzzy programming [6] based on the Ritter partitioning procedure [3] and the Dantzig-Wolfe decomposition principle [1] in order to efficiently derive an satisficing solutions for DMs to (1).

3   Interactive Fuzzy Programming

The interactive fuzzy programming proposed by M. Sakawa et al. [6] is a two-phase interactive decision making approach as follows.

Phase 1:

Considering the ambiguity or fuzziness of the decision makers' judgements on each objective functions $z_i(x)$, $i = 0, 1, \ldots, k$ in (1), it seems natural to introduce such fuzzy goals for objective functions as "$z_i(x)$ should be subjectively less than or equal to a certain value."

Then, the block angular decentralized two-level linear fractional programming problem (1) can be interpreted as

$$
\left.
\begin{array}{ll}
\underset{\text{DM0}}{\text{maximize}} & \mu_0(c_0 x) \\
\underset{\text{DM1}}{\text{maximize}} & \mu_1(c_1 x) \\
\quad\vdots & \quad\vdots \\
\underset{\text{DM}p}{\text{maximize}} & \mu_p(c_p x) \\
\text{subject to} & x \in X
\end{array}
\right\}
\tag{3}
$$

where $\mu_i(\cdot)$, $i = 0, 1, \ldots, k$ are membership functions which quantify fuzzy goals to consider the imprecise nature of DMs' judgement for objective functions in problem (1).

First, individual minima $z_i^{\min} = z_i(x^{i,\min})$ and maxima $z_i^{\max} = z_i(x^{i,\max})$ of objective functions $z_i(x)$ are calculated by solving the following problems.

$$
\underset{x \in X}{\text{minimize}} \; z_i(x), \quad i = 0, 1, \ldots, k
\tag{4}
$$

$$
\underset{x \in X}{\text{maximize}} \; z_i(x), \quad i = 0, 1, \ldots, k
\tag{5}
$$

Since these problems are transformed into block angular linear programming problems with coupling variables by the Charnes-Cooper variable transformation, we can use the Ritter partitioning procedure [3] to solve them.

After each of membership functions $\mu_i(z_i x)$, $i = 0, 1, \ldots, p$ is subjectively specified by the corresponding DMi, assuming that a hypothetical decision maker aggregating all DMj,

$j = 1, \ldots, k$ at the lower level by the minimum operator is on an equal position with DM0 at the upper level, the following problem (6) is solved.

$$\underset{x \in X}{\text{maximize}} \ \min \left\{ \mu_0(z_0(x)), \ \underset{j=1,\ldots,p}{\min} \mu_j(z_j(x)) \right\} \tag{6}$$

The problem (6) is equivalently rewritten as:

$$\left. \begin{array}{ll} \text{maximize} & v \\ \text{subject to} & \mu_i(z_i(x)) \geq v, \quad i = 0, 1, \ldots, k \\ & x \in X, \ v \in [0, 1]. \end{array} \right\} \tag{7}$$

In order to solve problem (7), the optimal value of $v = v^*$ is calculated by the combination of the bisection method and the phase one of the two-phase simplex method. In the above solution algorithm, since (7) is reduced to a block angular linear programming problem, we can apply the Dantzig-Wolfe decomposition principle [1]. Then, the optimal solution x^* corresponding to v^* is obtained by solving the following problem.

$$\left. \begin{array}{ll} \text{minimize} & z_0(x) \\ \text{subject to} & \mu_i(z_i(x)) \geq v^*, \quad i = 1, \ldots, k \\ & x \in X \end{array} \right\} \tag{8}$$

Since (8) is reduced to a block angular linear programming problem with coupling variables through the Charnes-Cooper variable transformation, we can apply the Ritter partitioning procedure [3].

If DM0 at the upper level is satisfied with the optimal solution to (6), *Phase 1* is terminated. Otherwise, after DM0 subjectively specifies the minimal satisfactory level $\hat{\delta}$ taking into account the ratio of the satisfactory degree of the aggregate decision maker at the lower level to that of DM0 at the upper level $\Delta = [\min_{j=1,\ldots,k} \mu_j(z_j(x))]/\mu_0(z_0(x))$, the following problem (9) is solved.

$$\left. \begin{array}{ll} \text{maximize} & \underset{j=1,\ldots,p}{\min} \mu_j(z_j(x)) \\ \text{subject to} & \mu_0(z_0(x)) \geq \hat{\delta} \\ & x \in X \end{array} \right\} \tag{9}$$

As in solving (7), problem (9) can be also solved efficiently by the Dantzig-Wolfe decomposition principle [1] and the Ritter partitioning method [3].

If the following two conditions hold simultaneously, *Phase 1* is terminated.

(1-1) The ratio of satisfactory degrees Δ corresponding to the optimal solution to (9) is in the closed interval $[\Delta_{\min}, \Delta_{\max}]$, where $\Delta_{\min}$ and $\Delta_{\max}$ denote the lower bound and the upper one of Δ, specified by DM0.

(1-2) DM0 at the upper level is satisfied with the optimal solution to (9).

Otherwise, the problem (9) will be repeatedly solved with updating $\hat{\delta}$ until the conditions (1-1) and (1-2) are fulfilled.

Phase 2:

First, ratios of the satisfactory degree of DM0 and that of each of DMj, $j = 1, \ldots, p$, $\Delta_j = \mu_j(z_j(x))/\mu_0(z_0(x))$ are calculated for the solution obtained in *Phase 1*. Then, DM0 specifies the maximal satisfactory level $\bar{\delta}_j$ for j such as $\Delta_j \notin [\Delta_{\min}, \Delta_{\max}]$, and solves the following problem.

$$
\left.
\begin{aligned}
\text{maximize} \quad & \min_{j=1,\dots,p} \mu_j(z_j(\boldsymbol{x})) \\
\text{subject to} \quad & \mu_0(z_0(\boldsymbol{x})) \geq \hat{\delta} \\
& \mu_j(z_j(\boldsymbol{x})) \leq \bar{\delta}_j, \quad j \in V \\
& \boldsymbol{x} \in X
\end{aligned}
\right\}
\tag{10}
$$

where $V = \{j \mid \Delta_j \notin [\Delta_{\min}, \Delta_{\max}]\} \subseteq \{1, \dots, p\}$. As in solving (7), the Dantzig-Wolfe decomposition principle [1] and the Ritter partitioning procedure [3] are applicable to solve problem (10).

If the following two conditions hold simultaneously, *Phase 2* is terminated.

(2-1) Ratios of satisfactory degrees Δ_j, $j = 1, \dots, p$ corresponding to the optimal solution to (10) are in the closed interval $[\Delta_{\min}, \Delta_{\max}]$.

(2-2) DM0 at the upper level is satisfied with the optimal solution to (10).

Otherwise, the problem (10) will be repeatedly solved with updating $\bar{\delta}_j$ for all $j \in V$ until the conditions (2-1) and (2-2) are fulfilled.

4 Conclusion

In this paper, focusing on a decision making situation where a single decision maker (DM0) exists at the upper level and p decision makers (DM1, $\dots$, DMp) exist at the lower level who can almost independently decide their strategies one another, we formulated it as a decentralized two-level linear fractional programming problem with block angular structure. For the formulated problem, we considered the application of the interactive fuzzy programming to derive a satisficing solution for the DMs and showed the applicability of the Dantzig-Wolfe decomposition principle and the Ritter partitioning procedure in the solution procedure of the interactive fuzzy programming.

References

[1] G.B. Dantzig and P. Wolfe, The decomposition algorithm for linear programming, *Econometrica* 29 (1961) 767-778.

[2] Y.-J. Lai, Hierarchical optimization: a satisfactory solution, *Fuzzy Sets and Systems* 77 (1996) 321-335.

[3] K. Ritter, A decomposition method for linear programming problems with coupling constraints and variables, *Report 739, Mathematics Research Center, University of Wisconsin*, 1967.

[4] M. Sakawa, I. Nishizaki and Y. Uemura, Interactive fuzzy programming for multi-level linear programming problems, *Computers & Mathematics with Applications* 36 (1998) 71-86.

[5] M. Sakawa, I. Nishizaki and Y. Uemura, Interactive fuzzy programming for multi-level linear programs with fuzzy numbers, *1998 Second International Conference on Knowledge-Based Intelligent Electronic Systems, Proceedings of KES'98* 1 (1998) 109-118.

[6] M. Sakawa and I. Nishizaki, Interactive fuzzy programming for decentralized two-level linear programming problems, *Proceedings of the Eighth International Fuzzy Systems Association World Congress* 1 (1999) 514-518.

[7] K. Shimizu, Y. Ishizuka and J.F. Bard, *Nondifferentiable and Two-Level Mathematical Programming*, Chapter 16, Kluwer Academic Publishers, Boston/London/Dordrecht, 1997.

[8] H.-S. Shih, Y.-J. Lai and E.S. Lee, Fuzzy approach for multi-level programming problems, *Computers and Operations Research* 23 (1996) 73-91.

KES 2002
E. Damiani et al. (Eds.)
IOS Press, 2002

Interactive decision making for a multiobjective 0-1 programming problem involving fuzzy random variable coefficients

Hideki Katagiri, Masatoshi Sakawa and Yoshihiro Sato
Graduate School of Engineering, Hiroshima University, Japan

Abstract. In this paper, we deal with a multiobjective 0-1 programming problem involving fuzzy random variable coefficients. Introducing fuzzy goals for the objective functions, we consider the problem to maximize the degrees of possibility that the objective function values satisfy the fuzzy goals, which becomes a multiobjective stochastic programming problem. Using an expectation model in stochastic programming, we formulate the problem as a multiobjective 0-1 linear fractional programming problem. In order to find a satisficing solution for a decision maker, we propose an interactive satisficing method based on the reference point method and show that the problem to be solved is reduced to a mixed 0-1 linear programming problem.

1 Introduction

In the classical mathematical programming, the coefficients of objectives or constraints in problems are assumed to be completely known. However, in real systems, they are rather uncertain than constant. In order to deal with such uncertainty, stochastic programming [1] and fuzzy programming [2, 3] were considered. They are useful tools for the decision making under a stochastic environment or a fuzzy environment, respectively.

Most researches in respect to mathematical programming take account of fuzziness or randomness. However, in practice, decision makers face with the situations where both fuzziness and randomness exist. For instance, in the case where some expert estimates coefficients of objective functions or constraints with uncertainty, they are not always given as random variables or fuzzy sets but as the values including both fuzziness and randomness. Fuzzy random variables [4, 5] are one of the mathematical concepts dealing with fuzziness and randomness simultaneously. Recently, several authors considered linear programming problems involving fuzzy random variables [6, 7, 8, 9]. In this research, we consider multiobjective 0-1 programming problem using the concept of possibility measures [10] and an expectation model in stochastic programming. In Section 2, we consider a multiobjective 0-1 programming problem with fuzzy random variable coefficients and formulate it as a multiobjective 0-1 programming problem where each objective function is the expected degree of possibility with respect to a fuzzy goal. Section 3 proposes an interactive satisficing method for the problem to obtain the satisficing solution for a decision maker. Since the problem involves the product of a continuous decision variable and a 0-1 variable, it becomes a nonlinear mixed 0-1 programming problem. Using some property, we transform the problem into a linear mixed 0-1 programming problem. Finally, in Section 4, we conclude this paper and discuss further research.

2　Formulation

Consider the following problem:

$$
\left.
\begin{array}{ll}
\text{minimize} & \tilde{c}_i(\omega)x, \quad i = 1, \dots, k \\
\text{subject to} & Ax \leq b, \ x \in \{0,1\}^n,
\end{array}
\right\}
\tag{2.1}
$$

where $x = (x_1, \dots, x_n)^t$ is a 0-1 decision vector and $\tilde{c}_i(\omega) = (\tilde{c}_{i1}(\omega), \dots, \tilde{c}_{in}(\omega))$ is a coefficient vector. Let A be an $m \times n$ matrix and b an $m \times 1$ vector. Each $\tilde{c}_{ij}(\omega)$ is a fuzzy random variable; namely it is a random variable whose actual value is a symmetric triangular fuzzy number $\tilde{C}_{ij}(\omega_{is})$ with the following membership function:

$$
\mu_{\tilde{C}_{ij}(\omega_{is})}(t) \;=\; \max\left\{0, 1 - \frac{|t - c_{ij}(\omega_{is})|}{\alpha_{ij}}\right\}, \ i = 1, \dots, k, \ j = 1, \dots, n
\tag{2.2}
$$

where ω_{is}, $s = 1, \dots, l_i$ denote outcomes of the event corresponding to the fuzzy random variable $\tilde{c}_{ij}(\omega)$. In other words, each ω_{is} is regarded as a scenario. Let p_{is} be the probability that each scenario ω_{is} occurs. We assume that $\sum_{s=1}^{l_i} p_{is} = 1$ holds. Each α_{ij} denotes the spread parameter of a fuzzy number. This type of fuzzy random variable is equivalent to a *hybrid number*, which was introduced by Kaufman and Gupta [11].

Since the coefficients of objective functions are symmetric triangular fuzzy random variables, each objective function becomes a single fuzzy random variable $\tilde{Y}_i(\omega)$ with the following membership function:

$$
\mu_{\tilde{Y}_i(\omega)}(y) \;=\; \max\left\{0, \ 1 - \frac{\left| y - \sum_{j=1}^{n} c_{ij}(\omega)x_j \right|}{\sum_{j=1}^{n} \alpha_{ij}x_j}\right\}, \ i = 1, \dots, k.
\tag{2.3}
$$

Considering the imprecision or fuzziness of the decision maker's judgment, for each objective function of problem (2.1), we introduce the fuzzy goal $\tilde{G}_i$ with the membership function expressed as

$$
\mu_{\tilde{G}_i}(y) = \left\{
\begin{array}{ll}
0, & y > h_i^0 \\[2mm]
\dfrac{y - h_i^0}{h_i^1 - h_i^0}, & h_i^1 \leq y \leq h_i^0 \\[2mm]
1, & y < h_i^1, \quad i = 1, \dots, k.
\end{array}
\right.
\tag{2.4}
$$

Since for each ω_{is}, the membership function $\mu_{\tilde{Y}_i(\omega_{is})}$ is regarded as a possibility distribution, the degree of possibility $\Pi_{\tilde{Y}_i(\omega_{is})}(\tilde{G}_i)$ that the objective function value satisfies the fuzzy goal is

$$
\Pi_{\tilde{Y}_i(\omega_{is})}(\tilde{G}_i) \;=\; \sup_{y} \min\left\{\mu_{\tilde{Y}_i(\omega_{is})}(y), \ \mu_{\tilde{G}_i}(y)\right\}, \ i = 1, \dots, k.
\tag{2.5}
$$

In this research, taking account of all scenarios, we set h_i^0 and h_i^1 as the following form:

$$
h_i^0 \;=\; \max_{s} \max_{x \in X} \sum_{j=1}^{n} \{c_{ij}(\omega_{is}) - \alpha_{ij}\}x_j, \quad i = 1, \dots, k,
$$

$$
h_i^1 \;=\; \min_{s} \min_{x \in X} \sum_{j=1}^{n} c_{ij}(\omega_{is})x_j, \quad i = 1, \dots, k,
$$

where $X \triangleq \{x | Ax \leq b, \ x \in \{0,1\}^n\}$. By using (2.3) and (2.4), the degree of possibility defined in (2.5) is represented as follows:

$$\Pi_{\tilde{Y}_i(\omega_{is})}(\tilde{G}_i) = \frac{\sum\limits_{j=1}^{n} \{\alpha_{ij} - c_{ij}(\omega_{is})\} x_j + h_i^0}{\sum\limits_{j=1}^{n} \alpha_{ij} x_j - h_i^1 + h_i^0}, \quad i = 1, \ldots, k, \ s = 1, \ldots, l_i.$$

Note that the above values depend on scenarios and hence they are considered as random variables. Consequently, the mean value is calculated and is expressed by

$$E[\Pi_{\tilde{Y}_i(\omega)}(\tilde{G}_i)] = \frac{\sum\limits_{s=1}^{l_i} p_{is} \left[\sum\limits_{j=1}^{n} \{\alpha_{ij} - c_{ij}(\omega_{is})\} x_j + h_i^0 \right]}{\sum\limits_{j=1}^{n} \alpha_{ij} x_j - h_i^1 + h_i^0}, \quad i = 1, \ldots, k.$$

As a result, problem (2.1) is interpreted as the following multiobjective linear fractional 0-1 programming problem:

$$\left.\begin{array}{ll} \text{maximize} & z_i(x) = E[\Pi_{\tilde{Y}_i(\omega)}(\tilde{G}_i)], \ i = 1, \ldots, k \\ \text{subject to} & Ax \leq b, \ x \in \{0,1\}^n. \end{array}\right\} \tag{2.6}$$

3   Interactive Decision Making Using an Expectation Model Based on a Possibility Measure

Since problem (2.6) has several objective functions, there does not generally exist the solution optimizing all functions. Therefore, in this section, we discuss the interactive decision making based on the reference point method [12] to obtain a Pareto optimal solution.

For each of the multiple conflicting objective functions, assume that the decision maker can specify the so-called reference point $\bar{\pi} = (\bar{\pi}_1, \ldots, \bar{\pi}_k)$ which reflects in some sense the desired values of the objective functions of the decision maker. Also assume that the decision maker can change the reference point interactively due to learning or improved understanding during the solution process. When the decision maker specifies the reference point $\bar{\pi} = (\bar{\pi}_1, \ldots, \bar{\pi}_k)$, the corresponding Pareto optimal solution, which is, in the minimax sense, nearest to the reference point or better than that if the reference point is attainable, is obtained by solving the following minimax problem:

$$\left.\begin{array}{ll} \text{minimize} & \max\limits_{1 \leq i \leq k} \{\bar{\pi}_i - z_i(x)\} \\ \text{subject to} & Ax \leq b, \ x \in \{0,1\}^n. \end{array}\right\} \tag{3.7}$$

By introducing the auxiliary variable λ, problem (3.7) is rewritten as follows:

$$\left.\begin{array}{ll} \text{minimize} & \lambda \\ \text{subject to} & \bar{\pi}_i - z_i(x) \leq \lambda, \quad i = 1, \ldots, k \\ & Ax \leq b, \ x \in \{0,1\}^n. \end{array}\right\} \tag{3.8}$$

In (3.8), $\bar{\pi}_i - z_i(x) \leq \lambda$ is equivalent to the following:

$$\sum\limits_{j=1}^{n} \alpha_{ij} x_j \lambda + h_i^1 \lambda - h_i^0 \lambda \geq \bar{\pi}_i Q_i(x) - P_i(x) \tag{3.9}$$

where

$$P_i(\boldsymbol{x}) = \sum_{i=1}^{l_i} p_{is}\left[\sum_{j=1}^{n}\{\alpha_{ij} - c_{ij}(\omega_{is})\}x_j + h_i^0\right] \quad \text{and} \quad Q_i(\boldsymbol{x}) = \sum_{j=1}^{n}\alpha_{ij}x_j - h_i^1 + h_i^0.$$

It should be noted that problem (3.8) is a nonlinear mixed 0-1 programming problem due to the product of a binary variable x_j and a continuous variable λ. However, it is transformed into a set of linear constraints by using the following theorem:

Theorem 1. *Let x_j be a 0-1 variable and λ a continuous variable. When $0 \leq \lambda \leq 1$, $y_j = x_j\lambda$ can be represented by the following linear inequalities: (1) $\lambda - y_j \leq 1 - x_j$, (2) $y_j \leq \lambda$, (3) $y_j \leq x_j$, (4) $y_j \geq 0$.*

This theorem can be easily proved in the same way as shown by H.L.Li [13]. Applying the above theorem to our problem, we transform the original problem (3.8) into the following linear mixed 0-1 programming problem:

$$\left.\begin{aligned}
&\text{minimize} \quad \lambda \\
&\text{subject to} \quad \sum_{j=1}^{n}\alpha_{ij}y_j + h_i^1\lambda - h_i^0\lambda \geq \bar{\pi}_iQ_i(\boldsymbol{x}) - P_i(\boldsymbol{x}), \\
&\qquad\qquad \lambda - y_j \leq 1 - x_j,\ y_j \leq \lambda,\ y_j \leq x_j,\ y_j \geq 0,\ i = 1,\ldots,k,\ j = 1,\ldots,n, \\
&\qquad\qquad A\boldsymbol{x} \leq \boldsymbol{b},\ \boldsymbol{x} \in \{0,1\}^n.
\end{aligned}\right\}$$

$$(3.10)$$

Given the reference point $\bar{\pi}$, the corresponding minimax problem becomes a linear mixed 0-1 programming problem. Therefore, the problem is solved by the branch-and-bound method and so on. From the above discussion, an interactive satisficing method for fuzzy random linear programming problems is as follows.

[An interactive satisficing method for a fuzzy random linear programming problem through an expectation model using a possibility measure]

Step 1: Calculate h_i^0 and h_i^1, $i = 1,\ldots,k$, and elicit the membership functions of fuzzy goals for the objective functions.

Step 2 Set the initial reference values $\bar{\pi}_i$, $i = 1,\ldots,k$ to 1.0.

Step 3: For the reference point $\bar{\pi}$, solve the corresponding minimax problem (3.10).

Step 4: If the decision maker is satisfied with the current level of the Pareto optimal solution with respect to the expected degree of possibility, stop. Then the current solution is the satisficing solution for the decision maker. Otherwise, ask the decision maker to update the current reference point by considering the current values of $\bar{\pi}_i$, $i = 1,\ldots,k$ and return to Step 3.

4   Conclusion

In this research, we have investigated multiobjective 0-1 programming problems with fuzzy random variable coefficients. Incorporating the expectation model and the concept of possibility measures, we formulate the problem as a multiobjective nonlinear 0-1 programming problem to maximize the expected degree of possibility. In future, we will try to consider the model minimizing the variance of degree of possibility.

Although we have assumed that the parameter of spread, α_{ij}, $i = i, \ldots, k$, $j = 1, \ldots, n$, are not dependent on senarios, the assumption that the parameters take different values dependent on the events is more realistic. In such a case, however, the proposed method in this research is not applicable directly and hence we will extend our method or consider another approach.

References

[1] S. Vajda, Probabilistic Programming, Academic Press (1972).

[2] M. Sakawa, Fuzzy Sets and Interactive Multiobjecive Optimization, Plenum, New York (1993).

[3] M. Inuiguchi and J. Ramik, Possibilistic linear programming: a brief review of fuzzy mathematical programming and a comparison with stochastic programming in portfolio selection problem, Fuzzy Sets and Systems **111** (2000) 3–28.

[4] H. Kwakernaak, Fuzzy random variable-1: definitions and theorems, Information Sciences **15** (1978) 1–29.

[5] M.L. Puri and D.A. Ralescu, Fuzzy random variables, Journal of Mathematical Analysis and Applications, **114** (1986) 409–422.

[6] G.-Y. Wang and Q. Zhong, Linear programming with fuzzy random variable coefficients, Fuzzy Sets and Systems **57** (1993) 295–311.

[7] M.K. Luhandjula and M.M. Gupta, On fuzzy stochastic optimization, Fuzzy Sets and Systems **81** (1996) 47–55.

[8] H. Katagiri and H. Ishii, Chance constrained bottleneck spanning tree problem with fuzzy random edge costs, Journal of the Operations Research Society of Japan **43** (2000) 128–137.

[9] H. Katagiri and H. Ishii, Linear programming problem with fuzzy random constraint, Mathematica Japonica **52** (2000) 123–129.

[10] L.A. Zadeh, Probability measure of fuzzy events, Journal of Mathematical Analysis and Applications **23** (1968) 421–427.

[11] A. Kaufman and M.M. Gupta, Introduction to Fuzzy Arithmetic: Theory and Applications, Van Nostrand Reinhold Company (1985).

[12] A.P. Wierzbicki, The use of reference objectives in multiobjective optimization, in: G. Frande and T. Gal (eds.), Multiple Criteria Decision Making: Theory and Application, Springer-Verlag (1980).

[13] H.-L. Li, A global approach for general 0-1 fractional programming, European Journal of Operational Research **73** (1994) 590–596.

KES 2002
E. Damiani et al. (Eds.)
IOS Press, 2002

1237

Fuzzy Multiple Decision Maker-Multiple Objective Programming Problems Using the Weighted Minimax Method

Hitoshi Yano

School of Humanities and Social Sciences, Nagoya City University, Nagoya 467-8501 Japan

Abstract. In this paper, we focus on fuzzy multiple decision maker-multiple objective programming problems, where each of multiple decision makers has his/her own multiple objectives which conflict each other. In order to deal with such problems, two kinds of extended Pareto optimal concepts are introduced and the weighted minimax problem is formulated to obtain the candidate of the agreeable solution from among extended Pareto optimal solution set. Using the weighted minimax method, an interactive algorithm is proposed to obtain the agreeable solution through the interaction between multiple decision makers.

1 Introduction

Recently, in order to deal with the real-world decision problems, many kinds of methods for multiple objective programming problems [3, 4] have been proposed to obtain a compromise or satisfactory solution, where a *single* decision maker is involved and he/she has his/her own multiple, noncommensurable and conflicting objective functions. However, in most real-world complicated decision situation, not a single decision maker but *multiple* decision makers are often involved. They may have their own multiple objective functions and each of them may have his/her own inherent point of view for his/her objective functions. In order to handle such multiple decision maker-multiple objective decision problems (MDMOP), many types of group decision support systems (GDSS) have been investigated [2], which can be regarded as interactive computer-based systems to communicate between multiple decision makers, aggregate multiple decision makers' preference, and find out the agreeable solution.

In this paper, we propose an interactive algorithm to obtain the agreeable solution of multiple decision makers in MDMOP on the basis of multiple objective programming techniques [3, 4], In the proposed algorithm, considering the vague nature of human subjective judgement in MDMOPs, it is assumed that each of multiple decision makers has a fuzzy goal for each of his/her own objective functions. It is also assumed that, in MDMOP, each of the decision makers may have the different decision power to reach the agreeable solution.

After eliciting the membership function for each of the objective functions from each of the decision makers in his/her subjective manner, he/she derives his/her own satisfactory solution from among his/her partial M-Pareto optimal solution set which is defined in his/her own membership space. Each of the satisfactory solutions of multiple decision makers is adopted as the ideal point for each of the decision makers in MDMOP. However, such satisfactory solutions can not be usually obtained simultaneously. In order to obtain the agreeable solution, the weighted minimax problem is formulated, where their own satisfactory solutions are used as ideal points, and the decision makers set their decision powers corresponding to the power relationship in the decision situation. Updating the weighting parameters in the weighted minimax problems, the corresponding candidate of the agreeable solution is obtained from among total M-Pareto optimal solution set.

2 Fuzzy Multiple Decision Maker-Multiple Objective Programming Problems

We focus on the following multiple decision maker-multiple objective programming problem (MDMOP) where each of multiple decision makers ($DM_i, i = 1, \cdots, p$) has his/her own multiple objective functions $f_{ij}(\boldsymbol{x}), j = 1, \cdots, k_i$ which conflict each other, and all of the objective functions are defined in the common feasible region X.

[MDMOP]

$$DM_1 : \quad \min \quad (f_{11}(\boldsymbol{x}), \cdots, f_{1k_1}(\boldsymbol{x})) \tag{1}$$

$$\cdots\cdots\cdots\cdots$$

$$DM_p : \quad \min \quad (f_{p1}(\boldsymbol{x}), \cdots, f_{pk_p}(\boldsymbol{x}))$$

$$\text{subject to} \quad \boldsymbol{x} \in X = \{\boldsymbol{x} \in R^n \mid g_j(\boldsymbol{x}) \leq 0, \ j = 1, \cdots, m\}$$

By considering the vague nature of human subjective judgement, it is quite natural to assume that all of the decision makers may have fuzzy goals [3, 6] for their own objective functions. Through the interaction with the decision makers, these fuzzy goals can be quantified by eliciting the corresponding membership functions which are denoted by $\mu_{ij}(f_{ij}(\boldsymbol{x})), i = 1, \cdots, p, j = 1, \cdots, k_i$ respectively. Then, the corresponding fuzzy multiple decision maker-multiple objective programming problem (FMDMOP) can be expressed as the following form:

[FMDMOP]

$$DM_1 : \quad \max \quad (\mu_{11}(f_{11}(\boldsymbol{x})), \cdots, \mu_{1k_1}(f_{1k_1}(\boldsymbol{x}))) \tag{2}$$

$$\cdots\cdots\cdots\cdots\cdots\cdots$$

$$DM_p : \quad \max \quad (\mu_{p1}(f_{p1}(\boldsymbol{x})), \cdots, \mu_{pk_p}(f_{pk_p}(\boldsymbol{x})))$$

$$\text{subject to} \quad \boldsymbol{x} \in X$$

In order to deal with FMDMOP, we first introduce the following extended Pareto optimal concept which is defined in membership space for all of the decision makers.

[Definition 1]
In FMDMOP, $\boldsymbol{x}* \in X$ is said to be a total M-Pareto optimal solution for all of the decision makers, if and only if there does not exist another $\boldsymbol{x} \in X$ such that $\mu_{ij}(f_{ij}(\boldsymbol{x})) \geq \mu_{ij}(f_{ij}(\boldsymbol{x}*)), i = 1, \cdots, p, j = 1, \cdots, k_i$, with strict inequality holding for at least one i and one j.

Now, consider the decision situation in FMDMOP where only one decision maker (DM_i) is involved. Then, FMDMOP can be reduced to the following usual multiple objective programming problem for single decision maker DM_i.

[FMOP$_i$]

$$\max_{\boldsymbol{x} \in X} \ (\mu_{i1}(f_{i1}(\boldsymbol{x})), \cdots, \mu_{ik_i}(f_{ik_i}(\boldsymbol{x}))) \tag{3}$$

In FMOP_i, the partial M-Pareto optimal concept for the decision maker DM_i can be defined in membership space as follows.

[Definition 2]
In FMOP_i, $\boldsymbol{x}* \in X$ is said to be a partial M-Pareto optimal solution for the decision maker DM_i, if and only if there does not exist another $\boldsymbol{x} \in X$ such that $\mu_{ij}(f_{ij}(\boldsymbol{x})) \geq \mu_{ij}(f_{ij}(\boldsymbol{x}*))$, $j = 1, \cdots, k_i$, with strict inequality holding for at least one j.

In the following, denote the partial M-Pareto optimal solution set for the decision maker DM_i as $X_i^P, i = 1, \cdots, p$, and the total M-Pareto optimal solution set for all of the decision makers as X_0^P. In FMOP_i, the decision maker DM_i can obtain his/her own satisfactory solution from among X_i^P by applying multiple objective decision making methods [3, 4]. If all of the decision makers $\text{DM}_i, i = 1, \cdots, p$ obtain their own satisfactory solution $\boldsymbol{x}_i^s$ from among X_i^P,

$$(\mu_{i1}(f_{i1}(\boldsymbol{x}_i^s)), \cdots, \mu_{ik_i}(f_{ik_i}(\boldsymbol{x}_i^s))), i = 1, \cdots, p \tag{4}$$

can be regarded as an ideal point in membership space for all of the decision makers, which reflects their own preferences. Unfortunately, such an ideal point is not usually attainable simultaneously because the membership functions of the decision makers conflict each other. In the next section, using the weighted minimax method, we propose an interactive algorithm to obtain the agreeable solution in FMDMOP, which is, in a sense, close to the satisfactory solutions $\boldsymbol{x}_i^s, i = 1, \cdots, p$.

3 An Interactive Algorithm for FMDMOP

After obtaining the satisfactory solutions $\boldsymbol{x}_i^s, i = 1, \cdots, p$ for FMOP_i, the decision makers have to find out their agreeable solution which reflects not only the preferences of the decision makers for their membership functions but also the power relationship between decision makers. In this paper, we assume that all of the decision makers find out their agreeable solution from among the total M-Pareto optimal solution set X_0^P. Then, the candidate of the agreeable solution is obtained by solving the following weighted minimax problem.

[Weighted Minimax Problem]

$$\min_{\boldsymbol{x}\in X} \ x_{p+1} \tag{5}$$
$$\text{subject to} \quad w_{ij}\{\mu_{ij}(f_{ij}(\boldsymbol{x}_i^s)) - \mu_{ij}(f_{ij}(\boldsymbol{x}))\} \leq x_{p+1}, \quad i = 1, \cdots, p, j = 1, \cdots, k_i \tag{6}$$

where $w_{ij} \geq 0, i = 1, \cdots, p, j = 1, \cdots, k_i$ are weighting parameters specified by the decision makers $\text{DM}_i, i = 1, \cdots, p$. In order to cope with the degree of influence of each of the decision maker in the decision making processes, we introduce the concept of the decision powers which are defined by the weighting parameters w_{ij} as follows:

$$C_i = \sum_{j=1,\cdots,k_i} w_{ij}, \ i = 1, \cdots, p, \tag{7}$$

where $C_i, i = 1, \cdots, p$ are decision powers of the decision maker DM_i in FMDMOP. It is easily understood that if only one decision power C_i is positive and all of the others is zero, i.e., $C_\ell = 0, \ell = 1, \cdots, p, \ell \neq i$, then it is clear that the weighting parameters $w_{\ell j} = 0, j = 1, \cdots, k_\ell, \ell \neq i$ and the optimal solution of the weighted minimax problem for any weighting parameters $w_{ij} \geq 0, j = 1, \cdots, k_i$ always becomes the satisfactory solution $\boldsymbol{x}_i^s$. If all of the decision makers are equal in FMDMOP, the decision powers should be set as follows.

$$C_1 = C_2 = \cdots\cdots = C_p > 0 \tag{8}$$

The relationships between the optimal solutions of the weighted minimax problem and the total M-Pareto optimal solutions can be characterized by the following theorem.

[Theorem 1]

If $\boldsymbol{x}* \in X$ is a unique optimal solution of the weighted minimax problem for some weighting parameters $w_{ij} \geq 0, i = 1, \cdots, p, j = 1, \cdots, k_i$, then $\boldsymbol{x}* \in X$ is a total M-Pareto optimal solution.

(Proof)

Assume that $\boldsymbol{x}* \in X$ is not a total M-Pareto optimal solution, then there exists $\boldsymbol{x} \in X$ such that $\mu_{ij}(f_{ij}(\boldsymbol{x})) \geq \mu_{ij}(f_{ij}(\boldsymbol{x}*)), i = 1, \cdots, p, j = 1, \cdots, k_i$ with strict inequality holding for at least one i and one j. This implies that

$$\mu_{ij}(f_{ij}(\boldsymbol{x}_i^s)) - \mu_{ij}(f_{ij}(\boldsymbol{x})) \leq \mu_{ij}(f_{ij}(\boldsymbol{x}_i^s)) - \mu_{ij}(f_{ij}(\boldsymbol{x}*)), i = 1, \cdots, p, j = 1, \cdots, k_i.$$

Then it holds that

$$\max_{\substack{i=1,\cdots,p \\ j=1,\cdots,k_i}} \{w_{ij}\{\mu_{ij}(f_{ij}(\boldsymbol{x}_i^s)) - \mu_{ij}(f_{ij}(\boldsymbol{x}))\}\} \leq \max_{\substack{i=1,\cdots,p \\ j=1,\cdots,k_i}} \{w_{ij}\{\mu_{ij}(f_{ij}(\boldsymbol{x}_i^s)) - \mu_{ij}(f_{ij}(\boldsymbol{x}*))\}\}$$

which contradicts the fact that $\boldsymbol{x}* \in X$ is a unique optimal solution of the weighted minimax problem.

Now given the total M-Pareto optimal solution $\boldsymbol{x}* \in X_0^P$ corresponding to some weighting parameters which are specified by all of the decision makers, if they are able to compromise at $\boldsymbol{x}*$, then $\boldsymbol{x}*$ can be regarded as one of the agreeable solutions. However, at least one of the decision makers can not compromise at $\boldsymbol{x}*$, they have to find out the agreeable solution from among the total M-Pareto optimal solution set through the negotiation between them. In order to help the decision makers update their weighting parameters of the weighted minimax problem according to their own preferences, the trade-off information between membership functions seems to be very useful. Such trade-off rates between membership functions are easily obtained explicitly under some appropriate conditions [1].

[Theorem 2]

Let us assume that the weighted minimax problem has a unique optimal solution $\boldsymbol{x}* \in X$ satisfying the conditions [1] that, $\boldsymbol{x}*$ is a regular point, the second-order sufficiency conditions are satisfied at $\boldsymbol{x}*$, and there are no degenerate constraints at $\boldsymbol{x}*$. Also assume that the constraints (6) for the indices (i_1, j_1) and (i_2, j_2) are active, *i.e.*,

$$w_{ij}\{\mu_{ij}(f_{ij}(\boldsymbol{x}_i^s)) - \mu_{ij}(f_{ij}(\boldsymbol{x}*))\} = x_{p+1}^*. \tag{9}$$

Then it holds [5] that

$$\frac{\partial \mu_{i_1 j_1}(f_{i_1 j_1}(\boldsymbol{x}*))}{\partial \mu_{i_2 j_2}(f_{i_2 j_2}(\boldsymbol{x}*))} = -\frac{w_{i_2 j_2}\lambda_{i_2 j_2}^*}{w_{i_1 j_1}\lambda_{i_1 j_1}^*}, \tag{10}$$

where $\lambda_{i_1 j_1}^*, \lambda_{i_2 j_2}^* > 0$ are the Lagrangian multipliers corresponding to the constraints (6).

In FMDMOP where many decision makers are involved, it may be difficult to obtain the agreeable solution efficiently by updating their weighting parameters of the weighted minimax problem. In order to reach an agreement efficiently in such decision situations, we extend the agreeable solution concept as follows, which depend on the majority parameter $q(0 \leq q \leq p)$.

[Definition 3]
$x* \in X_0^P$ is said to be an extended agreeable solution in FMDMOP, if and only if at least $q(0 \leq q \leq p)$ decision makers accept $x* \in X_0^P$ as one of the agreeable solutions, where q is the majority parameter.

The majority parameter q is specified by all of the decision makers in their subjective manner. Usually, the majority parameter q is set as p.

Now, we can construct an interactive algorithm based on the weighted minimax problem to obtain an extended agreeable solution from among the total M-Pareto optimal solution set.

[Interactive algorithm]
[Step 1] Each of the decision maker (DM$_i$) obtains his/her own satisfactory solution $x_i^s \in X_i^P, i = 1, \cdots, p$ for the usual multiple objective programming problem FMOP$_i$ by applying multiple objective programming techniques [3, 4].
[Step 2] Set the decision powers $C_i, i = 1, \cdots, p$ which reflect the power relationship between the decision makers.
[Step 3] Set the initial majority parameter $q = p$.
[Step 4] Set the initial weighting parameters $w_{ij} = C_i/k_i, i = 1, \cdots, p, j = 1, \cdots, k_i$.
[Step 5] Solve the weighted minimax problem and obtain the corresponding total M-Pareto optimal solution $x* \in X_0^P$ and the trade-off information between the membership functions.
[Step 6] If more than q decision makers can compromise at $x*$, then stop. $x*$ is the extended agreeable solution. Otherwise, go to Step 7.
[Step 7] If all of the decision makers agree to change the value of the majority parameter q in order to obtain the extended agreeable solution as soon as possible, go to Step 8. Otherwise, considering the current values of the membership functions and trade-off information between membership functions, the decision makers who do not accept their current membership function values update their own weighting parameters, and go to Step 5.
[Step 8] Set the majority parameter q. Corresponding to the majority parameter q, each of the decision makers sets his/her own weighting parameters, and go to Step 5.

4 Conclusions

In this paper, we propose an interactive algorithm to obtain the extended agreeable solution for fuzzy multiple decision maker-multiple objective programming problems, which reflect the preferences of the decision makers for each of their membership functions and the power relationship between them.

References

[1] A.V.Fiacco, *Introduction to Sensitivity and Stability Analysis in Nonlinear Programming*, Academic Press, 1983.

[2] N.F.Matsatsinis and A.P.Samaras, " MCDA and Preference Disaggregation in Group Decision Support Systems" *European Journal of Operational Research*, Vol.130, 2001, pp 414-429.

[3] M.Sakawa, *Fuzzy Sets and Interactive Multiobjective Optimization*, Plenum Press, New York, USA, 1993.

[4] R.E.Steuer, *Multiple Criteria Optimization: Theory, Computation, and Application* , Wiley, 1986.

[5] H.Yano and M.Sakawa, " Trade-off Rates in the Weighted Tchebycheff Norm Method" *Large Scale System*, Vol.13 No.2, 1987, pp.167-177.

[6] H.-J.Zimmermann, *Fuzzy Set Theory and Its Applications*, Kluwer Academic Publishers, Boston, 1991.

KES 2002
E. Damiani et al. (Eds.)
IOS Press, 2002

Nash-Stackelberg equilibrium solutions for noncooperative two-level four-person games

Ichiro Nishizaki, Masatoshi Sakawa and Yoshinobu Fujita
Graduate School of Engineering, Hiroshima University,
1-4-1, Kagamiyama, Higashi-Hiroshima, 739-8527 Japan

Abstract. In this paper, we consider a noncooperative two-level four-person game where there are two players at the upper level and the other two players at the lower level. Each of the two players at the upper level first chooses a decision and then the corresponding player at the lower level makes a decision with full knowledge of the decision of the player at the upper level. The two players at the upper level make decisions simultaneously or they do not know decisions of the opponents even if they make decisions one by one. For such a noncooperative two-level four-person games, we suppose the two players at the upper level to be in Nash equilibrium and a pair of the players at the upper and the lower levels to be in Stackelberg equilibrium, and define a Nash-Stackelberg equilibrium solution. Especially, for a two-level four-person bimatrix game, we develop a computational method for obtaining the Nash-Stackelberg equilibrium solution.

1 Introduction

In this paper, we deal with two-level four-person games where there are two players at the upper level and the other two players at the lower level. We assume that each of the players does not have a motivation to cooperate one another and there is no communication among them. Each of the two players at the upper level first chooses a decision and then the corresponding player at the lower level makes a decision with full knowledge of the decision of the player at the upper level as well as their objective functions and the constraints of the problem. Therefore, the player at the lower level chooses a decision so as to minimize/maximize the objective function of self. On this assumption, the player at the upper level also makes a decision such that the objective function of self is minimized/maximized. The two players at the upper level make decisions simultaneously or they do not know decisions of the opponents even if they make decisions one by one. Then, it is natural that each of the two players chooses a decision miniming/maximing the objective function for a given decision of the opponent.

While computational methods for Nash equilibrium solutions and those for Stackelberg equilibrium solutions have been developed independently [1, 3, 4, 5, 6, 7], computation of solutions to the noncooperative two-level four-person games with Nash equilibrium between the two players at the upper level and Stackelberg equilibrium between a pair of the two players at the upper and the lower levels have so far been not studied. We define Nash-Stackelberg equilibrium solutions for the noncooperative two-level four-person games, and especially, for a two-level four-person bimatrix game which is a special case of the noncooperative two-level four-person game, we develop a computational method for obtaining the Nash-Stackelberg solution by using the scheme of the genetic algorithms.

2 Two-level four-person bimatrix games

2.1 Nash-Stackelberg equilibrium solutions

In this section, we deal with a two-level four-person bimatrix game which is a special case of the noncooperative two-level four-person game and develop a computational method for obtaining Nash-Stackelberg equilibrium solutions by using the necessary conditions and the scheme of the genetic algorithms [2].

Let Players 1-1 and 2-1 denote the players at the upper level and Players 1-2 and 2-2 the players at the lower level corresponding to Players 1-1 and 2-1, respectively. Let $K \triangleq \{1,\ldots,k\}$ denote a set of pure strategies of Player 1-1, $L \triangleq \{1,\ldots,l\}$ a set of pure strategies of Player 1-2, $M \triangleq \{1,\ldots,m\}$ a set of pure strategies of Player 2-1, and $N \triangleq \{1,\ldots,n\}$ a set of pure strategies of Player 2-2. Let e_x, e_y, e_u and e_v denote k-, l-, m- and n-dimensional vectors whose elements are all one, respectively. Then, mixed strategies of Players 1-1, 1-2, 2-1 and 2-2 can be represented by $x = (x_1,\ldots,x_k)^T \in X \triangleq \{x \in \mathbb{R}^k \mid x \geq 0,\ e_x^T x = 1\}, y = (y_1,\ldots,y_l)^T \in Y \triangleq \{y \in \mathbb{R}^l \mid y \geq 0,\ e_y^T y = 1\},\ u = (u_1,\ldots,u_m)^T \in U \triangleq \{u \in \mathbb{R}^m \mid u \geq 0,\ e_u^T u = 1\}$ and $v = (v_1,\ldots,v_n)^T \in V \triangleq \{v \in \mathbb{R}^n \mid v \geq 0,\ e_v^T v = 1\}$, respectively. The superscript T represents transposition of vectors or matrices.

Let A and B denote payoff matrices between Players 1-1 and 1-2, C and D payoff matrices between Players 2-1 and 2-2, and E and F payoff matrices between Players 1-1 and 2-1. The payoff matrices A, B, C, D, E and F are $(k \times l)$, $(k \times l)$, $(m \times n)$, $(m \times n)$, $(k \times m)$ and $(k \times m)$ matrices and we assume that all elements of the matrices are nonnegative. Then, the two-level four-person bimatrix game can be formulated as

$$\left.\begin{array}{ll} \text{maximize} & f_{11}(x,y,u) = x^T A y + x^T E u \\ \text{subject to} & e_x^T x - 1 = 0,\ x \geq 0 \\ & \quad \text{maximize} \quad x^T B y \\ & \quad \text{subject to} \quad e_y^T y - 1 = 0,\ y \geq 0, \end{array}\right\} \qquad (1)$$

$$\left.\begin{array}{ll} \text{maximize} & f_{21}(x,u,v) = u^T C v + x^T F u \\ \text{subject to} & e_u^T u - 1 = 0,\ u \geq 0 \\ & \quad \text{maximize} \quad u^T D v \\ & \quad \text{subject to} \quad e_v^T v - 1 = 0, v \geq 0. \end{array}\right\} \qquad (2)$$

For a given pair of strategies x and u, we assume that an optimal solution to the problem

$$\left.\begin{array}{ll} \text{maximize} & x^T B y \\ \text{subject to} & e_y^T y - 1 = 0,\ y \geq 0 \end{array}\right\} \qquad (3)$$

in the constraints of problem (1) is unique and let $y^*(x,u)$ denote the optimal solution to problem (3).

Similarly, a given pair of strategies x and u, we assume that an optimal solution to the problem

$$\left.\begin{array}{ll} \text{maximize} & u^T D v \\ \text{subject to} & e_v^T v - 1 = 0,\ v \geq 0 \end{array}\right\} \qquad (4)$$

in the constraints of problem (2) is unique and let $v^*(x,u)$ denote the optimal solution to problem (4).

Because optimal solutions to problems (3) and (4) are called Stackelberg equilibrium solutions and solutions equilibrating objective functions f_{11} and f_{21} of Players 1-1 and 2-1 are called Nash equilibrium solutions, we call solutions optimizing problems (1) and (2) simultaneously by Nash-Stackelberg equilibrium solutions. Thus, a Nash-Stackelberg equilibrium solution is also a solution (x^*, y^*, u^*, v^*) satisfying

$$f_{11}(x^*, y^*, u^*) = \min_x f_{11}(x, y^*(x, u^*), u^*), \tag{5}$$

$$f_{21}(x^*, u^*, v^*) = \min_u f_{21}(x^*, u, v^*(x^*, u)). \tag{6}$$

2.2 *Necessary conditions*

By using the Kuhn-Tucker conditions for problems (1) and (2), we derive necessary conditions that a four-tuple of strategies of the four players be a Nash-Stackelberg equilibrium solution. Because the Kuhn-Tucker conditions of the maximization problem in the constraints of problem (1) are necessary and sufficient conditions that a strategies y is a rational response of Player 1-2, the following problem substituting the Kuhn-Tucker conditions for the maximization problem

$$\left. \begin{array}{ll} \text{maximize} & x^T A y + x^T E u \\ \text{subject to} & e_x^T x - 1 = 0, \ x \geq 0 \\ & B^T x - (x^T B y) e_y \leq 0 \\ & e_y^T y - 1 = 0, \ y \geq 0 \end{array} \right\} \tag{7}$$

is equivalent to problem (1). The Lagrange function of problem (7) is expressed as

$$L_{11}(x, y, u; \lambda) = -(x^T A y + x^T E u) + \lambda_1 (e_x^T x - 1) + \lambda_2^T (-x)$$
$$+ \lambda_3^T (B^T x - (x^T B y) e_y) + \lambda_4 (e_y^T y - 1) + \lambda_5^T (-y) \tag{8}$$

and the Kuhn-Tucker conditions

$$\left. \begin{array}{l} -Ay - Eu + \lambda_1 e_x - \lambda_2 + B\lambda_3 + (\lambda_3^T e_y)By = 0 \\ -A^T x - (\lambda_3^T e_y)B^T x + \lambda_4 e_y - \lambda_5 = 0 \\ e_x^T x - 1 = 0, \ x \geq 0 \\ B^T x - (x^T B y) e_y \leq 0 \\ e_y^T y - 1 = 0, \ y \geq 0 \\ -\lambda_2^T x + \lambda_3^T (B^T x - (x^T B y) e_y) - \lambda_5^T y = 0 \\ \lambda_2 \geq 0, \ \lambda_3 \geq 0, \ \lambda_5 \geq 0 \end{array} \right\} \tag{9}$$

are necessary optimality conditions for problem (7).

Similarly, from the following problem substituting the Kuhn-Tucker conditions for the maximization problem in problem (2)

$$\left. \begin{array}{ll} \text{maximize} & u^T C v + x^T F u \\ \text{subject to} & e_u^T u - 1 = 0, \ u \geq 0 \\ & D^T u - (u^T D v) e_v \leq 0 \\ & e_v^T v - 1 = 0, \ v \geq 0, \end{array} \right\} \tag{10}$$

we have the necessary conditions

$$\left. \begin{array}{l} -Cv - F^T x + \mu_1 e_u - \mu_2 + D\mu_3 - (\mu_3^T e_v)Dv = 0 \\ -C^T u - (\mu_3^T e_v)D^T u + \mu_4 e_v - \mu_5 = 0 \\ e_u^T u - 1 = 0, \ u \geq 0 \\ D^T u - (u^T D v) e_v \leq 0 \\ e_v^T v - 1 = 0, \ v \geq 0 \\ -\mu_2^T u + \mu_3^T (D^T u - (u^T D v) e_v) - \mu_5^T v = 0 \\ \mu_2 \geq 0, \ \mu_3 \geq 0, \ \mu_5 \geq 0, \end{array} \right\} \tag{11}$$

where $(\mu_1, \mu_2, \mu_3, \mu_4, \mu_5)$ are Lagrange multipliers. Therefore, the conditions (9) and (11) are necessary conditions that a four-tuple of strategies (x,y,u,v) be a Nash-Stackelberg equilibrium solution of the two-level four-person bimatrix game. However, because problems (7) and (10) are not convex programming problems, the conditions (9) and (11) are not sufficient conditions.

To obtain Nash-Stackelberg equilibrium solutions, we first find four-tuples of strategies (x,y,u,v) satisfying the conditions (9) and (11) and then verify that they are Nash-Stackelberg equilibrium solutions. In this paper, we develop a computational method implemented the above procedure on the basis of the scheme of the genetic algorithms.

3 A computational method based on the genetic algorithms

Because the constraints including mixed strategies (x,y,u,v) of the players in the necessary conditions (9) and (11) are

$$
\left.
\begin{aligned}
&e_x^T x - 1 = 0 \\
&B^T x - (x^T B y)e_y \leq \mathbf{0} \\
&e_y^T y - 1 = 0 \\
&e_u^T u - 1 = 0 \\
&D^T u - (u^T D v)e_v \leq \mathbf{0} \\
&e_v^T v - 1 = 0 \\
&x \geq \mathbf{0},\ y \geq \mathbf{0},\ u \geq \mathbf{0},\ v \geq \mathbf{0},
\end{aligned}
\right\}
\tag{12}
$$

we employ a representation of an individual s by not the binary implementation but the floating point implementation and each of individuals s made up of (x,y,u,v) is generated so as to satisfy the constraints (12). After the vectors x and u satisfying the conditions (12) are randomly generated, the vectors y and v are determined so as to be rational responses for x and u, respectively. We generate N individuals as an initial population.

We evaluate individuals in our artificial genetic system through two stages. At the first stage, we introduce artificial variables z_1, z_2, z_3, z_4, z_5 and z_6 to the necessary conditions (9) and (11) and formulate the following minimization programming problem

$$
\left.
\begin{aligned}
\text{minimize}\quad & w_0 = z_1^T e_x + z_2^T e_y + z_3 + z_4^T e_u + z_5^T e_v + z_6 \\
\text{subject to}\quad & -Ay - Eu + \lambda_1 e_x - \lambda_2 + B\lambda_3 + (\lambda_3^T e_y)By + z_1 = \mathbf{0} \\
& -A^T x - (\lambda_3^T e_y)B^T x + \lambda_4 e_y - \lambda_5 + z_2 = \mathbf{0} \\
& -\lambda_2^T x + \lambda_3^T(B^T x - (x^T By)e_y) - \lambda_5^T y + z_3 = 0 \\
& -Cv - F^T x + \mu_1 e_u - \mu_2 + D\mu_3 - (\mu_3^T e_v)Dv + z_4 = \mathbf{0} \\
& -C^T u - (\mu_3^T e_v)D^T u + \mu_4 e_v - \mu_5 + z_5 = \mathbf{0} \\
& -\mu_2^T u + \mu_3^T(D^T u - (u^T Dv)e_v) - \mu_5^T v + z_6 = 0 \\
& \lambda_2 \geq \mathbf{0},\ \lambda_3 \geq \mathbf{0},\ \lambda_5 \geq \mathbf{0},\ \mu_2 \geq \mathbf{0},\ \mu_3 \geq \mathbf{0},\ \mu_5 \geq \mathbf{0} \\
& z_1, z_2, z_3, z_4, z_5, z_6 \geq \mathbf{0}.
\end{aligned}
\right\}
\tag{13}
$$

In problem (13), because (x,y,u,v) are parameters given by an individual s in the artificial genetic system, problem (13) is a linear programming problem. If the objective function value w_0 becomes zero, the individual s satisfies the necessary conditions (9) and (11).

In the second stage, we examine sufficiency of the individual satisfying the necessary conditions (9) and (11). To do so, for the individual $s^* = (x^*, y^*, u^*, v^*)$ such that the objective function value w_0 is zero, we solve problem (7) with the fixed value of $u = u^*$ and problem (10) with the fixed value of $x = x^*$. If the obtained optimal solutions are respectively equal to (x^*, y^*) and (u^*, v^*), the individual s^* corresponds to a Nash-Stackelberg equilibrium solution.

To solve problems (7) and (10), we also employ the scheme of the genetic algorithms and compute values w_1 and w_2:

$$w_1 = \max_{x,y}(x^T A y + x^T E u^*) - (x^{*T} A y^* + x^{*T} E u^*) \tag{14}$$

$$w_2 = \max_{u,v}(u^T C v + x^{*T} F u) - (u^{*T} C v^* + x^{*T} F u^*). \tag{15}$$

Using the values w_0, w_1 and w_2, we define the following fitness function $f(s)$ because the smaller the values w_0, w_1 and w_2 become, the better the individual s grow.

$$f(s) = \begin{cases} 1/(w_1 + w_2 + 1), & w = 0 \\ 1/(100(w_0 + 1)), & w \neq 0. \end{cases} \tag{16}$$

To keep appropriate levels of competition in the artificial genetic system, the fitness value should be scaled and then we employ the linear scaling. Furthermore, because there may exist multiple equilibrium solutions, to search widely, we use the sharing function which gives smaller weights for many individuals being in the same neighborhood and larger weights for isolating individuals. As a reproduction operator, we adopt the elitist-roulette selection which is a combination of the elitism and the roulette selection.

For crossover, we employ the single arithmetical crossover and the whole arithmetical crossover. For mutation, we employ the uniform mutation and the boundary mutation. Each type of crossover and each type of mutation is selected at the same probability.

4　Conclusions

In this paper, we have developed a computational method for obtaining the Nash-Stackelberg equilibrium solution of the two-level four-person bimatrix game. The computational method is based on the scheme of the genetic algorithms. Our artificial genetic system, individuals are evaluated through two stages. We verify whether or not individuals satisfy the necessary conditions to be Nash-Stackelberg equilibrium solutions at the first stage, and the sufficient conditions are examined at the second stage. We have carried out computational experiments and verified the feasibility and efficiency of the proposed methods.

References

[1] O.L. Mangasarian and H. Stone, Two-person nonzero-sum games and quadratic programming, *Journal of Mathematical Analysis and Applications*, **9**, 348–355, 1964.

[2] Z. Michalewicz, *Genetic Algorithms + Data Structures = Evolution Programs,* Third, revised and extended edition, Springer-Verlag, Berlin, 1996.

[3] J. F. Nash, Noncooperative Games, *Annals of Mathematics*, **54**, 286–295, 1951.

[4] Owen, G., *Game Theory*, third edition, Academic Press, San Diego, 1995.

[5] T. Parthasarathy and T. E. S. Raghavan, *Some Topics in Two-Person Games,* American Elsevier Publishing Company, New York, 1971.

[6] K. Shimizu, Y. Ishizuka and J.F. Bard, *Nondifferentiable and Two-Level Mathematical Programming,* Kluwer Academic Publishers, Boston, 1997.

[7] M. Simaan and J. B. Cruz Jr., On the Stackelberg strategy in nonzero-sum games, *Journal of Optimization Theory and Applications,* **11**, 533–555, 1973.

KES 2002
E. Damiani et al. (Eds.)
IOS Press, 2002

Computational methods for two-level integer programming problems with fuzzy parameters through genetic algorithms

Keiichi Niwa[†], Ichiro Nishizaki[‡] and Masatoshi Sakawa[‡]
[†] *Faculty of Economics, Hiroshima University of Economics*
5-37-1, Gion, Asaminami-ku, Hiroshima, 731-0192, Japan
[‡] *Graduate School of Engineering, Hiroshima University*
1-4-1, Kagamiyama, Higashi-Hiroshima, 739-8527, Japan

Abstract. From the observation that possible values of parameters involved in objective functions and constraints of mathematical programming problems are often only imprecisely or ambiguously known to experts, in this paper, we consider two-level integer programming problems with fuzzy parameters represented by fuzzy numbers. A computational method based on genetic algorithms for obtaining the Stackelberg solution to a two-level integer programming problem with fuzzy parameters is developed. Computational experiments are carried out in order to demonstrate the efficiency of the proposed computational method.

1   Introduction

In this paper, we consider two-level programming problems in which there are two decision makers; the leader at the upper level and the follower at the lower level. We consider a situation that decision variables of the leader and the follower are integer variables, and between the leader and the follower, there is no motivation to cooperate with each other. For such two-level integer programming problems, assuming that the follower determines a decision such that an objective of the follower is optimized with full knowledge of a decision of the leader, the leader specifies the decision so as to optimize an objective of the leader. In this paper, we refer to this solution as a Stackelberg solution.

From the observation that possible values of parameters involved in objective functions and constraints of mathematical programming problems are often only imprecisely or ambiguously known to experts, in this paper we formulate two-level integer programming problems involving fuzzy parameters and present a computational method for obtaining Stackelberg solutions. These fuzzy parameters, reflecting the experts' vague or fuzzy understanding of the nature of parameters in the problem-formulation process, are assumed to be characterized as fuzzy numbers [1].

Suppose that the leader chooses a decision and then if possible, it would be desirable for the follower to fix fuzzy parameters at some value so as to minimize the objective function of the follower under the constraints.

From such a point of view, in this paper we assume that the leader makes a decision optimizing the objective function of the leader, supposing that the follower responds to the precedent decision of the leader such that the objective function of the follower is optimized by controlling not only the integer decision variables of the follower but also parameters involved in the two-level integer programming problems.

In the 1970's, genetic algorithms (GAs) were proposed by Holland [2], as a new learning paradigm that models a natural evolution mechanism. GAs have recently attracted considerable attention in a number of fields as a methodology for optimization, adaptation and learning [3, 5]. From such a viewpoint, we present the computational method using GAs to obtain Stackelberg solutions to the above mentioned problems. We perform computational experiments to demonstrate feasibility of the proposed computational method.

2 Two-level integer programming problem with fuzzy parameters

We consider a two-level integer programming problem as follows:

$$\left.\begin{array}{ll} \underset{x}{\text{minimize}} & z_1 = \tilde{c}_1 x + \tilde{d}_1 y \\ \text{subject to} & \underset{y}{\text{minimize}} \quad z_2 = \tilde{c}_2 x + \tilde{d}_2 y \\ & \text{subject to} \quad \tilde{A} x + \tilde{B} y \leq \tilde{b} \\ & \qquad x_i \geq 0, x_i \in Z, \ i = 1, \ldots, n_1 \\ & \qquad y_j \geq 0, y_j \in Z, \ j = 1, \ldots, n_2 \end{array}\right\} \quad (1)$$

where $\tilde{c}_i$, $i = 1, 2$ are n_1-dimensional row coefficient vector of the fuzzy parameters; $\tilde{d}_i$, $i = 1, 2$ are n_2-dimensional row coefficient vector of the fuzzy parameters; $\tilde{A}$ is an $m \times n_1$ coefficient matrix of the fuzzy parameters; $\tilde{B}$ is an $m \times n_2$ coefficient matrix of the fuzzy parameters; $\tilde{b}$ is an m-dimensional column constant vector of the fuzzy parameters; x is an n_1-dimensional column integer decision variable vector of the leader; and y is an n_2-dimensional column integer decision variable vector of the follower; Z is a set of integers. We deal with a real fuzzy number M whose membership function $\mu_M(x)$ is defined[4].

We introduce the α-level set of the fuzzy numbers $\tilde{c}_1$, $\tilde{d}_1$, $\tilde{c}_2$, $\tilde{d}_2$, $\tilde{A}$, $\tilde{B}$ and $\tilde{b}$ defined as the ordinary set $(\tilde{c}_1, \tilde{d}_1, \tilde{c}_2, \tilde{d}_2, \tilde{A}, \tilde{B}, \tilde{b})_\alpha$ in which value of each membership function exceeds the level α.

Now suppose that the leader considers that the membership function value of each of the fuzzy numbers involved in the two-level integer programming problem should be greater than or equal to some degree α.

Then, for such a degree α, the two-level integer programming problem (1) can be interpreted as the following nonfuzzy two-level integer programming problem which depends on a coefficient vector $(c_1, d_1, c_2, d_2, A, B, b) \in (\tilde{c}_1, \tilde{d}_1, \tilde{c}_2, \tilde{d}_2, \tilde{A}, \tilde{B}, \tilde{b})_\alpha$ [4].

$$\left.\begin{array}{ll} \underset{x}{\text{minimize}} & z_1 = c_1 x + d_1 y \\ \text{subject to} & \underset{y}{\text{minimize}} \quad z_2 = c_2 x + d_2 y \\ & \text{subject to} \quad A x + B y \leq b \\ & \qquad x_i \geq 0, x_i \in Z, \ i = 1, \ldots, n_1 \\ & \qquad y_j \geq 0, y_j \in Z, \ j = 1, \ldots, n_2. \end{array}\right\} \quad (2)$$

Observe that there exist an infinite number of such a problem (2) depending on the coefficient vector $(c_1, d_1, c_2, d_2, A, B, b) \in (\tilde{c}_1, \tilde{d}_1, \tilde{c}_2, \tilde{d}_2, \tilde{A}, \tilde{B}, \tilde{b})_\alpha$ and the values of $(c_1, d_1, c_2, d_2, A, B, b)$ are arbitrary for any $(\tilde{c}_1, \tilde{d}_1, \tilde{c}_2, \tilde{d}_2, \tilde{A}, \tilde{B}, \tilde{b})_\alpha$ in the sense that the degree of all of the membership functions for the fuzzy numbers in the problem (2) exceeds the level α. Suppose that the leader chooses a decision $\hat{x}$ and then if possible, it would be desirable for the follower to choose $(c_2, d_2, A, B, b) \in (\tilde{c}_2, \tilde{d}_2, \tilde{A}, \tilde{B}, \tilde{b})_\alpha$ so as to minimize the objective function of the follower under the constraints.

From such a point of view, for a certain degree α, it seems to be quite natural to interpret the follower's problem for obtaining a rational reaction as the following nonfuzzy problem:

$$\begin{aligned}
&\underset{y, c_2, d_2, A, B, b}{\text{minimize}} && c_2\hat{x} + d_2 y \\
&\text{subject to} && A\hat{x} + By \le b \\
& && y_j \ge 0, y_j \in Z, \ j = 1, \ldots, n_2 \\
& && (c_2, d_2, A, B, b) \in (\tilde{c}_2, \tilde{d}_2, \tilde{A}, \tilde{B}, \tilde{b})_\alpha.
\end{aligned} \right\} \quad (3)$$

From the properties of the α-level set for the matrices of fuzzy numbers $\tilde{A}$ and $\tilde{B}$ and the vectors of fuzzy numbers $\tilde{c}_2, \tilde{d}_2, \tilde{b}$, the feasible regions for c_2, b_2, A and B can be represented respectively by the closed intervals $[c_{2\alpha}^L, c_{2\alpha}^R]$, $[d_{2\alpha}^L, d_{2\alpha}^R]$, $[A_\alpha^L, A_\alpha^R]$, $[B_\alpha^L, B_\alpha^R]$ and $[b_\alpha^L, b_\alpha^R]$.

Thus, we can obtain an optimal solution to the problem (3) by solving the following integer programming problem:

$$\begin{aligned}
&\underset{y}{\text{minimize}} && c_{2\alpha}^L \hat{x} + d_{2\alpha}^L y \\
&\text{subject to} && A_\alpha^L \hat{x} + B_\alpha^L y \le b_\alpha^R \\
& && y_j \ge 0, y_j \in Z, \ j = 1, \ldots, n_2.
\end{aligned} \right\} \quad (4)$$

On the assumption the problem (1) can be rewritten as follows:

$$\begin{aligned}
&\underset{x}{\text{minimize}} && \tilde{c}_1 x + \tilde{d}_1 y \\
&\text{subject to} && \underset{y}{\text{minimize}} && c_{2\alpha}^L x + d_{2\alpha}^L y \\
& && \text{subject to} && A_\alpha^L x + B_\alpha^L y \le b_\alpha^R \\
& && && x_i \ge 0, x_i \in Z, \ i = 1, \ldots, n_1 \\
& && && y_j \ge 0, y_j \in Z, \ j = 1, \ldots, n_2.
\end{aligned} \right\} \quad (5)$$

Similarly, if possible, it would be desirable for the leader to choose $(c_1, d_1) \in (\tilde{c}_1, \tilde{d}_1)_\alpha$ so as to minimize the objective function of the leader under the constraints. From such a point of view, it seems to be quite natural to have the two-level integer programming problem for obtaining a Stackelberg solution as the following nonfuzzy problem:

$$\begin{aligned}
&\underset{x, c_1, d_1}{\text{minimize}} && c_1 x + d_1 y \\
&\text{subject to} && (c_1, d_1) \in (\tilde{c}_1, \tilde{d}_1)_\alpha \\
& && \underset{y}{\text{minimize}} && c_{2\alpha}^L x + d_{2\alpha}^L y \\
& && \text{subject to} && A_\alpha^L x + B_\alpha^L y \le b_\alpha^R \\
& && && x_i \ge 0, x_i \in Z, \ i = 1, \ldots, n_1 \\
& && && y_j \ge 0, y_j \in Z, \ j = 1, \ldots, n_2
\end{aligned} \right\} \quad (6)$$

From the properties of the α-level set, the problem (6) is written as follows:

$$\begin{aligned}
&\underset{x}{\text{minimize}} && c_{1\alpha}^L x + d_{1\alpha}^L y \\
&\text{subject to} && \underset{y}{\text{minimize}} && c_{2\alpha}^L x + d_{2\alpha}^L y \\
& && \text{subject to} && A_\alpha^L x + B_\alpha^L y \le b_\alpha^R \\
& && && x_i \ge 0, x_i \in Z, \ i = 1, \ldots, n_1 \\
& && && y_j \ge 0, y_j \in Z, \ j = 1, \ldots, n_2
\end{aligned} \right\} \quad (7)$$

We propose a computational method through GAs in order to solve the problem (7).

3 Computational method through GAs

In order to obtain Stackelberg solutions to the reduced problem (7), we propose a computational method through GAs. In this section, we describe fundamental elements of GAs, which are a coding procedure, a decoding procedure and genetic operators, used in the proposed algorithm.

3.1 Coding and decoding

In the proposed method, integer decision variables x and y of the leader and the follower are expressed as double strings [5], in which each cell of the upper string corresponds to an index of a variable and each cell of the lower string corresponds to a value of a variable in afrificial genetic systems and the proposed method has nested structure. The double strings consist of two parts of strings for a decision x of the leader and for decisions y of the follower. We call the GAs operating to decisions x of the leader the upper level GAs and for the GAs operating to decisions y of the follower the lower level GAs.

We adopt a decoding procedure using reference solutions proposed by Sakawa et al.[6] in the lower level GAs. In this procedure, an initial set of reference solutions is generated by using GAs based on penalty technique. In order to generate only feasible solution from individuals, the reference solutions are used. Moreover, some reference solutions are updated for every generation.

3.2 Genetic operators

We employ the following genetic operators in order to effectively solve the two-level integer programming problems with fuzzy parameters using GAs with double strings.

After feasible solutions are generated by using the decoding procedures, by using values of the objective functions at both levels, fitness of each individual in the upper level and the lower level GAs are computed. As a reproduction operator, we employ a combination of the elitist expected value selection. We use crossover operator which is a revised version of PMX [5] to deal with double strings. For the lower string of double strings, mutation of bit-reverse type is adopted and, for the upper string, another genetic operator, an inversion, is employed.

3.3 Procedure for obtaining Stackelberg solutions

We present a computational method through GAs with the above mentioned structure of a string and decoding procedure as follows:

Procedure for obtaining Stackelberg solutions

Step 1 Generate N_1 initial individuals in the upper level GAs for the decision variable x of the leader at random.

Step 2 For each individual with respect to x, the following procedure is repeated.

Step 2-1 Generate N_2 initial individuals for the decision variable y of the follower by using GAs based on penalty technique in order to obtain feasible solutions.

Step 2-2 Evaluate each individual for the decision variable y with given x by using the decoding procedures.

Step 2-3 Go to Step 3 after the procedure of the lower level GAs is repeated until attaining the termination condition, which is the number of generations specified in advance.

Step 2-4 Apply the genetic operators to each individual in the lower level GAs and return to Step 2-2.

Step 3 Evaluate each individual for the decision variable x with an individual $y(x)$, which is thought as a rational reaction.

Step 4 If the termination condition of the upper level GAs is attained, regard a solution with the best fitness as a Stackelberg solution and the algorithm stops; otherwise go to Step 5.

Step 5 Apply the genetic operators to each individual in the upper level GAs and return to Step 2.

4 Conclusions

We have considered two-level integer programming problems with fuzzy parameters, and have developed a computational method through genetic algorithms with double strings so as to obtain Stackelberg solutions to the problems. To demonstrate the feasibility of the proposed method, computational experiments have been carried out.

References

[1] D. Dubois and H. Prade, "Operations on fuzzy numbers," *International J. of Systems Science,* vol. 9, pp. 613–626 (1978).

[2] J.H. Holland, *Adaptation in Natural and Artificial Systems,* MIT Press, Cambridge (1992).

[3] Z. Michalewicz, *Genetic Algorithms + Data Structures = Evolution Programs,* Third, revised and extended edition, Springer-Verlag, Berlin (1996).

[4] M. Sakawa, *Fuzzy Sets and Interactive Multiobjective Optimization,* Plenum Press, New York (1993).

[5] M. Sakawa, K. Kato, H. Sunada and T. Shibano, "Fuzzy programming for multiobjective 0-1 programming problems through revised genetic algorithms," *European Journal of Operational Research,* vol. 8, pp. 149–158 (1997).

[6] M. Sakawa,K. Kato, "Integer Programmings through Genetic Algorithms with Double Strings Based on Reference Solution Updating," IECON-2000, pp. 2744-2749 (2000).

KES 2002
E. Damiani et al. (Eds.)
IOS Press, 2002

Clustering using Small World Structure

Yutaka Matsuo   National Institute of Advanced Industrial Science and Technology,
Aomi 2-41-6, Koto-ku, Tokyo, 113-0064 JAPAN

Abstract. Small world topology has been receiving much attention recent years. Many real-world graphs actually have a small world topology. In a small world graph, a shortcut plays an important role to make the world small, which means if shortcuts are eliminated, the graph is separated into clusters. In this paper, I propose a new clustering algorithm by eliminating "shortcuts" and show promising results.

1 Introduction

In the 1960s, Stanley Milgram showed that any two randomly chosen individuals in the United States are linked by a chain of six or fewer first-name acquaintances, known as "six degrees of separation." Watts and Strogatz defined what is small world, and showed some networks have small world characteristics [11]. Since their introduction, small-world networks and thir properties have received considerable attention. Numbers of networks are shown to have a small-world topology. Examples include social networks such as acquaintance networks and collaboration networks, technological networks such as the Internet, the World-Wide Web, and power grids, and biological networks such as neural networks, foodwebs, and metabolic networks. (For reference, see [4].) Matsuo et. al. showed that word co-occurrence in a technical paper also consists a small world graph [8].

In a "small world" graph, nodes are highly clustered yet the path length between them is small. Although some recent works have proposed different definitions of "small world" (for example, [6]), one by Watts and Strogats is appropriate to grab idea of node distance and clusters. They define the following two invariants [11]:

- The *characteristic path length*, L, is the path length averaged over all pairs of nodes. The path length $d(i, j)$ is the number of edges in the shortest path between nodes i and j.

- The *clustering coefficient* is a measure of the cliqueness of the local neighborhoods. For a node with k neighbors, then at most $_kC_2 = k(k-1)/2$ edges can exist between them. The clustering of a node is the fraction of these allowable edges that occur. The clustering coefficient, C is the average clustering over all the nodes in the graph.

Watts and Strogatz define a small world graph as one in which $L \geq L_{rand}$ (or $L \approx L_{rand}$) and $C \gg C_{rand}$ where L_{rand} and C_{rand} are the characteristic path length and clustering coefficient of a random graph with the same number of nodes and edges.

In this paper, I propose a new method for detecting clusters based on the small world structure. In a small world network, shortcuts (or weak ties) play an important role to connect clusters (or communities). I eliminate a given number of edges from the graph to make C and L large, which means nodes are clustered and clusters are separated.

A cluster corresponds to a community in social networks, a Web page community in WWW, and a topic in a technical paper. Finding these clusters is an essential task in data exploration, and useful for understanding the overview of data.

2 Related Works

Clustering is an important data exploration task in chance discovery[10] as well as in data mining. The first hierarchical clustering dates back to 1951 by K. Florek, and since then there have been numerous modifications. One of the most widely used clustering methods is the single linkage clustering. It is a kind of hierarchical clustering, and its cluster relationships can be represented by a rooted tree, called dendrogram. A cluster is produced by cutting edges of the dendrogram with a threshold. However, application of only one threshold for all clusters would produce many too small clusters and a few large clusters.

To tackle this problem, a clustering method based on a linkage graph is proposed in [5] to cluster protein sequences into families. They formulate clustering as a kind of graph partitioning problem [2] of a weighted linkage graph and find *minimal cut* with consideration of balancing the size of clusters. The graph partitioning problem is of interest in areas such as VLSI placement and routing, and efficient parallel implementations of finite element methods, e.g., to balance the computational load and reduce communication time. [3] develops an algorithm to find communities on the Web by maximum flow / minimum cut framework. This algorithm performs well in practice i.e., under the condition that one couldn't have rapid access to the entire Web.

Another approach focuses on the *betweenness* of an edge in a linkage graph as the number of shortest paths between pairs of nodes that run along it [4]. An edge in a graph is iteratively removed if the betweenness of the edge is highest in order to find communities in social and biological networks.

3 Small World Clustering

I formalize my clustering algorithm, called *Small World Clustering*, as an optimization problem. It differs to the conventional graph partition problem [2] in that I give the number of edges to prune, instead of the number of clusters to be partitioned into. I intend to use this clustering to grab the overview of data, rather than to assign tasks into a given number of computers, thus to give the number of clusters is burdensome to a user.[1]

Definition 3.1 (Small World Clustering)
Given a graph $G = (V, E)$ and k where V is a set of nodes, E is a set of edges, and k is a positive integer, *Small World Clustering (SMC)* is defined as finding a graph G' such that k edges are removed from G so that

$$f = aL_{G'} + bC_{G'}$$

is to be maximized. ($L_{G'}$ and $C_{G'}$ are L and C for graph G' respectively, and a and b are constants.)

To deal with a disconnected graph, I extend the definition of L as follows.

---

[1] I can modify my algorithm so that it can handle the number of clusters.

Definition 3.2 (Extended path length)
An *extended* path length $d'(i, j)$ of node i and j is defined as follows.

$$d'(i, j) = \begin{cases} d(i, j), & \text{if } (i, j) \text{ are connected,} \\ n, & \text{otherwise.} \end{cases} \tag{1}$$

where n is a number of nodes in G.

The problem of finding an optimal connection among all possible pairs of nodes of a graph has been proven to be NP-complete. Therefore I consider an approximate algorithm for SWC as follows.

1. Prune an edge which maximize f iteratively until k edges are pruned.

2. Add an edge which maximize f. If an edge to be added is the same as the most previously pruned one, terminate.

3. Prune an edge which maximize f. Go to 2.

The second and third proceedures are optional. If clustering needs to be finished rapidly, these proceedures can be skipped. However in some cases, they provide a little better solution.

4 Examples

I show an example of SWC applied to a word co-occurrence graph. A word co-occurrence graph is constructed as follows [8];

1. pick up n frequent words as nodes,

2. calculate a Jaccard coefficient[2] for each pair of words, and add an edge if the coefficient is larger than a given threshold.

A word co-occurrence graph is shown to have small world characteristics [8]. Thus, the characteristics is utilized for clustering by SWC.

Fig. 1 is a word co-occurrence graph derived from a technical paper [9] with the single linkage clustering. I can see a big cluster, one little cluster, and six isolated nodes. Resulting in a big cluster and many isolated nodes is very common when single linkage clustering is applied. It is very difficult to grab the 'meaning' behind each cluster.

Fig. 2 shows a graph derived from the same paper but by SWC[3] instead of the single linkage clustering (with the same number of nodes and links). Four big clusters, three pairwise nodes and one single node are extracted. Because the nodes in a cluster is well connected (as C should be high) and clusters are properly separated (as L should be high), it is easier to grab the meaning of cluster; for example, the left upper cluster is words related to "extract author's basic concept," the center upper cluster is about "existing retrieval method", the left lower cluster is "the key concept of this paper", and the right big cluster is about "the proceedure of the new algorithm."

In the result graph, a lot of complete subgraphs (or cliques) and stars emerge. These are two frequent types of subgraphs when small worlds are generated artificially [7].

---

[2]The Jaccard coefficient is the number of sentences that contain both words divided by the number of sentences that contain either words.

[3]Constant a is set 1, and b is set 100.

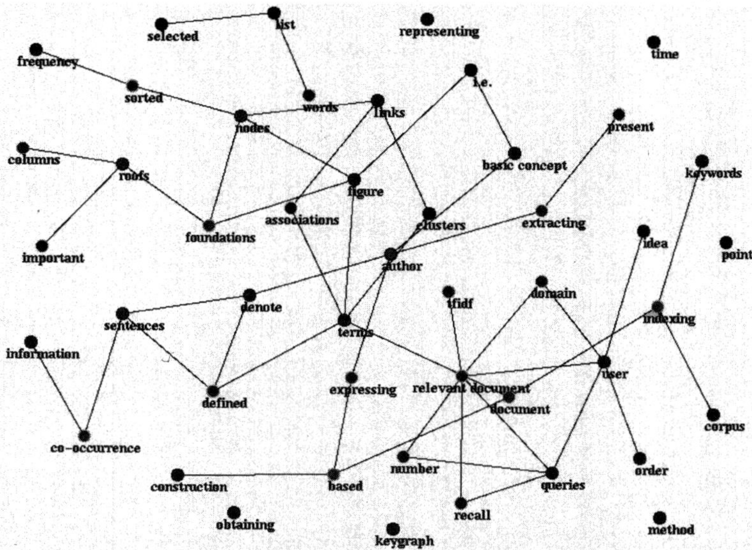

Figure 1: A word co-occurrence graph with a single linkage clustering. $C = 0.201$, $L = 12.1$.

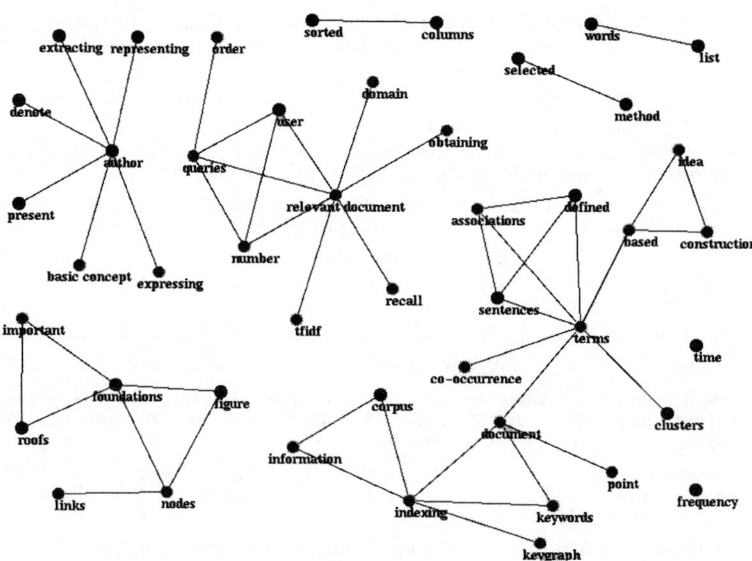

Figure 2: Clusters obtained by SWC. $C = 0.689$, $L = 18.3$

5   Discussion and Future Work

There is no consensus among the reserchers as to what constitutes a cluster. There is only some intuitive understanting: the intuitive idea behind clustering consists in condensing a subgraph into a single node, where the choice of the cluster is application-dependent [1].

In data exploration, a user will find meaning of a cluster from data. From chance discovery point of view, if a user can understand the meaning easily, the clustering algorithm is preferable. In the area of exploratory data analysis (EDA), data visualization as well as data clustering is an essential task. It is also very important to make a user understand the meaning of clusters by visualization.

In this paper, I show an example of clustering using a word co-occurrence graph. Quantitative evaluation is an ongoing research. However, there is a technical problem ahead; a problem of node duplication. For example, assume there are two well-connected subgraphs A and B. If A and B share a node, say v, then two subgraphs are connected by node v. In this case, I want to duplicate node v rather than eliminate node v or remove links from v. One of the interesting aspect of small world lies in this point; one belongs to multiple communities, thus bridges communities. One is a member of community, and at the same time, is a bridge to other community. (This is true for a word co-occurrence graph; A word sometimes has more than one meaning.) Therefore to extract communities precisely may require node duplication. When and how to duplicate nodes is one of the future works.

References

[1] M. Ancona, W. Cazzola, E. Martinuzzi, P. Raffo, and I.B. Vasian. Clustering algorithms for the optimization of communication graphs. In *Proc. 4th Conf. Italo-Latino American of Industrial and Applied Mathematics*, 2001.

[2] Per-Olof Fjällström. Algorithms for graph partitioning: A survey. *Computer and Information Science*, 3, 1998.

[3] Gary William Flake, Steve Lawrence, and C. Lee Giles. Efficient identification of Web communities. In *Proc. ACM SIGKDD-2000*, pages 150–160, 2000.

[4] Michelle Girvan and M. E. J. Newman. Community structure in social and biological networks. *submitted to the Proceedings of National Academy of Sciences*.

[5] Hideya Kawaji, Yosuke Yamaguchi, Hideo Matsuda, and Akihiro Hashimoto. A graph-based clustering method for a large set of sequences using a graph partitioning algorithm. *Genome Informatics*, 12:93–102, 2001.

[6] M. Marchiori and V. Latora. Harmony in the small-world. *Physica A*, 285:539–546, 2000.

[7] N. Mathias and V. Gopal. Small worlds: How and why. *Physical Review E*, 63(2), 2001.

[8] Yutaka Matsuo, Yukio Ohsawa, and Mitsuru Ishizuka. KeyWorld: Extracting keywords from a document as a small world. In *Proceedings the Fourth International Conference on Discovery Science (DS-2001)*, 2001.

[9] Y. Ohsawa, N. E. Benson, and M. Yachida. KeyGraph: Automatic indexing by co-occurrence graph based on building construction metaphor. In *Proc. Advanced Digital Library Conference (IEEE ADL'98)*, 1998.

[10] Yukio Ohsawa. Chance discoveries for making decisions in complex real world. *New Generation Computing*, to appear.

[11] D. Watts and S. Strogatz. Collective dynamics of small-world networks. *Nature*, 393:440–442, 1998.

KES 2002
E. Damiani et al. (Eds.)
IOS Press, 2002

Extracting Characteristic Sentences from Related Documents

Naoaki Okazaki *[†]   Yutaka Matsuo [‡]   Naohiro Matsumura *[†]   Hironori Tomobe *
Mitsuru Ishizuka *
Hongo 7-3-1, Bunkyo-ku, Tokyo, 113-8656 Japan.

Abstract.
More and more information is available recently. To find a chance i.e., an important event for decision-making, we have to be prepared for the chance. Recent progress of automatic summarization may contribute to Chance Discovery in that it helps a user read a lot of documents easily and be prepared for the chance. In this paper, we develop a new method for multi-document summarization which extracts a set of characteristic sentences that maximizes the coverage of an original content and minimizes the redundancy of a summary. On top of the summary result, we provide a word cooccurrence graph and show why the result is obtained.

1 Introduction

Text summarization is the process of distilling the most important information from a source (or sources) to produce an abridged version for a particular user and task [1]. Although the measure of 'importance' in this definition varies from one user or task to another, most of the current researches hardly pay attention to that.

From chance discovery point of view, a summary should be useful for a particular user to make decisions, or at least to understand the content for making decisions. Importance of a sentence or a set of sentences is dependent on who makes the decision (ie. a user) and what kind of decision one makes (ie. a task). We are aiming at user-task-dependent summarization, however, in this paper we show a general-purpose summarization system. The important feature of our system is that we provide a word cooccurrence graph of the original documents on top of the summary result and show a user why the summary is obtained. The explanation of the summary result may help a user convince the summary result, and attract new interests. It makes a great benefit in a user's decision-making (even if the algorithm is currently general-purpose).

To enable the visual presentation of a summary result, our method is based on a graph representation. We transform the summarization task into an edge covering problem on the graph. Because the edge covering problem is NP-complete, we use a fast hypothetical reasoning solver to obtain the result immediately.

---

*Graduate School of Information Science and Technology, University of Tokyo
[†]Japan Science and Technology Corporation
[‡]Cyber Assist Research Center, National Institute of Advanced Industrial Science and Technology

The rest of the paper is organized as follows: In the following section, we overview the summarization. Then we show how to compile the summarization into a hypothetical reasoning problem. In Section 4, the examples are shown. We discuss the future works and conclude this paper.

2 Summarization overview

Considering the human's process of summarization, we (1)understand the content of a passage, (2)pick up sentences or phrases that are considered as significant, and (3)synthesize an outputting summary together with splicing the extracted textual segments. Since (2) is relatively easier for computer to deal with than (1) and (3), extracting significant units has researched since 1950s [2, 3] and been the basis of automatic summarization.

Extracting significant units is a method of evaluating textual units in the source document(s). Most of the current research is not necessary for catching on what the text actually said to estimate the significance, but takes advantage of the surface phenomenon behind a text.

The extension of single-document summarization to collections of related documents is called multi-document summarization. We frequently meet related documents, for example, a collection of documents retrieved from a search engine with some queries, messages on an internet discussion board or mailing list, etc. As the related documents have some similar contents or expressions, extracting the significant textual units often results in a redundant summary. Multi-document summarization should therefore be capable of identifying common and different parts and removing redundancy in a summary.

3 Extracting the best combination of sentences

3.1 Formulation of extracting characteristic sentences

Based on the above discussion, we created a system extracting a set of sentences from related multi-documents. It uses a word cooccurrence graph to extract the best combination of sentences. We formulate this multi-summarization problem as follows.

First, we make a word cooccurrence graph from documents. Figure 1 shows a word cooccurrence relation between terms in a set of articles about "hybrid car." A node stands for a term, and we link nodes when a pair of terms is appeared in the same sentences more than twice.

What kind of sentences is characteristic in the graph? Each sentence in a document presents the relations between terms [4]. That is equivalent in the graph to covering several links. As a consequence, we should choose a set of sentences that covers as many links as possible in the graph. It is useless to pick up a sentence which covers the same links as the previously selected sentence does. Therefore we obtain an edge covering problem defined as the following optimization problem;

$$\min f = \sum_{i \in K} cost_i x_i \qquad (1)$$

where K is a set of links, $cost_i$ is a penalty cost when link i is not included in the summary, and x_i is a 0–1 boolean variable whether link i is included(1) or not(0).

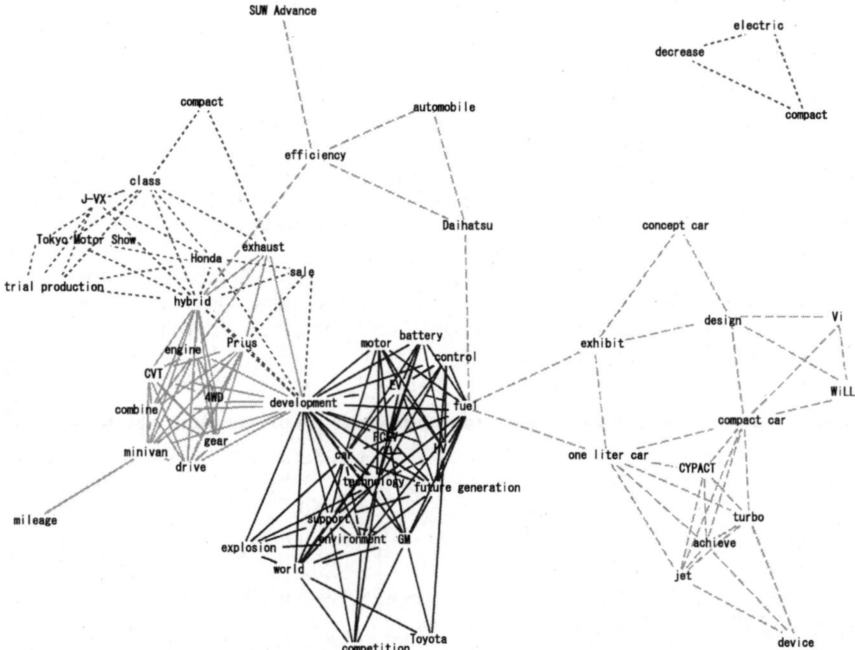

Figure 1: A word cooccurrence graph of a set of news articles. The source articles are a set of news articles about hybrid car written by the Mainichi newspaper(originally written in Japanese). The distance between nodes(terms) is about in inverse proportion to the times of cooccurrence. A line style corresponds to an article.

3.2　Transformation of the optimization problem into cost-based hypothetical reasoning

We solve the optimization problem by applying cost-based hypothetical reasoning as follows. We denote k as the total number of links and m as the total number of sentences, We define goal G to represent *"All links are taken into consideration"* as follows.

$$G \leftarrow x_1, x_2, ..., x_k \tag{2}$$

A hypothesis h_{s_j} represents *"sentence j is selected"* and has the corresponding cost with its length. For example, if sentence 1 has link#13, link#220, link#223, then we get the following rules.

$$x_{13} \leftarrow h_{s_1}, x_{220} \leftarrow h_{s_1}, x_{223} \leftarrow h_{s_1} \tag{3}$$

For unselected link i, on the other hand, we introduce a hypothesis h_{emp_i} to represent *"sentence i is not included in the summary"* and the following rules.

$$x_i \leftarrow h_{emp_i} (i = 1, ..., k) \tag{4}$$

We annotate h_{emp_i} with penalty cost 1.

At last we can decide a set of sentences by generating knowledge base and finding a combination of sentence that proves goal G. We use a fast hypothetical reasoning method [5]

Toyota was the first to develop the hybrid car and started selling the Prius in December of last year. October of last year at the Tokyo Motor Show, Honda presented the J-VX, a 1000cc class experimental hybrid engine car.

On the 19th, Toyota and General Motors (GM) announced at the same time their plans to co-operate in automobile technology advancement for environmentally friendliness in automobiles with the prospect of making next generation low-pollution vehicles such as the awaited Fuel Cell Electric Vehicle (FCEV). They will collaborate in the research of chemical reactions of hydrogen from fuel and oxygen from air to generate power, plus a wide variety of technologies which should bring about the production of more hybrid vehicles (HV) like the FCEV which join a gasoline engine and an electric motor, and electric vehicles (EV) whose motors are run on a storage battery.

Up to now, the only one to combine gasoline and electricity has been the Prius' 1500cc engine, but for a minivan that needs more power, they revealed to newspapers the plans for the first hybrid with four-wheel drive, a Continuously Variable Transmission (CVT) vehicle which combines a 2400cc engine and motor.

Nissan has also achieved CYPACT, a three-liter compact car with a jet-fueled diesel turbo engine

Figure 2: An example of summary. The source is a collection of 4 articles about "hybrid car" in the Mainichi. (Translated from Japanese for purposes of illustration)

which solves a hypothetical reasoning problem quickly by transforming the problem into two continuous optimization problems.

3.3   Implementation

First, we analyze the source text into morpheme and identify part of speech of each term by using Chasen [6]. Sorting nouns and verbs from terms, we enumerate cooccurrence relations between the terms in the same sentence. And then we make and solve a summarization problem described above.

We participate in a competition of summarization, TSC(Text Summarization Challenge) [7] task organized by NTCIR-3 project and we used a collection of the Mainichi articles for an experiment. Current system is only for Japanese news paper articles because of the morphlogicalanalysis, but the core algorithm can be easily applied to other languages.

4   Discussion and Conclusion

Figure 2 and 3 are two examples of summaries. The source articles are omitted due to limitations of space. Figure 2 is produced from Figure 1, a word cooccurrence graph in a collection of articles about "hybrid car." As can be seen from the summary, our system depicts various efforts of makers toward bybrid cars. Several brand names in Figure 1 indicate hybrid cars, ie. *Prius*, *J-VX*, and *SUW Advance*, which are located near 'hybrid' node. *WiLL* and *CYPACT* are located, on the other hand, far from 'hybrid' node. They are not actually hybrid cars.

Figure 3 is a summary of the article collection about earning gold medals of Japanese athletes. In addition to prompt reports, we must not miss that it includes some anecdotes about the win. It is not novel in the summary that Japanese athletes won the games because the ar-

> On the 10th, day four of the eighteenth winter olympics in Nagano, male speed skater Shimizu won the first gold medal for Japan in the men's 500 meter competition held at Nagano city's M-Wave. From this competition on, scores are determined by the combined event times of two races, skating on the outside and inside courses.
> On the 11th, women's freestyle skiing and mogul competitions were held. Satoya, on a foreign expedition, was shaken at Masaaki's illness when she was called just before. She said with strong conviction, "I skiied for him as well as for myself."
> On the 15, day nine at the large hill (120m) ski jump in Hakuba village, Kazuyoshi Funaki (of Descente) won the gold, and Masahiko Harada (of Snow Brand) took third place.
> On the 15th at the individuals large jump competition, Harada, the 25th jumper in his second jump, jumped 136 meters, but his result as combined with his first jump were not immediately displayed on the electrical scoreboard, but they were finally shown about ten minutes after finishing, after the calculation and confusion over the reporting of the winners.
> The Japanese ski jump team won their first olympic competition, and Japanese athletic teams have secured one-hundred gold medals in all in the summer and winter olympics.

Figure 3: Another example of summary. The source is a collection of 7 articles retrieved with queries, "Nagano Olympic, Japan, gold, win" in the Mainichi. (Translated from Japanese for purposes of illustration)

ticles were collected intentionally with queries, *"Nagano Olympic, Japan, gold, win."* These queries often appear in the same sentence and have close cooccurrence ralations. Because our summarization strategy tends not to bring such same cooccurrence relations into a summary, it chose instead some secret stories of which some users might not know.

In conclusion, we have developed a new summarization algorithm which solves the transformed edge covering problem. By showing a word cooccurrence graph, we can make the summary more understandable for a user. We are going to extend our method to deal with user- and context-dependent summarization system.

References

[1] Mani, I. *Automatic Summarization.* John Benjamins Publishing Company, 2001.

[2] Luhn, H. P. The automatic creation of literature abstracts. *IBM journal of Research and Development*, Vol. 2, No. 2, pp. 159–165, 1958.

[3] Salton, G. *Automatic Text Processing.* Addison-Wesley, 1989.

[4] Halliday, M.A.K, Hansa, R, *Cohesion in English*, Langman, 1976.

[5] Matsuo, Y., Ishizuka, M. Two Transformation of Clauses into Constraints and their Properties for Cost-based Hypothetical Reasoning, PRICAI-02, to appear.

[6] Chasen's Homepage: `http://www.chasen.org/`

[7] TSC2's Homepage: `http://lr-www.pi.titech.ac.jp/tsc/index-en.html`

1262

KES 2002
E. Damiani et al. (Eds.)
IOS Press, 2002

Utilizing Fault Cases for Supporting Fault Diagnosis Tasks

Yoshikiyo Kato[†], Takahiro Shirakawa[‡] and Koichi Hori[†]

[†]*Department of Advanced Interdisciplinary Studies, University of Tokyo*
4-6-1 Komaba, Meguro-ku, Tokyo 153-8904, Japan
[‡]*Department of Aeronautics and Astronautics, University of Tokyo*
7-3-1 Hongo, Bunkyo-ku, Tokyo 113-8656, Japan
{yoshi,shirakawa,hori}@ai.rcast.u-tokyo.ac.jp

Abstract. When building large and complex systems, like satellites, all sorts of risks have to be managed if they were to be successful. Although there have been various techniques for fault diagnosis, applying them requires both deep domain knowledge and extensive efforts of domain experts. In this paper, we present an approach to support fault diagnosis at low cost. The approach utilizes fault cases experienced during testing of the system. We show the effectiveness of the approach by applying it to a case taken from a satellite development project.

1 Introduction

During the development process of a large and complex system, all sorts of issues or concerns have to be managed to achieve a successful result. Especially for aerospace systems like launch vehicles and spacecrafts, because of its limited opportunity for operation and enormous costs involved, failure of the entire mission resulting from a small oversight is not affordable. Therefore, it is important to manage risks during the development process.

To avoid risks associated with the design and implementation of artifacts, many reliability analysis methods, such as FTA or FMEA, have been employed in the development process. One of the problems with these methods is that the effectiveness of any method relies on experts' knowledge and their extensive efforts on the analysis. As a system to be developed becomes larger in its scale and the complexity grows, it becomes much harder for experts to cover the entire system. To overcome this problem, many computerized methods have been proposed [4, 7].

Whenever one tries to employ a computerized method, knowledge have to be represented in some kind of formalism, and there is always the problem of cost associated with formalizing knowledge. It has been reported that there is a tremendous cost associated with capturing design rationales formally [1, 5, 6]. Similar problem is recognized in the domain of knowledge management as *capture bottleneck* [3]. Capture bottleneck problem in knowledge management domain refers to the lack of enough incentives to the users to invest time for sharing their knowledge or resources. The problem of formalizing knowledge for computerized reliability analysis methods is similar to the capture bottleneck problem in its nature.

Motivated by the capture bottleneck problem, we propose the concept of *knowledge recycling*. During development process of a large and complex system, huge amount of information is produced, whether it is formal or informal. Buried in pile of such information are *knowledge fragments*. A knowledge fragment is a piece of information which reflects deliberation, reasoning, or experience of developers. An argument behind a design decision, appearing in e-mail communications between developers, or a test report describing an anomaly observed during a test and its cause are examples of knowledge fragments. With additional cost on developers, we can collect knowledge fragments, put them together, and provide useful information which supports critical decision-making in the development or operation of the system. Besides, accumulated knowledge fragments may serve not only a single project, but also projects in the future.

In this paper, we describe a method to utilize fault cases experienced during development for supporting fault diagnosis tasks in operation. We explain how the method is applied to support fault diagnosis tasks in satellite operation through a case of applying the method. Finally, we give discussion and conclusions.

2 Fault Case as Knowledge Fragment

In the course of development of a large and complex system, many tests are conducted on the system and its subsystems to verify that they are working as expected. The system does not always behave as expected. In such a case, the reason why it does not work is investigated, and the problem gets corrected. Information collected on such *fault cases* are precious resources for understanding the actual behavior of the system, and can play a key role in diagnosis task. To utilize fault cases experienced in testing of the system, we use use ontology to capture and index fault cases experienced during the development and use that information for supporting fault diagnosis during the operation of the system. Then, based on captured fault cases, we compute *candidate failure paths* in the functional structure model, which represents functional dependencies among components of the target system. In the following sections, we explain how we represent fault case information, the ontologies we use, and the functional structure model, which is the basis for diagnosis support.

2.1 Representation of Fault Case Information

The schema of fault case information is shown in Table 1. Fault case information includes the component at which fault has occurred, the cause of the fault, functions affected by the fault, and description of the fault. Although it seems expensive to describe fault cases in this form, as test reports are produced anyway in any decent engineering process, we assume the additional cost is not significant.

2.2 Function Ontology and Component Ontology

Ontologies are hierarchically organized taxonomies of a domain and provide meaning of terms and relationship between them. We use two kinds of ontology: component ontology and function ontology. Component ontology defines components that form the target system. Function ontology defines functions that components requires and provides. Functions

Item	Description
Label	The label of the fault case.
Affected Component	The component which failed in the fault case.
Affected Function	The function of the affected component which was impaired in the fault case.
Cause Function	The function of the cause component on which the affected component depended to be operational.
Details	The description of details of the fault case.

Table 1: The schema of fault case information. Components and functions to be specified in the schema are defined in component ontology and function ontology, respectively.

required and provided by a component are defined as slots of the component class in the component ontology.

We primarily use these ontologies as the constructs for the functional structure model described in the next section. Components in the functional structure model are instances of component classes of the component ontology. Besides using it in building functional structure model, ontologies are used to associate fault case information with design documents or design rationale. In this way, besides fault cases, users can refer to the richer information related to the fault case in consideration, such as decisions or arguments about the component's design.

2.3 *Functional Structure Model*

Functional structure model describes the connection between components of the system, and dependencies of functionalities among components (Figure 1). Components used in the model are defined in component ontology. Likewise, functions that the components require and provide are defined in function ontology. For each component, functions required for the component to be operational, and functions it provides once it is operational, are defined as slots of a component class.

We intend to use the functional structure model as a basis for inferencing possible failure paths from fault cases experienced during the development and use that information for supporting fault diagnosis during the operation of the system. From a functional structure model, range of effects of a failure of a component can be computed from the function dependency information. Presenting fault cases along with the components identified to be affected by the failure will help the process of fault diagnosis.

3   Supporting Fault Diagnosis of Satellites

Fault diagnosis task of satellites is unique in a sense that, when some kind of anomaly is observed in the behavior of a satellite, it cannot be stopped and directly inspected. Consequently, operators on the ground have to guess what is happening out in the space. As a matter of course, such guesses are based on information available at hand: i.e. house-keeping data which is sent from the satellite to the ground station as telemetry data, and consists of parameters which indicate the status of the satellite.

In our approach, we support fault diagnosis task by presenting to the operator of the satellite candidate *failure paths* of the observed anomaly. An anomaly in system behavior indicates a risk of system failure. The proposed method does not support discovering risks. Rather, it provides possible explanations to the perceived risk, and the explanations are generated on

Label	Temperature dependency of secondary battery	Inability of DC/DC converter to boost up voltage	Inability of receiver to receive commands
Affected Component	Secondary battery	DC/DC converter	FM receiver
Affected Function	Power supply	Power supply	Command reception
Cause Function	Temperature dependency	Power supply	Power supply
Details	When the temperature of a battery becomes low, the voltage provided by the battery becomes low.	When the input voltage becomes low, DC/DC cannot provide enough boost up to the output voltage.	There were some cases where commands could not be received when the input voltage to FM receiver becomes low.

Table 2: Examples of fault case information.

the fly based on fault cases experienced so far. The more fault cases are accumulated, the more variation for explanations there are.

We built a tool for modeling functional structure model (Figure 2). In the tool, the operator can specify a function, which is impaired, from function ontology, and the tool indicates those components in the model which are related to the specified function. Then, s/he specifies one of the indicated components. From the specified component in the model, the tool computes every possible path, traversing the functional dependency links between components. Stored fault case information is related to the components or functional dependency links in the model via ontology. When presenting failure paths, the tool ranks them so that paths with more fault cases presented first. Given the list of candidate failure paths, the operator can systematically investigate the cause of the failure, referencing fault case information along the paths.

3.1 Initial Results

We have applied the system to a case of failure experienced in an experiment conducted as a part of development process of a satellite. In the experiment, a model of the satellite was put on a balloon and communication between ground station and the satellite was tested. In the middle of the experiment, the satellite seemed to stop accepting commands from the ground station. From documents produced before the experiment, we extracted about 20 fault cases that relate to components of communication subsystem. We show some of the extracted fault cases in Table 2. Even with such a small number of fault cases, we could effectively show some candidates of the failure cause, which were pointed out in the analysis by developers after the incident. The result indicates that the method can be useful in supporting systematic pursuit of possible cause of a problem during operation of a system. However, the method was applied to a relatively small scaled problem, and we still need to investigate the scalability of the method.

4 Discussion and Conclusions

In this paper, we presented an approach to collect fault cases experienced during development of a large and complex systems, and to utilize them for supporting fault diagnosis tasks. Behind the approach is a motivation to overcome the capture bottleneck problem prevalent in design rationale and knowledge management domain. We presented the concept of knowledge recycling, to capture knowledge from now wasted information at low cost. We illustrated the usefulness of the method by applying it to a case experienced in a real satellite development project.

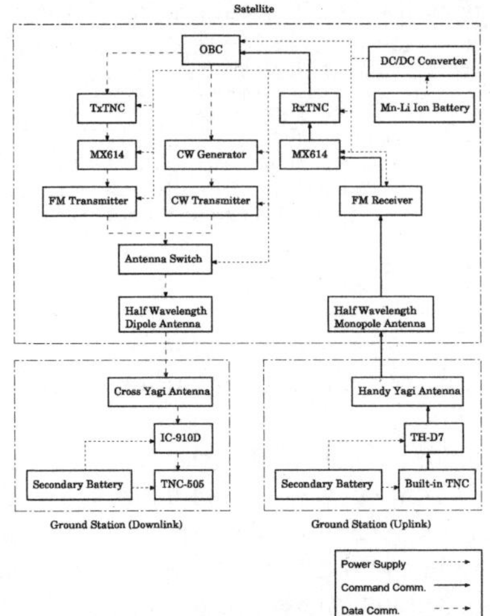

Figure 1: A functional structure model of the communication subsystem of the CubeSat satellite. Directed edges denotes that the source object provides a function to the destination object.

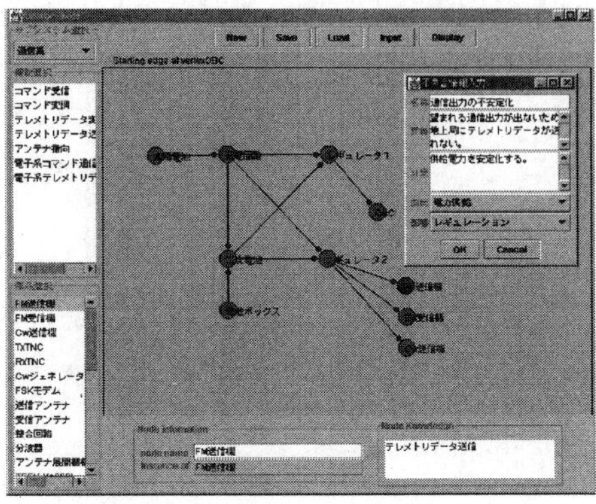

Figure 2: A screenshot of a tool for building functional structure models.

The method was tested in a small scaled problem, and we still need to evaluate it with larger scale problems. As functional structure models are represented as 'flat graphs', it will be cumbersome to represent a system with multiple subsystems. By adopting hierarchical graphs, the representation can be more scalable. There is also the problem of combinatorial explosion when computing candidate failure paths with larger scale models. One way to avoid this problem is to limit the computation at certain levels, when models are represented as hierarchical graphs. More sophisticated ranking mechanism may be needed to present up front the 'more interesting' failure paths to the user.

We are planning to apply the method in a real-life satellite development project.

References

[1] E.J. Conklin and K.C. Burgess Yakemovic, A process-oriented approach to design rationale, Human-Computer Interaction **6** (1991) 357–391.

[2] W.S. Lee, D.L. Grosh, F.A. Tillman and C.H. Lie, Fault tree analysis, methods, and applications – A review, IEEE Transactions on Reliability **34** (3) (1985) 194–203.

[3] E. Motta, S.B. Shum and J. Domingue, Ontology-driven document enrichment: principles, tools and applications, International Journal of Human Computer Studies **52** (2000) 1071–1109.

[4] C. Price and N. Taylor, Multiple fault diagnosis from FMEA, Proc. of AAAI-97/IAAI-97 (1997) 1052–1057.

[5] F.M. Shipman and C.C. Marshall, Computer-Supported Cooperative Work **8** (4) (1999) 333–352.

[6] S.B. Shum and N. Hammond, Argumentation-based design rationale – what use at what cost, International Journal of Human Computer Studies **40**(4) (1994) 603–652.

[7] K.K. Vemuri, J.B. Dugan, and K.J. Sullivan, A design language for automatic synthesis of fault trees, 1999 Proc. Annual Reliability and Maintainability Symposium (1999) 91–96.

Mining and Characterizing Opinion Leaders from Threaded Online Discussions

Naohiro Matsumura [1,2]   Yukio Ohsawa [1,3]   Mitsuru Ishizuka [2]

[1] *PRESTO, Japan Science and Technology Corporation,*
2-2-11 Tsutsujigaoka, Miyagino-ku, Sendai, Miyagi, 983-0852 Japan
[2] *Graduate School of Engineering, The University of Tokyo,*
7-3-1 Hongo, Bunkyo-ku, Tokyo, 113-8656 Japan
[3] *Graduate School of Business Science, University of Tsukuba,*
3-29-1 Otsuka, Bunkyo-ku, Tokyo, 112-0012 Japan

Abstract. Activating communities on the Internet is catching attentions of web site's designers because the activity affects the growth of communities. In this paper, as a trigger of activation, we aim at mining and characterizing opinion leaders from threaded online discussions on the Internet. Then we try to understand the relations between opinion leaders and their characteristics by using correspondence analysis.

1 Introduction

Communication places on the Internet, such as BBS and chat rooms, are used for the sake of gathering people into particular web sites [2]. The aim of setting such places is to make communities where people share the common context [1] by activating the interaction among people. However, it is not always easy to activate in fact because of the lack of triggers of topics. In other words, we might control the activation of a community if we could throw fascinating topics into the community. Here we focus on "opinion leader" [4] who are sensitive to the trend and having a great influence on peoples' decision making. We believe that opinion leaders can provide fascinating topics which trigger the activation of the community. In this paper, we aim at mining and characterizing opinion leaders from threaded online discussions on the Internet. Then, we try to understand the relations between opinion leaders and their characteristics by using correspondence analysis.

2 Influence Diffusion Model

Influence Diffusion Model (IDM) is a method for discovering influential comments, people, and terms from threaded online discussions, such as BBS [3]. One of the features of threaded online discussions is that communications between people are done by exchanging comments. The first assumption of IDM is that the relations of comments, called *comment-chain*, show the flow of influence. For example, if comment C_y replies to comment C_x, it is considered that C_y is affected by C_x. Similarly, if person Y replies to a comment of person X, Y is considered to be affected by X. In these cases, the influence diffuses from C_x to C_y / from X to Y. In this way, the influence diffuses throughout the comment-chains. Another feature of threaded online discussions is that comments are written by natural language composed of terms. The second assumption of IDM is that people's idea is expressed and propagated by the medium of terms. Based on these assumptions, the process of diffusing influence is defined as follow.

Definition 1. *In text-based communication, influence diffuses along the comment-chains by medium of terms, i.e., words or phrases.*

According to **Definition 1**, the influence is defined by the degree of terms propagating throughout the comment-chains. For example, If C_y replies to C_x, the influence of C_x onto C_y, $i_{x,y}$, is defined as

$$i_{x,y} = \frac{|w_x \cap w_y|}{|w_y|}, \tag{1}$$

where w_x and w_y are the set of terms in C_x and C_y respectively, and $|w|$ denotes the count of w. In addition, if C_z replies to C_y, the influence of C_x onto C_z through C_y, $i_{x,z}$, is defined as

$$i_{x,z} = \frac{|w_x \cap w_y \cap w_z|}{|w_z|} \cdot i_{x,y}, \tag{2}$$

where w_z are the terms in C_z.

It is considered that the more a comment affects other comments, the more the influence increases. And the same can be applied to the influence of people/concepts. The influence of a subject (including a comment, person or a term) then comes to be measurable.

Definition 2. *The influence of a subject (a comment, person or a term) to the community is measured by the sum of influence diffused from the subject to all other members of the community.*

Applying **Definition 2** to C_x, the influence (here after, let us skip "to other members of the community") is measured by the sum of influence diffused from C_x, i.e., $i_{x,y} + i_{x,z}$ if the community has three members x, y and z.

3 Mining Opinion Leaders

In IDM, the influence of a person X is defined as the sum of influence of X's comments. Here we apply IDM to Yahoo!Japan's BBS discussing about the clothing of UNIQLO.com [1]. The top 5 people in the order of values of diffusing influence (D_X) are listed in Table 1.

Table 1: The top 5 people in the order of diffusing influence.

Ranking	*Member ID*	D_X
1	M011	36.09
2	M002	3.949
3	M004	3.340
4	M010	2.985
5	M021	2.841

---

[1] http://www.uniqlo.com

Let me introduce each people in Table 1. The top-rank people, M011, was a person that posted comment #1 which is no doubt the most influential comment because it is the beginning of the comment-chains in the BBS. The second-rank people, M002, and the fourth-rank people, M010, were the staffs of UNIQLO (M002 is male and M010 is female). They frequently offered the hot information about new clothing, hot-selling clothing or advices on dressing. The third-rank people, M004, posed comments which often raised argumentive topics. The fifth-rank people, M021 who had been an UNIQLO enthusiast for four years, had been also offering much information from customer's point of view. All of them actively offered influential comments that catch participants' interest and caused constructive discussions. Therefore, we believe that they were suitable for opinion leaders.

4   Characterizing Opinion Leaders

The characteristics of opinion leaders are different from each other. For example, one might have a great influence on the color of fleece, and another might have on the style of fleece. Here we regard one's influential terms as his/her characteristics. The influential terms for each opinion leaders in Table 1 are listed in Table 2.

Table 2: Opinion leaders and their characteristics (influential terms).

Member ID	*Characteristics (Influential Terms)*
M011	UNIQLO, cardigan, jam, advice, sneaker, scarf, return, wear, fleece, fashionable, jacket ...
M002	UNIQLO, wear, fitting, fleece, advice, shape, sleeve, sweater, design, color, catalogue ...
M004	touch, jeans, pants, color, UNIQLO, return, clerk, fitting, cheap, size, shape, fleece, skirt ...
M010	cute, fleece, coverall, slipper, UNIQLO, beige, T-shirt, beige, demin, size, pink, color ...
M021	fleece, spring, slipper, color, cheap, full zipper, winter, walk, the Internet, UNIQLO, bag ...

5   Correspondence Analysis

For understanding the relations between opinion leaders and their characteristics, we employ correspondence analysis [6] to visualize the relations as a two-dimentional positioning map. We skip the details of correspondence analysis because it is beyond the scope of the paper. Figure 1 shows the positioning map for the data in Table 2. By seeing Figure 1, we can clearly understand the characteristics of opinion leaders and their terms as follows:

- M002 and M004 have similar characteristics, which are price, jeans, catalogue, T-shirt, basic, etc.

- M021 has the characteristics, which are spring, winter, the Internet, weekend, etc.

- M010 has the characteristics, which are beige, color, pink, cheap, etc. Note that M010 is in the position among M002, M004, and M0021.

- M011 has the characteristics, which are jacket, scarf, cardigan, etc.

- M002, M004, M0010 and M0011 have common characteristics, which are return, advice, favorite, etc.

- All the opinion leaders have common characteristics, which are UNIQLO, fleece, etc.

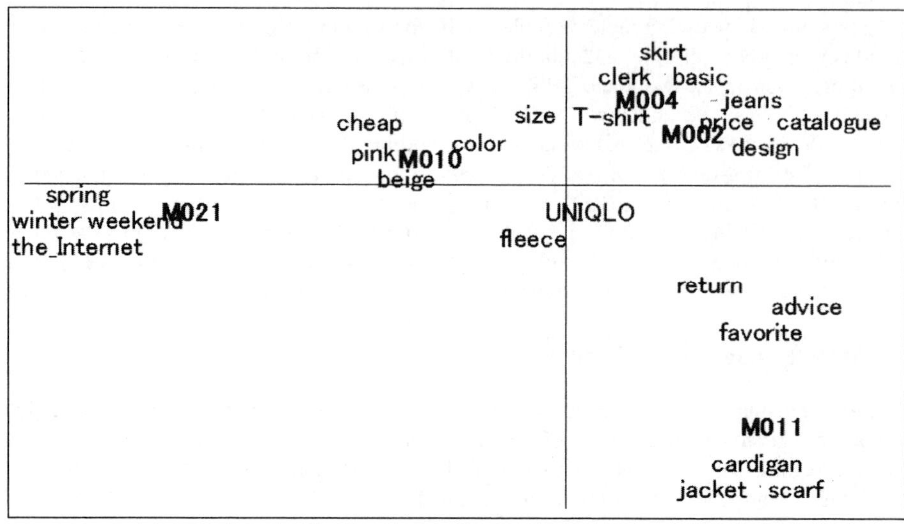

Figure 1: Positioning map of opinion leaders and their characteristic terms. For ease of understanding, moderate number of terms are shown.

6   Conclusions

The results mentioned in this paper help us understand the opinion leaders and their characteristics. As the future strategy for activating a community, for example, we plan to let a opinion leader whose characteristics are related to the community's current topic provide some topics for activating the community if the community becomes stagnant.

References

[1] Yukio Ohsawa: Chance Discovery for Making Decision in Complex Real World, *New Generation Computing*, Vol. 20, No. 2, 2002.

[2] Naoto Ishikawa: Internet Community Strategy, SoftBank Publishing, 2001. (in Japanese)

[3] Naohiro Matsumura, Yukio Ohsawa, and Mitsuru Ishizuka: Influence Diffusion Model in Text-based Communication, *WWW02*, 2002. to appear

[4] E.M. Rogers: Diffusion of Innovations, The Free Press, 1962.

[5] S. Kiesler, J. Siegel, and T.W. McGuire: Social Psychoogical Aspects of Computer-Mediated Communication, *American Psychologist*, Vol. 39, pp. 1123–1134, 1984.

[6] Miyagawa Masami: Graphical Modeling, Asakura Publisher, 1997 (in Japanese)

KES 2002
E. Damiani et al. (Eds.)
IOS Press, 2002

How Can We Facilitate Concept Articulation in Purchasing?

Hiroko Shoji,[†][‡]   Mikohiko Mori[†]   and   Koichi Hori[†]
[†]*University of Tokyo, 4-6-1 Komaba, Meguro-ku, Tokyo 153-8904, Japan*
[‡]*Kawamura Gakuen Women's University, 1133 Segedo, Abiko, Chiba 270-1138, Japan*
`{hiroko, mori, hori}@ai.rcast.u-tokyo.ac.jp`

Abstract. This study works on realizing the interactive system which can effectively promote the customer's decision-making during shopping online. Specifically, it is targeted at the "purchasing as concept articulation" which provides gradual clarification of their needs of initial vagueness, which has not been handled in traditional online shopping systems. This paper describes the way of presenting information to effectively help the customers in their concept articulation process, and the interaction design framework which enables it.

1 Introduction

The "era of products in excess" requires the marketing strategy which puts emphasis on emotional elements which have a great influence on the customer's decision-making [3]. Online shopping, which is recently getting more attention, also needs not only to try to have a wider selection of products and/or reduce the prices but also to present the customers with information in a manner appealing to their sensibility. Purchasing is usually assumed to be buying what we want, however, our wants are often not determined until actually shopping around as Underhill pointed out in his book[7]. The authors call such type of purchasing *purchasing as concept articulation*, where the shop environment and/or interaction with a salesclerk allows the customers whose initial requirement or image of their wants is vague to convince themselves of what they truly want and then reach a decision. Although purchasing as concept articulation is often observed in the real world, existing online shopping sites only deal with purchasing as problem solving where the customers have clear requirements.

Therefore, this study aims to build the online shopping system to aid in purchasing as concept articulation. As part of this study, the analysis of human behavior in the actual purachase activities has shown that aiding the customers both in their *conception* and in their *conviction* are important to faciliate purchasing as concept articulation. The spatial-arrangement style of information presentation is expected to be useful for the former, while the scene information presentation is for the latter. Therefore, this study has created S-Conart system which helps the customers in their conception through the spatial-arrangement style interface and in their conviction through the scene information presentation. The user study as part of this research has confirmed that the spatial-arrangement style of information presentation is effective in the support for the customers' perception and realization. The effectiveness of the scene information presentation in the support for their conviction should be addressed in the future.

2   Analysis of human behavior in the actual purchase activities

2.1   *Two types of purchasing*

This study started with observing customer's behavior in actual shops. A detailed observation of actual purchase activities has shown that purchasing can be roughly divided into a problem-solving type and a concept-articulation type.

Purchasing as problem solving With purchasing as problem solving, the customer initially has a clear idea of what desired product is like and/or what functionality it requires, and searches for the items which meet his/her requirements. That is, purchasing as problem solving means that the customer has previously determined what to buy. The customer who follows purchasing as problem solving searches for the products meeting his/her requirements to discover solution candidates, balances between them (if there are more than one), evaluates them, and then decides whether to buy them.

Purchasing as concept articulation With purchasing as concept articulation, on the other hand, the customer initially has a unclear requirements for his/her needs and gradually builds up a concrete image of target products through the interaction with a salesclerk. That is, purchasing as concept articulation means that the customer determines what to buy after due consideration in the shop. The customer who follows purchasing as concept articulation starts with vague requirements of his/her own, perceives and realizes underlying requirements with a trigger of some information provided while looking around various products, understands what true requirements are, satisfies himself/herself of products matching them, and then makes a decision on whether to buy those products. The customer doesn't perceive his/her true requirements until he/she looks at the products.

The actual purchase activities use either problem-solving or concept-articulation types according to circumstances. Also, both of them are sometimes observed to be mixed in each individual purchase.

2.2   *Role of communication*

This study considers the information presentation to effectively support purchasing as concept articulation which provides gradual clarification of the customer's requirements of initial vagueness, and aims to apply it to online shopping systems. In order to serve as a reference to this human-computer interaction design, the authors have observed the communication between the customer and salesclerk in actual purchase activities, and investigated what change in the customer's mental world the communication with the salesclerk causes. As a result, the customer's decision-making process during purchasing as concept articulation has demonstrated that the appropriate information presented in a timely manner causes the change of the customer's viewpoint, which in turn triggers a change of the search goal itself in many cases. This interaction has proven to be effective in the decision-making during purchasing as concept articulation [4].

Findings obtained through this analysis are summaried as follows:

1. Salesclerk's reaction is roughly divided into two types, i.e. expected reaction and unexpected reaction.

Expected reaction Salesclerk's role is considered to be presenting the solutions (products) to meet the customer's requirements. The reaction to fill this role is called expected reaction. It is often useful for purchasing as problem solving.

Unexpected reaction The reaction which presents information from a different viewpoint than the customer's current thought is called unexpected reaction. It is often useful for purchasing as concept articulation. This is because a new viewpoint presented by unexpected reaction in a timely manner causes the search goal itself to change to be more suited for the customer's potential requirements, and allows the customer to have a clearer image of his/her own requirements.

2. The unexpected reaction is useful as a trigger to the concept articulation, meaning that it is useful for helping the customer in purchasing as concept articulation. There are two main features useful for facilitating the concept articulation, as follows:

Support for conception The unexpected reaction sometimes causes the change of the customer's viewpoint, which in turn triggeres the change of the search goal itself, resulting in the promotion of his/her decision-making. Skillful salesclerks can use the unexpected reaction appropriately to facilitate the customer's conception.

Support for conviction The customer who has noticed a new viewpoint needs conviction in order to accept the viewpoint smoothly. Skillful salesclerks can facilitate the customer's conviction through their good demonstration of concrete use scenes and/or usage of possible products, and others.

3   S-Conart: Concept Articulator for Shoppers

3.1   Overview of S-Conart

Based on the findings obtained through the analysis of human behavior in actual purchase activities, this study considered an effective way of information presentation to help the customer in purachasing as concept articulation, and created an experimental system named S-Conart (Concept Articulator for Shoppers). The authors are developing a system with special emphasis on the appropriate information presentation for facilitating the customer's concept articulation instead of replacing human communication with Human-Computer Interaction (HCI) as is [5].

As previously mentioned, the analysis of human behavior in actual purchase activities has shown that the support for conception and the support for conviction are both important to help the customer in purchasing as concept articulation. S-Conart attempts to achieve these two types of support using the following approach.

Support for conception with spatial-arrangement style of information presentation The findings from the study of creativity support suggests that the spatial-arrangement style of information presentation is useful for facilitating the customer's conception[1][6]. Therefore, S-Conart implements the support for the customer's conception using spatial-arrangement style of information presentation based on the Multi-Dimensional Scaling Method (MDS) (Figure 1).

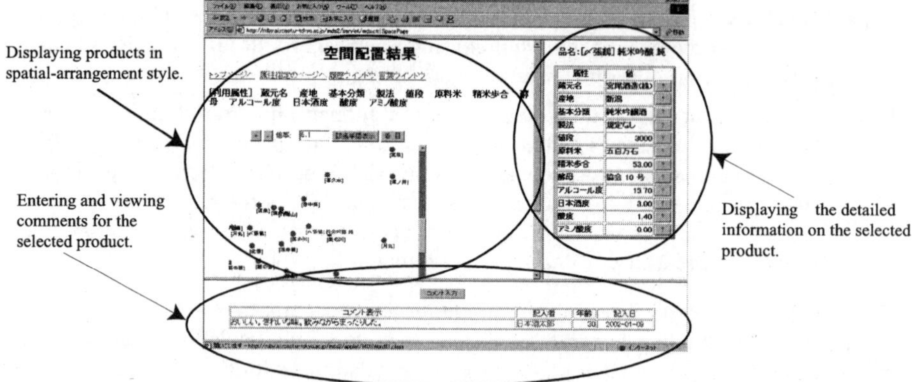

Figure 1: Product information presentation screen with spatial-arrangement style

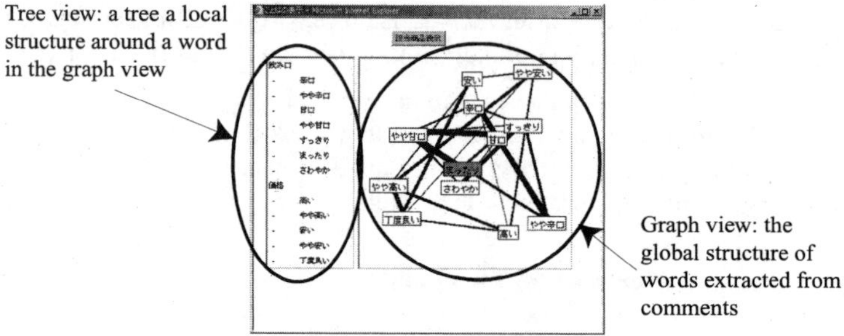

Figure 2: Words window

Support for conviction using scene information The information on image and/or use scenes of products (herein called scene information) has proven to be effective in the concept formation process [2]. Therefore, S-Conart implements the two functions shown below, presenting scene information suited for the user's current thought to facilitate his/her concept articulation. (1) Facilitates the user's conviction by allowing for browsing the comments which contains scene information on the products (Figure 1). (2) Facilitates the user's concept articulation with both graph and tree styles of presentation of words extracted from the comments on all the products (Figure 2).

3.2   User Studies

Through the purchasing experiment using S-Conart, an experimental system built as part of this study, user studies examined the effectiveness of spatial-arrangement style of information presentation in the support for the user's conception [5]. List style of information presentation was used for comparison. The result confirmed that spatial-arrangement style of information presentation is effective in prompting the user to conception of a new viewpoint. Therefore, using the spatial-arrangement style appropriately enables the support for purchasing as concept articulation. In contrast, the list style of information presentation is useful for purchasing as problem solving. The effectiveness of the scene information presentation in the support for the customer's conviction is currently under examination and will be reported in the future.

Through this analysis, this study argues that changing the way of presenting information provided by the system can bring a change to the customer's (or user's) mental world, which is equivalent to that by actual salesclerk's reaction based on his/her strategic knowledge with the exception of its form.

4 Conclusion

This paper considered the interaction design which focuses on the issue of the change in the human mental world. It dealt with purchasing as a concrete example and described the effectiveness of the interaction in purchasing as concept articulation, which faciliates the customers' conception of a new viewpoint and helps their conviction which enable a smooth acceptance of the new viewpoint.

In the past, researchers and engineers in the field of engineering have been making efforts to realize novel and innovative functionalities or systems. However, there is no denying that their devotion sometimes lead to the "system first" way of thinking. The authors think we should now get out of this thinking and pay attention to the fact that even simple mechanisms or tools could create a rich human-computer interaction. One of the clues to this turnover is the effectivenss of spatial-arrangement style of information presentation described in this paper. Designing systems leveraging advantages of computer (such as fast search and high-capacity memory) after long series of these basic and microscopic analyses may create a new style of human-computer interaction which is totally different from traditional ones.

References

[1] Hori, K., Concept space connected to knowledge processing for supporting creative design, Knowledge-Based Systems, Vol.10, No.1, pp.29-35, (1997).

[2] Ishino, Y., Hori, K. and Nakasuka, S.: Concept development of consumer goods utilizing strategic knowledge, *Knowledge-Based Systems* Vol.13, pp.417-427, (2000).

[3] Peppers, D. and Rogers, M.: *The One to One Future*, Doubleday, (1993).

[4] Shoji, H. and Hori, K.: Chance Discovery by Creative Communicators Observed in Real Shopping Behavior, T. Terano et al.(Eds.), JSAI2001 Workshops, LNAI2253, pp.462-467, (2001).

[5] Shoji, H. and Hori, Strategy Emergence from Human-Computer Interaction, J. S. Gero and K. Hori(eds), Strategic Knowledge and Concept Formation III, pp.87-99, (2001).

[6] Sugimoto, M., Hori, K. and Ohsuga, S.: A method to assist building and expanding subjective concepts and its application to design problems, Knowledge-Based Systems, Vol.7, No.4, pp.233-238, (1994).

[7] Underhill, P.: *Why We Buy: The Science of Shopping*, Touchstone Book, (1999,2000).

KES 2002
E. Damiani et al. (Eds.)
IOS Press, 2002

Toward a Chance Discovery-Oriented Recommender System: A Prototype

*1 Makoto Mizuno, *2 Hiroko Shoji, *3 Yukio Ohsawa, *4 Yutaka Matsuo,
*5 Naohiro Matsumura, *6 Yuzo Miyake
*1 *Hakuhodo Inc., 3-4-1 Shibaura, Minato, Tokyo 108-8088, Japan*
*2 *Kawamura Gakuen Women's University, 1133 Sagedo, Abiko, Chiba 270-1138,
Japan*
*3 *University of Tsukuba, 3-29-1 Otsuka, Bunkyo, Tokyo 112-0012, Japan*
*4 *National Institute of Advanced Industrial Science and Technology
Aomi 2-41-6, Koto, Tokyo 135-0064, Japan*
*5 *University of Tokyo, 7-3-1 Hongo, Bunkyo, Tokyo 113-8656, Japan*
*6 *Kozo Keikaku Engineering Inc., 4-5-3 Chuo, Nakano, Tokyo 164-0011, Japan*

Abstract: Existing recommender systems are based on the concept of 'likelihood' and are not necessarily designed for experienced consumers. What we propose is a Chance Discovery-oriented system. A preliminary experiment suggests that this system holds promise although there are some remaining issues that must be addressed.

1. Introduction

There is an emerging trend in marketing that some leading marketing scholars have referred to as 'reverse marketing' [1] or 'customerization' (note that the term is not 'customization') [2]. Both terms are based on the observation that a significant power shift from manufactures and distributors to customers is taking place, resulting in a transition from persuasion to recommendation.

With the penetration of on-line shopping, a variety of recommender systems have been developed and put into practical use [3]. Most of these systems adopt one or more of the following fundamental approaches:

- *Commonsense*: If knowledge of the domain of interest is well established, one can recommend the option that is 'best' in the conventional sense for a certain occasion.
- *Individual preference analysis*: Using a user's purchase records or alternative data, one can recommend the option that is statistically predicted to be best preferred.
- *Clustering*: Based on similarities between a user and other users, one can recommend the option that has been best preferred by his or her peers.

While each of the above approaches has distinctive features, all share the common concept of 'likelihood' (i.e., the recommended option is most likely to be preferred in general, personally, or among peers). A likelihood-oriented recommender system may be effective for novice consumers since it can complement elementary knowledge of the domain. However, it would be less effective for experienced consumers since they may already know the options that are recommended to them, whether they have already purchased them or not. Such customers would seek something new.

An alternative concept for recommender systems is 'Chance Discovery.' The original definition of it was "the awareness on and explanation of the significance of a chance, especially if the chance is rare and its significance has been unnoticed [5]." We propose applying this concept to recommender systems for experienced consumers.

Our main purpose is to offer consumers surprising or exciting options.   In other words, the systems should answer to 'variety seeking,' which is the tendency for consumers to seek something new [4].   The tendency seems to be more prominent among experienced consumers.

In the remaining part of this paper, we will report on the experimental development of a Chance Discovery-oriented recommender system.   First, we will briefly discuss the potentiality of Chance Discovery, particularly KeyGraph, for building recommender systems.   Second, we will propose a prototype system designed to support consumers in their selection of wines.   Next, we will evaluate whether or not the system will work as expected, based on the preliminary experiment.   Finally, we will discuss the limitations and future directions of our research.

2. KeyGraph as an Engine

The concept of 'Chance Discovery' is embodied in KeyGraph, which has been successfully applied to tasks such as the extraction of keywords from scientific papers [6] and the detection of risky faults based on earthquake data [7].   KeyGraph deals with data as a set of 'sentences,' each of which are comprised of 'words.'   It is capable of extracting 'foundations,' which are highly frequent concurrences of words, as well as 'roofs,' which are not so frequent but critical bridges that tie foundations.   The links between foundations and roofs are called 'columns,' and are interpreted as signs of a 'chance.'

Figure 1. Graph of wine items and foods drawn using KeyGraph

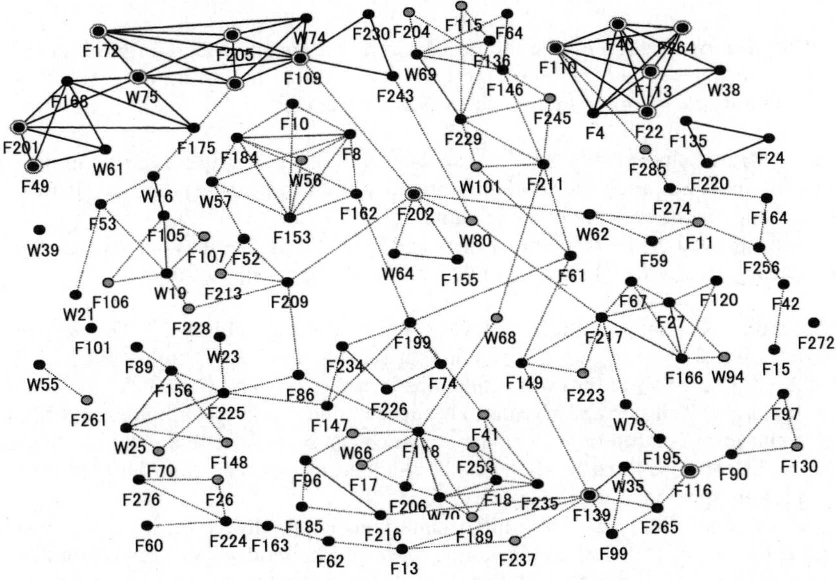

In this figure, nodes marked with W are assigned to wine items, and nodes marked with F to foods. Black nodes are the wines/foods frequently endorsed by experts.   When these are frequently endorsed concurrently with one another, they are linked by black lines to form 'foundations.'

Gray nodes represent 'roofs,' which are rarely endorsed but are concurrent with 'foundations,' and broken lines represent 'columns,' which are links between black and gray nodes.

In this paper, we propose to use an extension of KeyGraph as an engine for Chance Discovery-oriented recommender systems. When applying KeyGraph to the selection of wine, two types of data can be used. In both types of data, the items of wine are coded by region, type of wine, and price tier, such as 'Bordeaux: Red-Full Body: High Price,' to limit the number of wines to a tractable level. The first type of data is disaggregate purchase records, in which a shopping trip is treated as a sentence, and each purchased wine as a word. KeyGraph extracts the frequently co-purchased items as foundations and also extracts the not so frequently purchased items that bridge several foundations as columns. The latter items are interpreted as being purchased for 'a change of pace,' to give variety to consumption experiences.

Another type of data is the expert knowledge of recommended combinations of wines and foods, in which a sentence contains several wines and foods that should be enjoyed together. The foundations extracted here may be 'common sense' and familiar to experts, while the columns may be 'chances' that even experts are not aware of. We applied KeyGraph to both types of data and found that analysis of the latter produced more convincing results in this case. As shown in Figure 1, only three foundations containing more than one wine/food were formed, as most of the frequently endorsed wines/foods were not very concurrent with one another. This suggests that there are very few examples of 'common sense' and very many 'hidden chances' in wine-food combinations.

3. WineNavi as a Visualized Interface

To make the information derived from KeyGraph more approachable for consumers, we developed an interface called 'WineNavi'. Instead of importing outputs directly from KeyGraph, we modified them in the following manner:

- First, we formed clusters by taking into account all links drawn by KeyGraph in which each wine/food was assigned to either none or one of the clusters, and labeled each cluster based on our subjective interpretation.
- Next, we removed the links within clusters and redrew only the links between clusters. Consequently, wines/foods are connected not directly but hierarchically.

Four windows compose the interface of WineNavi (see Figure 2): Global View (lower left), Local View (right), List Display (upper left) and Item Display (middle left). Global View is used to overview the relations between clusters. Local View focuses on the highlighted cluster and its adjacent clusters. In Local View, wines and foods within a cluster are placed on a round table, which can be rotated freely. List Display shows the list of wines and foods within a specified cluster, and Item Display shows detailed information on a specified wine or food.

Users of this software can shift their viewpoints freely. If they search recommendations of standard combinations, they will be guided to an appropriate and conventional cluster of interest. Alternatively, they can move to the adjacent clusters to discover somewhat unusual but potentially connected wines or foods, thus broadening their experiences. This is the process of Chance Discovery for consumers who seek a variety in their selection of wines.

Figure 2. The interface of WineNavi

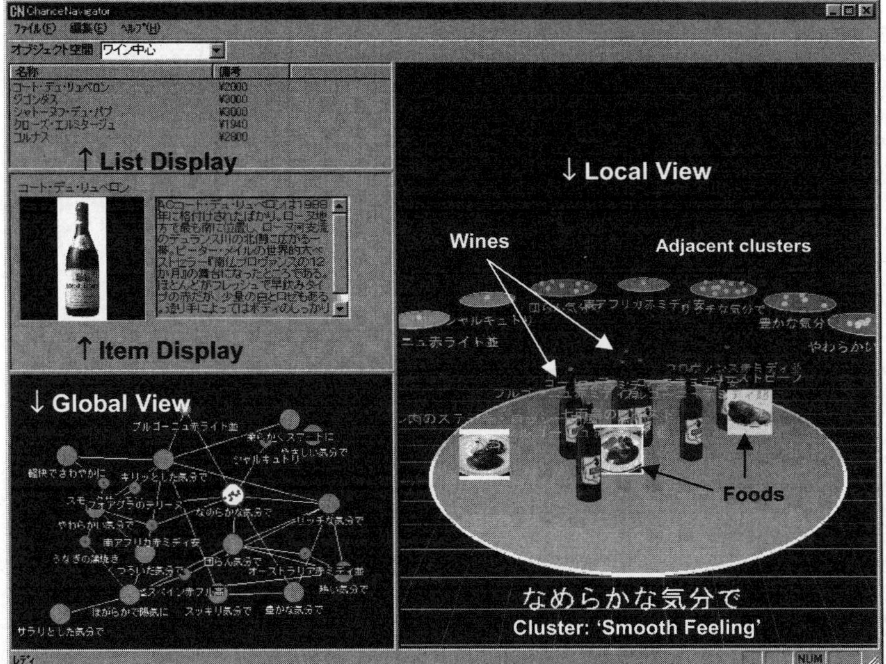

4. User Evaluation

We performed a preliminary experiment to evaluate the prototype of WineNavi. About 10 people were asked to use WineNavi and evaluate it verbally. Although many participants agreed that the interface was 'cool,' it was unclear as to whether or not it could produce the expected effect – Chance Discovery for consumers. One participant, however, reported a supportive experience: First, two relatively big clusters respectively labeled 'soft feeling' and 'smooth feeling' attracted his attention. The former cluster involved many white wines produced in Germany or France and boiled crabs; the latter, some Bourgogne red wines and meat dishes like roast beef. Interestingly, the two clusters were not directly linked but connected by a cluster with the element: 'South-Africa: Red-Medium: Low-Price.' This linkage suggested that this wine (category) could be served with foods in seemingly unconnected clusters and broaden his experiences by smoothly bridging those clusters. Indeed the participant admitted to having bought this wine after the experiment.

The following comments were collected from other participants:

- Automated navigation is needed. At least in the beginning, users should be led to options they are likely to prefer.
- The relationship between Global and Local View is a little confusing because the clusters are positioned in different ways.
- The meanings of clusters -- e.g., why they are grouped together, why they are labeled

in a certain way -- are sometimes not comprehensible to users.

The first two comments are understandable and can be easily addressed in the next stage of the development.   On the other hand, the third comment indicates an essentially difficult problem, since the results produced by Chance Discovery often contradict common sense and hence are less acceptable to the average consumer.   In other words, evaluation of the results is dependent on the users' flexibility and imagination.

In any case, this preliminary experiment inhibits generalization.   It is necessary for us to perform a more formal experiment in order to judge the validity of the system in a more systematic manner.   Moreover, if possible, we should aim to compare this system with traditional recommender systems based on established data mining tools.

5. Discussion

Traditional recommender systems may be useful for novice consumers but not for experienced consumers, who are bored with mundane recommendations and are seeking more variety.   A more novel recommender system for such consumers can be developed based on the concept of Chance Discovery, or, more specifically, by incorporating KeyGraph as an engine.   We therefore propose a prototype of a Chance Discovery-oriented recommender system named WineNavi, which is designed to support consumers in their selection of wines.   The results of a preliminary experiment suggest that this direction is promising.

The prototype will be extended in the following ways: Firstly, automated navigation, either partial or total, will facilitate the users' search process.   Secondly, personalization may be productive; for instance, shared knowledge may be combined with personal experiences.   Thirdly, there is room for improvement in our procedure for clustering and positioning items in a display space to make the outputs more comprehensible.   Lastly, there may be a different and more robust/practical way to extend the prototype.   If we implement this system on the Internet, some of the 'cool' features may have to be sacrificed because of the heavy volume of the graphics: That is, we may have to reduce the system to the core function of Chance Discovery for the benefit of consumers.

References

[1] M. Sawhney and P. Kotler, Marketing in the Age of Information Democracy. In: D. Iacobucci (ed.), Kellogg on Marketing. Wiley, New York, pp.386-408.

[2] J. Wind and A. Rangaswamy, Customerization: The Next Revolution in Mass Customization. *Journal of Interactive Marketing* **15** (2001) 13-32.

[3] H. Kautz (ed.), Recommender Systems. Technical Report WS-98-08, AAAI Press, California, 1998.

[4] L. McAlister and E. Pessemier, Variety Seeking Behavior: An Interdisciplinary Review. Journal of Consumer Research **9** (1982) 311-322.

[5] Y. Ohsawa: Chance Discoveries for Making Decisions in Complex Real World. *New Generation Computing* **20** (2002).

[6] Y. Ohsawa, N. E. Benson and M. Yachida, KeyGraph: Automatic Indexing by Co-occurrence Graph based on Building Construction Metaphor. Proc. Advanced Digital Library Conference (IEEE ADL'98), (1998) 12-18.

[7] Y. Ohsawa and M. Yachida, Discover Risky Active Faults by Indexing an Earthquake Sequence. Proc. International Conference on Discovery Science (DS'99)(1999).

KES 2002
E. Damiani et al. (Eds.)
IOS Press, 2002

An Approach to a Knowledge Reconstruction Engine for Supporting Event Planning

Shigeki AMITANI, Mikihiko MORI and Koichi HORI
{amitani, mori, hori}@ai.rcast.u-tokyo.ac.jp
Research Center for Advanced Science and Technology (RCAST),
The University of Tokyo, 4-6-1 Komaba, Meguro-ku, Tokyo, 153-8904, JAPAN
Tel: +81-3-5452-5289 Fax: +81-3-5452-5312

Abstract. The main goal of our current research is to establish a methodology to articulate the gap between the event designers' intention and the event visitors' mental impression, which is the first step for "a knowledge reconstruction system". At the actual events, the visitors' interactions with the event objects were observed and the verbal reports (protocol data) were recorded. From analysis of these data, a lot of gaps between the planners and the visitors are found as well as unexpected mental processes. This analysis provides "chances" to create new ideas with planners. In this paper, we are going to propose "a supporting system for knowledge reconstruction" as a tool for "Chance Discovery" by applying the framework of creativity support to the actual event designing work.

1. Introduction

Every year, event-planning companies hold various events. So far event planning is conducted with implicit knowledge of the experienced planners and the visitors' impression of the events is usually measured only by questionnaires. In the actual situation, it is said that the planners cannot obtain adequate and proper knowledge for the future planning from the statistical data from questionnaires because the data is information "without context where knowledge was produced". That means the planners are unable to evaluate the event they designed. Planners need to know what visitors to the event actually feel and how they behave when they are at the event booth to construct the strategies for next event planning. This is because knowledge itself cannot be applied to the real situation if it does not have the context where it was produced [1]. Qualitative data and quantitative data should complement with each other.

In this paper, we are going to propose a methodology to articulate gaps between planners' intention and visitors' impression. Our methodology was applied to the actual events, - "Tokyo Motor Show 2001" and "World PC Expo 2001", in cooperation with Dentsu Inc. The visitors' interactions with the event objects were observed and the verbal reports (protocol data) were recorded. The results of this experiment and, furthermore, the image of "a knowledge reconstruction system" is described. From analysis of the data, we obtained the prospect that our microscopic and detailed approach is useful and effective toward event planning in the real world.

2. What is "Knowledge Reconstruction"?

Though the importance of knowledge has been claimed since middle of the 1980s, the main concern of business theory is how to obtain and accumulate established knowledge. Little research has been conducted on how innovative knowledge is created [2]. Nonaka [2] claims that there are four modes of knowledge transition: Socialization, Externalization,

Combination and Internalization. Though a lot of companies have attempted to apply this theory to their actual works, it does not seem to be successful. Nonaka theory is proposed as a theory and no method has been indicated. In addition, this is the theory for "transition of knowledge mode", not the theory for "manipulation of knowledge mode by the user". In our research we are going to apply Nonaka's theory to the actual situations. We are going to propose a methodology and a system to support to apply this theory to the actual situations (Fig 1). The italic phrases are what we are going to provide.

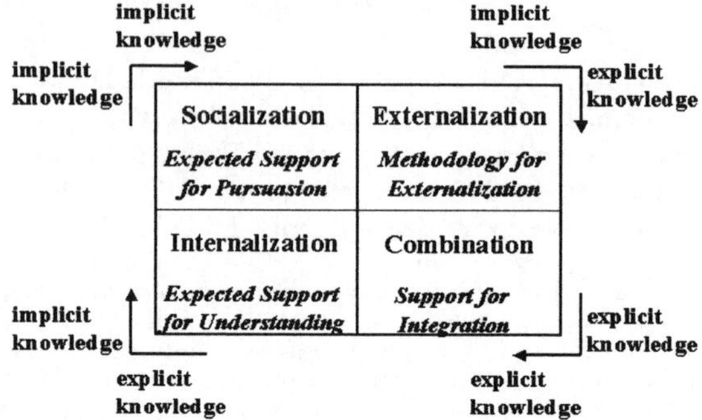

Fig 1 A Cyclic Process of Knowledge Reconstruction

Our system aims to "use in the real world". Knowledge cannot be separated from actual contexts to utilize the knowledge, i.e., knowledge management. That means that it is necessary to preserve knowledge together with "the real context". Event planners need to know how the knowledge was produced. It is necessary for planners to obtain and understand knowledge, to integrate and create innovative knowledge, and to apply knowledge to actual event planning. This is what we call "knowledge reconstruction". Knowledge must be dynamically integrated and innovated. We are going to propose "a supporting system for knowledge reconstruction". At the current state, our research obtained such perspective that our approach can be driving force of this cyclic process. In the following sections, the experiments at actual events and an example of findings are described. Moreover, the example that the planner hit upon an innovative idea with providing the results of our investigation is also described. The system image is described in section 4.

3. Experiment

The experiment of our methodology was conducted to extract the actual cognitive processes at event sites of "World PC Expo 2001 (WPC: 19-22, Sept., 2001)" and "Tokyo Motor Show 2001 (Motor Show: 26, Oct. - 7, Nov., 2001)" held at Makuhari Messe in Japan, to see if our methodology works. Three booths for WPC and one booth for the Motor Show were selected for the experiment in cooperation with the event organizers.

To investigate what the designers' intentions are and how they implement the event objects to express their intentions, we had interviews with the planners in advance.

To collect protocol data, two wearable computers were prepared[1]. Because of the sponsors' intention, the normal digital video camera was adopted for collecting protocol data. 9 subjects (one person for 3 sessions + one pair for 3 sessions) at WPC and 12 subjects (one person for each session) at Motor Show were employed. Subjects are asked to look around the designated booth(s). After visiting the booth, the procedure in the interview room is:

---

[1] In cooperation with MIT Media Lab and Intelligent Cooperative Systems Laboratory at Research Center for Advanced Science and Technology, the University of Tokyo. And a normal digital video camera was also prepared

1. *Retrospective reports with visual aid the subject recorded*: The subjects were asked to report "what you look at", "what you think about it" and "what you do" along with the VTR as an memory aid.
2. *Questions about the subjects' impression on the event objects*: The questions were made based on the interviews with the planners and the planning papers. This is to investigate how the planners' intention and the visitors' impression match or mismatch with each other.
3. *Keyword questionnaires*: The keywords are also extracted from the interviews with the planners and the planning papers. This is to investigate what keywords the planners presented were impressive for visitors.

4. Results and Discussions

Though there are a lot of findings that surprised the planners, in this paper we are going to show one example of a lot of unexpected findings that are beyond the planners' expectation at Motor Show. These findings devote to create new knowledge, that is "knowledge reconstruction". One example of "the effect of the other visitors" at Motor Show is described.

The other visitors can provide a context that raises the degree of satisfaction of the visitor. The following report was obtained:

> *A companion took a picture with a family. Both of the companion and the child smiled. My (= the subject's) children also like cars. They would be delighted if I took them here. That is a good idea.*

This observed data was reported to a planner and he hit upon a new strategy:

> *By inviting families that are customers of the company, the other visitors will feel in a way mentioned above. Moreover, the invited family will also feel better because they feel "they are invited as special guests" and this family can enjoy being a customer of the company, which will be great benefit to the company, too.*

This is a good example of "knowledge reconstruction". We call it "reconstruction" because implicitly they might know the follows:
- A visitor is affected by another visitor at an event site.
- Customers are delighted if they are invited as special guests.
- If customers like the company, it is beneficial to the company.

But these pieces of information have not been connected. That means the knowledge obtained through this analysis can support the event designing if it is properly reported to the planners. We are going to propose a supporting system that provides "knowledge for strategies with real context".

5. Toward a Supporting System for Knowledge Reconstruction

As our investigation externalized a lot of knowledge successfully, we are going to propose the supporting system for the transition among modes in Fig 1, i.e. knowledge reconstruction. Each mode is supported in the following way:

Externalization: This mode is supported by the methodology we proposed in the former section.
Combination: From the data obtained with the analysis, this mode is supported by "exhaustive search" and "presentation of contexts" Based on the conclusion of Hori and

Yamamoto et al., the spatial representation is adopted.
Internalization: It is expected that this system promote the user's understanding of the knowledge.
Socialization: For example, "pursuasion of clients" is expected by using this system.
To accomplish this aim, our system should be implemented with the following features:

- **Accumulating the knowledge:** the system should accumulate the data of the analysis
- **Browsing and reconstruct the knowledge:** the system should present the knowledge with the real context to promote and amplify the user's reflection [3][4]. And moreover, it should foster the discovering and creating process of knowledge.

We adopt spatial representation to present data to the planner. Hori [5] conducted the experiment for the effect of the spatial representation on the conceptual design. The system named AA1 presents words, which represents the concept the user vaguely conceives, on the two-dimensional space, the user can change the location of the words on the space. This action helps the user clarify his/her concept gradually and such phenomenon was observed as the user came to generate new concept by looking at the blank area on the space.

The image of the system is shown in Fig 2.

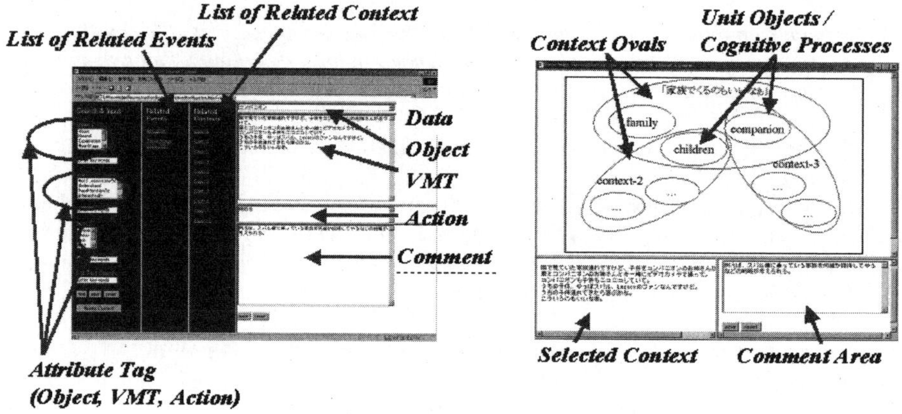

Fig 2 A Snap Shot of the System

The attribute set of "Object-VMT (Visitors' Mental Transition)-Action", i.e., the transition of the visitors' cognitive process "what you percept" - "what you think" - "what you do" are recorded together with the "real context" data. These attributes are composed of "unit object" such as "family", "companion" and "unit cognitive process" such as "Remember", "FeelStarting" are currently being defined to code the data [6][7]. The user can save the raw data with attaching attributes tag onto it. The data can be refered by searching with the attribute tags or by full text search. The user can add new attributes freely.
The expected interaction to reconstruct knowledge: The situation is that the planner would like to browse the existing contexts to create an innovative one.

1. The planner selects attribute(s) from the lists. e.g.) "how could the visitors feel something is starting?"
2. The system exhaustively searches contexts along with the selected attribute. e.g.)"FeelStarting"
3. Then the system lists the contexts that contain the selected attribute tags, so that the planner can refer to the real contexts. The related events are shown on the "Related

Events" list and the related contexts on the "Related Contexts" list (the third column of the left figure).

4. By clicking the button "Relate Context" (left most bottom of the left figure), the contexts are extracted along with the units and all of the contexts (shown as ovals in the right figure) are shown on the space. It shows the relations among contexts that the planner might not notice. Moreover, it is expected that the planner hit upon innovative idea by looking at the blank or combining the parts of existing contexts [8].

5. And it is also expected that the system automatically generate the new contexts by combining the existing Objects, VMTs and Actions.

As for the automatic generation of the context, Ogata [9] developed a basic framework for narrative conceptual structure generation. This is a system based on narrative techniques and strategies. By changing the strategies, that is, tuning the strategies for a certain domain, it is possible to generate a certain structures suitable for the domain.

6. Conclusion

In this paper, we proposed a real application of the methodology to investigate the cognitive processes at event site toward "knowledge reconstruction system". The microscopic investigation was conducted to understand the cognitive processes and their transitions. Through protocol analysis, the following knowledge is found with "the real context", which has not been obtainable from traditional statistical investigation.

- Effect of the other visitors (in our result, for example, "effect of a family")
- Effect and difference of understanding of congestion between designers and visitors

These knowledge with the real contexts are useful to establish new strategies, although the statistic data lacks hence they are discarded. That is why we propose "knowledge reconstruction system" that provides with the planners "knowledge with real context".

From now on, we are going to develop a knowledge reconstruction system and evaluate it at the actual event-planning phase.

Acknowledgement

The authors gratefully acknowledge the generous assistance of Ms. Shoji, Mr. Shibata and Mr. Kanazaki for experiments and discussions. With thanks to Ms. Ueoka in RCAST and Mr. Clarkson at MIT Media Lab for wearable computers.

References

[1] Gerhard Fischer, Jonathan Ostwald: "Knowledge Management: Problems, Promises, Realities, and Challenges", IEEE Intelligent Systems, Vol.16, No.1, pp. 60-72, January/February 2001

[2] Ikujiro Nonaka, Hirotaka Takeuchi: "The Knowledge-Creating Company : How Japanese Companies Create the Dynamics of Innovation", Oxford University Press, 1995

[3] Schoen, D A: "The Reflective Practitioner: How Professionals Think in Action", Basic Books, NY, 1983

[4] Donald A. Norman: "Things That Make Us Smart: Defending Human Attributes in the Age of the Machine", Addison Wesley Publishing Company, 1994

[5] Hori, K.: "A System for Aiding Creative Concept Formation", IEEE Trans. Systems, Man, and Cybernetics, Vol.24, No.6, pp.882-894, 1994

[6] M. Suwa and T. Purcell and J. Gero: "Macroscopic analysis of design processes based on a scheme for coding designers' cognitive actions", Design Studies, Vol.19, No.4, pp.455--483, 1998

[7] Shigeki Amitani, Koichi Hori: "Supporting Musical Composition by Externalizing the Composer's Mental Space", Journal of Information Processing Society of Japan, Vol.42, No.10, pp.2369-2378, 2001

[8] Hori, K.: "A Model to Explain and Predict the Effect of Human-Computer Interaction in the Articulation Process for Concept Formation", Information Modelling and Knowledge Bases, Vol.7, pp.36--43, IOS press, 1996

[9] Takashi Ogata, Koichi Hori, Setsuo Ohsuga: "A Basic Framework for Narrative Conceptual Structure Generation Based on Narrative Techniques and Strategies", J. of JSAI, Vol.11, No.1, pp.148-159, Jan., 1996

KES 2002
E. Damiani et al. (Eds.)
IOS Press, 2002

Visual Representation of Scene Information for Shopping as Concept Articulation

Mikihiko Mori[†]   Hiroko Shoji[† ‡]   Koichi Hori[†]
Research Center for Advanced Science and Technology, The University of Tokyo,
4-6-1 Komaba, Meguro-ku, Tokyo 153-8904 Japan

Abstract. We have built the online-shopping experiment system, S-Conart to analyze the effect of forms of information representation in its user interface. Through the results of the analysis, we hypothesize that information representation in situation information of using bought merchandise effectively promote decision-making in purchase. This paper is written about added interfaces to S-Conart in order to confirm the hypothesis. We explain the concept of the scene tree and network interfaces that use situational words, describe the implementation of the interfaces, and illustrated the system.

1   Introduction

We study to shape interactive systems for online-shopping that advances decision-making of sales clients for shopping to discover sales chances[3, 5]. We developed the online-shopping system named S-Conart (Concept Articulator for Shopping) that represented spatial arrangement of merchandise that used relationship between them by using their attributes. The experimental data set was run with Japanese sake as merchandise.

In this paper, we describe new concepts with expressions of scene information for decision-making support of shopping, where a scene denotes situation of a client who will use items of merchandise, and describe visual representation of the scene information with examples of execution and internal representation of the scene information for S-Conart.

2   Concept Articulating Supports for Clients

From both analyses of sales clerk's behavior in real shops[3] and clients behavior in S-Conart[4], the following perspectives is necessary for online-shopping systems in case of articulating concept of merchandise:

Awareness of merchandise attributes Showing arrangements of merchandise causes to make client get the laws of the arrangements for expected items of merchandise or unexpected ones and to make be aware particular attributes.

[†]Research Center for Advanced Science and Technology, The University of Tokyo
[‡]Faculty of Education, Kawamura Gakuen Women's University

Satisfaction of merchandise concepts Providing enough scene information on client's side causes to make him/her mental transition, where the scene information means description of scenes, the scene information is expressed by words or phrases. Therefore, throughout the process of the transition, the scenes gives him/her satisfaction to desire for buying nice items.

We note that both perspectives are important to make clients decide to buy from merchandise. The first perspective has been discussed in the previous paper[4] that some clients had set up more detailed scenes on which they would use the merchandise that they were selecting than the ones in the experimental settings. On the other hand, a excellent clerk provided sufficient scene information to give clients satisfaction, which meets the second perspective. The support of satisfaction-making for an online-shopping system is the new perspective for this paper.

We propose the following expressions of scene information for the satisfaction-making support:

Comments of each item of merchandise Through presenting comments of items of merchandise, clients can learn their criterion and can know distinctions of items respectively. Therefore, they are inspired to recognize concept of each item by the comments.

A scene tree expression The tree is a personal viewpoint for merchandise. A node of the tree is correspondent to a scene on the client. The tree is changed when the viewpoint of a client changes to another or when another client uses it. Searching the tree from the root to a leaf is considered finding out more detailed concept or discovery of next scenes as time rolls on.

A scene network expression The network expresses relations of any scenes in any viewpoints. The network can has a client getting the general viewpoint. Here, a relation between scenes has strength and weakness.

A child-parent relationship of scenes in the tree need not be fixed to particular conceptual structure, and also it does not have the intention of eliminating a item's name as words or a phrase for a scene, but the relation that is contrary to client's intuition is not desirable.

3   Presentation of Scene Information

We will adopt the following hypothesis stood on the findings of the behavior both the real-shop's clients and the S-Conart's clients: scene information inspires clients to decide for the purpose of merchandise selection. On the basis of the hypothesis, we add new visual interfaces that give clients scene information for decision-making support for shopping. After shopping, it is expected that the clients understand merchandise's concept and that the merchandise give them satisfaction about the shopping.

As related works, Niwa et al. proposed the system that has two views[2]. These views cooperated to search dictionaries and logs of news papers. Hirata et al. presented concept associative representation by a tree and a map[1]. The tree placed a phrase as the top node which was searched by a user. The map that centered a searched phrase put associated phrases. However, these methods are used the only personal viewpoint and do not consider the general viewpoint.

3.1   Visual Representation of Scene Information

We mainly supported awareness of attributes of merchandise in the old S-Conart, namely decision-making support in online-shopping by the spatial arrangement. However, it is important to use words or phrases representing scenes in the situation by reason of describing above, and showing a scene is more natural than to only show attributes of merchandise.

Current S-Conart provides scene information to support satisfaction of merchandise concept explained above. We add comment view and input form of selected item in old S-Conart window named merchandise window (See Figure 1). And also we set up new window that shows a scene tree representation and a scene network representation, which is named word window (See Figure 2).

The scene network representation crystallizes the scene network expression. A node means a scene. The label of the node represents a scene name as words or a phrase that is framed by a thin box. An arc between nodes means a relationship between scenes. The relationship represents various thickness of a line by its strength. Each scene can drag to move when you want. If you select scenes and click to show related items, the representation of spatial arrangement in the merchandise window is changed to show related items highlighting so as to provide relationship between items and scenes.

The scene tree representation crystallizes the scene tree expression. The tree representation draws like the folder-tree view in which one node corresponds to one icon of folder- or document-like representation. A node also means a scene, and the label of the node also represents a scene name. The node can fold to make children disappear or unfold to make them appear in case of the node have children.

When a node in the tree is selected, the node in the network is selected synchronously. The visible nodes in the network are only the nodes shown in the tree in order to show legible even if it shows an enormous amount of words. If you fold a node at the tree, the node of same scene name and its children will disappear in the network. At first time of displaying the word window, S-Conart shows only top nodes in the tree.

3.2   Internal Representation of Scene Information

To present scene information, we have to choose words or a phrase as each scene. We use words in the comments of each item of merchandise. It is simple ways on this system to make scene names:

1. All comments in all items of the merchandise is applied morphological analysis to make a word set.

2. Edit the word set to eliminate meaningless words or miss-analyzed words. Make words or a phrase as a scene if you need.

The network is made of relations between scenes. A relation between two scenes is generated from cooccurrence of both of scene names by words or phrases. Cooccurrence is calculated by way of counting both of scene names occurred in the same comment. Then, the counter of cooccurrence is normalized by the counter at its maximum.

The tree is founded on the scene on which a client have stayed from the network. A new tree is generated as follows:

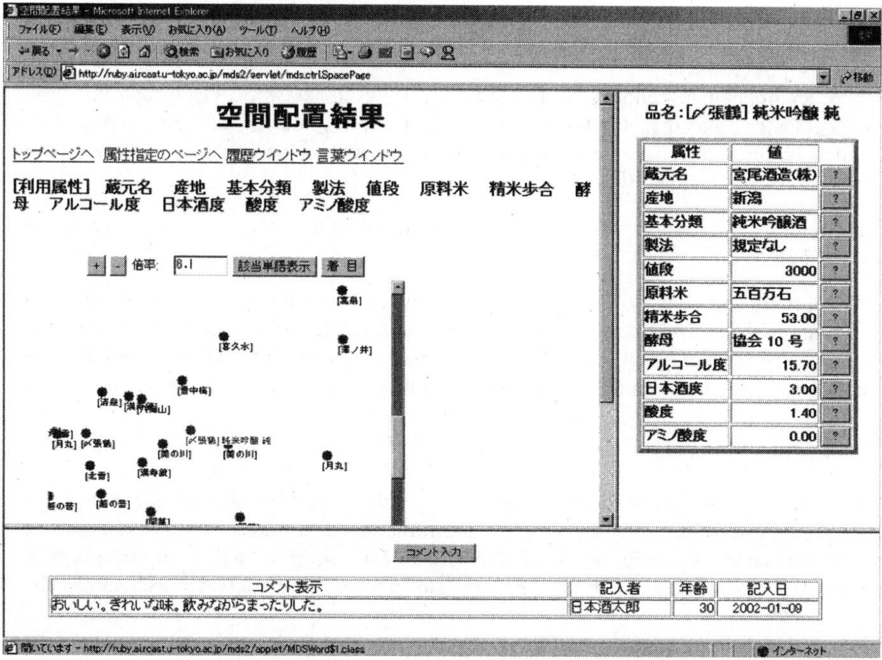

Figure 1: A merchandise window: items in spatial arrangement (the upper left frame), attributes of an item (the upper right frame) and comments of the item (the lower frame).

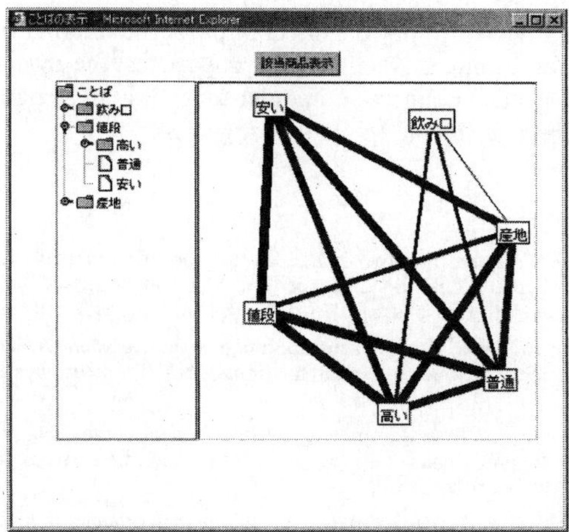

Figure 2: A word window: a scene tree (the left frame) and a scene network (the right frame).

1. The system recognizes the nodes which are selected by a client for the purpose of reconstruction of the tree.

2. A_u is an arc, and A_v denotes the arc adjacent to A_u in the network. The weight value that A_v gets is calculated by the law $W_v = W_u/2$ where W_u and W_v are the weights of A_u and A_v respectively. The weight spreads from a selected node until all arc are weighted in breadth-first search order.

3. The total weight for an arc is calculated by multiplying the cooccurrence at the arc by the sum of all of the weights.

4. A minimum spanning tree is made from the network, which uses the total weights. The system considers a minimum spanning tree whose tops are the selected nodes as the scene tree. In the time of generating a tree, it can make the tree with about three or four top nodes, no many but nothing at least, at a time like excellent clerks' introduction.

4 Conclusion and Future Works

In order that a client inspires himself/herself to purchase merchandise in online-shops, an online-shopper make the client recognize items of merchandise and satisfy them as purchasing in a real shop. It is one of effective technique for which the online-shopping system remind the concrete scene when the items are used and the system gives the client satisfaction.

We suggested the representation of scene information to make S-Conart provide it so as to be recognized and satisfied on shopping for clients. The scene information expressed a scene tree and a scene network. The scene tree expressed client's particular local situation that is difference in the case. On the other hand, The scene network expressed overviews of relationships between scenes not to be dependent on a situation. Thus, we expect to change client's view of items inspired by scene information.

For the future works, we will verify the effect of the scene information in term of recognition and satisfaction for the client. Also, we will propose that the arrangement of the scene network will effect the spatial arrangement in order to be easy to recognize relationship between the spatial arrangement and the network arrangement.

References

[1] Takashi Hirata, Harumi Murakami and Toyoaki Nishida: Support for Community Knowledge Sharing with Associative Representation and Talking-Virtualized-Egos Metaphor, Transactions of the Japanese Society for Artificial Intelligence, Vol. 16, No. 2, pp. 225–233, 2001 (in Japanese).

[2] Yoshiki Niwa, Makoto Iwayama, Toru Hisamitsu, Shingo Nishioka, Akihiko Takano, Hirofumi Sakurai and Osamu Imaichi: Interactive Document Search with DualNAVI, Proceedings of NTCIR'99, pp. 123–130, 1999.

[3] Hiroko Shoji and Koichi Hori: Toward Improving the Interface for Online-shopping —Suggestions from the Analysis of Real Shopping Behavior—, Journal of Information Processing Society of Japan, Vol. 42, No. 6, pp. 1387–1400, 2001 (in Japanese).

[4] Hiroko Shoji and Koichi Hori: Strategy emergence from human-computer interaction, J. S. Gero and K. Hori(eds), Strategic Knowledge and Concept Formation III, pp.87–99, 2001

[5] Hiroko Shoji, Mikihiko Mori and Koichi Hori: Creative Communication for Concept Articulation in Shopping, Proceedings of The 6th World Multiconference on Systemics, Cybernetics and Informatics, 2002 (to appear).

KES 2002
E. Damiani et al. (Eds.)
IOS Press, 2002

User's interests change as Chance Discovery

Akinori Abe
NTT MSC
No. 43000, Jalan APEC, 63000 Cyberjaya Selangor Darul Ehsan, Malaysia
ave@cslab.kecl.ntt.co.jp

Abstract

Recently, the necessity to the assistance of creative work has been increased. Creative works are usually regarded as specialized skills. However, from the viewpoint of chance discovery, creative work can be thought of as chance. This is because creative means novel (not rare) and creative work comes from nothing or from something relative but whose relation was hidden.

Therefore, in this paper, we regard chance as creative work, then we will show and model one of the process to conceive chance that is creative work.

1. Introduction

Recently, the necessity to the assistance of creative work has been increased. As to creative works, we usually expect artistic works by professionals. However, in this paper, we expand the definition of creative work to daily works that need some potential senses. First, we show some of definitions of "creative".

Carl Rogers defined creative process as follows [7].

> The creative process is the emergence in action of a novel relational product, growing out of the uniqueness of the individual on the one hand, and the materials, events, people, or circumstances of his life on the other.

Also, in DynaGloss [2], collective creativity is defined as follows.

> A term describing the phenomenon where concepts emerge in people's mind through interacting with knowledge in the world — external representations, with other people, or with computer systems. Designers evolve artifacts by externalizing such concepts and by sharing them with other people. Though creative individuals are often thought of working

in isolation, the role of interaction and collaboration with other individuals is also critical. Thus, creative activity grows out of the relationship between an individual and the world of his or her work, and out of the ties between an individual and other human beings.

Our definition of creative work will be similar to collective creative.

> Creative means novel and creative work comes from nothing or from something relative but whose relation to the work was hidden. Actually, for creativity, a sort of intuition or special sense will be necessary. An intuition or special sense will emerge with some stimulus like information (background knowledge). In emergence, knowledge will work as a catalyst.

As shown in our definition, we need various types of background knowledge to unveil our intuition or special sense for creativity which are hidden in our mind.

From the computational viewpoint, computers can easily deal with a large amount of information. Furthermore, recently internet has become easy to be accessed for everyone, as a result, it is quite easy to search necessary information on the internet. Actually, DSIU [3] that is a question and answer system obtains the necessary knowledge from the internet, then creates missing or new knowledge to make inferences. Similarly, as shown in CYC [4], a large amount of organized information seems to help us in common sense reasoning. In addition, a large amount of knowledge is thought of as a help to achieve creative work. As to creative work assistance, Shibata and Hori proposed creative thinking supporting system [8] which shows information and related information to stimulate the user's thinking. Anyway, due to the widespread internet, it is easy to deal with a large amount of information. Accordingly, it seems to be slightly easy to stimulate the user's creative thinking by the way of internet.

Anyway, creativity will emerge with some stimulus like background knowledge. We also defined that creativity hide in out mind, and creativity needs to be conceived. Ohsawa defined chance as follows[6]:

> A chance (risk) is a new event/situation that can be conceived either as an opportunity or a risk.

In this sense, the creative work can be thought of as chance. This is because creative means novel (not rare) and creative work comes from nothing or from something relative but whose relation was hidden. For example, in [9], Sunaga showed one example for creativity. That is, in the night, he saw something strange that is motorcycle with bag. At first, we could not recognize it as motorcycle. But, after he recognized that was motorcycle with bag, he obtained the new design idea for motorcycle. Something strange stimulated his creativity. Something strange can be regarded as something relative but whose relation (to motorcycle) was hidden. In addition, Shibata and Hori discussed creative work assistance from the viewpoint of chance discovery.

In this paper, we regard chance as creative work, then we will show and model one of the process to conceive chance that is creative work.

2. Creative work as chance

In general, creative works are regarded of as philosophical thinking, drawing paintings, poem writing, melody composing etc. Sometimes they come from nothing like an oracle during sleeping, walking, swimming, dining... However, usually a certain seed hides behind the work. As shown in definition by us, for creativity, we think collaboration with organized knowledge is important. The problem is how to obtain organized knowledge. Indeed, it will be easy to obtain knowledge on the internet. However, it is slightly difficult to catch up with the changing information. One of the powerful tool to catch up with will be a search engine.

This section shows one of the creative process by the example of writing technical paper. In fact, we show the process by using a search engine.

2.1 Non-creative search result

Let the situation be where we do not have enough reference for writing paper, and proper book shops and library are not close to us but we have an internet-connected computer. The easy solution is to consult with the internet. Suppose we want to write a paper on the "effects of humour in communication". It is natural to input words 'effect', 'humor', and 'communication' to search engine site, for example, `www.yahoo.com`. Then the result is as follows.

1. Humor

   ... separately from the purposes of humor, but, as will be seen ... to the success of humorous communication. ... order to achieve a humorous effect and bring the audience ...

   http://www.brown.edu/Departments/Anthropology/...

2. Jest for the Health of It! – "Humor Skills..." by Patty Wooten

   ... Humor and Laughter Effect the Body. ... field of psychoneuroimmunology which defines the communication links and relationships between our emotional ...

   http://www.jesthealth.com/artantistress.html
   More Results From: www.jesthealth.com

3. Encyclopedia.com - Results for humor

   ... change, perceived funniness, and the effect of humor stimuli. ; Behavioral Medicine ... Why do we laugh?(Humor) ; Communication World Fatt, James PT ...

   http://www.encyclopedia.com/articles/22511.html

4. Refereed Publications

   ... transitory dominance as a communication variable affecting humor ... suffer as factors in humor appreciation. Journal of ... questions and its effect on the learning ...

   http://www.joannecantor.com/refereed.htm

Actually, they are useful to write papers about "the effects of humor in communication". However, we cannot get more information from the results. That is, this result is not suitable to creative works.

2.2 Creative search result

Let's see another situation. Suppose we are going to write a paper on "the usage of language in advertisement". If we input 'advertisement' and 'language' to `www.yahoo.co.jp`, the result is not satisfactory. If we input the words to `www.yahoo.com`, the result is also unsatisfactory. That is, no proper result is listed from the search engine. Since some of the advertisements will be shown in (multi)media, we change the keyword from 'advertisement' to 'media'. If we input 'media (メディア)' and 'language (ことば)' to `www.yahoo.co.jp`, the result includes the following lists.

5. What wrong with our literary style?[†1]....
   http://www.ipsj.or.jp/members/SIGNotes/Jpn/33/
   1998/034/article001.html

6. This is not a clock.....
   http://duck.modern.tsukuba.ac.jp/develop/develop-
   5/develop-5-0.html

Actually, these page are not what we want to refer to, but these results give us another viewpoint of the theme for "language in media or advertisement". Then we can research for another type of research on language or we can include another viewpoint in our research.

The case in humour returns proper results, but the case in media language does not return proper results. However, the case in media language has possibility to extend or change our interest to our potential interests. In addition, the site shown in [5.] is page for SIG (Special Interest Group: like workshop) on information media, therefore, we can obtain another papers. Thus, the failure of searching sometimes provide us better direction.

We regard creative work as chance. As shown above, creativity is something novel and potential. The creativity appears by a certain stimulus. In the above case, slightly different research area can be a stimulus. Actually, as Shibata and Hori pointed out, similar area will be also stimulus, however, it cannot change the user's interest. In fact, when we are in deadlock to have a new idea or thought, the change of viewpoint or interest works well.

The next section will show modeling of creativity process as chance discovery process.

3. Chance discovery process

3.1 Chance discovery process modeling

Shibata and Hori showed that related problems and ideas stimulate the user's creative thinking. Indeed, it is true, however, we showed another process of creative thinking. The process is show in Fig. 1. This figure shows the thinking process modeling shown in the previous section.

The process shown in the right side is that shown in the case of humour. The expected or wanted results are shown. In this case, normal (expected) result will be created. Anyway, usually, we want this type of search result.

On the contrary, the process shown in left side is the process shown in the case of media language. No results

---

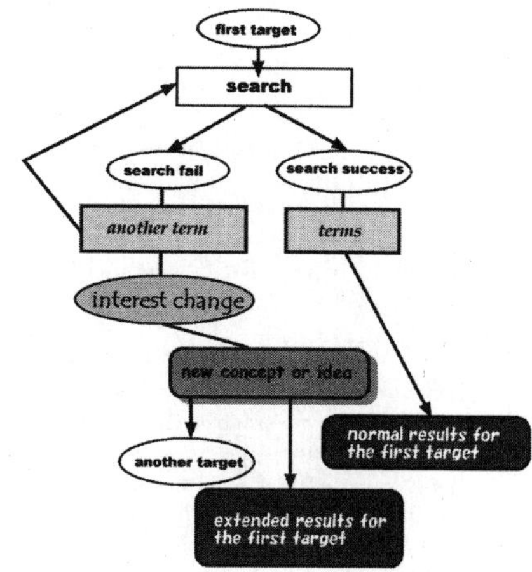

Figure 1. Chance discovery process

or unexpected results are shown. Actually, we do not want to have unexpected results. Indeed, 80% of these types of results will be useless or waste of time. However, the interesting case is the process shown in left side. In this case, indeed the searching fails, that is, the result is far from our expectation or desire. However, the user can find a new interest or viewpoint to do better or another research. We mean that slightly different result simulates our hidden creativity. Slightly different or unexpected result will be a catalyst to stimulate our hidden creativity or desire.

We think this type of unpredicted interest change is chance. By this interest change, we can go into the new concept or change our viewpoint. Of course, we can start a new research.

3.2 How to set search words?

Actually, we show the only successful example in the previous section. Of course we do not always succeed in obtaining proper site to stimulate our creativity. For example, if we input 'media' and 'language' to www.yahoo.com, the result is not satisfactory. In addition, as to very specialized field like 'narratology', the search result will be expected. For creativity, ambiguous (has multiple meanings) or non-specialized search words seem to be better.

However, if we input words 'humor' and 'communication' to www.yahoo.com, then the result is as follows.

7. Assessment Staff

   ... breadth of subject knowledge, sense of humor, volumes of practical and useful materials ... her BS Degree in Speech Communication from Oregon State University. She ...

   http://www.nwrel.org/assessment/Staff.asp?odelay=0&d=0

8. Humor and Respect

   ... must be tortured and die before we reach out for understanding and communication? Homosexuals are real people, with real hurts, desires, and needs. Somehow ...

   http://members.aol.com/graceeaca/chapter14.html

Actually, this result does not stimulate our creativity. In fact, the coverage of the field seems to be larger, however, the field is quite close to the result from three keywords. Finally, if we input only word 'communication' to www.yahoo.com, then the result will be overflowed. Accordingly, the result is useless or sometimes harmful.

The search engine is easy and useful, however, for proper results, we need to select keyword. If we need keen result, we will input a lot of keyword to restrict the result. This will be a real usage. Thus, it is quite difficult to obtain result that will stimulate our hidden creativity.

Moreover, if the searched site has various sorts of papers, it will be help to stimulate our thinking from the other viewpoint. For example, when we search a certain theme, if we can find the site like http://syass.kwansei.ac.jp/kiyou/ (database of the papers of Kwansei university), it is very helpful to stimulate or change our interest. Hence it has various paper from unexpected field. Actually, this case is ideal or lucky case, and it is quite difficult to decide the keyword. If we provide some database for such an archive, there might be some possibilities to reach such proper sites.

In this paper, we cannot show a generalized method to select proper keywords to have the proper search result that stimulates our hidden creativity. However, for creativity, we found some of importance in selection of search words:

- properly ambiguous (has multiple meanings) words

- non-specialized words

- non-common used words

4. Conclusions

This paper showed chance as creative work, and shows the process of chance discovery. This formalization comes from author's experience. The author works in abroad and does not bring enough references and there is not proper libraries around office and house. When he wrote paper, he naturally used an internet search engine like yahoo. Actually, each search engine has its own feature, therefore we must know their feature to obtain proper results. This task will be a sort of craftsman performance. Sometimes search engine returns different result from our intention. Usually, such wrong results are useless, however, in some cases, the wrong result leads us novel thinking or confirm our original target. We modeled this type of process as a chance discovery process.

Nara and Ohsawa proposed double helical model of chance discovery process [5]. Their model is similar to ours. Our model is based on sequential processing, on the contrary, their model presumes parallel processing. Actually, we sometimes think in a parallel way, but our model does not deal with parallelism. This is because we asuume the model to write only one paper. Anyway, we must consider the model of doing multiple creative works at the same time. In the next paper, we should take the parallelism in consideration.

In this paper, we could not show a generalized method to select proper keywords to have the proper search result that stimultes our hidden creativity. We only gave some hints for keywords. In the future we should find generalized method to select proper keywords.

In another paper, we proposed context change in abduction to generate homour phrase [1]. Interest change can be caused by a context change. In the future paper, we will formalize the shown creative process by abduction.

References

[1] Abe A.: Applications of Abduction, *Proc. of ECAI98 Workshop on Abduction and Induction in AI*, pp. 12–19 (1998)

[2] DynaGloss: http://Seed.cs.colorado.edu/

[3] Fujimoto K. and Matsuzawa K.: Intelligent systems using web-pages as knowledge base for statistical decision making, *New Generation Computing*, Vol. 17, No. 4, pp. 349–358 (1999)

[4] Guha R. V. and Lenat D. B.: Cyc: A Midterm Report, *AI Magazine*, Vol. 11, No. 3, pp. 33-59 (1990)

[5] Nara Y. and Ohsawa Y.: Understanding Internet Users on Double Helical Model of Chance-

Discovery Process, *New Generation Computings*, Vol. 21, No. 1 (2002) (to appear)

[6] Ohsawa Y.: Chance Discovery for Making Decision in Complex Real World, *New Generation Computings*, Vol. 20, No. 2, pp. 143–163 (2002)

[7] Rogers C.: Toward a Theory of Creativity, in Creativity and Its Cultivation (Harold Anderson eds.), Harper & Row (1959)

[8] Shibata H. and Hori K.: An Approach to Support Long-Term Creative Thinking and Its Feasibility, *Post-Proc. of Joint JSAI Worlshop (LNAI 2253)*, pp. 455–461 (2001)

[9] Sunaga T., Matsuura H., Matsuzawa K., Hori K., Abe A.: An Invitation to Language Sense processing Engineering — Creator or Assistant?, *J. of JSAI*, Vol. 15, No.3, pp. 456–447 (2000) (in Japanese)

KES 2002
E. Damiani et al. (Eds.)
IOS Press, 2002

Self-Organizing Map
with Limited Scope Learning

Masahiro MICHIHATA, Tsutomu MIYOSHI, Hiroshi MASUYAMA
Information and Knowledge Engineering, Tottori University
Tottori-shi Koyama-chou Minami 4-101, Tottori, Japan

Abstract: Self-Organizing Map (SOM) is a kind of neural networks that learns without supervision. In this paper, we proposed "Limited Scope Learning" on SOM. This technique is able to get the feature map that is shown by distributed expression, i.e., at the learning time, more than one winners are selected out of the whole map. In the case that the troubled nodes exist on the map, the degree of node fault tolerance will be improved by using this method rather than the conventional technique.

1. Introduction

Self-Organizing Map (SOM) is a kind of neural networks that learns without supervision. It is necessary for process of self-organizing to require the following general rules.
1. To specify the position of winner neuron, which is the best fit for input data.
2. To adjust connection weight of the winner and its neighbor neurons to the input data.
On conventional SOM, only one winner is selected from all nodes of output-layer. This is, however, undesirable in respect of information expression or the degree of node fault tolerance. In this paper, we propose "SOM with Limited Scope Learning" that separate all neurons of output-layer by Limited Scope, and select winners from each area.

Hereafter, Chapter 2 explains SOM and Chapter 3 describe the "Limited Scope Learning." And it indicates about experiments and its result at Chapter 4, and Chapter 5 describes a conclusion.

2. Self-Organizing Map (SOM)

Self-Organizing Map (SOM) usually provides a topology-preserving mapping from the high-dimensional space to two-dimensional map. SOM groups similar input data, which are nearby each other in the input space. And input data are mapped to nearby map units. SOM can thus serve as a clustering tool as well as a tool for visualizing high-dimensional data [1]-[3].

SOM requires two layers of units: the first is an input layer, that contains units for each element in the input data, and the second is an output layer or grid of units that is fully connected with those at the input layer. When input patterns are presented to the input layer, the output units compete with each other. The winner is the output unit whose incoming connection weights are the closest to the input pattern. In learning process, connection weights of the winner are then adjusted, i.e. move in the direction of the input pattern. That is, SOM configures the output units into a topological representation of the original data, through a process called self-organization.

3. Limited Scope Learning

On conventional SOM, only one winner is selected from all nodes of output-layer, so information expressed as local expression. Therefore following problems are showed:

1. Information expression efficiency is low, because certain information is represented to certain domain of units by one to one correspondence.

2. The degree of node fault tolerance is low, because, in case that the domain of nodes breaks down, the information that they were expressing is completely lost.

However, originally, a neural network is a parallel distribution mechanism and a distributed expression is greatly excellent in respect of information expression or the degree of node fault tolerance.

Therefore we proposed "Limited Scope Learning." This technique is able to get SOM map that is shown by distributed expression. At the learning time, a winner is not selected out of the whole map, but selected within each "Limited Scope." We suppose, that two or more winners will exist by using this technique, so information will disperse on the map. Thus it is expected to solve above-mentioned problems.

3.1 Learning Algorithm of proposed technique

Proposed algorithm is summarized the following:

(1) Calculate a distance between the input data (vector) and the weights vector of each output unit. And search the closest output units within the "Limited Scope" around each node, it is the winner node of the scope.

$$d_{ij}(t) = \sum_{i=0}^{n-1} (x_i(t) - w_j(t))^2$$

(2) Adjust weights vector of the winner unit and neighborhood units of the winner, to move closer to input vector.

$$w_j(t+1) = w_j(t) + h(t)(x_i(t) - w_j(t))$$

(3) Repeat (1)-(2) while learning time.

(4) As learning progress, the size of the neighborhood around the winner gets smaller. Then, initially large number of output units will be updated. And, as the learning progress, the number of updated units gets less and less. Finally only one unit will be updated at the end of the learning. Similarly, the learning rate will decrease as the process progresses. But the size of Limited Scope will not be changed.

Where t stands for a learning times, $d_{ij}(t)$ stands for a distance between the input vector and the weights vector of each output unit, $x_i(t)$ stands for a input value, $w_j(t)$ stands for a connection weights value, and $h(t)$ stands for a learning rate.

4. Experiments

We experimented how influence may appear in the map that formed by proposed technique.

1. Check whether the self-organization of the map can be carried out by proposed technique.

2. Investigate whether the degree of node fault tolerance improved by the proposed technique as compared with the conventional technique.

3. Verify whether it is safe to discernment of data, because the map will be visually complicated.

The conditions in experiments are shown in Table1. The used data were assumed sets of points scattered at center and among it on 3-dimensional space. Then those scattered points are regular random number.

$$x_i = \sqrt{-2\sigma^2 Rand_1} \sin 2\pi Rand_2 + \mu_i$$

Where σ^2 stands for the dispersion and sets to 0.1, $Rand_1$ and $Rand_2$ stand for the random number, and μ_i stands for the center of cluster. The data are classified into 8 clusters, which are shown in Table2. Those data were 50 points in each class, in total 400 points as learning data, and 1000 points each, in total 8000 points as estimated one.

Table 1: the conditions of experiment

	Conventional technique	Proposed technique
Size of Map	20x20	
Number of learning data	400	
Number of estimated data	8000	
Learning time	1000	
First rate of learning	0.8	
First range of neighborhood	20	2-20
Size of Limited-Scope	Whole of Map	2-20

Table 2: cluster of experiment data

Cluster	Center	Cluster	Center
A	(0.0, 0.0, 0.0)	E	(5.0, 5.0, 0.0)
B	(5.0, 0.0, 0.0)	F	(5.0, 0.0, 5.0)
C	(0.0, 5.0, 0.0)	G	(0.0, 5.0, 5.0)
D	(0.0, 0.0, 5.0)	H	(5.0, 5.0, 5.0)

4.1 Experiment 1: Self-Organizing

Fig.2 shows one of the maps formed by proposed technique. "#" stands for the border line that the nodes are too few selecting as winner to assign any cluster. In Fig.2, it is showed that each cluster organized itself into their domain. Therefore it is proved that the self-organization of the map can be carried out by proposed technique. And it is proved that the map formed by proposed technique is represented by distributed expression, since each cluster divided into two or more groups.

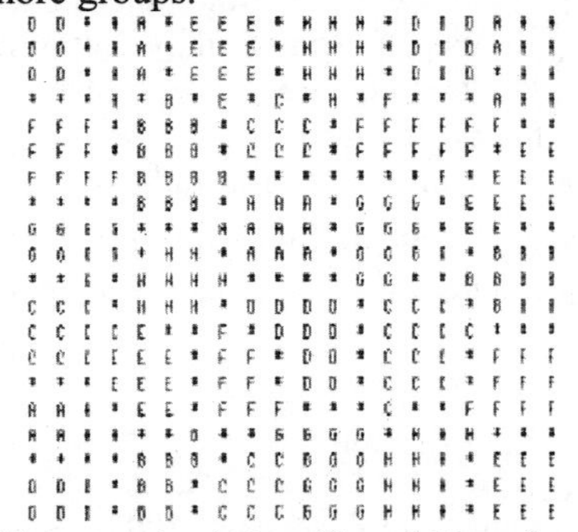

Fig 2: map with proposed technique (the neighbor of competition: 10)

4.2 Experiment 2: the degree of node fault tolerance

On the node-fault-tolerances, we confirmed there are some differences would be in the conventional technique and proposed one, in the state where some output units break down. We investigate three kinds of distribution of a fault node as following:

1.Random.   2.Single burst. (It gathers in one place.)   3.Plural burst. (It gathers in two or more places.)

And calculate following values to techniques.

$$t_{node} = \frac{\sum N^{(Troubled)}}{\sum N^{(All)}} \times 100\% \qquad (1)$$

$$d^{(Average)} = \frac{\sum d_c}{D} \qquad (2)$$

Where t_{node} stands for the rate of fault node on the map, it is expressed with a formula (1), $d^{(average)}$ stands for the average distances between input data and winner node, it is expressed with a formula (2), d_c stands for the distance between input data and winner node, D stands for the number of input data. If $d^{(average)}$ is large, it can be said that a map cannot reflect the feature of input data.

At Fig.3, the x-axis represented the $d^{(average)}$, and the y-axis represented the t_{node}. In Fig.3, when the fault node exists at random, it is almost same to the conventional technique and the proposed one, however, when the fault node gathers, the difference appears notably as the rate of fault node become large, i.e., proposed technique is better. Because of distributed expression, even if fault nodes gather in certain part, other clusters are able to make up for fault node.

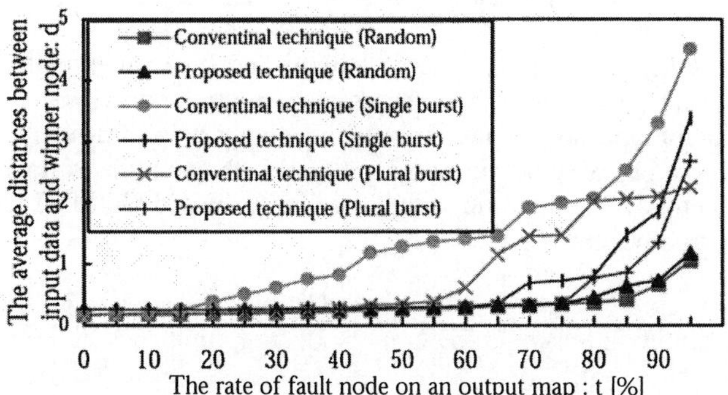

Fig 3: The degree of the node-fault-tolerances

4.3 Experiment 3: Visual Understanding

The map formed by proposed technique is shown by distributed expression, so it is complicated compared with the map by conventional technique. Therefore we examined about visual understanding of the map.

On proposed technique, the map is formed in response to the influence of "Limited Scope" size or width. So we considered the following three cases. Where l stands for the length of one side of the map that is a square-like, and N_c stands for the width of "Limited Scope".

[Case 1: $N_c \geq 2l$] Each class consists of a single cluster. That is, the map is expressed with local expression as well as conventional technique.

[Case 2: $l \leq N_c \leq 2l$] It exists both classes whose cluster is single and classes that have two or more cluster.

[Case 3: $N_c \leq l$] Each class consists of two or more clusters. That is, the map is expressed with distributed expression.

At Fig.4, the x-axis represented N_c, and the y-axis represented the number that map was divided. It can be said in feature map that the less division, the more simple or the better understanding. The number of division has decreased so that N_c is large. In Fig.4, simple natures increase notably at $N_c \geq 8$.

In respect of the improvement in the degree of node fault tolerance, it is desirable that the map is expressed with distributed expression, that is, each class must consist of two or more clusters. In order to obtain two or more clusters for all classes, it is necessary to set the width of "Limited Scope" as the range of $N_c \le l$. In this experiment, $N_c < 20$.

Therefore, in case that N_c set to $8 \le N_c \le 20$, it is expected to get simple and distributed expression map.

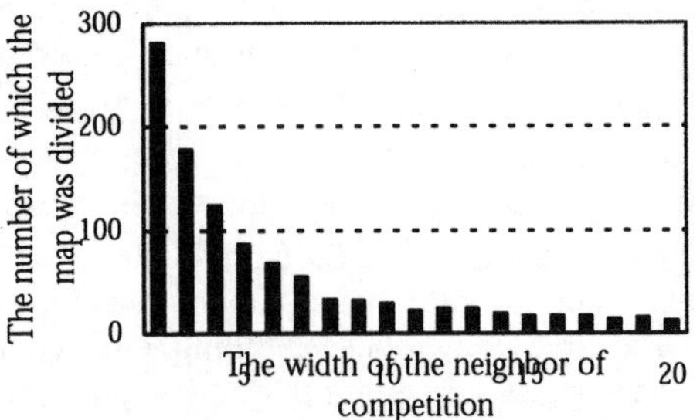

Fig 4: The number of which the map was divided

In conventional SOM, the input data are distinguished by winner node position in the map. In proposed SOM, the map whose expression mode is distributed, there could be many nodes to win for one input data. As making comparison between Fig.5 and Fig.6, each output is different not only output position but also the number of node groups. It can be said, that parameters of distinction is increase in proposed SOM. That is to say the distinction of input data becomes easier.

Fig 5: example 1 of output for cluster "A"   Fig 6: example 2 of output for cluster "A"

5. Conclusion

In this paper, we proposed "Limited Scope Learning". From the result of experiments, we understand the followings:
* SOM is able to self-organize by using proposed technique,
* the map is represented by distributed expression,
* node fault tolerance is improved especially burst fault,
* the discrimination of input data is improved.

We considered the followings as future work:
* Optimum value for the neighbor of competition,
* Shorten learning time.

References
[1]   T.Kohonen: "Self-Organizing Maps", Springer (1997).
[2]   R.Hecht-Nielsen: "Neurocomputing," Addison-Wesley Pub. Co. (1992).
[3]   Stephen T.Welstead: "NEURAL NETWORK AND FUZZY LOGIC APPLICATIONS IN C/C++" John Wiley & Sons, Inc. (1994).

E. Damiani et al. (Eds.)
IOS Press, 2002

Steepest Ascent Training
of Support Vector Machines

Shigeo Abe, Youichi Hirokawa, and Seiichi Ozawa
Graduate School of Science and Technology, Kobe University
Rokkodai, Nada, Kobe, Japan

Abstract

Since the training of support vector machines needs to solve the dual problem with the number of variables equal to the number of training data, training becomes slow when the number of training data is large. To speed up training the Sequential Minimal Optimization (SMO) technique has been proposed, in which two data are optimized simultaneously. In this paper, we propose to extend SMO so that more than two data are optimized simultaneously. Namely, we select a working set including variables, solve the equality constraint for one variable included in the working set, and substitute it into the objective function. Then we solve the subproblem related to the working set by calculating the inverse of the Hessian matrix. If the Hessian matrix is singular, we solve the problem for the subset of the working set that is not singular. We evaluate our method for the five benchmark data sets and show the speed-up of training over SMO.

1 Introduction

The high generalization ability of support vector machines compared to other methods has been shown for many applications but the major problem is slow training especially when the number of training data is large. Therefore, many fast training methods have been proposed [1]. Conventional learning algorithms developed for perceptrons are adapted to be used in the feature space [1, 2]. Since the kernel-Adatron algorithm [2] does not consider the equality constraint, the optimality for the bias term is not guaranteed. Sequential Minimal Optimization is a training method developed by Platt [3]. It optimizes two data at a time.

In this paper we discuss the training algorithm that is based on steepest ascent. The algorithm reduces to SMO when two data are optimized simultaneously. Namely, we select a working set that includes support vector candidates, solve the equality constraint for one variable included in the working set, and substitute it into the objective function. Then we solve the subproblem related to the working set by calculating the inverse of the Hessian matrix.

In Section 2, we explain the decomposition technique [4] used for training, and in Section 3 we discuss the steepest ascent method. Finally in Section 4, we compare performance of the steepest ascent method with that of SMO and the primal dual interior-point method.

2 Decomposition Technique

Because of the space limitation we only show the dual problem of the support vector training. Namely, find $\alpha_i\ (i = 1, \ldots, M)$ that maximize

$$Q(\alpha) = \sum_{i=1}^{M} \alpha_i - \frac{1}{2} \sum_{i,j=1}^{M} \alpha_i\,\alpha_j\,y_i\,y_j\,H(\mathbf{x}_i, \mathbf{x}_j) \tag{1}$$

subject to the constraints

$$\sum_{i=1}^{M} y_i\,\alpha_i = 0, \qquad 0 \le \alpha_i \le C. \tag{2}$$

Here, $\mathbf{x}_i\ (i = 1, \ldots, M)$ are m-dimensional inputs which belong to either Class 1 or 2 and the associated labels are $y_i = 1$ for Class 1 and -1 for Class 2, α_i are the Lagrange multipliers associated with $\mathbf{x}_i$, and C is the upper bound for α_i. If we use the square sum of slack variables, we replace $H(\mathbf{x}_i, \mathbf{x}_j)$ with $H(\mathbf{x}_i, \mathbf{x}_j) + \delta_{ij}/C$ and delete the upper bound C for α_i [1].

To reduce the number of variables to be processed at once, Osuna et al. [4] propose to decompose the problem into two. Let the index set $\{1, \ldots, M\}$ be partitioned into two sets B and N, where $B \cup N = \{1, \ldots, M\}$ and $B \cap N = \phi$. Then decomposing $\{\alpha_i | i = 1, \ldots, M\}$ into $\alpha_B = \{\alpha_i | i \in B\}$ and $\alpha_N = \{\alpha_i | i \in N\}$ we define the following subproblem. Fixing α_N, maximize

$$
\begin{aligned}
Q(\alpha_B) \;=\; & \sum_{i \in B} \alpha_i - \frac{1}{2} \sum_{i,j \in B} \alpha_i \alpha_j y_i y_j H(\mathbf{x}_i, \mathbf{x}_j) - \sum_{\substack{i \in B \\ j \in N}} \alpha_i \alpha_j y_i y_j H(\mathbf{x}_i, \mathbf{x}_j) \\
& - \frac{1}{2} \sum_{i,j \in N} \alpha_i \alpha_j y_i y_j H(\mathbf{x}_i, \mathbf{x}_j) + \sum_{i \in N} \alpha_i
\end{aligned}
\tag{3}
$$

subject to the constraints

$$
\sum_{i \in B} y_i \alpha_i = - \sum_{i \in N} y_i \alpha_i, \qquad 0 \le \alpha_i \le C \quad \text{for} \quad i \in B.
\tag{4}
$$

3　Steepest Ascent Method

Now consider solving the subprogram for α_B. Solving the equality in (2) for $\alpha_s \in \alpha_B$, we obtain

$$
\alpha_s = - \sum_{\substack{i=1 \\ i \neq s}}^{M} y_s y_i \alpha_i.
\tag{5}
$$

Substituting (5) into (1), we can eliminate the equality constraint. Let $\alpha_{B'} = \{\alpha_i | i \neq s, i \in B\}$. Now since $Q(\alpha_{B'})$ is quadratic, we can express the change of $Q(\alpha_{B'})$, $\Delta Q(\alpha_{B'})$, as a function of the change of $\alpha_{B'}$, $\Delta \alpha_{B'}$, by

$$
\Delta Q(\alpha_{B'}) = \frac{\partial Q(\alpha_{B'})}{\partial \alpha_{B'}} \Delta \alpha_{B'} + \frac{1}{2} \Delta \alpha_{B'}^t \frac{\partial^2 Q(\alpha_{B'})}{\partial \alpha_{B'}^2} \Delta \alpha_{B'}.
\tag{6}
$$

Considering that

$$
\frac{\partial \alpha_s}{\partial \alpha_i} = - y_s y_i,
\tag{7}
$$

we derive the partial derivatives of $Q(\alpha_{B'})$ with respect to α_i $(i \neq s, i \in B')$:

$$
\frac{\partial Q(\alpha_{B'})}{\partial \alpha_i} = 1 - y_s y_i - \sum_{j=1}^{M} \alpha_j y_i y_j H(\mathbf{x}_i, \mathbf{x}_j) + \sum_{j=1}^{M} \alpha_j y_j y_i H(\mathbf{x}_s, \mathbf{x}_j).
\tag{8}
$$

Using (8), the second partial derivatives of $Q(\alpha_{B'})$ with respect to α_i and α_j $(i, j \neq s, i, j \in B')$ are

$$
\frac{\partial^2 Q(\alpha_{B'})}{\partial \alpha_i \partial \alpha_j} = y_i y_j \left(-H(\mathbf{x}_i, \mathbf{x}_j) + H(\mathbf{x}_i, \mathbf{x}_s) + H(\mathbf{x}_s, \mathbf{x}_j) - H(\mathbf{x}_s, \mathbf{x}_s) \right).
\tag{9}
$$

Since $-\partial^2 Q(\alpha)/\partial \alpha_{B'}^2$ is expressed by the product of a transposed matrix and the matrix, it is positive semi-definite. Assuming that it is positive definite and neglecting the bounds, $\Delta Q(\alpha_{B'})$ has the maximum at

$$
\Delta \alpha_{B'} = - \left(\frac{\partial^2 Q(\alpha_{B'})}{\partial \alpha_{B'}^2} \right)^{-1} \frac{\partial Q(\alpha_{B'})}{\partial \alpha_{B'}}.
\tag{10}
$$

Assume that $\alpha_i (i = 1, \ldots, M)$ satisfy (2). Then from (2) and (10), we obtain the correction of α_s:

$$
\Delta \alpha_s = - \sum_{i \in B'} y_s y_i \Delta \alpha_i.
\tag{11}
$$

For α_i $(i \in B)$, if

$$
\alpha_i = 0, \quad \Delta \alpha_i < 0, \quad \text{or} \quad \alpha_i = C, \quad \Delta \alpha_i > 0,
\tag{12}
$$

we cannot modify α_B, since this will violate the lower or upper bound. If this happens, we delete these variables from the working set, and repeat the above procedure for the reduced working set.

If (12) does not hold for any variable in the working set, we can modify the variables. Let $\Delta\alpha_i'$ be the maximum or minimum correction of α_i that is within the bounds. Then if $\alpha_i + \Delta\alpha_i < 0$, $\Delta\alpha_i' = -\alpha_i$. And if $\alpha_i + \Delta\alpha_i > C$, $\Delta\alpha_i' = C - \alpha_i$. Otherwise, $\Delta\alpha_i' = \Delta\alpha_i$.

We modify α_B by

$$\alpha_B^{\text{new}} = \alpha_B^{\text{old}} + r\,\Delta\alpha_B, \quad r = \min_{i \in B} \frac{\Delta\alpha_i'}{\Delta\alpha_i}, \tag{13}$$

where $0 < r \leq 1$. It is clear that the obtained α_B^{new} satisfies the equality constraint. In addition, $Q(\alpha_B^{\text{new}}) \geq Q(\alpha_B^{\text{old}})$ is satisfied.

If $\partial^2 Q(\alpha_{B'})/\partial\alpha_{B'}^2$ is singular, we can use the pseudo-inverse. But since it is time consuming, we delete the variables that cause singularity from the working set, and if the number of reduced variables is more than 1, do the above procedure for the reduced working set. Consider deleting the variables by using the symmetric Cholesky factorization in calculating (10). In factorizing the ith column of $\partial^2 Q(\alpha_{B'})/\partial\alpha_{B'}^2$, if the value of the ith diagonal element of the lower triangular matrix is smaller than the prescribed small value, we discard the ith column and row of $\partial^2 Q(\alpha_{B'})/\partial\alpha_{B'}^2$, and proceed to the $(i+1)$st column. To simplify this procedure, in the following simulations, we stop factorization and delete the variables corresponding to the ith to $(|B'|)$th diagonal elements.

Let V be the index set of support vector candidates. At the start of training, V includes all the indices, i.e., $V = \{1, \ldots, M\}$. Let assume that the maximum working set size $|B|_{\max}$ is given. We change the working set size according to the change of $|V|$ as follows:

$$|B| = \begin{cases} |B|_{\max} & \text{for} \quad |V| \geq |B|_{\max}, \\ \max\{2, |V|\} & \text{for} \quad |V| < |B|_{\max}. \end{cases} \tag{14}$$

We call the corrections of all the variables α_i $(i \in V)$, one epoch of training. At the beginning of a training epoch, we select one variable α_i $(i \in V)$, and set $B = \{i\}$ and $V = V - \{i\}$. Then we randomly select $(|B| - 1)$ variables α_i $(i \in V)$, where $|B|$ is given by (14), and delete and add the associated indices from V and to B, respectively. We calculate the variable vector α_B^{new}, and iterate the above procedure until V is empty.

For each calculation of α_B^{new}, we check if $|r\,\Delta\alpha_i| < \varepsilon_{\text{var}}$ hold for all i $(i \in B)$, where ε_{var} is a tolerance of convergence for variables. At the end of a training epoch, we calculate the bias terms b_i for $\alpha_i (i \in B)$ that are within the bounds, i.e., $0 < \alpha_i < C$ and their average b_{ave} by

$$b_i = y_i - \sum_{j \in B} \alpha_j\, H(\mathbf{x}_j, \mathbf{x}_i), \quad b_{\text{ave}} = \frac{1}{N_b} \sum_{\substack{i \in B \\ 0 < \alpha_i < C}} b_i, \tag{15}$$

respectively, where N_b is the number of α_i $(\in V)$ that satisfy $0 < \alpha_i < C$.

To accelerate training, for N_{ite} consecutive iterations, where N_{ite} is the positive integer, we confine calculations within the support vector candidates [3]. Namely, at the end of the training epoch we delete the index i with $\alpha_i = 0$ from V, and proceeds training.

If this does not happen and $|r\,\Delta\alpha_i| < \varepsilon_{\text{var}}$ hold for all α_B's in the training epoch, and if

$$|b_{\text{ave}} - b_i|/|b_{\text{ave}}| \leq \varepsilon_b \tag{16}$$

hold for all i $(i \in V, 0 < \alpha_i < C)$, we check whether there are a new candidate $\mathbf{x}_i$ $(i \notin V)$ of support vectors that satisfy

$$y_j \left(\sum_{i \in V} \alpha_i\, y_i\, H(\mathbf{x}_i, \mathbf{x}_j) + b_{\text{ave}} \right) < 1. \tag{17}$$

If there is none, we stop training. Otherwise, we add the indices to V and proceed training. At the end of every $(I_{\text{ite}} + 1)$st training epoch, we check (17) and add the support vector candidates if there are some.

To further speed up training, we impose

$$(Q^{(n+1)}(\alpha) - Q^{(n)}(\alpha))/Q^{(n)}(\alpha) \leq \varepsilon_q, \tag{18}$$

where ε_q is a small positive parameter. If this criterion is applied at the early stage of convergence, the calculation is terminated before the solution reaches the optimal solution. Therefore, we apply the above criterion after $N_q\,(>1)$ epoch of training.

To reduce the number of calculations of $H(\mathbf{x}_i, \mathbf{x}_j)$, we prepare an $N_M \times N_M$ matrix H for $H(\mathbf{x}_i, \mathbf{x}_j)$. When we calculate $H(\mathbf{x}_i, \mathbf{x}_j)$ where α_i and α_j are support vector candidates, i.e., i , $j \in V$, we store it into H in the similar way as the cache memory does [3].

4   Performance Evaluation

We used a SUN UltraSPARC-IIi (335MHz) workstation for the thyroid, blood cell, and hiragana data sets listed in Table 1 [5]. The input ranges were globally scaled into $[-1, 1]$. We used polynomial kernels $(1 + \mathbf{x}^t\mathbf{x}')^d$ and RBF kernels $\exp(-\gamma\|\mathbf{x} - \mathbf{x}'\|)$. The parameters used for simulations are as follows: $C = 5000$, $\varepsilon_b = 10^{-2}$, $\varepsilon_{var} = 10^{-6}$, $\varepsilon_q = 10^{-7}$, $N_{ite} = 10$, and $N_q = 50$. Since for the thyroid data and blood cell data, $N_q = 50$ was too small we set as follows: for the thyroid data except for $d = 4$ and the blood cell data, we set $N_q = 100$, and for the thyroid data with $d = 4$ we set $N_q = 600$. We used pairwise classification with resolution of unclassifiable regions by membership functions [6]. To compare the calculation time with the quadratic programming technique, we used the software developed by London University(http://svm.cs.rhbnc.ac.uk/) [7], which uses the primal dual interior-point method (LOQO). The working set size was set to be 50.

Table 1: Feature of benchmark data

Data	Inputs	Classes	Train.	Test
Thyroid	21	3	3772	3428
Blood cell	13	12	3097	3100
Hiragana-50	50	39	4610	4610
Hiragana-105	105	38	8375	8356
Hiragana-13	13	38	8375	8356

Table 2 shows the result for the different training methods. In the table, columns LOQO, SMO, SAM (steepest ascent method), SMO-Q, and SAM-Q show the training time and in the brackets the speedup ratios to SMO, where Q indicates the square sum of slack variables. There is not so much difference between the linear sum of slack variables and the square sum. Since the implementation of SMO is different from that in [3], only comparison of the training time with that of SAM is meaningful. In [3], SMO was shown to be faster than the projected conjugate gradient method with the working set size of 500. But, comparing LOQO and SMO, LOQO outperformed SMO for all the data sets evaluated. For LOQO also the working set size affected the training time considerably. For instance, the training time by LOQO for blood cell data with $d = 4$ and working set size of 500 was 936 seconds. Thus the optimal selection of the working set size for the conventional optimization techniques is important.

Comparing SMO and SAM, there is not so much difference for the hiragana data but for the thyroid and blood cell data training by SAM is much faster for most cases. SAM is comparable with LOQO for the hiragana data, but for the thyroid data, LOQO is much faster. This means that when class overlap exists SAM becomes slower than LOQO with the optimal working set size.

Table 2: Performance comparison of training methods

Condition	LOQO (s)	SMO (s)	SAM (s)	SMO-Q (s)	SAM-Q (s)
Thyroid ($d = 4$)	**14 (145)**	2032 (1)	109 (19)	2362 (0.86)	134 (15)
Thyroid ($\gamma = 10$)	**108 (37)**	4044 (1)	424 (9.5)	4309 (0.94)	390 (10)
Blood Cell ($d = 4$)	**19 (18)**	338 (1)	31 (11)	353 (0.96)	31 (11)
Blood Cell ($\gamma = 10$)	185 (1.3)	238 (1)	**165 (1.4)**	235 (1.0)	**165 (1.4)**
Hiragana-50 ($d = 2$)	143 (1.4)	202 (1)	**117 (1.7)**	215 (0.94)	124 (1.6)
Hiragana-50 ($\gamma = 0.1$)	288 (1.1)	321 (1)	**194 (1.6)**	333 (0.96)	199 (1.6)
Hiragana-105 ($d = 2$)	**318 (2.1)**	671 (1)	394 (1.7)	698 (0.96)	414 (1.6)
Hiragana-105 ($\gamma = 0.1$)	1889 (1.2)	2262 (1)	**1479 (1.5)**	2316 (0.98)	1525 (1.5)
Hiragana-13 ($d = 2$)	99 (1.4)	136 (1)	**93 (1.5)**	147 (0.93)	104 (1.3)
Hiragana-13 ($\gamma = 1$)	**181 (1.8)**	325 (1)	190 (1.7)	333 (0.98)	195 (1.7)

5　Conclusions

We discussed training of support vector machines by the steepest ascent method by calculating the inverse of the Hessian matrix for a working set. If the Hessian matrix is singular, we solve the problem for the subset of the working set that is not singular. We evaluate our method for the five benchmark data sets and show the speed-up of training over SMO.

References

[1] N. Cristianini and J. Shawe-Taylor. *An Introduction to Support Vector Machines and Other Kernel-based Learning Methods.* Cambridge University Press, 2000.

[2] T.-T. Frieß, N. Cristianini, and C. Campbell. The kernel-Adatron algorithm: a fast and simple learning procedure for support vector machines. *Proc. ICML'98*, pp. 188–196, 1998.

[3] J. C. Platt. Fast training of support vector machines using sequential minimal optimization. In B. Schölkopf, C. J. C. Burges, and A. J. Smola, Eds., *Advances in Kernel Methods: Support Vector Learning*, pp. 185–208. MIT Press, 1999.

[4] E. Osuna, R. Freund, and F. Girosi. An improved training algorithm for support vector machines. *Proc. NNSP'97*, pp. 276–285, 1997.

[5] S. Abe. *Pattern Classification: Neuro-fuzzy Methods and Their Comparison.* Springer-Verlag, 2001.

[6] T. Inoue and S. Abe. Fuzzy support vector machines for pattern classification. *Proc. IJCNN'01*, vol. 2, pp. 1449–1454, 2001.

[7] C. Saunders, M. O. Stitson, J. Weston, L. Bottou, B. Schölkopf, and A. Smola. Support vector machine reference manual. Technical Report CSD-TR-98-03, Royal Holloway, University of London, 1998.

KES 2002
E. Damiani et al. (Eds.)
IOS Press, 2002

Comments on Using MLP and FFT for Fast Object/Face Detection

HAZEM M. EL-BAKRY

Faculty of Computer Science &Information Systems
Mansoura University - Egypt
helbakry1@hotmail.com

ABSTRACT- Recently, fast neural nets for object/face detection are presented in [1-3]. The speed up factor of these networks based on cross correlation in frequency domain between the input image and the weights of the hidden layer. But, these equations presented in [1-3] for conventional and fast neural nets as well as speed up ratio are not valid for many reasons presented here. In this paper, a correct formula for the computation steps required for conventional, fast neural nets and speed up ratio is introduced. Practically, simulation results show that only in case of the input image is symmetric or the weights are symmetric, neural nets presented in [1-3] are faster than conventional neural nets.

1. Introduction

The authors in [1-3] have proposed a multilayer perceptron (MLP) algorithm for fast object/face detection. They claimed that applying cross correlation in frequency domain between the input image and the neural weights is much faster than object detection using conventional neural nets. They introduced formulas for the number of computation steps needed by conventional and fast neural nets. Then they deduced an equation for the speed up ratio. Unfortunately, these equations contain many errors, which leads to invalid speed up ratio. The main objective of this paper is to correct the formulas and equations given for conventional, and fast neural networks. Finally, the number of floating point operations required by conventional neural nets is proved to be less than that needed for fast neural nets proposed in [1-3]. Simulation results using Matlab confirm such approval. In section 2, fast neural nets for object/face detection are described. Comments on conventional neural nets, Fast neural nets, and the speed up ratio of object/face detection are presented in section 3.

2. Fast Neural Nets for Human Face Detection

In [1-3], a fast algorithm for object/face detection based on two dimensional cross correlations that take place between the tested image and the sliding window (20x20 pixels) was described. Such window is represented by the neural net weights situated between the input unit and the hidden layer. The convolution theorem in mathematical analysis says that a convolution of f with h is identical to the result of the following steps: let **F** and **H** be the results of the Fourier transformation of f and h in the frequency domain. Multiply **F** and **H** in the frequency domain point by point and then transform this product into spatial domain via the inverse Fourier transform. As a result of this, these cross correlations can be represented by a product in frequency domain. So, by using cross correlation in frequency domain, speed up in an order of magnitude can be achieved during the detection process [1-3].

In the detection phase, a sub image **I** of size mxn (sliding window) is extracted from the tested image which has a size **PxT** and fed to the neural network. Let X_i be the vector of weights between the input sub image and the hidden layer. This vector has a size of mxn and can be represented as mxn matrix. The output of hidden neurons **h(i)** can be calculated as follows:

$$h_i = g\left(\sum_{j=1}^{m} \sum_{k=1}^{n} X_i(j,k)I(j,k) + b_i \right) \tag{1}$$

where g is the activation function and **b(i)** is the bias of each hidden neuron **(i)**. Equation 1 represents the output of each hidden neuron for a particular sub-image **I**. It can be obtained to the whole image **Z** as follows:

$$h_i(u,v) =$$

$$g\left(\sum_{j=-m/2}^{m/2} \sum_{k=-n/2}^{n/2} X_i(j,k) \; Z(u+j,v+k) + b_i \right) \tag{2}$$

Equ.2 represents a cross correlation operation. Given any two functions **f** and **d**, their cross correlation can be obtained by:

$$f(x,y) \otimes d(x,y) =$$

$$\left(\sum_{m=-\infty}^{\infty} \sum_{n=-\infty}^{\infty} f(m,n) d(x+m,y+n) \right) \tag{3}$$

Therefore, equ. 2 may be written as follows:

$$h_i = g\left(X_i \otimes Z + b_i \right) \tag{4}$$

where h_i is the output of the hidden neuron **(i)** and $h_i(u,v)$ is the activity of the hidden unit **(i)** when the sliding window is located at position **(u,v)** and $(u,v) \in [P-m+1, T-n+1]$.

Now, the above given cross correlation can be expressed in terms of Fourier Transform:

$$Z \otimes X_i = F^{-1}\left(F(Z) \bullet F^*\left(X_i \right) \right) \tag{5}$$

Hence, by evaluating this cross correlation, a speed up ratio can be obtained compared to conventional neural networks. Also, the final output of the neural network can be evaluated as follows:

$$O(u,v) = g\left(\sum_{i=1}^{q} w_o(i) \, h_i(u,v) + b_o \right) \tag{6}$$

O(u,v) is the output of the neural network when the sliding window located at the position **(u,v)** in the input image **Z**.

The authors in [1,2,3] analyzed their proposed fast neural network as follows: For a tested image of NxN pixels, the 2D-FFT requires $O(N^2 \log_2 N^2)$ computation steps. For the weight matrix X_i, the 2D FFT can be computed off line since these are constant parameters of the network independent of the tested image. The 2D FFT of the tested image must be computed. As a result, q backward and one forward transforms have to be computed. Therefore, for a tested image, the total number of the 2DFFT to compute is $(q+1)N^2 \log_2 N^2$. Moreover, the input image and the weights should be multiplied in the frequency domain. Therefore, computation steps of (qN^2) should be added. Finally, a total of $O((q+1)N^2 \log_2 N^2 + qN^2)$ computation steps must be evaluated for fast the neural algorithm.

Using sliding window of size nxn, for the same image of NxN pixels, qN^2n^2 computation steps are required when using traditional neural networks for the face detection process. The theoretical speed up factor η can be evaluated as follows [1]:

$$\eta = O\left(\frac{qn^2}{(q+1)\log^2 N}\right) \tag{7}$$

3. Comments on Fast Neural Net Presented for Object/ Face Detection

The speed up factor introduced in [1] and given by equ. 7 is not correct for the following reasons:

1- The number of computation steps required for the 2D FFT is $O(N^2\log_2 N^2)$ and not $O(N^2\log^2 N)$ as presented in [1,2]. Also, this is not a typing error as the curve in Fig. 2 in [1] realizes equ.8, and the curve in Fig. 5 in [2] realized equ. 14 in [2] which is given by:

$$\eta = O\left(\frac{qn^2}{(q+1)\log^2 N}\right) \tag{8}$$

2- Moreover, the speed up ratio presented in [1] not only contains an error but is also not precise. This is because for fast neural nets, the term $(6qN^2)$ corresponds to complex dot product in frequency domain must be added. Such term has a great effect on the speed up ratio. Adding only $q N^2$ as stated in [2] is not correct since a one complex multiplication requires six real computation steps.

3- Furthermore, for conventional neural nets, the number of operations is $(q(2n^2)(N-n+1)^2)$ and not (qN^2n^2). The term n^2 is required for multiplication of n^2 elements (in the input window) by n^2 weights which results in another new n^2 elements. Adding these n^2 elements, requires another n^2 steps. So, the total computation steps needed for each window is $(2n^2)$. The search for a face is done using a window (with nxn elements) in an image, and the search process is done at each pixel in the image. So, the search is repeated $(N-n+1)^2$ times and not N^2.

4- Before applying cross correlation, the FFT2 of the weight matrix must be computed. Because dot product is done in the frequency domain, the size of weight matrix should be increased to be the same as the size of the input image (under test). Computing the FFT2 of the weight matrix off line as stated in [1-3] is not true. Because, this puts a restriction on all of the images under test to have the same size and as a result, the image under test has only a one fixed size. This means that, the testing time for an image of size 50x50 pixels is the same as that image of size 1000x1000 pixels and of course this is unreliable. So, another number of complex computation steps to perform FFT2 for (NxN) matrix should be added to the complex number of computation steps (σ) required for fast neural nets as follows:

$$\sigma=((2q+1)(N^2\log_2 N^2) + 6qN^2) \tag{9}$$

This will increase the computation steps required for fast neural nets especially when q is more than one neuron.

5-It is not valid to compare complex number by another real one directly. The number of computation steps given in [1,2,3] which is required for conventional neural nets is real while that is required for fast neural nets is complex. To obtain the speed up ratio, the authors in [1,2,3] have divided the two formulas directly without converting the number of computation steps required by fast neural nets into a real one. It is known that the two dimension Fast Fourier Transform requires $N^2/2\log_2 N^2$ complex multiplications and $N^2\log_2 N^2$ complex additions. Every complex multiplication is realized by six real floating point operations and every complex addition is implemented by two real floating point operations. So, the total number of computations steps required to obtain the 2D-FFT of an NxN image is:

$$\rho=6(N^2/2\log_2 N^2) + 3(N^2\log_2 N^2) \tag{10}$$

which may be simplified to:

$$\rho=5(N^2/2\log_2 N^2) \tag{11}$$

6- Also, a number of computation steps equal to $N\log_2 N$ is required to create butterflies complex numbers $(e^{-jk(2\Pi/N)})$, where $0<K<L$. These complex numbers are multiplied by the input values. So, the total number of floating point computation steps required for fast neural nets is:

$$\sigma=((2q+1)(5N^2\log_2 N^2) + N\log_2 N + 6qN^2) \tag{12}$$

7- For fast neural nets, to give a correct face detection result as conventional neural nets, either the weights or the input image must be symmetric. It is very complex to allow the weights to be symmetric in the required form which is needed to be as follows:

$$W = \begin{bmatrix} w & w \\ w & w \end{bmatrix} \tag{13}$$

Adding this constrain to the learning rules will cause many well known problems during the training process of the neural network. The probability that the network will not trained is very high.The solution is to convert the input image into the required symmetric form as shown in Fig. 1. As the input image has a dimension of (N), the new symmetric image will have a length of (2N). In this case, the number of computation steps required for fast neural nets can be calculated as follows:

$$\sigma_{2N}=((2q+1)(5(2N)^2\log_2(2N)^2) +(2N)\log_2(2N) + 6q(2N)^2) \tag{14}$$

But, converting the non-symmetric input image into a symmetric one will slow down those fast neural nets compared to conventional neural nets. For symmetric input images, a comparison between the time taken by fast and conventional to manipulate symmetric images with different sizes is shown in table (1) using 200 MHz processor and Matlab. It is also shown that, the time taken by fast neural nets for (2N) images (for example 800x800 pixels) is more than that taken by conventional neural nets for (N) images (400x400 pixels). This proves that converting a non-symmetric input image into a symmetric one drops the speed of fast neural nets to be less than conventional neural nets.

8- Images are tested for the presence of a face (object) at different scales by building a pyramid of the input image which generates a set of images at different resolutions. The face detector is then applied at each resolution and this process takes much more time as the number of processing steps will be increased. In [1-3], the author stated that the Fourier transforms of the new scales do not need to be computed. This is due to a property of the Fourier transform. If f(x,y) is the original and g(x,y) is the sub-sampled by a factor of 2 in each direction image then:

$$g(x, y) = FT(2x,2y) \tag{15}$$

$$F(u, v) = FT(f(x, y)) \tag{16}$$

$$FT(g(x, y) = G(u, v) = \frac{1}{4}F(\frac{u}{2},\frac{v}{2}) \tag{17}$$

This implies that we do not need to recompute the Fourier transform of images, as it can be directly obtained from the original Fourier transform. Then, the author claimed that the processing needs $O((q+2)N^2\log^2 N)$ additional number of computation steps. Thus the speed up ratio will be [1]:

$$\eta = O\left(\frac{qn^2}{(q+2)\log^2 N}\right) \tag{18}$$

Of course this is not correct, because the inverse of the Fourier transform is required to be computed at each neuron in the hidden layer (for the result of dot product between the Fourier transform of input image and the weights of the neuron, the inverse of the Fourier transform must be computed). So, the term (q+2) presented in [1] should be (3q+1) because the inverse FFT2 must be done at each neuron in the hidden layer.

4. Conclusion

I have shown that the equations given in [1-3] for conventional, fast neural nets and speed up ratio contain errors. The reasons for these errors have been presented and then corrected. Also, only under the constrain that the input image is symmetric or the weights are symmetric, fast neural nets (previously proposed by other authors [1-3]) have been proved to be faster than conventional neural nets. Otherwise, conventional neural nets are faster than fast neural nets presented in [1-3]. This approval has been confirmed by Matlab simulation results.

References

[1] S. Ben-Yacoub, "Fast Object Detection using MLP and FFT, " IDIAP-RR 11, IDIAP, 1997.

[2] Beat Fasel, "Fast Multi-Scale Face Detection, ", IDIAP-Com 98-04, 1998

[3] S. Ben-Yacoub, B. Fasel, and J. Luettin , "Fast Face Detection using MLP and FFT" in Proc. Second International Conference on Audio and Video-based Biometric Person Authentication (AVBPA'99)", 1999.

Table 1: A comparison between the time taken (in seconds) by conventional and fast neural nets to manipulate images with different sizes using Matlab and 133MHz processor.

Image size	Conventional Neural Nets	Fast Neural Nets	Image size	Conventional Neural Nets	Fast Neural Nets
100x100	.44	.11	500x500	7.20	2.86
200x200	1.32	.55	600x600	10.50	4.61
300x300	2.53	.99	700x700	13.56	6.68
400x400	4.17	1.7	800x800	27.08	12.97

Fig. 1. Image conversion from non-symmetric to symmetric one.

KES 2002
E. Damiani et al. (Eds.)
IOS Press, 2002

Writing Style Variation Absorption for a Hybrid Neuro-Markovian On-line Handwriting Recognition System

Haifeng LI, Thierry ARTIERES, Patrick GALLINARI, Bernadette DORIZZI[*]
Computer Science Lab., University Paris 6, 8 Rue du Capitaine Scott, 75015 Paris, France
[*]*National Telecommunication Institute, 9 Rue Charles Fourier, 91011 Evry, France*

Abstract. Hybrid neuro-markovian methods have been broadly used in recent years in many human-computer interfaces. Usually, a left-to-right topology architecture is applied and the transition probabilities are assigned in advance. However, this kind of architecture has appeared to lack the capacity for modelling the immense variability of the cursive handwriting styles of many writers. In this paper, two approaches are studied to improve the system performance through augmenting the variation absorption ability at HMM chain level and state level, without involving the difficult task of transition probability estimation. Evaluated on the UNIPEN database, the suggested methods are approved efficient, simple and universal for various other kinds of applications.

1. Introduction

Being one of the most natural and practical communication medias, handwriting becomes increasingly important for human-computer interfaces. However, the problem of handwriting recognition is still far from a perfectly resolved one for many reasons, among which the attached-character segmentation and the handwriting style variation are the most difficult obstacles. The hidden Markov models (HMM) and artificial neural networks (ANN) have been applied to this problem and achieved great success [1,2,3].

In our former works, a hybrid approach of these two techniques was used in speaker verification and on-line handwriting recognition [4,5,6]. In this approach, neural networks are used to create a prediction of the current signal frame according to a frame context. Each network is trained to represent a special signal segment that is called a primitive. Primitives may be phonemes in speech or strokes in handwriting. The different occurrence orders of primitives represent different pattern classes and different styles within a class. Based on this neural auto-regression at signal primitive level, HMMs are used as high-level models (letter model, word model, speaker model, *et al.*). The topology architecture of HMMs should model the intra-class variations. However, a predefined static structure is usually used for practical reasons. Obviously, it could not be powerful enough to model all the variations. For example in an on-line handwriting recognition case, great style variation is often observed for letters 'b', 'f' and 'r', and some styles may be completely different [7].

In this paper, we focus on improving the variation absorption ability for a better performance. In Section 2, the overview of the neuro-markovian approach is given. In Section 3, our general idea is discussed and resumed theoretically. In Section 4 and 5, two methods are studied separately at the HMM chain level and the state level. Experimental results are given in Section 6. Finally, our conclusion is drawn in Section 7.

2. System Overview

This system has been designed for recognizing isolated handwritten words. It contains two layers: a letter layer and a word layer. Here, we focus on the letter layer where the handwriting signal is modeled. Each "letter-model" is a left-to-right (Bakis) HMM chain of m states [6] (Fig-1). The only authorized transitions are from one state to itself or to the state that fellows. All the transition probabilities are assumed identical. Such architecture makes states focus on modelling the successive parts of a letter (handwriting primitives). A word model may be easily built by concatenating together all HMM chains of its letters. In practice, m is empirically set to 7.

Fig-1: HMM left-to-right (Bakis) architecture of a letter-model

In HMM chain, emission probability densities are modeled by multi-layer perceptron neural networks (MLP) that work in a predictive way. The main advantage lies in the automatic modelling of signal dynamics. This idea bases on the assumption that a frame dependents on its preceding frames, as described in the following equation:

$$o_t = F_s\left(o_{t-1},...,o_{t-p}\right) + \varepsilon_t \tag{1}$$

Where o_t is the t-th frame of a signal, p is the auto-regression order, F_s is the prediction function and ε_t is a residual error. This hypothesis means that:

$$p_s\left(o_t \mid o_{t-1},...,o_{t-p}\right) = p_\varepsilon\left(\varepsilon_t\right) \tag{2}$$

Here, a MLP is used to approximate the prediction function F_s , so it is also called a prediction neural network (e.g. PNN). By assuming the residual ε_t to be a white Gaussian noise, we have:

$$-\log\left(p_s\left(o_t \mid o_{t-1},...,o_{t-p}\right)\right) \approx \left\| o_t - F_s^{NN}\left(o_{t-1},...,o_{t-p}\right)\right\|^2 \tag{3}$$

Where F_s^{NN} is the prediction function realized by a PNN. In our experiments, each PNN has 15 output units according to the 15 features in a frame, 8 hidden units and $(p*15)$ input units with $p=2$, implying that it considers a 2-precedent-frame context $\{o_{t-1}, o_{t-2}\}$. The training algorithm is Viterbi-based and has been described in detail elsewhere [5, 6].

3. General Idea

According to our primary studies and other similar works, writing style is modeled mainly by the HMM architecture. Nevertheless, the only HMM chain in a letter-model fails to model different styles. Augmenting the architectural complexity is a natural way to solve this problem while tying HMM states to the handwriting primitives [7]. Meanwhile, one also expects that improving the modelling capability of the HMM states may bring the same effect [8]. We interest in both ideas and will compare them latterly.

The general idea of modelling different handwriting styles is very simple. Suppose that there are J different styles for a letter l: $ST_1, ST_2,..., ST_J$. The letter-model is thereby obliged to model all these styles through a mixture of J sub-models. Therefore, the probability that this letter-model l produces an observation frame sequence X_1^T is given by:

$$P\left(X_1^T \mid l\right) = \sum_{j=1}^{J} P\left(X_1^T \mid ST_j, l\right) P\left(ST_j \mid l\right) \tag{4}$$

Where, $P(ST_j \mid l)$ is the priori probability of style STj, and $P(X_1^T \mid ST_j,l)$ is the conditional emission probability. In a simplified form, the coefficients $W_1, W_2,..., W_J$ are used to replace the priori probabilities, so we have:

$$P(X_1^T \mid l) = \sum_{j=1}^{J} W_j \bullet P(X_1^T \mid ST_j, l) \tag{5}$$

These two formulas form the general principle of the mixture of different style models.

4. Mixture of HMM Chains for Letter-Model

Firstly, we consider the augmentation of the architecture complexity for the letter-models. We intend to use more HMM chains to model different writing styles. Such a mixture of HMM chains is derived directly from Equation (4). As it is shown in Fig-2, each new letter-model has J parallel HMM chains. A pseudo-initial state and a pseudo-finish state are also introduced in this schema. In a simple form, we let all the HMM chains have the same number of states and all the letter-models have the same number of HMM chains.

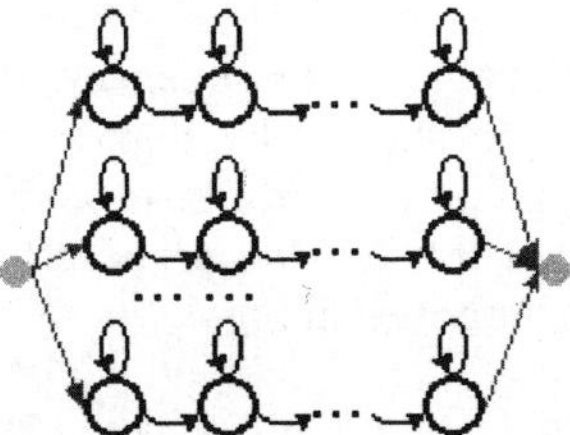

Figure 2: Mixture of HMM chains in a letter-model

All these J Bakis HMM chains are parallel and may work in competition or fusion. In a fusion form, the probability of a frame sequence is a weighted sum of all HMM chain probabilities, exactly as in Equation (4) or (5). Each $P(X_1^T \mid ST_j,l)$ can be easily obtained by the Viterbi algorithm [5]. In a competition form, the things become as simple as this:

$$P(X_1^T \mid l) = \max_{j=1,2,...,J} \left\{ P(X_1^T \mid ST_j,l) \right\} \tag{6}$$

Once the letter-model architecture and the mixture mechanism are defined, a word-model is then constructed easily by a simple concatenation of all the letter-models in the word. Though we concentrate in character recognition now, we still let all the HMM chains work in a no-synchronized manner. The reason is quite clear that letting them compete beyond letter boundaries conduct surely a more powerful word recognition system.

This a mixture of HMM chains is in fact at a high level – the letter level. The advantages are the simplicity and the generality, because it can model globally different writing styles. The drawback is also clear that there are surely many redundant states.

5. Mixture of Primitive Models for HMM State

The second method however, focuses on the fusion of the primitive models at a lower level – the state level. The idea concerns modelling only those special strokes in the different styles and sharing the common primitives. Such a method may be less global, but it can model more precisely and efficiently. In the reference system, we replace the only PNN in each HMM state by a mixture of J PNNs, like what is shown in Fig-3.

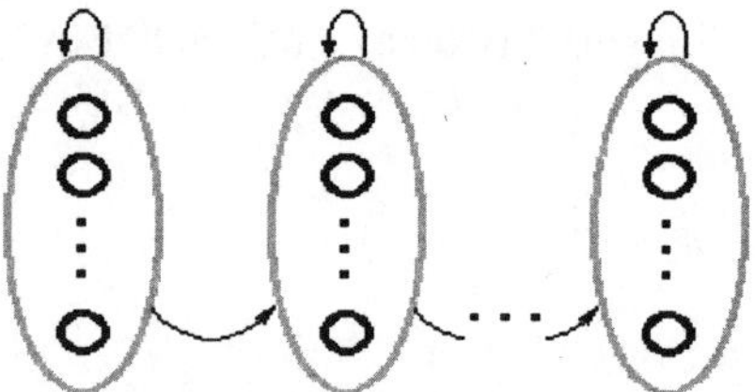

Fig-3: General schema of the mixture of primitive models at the state level

For a letter l, the model at the state s ($s=1\sim m$) is now a mixture of J PNNs: $PR_{(1,s)}, PR_{(2,s)}, ..., PR_{(J,s)}$, the probability of this state for an observation frame $x(t)$ in the input signal X_1^T can then be calculated by:

$$P(x(t)|s,l) = \sum_{j=1}^{J} P(x(t)|PR_{(j,s)},s,l) P(PR_{(j,s)}|s,l) \tag{7}$$

Therefore, the probability of this letter-model for the input signal is estimated by:

$$P(X_1^T|l) = \max_{\{S_1^T\}} \{P(X_1^T|S_1^T,l)\} = \max_{\{S_1^T\}} \left\{ \prod_{S_1^T:t=1\sim T} \{P(x(t)|s(t),l)\} \right\} \tag{8}$$

Where, $\{S_1^T\}$ is the set of all possible state sequences. Just the same as in the former method, the mixture of PNNs can be realized by fusion or competition. Besides, a new problem has to be resolved in the mixture of primitive models – what kinds of transitions among the PNNs may be allowed. Therefore, we propose two types of topologies: a free architecture and a trajectory one, as described in Fig-4.

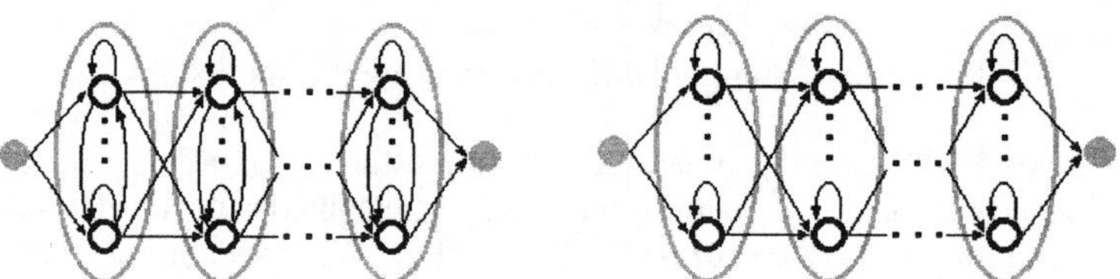

Fig-4: Free architecture (left) and trajectory one (right) for the mixture of PNNs

In the free structure, all kinds of transitions within a state are allowed. Between two successive states, a model is connected to all the PNNs in the following state. Each moment, the best model is chosen from all these $2J$ models. Such a structure is then able to model the variance of only one frame length.

In the trajectory structure, the transitions among the PNNs within a state are forbidden. Consequently, some kinds of variation may be ignored, and the primitives become longer and thus more stable. This structure is expected to be more robust.

For both methods, only small modifications have to be made in the training and recognition algorithms of the reference system [6]. Here, we don't discuss them in detail.

6. Experimental Results

Our experiences were realized on the UNIPEN handwriting corpus [9]. We used 40000 handwritten characters of multi-writers in an on-line form. After necessary pre-processing, 15 features were extracted on each sampling point, the same as [6]. Neither normalization nor slant correction was performed to the variation such that the proposed methods may be evaluated to the best. 30000 were used in the training database and other 10000 for testing the system.

Table-1 shows the performances of the proposed methods. The mixture order J varies from 1 to 3. It's clear that the best value may be much larger according to the obtained results. When $J=1$, we return to the reference system.

We notice that the mixture of primitives always creates a better result than the mixture of HMM chains. According to us, it is due to the larger structural liberty that allows a more powerful modelling at the same system complexity level. The reason why the trajectory structure acts better than the free structure is due to a better compensation between the modelling capacity and the discrimination. As a 3-dimensional Viterbi algorithm should be used for the trajectory architecture, this method appears very time consuming.

Here, the results about the fusion methods are not mentioned, because no improvement is brought to the performance. The reason is that the fusion losses too much of the discrimination.

Table-1: Comparison of the performance of the proposed methods

	Mixture of HMM Chains	Mixture of P. M. with Free Structure	Mixture of P. M. with Trajectory Structure
$J = 1$	75.2 %	75.2 %	75.2 %
$J = 2$	78.0 %	78.3 %	79.9 %
$J = 3$	79.6 %	79.7 %	81.3 %

7. Conclusions

We studied several methods of writing style variation absorption for a hybrid neuro-markovian on-line handwriting recognition system. More primitive models were used for modelling precisely the different handwritings. Two types of model mixture were proposed: at the letter level, a mixture of HMM chains and at the state level, a mixture of primitive models. Two possible topology architectures were also introduced, such as a free structure and a trajectory one. Effectiveness of our proposals was approved by the experiments realized on the wildly used UNIPEN database.

References

[1] J. Hu, M.K. Brown and W. Turin, HMM Based On-Line Handwriting Recognition. In IEEE Transactions on Pattern Analysis and Machine Intelligence, Vol.18, No.10, October 1996, pp.1039-1045.

[2] I. Guyon, Application of Neural Networks to Character Recognition. In *Character & Handwriting Recognition* --- World Scientific Series in Computer Science, Vol.30.

[3] G. Seni, N. Nasrabadi and R. Srihari, An On-line Cursive Word Recognition System. In Proceedings of IEEE Conference on Computer Vision and Pattern Recognition, Seattle USA, June 1994, pp.404-410.

[4] T. Artières and P. Gallinari, Multi-state Predictive Neural Networks for Text-Independent Speaker Recognition. In proceedings of EUROSPEECH'95, 1995.

[5] S. Garcia-Salicetti, B. Dorizzi, P. Gallinari, *et al.*, A Neural Predictive Approach for On-line Cursive Script Recognition. In Proceedings of ICASSP'95, 1995.

[6] T. Artières, B. Dorizzi, P. Gallinari, *et al.*, From Character to Sentences: a Hybrid Neuro-Markovian System for On-line Handwriting Recognition. In *Hybrid Methods in Pattern Recognition*. H. Bunke, A. Kandel (eds), World Scientific Publ. Co. Parution courant 2001, pp.1-27.

[7] J.J. Lee, J.W. Kim and J.H. Kim, Data Driven Design of HMM Topology for On-Line Handwriting Recognition. In Proceedings of IWFHR'2000, Amsterdam, September 2000, pp.239-249.

[8] S. Garcia-Salicetti, Une Approche Neuronale Prédictive pour la Reconnaissance de l'Ecriture Cursive. PhD thesis, Université Paris 6, 1996.

[9] I. Guyon, I. Schomaker, L. Plamondon, *et al.*, UNIPEN Project of On-Line Data Exchange and Recognizer Benchmarks. In Proceedings of ICPR'94, Israel, October 1994, pp.29-33.

KES 2002
E. Damiani et al. (Eds.)
IOS Press, 2002

Neural Network Approximation of Stochastic Processes: A Recursive Algorithm

Paolo CRIPPA, and Claudio TURCHETTI
Dipartimento di Elettronica e Automatica, University of Ancona,
via Brecce Bianche, I-60131 Ancona, ITALY

Abstract. Artificial Neural Networks (ANNs) must be able to learn by experience from environment. Some neural networks are capable of approximating not only deterministic (non-random) functions but also stochastic processes. These networks, named Stochastic Neural Networks represents a generalisation of the usually defined neural networks. From an application point of view such a class of networks is more adherent to the real world whose nature is essentially stochastic. In this paper an original recursive training algorithm for the approximation of a large class of stochastic process has been presented. The methodology is based on the canonical representation of non-stationary stochastic processes by means of Brownian motion processes.

1 Introduction

One attracting property of neural networks is their capability in approximating not only arbitrary non-random input-output mappings [1, 2, 3, 4] but also stochastic processes or, more in general, input-output transformations of stochastic processes [5, 6]. This achievement is of utmost importance because neural networks (biological or artificial) operates in the real world where all the signals are undoubtedly stochastic. This occurs because either the noise could be superimposed to a non-stochastic signal or the signal could be inherently stochastic being unidentifiable a predictable part of it. The aim of this work is to show a recursive algorithm based on a recent model for neural processing in which the network acts as a universal approximator of stochastic processes.

2 Canonical Representation of Stochastic Process

Let us consider a stochastic process (SP) $\varphi(t) = \varphi(t, \omega)$ or $\{\varphi(t), t \in T\}$ such that for every $t \in T$ the random variable (RV) $\varphi(\omega)$, defined on a fixed probability space $\{\Omega, \mathcal{B}, \mathcal{P}\}$, satisfies the conditions $E\{\varphi(t)\} = 0$, $E\{|\varphi(t)|^2\} < \infty$ where $E\{\cdot\}$ represents the expectation of a RV. The set of all such RVs forms a real Hilbert space $L^2(\Omega)$ [7]. We consider a class of real-valued random processes $\{\varphi(t), t \in T\}$ whose autocorrelation function $R_{\varphi\varphi}(t, s) = E\{\varphi(t)\varphi(s)\}$ admits the representation

$$R_{\varphi\varphi}(t, s) = \int_0^{+\infty} \boldsymbol{\Phi}^T(t, \lambda) \cdot \boldsymbol{\Phi}(s, \lambda)\, dF(\lambda) \tag{1}$$

where $\boldsymbol{\Phi}(t, \lambda) = [g_1(t, \lambda), g_2(t, \lambda)]^T$ is a family of real-valued functions of the variable λ that depend on the parameter $t \in T$, and $F(\lambda)$ is a monotone non-decreasing function defining a Stieltjes measure on the measurable sets $\Delta\lambda = [\lambda, \lambda + \Delta\lambda)$ of $\mathbb{R}^+$. Processes of such a kind constitute a wide class that includes all the stationary processes other than many non-stationary processes. It is well known from the SP theory that $\varphi(t)$ admits the following canonical representation:

$$\varphi(t) = \int_0^{+\infty} \mathbf{\Phi}^T(t, \lambda) \cdot \mathbf{dy}(\lambda) \tag{2}$$

where the integral is a stochastic integral and the components of $\mathbf{y}(\lambda) = [y_1(\lambda), y_2(\lambda)]^T$ are orthogonal increments (OI) processes such that $E\left\{|dy_k(\lambda)|^2\right\} = dF(\lambda)$, $k = 1, 2$. However even if this representation is useful from a theoretical point of view, it is not easily implementable by neural networks because only these OI processes are allowed. Instead, as the Brownian Motion (BM) processes are straightforward to generate, it is more convenient to derive a representation in terms of such processes. To this end it can be shown that the following canonical representation holds [5, 6]:

$$\varphi(t) = \int_0^{+\infty} c(\lambda)\,\mathbf{\Phi}^T(t, \lambda) \cdot \mathbf{d\nu}(\lambda) = \int_0^{+\infty} \mathbf{\Psi}^T(t, \lambda) \cdot \mathbf{d\nu}(\lambda) \tag{3}$$

where the components of $\mathbf{\nu} = [\nu_1(\lambda), \nu_2(\lambda)]^T$ are independent BM processes for which $E\left\{|d\nu_k(\lambda)|^2\right\} = dG(\lambda) = \sigma^2 d\lambda$ and $c(\lambda)$ is a non-negative function such that $F(\lambda) = \int_0^\lambda [c(\xi)]^2\,dG(\xi)$.

3 Recursive Learning Algorithm

Let us consider a stochastic process of such a kind

$$\varphi_n(t) = \int_0^{+\infty} \mathbf{\Phi}_n^T(t, \lambda) \cdot \mathbf{dy}(\lambda) \quad \text{with} \quad \mathbf{\Phi}_n^T(t, \lambda) = \sum_{i=1}^n \mathbf{a}_i^T u_i(t)\, u_i(\lambda) \tag{4}$$

where $u_i(\cdot)$ are approximate identity functions and $\mathbf{a}_i^T$ are vectors of weights. Networks of such a kind are able to approximate a wide class of functions with high accuracy [6]. By using the transformation (3), it also results

$$\varphi_n(t) = \int_0^{+\infty} \mathbf{\Psi}_n^T(t, \lambda) \cdot \mathbf{d\nu}(\lambda) \quad \text{with} \quad \mathbf{\Psi}_n^T(t, \lambda) = \sum_{i=1}^n \mathbf{b}_i^T u_i(t)\, u_i(\lambda)\,. \tag{5}$$

The transformation mapping $\varphi(t) \to \mathbf{\Phi}^T$ (or $\varphi(t) \to \mathbf{\Psi}^T$) preserves the distances, i.e.

$$d(\varphi, \varphi_n) = E\left\{|\varphi(t) - \varphi_n(t)|^2\right\} = \int_0^{+\infty} \left|\mathbf{\Psi}^T(t, \lambda) - \mathbf{\Psi}_n^T(t, \lambda)\right|^2 G(d\lambda) = d(\mathbf{\Psi}, \mathbf{\Psi}_n)\,. \tag{6}$$

Therefore an algorithm that minimizes the distance $d(\mathbf{\Psi}, \mathbf{\Psi}_n)$ in the space of non-random functions $\mathbf{\Psi}$ also guarantees the convergence of the stochastic process $\varphi_n(t)$ to $\varphi(t)$. This distance can be viewed as an *error function* $\mathcal{E}_n(t)$ that depends on the time t:

$$\mathcal{E}_n(t) = R_{\varphi\varphi}(t, t) + R_{\varphi_n\varphi_n}(t, t) - 2\int_0^{+\infty} \mathbf{\Psi}^T(t, \lambda) \cdot \mathbf{\Psi}_n^T(t, \lambda)\, G(d\lambda) \tag{7}$$

Thus the set of weights in (5) can be obtained integrating this error function over the time interval $[0\ T]$ by considering time-limited signals, i.e. $\varphi(t) = 0$ for $t \notin [0\ T]$, and minimizing the quantity

$$\mathcal{M}_{\mathcal{E}_n} = \int_0^T |\mathcal{E}_n(t)|^2\, dt \tag{8}$$

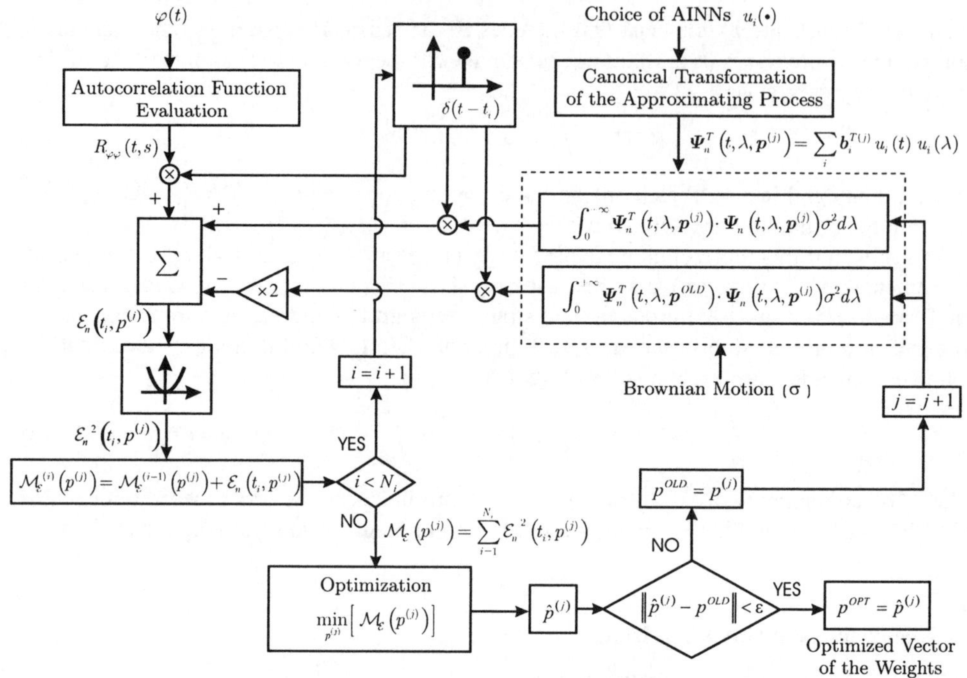

Figure 1: Flow chart of the recursive learning algorithm .

by using the recursive algorithm summarized in the flow chart depicted in Fig. 1. The algorithm has been implemented in C++ language on a Pentium III PC.

4   Application Example

As an application example of the recursive algorithm we refer to the signal

$$\varphi(t) = \chi(t)\, t\, \exp(-\omega_0 t) \tag{9}$$

where ω_0 is a constant and $\chi(t)$ is wide-sense stationary process with zero mean value and autocorrelation function given by $E\{\chi(t)\chi(s)\} = R_{\chi\chi}(\tau) = \exp(-\alpha|\tau|)$ with $\tau = t - s$. The process $\varphi(t)$ is non-stationary being its covariance function given by

$$R_{\varphi\varphi}(t, s) = E\{\chi(t)\chi(s)\}\, t\, \exp(-\omega_0 t)\, s\, \exp(-\omega_0 s). \tag{10}$$

By using the spectral theory of stationary processes it is straightforward to show that it results

$$R_{\varphi\varphi}(t, s) = \int_0^{+\infty} [g_1(t, \lambda)\, g_1(s, \lambda) + g_2(t, \lambda)\, g_2(s, \lambda)]\, F(d\lambda) \tag{11}$$

where $g_1(t, \lambda) = \cos(\lambda t)\, t\, \exp(-\omega_0 t)$, $g_2(t, \lambda) = \sin(\lambda t)\, t\, \exp(-\omega_0 t)$, and $F(d\lambda) = 2\,\alpha\, d\lambda\,/[\pi(\alpha^2 + \lambda^2)]$. Thus, by considering the Brownian motion $G(d\lambda) = \sigma^2 d\lambda$, the components of the function $\boldsymbol{\Psi}^T(t, \lambda) = [h_1(t, \lambda),\, h_2(t, \lambda)]^T$ to be approximated are given by

$$h_1(t, \lambda) = \frac{1}{\sigma}\sqrt{2\alpha/[\pi(\alpha^2 + \lambda^2)]}\cos(\lambda t)\, t\, \exp(-\omega_0 t)\,, \tag{12}$$

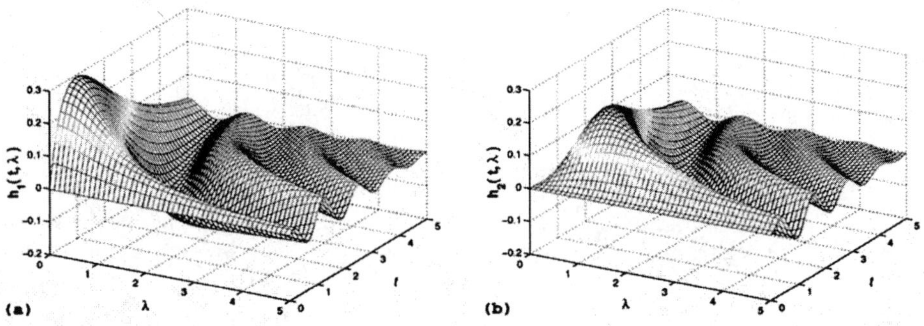

Figure 2. Target functions $h_1(t, \lambda)$ and $h_2(t, \lambda)$ given by (12)–(13).

$$h_2(t, \lambda) = \frac{1}{\sigma} \sqrt{2\alpha / [\pi(\alpha^2 + \lambda^2)]} \sin(\lambda t) \, t \, \exp(-\omega_0 t) . \tag{13}$$

From the above methodology, approximating the functions h_1 and h_2 it is equivalent to determining both the parameter σ of the Brownian motion G and the decay constant ω_0. Figs. 2(a)–(b) report the functions $h_1(t, \lambda)$ and $h_2(t, \lambda)$ to be approximated as functions of λ and t. Figs. 3(a)–(h) report the approximating functions $\widetilde{h}_1(t, \lambda)$ and $\widetilde{h}_2(t, \lambda)$ as functions of λ and t at different simulation epochs. The outputs of the recursive approximation, i.e. the final values of the functions $\widetilde{h}_1$ and $\widetilde{h}_2$ are displayed in Fig. 3(g) and Fig. 3(h), respectively. As you can see, there is an excellent agreement with the target functions h_1 and h_2 in Figs. 2(a)–(b) that confirms the validity of this approach.

5 Conclusion

In this paper a recursive algorithm for the approximation of a wide class of stochastic processes via neural networks has been implemented in C++ language. As an application example, the algorithm, based on the AINNs theory, has been used to approximate a non-stationary SP. The results confirmed the validity of the methodology.

References

[1] T. Poggio, and F. Girosi, "Networks for approximation and learning," *Proceedings of IEEE*, vol. 78, No. 9, pp. 1481–1497, 1990.

[2] K. Funahashi, "On the approximate realisation of continuous mappings by neural networks," *Neural Networks*, vol. 2, No. 3, pp. 183–192, 1989.

[3] G. Cybenko, "Approximation by superposition of sigmoidal function," *Math. Control, Systems, Signal*, vol. 2, No. 4, pp. 303–314, 1989.

[4] K. Hornik, M. Stinchcombe, and H. White, "Multilayer feedforward networks are universal approximators," *Neural Networks*, vol. 2, No. 5, pp. 395–403, 1989.

[5] M. R. Belli, M. Conti, P. Crippa, and C. Turchetti, "Artificial neural networks as approximators of stochastic processes," *Neural Networks*, vol. 12, No. 4-5, pp. 647–658, 1999.

[6] C. Turchetti, M. Conti, P. Crippa, and S. Orcioni, "On the approximation of stochastic processes by approximate identity neural networks," *IEEE Trans. Neural Networks*, vol. 9, n. 6, pp. 1069–1085, 1998.

[7] J. L. Doob, *Stochastic processes. J. Wiley & Sons*, New York, USA, 1990.

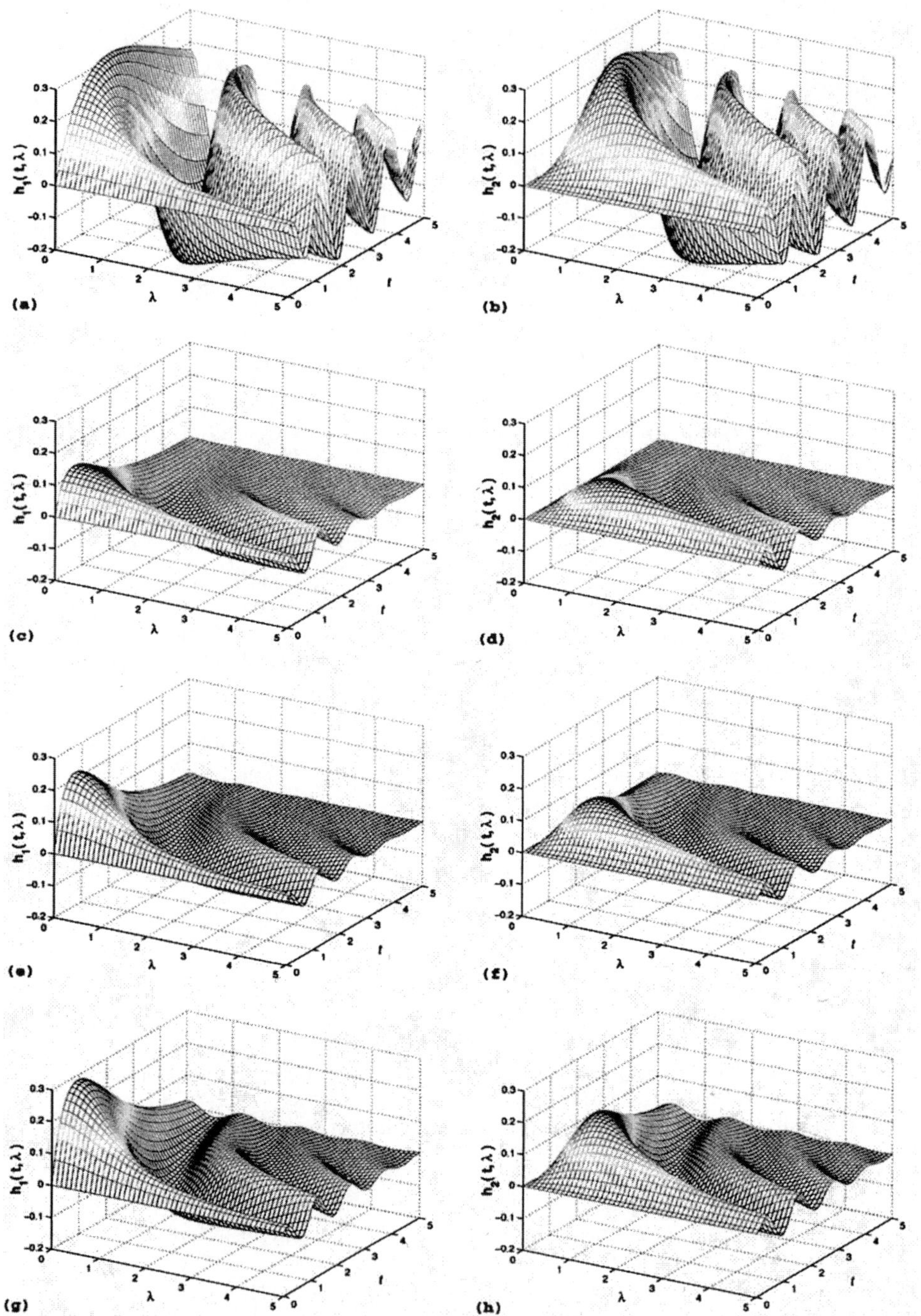

Figure 3: Recursive algorithm outputs $\widetilde{h}_1(t, \lambda)$ and $\widetilde{h}_2(t, \lambda)$ at different epochs in the approximation of functions $h_1(t, \lambda)$ and $h_2(t, \lambda)$ given by (12)–(13); (a), (b) ((g) ,(h)) are the starting (final) functions.

KES 2002
E. Damiani et al. (Eds.)
IOS Press, 2002

Interpersonal Cognition
in Anonymous Community

AZECHI Shintaro and MATSUMURA Ken'ichi
Hokkaido Tokai University, Japan
Synsophy Project, CRL, Japan

Abstract. A social psychological research indicates how people in anonymous community percept numbers of other group members. First, people in anonymous community tends to consistently recognize that the numbers of the members is fewer than the numbers of the statements in the discussion. Second, people think there are much members who have same attitude with their own more than who have the different one. We discuss these findings from the view ponint of social cognition and propose how do we solve the problems caused by the 'outgroup bias'.

1 Background

There are many anonymous community in the internet, for example, a huge community called "2channel (Ni-Channel)[1]" in Japan that is thread styled bulletin boards accessed by more than one million people per a day. In other many communities mediated by bulletin board systems or Public Opinion Channel [2], people also enjoy discussion with anonymity.

Why such a many people need anonymous communities? One main reason is that anonymity enhances exchange of information and interactions [1]. When people live in the real world, they are often needed to stop to express their own opinions because they are not appropriate to the social situation. For example, when someone has excellent opinion about protection of whales, an authority express that whaling is indispensable to manage marine products, s/he might be easily discouraged to express their own opinion. On the other hands, in an anonymous community, everyone don't have to take care of social situations or authorities because no one can know who is s/he. One can express her or his opinion without fear to be burden by other people's social pressures.

On the other hands, there are troubles in anonymous community. One most serious trouble is hostility to other community members. Some people think they are continuously blamed by someone who has counter opinion against them. They regard that there are few people against them although the few people describes their unacceptable opinion repeatedly with abusing the anonymity. They easily ignore people who have opposite opinions regarding as a minority, and sometimes they purpose to enclose their opinion. Finally, the anonymous community that was supposed to enhance discussions becomes suppressive field preventing free discussion.

---

[1]http://www.2ch.net/

2 Psychological Processes in Anonymous Community

There are two biased social cognition processes that causes such a trouble in anonymous community. One is 'bias for outgroup' and the other is 'false consensus effect'. The purpose of our study is to know how people in the anonymous community recognize other members.

2.1 Bias for Outgroup

This well-known bias is defined as follow: a group member regards members of her or his own group (ingroup members) as various good people: s/he regards members of other group (outgroup members) as uniform bad people. Linville [4] suggests that the difference of complexity of representations between for ingroup and for outgroup causes this biased cognition. This 'group' is formed easily when a person recognizes some salient characteristics in other people, such as gender, ethnicity and attitude differences.

People in anonymous community is supposed to be affected this bias. They think community members who have opposite opinion against them. They also think the opposite opinions are in a pattern. Finally they think that a few people repeatedly express patterned opinion against them.

2.2 False Consensus Effect

False consensus effect is defined as a biased cognition that people believe they are socially majority [3]. Even when they know they are socially minority, they think the difference of the population between majority and minority is smaller than the real difference.

In anonymous community, people are also affected by false consensus effect. Even when members think they are minority in the community, they recognize the number of their supporter is larger than real populations.

3 Hypotheses

How the members of anonymous community recognize the population of other memberswho have opinion agree with versus opposite against them? We planned an experiment to test following two hypotheses.

1. Members in an anonymous community generally recognize the size of community is smaller than the numbers of statements through the discussion.

2. Members recognize that the population of members who has the same attitude with their own is greater than the real one.

4 Method

4.1 Participants

83 undergraduate students at Hokkaido Tokai University who take "Cognitive Psychology" class participate this experiment for the credit.

4.2 Procedures and Materials

Participants are instructed to solve a task in a questionnaire that contains log of an anonymous bulletin board and some questions about the log.

Questionnaire has three parts. The first part asks participant's age, gender and attitude for whaling. Participant are needed to describe they approve with or oppose to whaling in four point scale ('Approval', 'Rather Approval', 'Rather Opposite' and 'Opposite'). Participants are divided into approvers who checked 'Approval' and 'Rather Approval' and opposites who checked 'Opposite' and 'Rather Opposite'.

The second part, participants are demanded to read log of anonymous bulletin board where people discuss about whether whaling is proper to do. The numbers of statements are 18.

There are two types of the log and participants read either one. One types, named as 'ap12op6', has 12 approval statements for whaling and 6 the opposite ones. In this condition, participant who has approval opinion for the whaling is majority. The other, named as 'ap6op12', contains vice versa 6 approval statements and 12 opposite ones. In this condition, participants who have approval opinion should feel they are minority.

After reading the log, participants are needed to reply to two questions. 1) How many people describe the statements? 2) How do you infer the rate of the people who approve versus oppose the whaling?

The third part, participants are demanded show their attitude for whaling again.

5 Results

There is no significant effect by participant's age (M=21.11) and gender (Female: 33 persons, Male: 50 persons). 4 participants did not complete all the task, so that finally 79 replies are analysed.

Attitude for whaling of participants between pre and post engaging the task is little changed. Peason's correlation coefficient between pre and post attitude is .77 ($p < .00$).

33 participants (41.8%) agree to do whaling and 46 participants (58.2%) are against it. Pro-whaling is minority and anti-whaling is majority in this participant universe. This tendency is not match to Japanese general distribution of attitude for whaling, over 60% people approve of wharing[2].

5.1 Estimated Size of the Community

The average of the response to the question one, "How many people describe the statements?", is about 8 (M=7.94, SD=3.57). No participant rates over 18 that is numbers of the statements.

5.2 Outgroup Cognition in Anonymous Community

Participants describe the rate not complying with percentile but like "approval : opposite = 8 : 9", so that we calculate the index number (IN) indicating how the participant recognize the rate of approver with percentile.

---

[2] http://www.jfa.maff.go.jp/whaling/kentoukai/rep.html

$$IN = \frac{rate\ of\ approval\ ones}{rate\ of\ approval\ ones + rate\ of\ opposite\ ones} \tag{1}$$

Table 1 shows IN broken down by participant's attitude for whaling and type of the task.

Table 1: IN broken down by participant's attitude and task type

Version of Task	Attitude of Participant for Whaling	
	Approval (N)	Opposite (N)
Ap12-Op6	.64 (17)	.56 (24)
Ap6-Op12	.40 (16)	.37 (22)

ANOVA is applied to analyze the table. The main effect of participant's attitude for whaling is significant ($F(1, 78) = 3.96, p = .050$). The main effect of type of the task is also significant ($F(1, 78) = 50.26, p < .00$). There is no interactive effect between attitude and type of the task.

6 Discussion

6.1 *Estimated Size of the Community*

A consistent tendency is found out. People think that the number of the people who engage in discussion is smaller than the number of statements. It indicates that people think only a few people describe their opinion in many times. No one notice the two possibilities, 1) there are so-called 'ROM (Read Only Member)'s who don't describe any comment but only read the statements are join there, 2) every statement is described by different person.

This tendency suggests that people suppose the group size of the community smaller then the number of statements there. In fact, because there are many ROMs, the group size of the anonymous community is much larger than the number of statements. We consider that this biased perception of members for the community size is a salient characteristic of the anonymous community.

6.2 *Outgroup Cognition in Anonymous Community*

In both task conditions whether approver is majority versus minority, participants who approve for whaling estimate the rate of approver is larger than the rate percepted by opposite participants. This result shows that participants consistently estimate the ingroup size rather large. So that participants are supposed to regard one outgroup member describe more time than one ingroup member. This is a evidence that bias for outgroup is occurred.

Is this result indicating false consensus effect? It is unclear. The result seems be reflect false consensus effect because participants consistently think them as rather major than counter group members think so. But to assess false consensus effect, we are needed to know how participants see the distribution of opinions in the real society. There are complex levels of false consensus effect. We discuss about this problem of in section7.1, future works.

7　Perspectives

7.1　How Control Bias in Anonymous Community

This study explains that member of anonymous community recognize the population of community members with biased way. We have already suggested that this biased cognition make some serious troubles of anonymous community. How can we solve these problems?

One possible solution is that a system which mediates anonymous community should be implemented functions to show users the actual conditions of the community. To solve misunderstanding of users that a few people smaller than statements is there, indicating actual community size seems to be effective. Indicating the distribution of opinion in the community in real time also affects to solve outgroup bias, because members can know how their opinions are accepted by other members, and know how much they must engage to persuade other members. Those two considerable functions would solve the problem of anonymous community and improve the efficacy for free discussion to enhance user's motivation.

7.2　Future Works

Unfortunately, we can not indicate how false consensus effect affects member's interpersonal cognition in anonymous community. There are complex level of false consensus, that is, 1) social level that participant's belief about distribution of opinions, 2) community level that how participants infer distribution of members who have approval versus opposite opinion, 3) descriptive level how participants We are planning experiments that make clear those three levels of false consensus effect and the relationship the effect and bias for outgroups.

References

[1] Azechi, S., and Matsumura, K., Motivation for Showing Opinion on Public Opinion Channel: A Case Study, IEEE KES2001 proceedings **2** (2001) 344–347.

[2] Fukuhara, T., Matsumura, K., Azechi, S., Fujihara, N., Terada, K., Yamashita, K., and Nishida, T., Creating city community consanguinity: use of public opinion channel in Digital Cities, In Tanabe, M., van den Besselarr, P., and Ishida, T. (eds): Digital Cities II: Computational and Sociological Approarches (in press).

[3] Ross, L., Greene, D., and House, P., The 'false consensus effect': An egocentric bias in social perception and attribution processes, Journal of Experimental Social Psychology **13** (1977) 279–301.

[4] Linville, P. W., The complexity extremity effect and age-based stereotyping, Journal of Personality and Social Psychology **42** (1982) 192–211.

KES 2002
E. Damiani et al. (Eds.)
IOS Press, 2002

Conversational Contents
Making a Comment Automatically

Hidekazu Kubota [†]   Koji Yamashita [†,†]   Toyoaki Nishida [‡]

[†] *Department of Information and Communication Engineering,*
School of Engineering, The University of Tokyo
[††] *Synsophy Project, Communications Research Laboratory*
[‡] *Information and Communication Engineering,*
Graduate School of Information Science and Technology, The University of Tokyo

Abstract. This paper presents a method for transforming a text form into a conversation form by inserting different comments in a documented text. A dialogue documented in a fixed text form lacks a diversity of the original conversation. In order to evolve the stored knowledge, diverse reevaluations of the knowledge are essential. We propose the concept of *Conversational Contents* that enables people to keep discussing a knowledge product by talking with the very product. Our experimental result called POC caster inserts different comments into a text in order that people can focus on many sides of the text. The effects of different comments upon understanding and studying a text are discussed.

1. Introduction

A great deal of dialogue is documented in a text form, for example, interviews, proceedings, mailing lists and log files of BBS, which is useful knowledge resource to us. In order to evolve such knowledge, diverse reevaluations are essential. The past knowledge must be evaluated and organized again and again to adapt new situations. What seems to be lacking in the present text-formed knowledge resource is a means to keep studying and discussing the knowledge diversely. In conversation, we often discuss an old issue with asking again, supposing new hypotheses and connecting novel associations. On the other hand, an issue described in a text form provides only the past ways of thinking. At this point, the problem which people gets confused about is that a dialogue documented in a fixed text form lacks a diversity of the original conversation.

In this paper, we propose Conversational Contents that enables people to keep discussing a knowledge product by talking with the very product. The term 'discuss' implies that people studies, reevaluates and reorganizes knowledge in some kind of discussion. Sentences in a conversational form are exchanged between a user and Conversational Contents. A user can retrieve, evaluate and operate knowledge by dialogue.

Our first study of Conversational Contents is how to transform the contents from a text form into a conversational form. We propose a transform method that inserts different comments into the original text. We suppose that the diverse comments in a conversational form enable people to focus on many side of the text. POC caster is our early result that is composed of a main caster agent and an announcer agent. By using POC caster, we have an experiment about understanding of such a transformed conversation. In discussion, we study the effects of the transformation on understanding and studying a text to progress Conversational Contents.

2. Conversational Contents

We often explore the knowledge contents by annotating our processes of thinking. In a classroom, a student studies the contents of a textbook in a conversation with a teacher or other students. What is important in learning more is an annotation of the textbook that reminds her/him of what was discussed and how s/he explored the knowledge. The same may be said of a teacher. The teacher practically learns how to instruct the contents of a textbook by discussing with students again and again. So the annotation of a teaching note that reminds her/him of how to evolve the knowledge in the class is useful. The question we have to ask here is how to revise the annotations. Such annotations lack a diversity of the original lively discussion because only the old process of thinking is fixed into the annotation documented in a text form. Consequently, a sophisticated skill must be needed to have a rethink freely on the fixed annotation.

Conversational Contents is our answer to the lack of a diversity. Conversational Contents enables people to explore the past knowledge contents by talking with the very contents. In the aforesaid example of a classroom, Conversational Contents is pedagogical contents that can talk about itself and adopt new annotations of students and a teacher. A student can learn from the pedagogical contents adapted to the each student's progress and ask the contents again about a problem until s/he satisfies the explanation. Conversational Contents is also helpful for a teacher to reevaluate and reorganize the pedagogical contents because teacher can learn from the adopted contents about how students explored the content. Thus, the knowledge in a classroom evolves in the discussion mediated by Conversational Contents.

Know-How base is another application of Conversational Contents. When the past case study is reevaluated again and again in the new context, it is difficult to discuss about Know-How in a text form. Conversational Contents that can play back the past talk vividly seems to be of great use to make a comment. In addition to supporting knowledge evolution in a group, Conversational Contents applies to thinking support for individual. Since the human process of thinking often implies a contradiction that seems not to be resolved, we must be keeping company with such a contradiction. Conversational Contents help us to tackle a maze by recording exactly the process of wandering.

Fig.1 shows the concept of Conversational Contents. The Conversational Contents consists of a dialogue database (DB), conversational agents, generation rules and a DB reorganizer. The memory of Conversational Contents is a Dialogue DB that stores dialogue such as e-mails, log files of BBS and proceedings in a text form. It is appropriate to use human-like conversational agent as an interface between a user and Conversational Contents because it is easy for humans to talk with a character similar to themselves. Each agent has a generation rule that compiles text-formed dialogue into conversational explanation. In the Conversational Contents, the agent has an individual generation rule and the common memory. The reason why the agent doesn't have an individual memory is that we are not concerned with the multi-agents computing but interested in the very contents. Since the gathered utterances in dialogues are connected weakly and not separated clearly, Conversational Contents stores the utterances as they are into a database and compiles them into diverse explanations by applying different generation rules. All the explanations and comments discussed on Conversational Contents are adopted into dialogue DB by a DB reorganizer.

The processes of evolving knowledge are as follows. When a user says a short comment on some issues (process (1)), the agents listen the utterances (process (2)) and DB organizer records them (process (3)). The agents query about these utterances (process (4)), with the result that they retrieve related dialogues (process (5)). After the generation rules are applied into the retrieved original dialogues (process (6)), the agents utter the revised dialogue (process (7)) and DB organizer records them (process (8)). Finally, DB organizer reevaluates all the discussion on Conversational Contents and reorganizes the dialogue DB (process (9)).

Our first discussion of Conversational Contents is about the generating rules. The primary question of developing Conversational Contents is how people understand dialogues. In case of transforming a text form into a conversational form, no one knows what will change in understanding of the text. In the next section, we discuss a change in an understanding of a text if we apply simple rules to transform the text into a conversational form.

3. Method of transforming a text form into a conversation form

Commenting on a text is one of the simplest methods for transforming a text form into a conversational form. Fig.2 shows an example of such a simple transform. Consider two agents that introduce the original text in turn. At first, the original text form is divided into two parts by a period. A conversation proceeds as follows. (1) Agent I says the former part of the text. (2) Agent II comments on the former text. (3) Agent I says the latter part of the text.

Such a rule is surely simple, however, the results are quite deep. When Agent II says "How do they communicate?" instead of "What is the change?", the focus of the conversation changes from 'what' to 'how' and from 'change' to 'communication'. A new comment can focus on a particular part of the original text. While the focus removes ambiguity of the text, it changes a nuance of the text. It seems to be enough for the first study to discuss the complex changes of understanding of the text by using simple transformation.

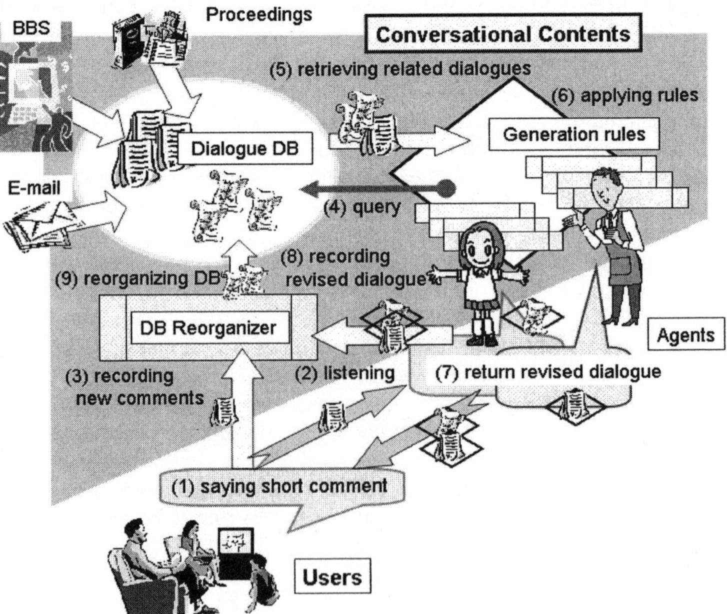

Fig. 1: Concept of Conversational Contents

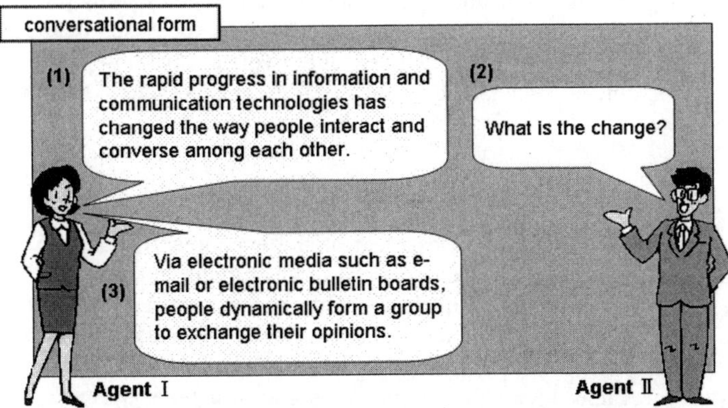

Fig. 2: Simple transform from a text form to conversational form

We propose POC caster [Kubota2002] that is a broadcasting agent system using a method for transforming a text form into a conversation form. The background of POC caster is Public Opinion Channel (POC) that is an interactive broadcasting system supporting community knowledge creation [Azechi 2000]. In POC, the community members evolve their knowledge by exchanging POC cards. A POC card is composed of one title line and two or three sentences like as general BBS. POC caster plays a role of a broadcaster in POC that introduces POC cards through a conversation. Dialog DB stores POC cards posted by community members. POC caster agent consists of a main caster agent and an announcer agent. Generaion rules of POC caster is dividing the original text and commenting on a text. DB organizer is not included in POC caster because we concentrate the generation rules on this system.

We conducted an experiment in understanding of a text by using POC caster [Kubota 2002]. 24 examinees listen and compare two types of stimulus sentences. One is the original text formed sentences about various fields, and another is conversational form of the original text. Two generation rules applied to the original text appears in Table 1.

Table 1: Two generation rules

Type of the rule	(A) Leading details	(B) Simple response
Modality of the former sentence	Introduction	No use
Inserting comment	"What is that?"(*1), "Tell me more details."(*2) and so on.	"Yes."(*3)
Function of the comment	The comment makes a listener to expect detailed continuation of the former sentence.	The comment divides the long text into two short parts.

(*1) The proper sentence is "Sore ha nandesuka?" in Japanese.

(*2) The proper sentence is "Motto kuwashiku oshiete kudasai." in Japanese.

(*3) The proper sentence is "Hai.", "Un." and so on in Japanese.

Rule (A) is applied when the original text can be divided into the former introduction part and the latter detailed part by a particular period. Rule (A) inserts a comment between the former and the latter sentences that makes a listener expect detailed continuation. At first, (1) the announcer agent (Agent I) says the former introduction part, (2) the main caster (Agent II) inserts a comment type (A) randomly from the list in Table 1 and (3) the announcer agent says the latter detailed part (like as the processes shown in Fig. 2). In a similar fashion, Rule (B) is applied when the original text is composed of more than three sentences. Rule (B) divide the original long text into two short parts by some period. Rule (B) gives a listener time to understand a long text. Examinees rate understandability of stimulus sentences on a scale ranging from 1 to 7.

4. Discussion

It was found from the subjective assessment of the experiment [Kubota 2002] that the conversational form is more understandable than the original text form and Rule (A) is better than (B) in the conversational form. This result suggests that even a simple conversational form can revise the contents in a text form. It was also found from the result that the general comments can even change understanding of the text largely. The general comments listed on Table 1 can be applied to broad situations because they have no reference to the original text. We will now expand these rules into the comments that have references to the former and the latter part of the original text.

In the example above (see Fig.2), "What is the change?" or "How do they communicate ? " refer to 'change' or 'communication' in the former part of the text. This repetition of the former part seems to emphasize the background of the text. "How does the electronic media effect on people?" is an example of referring the latter. This comment includes the common words 'electronic media' that makes a listener expect the common topic in the next sentence. It is inferred from these ideas that the different comments which refer the former part, the latter part or no part enable people to reevaluate the same contents diversely.

5. Conclusion

We are studying Conversational Contents that enables people to talk with a knowledge product. In order to enable people to discuss about the knowledge in a text form diversely, we proposed a method for transforming a text form into a conversation form by inserting different comments in a documented text. So far, we have seen that inserting simple comments can revise the documented text. Three types of reference in the comments remain as a matter to be discussed further.

References

[Azechi 2000] S. Azechi, N. Fujihara, S. Kaoru, T. Hirata, H. Yano, and T. Nishida, Public Opinion Channel: A Challenge for Interactive Community Broadcasting, In T. Ishida (ed.): Digital Cities: Experiences, Technologies and Future Perspectives, Lecture Notes in Computer Science, Springer-Verlag, 2000, 427-441.

[Kubota 2002] H. Kubota, K. Yamashita, T. Fukuhara, T. Nishida, POC caster: Broadcasting Agent Using Conversational Representation

for Internet Community, Transactions of the Japanese Society for Artificial Intelligence, Vol.17, No.3, 313-321.

Forming a Sense of Reality
from the Viewpoint of Cognitive Psychology

Koji YAMASHITA
Synsophy Project, Communications Research Laboratory, JAPAN
koji@crl.go.jp

Abstract. This paper presents a discussion of sense of reality from the viewpoint of cognitive psychology. Discussions that address the question of cognitive processes in developing a sense of reality are presented with emphasis on the reality of objects and the reality of relations and their interactions with bottom-up and top-down processing. Some factors influencing the formation of a sense of reality are reviewed. Finally, it is suggested that communication plays a significant role in forming such a sense.

1. Introduction

What is reality? How do we sense that there is a reality? Where does the sense of reality come from? Recent advances in communications and information technology have altered how we use the word 'reality'. For example, we can now speak of virtual reality (VR), augmented reality (AR), and mixed reality (MR). The question, however, is to what extent such technologies, or the 'worlds' created with them, convey a sense of reality.

This paper focuses on the formation process of the sense of reality, and presents perspectives to support the reality-sense formation. In section 2, we consider what reality is by focusing on changes and enhancements being made the world we live in. In section 3, we discuss the reality-sense formation process and point out some factors affecting it. The importance of communication with objects and others in reality formation is suggested.

2. What is Reality?

2.1 Enhancement of Reality

Given that cognition of a 'real world' is what could be called reality-sense formation in a broad sense, it can be said the environmental image which we perceive and experience through our sensory organs, such as the eyes, ears, mouth, nose, and skin, is a starting point of reality-sense formation. We term this image *primary reality*. Humans have long enhanced this primary reality by developing various informational technologies. These technologies included language, writing, typographic printing, mass media, the computer, and the Internet, innovations that have greatly affected our lives. Using language, for instance, we can express our individual environmental images, and thus share a reality beyond one person. Writing has allowed us to conceptualize and think in the abstract, and this also has meant being able to convey a sense of reality beyond the time and space of the moment. Our reality could be easily conveyed by typographic printing, and with the appearance of mass media (newspaper, radio, and television), a common reality, as well as, vicarious environments, became possible. Since the mid 20th century, the advent of the computer and the Internet has produced cyber space and the world of virtual reality.

These changes may have enhanced *primary reality*, which remained perceivable under the invisible abstract, past or future world, or world beyond time and space. However, although it seems that reality has been changed and enhanced, have these changes actually affected the nature of reality or our sense of reality? We will discuss this issue in the following section.

2.2 Two kinds of Reality

When we use a video conference system, we want to see and hear each other as well as we could in face-to-face communication. In the case of realistic images, one supposes the visual reality of the object. In the case of sound, one assumes acoustic reality. For example, in virtual reality, the most important problems are "how does it look?" and "how does it sound?" This is a sensory-centered view, i.e., one that looks on a stimulus as existing in the outer world as an object, and it can be called *Reality of Object*. This *Reality of Object* viewpoint aims to formulate a sense of reality by improving fineness of inputted sensory information (VR), and by adding information to the object (AR).

Even if we look at an object directly, however, we cannot feel the object as being real if we cannot acknowledge a connection with the object. That is, it seems that the sense of reality can be formulated only when we feel a connection with the object (know it is there at a sensory level is not sufficient). We call this *Reality of Relation*. The *Reality of Relation* makes a meaning for the perceived object, and is defined in terms of knowledge and contextual information about the object. For instance, although text-based communication such as MUD and ELISA has much less *Reality of Object* than a video conference system, it does not always have less *Reality of Relation*. In fact, text gives us not only literal information but also the sender's background knowledge or contextual situation.

As described above, it is clear that there are a *Reality of Object* that is sensory-centered and a *Reality of Relation* that is meaning-centered. Although it appears that recent progress has enlarged reality, it seems there are no changes in the nature of reality. When we consider human reality-sense formation, we need to take into account these two realities.

3. Formation Process of Reality Sense

3.1 As an Information Processing System

In cognitive psychology, the human being is viewed as an information processing system or a processor of information. That is, cognitive psychology has the view that information in the external world is inputted via sense organs, processed somehow and outputted as a response. Thus, it assumes place or process for judging what inputted information is and deciding what response is output. This problem solving process is an internal process to perceive information selectively from surrounding environment, to process that information, and to decide the next response.

Two kinds of information are always used in personal internal activities: bottom-up and top-down information (Figure 1). Bottom-up information is information inputted through sensory

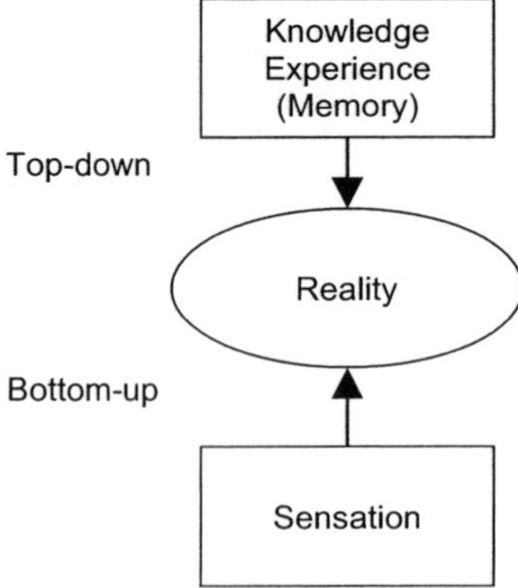

Figure 1: Reality formation process

organs, processing using this information is called bottom-up processing (or data-driven processing). Bottom-up processing begins from low-level analysis of physical features of objects, and advances upward toward the construction of a final interpretation. On the other hand, top-down information means information stored in our memory from past experience such as knowledge, concepts, and so on, and processing using this information is called top-down processing (or conceptual-driven processing). Top-down processing begins from high-level processing that produces anticipation and hypothesis-related interpretation and evaluation of the inputted information, and moves downward. Both processings are related through control flow of information processing and comprised of various cognitive activities, for example, memory, language comprehension, and thought. They proceed in parallel and interact [1].

Taking language comprehension for instance, we can see that there are several levels of analysis in the interaction between top-down and bottom-up processing. Low-level physical analysis of the object includes extracting visual features of characters and acoustic cues of utterances. High-level processing includes syntactic analysis and semantic analysis. Reader and listener may anticipate outcomes by using spelling, phonetic constraint, grammatical knowledge, contextual knowledge, and experience gained in the outer world. Thus, top-down and bottom-up processing interact to make possible reading of a text and a conversation with someone else. Accessibility of contextual information and the nature of stimulus determine how far top-down processing accounts for the whole processing. Where we can use contextual information, the expected letters or words fill gaps caused by losses. In that context, top-down processing provides a shortcut to the end, and there is no need to analyze the message completely. In the comprehension of conversation, top-down processing plays a large role, because spoken language is of poorer quality compared with written text. It is very difficult for us to understand spoken language out of context.

It is expected that bottom-up and top-down processing interact with each other in formulating a sense of reality. As pointed out in the above section, bottom-up processing is analysis of physical features of the object, and this is related to *Reality of Object*. On the other hand, top-down processing using past experience or contextual knowledge is involved with *Reality of Relation*. By arousing our reality-sense formation by increasing and elaborating bottom-up information through sensory organs directly, only part of reality is affected. This suggests that we cannot get the complete picture of reality-sense formation unless we examine the interaction between bottom-up processing (*Reality of Object*) and top-down processing (*Reality of Relation*). In the next section, we will address the issue in this regard.

3.2 Some Factors which have Influence on Reality-sense Formation

A sense of reality is formed by the interaction between bottom-up and top-down processing. Certain factors affect this formation in each of the two processing stages. The bottom-up processing is an analysis of the physical features of objects, that is, perception itself. It needs to pay attention that perception is defined as an "activity and process in which the organism recognizes directly and viscerally an event in the external world or itself". According to this definition, perception occurs not only in the external world but also in the internal world, that is, it can be said that perceptual experience arises by imagining or thinking.

Perceptual experience has two conditions: that the stimulus (i.e. object) exists in the outer or inner world, and that the stimulus goes through a sensory organ.

There are, however, some exceptions, for example, Phantom limb or sensory deprivation. Phantom limb is a phenomenon in which the affected person feels a body part that has been amputated as if it were still attached, or experiences aching pain, feeling of numbness, or sense of muscular movement. Perception of the Phantom limb occurs without the object being present. Sensory deprivation is a phenomenon in which hallucinations can be induced by decreasing or obliterating stimuli from the behavioral environment (Figure 2 [2]).

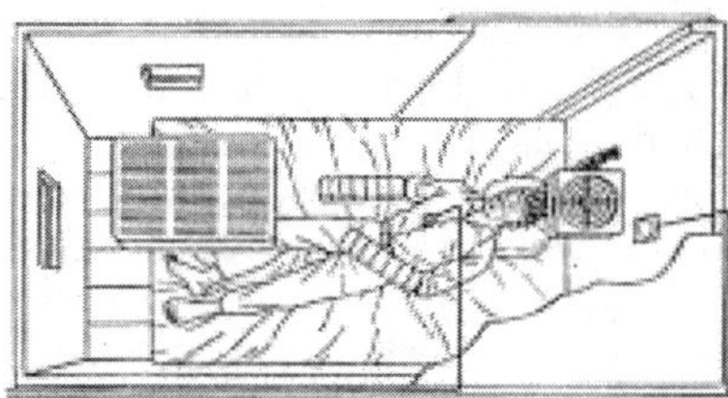

Figure 2: Sensory Deprivation Experiment

Figure 3: An Ambiguous Figure

On the other hand, the same perception is not guaranteed even if the same stimulus passes through a sensory organ. When we look at the ambiguous figure (Figure 3 [3]), our perceptual experience depends on what we pay attention to in the figure. Moreover, visual illusions occur where the geometrical nature of the perceived plane figure (size, length, direction, angle, curvature, etc) is systematically and significantly different from the actual one. This phenomenon suggests that our visual world is affected by the relationship between the stimulus and the surroundings. It seems that these factors mainly influence bottom-up processing.

$$\text{A} \quad \text{B} \quad \text{C}$$
$$\text{12} \quad \text{13} \quad \text{14}$$

Figure 4: Contextual Information

Knowledge of the object and contextual information are important for top-down processing. In Figure 4 [4], although the central figures are the same in the upper and lower lines, they may be perceived differently, that is, the central figure is perceived as the letter "B" in the upper line, and the number "13" in the lower line. Another example is a belief bias that produces inaccurate results by relying on intuition and prior experience rather than thinking logically. Although there are many other examples showing the effect of knowledge and context, the most famous and broadly applicable explanation for them is schema theory [5]. A schema is a set of prior expectations on particular topics, events, scenarios, and actions. Using schema theory, Bartlett explained that when we comprehend and remember, we tend to reconstruct an original story corresponding to existing knowledge and past experience. The schema are very useful, although they can produce problems if the event does not correspond to existing schema. Either little is remembered, or the information is biased, to make it fit the existing schema. It follows from the above considerations that both bottom-up and top-down processing play an important role in sense of reality formation, and we have to examine the interaction of both processings in detail.

3.3 Disorder of Reality Formation

If we can not distinguish a real thing from an imagined thing, what happens? The disorder of the reality-sense formation is one of the symptoms of schizophrenia. Schizophrenia is the label applied to a group of disorders characterized by severe personality disorganization, distortion of reality, and an inability to function in daily life. Schizophrenia sufferers are unable to link their thoughts and feelings to real life. They suffer from delusions and withdraw increasingly from social relationships into a life of imagination.

We all have experienced situations similar to those of schizophrenia sufferers. Reality monitoring is the ability to distinguish between externally derived memories that originate from perceptions, and internally derived memories that originate from imagination [6], and is a crucial aspect of competence in everyday life. External memories represent events that really occurred, objects that were perceived, actions that were performed, and words that were spoken or written. Internal memories are of events that have only been imagined, actions that were planned, considered or intended, and words that were thought but never uttered.

4. Importance of Communication

To differentiate between a real thing and an imagined thing, and to improve or facilitate the reality-sense formation process, requires communications; 1) repeated interaction with the object and 2) interaction with other persons. These assure us of the fact and lets us avoid various biases. It is probable that conversational agents can assist our reality-sense formation process. There are two effects in conversation itself, that is, clarification of current topic and offering of information. Interaction with an object could be facilitated by conversing with an agent, i.e., asking various questions about the object. If agent can provide related information about the object efficiently, this would be a good way to process contextual information. It will be important to examine the reality-sense formation process through conversation with agents.

5. Concluding Remarks

This paper argued for the formation of a sense of reality from the viewpoint of cognitive psychology. We proposed two kinds of reality as comprising what one would call 'the nature of reality'; that is, Reality of Object and Reality of Relation. We then examined the interaction between these realities and bottom-up and top-down processing and suggested a role for communication in the reality-sense formation process.

References

[1]   Norman, D. A., and Rumelhart, D. (1975). Exploration in cognition. San Francisco: Freeman.
[2]   Heron, W. (1957). The pathology of boredom. Scientific American, January.
[3]   Attneave, F. (1971). Multistability in perception. *Scientific American*, **225**, 62-71.
[4]   Bruner, J.S., and Minturn, A.L. (1955). Perceptual identification and perceptual organization. *Journal of General Psychology*, **53**, 21-28.
[5]   Bartlett, F. C. (1932). Remembering. Cambridge: Cambridge University Press.
[6]   Johnson, M. K., and Raye, C. L. (1981). Reality monitoring. *Psychological Review*, **88**, 67-85.

KES 2002
E. Damiani et al. (Eds.)
IOS Press, 2002

POC Communicator: A System for Collaborative Story Building

Tomohiro FUKUHARA[1], Toyoaki NISHIDA[2], and Shunsuke UEMURA[3]
[1]*Synsophy Project, Communications Research Laboratory*
Kyoto 619-0289, Japan
E-mail: tomohi-f@acm.org
[2] School of Engineering, The University of Tokyo
Tokyo 113-8656, Japan
E-mail: nishida@kc.t.u-tokyo.ac.jp
[3] Graduate School of Information Science, Nara Institute of Science and Technology
Nara 630-0101, Japan
E-mail: uemura@is.aist-nara.ac.jp

Abstract. We propose a system called *POC Communicator* for creating realistic contents collaboratively. Creating contents that bring us reality is important for learning and knowledge sharing among people. We propose a notion of *collaborative story building* for adding reality to contents. We implemented a prototype system based on this notion. Design concept and an overview of the system are described.

1 Introduction

Reality is an important factor for human's understanding. We feel reality in various contents such as movies, novels, and grandfather's stories. An important thing is that we often learn lessons from them. For example, when we listen to experiences of earthquake victims, we understand their sorrow, and learn ways to avoid risks in a disaster area. On the other hand, it is difficult to feel reality in news articles that describe only facts of the earthquake. There is a big difference between the victims' stories and the news articles. More reality we feel, more lessons we learn. Thus, creating contents that bring us reality is important for learning and knowledge sharing among people.

One of issues on communicative reality is how to create realistic contents. Creating realistic contents requires much skills. For example, writing or telling our experiences detailedly and expressively is difficult except for professional scenario writers or storytellers who know techniques to attract readers or audience. Although there are many realistic stories around us, it is difficult for us to write and tell our stories to others because of lack of those techniques.

We propose a notion called collaborative story building for facilitating group members to create realistic contents collaboratively. We can tell our experiences realistically when we tell them in cooperating with people who had the same experiences. This is because they can collaborate to tell a story by adding details or complementing information mutually. We implemented a prototype system called POC Communicator based on this notion.

This paper is organized as follows. Section 2 describes our definition of communicative reality, the notion of collaborative story building, and the design concept of a prototype system. Section 3 describes an overview of the system.

2 Collaborative story building for adding reality to contents

2.1 Communicative reality

Following is our definition of "Communicative Reality (CR)".

Communicative Reality (CR)
CR is reality of contents exchanged through communication between a human and a *contents provider* that has an intention for transmitting information realistically.

In this definition, the contents provider might be a human or an artifact that provides contents statically or dynamically. For example, books and recordings are examples of static contents providers[1]. Conversational agents and interactive web pages are examples of dynamic ones.

2.2 Collaborative story building

We propose a notion of collaborative story building for facilitating group members to write realistic contents. Our hypothesis is that telling a story collaboratively in a group makes the story more realistic than telling it alone. Suppose a group such as a family, a sport club, or a class in a school, members share feelings and memories about common events to them. They remember those events and tell various episodes realistically even though they meet again after long years. Takatori reported that recalling things collaboratively works better than recalling them alone[1]. This is because humans recall common events to them correctly by complementing information mutually. Thus, we consider that telling stories collaboratively enable them realistic.

2.3 Design concept of a prototype system

Followings are points for designing a prototype system.

One can write a fragment of a story.
For lowering cognitive burdens for writing a story, one can write a fragment of a story. A *story* here is a writing of one's experience. Writing such story requires motivation and a cognitive load. We consider that writing a *message*, which is a fragment of a story, lowers cognitive burdens of users. A message consists of a title, a picture, and a comment to the picture (Figure 1).

Fragments of a story are shared in a group.
For facilitating group members to start writing a story, messages are shared and utilized in a group. By this policy, one can reuse a message by editing title and comment of the original one.

---

[1]Note that the nature is not the contents provider because it has no intention for transmitting information.

Figure 1: A message consists of a title, a picture, and a comment to the picture.

Figure 2: Conceptual image of the collaborative story building. Group members share messages and stories, and create and edit stories mutually.

Stories are edited mutually.

For evolving stories in a group, stories are edited mutually. Group members can edit not only their own stories but also other stories contributed by others.

Figure 2 shows the conceptual image of the collaborative story building. In the figure, a message, which is posted by one of group members, becomes a component of a story for other members. By sharing and editing messages and stories mutually, the group evolve their stories mutually.

2.4   Previous work

World Wide Web (WWW) has limitations on collaborative story building. This is because most Web pages don't allow people to edit them directly. Although WikiWikiWeb[2] and CoWeb[3] allow people to edit Web pages directly, it is difficult to find relationship among Web pages. For facilitating collaborative story building, a system should lower burdens for creating information. Furthermore, the system should facilitate users to understand relationship among pieces of information.

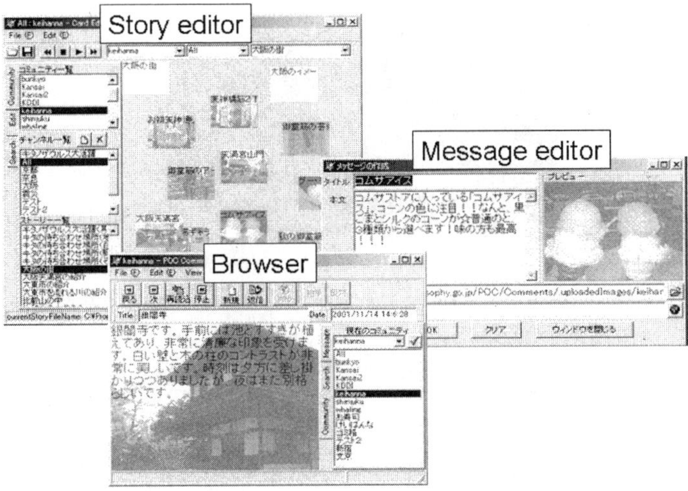

Figure 3: Screen image of POC Communicator. In the message editor window, users can compose and edit a message. In the story editor window, users can edit stories by arranging messages. In the browser window, users can browse messages and stories passively.

3   POC Communicator

We implemented a prototype system called POC Communicator. POC Communicator is the client software of the *Public Opinion Channel (POC)* system[4]. In POC Communicator, a story consists of several messages. Users can create a story by adding messages to a story.

3.1   Overview

Figure 3 shows an overview of the system. The system consists of (1) message editor window where users can create a message, (2) story editor window where users can create and edit a story, and (3) browser window where users can browse messages and stories.

Messages and stories are stored on a server. When a user retrieves messages on the server, the server returns messages. The user can incorporate those messages into his or her stories, and upload them to the server. Stories on the server are available for users. Users can download existing stories, and edit and upload them again.

3.2   Functions

Major functions of the system are follows.

Automatic message playing function

The browser window displays messages automatically. This function helps a user to recall his or her stories. When a user hits upon an idea of a story, s/he can create a new story

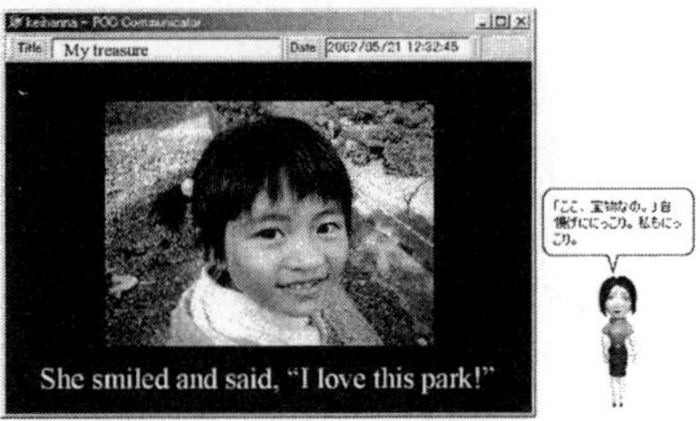

Figure 4: A screen image of a story. An interface agent explains the picture with her Text-to-Speech system.

instantly by incorporating the message into the story. This is done by a dragging the message from the browser window and dropping it to the story editor window.

Story editing function

Users can edit not only their own stories but also others' stories directly. A story is displayed on the story editor window. In the window, users can edit a story by (1) adding or removing messages to/from the story, and (2) editing messages contained in the story.

Story playing function

Users can browse a story as a picture-story show in the browser window. An interface agent introduces the story with her Text-to-Speech system. Figure 4 shows a scene of playing a story.

4   Conclusion

We proposed a notion of collaborative story building for facilitating group members to create realistic contents collaboratively. Our hypothesis is that a group can create realistic stories through writing and editing them collaboratively. We will verify this hypothesis through an experiment using the prototype system.

References

[1] Takatori,K. A role of communication in recalling, Japanese Journal of Educational Psychology, Vol.28,pp.108-113, 1980. [in Japanese].

[2] Leuf,B. and Cunningham,W. Wiki Way: The quick collaboration on the Web, Addison-Wesley Publishing, 2001.

[3] Guzdial,M. Collaborative websites supporting open authoring, The Journal of the Learning Sciences, 1998.

[4] Fukuhara,T., Matsumura,K., Azechi,S., Fujihara,N., Terada,K., Yamashita,K., and Nishida,T. Creating city community consanguinity: Use of public opinion channel in digital cities, in: Tanabe,M., Besselaar,P., and Ishida,T.(eds.): Digital Cities II: Computational and Sociological Approaches, Lecture Notes in Compute Science, Springer-Verlag (to appear).

KES 2002
E. Damiani et al. (Eds.)
IOS Press, 2002

The motivations to get and send information

Ken'ichi MATSUMURA

Synsophy Project, Keihanna, Human Info-Communication Research Center, CRL, Japan
Department of Social Psychology, Graduate School of Human Sciences, Osaka University,
Japan

Abstract. In this paper, we proposed the psychological scales for motivations to send and get information, and behavior of sending and getting information. We researched whether users' motivations to send and get information could lead to posting information and access to their community. The result indicated that users with high motivation to send information posted more messages than users with low motivation and that users with low motivation to access to network communities more frequently than users with high motivation.

1. Introduction

In this paper, we propose the scale for motivations to send and get information. This scale is being developed to estimate the network communication tool from human's psychological view.

Fishbein & Aijen proposed "theory of reasoned action" [1]. They said that when humans behave, at the beginning they form an intention to behave. Therefore when we construct the scale, we need to separate behavior and intention. The purposes of this study are to construct the scale for motivations to send and get information, and to examine whether those motivations will lead to behavior of sending and getting information.

2. The two types of reality in the network community

In network community, two types of realities will be formed from information sent by community's members. One of them is the reality as a knowledge which a member can get from others' utterances. In other words, that is the reality for objects. This reality is formed by understanding others' opinion or information from others. The reality for objects will make community members feel usage of tools. Users' feeling usage of tools will facilitate them to use them. Therefore, we can say that this reality influences members' behavior of getting information

The other is the reality about users' cognition to others' having opinion. In other words, this is the reality for phenomena. For example, that is community members' cognition to distribution of opinions in network community. This reality is formed from members' cognition to which opinions others have. This reality could reduce community members' sending information, because they might have different opinions from others [2]. This means that the reality for phenomena influence community members' behavior of sending information. As has been noted, we expect that two types of reality being in network community will influence community members' behavior. Therefore, it is valuable to estimate network community tools from viewpoints of users' motivations.

3. The estimation for Public Opinion Channel (POC)

The Public Opinion Channel (POC) is an automatic community broadcasting system that elicits opinions from community members, generates a story from them, and broadcasts the story within the community [3]. Therefore, it is very important that community member post messages to POC community. And when we test the effect of POC, we need to research whether POC could support members' to send and get information.

4. Motivations to send and get information

The purpose of this investigation was to construct the scales for motivations to send and get information. In this study, we aim to examine the relations between information behavior and motivation.

4-1. Method

4-1-1. Participants

We distributed questionnaire to 120 participants. (67 females and 53 males, Mean of age : 23.9) We distributed questionnaire to 80 subjects who visited our laboratory to participate in another experiment and requested them to answer the questionnaire. And we emailed questionnaire to 40 participants and requested them to answer that, in the same time, requested them to join the experimental network community.

4-1-2. Questionnaire Items

Questionnaire items included 56 items. They were consisting of 25 items about motivations to get or send information, 10 items about competitive achievement motivation, the behavior of sending information, and the behavior of getting information. Each item was accompanied by 5-point scale.

Table1. Result of principal factor analysis with varimax rotation

Items	Factor1	Factor2	Factor3	Factor4	h^2
I study or work hard not to be inferior to others.	**.742**	.005	-.164	-.094	.60
I'm grad to win competition with others.	**.714**	-.005	.142	.099	.55
I want to do everything better than others.	**.682**	.135	.277	.129	.58
I strongly want to be superior to others.	**.659**	.093	.077	-.047	.46
I like to chat.	-.070	**.799**	.085	.120	.69
I want somebody to listen in my word.	.291	**.750**	-.156	.265	.76
I like to have anybody listen in my word.	.227	**.677**	.167	.025	.63
I want to answer the question somebody asks me.	-.278	**.442**	.066	.393	.48
I want to know a lot of information.	.056	-.051	**.751**	.038	.57
I want to grow from leaning various things.	.023	.070	**.562**	.313	.45
I want to get information that nobody knows.	.296	.032	**.545**	.035	.40
I am interested in news about trends.	-.025	.100	**.431**	.017	.35
I am happy to have a response to information I send.	.004	.040	-.040	**.658**	.48
I want to go where somebody recommend.	-.149	.009	.274	**.577**	.48
I want to inform the information I know to somebody.	.085	.134	.101	**.501**	.30
I want to do something that makes everybody grad.	.180	.257	.099	**.434**	.35
Eigenvalue	4.12	2.68	2.03	1.58	8.12
Variance explained	20.61	13.38	10.17	7.88	52.03

4-2. Result

4-2-1. Factor analysis

We conducted a factor analysis to confirm the 25 items about motivations structure of all 120 participants. These items were analyzed using a principal factor analysis with varimax rotation, and four factors were left with eigenvalues greater than 1.0 (Table1). Based on this result, we named factor1 as competitive achievement motivation, factor2 as motivation to send information, factor3 as motivation to get information, and factor4 as motivation to keep human-relations. We considered the average score of each factor's variables per participants as their factor's scores. We focused on the motivations to send and get information, because the purpose of this study was researching whether they would lead to the behavior of sending and getting information.

4-2-2. Motivation to send information

It was hypothesized that people with high motivation to send information, would send much information than those with low motivation.

Participants were distributed of three groups according to the score of motivation to send information. 33 participants having score of motivation to send information higher than 4.75 (third quartile) were in high motivation group, 42 participants having score lower than 3.75 (first quartile) were in low motivation group, the others were in medium motivation group. Table2 shows the means of the score about motivation to send information and two behaviors of sending information by motivation categories. To test the hypothesis, an ANOVA for motivation category was run.

In this analysis, there was a significant difference on the behavior of sending information to others ($F(2, 117) = 25.155$, $p < .01$), but there was no significant difference on the behavior of sending information on the internet ($F(2, 117) = 0.535$, n.s.). These results revealed that people with high motivation to send information led to behavior of sending information to others more frequently than people with medium and low motivation.

Table2. Means of the score about the behavior of sending information on the internet and the behavior sending information to others by motivation categories.

	N	Motivation to send information	The behavior of sending information on the internet	The behavior of sending information to others
High	33	4.72	1.93	4.12 a
Medium	45	3.98	2.12	3.48 b
Low	42	3.06	1.86	3.10 c

Notes. For each behavior, means with different letters are significantly different at $p < .05$.

4-2-3. Motivation to get information

By the same way, participants were distributed to three groups, according to motivation to get information, and ANOVA was run. There was a significant difference on the behavior of getting information from others between high motivation group and low motivation group ($F(2, 117) = 3.65$, $p < .05$), and on the behavior of getting information with watching TV ($F(2, 117) = 5.304$, $p < .01$). People with high motivation get information from others and watching TV more frequently than participants with low motivation. This showed that the motivation to get information measured by this scales led to the behavior of getting information from others and TV.

Table3. Means of the score about the behavior of getting information on the internet, the behavior of getting information from others, and the behavior of getting information with watching TV by motivation categories.

	N	Motivation to get information	The behavior of getting information on the internet	The behavior of getting information from others	The behavior of getting information with watching TV
High	31	4.85	4.37	3.60 a	3.53 a
Medium	55	4.25	4.12	3.36	3.02 a
Low	34	3.31	3.91	3.07 b	2.87 b

Notes.   For each behavior, means with different letters are significantly different at $p < .05$.

5. Motivation and behavior in network community

We examined whether community members' motivations to send and get information would lead to send and get information on network community.

5-1. Procedure

Participants were 40 members living in Kansai Area in Japan. They were distributed to 4 groups. 10 members were asked to use BBS on internet for four weeks, and 10 members for two weeks. 10 members were asked to use POCviewer, and 10 members for two weeks. The members of our experimental community spent two weeks or four weeks using the POCviewer or BBS to talk about domestic topics in the Kansai area of Japan. But one of members in POC community never accessed to POC community, so he was excluded from analysis. Therefore, subjects for analysis were 39 members (20 females and 19males, mean of age is 27.54).

5-2. Results

5-2-1. Sending information

We distributed 40 participants to 3 groups according to percentiles of their motivation to send information. We considered participants' posting of message to each community as the behavior of sending information and calculated the averages number of a member's posting messages in a day. To test the validity of scale for motivation to send information, we compared high motivation category and low motivation category on the average number of posting messages in a day by ANOVA. In this analysis, there were significant differences on the number of posting messages ($F(1, 22) = 7.54$, $p < .05$). This result indicated motivation to send information led to sending information in network community.

We distributed participants in POC community to high motivation or low motivation groups by mean of motivation to send information. Participants with high motivation didn't differ from those with low motivation on the number of posting messages in a day ($t(17) = 1.31$, *n.s.*). On the other hand, in BBS community, there was a significant difference ($t(18) = 2.49$, $p < .05$). These results indicated that community members with low motivation to send information posted as much as members with high motivation in POC community, but members with low motivation couldn't post as much as members with high motivation in BBS community.

5-2-2. Getting-information

In the same way, we distributed participants to 3 groups according to percentiles of their motivation to get information. We considered the frequently of members' access to community as behavior of getting information, and calculated frequency of members' access in a day. We compared high motivation category and low motivation category by ANOVA. In this analysis, there was a significant difference ($F(1, 16) = 5.33$, $p < .05$). This result indicated that members with low motivation to get information accessed to their community more frequently than other members with high motivation to send information.

We distributed participants in POC community to high motivation or low motivation groups by mean of motivation to send information. Both in POC community and BBS community, we could find no significant difference on mean of frequency of members' access to the community in a day (POC: $t(17) = -0.10$, n.s., BBS:$t(18) = -1.45$, n.s., see table5). We expected members with high motivation to get information would frequently access to their community. But they didn't do that. That might indicate that they got information from friendly others or TV, and they might not need to access to their community to get information.

Table4. Means of a member's posting messages in a day

Sending information	All	POC	BBS
High motivation	0.64	0.65	0.59
Low motivation	0.40	0.45	0.40

Table5. Means of number of access to their community per participant in a day

Getting information	All	POC	BBS
High motivation	0.67	0.43	0.91
Low motivation	1.24	0.89	1.46

6. Conclusion

In this paper, we proposed the scale for motivations to send and get information. The motivations measured by this scale make it possible to expect user's behavior of sending information in network community. But we have to improve the scale for motivation to get information. Because people with low motivation access to community more frequently than those with high motivation. This result was the opposite of our expectation. But people with high motivation to get information might acquire necessary information from others in their daily life. Therefore, we can think that they didn't need to get information by POC or BBS.

In the future, we need to elaborate the scale for motivations to send and get information. In particularly, we should consider the scales for motivation to get information and the behavior of getting information.

References

[1] Fishbein, M., & Ajzen, I. (1975): Belief, attitude, intention and behavior : *An introduction to theory and research* .

[2] Matsumura, K. (2001): Consensus formation process in network community: *Knowledge-based intelligent information engineering systems & allied technologies*, KES'2001, 348-352

[3] Azechi, S., Fujihara, N., Kaoru, S., Hirata, T., Yano, H., and Nishida, T. (2000): "Public Opinion Channel: A challenge for interactive community broadcasting," In: Ishida, T., and Isbister, K. (eds.), *Digital Cities: Experiences, Technologies and Future Perspectives*, Lecture Notes in Computer Science, 1765, Springer-Verlag, pp. 427-441

KES 2002
E. Damiani et al. (Eds.)
IOS Press, 2002

Application of POC System
to Education

Nobuhiko FUJIHARA [*], Mikiya TANIGUCHI [*], Tomohiro FUKUHARA [**],
and Toyoaki NISHIDA [**,***]
[*] *Naruto University of Education /* [**] *Synsophy Project /* [***] *The University of Tokyo*
[*] *748 Takashima, Naruto, Tokushima, 772-8502, JAPAN*
fujihara@naruto-u.ac.jp

Abstract. We discuss the application of POC system to education. One of the most
straightforward ways is to use it for sharing information among members of a learning
community. POC system activates the potential that the volumes of information stored
worldwide embody and enables us to find new relations among seemingly unconnected
pieces of information. Community members would be stimulated to think creatively. POC
system seems well suited to divergent thinking activities. We introduce a program we are
currently developing called *degi-shibai*, which is intended for use in art education. The
possibility of using POC Communicator, which is a client software of POC system, as a
tool for creative expression, not only for communication purposes, is discussed. We also
introduce a collaboration model for expression and thinking, a so-called TEC (Thinking,
Expression, and Communication) cycle. We discuss how the model can be a useful tool to
analyse the effects of communication on the creative process.

1. Introduction

Information technology has become firmly nestled into the infrastructure of the educational
community. Recently, the worldwide proliferation of the Internet has accelerated the
application of information technology to education as a means of communication. In this
paper, we'll discuss the possibility of application of POC (Public Opinion Channel) system,
which we have proposed as a tool for communication through the Internet, to educational
communities, such as on school campuses and in the classroom.

POC system is tool that automatically broadcasts information to a community of users
[1, 2]. In a POC community, members of the community provide information through the
Internet. POC system collects pieces of information and edit as a story and broadcast it.
Members can reply to other members' information. POC system works as a circulator of
information in a community. Thus, when POC system is applied to a learning community, we
can expect that it should facilitate communications among members and support the learning
experience of them. Some people may point that this kind of function has already
implemented by other media such as the World Wide Web (WWW) and mailing lists. Indeed,
other media could be used to support the communications in a learning community. However,
some of the features of POC system may produce different effects. For example, POC system
will receive broadcasts automatically without the need for users to voluntarily access
information. Another feature is a high degree of anonymity that masks the identities of the
senders in a POC Community. If POC system is applied to situations where its features
facilitate communications, the realization of a communicative environment would ensue, and
POC system would thus support the learning experience of the community members.

Therefore, what we have to consider is to explore the situations where the features of POC system work best.

We will posit that POC system be used as an idea generation tool in divergent thinking situations, such as brainstorming, rather than in a convergent thinking activity that seeks to acquire a conclusion. In addition, we will discuss the possibility of using POC Communicator, which is a client software of POC system, to support the expression of young learners as well as to support their learning community. We will introduce a *degi-shibai* program, which is educational practice in art education.

2. Application of POC system to education

2.1. POC system as a tool for sharing information in a community

One of the simplest and most straightforward applications of POC system is to use it to share information among members of a research group. In a POC community, members introduce information related to their research. Members are able to see many pieces of information all together, although they were introduced by different people and in different contexts. POC system provides members with opportunities to discern important relations among separate pieces of information, which at a glance seem to be unrelated, and to use the knowledge gained to generate new ideas. When information is kept in static databases, such as the computer hard disks and the WWW, it often becomes dead storage. On the other hand, POC system activates the potential of this often-static information when it is appropriately disseminated. Information is always circulated in a POC community. POC system creates an environment where community members can use ideas developed in the past more actively. This type of environment would stimulate the creativity of members connected by POC.

POC system can be used in a laboratory, university, a school, and other learning environment to announce information related to everyday life, upcoming events, and so on. Some may remark that mailing lists and the WWW can be used to broadcast these kinds of announcements and that POC system is not needed. However, POC system does have its merits. One of the more remarkable merits is POC system can display information without the user having to voluntarily access a database, such as mailing lists. In other words, POC system can play the role of a digital secretary that reminds users of their schedule. In order to realize this function, we note of the following two issues. First, a given piece of information should be presented at an appropriate time. When the database of POC system contains a lot of information, each piece of information is not presented so frequently. Probabilistically, a chance that appropriate information is presented just before a given event is decreased. POC system should have a function that information is presented in proper time. Second, the problem of habituation should be resolved. When information is presented repeatedly in a monotonous fashion, people adapt to this situation and gradually pay less and less attention. POC system, which broadcasts information repeatedly, may be recognized as "a ground", not as "a figure" of our recognition. That is, the information just blends in to the background of the user's environment. To keep this from happening, some improvements are required. For example, it may be effective to rate the importance of the information being broadcast and to display some notice when very important information is presented.

The functions to present information in an appropriate time and to display a notice when important information is broadcast have yet to be into POC system. These functions may facilitate communication among members and serve an important role in creating a realistic environment for the learning community. With taking cases of other communities as well as a learning community into account, we have to discuss whether the functions are necessarily required.

2.2. POC system as a tool for a discussion in a community

POC system could be used as a tool for a discussion among learning community members. Community members exchange ideas in the POC community. However, there are some inherent difficulties involved in participating in a discussion through POC system. Basically, POC system treats information senders as anonymous ones. As information is presented repeatedly, audiences are not aware of the time sequence and the causal relationship of the various pieces of information. In addition, members cannot respond to information synchronously. When a member asks a question, they may not necessarily get a response to that particular query. To hold a discussion, those participating need to understand what is on each other's mind and develop and understanding of their sense of logic. The thinking process of a participant would be disrupted unless their questions are responded to quickly. Perhaps the characteristics of POC system may not be conducive to the communicative realities required for a discussion. Especially, trying to use a POC system as a tool for convergent thinking might prove disappointing. Rather, POC system seems to be better suited to divergent thinking processes, such as brainstorming.

2.3. Another application of POC system to education: a tool for expression in art education environment

So far, we have considered POC system as a tool to support a learning community. In this section we will propose using POC communicator, which is a client software of POC system, as a tool to support an individual.

We have been developing a *degi-shibai*[1] program, which is an educational practice used in art education classes, especially with primary school pupils. In the program, pupils use a digital camera and Microsoft PowerPoint to create stories in a picture-card-show format (Figure 1). Pupils can create stories in their own way, or they can use one of the following formats:

- *Fixed-point observation*
  Pupils take pictures of the same scene or object at regular time intervals.

- *Walk around my town*
  Pupils walk around their own towns and take pictures of the scenery and objects that pique their interest.

- *Journey of my favorites*
  Pupils take pictures of their favorites objects (e.g., a stuffed animal or toy), in a setting that they choose.

One aim of a *degi-shibai* is, of course, to have pupils enjoy creative activities. Moreover, it is expected that pupils will develop new insights of the subjects of their pictures, their environments, and even themselves. To take pictures of things that appeal to them as well as pictures of aspects of their everyday life would facilitate their meta-cognition that relates to the

---

[1] The term *degi-shibai* is a word coined by the authors. *Degi* is a compound prefix of the words design and digital. *Shibai* means a play in Japanese. And in Japanese, a picture-card show is called "*kami* (paper) – *shibai*." In this educational practice, art educators teach pupils to create picture-card-show like stories with digital cameras in art education classes. So, we call it *degi-shibai*.

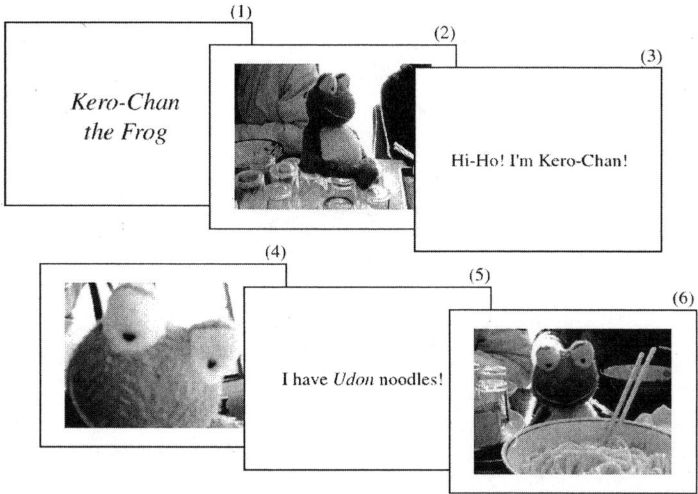

Figure 1. "*Kero-chan the frog*", an example of *degi-shibai* stories.
(These are just a top part of the story.)

subject of their pictures. Their choice of subject matter often reflects their thoughts, emotions, and inner self. A *degi-shibai* may give pupils the opportunity to observe their environments and themselves objectively.

Presently, we use Microsoft PowerPoint to make *degi-shibai* stories. Instead, we propose using POC Communicator to make *degi-shibai*. In this case, pupils would make stories by assembling images as if they are editing a TV program. They can express their emotions and thoughts as they create this kind of movie. In fact, the making of TV programs and movies is indeed a very difficult creative activity that requires a lot of skill. Probably most pupils would not find making movies an easy task. Such an application of POC Communicator to art education classes makes the pupils learning experience more enjoyable. POC Communicator expands the opportunities for pupils to express themselves creatively.

Furthermore, pupils can make *degi-shibai* stories collaboratively. POC Communicator affords them a place to collaborate. They may show their *degi-shibai* to each other through POC Communicator and get responses in the form of feedback. This kind of feedback may activate their meta-cognitive thinking. That is, the feedback may remind them of their own emotions and way of thinking. As a result, pupils can make improvements to their work by taking the feedback they receive into account.

Certain conditions have to be fulfilled for such interaction to work well. We have considered carefully what conditions are required to facilitate pupils' creations. For this purpose, we consider a TEC (Thinking-Expression-Communication) cycle as a model of collaborative creations (Figure 2). The TEC cycle hypothesizes that the creative thoughts of a pupil are facilitated by the externalization of their ideas (expression) and by the ongoing interactions with other people (communication). This model gives us a very simple framework. It is useful to clarify issues and problems related to collaborative creations. For example, this model suggests that creations, which are mainly composed by expression, can be started by using any of the above factors. Some can start creations from just thinking about what they desire to make. However, the model shows that talking with others about their own town or their favorites something can be a starting point. Thus, the creative process can begin without there being a clear idea of what a story is going to be about. The TEC cycle can be used to

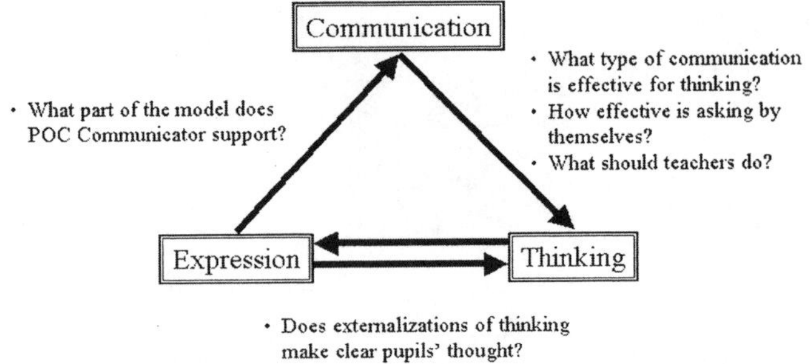

Figure 2. The TEC (Thinking-Expression-Communication) cycle.

analyse the effect of POC Communicator when used in the creative process. Perhaps, such an analysis would reveal issues concerning POC Communicator and our communication skills that require some improvement.

We have mainly discussed here the application of POC Communication to a collaborative art-education environment. Needless to say, POC Communicator can be applied to other creative collaborations, such as brainstorming. For example, POC Communicator can be used to facilitate brainstorming activity where the desired outcome is not with stated in. Such a case can be analysed by using the TEC cycle.

3. Future works

In this paper, we consider the application of POC system to education. We proposed that POC system is suitable for divergent and creative thinking activities. In addition, we introduced the *degi-shibai*, as an art education program that we feel is well suited to take advantage of the merits of POC Communicator.

Most recently, we started practical *degi-shibai* program in a junior high school. Our hypothesis of the TEC cycle will be examined through this practice and hopefully the communicative reality we expect will emerge and we can then study the effect of this type of collaborative communication on the creation process. We are now using Microsoft PowerPoint to make *degi-shibai* stories. In the near feature, we will introduce POC Communicator to this process. We believe the effectiveness of POC Communicator when used in a learning community will become apparent when we explore the creation of a *degi-shibai* project using POC Communicator.

References

[1]   T. Nishida (ed.), Dynamic Knowledge Interaction, CRC Press, 2000.
[2]   T. Fukuhara *et al.*, Creating City Community Consanguinity: Use of Public Opinion Channel in Digital Cities, in: Tanabe, M., Besselaar, P., and Ishida T. (eds.): Digital Cities II: Computational and Sociological Approaches, Lecture Notes in Computer Science, Springer-Verlag, to appear.

KES 2002
E. Damiani et al. (Eds.)
IOS Press, 2002

An Active-Affordance-based method for Communication between Humans and Artifacts

Kazunori Terada* Toyoaki Nishida**
*Synsophy Project, Communications Research Laboratory
2-2-2, Hikaridai, Seika-cho, Soraku-gun, Kyoto, 619-0298, Japan
kazuno-t@crl.go.jp
**The University of Tokyo
7-3-1, Hongo, Bunkyo-ku, Tokyo, 113-8656, Japan
nishida@kc.t.u-tokyo.ac.jp

Abstract.
The development of computer technology has created artifacts that have more complex and intelligent functions. Such artifacts need more sophisticated interfaces than primitive artifacts. In this paper, we discuss which characteristics are appropriate for interfaces with artifacts and propose a concept of *active affordance*. We describe an *autonomous mobile chair* that we built as a test bed for active affordance. We also describe a experiment that we performed with a real robot that show the validity of our proposed method.

keywords: *active affordance, human-artifact communication, embodiment*

1 Introduction

Artifacts have been created as a way of extending our abilities beyond our own bodies. We manipulate an artifact in the way that brings out the function which is peculiar to that artifact. When we use an artifact we need to convey our intention to it through some communications channel, such that the artifact 'understands' our intention and responds with some appropriate behavior. A human communicates with an artifact to convey his of her intention and control the artifact in the desired way. In the extreme case, we might expected to treat an artifact as if it ware a limb or set of limbs.

Communication is even needed in the case of the manipulation of a pair of scissors, a rather primitive tool. The intention of cutting a sheet of paper is generated, and we apply our hand to move the handle of the scissors. As a result, we cut the paper as we intended. In this case, the handle of the scissors acts as the interface where our intention is translated into functional motion. In the case of primitive tools such as a pairs of scissors, the manipulation of the artifact is directly related to the function.

However, the development of computer technology giving many artifacts more complex and intelligent functions. Such artifacts need more sophisticated interfaces than those of primitive artifacts. In this paper, we discuss the appropriate form for interfaces with artifacts.

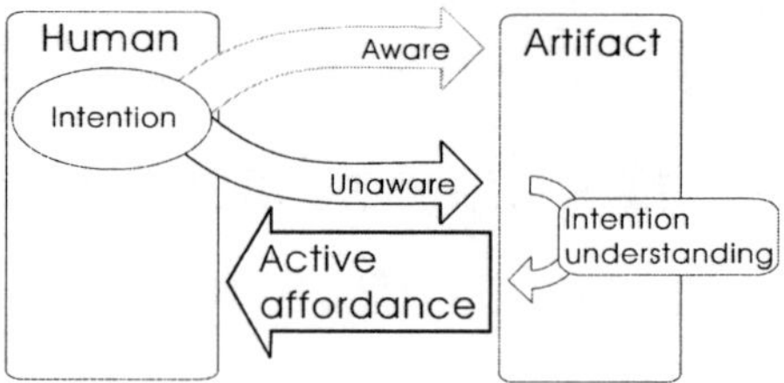

Figure 1: Communication between human and artifact

2 Active artifacts

Artifacts have been evolving for a long time and continue to evolve. In recent times some artifacts have been equipped with sensors and actuators, for example, doors becomes an automatic door. We call such an artifact an *active artifact*. The characteristic of the active artifact is that it realizes its function autonomously. It is expected that active artifact will improve our life. Actuators are capable of reducing the load of our everyday work.

We explain the characteristics of the active artifact by comparing the humanoid paradigm with the active-artifact paradigm. In recent years, there has been much effort toward the design of humanoid robots [4][2]. It is considered that humanoid robots may be most suitable artifact for the labor of our everyday work, because our living areas are designed with human usage. However, humanoid robot also have to use some tools that are required for task execution like human . If we want to take a rest, we request that a humanoid robot bring a chair. Then the humanoid robot bring the chair for us but the humanoid robot itself does not become a chair. However, in the active-artifact paradigm, the chair itself would come to us. We do not need a task-mediator such as a humanoid robot.

If the artifact is passive and does not move, we ourselves need to do everything for the task execution. However, if the artifact is active, the task is collaboratively executed by the human and the artifact. The main issue in designing an active artifact is working out how to communicate with it. We will discuss this issue in the next section.

3 Communication between Human and Artifact

The purpose of human-human communication is mutual comprehension and the knowledge sharing. On the other hand, the purpose of human-artifact communication is task sharing.

3.1 Model of communication

Figure 1 shows our model of human-artifact communication and the concept of active affordance we propose. The communication can be divided into the following parts.

- Conveyance of intention from the human to the artifact, which includes

    - aware communications, and.

    - unaware communications; and

- actions from the artifact to the human.

While human-human communication is bidirectional, human-artifact communication is mono-directional. A human has an intention when he is going to carry out some task. The intention is conveyed to the artifact through the communications channel. There are two modes for the communications channel; 1) aware communications, the means of which include natural language, sign language, and gesture and 2) unaware communication channel, which refers to nonverbal behavior other than gesture. Note that, in this paper, gesture is considered verbal because some gesture has articulation or grammar. If an artifact has a modality for aware communications, a human is able to use the channel for aware communications by utilizing the methods for human-human communication.

Some psychological researchers have concluded that more than 65 percent of the information exchanged during a face-to-face interaction between humans is expressed nonverbal [1]. The unaware communications channel is important the human-human communication, and should also be important in human-artifact communication. Unaware communications is used as a means for conveying human intention in communications between humans and artifacts. An intelligent active artifact is able to completely understand the user's intention.

3.2　Active Affordance

If the artifact can comprehend the user's intention, the artifact then to complement the user's actions that the user intend to do. We call such an action as *active affordance*. The concept of affordance was introduced by the psychologist J.J.Gibson [3]. Affordance refers to the possibilities for the action that available in the environment or the object, and which are revealed by interaction between the human and the environment. For example, the affordance of a chair is such that it allows to a human to sit on the chair, but is not manifested until the human generates the action of siting down.

Although affordance in the original concept is realized by the user's action [5], active affordance is realized by the artifact's action. Affordance which is not found by a human is meaningless. The reasons for affordance not being found are as follows.

- The function of the artifact is unknown.

- A user does not know how to use the artifact (the user is, however, conscious of his intention).

- A user is not even conscious of his intention.

Active affordance solves these problems.

4　Our implementation of the comprehension of intention

Our use of the aware communications channel and the unaware communications channel as means for the 'comprehension' of intention must be appropriate. In this work, we use the following methods.

4.1　Channel for aware communications

We use gesture as one means for aware communication. We thus compare the user's motion with a set of predetermined gesture pattern.

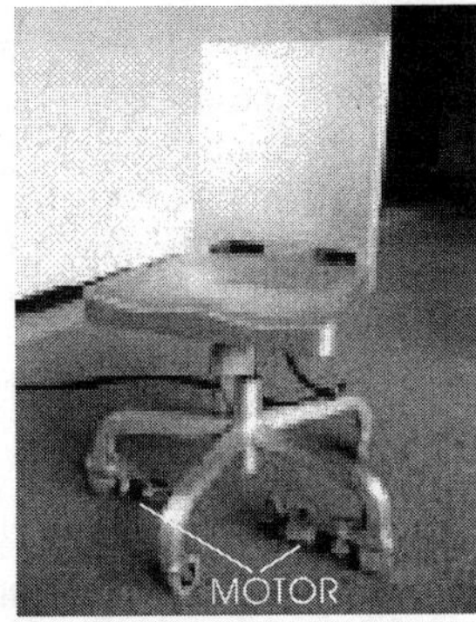

Figure 2: The autonomous mobile chair.

4.2   Channel for unaware communications

The user's intention is revealed in unconscious motion. The action generated when a human manipulates some artifact varies according to the artifact's physical properties and functions. The artifact might take advantage of the peculiarities of the various forms of motion to detect the user's intention. We use following heuristics to realize the unaware communication.

- Physical contact always occurs in object manipulation, and indicates a critical state.

- The distance between the surfaces of user's body and the artifact is reduced by the action of reaching.

- This reduction of distance indicates that the user intends to manipulate the given artifact.

5   Experimental results

To show the validity of our proposed method, we performed a preliminary experiment using a real robot. The scenario of this experiment is

1. the subject calls the autonomous mobile chair by a beckoning gesture, and

2. the autonomous mobile chair moves to the subject.

In this experiment, comprehension of intention through unaware communication was not implemented.

5.1   Autonomous mobile chair

We have built an autonomous mobile chair as an example of an active artifact. We remodeled some parts of an aluminum chair to allow it to move around (See Figure 2). The autonomous mobile chair has two powered wheels so that it can move around and is equipped motion-capture system made by Ascension Technology, which enables to measurement of the position and orientation of the chair's body. The motion-capture system employs pulsed-DC

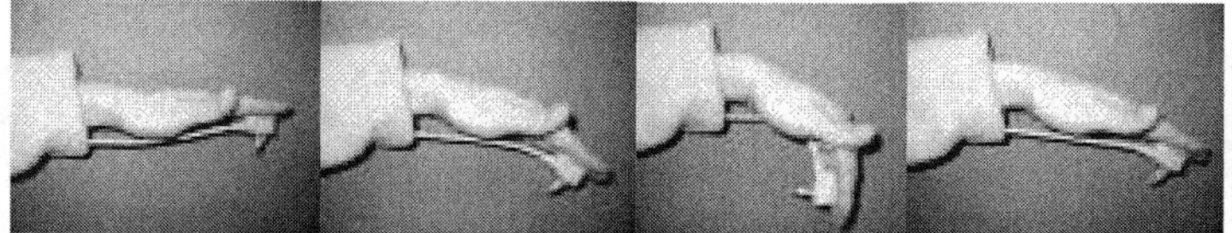

Figure 3: Beckoning gesture.

magnetic-field transmission and sensing technology to measure the positions and orientations of miniaturized sensors that are attached to the measuring equipment. The autonomous mobile chair is controlled by a Linux PC to which it is connected via RS232C cable. A subject in this experiment also has to a carry motion sensor so that the autonomous mobile chair is able to determine the reaching point.

5.2 Gesture recognition

A magnetic sensor is attached on the subject's fingertip to recognize a beckoning gesture. The position and orientation of the fingertip is measured by means of this sensor. Gesture recognition is performed by applying simple rules;

1. height of the hand is between 110 cm to 150 cm, and

2. the angular velocity is greater than 7 rad/sec.

Figure 3 shows the magnetic sensor and a beckoning gesture.

5.3 Generation of behavior

We utilize a utility function in order to control the autonomous mobile chair [6]. The utility function is widely used in a research in autonomous agent. The utility function is calculated by *dynamic programming*.

5.4 Comprehension of intention and generation of behavior

Figure 4 shows an experiment in the comprehension of intention through aware communication and generation of behavior. The subject generates a beckoning gesture: palm down, the hand flaps at the wrist. The gesture is perceived by the autonomous mobile chair through its motion capture system. The chair then moves to the subject.

6 Conclusion

In this paper, we discussed an appropriate interface for artifacts and proposed the concept of *active affordance*. Active affordance is valid in the following cases:

- where the the functions of the artifact are unknown;

- where the user does not know how to use the artifact;

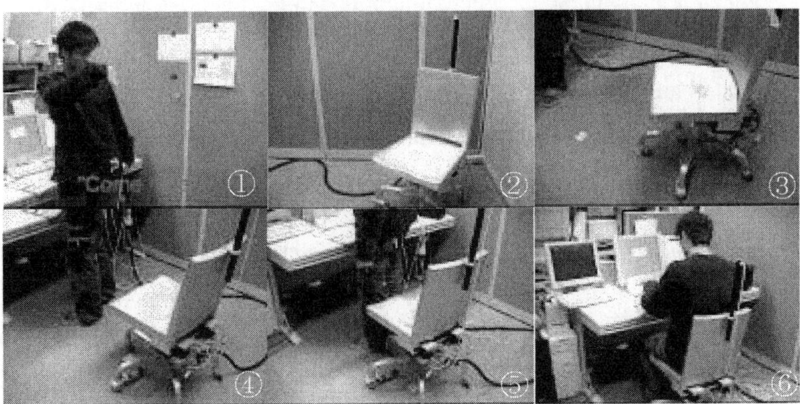

Figure 4: Comprehension of intention and generation of behavior

- where the user is not conscious of even his intention.

We built an *autonomous mobile chair* as an example of an active artifact. We gave a result of a preliminary experiment in the comprehension of intention and the generation of behavior. We plan to achieve another experiments for comprehension of intention through unaware communication to show the validity of active affordance.

References

[1] Michael Argyle. *Bodily Communication*. Methuen & Co., 1988.

[2] Rainer Bischoff and Tamhant Jain. Natural communication and interaction with humanoid robots. In *Second International Symposium on Humanoid Robots*, pages 121–128, 1999.

[3] James J. Gibson. *The Ecological Approach to Visual Perception*. Houghton Mifflin Company, 1979.

[4] Kazuo Hirai, Masato Hirose, Yuji Haikawa, and Toru Takenaka. The development of honda humanoid robot. In *IEEE International Conference on Robotics and Automation*, pages 1321–1326, 1998.

[5] Donald A. Norman. *The psychology of everyday things*. Basic Books Inc., 1988.

[6] Kazunori Terada and Toyoaki Nishida. Active artifacts: for new embodiment relation between human and artifacts. In *Intelligent Autonomous System 7 (IAS-7)*, pages 333–340, 2002.

KES 2002
E. Damiani et al. (Eds.)
IOS Press, 2002

Key Words Extraction for Image Search of WWW

Seiji Ito, Yasue Mitsukura, Minoru Fukumi and Norio Akamatsu

The department of Information Science & Intelligent Systems
Faculty of Engineering, University of Tokushima Japan
2-1, Minami-josanjima, Tokushima, 770-8506 Japan
e-mail: {seiji,mitsu,fukumi,akamatsu}@is.tokushima-u.ac.jp

Abstract

In this paper, key words in the image are extracted by using neural network.

As images preprocessing, objective images are shaded off color and segmented by maximin-distance algorithm. Small regions are integrated into a near region. Thus, objective images are segmented into some region. After this images preprocessing, key words in the image are extracted by using neural network. Grid is considered to segmentations decompozed image to classify the weight of important key words. By using procedure, key words in the image are extracted objectively.

Keywords : Median Filtering, Maximin-Distance Algorithm, Integration of Small Region, Grid

1 Introduction

The rapid permeation of information tecnology, such as the Internet, into our society is increasing the demand for imformation system development. However, there is the enormous quantity of information on the Internet. Therefore, it's very difficult to search necessary information exactly. Futhermore, it is the same when an image in Internet is sought out. In the case of retrieving images, a file name given to an image is sought in many case. Therefore, it is very difficult to obtain desired information. In this paper, it pays attention to the way of giving the name of images to search necessary images. Key words are then extracted from the images automatically by using neural networks (NN)s. That is automatic key word extraction for key word search is proposed. There have been another automatic key word extraction methods [1]-[3], but there is no extraction for difference in how to give a key word by human. For example, given an image with sky and sea, there are some human called "This is a sea image." and there is a human "This is a sky image.". This paper proposes an appropriate key word extraction method. Objective images are used only outdoor scenes in this paper. Only grid points are segmented and are recognized by Neural Network (NN) [5]. The evaluation method is performed by comparison with a questionnaire result.

2 Preprocessing of image

First of all, preprocessing is done to make an image easy to handle. Fig.1 shows the process of preprocessing. Median filtering is used as a preprocessing.

2.1 Median Filtering

First, images out of focus are made from original images by using median filtering. As for original images, edge is divided into some segmentation. On the other hand, images out of focus is not divided into any segmentation. Fig.2 shows regional segmentation after the procedure of median filtering. However, this regional segmentation is obtained using a method of S of HSI color system. Comparing these figures, the image after a median filtering is roughly divided rather than the original image.

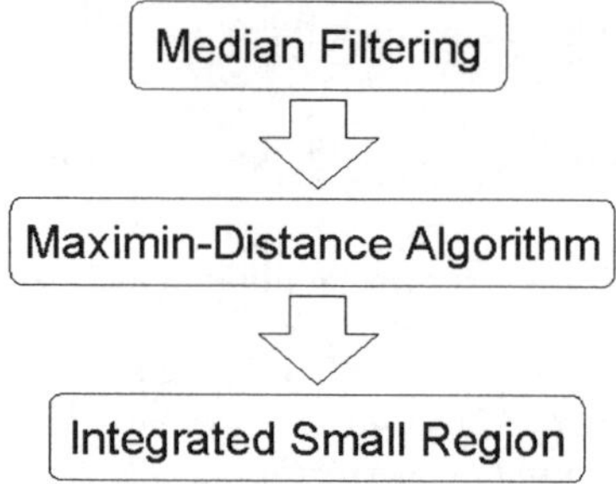

Fig. 1: Process of preprocessing an image

(a) An original image

(b) Regional division result of an original image

(c) Regional division result of an image after median filtering

Fig. 2: Regional division result of the original image and the image after median filtering

2.2 Maximin-Distance Algorithm

Image pixels with similar values after preprocessing of median filtering are classified into the same cluster by using a Maximin-Distance Algorithm [6]. The Maximin-Distance Algorithm is as it were the longest point is searched from existing cluster. If this distance is greater than a threshold, this point is set as the center of a new cluster. The detailed algorithm is as follows.

1. The cluster which sets $\mathbf{x}_1$ as a cluster center is made.

2. $D_i = \min\limits_{j} d(\mathbf{x}_i, \bar{Z}_j)$ is calculated to each $\mathbf{x}_i$. However, $\bar{Z}_i$ is set as the center of cluster Z_i, $D_i(X, Y)$ (Distance between a pixel X and a pixel Y) is computed as follows:

$$D_i = \sqrt{(X_r - Y_r)^2 + (X_g - Y_g)^2 + (X_b - Y_b)^2}$$

   Where, X_r, X_g, X_b means RGB of a pixel X, Y_r, Y_g, Y_b means RGB of a pixel Y.

3. $l = \max\limits_{i} D_i$ is calculated. Then, i is set as k.

4. If $l/\mathrm{MAX} > r$, then $\mathbf{x}_k$ is set as the center of a new cluster, and back to step2, where, MAX is given by $\mathrm{MAX} = \max\limits_{ij} d(\bar{Z}_i, \bar{Z}_j)$ and r is parameter.

5. If $l/\mathrm{MAX} \leq r$, then this algorithm is finished.

Fig.3 shows the result of regional segmentation obtained by using the Maximin-Distance Algorithm. From this figure, it is thought that this regional segmentation method by using Maximin-Distance Algorithm is better than the result of Fig.2. From this result, the Maximin-Distance Algorithm is used as preprocessing.

Fig. 3: Regional segmentation obtained by using the Maximin-Distance Algorithm

2.3 Integration of Small Region

In Fig.3, there are a lot of small-sized regions. In this case, the image is divided into big regions by integration of small regions [4]. In this paper, a small region is defined as 0.5% or less area of a picture. The integration algorithm of small regions is summarized as follows:

1. Objective pixel is defined by the maximum outline pixel of small region.

2. In the 8-neighbor distance pixel of objective distance, regional information of another region is obtained.

3. In the obtained regional information, the most region information is selected.

4. Objective regions are integrated into selected region.

5. This algorithm is continued in all of small region.

Fig.4 shows the result of the integration of small regions.

Fig. 4: The result of the integration of small regions

3   Key words extraction method by using grid point

After doing the above procedure, key words are extracted. It is thought that an objective region with a larger area tends to become a key word. Therefore, when key words are extracted, each region is weighted using grid points. By using the weights, the key words in an image is added. For the extracted key words, we count the number of grids. The number of grids in the key word is defined as points in the image. Fig.5 shows in a grid points example. Key words with the largest number of points are defined as key word in the image. There are grids at intervals of 30 pixels. A key word is learned by using NN. The inputs of NN are the average and variance value of RGB, HSI in the key word. Table1 shows the structure of NN. In this NN, 10 key words (Bule sky, Rock, Cloud, Sandy beach, Sea, Mountain, grassland, Sun, Tree, and Snow) can be recognized.

Table 1: Structure of NN

Unit	The number of units	
Input layer	12	Average and variance of HSI, RGB
Hidden layer	11	
Output layer	10	The number of key words

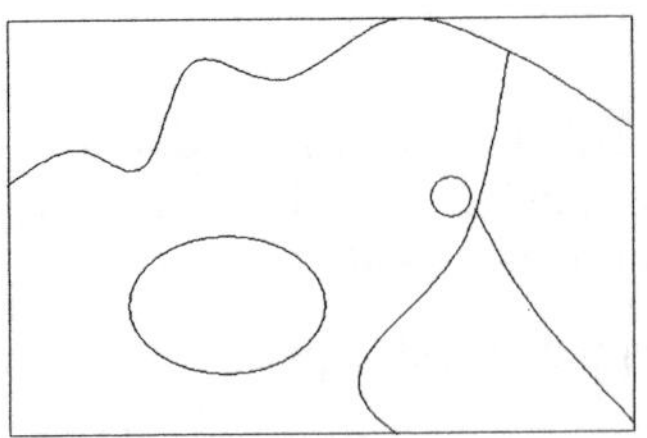

An image of regional segmentation.

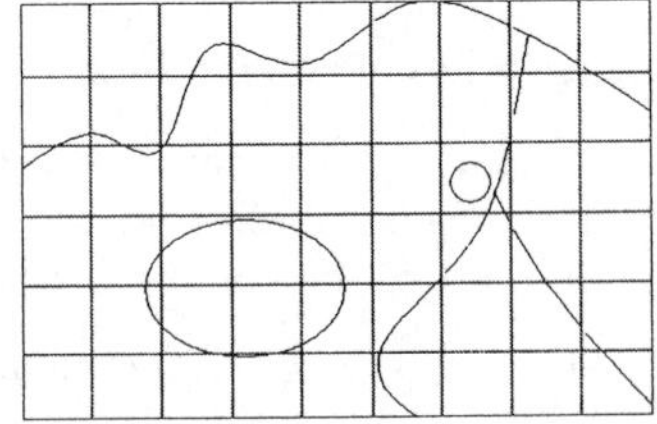

An image of regional segmentation with grid points.

Fig. 5: An example of grid point

4   Computer simulation

In order to show the effectiveness of the proposed method, computer simulations are done. Learning data are 100 images (about 10 images in a key word). This learning images are picked out by hand. The questionnaire is done. The phrase is as follows: "which is better key word in test images do you think?" This phrase is used as an evaluation method. Table2 shows simulation results. Simulation results are shown in Table2. In Table2, key words of top 3 are described. However, a simulation result of image4 is not included at all. Futhermore, considering the ranking, simulation results are not almost accorded to questionnaire. Therefore, it is not only thought the size of segmentation, but feature weighting have to be thought.

Table 2: Results of computer simulations

Image1	First	Second	Third	Image2	First	Second	Third
Questionnaire	Sea	Mountain	Rock		Rock	Sea	Blue sky
Rsult of simulation	Rock	Mountain	Sea		Sea	Sandy beach	Cloud
Image3				Image4			
Questionnaire	Blue Sky	Grassland	Tree		Snow	Sun	Tree
Rsult of simulation	Grassland	Blue sky	Tree		Blue sky	Tree	

5   Conclusion

This paper proposed the key word extraction for key word search. As for image preprocessing, this paper performs median filtering, Maximin-Distance Algorithm, and Ingegration of Small Regions. Therefore, a preprocessed image is extracted key words by using NN. For higher accuracy, the new method is developed in future work.

References

[1] M.Mukunoki, M. Minoh, K. Ikeda "A Retrieval Method of Outdoor Scenes Using Object Sketch and an Automatic Index Generation Method", IEE D- II Vol.J79-D- II NO.6, pp.1025-1033,1996, in Japanese

[2] A.Yamamura, M.Hagiwara "Recognition of Scenery Images considering Positional Relation using Fuzzy Interface Neural Networks", T.IEE Japan Vol. 122-C No.3, pp506-511, 2002, in Japanese

[3] H. Iyatomi, M. Hagiwara "Knowledge Extraction from Scenery Images and the Recognition Using Fuzzy Inference Neural Networks", IEE D- II Vol.J82-D- II NO.4, pp.685-693, 1999, in Japanese

[4] S. Sakaida, Y. Shishikui, Y. tanaka, and I. Yuyama "An Image Segmentation Method by the Region Integration Using the Initial Dependence of the K-Means Algorithm", IEE D- II Vol.J81-D- II NO.2, pp 311-322, 1998, in Japanese

[5] N. baba "Foundation and Application of Neural Net", Kyouritsu-syuppan, 1994 in Japanese.

[6] M. Nagao "Multimedia Infomation - Infomational Organization", Iwanamisyoten, 2000 in Japanese

[7] http://www.google.com/

KES 2002
E. Damiani et al. (Eds.)
IOS Press, 2002

A Proposal of Emotional Detection System from Speech Data

Hideaki Sato, Yasue Mitsukura, Minoru Fukumi, and Norio Akamatsu
Guraduate school, University of Tokushima
2-1 Minami-Josanjima Tokushima, 770-8506 Japan
{hidehide, mitsu, fukumi, akamatsu}@is.tokushima-u.ac.jp

Abstract

In this paper, prosodic characteristics are obtained from 4 kinds of emotional speeches that are neutral, angry, sad, and joyful speech, and are analyzed. Furthermore, each emotional speech is classified by using a neural network. In prosodic characteristics, the pitch has characteristics of frequency domain and the length of utterance time has characteristics of time domain. These elements are considered to be the most important elements. A prosodic characteristics extraction method is employed a special integrated circuits (IC) , which combine analog and digital processing. We have already gotten a patent for this IC.

key words : classification of emotional speech, prosodic characteristics, neural networks

1   Introduction

In this paper, an emotional speech classification system is established that uttered from various emotional states. The pitch and utterance time length are used as characteristic parameters which obtained from emotional speeches. It is difficult to detect pitch fast and accurately by software. Therefore, as a first step, pitch is detected through our proprietary integrated circuits (IC) which combined analog and digital. As a second step, four parameters (1.pitch pattern, 2.variance, 3.the whole utterance time length, 4.the utterance time length for each syllable) are extracted, and then emotional speeches are analyzed and classified by learning these parameters by a NN. Furthermore, we have carried out computer simulations for four kinds of emotional speeches. By using this system, it should become possible to make a robot take actions after one understands the emotion of the speaking human. Furthermore, we expect that it will generate an emotional speech from a synthetic voice by giving the information of the pitch change, utterance time length.

2   Characteristics of speech

Characteristics of speech[1] contain 3 elements; that is prosody, phoneme, and vocal quality. Prosodic characteristics include pitch, stress, and tempo information[2]. The accent and the intonation are affected by a pitch structure. Furthermore, pitch level is related to fundamental frequency. The amplitude is decided on an stress structure. The rhythm is related to a temporal architecture. By the way, the role of phoneme is discrimination of language information such as vowels and consonants. Vocal quality shows the difference between men and women which cannot express it by only pitch level, and characterizes individual characteristics such as husky and clear voices. As mentioned above, there are many characteristics of speech. In these elements, changes in the pitch and tempo have relations with emotion, so that the most important element is prosodic characteristics[3], [4]. In this study, the pitch pattern and the length of utterance time in prosody are used for an emotional classification. The reason for using a pitch and the length of utterance time is as follows:

1.　The pitch is the frequency domain parameter obtained from vibration of the vocal cords. Therefore, pitch cannot be easily influenced by the external state when speech material is taken in.

2.　The length of utterance time is uniquely decided by measurement how long it takes to finish.

Human being has various kinds of emotions and degree in one emotion is different. In this paper, emotional speech classification is performed to basic emotions of anger, sadness, and joy, in addition to neutral. It isn't taken into consideration about the emotional degree such as "very angry" and "a little angry". The number of subjects is 2 men and the emotional expression word is "KIMURA". Emotions are neutral, anger, sadness, and joy. 64 emotional speeches can be obtained and are used for simulations. Sampling rate is 20,480 Hz. Quantization is 16 bits and mono-channel is used.

3　Prosodic characteristics extraction method

The pitch is one of the very important elements for emotional speech. There are many detection methods such as a cepstrum analysis, autocorrelation function, and so on. However, their methods have been not established. Because, it is not rare to be wrongly extracted by the influence of noises, as the double pitch and the half-pitch. From these reasons, pitch is detected through our proprietary IC combined analog and digital processing. The IC is developed by the authors. Two waveforms that show a speech and a pitch information obtained through the developed IC are illustrated in Fig.1. The IC developed by us extracts the pitch based on speech waveform characteristics and has taken out the patent. Thus, the algorithm about our IC is shown on Web site[5] in detail.

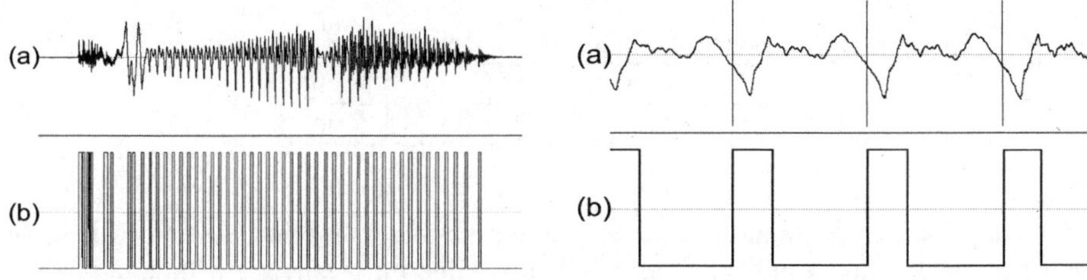

Fig. 1: Speech and pitch waveform (left:whole, right:close-up of a part)
(a) : speech waveform ,　　(b) : pitch waveform

The pitch pattern is detected accurately as shown in Fig.1. In Fig.1(b), the up point is defined as a point where moved from minus domain to plus domain and the down point is defined as a point where moved from plus domain to minus domain. The pitch is obtained by calculating the time between each up point in the pitch waveform. The pitch changes with time in each emotion, as illustrated in Figs.2, 3, 4, and 5. The normalization of the pitch was performed to each pitch pattern by the median value of their pitches. The purpose of the normalization is to consider the feature of the form of pitch patterns without distinguishing the height difference among pitches of individuals. The normalization of the time was performed to extract 12 pitches from each area which divided the length of utterance time into three equal parts. These extracted pitches contain 75-100% information for each area. The variance calculated from these pitches is also added as a characteristic parameter.

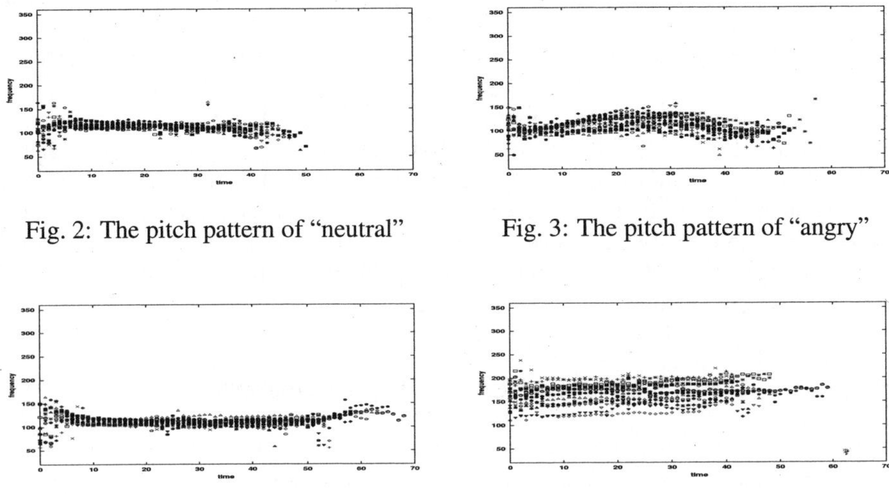

Fig. 2: The pitch pattern of "neutral"   Fig. 3: The pitch pattern of "angry"

Fig. 4: The pitch pattern of "sad"   Fig. 5: The pitch pattern of "joyful"

The variance of each pitch pattern is calculated and is shown in Table 1. The pitch pattern of "angry" has the largest change. The variance of "neutral" and "sad" have a similar value. From this Table 1, it is predicted that classification of "neutral" and "sad" is difficult to do with only pitch pattern and variance. Therefore, the length of utterance time is obtained as another parameter.

Table 1: The variance of pitch pattern of each emotional speech

	Neutral	Angry	Sad	Joyful
Variance	0.164	0.273	0.174	0.065

It is clear that the beginning of utterance is the first up point and the end of it is the last down point, as shown in Fig.1. It is easy to decide on the length of utterance time. Moreover, the length of each syllable is extracted as characteristic parameters. The procedure of extraction is the following. First of all, FFT is calculated on all of up points, and then the difference is obtained from adjacent up points. The vowels comparatively have periodicity and the consonants don't have it. The boundaries between syllables are decided according to this feature. Table 2 shows the result. Numerals in Table 2 show summation of each emotion and numerals in parentheses show the rate for the total time. Time of "joyful" is short, and time of /RA/ in "sad" is long.

These characterisitic parameters are used for neural network inputs. NN outputs are intended to indicate the emotion through learning.

4   Computer simulations

The NN has a 4-layered structure and the back propagation (BP) method is applied to a NN learning. The number of input units is 41. The input signals are pitches, a variance of these

Table 2: Summations of the utterance time in each syllable(proportion)

	Neutral	Angry	Sad	Joyful
KI	2.107(0.30)	2.353(0.28)	2.259(0.26)	1.502(0.27)
MU	1.816(0.26)	2.095(0.25)	1.695(0.19)	1.244(0.22)
RA	3.140(0.44)	3.932(0.47)	4.795(0.55)	2.904(0.51)
Total	7.063	8.380	8.749	5.650

pitches, the length of utterance time, and the length of each syllable described in the section 3. The number of hidden-1 units, hidden-2 units and output units is 10, 3, and 4, respectively. Each output unit corresponds to each emotion. 32 samples (8 samples per one emotion) are used for learning, and the other 32 samples (8 samples per one emotion) are used for recognition test. The simulations were executed ten times, and the result of classification shown in Table 3 is average values. Recognition accuracies of neutral, angry, and joyful emotion are considered to be very good results. However, accuracy of "sad" emotion is not very good. As the reason, it is conceivable that the important information expressing sadness was lost by normalization of time.

Table 3: The results of emotional speech classification
(a) Classification of 2 kinds of emotions

Emotion	Neutral	Angry	Neutral	Sad	Neutral	Joyful
Accuracy (%)	100.0	100.0	84.4	81.3	87.5	97.5
Emotion	Angry	Sad	Angry	Joyful	Sad	Joyful
Accuracy (%)	96.9	100.0	95.0	97.5	72.9	100.0

(b) Classification of 3 kinds of emotions

Emotion	Neutral	Angry	Sad	Neutral	Angry	Joyful
Accuracy (%)	90.6	87.5	68.8	87.5	85.0	90.5
Emotion	Neutral	Sad	Joyful	Angry	Sad	Joyful
Accuracy (%)	79.2	75.0	100.0	87.5	80.5	82.5

(c) Classification of 4 kinds of emotions

Emotion	Neutral	Angry	Sad	Joyful	
Accuracy (%)	82.5	96.3	77.5	81.3	

In order to analyze emotional speeches themselves, 5-layers NN with identical structure is applied for this simulation. This method is useful for compressing multidimensional data into two or three dimensional data. The same data is used in units of an input layer and teacher signals. The number of input and output units is 41, the number of hidden-1 and hidden-3 units is 10, and the number of hidden-2 units is 3. All of sample data are used for learning and testing. Fig.6 illustrates the data obtained from hidden-2, Figs.7, 8, and 9 are shown that the relationship between three units. From these figures, angry and joy are relatively separated from other emotions.

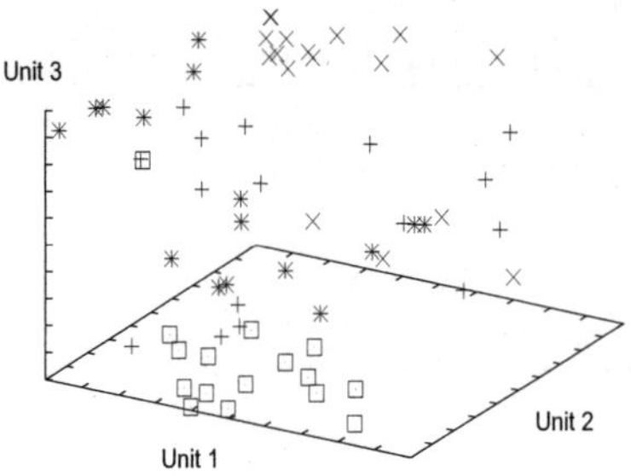

Fig. 6: Output values of units in the hidden 2 layer

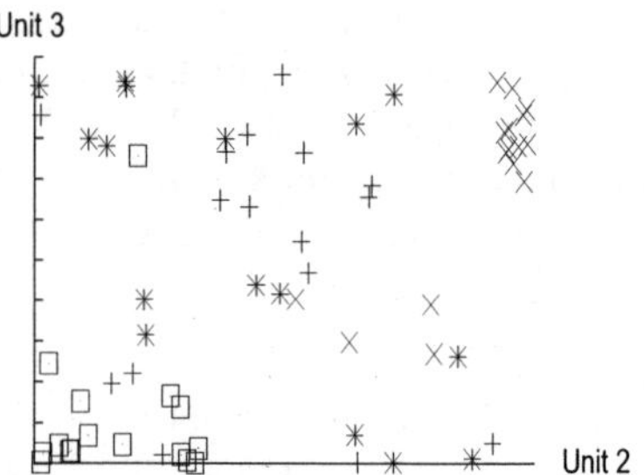

Fig. 7: The relationship between the unit 1 and unit 2

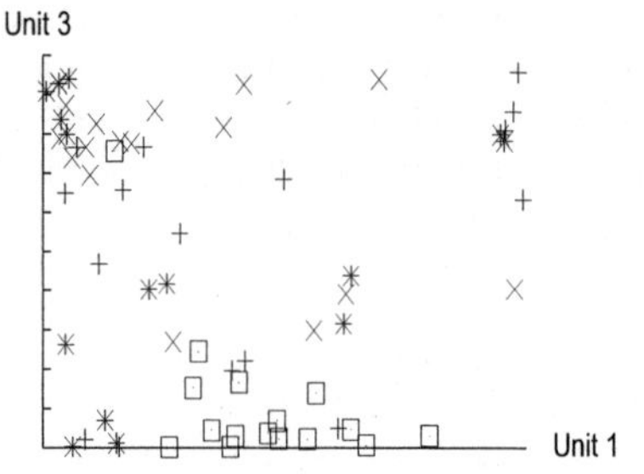

Fig. 8: The relationship between the unit 1 and unit 3

Fig. 9: The relationship between the unit 2 and unit 3

(Emotion : + "neutral", × : "angry", ∗ : "sad", □ : "joyful")

5 Conclusion

In this paper, the pitch pattern and the length of utterance time among prosodic characteristics were considered as important elements of emotional speech. These parameters are extracted by using the IC which is proposed in our lab., and are used for simulations. Finally, computer simulations are done. From these simulation results, it is clear that the pitch and the utterance time have characteristics of emotion, so that emotional speeches contain particular features yielded by emotions and can be classified.

References

[1] Tanetoshi Miura. *The auditory sense and speech*. The IEICE Japan, 1998. (in Japanese).

[2] Cecile Pereira and Catherine Watson. Some Acoustic Characteristics Of Emotion. *Proc. ICSLP*, Vol. 3, , 1998.

[3] A.Paeschke and W.F.Sendlmeier. Prosodic characteristics of emotional speech : Measurements of fundamental frequency movements. *ISCA*, SEPTEMBER 2000.

[4] Roddy Cowie and Ellen Douglas-Cowie. Automatic statistical analysis of the signal and prosodic signs of emotion in speech. *Proc. ICSLP*, MARCH 1996.

[5] Japan Patent Office, URL http://www.jpo.go.jp/index.htm , number:7297.

Learning Correspondences of Syntax Structures From Bilingual Corpora

Toshiki HIRANO
Kazuhiko TSUDA
Graduate School of Systems Management, University of Tsukuba
3-29-1, Otsuka, Bunkyo-ku, Tokyo 112-0012, Japan

Abstract. This paper proposes a machine translation method to translate similar sentences based on correspondences of their syntax structures and similarity of their words. Then this paper describes the one of the essential technique for this translation method, which is to learn and store correspondences of syntax structures from bilingual corpus. We are under development of a method to extracts correspondences of syntax structures between English and Japanese to determine which part of new sentence need to be changed by comparing similar sentences. We conducted sampling evaluation of this method using similar sentences in bilingual corpus of an automobile regulation and have confirmed the effectiveness of the method. As future work, we will systemize this method and conduct experiments in larger scale.

1. Typing Area

Many kinds of English-Japanese translation software are in the market. However, their translation qualities are not enough for human being to read through it without suffering stress. The main reason is that the conventional English-Japanese machine translation systems fully rely on machines to generate the translation.

To address this problem, we study a method to generate a translation by retrieving similar sentences from bilingual corpus and changing the word(s) that need to be changed. This method enables the generation of reader-friendly, natural translations by diverting existing human translated sentences. One of the essential techniques to accomplish this method is to learn and store correspondences of syntax structures from bilingual corpus. In chapter 2, we outline the existing techniques for machine translation and their problems. Chapter 3 describes our proposed method and chapter 4 describes the sampling evaluation we conducted this time.

2. Existing Machine Translation Methods and Problems

This chapter summarizes the existing machine translation methods and their problems.

2.1 Rule-based machine translation

This is the method that is used in many kinds of translation software in the market. The translation quality is much improved in recent years; however, it is still difficult to translate long sentences with complicated syntax structures. This problem indicates the limit of storing vast number of rules manually without losing consistency. Another problem of this method is that the generated translations are machinelike and not reader-friendly. This indicates the difficulty to generate natural translation by fully relying on machines.

2.2 Corpus-based syntactic parsing

In the past decade, corpus-based method becomes the mainstream of syntactic parsing method. This stream is brought by the progress of probabilistic language models such as N-gram model, hidden Markov model, and probabilistic context-free grammar. Many kinds of parser with high accuracy have come into practical use in recent years.

2.3 Translation memory based translation support system

Actual translation work, such as translation project of computer software manuals, includes many repetitions of similar sentences. When upgrading software, only small parts of sentences are changed. To support this kind of translation work, several kinds of translation support tools are in practical use [TRADOS]. The translation memory is a kind of corpus, which stores parallel translations in the database. In actual translation work, the tool retrieves the similar sentence and human translators make changes to the retrieved translation based on the new original sentence. The newly translated parallel translation is stored in the translation memory for future use. The limitation of this kind of tools is that they rely on human translators to determine which part of the retrieved translation needs modification. Therefore, these tools are only translation "support" tools.

3. Machine Translation Using Syntax Correspondences

As described in previous chapter, it is difficult to generate reader-friendly translation by the method used in conventional machine translation systems. To address this situation, we are developing a method to extracts parallel translations that are similar to the sentence to be translated from bilingual corpus and automatically change the parts need to be changed in new sentence. To accomplish this method, the differences between similar original sentences need to be reflected to the translation. To do this, we extract correspondences of syntax structures between English and Japanese to automatically determine the parts need to be changed in new sentence. Figure 1 shows the outline of this method.

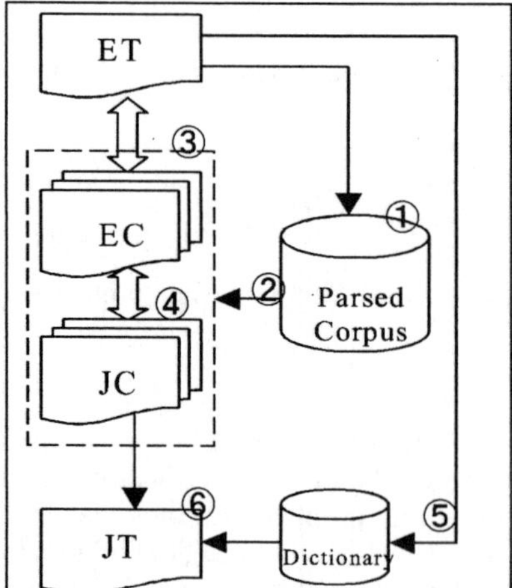

Figure 1. Outline of machine translation method using syntax correspondences

The step numbers in Figure 1 indicate that the following processes are executed:

1. Construct a parsed bilingual corpus of EC and JC.
2. Extract parallel translations (EC and JC) that have the same syntax structure with ET.
3. Compare each word between EC and ET and select the most similar EC.
4. Determine which parts of JC need to be changed based on the difference between EC and ET.
5. Pick up dictionaries and decide the terms to be used in the translation.
6. Generate translation by replacing the words determined in step 4 with the terms selected in step 5.

The procedures to accomplish this method are described below:

1) Construct a parsed bilingual corpus

We construct the parsed bilingual corpus by manually modifying the result of English and Japanese parsers. This is to neglect the accuracy of parsers so that we can conduct the experiments based on the idea that the outputs from parsers are always accurate.

2) Make correspondences of words in parallel translations

To determine the parts to be changed in translation, the correspondences of words between EC and EJ must be made. The correspondences can also be used to generate dictionary of technical terms. Moreover, this can be applied as a tool to check disunion of translation terms by finding that one word corresponds to n translation terms.

3) Compare syntax trees

Compare syntax trees between EC and ET. We use parsers that handle parenthetical expression separately. In this way, we simplify the complex sentences and increase the rate of correspondences. We also have chosen the automobile regulation as the material, which use a lot of similar expressions repeatedly.

4) Measure similarities of sentences

We measure similarities of sentences based on similarities of both syntax trees and words. We score the similarity points with weighting (correspondences of syntax trees with heaviest weight, then correspondences of verbs, and similarity of other words with the lightest weight). We especially pay notice on verbs to enable the use of case grammar as one of measure of determination.

5) Select terms for translation

In the case of English to Japanese translation, most of the semantic correspondences of terms are one to many. In our method, "dedicated dictionary" is searched preferentially to reduce the load on normal dictionaries. The "dedicated dictionary" is the dictionary generated based on the correspondences of words between EC and JC of the targeted material of translation. It covers most of words appear in the material and main parts of speech, which are nouns, verbs, adjectives, and adverbs, are all included. Since it is the dedicated dictionary for the targeted material, most of the words in the dictionary should correspond in one to one relations. That is, the words appeared in the corpus must be corresponded and the selections of translation terms are only required for newly appeared words.

4. Evaluation

This chapter describes the result of sampling evaluation. The data used in this evaluation is English - Japanese parallel translation of automobile regulation. The reason we select the automobile regulation is that it contains a lot of repeated similar sentences. In this evaluation, we extract similar sentences manually and use them as the samples to evaluate our method.

Here we show one example from samples. Table 1 contains five sentences with very similar syntax structures. The sample syntax trees of English and Japanese are shown in Figure 2. The correspondence of each word is indicated by upper and lower cases of alphabet. That is, the English word in "A" node corresponds to the Japanese word in "a" node.

As a result of automatic generation of translation using these information, the sentence shown in Table 2, which is "A stop lamp with two levels of illumination (S2) approved in accordance with the 02 series of amendments to Regulation No. 7." is translated into the Japanese sentence "規則 No.7 の 02 改訂シリーズに基づいて認可された 2 つの照射レベルを有するストップランプ(S2)". The generated sentence is appropriate and reader-friendly, and we confirm the effectiveness of our method.

In future, we systemize this process and conduct experiments with larger volume of data.

Table 1. One example of sample data used for evaluation

A rear fog lamp (F) approved in accordance with Regulation No. 38 in its original form;	規則 No.38 の初版に基づいて認可されたリアフォッンプ(F)
A rear direction indicator lamp of category 2a approved in accordance with the 01 series of amendments to Regulation No. 6;	規則 No.6 の 01 改訂シリーズに基づいて認可されテゴリー2a の後部方向指示灯
A red rear position lamp (R) approved in accordance with the 02 series of amendments to Regulation No. 7;	規則 No.7 の 02 改訂シリーズに基づいて認可され赤色リアポジションランプ(R)
A reflex-retro reflector of class IA approved in accordance with the 02 series of amendments to Regulation No. 3;	規則 No.3 の 02 改訂シリーズに基づいて認可されラス 1A のリフレックスレトロリフレクター
A reversing lamp (AR) approved in accordance with Regulation No. 23 in its original form;	規則 No.23 の初版に基づいて認可されたリバースプ(AR)

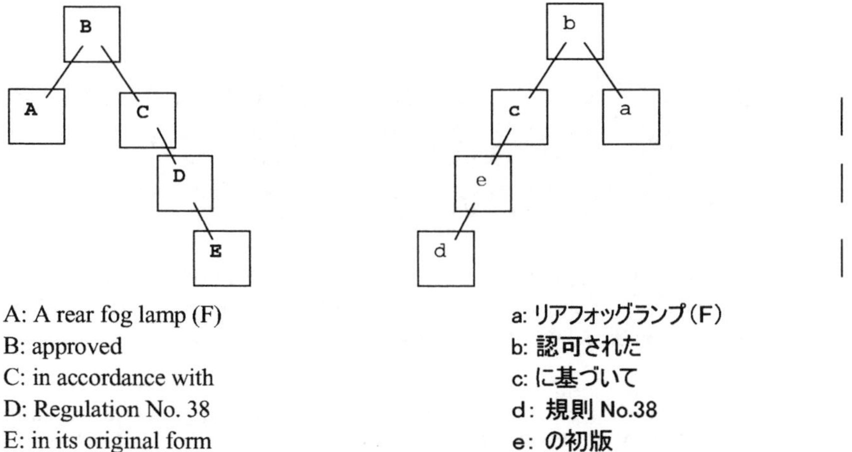

A: A rear fog lamp (F)
B: approved
C: in accordance with
D: Regulation No. 38
E: in its original form

a: リアフォッグランプ（F）
b: 認可された
c: に基づいて
d: 規則 No.38
e: の初版

Figure 2. Syntax trees of the sample data

Table 2. Example of processed data

A stop lamp with two levels of illumination (S2) approved in accordance with the 02 series of amendments to Regulation No. 7.	規則 No.7 の 02 改訂シリーズに基づいて認可された 2 つの照射レベルを有するストップランプ（S2）

5. Conclusion

In this paper, we have proposed an English to Japanese translation method using similarities of sentences to generate user-friendly translation. This method can be smoothly applied to Japanese to English translation since it holds correspondences of syntax structures. Furthermore, our method should be easily applied to the translation of other languages if the correspondences of syntax structures are obtainable.

In future, we systemize this process and conduct experiments with larger volume of data. We plan to apply this method to Japanese to English translation of computer software and hardware documentations to confirm the further applicability of our method.

References

[1]   Y. Matsumono, H. Ishimoto, and T. Utsuro, Structural Matching of Parallel Texts, *ACL-93: 31st Annual Meeting of the Association for Computational Linguistics*, pp.23-30, 1993

[2]   W.A. Gale, and K.W. Church, Identifying word correspondences in parallel texts, *Proc. 4th DARPA Speech and Natural Language Workshop*, pp.152-157, 1991

[3]   P.F. Brown, J.C. Lai, and R.L.Mercer, Aligning sentences in parallel corpora, Proc. 29th Annual Meeting of the Association for Computational Linguistics, pp.177-184, 1991.

[4]   http://www.trados.com

KES 2002
E. Damiani et al. (Eds.)
IOS Press, 2002

Remote Consultation Business Model on Steel Material Selection Using KNOW-HOW and KNOW-WHO

Masakazu TAKAHASHI[†1,†2], Akio FUJI[†1] and Kazuhiko TSUDA[†2]

[†1] *Isikawajima-Harima Heavy Industries Co. , Ltd., 2-16-3 Toyosu, Koto-ku, Tokyo, Japan*
[†2] *Tsukuba University, 3-29-1 Otsuka, Bunkyo-ku, Tokyo, Japan*

Abstract. This paper proposes Best Steal Material Selection System (below, BSMSS) and a business model of remote consultation that uses BSMSS. BSMSS is the expert system that uses databases of KNOW-HOW (the knowledge on selecting steel materials) and KNOW-WHO (the knowledge of contacts to experts), and is accessed via the Intranet and the Internet. The business model realizes best steel material selection for small companies or appropriate insurance rate setting against steel structures.

1. Backgrounds of Remote Consultation Business Model

Our daily life is surrounded by many kinds of steel structures such as boilers, plants, land machinery, and bridges. These steel structures are used over the long term as infrastructures and they need to be highly reliable. To build highly reliable steel structures, we need to select appropriate steel materials as well as appropriate methods. The knowledge to select appropriate steel materials is accumulated by small groups of researchers within steel structure makers, and is mainly used on consultation services for selling the products of each maker. Additionally, the steel material selection requires considerations on various conditions (such as use environment, installation location, legal imperatives, request from customers, and destination of export), and researchers deal with them individually. For this reason, it has been rare case that knowledge on steel material selection is used outside the steel structure makers. Steel structures are essential to our daily life. We assume that the knowledge of selecting appropriate steel materials must be demanded not only by steel structure makers, but also by others such as medium and small companies that do not hold expert researchers and engineers, or insurance companies that provide insurance against steel structures. This paper proposes a business model of remote consultation on steel material selection using databases of KNOW-HOW (the knowledge on selecting steel materials) and KNOW-WHO (the knowledge of contacts to experts) that can be accessed via the Internet.

2. Problems and Solutions on Steel Material Selection

This chapter describes the problems on conventional methods for selecting appropriate steel materials and the solutions.

1) Insufficient investigation on use conditions

Selecting appropriate steel materials requires different investigations depending on various use conditions such as use environment, installation location, legal imperatives, request from customers, and destination of export However, designers tend to design the steel structures based on insufficient investigations and this results in problems after start using the structures.

2) Improper reuse of design drawings

Design drawings of steel structures are often reused for other steel structures. However, only design information is inherited in repeated use of design drawings and the process information (for example, why this design is adopted) disappears. Consequently, design drawings are reused improperly and it causes problems later time.

3) Lack of knowledge about steel materials

In conjunction with the problem 2), reusing design drawings often results in reusing the steel materials adopted on the drawings as well. This makes designers to select older steel materials even when cheaper and more durable materials are available. This problem can be avoided by asking expert researchers for advice. However, the number of expert researchers is limited and it is difficult for them to deal with every case. We address this problem by constructing a database of experts' knowledge (KNOW-HOW) on selecting steel materials. But the KNOW-HOW database is inadequate to resolve complicated cases. We also construct a database of experts' contacts (KNOW-WHO) to allow designers to contact experts for more detailed advice when needed. We propose a business model of remote consultation on steel material selection by providing the environment to access these KNOW-HOW and KNOW-WHO databases via the Internet.

3. Outline of Best Steel Material Selection System

This chapter describes the outline of Best Steel Material Selection system (below, BSMSS) that consists of KNOW-HOW and KNOW-WHO databases.

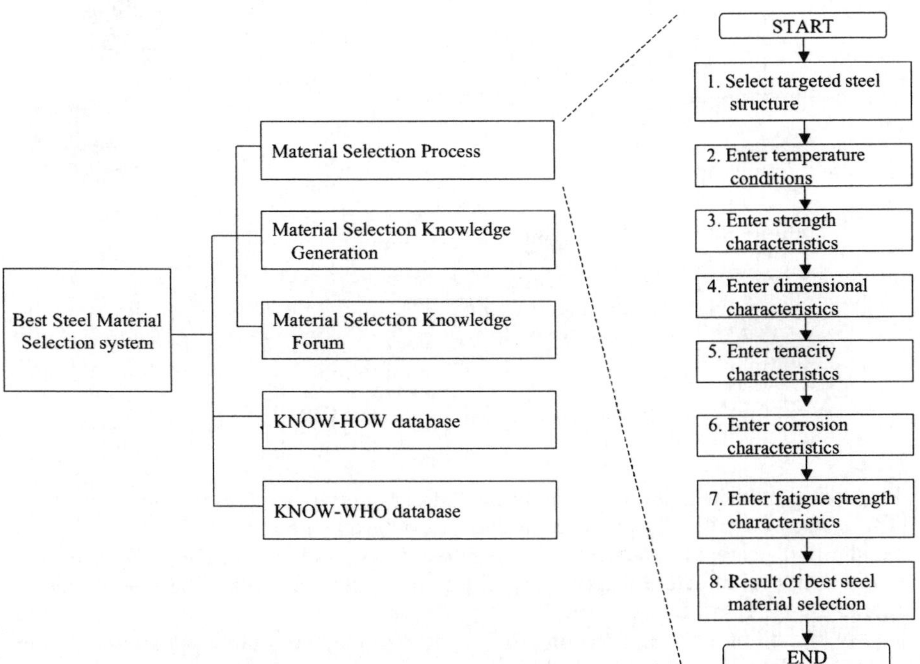

Figure 1. Functional configuration of BSMSS

Figure 1 illustrates the functional configuration of the BSMSS. The system implements the Material Selection subsystem separately from KNOW-HOW and KNOW-WHO databases. This is because that different knowledge is required for selecting materials for different steel structures. Designers may obtain only an insufficient answer as a result of selecting steel materials using KNOW-HOW database and the Material Selection subsystem. If this is the case, the BSMSS recommends experts from KNOW-WHO database for more concrete consultation. The Material Selection Knowledge Generation subsystem extracts knowledge of steel material selection and registers it with KNOW-HOW database. The Material Selection Knowledge Forum is for exchanging opinions between designers and experts about steel material selection on the Web. KNOW-HOW and KNOW-WHO information is automatically extracted from this forum: knowledge of selecting steel materials is extracted and registered with KNOW-HOW database, and the contact information of the experts who offer the opinion is registered with KNOW-WHO database. The material selection processes are independent from targeted steel structures. Therefore, the system allows designers to select the targeted steel structure first, then the selection knowledge for the structure is retrieved from KNOW-HOW database.

Figure 2 shows the configuration of the BSMSS. The system can be accessed both from inside and outside of our company via our Intranet or the Internet.

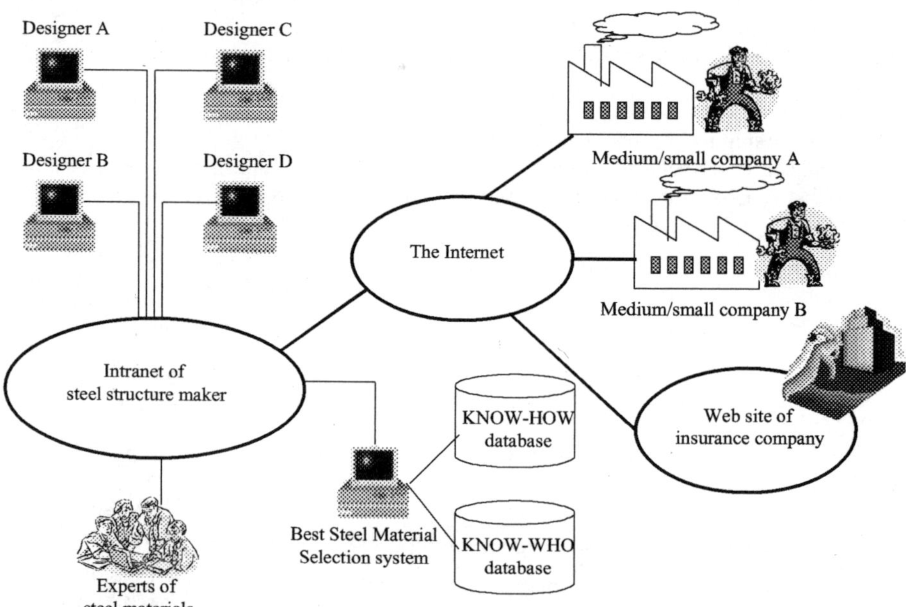

Figure 2. System configuration of BSMSS

4. Proposal of a Business Model Using the Best Steel Material Selection System

This chapter describes the proposed business model using the BSMSS.

As the first phase of constructing the business model, we release the BSMSS on the Intranet of a steel structure maker. This allows every designer within the steel structure maker to use the BSMSS (see Figure 3).

As the second phase of constructing the business model, we connect the Intranet to the Internet to release the system outside the company. This enables us to provide remote consultation services to other companies, such as and small companies that do not hold experts for selecting steel materials (broken line part in Figure 3).

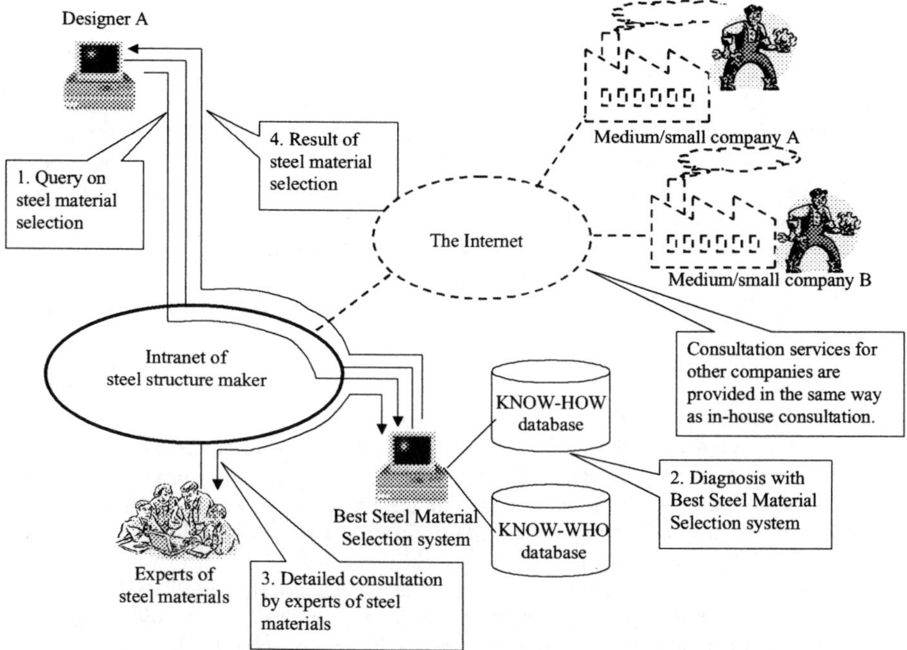

Figure 3. A business model using BSMSS for other makers

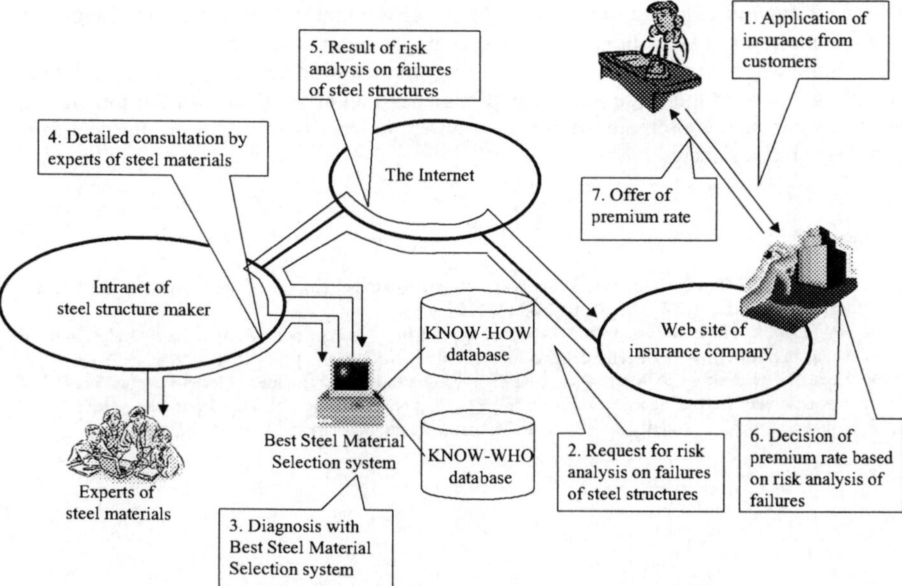

Figure 4. A business model for premium rate decision under cooperation with insurance companies

As the third phase of constructing the business model, we register our system with the Web site of insurance companies and cross-link with them. This cross-link provides information from insurance companies to steel structure makers; for examples, information of the steel structures that are covered by insurance, or risk information regarding the safety of the products. Insurance companies decide premium rates (insurance prices) based on the risk analysis of the products. The Japanese insurance companies calculate the premium rates of steel structures based on the failure probability of the structure of the past. This sometime results in inappropriate premium rates. The BSMSS enables to precisely determine the risk on steel structures, and accomplishes fair price setting for both insurance companies and the customers. Using our system, the insurance company can decide fair premium rates, which normally lowers the rates, and can enhance the competitiveness against the other insurance companies.

By following these steps, we realize the remote consultation business for other industries.

5. Conclusion and Future Work

This chapter concludes our discussion by describing advantages of our BSMSS.

Generally, steel structure makers use materials of various steel makers and are familiar with the characteristics of each maker's materials. They can select steel materials in consideration of the characteristics of each maker's materials as well as the specification of materials.

Additionally, steel structure makers have knowledge on long-term use of steel materials since they have maintained steel structures for a long time (for one decade to several decades). On the other hand, steel makers normally hold steel material testing data only for several years. Selecting steel materials in consideration of long-term use is only available to steel structure makers with the long-term knowledge. Our BSMSS is based on the database of knowledge such as the characteristics of each maker's materials and long-term use of materials, and this accomplishes the appropriate material selection.

As a future work, we plan to link the BSMSS with condition monitoring system for steel structures to monitor their operating conditions and soundness. Based on the monitoring functions, we consider a remote maintenance system to warn the occurrences the problems before they actually occur.

References

[1] J. Conklin and M. Begeman, gIBIS: A Tool for All Reasons. Journal of the American Society for Information Science, Vol.40, No.3, pp.200-213, 1989.
[2] N. Fujino and K. Himeno, Business Models for Supply Chain Management: Analysis and Design, Journal of the Japan Society for Management Information, Vol.10, No.3, 2001, pp.3-20 (in Japanese)
[3] M. Ueda and H. Matsuo, A Method for Classifying Internet-Related Business Model Patterns, Journal of the Japan Society for Management Information, Vol.10, No.3, 2001, pp.71-84 (in Japanese)
[4] P. Timmers, Business Models For Electronic Markets, Electronic Markets, Vol.8, No.2, 1998

Real-time Robust Recognition of Human Using Illuminance-depending Gazing GA

Mamoru MINAMI, Hidekazu SUZUKI and Masato MIURA
Faculty of Eng., Fukui University, 3-9-1, Bunkyo, Fukui 910-0017, Japan

Abstract.   Many trials to realize human's visual recognition by image processing have been carried out. Recognition and tracking ability of human are expected to be applied for a security system and a field of ITS (Intelligent Transport System). This paper aims at construction of real-time recognition system with robustness against changing of illumination. We employ Genetic Algorithm (GA) and a gray-scale image termed here as raw-image to execute recognition process by a model-based matching method in real time. Proposed recognition method can be divided two main functions, that is, global and local searching, which is switched depending on the matching degree of a human in the raw-image and a model to be matched to the human. Furthermore, we improved the recognition performances, to shorten recognition time and to raise the accuracy, by adding gazing operation in local searching, which is inspired from gazing action of human. Moreover, in order to improve the recognition system to be robust against lighting condition varieties, we propose an Illuminance-depending Gazing GA, which changes the threshold value automatically based on the brightness value of the image. We performed experiments to recognize a walking human to confirm the effectiveness of our proposed recognition system in an indoor environment.

1 Introduction

So far, many trials to realize human's visual recognition by image processing have been carried out. Human recognition and tracking are expected to be applied for a security system of buildings and a field of ITS (Intelligent Transport System) to avoid traffic accident harming humans. For example, in a field of ITS, an image recognition systems mounted on a car to support human's visual recognition process and, if possible, to replace it for automatic driving system whose technology consists of lane recognition, obstacle detection, road signs detection, walker recognition, and so on. This research aims at construction of recognition system of human in real time. Here, the real-time recognition means that an object should be recognized in the images input within video rate, i.e., 33 [ms].

For image recognition purposes, many researchers [1]-[4] have employed a binary image and some edge extraction methods that require several preprocessing steps. The various filtering stages seem to be a time-consuming process, and to us, these are not convenient for real-time recognition. Then, we use directly the unprocessed gray-scale image termed here as raw-image. Basically, this research is based on a model-based pattern-matching method. We employ a model designated as surface-strips model for the recognition purposes of a human considered here as the target. An objective function, whose computation is based on the configuration of the surface-strips model, is used to evaluate the extent to which the surface-strips model matches with the object being imaged, by changing the recognition problem into an optimization problem. Therefore, We use a Genetic Algorithm (GA) in the image recognition, because of its high performance of optimization.

GA is well known as a method for solving search and parameter optimization problems [5]. Moreover, to use the GA process in real time, i.e., to extract its position from the consecutively input images, we used the GA such a way that every input image is evaluated only one time by target-model-based fitness function, which we called Step GA [6]. Furthermore, in order to increase the tracking performance to a target human, here we employ hybrid searching method, which is a localized search technique of a GA combined with a global GA process, that is, conventional GA without loosing real-time nature of Step GA. Using this localized search technique, we confirmed in previous researches that the hand-eye manipulator can catch a fish swimming in a pool with a net attached at the hand [7].

To make our recognition process to be robust against illuminance changing of the environment while keeping the tracking performance of the Step-GA, the switching of the local search process should be determined depending on illuminance level of the input image. In this report, in order to perform the recognition of a human in spite of changing of the lighting condition, we

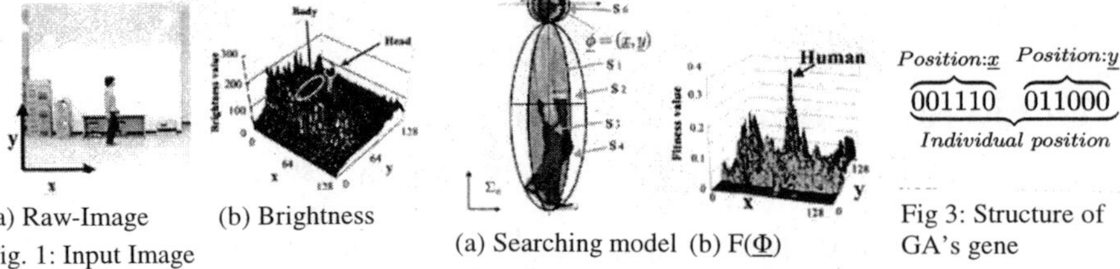

(a) Raw-Image　　　(b) Brightness

Fig. 1: Input Image

(a) Searching model　(b) F(Φ)

Fig. 2: Searching model and F (Φ)

Fig 3: Structure of GA's gene

propose an illuminance-depending gazing GA, which changes the threshold value to determine the gazing area automatically based on the brightness level of the image. We performed experiments to recognize a human to confirm the effectiveness of our proposed recognition system.

2　Recognition Method Using Model-based Matching

Consider the 2-D raw-image of a target human shown in Fig.1(a), its corresponding 3-D plot is shown in Fig.1(b). In this figure, the vertical axis represents the image brightness values, and the horizontal axis, the image plane. To search for a such target human in the raw-image, a geometrical human model as shown in Fig.2(a) is used. In this research, a geometrical model, which is composed of an internal surface and contour-strips, is employed. The internal surface approximated the most the 2-D top surface of the target. Such model is designated as surface-strips model. In Fig.2(a), S_1, S_3, S_5 denotes the inside surface of the model, and S_2, S_4, S_6 denotes the contour-strips. Also, the combination is designated as S. When the position of surface-strips model S is defined as a function of $\phi = [x, y]^T$, which designates the position of the origin of the model, then S moves in the camera frame and a set of x-y coordinates in the moving model is expressed as $S(\phi)$. And moreover, the brightness distribution of raw-image corresponds to the area of the moving model is expressed as $p(\tilde{r}), \tilde{r} \in S(\phi)$, then the evaluation function $F(\phi)$ representing matching degree of the model and the raw-image, of the moving surface-strips model is given as follows in Eq.(1).

$$F(\phi) = \sum_{i=1,3,5} | \sum_{\tilde{r} \in \underline{S}_i(\phi)} p(\tilde{r}) - \sum_{\tilde{r} \in \underline{S}_{i+1}(\phi)} p(\tilde{r})| \tag{1}$$

This function means the integrated brightness difference of the input raw-image between the one of the internal surface and the one of the contour-strips of the surface-strips model. The filtering result by using Eq.(1) of the surface-strips model-based fitness function, with respect to Fig.1(a) is shown in Fig.2(b). The filtering result in Fig.2(b) has a peak corresponding to the target human in the raw-image. An evaluation using the surface-strips model means that $F(\phi)$ takes into account the differentiation between an object signal and the background brightness. As long as we can set such an environment that the highest value of $F(\phi)$ is obtained only if S fits to the target object being imaged. Then the problem of recognition of a human and detection of its position is converted to a searching problem of ϕ such that maximize $F(\phi)$. $F(\phi)$ is used as a fitness function of GA.

GA is well known as a parallel search and parameters optimization algorithm. The GA is viewed as an optimization method since the iterative evolution process of the potential solutions toward better solutions is equivalent to the process of optimizing the objective function used as a fitness function in GA search. The term "parallel", in "parallel search" above is related to the implicit parallelism of GA. Implicit parallelism is explained in Goldberg (1989, pp. 40-41) [5]. The GA operates with a population of searching binary strings designated as individuals, shown in Fig.3, considered to be the potential solutions to a given problem. The search by a GA for the solution is performed through an evolution process, from generation to generation.

To recognize a target in a dynamic image, the recognition system must have real-time nature, that is, the searching model must converge to a human in the successively input raw images. We have proposed a new idea of an evolutionary recognition process for dynamic image, in which the GA is applied only one time to the newly input raw image. Therefore every input image is evaluated only one time, we named it as "Step-GA" [6].

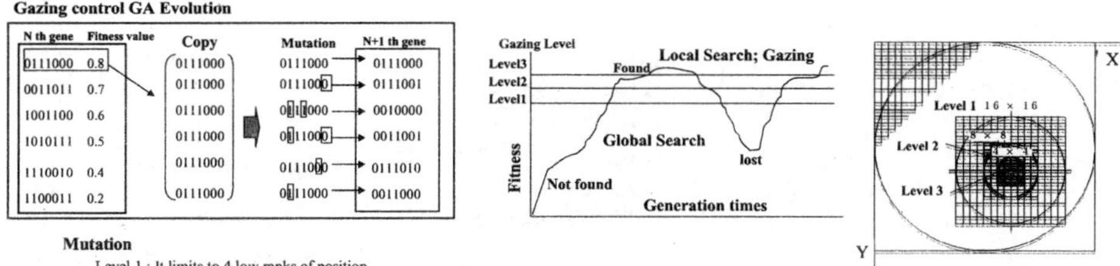

Fig. 5: Gazing switching   Fig. 6: Area switching

Fig. 4: Searching method of Gazing GA

3 Illuminance-depending Gazing GA

In order to increase the tracking performance in dynamic images, here we employ a localized search technique of a GA, combined with a global GA process. Like you and me, if we were to track a moving target with our eyes, at a certain degree we do not look at the surrounding and just focus our attention on the target, that is gazing it. Here, we think of how to propose a similar action as the human being to real-time recognition by machine. In the reproduction of conventional GA, the genes of a subsequent generation are obtained via selection and mutation in a probability based procedure to maintain exploration of the search domain. Contrarily to the global GA search, the reproduction of the proposed algorithm describing the localized search technique of GA is represented in Fig.4. In this figure, let us assume that the highest fitness value of an $N-th$ generation has reached a certain value where the target's position is in a range of acceptable recognition. The individual that has the highest fitness value is positioned on the place indicated by the gene of ranking 1 by sorting. Then what we denoted as pure selection is performed, that is, the genes of the individual of ranking 1 are selected to be copied s times, thus making an intermediate population of s identical individuals, for a population of s individuals. Next, in the reproduction process, except that ranking 1 individual, a mutating operation is gradually performed on the lower level bits of the genes of the other individuals in order to increase the fitness value and at the same time to obtain better positional results, that means concentrated searching in a vicinity of the top gene.

The switching process between the global GA search and the local GA search technique can also be described by the curve shown in Fig.5. This figure indicates the global search of GA that proceeds to the search in order to find a target, and once a target is found, the system switches to the local GA search as a gazing action, in order to perform an intensive localized search so as to fasten and improve the recognition results of the target. When the GA happens to loose the target, which means the fitness value decreases under the level of local search, the GA search returns to the global search state as shown in Fig.5. Level 1-3 in Fig.5 correspond to the gazing area around the position designated by gene of ranking 1, and the area is depicted in Fig.6 In this gazing control, we also have to consider the time used for the restricted area searching. To maintain the real-time recognition, the restricted searching is executed in the next newly input image. In practice, once the global GA has achieved the stage of the detection of a target, after reaching a certain fitness value as a threshold value, it switches to the local GA that performs the fine and fast recognition of the target at the *next* iteration. The determination of this threshold values, which are shown in Fig.5 as "Level 1,2,3 ", is environment dependent.

Therefore, it is important to determine the threshold value depending on an illuminance changing to make the system to be robust against lighting condition varieties. Figure 7 denotes a transition of the threshold value of local search based on the illuminance of three stationary images (a) $\sim$ (c) taken by a CCD camera in difference lighting condition. In the Fig.7, A $\sim$ C represent maximum fitness values of each image (a) $\sim$ (c) and mean value of sampled brightness value in the input image represents a illuminance of the image. Then, a line connected to A $\sim$ C is a function, which indicates a relation between the fitness value and the corresponding illuminance. Moreover, this function shows that the darker the illuminance of the input image changes, the lower the maximum fitness value becomes. In order to perform the recognition of a target human depending on its lighting environment, we propose an Illuminance-depending Gazing GA, which changes the threshold value of local search automatically based on the illuminance level of the input image by using a relation of Fig.7, which determine the gazing threshold proportionally to an illuminance of environment in the input image.

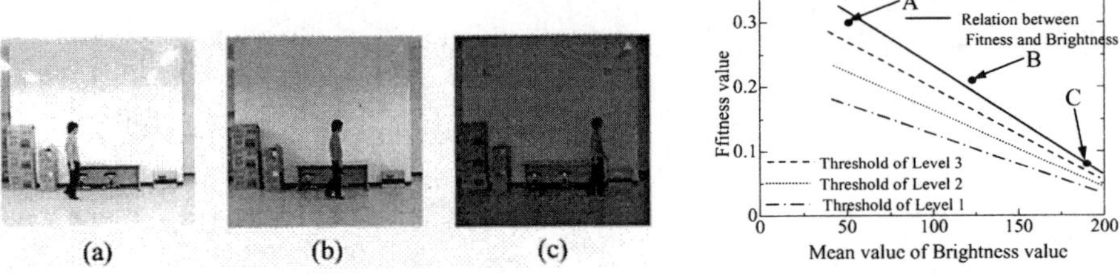

Fig. 7: Relation between Intensity of Object and Fitness value

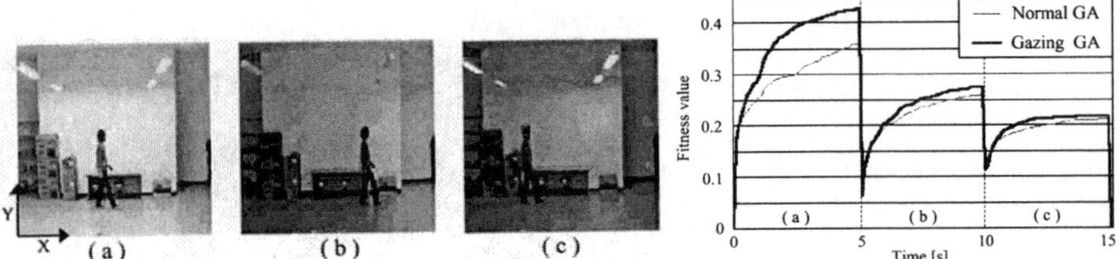

Fig. 8: 3 Input Images

Fig. 9: Recognition Result of Normal GA Search and Gazing GA Search

4 Recognition Experiment of Human

[Convergence Speed and Adaptation for Illuminance Changing]

Experiments of recognition and tracking of a human in indoor environment using raw images have been performed to appraise the performances of the proposed scene recognition method. First, recognition experiment has executed using three fixed images of different illuminances as shown in Fig.8. In this experiment, the images have been input into the recognition system at 0, 5 and 10 [s], as a step-input manner. Moreover, this experiment has been repeated for 100 times. Fig.9 shows the mean value of the best fitness values of both normal GA and gazing GA versus the time used for the recognition, for comparison between the results of the normal GA (Conventional GA) and the results of the gazing GA (illuminance-depending). For all images, the fitness value of the Illuminance-depending Gazing GA search method increases quickly and higher than the one of the normal GA. These results indicate, at first, it can shorten the time for the recognition, that means proposed system improves the dynamical response of the recognition, relating to the improvement of the tracking performance to moving object. The second is the improvement of the recognition accuracy pointed out by the higher fitness value, which means closer matching of the searching model to the target human in the input images. These results show the effectiveness of the proposed search technique for real-time recognition having an adaptation for illuminance changing of the environment.

[Real-time Recognition of Human in Dynamic Scene]

Furthermore, we would like to confirm the validness of the proposed method while using dynamic scene. In the real-time tracking experiment, the images are obtained from a camera fixed in a room. This experiment has been performed at night, and we made a situation being changed the illuminance by putting on and off the light intentionally. Figure 10 presents a sequence of the tracking results observed by the CCD camera, when a human is walking. The white and connected two circle is representing a position of the detected human. Here, we change the searching area of gazing GA with 3 steps (level 1,2,3), and the calculation time is 30 [ms] for Step-GA recognition, by the computer DELL optiplex GX1 (CPU:Pentium2, 400MHz). The experimental results have confirmed the effectiveness of the proposed method by showing the robustness against an illuminance changing in indoor environment.

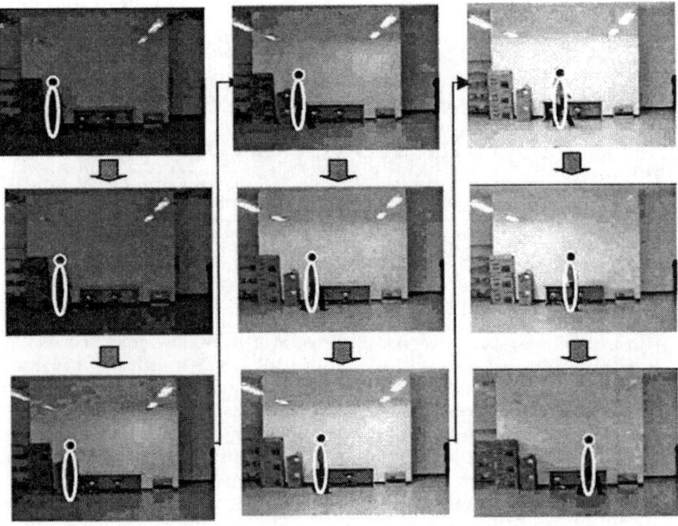

Fig. 10: Recognition Result of Indoor Environment

5   Conclusion

We have proposed in this paper a vision related technique for a real-time recognition of a human, which utilizes the global search feature of a genetic algorithm (GA) together with a local search technique of the GA, and also the unprocessed gray-scale image called here as raw image. And moreover, in order to perform the recognition of a target human adapting to changing of the lighting environment, we propose an Illuminance-depending Gazing GA, which changes the threshold value to change local search area automatically based on the brightness value of the input image, i.e., illuminance, by using a relation that the threshold value is proportional to an illuminance of environment in the input image. Results of the experiment to track a target human whose position and the illuminance are changed by stepping manner, have shown the effectiveness of proposed GA search method. Furthermore, we confirmed the validness of the proposed method for the dynamic scene. The experimental results have confirmed the effectiveness of the proposed method by showing the robustness against the illuminance changing in indoor environment.

References

[1] T. Fukuda and K. Shimojima "Intelligent Control for Robotics," Computational Intelligence, pp.202-215, 1995.

[2] G. Ao, H. Akazawa, M. Izumi and K. Fukunaga, "A Method of Model-Based Object Recognition," Japan/USA Symposium on Flexible Automation, vol.2, pp. 905-912, ASME, 1996.

[3] T. Nagata and H. Zha, "Recognizing and Locating a Known Object from Multiple Images," IEEE Transactions on Robotics and automation, vol. 7, No. 4, pp. 434-447, 1991.

[4] K. Sumi, M. Hashimoto and H. Okuda, "Three-level Broad-Edge Matching based Real-time Robot Vision," in Proc. IEEE Int. Conf. on Robotics and Automation, pp. 1416-1422, 1995.

[5] D.E.Goldberg, "Genetic algorithm in Search, Optimization and Machine Learning. Reading", Addison-Wealey, 1989

[6] M.Minami, J.Agbanhan and T.Asakura: "Manipulator Visual Servoing and Tracking of Fish Using Genetic Algorithm", Int. J. of Industrial Robot, 29-4, pp.278-289, 1999

[7] M.Minami, H.Suzuki, J.Agbanhan, T.Asakura: "Visual Servoing to Fish and Catching Using Global/Local GA Search", Int. Conf. on Advanced Intelligent Mechatronics, Proc., 2001, pp.183-188.

KES 2002
E. Damiani et al. (Eds.)
IOS Press, 2002

Knowledge Sharing based on Dependability in Multi-Robot Exploration

Futoshi Kobayashi, Shigeru Shinyama and Fumio Kojima
Graduate School of Science and Technology, Kobe University
1-1 Rokkodai, Nada, Kobe 657-8501, Japan

Abstract. In this paper, we consider the problem of sharing knowledge in multi-robot exploration. We use the belief measure as knowledge of explored information in each robot for exploring an unknown environment. Each robot has grid maps for storing explored knowledge and for sharing knowledge explored by other robots. Here, each robot calculates the dependability for other robots by using explored knowledge in multi-robot exploration. Then, multiple robots share knowledge considering the dependability for other robots. The effectiveness of our approach is demonstrated by a real experiment for the case of two mobile robots.

1   Introduction

Recently, some research works with multiple robots are studied in various field because various types of robots are developed and produced. In research works, exploration of an unknown environment with multiple robots have received much attention. The use of multiple robots for exploration have several advantages over single robot systems. Firstly, multiple robots can explore an environment faster than a single robot. Secondly, using some cheap robots can be expected to be more fault-tolerant than using one powerful and expensive robot. Finally, multiple robots can localize themselves more efficiently by exchanging sensing information about the environment. Though the exploration problem has been dealt with for single robot[1], there are few approaches for multiple robot. Rekleitis et al. proposed a multi-robot exploration of reducing the odometry error during exploration[2], Burgard et al. consider a collaborative multi-robot exploration in order to minimize the overall exploration time[3], Singh et al. proposed a technique for heterogeneous robots[6], and Kurabayashi et al. proposed an Intelligent Data Carrier (IDC) for cooperation of multiple autonomous mobile robots[7]. Moreover, we have developed a sharing method of explored knowledge for multi-robot exploration[8]. In this method, each robot implements the temporal knowledge sharing for sensor values which measured by itself and the spatial knowledge sharing considering the dependability for other robots. However, the difference between the belief measure of itself and the belief measure of other robots is not considered, because the dependability is calculated by the distance between robots.

In this paper, we consider the spatial knowledge sharing based on the dependability for other robots in multi-robot exploration. Here, the dependability is determined by the difference of the belief measures among robots and the explored area by each robot. The robots have grid maps expressed by the belief measures for explored knowledge. The robots have not a common map, but a map for sharing sensing information. For showing the effectiveness, our approach is implemented on real robots in real environment.

2　Knowledge Sharing based on Dependability

2.1　Expression of Knowledge by Belief Measure

In order to explore an unknown environment, we use grid maps to represent the environments. The concept of grid maps is to use a grid of equally spaced cells and to store in each cell the degree of confidence that this cell is occupied by an obstacle. In this paper, we represent the belief measure[4, 5] as the degree of confidence. Here, we define the 2 types of sets as follows:

O:　Cell without Obstacle,
N:　Cell with Boundary of Obstacle.

Accordingly, each cell has the belief measures $Bel^r_{x,y,t}(O)$, $Bel^r_{x,y,t}(N)$ and $Bel^r_{x,y,t}(O \cup N)$. Here, x and y represents the location of the cell, and r represents the number of the mobile robots. The belief measures are expressed by the basic assignments $m(A)$ which is interpreted either as the degree of evidence supporting the claim that a specific element of X belongs to the set A but not to any special subset of A, or as the degree to which we believe that such a claim is warranted. The belief measures are expressed by the basic assignments as follows:

$$\begin{aligned}
Bel^r_{x,y,t}(O) &= m^r_{x,y,t}(O) \\
Bel^r_{x,y,t}(N) &= m^r_{x,y,t}(N) \\
Bel^r_{x,y,t}(O \cup N) &= m^r_{x,y,t}(O \cup N) + m^r_{x,y,t}(O) + m^r_{x,y,t}(N)
\end{aligned} \tag{1}$$

Here, $Bel^r_{x,y,t}(O \cup N)$ is defined as 1 by the axiom of the belief measure.

2.2　Temporal Knowledge Sharing

In this subsection, we explain the knowledge sharing method of explored knowledge in each robot. When the evidences $m_{x,y}(\{O, N, O \cup N\})$ are obtained from sensor values, $m^r_{x,y,t}(\{O, N, O \cup N\})$ are combined to obtain a joint basic assignment $m^r_{x,y,t+1}(\{O, N, O \cup N\})$ from the Dempster's rule of combination[4] as follows:

$$\begin{aligned}
m^r_{x,y,t+1}(O) &= \frac{m^r_{x,y,t}(O)m_{x,y}(O) + m^r_{x,y,t}(O)m_{x,y}(O \cup N) + m^r_{x,y,t}(O \cup N)m_{x,y}(O)}{1 - m^r_{x,y,t}(O)m_{x,y}(N) - m^r_{x,y,t}(N)m_{x,y}(O)} \\
m^r_{x,y,t+1}(N) &= \frac{m^r_{x,y,t}(N)m_{x,y}(N) + m^r_{x,y,t}(N)m_{x,y}(O \cup N) + m^r_{x,y,t}(O \cup N)m_{x,y}(N)}{1 - m^r_{x,y,t}(O)m_{x,y}(N) - m^r_{x,y,t}(N)m_{x,y}(O)} \\
m^r_{x,y,t+1}(O \cup N) &= 1 - m^r_{x,y,t+1}(O) - m^r_{x,y,t+1}(N)
\end{aligned} \tag{2}$$

Here, when the robot can detect the obstacle by its sensors, the evidences $m_{x,y}(O)$, $m_{x,y}(N)$, and $m_{x,y}(O \cup N)$ for corresponding cells are 0.05, 0.0, and 0.95, respectively. On the contrary, when the robot cannot detect, the evidences $m_{x,y}(O)$, $m_{x,y}(N)$, and $m_{x,y}(O \cup N)$ are 0.0, 0.05, and 0.95, respectively.

2.3   *Spatial Knowledge Sharing based on Dependability*

In our previous research, the dependability for another robot is determined by the distance between the robot and another robot. In this paper, we propose a novel method of determining the dependability for another robot by the belief measure. Consequently, we consider the sensing ability and the motion ability of each robot.

The dependability of the robot r for an other robot r' is calculated as follows:

$$p_{rr'}(t+1) = p_{rr'}(t) + a \cdot s(t) * p_{rr'}^{berief}(t) + b \cdot p_{rr'}^{renual}(t) + c \cdot p_{rr'}^{difference}(t). \qquad (3)$$

where, $p_{rr'}(t)$ represents the dependability for the robot r', and a, b and c represent the coefficients. In the first term, $p_{rr'}^{berief}(t)$ is determined by the difference of the belief measure as expressed by Eq. (4).

$$p_{rr'}^{belief}(t) = -\frac{\displaystyle\sum_{(x,y)\in K^2} \left\{ Bel_{x,y,t}^{r}(Q) - Bel_{x,y,t}^{r'}(Q) \right\}}{k} \qquad (4)$$

where, Q represents the set with higher belief measure among O and N, K^2 represents cells which are updated by the robot r and are explored by the robot r', and k represents the number of cells which belongs to K^2. Then, $s(t)$ is calculated by Eq. (5).

$$s(t) = \frac{1}{1 + \exp\left\{(-t+\alpha)/\beta\right\}} \qquad (5)$$

where α and β represent the fixed number defined in advance. By using the function $s(t)$, the difference of the belief measure is not affect the dependability in the beginnings of exploration. In the second term, $p_{rr'}^{renewal}$ is determined by the explored area by robot r' as expressed by Eq. (6).

$$p_{rr'}^{renual}(t) = renual/n \qquad (6)$$

where $renual$, n represent the number of explored cells by robot r' and the number of whole cells, respectively. In the third term, $p_{rr'}^{difference}(t)$ is calculated by the number of cells dif which knowledge of robot r is differ from knowledge of robot r' as follows:

$$p_{rr'}^{difference}(t) = dif/n \qquad (7)$$

By using the dependability, the basic assignments $m_{x,y,t}^{\prime r}(O, N, O \cup N)$ after sharing information between robots are calculated from the Dempster's rule of combination as follows:

$$m_{x,y,t}^{\prime r}(O) = \frac{\begin{aligned}&m_{x,y,t}^{r}(O)\left\{p_{rr'}(t)\cdot m_{x,y,t}^{r'}(O)\right\} + m_{x,y,t}^{r}(O)\left\{p_{rr'}(t)\cdot m_{x,y,t}^{r'}(O\cup N)\right\}\\&\qquad\qquad +m_{x,y,t}^{r}(O\cup N)\left\{p_{rr'}(t)\cdot m_{x,y,t}^{r'}(O)\right\}\end{aligned}}{1 - m_{x,y,t}^{r}(O)\cdot m_{x,y}(N) - m_{x,y,t}^{r}(N)\cdot m_{x,y}(O)}$$

$$m_{x,y,t}^{\prime r}(N) = \frac{\begin{aligned}&m_{x,y,t}^{r}(N)\left\{p_{rr'}(t)\cdot m_{x,y,t}^{r'}(N)\right\} + m_{x,y,t}^{r}(N)\left\{p_{rr'}(t)\cdot m_{x,y,t}^{r'}(O\cup N)\right\}\\&\qquad\qquad +m_{x,y,t}^{r}(O\cup N)\left\{p_{rr'}(t)\cdot m_{x,y,t}^{r'}(N)\right\}\end{aligned}}{1 - m_{x,y,t}^{r}(O)\cdot m_{x,y}(N) - m_{x,y,t}^{r}(N)\cdot m_{x,y}(O)}$$

$$m_{x,y,t}^{\prime r}(O \cup N) = 1 - m_{x,y,t}^{\prime r}(O) - m_{x,y,t}^{\prime r}(N) \qquad (8)$$

Accordingly, each robot acquire the grid map for the unknown environment by sharing the sensing information of other robots.

3   Exploration by Multi Khepera Robots

Our approach has been implemented on real robots and in real environment. In this experiment, we use two Khepera robots equipped with 8 Infra-red proximity sensors which can measure the distance between a robot and an obstacle in an environment as shown in Figure 1. The diameter of the Khepera robot is $52.5mm$, the size of the environment to be explored in this experiment is $400 \times 400mm^2$, and the size of a grid cell is $2 \times 2mm^2$.

For exploring the environment, each robot implements three process, self localization, sensing, sharing and motion. In self localization process, each robot calculates the current position and the direction by encoder data, and updates the belief measures of cells occupied by itself in temporal knowledge sharing. In sensing process, each robot measures the distance by IR proximity sensors and the belief measures of cells in front of IR proximity sensors sensors are updated in temporal knowledge sharing. In sharing process, each robot gets exploring information from other robots, calculates the depedability for other robot as shown in Figure 2 and updates the belief measures of cells considering the dependability in spatial knowledge sharing. In motion process, each robot moves along the right side of obstacles and avoid obstacles.

Figures 3 and 4 show the individual grid map of each robot for $Bel(O)$ and $Bel(N)$, respectively. Figures 5 and 6 show the shared grid map of each robot for $Bel(O)$ and $Bel(N)$, respectively. The shading of gray for each cell in these figures indicates belief measures. As shown in each sharing grid map, cells that the robot measures by its sensors is dark, and cells that other robots measures is light. By this experiments, two robots can share their sensing information efficiently and explore the unknown environment.

4   Summary

We have proposed a knowledge sharing method with the belief measure for multi-robot exploration. The proposed knowledge sharing method consists on two types of knowledge sharing, such as the temporal knowledge sharing and the spatial knowledge sharing. In this paper, we proposed the dependability for other robots in the spatial knowledge sharing. The dependability for other robots is calculated by the difference of the belief measure among robots and the area explored by robots. Our approach has been implemented on real robots. The experiment in this paper presents that our approach can share the knowledge of each robot.

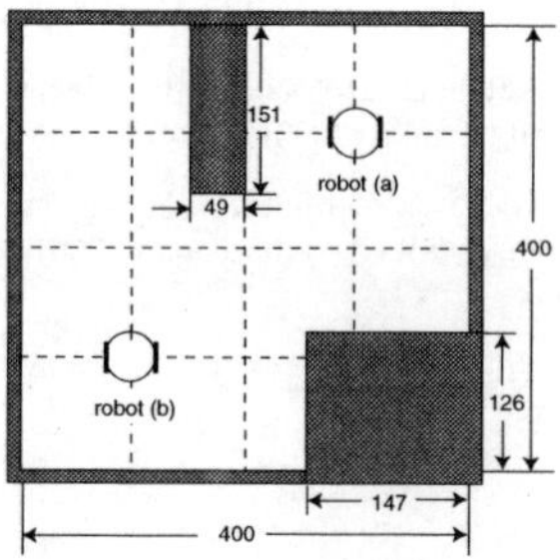

Figure 1: Exploration Environment

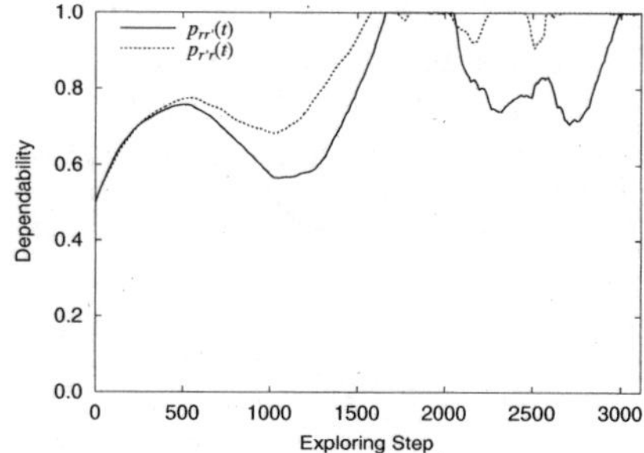

Figure 2: Dependability

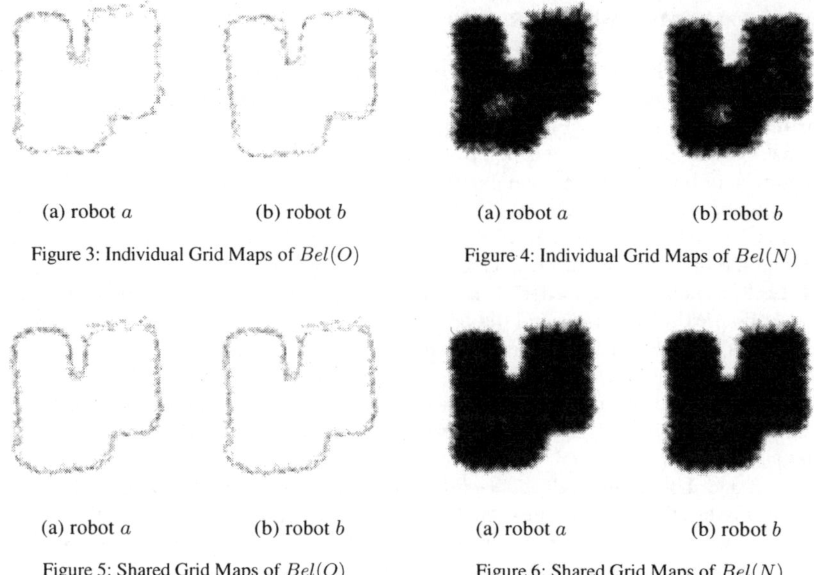

(a) robot a (b) robot b (a) robot a (b) robot b

Figure 3: Individual Grid Maps of $Bel(O)$ Figure 4: Individual Grid Maps of $Bel(N)$

(a) robot a (b) robot b (a) robot a (b) robot b

Figure 5: Shared Grid Maps of $Bel(O)$ Figure 6: Shared Grid Maps of $Bel(N)$

References

[1] H. Gonzalez-Banos and J.C. Latombe, Planning Robot Motions for Range-Image Acquisition and Automatic 3D Model Construction, Proc. of AAAI Fall Symposium (1998).

[2] I.M. Rekleitis, G. Dudek, and E.E. Milios, Graph-Based Exploration using Multiple Robots, Proc. of 5th International Symposium on Distributed Autonomous Robotic Systems (2000) 241–250.

[3] W. Burgard, M. Moors, D. Fox, R. Simmons, and S. Thrun, Collaborative Multi-Robot Exploration, Proc. of the 2000 IEEE International Conference on Robotics and Automation (2000) 476–481.

[4] A.P. Dempster, Upper and lower probabilities induced by a multi-valued mapping, Ann. Math. Stat **38** (1967) 325–339.

[5] G. Shafer, A Mathematical Theory of Evidence, Princeton Univ. (1976).

[6] K. Singh and K. Fujimura, Map Making by Cooperative Mobile Robots, Proc. of IEEE International Conference on Robotics and Automation (1993) 254–259.

[7] D. Kurabayashi and H. Asama, Knowledge Sharing and Cooperation of Autonomous Robots by Intelligent Data Carrier System, IEEE International Conference on Robotics and Automation (2000) 464–469.

[8] F. Kobayashi, S. Sakai, and F. Kojima, Knowledge Sharing Using Belief Measure in Multi Robot System, Proc. of Knowledge-Based Intelligent Information Engineering Systems & Allied Technologies (2001) 535–539.

Communication of A Partner Robot Based on Perceiving-Acting Cycle

Naoyuki Kubota* Daisuke Hisajima*
Fumio Kojima Toshio Fukuda*****

** Fukui University, 3-9-1 Bunkyo, Fukui 910-8507, JAPAN*
***Graduate School of Kobe University, 1-1 Rokkodai, Nada, Kobe 657-8501, JAPAN*
****Graduate School of Nagoya University, 1 Furo, Chikusa, Nagoya 464-8603, JAPAN*

Abstract. This paper deals with communication of a partner robot interacting with a human. We focus on communication based on the concept of perceiving-acting cycle of ecological psychology, and apply modular neural networks and fuzzy inference for behavior learning of a partner robot developed by us. In this paper, we discuss how to form the communication of the partner robot with the human through navigation experiments.

1. Introduction

Recently, various types of pet robots, partner robots, and amusement robots have been developed as the next generation of electronic toys. Like animals, these robots require several capabilities such as perceiving, acting, communicating, and surviving. Especially, the communication capability plays a very important role in order that a human grows the friendly ship with a pet robot. When a child begins to play with a pet robot, the child would try to have contact with the pet robot in various ways. And then, the child searches for causal relationship between the sensory inputs and motion outputs of the pet robot. Figure 1 shows the communication between a human and a robot. If the structure of expression exists in the action patterns of the robot related with a human contacting pattern, the human can learn the structure. This indicates that the human gradually succeeds to find the boundaries of the action patterns of the robot if the robot can form the boundaries of action patterns. Then, we discuss the perceptual system and the action system for a partner robot to generate specific action patterns related with human specific contacting patterns. The perceptual system doesn't extract all features of the object, but picks up the specific information of the object according to the spatio-temporal context of the facing situation. Consequently, the perceptual system does not construct a complete world model, but makes ready beforehand for a next specific perception. That is, the perceptual system picks up the specific information from sensory inputs, using several functions based on internal state. At the same time, the action system generates specific motion outputs using mapping functions based on the specific information related with the human contacting patterns. Furthermore, the motion outputs of the robot construct the spatio-temporal context for human specific perception with the dynamics of the environment. In the above discussion, the perceptual system and the action system restrict each other through interaction with the environment. In ecological psychology, this is called perceiving-acting cycle [1-2]. Next, we consider the mechanism for the communication of the partner robot based on the perceiving-acting cycle.

In most of pet robots, their behavior patterns and communication forms are designed beforehand. We think the mechanism to enrich the relationship between a human and robot is not the architecture for realizing the given behavior patterns, but the architecture for learning the interrelation between the human and robot. Consequently, the communication of a robot with a human requires the continuous interaction with the human, because the human tries to find out the causal relationship between human approach and the robotic behavior, and furthermore, the human tries to find more complicated relationship according to the found relationships. Therefore, the robot needs to accumulate its perceptual system and action system through interaction with the human.

On the other hand, cybernetics has been discussed as the theoretical study of communication and control processes in biological, mechanical, and electronic systems [3-5]. The traditions of cybernetics can be discussed from three different viewpoints; Wiener's cybernetics, Turing's cybernetics, and McCulloch's cybernetics [5].

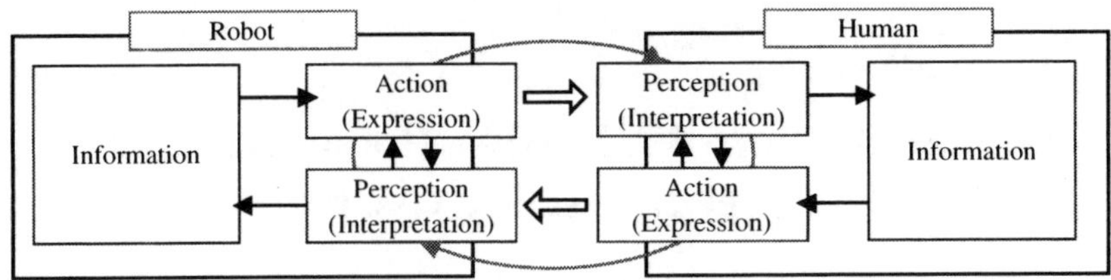

Figure 1 Communication between a human and a partner robot

Wiener's cybernetics is the study of control systems based on the concept of feedback [3]. The computational model for a system is based on the functions, not symbolic representation and manipulation. To contrast, Turing's cybernetics is the study of the intelligence in calculations and machines based on computability [6]. A Turing machine is a theoretical model of a computer. McCulloch's cybernetics is the study of neuroscience. McCulloch and Pitts suggested a mathematical model of a single neuron as a binary device performing a simple threshold logic [7]. The brain is a network of neurons and this is considered as the first model of connectionism. Furthermore, the McCulloch's tradition in cybernetics led to the development of second-order cybernetics [5]. However, we need the concepts and methodologies of these traditions in order to realize communication between a human and a partner robot, because we believe that robotic intelligence emerges from various information processing related to intelligence. Consequently, we should consider an entire structure of intelligence for information processing. We have proposed the concept of structured intelligence [8]. Structured intelligence emphasizes the importance of the close linkage among perception, decision making, and action. In this paper, we discuss communication of a partner robot with structured intelligence based on the perceiving-acting cycle. Modular neural networks (MNN) are used as one method for realizing the perceiving-acting cycle. Furthermore, we conduct some experiments about communication between human and a partner robot.

This paper is organized as follows. Section 2 explains the concept of structured intelligence for partner robot. Section 3 proposes the control algorithm of a partner robot using MNN. Section 4 shows an experiment result.

2. Structured Intelligence for A Partner Robot

The concept of structured intelligence includes three functions: intuitive inference, logical inference, and self-consciousness. Since it is difficult to define consciousness, we focus on the functional features of consciousness. Self-consciousness is consciousness toward a self, and this is regarded as a high level of consciousness. We can take various behaviors unconsciously, whereas we can take logical behaviors when the degree of self-consciousness is high. Furthermore, in our view, self-consciousness perceives the internal state, while the consciousness perceives the external state. In this sense, consciousness can be used in the same role of perception, and therefore, consciousness is related with perceiving-acting cycle. Furthermore, self-consciousness combines and selects the outputs from the intuitive and logical inference. We consider that intuitive inference is realized by neural computing based on numerical processing, while logical inference is realized by production rules and fuzzy rules based on symbolic processing.

2.1 Perceiving-Acting Cycle for A Partner Robot

Figure 2 shows the coupling of perceptual system and action system corresponding to perceiving-acting cycle. Each module of perception system or action system is a specific perception module or action module. When the perceiving-acting cycle forms a coherent relationship with the environment, the specific perceptual information generates the specific action outputs like reactive motions. We consider the couple of perceptual system and action system situated to the facing environment as one phase of the perceiving-acting cycles (Figure 2). In a phase of perceiving-acting cycle, the perceptual system picks up the specific information. To persist perceiving-acting cycle or to connect to another phase, the perceptual system makes ready beforehand for a next specific perception. The self-consciousness searches the phase situated to the facing environment, and updates the phase of perceiving-acting cycles.

2.2 Logical Inference using Fuzzy Computing

We have proposed multi-objective behavior coordination as a logical inference model. A robot has a set of

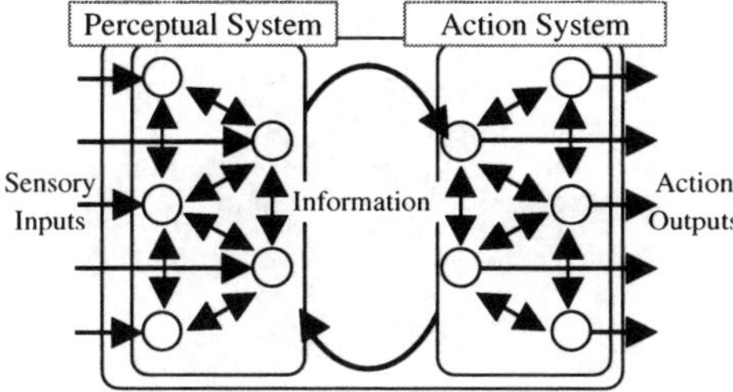

Figure 2 Coupling of perceptual system and action system

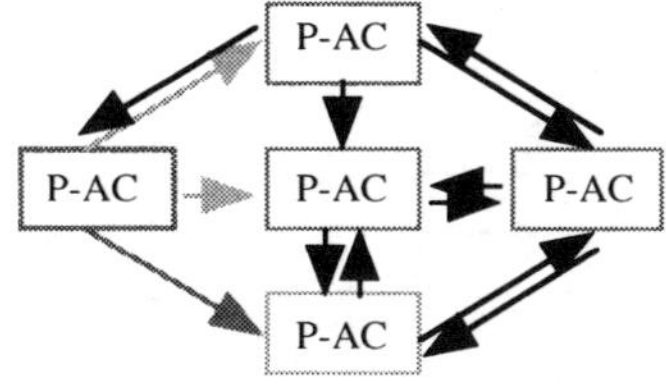

Figure 3 Phase transition of perceiving-acting cycles

primitive objective-based behaviors written by fuzzy rules. A behavior weight is assigned to each behavior and the kth action output is calculated by

$$y_j = \frac{\sum_{k=1}^{b}\left(wgt_k(t)\sum_{i=1}^{r}\mu_{k,i}\cdot w_{k,i,j}\right)}{\sum_{k=1}^{b}\left(wgt_k(t)\sum_{i=1}^{r}\mu_{k,i}\right)}$$

(1)

where r and $m_{k,i}$ are the number of fuzzy rules and firing strength of the ith fuzzy rule, b and $wgt_k(t)$ are the number of behaviors and a behavior weight of the kth behavior over the discrete time step t, respectively. By updating the behavior weights, the robot can take a multi-objective behavior according to the time series of perceptual information. This method can be considered as a mixture of experts, if the behavior coordination mechanism is considered as a getting network.

2.3 Intuitive Inference using Neural Computing

A robot makes decisions using logical inference in unknown situations, and the behaviors of the robot are improved gradually and refined through learning. As the result, the robot takes actions using intuitive inference. This indicates the robot acquires the specific perceptual systems and action systems suitable to facing environments using logical inference models as behavioral knowledge. In general, the refined intuitive inference takes a little time, while logical inference based on the multi-objective behavior coordination takes considerable time. We have proposed MNNs as an intuitive inference model related to consciousness. Each NN performs the decision making and state prediction simultaneously. That is, inputs to NN are the current perceptual information and action output, and the outputs predict perceptual information, action outputs, and connectability values to other NNs. The output of each NN is calculated as follows,

$$Y_p^l = S\left(\sum_{q=1}^{N_{l-1}} W_{p,q}^l \cdot Y_q^{l-1} - \theta_p^l\right)$$

(2)

where Y_p^l is an output of the pth neuron in the lth layer, S is a sigmoid function, $W_{p,q}^l$, θ_p^l, and N_l are a weight parameter between pth neuron of the lth layer and qth neuron of the $(l-1)$th layer, threshold of the pth neuron, and the number of neuron of the lth layer, respectively. The inputs are the current perceptual information $I(t)$ and action $O(t)$, and the outputs are predicting perceptual information $I'(t+1)$, action outputs $O(t+1)$, connectability vector $C(t+1)$ composed of the values of connectability to the sth NN; $C_s(t+1)$.

2.4 Learning of Modular Neural Networks

In this section, we consider learning for perceptual system and action system of a partner robot. The communication of a partner robot requires the continuous interaction with the human, because the human tries to find out the relationship between human approach and the robotic behavior, and furthermore, the human tries to find more complicated relationship according to the found relationships. Therefore, the robot needs to accumulate its perceptual system and action system through interaction with the human. Consequently, its perception and action are based on the current perceptual system and action system, while the learning of the perceptual system and action system is based on the current relationship of perception and action. Therefore, the

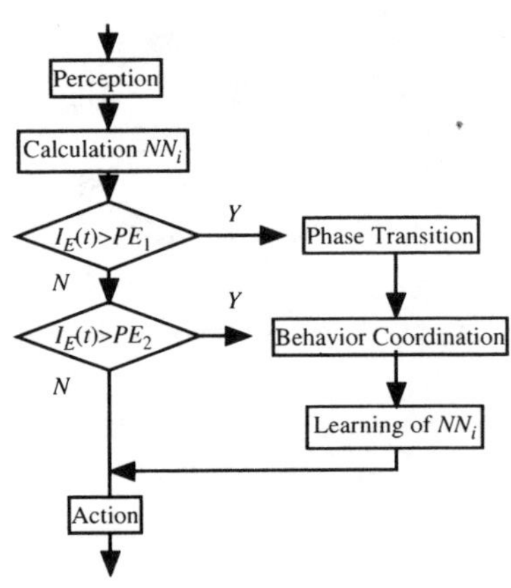

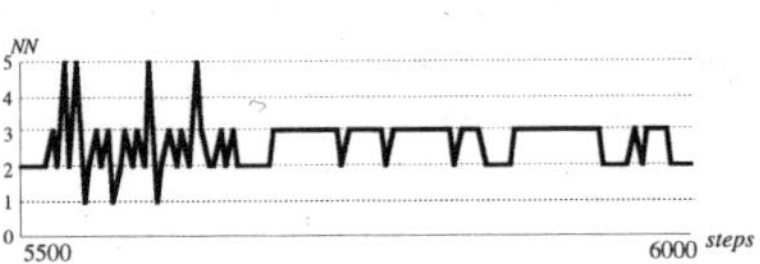

Figure 5 A Partner Robot

Figure 4 Learning procedure of MNN

Figure 6 Phase transition of robot

learning structure for accumulating perceptual system and perceptual system is required.

In general, self-consciousness is activated when the predicting perceptual information is different from actual perceptual information. When the predictive difference is high, the robot selects other NNs suitable to the facing situation according to the connectability values. The MNNs are trained using the outputs of the multi-objective behavior coordination through interaction with the environments. Consequently, the inference system of multi-objective behavior coordination is partially replaced by MNNs composed of compact NNs. Figure 4 shows the procedure of the phase transition and learning. After perception, one NN is calculated, and then, the predictive difference is calculated by the following equation.

$$I_E(t) = |I(t) - I'(t)| \tag{3}$$

If the $I_E(t)$ is larger than a given threshold PE, the robot selects the NN corresponding to the highest connectability value in $C(t)$. In this way, a phase of perceiving-acting cycle suitable to the facing situation is selected. The learning of each NN is done by the back-propagation learning algorithm [2]. A weight parameter is updated by

$$\Delta W_{p.q}^I = -\eta \cdot \frac{\partial E}{\partial W_{p.q}^I} \tag{4}$$

where E and η are the squared error between the outputs of NN and the desired outputs and the learning rate. Here the outputs of the multi-objective behavior coordination are used as training data in the learning of each NN. Consequently, the inference system of multi-objective behavior coordination is partially replaced by MNN composed of compact NN. The teaching signals for the connectability values are generated using the results of the feedforward calculation of all NN. The desired output of neuron corresponding the NN with the least error is 1 as the teaching signal for the connectability value $C(t)$. The other connectability values are 0. To summarize, the robot using MNNs does not perform the inference of all possible behaviors, but performs only the inference of behaviors based on the predictive difference.

3. Experimental Results

We are developing partner robots (Figure 5). The partner robot uses the body of all-in-one personal computer (PC) familiar with various people. While we can use this robot as a standard PC, this robot can be used as a pet robot. We aim to develop a mobile PC that we want to continue to use as a partner. Two CPUs are used for PC and robotic behaviors. The robot has two servo motors, four ultrasonic sensors, and four light sensors. Therefore, the robot can take collision avoiding, human approaching, line tracing behaviors, and others. The robot should have motion patterns easy for the human to understand, but such motion patters can not be prepared beforehand.

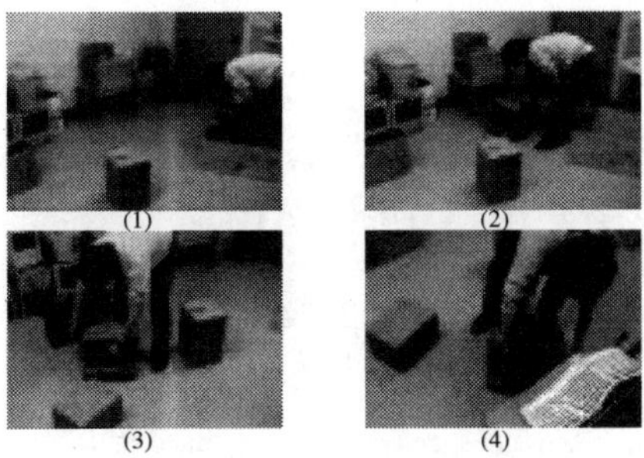

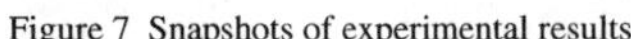
(1) (2)
(3) (4)

Figure 7 Snapshots of experimental results

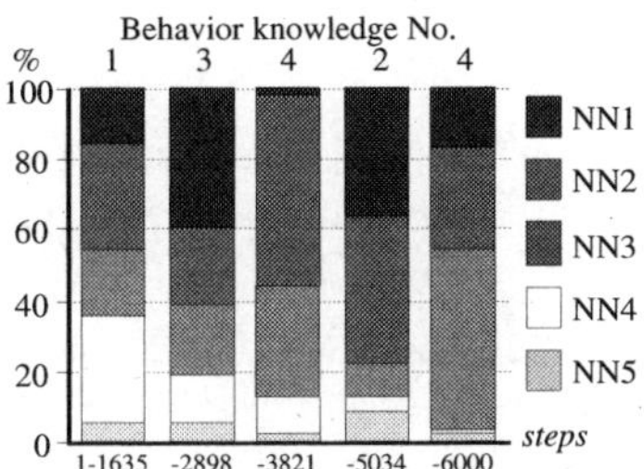

Figure 8 The rate of NNs used in each behavior knowledge

Therefore, the robot should generate or prepare some motion patters. To simplify, we prepare 4 different types of fuzzy rule sets as behavior knowledge of the robot in this experiment. The fuzzy rule set is switched if one NN is used for 500 steps through the task of the human navigation. Figure 7 shows some snapshots of experimental results of a partner robot. The MNNs of the robot are trained gradually using the fuzzy rules selected as behavior knowledge. The human makes contact with the robot in various ways, and then the robot gradually learns some of the contacting patterns using MNNs (Figure 7 (1) and (2)). As a result of this interaction, the human could find the causal relationship between the human contacting pattern and its corresponding expression of the robot (Figures 7 (3) and (4)). Figure 6 shows the transition of NNs used for the navigation by the human. The partner robot uses the NNs properly according to the contacting patterns of the human. When the predictive error is high, he robot searches for an appropriate phase while switching the NN frequently.

Figure 8 shows the rate of NNs used in each behavior knowledge. In this result, the main NN used for each behavior knowledge is differentiated. This indicates the role of each NN is specialized. The robot does not learn all motion input-output patterns of behavior knowledge, but the robot partially learns some relationship of behavior knowledge corresponding the human contacting patterns. By switching behavior knowledge, the robot succeeded to find the motion output patterns suitable for the navigation by the human.

4. Concluding Remarks

This paper discussed the form of communication between human and a partner robot from the viewpoint of ecological psychology. Modular neural networks and fuzzy inference are applied for behavior learning of the partner robot through interaction with the human. Experimental results show that the human could find the causal relationship between the human contacting pattern and its corresponding expression of the robot and succeed to navigate the robot to a target point by using the found relationships. However, these interrelations are not designed beforehand, but the interrelations have been built according to the interaction between the human and the robot. As future work, we must discuss the relationship among communication, consciousness, and learning through interaction with a human.

References

[1] M.T.Turvey and R.E.Shaw, Ecological Foundations of Cognition I. Symmetry and Specificity of Animal-Environment Systems, *Journal of Consciousness Studies* 6, No.11-12, pp.95-110, 1999.

[2] R.E.Shaw and M.T.Turvey, "Ecological Foundations of Cognition II. Degree of Freedom and Conserved Quantities in Animal-Environment Systems," Journal of Consciousness Studies 6, No.11-12, pp.111-123, 1999.

[3] N.Wiener, *Cybernetics*, John Wiley and Sons, New York, 1948.

[4] W.R.Ashby, *An Introduction to Cybernetics*, Chapman & Hall, London, 1956, Internet (1999): http://pcp.vub.ac.be/books/IntroCyb.pdf.

[5] S.A.Umpleby and E.B.Dent, The Origins and Purposes of Several Traditions in Systems Theory and Cybernetics, *Cybernetics and Systems*, Vol.30, pp.79-103, 1999.

[6] A.M.Turing, Computing Machinery and Intelligence, *Mind*, Vol.59, pp.433-466, 1950.

[7] J.A.Anderson and E.Rosenfeld, *Neurocomputing*, The MIT Press, 1988.

[8] T.Fukuda and N.Kubota, An Intelligent Robotic System Based on A Fuzzy Approach, *Proceedings of IEEE*, 1999, Vol.87, No.9, pp.1448-1470, 1999.

KES 2002
E. Damiani et al. (Eds.)
IOS Press, 2002

Motion Intelligence of a 1-link Mobile Manipulator Using GA evolved in Time Domain

Mitsunori KOBATA, Mamoru MINAMI and Atsushi TAMAMURA
Faculty of Eng., Fukui University, 3-9-1, Bunkyo, Fukui 910-8507, Japan

Abstract. We propose a method to acquire a kind of machine intelligence to utilize unknown dynamics, which is designated here as "motion intelligence". The motion intelligence is defined here as an ability that the machine can find by itself a way to use its own but unknown dynamics for extracting a desired motion. We propose a causality-based learning strategy, i.e., the way that a machine uses effects of the dynamical interferences and friction non-linearity to achieve an objective motion. Here, the desired motion that simulates a human's motion to make caster-chair travel by swinging the body, is traveling of a 1-link mobile manipulator by using interfering motion of the mounted link, which does not possess driving motors nor brakes for mobile base. Further considering the motion intelligence of the human to let the caster-chair go forward looks to be realized in sequential trials in time domain, we propose a method to achieve motion intelligence using a GA evloved in time domain. We confirmed by simulations and real experiments that the 1-link mobile manipulator could find effective motion of the link that makes it travel forward.

1 Introduction

As a challenging trial to give robots an adapting ability to unknown surroundings, Brooks's Subsumption Architecture [1] is a remarkable proposal. Brooks built in plural reflex nerves between sensors and actuators in the robot controller, and he made a robot that has adapting ability for the unknown surroundings through interacting combinations of plural reflex nerves and surroundings. As an another way to realize an behavior-based artificial intelligence, a method combined an optimization by Genetic Algorithm (GA) with learning by neural network (N.N.) is proposed [2], further it is identified that the learning ability is better than the method simply evolving coefficients of N.N. [3]. Furthermore, D. Floreano realized an evolving robot having high ability of adaptation to environments by representing the changing process of synapse by "Hebb rule" and evolving its combination by GA [4].

The occurred motion caused by some input is explained by the causality, putting it concretely in another words, a motion result is a reflection of an equation of motion and input. Contrarily, through the resulted motion governed by the causality, the input can be evaluated based on the resulted motion. Here, since we use only the causality, a machine can make efforts to optimize the motion through input by evaluating noteworthy output of the dynamics even if the dynamic model is unknown. Behavior-based AI [5] is thought to be a methodology to achieve intelligence in machine. However, the research direction of AI seems so far to be concentrated on kinematics, as the result of it, the researches to use the dynamics aggressively for the intelligent motion are very sparse if not none. We called such intelligence concentrated on the dynamical motion as "Motion Intelligence", which is a part of behavior-based robotics. As an example of motion intelligence, human can find a motion pattern in trial and error to make travel the sitting chair with caster by swinging the upper body without touching one's foot to the floor, without considering the equation of motion. This is thought that human improves the input based on the causality described above, so it can be said "Causality-based Motion Intelligence".

In this paper, we consider the intelligent motion by using the 1-link mobile manipulator, which simulates a human to make travel the sitting chair by swinging motion of the upper body mentioned above, and discuss how to realize such ability by a machine itself. A machine learning method proposed in this paper improves its motion by using the evaluation of the input through the motion result reflecting dynamical interferences, nonlinear friction and input, in which evolution of genetic algorithm (GA) is utilized to search for a best input for objective motion, that is, in this presentation, going forward with the conditions of not having driving motor for mobile robot nor break. That is, we gave the mobile manipulator a situation similar to a man sitting on a chair with caster, and let the machine find by itself how to go forward without any knowledge of dynamics.

So far, we confirmed that the machine could find by itself a swinging way to make travel to the desired direction by using interfering motion of the mounted link, in which the searching of

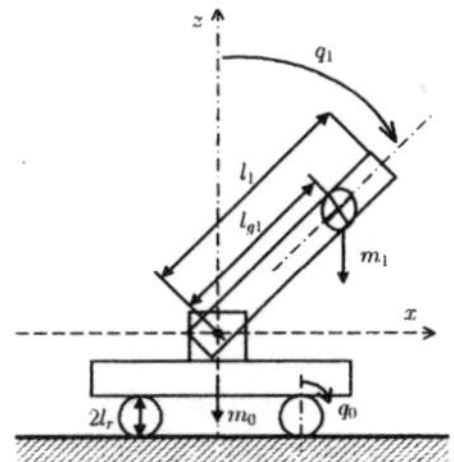

Fig. 1: mobile manipulator

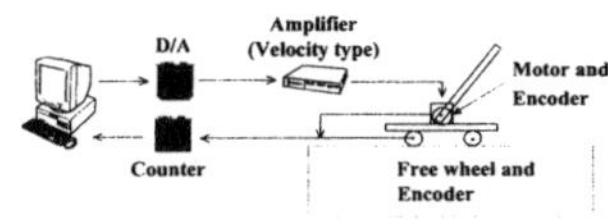

Fig. 2: Experimental system

Fig. 3: Searching flow chart

motion was executed by discrete trials of motions whose initial angles are all same. Contrarily, human acquires the motion intelligence for attaining the purpose in continuous motion, for example, as you see human searches continuously how to use one's body to magnify the amplitude of a swing. We thought that such motion intelligence can be used for a robot, and it is effective to improve the motion. In this presentation, new approach of a GA to realize motion intelligence in continuous time domain is proposed. The objective of motion is to make the mobile manipulator travel by the swinging motion of the mounted manipulator. We confirmed that the proposed method could find how to use the mounted link to travel during continuous motion, it means that the 1-link mobile manipulator got the motion intelligence in continuous time domain similar to humans.

2 Model and Control System

For modeling, we assumed that the wheel does not slip and the floor is flat. Further, to discuss the motion intelligence while avoiding the non-holonomic character of mobile robot, it is assumed that the mobile robot travels without steering. The wheel's angle is represented by q_0 and link's angles of mounted manipulator by $q_i(1, 2, \cdots, n)$, then the equation of motion [6] is,

$$M(q)\ddot{q} + h(q, \dot{q}) + g(q) + d(\dot{q}) = \tau, \tag{1}$$

where, $q = [q_0, q_1, \cdots, q_n]^T$, $\tau = [\tau_0, \tau_1, \cdots, \tau_n]^T$, $M(q)$: the inertia matrix, $h(q, \dot{q})$: the vector of coriolis force and centrifugal force, $g(q)$: the gravity. Besides, the frictional force $d(\dot{q}) = [d_0, d_1, \cdots, d_n]^T$ is,

$$d_i = \mu_{d_i}\dot{q}_i + \mu_{si}sign(\dot{q}_i)exp(-a_i \mid \dot{q}_i \mid), \tag{2}$$

where μ_{di} is viscous friction coefficient and μ_{si} is static friction coefficient, and $sign(\dot{q})$ denotes the sign of $\dot{q}$.

In this paper, we simulate a motion of human to make travel the sitting chair with caster by swinging the upper body, by the 1-link mobile manipulator. The degrees of freedom of this motion are two, then $i = 0, 1$, and $\tau_0 = 0$, since the mobile robot does not have the driving power nor brake. We show the structure of 1-link mobile manipulator used for the experiments and simulations in Fig.1. The components in eq.(1) in case of the 1-link mobile manipulator are as follows: $M = (m_{ij})$, $m_{11} = (m_0 + m_1)r^2$, $m_{12} = m_1\ell_g rcosq_1$, $m_{21} = m_1\ell_g rcosq_1$, $m_{22} = m_1\ell_g^2 + I_1$, $h = [-\dot{q}_1^2 m_1\ell_g rsinq_1, 0]^T$, $g = [0, -m_1 g\ell_g sinq_1]^T$.

To find a motion that human makes the sitting chair travel corresponds to find $\tau_1(t)$ that make $\dot{q}_0$ maximize for the 1-link mobile manipulator. Here, $\tau_0 = 0$ means that the mobile robot is not driven by motor, that is, under actuated, therefore the equation of motion of the mobile robot concerning q_0 represents a non-holonomic constraint [7],[8]. Despite that the non-holonomic constraint condition of wheel's rotating motion with steering is removed in the above discussion, this dynamic system is non-holonomic again, then the traveling of human sitting on a chair with caster is also non-holonomic. Therefore, to find $\tau_1(t)$ that maximize $\dot{q}_0$ can be said in another words, to optimize an input that maximizes one of outputs of non-holonomic constraints. The try and error of human is an endeavor to optimize the input by using nothing but the motion results based on the causality, in this paper, we discuss a method to find $\tau_1(t)$ that maximizes the traveling motion merely by $\dot{q}_0$, which is the motion result. The genes of GA defines the desired trajectory of mounted manipulator $q_{1d}(t)$, and $\tau_1(t)$ is calculated as the output of PD controller using $q_{1d}(t)$ stated in the next section. By searching $q_{1d}(t)$ maximizing $\dot{q}_0$, $\tau_1(t)$ is optimized indirectly. One gene means one trial of sample motion, and the genes are evolved so as to maximize $\dot{q}_0$.

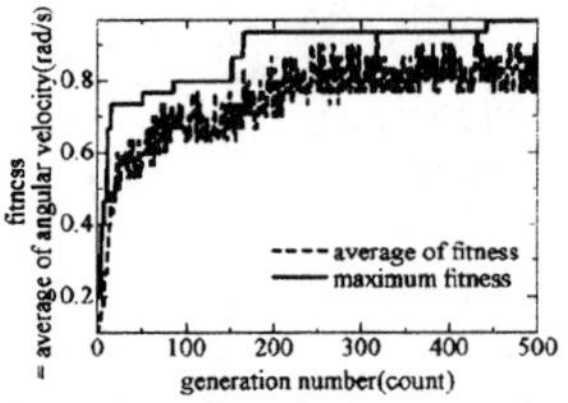

Fig. 4: Simulation results (Evolution process)

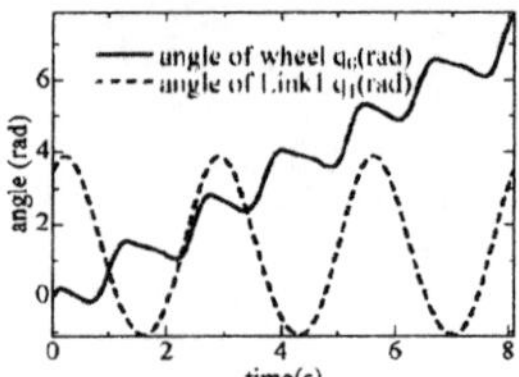

Fig. 5: Simulation results (Traveling motion)

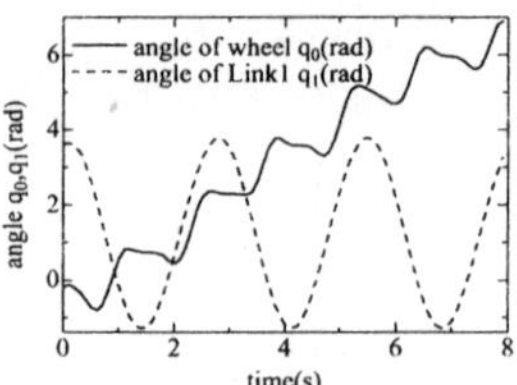

Fig. 6: Experimented traveling motion

The equation of motion of the actuator to drive the mounted link is given as,

$$J_m \ddot{q}_m + D_m \dot{q}_m + R\tau_1 = \tau_m, \tag{3}$$

where, D_m is viscous friction coefficient, R: reduction gear ratio, $R\tau_1$: load torque of link through reduction gear, τ_m: output torque of motor. The τ_1 in eq.(3) is a component of $\boldsymbol{\tau}$ in eq.(1), and is the output torque of the reduction gear unit.

Next, we described the relation between the motor and servo amplifier. The motor is driven by the servo system of velocity control type. Denoting that L: torque constant, R_m: electric resistance, K_s: induced voltage coefficient, K_F: velocity feed back gain, V_{ref}: input voltage to amplifier, then the relation of these variables are,

$$V_{ref} = \frac{RR_m}{LK_m}\tau_1 + \frac{K_F}{R}\dot{q}_1. \tag{4}$$

τ_1 is decided by PD control described in eq.(5) using desired angle of mounted link $q_{1d}(t)$.

$$\tau_1 = d_p(q_{1d} - q_1) + d_d(\dot{q}_{1d} - \dot{q}_1) \tag{5}$$

Furthermore, the velocity command voltage to the amplifier is given by eq.(4) to generate the τ_1.

3   Causality-based Motion Intelligence

The composition of the experimental hardware system is shown in Fig.2. The mounted link with one degree of freedom is driven by servo motor through the velocity type amplifier. Just displacement angle of the vehicle's wheel is measured by the encoder installed at the mobile robot shown in Fig.2. By inputting V_{ref} given by eq.(4) as a velocity command to D/A converter, the motor of the link is driven. The rotated angles, q_0 and q_1, measured by encoders are sent back to the computer through counter board.

[Desired trajectory searching by GA]

Here we discuss a method to acquire a desired trajectory to maximize the objective traveling motion by GA, in which the desired periodic trajectory is represented using the Fourier Series. The periodic function $c(t)$ with time cycle T is defined as,

$$c(t) = \frac{a_0}{2} + \sum_{n=1}^{\infty} \{a_n \cos(\frac{2n\pi}{T}t) + b_n \sin(\frac{2n\pi}{T}t)\}, \tag{6}$$

where $a_0/2$ is average value of the function, and $a_n, b_n (n = 1 \sim \infty)$ denote Fourier coefficients of n-th term. Since the summation of $1 \sim \infty$ is impossible for the computer, the $c(t)$ is approximately defined by the summation to m-th term. An individual of a GA is composed of $(2 + 2m)$ genes as $T, a_0, a_1, b_1, \cdots, a_m, b_m$. The j-th individual in i-th generation denoted by $\boldsymbol{g}_j^i$ is defined as,

$$\boldsymbol{g}_j^i = [T_j^i, a_{0j}^i, a_{1j}^i, b_{1j}^i, \cdots, a_{mj}^i, b_{mj}^i]. \tag{7}$$

The periodic function determined by eq.(6) and (7) defines the desired trajectory of the link, $q_{1d_j^i}(t)$ of j-th individual in i-th generation, and the wheel's angle of the mobile robot caused by $q_{1d_j^i}(t)$ is denoted by $q_{0j}^i(t)$. The function evaluating the results of the interfered motion controlled by eq.(5), that is, fitness function f_{1j}^i of j-th individual in i-th generation, is given as,

$$f_{1j}^i = \frac{1}{3T_j^i}(q_{0j}^i(3T_j^i) - q_{0j}^i(0)). \tag{8}$$

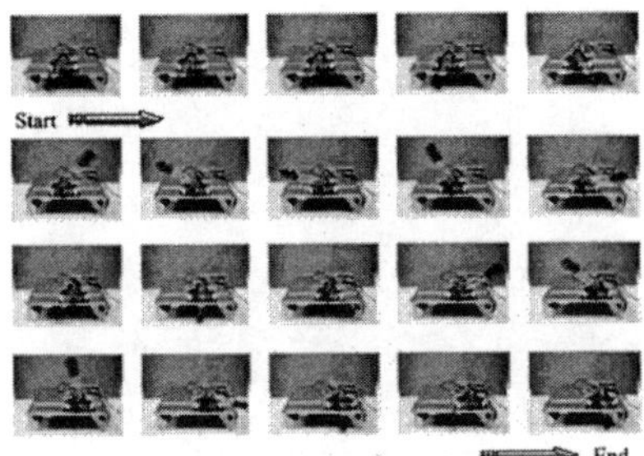

Fig. 7: Photographs of traveling

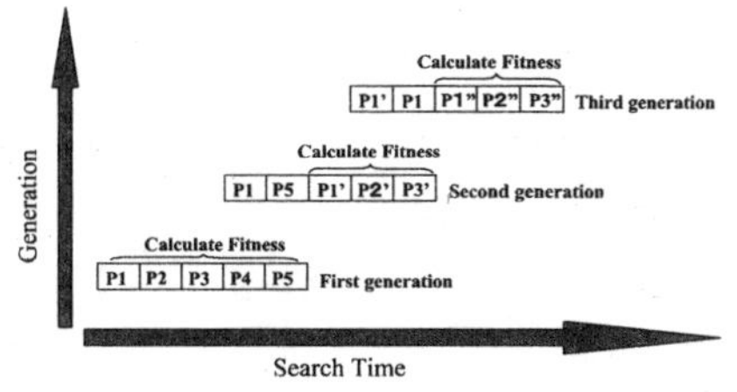

Fig. 8: GA system in continous motion

The f_{1j}^{i} represents the average angular velocity of wheel during period of $3T_{j}^{i}$.

[Searching simulation]

The flow chart of the simulation searching for $q_{1d_{j}}^{i}(t)$ that maximizes f_{1j}^{i} is shown in Fig.3. Population number p=10, 2-points crossover rate $c = 40(\%)$, selecting rate $s = 20(\%)$ and mutation rate $z = 5(\%)$, which are parameters of GA process. In order that the GA can search the optimum solution, the parameters mentioned above should be set appropriately. The above parameter values obtained through preliminary experiments preserve the fastest convergence speed in GA process to solution. In the following simulations and experiments, the length of the gene is 8 bits and m=2, that is, $T, a_0, a_1, b_1, a_2, b_2$, therefore, the binary length of chromosome is 48 bits.

The fitness values of the best individual in GA, versus the generation times is shown in Fig.4, and motion result caused by the swinging motion of the link, which is got by GA searching after 500 generations, is shown in Fig.5. The dotted line in Fig.5 is the desired angle of the link, and the solid line denotes the angle of the wheel, and the figure shows how the mobile robot travels. The fitness of selected gene got by searching is $f_1 = 0.9661$, and Fourier coefficients are $T = 2.70$, $a_0 = 1.26$, $a_1 = 2.46$, $b_1 = 0.56$, $a_2 = -0.02$, $b_2 = 0.02$. From Fig.4, it is comprehensive that the fitness is increased in every generation monotonously, that is, the swinging method of the link to travel faster is being found. The monotonous increasing is owing to the elitist model preservation strategy in GA. The increasing of fitness (solid line) together with the mean value of all individuals (broken line) means that the searching is executed systematic to exploit the best solution.

[Traveling experiments]

Using the desired angle function of the link obtained by the simulation, the traveling is confirmed by the experimental 1-link mobile manipulator. The result is shown in Fig.6, and the photographs took at every 0.3[s] successively are shown in Fig7. The mobile manipulator traveled to right hand side. Comparing Fig.4,Fig.5(Simulation result) and Fig.6, the both results are consistent.

4 Real Time Implementation

So far we discussed how to let the machine find by itself to travel forward using it's own dynamics such as dynamical coupling and nonlinear friction. However searching action for every individual of GA is not executed in continuous time, that is the state variables are initialized to identical values for each individual, meaning every motion trials have the same initial states. And this is far from how humans and animals are using motion intelligence. Therefore, we want to discuss in this section how a machine achieves motion intelligence in continuous trial using the same 1-link mobile manipulator.

[Simulation in continuous motion]

Continuous motion is defined here as successive motion, which succeeds the end state variables of motion of the previous individual corresponding to previous motion trial with some time period, to initial state variables of the following individual between each motion made by inputs generated by successive individuals of GA. The searching system in continuous motion is shown Fig.8. The P1 to P5 in the first generation in the figure correspond to five individuals and five different motion results whose inputs are determined the individuals. The connection of the motion at the end state of previous individual and the initial state of following individual means the motion is continuous. The flow chart of the continuous searching simulation for desired angle function that maximizes f_{1j}^{i} is shown in Fig.9. At first, calculation of the fitness of each individual is done at the time when each trial motion is executed. Then, select-

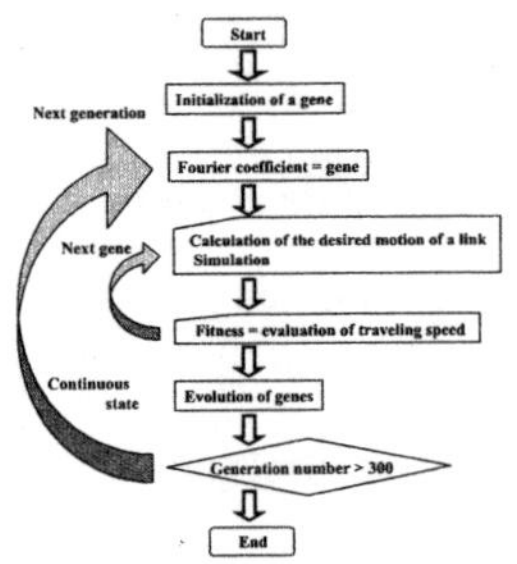

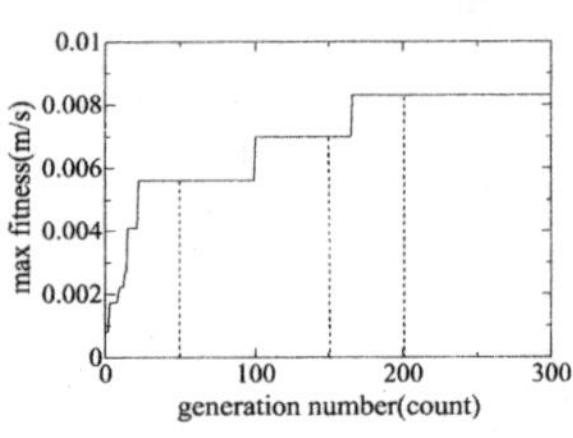

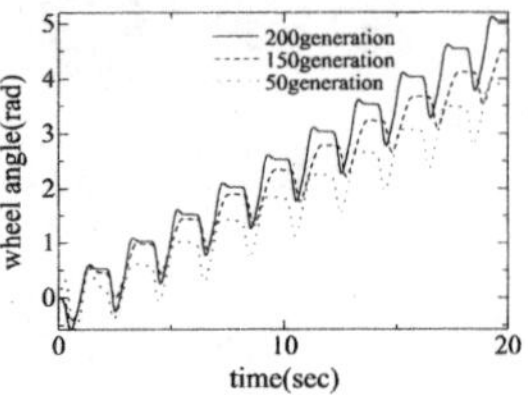

Fig. 9: Searching flow chart in continous motion

Fig. 10: Evolution process of of traveling motion in continous motion

Fig. 11: Evolution process fitness in continous motion

ing, crossover and mutation are carried out in a short time at the end of the first generation. The two individuals (P1, P5) with the highest and the second fitness value survives to second generation, when selecting rate is set to be 40(%). The new motion in the second generation defined by new individual(P1',P2',P3') examined and the previous fitness including (P1,P5) are evaluated altogether. The highest two individuals (P1',P1) are preserved in the third generation and (P1",P2",P3") are generated by the procedure of conventional GA.

Input torque of the link is determined directly in the section using Fourier Series as the following equation,

$$\tau_1 = \frac{a_0}{2} + \sum_{n=1}^{m} (a_n cos(\frac{2\pi n}{T}t) + b_n sin(\frac{2\pi n}{T}t)) \tag{9}$$

The fitness values of the best individual in GA, versus the generation times is shown in Fig.10. And the motion results of wheel caused by swinging motion of the link are shown in Fig.11. The dashed line in Fig.11 is angle of the wheel caused by the individual of best fitness at 50 generation, and the dotted line denotes the one at 150 generation, and the solid line at 200 generation, and the figure shows how the mobile robot travels when the motions corresponding to each best individual are examined during 20 seconds. Using elite preservation strategy which reserve individual having highest fitness acquired by the continuous searching, the fitness increase monotonously as shown in Fig.10. The angle of the wheel increase whenever a generation increases, it indicates the motion intelligence of 1-link mobile manipulator is evolved in time domain. Moreover, motion intelligence in sequential motions in real time is confirmed by simulation.

5   Conclusion

In this paper, we proposed a causality-based learning system, and the effectiveness is examined through real experiments. The motion intelligence is defined here as the ability to use the unknown dynamics for the objective motion. To confirm whether the causality-based learning can realize the motion intelligence, traveling experiment of a 1-link mobile manipulator not having driving power and brake, is executed. As the result of it, it is identified that the machine can find an effective procedure written as a time function to use the unknown dynamics for the objective motion, i.e., traveling, while searching in continuous motion. These results show that the motion intelligence to use unknown dynamics is realized by proposed causality-based learning system.

References

[1] Brooks, R. A. :"New Approaches to Robotics", Science, Vol.253, September 1991, pp.1227-1232. [2] D.E.Goldberg :Genetic Algorithms in Search, Optimization and Machine Learning.Reading:Addison-wesley, 1989. [3] R. K. Belewnd, M.Mitchell, editors: Adaptive Individuals in Evolving Populations Models and Algorithms, Addison-wesley, Redwood City,CA1996. [4] D.Floreano and F.Mondada.Evolutionary neurocontrollers for autonomous mobile robots. ;Neural Networks, 11:1461-1478(1998) [5] R.C. Arkin, Behavior-based Robotics, MIT Press, (1998) [6] M.Minami, T.Asakura, N.Fujiwara, K.Kanbara: Inverse Dynamics Compensation Method for PWS Mobile Manipulators; Int. J. of JSME, 40-2 C, pp.291-298, 1997. [7] F.Nakamura :"Nonholonomic Robot Systems Part 4, Motion under Dynamical Nonholonomic Constraints", Journal of the Robotics Society of Japan, vol.11, No.7, pp.999-1005, 1993. [8] F.Nakamura, T.Iwamoto and K.Yosimoto :"Control of Nonholonomic Mechanisms with Drift", Journal of the Robotics Society of Japan, vol.13, No.6, pp.830-837, 1995.

KES 2002
E. Damiani et al. (Eds.)
IOS Press, 2002

Perceptual System of Multiple Robots for Quasi-Ecosystem

Naoyuki KUBOTA[*], **Masanori MIHARA**[*] **and Fumio KOJIMA**[**]

[*] *Dept. of Human and Artificial Intelligent Systems, Fukui University*
3-9-1 Bunkyo, Fukui 910-8507, JAPAN

[**] *Dept. of System Function Science, Kobe University*
1-1 Rokkodai-cho, Nada-ku, Kobe 657-8501, JAPAN

Abstract. This paper deals with perceptual system and target selection of multiple robots in a quasi-ecosystem which is composed of insects and plants. In this ecosystem, the plants become easy to be eliminated as the population size of the insects increases. Therefore, multiple robots are introduced to remove the insects from the quasi-ecosystem. Each robot has typical types of behavior rule sets that are given beforehand, but the robot must select its suitable aim in order to adapt to the current state of the quasi-ecosystem. In this paper, we discuss the perceptual system for insect removing and plant reaping behaviors through several computer simulations.

1 Introduction

Intelligent technologies have been applied to various engineering applications. Especially, intelligent robots have been used not only in industry, but also, in agriculture and ecological systems [1]. Furthermore, quasi-ecosystems in artificial life have been studied mainly with the population dynamics and evolutionary dynamics [2,3]. The essentials of the quasi-ecosystems are on the adaptability and redundancy of species. In this paper, we deal with multiple robots for maintaining a quasi-ecosystem model composed of two species of insects and plants, and we discuss the adaptability of multiple robots. The relationship between two species in a quasi-ecosystem is a kind of parasitism on discrete cell space. We assume the plants become easy to be eliminated as the increase of the population size of the insects. Consequently, multiple robots are introduced to maintain the numerical balance of plants and insects. In this kind of ecosystem with complicated relationships, the robots must acquire strategies using only local observation to maintain the numerical balance according to unknown population dynamics. Therefore, the robot should find or extract essential information from sensing information. In our previous study, we have discussed how to share decision rules for robotic behaviors [5,7]. The decision rules are described by a set of simple if-then rules to move toward a good situation and to remove insects based on local observing information. Furthermore, the decision rules are updated among multiple robots by using evolutionary computation [8,9]. In this paper, we propose multi-objective behavior coordination for robots in the quasi-ecosystem, and discuss the role of the perceptual system through several computer simulations.

This paper is organized as follows. Section 2 describes a quasi-ecosystem composed of plants and insect. Section 3 proposes multi-objective behavior coordination of robots for maintaining the balance of insects and plants. Section 4 shows simulation results of the quasi-ecosystem with multiple robots. Finally, section 5 concludes this paper and discusses future works.

2 A Quasi-Ecosystem

A quasi-ecosystem composed of plants and insects is often approximated into 2-dimensional discrete cell space. We apply cellular automata (CA) on the 2-dimensional cell space [4]. Figure 1 shows an example of cell space. An automaton is connected in a feedback loop with the environment including its neighboring automata. CA is defined by $\{S, N, n, g\}$ where S is a set of the state, N is a neighborhood, n is the size of N ($n=|N|$), and g is a transition function. The state of each cell is decided according to the state of neighborhood by using the following mapping,

$$g: S^n \rightarrow S$$

(1)

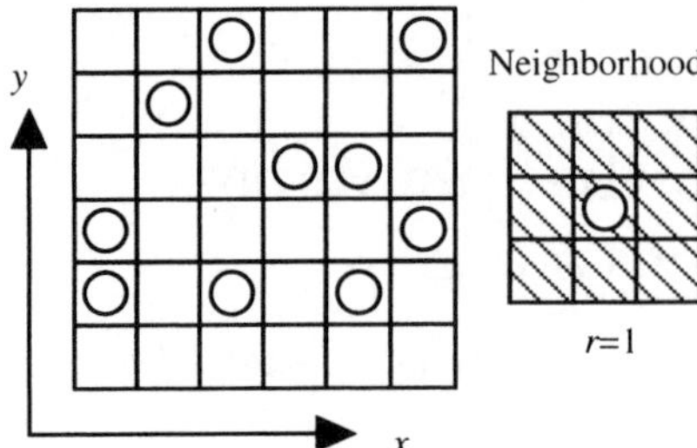

Neighborhood

$r=1$

Table 1 Transition rules of CA for plants

Number of living plants	State
0, 1	No change
2, 3	Birth
4 ~ 8	Death

Fig.1 2-D Cellular automaton

Based on this mapping, the state of a cell $c(x, y, t)$ on the 2-D cell space is updated in the following,

$$c(x,y,t+1)=g\{c(x-d,y-d,t),\cdots,c(x,y,t),\cdots,c(x+d,y+d,t)\}$$

(2)

where t is discrete time step and d is the neighborhood range called Moore neighborhood (see Fig.1). In this way, CA evolves according to the state of neighborhood by the transition function. Here we use simple transition rules based on the number of alive plants in the neighborhood shown in Table 1 [6].

Insects also live on the same discrete cell space. It is assumed that an insect has a lethal age and it is removed when its age reaches to the lethal age (L-age). Furthermore, the insect over the breeding age (B-age) can breed one if a plant exists in the neighborhood of the insect. Here if many plants exist in its neighborhood, the insect randomly selects one of them and breeds. Consequently, the change of the population sizes depends largely on the L-age and B-age of insects. Furthermore, we assume that a plant is removed owing to disease by viruses if there is no insect in its range of $u \times u$. Consequently, the numerical balance of both species in the cell space must be maintained. The cell space is divided into 20×20. Initial location of plants and insects is randomly decided according to the probabilities at 0.35 and 0.2, respectively [10].

3 Multiple Robots in Quasi-Ecosystem

3.1 Multi-Objective Behavior Coordination

To maintain the numerical balance of plants and insects, multiple robots are introduced into this quasi-ecosystem (Fig.3). First of all, we explain the structure of a robot. The robot has two types of sensors to detect insects and plants and the number of its sensing directions is 8 neighborhood (Fig.4). In addition, the robot has sensors to detect other robots or obstacles, and sensors for estimating the self-location. Therefore, the robot can take behaviors such as target tracing, collision avoiding, plant reaping, and insect removing. Furthermore, each robot has a parameter for the energy state, $Eng(t)$. The energy of each robot is reduced according to the actions. The robot obtains some energy according to the amount of insects and plants when reaching at a target point. The environment includes some target points with different reward energy functions. We assume that the robot knows the location of all target points, but the robot does not know their reward energy functions. Consequently, a robot might obtain bad rewards, i.e., penalties, if the robot gets too much plant that causes the elimination of the plant species. Therefore, the rth robot should estimate the reward $f_{r,s}$ at the tth target point as follows,

$$f_{r,s} \leftarrow (1-\alpha)f_{r,s} + \alpha \cdot F_s(I_r, P_r)$$

(3)

where $F_s(I_r, P_r)$ is a reward energy function, I_r, P_r, and a are the numbers of the removed insects and reaped plants, and coefficient, respectively. And then, the obtained $F_s(I_r, P_r)$ is added to $Eng(t)$. Therefore, based on the estimated rewards, the robot must remove insects and get plants suitable to the state of its surrounding environment. If the energy of a robot is less than the given threshold, the robot stops without taking any actions. In such a case, the robot can receive some energy from another robot with enough energy in the communication range. For example, if there are many insects in the current environment, the robot should remove many insects, while the robot should not remove if there are few insects. Basically, this numerical balance of quasi-ecosystem is kept by this removal behavior of robots.

The robot has sets of simple if-then rules for plant reaping and insect removing behaviors. The if-then rule concerning insects is described in the following.

If x_1 is $X_{i,1}$ and ... and x_m is $X_{i,m}$ **Then** y is $W_{i,1}$

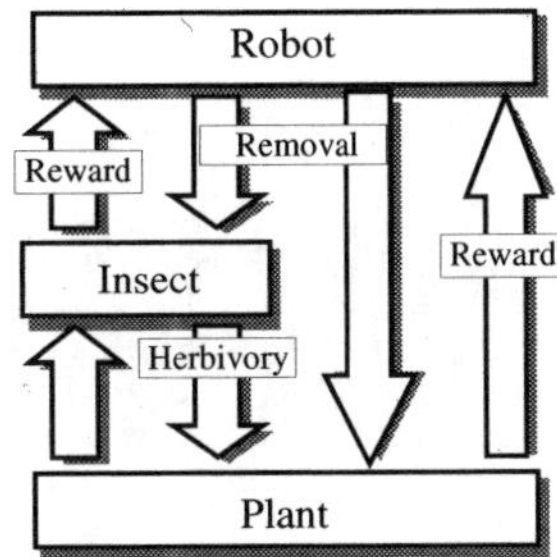

Fig.3 Relationship among plants, insects, and robots

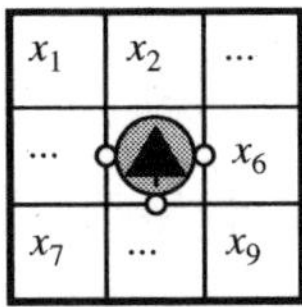
(a) Sensing range

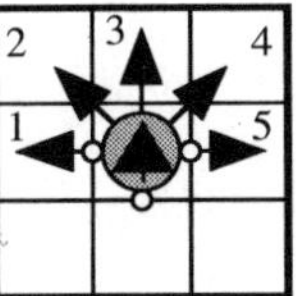
(b) Moving directions

Fig.4 Sensing range and moving directions of a robot in cell space

where $X_{i,j} \in \{0, 1, \#\}$, $W_i \in \{0,1, \dots , 6\}$, and "#" indicates a wild card character. If the condition part is satisfied, the consequent part is used as the output of the robot. Here x_j ($j=1, 2, \dots , 9$) indicates the state of the surrounding cell corresponding to the number in Fig.4. The output characters of W_i correspond to the removal behavior (0), moving directions (1~5), stop (6) shown in Fig.4. The plant reaping behavior is also described using simple if-then rule.

Furthermore, a robot can have fuzzy rules for avoiding collisions with other robots or obstacles [5]. In general, a fuzzy if-then rule based on simplified fuzzy inference is described as follows,

If x_1 is $A_{i,1}$ and ... and x_m is $A_{i,m}$ **Then** y_1 is $w_{i,1}$ and ... and y_n is $w_{i,n}$

where $A_{i,j}$ and $w_{i,k}$ are is a symmetric triangular membership function for the jth input and a singleton for the kth output of the ith rule; m and n are the numbers of inputs and outputs, respectively. Fuzzy inference is generally described by,

$$\mu_{Ai,j}\left(x_j\right) = \begin{cases} 1 - \dfrac{\left|x_j - a_{i,j}\right|}{b_{i,j}} & \left|x_j - a_{i,j}\right| \le b_{i,j} \\ 0 & otherwise \end{cases} \tag{4}$$

$$\mu_i = \prod_{j=1}^{m} \mu_{Ai,j}\left(x_j\right) \tag{5}$$

$$y_k = \frac{\sum_{i=1}^{R} \mu_i w_{i,k}}{\sum_{i=1}^{R} \mu_i} \tag{6}$$

where $a_{i,j}$ and $b_{i,j}$ are the central value and the width of the membership function $A_{i,j}$; R is the number of rules. Outputs of the mobile robot are steering angle and its velocity.

Based on the above behaviors, the robot should take an action from the inference results of various behaviors, that is, the robot requires a perceptual system for recognizing the current situation. We have proposed a multi-objective behavior coordination mechanism [7]. A behavior weight is assigned to each behavior. By extending eq.(6), the output is calculated by

$$y_k = \frac{\sum_{h=1}^{B}\left(wgt_h(t)\sum_{i=1}^{R} \mu_{h,i} \cdot w_{h,i,k}\right)}{\sum_{h=1}^{B}\left(wgt_h(t)\sum_{i=1}^{R} \mu_{h,i}\right)} \tag{7}$$

where B and $wgt_h(t)$ are the number of behaviors and a behavior weight of the hth behavior over the discrete time step t, respectively. By updating the behavior weights, the robot can take a multi-objective behavior according to the time series of perceptual information. Consequently, the recognition of the current situation is performed by assigning behavior weights suitable to the current situation. This method can be considered as a mixture of experts, because the behavior coordination mechanism is considered as a gating network.

3.2 Target Selection

The environment of the quasi-ecosystem includes 9 target points for multiple robots. Each target point gives

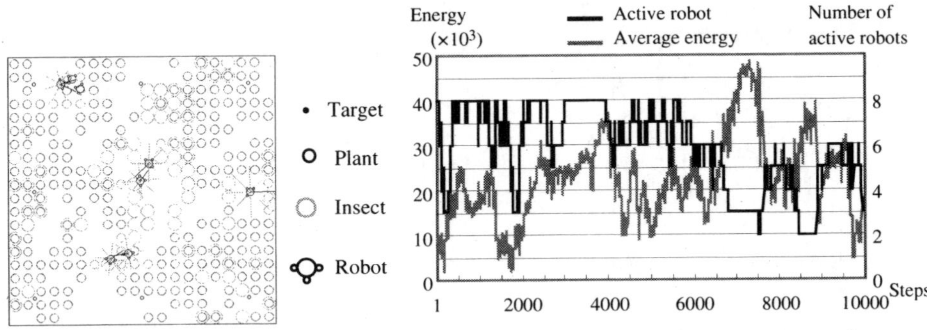

Fig.5 Quasi-ecosystem with multiple robots

Fig.6 The change of average energy and
the number of active robots

different energy rewards to a robot according to its reward functions based on the number of insects and plants the robot has. Therefore, the robot should find the best target point according to the current state of the quasi-ecosystem. Basically, the probability $p_{r,s}$ for the rth robot to select the sth target is calculated by,

$$p_{r,s} = \frac{\exp(f_{r,s} \, / \, T)}{\sum_{i=1}^{M} \exp(f_{r,i} \, / \, T)}$$

(8)

where M and T are the number of targets and the temperature, respectively. Furthermore, the robot should select suitable behaviors from all possible behaviors to obtain high rewards. To do so, the robot decides the behavior dimensions or the validities of behaviors used in the multi-objective behavior coordination when the robot selects a next target point. Here the robot estimates the value of each target and selects the behavior sets like reinforcement learning methods.

4 Computer Simulation

This section shows several simulation results of the robotic behaviors in the quasi-ecosystem. The cell space is divided into 20×20. Initial location of plants and insects is randomly decided according to the probabilities at 0.35 and 0.02, respectively. The number of target points (M) is 9. Here a robot can makes decisions 20 times per step of the plants and insects. The total number of decision times for a robot is 10000. We conduct simulations where the numbers of initial robots is set at 8. The numbers of if-then rules for insect removing and plant reaping behaviors are 10 and 10, respectively. Figure 5 shows a snapshot of the quasi-ecosystem including multiple robots. In the figure, large, medium, and small circles are plant, insect, target point, respectively. The robot can move without considering the cell space. The sensing ranges for the collision avoiding behavior of the robot are drawn from the robot. The full line is used if the robot detects other robot. Otherwise, the dotted line is used.

Figure 6 shows the the change of average energy and the number of active robots. First, the robot has enough energy to move, but the energy gradually decreases. If the robot finds a good target point, the robot gains much energy by estimating the reward of the good target point. Table 2 shows the number of the selected target points during 10000 steps. The most frequently used target point is T2, and the second is T6. At these target points a robot can gain much energy. However, if all of robots select these two target points, the target points would be much crowded. Therefore, some of the robots select other good target points such as T8 and T9. In this way, the negotiation of the target selection among robots can be performed through the estimation of rewards. Furthermore, the target-dependet reward function enables a robot to select behaviors suitable to the current situation by its perceptual system. The perceptual system controls the behavior weights and behavior validity according to the sensing information and rewards.

5 Summary

This paper proposed multi-objective behavior coordination and target selection for multiple robots in a quasi-ecosystem. Furthermore, simulation results show the perceptual system can control the behavior weights and behavior validity suitable to the current state of the quasi-ecosystem in order to maintain numerical balance between plants and insects as well as to gain the energy. However, this paper only shows simulation results,

Table 2 The number of targets selected by robots during 10,000step

	R 1	R 2	R 3	R 4	R 5	R 6	R 7	R 8	Total
T 1	21	7	10	29	12	15	23	28	145
T 2	<u>71</u>	<u>41</u>	<u>56</u>	<u>70</u>	<u>35</u>	<u>41</u>	16	<u>80</u>	<u>410</u>
T 3	29	2	<u>29</u>	<u>67</u>	18	40	49	27	261
T 4	42	12	21	32	<u>37</u>	<u>54</u>	<u>57</u>	54	309
T 5	31	14	22	19	17	19	34	54	210
T 6	<u>72</u>	28	25	47	24	29	39	<u>70</u>	<u>334</u>
T 7	33	10	22	29	18	11	20	24	167
T 8	22	22	19	7	6	19	<u>45</u>	11	151
T 9	0	<u>32</u>	12	10	3	28	0	<u>40</u>	125

and therefore, we have many future works for this study. We will discuss the evolution and learning of if-then rules. Furthermore, we intend to discuss the dynamical relationship between plants, insects, and robots in a quasi-ecosystem in detail.

Acknowledgements

This research is partially supported by the Foundation of the Fusion of Science and Technology, Japan.

References

[1] I.Kelly, O.Holland, C.Melhuish, SlugBot: A Robotic Predator in the Natural World, The 5th International Symposium on Artificial Life and Robotics (AROB 5th), pp 470-475(2000).

[2] M.P.Hassell, The Dynamics of Arthropod Predator-Prey Systems, Princeton University Press, Princeton (1978).

[3] E.Teramoto, K.Kawasaki and N.Shigesada, Switching effect of predation on competitive prey species. Journal of Theoretical Biology 79: pp.303-315 (1979).

[4] C.G.Langton. Artificial Life -An Overview, The MIT Press (1995).

[5] T.Fukuda and N.Kubota, An Intelligent Robotic System Based on A Fuzzy Approach, Proceedings of The IEEE, Vol.87, No.9, pp.1448-1470 (1999).

[6] N.Kubota, M.Ogishi, and F.Kojima, Learning of Multiple Robots in Quasi-Ecosystem, Proc. (CD-ROM) of 2000 26th Annual Conference of the IEEE Industrial Electronics Society, pp.2105-2110 (2000).

[7] N.Kubota, S.Yamaji, F.Kojima, and T.Fukuda, Behavior Learning of Human-Friendly Robots by Symbolic Teaching, Machine Intelligence and Robotic Control, Cyber Scientific, Vol.1, No.2, pp79-86(1999).

[8] D.E.Goldberg, "Genetic Algorithms in Search, Optimization, and Machine Learning", Addison Wesley (1989).

[9] T.Fukuda, N.Kubota, and T.Arakawa, GA Algorithms in Intelligent Robots, Fuzzy Evolutionary Computation (edited by W.Pedrycz): Kluwer Academic Publishers, pp.81-105 (1997).

[10] N.Kubota, M.Mihara, and F.Kojima, Evolutionary Robotics for Quasi-Ecosystem, Proc.CEC2001, pp.115-120, 2001

Author Index

Abbattista, F.	1103	Berthouze, N.	785
Abe, A.	1291	Bevo, V.	26
Abe, S.	1301	Beydoun, G.	458
Abolhassani, H.	801	Bhalla, S.	785
Abran, A.	26	Billhardt, H.	955
Adachi, Y.	765,770	Bisci, D.	229
Aguilar-Ruiz, J.S.	260,275	Blonda, P.	438
Ahmet, K.	311	Bonifacio, M.	448
Ahriz, H.	189	Bonini, L.	75
Aizawa, T.	688	Bosc, P.	3
Akamatsu, N.	343,348,353,364,	Boscolo, A.	1182,1187
	1357,1362	Bouchon-Meunier, B.	879
Akashi, T.	343	Bouquet, P.	448
Akutsu, T.	1048	Bremond, F.	807
Alexandre, A.	1177	Brinkers, M.	922
Al-Jadir, L.	458	Buglione, L.	26
Alonso-Betanzos, A.	1018	Cabrero-Canosa, M.	1023
Al-Rawi, K.R.	522	Cairó, O.	286
Amitani, S.	1281	Carbo, J.	131
Amorim, R.	1028	Carrascal, A	945
Amy, B.	433	Carreiro, P.	250
Ancona, N.	438	Casadio, R.	464
Aoba, M.	219	Castellano, G.	443
Apolloni, B.	296	Castellanos, J.	527
Arana, I.	156,189	Castiello, C.	443
Ardissono, L.	502	Cebreiro, B.	1028
Ardizzone, E.	609	Chan, H.Y.	407
Arguello, M.	1013	Chan, I.-K.	663
Arquero, A.	522	Chang, C.-T.	838,858
Arsene, C.	427	Chang, J.-F.	1132
Artieres, T.	1311	Chang, S.-Y.	838
Aruga, R.	60	Chang, T.M.	653
Azechi, S.	1321	Chashikawa, T.	224
Baba, N.	374	Chen, G.	326,552
Babic, A.	955	Chen, H.	801
Babij, T.M.	1137	Chen, T.-Y.	422,1127
Bai, Y.	136,1110	Chen, W.-C.	678
Balducelli, C.	45	Chen, Y.H.	407
Baratti, R.	848	Chen, Y.-W.	1167,1172
Bargiela, A.	427	Cheng, S.S.	407
Barro, S.	1028	Chien, B.-C.	648
Bassis, S.	296	Chrétien, J.	542
Belhaire, E.	1177	Chu, S.-C.	1132,1152
Benfenati, E.	542	Compatangelo, E.	156,306
Bento, C.	250	Constantin, N.	1120

1404

Conti, V.	1162
Corallo, A.	176
Corchado, E.	245
Corsini, P.	885
Crespo, J.	955
Crippa, P.	1316
Cucchiara, R.	166
Cuel, R.	448
Cunko, K.	184
Czyz, J.	1207
D'Addabbo, A.	438
Dadunashvili, S.	65
Damarapu, S.	992
Damiani, E.	v,176
Day, R.O.	477
Deravi, F.	1217
Des, J.	1013
Desharnais, J.-M.	26
Dharmar, C.	1002
Di Lascio, L.	55
Di Stefano, L.	115
Díaz, F.	517
Díaz, M.A.	532
Dillon, T.	594
Ding, L.	311
Dixit, S.	987,1007
Doeben-Henisch, G.	572
Dorizzi, B.	1311
Dümcke, K.	16
Dumitriu, L.	537
Dzakpasu, R.	927
El-Bakry H.M.	1306
Elia, G.	176
Endo, S.	718
Enk, S.	70
Erasmus, L.	572
Fairhurst, M.C.	1217
Fanelli, A.M.	443
Fang, T.M.	407
Fariselli, P.	464
Fazekas, M.	950
Fernández García de la Rocha, J.G.	1033
Fernandez-Lopez, M.	131
Fernández-Morante, C.	1028
Ferreira, J.L.	250
Ferrer-Troyano, F.J.	260,275
Fiorotto, V.	1182
Fu, H.C.	407
Fuji, A.	1372
Fujihara, N.	1346
Fujii, S.	1043,1048
Fujimoto, N.	817
Fujita, Y.	1242
Fukuda, T.	1387
Fukuhara, T.	1336,1346
Fukuhara, Y.	1157
Fukumi, M.	343,348,353,364, 1357,1362
Fukuta, Y.	348
Fyfe, C.	240,245
Gaito, S.	296
Gallanti, M.	229
Gallinari, P.	1142,1311
García-Remesal, M.	955
Garda, P.	1177
George, J.	1217
Gierl, L.	16,21,453
Giménez, V.	512,517
Gini, G.	542
Giraldez, R.	260,275
Gisolfi, A.	55
Giumelli, M.	542
Gomes, P.	250
Gómez, P.	517
Gong, P.	900
Gonzalo, C.	522
Gonzalo, R.	527
Goy, A.	502
Grana, C.	166
Granado, B.	1177
Guijarro-Berdiñas, B.	1018
Hagihara, K.-I.	817
Hamabe, R.	557,567
Hamamoto, Y.	121
Han, Y.	240
Haque, S.	992,997,1002
Hara, K.	1038
Harada, K.	728
Hasebrook, J.	572
Hasegawa, M.	785
Hashimoto, N.	623
Hata, Y.	600,604,614,618
Hatano, T.	1197
Havran, C.	1207
Hayashi, I.	105
Hernández, C.	527
Hernández-Pereira, E.	1023
Hirano, T.	1367
Hiraoka, S.	567
Hirokawa, Y.	1301
Hirose, A.	638,643
Hisajima, D.	1387

Ho, T.B.	547	Kawaoka, T.	60
Holland, M.	502	Kawasaki, S.	547
Honda, K.	235	Kerre, E.	326,552
Hong, T.-P.	668,683	Khoo, I.	214
Hori, K.	1262,1271,1281,1286	Khosla, R.	791
Housel, T.	291	Kiem, H.	32
Hsu, C.-H.	1137	Kikuchi, Y.	734
Hu, B.	156	Kim, I.C.	40,589
Huang, H.-C.	412,422	Kim, S.	469
Huang, K.C.	1147,1152	Kimura, N.	853
Hupet, L.	1207	Kiso, S.	121
Huynh, V.-N.	1063,1068,1073	Kitamura, Y.T.	604
Hwang, G.-H.	663	Kobashi, S.	600,604,614,618
Hwang, G.-J.	658,663	Kobata, M.	1392
Hyvärinen, A.	397	Kobayashi, F.	1382
Ichihashi, H.	235	Kojima, F.	1382,1387,1397
Ichimura, H.	1048	Kojima, M.	85
Ichimura, T.	688,698,703,708,723	Kolchanov, N.A.	487
Ikeda, T.	204	Komatsu, N.	1202
Ikegami, T.	728	Kondo, H.	562
Imawaki, S.	618	Kondo, K.	600,604,614,618
Inuzuka, N.	754	Kondo, T.	1202
Ionescu, F.	199	Koono, Z.	801
Ishida, Y.	713,739	Kortelainen, J.	960
Ishigaki, H.	614	Kouda, T.	562
Ishii, N.	744,749,759,765,780	Kovalerchuk, B.Y.	487
Ishikawa, M.	618	Kubota, H.	1326
Ishikawa, T.	562	Kubota, N.	1387,1397
Ishimaru, D.	638	Kukkurainen, P.	969
Ishizuka, M.	1257,1267	Kunifuji, S.	827
Ito, J.	895	Kunstic, M.	184
Ito, S.	827,1357	Kuramoto, I.	817
Iwade, S.	85	Kurano, M.	255
Iwata, J.	1043	Kuroda, C.	853
Jain, L.C.	105,402,417,1147	Kuroe, Y.	623
Jain, R.K.	321	Kyuragi, H.	1197
Jajoo, R.	997	L'Abbate, M.	497
Jannach, D.	502	Lai, C.-M.	653
Jedrzejowicz, J.	93	Lama, M.	1028
Jedrzejowicz, P.	93	Lamont, G.B.	477
Jesus Oliveira, E.M.	301	Lazkano, E.	932
Jevtic, D.	184	Lazzerini, B.	885
Jezic, G.	126	Lee, D.D.	387
Jimbo, T.	759	Lee, H.-S.	80
Jurica, P.	922	Lee, J.	1207
Karagiannis, D.	316	Lee, S.-L.	658
Katagiri, H.	1222,1232	Lepera, G.	45
Kato, K.	1222,1227	Li, H.	1142,1311
Kato, Y.	1262	Li, Z.	194
Kawaguchi, M.	759	Liao, C.-H.	663
Kawamae, N.	890	Liao, L.	469

Lien, C.-H.	678
Lin, J.-Y.	648
Lin, N.-Y.	663
Lin, W.-Y.	673
Lo, W.-S.	683
Loia, V.	55,331
Lopez-Alonso, V.	940
Lopez-Campos, G.	940
Lops, P.	1103
Lovrek, I.	126
Ludermir, T.B.	301
Luukka, P.	974,982
Ma, L.	1142
Mabuchi, K.	614
Machida, K.	1197
Maeda, A.	812
Maeda, K.	614
Maeda, T.	105
Malchiodi, D.	296
Mangiavacchi, C.	1187
Manrique, D.	945
Maojo, V.	955
Marcelloni, F.	885
Martelli, P.L.	464
Martín, F.	955
Martín, V.	955
Martínez, A.	527
Martínez, D.	1013
Martínez, E.	522
Martínez, R.	517
Martin-Merino, M.	873
Martín-Sanchez, F.	940
Marui, E.	1098
Mase, K.	827
Massey, L.	161
Masuyama, H.	1296
Matsui, M.	600
Matsumoto, H.	853
Matsumoto, K.	1048
Matsumura, K.	1321,1341
Matsumura, N.	1257,1267,1276
Matsunaga, N.	121
Matsuo, T.	843
Matsuo, Y.	1252,1257,1276
Matsushita, K.	744
Mayers, A.	26
Mazzetti, A.	75,229
Mechefske, C.K.	50
Mei, X.	1127
Meisel, H.	306
Mera, K.	688,708
Michihata, M.	1296
Mihara, M.	1397
Miki, Y.	708
Minami, M.	1377,1392
Mingo, L.F.	532
Mira, J.	1013
Mital, D.P.	987,992,997,1002,1007
Mitani, Y.	121
Mitsukura, Y.	348,364,1357,1362
Miura, M.	1377
Miyagi, H.	718
Miyake, Y.	1276
Miyamoto, S.	1078
Miyoshi, T.	1296
Mizuno, M.	1276
Mizuno, T.	85,1043
Mola, M.	115
Molan, G.	110
Molan, M.	110
Monasterio, F.	512
Monirul Islam, Md.	11
Moreno, L.	940
Moret-Bonillo, V.	1023
Mori, M.	1271,1281,1286
Mori, T.	623
Mosier, K.	987
Mosqueira-Rey, E.	1033
Moura Pires, F.	507
Munemori, J.	822,1053,1058
Muñoz, A.	873
Murai, T.	1078,1083
Murakami, T.	693
Murase, K.	11
Murayama, M.	582
Nachtegael, M.	326,552
Nafalski, A.	321
Nagamoto, K.	1048
Nagamune, K.	604
Nagao, K.	744
Nakada, K.	1048
Nakagami, J.-I.	255
Nakamori, Y.	812,1068,1073
Nakamura, T.	698
Nakamura, Y.	582
Nakanishi, M.	1197
Nakano, T.	754
Nakao, Z.	1172
Nakata, M.	1078,1083
Nakayama, S.	1038
Nasu, Y.	614
Neagu, D.	1088,1093

Nemoto, I.	633
Neri, G.	115
Nguyen, D.D.	547
Nishida, T.	5,1326,1336,1346,1351
Nishizaki, I.	1227,1242,1247
Nishizawa, H.	775
Niskanen, V.A.	965
Nitta, T.	628
Niwa, K.	1247
Noda, J.	817
Oeda, S.	723
Ogaji, S.	141,171
Oh, C.H.	235
Ohsawa, Y.	1267,1276
Ohta, Y.	582
Okamoto, T.	713
Okazaki, N.	1257
Okazaki, Y.	1197
Okuya, K.	775
Olarte, J.G.	286
Orihara, R.	693
Orlov, Y.L.	487
Ozaki, M.	765,770
Ozawa, S.	1301
Pachter, R.	477
Paiva, P.	250
Palade, V.	1088,1120
Palencia, V.	532
Pan, J.-S.	402,412,417,422,
	1127,1147
Pasquariello, G.	438
Pecheanu, E.	537
Pedrycz, W.	331
Pereira, F.C.	250
Pérez, M.	532
Peri, D.	609
Petrovic, S.	336
Piclin, N.	542
Pilato, G.	1162
Pimentão, J.P.	507
Pinna, A.	1177
Pintore, M.	542
Pirrone, R.	609
Poggio, T.	392
Poluzzi, R.	1192
Pozdnyakov, M.A.	487
Prim, M.	1115
Qu, R.	336
Raffone, A.	910
Rakhlin, A.	392
Rao Vemur, V.	7
Resconi, G.	1063,1068,1083
Resta, M.	369
Reyes Salgado, G.	433
Riera, A.	1028
Rifqi, M.	879
Ríos, J.	945
Riquelme, J.C.	260,275
Rodríguez-Baena, D.S.	260,275
Rodríguez-Pedrosa, J.	955
Roig, J.	1115
Rossi, C.	945
Roy, N.K.	321
Rubio, M.	517
Russo, D.	1182
Saastamoinen, K.	974,982
Sagara, T.	557
Saito, K.	1197
Saitou, A.	547
Sakano, H.	146,151
Sakawa, M.	1222,1227,1232,
	1242,1247
Sakthivel, P.	796
Sampath, S.	141,171
Sánchez, E.	1028
Sanchez-Miralles, A.	265
Santangelo, A.	55
Sanz-Bobi, M.A.	265
Sasaki, H.	749
Sasaoka, H.	843
Satalino, G.	438
Sato, A.	146,151
Sato, F.	85
Sato, H.	1362
Sato, Y.	1083,1232
Satoh, T.	204
Saul, L.K.	387
Savi, A.	1192
Sazaklis, G.	1212
Schäfer, R.	502
Schmidt, R.	16,21,453
Seco, N.	250
Semeraro, G.	1103
Servida, A.	848
Sessa, S.	331
Seung, H.S.	387
Shahjahan, Md.	11
Shankar, S.	987,1007
Shieh, C.-S.	422
Shigematsu, S.	1197
Shigenobu, T.	1053
Shimizu, K.	734

Shimizu, T.	85
Shimmin, J.	214
Shimooka, T.	734
Shinyama, S.	1382
Shirakawa, T.	1262
Shiratori, N.	728
Shoji, H.	1271,1276,1286
Sierra, B.	932
Silva, A.	955
Simeoni, R.	502
Singh, R.	141,171
Sinkovic, V.	126
Soler, V.	1115
Son, D.M.	32
Son, N.H.	32
Sorbello, F.	1162
Sousa, A.	955
Sousa, P.A.C.	507
Sridharan, D.	796
Srinivasan, S.	992,997,1002
Srivatsa, S.K.	796
Stanojevic, M.	381
Stefan, M.	270
Stefanescu, D.	537
Stefanoiu, D.	199
Steffen, D.	16
Steiger-Garçao, A.	507
Sterling, G.	594
Stoop, R.	905
Su, Y.-M.	668
Suárez-Romero, J.A.	1018
Sueda, N.	693
Suenaga, T.	146,151
Suganuma, S.	1073
Sugiyama, K.	812
Suka, M.	698
Suksmono, A.B.	638
Sumi, Y.	827
Sun, S.	1127
Suto, H.	1197
Suzuki, H.	1377
Suzuki, T.	614
Taboada, M.	1013
Tabus, I.	199
Tadokoro, S.	353
Tajima, N.	698
Takahashi, M.	1372
Takefuji, Y.	209,219,224,280, 1157
Takeoka, S.	770
Takeuchi, H.	780
Tamamura, A.	1392
Tamashiro, H.	1167
Tanaka, M.	427
Taniguchi, M.	1346
Tao, Y.-H.	668
Tarantino, C.	438
Tate, S.	280
Tateyama, T.	1167
Terada, K.	1351
Testoni, G.	848
Thiel, U.	497
Thomopoulos, S.C.A.	1212
Thonnat, M.	807
Timusk, M.A.	50
Tolonen, Y.	960
Toma, N.	718
Tomašević, V.	381
Tomb, J.-F.	469
Tomimatsu, K.	567
Tomita, J.	204
Tomita, S.	863
Tomobe, H.	1257
Torres, C.	512
Traphöner, R.	492
Tronci, S.	848
Tseng, C.L.	407
Tseng, J.C.R.	658
Tseng, M.-C.	673
Tseng, S.S.	678
Tsuda, K.	1367,1372
Turchetti, C.	1316
Tyukin, I.	917
Ueda, K.	121
Uemura, S.	1336
Ulian, M.	1187
Umano, M.	105
Umeno, M.	759
Ursu, M.F.	98
Usmanji, P.	791
Vago, D.	1192
Valasek, M.	270
Valente, A.	291
Valentini, G.	482
Van der Weken, D.	326,552
Van Leeuwen, C.	900,910,917,922
Vandendorpe, L.	1207
Verleysen, M.	1207
Viarani, E.	115
Vicoli, G.	45
Vila, J.	1028
Vishnevsky, O.V.	487

Vitabile, S.	1162	Yanagida, T.	604
Vityaev, E.E.	487	Yang, T.-C.	402,1147,1152
Vraneš, S.	381	Yano, H.	1237
Vu, V.-T.	807	Yasuda, H.	121
Wagner, C.	905	Yasuda, M.	255
Wang, F.-H.	402,412,417	Yeo, G.	392
Wang, S.	1073	Yonezawa, Y.	703
Wang, S.-L.	683	Yoshida, K.	688,698,723,1043,1048
Wang, Y.-F.	858	Yoshida, Y.	255
Watabe, H.	60	Yoshie, M.	688
Watanabe, Y.	739	Yoshiike, N.	209
Woitsch, R.	316	Yoshino, T.	822,1053,1058
Wu, P.-Y.	1132	Yoshioka, T.	775
Yamaba, H.	863	Yue, Y.	311
Yamada, K.	85	Yuizono, T.	1038
Yamada, K.	718	Zambetta, F.	1103
Yamamoto, H.	1098	Zanker, M.	502
Yamashita, K.	703	Zeng, X.-Y.	1167,1172
Yamashita, K.	1326,1331	Zhang, L.	562
Yamashita, T.	688,703,708,723	Zhang, Y.	136,1110
Yamashita, Y.	833	Zhang, Z.	194
Yamauchi, K.	780	Zhou, L.	868
Yamawaki, S.	358	Zimmer, R.	98
Yamazaki, Y.	1202	Zochowski, M.	927
		Zydallis, J.B.	477

Intelligent Knowledge Management

Edited by

Ajith Abraham
Maumita Bhattacharya
Lakhmi Jain

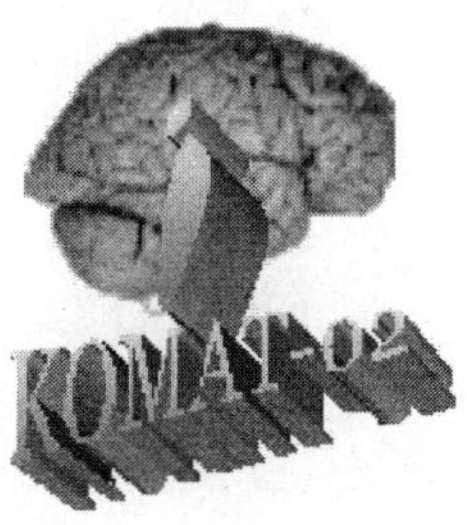

IKOMAT'02
2002 International Workshop on Intelligent Knowledge
Management Techniques

In conjunction with KES'02

16-18 September 2002, Podere d'Ombriano, Crema, Italy

Preface

Since its recent inception as a structured field, Knowledge Management (KM) has fast been recognised as the most essential tool for universal knowledge workers. The concept of KM, starting with its deep association with corporate information management, still carries multiple, even conflicting interpretations. The most popular one being a structured field that encompasses processes and techniques for knowledge discovery, indexing, organisation, and fusion. Where the classical approach to knowledge management tends to rely on techniques like concept maps, hypermedia and object-oriented databases, artificial intelligence techniques for core KM activities like knowledge discovery, organisation, and knowledge fusion are rapidly gaining popularity. In the evolved scenario, KM may be interpreted as a field that deals with acquisition, storage and application of knowledge for a range of knowledge intensive tasks – whether that be decision support, learning or research support. A very recent trend is the fast emergence of a second generation of Knowledge Management. This is quite interesting and has two thrusts. The first thrust is the people-centric focus on how knowledge really is used by people to handle situations effectively (sense making, decision making, execution, and monitoring). The second thrust involves the application of intelligent inanimate processes – essentially the major theme of IKOMAT' 2002.

IKOMAT'02, the First International Workshop on Intelligent Knowledge Management Techniques, took place in Crema (Italy), September 16-18, 2002. The theme of the workshop was "Intelligent Knowledge Management" using computational intelligence, knowledge based systems, artificial intelligence paradigms, modern heuristics and so on. The papers presented here reflects the aim of IKOMAT'02 to offer a premier technical forum to researchers/developers and practitioners of computational intelligence and knowledge management techniques. IKOMAT'02 attracted 57 full papers from over 23 countries and each paper was peer reviewed by at least two independent referees. The papers have been a good mix of contributions by researchers from both academic and industrial background, including few interesting documentations of major real life projects. Based on the evaluation process and the recommendation of the reviewers, 34 papers were finally included in the workshop program and the major topics are as follows:

- Knowledge Management, Knowledge Engineering and Related Systems

- Fuzzy Systems in Knowledge Management

- Knowledge Management Using Connectionist Paradigms and Adaptable Systems

- Evolutionary Computation in Knowledge Discovery

- Ontology, Data Mining and Image Processing

We express our sincere thanks to the authors, technical sponsors and other organizations, for their support, which is behind the successful realization of IKOMAT' 02. The workshop was technically co-sponsored by The Knowledge Based Intelligent Systems Centre, Australia. We are deeply indebted to the members of the International technical committee and all those in the community without whose help IKOMAT'02 would not have been possible. We would like to express our gratitude to all the reviewers for the tremendous service by critically reviewing the papers within the stipulated deadline. We are grateful to Professor R.J. Howlett (KES Secretariat at the University of Brighton, UK) and Professor E. Damiani (General Chair of KES'02) for the support and timely advices, making IKOMAT'02 a reality. Our special thanks to IOS Press, Netherlands for their excellent cooperation to produce this important scientific work. Last but not the least, we would like to express our gratitude to our colleagues at Monash University, Australia, Department of Computer Science and Software Engineering, The University of Melbourne, The KES Secretariat, University of Brighton, UK and The Department of Information Technology, University of Milan, Italy, for supporting us in the organization of IKOMAT' 02.

Ajith Abraham, Maumita Bhattacharya and Lakhmi Jain (Editors)
June 2002

IKOMAT'02 Organizing Committee

Workshop Chairs

Honorary Chair

Lakhmi Jain
Knowledge-Based Intelligent Engineering Systems Centre
University of South Australia, Adelaide
Mawson Lakes, South Australia 5095, Australia
Phone +61 8 8302 3315, Fax +61 8 8302 3384
Email: L.Jain@unisa.edu.au, Web: http://www.kes.unisa.edu.au

Organizing Chairs

Maumita Bhattacharya
School of Computing and Information Technology
Monash University (Gippsland Campus), Victoria 3842, Australia
Phone: +61 3 9902 6396, Fax: +61 3 9902-6842
Email: M.Bhattacharya@infotech.monash.edu.au

Ajith Abraham
School of Business Systems, Faculty of Information Technology,
Monash University (Clayton Campus), Victoria 3168, Australia
Phone: +61 3 9905 9766, eFax: +1 (509) 691 2851
Email: ajith.abraham@ieee.org, Web: http://ajith.softcomputing.net

Program Chairs

Frada Burstein
School of Information Management and Systems
Monash University (Caulfield Campus), Victoria 3145, Australia
Phone: + 61 3 9903 2011, Fax: +61 3 9903-2005
Email: Frada.Burstein@sims.monash.edu.au,
Web: http://www.sims.monash.edu.au/staff/fb/

Baikunth Nath
Department of Computer Science and Software Engineering
The University of Melbourne
Melbourne, Victoria 3010, Australia
Phone: +61 3 8344 9316
Email: Baikunth@unimelb.edu.au, Web: http://www.gscit.monash.edu.au/~bnath/

IKOMAT'02 is technically co-sponsored by

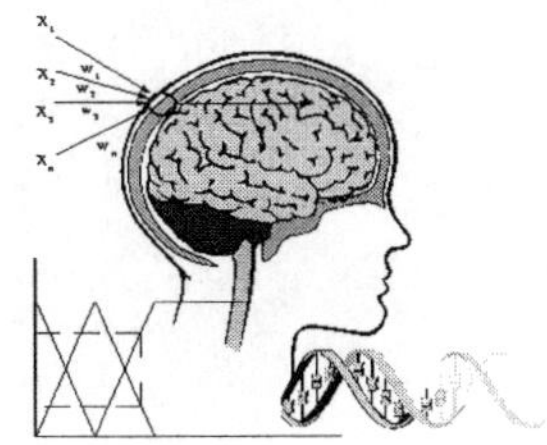

Knowledge-Based Intelligent Engineering Systems Centre
University of South Australia, Adelaide
Mawson Lakes, South Australia 5095, Australia
Web: http://www.kes.unisa.edu.au

International Technical Committee

José Manuel Benítez, University of Granada, Spain
Lakhmi Jain, KES Center, University of South Australia, Australia
Sankar K Pal, Indian Statistical Institute, India
Kate Smith, Monash University, Australia
Rajkumar Roy, Cranfield University, United Kingdom
Frada Burstein, Monash University, Australia
Nada Lavrac, Jozef Stefan Institute, Slovenia
Manuel Grana Romay,University of the Basque Country, Spain
Baikunth Nath, The University of Melbourne, Australia
Janos Abonyi, University of Veszprem, Hungary
Gleb Beliakov, Deakin University, Australia
Vijayan Asari, Old Dominion University, USA
Alvaro del Val, Universidad Autónoma de Madrid, Spain
Saeid Belkasim, Georgia State University, USA
Mario Köppen, Fraunhofer IPK-Berlin, Germany
Ajith Abraham, Monash University, Australia
Maumita Bhattacharya, Monash University, Australia
Xiong Wang, California State University, USA
Tanja Urbancic, Jozef Stefan Institute, Slovenia
Morshed U Chowdhury, Deakin University, Australia
Ashish Ghosh, Indian Statistical Institute, India
Xiao Zhi Gao, Helsinki University of Technology, Finland.
Costa Branco P J, Instituto Superior Technico, Portugal

IKOMAT'02 Additional Reviewers

Fabio Abbatista, Universita di Bari, Italy
Katrin Franke, Fraunhofer IPK-Berlin, Germany
Zorica Nedic, University of South Australia, Australia
Son Kuswadi, Tokyo Institute of Technology, Japan
Bhanu Prasad, Georgia South-western State University, USA
Yanqing Zhang, Georgia State University, Georgia
Emma Regentova, University of Nevada, Las Vegas, USA
Ninan Sajith Philip, Cochin University of Science and Technology, India
Chris Cornellis, University of Ghent, Belgium
Bernadette Garner, Monash University, Australia
Arpad Kelemen, University of Memphis, USA
Lance Chambers, Australia
Nicoletta Del Buono, Università degli Studi di Bari
Shonali Krishnaswamy, Monash University, Australia
Hepu Deng, RMIT, Australia
Sergio Viademonte, Monash University, Australia

Contents

Preface 1413
IKOMAT'02 Organizing Committee 1415

Knowledge Management, Knowledge Engineering and Related Systems

Managing Business Intelligence in a Virtual Enterprise Model: A Case Study and
Knowledge Management Lessons Learned, *M. Jermol, N. Lavrač and
T. Urbančič* 1419
Knowledge Engineering for Real Time Intelligent Control, *J.M. Evans,
E.R. Messina, J.S. Albus and C.I. Schlenoff* 1424
A Modelling Framework for Capturing the Design Rationales of Problem Solving
Processes, *K. Seta and O. Kakusho* 1429
Multimedia Platform to Support Knowledge Processes Anywhere and Anytime,
M. Mesenzani, T. Schael and S. Albolino 1434
κ-ShaRe: An Architecture for Sharing Heterogeneous Conceptualisations,
E. Compatangelo and H. Meisel 1439

Fuzzy Systems in Knowledge Management

Use of Fuzzy Sets to Measure the Degree of Inclusion between Vague Concepts in
Semantic Nets, *M-N. Omri and M.A. Mahjoub* 1444
Similarity Measures for Fuzzy Values, *J. Dvořák and M. Šeda* 1449
Rule-based Valid-Time Expert Systems with Multicriteria Fuzzy Sets, *I.V. Filis,
C.P. Yialouris and A.B. Sideridis* 1454
Possibilistic Abduction Versus Fuzzy Pattern Matching, *C. Segal* 1461
SAR Images Classification using Fuzzy Subsethood Operator, *G. Angiulli,
V. Barrile and M. Versaci* 1466

Knowledge Management using Connectionist Paradigms and Adaptable Systems

GCS Networks, Views, and Path Planning, *C. Baroglio* 1471
An Artificial Neural Network Algorithm for Track Reconstruction in the ALICE
Inner Tracking System, *A. Pulvirenti, A. Badalà, R. Barbera, G. Lo Re,
A. Palmeri, G.S. Pappalardo and F. Riggi* 1476
AppART + Growing Neural Gas = High Performance Hybrid Neural Network for
Function Approximation, *L. Martí, A. Policriti, L. García and R. Lazo* 1483
Qualitative Traffic Analysis using Image Processing and Time-Delay Neural Network,
S. Navabzadeh Razavi and M. Fathy 1488
Adaptive Neuro Fuzzy Inference System Applied to EEG Signals for Estimating
Movement-Related Potentials, *D.D. Ben Dayan Rubin, G.F. Inbar and
S. Cerutti* 1497

1418

Support Vector Machine and Neural Network with Others for Two Dimensional
Classification Problem: An Empirical Study, *A.B.M. Shawkat Ali* — 1502
Forecasting of North Northeast Brazil Rainfall Anomalies: A Hierarchical
Neuro-Fuzzy Model Application, *F.J. de Souza and M.L. Velloso* — 1507
Lattice-Gas Cellular Automata Approach for Fluid Flow in Porous Media,
S.N. Khotimah, I. Arif and H.L. The — 1512
A Case-based System for Asthmatic Patient Health Care, *I. Sefion, M. Gailhardou
and A. Ennaji* — 1518

Evolutionary Computation in Knowledge Discovery

A New Heuristic Genetic-based Algorithm for Constrained Multicast Tree
Construction, *S.M. Youssef, M.A. Ismail and S.A. Bssiouny* — 1523
Boundaries Detection based on Polygonal Approximation by Genetic Algorithms,
W. Barhoumi and E. Zagrouba — 1529
Skill-based Resource Allocation using Genetic Algorithms and Ontologies,
K. Nammuni, J. Levine and J. Kingston — 1534

Ontology, Data Mining and Image Processing

Re-usable Knowledge: Development of an Object Oriented Industrial KBS and a
Collaborative Domain Ontology, *P. Crowther, G. Berner and R. Williams* — 1539
Ontology-based Communication Forum, *M. Mach, P. Macej and J. Hreno* — 1544
Relational Text Mining and Visualization, *B. Kovalerchuk* — 1549
Content-based Image Retrieval Based on Color Histogram and Discrete Cosine
Transform, *G. Sorwar, M. Murshed and L. Dooley* — 1555
A Variable Pattern Selection Algorithm with Improved Pattern Selection Technique
for Low Bit-Rate Video-Coding Focusing on Moving Objects, *M. Paul,
M. Murshed and L. Dooley* — 1560
Content Based Image Retrieval using Discrete Wavelet Transform, *X. Hong,
S. Belkasim and O. Basir* — 1565
Causal Possibilities Model Structures, *L.J. Mazlack* — 1571

Author Index — 1577

KES 2002
E. Damiani et al. (Eds.)
IOS Press, 2002

Managing Business Intelligence in a Virtual Enterprise Model: A Case Study and Knowledge Management Lessons Learned

Mitja JERMOL (1), Nada LAVRAČ (2) and Tanja URBANČIČ (2,3)
(1) Perenič Consulting, Mestni trg 8, 1000 Ljubljana, Slovenia
(2) J. Stefan Institute, Jamova 39, 1000 Ljubljana, Slovenia
(3) Nova Gorica Polytechnic,Vipavska 13, p.p. 301, 5001 Nova Gorica, Slovenia

Abstract. This paper presents the need for knowledge management in a virtual enterprise model for networking of international expert teams from academia and business in the area of data mining and decision support. The knowledge management aspects of the business intelligence as implemented in the virtual enterprise model are analysed from the point of view of (a) the need for the construction and management of a knowledge map of the available tools, expertise and collaborative work procedures, (b) the cognitive authority aspect in collaborative work management, as well as (c) the network intelligence aspect of the virtual enterprise endeavour. The paper concludes by discussing the value network aspect and subsequent knowledge spaces, as well as the potential of AI methods of data mining and decision support in knowledge management of a virtual enterprise.

1. Introduction

Research and development projects as well as partnership networks, such as EU funded European Networks of Excellence; support the collaboration of international expert teams from academia, business and industry. The EU funded project "Data Mining and Decision Support for Business Competitiveness: A European Virtual Enterprise" (IST-1999-11495 project Sol-Eu-Net, http://soleunet.ijs.si), used as a case study in this paper, aims at forming a dynamic network of expert teams with long term experience in data mining (DM) [1] and decision support (DS) [2] whose functionalities are complementary and oriented towards solving difficult practical problems. The organizational model of this partnership is a virtual enterprise model [3],[4], in which the participating partners join their efforts and expertise in developing methods, problem-solving protocols and practical DM and DS solutions, aimed at increasing their visibility and success in the market. The novelty of this virtual enterprise model lies in a flexible association of academic institutions and business entities which (although having different motivations for this partnership) share the main objective of promoting and selling advanced services offered by the pool of partners.

In knowledge intensive services, success critically depends on recognizing partners' expertise, tools and skills as marketable knowledge assets. The virtual enterprise has to solve the problem of efficiently storing, updating, sharing, promoting and transferring knowledge. In addition to technological solutions, organizational, economic, legislative, psychological and cultural issues have to be addressed as well [5]. Appropriate knowledge management [6] leads to a quick recognition of a business opportunity and timely response (e.g., a business offer), so that geographic dispersion of clients and expert teams need not necessarily be a limit to successful business operations.

The partnership model developed in the RTD project Sol-Eu-Net aims at alleviating the problem of partnership discontinuation after the end of the EU funding period. A virtual enterprise (VE) model has been proposed to support business activities of a pool of DM and DS experts, following the VE definition as a temporary aggregation of core competencies and associated resources collaborating to address a specific situation, presumed to be business opportunity [7]. This model includes protocols and standards for partner collaboration in distributed DM and DS projects, and many other issues, including protocols for information and knowledge management, which are in the main scope of interest of this paper.

In this paper, the business intelligence as implemented in the virtual enterprise model is analysed from the point of view of the need for a knowledge map tools, expertise and collaborative work procedures (Section 2), the cognitive authority aspect in collaborative work management (Section 3), as well as the network intelligence aspect of the virtual enterprise endeavour (Section 4). The paper concludes with discussing the value network aspect and subsequent knowledge spaces and the potential of AI methods of data mining and decision support in knowledge management of a virtual enterprise.

2. Knowledge Map as a Tool for Successful Virtual Enterprise Management

The network of competences of the Sol-Eu-Net virtual enterprise has three levels: strategic, management and organisational level. For the RTD project, the strategic level was executed by the project coordinators, guided by the project technical annex as well as the evaluators' guidelines. The problem of the virtual enterprise strategic management remains an open challenge for the virtual enterprise formed as a result of the RTD project. The second level is formed by the DS and DM "core competence" partners, while on the third, physical level, there are partners supporting traditional business functions like marketing, accounting, finance, information infrastructure, etc. The basic asset in a knowledge based enterprise like Sol-Eu-Net is not just knowledge, which resides in the heads of individuals, but also group and team knowledge, organisation knowledge and, because of a highly dispersed and multicultural enterprise, the cultural knowledge.

In a virtual enterprise, the added-value and the main strength of each partner is due to linking his expertise with supplementary and complementary knowledge of other partners. Moreover, also the most important strength of the whole virtual enterprise lies in the range of competencies the partners can offer together. The wider is the network of knowledge the more competitive is the enterprise. But this can be fully exploited only if there is well structured and updated information about the competencies available to the managers that have to respond to particular business opportunities as well as to clients looking for a special kind of products and services. This is why a "catalogue" of each partner's knowledge has to be made accessible to other partners, in particular to the strategic decision makers, managers of the virtual enterprise and the marketing people.

To be able to manage the virtual enterprise, an appropriate knowledge map has to be formed, in order to provide "a visual representation of a knowledge domain according to criteria that facilitate the location, comprehension or development of knowledge" [8]. Building the knowledge map of partners core competencies, additional expertise, tools and procedures in terms of explicit knowledge will cover the most of the knowledge management and business needs.

In Sol-Eu-Net virtual enterprise, several protocols have been developed for gathering information from the partners. In practice, this is done by filling-in the electronic forms, which are then used to update and upgrade the Sol-Eu-Net database. As a result, there are several repositories available: Repository of DM and DS tools, Repository of DM and DS solved problems, Sol-Eu-Net on-line library and Repository of educational modules. How to scan and put on the map other types of knowledge like tacit, group, organisational and cultural knowledge remains one of the main challenges for future work. Before a better and more systematic way is established, the Sol-Eu-Net partners have been trying to systematically gather the lessons learned in any form, mostly in the form of papers and reports, just to keep the organizational memory up-to-date and at disposal to the other partners and the professional communities as a whole. Considerable efforts have been devoted to the design of a systematic and powerful way of capturing the lessons learned in a way that would, at a later stage, enable their inclusion into the knowledge map. A current effort in this direction is the gathering of the lessons learned in solving particular DM and DS problems as part of the DM and DS solved problems repository.

3. Cognitive Authority

When products and services of a virtual enterprise are highly knowledge intensive, it is difficult to decide whether the partners connected in such an enterprise are capable to respond to a business opportunity successfully or not. Unlike in businesses that deal with well-defined material products, it is usually not sufficient just to look into a catalogue and find with a simple search if the enterprise has at disposal what a particular client needs. Problems that clients have are typically expressed in the language of end users, not in the language of methods and tools that should be applied. Besides, in typical applications that are very similar to the previously solved cases stored in such a catalogue, considerable level of knowledge and experience is needed to see the similarities between a new problem and an already solved problem. The problem has to be "translated" into the language of problem solvers and checked against the cases that might look very different at a first sight, but bear some similarities when looked at by a highly skilled professional. This indicates that there is a missing link between the knowledge map and a problem description that has to be overcome by a Cognitive Authority [9]. Similarly, when dealing with knowledge based virtual enterprises involved in co-operation trough an extensive network of knowledge resources then we are facing a knowledge overload problem: which partners to choose and how to manage knowledge and the processes become the task for a Cognitive Authority.

We found out in Sol-Eu-Net that a central Cognitive Authority can be replaced by a network of DM and DS authorities: when a new business opportunity arises, Sol-Eu-Net partnership uses electronic means (a workgroup support system ZENO) for discussing a new problem, trying to find out the best methods, techniques, tools and expert teams to develop a problem solution, using the collaborative DM and DS methodology being developed in the course of the project. We also realized that there is a strong dependency between Cognitive Authority and the management of business intelligence, especially when dealing with the management of collaborative work in highly dynamic business environment. Cognitive Authority is strongly related to the notion of Network Intelligence (NQ) described below.

4. Network Intelligence of a Successful Virtual Enterprise

In traditional enterprises the culture of the company is mainly influenced by organisational behaviour, characterised by long-term involvement, very specific rules, top-down decisions and segmentation. On the other hand, in virtual enterprises the form of work is networking. Because of the unbounded short-term relations between partners, each individual is expected to share most of her/his knowledge resources. The culture of the enterprise and the motivation of the individual have to be shifted to the network culture. This is why knowledge management must tackle the potential for deeper social fragmentation that is inherent in technological networks. For this reason Joy Palmer has developed the concept of network intelligence, defined as "the capacity for connecting to others" [10].

Human intelligence (IQ) is a familiar and well understood notion. We are becoming increasingly aware of the influence the emotional intelligence (EQ) has on the business relations, but the network intelligence (NQ) concept of sharing is remote to us, since "NQ gives us the ability to make sense of experience beyond the narrow confines of specific and fixed identity. It enables us to build human networks based on group phenomena such as dialogue, mutuality, and trust" [10]. As such, NQ, in the same way as EQ, looks far beyond the IQ.

In Section 2 we argued that the main strength of the whole virtual enterprise lies in the range of competencies the partners are able to offer jointly. Here the key is actually in the word "jointly". In the Sol-Eu-Net virtual enterprise, there is full awareness of the fact that the advantage of having several partners is not just having a bigger collection of tools, methods and solved cases. Rather, this is amplified by active exchange of knowledge and experience at joint seminars and workshops, and mostly by developing and implementing different procedures for collaborative work, including collaborative solving of DM and DS problems. In the Sol-Eu-Net virtual enterprise many partners are active researchers. Their work culture is lined with the research community culture in which sharing ideas and knowledge is of vital importance. However, when faced with the commercial exploitation of research results, academic and commercial partners establish a principal-agent relationship, facing many caveats to success [11]. The problem is well known, and it can be to a certain extent solved by carefully elaborated contracts, providing clear solutions in terms of IPR. However, there will never be possible to regulate everything by contracts. Here, NQ turns out to be the crucial factor that can even mean success or failure of a virtual enterprise. The Sol-Eu-Net virtual enterprise has already gone through several phases, looking for the best ways to deal with these issues, and more are to be faced in the future. Therefore, continuous development and evaluation of the model of operation is one of the important tasks that has to take into the account the fact that developing NQ is essential, but a very long lasting process. We found out that NQ as well as EQ cannot be taught or learned but it has to be carefully thought of and developed to raise the network culture. The individuals working in networked enterprises, as well as future networked global citizen needs to be raised up with network culture values such as identity, diversity and self-respect.

5. Conclusions, Revealing the Potential of AI Methods and Tools in Knowledge Management of a Virtual Enterprise

As pointed out in Section 2, it is much easier for such a virtual enterprise to compete in the market if its functionality is built upon a Knowledge Map. A Knowledge Map enables decision makers to estimate quickly if the consortium networked in a virtual enterprise is capable to respond to a particular business opportunity. In the most optimistic scenario, the knowledge map leads towards a positive answer in a relatively direct manner.

The traditional corporate value chain from the industrial age has been transformed into the information chain in the digital age. In the "networked age", the value chain further evolves to a value network, or web, in which companies engage in multiple two-way relationships. All three business models have a common ground: fixed relations. A Value Network is in fact forming a "business layer" on top of the fixed or flexible web of partners. In an ultimate liberal economy each individual or organisation would form one of the nodes in the web and be presented by his/her expertise and tools, referred to as a Knowledge Space.

A virtual enterprise is actually a one-path business process or an "added-value" flux over the web of Knowledge Spaces. So the web is not formed from nodes of individuals or organisations anymore but rather from Knowledge Spaces and functions. Since knowledge and beliefs are changing permanently, the web becomes highly complex and self-organized. To manage business processes in such an environment, a suitable knowledge management model needs to be developed based on few rules and a lot of freedom.

Above-mentioned theses and experiences have been learned from the Sol-Eu-Net case. In our experience, known knowledge management models are too focused on the core competencies and subsequent processes what make the enterprise too rigid. In future development we will try to implement the above-mentioned paradigm together with the "real business" motivation that makes a Value Network effective.

Sol-Eu-Net virtual enterprise has a strong competitive advantage in mastering DM and DS methods: besides solving the problems of clients, they can contribute also to better knowledge management of the virtual enterprise itself. For example, when faced with a new problem to be solved, advanced technologies of web mining, mastered by the Sol-Eu-Net partners, make it possible to find complementary expertise needed but unavailable within the current network. A virtual enterprise with technical aid of these technologies and with the developed NQ will be able to recognize potential partners and invite them to cooperate whenever appropriate. Furthermore, DM can be used to mine the project databases and the knowledge map to uncover some hidden regularities. For instance, the database of business participants of education and training courses can be used to find subgroups of individuals with common interests, to whom to address a marketing campaign promoting our expertise, education offer and business solutions. This is a step to be made in the future.

References

[1] Fayyad U, Piatetski-Shapiro G, Smith P, and Uthurusamy R (eds.) Advances in Knowledge Discovery and Data Mining. MIT Press, Cambridge, MA, 1996.

[2] Mallach EG. Understanding Decision Support Systems and Expert Systems. Irwin, 1994.

[3] Camarinha-Matos LM, Afsarmanesh H, Rabelo R. Supporting agility in virtual enterprises. Proceedings of PRO-VE 2000 – 2nd IFIP Working Conference on Infrastructures for Virtual Enterprises. Florianopolis, Brasil, 4-6 Dec. 2000.

[4] Camarinha-Matos LM, Afsarmanesh H, Rabelo R (eds). E-Business and Virtual Enterprise: Managing Business-to-Business Cooperation. Kluwer Academic Publishers, 2000.

[5] McKenzie J, van Winkelen C. Exploring E-collaboration Space. Henley Knowledge Management Forum, 2001.

[6] Smith RG, Farquhar A. The Road Ahead for Knowledge Management: An AI Perspective. AI Magazine, Vol. 21, No. 4, 17-40, 2000

[7] Goranson HT. The Agile Virtual Enterprise: Cases, Metrics, Tools, Quorum Books, 1999.

[8] Eppler, Martin J.: Knowledge Management Terminology. Guide, St. Gallen; mcm institute, 1999, 07/2000. URL: <http://www.netacademy.org/netacademy/publications.nsf/all_pk/1617> [03/27/2002].

[9] Wilson, P. Second-Hand Knowledge: An Inquiry Into Cognitive Authority (Chapters 1 & 2, pp. 1-37), Westport, Conn.: Greenwood Press, 1983

[10] Palmer, J, The Human Organization Journal of Knowledge Management, Vol. 1, Number 4, June 1998,

[11] Lavrač N., Urbančič T., Orel A., Virtual Enterprise for Data Mining and Decision Support: A Model for Networking Academia and Business, Proceedings of the Third IFIP Working Conference on Infrastructures for Virtual Enterprises (PRO-VE 2002) (in press).

KES 2002
E. Damiani et al. (Eds.)
IOS Press, 2002

Knowledge Engineering for Real Time Intelligent Control

John M. Evans, Elena R. Messina, James S. Albus, Craig I. Schlenoff
john.evans, elena.messina, james.albus, craig.schlenoff @nist.gov
Intelligent Systems Division
National Institute of Standards and Technology
Gaithersburg, MD 20899-8230

Abstract. The key to real-time intelligent control lies in the knowledge models that the system contains. Three main classes of knowledge are identified: parametric, geometric/iconic, and symbolic. Each of these classes provides unique perspectives and advantages for the planning of behaviors by the intelligent system.

1. Introduction

The concept of intelligence in control applies to a variety of approaches to extending classical control theory that include learning, non-linear control, model-based control, and, in general, control of complex systems that will "do the right thing" when confronted with unexpected or unplanned situations [1]. It can be said that all "intelligent" systems have some knowledge of the system to be controlled or that they use some model of the system in calculating control outputs. In fact, the American Heritage Dictionary defines intelligence as "the capacity to acquire and apply knowledge."

Creating, capturing, and using the knowledge of the system to be controlled is one branch of what is known as knowledge engineering. The real-time aspects of control make this

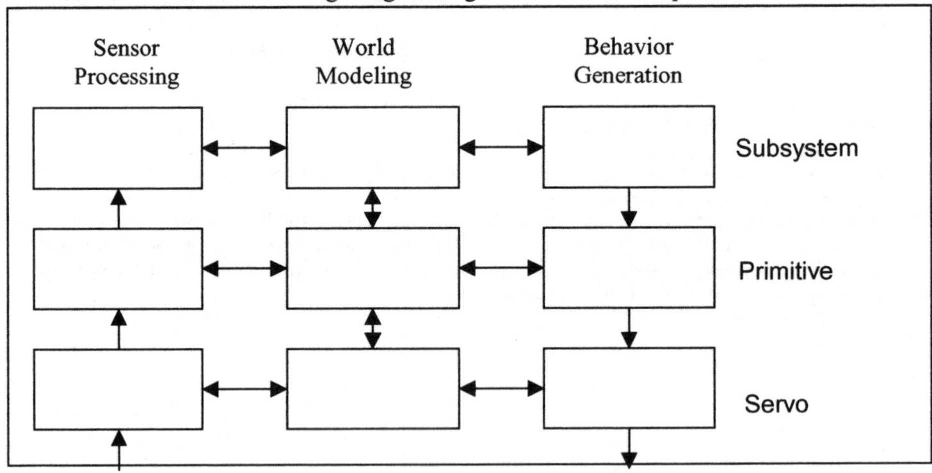

Figure 1: General Framework for an Intelligent Control System

problem domain uniquely different than other knowledge engineering problems. However, as we will discuss, intelligent control requires several different types of "knowledge", and the highest levels of control require the same symbolic knowledge as ontologies, expert systems, or logic systems.

2. Classes of Knowledge

A general framework for a model-based control system is shown schematically in Fig. 1. This framework shows a hierarchical control structure with a world model hierarchy explicitly interspersed between the sensor processing hierarchy and the behavior generation or task decomposition hierarchy, allowing for model-based perception and model-based control [2], [3]. Example labels for three of the levels (subsystem, primitive, and servo), per [2] are shown.   This paper presents an overview of the types of data needed for the world model hierarchy.

We argue that there are three distinctly different classes of knowledge in such a control hierarchy: system parameters at the servo level; maps and images at the middle levels, and symbolic data at the highest levels. We will consider each of these below.

We can further distinguish knowledge that is learned or acquired, which we will call *in situ* knowledge, from knowledge that is pre-programmed or referenced from an outside database, which we will call *a priori* knowledge. This provides a framework for considering learning and adaptive control.

2.1 Parametric Level Knowledge

The lowest levels of any control system, whether for an autonomous robot, a machine tool, or a refinery, are at the servo level, where knowledge of the value of system parameters is needed to provide position and/or velocity and/or torque control of each degree of freedom by appropriate voltages sent to a motor or a hydraulic servo valve. The control loops at this level can generally be analyzed with classical techniques and the "knowledge" embedded in the world model is the specification of the system functional blocks, the set of gains and filters that define the servo controls for a specific actuator, and the current value of relevant state variables. These are generally called the system parameters, so we refer to knowledge at this level as parametric knowledge. Fig. 2 shows a traditional PD servo control for a motor of a robot arm.

2.2 Iconic or Geometric Level Knowledge

Above the servo level are a series of control loops that coordinate the individual servos and that require what can be generally called "geometric knowledge," "iconic knowledge," or "patterns."   Iconic knowledge includes maps, images, models of the kinematics of the machines being controlled, and knowledge of the spatial geometry of parts or other objects

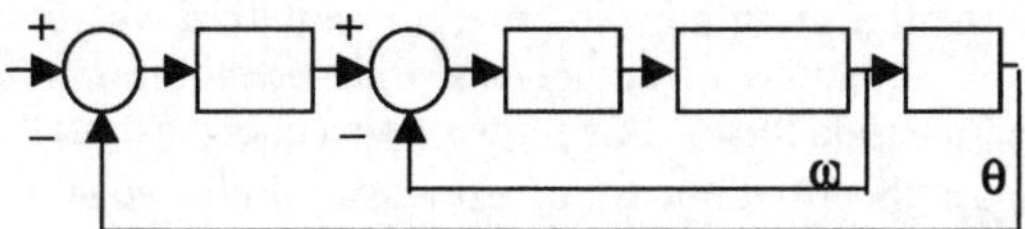

Figure 2: PD Servo Control

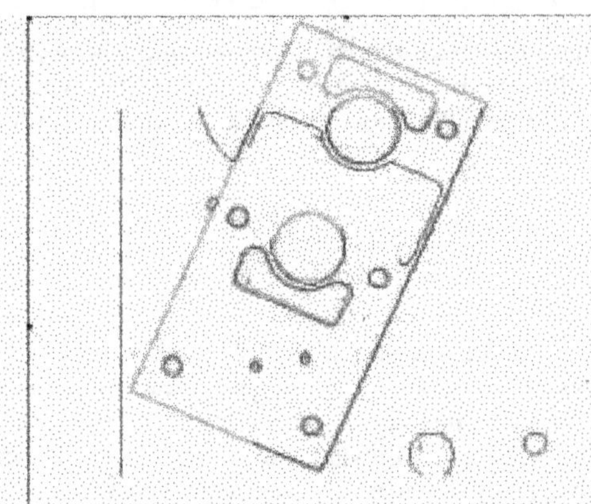

Figure 3: Part Pose Computation

that are sensed and with which the machine interacts in some way. This is where objects and their relationship in space and time are modeled as to represent and preserve those spatial and temporal relationships, as in a map, image, or trajectory.

For industrial robots, machine tools, and coordinate measuring machines, the first level above the servo level deals with the kinematics of the machine, relating the geometry of the different axes to allow coordinated control. Linear, circular and other interpolation and motion in world or tool coordinates are enabled by such coordination. The "knowledge" here may be the kinematic equations or Jacobian coefficients that define the geometric relationships of the axes, or the mathematical routines for interpolation or coordinate transformations. It is at this level that systematic multi-dimensional geometric errors such as non-orthogonality of axes of a machine tool and Abbe offset errors are considered [4]. Fig. 3 shows an investigation of fixtureless inspection, in which a part is placed on the table of an inspection machine without a fixture and the pose of the part is determined by matching an image of the part (dark edges) with a predicted image derived by rotating and translating a CAD model of the part (light edges) [5].

2.3 Symbolic Knowledge

At the highest levels of control, knowledge will be symbolic, whether dealing with actions or objects. It is at this level that a large body of relevant work exists in knowledge engineering for domains other than real-time control, such as formal logic systems or rule based expert systems. Whether the knowledge is represented in terms of mathematical logic, rules, frames, or semantic nets, there is a formal linguistic structure for defining and manipulating and using the knowledge. A good presentation of different concepts of knowledge representation is found in Davis [6].

An example of a formal description of a solid model of a part is shown in Fig. 4. A block is being described using International Standards Organization Standard for the Exchange of Product Model Data (STEP) Part 21 [7]. Note the fundamentally different nature of this linguistic representation from a geometric representation where, for example, a block might be represented by equations of six planes with bounding curves and a coordinate transformation matrix to position the block within a given coordinate system.

Linguistic representations provide ways of expressing knowledge and relationships, and of manipulating knowledge, including the ability to address objects by property. Tying symbolic knowledge back into the geometric levels provides the valuable ability to identify

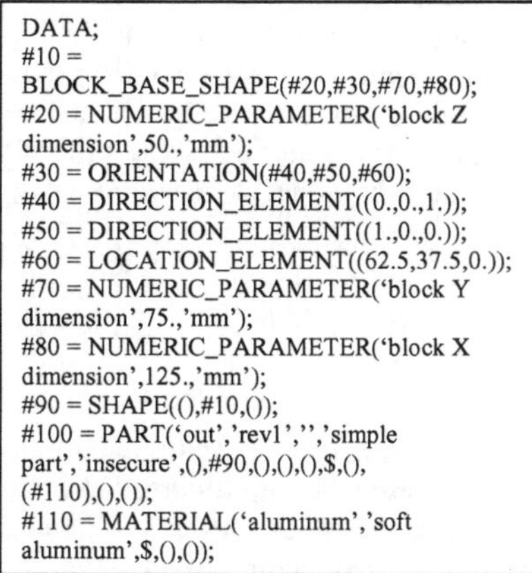

```
DATA;
#10 =
BLOCK_BASE_SHAPE(#20,#30,#70,#80);
#20 = NUMERIC_PARAMETER('block Z
dimension',50.,'mm');
#30 = ORIENTATION(#40,#50,#60);
#40 = DIRECTION_ELEMENT((0.,0.,1.));
#50 = DIRECTION_ELEMENT((1.,0.,0.));
#60 = LOCATION_ELEMENT((62.5,37.5,0.));
#70 = NUMERIC_PARAMETER('block Y
dimension',75.,'mm');
#80 = NUMERIC_PARAMETER('block X
dimension',125.,'mm');
#90 = SHAPE((),#10,());
#100 = PART('out','rev1','','simple
part','insecure',(),#90,(),(),(),$,(),
(#110),(),());
#110 = MATERIAL('aluminum','soft
aluminum',$,(),());
```

Figure 4: STEP Representation of a Block

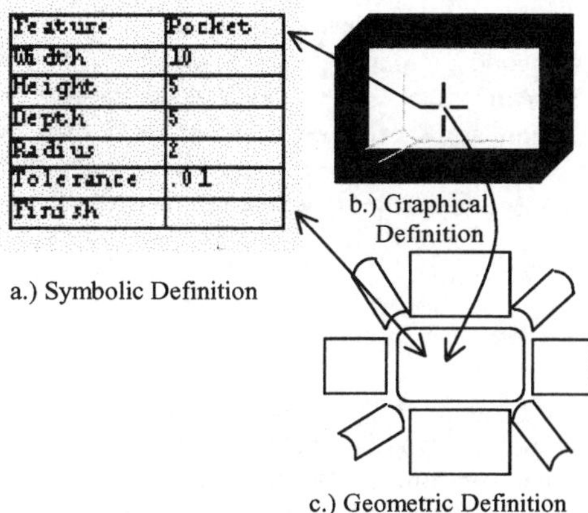

Feature	Pocket
Width	10
Height	5
Depth	5
Radius	2
Tolerance	.01
Finish	

a.) Symbolic Definition

b.) Graphical Definition

c.) Geometric Definition

Figure 5: Pocket Feature

objects from partial observations and then extrapolate facts or future behaviors from the symbolic knowledge. In the manufacturing domain, using a feature-based representation (which is symbolic) is reasonable at the generative planning level (Fig. 5a). The geometric representation of each edge and surface that comprise a feature (Fig. 5c) can be tied to the feature definition in order to facilitate calculations for generating the tool paths. Graphical primitives (Fig. 5b) that relate to the geometry can also be tied to features to easily let users pick a feature by selecting on a portion of it on the screen.

STEP Part 21 files are one or many ways of representing symbolic information. Different representation techniques often offer different advantages. For example, as is the case in almost any planning and control system, it is often advantageous to be able to reason over information that is represented. This includes being able to infer information that may not be explicitly represented, as well as the ability to pose questions to the knowledge base and receive answers in return. One way of enabling this functionality is to represent the symbolic information in the world model in a logic-based, computer-interpretable format, such as in the Knowledge Interface Format (KIF) representation [8].

Through the use of an inference engine or theorem prover, information represented in this format could be queried, and logically-proven answer could be returned. As an example, a manufacturer may want to know whether a given set of fixture positions is suitable to fully inspect a part. Assuming that the necessary inspection points, access volumes, and machine capabilities are represented in KIF, the manufacturer could enter in the fixture positions and the system could logically-prove whether those positions are sufficient to fully inspect the part. Future work will be exploring this area in more detail via the implementation of logic-based ontologies to represent the symbolic information in the control hierarchy.

3. Control With Multiple Levels of Knowledge

The most significant and complex autonomous mobile robot built to date is the Army's Experimental Unmanned Vehicle (XUV) being developed for scout missions (reconnaissance, surveillance, and target acquisition (RSTA) missions). The architecture for this vehicle is called 4D/RCS, merging the work of Dickmanns in Germany on road following [9] and the work of Albus at NIST [3]. Both use data from multiple sensors to build a world model and then use that model for planning what the vehicle should do.

The Army XUV has successfully navigated many kilometers of off road terrain, including fields, woods, streams and hilly terrain, given only a few way points on a low resolution map by an Army scout. The XUV used its on-board sensors to create high definition multi-resolution maps of its environment and then navigated successfully through very difficult terrain. Over the next several years, symbolic knowledge will be added to enable tactical behaviors and human-machine interaction.

4. Conclusion

No single type of knowledge representation is adequate for all purposes. Davis [6] argues that representation and reasoning at the symbolic level are inextricably intertwined, and that different reasoning mechanisms, such as rules and frames, have different natural representations that must be integrated in a representation architecture to achieve the advantages of multiple approaches to reasoning.

We would go further and argue that there is a requirement for integrating iconic and parametric knowledge with multiple types of symbolic knowledge and that, as Davis argues, there is a basic need for a representational architecture to provide a basis for intelligent control, which we have presented above in Figure 1.

For example, with the ability for the Army XUV to not only sense when there is an obstacle in its path, but to also be able to compare that sensed data (possibly represented in an occupancy grid) with a priori knowledge of obstacles (possibly represented in KIF in a symbolic world model), the XUV can make more informed decisions about the best action to take, taking into consideration the type of obstacle it encounters. If the obstacle is deemed to be a boulder, the XUV must take evasive maneuvers. However, if it is simple tumbleweed, the XUV may be able to drive through it. This is the essence of intelligent control.

References

[1] Antsaklis P. J., "Defining Intelligent Control", Report of the Task Force on Intelligent Control, P.J Antsaklis, Chair, IEEE Control Systems Magazine, pp. 4-5 & 58-66, June 1994.

[2] Albus, J., "4-D/RCS: A Reference Model Architecture for Demo III," NISTIR 5994, Gaithersburg, MD, March 1997.

[3] Albus, J.S., Lumia, R., Fiala, J., Wavering, A., "NASREM -- The NASA/NBS Standard Reference Model for Telerobot Control System Architecture", Proceedings of the 20th International Symposium on Industrial Robots, Tokyo, Japan, October 4-6, 1989.

[4] ANSI/ASME B5 TC52 Committee, "Interim Revision of Methods for Performance Evaluation of Computer Numerically Controlled Machining Centers." ASME standard B5.54, Version 5.0, May 2001.

[5] Messina, E., Horst, J., Kramer, T., Huang, H.M., Tsai, T.M., Amatucci, E., "A Knowledge-Based Inspection Workstation," Proceedings of the IEEE Intn'l. Conference on Intelligence, Information, and Systems, Bethesda, MD., Oct. 31-Nov. 3, 1999.

[6] Davis, R., et al. 1993. "What is in a Knowledge Representation?" AI Magazine, Spring 1993.

[7] ISO 10303-21:1994; Industrial automation systems and integration – Product data representation and exchange - Part 21: Clear Text Encoding of the Exchange Structure; ISO; Geneva, Switzerland; 1994.

[8] Gensereth M., Fikes R. 1992. Knowledge Interchange Format. Stanford Logic Report Logic-92-1, Stanford Univ. http://logic.stanford.edu/kif/kif.html .

[9] Dickmanns, E.D., "A General Dynamic Vision Architecture for UGV and UAV." Journal of Applied Intelligence, **2**, p.251, 1992.

KES 2002
E. Damiani et al. (Eds.)
IOS Press, 2002

1429

A modeling framework for capturing the design rationales of problem solving processes

Kazuhisa SETA[*1] and Osamu KAKUSHO[*2]
*1 Osaka Prefecture University, 1-1, Gakuen-cho, Sakai, Osaka, 599-8531 Japan
*2 Hyogo University,2301, Shin-zaike, Hiraoka-machi, Kakogawa, Hyogo, 675-0101 Japan

Abstract. Many researchers aim at developing the computer assisted knowledge management systems which adequately support creating and building up the sharable knowledge assets in an organization. In general, the steps towards computer assisted knowledge management are roughly divided into two phases, that is, modeling phase and using phase. Many research projects focus on the latter phase and aim at achieving the knowledge creation support, however, we think a matter that must be settled first is to provide a framework which can capture the helpful information for knowledge sharing/inheritance. The evolution such as the improvement of existing business models called "as is model" to desired models called "to be model" is not carried out without understanding the design rationales of the as is model. We will present a framework for capturing the design rationales of the business workflow to achieve effective sharing of the problem solving knowledge.

1. Introduction

Many researchers aim at developing the computer assisted knowledge management systems which adequately support creating and building up the sharable knowledge assets in an organization [1][4][7]. In general, the steps towards computer assisted knowledge management are roughly divided into two phases, that is, modeling phase and using phase. Basically, business processes and their associated documents/information are modeled and stored in the former phase and the models are used by the organization members who need for performing their task in the latter phase. Based on this fundamental framework, many systems such as business process modeling, work flow management and document management were built and run in many business areas. However, some problems are becoming clear through practical experience. That is, only modeling and storing the business flow in a repository are not sufficient to achieve effective sharing/inheritance of the knowledge. The essential reason is that the design rationales of the business logic are not explicitly represented in the model. As the environment which surrounds a business process today is changing with dizzy vigor, it is necessary for the business model to be adapted to the change promptly. The evolution, i.e. improvement of existing business models called "as is model" to desired models called "to be model", is not carried out without understanding the design rationales of the as is model. Many research projects focus on the using phase and aim at achieving the knowledge creation support, however, we think a matter that must be settled first is to provide intelligent techniques and frameworks for capturing the helpful information for knowledge sharing/inheritance of the problem solving knowledge.

The purpose of this paper is to present a framework for capturing the design rationales of the business workflow and the information for understanding them to ensure effective sharing/inheritance of the problem solving knowledge. Firstly, we will examine a product model which stands for what the problem solving model should be. Secondly, we will describe a process model which

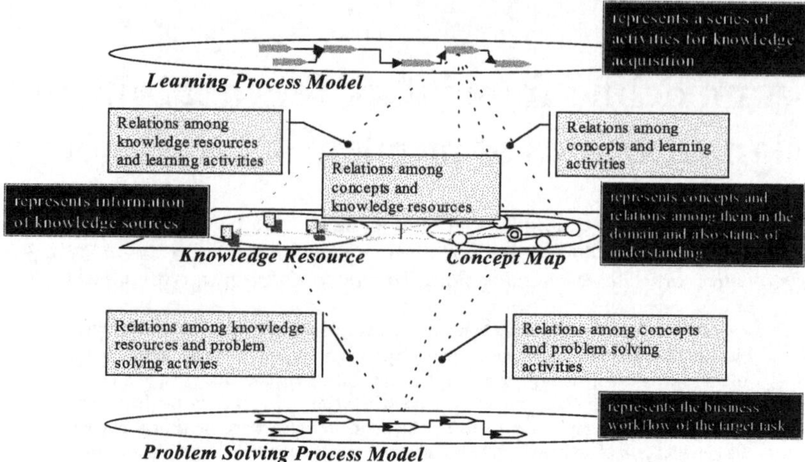

Figure 1. A framework for representing problem solving model

captures how we construct the model. And then, we will briefly explain a modeling system based on the framework.

2. A product model for capturing the design rationales of problem solving processes

2.1 Design principle

To enable effective sharing and inheritance of problem solving knowledge, we must represent the helpful information to understand the design rationales of the problem solving processes. Representing them in a model ensures that the others can easily understand the reason why each problem solving process is combined in such a way (problem solving context), that is, the roles, functions, dependencies, antecedents and criterions of the problem solving processes. Needles to say, this information is much useful when we try to redesign or evolve the problem solving processes according to the changes of the real world. The framework of knowledge management must have the capability for capturing these information.

2.2 Construction of a model

Figure 1. shows our framework designed for representing problem solving knowledge. The framework is composed of knowledge map and task flow model. Furthermore, knowledge map is divided into concept map and knowledge resource, and task flow model is divided into problem solving process model and learning process model. We will briefly explain each of them.

- *Knowledge Resource (KR)* captures knowledge sources, such as web pages, electronic texts, human resources and so forth, useful for understanding and performing the target task.
- *Concept Map (C-Map)* which stands for the antecedent of the task domain captures the concepts and relationships among them using basic semantic links such as "is-a", "part-of." Each concept expresses the meaning, roles and manners of the problem solving processes and objects processed by the problem solving processes.
- *Problem Solving Process Model (PSPM)* captures a problem solving workflow that solves the target problems. Performing the problem solving activities changes the status of the real world.

- *Learning Process Model (LPM)* captures a series of learning activities which is recognized as a problem solving of acquiring the knowledge required for performing the task. By following this model we can understand the design rationales of the problem solving process model.

Each of these models is connected each other. Intuitively, the relation between C-Map and KR is that nodes on C-Map play a role of annotation of connected nodes on KR. Roughly speaking, each node on KR, such as web pages, human resources, is for humans to refer for their problem solving. C-Map is represented both human and machine readable manner, so humans/computers can manage KR through the C-Map. The relation between PSPM and C-Map is as follows. Each node on C-Map is referred by the activities in PSPM that need for running. Furthermore, the relation between LPM and C-Map is that the concepts and relations among them represented in C-Map are linked to the corresponding nodes in LPM. Therefore, LPM forms a series of activities for effective learning of the concept and relations in C-Map. By representing problem solving knowledge based on the framework, not only the problem solving methods but also the design rationales behind them could be effectively shared/inherited among organization members. This means that they can evolve according to the requirement and the changes occurred in the real world.

3. A process model for designing problem solving model

The framework we described in section 2 is a meta model of problem solving knowledge. That is, it represents what the problem solving model should be. Furthermore, we also need to provide a process model with the model creators so that they can smoothly construct the accurate problem solving plans.

If we suppose the existence of the "as is model", it is not so difficult to model the problem solving processes without any support. Our research on knowledge management, however, aims at supporting the problem solving processes design from scratch even if there is no reference model. We think this goal is especially important, because we cannot expect no longer the existence of the reference models especially in the world of IT revolution. There is no doubt that preparing the process model which supports the problem solving processes design is a key technique towards knowledge management. Actually, the process model would add a striking effect when someone who is a novice about the domain of problem solving tries to design the problem solving processes from scratch.

In general, if we are not yet sufficiently experienced in a work and asked to solve some problems, we might iterate the spiral processes depicted as Figure 2, that is, the processes of investigating where available KRs are, what kinds of knowledge we can acquire from them and acquiring the knowledge on the target world, the processes of designing effective problem solving/learning processes and the processes of performing the workflow. The process model must have the capability to adequately capture these processes.

We present a Knowledge Map (K-Map) centered process model depicted in Figure 3. It captures the processes of the business workflow design. In this figure, center, above left and below right stand for the processes of modeling K-Map, LPM and PSPM, respectively. Each of them stands for that each model is gradually detailed and refined by spirally repeating the specification and revising processes.

In general, the methods of problem solving and/or their design processes much depend on what kinds of knowledge resources/tools

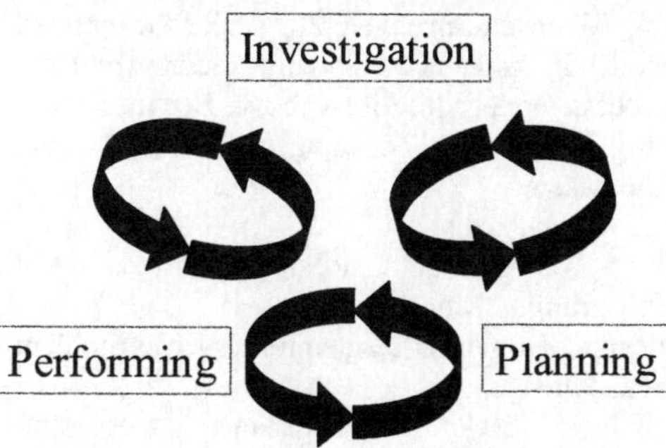

Figure 2. A process model for problem solving

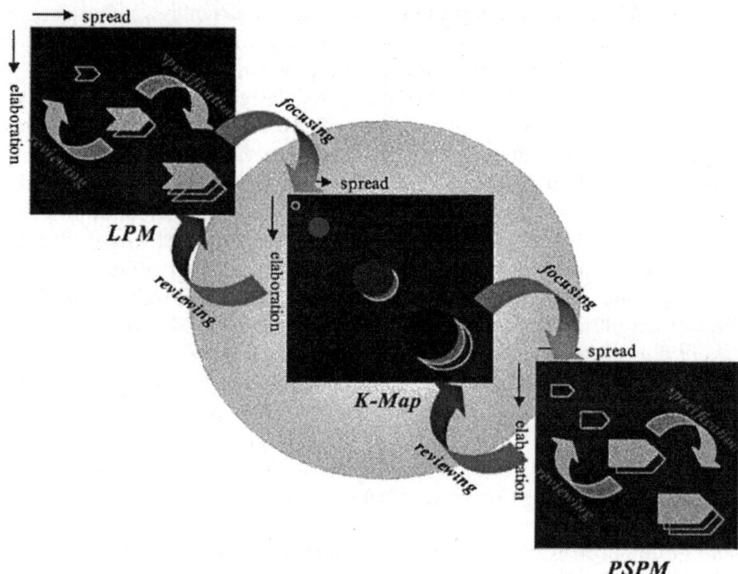

Figure 3. A K-Map centered process model

are available in the organization. In this process model, by referring such information represented in the KR, users can understand the meaning of concepts and relations among them and express their understanding as a C-Map. Based on the user's model of understanding represented as the C-Map, furthermore, they can adequately arrange, reform and execute the LPM for acquiring the knowledge required for performing the task. This process also encourages for users to review the K-Map. And similarly, based on the links between C-Map and PSPM it also does to review the PSPM. Consequently, the model supports users' work of incremental design of problem solving process plan as a whole.

4. Advantages of the framework

It is quite time consuming work for users to describe/design the valid, sharable and inheritable models. To lighten the loads of users, we must develop a "modeling methodology aware tool. " This means that the tool knows the structures of the product and how it can be constructed. It enables the tool to provide useful information for users' work. We have been developing such a tool entitled Kassist. The Kassist, which is an environment for modeling/designing and sharing/inheriting problem solving knowledge, employs the product and process models described above. Because of the space limitation, we will concentrate on describing the functional advantages of Kassist that are realized by laying the framework on the basis of the system.

When we develop Kassist, we set the following two main goals.

a. The system can support novice workers' design process so that they can make effective and efficient problem solving processes that include learning activities.

b. The system can support effective use of experiences of organization members, such as problem solving strategy or know-how and know-what knowledge.

The reason why we define these goals are as follows. Most knowledge management tools suppose the existence of the reference models, however as you know, there is no reason to believe

especially in the industrial world of information technology. The time when people should just be giving even the routine work has already passed away. They must arrange/redesign the adequate workflows from scratch to solve the problem even in a strange domain. So, we must attain the goals to support their works. Kassist has the following advantages based on the framework.

Modeling support:
- Based on the process model, the system can support users' work to design problem solving plan even in case that there is no expert to turn to.

Supporting for sharing and inheritance of problem solving knowledge:
- The system has basic functionality which ordinary workflow systems have, such as KR retrieval to perform the task, scheduling management and so on. So, they can solve the problems by referring the shared models.
- The system can support the learning processes for understanding the design rationales of the problem solving methods.

Supporting for evolving the problem solving processes:
- If the problem solving context changes, we need to adapt the model. Because the product model captures the problem solving context, the system can understand the scope needed to review by tracing the propagation through the links among the models and report the helpful information to users.
- We can understand the C-Map as a valid domain ontology sufficient for performing the task. An ontology is a specification of a conceptualisation in the domain [2]. Thus, based on the model, users and the system can collaboratively retrieve the useful information from the WWW using the semantic information. With an outcome of the semantic web research [3][8], users might get the useful and high quality information for making a "to be model."

5. Concluding Remarks

In this paper, we concentrated on presenting an outline of the modeling framework for enabling effective knowledge sharing. To make a generic framework for knowledge management, we presented a domain independent framework in this paper. We are currently implementing Kassit based on design principle presented in this paper. Our work on task ontology and user modeling [5][6] also taught us the importance of embedding the task/domain dependent ontology in the system to realize the more effective support capturing the contents of the target world. This subject will be addressed more carefully in the future work.

References

[1] Daniel O'Leary: "Special Issues on Knowledge Management," IEEE Intelligent Systems, vol. 13, No. 3, (1998).

[2] Dieter Fensel: Ontologies: A Silver Bullet for Knowledge Management and Electronic Commerce, Springer, (2001).

[3] Dieter Fensel et. al.: "Special Issues on the Semantic Web, " IEEE Intelligent Systems, vol. 16, No. 2, (2001).

[4] I. Nonaka, and H. Takeuchi: "The Knowledge-Creating Company," Oxford, UK, Oxford University Press, (1995).

[5] K. Seta, M. Ikeda, O. Kakusho and R. Mizoguchi: "Capturing a Conceptual Model of Problem Solving for End-user Programming," Proc. of the Sixth International Conference on User Modeling (UM-97), Sardinia, Italy, pp.203--214. (1997).

[6] M. Ikeda, K. Seta and R. Mizoguchi: "Task Ontology Makes It Easier To Use Authoring Tools," Proc of International Joint Conference on Artificial Intelligence '97 (IJCAI-97), Nagoya, Japan, pp.342--347 (1997).

[7] Schreiber, et. al: "Knowledge Engineering and Management: "The CommonKADS Methodology," The MIT Press, (1999).

[8] SemanticWeb.org: http://www.semanticweb.org/

KES 2002
E. Damiani et al. (Eds.)
IOS Press, 2002

Multimedia platform to support knowledge processes anywhere and anytime

Maurizio Mesenzani[1], Thomas Schael[2], Sara Albolino[3]
c/o Butera e Partners, Piazza Giovine Italia 3, Milano, Italia

Abstract. The paper contains a presentation of the work done during the past years in two ESPRIT Projects in the knowledge management area: Klee&Co (Knowledge and Learning Environments for European & Creative Organisations, started in 1998 and ended in 2000) and MILK (Multimedia Interaction for Learning and Knowing, started in early 2002 and expected to be closed by 2004). Klee&Co developed an approach and a web-based prototype allowing users to view knowledge in its context: the main feature of the Klee&Co system is the "view with context". MILK is going to develop a system supporting users in any working situation, even mobile and social situations. It means a strong effort in interface and system design in order to create differentiate technological platforms for PC environment, social environment, mobile environment. To get the expected result, it is important to start from the observation of users' needs and from the definition of a clear knowledge management strategy. The seductive design approach of Klee&Co and MILK will be presented to show the relevance of the integration between users, technology developers and designers visions for a successful knowledge management solution.

1. Introduction

The paper is focused on the dynamic idea of knowledge management, as a way to enable learning, communication and networking processes between and within communities of practice. The hypothesis is that "inspirational knowledge" is the driver to enact a network of knowledge-intensive communities, where people and organisations get value from relations and knowledge exchange. In order to do that, we developed a solution where "knowledge objects" can be visualised in their knowledge context. In this paper we present two different European research & tech. dev. projects: Klee&Co[4] project (1999-2000) and MILK[5] (2002-2004) project.

Klee&Co was aimed at implementing a knowledge management solution to support professional work and business processes in innovative environments (design companies). The goal was to develop a system able to support working processes providing the "right knowledge at the right time". Starting from the Klee&Co output, the challenge of the MILK project will be to define an integrated and comprehensive knowledge management solution supporting professional work in any situation.

2. Knowledge management theories

Near 80% of worldwide leading companies have some knowledge management efforts under way. Some of them have already established a clear pattern of knowledge management responsibilities within their organisation chart (Chief Knowledge Officers,

---

[1] Maurizio Mesenzani is researcher and senior consultant in Irso - Butera e Partners. He is leading knowledge management projects in public and private companies. From 1997 to 2000 he has been working in Andersen Consulting. Email: mesenzani@irso-bep.it

[2] Thomas Schael is Managing Director Irso and manager of knowledge management practice Butera e partners. Former Project Manager in Klee&Co project (ESPRIT project 28842) , Project Manager in MILK project (IST project 2001-33165). Email: schael@irso-bep.it

[3] Sara Albolino is a researcher and consultant in Irso - Butera e Partners, expert in the knowledge management area and work practices and community of practices analysis. She is collaborating with University of Milan Bicocca in the Social Science Department. Email: albolino@irso-bep.it

[4] Klee&Co (ESPRIT Project 28842) is the acronym of Knowledge and Learning Environments for European & Creative Organisations. Partners: Irso, University of Milan Bicocca, Domus Academy, Philips Design, Xerox Professional Services.

[5] MILK (IST Project 2001-33165) is the acronym of Multimedia Interaction for Learning and Knowing. Partners: Irso, Butera e Partners, Orbiteam, University of Milan Bicocca, Domus Academy, Fraunhofer Institute, PictureSafe, Xerox Research Centre, Xerox Professional Services.

knowledge managers, knowledge champions, ...)[6][1]. A knowledge management programme is often related to organisational development, change management and human resources management issues, such as training and continuous learning. This is due to the fact that knowledge management and organisational learning are the two faces of the same coin. A knowledge-based innovative organisation is taking care of individual learning and knowledge sharing processes and is a "community of communities", where people are used and encouraged to launch new ideas to innovate processes, practices and products.

2.1 Tacit and explicit knowledge

There are two different kinds of knowledge to be managed: tacit knowledge ("kept in people mind") and explicit knowledge (memorised through ICT[7] and paper supports). A knowledge management system has to manage two different kind of knowledge: tacit knowledge ("kept in people mind") and explicit knowledge (memorised through ICT[8] and paper supports). The analysis of knowledge creation and transfer processes is based on Nonaka and Takeuchi model [2]. Knowledge increasing the overall organisational value is not an object stored into a database neither a personal set of competencies and experiences (hidden in people mind). The relevant knowledge for organisations is collectively shared by members and can be effectively used in business processes. So the knowledge creation process is based on the transformation of tacit and/or explicit knowledge into new knowledge.

2.2 Knowledge management and communities of practice

A community of practice is as *"an informal aggregation of people who share work practices and common experiences"*[3]. The community is the place where social processes are enacted; through these processes people acquire skills and methodologies to act their organisational role and to develop themselves in a lifelong perspective.

A community is a strategic tool for the enhancement of knowledge and learning processes [4]. In a community of practice people are consciously or unconsciously teaching and learning at the same time. Communities around topics and contents grow up spontaneously and should be recognised and cultivated to make them additional arms of the organisation. They have common goals and repertoires (i.e. shared archives), they are based on spontaneous participation and informal behaviour codes, and they are social networks among people sharing the same interests. Social exchanges and relations among people permit to mix the level of expertise and to differentiate skill development.

3. Knowledge Management to increase the value of organisations

Klee&Co and MILK start from the assumption that knowledge management increase the overall organisational value [5], in terms of performance and assets. They are based on the adoption of communication-intensive platforms, combining the management of objects with the management of people profiles (people experiences and competencies) and relations between people and objects (object to object, object to person, person to person). In Klee&Co and MILK, knowledge management means a way to develop knowledge and communication circles and networking within and across communities: this is the reason why it is so important to understand working practices and organisational culture in order to identify users needs and behaviour. Communication flows in business processes are often hidden for the organisation and use different channels. A knowledge management system should capture the knowledge exchanged within communication processes at the scope to move tacit to explicit knowledge and to build an organisational memory. The results are modified and improved business processes, this is the reason why several knowledge management projects are named and assume the aspect of change management or BPR[9] projects.

---

[6] *Beyond Knowledge Management: new ways to work and learn* - Research report, Conference Board, 2001.
[7] Information and Communication Technology
[8] Information and Communication Technology
[9] Business Process Reengineering

4. Klee&Co and MILK solution

In this section of the paper the adopted solution will be presented in terms of approach and main features.

4.1 Our approach: the "Seductive Design"

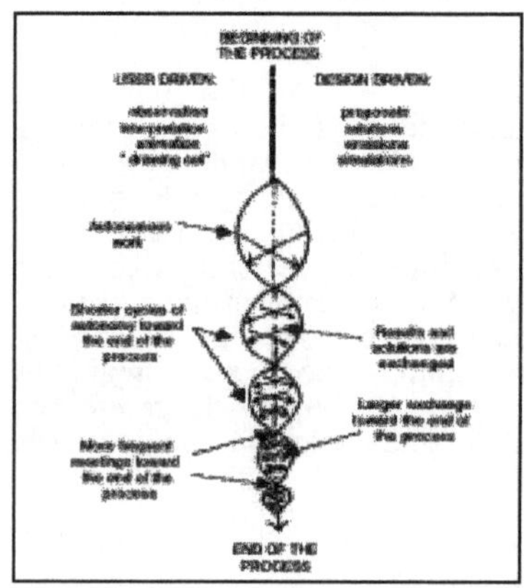

Figure 1 The seductive approach

The approach chosen to design the Klee&Co and the MILK solution are strongly focused on the need to integrate three different visions coming from different subjects involved in system design [5]: users, technology developers and designers. An effective knowledge management system has to be understood, used and useful for *users*: it means that they should be involved in the early stage of system development, the system could be tailored on business processes, working needs and working culture of future users. *Technology developers* should be involved in the programming and system integration activities in order to translate users needs into system functionalities. *Designers* involvement in the system development is an innovative method to enlarge system's capabilities from the beginning.

4.2 Our solution: content, context and community in different working situations

The inspirational side of knowledge management and the dynamic idea of knowledge management have been turned into the Klee&Co prototype in the "view with context" solution. Information and knowledge are presented in their context and the visualisation changes according to the different working situation they support ("the right knowledge in the right time"). Once a document is launched, the system conducts a background search aimed to find related documents and related people. The Klee&Co prototype is a web-based environment where added value functionalities have been combined with basic document management features. In order to reduce users effort, in Klee&Co the uploading process is supported through a keyword extraction tool and through automatic abstracting technologies. Once the user identifies the file he wants to upload, the system produces an abstract and extracts a set of keywords. These indexes are used by the system to build relations among objects. In addition to automatic relations, Klee&Co has been designed to track users' defined relations.

4.3 Technical aspects

The Klee&Co prototype is an environment where added value functionalities have been combined with basic document management features.

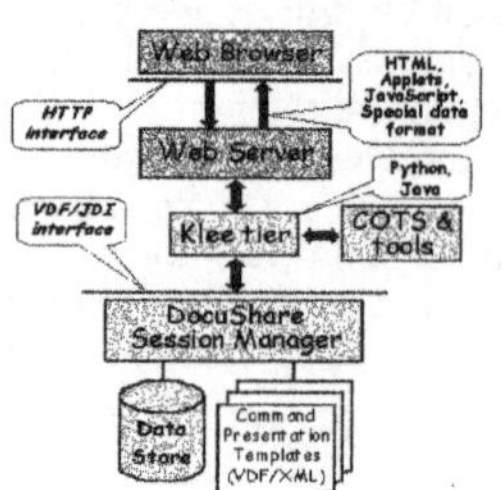

Figure 2 The system

The document management system is Docushare[10]. Users enter in the system through a browser, they get to the web server and enter the Klee tier. The system architecture shows (see figure 2) that the Klee&Co tier is made of several components, supporting the basic processes of a knowledge management system. Klee&Co is supporting the visualisation process in order to build the "view with context": once a document is launched, the system conducts a background search aimed to find related documents and related people. Documents and objects (people profiles, mail messages, tags…) are archived in the document management system and are indexed through a thesaurus combining institutional keywords with users' chosen terms.

---

[10] Docushare is a document management product owned and distributed by Xerox Corp.

The uploading process is supported through a keyword extraction tool and through automatic abstracting technologies. Once the user identify the file he wants to upload, the system produces an abstract and extract a set of keywords, that users can accept, refuse and weight using a ranking tool. These indexes are used by the system to build relations among objects. In addition to automatic relations, built by the system using indexes, Klee&Co has been designed to track users' defined relations, this is a way to capture tacit knowledge: when looking at a document, users can say that the document is related to a different one, archived in a separate folder or in a different place in the document management system, through a command they can explicit this relation and make it available for other users. The system is also able to detect who is on line, matching people profiles with event detection tools: once the users log into the system they are recognised as on line and presented when their name appears in the related people framework. Through this framework, users can activate communication sessions, by clicking on people's icons: in example they can launch a chat session, a tele/video conference, or they can just send an e-mail. MILK will extend these functionalities incorporating dynamic profiling features to automatically update people's and objects' profiles. The picture presents the MILK architecture, presenting the different way to access the system and the interaction managers tools: an interface will be designed for personal PC environment, focusing on the results of Klee&Co, an interface will be designed for mobile devices A KM engine will enrich the functionalities of the document management system. The personal mobile environment will be based on palm laptops, mobile phones, over GPRS/WAP technologies, the social environment will be based on wall screen technologies. Depending on users' authentication, the system will choose the proper interface. Under a human-computer interaction viewpoint the matter is to design the right interface tailored on the working environment, presenting commands, functionalities and information according to the users needs detected with ethnographic methods. The design of a knowledge system based on "the right knowledge at the right time" starts from the analysis of typical working situations. The hypotheses of working scenarios show different situations:

- *PC Environment:* the person works at his workstation using a PC. He needs to get information tailored on his preferences and related to the project or customer is working on (e.g. client data, project plan, personal agenda…).
- *Social Environment:* people work together in group work or meeting sessions. They need information tailored on projects and customers, stored in the intranet.
- *Mobile Environment:* he needs synthetic and specific information and data about the client organisation and the activities and projects he or his colleagues are realizing with him.

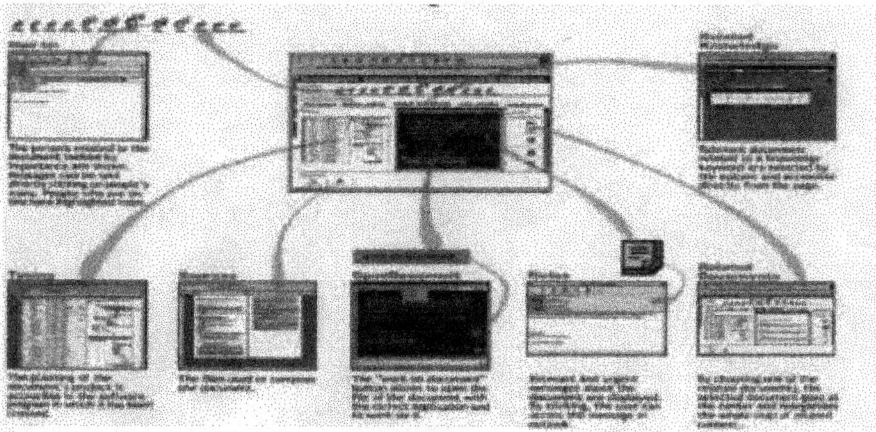

Figure 3 - Klee&Co: The view with context

2. Lessons learned and future challenges

Looking back at the work done, these are the main findings:
- An effective knowledge management solution should start from the comprehension of users needs and behaviour.
- An effective knowledge management solution is not only a "set of software tools", it is the joint combination of processes, organisation, culture and technology.
- Knowledge in context is the way to effectively support knowledge sharing and knowledge creation processes
- Users can work in different situations, so the system should be able to present the same contents in different ways.

All these points confirm that a knowledge management solution is a combination of top-down and bottom-up perspectives and implies a cultural change to the way people use to exchange their knowledge among themselves. In terms of future perspectives, adding "mobile and social working situations" features to a knowledge management system is giving strong opportunities to the development of knowledge processes within organisations. This is also opening a new door towards new ways of working. The MILK social environment is completely changing the space/content perception of an organisation, affecting logistics, people management, content management and visualisation, giving to individuals and groups emerging capabilities that right now we can't even imagine. Same reasoning can be applied to the MILK mobile environment: mobile communication is growing a lot, in terms of services, protocols, devices and related components. It means that there is a strong need for people to communicate all during the day (working day or not). Giving employees the possibility to combine communication with knowledge management means to create a strong value opportunity for the company and, at the same time, for the society.

References

[1] *Beyond Knowledge Management: new ways to work and learn* - Research report, Conference Board, 2001.
[2] Nonaka I., Takeuchi H. (1995), *The Knowledge Creating Company*, Oxford University Press, New York.
[3] Wenger E., *Communities of practice, learning meaning and identity*, Cambridge University Press, New York.
[4] Butera F., "L'organizzazione a rete attivata da cooperazione, conoscenza, comunicazione, comunità: la R&S", *Studi Organizzativi*, 1999, n.2.
[5] Davenport T., Prusak L., *Working Knowledge: how organizations manage what they know*, Harvard College, USA, 1998.
[6] Agostini A., De Michelis G., Susani M., "From user participation to user seduction in the design of innovative user-centered system", in R. Dieng, et al. (Eds.), Designing Cooperative Systems: The Use of Theories and Models. Proceedings of the 5th International Conference on the Design of Cooperative Systems (COOP'2000, Sophia Antipolis, France, 23-26 May 2000), IOS Press, 2000, pp. 225-240.

$\mathcal{K}$–ShaRe: an architecture for sharing heterogeneous conceptualisations

Ernesto Compatangelo   Helmut Meisel
Department of Computing Science, University of Aberdeen, AB24 3UE Scotland, UK
e-mail {compatan, hmeisel}@csd.abdn.ac.uk

Abstract We propose an architecture for the development of an intelligent knowledge management environment based on description logics. This environment addresses the problem of heterogeneous reasoning underpinning knowledge analysis, sharing, and reuse. We explicitly focus on selectively variable expressive power, hybrid reasoning about expressive knowledge, and heuristic lexical reasoning, with major emphasis on reasoning at the conceptual level. We outline the usage of emulators in semantic analysis, describing how conceptual reasoning jointly extends and constrains how the description logic is used for the purpose of semantic reasoning. Most critical portions of the $\mathcal{K}$–ShaRe architecture have been either implemented as part of our CONCEPTOOL support system or as standalone functionalities.

1   Introduction

Knowledge Management (KM) develops formal frameworks for knowledge construction, access and reuse. KM approaches are based on the concept of a "knowledge lifecycle", which encompasses acquisition, modelling, retrieval, publishing, maintenance and reuse. All these activities are considered as the explicit research challenges of the Advanced Knowledge Technologies (AKT) Consortium, which aims at producing an integrated approach to the knowledge lifecycle that encompasses a multi-disciplinary range of viewpoints [13].

The management of conceptualisations (e.g. ontologies and, more widely, declarative knowledge bases) in the AKT framework needs substantial intelligent support (i.e. logic deductions or heuristic inferences) to combine knowledge from heterogeneous sources.

We have been providing this support in terms of two core KM functionalities, namely:

- *Analysis functionalities*, which offer verification, validation and augmentation services for a declarative Knowledge Body (KB);

- *Sharing functionalities*, which offer knowledge translation and alignment services that enable the same interpretation of (parts of) one or more KBs by different humans or computer agents.

Analysis implies that inferences and their results are meaningful at the "conceptual level", where knowledge is modelled using a variety of primitives such as entities, relationships, functions, goals, etc. It contributes to enable knowledge sharing by highlighting impediments such as intra-schema and inter-schema inconsistencies. Sharing implies the same interpretation of (parts of) a KB by different humans or computer agents. Shared knowledge can be reused in contexts other than the original one, leading to a reduction of KB development costs and times as well as to an increase in KB quality and size.

2 $\mathcal{K}$–**ShaRe: motivations and rationale**

Current support to analysis and sharing during lifecycle management is based on a wide variety of structured knowledge models, each enabling different automated reasoning capabilities. Some models, such as UML, represent knowledge at the conceptual level. Unfortunately, most of them have an ill-defined semantics and thus do not enable any semantic deduction. Frame-based models like the Protégé one [11] represent knowledge at the epistemological level (i.e. they use the two generic primitives class and role). These models enable automated inferences like class membership, but not deductions like subsumption. Finally, models based on Description Logics (DLs) like DAML+OIL/OIL [3] broaden the spectrum of frame-based inferences with a whole set of specialised deductions such as subsumption.

Similarly to DAML+OIL/OIL, our core KM architecture, denoted as $\mathcal{K}$–ShaRe (and pronounced *key-share*) is also partially based on DLs. However, as we aim at providing a broad range of analysis and sharing functionalities for heterogeneous, expressive knowledge, our reasoning approach cannot be limited to DL-based deductions only. Consequently, our $\mathcal{K}$–ShaRe architecture has been conceived to provide the following additional features:

- *Reasoning at the conceptual level.* Different rules are used to define links (e.g. hierarchies) within different categories of concepts such as entities, relationships, and functions. Correspondingly, different deductions should hold for different categories.

- *Reasoning with a selectively variable expressive power.* It should be possible to ignore selected portions of the expressive power while reasoning about a KB. This could lead to further deductions which would not be otherwise highlighted.

- *Hybrid reasoning with an expressive knowledge model.* In most cases, the expressivity of a KB is wider than that processable within a single reasoning framework. Different inferential engines should be then used, and their results should be "fused" together.

- *Reasoning at the linguistic level.* The results of lexical queries (e.g. about lexical synonymity or subsumption between terms) can be then combined with those of conceptual deductions in a methodologically disciplined way.

- *Heuristic reasoning about potential correlations.* Groups of concepts or attributes whose names (structures) have terms (components) in common should be suggested as candidates for the introduction of explicit relationships between them.

Although most of the above issues have been *separately* addressed by different knowledge representation and reasoning frameworks, to the best of our knowledge they have not been *jointly* addressed. This motivates our $\mathcal{K}$–ShaRe architecture, which introduces a set of (partially and fully) automated KM functionalities that integrate different reasoning paradigms.

3   **The** $\mathcal{K}$–**ShaRe architecture in practice**

A graphical view of the $\mathcal{K}$–ShaRe modular architecture is shown in Figure 1, where each shaded block copes with one or more of the reasoning features listed in Section 2. These features are being implemented in an Intelligent Knowledge Management Environment (IKME) called CONCEPTOOL as independent but linkable automated functionalities [1]. An overview of each functionality is outlined in the following sub-sections, where a brief comparison with existing intelligent technologies for knowledge management is also discussed.

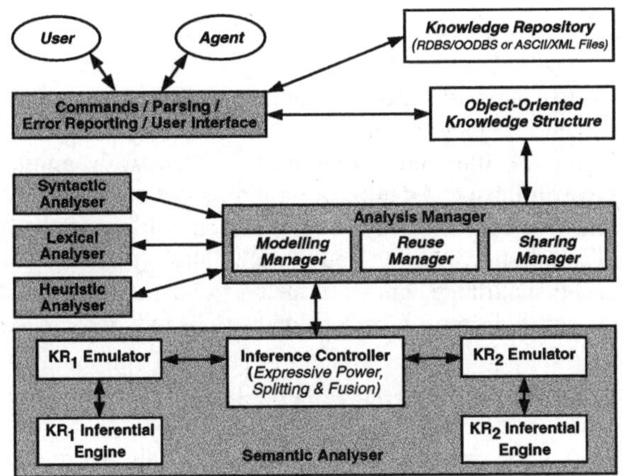

Figure 1: The $\mathcal{K}$–ShaRe functional architecture

3.1 Reasoning at the conceptual level

CONCEPTOOL currently uses an expressive Entity-Relationship (ER) modelling language called $\mathcal{EDDL}_{EER}$, which includes entity identifiers and attributes, entity and relationship hierarchies, attribute fillers, and full coverage and disjointness assertions. The semantics of $\mathcal{EDDL}_{EER}$ is formalised by its mapping into the $\mathcal{SHIQ}$ DL, which is interpreted by the iFaCT terminological reasoner [5]. The rewrite rules that define this mapping allow $\mathcal{EDDL}_{EER}$ KBs to be semantically analysed using the automated deductive services provided by iFaCT. This behaviour is controlled by the *semantic analyser* shown in Figure 1, where an *inference controller* selects the right reasoner. In the current version of CONCEPTOOL, only a $\mathcal{SHIQ}$ *emulator* has been implemented which interacts with the iFaCT *inferential engine*.

Rewrite rules emulate those portion of the $\mathcal{EDDL}_{EER}$ expressive power which are not directly available in $\mathcal{SHIQ}$ (e.g. entity identifiers, fillers, relationships). Moreover, they are designed to generate all the expected conceptual deductions. However, some skews between conceptual analysis and DL reasoning cannot be overcome by using rewrite rules only.

For instance, subsumption between relationships conceptually differs from subsumption between entities. More specifically, (i) the same arity and (ii) equal or hierarchically related relationship roles are necessary conditions for a superclass/subclass link between two relationships. Unfortunately, a representation of relationships as $\mathcal{SHIQ}$ classes does not provide inferences in accordance with such constraints. In fact, there can be wrong subsumptions between binary and ternary relationships, as well as between two relationships with one link to the same entities. In order to prevent this, CONCEPTOOL reproduces the expected conceptual deductions (i) by modifying the way $\mathcal{EDDL}_{EER}$ concepts are rewritten in $\mathcal{SHIQ}$ for the purpose of reasoning and (ii) by checking whether the above conditions are fulfilled.

Existing systems for intelligent conceptual modelling (e.g. the i•COM tool [4]), which are also using a DL-based approach, do not recognise the above characteristics of relationship subsumption. Moreover, the expressive power of i•COM does not support identifiers, multivalued attributes, fillers and generic cardinality constraints in relationships.

3.2　Reasoning with a selectively variable expressive power

The inference controller allows to decrease the expressive power *subject to reasoning* by selecting which differences should be ignored. For instances, attribute names only could be used, ignoring their multiplicities and their domains. Alternatively, minimum and maximum multiplicities can be overridden to default to a certain couple of values and/or domains can be overridden to default to a predefined class and/or attribute fillers can be ignored. Moreover, relationships and/or assertions can be ignored as well. Finally, preferred aliases (if any) can be used in place of the actual attribute names. To the best of our knowledge, no other intelligent knowledge management environment supports deductions with tunable expressive power.

3.3　Hybrid reasoning with an expressive knowledge model

If a knowledge model is too expressive to be analysed within the framework of DLs, then other reasoning paradigms must be jointly used. For instance, using a class-centred knowledge model with complex local numeric constraints, the inference controller splits a KB in two homogeneous emulators (e.g. a DL one and a constraint one), which then interact with two different inferential engines (e.g. a DL terminological classifier and a constraint solver). Deductions or inferences from these two engines are then *fused* by the inference controller to produce an overall result. Although not integrated in CONCEPTOOL yet, this approach has been already developed as a standalone semantic analyser with an inference controller [6].

Several approaches have been introduced in the past which deal with hybrid reasoning (e.g. about knowledge which integrates DLs and Horn Rules [7]). However, these approaches do not address the problem of reasoning with highly expressive heterogeneous knowledge.

3.4　Linguistic (lexical) and heuristic reasoning

A lexical analyser uses a source of external knowledge to reason about the terminology of an ontology. It proposes taxonomic links or detects lexical synonyms (e.g. the term CLIENT is a synonym for the term CUSTOMER). The lexical database WordNet [9] can be used to perform this kind of reasoning, as in the articulation of ontology interdependencies [10].

The existing kinds of heuristic analysis include string-substring matching or the detection of cycles (e.g. a MAMMAL is-a LIVING_THING and a LIVING_THING is-a MAMMAL). The *heuristic analyser* in Figure 1 proposes correlations between concepts with lexical or structural similarities, as in the Chimaera [8] system, which uses string-substring matching.

In contrast with the above approaches, $\mathcal{K}$–ShaRe jointly uses both kinds of analysis.

3.5　Supporting knowledge sharing and reuse

The $\mathcal{K}$–ShaRe architecture provides functionalities to support the sharing and the reuse of declarative knowledge (e.g. ontologies). Currently, the *sharing manager* in CONCEPTOOL supports the alignment of two ER KBs using an articulation schema and a corresponding set of mappings between concepts. A detailed description of this functionality is presented in [2].

The PROMPT [12] and Chimaera [8] systems currently support ontology reuse through knowledge merging. However, they are both based on expressive frame models which do not enable DL-like automated subsumption or other extensive reasoning capabilities.

4   Conclusion

In this paper, we have introduced the $\mathcal{K}-$ShaRe architecture, which addresses the problem of automated reasoning during modelling, analysis, sharing and reuse of conceptual knowledge.

We have highlighted that this architecture cannot rely on a DL-based approach alone. In fact, DL reasoning fails to differentiate between the different kinds of deductions which apply to distinct categories of concepts such as entities, relationships, functions, etc. Therefore, we have advocated the joint usage of different kinds of (conceptual, linguistic, and heuristic) reasoning, together with the modular variability of the "reasonable" expressive power, as a key issue in order to cope with the complexity of the knowledge lifecycle. We have also briefly discussed how existing systems only provide some of the $\mathcal{K}-$ShaRe functionalities.

Acknowledgements

This work is supported by the EPSRC under grants GR/R10127/01 and GR/N15764. Derek Sleeman and Peter Gray provided useful suggestions during the definition of our framework.

References

[1] E. Compatangelo and H. Meisel. Intelligent support to knowledge management: conceptual analysis of schemas and ontologies. Submitted for publication.

[2] E. Compatangelo and H. Meisel. Intelligent support to knowledge sharing through the articulation of class schemas. Submitted for publication.

[3] D. Fensel et al. OIL: An Ontology Infrastructure for the Semantic Web. *Intelligent Systems*, 16(2):38–45, 2001.

[4] E. Franconi and G. Ng. The i•com Tool for Intelligent Conceptual Modelling. In *Proc. of the 7th Intl. Workshop on Knowledge Representation meets Databases (KRDB'00)*, 2000.

[5] I. Horrocks, U. Sattler, and S. Tobies. Practical reasoning for expressive description logics. In *Proc. of the 6th Intl. Conf. on Logic for Programming and Automated Reasoning (LPAR'99)*, number 1705 in Lecture Notes In Artificial Intelligence, pages 161–180. Springer-Verlag, 1999.

[6] B. Hu, E. Compatangelo, and I. Arana. Coordinated reasoning with inference fusion. Submitted for publication.

[7] A. Y. Levy and M. C. Rousset. Carin: a representation language integrating rules and description logics. In *Proc. of the 12th European Conf. on Artificial Intelligence (ECAI'96)*, pages 323–327, 1996.

[8] D. L. McGuinness et al. An Environment for Merging and testing Large Ontologies. In *Proc. of the 7th Intl. Conf. on the Principles of Knowledge Representation and Reasoning (KR'2000)*, 2000.

[9] G. A. Miller. WordNet: a Lexical Database for English. *Comm. of the ACM*, 38(11):39–41, 1995.

[10] P. Mitra, G. Wiederhold, and M. L. Kersten. A Graph-Oriented Model for Articulation of Ontology Interdependencies. In *Proc. of the VII Conf. on Extending Database Technology (EDBT'2000)*, Lecture Notes in Computer Science, pages 86–100. Springer-Verlag, 2000.

[11] N. F. Noy, R. W. Fergerson, and M. A. Musen. The knowledge model of Protégé-2000: Combining interoperability and flexibility. In *Proc. of the 12th Intl. Conf. on Knowledge Engineering and Knowledge Management (EKAW'2000)*, 2000.

[12] N. F. Noy and M. A. Musen. PROMPT: Algorithm and Tool for Automated Ontology Merging and Alignment. In *Proc. of the 17th Nat. Conf. on Artificial Intelligence (AAAI'00)*, 2000.

[13] The Advanced Knowledge technologies (AKT) Consortium. The AKT Manifesto, Sept. 2001. Available from http://www.aktors.org/publications.

KES 2002
E. Damiani et al. (Eds.)
IOS Press, 2002

Use of Fuzzy Sets to Measure the Degree of Inclusion between Vague Concepts in Semantic Nets

M-N. OMRI & M. A. MAHJOUB
IPEIM, Department of Mathematical and Data Processing,
Preparatory Institute of Engineering Studies of Monastir
Kairouan Road, 5019 Monastir, Tunisia

Abstract. The approach described here present a method to measure the degree of inclusion between vague Concepts based on the degree of inclusion between fuzzy properties and the degree of inclusion between fuzzy Goals, in a fuzzy Semantic Network. The set of properties, based on the fuzzy sets theory, describe vague Concepts (Goals and Objects). With the learning of new Objects and new Goals, from the requests of users, our system improves its representation schema, which implies the valuation of the relation between two vague Goals and between two vague Objects.

1. Introduction

Using natural synonyms or words which are not in the system's dictionary makes a request not understandable to a computer, because the existing computer-based systems require to present user's requests in a very precise special formal language. From the user viewpoint, this requirement is a big drawback, and then, it is desarable to make computer systems understand natural language requests. To make this system[1] understand natural language Objects and Goals, it is therefore desirable, for each such Object and Goal, to find the standard Object and Goal that present the most general structure which allow to identify the original ones.

One of the main problems with understanding natural language requests is that natural language is often ambiguous (fuzzy). It is therefore natural to use *fuzzy logic* as a tool to interpret fuzzy users' Objects and Goals. We have successfully used the formalism of fuzzy Semantic Networks[2] as a knowledge base for designing such a system.

This system interprets an unknown word by using the links created between this new word and known words[3,4]. The main link is provided by the context of the request, when user's request is confused with an unknown Goal applied to a known Object denoted in the Semantic Network, the system infer that this new Goal corresponds to one of the known Goal[5].

Both for Objects and for Goals, we must therefore be able to describe the degree of inclusion. The main Objective of this paper is to describe how these degrees of inclusion can be defined.

2. Fuzzy Goals and Fuzzy Objects

In fuzzy logic, there has been a lot of research into the notion of inclusion degree. However, so far, researchers have mainly analyzed the inclusion degree between fuzzy sets, while we need a somewhat more sophisticated notion: inclusion degree between Objects and between Goals. To extend definitions of inclusion degree from fuzzy sets to fuzzy Objects and Goals, we must first analyse how fuzzy Objects and Goals can be described in terms of fuzzy sets.

How does a user formulate his or her Goals? For example, how do we describe a Goal when we look for a house to buy? A natural Goal is to have a house not too far away from work, not too expensive, in a nice neighborhood, etc. In general, to describe a Goal:

- we list *properties P_1, ..., P_n* (in the above example, distance, cost, and neighborhood quality), and
- we list the desired (fuzzy) values $V_1,...,V_n$ of theses properties (in the above example, these values are, correspondingly, "not too far", "not too expensive", and "nice".

Each of the fuzzy values V_i like "not too far" can be represented, in a natural way, as a fuzzy set formed by couples (v_i, d_i), where v_i (i.e., far) is a possible synonym of V_i with a certain degree (i.e., 0.9). Similarly, an Object can be described if we list the properties and corresponding values. At first glance, it may seem that from this viewpoint, a description of an Object is very much alike the description of a Goal, but there is a difference.

When describing a property of an Object, we may want to distinguish between the expert value of this property and a user value. For example, when we describe a sports car, "high" may be a necessary value of the attribute "speed" but "very high" is a possible value. Correspondingly, for each property, we may have two different fuzzy values: the expert's and the user's values. In what follows, we first describe the structure of Expert's and User's fuzzy properties which define Goals and Objects, in the Semantic Network. We next explain how the degree of inclusion between two Goals and between two Objects are calculated.

3. Fuzzy Properties

As we have mentioned before, a natural way to describe a fuzzy goal is to describe the (fuzzy) values $V_1,..., V_n$ of the desired properties $P_1,..., P_n$. Each property P_i is described by two types of values: Expert's value and User's value. Let:

- the values $v_1,..., v_m$ correspond to the possible values of V_i, relatively to the property P_i and the degrees $d_1,..., d_m$ correspond to the possible degrees associated by the Expert to v_i, $i \in [1,n]$.
- the values $v_1,..., v_m$ correspond to the possible values of V_i, relatively to the property P_i and the degrees $f_1,..., f_m$ correspond to the possible membership functions obtained from responses associated by the User to v_i, $i \in [1,n]$.

4. Degree of Inclusion Between Vague Concepts

Many researchers have suggested that vague, non-precise observations should be described by fuzzy sets. Fuzzy set theory originated by Zadeh[6] relies on ordering relations that express degree of membership of an object in a set. Recently Pawlak[7] has proposed rough set methodology as a new approach in handling classificatory analysis of vague concepts. In this methodology any vague concept is characterized by a pair of precise concepts called the lower and the upper approximations. Several studies have already been conducted about combinations of rough and fuzzy sets. Dubois and Prade[8,9], or Kuchneva[10] introduced the notions of rough fuzzy sets and fuzzy rough sets. Bouchon, Yager, and Zadeh[11] the uncertainty in knowledge bases to treat the vague concepts.

A general description of the degree of inclusion is given in[12]. In this paper, we show how some of these definitions can be extended to describe the degree of inclusion between fuzzy Goals and between fuzzy Objects.

5. Fuzzy Degree of Inclusion : General definition

Let X be a non-empty Universe of discourse.

Definition 1. *A mapping D_C which assigns a number $D_C(A,B) \in [0,1]$ to every pair of fuzzy subsets of X is called a fuzzy degree of inclusion if the three following conditions are satisfied:*

- *$D_C(A,A)=1$;*
- *$A \subset B$ if, and only if, $D_C(A,B)=1$;*
- *$0 \leq D_C(A,B) \leq 1$.*

If A and B are two fuzzy sub-sets in a universe X, the inclusion of A in B is defined by:

$$A \subset B \Leftrightarrow f_A(x) \leq f_B(x)$$

A degree of inclusion is defined by:

$$D_C(A,B) = \varepsilon \frac{\int_X f_{A \cap B}(x)\,dx}{\int_X f_A(x)\,dx} + (1-\varepsilon) \frac{\sum_{x \in \text{sup } p(A \cap B)} \min(f_A(x), f_B(x))}{\sum_{x \in \text{sup } p(A)} f_A(x)}$$

with: $\varepsilon=0$ if supp(A) is countable, $\varepsilon=1$ else.

As a result, we get the following two fuzzy degree of inclusion: in the case where the Universe X is discreet, then the degree of inclusion is defined by the following definition:

$$D_C(A,B) = \frac{\sum_{x \in X} f_{A \cap B}(x)}{\sum_{x \in X} f_A(x)} \qquad (1)$$

In the case where the membership functions of A and B, f_A and f_B, are continuous, we have:

$$D_C(A,B) = \frac{\int_{x \in X} f_{A \cap B}(x)\,dx}{\int_{x \in X} f_A(x)\,dx} \qquad (2)$$

6. Degree of Inclusion Between Fuzzy Goals

As it is mentioned before, each Goal is described by two types of fuzzy values (expert's value and user's value). Then, to determined the degree of inclusion between two Goals, it is natural to calculates this degree from the degree of inclusion between the corresponding fuzzy values.

6.1. Degree of Inclusion Between Fuzzy Values

We consider two fuzzy properties P_1 and P_2 defined on the same Universe. Let:
- the values $v_{11},..., v_{1p}$ correspond to the possible values V_1 of the property P_1 and the degrees $d_{11},..., d_{1p}$ their associated possible degrees.
- the values $v_{21},..., v_{2p}$ correspond to the possible values V_2 of the property P_2 and the degrees $d_{21},..., d_{2p}$ their associated possible degrees.

Intuitively, the value V_1 is included in the value V_2 if each component of V_1 is included in the corresponding component of V_2, i.e., if v_{11} is included in v_{21}, and v_{12} is included in v_{22}, etc. For each i, the degree to which v_{1i} is included in v_{2i} is characterized by the value $D_\subset(v_{1i}, v_{2i})$. If we use the formula given in the general definition to represent "$\subset$", then we get the following expression for the inclusion degree $D_\subset(V_1, V_2)$ of V_1 in V_2:

$$D_\subset(V_1,V_2) = \frac{\sum\limits_{p \in P} f_{V_1 \cap V_2}(p)}{\sum\limits_{p \in P} f_{V_1}(p)} \tag{3}$$

To represent the intersection between V_1 and V_2, we use the t-norm min and we get the following expression for the membership function of the intersection $f_{V_1 \cap V_2}$:

$$f_{V_1 \cap V_2} = \min_{1 \le i \le n}\left(d_{1i}, d_{2i}\right) \tag{4}$$

We can now apply the formula (4) to the formula (3), it results that the degree of inclusion of V_1 in V_2 is given by:

$$D_\subset(V_1,V_2) = \frac{\sum\limits_{i=1}^{n} \min\limits_{1 \le i \le n}(d_i, d_i)}{\sum\limits_{i=1}^{n} d_i} \tag{5}$$

6.2. Degree of Inclusion Between Fuzzy Properties

The degree of inclusion between two fuzzy properties is calculated according to the two degrees of inclusions calculated previously, we have two fuzzy values: expert's value and user's value. Thus, for each property, instead of a single inclusion's degree, we have two inclusion's degrees $D_\subset(V_1^E, V_2^E)$ and $D_\subset(V_1^U, V_2^U)$ corresponding to expert's and user's values. If we want a simple degree for each property, we must combine these two degrees. A natural combination is an arithmetic average:

$$D_\subset(P_1,P_2) = \frac{D_\subset(V_1^E, V_2^E) + D_\subset(V_1^U, V_2^U)}{2} \tag{6}$$

6.3. Degree of Inclusion Between Fuzzy Goals

As it is shown in the previous paragraphs, a Goal of an Object in the Semantic Network is defined by its properties. Let:
- the properties $P_1, P_2,..., P_n$ defines the Goal G_1, and
- the properties $P_1', P_2',..., P_n'$ defines the Goal G_2.

If we use min to represent "and", then we get the following expression for the inclusion degree $D_\subset(G_1, G_2)$ between the Goals G_1 and G_2.

$$D_\subset(G_1, G_2) = \min_{1 \leq i \leq n}(D_\subset(P_i, P_i')) \tag{7}$$

7. Degree of Inclusion Between Fuzzy Objects

As we have mentioned before, an Object in the Semantic Network is defined by its Goals as a properties. Let:

- the Goals $G_1, G_2, ..., G_n$ defines the Object O_1, and
- the Goals $G_1', G_2', ..., G_h$ defines the Object O_2.

If we use the same principale used in the previous paragraph, we get the following expression for the inclusion degree $D_\subset(O_1, O_2)$ between O_1 and O_2.

$$D_\subset(O_1, O_2) = \min_{1 \leq i \leq n}(D_\subset(G_i, G_i')) \tag{8}$$

8. Conclusion

In this paper, we use the fuzzy sets theory to describe the Goals and Objects, in a Semantic Network. This description is based on two types of fuzzy properties: Expert's propertiy and User's property. We have calculated the different degree of inclusion between Goals and between Objects. Both for Goals and for Objects, the degree of inclusion is defined based on degree of inclusion between fuzzy properties.

References

[1]   Omri, M.N., Tijus C.A. and Bouchon-Meunier, B. Fuzzy Sets System for User's Assistance: How Sifade Diagnoses User's Procedure. *The Second World Congress on Expert Systems: "Moving towards Expert Systems Globally in the 21st Century". Macmillian ed. CD Rom.* Estoril, Lisbonne, (1993).

[2]   Omri, N.M., Tijus, C.A., Poitrenaud, S. & Bouchon-Meunier, B. Fuzzy Sets and Semantic Networks for On-Line Assistance. *Proceedings of the Eleven IEEE Conference on Artificial Intelligence Applications.* Los-Angeles, (1995).

[3]   Omri, N.M., Fuzzy Interactive System to Use Technical Divices: SIFADE. *Ph. D. Dissertation, University Paris VI, Paris, (1994).*

[4]   Omri, M.N. & Chouigui N. Linguistic Variables Definition by Membership Function and Measure of Similarity. *14th International Conference on Systems Science.* Wroclaw, Poland, (2001).

[5]   Omri, M.N. & Chouigui N. Measure of Similarity Between Fuzzy Concepts for Identification of Fuzzy User's Requests in Fuzzy Semantic Networks. *International Journal of Uncertainty, Fuzziness and Knowledge-Based Systems.* Vol. 9, No. 6, 743-748, (2001).

[6]   Zadeh L.A. Fuzzy Sets. *Information and Contro.* 8, 338-353, (1965).

[7]   Pawlak, Z. Rough Sets. *International Journal of Computer and Information Sciences.* 11, 341-356, (1982).

[8]   Dubois D. and Prade H. Rough Fuzzy sets and Fuzzy Rough Sets. *International Journal of general Systems.* 17, 191-209, (1990).

[9]   Dubois D. and Prade H. Vagueness and typicality in class hierachies. *Proceedings 3rd IFSA Congress,* Seattle, (1989).

[10]   Kuchneva, L. I. Fuzzy Rough Sets: Application to Feature Selection. *Fuzzy Sets and Systems.* 51, 147-153, (1992).

[11]   Bouchon B., Yager R.R., and Zadeh L.A. (eds). Uncertainty in Knowledge Bases. *Lecture Note in Computer Science, 521, Spring Verlag,* (1991).

[12]   Zadeh L.A. Pruf-a meaning representation language for naturel languages. *InternationalJournal of Man-Machine Studies,* 10:395-460, (1978).

KES 2002
E. Damiani et al. (Eds.)
IOS Press, 2002

Similarity Measures for Fuzzy Values

Jiří DVOŘÁK and Miloš ŠEDA
*Institute of Automation and Computer Science, Faculty of Mechanical Engineering,
Brno University of Technology, Technická 2, 616 69 Brno, Czech Republic*

Abstract. The concept of similarity is the key notion in many fields (e.g. case-based reasoning, multimedia databases, expert systems). In this paper, we study selected similarity measures for fuzzy values. We distinguish measures sensitive and non-sensitive to a distance between fuzzy values. We compare the properties of these measures and propose some combined intersection-sensitive measures and a distance-based measure for fuzzy numbers.

1 Introduction

Retrieval of similar objects is an important task in solving many problems. For example, in case-based reasoning, previously solved problems similar to a given problem are searched for and their solution is then adapted to the new problem. The other related areas are multimedia databases and knowledge-based systems.

A variety of similarity measures, together with interesting properties, have been proposed in the literature. In [7], a broad review of the existing similarity measures can be found. An object description is often only vague and thus fuzzy set-based techniques must be used to represent and match the imprecise and uncertain values. In our paper we study selected similarity measures between fuzzy values and propose some new approaches.

A similarity relation can be defined as a mapping $sim : U \times U \to [0, 1]$, satisfying the three following properties [3]:

$$\forall x \in U, \; sim(x, x) = 1, \quad \text{(reflexivity)} \tag{1}$$

$$\forall x, y \in U, \; sim(x, y) = sim(y, x), \quad \text{(symmetry)} \tag{2}$$

$$\forall x, y, z \in U, \; sim(x, y) \otimes sim(y, z) \leq sim(x, z), \quad (\otimes - \text{transitivity}) \tag{3}$$

where $\otimes$ is a binary operation on the unit interval with some additional properties. This operation is usually a t-norm operation, i.e. $\otimes$ is a non-decreasing operation on [0, 1], satisfying associativity, commutativity, 1 being the neutral element and 0 being the absorbent element ($1 \otimes a = a$, $0 \otimes a = 0$).

While the properties $0 \leq sim(x, y) \leq 1$, reflexivity and symmetry are usually assumed (but there exist also non-symmetric similarity measures), extended transitivity does not seem always compulsory. Sometimes a stronger property than reflexivity is required for similarity relations:

$$\forall x \in U, \; sim(x, y) = 1 \Leftrightarrow x = y, \quad \text{(separativity)} \tag{4}$$

Fuzzy values can represent linguistic values, inexact numbers or inexact intervals. Membership functions can also represent special user similarity functions which determine what is considered as similar and what not. Such a function $\mu_y(x)$ corresponds to the function $sim(x, y)$ and determines the fuzzy set of values x similar to the value y. A frequently used type of fuzzy value is a fuzzy number. A fuzzy number A is a normal convex fuzzy set in the universe of real numbers which is represented by 4-tuple (a_1, a_2, a_3, a_4) and a piecewise continuous membership function with the following properties: (i) $a_1 \leq a_2 \leq a_3 \leq a_4$; (ii) $\mu_A(x) = 0$ for $x \leq a_1$ or $x \geq a_4$; (iii) $\mu_A(x) = 1$ for $a_2 \leq x \leq a_3$; (iv) $\mu_A(x)$ is increasing on $[a_1, a_2]$ and decreasing on $[a_3, a_4]$. A fuzzy number can be also represented as a function pair $[L_A(\alpha), R_A(\alpha)]$, where $L_A(0) = a_1$, $R_A(0) = a_4$, and $L_A(\alpha) = \inf\{x | \mu_A(x) \geq \alpha\}$, $R_A(\alpha) = \sup\{x | \mu_A(x) \geq \alpha\}$ for $\alpha \in (0, 1]$.

2 Similarity Measures Non-Sensitive to a Distance between Fuzzy Values

Similarity measures for fuzzy values are mostly constructed only for fuzzy sets with non-empty intersection and are not sensitive to a "distance" between fuzzy sets. It is desirable that an intersection-oriented similarity measure should have the properties (2), (4) and the property

$$sim(A, B) = 0 \Leftrightarrow A \cap B = \emptyset \tag{5}$$

A very simple similarity measure, used often e.g. for matching in fuzzy expert systems, is the measure [2, 4]

$$sim(A, B) = T_{A,B} = \sup_{x \in U} \min\{\mu_A(x), \mu_B(x)\} \tag{6}$$

This measure satisfies the properties (2) and (5), but not the properties (1) and (4). If A is not a normal fuzzy set, then $T_{A,A} < 1$. We may also have $T_{A,B} = 1$ for $A \neq B$. For example, let $A = (a_1, a_1, a_2, a_2)$ and $B = (a_2 - \varepsilon, a_2 - \varepsilon, b, b)$ be trapezoidal fuzzy numbers where ε is a small positive number. A and B are therefore almost disjoint crisp intervals, but $T_{A,B} = 1$. The measure $T_{A,B}$ is also not sensitive to the area of intersection. Another measure with a wrong sensitivity to the area of intersection is the measure $L_{A,B}$ [5]:

$$sim(A, B) = L_{A,B} = 1 - \sup_{x \in U} |\mu_A(x) - \mu_B(x)| \tag{7}$$

This measure satisfies the properties (2) and (4), but not the property (5). If $A \cap B = \emptyset$, A and B are not normal fuzzy sets, then $L_{A,B} > 0$. We may also have $L_{A,B} = 0$ for $A \cap B \neq \emptyset$. For example, let $A = (a_1, a_1, a_2, a_2)$ and $B = (a_1 + \varepsilon, a_1 + \varepsilon, a_2 + \varepsilon, a_2 + \varepsilon)$ be trapezoidal fuzzy numbers where ε is a small positive number. A an B are therefore almost equal crisp intervals, but $L_{A,B} = 0$.

Karacapilidis and Pappis [5] have also proposed the next two similarity measures for fuzzy sets in a discrete universe. For a continuous universe they can be modified as follows (see also [6]):

$$sim(A, B) = M_{A,B} = \frac{area(A \cap B)}{area(A \cup B)} \quad \text{where} \quad area(A) = \int_U \mu_A(x)dx \tag{8}$$

$$sim(A, B) = S_{A,B} = 1 - \frac{\int_U |\mu_A(x) - \mu_B(x)|dx}{\int_U (\mu_A(x) + \mu_B(x))dx} \tag{9}$$

These measures satisfy the properties (1), (2) and $A \cap B = \emptyset \Rightarrow sim(A, B) = 0$. If A and B are fuzzy numbers, then the properties (4) and (5) hold as well.

Chen et al. [2] have proposed the measure W for fuzzy sets in a discrete universe, which gives some counter-intuitive results for the fuzzy sets with empty intersection. The repaired measure for the continuous universe has the form (see also M_2 [6]):

$$sim(A, B) = W_{A,B} = 1 - \frac{1}{m(supp(A) \cup supp(B))} \int_U |\mu_A(x) - \mu_B(x)| dx \qquad (10)$$

where $m(supp(A))$ is the measure (size) of the crisp set $supp(A)$. If $supp(A) = [a1, a2]$, then $m(supp(A)) = a_2 - a_1$. This measure satisfies the properties (1) and (2), but not the property $A \cap B = \emptyset \Rightarrow sim(A, B) = 0$. If $A \cap B = \emptyset$, A and B are not crisp intervals, then $W_{A,B} > 0$. Moreover, $W_{A,B}$ may be almost 1 for sets with empty intersection. For example, let A and B be fuzzy numbers with $supp(A) = [a, a+1]$, $supp(B) = [b-1, b]$, $a+1 < b-1$, and membership functions

$$\mu_A(x) = \begin{cases} 1 - (x - a)(1 - \varepsilon)/\varepsilon & a \leq x \leq a + \varepsilon \\ \varepsilon - (x - a - \varepsilon)\varepsilon/(1 - \varepsilon) & a + \varepsilon \leq x \leq a + 1 \end{cases} \qquad (11)$$

$$\mu_B(x) = \begin{cases} (x - b + 1)\varepsilon/(1 - \varepsilon) & b - 1 \leq x \leq b - \varepsilon \\ \varepsilon + (x - b + \varepsilon)(1 - \varepsilon)/\varepsilon & b - \varepsilon \leq x \leq b \end{cases} \qquad (12)$$

where ε is a small positive number. Then $W_{A,B} = 1 - \varepsilon$ and therefore, for small ε, it would mean that A is almost equal to B.

None of the measures (6), (8), (9), (10) is sensitive to the position of A in B, if $A \subseteq B$. Let A' be defined as follows: $\mu_{A'}(x) = \mu_A(x + \Delta)$ where Δ is such that $A' \subseteq B$. Then $sim(A, B) = sim(A', B)$. This circumstance can be removed by means of some combination with a distance-based similarity.

Among the compared measures, the measures M (8) and S (9) are the best. But in the case of fuzzy numbers it may not be of advantage that these measures, in contrast to the measure T (6), are not sensitive to the height of the intersection of the given sets. This drawback can be removed by means of the following combined measures:

$$sim(A, B) = v_T T_{A,B} + v_M M_{A,B}, \quad sim(A, B) = v_T T_{A,B} + v_S S_{A,B} \qquad (13)$$

where $v_T, v_M, v_S > 0$ and $v_T + v_M = 1$, $v_T + v_S = 1$. If A, B are fuzzy numbers, then these measures satisfy the properties (1), (2) and (5). The measures (13) partly solve another problem of the measures (8) and (9). Let $A = (a - \varepsilon, a, a, a + \varepsilon)$ and $B = (b_1, b_2, b_3, b_4)$ be trapezoidal fuzzy numbers, where ε is a small positive number and $a \in [b_1, b_4]$. Then $\lim_{\varepsilon \to 0} T_{A,B} = \mu_B(a)$, while $\lim_{\varepsilon \to 0} M_{A,B} = 0$ and $\lim_{\varepsilon \to 0} S_{A,B} = 0$, which is counter-intuitive, because the similarity between the crisp value a and the fuzzy value B should not be equal to zero.

3 Distance-Sensitive Similarity Measures

Intersection-based measures that assign zero to fuzzy sets with empty intersection are proper in situations where such fuzzy values should be rejected as quite dissimilar. But in practice there are often situations, when the system does not find any fuzzy value with non-empty intersection and we wish to know at least the "nearest" fuzzy value. Then the distance-based similarity measure should be used.

The distance-based similarity measure can be defined in this way:

$$sim(x, y) = \Phi\big(dist(x, y)\big) \tag{14}$$

where Φ is a decreasing function satisfying $\Phi(0) = 1$ and $dist$ is a distance relation (usually normalized, i.e. having values from $[0, 1]$). The distance relation should satisfy the metric properties (non-negativity, symmetry, triangle inequality and the property $dist(x, y) = 0 \Leftrightarrow x = y$). If it holds only $dist(x, y) = 0 \Leftarrow x = y$, then the relation $dist$ is a pseudometric.

A frequently used distance-based similarity measure is:

$$sim(x, y) = 1 - dist(x, y) \quad \text{(the distance should be normalized)} \tag{15}$$

It is easy to prove that, if $dist$ is a pseudometric, then this similarity measure satisfies the similarity properties (1)-(3), where $\otimes$ is Lukasiewicz operation, i.e. $\max(0, x + y - 1)$. If $dist$ is a metric, then the property (4) holds as well.

The similarity measure proposed by Xu et al. [9] has the form (15), where the non-normalized distance is based on centres of gravity (general mean values). Let $U = [c, d]$. After normalizing, this pseudometric has the form:

$$dist(A, B) = \frac{|mean(A) - mean(B)|}{d - c} \quad \text{where} \quad mean(A) = \frac{1}{area(A)} \int_U x\mu_A(x)dx \tag{16}$$

It is obvious that the representation of fuzzy set by only one crisp number is not sufficient for determining the distance between fuzzy sets. For example, let $A = (a_1 - \alpha, a_1, a_2, a_2 + \alpha)$, $A' = (a_1 - \alpha', a_1 - \alpha'', a_2 + \alpha'', a_2 + \alpha')$ and $B = (b_1, b_2, b_3, b_4)$ be trapezoidal fuzzy numbers. Then $dist(A', B) = dist(A, B)$. To remove this obstacle we propose the following function defined for fuzzy numbers in the universe $U = [c, d]$:

$$dist(A, B) = v_1 dist_1(A, B) + v_2 dist_2(A, B) \tag{17}$$

where v_1, v_2 are positive normalized weights (i.e. $v_1 + v_2 = 1$), the distance $dist_1$ is given by (16) and $dist_2$ is defined in this way:

$$dist_2(A, B) = \frac{1}{d - c} \int_0^1 [\max\{0, L_B(\alpha) - R_A(\alpha)\} + \max\{0, L_A(\alpha) - R_B(\alpha)\}]d\alpha \tag{18}$$

The function (17) does not satisfy all metric axioms. It is non-negative, symmetric and it holds $dist(A, A) = 0$. But if the similarity property (3) is not required, the triangle inequality for $dist$ is not necessary.

The described distance-based measures do not assume the intersection of fuzzy sets. Moreover, when a similarity measure is based on (16), then fuzzy sets with non-empty intersection can have the same similarity as fuzzy sets with empty intersection. Therefore it is necessary to combine these measures with intersection-sensitive measures into a hybrid measure:

$$sim(A, B) = v_I sim_I(A, B) + v_D sim_D(A, B) \tag{19}$$

where sim_I is an intersection-sensitive measure, sim_D is a distance-based measure, and v_I, v_D are positive normalized weights. If sim_I and sim_D satisfy properties (1), (2) or (2), (4), then also sim satisfies these properties.

4 Conclusion

In this paper we are concerned with similarity measures between fuzzy values defined in a continuous universe. We study properties of selected similarity measures and propose some new approaches or modifications of existing ones. The next problem we study is the accumulation of local similarity measures of object attributes into a global similarity measure of the whole object.

A global similarity function can be composed from local similarity functions by a non-decreasing composition function. According to Richter [8] we can accumulate intersection-sensitive measures satisfying the property (5) by means of some t-norm, while distance-based and hybrid measures can be accumulated by means of some t-conorm. A t-conorm is a non-decreasing operation on $[0, 1]$, satisfying associativity, commutativity, 0 being the neutral element and 1 being the absorbent element. Similarities of different types can be combined e.g. as follows: $\lambda\alpha + (1 - \lambda)\beta$ or $\lambda \min(\alpha, \beta) + (1 - \lambda) \max(\alpha, \beta)$, where $0 \leq \lambda \leq 1$. Other pairs of aggregation operators corresponding to fuzzy AND and fuzzy OR are proposed in [1] as follows: $M(\mathbf{x}) = 1 - d(\mathbf{1}, \mathbf{x})$ (fuzzy AND), $N(\mathbf{x}) = d(\mathbf{0}, \mathbf{x})$ (fuzzy OR), where $\mathbf{x} \in [0, 1]^n$, $\mathbf{1} = (1, 1, \ldots, 1)$, $\mathbf{0} = (0, 0, \ldots, 0)$ and d is a (pseudo)metric normalized to 1.

Acknowledgments

This paper has been supported by the research project CEZ: J22/98: 261100009 *Non-Traditional Methods for Investigating Complex and Vague Systems* and the project MSM 262100024 *Research and Development of Mechatronic Systems*.

References

[1] G. Beliakov. Definition of General Aggregation Operators through Similarity Relations. *Fuzzy Sets and Systems* **114** (2000) 437–453.

[2] S.-M. Chen, M.-S. Yeh, and P.-Y. Hsiao. A Comparison of Similarity Measures of Fuzzy Values. *Fuzzy Sets and Systems* **72** (1995) 79–89.

[3] D. Dubois, F. Esteva, P. Garcia, L. Godo, R.L. de Mántaras, and P. Henri. Fuzzy Set-Based Models in Case-Based Reasoning. *Lecture Notes in Artificial Intelligence* **1266** (1997) 599–610.

[4] D.H. Hyung, Y.S. Song, and K.M. Lee. Similarity Measure between Fuzzy Sets and between Elements. *Fuzzy Sets and Systems* **62** (1994) 291–293.

[5] N. Karacapilidis and C. Pappis. Computer-Supported Collaborative Argumentation and Fuzzy Similarity Measures in Multiple Criteria Decision Making. *Computers & Operations Research* **27** (2000) 653–671.

[6] T. W. Liao. Classification and Coding Approaches to Part Family Formation under a Fuzzy Environment. *Fuzzy Sets and Systems* **122** (2001) 425–441.

[7] T. W. Liao, Z. Zhang, and C.R. Mount. Similarity Measures for Retrieval in Case-Based Reasoning Systems. *Applied Artificial Intelligence* **12** (1998) 267–288.

[8] M.M. Richter. Perspectives on the Integration of Fuzzy and Case-Based Reasoning Systems. Technical Report, International Computer Science Institute, University of California at Berkeley, March 1997.

[9] M. Xu, K. Hirota, and H. Yoshino. Fuzzy Theoretical Approach to Case-Based Representation and Inference in CISG. *Artificial Intelligence and Law* **7** (1999) 259–272.

KES 2002
E. Damiani et al. (Eds.)
IOS Press, 2002

Rule-Based Valid-Time Expert Systems with Multicriteria Fuzzy Sets

I. V. Filis, C. P. Yialouris, A. B. Sideridis

Informatics Laboratory, Agricultural University of Athens,
75 Iera Odos, 118 55 Athens, Greece

Abstract. This paper proposes a Rule Based Expert System based on the Relational Data Model with its rules based on the Object-Attribute-Value (O-A-V) triplet representation model with respect to the valid time of each one rule. The values of the O-A-V triplets are related to fuzzy sets so these become fuzzy values, and each one fuzzy value is able to depend on one or more criteria affecting it. Therefore, it is a Rule-Based Valid-Time Expert System with Multicriteria Fuzzy Sets. The development of the system on a relational database is also presented.

1. Introduction

Expert Systems (ES) are computer programs designed to represent human expertise in a specific domain [10]. A Rule-Based Expert System (RBES) is an ES that represents knowledge by means of rules, aiming to help a no expert user to solve real-world problems. The elements of a RBES are the Knowledge Base (KB) containing the expert's knowledge, the inference engine which decides the appropriate rules to be fired in order to suggest the best solution(s) and the user interface which is the contact between the ES and the end user. The KB of a RBES has as a fundamental component, the production rule, which in the simplest case has the form:

IF <assertionIf> THEN < assertionThen>.

Since a rule is a piece of knowledge, and Valid Time (VT) of a piece of knowledge is the time during which this knowledge is true in the real world [8], VT of a rule is the time interval during which, the rule we are referring to, is true or exists. As an example of the above definitions is the rule 'R' which is acceptable in our days (VT):

R: "IF a man is middle-aged THEN he has to work in order to live"

But what is the meaning of the word 'middle-aged'? Am I a middle-aged man? One could consider ages from 20 to 60 but since I am in this interval, I consider from 40 to 65.

To resolve this argument we use the membership grade of each crisp value, which derives from a membership function in order to define the linguistic value 'middle-aged'. A membership function is a function that assigns to each possible individual in the universe of discourse, a value representing its grade of membership in the fuzzy set [17]. Larger values denote higher degrees of set membership A fuzzy set is a set with the capability to express gradual transition from membership to non-membership and vice versa, and it is defined by the membership function.

Therefore, we can use the membership function 'A_m' to define a middle-aged man as follows:

$$A_m(x) = \begin{cases} 0 & \text{when either } x \leq 20 \text{ or } \geq 60 \\ (x-20)/15 & \text{when } 20 < x < 35 \\ (60-x)/15 & \text{when } 45 < x < 60 \\ 1 & \text{when } 35 \leq x \leq 45 \end{cases}$$

By resolving the membership function for x=30 we get that a 30 years old man is a member of the set of middle-aged men, having a membership grade $\mu(30)=0.66$, or in a more meaningful way, 'a 30 years old man belongs by 66% in the set of middle aged men'.

Up to now, according to the rule 'R', which is 'valid' nowadays, and according to the membership function 'A_m', we have concluded that a '30 years old man' is 'middle-aged' so 'he has to work in order to live'.

Nevertheless, is this rule correct for a man born and growing old in Africa? Is this rule correct for a man born one thousand years before? Alternatively, is this rule correct for a man born one thousand years before in Africa? All these are criteria (or a combination of criteria) that one has to take into account before reaching to a conclusion.

We must note that these criteria do not affect the assertion (assertion IF) of the rule 'R' directly, but the membership grade that affects the assertion of the rule. We must also note that there is a great difference between the time criterion we have used in the above question (Is this rule correct for the men born one thousand years before?) and the Valid-Time definition we have introduced at the beginning of this section. The first is a criterion that affects the rule while the latter is the time period during which the rule is valid. If our knowledge about the 'middle-aged man' definition changes later, we shall define a new rule 'R1' with a new membership function and new criteria, keeping the old one as historical knowledge.

In order to represent the above example (rule 'R') with all the mentioned components (validity of time, fuzziness, multi criteria) we introduce an ES based on a Relational Data Model, which is a combination of the structural part (in which the DB is a collection of relations), the integrity part (in which are described primary and foreign keys), and the manipulative part (in which is used relational algebra and relational calculus) [7].

Summarizing the above, this is an Expert Database System [15] with multicriteria fuzzy sets aiming to be used in a wide range of decision making problems where there is no a step by step process [11], according to the KB of the system.

This paper is organized as follows: In Section 2 we expound the components and the methodology of a Rule-Based Valid-Time Expert System with Multicriteria Fuzzy Sets, while in Section 3 we implement the methodology by describing the development of the proposed ES on a relational database. In Section 4 we conclude about the integration of a RBES with Fuzzy Logic and multi criteria affecting the fuzzy sets, and the use of this integration in Relational Data Model.

2. Methodology

A Rule is a conditional statement of two parts. The first part, comprised of one or more IF clauses, establishes the conditions of the rule. The second part, comprised of one or more THEN clauses, establishes the actions of the rule that have to be undertaken, and finally, the first part applies the second. The clauses of both parts, are represented in Object – Attribute – Value (O-A-V) triplets, since this representation easily fits into any RBES development tool [16].

The KB of all RBES, share the same most fundamental component, the production rule, which in the simplest case has the following form:

IF < assertion_i1 >	THEN < assertion_t1 >
AND / OR < assertion_i2 >	AND / OR < assertion_t2 >
AND / OR ...	AND / OR ...
AND / OR < assertion_im >	AND / OR < assertion_tm >

In our system the assertions 'assertion_i1, assertion_i2, ..., assertion_im' of the 'IF' part of the rule (the antecedents), are kept in the 'Condition' table and these are represented as O-A-V triplets. The Value (V) of each condition may be either a single value, or the conclusion of another existed rule.

The assertions 'assertion_t1, assertion_t2, ..., assertion_tm' of the 'THEN' part of the rule (the consequents), are kept in the 'Conclusion' table and these may be either a single value, or a single condition. We are able to use a single condition as a conclusion to support the forward chaining inference engine. If there is a need for a reaction as an entailment for a specific conclusion, we use the 'Reaction' table.

Since a rule consists of condition(s), conclusion(s) and possibly reaction(s), the three above tables are related to the 'Rule' table, with intermediate tables by 'many to many' relationships. The Validity of Time (VT) is performed at rule level, so using the appropriate fields, a rule is valid in a specific time interval. Another approach of the VT is to affect each one condition instead of the rule. This helps when a condition changes while all the others (both conditions and conclusions) remain unchanged and we don't want to make a new rule for just one change. This approach is helpful in legal KBs [8], but the procedures of knowledge entry and maintenance are difficult since we must enter valid time interval for each one condition.

There are also fields in the related tables to represent the essential attributes of each rule. More specifically, at condition level, in order to express the uncertainty of the knowledge contained in the knowledge base [14] we have included the missing factor, the weight factor and the user's certainty factor [13], as well as the order of every condition within a specific rule. At conclusion and reaction levels, we have included the confidence factor and the order within the specific rule, as before.

The described system up to now is a simple RBES on a relational DB. In more complex ES, the values of the condition may be vague. For example in the condition: 'The tank has water level that is low', the value 'low' can be represented by a discrete fuzzy set defining vaguely the 'low' for the specific condition which is part of a specific rule.

A fuzzy relational data model, as an extension of classical relational model, incorporates the impreciseness in data values and their associations [2]. A Fuzzy Relational DataBase (FRDB) represents imprecise attribute values and close domain elements with possibility distributions and closeness relations respectively [4]. Buckles et al [3] proposed one of the earliest versions of Fuzzy Relational Database System (FRDBS) by merging the theory of fuzzy set and Relational Database System (RDBS). Chiang et al [5], following some works, defined an extended fuzzy relational model by the first-order logic.

In our system, we did not use the pre-described fuzzy relational data model, but the values of the O-A-V (condition) triplets are used as separate fuzzy sets related to one or more conditions. Therefore, in order to represent the vagueness of the values, we use a table to store values' discrete fuzzy sets. From the two alternative ways used to represent a membership function (continuous and discrete), we have used the later one.

The next step we introduce is a way to handle the fuzzy values in miscellaneous situations (criteria). At the last example what is the 'low' for a refinery's oil tank, and for a glass of water? In order to represent the shifting of the fuzzy sets under various criteria, we use a table to store the criteria and another one to relate it with the discrete fuzzy sets.

The proposed system could be easily confused with the Active Database Management System (ADBMS) which is standard 'passive' DBMS with the capability to react to events (not only internal that were raised by control or database operations, but external events also) of a prescribed nature [6]. This capability is described in terms of a set of Event-Condition-Action (ECA) rules.

An ADBMS acts as follows: The database system is able to monitor a special situation represented by an event and one or more conditions. When the event occurs and the conditions evaluate to true, the corresponding actions are executed. Hence, active database systems can recognize specific situations and react to them without direct explicit user or application requests [12].

The advantage of maintenance of ADBMS is that when rule requirements are changing, only ECA-rules (and not whole application programs) have to be updated in accordance with possible new requirements. A problem arises when, if one wants to delete or temporary disable the production rule (since one production rule corresponds to a set of ECA rules), he/she should perform the same operation to all related ECA rules. However, this requires special care since the user might forget some of the ECA rules and the rule base would then become inconsistent [1].

The differences between our system and an ADBMS are the following:

- The ADBMS acts on events, while our system first concludes and then reacts according to the conclusion.
- As a consequence, our system has reasoning abilities to explain the 'how' and 'why' of the conclusion.
- The conclusions in our system come from a separate module, the inference engine (not described in this paper) giving the ability to infer either forward or backward.

3. Implementation

In order to describe the implementation procedure of the above we present the development of the ES into three steps. Firstly we introduce the Relational Rule Based Expert System (RRBES), which is a RBES based on the Relational Data model, then we extend it presenting a Fuzzy RRBES and finally in order to handle fuzzy values under multiple criteria we present a Multicriteria FRRBES.

The Entity Relationship model of these three steps is presented in Figure 1, divided in three sections.

3.1 A Relational Rule Based Expert System (RRBES)

The Section 1 of Figure 1 is the Relational Rule Based Expert System (RRBES), which is a RBES based on the Relational Data model.

We have used the Object Attribute Value (O-A-V) representation model, using the "has-a" and "is-a" links between them [9]. The RRBS model works as following:

An Object ('Object' table) has attribute(s) ('Attribute' table), every attribute is valued ('Value' table), and every single combination of them ("O-A-V" triplet) constitutes a condition ('Condition' table). Rules ('Rule' table) and conditions are basic elements of the system, which when are related each other (in 'RuleCond' table) build the 'IF' part of the "IF P THEN Q" statement, while when the rules are related to conclusions ('Conclusion' table relation to 'RuleConcl' table) they build the 'THEN' part of the "IF P THEN Q" statement. In every Rule we may (or not) act in response ('Reaction' table relation to 'RuleReact' table).

As we have mentioned, one condition (in 'Condition' table) is a relation between an object that has an attribute with a value. Sometimes, an attribute's value may not be a strict value but when this is the conclusion of another rule, we use the reference to the Conclusion table ('v_cl_id' field). For example, in the condition "The hives have honey production that is less than normal honey production", the value derives from another rule, which defines the meaning of 'less than normal honey production'.

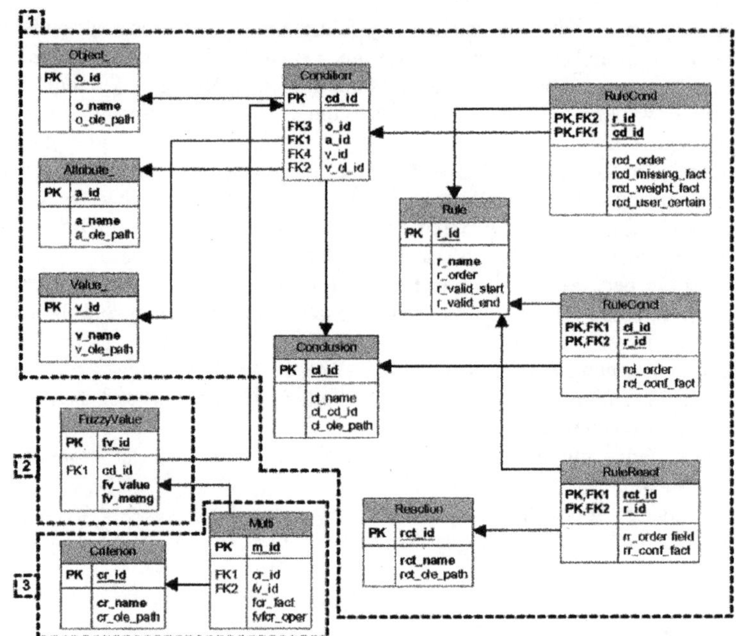

Figure 1   Entity Relational model of the Multicriteria Fuzzy Relational Rule-Based Expert System

In the 'Rule' table, apart from the rule name, we have used a time interval during which, the rule is valid ('r_valid_start' and 'r_valid_end' fields). It is very significant since a rule may be dropped during the time and a new rule may be born using similar conditions, reaching in the same or similar conclusion. This feature determines the temporality of our RRBES. We have also included the order of rule ('r_order' field) in order to represent the rules (optionally) in a hierarchical format.

A rule consists of at least one condition, and a condition takes part in one or possibly more rules. To represent this 'many-to-many' relationship between the 'Rule' and 'Condition' tables, we have used the 'RuleCond' table, with the appropriate fields to define the order of every condition within a specific rule, the missing factor, the weight factor and the user's certainty factor).

In 'Conclusion' table, we have used two different ways to keep the conclusion name, either by using a simple string field (cl_name field) or by using a reference to an existing condition (cl_cd_id).

Like Rules and Conditions, a rule includes at least one conclusion, and a conclusion takes part in one or maybe more rules. To represent this 'many-to-many' relationship between the 'Rule' and 'Conclusion' tables, we have used the 'RuleConcl' table, with the appropriate fields (order and confidence factor).

Another table we have used is the 'Reaction' table, which informs the user how to react when the user has reached in a conclusion within a rule.

The link between the 'React' and the 'Rule' table is the 'RuleReact' table, which represents a many-to-many relationship between them. The order of every reaction ('rr_order' field) and the confidence factor ('rr_conf' field) within a specific rule that we have included in this table, act similarly to the fields of the 'RuleConcl' table.

3.2 A Fuzzy Relational Rule Based Expert System (FRRBES)

In the previous example, (RRBES) we used the condition "The colony has location that is near the pinewood". It means that the object 'The colony' has the attribute 'location' that is valued 'near the pinewood'. Nevertheless, what is the meaning of the value 'near the pinewood'? One could say 5 km is near while the other could say 1 km is near. Both of them are correct since this distance is 'near' for bees. But the second opinion is 'more near' than the other.

In order to resolve this type of conflicts, we propose an extension to our RRBES (Section 2 of Figure 1) to handle fuzzy values, creating a new table, the 'FuzzyValue' table. In this table we have included a link field to relate the table to 'Condition' table ('cd_id' field), and the fuzzy sets (many records having the same 'cd_id' field) consisted of the value ('fv_value') and grade of membership ('fv_memg') fields.

We must note that we have related the new 'FuzzyValue' table to the 'Condition' table and not to the 'Value' table, because the fuzziness of the value depends on the attribute of the object which it comes from.

3.3 A Multicriteria Fuzzy Relational Rule Based Expert System (MFRRBES)

In the fuzzy extension of the RRBES we used the fuzzy sets to handle the linguistic variables of a condition. In this way we were able to conclude about the truth of the linguistic variable 'near the pinewood' according to the corresponding membership grade. Nevertheless, how can we handle the conditions when there are criteria that affect them? For example in the RRBES we have used a rule to define the 'reduced honey production near the pinewood'. There are more criteria that one could take into account in order to make the rule more complete, as 'the times that the beekeeper transports his colony', 'the season of the year' or 'the existence of water near the colony'. These criteria can be included in the rule, so the rule will be more complete having more conditions. But what can we do for the criterion of 'attitude'? We used the example of 'the colony is 2 km distance from the pinewood', with the membership grade '0.7' but we don't know if this distance is horizontal or inclined. The membership grade should not be the same for an inclined bee flight.

In order to resolve this kind of conflicts, we propose an extension in our FRRBES (Section 3 of Figure 1) to handle fuzzy values under multiple criteria, creating the 'Criterion' table to keep all the criteria and the 'Multi' table that relates the 'Criterion' table to 'FuzzyValue' table.

We have included in the 'Multi' table a Fuzzy Criterion factor ('fcr_fact' field) which interacts with the membership grade of the 'FuzzyValue' table. This means that the membership grade of a value in the 'FuzzyValue' table increases and decreases alternately, according to the Fuzzy Criterion factor of the criterion that affects the fuzzy value. The operator (min, max, average, etc) for that fluctuation is stored in ('fvfcr_oper' field).

4. Conclusions

In this paper, we have tried to integrate many techniques in order to build a formal Rule Based Expert System based on the mostly used Database model, the Relational Data Model.

One critical point is the use of Valid Time knowledge, a very helpful point for changing Knowledge Bases in which we want to keep track of the changes for advisable purposes.

We have also tried to use Fuzzy conditions and conclusions in our rules, but since most Expert Systems use some kind of fuzziness, we have tried to relate our Fuzzy conditions and conclusions to any criteria able to affect the fuzzy values.

This ES is suitable for most of the cases where humans and machines need to communicate using linguistic terms in order to make a decision. It depends on the Knowledge Engineer to acquire the knowledge from the expert, to format it in the proper format so to be stored in our Knowledge Base.

References
[1] N. Bassiliades, I. Vlahavas, Processing production rules in DEVICE, an active knowledge base system, Elsevier, Data & Knowledge Engineering 24 (1997) 117-155.
[2] T. K. Bhattacharjee, A. K. Mazumdar, Axiomatisation of fuzzy multivalued dependencies in a fuzzy relational data model, Elsevier, Fuzzy sets and systems, 96 (1998) 343-352.
[3] B. P. Buckles and Petry F. E., A Fuzzy Representation of Data for Relational Databases, Elsevier, Fuzzy Sets and Systems, vol 7, 1982, 213-226.
[4] G. Q. Chen, Vandenbulcke J. and Kerre E. E., A step towards the theory of fuzzy database design, Proceedings of the 4th World Congress of International Fuzzy Systems Association (IFSA '91), Brussels, 44-47, 1991.
[5] D. Chiang, l. R. Chow, N. Hsien, Fuzzy information in extended fuzzy relational databases, Elsevier, Fuzzy sets and systems, 92, 1997, 1-20.
[6] A. Koschel, P. C. Lockemann, Distributed events in active database systems: Letting the genie out of the bottle, Data & Knowledge Engineering, 25 (1998) 11-28.
[7] M. Levene, G. Loizou, A guided tour of Relational Databases and beyond, Springer, 1999.
[8] N. A Lorentzos., C. P. Yialouris, A. B. Sideridis, Time-evolving rule-based knowledge bases, Data & Knowledge Engineering, 29, 1999, 313-335.
[9] R. E. Plant, Nicholas D. Stone, Knowledge-Based Systems in Agriculture, McGraw-Hill Inc, 1991.
[10] D. Saint-Joan, J. Desachy, A Fuzzy Expert System for Geographical Problems: An Agricultural Application, IEEE, 1995.
[11] Y. Siskos, A. Spyridakos, Intelligent multicreteria decision support: Overview and perspectives, European Journal of Operational Research, 1999.
[12] C.W. Tan, A. Goh, Implementing ECA rules in an active database, Knowledge-Based Systems 12 (1999) 137–144.
[13] G. Vouros, Representing, adapting and reasoning with uncertain, imprecise and vague information, Pergamon, Expert Systems with Applications, 19, 2000, 167-192.
[14] V. C. Yen, Rule selections in fuzzy Expert Systems, Expert Systems with Applications, 1999.
[15] H.Yang, A Simple Coupler to Link Expert Systems with Database Systems, Pergamon, Expert Systems with Applications, Vol 12, No. 2, 1997, 179-188.
[16] C.P. Yialouris, H.C. Passam, A.B. Sideridis, C. Metin, "VEGES – A multiningual Expert System for the diagnosis of pests, diseases and nutritional disorders of six greenhouse vegetables", Computers and electronics in agriculture, 1997.
[17] L. A. Zadeh, "Fuzzy sets, Inf. And control", 1965.

KES 2002
E. Damiani et al. (Eds.)
IOS Press, 2002

Possibilistic Abduction versus Fuzzy Pattern Matching

Cristina SEGAL
"Dunarea de Jos" University of Galati, Romania
tel:036.468061 fax:036.461353
E-mail: cristina.segal@ugal.ro

Abstract. This paper presents an analogy between possibilistic abduction and fuzzy pattern matching in a context of diagnosis problems. The diagnosis problem is presented as a causal graph that captures the knowledge about the prototypical manifestations of the possible disorders from the system, described through their prototypical fuzzy values entailed by a disorder. The observations are also fuzzy due to the imprecision of measurement or to their linguistical nature. The proposed analogy offers a new approach to the abductive method from the fuzzy pattern matching point of view.

1. Introduction

Diagnosis problem consists of the identification of the disorders that may cause some observed manifestations. In the context of diagnosis the possibilistic abduction is a very new approach used in particular for relational abductive diagnosis [1] [3] [4]. The goal of this paper is to prove that a diagnosis problem may be reduced to a pattern matching problem and also solved by the specific formalism. This paper has 4 sections. Section 2 briefly presents what fuzzy pattern matching is in the existing literature. The third section consists of a short presentation of a model of a relational possibilistic representation of a diagnosis problem. In section 4 it is proved, in a formal manner, how the relational diagnosis method from section 3 may be transformed into a fuzzy pattern matching problem.

2. A brief analysis of fuzzy pattern matching (upon Dubois and Prade [2])

In the existing literature the essential idea of fuzzy pattern matching consiss of a gradual evaluation of the overlapping between an object named *"pattern"*, denoted by P and a data, named *"datum"*, denoted by D, both entities belonging to the same discourse universe U. It is more appropriate proceed to fuzzy pattern matching in the case in which the two entities P and D are incompletely specified (the datum is pervaded with imprecision or uncertainty and in the pattern are included vague specifications). In the case of possibilistic pattern matching the two entities are defined by two fuzzy values. The characteristic function of the atom pattern, described by $\mu_P(u)$, expresses the compatibility of a value $u \in U$ with the pattern P. The data D is described through the possibility distribution $\pi_D(u)$ that expresses the measure in which it is possible for u to be the (unique) value of the attribute that describes the object modeled by the corresponding entity.

According to Dubois and Prade [2], in order to express the compatibility between a pattern P and a data D, two scalar measures are used:

- **The possibility degree of the fuzzy pattern matching:**

$$\Pi\ (P;D) = \sup_{u \in U} \min\ (\mu_P(u),\ \pi_D\ (u)) \tag{1}$$

- **The necessity degree of the fuzzy pattern matching:**

$$N\ (P;D) = \inf_{u \in U} \max\ (\mu_P(u),\ 1 - \pi_D\ (u)) \tag{2}$$

In other words, $\Pi(P;D)$ is the overlapping degree between the fuzzy subset of the values compatible with the concept expressed by P and the fuzzy subset of the possible value from D. The necessity degree $N(P;D)$ estimates to which extent it is necessary (certain) that the value to which refers D to be among those compatible with P. It may be said that $N(P;D)$ measures the inclusion degree of the set D in P. In the situation where the pattern and the datum are described by two vectors of attributes, the atomic measure of possibility and necessity mentioned in equations (1) and (2) are separately aggregated in order to obtain to global measure for the matching of the entire pattern and the entire datum. The semantical meaning of the atomic necessity and possibility measures is preserved by the following relation:

$$\Pi\ (P_1 \times ... \times P_n;\ D_1 \times ... \times D_n) = \min_{i=1,...n} \Pi\ (P_{ij},\ D_i) \tag{4}$$

$$N\ (P_1 \times ... \times P_n;\ D_1 \times ... \times D_n) = \min_{i=1..n} N\ (P_I,\ D_i) \tag{5}$$

where P_i and D_i are supposed to be defined on the same universe U_i and "$\times$" is the cartesian product of the two fuzzy sets.

3. The possibilistic relational model of the diagnosis problem (upon Segal [4])

Let us consider the following representation of the diagnosis problem: $D=\{d_1,...,d_n\}$ is the set of **all the possible disorders** and $M=\{m_1,\ ...,m_n\}$ denotes **the set of the n possible** manifestations. A disorder may be present or absent.

Each m_j is a couple (*manifestation_name, X_j*) where X_j is the linguistic (fuzzy) variable, which represents the set of fuzzy values that the fuzzy variable may take.

Example: m_j=(*liver firmness, X_j*) where X_j={*tender, normal, firm*}

For each disorder d_i there is a subset of fuzzy manifestations denoted by $M(d_i)^+$ which are (more or less) certainly caused by d_i, and a subset of fuzzy manifestations denoted by $M(d_i)^-$ that are (more or less) certainly not caused by the presence of $d_{i,}$. The set of the manifestations causally related to d_i is $M(d_i)=M(d_i)^+ \cup M(d_i)^-$. Each of its elements is a fuzzy subset $M(d_i)_j$ corresponding to the values of the manifestation m_j entailed by the presence of d_i. In fact $M(d_i)_j$ may be considered as a flexible constraint on the values of m_j in the presence of d_i. Beside the manifestations causally related to the disorder d_i there is a set of manifestations $M^{(o)}(d_i)=M-M(d_i)$ that seems to be unrelated with d_i, all that can be said about these manifestations is that it is unknown they follow or not from d_i or not. The existence of this set is due to the incompleteness of the causal relation. From the possibility theory point of view this situation has the following formulation: *if m_j is not causally related to d_i then all the values from its universe are equally possible:* $\forall m \in X_j, \mu_{M(d_i)_j}(m)=1$.

The **observation set Mo**, is a set of fuzzy sets because each observation mo_i is expressed by a fuzzy set Mo_i with the membership function μ_{Moi}. Taking into account that for each manifestation their values in the presence of a disorder can be different we must express the *presence level* for an observation according to their present value and their predicted one corresponding to the disorder causal related to this manifestation. The knowledge base of a diagnosis system for which the entities are modeled as above can be expressed as a collection $K=\{P_1,\ P_2,\ ...,\ P_n\}$. Each P_i is a set of rules P_{ij}, corresponding to all causal relations existing between a disorder d_i and the manifestation m_j, of the following type:

P_{ij}: *"if d_i is present then (more or less certain) m_j is $M(d_i)_j$"*

The causal relation between a disorder d_i and a manifestation m_j is expressed by a fuzzy relation $R_{ij}: \{d_i\} \times X_j \rightarrow [0,1]$ that describes an elastic constraint of the m_j values on all the possible couples (d_i, m), where $m \in X_j$. Knowing that the presence of d_i is certain and m_j is fuzzy, **the causal relation equation** is:

$$\forall d_i \in D, \forall m \in X_i, \mu_{R_{ij}}(m_j, d_i) = \mu_{M(d_i)_j}(m) \tag{6}$$

With the above-defined diagnosis problem at hand, an abductive method for generating the hypothesis set may be built. Taking into account the minimality property of an abductive explanation, in [4] Segal states that there are two types of disorder hypotheses that can be considered as explanations of the present observations:

- The set of the **relevant explanations** consisting of disorders those are consistent with the observations.

$$\mu_{D^\wedge}(d) = cons(Mo, M(d)) = \min_{M(d)_i \in M(d)} cons(Mo_i, M(d)_i) \tag{7}$$

where

$$cons(M(d)_i, Mo) = \max_{m \in X_i} (\min(\mu_{M(d)_i}(m), \mu_{Mo_i}(m))) \tag{8}$$

- The set of **pertinent explanations** consisting of disorders that are covers of the present manifestations

$$\mu_{D^\bullet}(d) = \min_{M(d)_i \in M(d)} incl(Mo_i, M(d)_i) \tag{9}$$

where the inclusion is expressed by one of the well known fuzzy operators:

$$incl(Mo_i, M(d)_i) = \min_{m \in X_i} (\mu_{Mo_i}(m) \rightarrow \mu_{M(d)_i}(m)) \tag{10}$$

4. Fuzzy pattern matching versus possibilistic abduction

The rest of this section is devoted to prove that the diagnosis problem formulated in the precedent chapters may be considered as a fuzzy pattern matching one. Let us consider a disorder d, for which we will compute its plausibility degree when an observation set Mo is given. From section 3 it is known that for a disorder d there is a set $M(d)$, that describes the prototypical manifestation of d. The set $M(d)$ may be considered as being a "tree like" pattern, whose leaves correspond to its attributes. In other words the leaves are the prototypical manifestations caused by this disorder. This pattern describes the prototypical restrictions of the values of its attributes. Each attribute is characterized by a fuzzy subset $M(d)_j$, that may be considered as an flexible constraint of its values in the presence of d. The membership function of the fuzzy set $M(d)_j$ ($\mu_{M(d)_j}$) expresses the compatibility degree between

the value $m \in X_j$ of the manifestation m_j and the appearance of the disorder d. The interpretation of the limit values of this function is the following:

- If $\mu_{M(d)_i}(m) = 1$ then the value m is totally compatible with the concept defined by d,

- $\mu_{M(d)_i}(m) = 0$ expresses the complete incompatibility between the value m and the concept defined by disorder d.

If the set $M(d)$ is considered as a vector of fuzzy attributes, thus a fuzzy pattern, it may be extended by including other manifestations that are not causally related to d in this set. As mentioned in section 3, when a manifestation m_k is not causally related to the disorder d, the possibility distribution of the values of the manifestation $\dot{m}_k$ in the presence of the disorder d is one over its entire universe: $\forall m \in X_k$, $\mu_{M(d)_k}(m) = 1$. If this manifestation is included as an attribute in the pattern that describe the disorder d expressed by the set $M(d)$ then the above remark is consistent with the interpretation of the fuzzy value of the attribute k, value described by the characteristical function $\mu_{M(d)_k}(m)$. This function expresses the compatibility of a value $m \in X_k$ with the pattern P. In this case it may be consider that in the situation in which there is not a causal relation between d and the manifestation m_k, any value of this manifestation is perfectly compatible with the disorder d: $\mu_{M(d)_k}(m)=1$.

The set *Mo* is the collection of current data (observations), expressed as a set of fuzzy subsets Mo_i type. These fuzzy values of the manifestation have their source in the imprecision of the measurement or in their linguistical nature. Thus, a value $\mu_{Moi}(m)$ expresses the possibility for m to be the unique value of the attribute j, that describes the manifestation. *Mo* is a subset of the set of the manifestations of the system. For all the other manifestations of the system, (*M-Mo*), nothing can be said about their values because they are not yet observed and their possible values cannot be predicted.

With the extended version of the set of attributes of a disorder d described as above, it may say that the attributes of the datum modeled by *Mo* can be found in each of the pattern that describes the disorders from the system, because these patterns contain all the manifestations from the system. For each manifestation m_i from *Mo* there may be built pairs of type *(measured fuzzy value, predicted fuzzy value entailed by each disorder d)* or in a symbolical representation, *(M_i, M(d)_i)*. Thus for each disorder $d \in D$, the expressions of the possibility pattern matching degree, $\Pi(M(d),Mo)$, and necessity pattern matching degree, $N(M(d),Mo)$ may be computed. Considering the equation (4) and (5), these expressions are:

$$\Pi\,(M(d);Mo) = \min_{i=1,\ldots n}\Pi\,(M(d)_i,Mo_i) \tag{11}$$

$$N\,(M(d);Mo) = \min_{i=1,\ldots n}N\,(M(d)_i,Mo_i) \tag{12}$$

where $M(d)_i$ and Mo_i are supposed to be defined on the same domain X_i, and n is the crisp cardinality of the set *Mo*. The atomic pattern matching degrees are given by:

$$\Pi\,(M(d)_i,Mo) = \max_{m\in X_i}(\min(\mu_{M(d)_i}(m),\mu_{Mo_i}(m))) \tag{13}$$

$$N\,(M(d)_i,Mo) = \min_{m\in X_i}(\max(\mu_{M(d)_i}(m),1-\mu_{Mo_i}(m))) \tag{14}$$

Having as starting point the above equation we will analyze the implications for the inclusion in the pattern of each disorder d of the manifestations that are not causally related with this disorder. Thus, if m_l is a manifestation that is not causally related to a disorder d, then it is well known that $\forall m \in X_l, \mu_{M(d)_l}(m)=1$ and so for these manifestations the atomic measures of fuzzy pattern matching derived from the equation (13) and (14) are:

$$\Pi\,(M(d)_l,Mo) = max_{m\in X_l}(\min(1,\mu_{Mo_l}(m))) = 1 \tag{15}$$

$$N\,(M(d)_l,Mo) = \min_{m\in X_l}(\max(1,1-\mu_{Mo_l}(m))) = 1 \tag{16}$$

If the fuzzy set Mo_i is normalized $max_{m\in X_i}\,\mu_{Mo_i}(m)=1$. Obviously, the necessity and the possibility measures from the equations (15) and (16) are one because if we say that a manifestation is not causally related to a disorder, it means that in fact we ignore the existence of a causal relation (more details about the interpretation of complete ignorance are given in section 1). Knowing that "min" operator aggregates the atomic fuzzy pattern matching measures, the existence of this ignorance situation modeled by the equations (15) and (16) does not influence the general fuzzy pattern matching result. Consequently, the following fuzzy pattern matching measures may be deduced:

$$\Pi\,(M(d);Mo) = \min_{i=1,\ldots n}\Pi\,(M(d)_l,Mo_l) = \min_{M(d)l\in M(d)}\Pi\,(M(d)_i,Mo_i) \tag{17}$$

$$N\,(M(d);Mo) = \min_{i=1,\ldots n}N\,(M(d)_i,Mo_i) = \min_{M(d)i\in M(d)}N\,(M(d)_i,Mo_i) \tag{18}$$

The equivalence between the equations (17) and (7), expresses that the consistency of two fuzzy sets is similar to the measure of possibility fuzzy pattern matching of this sets, because the latter estimates the overlapping degree (as intersection) of the two fuzzy sets. On the other hand, the equivalence of the equation (18) with the equation (9) may be used in evidence. The equation (17) expresses to which extent it is necessary to have the value referred by Mo_i among those compatible with d. In other words $N(M(d)_i, Mo_i)$ is the inclusion degree of the set Mo_i in the set $M(d)_i$. The implication operator allowed for this necessity degree is the one derived from the crisp implication

$(a{\rightarrow}b \Leftrightarrow \neg a \vee b)$, being a very restrictive one because it computes the inclusion degree at the core level. In the equation (18) the choice is made for the implication operator Dienes-Rescher, and then the equivalence of the two equations is evident.

Therefore the following conclusion may be stated:

- The *relevance measure* of a disorder d as explanation for the observation set is the same as the pattern matching possibility degree;
- The pattern matching *necessity degree* is a particular case of the *pertinence measure* of a disorder d as explanation for the observation set.

As in the case of pertinence and relevance measures it is obvious that the necessity and possibility degree of the fuzzy pattern matching may be used as a rank order criterion for expressing the compatibility between the datum (observation set) and all the existing patterns (the attributes of each disorders from the system). In this case we do not aim to establish a classification of the datum w.r.t. the existing patterns, but a similarity of the different datum with the existing patterns. It may be put in evidence the following situations:

i. None of the existing patterns is not compatible with the observations:
$$\Pi\,(M(d);Mo) = \mathrm{N}(M(d);Mo) = 0;$$
ii. None of the existing patterns does not include the observations, but there are patterns that overlap somehow the observations:
$$\Pi\,(M(d);Mo){>}0 \text{ and } N(M(d);Mo)= 0;$$
iii. One or many patterns are similar (more or less) with the observation set:
$$\Pi\,(M(d);Mo) > 0 \text{ and } \mathrm{N}(M(d);Mo) > 0.$$

It may be considered that the first situation appears because of the incorrectness of the knowledge bases generated either by ignorance, or by an inadequate estimation of the prototypical fuzzy intervals of the pattern attributes. The second situation is generated by a lack of measurement precision or by the use of too restrictive precision for the prototypical attributes. The third situation is the most common one. In this case, in order to evaluate the result of the fuzzy pattern matching an order relation must be defined based on the necessity and possibility degree defined above.

5. Conclusion

In conclusion it may be stated that the discussion from this paper proves in a formal and also intuitive manner that the results obtained by the fuzzy matching of the pattern (that describes the causal relations existing between a disorder and the manifestations from a system) and the datum (that models the observation set) is a particular case of the result obtained through the abductive relational possibilistic method presented in section 1, for generating the plausible explanation of the current observations set.

References

[1] Cayrac, D., Dubois, D., Prade, H., Handling Uncertainty with possibility theory and fuzzy sets in a satelite fault diagnosis application, IEEE Trans. On Fuzzy Sets and Systems, Vol 4, pp.251-269, 1996

[2] Dubois, D., Prade, H. C. Testemale, "Weighed fuzzy Pattern Matching", Fuzzy Sets and Systems 28, pp.313-331, 1988

[3] D. Dubois, M. Grabisch, H. Prade, "A general approach to diagnosis in a fuzzy setting", 8th *Int. Fuzzy System Association World Congress, Taipei,(IFSA'99) august, 1999*, Proceedings, pp.680-684

[4] C. Segal "Using possibility theory in abductive diagnostic problem, IPMU 2000, pp.1554-1560

KES 2002
E. Damiani et al. (Eds.)
IOS Press, 2002

SAR Images Classification Using Fuzzy Subsethood Operator

Giovanni ANGIULLI, Vincenzo BARRILE, Mario VERSACI
Dipartimento di Informatica, Matematica, Elettronica e Trasporti
Università degli Studi "Mediterranea" di Reggio Calabria
89100 Reggio Calabria - Italy

Abstract. The availability of synthetic aperture radar (SAR) images makes practical many geophysical applications. One of them is the classification of terrain structures. In the past years a number of techniques have been developed to classify ground terrain types from SAR images. The conventional approach to classification, which assigns a specific class for each region in an image, is often inadequate because the area covered by each region may embrace more than a single class. Fuzzy set theory, which has been developed to deal with imprecise information, can provide a more appropriate solution to this problem. In this work has been developed a new supervised fuzzy classification method based on the use of *Subsethood Operator*. Results of classifying a SAR image are presented and their accuracy is analysed and compared with other standard classification techniques.

1. Introduction

Classification of earth regions from synthetic aperture radar (SAR) imagery represents an important developing application of microwave remote sensing. Advances in microwave technology have provided an improved measurement capability, allowing development of coherent system employing multiple frequencies and polarizations. In addition, all weather, day and night operation and the ability to penetrate foliage or surface features, give synthetic aperture radar advantages over infrared and optical systems.

To make full use of collected measurements, however, and to reduce the requirement for photo interpretation, a quantitative procedure for systematic classification is needed. A number of supervised and unsupervised classification techniques have been applied to the problem of SAR image classification [1-2]. However, the conventional approach of terrain image classification which assigns a specific class for each region in a SAR image is often inadequate because the area covered by each region may embrace more than a single class.

More recently, fuzzy sets have been employed in the classification of land terrain covers [3]. Fuzzy sets theory, [4] which has been developed to deal with imprecise information, can provide a more appropriate solution to this problem. In fact, it provides useful concepts and tools to deal with imprecise information. It allows each region to have partial and multiple membership in several classes. A region's membership grade function with respect to a specific class indicates to what extent its properties are akin to that class. The value of the membership grade varies between 0 and 1. Closer the value is to 1 more that region belongs to that class. Partial membership allows that the information about more complex situations, such as cover mixture or intermediate conditions, can be better represented and utilized.

In this work a fuzzy classification method based on *Fuzzy Subsethood Operator* has been proposed for classification of SAR images. The paper is organized as follows: section 2 presents the basic on fuzzy sets theory and the definition of *Subsethood Operator*.

In section 3 we describe the employed algorithm. Section 4 shows the experimental results and a discussion on the usefulness of the proposed approach comparing it with other classification methods. Finally, in section 5 we illustrate the conclusions.

2. Geometrical Representation of Fuzzy Sets: Points in a Unit Hypercube

A fuzzy set can be considered as an abstract object that partially contains other objects. It can be represented by means of a function, called *membership function*, that maps objects in a space to the numbers between 0 and 1. A fuzzy set can also be a point in some space. This is the geometrical view of a fuzzy set or the view of sets as points. A continuous fuzzy set like the triangle or bell curve that stands for a variable defines a point in an abstract function space of set functions. The mind's eye cannot fully see these abstract spaces. But it can grasp the distance between two fuzzy sets as the length of the line segment that connects two points. It can grasp the neighbourhood of a fuzzy set as a ball or sphere that contains the fuzzy set as the point at the ball's centre. And it can grasp a changing or adapting fuzzy set as a point moving through the space. A discrete fuzzy set has the simplest geometry. It is a point in a fuzzy cube. A fuzzy cube is a unit hypercube that has the unit interval [0, 1] as each of its sides. The unit interval itself forms the simplest fuzzy cube or 1-D cube. It houses all truth-values of a fuzzy or multivalued logic. The unit square houses all fuzzy subsets of three objects and so on up to countable infinity. Nonfuzzy sets lie at the cube corners. A fuzzy cube contains all fuzzy subsets of a set X of n objects. The 2^n bivalent subsets of X lie at the 2^n corners of the n-cube [0, 1]^n. Consequently, a fuzzy set is representable in a cube as a point in which coordinates are the fuzzified values $f^A(x_j)$. In technical literature there are many operators which giving a membership measure of a fuzzy set to an assigned category. Here we have employed the *Subsethood Operator* defined by Kosko [5] as:

$$Subsethood(A, B) = 1 - \frac{\sum_{j=1}^{n} \max\left[0, f^A(x_j) - f^B(x_j)\right]}{M(A)} \tag{1}$$

where *M(A)* is a fuzzy set *A* measure defined as:

$$M(A) = \sum_{j}^{n} f^A(x_j) = \sum_{j}^{n} \left| f^A(x_j) - 0 \right| = \sum_{j}^{n} \left| f^A(x_j) - f^\varnothing(x_j) \right| = l^j(A, \varnothing) \tag{2}$$

and where $\varnothing$ is the empty set. The possibility that set B is contained into A is as high as great is the subsethood value *Subsethood(A,B)*. Next, we consider each well known category in an SAR image as a fuzzy set. They will be points lying in an unitary hypercube. An unknown zone in a SAR image will be represented by a point in this multidimensional space named *"unknown point"*. Using this approach, for the an assigned region in a SAR image, the unknown region membership to a well known category is reducible to a measure distance problem between fuzzy sets in a hypercube.

3. The proposed Subsethood Algorithm

In this section, the classification algorithm based on Subsethood operator will be examined in detail:

Step 1.   The image under analysis is transformed in 256 grey levels.

Step 2.   For the i-th set of pixels belonging to the class k we define the Q^k matrix as follows:

$$Q^k=[AV^k(i),\ SD^k(i),\ SK^k(i),\ KU^k(i)] \quad \text{for } i = 1\ldots m$$

where m represents the cardinality of the i-th set of pixels under analysis and $AV^k(i)$, $SD^k(i)$, $SK^k(i)$, $KU^k(i)$ are the values of average, standard deviation, skewness and kurtosis, respectively. For each Q^k we define $min(Q^k)$ and $max(Q^k)$ as:

$$min(Q^k)=[min(AV^k),\ min(SD^k),\ min(SK^k),\ min(KU)^k)]$$
$$max(Q^k)=[max(AV^k),\ max(SD^k),\ max(SK^k),\ max(KU^k)]$$

These parameters provide the range for each variable under study.

Step 3.   The fuzzy range for each class is obtained as:

$$\left[f\left(min\left(Q^k\right)\right) \quad f\left(max\left(Q^k\right)\right)\right]=\left[\frac{1}{1+e^{-A^k\left(min\left(Q^k\right)-C^k\right)}} \quad \frac{1}{1+e^{-A^k\left(max\left(Q^k\right)-C^k\right)}}\right]$$

in which the sigmoid functions are defined on $[min(Q^k),\ max(Q^k)]$ assuming value into $[0, 1]$. The pair $\left(A^k,\ C^k\right)$ is a vector that determines the shape and position of each sigmoid. In Table 3 are reported the obtained fuzzy range for each statistical parameter of the category under analysis. Fuzzy sets are defined lying into a unitary hypercube as reported in Fig.2 for the case of a fixed value of kurtosis.

Step 4.   Let X the unknown region in a SAR image. With $X=[AV_X\ SD_X\ SK_X\ KU_X]$ we indicate the statistical parameters vector associate to X. For analyze X steps 2 and 3 must be applied to $\mathbf{X}$, in particular we define:

$$f^k\left(\mathbf{X}(j)\right)=\frac{1}{1+e^{-A^k\left((\mathbf{X}(j))-C^k\right)}} \quad j=1\ldots 4$$

Step 5.   Now, is possible evaluate the *Subsethood Operator* for the unknown zone under analysis as:

$$Subsethood(\mathbf{X},k)=1-\frac{\sum_{j}max\left[0,\ f^k\left(\mathbf{X}(j)\right)-f\left(max\left(Q^k\right)\right)\right]}{M(\mathbf{X})} \quad j=1\ldots 4$$

where:

$$M(\mathbf{X})=\sum_{j=1}^{4}f^k\left(\mathbf{X}(j)\right)$$

represent the fuzzy set measure in the unitary hypercube.

Step 6.   Closer the value of *Subsethood*(**X**, k) is to 1, more the X region belongs to the class k. Vice versa, closer the value of *Subsethood(***X**, $k)$ is to 0, more the X region does not belong to the class k under consideration. However, for unknown transition region *Subsethood(***X**, $k)$ can assume any value contained in $[0, 1]$. Consequently, the association between region X and category k is obtained as:

$$\max_{k=1,\ldots,4} Subsethood(\mathbf{X},k)$$

4. Numerical Results

In order to evaluate the effectiveness of the procedure described in the last section we have analyzed a SAR image reported in Fig. 1. It consists of 256x310 pixels. In this image we can isolate two main classes or categories: *Sea* and *Natural Terrain*. Each category is characterized by specifically values of statistical parameter.

We have calculated these values for different samples representative of each category appropriately selected in the image under test. In the step 2 of our algorithm the shape of each set of pixels take into account is rectangular. For this work, we have considered 50 sets of pixel regarding the Sea and 500 sets of pixel for Natural Terrain. For each i-th set of pixels, we have calculated the values of average, standard deviation, skewness and kurtosis. These data are reported in Tables 1 and 2. The aim of the classification with the *Subsethood Operator* was to distinguish between the two classes. Table 4 shown the percentage of correct classifications obtained using the *Subsethood Operator* for different region samples in Fig. 1. Bayes and Gauss classificators has been employed also for this purpose. The results obtained using the developed approach are in good agreement with these standard techniques. However, the proposed technique improves SAR images classifications in the transition regions in which *Sea* and *Natural Terrain* are mixed between them.

5. Conclusions

In this paper a fuzzy approach for SAR images classification is developed. In particular, by means of the representation of fuzzy sets as points in a unit hypercube and using fuzzy *Subsethood Operator*, we are able to associate a selected region of a SAR image to a particular category. Experimental results show that the proposed method gives very promising results especially in the cases in which transition zones are considered.

Table 1 – Values assumed by statistical parameters for different samples representative of certain regions of Natural Terrain in Fig. 1

Average	Standard Deviation	Skewness	Kurtosis
3.919	0.403	-0.627	5.675
3.996	0.062	-16.031	258.003
3.804	0.474	-0.625	6.055
3.913	0.651	3.497	25.096
3.997	0.169	-0.491	34.894
3.786	0.483	-2.210	9.519
3.873	0.644	1.626	20.711
3.936	0.244	-3.579	13.812
3.724	0.567	0.057	2.491
3.875	0.331	-2.267	6.142

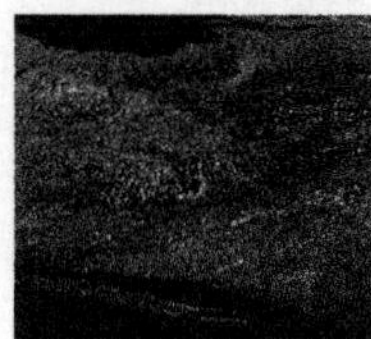

Figure 1 – Image exploited for our study that shows two typical classes: *Sea* and *Natural Terrain*.

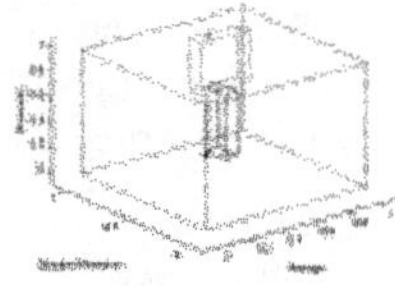

Figure 2 – Representation of the classes *Sea* (pearl grey box) and *Natural Terrain* (dark grey box) into a three dimensional hypercube corresponding to a fixed value of kurtosis.

Table 2 – Values assumed by statistical parameters for different samples representative of certain regions of Sea in Fig. 1

Average	Standard Deviation	Skewness	Kurtosis
197.786	33.517	-0.4594	2.611
198.786	28.150	-0.8454	4.665
203.852	34.550	-0.606	2.589
207.152	27.152	-0.454	2.729
167.301	30.956	-0.211	1.932
180.401	23.521	0.307	2.320
169.407	20.369	-0.430	2.410
180.922	34.785	0.062	2.171
192.208	24.438	-0.339	2.300
216.839	26.062	-0.787	3.046

Table 3 – Fuzzified ranges for statistical parameters of the Sea and Natural Terrain.

Statistical Parameters	Sea	Natural Terrain
Average	0.499 – 0.568	0.674 – 0.979
Standard Deviation	0.487 – 0.631	0.924 – 1.000
Skewness	0.177 – 0.700	0.445 – 0.864
Kurtosis	0.186 – 0.746	0.340 – 0.878

Table 4– Percentage of correct classifications for some regions samples in Fig. 1.

Type of Classifier	Sea regions	Natural Terrain regions	Regions with more Sea than Natural Terrain	Regions with more Natural Terrain than Sea
Bayes (AV)	100%	100%	75%	50%
Bayes (SD)	100%	100%	25%	100%
Gaussian	100%	100%	15%	100%
Subsethood	100%	100%	70%	100%

References

[1] A. Wacker *et al*, Minimum Distance Classification in Remote Sensing. Proc. 1st Canad. Symp. Remote Sens., Ottawa, 1972.
[2] A. Solberg *et al.*, A Markov Random Field Model for Classification of Multisource Satellite Imagery. IEEE Transaction on Geoscience and Remote Sensing, 34(1996) 100-113.
[3] Y.Chang Tzeng *et al*, A Fuzzy Neural Network to SAR Image Classification, IEEE Transaction on Geoscience and Remote Sensing, 36(1998) 301-307.
[4] L. A. Zadeh, Fuzzy Sets, Information Control, vol.8, pp.338-353, 1965.
[5] B. Kosko, Fuzzy Engineering, Prentice Hall International Editions 1997.

KES 2002
E. Damiani et al. (Eds.)
IOS Press, 2002

GCS networks, views, and path planning

Cristina BAROGLIO
Dipartimento di Informatica, Università degli Studi di Torino
corso Svizzera 185, 10149 Torino, Italy

Abstract. This paper presents a study of the application of Growing Cell Structure networks to learning map topology for performing path planning. The experimental framework is Webots, the well-known Khepera robot simulator. This work is the first step of a broader study aimed at merging symbolic planning together with path planning and control. GCS networks are a self-growing network model; the idea is to exploit the neighborhood structure that characterizes this model to support a potential field-like path planning. The use of GCS networks, w.r.t. other map representation methods, has the advantage that the agent builds its own representation of the environment. In the future this characteristic will be exploited for letting the agent acquire spatial knowledge about open environments, yet to be explored, and for tracking moving targets. I would like to thank my students P. Franceschina and A. Squarotti.

1. Introduction

Autonomous agents have a special appeal. They perform tasks by adapting to the conditions that they encounter. One of the capacities that a roaming, adaptive agent must have is the ability to acquire knowledge about its spatial environment. This means many things: first of all the agent should learn the topology of its world or, more precisely, of the free space where it can wander; such a knowledge should, then, be versatile, i.e. useful to perform tasks like "plan a path from here to there" but also "tell me where you are" or "find this kind of object". Such tasks, as well as many others, can be performed by reasoning at different abstraction levels. On one side there is the physical, continuous world, on the other an internal representation made of symbols. Many proposals have been done about what should fill the space in between the two extremes. One of the most successful suggests to use fuzzy logics (e.g. see [11]). Fuzzy rules have been used in many control tasks and interesting results about the relationships between fuzzy logics and other models have been obtained (for instance, a relationship between fuzzy rules and radial basis function neural networks has been underlined [12,4]).

In our work we have investigated the use of self-growing neural networks for acquiring a versatile spatial representation of the environment of a little Khepera robot. The spatial knowledge is captured by a variant of a Growing Cell Structure network, which is then used for self-localization and path planning. Path planning is obtained by exploiting a simple implementation of potential fields. Attractive potential is simply obtained by applying a graph-distance algorithm. Repulsive potential depends on the environment in the way that is described in Section 4. After a first set of experiments, we passed to a more sophisticate environment, where space is divided in rooms, each room having a name. Names are symbols. The new goal was to plan a path that follows a high-level, rough description given by an abstract oracle (e.g. a DyLOG symbolic planner [10]). So, for instance, if the oracle says: "go from the kitchen to the corridor and then to the living room", the Khepera agent will use the same spatial knowledge and will apply to it two different attractive potential functions: the first will allow it to reach the first sub-goal (the corridor), the second will allow it to reach the second sub-goal (the living room). The focus of this stage of the work is the planning task and not the development of control functions that make the robot follow the suggested path.

2. The simulator and the learning instances

Khepera is a little, two-wheeled, round robot, which can be equipped with a variety of devices (sensors for distance and light, a camera, ...). We do not own a real Khepera robot, we used its simulation (webots), which also allows to build worlds and add objects like balls, lights, etc. by means of a friendly gui. The simulator introduces some white noise (up to 10%) on sensor readings for making the setting closer to real world conditions. A webots world has a Cartesian reference system whose origin corresponds to the centre of the world; about rotation, the north direction corresponds to 0 degrees, west to 90 degrees, south to 180, and east to 270. Such a reference system is supplied to the user only, because in the simulation the robot has no way to know its position or orientation. In our experiments we used a single robot, equipped with distance sensors, in a big square world, divided in rooms by walls of the same color. We decided to use only straight walls because we do not want our robot to rely on peculiarities about the world, such as the shape or color of obstacles, as landmarks. The robot will try to acquire the free space topology. Since we exploit neural networks, the first step was to collect a set of instances for learning. If we want to learn the topology of the free space, we need to start from a set of points that belong to the free space. To this aim we lead our robot around the world to visit every room. In this phase we gave to the robot information about where it was located and how it was oriented, so that it could collect a set of instances, each of which contained position (used to build the network) and orientation. This choice is a simplification of what would happen in a non-simulated setting, where, supposing to take the start position of the robot as the origin (with orientation set to 0), the robot should compute its next positions by exploiting knowledge about its moves. Such a simplification helped us to avoid approximation problems and focus on the space representation problems (our target). For each recorded position we also store the sensor readings. Sensor readings will be useful in the path-planning phase. In order to exploit sensor readings we had to modify the neural network model in the way described in the next section. A little tuning had to be done in order to have the robot collecting sparse enough, uniformly distributed data.

3. The Growing Cell Structure Network model and Views

In this work we used Growing Cell Structure (GCS) networks; they belong to the family of *Growing Self-organizing* networks, which are vector-based models trained by competitive learning. Generally speaking, instead of defining the whole network in advance and, then, let it adjust its weights, one starts with a near-minimal network and lets it add (delete) neurons until it either has reached a predefined, maximum size or we gained the desired precision for the task at hand. The main characteristics of this approach are the evolution of the network structure through a *growth process* (which is also an advantage w.r.t. other approaches) and the use of *local statistical quantities* to decide when and where to insert/delete units. Another advantage is that these networks show a *fast training*, in particular in supervised learning (a factor of 30-100 over back-propagation has been observed for some problems). Nodes are organized in a neighborhood structure, where each node is connected to a set of topological neighbors. Connections respect the *Delaunay Triangulation* [2]. Among the applications we can find: topology learning [5], pattern classification [3], and clustering [5].

GCS[1] networks [1] have a network topology constrained to consist of k-dimensional simplices, where k is a predefined positive integer. The basic building block and also the initial configuration of each network is a k-dimensional simplex (e.g. a triangle, for $k=2$). Briefly, *node insertion* works as follows: learning instances are used to gather local error information at each node; after a predefined number of instances (*insertion step*) have been processed, the node with the maximum accumulated error is identified and the longest edge emanating from it is split, by inserting a new node. Afterwards the network is locally restructured for maintaining the k-simplex structure. See [1] for details. The *novelty* of our work stands in exploiting the knowledge captured by the neighborhood structure of GCS networks for path planning tasks.

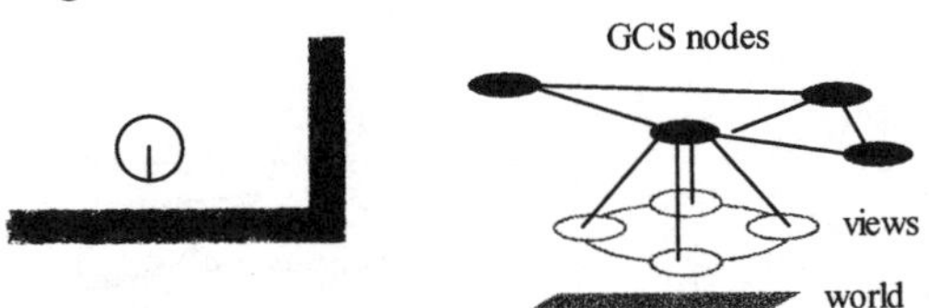

Figure 1 – Left: an agent facing a wall. Right: an example of the two-level representation: green nodes are GCS nodes, they capture locations in the free space, their connections capture its topology. Grey nodes are views: a set of views is associated to each GCS node, they store sensory data perceived at the node location.

To this aim, we modified the GCS nodes by adding *views*. Intuitively, each node represents a small *region* of the outer world; *views* are sensor readings at different orientations (what the robot perceives by "watching around" while standing at a point), which are added to nodes. Such an enriched network can be used for tasks where reasoning and sensing are combined, e.g. robot *position detection* (the robot "looks around", passes the information through the network, finds a set of candidate positions, and applies a disambiguation policy for restricting the set of candidates –this is a yet unpublished work–) and for tasks that involve navigation (e.g. turn off all the lights in the house). The association between nodes and views was initially based on the learning instances but during learning, nodes move in the input space and split producing new neurons; the problem is that it is not obvious how to change the associated views, the mathematical model of the sensors should be given. However, in a controlled environment the following procedure can also be adopted: let the network grow, then, for each neuron, move the robot to its position and let it perceive the world, associate the perceived views to the neuron. In different words: first learn the topology and afterwards anchor views to the nodes.

For the implementation, we modified the GCS C++ implementation of the *nno package*, developed at the *University of Bochum*. The best internal representations of maps were obtained by using the following values: learning step 0.006, neighbor learning step 0.0001, insertion step 20000 (the learning set contains about 500 instances, the number of learning epochs after which no improvement was obtained is 8200), maximum number of neurons 200, maximum number of connections 50, deletion disabled. Node deletion was disabled to avoid interruptions in the network structure. Some tuning of parameters is necessary to avoid connections that cross walls. Figure 2 shows an example of what GCS nets look like; each node is a region in the input map.

4. Potential fields

---

[1] GCS was the best performing unsupervised model, chosen after preliminary tests on our experimental setting.

Potential fields [6,7] are a powerful and widely used tool for robot path planning and control. Potentials are obtained by combining an attractive potential function that decreases towards the goal, and a repulsive potential function that keeps the agent far from obstacles.

In the work here described *attractive potential* is obtained in the following way: once the target position is known, it is passed through the network which returns the neuron that is the closest to it. The net discretizes the world into a set of nodes. The firing neuron will be the root of a *minimal spanning tree*, built by numbering nodes according to their minimal distance (tree distance) from the root. Now wherever the robot is placed, it can find the closest neuron to its current position (see Figure 2): the shortest path that joins such a node to the root will be the first approximation of the navigation path.

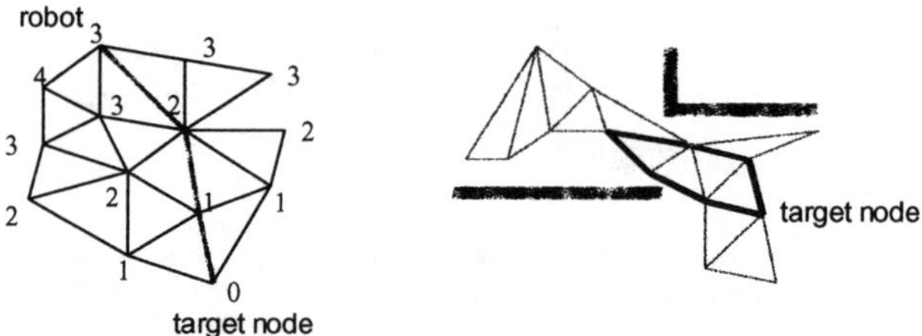

Figure 2 – Left: example of node distance measurement and of path built upon a piece of network. Right: example of alternative paths to a target node, red lines represent walls.

Nevertheless, the *neighborhood structure* of a GCS network is a graph, so a node at distance N+1 from the root may be connected to different nodes at distance N. As a consequence there may exist many paths with the same length, that join the root node and the robot position node. *Views* keep information about obstacles and are used by a heuristic that forces agents to prefer paths that are farther from obstacles. They correspond to repulsive potentials. Observe that we do not combine the two kinds of potential. Figure 2 shows a hypothetical example: we have two alternative paths with same length that lead to a target node. In this situation, the heuristic will choose the upper path because the lower passes too close to a wall (the robot knows that in that direction there is an obstacle and it may collide). While attractive potential depends on the goal, views depend on the map. When planning, visited nodes are marked, so in presence of local minima (the presence of obstacles can produce local minima), it is possible to perform backtracking in order to search for alternatives. In the worst case the whole graph is analised. Paths are sequences of segments, at execution time they can be smoothened by means of standard interpolation techniques.

5. Adding symbols

The next step was quite straightforward. Let's suppose to have a symbolic planner that exploits knowledge about the environment to build high-level plans (e.g. "you must reach the kitchen by going from the bathroom to the corridor and pass by the entrance"): we want our robot to automatically build a path that implements, at the GCS network level, the higher-level navigation plan. To this aim, we had to add to each neuron the name of a room, depending on the neuron's position. The planning procedure now works as follows: given a sub-goal, e.g. "go from kitchen to corridor", set as goal the neurons that are in the corridor and have neighbors in the kitchen. Build the attractive potential according to the current goal.

Build a path. Pass to the next sub-goal, taking as start position the end of the previously built path and continue until no sub-goal is left. This is just one of the possible uses of the above described representation

6. Results and future work

In this paper we have (briefly) described how a robot path-planning problem can be turned into a graph navigation problem where the graph that represents the topology of the free space is built by the agent itself through a learning process that exploits self-organizing networks. We have said that this representation can be augmented by adding information that the robot collects, while exploring its environment, and anchors to specific areas. We have also (briefly) described how such information can be used in a navigation task. By using the described knowledge structures, our robot was always able to quickly build satisfactory plans for reaching targets.

The idea of representing the free space as a graph is not new (e.g. see [8,9]), what is new is the way in which such a graph is acquired and, then, used. We believe that the incremental nature of GCS networks will be very useful in applications where an autonomous agent has to explore open environments (open means that we do not have a set of learning instances before the robot begins the exploration so learning occurs while the robot moves) and grow knowledge about it. In the future, besides deeper studies on view management, we mean both to integrate the developed system with a symbolic planner and to integrate it with controllers that will allow the agent to move along the obtained paths.

References

[1]   B. Fritzke, Growing cell structures - a self-organizing network for unsupervised and supervised learning, *Neural Networks*, (7)9, Pergamon Press, (1994) 1441-1460.

[2]   S. M. Omohundro, The Delaunay triangulation and function learning. Tr-90-001, International Computer Science Institute, Berkeley, 1990.

[3]   Fast learning with incremental radial basis function networks, *Neural Processing Letters*, (1)1, D Facto Publishing , (1994) 2-5.

[4]   C. Baroglio, A.Giordana, G. Lo Bello, and R. Piola, Learning Fuzzy Controllers from Examples, in Statistics and Machine Learning: The Interface, Wiley, (1997), 197-220.

[5]   B. Fritzke, A growing neural gas network learns topologies, G. Tesauro, D. S. Touretzky, and T. K. Leen eds, *Advances in Neural Information Processing Systems 7*, MIT Press, (1995) , 625-632.

[6]   O. Khatib, Real-time Obstacle Avoidance for Manipulators and Mobile Robots, *International Journal of Robotics Research*, (5) 1, 90-98, 1986.

[7]   D. E. Koditschek, Robot Planning and Control via Potential Functions, in *The Robotics Review 1*, Khatib O., Craig J. J an Lozano-Perez T. ed., MIT Press, 1989.

[8]   J. C. Latombe, Robot Motion Planning, Kluwer Acad. Publ. 1993.

[9]   G. Dudek, M. Jenkin, Computational Principles of Mobile Robotics, Cambridge Univ. Press, 2000.

[10] M. Baldoni, L. Giordano, A. Martelli, and V. Patti, Reasoning about Complex Actions with Incomplete Knowledge: A Modal Approach. In A. Restivo, S. Ronchi Della Rocca, L. Roversi, eds, Proc. of ICTCS'2001, LNCS2202 , Springer, (2001), 405-425.

[11] I. Bloch, A. Saffiotti, Why robots should use fuzzy mathematical morphology, Proc. of the 1st Int.ICSC-NAISO Congress on Neuro-Fuzzy Technologies, La Havana, Cuba, 2002.

[12] H.R. Berenji, A reinforcement Learning Based Architecture For Fuzzy Logic Control, Readings in Fuzzy Sets for Intelligent Systems, 13, Morgan Kaufmann, (1993), 368-380.

KES 2002
E. Damiani et al. (Eds.)
IOS Press, 2002

An Artificial Neural Network Algorithm for Track Reconstruction in the ALICE Inner Tracking System

A. Pulvirenti[1,2], A. Badalà[1], R. Barbera[1,2], G. Lo Re[1,2], A. Palmeri[1], G. S. Pappalardo[1], F. Riggi[1,2]

for the ALICE Collaboration.

1) INFN, Sezione di Catania - Italy
2) Dipartimento di Fisica e Astronomia dell'Università di Catania - Italy

Abstract. An artificial neural network algorithm for high transverse momentum track recognition in the ALICE Inner Tracking System (ITS) is presented. The model is based on the Denby-Peterson scheme but contains several original improvements to cope with the huge particle multiplicity expected at ALICE. Results concerning central PbPb events at LHC energy are shown and compared with those obtained with a full tracking algorithm based on the Kalman Filter.

1. Introduction

ALICE [1] is one of the four planned experiments at the CERN Large Hadron Collider (LHC) . Its main purpose is to study high energy heavy ion collisions (up to PbPb at 5.5 A TeV in the CM system) in order to observe the theoretically expected transition from normal hadronic matter to a plasma of "de-confined" quarks and gluons (QGP). Due to the unprecedented track density expected in PbPb collisions at LHC energy (more than $8 \cdot 10^4$ primary particles in the whole phase space) track finding and reconstruction in ALICE is a daunting task. This task is usually accomplished using the track information both from the Time Projection Chamber (TPC) and the Inner Tracking System (ITS) [2]. In this paper we present an artificial neural network algorithm [3] for high transverse momentum tracking in the ITS stand-alone mode, i.e. when the information from the TPC will not be available. This might be the case if the ITS should be used (without the TPC) together with other fast ALICE detectors for special purposes/studies. The paper is organized as follows. Section 2 contains a general description of the neural network used. Mapping of the ITS stand-alone tracking problem onto the neural network is discussed in Section 3. Boundary conditions for network creation and track reconstruction procedures are presented in Section 4 while results are shown in Section 5. Conclusions are summarized in Section 6, which also gives an outlook to future work.

2. The neural network schema

The network, which has been implemented, consists of the neural topology usually known as "associative memory" (or "Hopfield's network") [4], which is widely used for pattern recognition purposes. Using the definition of the so-called "energy function" of the Hopfield network it is possible to develop a variant of the neural scheme which allows to solve optimization problems. Such a variant is the application of an updating rule which introduces a stochastic choice of the activation, which is led by a "temperature" parameter. Due to its analogy to the annealing procedure commonly used during the production of pure silicon crystals, this method is called "simulated annealing" (SA) [5, 6]. However, a SA network usually has a very low speed, which is strong limitation in case of a huge amount of data such as the one expected to be collected at ALICE. Then, instead of the "pure" SA, we decided to adopt the so-called *MFT* approximation [7] of a Hopfield network where binary activations are replaced by real-valued ones described by a "logistic function" of the "temperature" parameter.

3. The track recognition procedure

This section describes the mapping of the neural schema proposed above onto a suitable track recognizer [8-10] together with some original improvements necessary to cope with the large track density expected at ALICE.

Experimental data coming from a tracking detector usually consists of a large ensemble of space points with no hints for identifying *which* ones belong to each particle's path and the result of a track finding procedure is just the resolution of these ambiguities. Due to the fact that the ALICE ITS consists of 6 sensitive layers (two inner layers of silicon pixel detectors, two intermediate layers of silicon drift detectors, and two outer layers of silicon strip detectors) then, 7-points polygonals should be obtained from a track finding procedure (including the primary vertex). Owing to this geometrical picture, a good method to map the tracking problem onto the neural network comes from the association of a neuron to a segment of such polygonal lines, i.e. a simple geometrical space segment linking two space points. Obviously, one must create many segments entering into and exiting from each space point and the neural network operation has to choose which one is more likely to be correct. In a more practical picture, the neural network will turn 'on' all correct segments and 'off' the others, in order to return, as its output, many chains of connected track segments going from the inner layer to the outer one. These chains will be the track candidates which have to be considered as the "guesses" for the particle trajectories. In order to take into account the correct time direction in the ensemble of points, an orientation must also be defined for each segment, following the convention to take the space points in the direction of the increasing distance from the primary vertex. Then, a neuron n can be represented by two indexes: a "tail" index and a "head" index (with $n_{ij} \neq n_{ji}$). Moreover, under the reasonable hypothesis of track continuity, non-zero weights will be allowed only for couples of neurons sharing a point. In general, there are two different possible non-zero weight configurations because of the neural orientation. When the tail of one unit coincides with the head of the other ($n_{ij} \rightarrow n_{jk}$), then their orientations lead to the definition of a path touching the three connected points in an ordered sequence. Such situations represent a possible guess for a couple of segments belonging to the same track, so they should be associated with an excitatory weight (positive value). In this case we can also define a 'quality parameter' which should give a different excitatory contribution according to some criteria related to the probability that the found sequence is likely to be a correct one. If the common point between two track segments is a head or a tail for both neurons a "competition" arises between these two units because the two neurons turn out to be incompatible with each other (bifurcations in the recognized tracks are not allowed). As a consequence, this situation requires a negative (and then inhibitory) weight which is kept constant in the whole neural network.

The final result of this construction scheme makes each neuron get two contributions from each updating cycle:

- a "gain" contribution, given by all sequenced neurons;
- a "cost" contribution, given by all competing neurons.

Neurons, which will remain active, are the ones that will collect large gain terms, due to well-aligned sequenced contributions, while the kinked units are very likely to be turned off at the end of the neural network operation. A sketch of the neural work flow is shown in figure 1:

1. the experimental points (represented by the small circles) are collected;
2. a lattice is created with many neurons, here represented by segments connecting two points; obviously, as it will be explained hereinafter, not all point couples will be connected;
3. at the end of the track finding procedure, the only active neurons will be the ones which will form the polygonal lines (track segments) connecting - in the correct order - all points of the same track; in a more realistic case, there could still be some wrong neurons turned on, but their activations will be much smaller than the ones of the correct neurons.

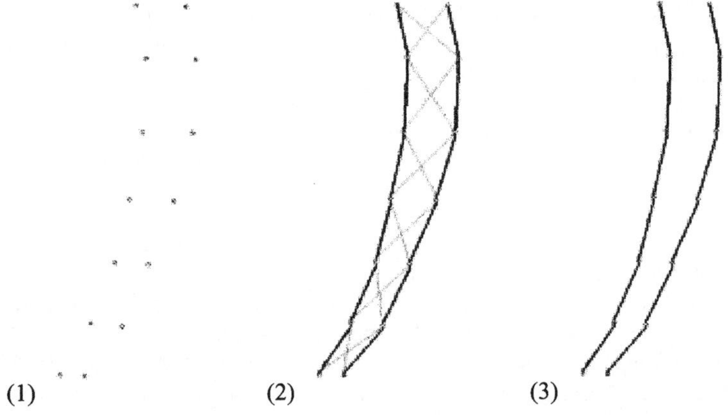

Figure 1: A sketch of the neural network work flow.

4. Boundary conditions for network creation and track reconstruction

The main source of wrong results from a neural track recognition procedure is generally related to the possible "confusion" due to a too large number of neurons starting from or ending to the same point. This is also a computational problem, because a too large number of neurons, in a simulation like the one which has been developed here, requires too much memory and CPU time to complete the work. This explains why some criteria are needed in order to select only the point pairs which look most likely to be correct. The first cut comes from the consideration that a 'useful' particle track (i.e., one whose physical parameters are recognizable with a good resolution) should contain as many points as possible, which, in the case of the ITS, means that they should be six (one per layer). The criterion implementing such an assumption can be defined by imposing that the two edges of a neuron must lie on adjacent layers. Other criteria follow from the physical behavior of a particle track lying in the high p_t range. If we can implement a weight which selects the well aligned track segments, then we can also create only those neurons which are well aligned along the radial direction going from the most-probable primary vertex position to the outermost ITS layer. This is performed, with two cuts:

- one on the difference in the polar angular coordinates;
- one in the curvature of the circle in the bending plane, passing through the primary vertex (found with the algorithm described in ref. [11]) and the two "candidate" points.

Another cut is defined looking at how much the two candidate points are likely to lie on the same helix passing through the primary vertex estimated above. For all of the above, the synaptic weight, which has been used throughout this paper, is defined by the following expression:

$$w_{ijkl} = A C_{ijl}\left(1 - \sin\theta_{ijl}^n\right)\delta_{jk} - B\left(1 - \delta_{ik}\right)\left(1 - \delta_{jl}\right) \text{ where } C_{ijl} = \frac{2\varepsilon}{d_{ij} + d_{jl}} \quad (1)$$

where θ_{ijl} is the space angle between two neurons and C_{ijl} is a quantity always larger than 1 whose value increases as the pairs of neurons are better disposed along a helix passing through the vertex. d_{ij} gives a quantitative estimation of how well the vertex and the neuron edges are fitted by the same helix. The left term is the excitatory one (in fact it is non zero only for sequenced neurons, according to the Kronecker symbol) and the other is the inhibitory one. Network creation cuts are very useful in simplifying the initial state of the network but can induce severe problems affecting the tracking efficiency. So, instead of performing a unique analysis step over all the space points, a set of curvature cuts has been defined and the network has been instructed to operate on the same ensemble of space points in many steps. At the end of each step, all used points are saved and removed from the arrays and the curvature cut is relaxed. This procedure allows to select at first the very well aligned (straight) tracks and then, after removing all the used space points, tracks with lower and lower transverse momentum (curved tracks).

In order to improve the speed of the algorithm two procedures have also been set up. According to the "adjacent layers" cut discussed above, all the arrays containing space points and neurons have been splitted with respect to the ITS layer index (from 1 to 6) thus avoiding to cycle over the whole array for each couple of points. Second, the whole ITS barrel has been logically divided into N sectors, according to the azimuthal angle, and the neural network algorithm runs on one sector at a time. This reduces the array population by a factor of N, so its contribution in speeding up the job is quite substantial. In the simulations described in the following the value $N=18$ has been used throughout.

Once the tracks have been recognized, space points belonging to them have then been fitted to a helix in order to determine the track parameters and their resolutions. The used fit function, expressed in cylindrical coordinates (ρ, φ, z), can be written as:

$$
\begin{cases}
\varphi(\rho) = \gamma_0 + \arcsin\left(\dfrac{C\rho + (1 + CD_t)D_t/\rho}{1 + 2CD_t} \right) \\[2ex]
z(\rho) = D_z + \dfrac{\tan\lambda}{C}\arcsin\left(C\sqrt{\dfrac{\rho^2 - D_t^2}{1 + 2CD_t}} \right)
\end{cases}
\tag{2}
$$

where C is one half of the track curvature, D_t and D_z values are the impact parameters in the transverse and longitudinal directions, γ_0 is the angle of initial transverse momentum direction in the bending plane, and λ is the track dip angle ($\tan\lambda = p_z / p_t$). The best-fit is done in two steps:

1. fitting the circular projection in the bending plane;
2. rewriting the equation (2-II) as a linear function:

$$
z = D_z + s\tan\lambda \qquad \text{where} \qquad s = \arcsin\left(C\sqrt{\dfrac{\rho^2 - D_t^2}{1 + 2CD_t}} \right)
$$

which is possible using the results obtained after the first step.

The first step is by far the most difficult one, due to its non-linearity. In order to perform it, we map the point coordinates into the dual space known as *Riemann sphere* where the circle fit reduces to a plane fit, which is exactly solvable. The procedure is explained in ref. [12].

5. Results

In order to evaluate the performances of the algorithm described so far, some tests have been made with simulated central PbPb collisions at LHC energy (5.5 A TeV in the CM system). For this purpose some classes simulating the neural network have been implemented in C++ within AliRoot [13], the object oriented ALICE simulation and reconstruction framework. Simulated events have been created both in the acceptance of the ALICE central detectors (pseudo-rapidity $|\eta|<0.9$ which corresponds to $45°<\theta<135°$) and in the acceptance of the whole ALICE (pseudo-rapidity $|\eta|<8$ which practically corresponds to $0°<\theta<180°$).

According to previous tracking studies in ALICE [14,15], tracking efficiency has been defined as the ratio of all "good" tracks over all "findable" tracks, where a "good" track is a track which satisfies the criteria required in order to accept it as a well recognized one and a "findable" track is every track which, in principle, can be recognized with the used criterion. In the present paper, two different criteria are used. A "hard"-er criterion defines as "findable" all tracks which leave a signal in *each* ITS layer; as a consequence, it marks as "good" all recognized tracks which contain *no* wrong assignments. A "soft"-er criterion allows for at last one wrong assignment in the "good" track and considers as "findable" also tracks which miss at last one ITS layer. Every recognized track which does not satisfy the "goodness" criterion is tagged as "fake". As an example, figure 2 shows, in the case of the "soft criterion", the "good" tracking efficiency (black symbols) and the "fake" track probability (white symbols) as functions of the particles' transverse momentum. Average values of the tracking efficiencies (for $p_t > 1$ *GeV/c*) are reported in table 1 together with those relative to the full tracking algorithm based on the Kalman filter [14].

For all good tracks which have been found by the neural network, track parameter resolutions have also been evaluated for p_t, γ_0, λ, D_t and D_z with the method cited in section 4. Standard deviations of the track parameter resolutions are reported in table 2 together with the values obtained with the Kalman filter algorithm.

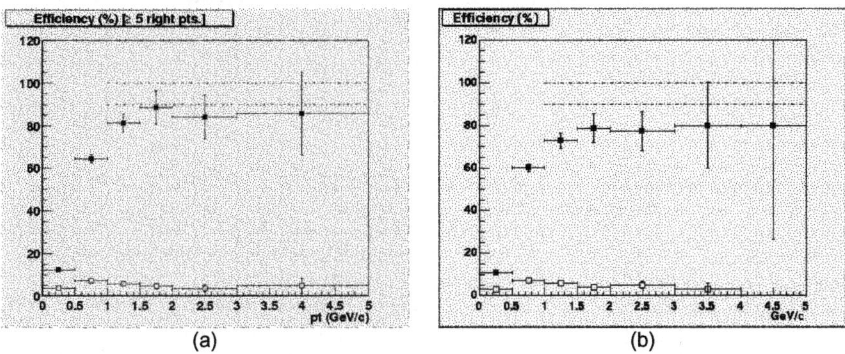

(a)   (b)

Figure 2: Neural tracking efficiency plots in the case of the "soft" criterion for mid-pseudorapidity (a) and full pseudo-rapidity (b) cases, respectively. Dotted lines are just to guide the eye for the efficiency levels of 90% and 100%.

Efficiency (%)	GOOD ($|\eta|< 0.9$)	FAKE ($|\eta|< 0.9$)	GOOD ($|\eta| < 8$)	FAKE ($|\eta| < 8$)
Neural - "hard"	84.7 ± 3.9	8.3 ±0.9	79.2 ±3.5	10.4 ±1.0
Neural - "soft"	83.3 ± 3.4	5.1 ± 0.6	79.8 ± 3.1	0.9 ± 0.5
Kalman - "hard"	87.5 ± 4.0	0.001 ± 0.001	83.0 ± 4.0	0.9 ± 0.5
Kalman - "soft"	87.0 ± 4.0	0.001 ± 0.001	83.0 ± 4.0	0.7 ± 0.4

Table 1: Average efficiencies for particles having a transverse momentum greater than 1 GeV/c with respect to both the "hard" and "soft" evaluation criterion (see text).

Track parameter (units)	Neural	Kalman (at p_t = 1 GeV/c)
p_t (%)	12.5 ± 0.3	1.57 ± 0.02
γ_0 (mrad)	2.45 ± 0.06	1.40 ± 0.08
λ (mrad)	3.04 ± 0.11	1.60 ± 0.08
D_t (µm)	139 ± 2	50 ± 2
D_z (µm)	398 ± 7	150 ± 2

Table 2: Resolutions of all track parameters with a comparison with those relative to the Kalman filter tracking. Kalman filter parameter resolutions have been calculated at p_t = 1 GeV/c.

It is worth stressing here that the number of space points used to perform the fit are quite different: an ITS stand alone tracking has only 6 points (plus the primary vertex) to work with, while the full Kalman filter tracking can exploit up to all the 160 space points registered in the ALICE Time Project Chamber [2]. This explains why, of course, track parameter resolutions obtained with the neural tracking are worse than those obtained with the Kalman filter.

6. Conclusions and outlook

In conclusion, in this paper we have exploited a modified Hopfield neural network to perform a track finding and reconstruction in the ALICE Inner Tracking System for high transverse momentum tracks ($p_t > 1$ *GeV/c*). Another future development line concerns a possible use of the neural network approach to perform a low transverse momentum tracking ($p_t < 0.2$ *GeV/c*) where Kalman filter efficiency begins to fall down.

References

[1] ALICE Collaboration, Technical Proposal, CERN/LHCC 95-71 (http://consult.cern.ch/alice/Documents/1995/01/abstract).

[2] ALICE Collaboration, Time Projection Chamber, Technical Design Report, CERN/LHCC 2000-001; ALICE Collaboration, Inner Tracking System, Technical Design Report, CERN/LHCC 99-12. Both documents can be found at the URL: http://alice.web.cern.ch/Alice/TDR/.

[3] Due to length limitations, the present paper is a condensate of the much longer internal report: A. Badalà, R. Barbera, G. Lo Re, A. Palmeri, G. S. Pappalardo, A. Pulvirenti, F. Riggi, ALICE Internal Note 2002-12 (http://edmsoraweb.cern.ch:8001/cedarnew/navigation.tree?cookie=966155&p_top_id=1505648327& p_top_type=P&p_open_id=1066276064&p_open_type=P). The reader is addressed to that document for any further details.

[4] J. A. Freeman and D. M. Skapura, Neural Networks: algorithms, application and programming techniques, Addison Wesley Ed., 1991.

[5] P. Salamon *et al.*, *Comp. Phys. Comm.* **49** (1988) 423-428.

[6] S. Kirkpatrick, C. D. Gelatt, M. P. Vecchi, *Science* **220** (1983) 671-680.

[7] C. Peterson and J. R. Anderson, *Complex Systems* **2** (1988) 59-89.

[8] G. Stimpfl-Abele and L. Garrido, *Comp. Phys. Comm.* **64** (1991) 46.

[9] C. Peterson, *Nucl. Instr. and Meth. in Phys. Res. A* **279** (1989) 537-545.

[10] B. Denby, *Comp. Phys. Comm.* **49** (1988) 429.

[11] A. Badalà, R. Barbera, G. Lo Re, A. Palmeri, G. S. Pappalardo, A. Pulvirenti, F. Riggi, Nucl. Instrum. and Methods Section A (2002); *ibidem* ALICE/ITS 2001-11 (http://edmsoraweb.cern.ch:8001/cedar/doc.info?cookie=966155&document_id=309560&version=1); *ibidem*, ALICE/ITS 2001-26 (http://edmsoraweb.cern.ch:8001/cedar/doc.info?cookie=966155&document_id=320645&version=1).

[12] A. Strandlie *et al.*, *Comp. Phys. Comm.* **131** (2000) 95.

[13] All details about Aliroot can be found in the official ALICE Off-line Project WWW pages located at the URL: http://AliSoft.cern.ch/offline/.

[14] A. Badalà, R. Barbera, G. Lo Re, A. Palmeri, G. S. Pappalardo, A. Pulvirenti, F. Riggi, Nucl. Instrum. and Methods Section A (2002), in press; *ibidem*, ALICE/ITS 2001-39 (http://edmsoraweb.cern.ch:8001/cedar/doc.info?cookie=966155&document_id=328788&version=2).

KES 2002
E. Damiani et al. (Eds.)
IOS Press, 2002

AppART + Growing Neural Gas = high performance hybrid neural network for function approximation*

Luis Martí[†,‡], Alberto Policriti[†], Luciano García[‡] and Raynel Lazo[‡]

[†]DIMI, Università degli Studi di Udine, Italy.   [‡]Fac. de Mat. y Comp., Universidad de La Habana, Cuba.

Abstract. We combine two notable streams of neural networks research: Adaptive Resonance Theory (ART) and Growing Neural Gas (GNG) networks. In particular we modify the AppART neural network formulation by introducing GNG based training features. The resulting neural network outperforms its original version as well as other neural models while maintaining the functional approximation properties and hybrid system conception.

1   Introduction

In this work we propose a neural model that combines two streams of connectionist research that has been recognized as notable achievements of modern artificial neural network theory: Adaptive Resonance Theory and Growing Neural Gas networks.

Adaptive Resonance Theory (ART) [1] is a theory of human cognition. ART networks have some features, such as match-based stable learning and intrinsic self-organization, that are appealing for constructing a hybrid (symbolic+connectionist) neural system. The search process involved in the production of a network output in ART networks is also of interest as it is a sort of hypothesis testing mechanism . Some of these features can also sensibly improve the high parameterization problem associated to Multi-Layer Perceptron (MLP) and Radial Basis Functions (RBF) networks [2].

Growing Neural Gas (GNG) [3] networks are intrinsic (growing) self-organizing RBF neural networks based on the Neural Gas [4] model. This models use cumulative errors associated to the input classes to determine where new classes should be inserted.

AppART [5] is an ART low parameterized neural model that incrementally approximates continuous-valued multidimensional functions from noisy data using biologically plausible processes. AppART performs a higher-order Nadaraya-Watson [6] regression and can be interpreted as an extension of the fuzzy logic's Standard Additive Model [7]. AppART allows the on-line insertion and extraction of fuzzy if-then rules. It also provides way of justifying network responses. AppART have been proved to have the universal and best approximation properties shared by RBF networks [2].

In this work we modify the original AppART formulation by introducing some GNG based training characteristics. The resulting neural network, which we named GasART, builds classes of similar inputs that have similar outputs. Each class has a desired output value assigned. The output of the network to a given input is the weighted mean of the degree of membership of the input to the classes stored and the desired output values assigned to each class. A match tracking mechanism induces the creation of more specific classes when the prediction of the network differs from the expected output at some degree. GasART keeps the general input propagation dynamics of AppART, while modifies the way new classes are

---

*The authors wish to thank the Dipartimento di Matematica e Informatica of the Università degli Studi di Udine for its support on the elaboration of this work.

Corr. author: L. M., DIMI, Univ. Udine. v. delle Scienze 208, Udine (UD) 33100 Italy; marti@dimi.uniud.it

inserted and the laws that control the way these classes are modified to fit the complexity of the training data. GasART will be described in the next section. After that, we show the results yielded when solving a well-known functional approximation benchmark problem: the approximation of the Mackey-Glass equation.

2   GasART dynamics and training

GasART has a layer of afferent or input nodes, *F1*, a classification layer, *F2*, a prediction layer, *P*, and an output layer *O*. The F2 layer stores classes of inputs. Its activation is a combined measure of the similarity of the input and the prototype of each class, and the size of the given class. Each class is represented by a Gaussian receptive field. The network output is obtained by propagating the output of the F2 layer through the P and O layers.

2.1   *Equations*

When an input $\mathbf{x} \in \mathbb{R}^n$ is presented to the input layer it is propagated to the F2 layer. F2 has N^* nodes, with N of them committed. Each committed node models a local density of the input space using Gaussian receptive fields with mean μ_j and standard deviation σ_j and has an associated cumulative error, e_j. A node is activated if it satisfies the match criterion. That means that the match function,

$$G_j = \exp\left(-\frac{\sum_{i=1}^{n}(x_i - \mu_{ji})^2}{2\sigma_j^2}\right), \ j = 1, \ldots, N,$$
(1)

must be greater than the F2 vigilance parameter, ρ_{F2}; according to this, the input strength of a node is computed as

$$g_j = \begin{cases} \frac{\eta_j G_j}{\sigma_j}, & \text{if } G_j > \rho_{F2} \\ 0 & \text{otherwise} \end{cases}, \ \rho_{F2} > 0,$$
(2)

where η_j is a measure of the node a priori activation probability. The activation of each node is then calculated normalizing the node's input strength,

$$v_j = \frac{g_j}{\sum_{l=1}^{N} g_l}.$$
(3)

The prediction and output layers conjointly calculate the prediction of the network. The P layer contains two types of nodes: *A* and *B*. There are as many A nodes as features are in the output vector $\mathbf{y} \in \mathbb{R}^m$ and only one B node. A and B nodes calculate its activations, a_k and b respectively, as weighted sums of the F2 activation vector

$$a_k = \sum_{j=1}^{N} \alpha_{kj} v_j, \ k = 1, \ldots, m,$$
(4)

$$b = \sum_{j=1}^{N} \beta_j v_j.$$

Nodes on the output layer are connected with one-to-one connections to a corresponding A node and to the B node. Their outputs,

$$o_k = \begin{cases} \frac{a_k}{b} & \text{if } b > 0 \\ 0 & \text{otherwise} \end{cases},$$
(5)

represent the most probable output value, $E(\mathbf{o}|\mathbf{x})$, expected after an input $\mathbf{x}$ have been presented.

2.2 Error handling and match tracking

After the presentation of an input, if no F2 node is active an uncommitted node must be committed. The task of detecting when an input is not sufficiently coded in F2 is accomplished by the F2 gain control, G_{F2}, that fires if no committed nodes are active. The signal

$$\Gamma_{F2} = \begin{cases} 1 & \text{if } \max_{j=1,\dots,N} v_j = 0 \\ 0 & \text{otherwise} \end{cases} \tag{6}$$

is used to commit an uncommitted node. It can also be used to offer an "I don't know" answer during the non-adaptive use of a network.

The E node computes the prediction error, ξ, measuring it as an absolute error,

$$\xi = |\mathbf{o} - \mathbf{y}| \;. \tag{7}$$

Incorrect predictions are detected by the output gain control, G_O. G_O compares the error measurement produced by the E node with a output error vigilance parameter, ρ_o,

$$\Gamma_O = \begin{cases} 1 & \text{if } \xi > \rho_o \\ 0 & \text{otherwise} \end{cases} \quad \rho_o > 0 \,. \tag{8}$$

If G_O remains inactive learning takes place in F2 and P. If G_O fires then a match tracking mechanism takes care of rising the F2 vigilance from its base value $\overline{\rho_{F2}}$, that is the minimum vigilance accepted. The purpose is to reset currently active categories that might be interfering in the calculation of an accurate prediction. A straight through approach could consist in raising the vigilance value as the minimum activation,

$$\rho_{F2} = \min_{j=1,\dots,N} v_j, \tag{9}$$

therefore deactivating the less active F2 node.

2.3 Learning

The learning rule used in F2 is based on the gated steepest descent learning rule [1].

First, the prediction errors obtained in this input presentation are added to the cumulative error of the nodes that had interfered in the production of the prediction. The amount of error added is modulated by the node activation, v_j,

$$e_j\,(t+1) = e_j\,(t) + \xi v_j \,. \tag{10}$$

After this, the cumulative category activation, η_j, is updated as

$$\eta_j\,(t+1) = \eta_j\,(t) + v_j \,, \tag{11}$$

in order to represent the amount of training that has taken place in the jth node. The use of η_j equally weights inputs over time with the intention to measure their sample statistics.

Among all active nodes the *best-matching node (BMN)* is determined as the one with maximum activation, v_j. A *second best-matching node (SBMN)* is selected, if its not bound with the BMN a link between them is established with $\kappa = 0$. If the link already exists its age is reset to zero. The center of the BMN receptive field is updated by allowing

$$\mu_{BMNi}\,(t+1) = \left(1 - \eta_{BMN}^{-1}v_{BMN}\right)\mu_{BMNi}\,(t) + \eta_{BMN}^{-1}v_{BMN}x_i \,, \tag{12}$$

All nodes bound to the BMN also modify their centers

$$\mu_{ji}\left(t+1\right) = \left(1 - \upsilon\eta_j^{-1}\upsilon_j\right)\mu_{ji}\left(t\right) + \upsilon\eta_j^{-1}\upsilon_j x_i . \tag{13}$$

but with a rate, $0 \le \upsilon \le 1$, that controls the amount of change. After this, the age of the links emanating from the BMN is incremented.

When a node is committed and after a training iteration, the standard deviations should be recomputed. They are set to the mean distance between the center of the node and the centers of the nodes bound it [3].

In the P layer α_{kj} is adapted to represent the corresponding cumulative expected output learned by each A node:

$$\alpha_{kj}\left(t+1\right) = \alpha_{kj}\left(t\right) + \varepsilon^{-1}\upsilon_j y_k , \tag{14}$$

where $\varepsilon > 0$, is a small constant. The weights of the B node, β_j, are updated in a similar way but tracking the amount of learning that have taken place in each F2 node.

$$\beta_j\left(t+1\right) = \beta_j\left(t\right) + \varepsilon^{-1}\upsilon_j . \tag{15}$$

2.4   *Creation of input classes*

GasART is initialized with two nodes committed ($N = 2$). The centers of these nodes could be either initialized to the first two inputs in the training set or to random values. Each node is bounded with each other with a link that indicates its topological vicinity. This bind has an age, κ, that is initialized to $\kappa = 0$ when a new bind is set up. A link between two nodes is removed if its age, κ, is bigger than a certain κ_{max}.

If the signal Γ_{F2} fires, then an uncommitted node is committed and N is incremented. The newly committed node is indexed by N and initialized with $\upsilon_N = 1$, $\eta_N = 0$ and $e_N = 0$. The two units with larger accumulated error are binded to the node N. Subsequently learning will proceed as normal.

3   Mackey-Glass equation approximation

We now focus on the problem of approximation the Mackey-Glass equation. We aim to study GasART performance as a function approximating algorithm and comparing it with other neural models like MLP, RBF and the original AppART.

The Mackey-Glass equation,

$$\frac{dx}{dt} = \frac{ax\left(t - \tau\right)}{\left[1 + x^c\left(t - \tau\right)\right]} - bx\left(t\right) , \tag{16}$$

is a time-delay differential equation that has been proposed as a model of white blood cell production [8]. The constant values are commonly set to $a = 0.2$, $b = 0.1$ and $c = 10$. The delay parameter τ determines the behavior of the system. For $\tau > 16.8$ the system produces a chaotic attractor. For our simulations we have chosen $\tau = 30$ and an input window of 6 time steps elements. A set of 4000 items was generated, 3200 were used as training set and the remaining 800 we used as test set. When applying AppART and GasART a voting strategy is used, with the number of voting networks was set to a 10 per cent of the size of the training set. In the case of MLP a backpropagation algorithm was used for fitting the network. For RBF network a hybrid training method (unsupervised in the hidden layer and supervised in the output layer) was used[1].

---

[1] Both of these training techniques are broadly discussed else where, *c.f.* [2])

Table 1: Mean square errors of the predictions of the Multi-Layer Perceptron (MLP), Radial Basis Functions network (RBF), AppART and GasART using the Mackey-Glass problem test set.

Neural model	Mean Square Error	Nodes needed
MLP	0.2406	1500
RBF	0.2173	1000
AppART	0.0262	698
GasART	0.0005	527

The results obtained[2] in table 1 clearly show that GasART not only makes a more accurate prediction by two orders of magnitude than AppART but also commits fewer nodes thus creating a more compact input space representation.

4   Concluding remarks

We have presented GasART. GasART combines ART and GNG networks. It modifies the learning laws of AppART to improve its accuracy and performance while keeping its main characteristics. Therefore, GasART incrementally approximates any function with any degree of accuracy from noisy training samples. In the benchmark tests carried out AppART outperformed neural models tested, generating compact knowledge representations. Because of the briefness of this communication we did not extended ourselves into solving more problems or testing other models, however more tests are needed. We are currently studying alternative ways of for making the merging of ART and GNG models even tighter and starting to apply GasART in bioinformatics problems.

References

[1] S. Grossberg, *Studies of Mind and Brain: Neural Principles of Learning, Perception, Development, Cognition, and Motor Control.* Boston: Reidel, 1982.

[2] C. M. Bishop, *Neural Networks for Pattern Recognition.* Oxford: Clarendon Press, 1995.

[3] B. Fritzke, "Fast learning with incremental RBF networks," *Neural Processing Letters*, vol. 1, pp. 2–5, 1994.

[4] T. M. Martinetz, S. G. Berkovich, and K. J. Shulten, "Neural-Gas network for vector quantization and its application to time-series prediction," *IEEE Transactions on Neural Networks*, vol. 4, pp. 558–560, 1993.

[5] L. Martí, A. Policriti, and L. García, "AppART: An ART hybrid stable learning neural network for universal function approximation," in *Hybrid Information Systems* (A. Abraham and M. Koeppen, eds.), (Heidelberg), pp. 93–120, Physica Verlag, 2002.

[6] E. A. Nadaraya, "On estimating regression," *Theory of Probability and Its Application*, vol. 10, pp. 186–190, 1964.

[7] B. Kosko, *Fuzzy Engineering.* New York: Prentice Hall, 1997.

[8] M. C. Mackey and L. Glass, "Oscillation and chaos in physiological control systems," *Science*, pp. 197–287, 1977.

---

[2]The results that do not have to deal with GasART are taken from [5]. In this work other neural networks were applied, GasART outperforms all of them.

KES 2002
E. Damiani et al. (Eds.)
IOS Press, 2002

Qualitative Traffic Analysis Using Image Processing and Time-Delay Neural Network

S. Navabzadeh Razavi
Tehran Municipality Computer Services Organization (TMCSO)
Iran University of Science and Technology (IUST)
navab@tmcso.com

M. Fathy
Iran University of Science and Technology (IUST)
mfathy@iust.ac.ir

Abstract. In this paper, we present an online, feature-based approach to estimate traffic qualitative parameters from a sequence of traffic images. Considering the factor of time and attempting to simulate the human behavior, a Time-Delay neural network is used to determine the traffic status through traffic lanes. The acquired frames are divided into a number of blocks based on number of lanes and road boundary coordinates, which are obtained automatically by another system. Two extracted principal features from each block of a lane will form the input vector of the neural network. The neural network will classify each lane into a level of traffic congestion based on trained data. Finally a description of traffic scene is obtained using descriptions of all lanes.

1. Introduction

Recently, traffic engineers are more eagered to find automated approaches for traffic surveillance. Several researchers have investigated the application of vision-based techniques and image processing in automated traffic monitoring and control.

According to the available literature, there dose not exist a system that can overcome all the inconvenience. Distortions due perspective, changing weather conditions, presence of shades and reflections, varied vehicles, shapes processing time, etc., seem still to be problems that each system have solved only partially.[4]

Many vision-based traffic surveillance systems try to emulate loop detectors. These techniques are older than the others. In some other techniques, systems are designed to monitor distinct lanes in order to monitor the traffic. Finally there have been researches carried out into locating and tracking individual vehicles through a traffic scene. [19], [21]. It is not possible to classify all algorithms into these three categories, but we can mention them as the most common techniques in designing a vision-based traffic monitoring system.

In loop emulation systems, knowledge extraction is carried out by processing groups of pixels at different location within the scene to give data on traffic counts, average speeds, headways, occupancy and vehicle lengths. This is the same type of data as can be obtained from an inductive loop or a pair of loops. [1], [4], [5].

"In traffic surveillance system based on lane monitoring, it is assumed that the traffic flow is taking place within individual lanes and that there is negligible lane changing. The algorithms aim to provide a way of monitoring traffic along a lane. For multi-lane roads the same technique is repeated for each lane in turn." [19] Researches such as [14], [19] fall into this category.

The principal role of traffic surveillance systems based on vehicle tracking is to locate each individual vehicle within the scene and to follow its path through a sequence of images. Thus the vehicle parameters in both space and time can be collected. Researches such as [3], [8], [11-12], [24] fall into this category.

The system presented in this paper falls in the "traffic surveillance systems based on lane monitoring" category.

Regarding to different traffic behavior in urban roads and high-speed roads such as motorways or freeways, the system is implemented to be used in both applications efficiently.

2. Constraints

The following constraints are assumed in designing of the current system:
1.   Motorized traffic surveillance cameras (capable to zoom, pan, tilt, etc.) which cause variable views of one traffic scene.
2.   Different vertical location of installed cameras which cause different depths of view
3.   Variable lighting conditions during daytime and night.
4.   Variable weather conditions, including rainy, snowy, foggy, dusty and etc.
5.   Possible unclear view of some roads or insufficient view of a traffic scene due to camera installation location.
6.   Disobeying driving rules that lead to violation and dictates restrictions in system design. Some of these restrictions are:
   - Driving not inside a lane
   - Vehicle stopping in forbidden or critical areas leading to local traffic disorder behind the vehicle
   - Presence of pedestrians in various segments of street neglecting pedestrian lines.
7.   Exclusive bus and emergency lanes that have different traffic conditions.
8.   Presence of signs, billboards and pedestrian transverse bridges in some zones of street leading to reduction of sight field in those zones.

Respecting existing facilities and above-mentioned restrictions and problems, this system is designed and implemented to have maximum possible efficiency and minimally affected by restrictions as well.

3. Algorithm Stages

3-1. Choosing suitable blocks of image for processing

In this stage, traffic images are received in a sequence of frames as input and unnecessary parts of each frame are removed. The first step is to separate the road boundary from its surrounding that is carried out by another system. Then the street itself is divided into several blocks to process the information of a minimum area with sufficient and suitable information. These blocks are square shaped and have an edge length equal to 50% of a lane width. There is no vertical gap between blocks and they are located exactly in the middle of the lanes; hence, the white lines of traffic lanes are mostly excluded from the processing images.

Because of different vertical position of installed cameras, invalid information of bottom of images especially in crossroads and insignificant visual information at the top of the frames, a constant bottom margin and varying top margin are ignored in processing. An instance of segmentation process is illustrated in Fig. 1.

Fig1. An instance of segmentation process

It is essential to have a constant number of blocks per lane to produce a constant number of input data for decision-making system. Therefore using a statistical consideration, 8 blocks per lane were selected for segmentation process.

In next step, a 3*3 median filter was used in each block to reduce noise without blurring the edges.

3-2. Feature Extraction

"There are two principal features of interest in the image of a traffic scene. The first one is the **distribution of vehicles through the scene**. The location of vehicles rather than exact location of an individual vehicle are. The other one is the **distribution of movement through the scene**. The relation between these two spatial distributions through the time describes the traffic behavior within the scene." [19]

In fact, what helps a person describe and analyse the traffic condition, is the ability to recognize the presence of vehicles and the quality of movements of the recognized vehicles and to analyse the traffic condition based on these data and previous trained knowledge.

The following aspects are considered in selecting an appropriate feature:
- The ability of working in varying lighting conditions
- The ability of working in different weather conditions
- The ability of working in various distances from camera

3-2-1. Vehicle Detector Feature

The feature which is used to determine the presence and absence of vehicles, is "the variance of equalized image of a block using the edge histogram of image of that block". It is important to note that edge images used in this application are thresholded. This feature will be discussed in more details later.

The presence of a vehicle in a block increases the number of edge pixels [5], [9], [10], [11], [14], [21], [24]. In the same way, the variance of block histogram increases in presence of a vehicle because of the contrast through the vehicle and existence of different shapes and edges in different directions which cause different gray values, shades and reflections.

The presence of a vehicle is detected indirectly by detecting the absence of vehicles. Logically, in case of not detecting the absence of vehicle, the presence of vehicle is detected.

To explain in more details, first consider the best case, in which there is no vehicle in the block. In this case, the edge pixels have somewhat equal values near zero. The cumulative distribution chart of edge histogram is nearly a horizontal line. Thus the pixels mapped by the use of this cumulative function will find equal gray values and the variances of the new histogram will have a very small value.

In this way, the variance range of blocks with no vehicle are quite distinct and in the other cases (which the variance is more than the distinct range), it can be inferred that a vehicle is present in the block. Figure 2 shows some results of applying this feature to a large number of frames in different conditions.

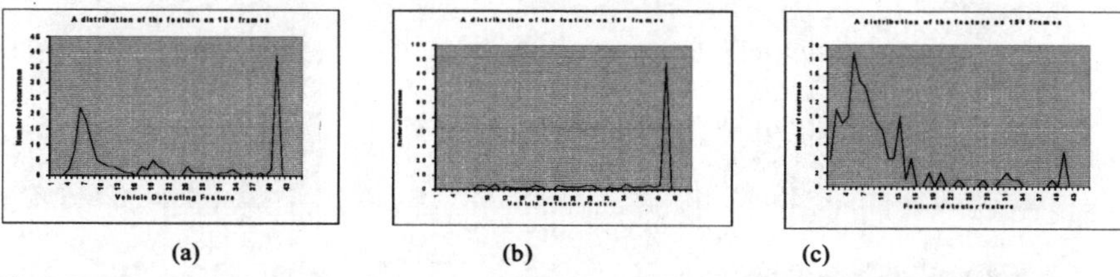

Fig 2: Results of vehicle detector feature experiments on 150 frames of normal weather traffic images. (a) A distribution of the feature in quite light traffic, (b) A distribution of the feature in heavy traffic, (c) A distribution of the feature in nearly light traffic (having some vehicles in some frames)

As it is showed in preceding and following figures, the chart areas related to non-appeared vehicle are quite distinct from that of a detected vehicle. In other way, the distinct areas of chart related to different status of feature are quite superposable in different experiments of different weather conditions.

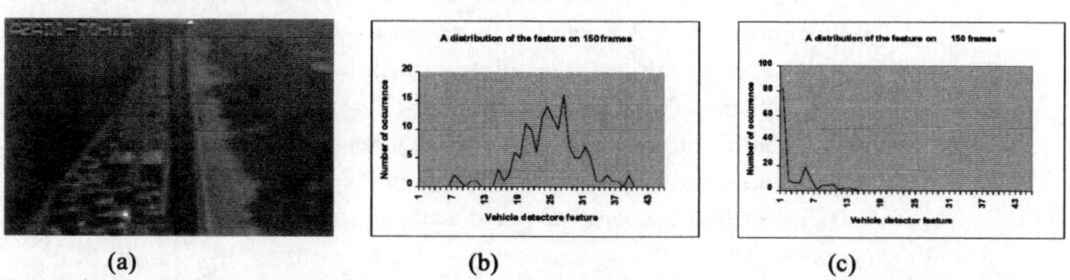

Fig 3: Results of vehicle detector feature experiments on 150 frames of adverse weather traffic images. (a) A sample frame (b) A distribution of the feature in heavy traffic (c) A distribution of the feature in quite light traffic

Finding an appropriate edge detection algorithm for this application is an important factor. Because the high rate of not detected edges in adverse weathers, especially foggy and dusty weathers is not acceptable in applicable systems. Therefore, different edge detectors were experimented on images of different weather and light conditions. The experimented algorithms were *Gaussian, Morphological, Soble, Marr-Hildreth and Rothwell*. According to the in following figures, *Rothwell* algorithm was selected as the best one for this application.

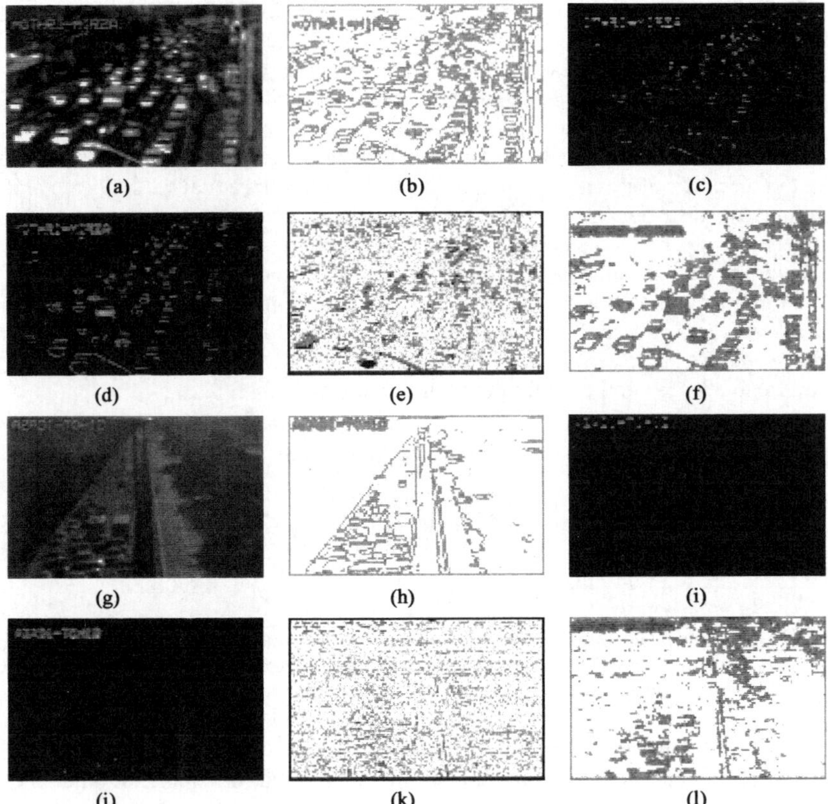

Fig 4. Results of several edge detection algorithms on sample images taken
through normal and adverse weather conditions. (a)A sample image of normal
weather condition. (d) A sample image of adverse weather condition.
(b, h) Results of applying *Rothwell* algorithm on sample image.
(d, j) Results of applying *Soble* algorithm on sample image.
(c, i) Results of applying *Morphlogic Min-Max* algorithm on sample image.
(k, e) Results of applying *Marr-Hildreth* algorithm on sample image.
(f, l) Results of applying *Canny* algorithms on sample image.

3-2-2. Motion Detector Feature

After receiving images and segmenting each image to blocks, noise reduction algorithm is implemented in each block. Next, the mean and variance of each block is calculated and after substitution in the following equation[22], the dimensionless amount of parameter λ will be in hand. This parameter will be processed in Lane Condition Decision step.

$$-\lambda = \left[\left((Var\,1 + Var\,2)/2 + (Mean1 - Mean2)^2/4 \right)^{2}\Big/_{(Var1\,*\,Var2\,)} \right]$$

Where Mean1, Var1 are mean and variance of preceding frame and Mean2 and Var2 are mean and variance of current frame respectively. Relative increase of λ means that negligible changes have occurred in current frame. This feature has a great influence in consistency and reducing sensitivity to noise in decision process.

The reason for inserting a power of 2 in mean factor and a multiplication operator in denominator of preceding equation is to obtain a dimensionless amount for λ.

It can be inferred that λ will somehow indicate the velocity of moving vehicle in block (with respect to suggested time intervals) either. If the vehicle moves faster, a considerable difference occurs between two frames during 0.5 second time interval. So λ will increas and vice versa.

The 0.5 second time interval between two adjacent frames is concluded from several experiments worked out on several images taken from urban roads. It is essential to use a time interval in which the normal movement of vehicles results in considerable change in λ. Figures 5.a through 6.b show cureves for efficiency of this feature.The charts have two distinct peaks for each mentioned states.

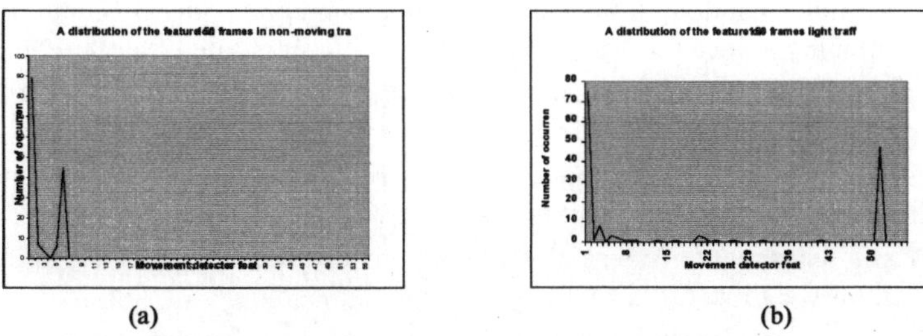

<table>
<tr><td style="text-align:center">(a)</td><td style="text-align:center">(b)</td></tr>
</table>

Fig 5 Results of experiments with λ feature in normal condition

(a) at non-moving traffic (b) at fluent traffic

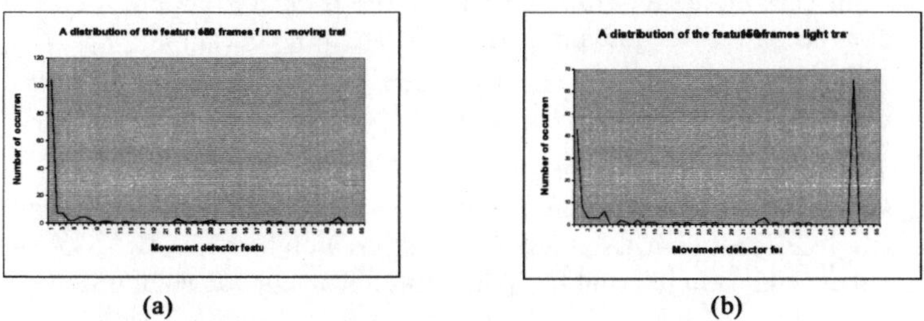

<table>
<tr><td style="text-align:center">(a)</td><td style="text-align:center">(b)</td></tr>
</table>

Fig 6 Results of experiments with λ feature in adverse weather condition

(a) at non-moving traffic (b) at fluent traffic

3-3. Determining Lane Traffic Status

It is essential to suggest a suitable rang of time for receiving information so as to guarantee a valid and correct decision making about qualitative status of traffic. In other hand, it is important to get results as soon as possible. The suggested time range should balance these two demands. A 10-second time range seems to be suitable regarding experimental and statistical studies and it is enough to respond to traffic requirements. A duration of 10 seconds and a fixed time interval of 0.5 seconds is used for image acquisition and decision phases. So there will be 16 input for each frame with respect to 2 features for each block and 8 blocks per lane.

Neural network is selected as a decision-making agent in this system because of intricated traffic conditions and lack of a distinct and evident structure for different traffic conditions [2], [6-7]. Ordinary neural networks have not the ability of working with dynamic inputs. Hence, a Time-Delay neural network is used as a powerful and efficient approach in this system.

The time delay neural network used in this system has following characteristics:

- Input Layer: An input layer with 16 feature during 10 consecutive frames of desired traffic scene.
- First Hidden Layer: A 5x8 matrix of transformed input data processed by the network activation function.
- Second Hidden Layer: A 5x5 matrix of transformed first hidden layer data processed by the network activation function.
- Output Layer: An array with 5 elements to describe 5 distinct traffic conditions of each lane.
- Learning Method: The back-propagation method is used to train the Time-Delay neural network.
- Activation Function: A Hyperbolic tangent function which is continuous and asymmetric is used for this purpose.
- Target Values: Because of using the *tanh function* as the activation function, the target values will fall into the range of [-1, +1].
- Initialization of Synaptic Weights: They are randomly selected from uniformly distributed number in a small range.
- Wight adjustment: Pattern-by-Pattern updating rather that batch updating is used for weight adjustment.
- Training Samples Presentation: The training samples are presented to the network randomly using uniform distribution.
- Coding Technique for Different Classes: The *Hamming* coding is used as output coding because of the maximum distance between codes.
- Stopping Criteria: The learning algorithm will stop working when the overall error of an epoch is equal or less than 0.01.
- Learning Rate: A learning rate equal to 0.05 was chosen.
- Training Set: To train the neural network, training sets with different sizes were employed. The training sets used in this research had 120, 270, 330, 440, 700 members and were selected from different traffic conditions of different roads.

3-4. Final Decision

After determining the traffic status of all lanes, it is time to have a judgement about the scene traffic status in a larger scale than an individual lane.

It is fully dependent to the end users' requirements that how to make a decision by use of lane descriptions. Some end users need to be informed about the traffic status of all individual lanes. In these cases, this stage of algorithm can be omitted and final reporting will be carried out by the result of preceding stage. In the current application, we use *voting* method for different statuses; the final result is the status, which has gained the most of the points in voting. Votes fall into a range of [-1, +1] according to the acceptable range of neural network outputs and there are 5 candidates (5 different traffic status). The votes of each candidate are added together to choose the largest-vote candidate as the final traffic status.

4. Results

Results of trained network with different sizes of training set are illustrated in table 1 and Fig. 7. The test data were obtained from 10 different trained-roads with 20 samples for each road and 10 different non-trained-roads with 20 samples each. The size of training data was varied during different experiments.

Table 1. Final results of network training.

AVERAGE PERCENTAGE OF SYSTEM RESPONSE ACCURACY IN NON-TRAINED ROADS	AVERAGE PERCENTAGE OF SYSTEM RESPONSE ACCURACY IN TRAINED ROADS	SIZE OF TRAINNG SETS
74%	85%	120
82%	92%	270
85%	94%	330
89%	95%	380
90%	99%	440
96%	99%	700

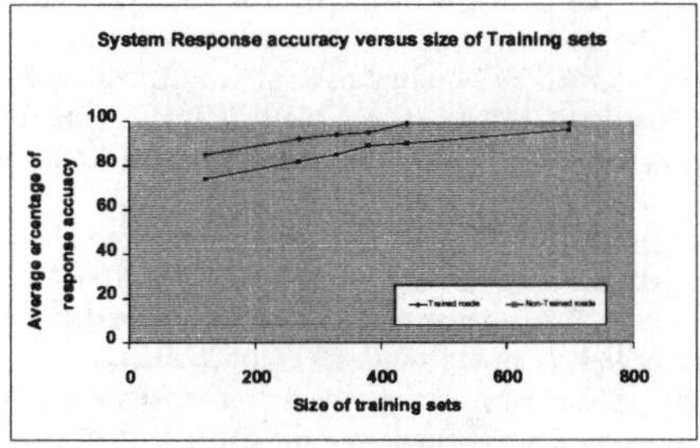

Fig 7. Response liability of system versus size of training set

5. Conclusion

In this paper we have presented a vision-based approach for traffic qualitative analysis. We tried to find a suitable approach for varying conditions in traffic which is based on previous experiments and tried to modify and enhance them in some parts and innovate some other parts in order to get a more applicable, efficient and accurate system. Using Time-Delay neural networks and a new feature for vehicle detection are some factors for getting a faster and more reliable system in different situations.

Considering different weather conditions, different lighting through day time and night, real-time response, using inconsiderable memory because of employing limited features instead of and image and experimenting the *Time-Delay neural network* in traffic applications are the most important features of this system.

6. Acknowledgement

I would like to thank *Tehran Municipality Computer Services Organization* for supporting the project. Without this support, the accomplishment of the project would have been impossible.

References

[1] S. H. Park, K.J. Jung, J. K. Hea, H. J. Kim, "Vision-based Traffic Surveillance System on the Internet", Proc. of IEEE third International conf. on Computational Intelligence and Multimedia Applications, 1999, pp.201-205.

[2] I. Ohe, H .Kawashima, M. Kojima, Y. Kanek, "A Method for Automatic Detection of Traffic Incidents Using Neural Networks", Proc. of Conf. on Road Traffic Monitoring and Control, 1995, pp.231-235.

[3] D. Beymer, P. McLauchlan, B. Cofman, J.Malik, "Real Time Computer Vision System for Measuring Traffic Parameter", IEEE Proc. of Computer Vision and Pattern Recognition, Puerto Rico, 1997, pp.495-501.

[4] A. Suto, A.Cipriano, "Image Processing Applied to Real Time Measurment of Traffic Flow", 28th South Eastern Symposium on System Theory, 1996, pp. 312-316.

[5] M. Fathy, M.Y. Siyal, "A Window-based image processing technique for quantitative and qualitative analysis of road traffic parameters", IEEE Trans. on vehicular technology, 1998, , pp. 591-895.

[6] K. Aoyama, "Next Generation Universal Traffic Management System (UTMS'21) in Japan", Proc. of Intelligent Transportation Systems, 1998, pp.649-654.

[7] T.H. Kang, B.G. Lee, D.S.H. Wang, "Applying Neural Network to Measure Traffic Congestion in the Urban Environment", IEEE International Conference on Intelligent Vehicles, 1998, pp. 676-680.

[8] J. Malik, S. Russell, "Measuring Traffic Parameters Using Video Image Processing, Intellimotion, Vol. 6, No. 1, 1997, pp.6-7, 12-14.

[9] M.Y. Siyal, M. Fathy, F. Dorry, "Neural -Vision based approach for Real-Time Road Traffic Application", IEE Electronic Letters, Vol. 33, 1997 , pp. 969-970.

[10] M. Fathy, M.Y. Siyal, "Image Processing Techniques For Real Time Qualitative Road Traffic Data Analysis", International Journal of Real Time Imaging, Academic Press, Vol. 5, 1999, pp.103-106.

[11] R. Cucchiara, M. Piccardi, P. Mello, "Image analysis and Rule-Based Reasoning for a Traffic Monitoring System", IEEE Trans. on Intelligent Transportation Systems, Vol. 1, No. 2, 2000, p.119-130.

[12] Y. K. Jung, Y.S. Ho, "Traffic Parameter Extraction Using Video-based Vehicle Tracking" IEEE Trans. on Intelligent Transportation Systems, 1999, pp.764-769.

[13] F. Jianping, A. K. Elmagarmid, "Statistical Approaches to Tracking-Based Moving Object Extraction" Proc. of Image Processing Conf, 1999, pp.375-381.

[14] Y, Iwasaki, "An Image Processing system to meseare vehicular queues and an adaptive traffic signal control by using the information of the qeues", Proc. of Intelligent transportation systems, 1998, pp.195-200.

[15] D.J. Dailey, T.W. Cathey, S. Pumrin, "An Algorithm to Estimate Mean Traffic Speed Using Uncalibrated Cameras", IEEE Trans. on Intelligent Transportation Systems, Vol. 1, No. 2, 2000, pp.98-107.

[16] H. Xu, C.M. Kwan, L. Haynes, D. Pryor, "Real Time Adaptive On-line Traffic Detection", Proc of IEEE International Symposium Intelligent Control, 1996, pp. 200-205.

[17] J. Rittscher, J. Kato, S. Joga, A. Blake, "A Probabilistic Background Model for Tracking", Engineering Science Department of Oxford University, 2000.

[18] J.M. Blsseville, "Image Processing for Traffic Management", Advanced Video-based Surveillance Systems, Kluwer Academic, 1999.

[19] N. Hoose, Computer Image Processing in Traffic Engineering, John Wiely & Sons Inc., 1991.

[20] G. L. Foresti, "Object Recognition and Tracking for Remote Video Surveillance", IEEE Trans. on Circutes and Systems for Viedeo Technology, Vol. 9, No. 7, 1999, pp.1045-1062.

[21] C. Nwagboso, "Smart Surveillance system for Intelligent transport system", IEEE Proc. of International Conf. on Intelligent vehicles, 1994, pp. 623-628.

[22] R. M. Haralic, L G Shapiro(1992), Computer and Robot Vision, Vol I, II Addison Weseley.

[23] E. Gose, R. Johnsnbaugh, S. Jost, Pattern Recognition and Image Processing, Prentice Hall, Inc., 1996.

[24] S. Kyo, T, Koga, K, Sakurai, S. Okazaki, "A Robust Vehicle Detecting and Tracking System for Wet Weather Condition using the IMAP-VISION Image Processing Board", IEEE/IEEJ/JSAI International Conference on Intelligent Transportation Systems (ITSC'99, Oct., 1999,), pp.423-428.

KES 2002
E. Damiani et al. (Eds.)
IOS Press, 2002

1497

*A*daptive*N*euro*F*uzzy*I*nference*S*ystem applied to EEG signals for estimating movement-related potentials

Daniel D. BEN DAYAN RUBIN*[§], Gideon F. INBAR[§], Sergio CERUTTI*
*Department Bioengineering, Politecnico di Milano, Milan, Italy
[§]Department of Electrical Engineering, Technion I.I.T, Haifa, Israel

Abstract. This study deals with recovering transient, trial–varying evoked potentials (EP), specifically the movement-related potentials (MRP), embedded within the background cerebral activity (EEG). An adaptive noise canceling technique that exploits a neuro–fuzzy system (*ANFIS*), enables tracking the MRP variability. The method is fully adaptive and no a-priori average templates for signal comparison are used. This framework consists of three main stages: *Preprocessing*: a discrete Laplacian spatial filtering is applied to obtain spatial decorrelation of the input channels. *Channel-selection*: various combinations of the inputs are tested to drive *ANFIS* to achieve the best estimation of the MRP: the most informative channel is chosen. *Noise-canceling*: the most informative channel is corrupted by noise produced by the other channels via an unknown process that is estimated using *ANFIS*.
Every stage of this method was tested with simulations to validate the analytical results before applying them to the real biological data. The processing method that is proposed here, enables evoked potential analysis on a single–trial basis.

1. Introduction

Two major problems are encountered in analyzing evoked potentials (EPs): (1) extremely low signal to noise ratio (SNR) with overlapping spectra of the evoked response embedded within the background EEG brain activity, ranging from 0dB to –20dB, depending on the type of evoked signals. (2) Possible vectorial signal summation, resulting in component overlap, which may cause partial or total occlusion of the desired component features. Usually, these field potentials are averaged to increase SNR and other phase lock EEG activity. The averaging methods do not take into account that in single epochs response activity may vary widely in both time course and scalp distribution. Single-trial analysis methods can avoid problems due to time and/or phase shifts and can potentially reveal richer information about event-related brain dynamics. On the other hand, these methods suffer from low SNR due to the background EEG activity, pervasive artefacts associated with blinks, eye-movements, and muscle noise, because non-phase locked background EEG activities often are larger than phase locked response components.

A set of 8 EEG leads is used to compute 4 spatially decorrelated channels by means of Laplacian Filtering. We assume the MRP signal is mainly located over one of the 4 channels that will be referred to as *primary* input in the sense of Widrow's *noise canceling* technique. Contrarily, the other three channels are supposed to carry mostly background activity (i.e. noise in our problem) and, therefore, can be used as *reference* inputs for noise cancellation from the *primary* input. An *ANFIS* network is used to adaptively estimate the relationship between the *reference* inputs and noise component of the *primary* input. We start by testing 4 different *noise canceling* models, each one with a different channel playing the role of *primary* input. Each model is adapted over two iteration cycles of *ANFIS*, thus obtaining a first approximation (mainly linear) of the respective noise canceling rule. The selection procedure chooses as representative channel the one that (at this early stage) has the largest component not explained by the *reference* inputs. After the selection only the chosen model is carried on with further trainings, in order to optimize the *noise cancellation*. The final residual is considered as single sweep estimate of the MRP, which is, thus obtained without any a-priori information or template, preventing distortions due to fitting procedures and enhancing sensitivity to the changes in the MRP. Both the last two stages, *channel selection* and *noise canceling* exploit *ANFIS* capabilities of tracking both non-linearity and linearity in multidimensional input spaces [3]. Since EEG activity could be subjected to different shapings, passing from one recording site to another, there are required tracking capabilities that should deal with linear, as well as with nonlinear transformations.

2. Methods

2.1 Experimental Setup

Two male subjects aged 26 and 29 years old, not suffering from neurological disorders, participated to the experiment. Micro-switches were placed under both right and left hand index fingers and both right and left foot toes. The subjects were told to press the four micro-switches randomly, self-pacing and as briefly as possible. Cortical potentials were recorded using electrodes (Ag-AgCl surface electrodes, circular, with a 6mm diameter) placed over Fp1, Fp2, F3, F4, C3, C4, T3 and T4, all referenced to an electrode placed over Cz. Cross impedances were kept strictly below 5kΩ. The on-off positions of the four micro–switches were recorded in order to synchronize events in the EEG with external events.

2.2 Discrete Laplace Filtering

This operation has been widely used in literature see [6, 1], it is meaningful for detection intrinsic properties of brain generators, achieving spatial decorrelation of the sensors. The result is proportional to either the current source or the current sink intensity. A 2-D interpolation of the data points was performed to obtain a regular grid (see Fig.1). The interpolation operation assumes that the propagation of the electric field is continuous and monotonically decreasing as the radius increases from the sensor sites. The sites F3, F4, C3, C4 are the center of a square grid which has on its edges 8 neighbor samples located over the horizontal-vertical and diagonal directions. The 8 samples are averaged and subtracted from each site (the center of each square grid) at every instant of time. The dimension of the filtering square grid, is chosen to minimize the distance between the interpolate samples placed on the square borders and the nearest neighboring recording site; this is done in order to better fit the real distribution of the electric field over the scalp. These four sensors sites F3, F4, C3, C4, from now on, will be labeled respectively 1, 2, 3, 4 channel [2].

2.3 Model of channel relationship

In the present study we assume that the information carried in different channels is useful for tracking the noise found in one of the channels. Let **Ch** be the set containing the channels

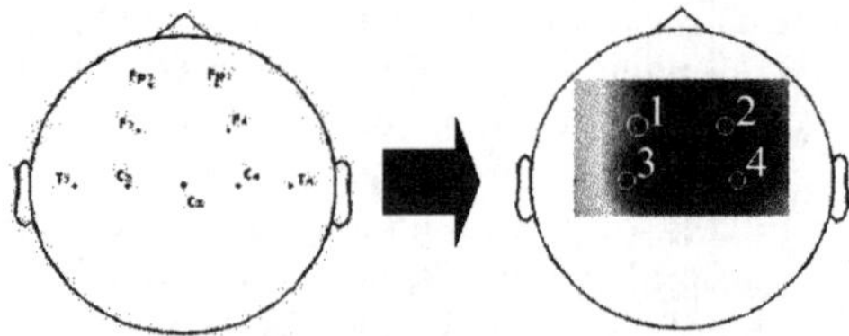

Fig. 1: electrode positions give the basis for the interpolation of the 2-D map over the surface scalp. The filtered-Laplace channels are white circled over the F3, F4, C3, and C4 positions that will be labeled 1, 2, 3, and 4 respectively.

used for the recording; let ch_i be the i-th element of the set; let $\mathbf{Ch}_i^*$ be the subset obtained from the set $\mathbf{Ch}$, removing channel ch_i; let $g(\bullet)$ be the unknown function that relates the elements of $\mathbf{Ch}_i^*$. Thus we can write:

$$ch_i = g_i(\mathbf{Ch}_i^*) + Signal, \tag{1}$$

where $g_i(\mathbf{Ch}_i^*)$ identifies the noise over the channel ch_i, and *Signal* is the MRP in this problem. We assume that (i) $\mathbf{Ch}_i^*$ is a set of decorrelated noise signals, which corresponds to the *reference* input set, while (ii) ch_i, that corresponds to the *primary* input, is assumed to be the signal that contains both the information regarding the function $g(\bullet)$, and the embedded signal to recover. Based on these observations, we processed the signals using the Widrow's '*adaptive noise canceling*' technique [7] in which the adaptive process describing and estimating the function $g(\bullet)$ is *ANFIS* [4]. The Widrow's filter tries to reproduce the closest replica of the additive noise placed in the primary input by means of iterative adaptive procedure and energy minimization. The estimation is performed using the *reference* input set as variable of a function $g(\bullet)$ that best relates that set to the noise placed over the *primary* input. The obtained replica is then subtracted from the *primary* input to obtain the embedded signal.

2.4 ANFIS

The synergism of fuzzy logic and neural networks has produced a functional system capable of learning and high level reasoning. It is an improved tool for determining the behavior of imprecisely defined complex dynamical systems. A detailed coverage of *ANFIS* can be found in [3], here some introductory concepts follow. *ANFIS* structure is based on a typical Sugeno fuzzy rule. The model can easily adapt itself to the surrounding context if it is placed into the framework of adaptive networks that can compute gradient vectors systematically. *ANFIS* combines two learning rules, the back-propagation method and the linear least-squares (LS) method, obtaining a *hybrid* learning algorithm, for an effective search of optimal parameters. If the output of an adaptive network or its transformation is linear in some of the network parameters, then these linear parameters can be identified by the well-known linear LS method. Accordingly, the *hybrid* approach converges much faster since it reduces the dimension of the search space of the original back-propagation method. The *ANFIS* algorithm is capable of tracking both highly nonlinear and linear characteristics of the input variable spaces. The fuzzy inference system was structured by 2 mfs (generalized bell mf) at each of the 3 input nodes. All the combinations of membership resulted in 9 rules that gave the basis to the weightings of corresponding crispy polynomial functions which were, accordingly, added to produce the output of the system.

2.5 Channel Selection Procedure

Once the four-signal output is derived from the Laplacian Filter, we need to obtain, via a heuristic way, an order of priority of these potential inputs and then we can use them accordingly. *ANFIS* can usually generate satisfactory results right after the first epochs of training [5]. Since the least-squares method is computationally efficient, we can construct *ANFIS* models for various combinations of inputs, train them with few applications of the hybrid method and then choose the one with the best performance and proceed for further training. Using the *noise canceling* technique, the system output serves as the error signal for the adaptive process. The system tries to reach the output trying to minimize the error. In this proposed adaptive method the input selection is done taking the channel that owns the greatest RMSE (Root Mean Square Error), in order to choose the most uncorrelated channel as the *primary* input. In this way we obtain the maximal possible information on the background EEG noise from the corresponding *reference* inputs. Contrarily, choosing the smallest RMSE, we would choose the primary input with the smallest prediction error, which is the closest replica of one of the reference inputs. In such case, the chosen channel

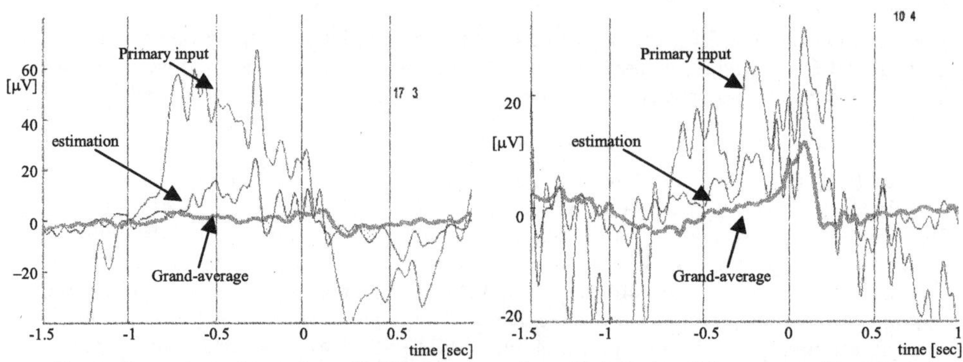

Fig. 2: Comparison of the estimated MRP, the primary input (*light line*), and the average of the same primary input (*thick line*). The numbers above each plot are, respectively, the serial number of the trial and the number of the relative primary input. The vertical line represents the instant in which the switch was pressed (i.e. the movement action).

would be the most correlated channel among the others.

2.6 Noise Cancellation Procedure

The selected *primary* input is presented to the *ANFIS* model, which is trained on the basis of the *reference* input set, $\mathbf{Ch}_i^*$ in (1), as variable of the function g. We choose to train the network for 1000 epochs, and, therefore, the estimated function $\hat{g}$ has a very fine-tuned estimation, where $\hat{g}$ estimates g in (1). Accordingly we can compute:

$estimated\ EP = ch_i - \hat{g}(\mathbf{Ch}_i^*)$, and for each trial this procedure is followed.

4. Results

In [2] simulations are presented to validate this method with analytical results. We will show results obtained applying this method to EEG sweeps. Two types of movements (left and right hand index movement) will be shown of the four types of movements (left and right both hands and feet) accounted in this study. The channel, over which the estimation is performed, may be different from trial to trial, even if the same type of movement is performed. This happens since the informative channel is selected according to the best possible estimation achievable among all the channels presented to *input selection* stage. For this reason the method estimates the MRP within different locations, not necessarily the one that overrides the physiological location of the corresponding MRP. In Fig.2 the estimation of two single trials is shown. In the plot on the left the selected channel was the third one and in the plot on the right it was the fourth one. The results are shown compared with both the non-denoised sweep and the Grand-average of the same selected channel.

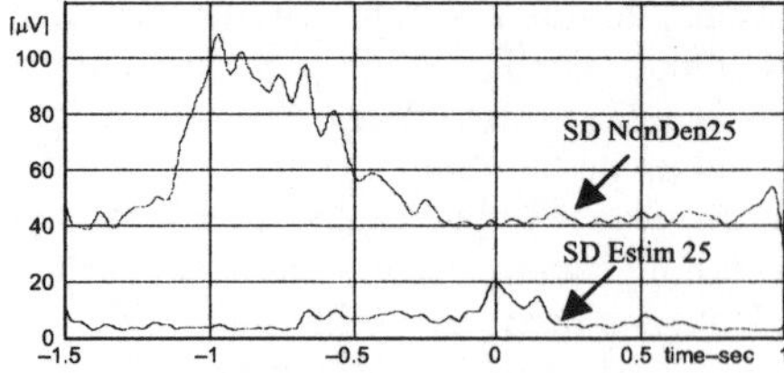

Fig. 3: left hand index movement of the first subject. Standard deviation (SD) of the averaged estimated signals each time (25) the 4th channel was chosen as primary input vs. the SD of the average non-denoised signals the very same times.

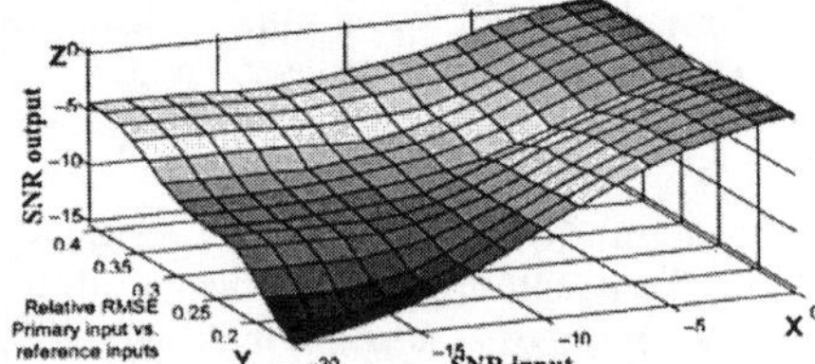

Fig. 4: Interpolation surface obtained from the data collected from all the trial sets of the two subjects. Estimation power: input (X), and output (Z) SNRs (i.e. both calculated with respect to the corresponding grand–average), and the relative RMSE (Y) between the primary input and the reference inputs.

We compare now the average of the estimated MRPs in those trial in which each channel was chosen as *primary input*, with the average of the non-denoised signals in the very same trials. We then compute the standard deviation (SD) of the two averages (Fig.3). The SD of the averaged estimated signals in all the trials remains below 20µV at its highest peak – in correspondence of the instant of time in which the switch is pressed (i.e. the MRP variability is maximal) – while the SD of the averaged non-denoised signals is in the range of $\{40,110\}$µV. We present the results for trials corresponding to the left hand index finger movement whenever the 4th channel was chosen.

5. Discussion and Conclusion

As earlier mentioned, the estimation power that *ANFIS* can achieve depends on the degree of mutual correlation between the input channels. Note that due to the intrinsic variability of the MRP and to the complete absence of a-priori knowledge on the averaged response, the resulting estimated signals cannot exactly correspond to the averaged response. Indeed the average response will serve us as the deterministic signal for quantifying the variance of estimated signals. Since ANFIS is capable of rejecting inputs not related to the estimation process, the correlation of one of the three inputs does not flaw the estimation, but indeed provides less information. We have computed the estimation power in terms of input SNR, output SNR (both SNRs calculated with respect to the corresponding Grand-averages), and the relative RMSE achieved among the channels during the input selection; in Fig.4 we plot a 3-D interpolated surface obtained using data collected from all the trial sets of the two subjects in study. The output represents the ratio between the estimation variance and the average response variance (over the selected *primary input*). Lower RMSE corresponds to lower estimation capabilities (i.e. the corresponding output is noisy) because there is not sufficient information regarding the corrupting noise among the channels. As the RMSE increases the estimation power increases even for low SNR levels of the input signal.

This present study opens a new and potentially useful window into complex event-related brain data that can complement other analysis techniques. Further research will be required to fully assess and to confirm the efficiency and limitations of the method here proposed. Other cross-validations as neurological tests might be performed. The use of electro-corticographic data in simulations for comparing the single trial MRP estimation presented in this study, is a plausible best approximation to the temporal dynamics of the unknown ERP brain generators. Only such data might be the effective prove of the efficiency and reliability of this work.

References

[1] C. Babiloni et al., Human Movement-Related Potentials vs Desincronization of EEG Alpha Rhythm : a High Resolution EEG study. *NeuroImage*, 10(6), : pp. 658–665, 1999.

[2] D. D. Ben Dayan Rubin. On the Application of *ANFIS* to the single–trial Movement Related Potential Analysis. M.Sc. Thesis. Technion I.I.T., Israel, Politecnico di Milano, Italy, 2000.

[3] J.–S.R. Jang. ANFIS: Adaptive–Network–Based Fuzzy Inference Systems. *IEEE Trans. on Systems, man and Cybernetics*, 23(03):665–685, 1993.

[4] J.–S.R. Jang. Adaptive Neuro–Fuzzy Inference Systems (ANFIS) for Noise Cancellation. *Proc. of the Internat. Joint Conf. of CFSA/IFIS/SOFT 1995*, Chiang and Lee, editors, pp. 127–132, World Scient., 1995.

[5] J.-S.R. Jang. Input Selection for ANFIS Learning. In *Proceedings of the IEEE International Conference on Fuzzy Systems*, New Orleans, 1996.

[6] J. Müller–Gerking, G. Pfurtscheller, H. Flyvbjerg: Designing Optimal spatial filters for single–trial EEG classification in a movement task, *Clin. Neurophys.*, 110: 787–798, 1999.

[7] B. Widrow and D. Stearns. Adaptive Signal Processing. Prentice–Hall, Englewood Cliffs, N.J., 1985.

KES 2002
E. Damiani et al. (Eds.)
IOS Press, 2002

Support Vector Machine And Neural Network With Others For Two Dimensional Classification Problem: An Empirical Study

A B M Shawkat Ali {Shawkat.Ali@infotech.monash.edu.au}
Gippsland School of Computing and Information Technology, Monash University, Victoria 3842, Australia.

Abstract. The performance analysis of most commonly used classifiers Support Vector Machine (SVM) and Neural Networks (NN) are compared with others six different classifiers on seventeen quite different, standard and extensively used datasets in terms of classification error rates and computational times. It is found that the average error rates for a majority of the classifiers are closes with each other but the computational times of the classifiers differ over a wide range. SVM algorithm based on statistical learning has the lowest average error rate and computationally it is faster than NN but computationally expensive than other classifiers. SVM and NN have also been considered with changing parameter values.

Keywords: SVM, NN, classification error, computationally expensive.

1. Introduction

Over the last decade classification of different size datasets is an important issue for data mining methodology. Now-a-days researchers in the machine learning and statistics communities are trying to build a better performance classifier. One study[11] discusses the comparison of prediction accuracy, the complexity and training time different of classification algorithms. The paper discusses the results of a comparison of twenty-two decision tree, nine statistical and neural network algorithms on thirty-two data sets in terms of classification accuracy, training time and (in the case of trees) number of leaves. Another study[14] compared the accuracy between several decision tree classifiers against some non-decision tree classifiers on a large number of datasets. Other studies that are smaller in scale include [15], [3], [4], [5], [9], and [8]. Our experiment compared mainly statistical classifier SVM and NN with others for different size of datasets. Most of datasets were from real life domains. Two classes divide all of the datasets and all of the attributes values have been considered numerical and categorical. Section 2 shortly describes each of the classification algorithms and section 3 gives the experiment setup. Section 4 explains the result and conclusions are given in section 5.

2. Algorithms Descriptions

2.1 ZeroR

ZeroR is the most primitive learner. It is simply predicts the majority class in the training data if the class is categorical and the average class value if it is numeric. Although it makes little sense to use this scheme for prediction, here it can be used to test other learners and served as baseline performance benchmark[19].

2.2 OneR

OneR is one of the simplest classification algorithms. It produces simple rules based on one attribute only[7]. It generates a one-level decision tree that indicates in the form of a set of rules that all test one particular attribute. It is a simple cheap method that often comes up with better rules for characterizing the structure in data[18]. It often gets reasonable accuracy on different tasks by simple looking at an attribute.

2.3 IBK

IBK is an achievement of the knearest-neighbors classifier. Each case is considered as a point in multi-dimensional space and classification is done based on the nearest neighbors. The value of 'k' for nearest neighbors can vary. This determines how many cases are to be considered as neighbors to decide how to classify an unknown instance. The time taken to classify a test instance with a nearest-neighbor classifier increases linearly with the number of training instances that are kept in the classifier. It needs a large storage requirement[18]. Its performance degrades quickly with increasing noise levels. It also performs badly when different attributes affect the outcome to different extents. One parameter that can be affect the performance of the IBK algorithm is the number of nearest neighbors to be used. By default it uses just one nearest neighbor.

2.4 j48.J48

j48.J48 is a top down decision tree classification algorithm. The algorithm considers all the possible tests that can split the data set and selects a test that gives the best information gain. For each discrete attribute, one test with outcomes as many as the number of distinct values of the attribute is considered. For each continues attribute, binary tests involving every distinct values of the attributes are considered. In order to gather the entropy gain of all these binary tests competently, the training data set belonging to the node in considerations sorted for the values of the continuous attribute and the entropy gains of the binary cut based on each distinct values are calculated in one scan of the sorted data. This process is repeated continuous for each attributes[9].

2.5 KernalDensity

KernalDensity algorithm works in very simple way as like Naive Bayes. The main difference is that, unlike Naive Bayes, it does not assume normal distribution of the data. Kernel Density tries to fit a arrangement of kernel functions. According to Beardah and Baxter[1], it estimates are similar to histograms but provide smoother representation of the data. They also illustrate some of the advantage of Kernel Density estimates for data presentation in archaeology. They show that it estimates can be used as a basis for producing contour plots of archeological data, which lead to a useful graphical representation of the data.

2.6 NaiveBayes

The NaiveBayes classification algorithm is created on Bayes rule, which is used to compute the probabilities and it is used to make predictions. It considers that the input attributes are statistically independent. It analyses the relationship between each input attribute and the dependent attribute to derive a conditional probability for each relationship[2]. These conditional probabilities are then combined to classify new cases. An advantage of this algorithm over some other algorithms is that it requires only one pass through the training set to produce a classification model. Naive Bayes works very well when tested on many real world datasets[17]. Naive Bayes can obtain results that are much better than other sophisticated algorithms. However, if a particular attribute value does not occur in the training set in conjunction with every class value, then Naive Bayes may not perform very well. It can also perform poorly on some datasets because attributes were treated, as through if they are not dependent, whereas in reality they are associated.

2.7 SVM

Support Vector Machine (SVM) is an elegant tool for solving pattern-recognition and regression problems. Over the past few years, it has attracted a lot of researchers from the neural network and mathematical programming community; the main reason for this being their ability to provide excellent generalization performance. Recently, Smola and Scholkopf[16] proposed an iterative algorithm called Sequential Minimal Optimization (SMO) for solving the regression problem using SVMs. The remarkable feature of the SMO algorithm is that they are fast as well as very easy to implement. SMO has been used here for training the datasets.

2.8 Neural Network

Neural networks are based on the function of biological neurons and perform two major tasks. One type of NN, the perceptron, uses a set of input data to predict an output. Another type of neural network clusters in with a method known as a self-organizing map or Kohonen network. However, NNs are capable of expressing a rich variety of non-linear decision surfaces. This is done through the structure of the networks, non-linear thresholds and the weight of the edges between the nodes. Here, neural network uses backpropagation (BP) algorithm to train the different datasets. It is capable of generating smooth nonlinear mappings between input and output variables by using BP algorithm. BP is one of many error-minimizing functions that tune weight to generate the desired mapping. For instance, it is minimizing the squared error cost function over a train set in this experiment. BP with momentum is advantageous when some training data are very different from the majority of the data[6].

3. Experimental Setup

In order to evaluate the performance of different classifiers, we have considered all the datasets from the UCI collection[13] and KDCentral[12]. The system configuration is Pentium III 933 MHz with 256 Mbytes. The name of the all datasets has been mentioned in Figure 2 and 3. Here we considered only the binary classes problems. We choose all the classifiers from Weka data mining tool. Weka is a collection of machine learning algorithms for solving real world data mining problems. It is written in Java and runs on almost any platform. Firstly we consider all the classifier with default setting and after that we set up the parameter value c is 0.5 and exponent is 0.5 for SVM* and training time is 100 for NN*. We also used 10-fold Cross-validation technique. Cross-validation is a method for estimating how good the classifier will perform on new data and is based on "re-sampling"[10]. Cross-validation is good for use especially when the datasets is small.

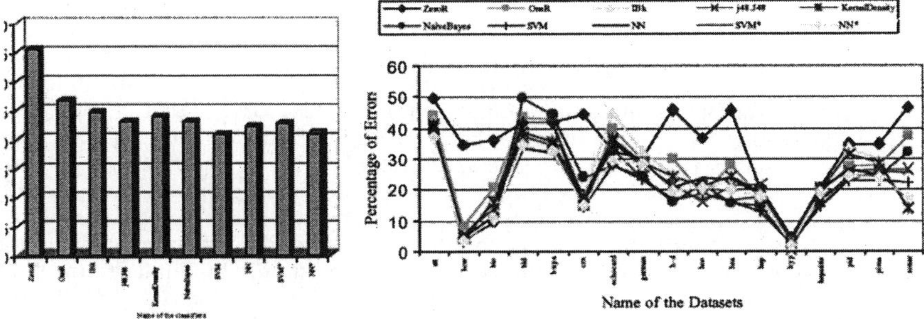

Figure 1: Performance curves for the individual
classifier on percentage of error.

Figure 2: Performance curve for the individual dataset
on percentage of error.

4. Results

A ranking of the classifiers in terms of average error rates as follows: SVM, NN*, NN, SVM*, NaiveBayes, j48.J48, KernalDensity, IBK, OneR and ZeroR. SVM has the lowest average error rate and zeroR the highest. The easiest datasets to classify is hyp; the error rates between 0.79 and 4.78. The most difficult dataset are bld, bupa, att and echocardiogram, where the minimum error rate is 27.88. Overall SVM shows the maximum number of minimum error rate for the individual dataset, second is NavieBayes. Figure 1 shows the individual performance for each classifier on percentage of error. Figure 2 shows the error percentage for each dataset on the basis of every classifier. Overall the fastest classifier is ZeroR and the slowest classifier is NN. ZeroR, OneR and SVM have showed zero time for testing the datasets. The modified SVM* has not performed well then SVM, but NN* showed the less error than NN default setting. It is noted that the training time of all the datasets have been showed 10 times less in figure 3 for NN for a good visual graph representation.　Figure 3 shows the training and testing time for each dataset on basis of classifier.

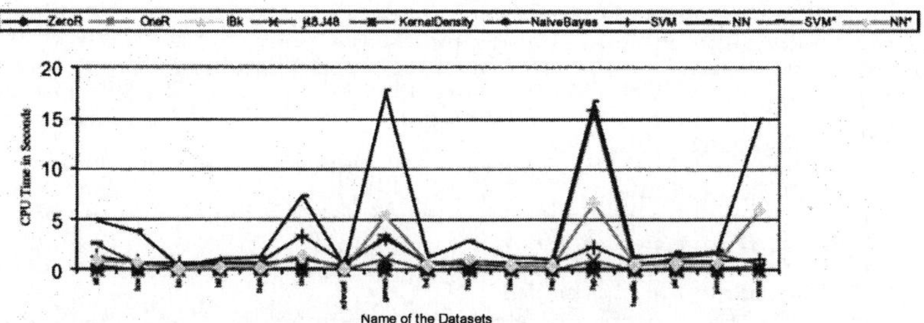

Figure 3: Curves represents the training time for individual dataset. Time for neural network
classifier is represented 10 times less for good visualization.

5. Conclusion

We have tried to find out the best classification algorithms and characteristics of datasets for binary class problem. Our results show that firstly no single algorithm can outperform any other when the performance measure is the expected generalization accuracy and secondly the average error rates of many classifiers are sufficiently similar that their differences are statistically insignificant. But SVM shows the minimum error rates among others. It is clear that if error rate is the sole criterion, SVM would be the method of choice. But one disadvantage of SVM is that it can't classify more than two classes problems at a time. We use over the experiment the Weka data mining tolls. Weka is more users friendly, much functionality, including classification, clustering, searching for association rules on applied datasets. Therefore, although a single algorithm cannot build the most accurate classifiers in all situations, some algorithms will perform better in specific domains.

6. References

[1] Beardah, C. and Baxter, M. (1996). *The Archaeological Use of Kernel Density, Estimates, Internet Archeology*, http://www.intarch.ac.uk.

[2] Brand, D. and Gerritsen, R. (1997). *Naive Bayes and nearest neighbor*, http://ww.dbmsmag.com/9807 m07.html.

[3] Brodley, C. E. and Utgoff, P. E. (1992). Multivariate versus univariate decision trees, *Technical Report 92-8*, Department of computer science, University of Massachusetts, Amherst, MA.

[4] Brown, D. E., Corruble, V. and Pittard, C. L. (1993). A comparison of decision tree classifiers with backpropagation neural networks for multimodal classification problems, *Pattern Recognition* **26**: 953-961.

[5] Curram, S. P. and Mingers, J. (1994). Neural networks, decision tree induction and discriminant analysis: an empirical comparison, *Journal of Operational Research Society* **45**: 440-450.

[6] Fauset, L. (1994). *Fundamentals of Neural Networks*. Prentice Hall, Englewood Cliffs, NJ.

[7] Holte, R. (1993). *Very Simple Classification Rules Perform Well on Most Commonly used data sets, Machine Learning*, Kluwer Academic Publisher, Boston, Vol. 11, pp.63-91.

[8] Ibrahim, R. S. (1999). Data Mining of Machine Learning Performance Data, *Master's Thesis*, RMIT, Australia.

[9] Joshi, K. P. (1997). Analysis of Data Mining Algorithms, http://userpages.umbc.edu/~kjoshi1/datamine/ proj_rpt.html

[10] Kohavi, R. (1995). *A study of Cross-Validation and Bootstrap for Accuracy estimation and Model selection, Max-Plank-Institute Proceedings* pp.1137-1145.

[11] Lim, T. and Loh. W. (2000). *A comparison of Prediction Accuracy, Complexity and Training Time of Thirty-Three Old and New Classification Algorithms*, Kluwer Academic Publishers, Machine Learning, 40, pp.203-229, Boston.

[12] Lim, T. S. (2000). Knowledge Discovery Central, Data Sets, http://www.KDCentral.com/.

[13] Merz, C. J. and Murphy, P. M. (1996). UCI Repository of Machine Learning Data-Bases. Irvine, CA: University of California, Department of Information and Computer Science, http://www.ics.uci.edu /~mlearn /MLRepository.html.

[14] Michie, D., Spiegelhalter, D. J. and Taylor, C. C. (eds) (1994). *Machine Learning, Neural and Statistical Classification*, Ellis Horwood, London.

[15] Shavlik, J. W., Mooney, R. J. and Towell, G. G. (1991). Symbolic and neural Learning algorithms: an empirical comparison, *Machine Learning* **6**: 111-144.

[16] Smola, J. and Scholkopf, B. (1998). A tutorial on support vector regression, *NeuroCOLT Technical Report TR 1998-030*, Royal Holloway College, London, UK.

[17] Witten, I. and Frank, M. (2000). *Data Mining: Practical Machine Learning Tool and Technique with Java implementation*, Morgan Kaufmann, San Francisco.

[18] Wolpert, D. and Macready, W. (1995). *No Free Lunch Theorems for Search* Santa Fe Institute, Technical report on., No. SFI-TR-95-02-010.

[19] Xintao, Wu. (2001). *Data Mining Using Weka System on Census-Income Dataset*, ITCS6163 Data Warehousing, SP.

KES 2002
E. Damiani et al. (Eds.)
IOS Press, 2002

Forecasting of North Northeast Brazil Rainfall Anomalies: A Hierarchical Neuro-Fuzzy Model Application

Flávio J. de Souza[1], Maria Luiza. Velloso[2]

Departments of Computer Engineering[1], Department of Electronics Engineering[2] - FEN
Uerj – Rio de Janeiro State University
Rua S. Francisco Xavier, 524
CEP 20550-013, Rio de Janeiro, Brazil
fjsouza@eng.uerj.br, mlfv@openlink.com.br

Abstract. Rainfall prediction for the rainy season is particularly important in north-northeast Brazil. In a severe year, a dry season disrupts the local subsistence agriculture and cause famine and migration. A novel hybrid system called Hierarchical Neuro-Fuzzy BSP (HNFB) was used to construct a parcimonious prediction model for forecasting rainfall indexes for the rainy season of this region. The HNFB system is based on the BSP partitioning (Binary Space Partitioning) of the input space and has been developed in order to bypass the traditional drawbacks of neuro-fuzzy systems: the reduced number of allowed inputs and the poor capacity to create their own structure. A great advantage of this modeling is the capacity for finding the relevant input lags in a lagged structure.

1. Introduction

The Political Northeast Region of Brazil (hereafter referred to as the Northeast) is a densely populated region located approximately between 1 and 18 $^\circ$ S and 35 and 47 $^\circ$ W. In this region, three different regimes have been identified throughout the year and represent annual precipitation amounts varying between 600 mm and 2000 mm: east part, southern part and northern part [1]. The northern part of Northeast is known for its semi-arid climate with its rainy season concentrated between February and May, and very large interannual variability that can reach values as high as 40% of the mean. North of the Northeast not only exhibits a high variability in the total amount of precipitation from year do year, but also a high spatial and temporal variability in the precipitation within its rainy season. Rainfall prediction is particularly important in this region for the rainy season. In severe years, a dry season may disrupt the local subsistence agriculture and cause famine and migration. Rainfall anomalies are standardized measures of rainfall representing deviation of norm or average calculated over a certain period. We construct an annual rainfall index in the rainy season by averaging over rainfall anomalies of 44 rain stations. This time series corresponds to the 1912-1991 period. Many works have identified strong relationship between the precipitation over the Northeast and Atlantic Ocean Sea Surface Temperature [1], [2], [3]. Then, as predictors we use an index for the pre-rainy season and an index compiled with the Sea Surface Temperature Anomalies in the tropical Atlantic in a lagged structure.

Neuro-fuzzy Systems [4],[5], are hybrid systems that match the learning capacity of artificial neural networks [6] with the linguistic interpretation power of the Fuzzy Inference Systems [7]. The basic idea of a Neuro-Fuzzy System is implementing a Fuzzy Inference System in a distributed parallel architecture in such way that the learning paradigms used in the neural networks can be employed in this hybrid architecture.

This work presents and applies a new Neuro-Fuzzy System called Hierarchical Neuro-Fuzzy BSP (HNFB) [8], [9] in annual rainfall indexes forecasting. The HFNB System is based on the Binary Space Partitioning (BSP) of the input space and was developed in order to bypass the traditional drawbacks of the neuro-fuzzy systems.

Section 2 briefly describes the BSP scheme. Section 3 presents the HFNB model, its basic cell, and its architecture. Section 4 describes an evaluation experiment and discusses the results.

2. BSP Partitioning

Neuro-Fuzzy Systems and the Fuzzy Systems perform a mapping of fuzzy regions of the input space into fuzzy regions of the output space. This mapping is made through fuzzy rules. The input and output variables of Fuzzy and Neuro-Fuzzy Systems are divided in some linguistic terms (for example: low, high) that are used by the fuzzy rules. The input space partitioning indicates how the fuzzy rules are related in the space. The most common partitioning is the fuzzy grid, which, in spite of being the simplest, limits the number of input variables. The HNFB System [8],[9] uses a recursive partitioning, called BSP, that aims to reduce this limitation, besides having a unlimited capacity to create and to expand its own structure.

The BSP Partitioning divides the space successively, in a recursive way, in two regions. An example of such partitioning, illustrated in Figure 1(a), shows that initially the space was divided into two parts in the vertical direction – x_2 (for example high and low).

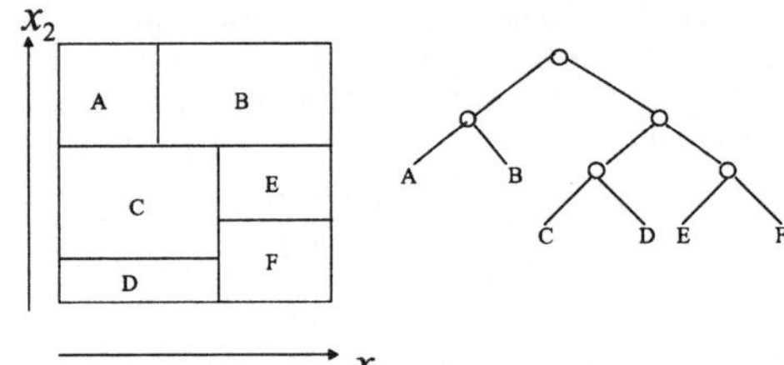

Fig. 1. (a) BSP Partitioning;   (b) BSP Tree referring to BSP partitioning.

In this example, the upper partition was subdivided in two new partitions A and B, according to the horizontal direction – x_1. The inferior partition, in turn, was subdivided successively, in the horizontal and vertical route, generating the partitions C, D, E and F. Figure 1(b) shows each final partition represented by letters in the BSP tree. Interior knots represent the existing intermediate partitions.

The BSP partitioning is flexible and minimizes the problem of the exponential growth of rules, since it only creates new rules locally, according to the training set. Its main advantage is to build automatically its own structure. This type of partitioning is recursive because and results in models with hierarchy in their structure and, consequently, hierarchical rules.

3. Hierarchical Neuro-Fuzzy BSP (HNFB) Model

An HNFB model may be described as a system that is made up of interconnections of HNFB cells. This cell is a neuro-fuzzy mini-system that performs binary fuzzy partitioning of the input space. The HNFB cell generates a precise (crisp) output after a defuzzification process. Figure 2 (a) illustrates the cell's defuzzification process and the concatenation of the consequents. In this cell, 'x' represents the input variable, while ρ and μ are the membership functions that generate the antecedents of the two fuzzy rules. Figure 2(b) illustrates the representation of this cell in a simplified manner.

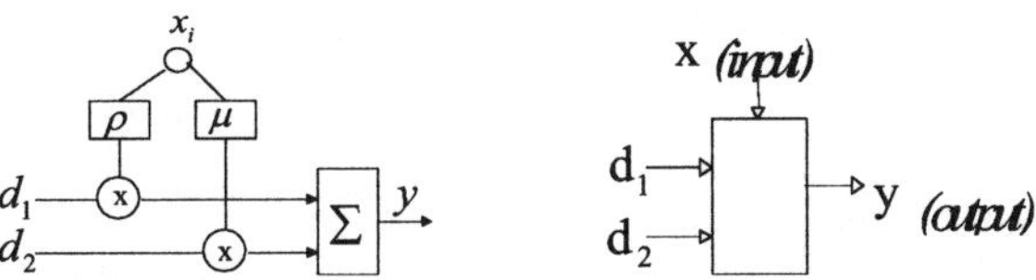

Fig. 2. (a) Interior of the Neuro-Fuzzy BSP cell. (b)Simplified Neuro-Fuzzy BSP Cell.

The linguistic interpretation of the mapping implemented by the HNFB cell is given by the following set of rules:

Rule 1: If $x \in \rho$ then $y = d1$.
Rule 2: If $x \in \mu$ then $y = d2$.

Each rule corresponds to one of the two partitions generated by BSP partitioning. When the inputs occur in partition 1, it is rule 1 that has a higher firing level. When they are incident to partition 2, it is rule 2 that has a higher firing level. Each partition can in turn be subdivided into two parts by means of another HNFB cell. Profiles of membership functions ρ and μ are presented in Figure 3 and are described as follows.

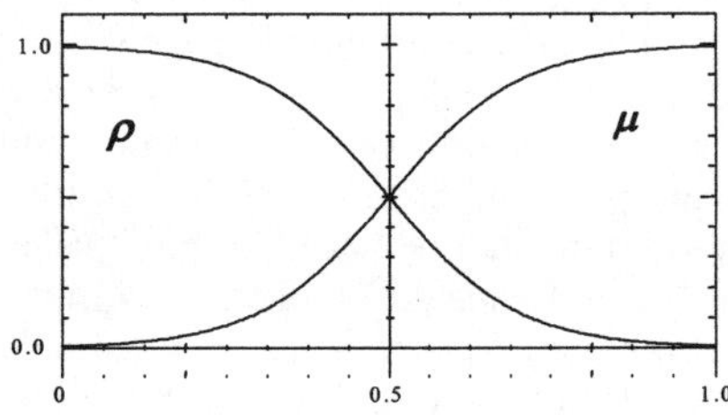

$$\mu(x) = \frac{1}{1 + exp(-ax + b)},$$
$$\rho(x) = 1 - \mu(x)$$

Fig. 3. Example of a profile of the BSP cell Membership functions:

The (crisp) output 'y' of an HNFB cell is given by the weighted average shown in equation (1).

$$y = \frac{\rho(x) * d1 + \mu(x) * d2}{\rho(x) + \mu(x)} \tag{1}$$

Due to the fact that the membership function ρ is the complement to 1 of the membership function μ, in the HNFB cell the following equation applies:

$$\rho(x) + \mu(x) = 1 \tag{2}$$

Therefore, the output equation is simplified as:

$$y = \rho(x) * d1 + \mu(x) * d2 \tag{3}$$

or

$$y = \sum_{i=1}^{2} \alpha_i \, *di \qquad (4)$$

where the αi's symbolize the firing level of the rules.

Each di of equation (4) corresponds to one of the three possible consequents: a singleton (di is constant), a linear combination of the inputs, and the output of a stage of a previous level.

The HNFB system has a training algorithm based on the gradient descent method for learning the structure of the model and consequently, the linguistic rules. The parameters that define the profiles of the membership functions of the antecedents and consequents are regarded as the fuzzy weights of the neuro-fuzzy system. Thus, the di's and parameters 'a' and 'b' are the fuzzy weights of the model. In order to prevent the system's structure from growing indefinitely, a parameter, named decomposition rate (δ), was created. It is an adimensional parameter and acts as a limiting factor for the decomposition process. Information on how this algorithm functions may be found in greater detail in [8],[9].

4. Experiments and Results

A Hierarchical Neuro-Fuzzy BSP (HNFB) model was used to construct a parsimonious prediction model for forecasting rainfall indexes of the rainy season based on annually historical data.The observational basis was monthly rainfall series of 44 rain gage stations of northern Northeast from January/1912 to December/1991 and Sea Surface Temperature Anomalies for $5°$ x $5°$ grid squares extending over the Atlantic Ocean.Two annual rainfall indexes were constructed accounting for the great space and time variability of tropical rainfall: a rainy season (February-May) index and a pre-rainfall (October-January) index. A third annual index was compiled based on SSTAs in January. In order to assess the effectiveness of the HNFB model, we carried out forecasting experiments on three indexes compiled and the results were compared with those obtained by a multilayer perceptron with one hidden layer and trained with the error back-propagation algorithm.

The lagged structure found in the HFNB model had 7 inputs and the lagged structure used in neural network model had 10 inputs. The best inputs of neural network model were found after many trials.

The forecasts were made with a one-step-ahead (one year ahead) prediction horizon. In order to evaluate the forecasting performance of the HNFB model, two metrics associated with prediction errors were used: the NRMSE (Normalized Root Mean Square Error) and the Theil's U coefficient. Table 1 shows the compared performances of HFNB and neural network models for training set and test set. It can be noticed that the performance of the HNFB model was higher for the test and the training set, when compared with that of the neural network. Furthermore, the input data used in the HNFB model were automatically.

Table 1 - Comparison of the HNFB model with neural network.

	Metric	Neural network	HFNB
Training Set	NRMSE	0.5780	0.38226
	U-Theil	0.3341	0.2852
Test Set	NRMSE	0.4484	0.31408
	U-Theil	0.3216	0.2253

Figure 4 shows the results for the test set in a graphical form. The dashed line represents the real data, the forecasts estimated by neural network are marked with triangles and the forecasts estimated by the HFNB model are marked with circles.

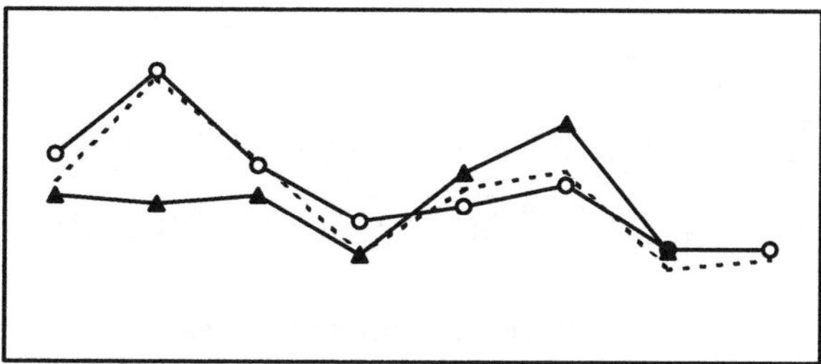

Fig. 4. Forecasts of the rainfall index. The dashed line represents the real data, the forecasts estimated by neural network are marked with triangles and the forecasts estimated by the HFNB model are marked with circles.

5. Conclusions

The objective of this paper was to introduce the HNFB model, a new neuro-fuzzy model, and present its application on time series forecasting. The HNFB model was developed with the purpose of improving the weaknesses of neuro-fuzzy systems. The BSP partitioning is flexible and minimizes the problem of the exponential growth of rules, since it only creates new rules locally, according to the   training set. Furthermore, its main advantage is to build automatically its own structure.The number of rules depends on the complexity of the problem. The results obtained by the HNFB model showed that it performs well in time series forecasting. In addition, this model allows the extraction of knowledge in a fuzzy rule-base format.

An interesting prospect for future work, which is already in the research stage, is the use of this model for classification systems that allow the extraction of knowledge in the format of nonhierarchical fuzzy rules for data mining applications.

References

[1] C. B. Uvo, C. A. Repelli, S. Zebiak and Y. Kushnir, A Study on the Influence of the Pacific and Atlantic SST on the Northeast Brazil Monthly Precipitation using Singular Value Decomposition. Technical Report, INPE-National Institute for Space Research, São Jose dos Campos, SP, Brazil, 1994.

[2] G. Hastenrath, – "Further Work on the Prediction of Northeast Brazil Rainfall Anomalies" – Journal of Climate, April 1993.

[3] S. Hastenrath, Prediction of Seasonal Rainfall in the North Northeast of Brazil Using Eigenvectors of Sea-Surface Temperature, International Journal of Climatology, Vol. 11, 711-143, 1991.

[4] S. K Halgamuge, M. Glesner, Neural Networks in Designing Fuzzy Systems for Real World Applications, Fuzzy Sets and Systems, N0.65, pp. 1-12, 1994.

[5] R. Kruse, D. Nauck, NEFFCLASS – A Neuro-Fuzzy Approach for the Classification of Data. Proceedings of the 1995 ACM Symposium on Applied Computing, Nashville.

[6] S. Haykin, A Comprehensive Foundation, Prentice Hall, New Jersey, 1999.

[7] J. Mendel, Fuzzy Logic Systems for Engineering: A Tutorial. Proceedings of the IEEE, Vol.83, n.3, pp.345-377, 1995.

[8] Flávio Joaquim de Souza, Modelos Neuro-Fuzzy Hierárquicos. Tese de Doutorado. Departamento de Engenharia Elétrica da Pontifícia Universidade Católica do Rio de Janeiro (1999) (in portuguese).

[9] F. J Souza, M.M.B.R Vellasco, M.A.C Pacheco, Hierarchical Neuro-Fuzzy QuadTree Models, Fuzzy Sets & Systems, (to be published), 2001.

KES 2002
E. Damiani et al. (Eds.)
IOS Press, 2002

Lattice-Gas Cellular Automata Approach for Fluid Flow in Porous Media

S.N. Khotimah, I. Arif, and H. L. The

Physics Department, Institut Teknologi Bandung, Jl. Ganesa 10 Bandung 40132, Indonesia

Abstract. The method of lattice-gas cellular automata was used to solve 2-dimensional fluid dynamics. If the flux between two stationary parallel plates is less than 0.2 then the velocity profile is parabolic and the flux is proportional to the quadratic of channel width. For flow in tortuous media, the formula for flux is the same as for the straight media if we use effective path length of the tortuous media. Constant volume rate of flow is also approved for incompressible flow if there is no source or sink along a pathway. For flux greater than 0.2 the lattice-gas result suggests that the fluid viscosity is dependent on the fluid velocity.

1. Introduction

Scientists developed new computational methods based on the concepts of cellular automata. Each automaton in its internal state will behave as a processor that receives inputs, processes the inputs, determines the next internal state and its outputs [1]. Therefore the computation method based on artificial neural network [2] can also be included as cellular automata method. This method enables to give solution in many problems that are difficult to be worked out conventionally.

In this paper, firstly we discuss fluid flow between two stationary parallel plates. This is the object of extensive study because parallel-plate flow is a simple model for flow through a crack or joint of a rock [3,4]. We used lattice-gas cellular automata to solve the fluid flow in this media. The lattice-gas results of steady state velocity profile and the volumetric flux were compared with the theory. We attempt to obtain the dependence of fluid viscosity on the fluid velocity for non-parabolic velocity profile. Secondly, flux was also measured in tortuous media to be compared to the straight media and the streamline patterns are shown.

2. Theoretical Background

The analytical solution for steady incompressible laminar fluid flow between stationary parallel plates has been known [5]. At a constant pressure gradient, velocity profile $u(y)$ for the laminar flow in Figure 1 is a parabolic curve, i.e.:

$$u(y) = \frac{1}{2\mu}\left(-\frac{dP}{dx}\right)\left[yH - y^2\right] \tag{1}$$

and the volumetric flux is proportional to the square of channel width.

$$q = \frac{H^2}{12\mu}\left(-\frac{dP}{dx}\right) = -\frac{H^2}{12\mu}\frac{\Delta P}{L} \tag{2}$$

where u is flow velocity in x-direction, H is channel width, $\Delta P = P(x = L) - P(x = 0)$ is pressure difference, L is channel length and μ is the dynamic viscosity of the fluid.

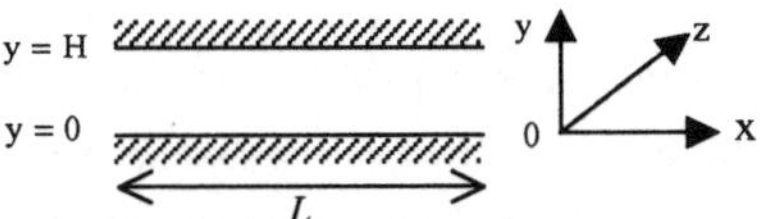

Figure 1. Two parallel plates at y = 0 and y = H

The continuity equation gives constant volume rate of flow along parallel plates:

$$H u = \text{constant} \tag{3}$$

For steady incompressible laminar flow in tortuous media, volumetric flux in equation (2) is still hold if the path length (L) is replaced by effective path length (L_e) [6]

3. The Lattice-Gas Cellular Automata

Lattice-gas automata model is introduced by Frisch, Hasslacher, and Pomeau in 1986 [7]. They show that this microscopic model simulates the Navier-Stokes equation. The automata in this model are distributed on the nodes of a periodic triangular lattice. The automata change their states simultaneously or by parallel iteration. The lattice-gas model is constructed of identical particles that move from node to node on a triangular lattice, colliding when they meet, and always obey the fundamental physical principles, i.e. conserving particle number and momentum. Particles of unit mass move with unit speed in the direction given by:

$$c_i = \left[\cos\left(\frac{2\pi i}{6}\right), \sin\left(\frac{2\pi i}{6}\right) \right] \qquad i = 1, 2, \ldots, 6. \tag{4}$$

Any particle that reaches a material point is reflected back with its negative velocity. This rule is used to fulfil the no-slip boundary conditions. Periodic boundary condition is applied so that the left end of the lattice is connected the right end.

The units for kinematics quantities are as follows. Length is in lattice units (l.u.), i.e. the distance between two nearest nodes. One time unit is the discrete time step (t.s.) required for the update of the entire automata. One mass unit (m.u.) is the mass of a single particle. Therefore, velocity is in lattice units per time step and volumetric flux has the same unit as velocity. Pressure is in (m.u.) x (l.u.)$^{-1}$ x (t.s.)$^{-2}$. Dynamic viscosity is in (m.u.)x(t.s.)$^{-1}$(l.u.)$^{-1}$.

4. Experimental Methods

4.1 Flux as a function of time and velocity profile

The volumetric flux is the average x-component of velocity per particle averaged over the entire lattice. This volumetric flux is calculated as a function of time steps. The lattice size for this 2-dimensional flow in Figure 1 is L (n_x automata) x H (n_y automata) and the density ρ is 2.4 particles per site. At initial condition t=0, a number input particles (2.4 x n_x x n_y) are distributed in the lattice randomly both their positions and velocity directions. Particles that leave the lattice at $x = L$ reenter the lattice at $x = 0$ and particles that leave at x=0 reenter at $x = L$. This condition maintains constant particle number in the lattice. A pressure gradient was created by changing the x-component of momentum particles at $x = 0$ each time step during the experiment and no forces are applied at $x = L$.

$$\Delta P = P(x = L) - P(x = 0) = -\frac{n_y f_x}{H} \tag{5}$$

where $n_y f_x$ is total change in the x-component of momentum from particles distributed in n_y automata at $x = 0$. f_x is the average change in the x-component of momentum at a single point at $x = 0$. Flux was plotted as a function of time until its value does not change significantly with time.

Velocity profile is then determined when system has achieved its steady condition. Average x-component of velocity per particle is presented as a function of y. In order to obtain better data quality, we need to average the results from many (6-12) runs with different initial conditions. The data are reported in the averages $\pm$ their standard deviations.

4.2 Volumetric flux as a function of channel width

For a constant pressure gradient, the volumetric fluxes in a steady-state condition are determined for several channel widths. (in $\sqrt{3}/2 = 0{,}866$ lattice units) : 20, 40, 60, 80, 120, and 160.

5. Results and Discussions

5.1 Velocity profile

Velocity profiles for steady parallel-plate flow are shown in Figures 2a-2b. In Figure 2a, the velocity profile is parabolic for the case $L = 128$, $H = 80\sqrt{3}/2$ and $f_x = 0.0125$. By applying equation (5) into equation (1), we found that the viscosity of the fluid was 0.315. For the case $L = 128$, $H = 120\sqrt{3}/2$ and $f_x = 0.0250$ in Figure 2b, this numerical experiment observed a non-parabolic profile for this greater pressure gradient.

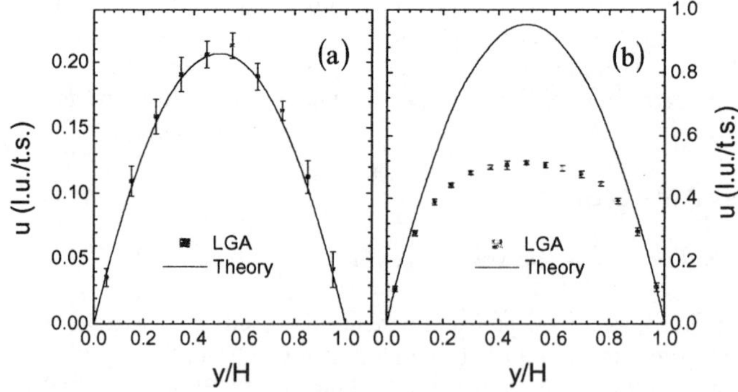

Figure 2. Numerical result for $u(y)$ compared to the theoretical parabolic profile for the case (a) $L = 128$, $H = 80\sqrt{3}/2$ and $f_x = 0.0125$, and (b) $L = 128$, $H = 120\sqrt{3}/2$ and $f_x = 0.0250$

The LGA result in Figure 2b suggests that the fluid viscosity is not constant. The viscosity of real fluid flows is generally a function of fluid velocity that caused by a non-linear effect. In this study, the viscosity for the case in Figure 2b may be approximated by:

$$\mu = \mu_0 + \alpha\, u^2 \left[1 - \left(\frac{r}{H/2} \right)^3 \right] \tag{6}$$

with the correlation coefficient to data is 0.99, μ_0 is the fluid viscosity obtained from equation (1) or for flow close to the parallel plates, $\alpha = 1 (\text{m.u.})(\text{t.s.})(\text{l.u.})^{-3}$ is a proportional constant, u is fluid velocity and it is a function of y, and r is distance from the center of channel width to the observed position in y direction so that $0 \le r \le H/2$.

5.2 Flux as a function of channel width

Volumetric flux as a function of channel width is given in Figures 3a and 3b for $f_x = 0.0125$ and $f_x = 0.0250$ respectively and $L = 128$. It is seen that the flux increases with the increment of channel width. When the flux is less than 0.2, LGA results have a good match with the theory. However, LGA results are lower than theory if the flux is more than 0.2. This result in agreement with the earlier results by Rothman [3].

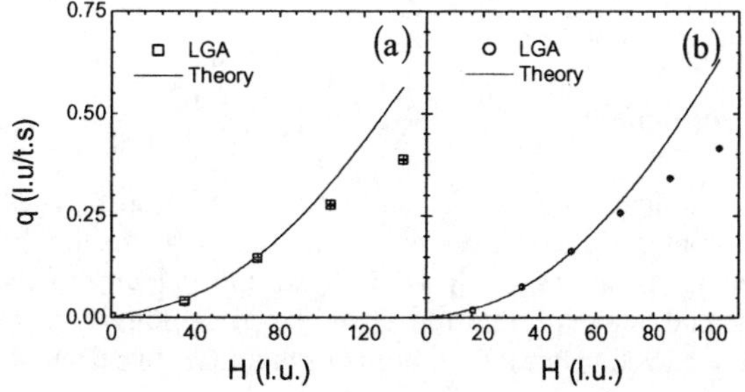

Figure 3. Flux as a function of channel width, for a pressure gradient created using (a) $f_x = 0.0125$ and $L = 128$ (b) $f_x = 0.0250$ and $L = 128$.

For the case $H = 80\sqrt{3}/2$, $L = 128$ and $f_x = 0.0125$, flux is proportional to the square of channel width (Figure 3a) and velocity profile is parabolic (Figure 2a). However, for the case $H = 120\sqrt{3}/2$, $L = 128$ and $f_x = 0.0250$, flux is less than the theory for laminar flow (Figure 3b) and velocity profile is not parabolic (Figure 2b).

5.3 Flux in tortuous media

A comparison for flux between straight and tortuous media is made in Figures 4a-4b. The flow in the straight medium ($H = 48\sqrt{3}/2$, $L = 240$) and the tortuous medium ($H = 48\sqrt{3}/2$, $L_e = 323$) are generated by the same pressure gradient. In this study, the flux is less than 0.2 (Figure 4b) therefore the equations for laminar flow are applicable.

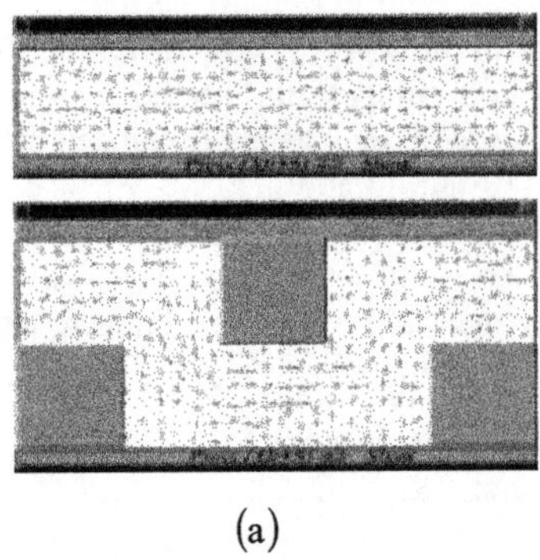

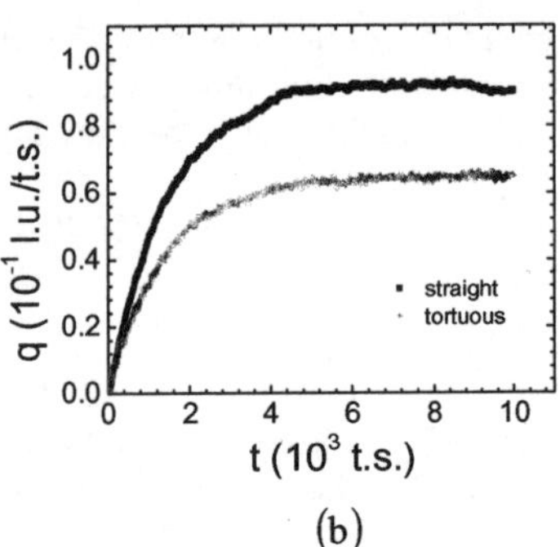

(a)   (b)

Figure 4. A comparison between straight and tortuous media with the same channel width.
(a) Flow in 2-D media. The dark regions represent impermeable material and
(b) Flux in straight and tortuous media.

The ratio for steady flux between the two media in Figure 4 is written as [6]:

$$\frac{q_{straight}}{q_{tortuous}} = \frac{L_e}{L}$$

$$\frac{0.091 \pm 0.009}{0.065 \pm 0.002} = \frac{323}{240} \cong 1.4$$

(7)

5.4 Constant volume rate of flow

For steady incompressible laminar flow, the continuity equation gives constant volume rate of flow along parallel plates if there is no source or sink. This means that the product of channel width to velocity is constant. Figure 5 shows the velocity of the flow increases when the size of channel width decreases. There are 4 regions with different sizes of channel width along the parallel plates. The results of this LGA experiment are:

$$80\sqrt{3}/2 \times (0.066 \pm 0.005) = 32\sqrt{3}/2 \times (0.167 \pm 0.007) = 64\sqrt{3}/2 \times (0.088 \pm 0.007)$$

$$= 80\sqrt{3}/2 \times (0.069 \pm 0.007) = \text{constant}$$

(8)

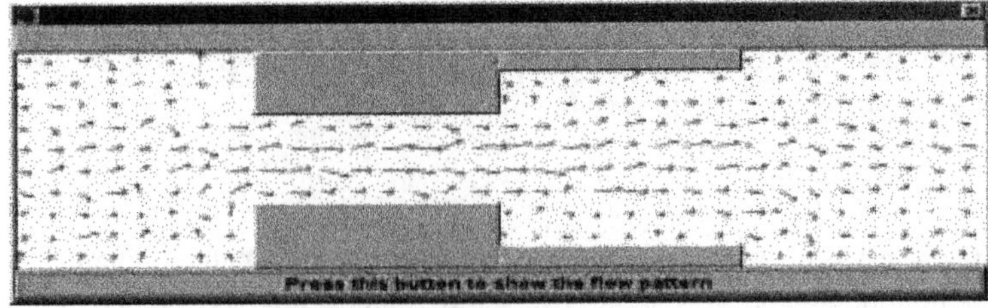

Figure 5. Steady incompressible flow along parallel plates with non-uniform channel width.

5. Conclusions

We have used the lattice-gas cellular automata to solve fluid flow in porous media. If the flux is less than 0.2 the Lattice-gas results are in agreement with the theory for steady laminar flow. Flux is inversely proportional to the effective path length in tortuous media. Constant volume rate of flow is also approved for incompressible flow if there is no source or sink along a pathway. If the flux is greater than 0.2 the lattice-gas results suggest that the fluid viscosity is dependent on the fluid velocity.

References

[1]   G. Weisbuch, Complex System Dynamics, An Introduction to Automata Network, Lecture Note Volume II, Santa FE Institute, Studies in the Science of Complexity, Addison-Wesley Publish. Co., 1991

[2]   R.C. Eberhart, and R.W. Dobbins, Neural Network PC Tools, Academic Press, 1990.

[3]   D. H. Rothman, Cellular-Automaton Fluids: A Model for Flow in Porous Media, *Geophys.*, **53**, 4, (1988) 509-518

[4]   S. Brown, Flow through Rock Joints: the effect of surface roughness. *J. Geophys. Res.* **92**(1987) 1337-1347

[5]   W. J. Duncan, A. S. Thom, and A. D. Young, *Mechanics of Fluids*, 2nded. ELBS and Edward Arnold Publ. Ltd., 1971.pp. 156-162

[6]   F.A.L. Dullien, Porous Media, Fluid Transport and Pore Structure, Academic Press, 1992

[7]   U. Frisch, B. Hasslacher, and Y. Pomeau, Lattice-gas automata for the Navier-Stokes equations, *Phys. Rev. Lett.*, **56** (1986) 1505-1508

KES 2002
E. Damiani et al. (Eds.)
IOS Press, 2002

A Case-Based System for Asthmatic Patient Health Care

Icham SEFION[1,2], Marc GAILHARDOU[2], Abdel ENNAJI[1]

[1] PSI Laboratory, *University of Rouen, pl. E. Blondel, 76821 Mt St Aignan, FRANCE*
[2] Alliance Médica, *100 route de Versailles, 78163 Marly le roi, FRANCE*

Abstract. Asthma is a distressing disease, affecting up to 7% of the French population and causing considerable morbidity and mortality. A medical decision support system can help physicians to control this chronic disease. Thanks to the health care network (RESALIS©) of Alliance Médica Society, asthma consultation data were collected. We chose Case-Based Reasoning paradigm to develop our medical decision support system. Intelligent data analysis methods have been used to determine the knowledge models for our system. A Self-Organizing Map has been used. A case represent a particular "profile" of the pathology associated to an adequate diagnosis and treatment. A classical classifiers are then used to design the system and a particular similarity metric (MVDM) is used in order to take into account the nature of the data (both numeric and symbolic description). A preliminary results are given.

1. Introduction

Asthma is a chronic inflammatory disease of the airways. In spite of the availability of effective drugs, this pathology remains insufficiently controlled. There are 2000 deaths per year in France out of which more than half would be avoidable. One estimates today at 2 500 000 the number of people in France, adults and children, suffering from asthma, which represent 5 to 7 % of the population. This alarming situation of public health shows that it is essential to improve the health care and quality of life of the asthmatic patients notably by providing a decision support system to physicians.

This work lies within the scope of the RESALIS© project which is an experiment of health care network aiming at improving quality of health care for asthmatic patients. When taking a decision for a patient (diagnosis and treatment prescription), physician [1] use both their experience and its academic knowledge. Our objective is then to improve the resort to the experience. After a complete overview on methods of medical decision support system, we chose the Case Based Reasoning paradigm. This well known technique of artificial intelligence attempts to solve a given problem by adapting established solutions to similar problems. In order to design a case based system, we should determine a knowledge model based on the use of intelligent data analysis methods and including a case model, a case indexing and a similarity metric. This paper is organised as follow. Section 2 present the context of this work, the nature of our data and some medical systems. Section 3 presents the methods that we used for intelligent data analysis including a self-organizing map and our similarity metric based on the MVDM method. Section 4 show our data analysis results obtained with clustering method, the evaluation of MVDM method and we discuss finally on the perspective of this work.

2. The experimental context

2.1 The RESALIS© Project

Our work lies within the scope of the RESALIS© project promoted by Alliance Médica. This project is a coordinated health care network which aims at facilitating access to health care for the asthmatic patients. An information system ensures the communication between physicians which respects the medical secrecy. A database, containing all the consultations of the patients included in the network (355 patients), is fed daily. This data constitute our raw material for the development of a decision support system for physicians.

2.2 The physician decision

When taking a decision for a patient, physician [1] use both their experience and its academic knowledge (FIG. 1) :

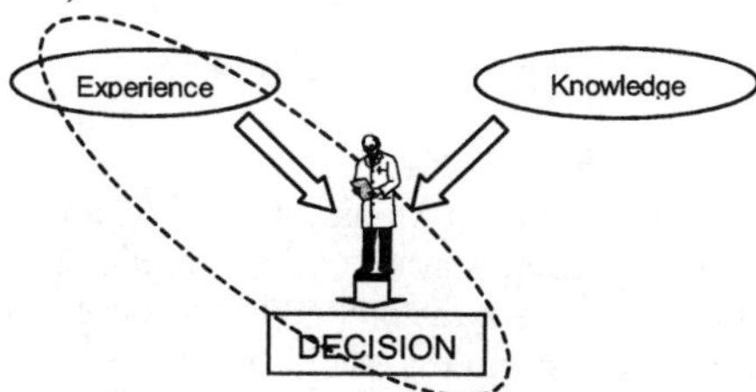

Figure 1 : the physician decision

We want to improve the resort to the experience with a medical decision support system.

2.3 The medical decision support

A complete overview on the principal methods of medical decision support system shows a broad use of rule based techniques. A rule-based system [5] breaks a problem down into a set of individual rules, each solves a part of the problem. Rules are combined together to solve a whole problem. To create these rules by hand, it's necessary to know how to solve the problem, and this task can be extremely complex and time consuming [4]. In this study, we chose the Case Based Reasoning (CBR) paradigm [2][3][4], CBR system differ fundamentally in that to use them, we do not need to know how to solve the problem, but only to recognize if we have solved a similar problem in the past. The major disadvantage of Neural Networks technology [6], in spite of their universal approximation property and, above all, their good generalization capabilities, is that an NN system functions as a "black box". No explanation or justification of any sort can be given by an Neural Network. Remember that CBR retrieves the most similar case and attempts to reuse the solution from case. We chose CBR technique for our medical decision support system.

3. Methods

3.1 The Case Based Reasoning

Case Based Reasoning is a technique of artificial intelligence that attempts to solve a given problem within a specific domain by adapting established solutions to similar problem[2].
We can describe CBR typically as a cyclical process comprising the four REs :

    1. REtrieve the most similar Case(s).
    2. REuse the case(s) to attempt to solve the problem.
    3. REvise the proposed solution if necessary.
    4. REtain the new solution as a part of a new case.

The following Figure 2 present the CBR cycle:

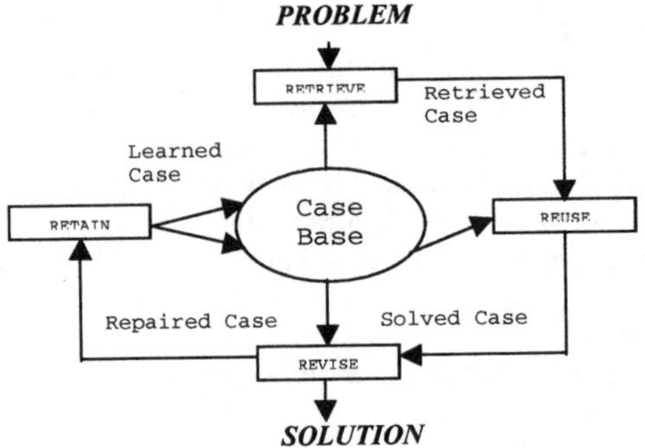

Figure 2 : the CBR cycle

3.2 Examples of CBR systems in Medicine

Several CBR systems were developed in Medicine [7]. CASEY [8] is a system that diagnoses heart failure. As input, it uses a patient's symptoms and produces a causal network of possible internal states that could lead to those symptoms. When a new case arises, CASEY tries to find cases of patients with similar but not necessarily identical symptoms. If the new case matches, then CASEY adapts the retrieved diagnosis by considering differences in symptoms between the old and new cases. PROTOS [9] was developed in the domain of clinical audiology. It learned to classify hearing disorders from descriptions of patients' symptoms, histories, and test results. PROTOS was trained with 200 cases in 24 categories from a speech and hearing clinic. After training, PROTOS had an absolute accuracy of 100%.

3.3 Case Representation

A case is a contextualized piece of knowledge representing an experience [2]. It contains the past lesson that is the content of the case and the context in which the lesson can be used. A case can be an account of an event, a story, or some record typically comprising.

- The *problem* that describes the state of the world when the case occurred
- The *solution* that states the derived solution to that problem

To design a Case Based Reasoning system, you should define a set of features for cases in your specific domain. A choice of variables for our cases was made by the physicians group of RESALIS©. We want to analyze the influence of variables choice on the data clustering with the Self-Organizing Map. With this method, we want also to estimate the nearness between cases in the case base. To resume, we used this method in order to represent efficiently a case in our CBR system.

3.3.1 Self-Organizing Map

The Self-Organizing Map (SOM) is a powerful neural network for analysis and visualization of high-dimensional data [10]. It maps non-linear statistical relationships between high-dimensional input data into simple geometric relationships on a usually two dimensional grid. The mapping roughly preserves the most important topological and metric relationships of the original data elements and, thus, inherently clusters the data. One can describe a SOM like a fishing net (FIG.3), whose meshs would be elastic, and whose nodes are the "neurons" of the map. The algorithm of training will deploy this net in the space of the data until it marries as well as possible the shape of the scatter plot.

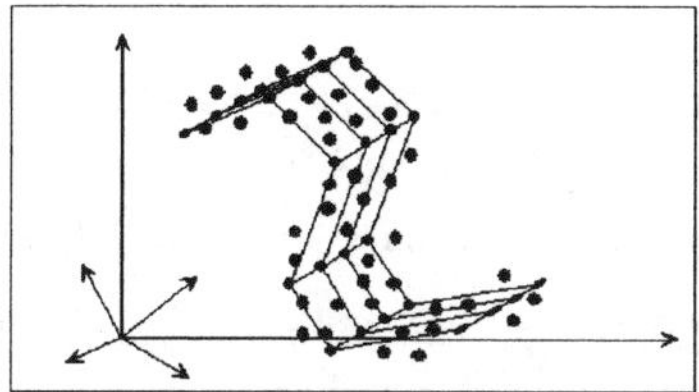

Figure 3 : a "fishing net" in the data space

Therefore, the SOM can be used at the same time both to reduce the amount of data by clustering, and for projecting the data nonlinearly onto a lower-dimensional display.

3.4 Case Retrieval

Remembering is the process of retrieving a case or a set of cases from case base. In general, two techniques are currently used in CBR tools : nearest neighbour retrieval and inductive retrieval. We chose the first method for its easiness to use. Two problem are similar if they are nearby in data space. You should calculate a similarity measure between cases to find similar case(s). To determine this distance, we chose to use the MVDM (Modified Value Difference Metric) technique [11].

3.4.1 MVDM method

It's a powerful new method for measuring the distance between values of features in domains with symbolic feature values. Using this method, a matrix defining the distance between all values of a feature is derived statistically, based on the examples in the training set. The distance δ between two values for a specific feature is defined in Equation 1 :

$$\delta(V1,V2) = \sum_{i=1}^{n} \left| \frac{C1_i}{C1} - \frac{C2_i}{C2} \right|^k \tag{1}$$

In the equation, V1 and V2 are two possible values for the feature. The distance between the values is a sum over all n classes. $C1_i$ is the number of times V1 was classified into category i, C1 is the total number of times value 1 occured, and k is a constant, usually set to 1. The total distance Δ between two instances is defined in Equation 2 :

$$\Delta(X,Y) = \omega_X \omega_Y \sum_{i=1}^{N} \delta(x_i, y_i)^r \tag{2}$$

where X and Y represent two instances (e.g., two consultations for the asthmatic patient health care), with X being an exemplar in memory and Y a new example. The variables x_i and y_i are values of the ith feature for X and Y, where each example has N features.

4. Results

4.1 Case Representation

Globally, one could show that the variables used for the analysis have a rather satisfactory capacity of discrimination taking into consideration our result. One could see that there were certain regularities within our data. However, we showed the presence of zones of recovery. This is explained partly by the presence of outliers and with the lack of homogeneity in the manpower of the various groups. Finally, following these various analyses for various configurations of variables, our choice for the representation of the cases remains that proposed by the doctors and the group of experts. It will be necessary however to eliminate the outliers which will be able to have a negative effect on the case retrieval.

4.2 Case Retrieval

We evaluated several methods for the similarity measure and we shown that the MVDM method was the best following our results presenting in table 1 :

Table 1 : recognition rate for several methods

	Euclidienne	**Manhattan**	**Chebychev**	**MVDM**
Recognition rate	59.0788 %	61.7628 %	63.4402 %	66.7772 %

The MVDM technique present a good performance for our type of medical data compared with other techniques. Our similarity measure will be based on this approach.

5. Conclusion and Perspectives

In this paper, we have presented our work about designing a case-based system for asthmatic patient health care. Case Based Reasoning was chosen for its advantages in medicine domain. We defined our case representation thanks to an intelligent data analysis: the self-organizing map. We shown that medical data contains noise such as outliers. Case representation was obtained after data analysis and discussion with physicians and experts in asthma disease. Next, we tested the MVDM technique for the similarity measure and we shown that this approach is the best for our application. A comparison of recognition rate of several methods shown that the MVDM method purpose the best result. These two aspects soon will be integrated in the case-based system. The Reuse phase is under development and should make it possible to adapt for example a treatment according to the age of a patient or a diagnosis following the international recommendations. Parallel to the designing of the system, the development of a user interface started. An UML specification (Unified Modeling Language) was made to facilitate this task of development but also for well the needs of the users. These potential users are general practitioners for whom the accent must be put on ergonomics, the speed and the ease of use while presenting to him at the screen relevant information for asthmatic patient health care.

References

[1] Degoulet P., Fieschi M., Informatique médicale, Paris : Masson, 1998.

[2] Kolodner J., Case Based Reasoning, San Mateo : Morgan Kaufmann Publishers, 1993.

[3] Aamodt A., Plaza E., Case Based Reasoning : Foundational Issues, Methodological Variations, and System Approachs, AI Communications, 1994;7 : pp39-59.

[4] Watson I., Applying Case-Based Reasoning : Technique for Enterprise Systems, San Francisco : Morgan Kaufmann Publishers, 1997.

[5] Fargeas X., Frydman F., Les Systèmes Expert en Médecine, Paris : Hermes, 1988.

[6] Ribert A., Structuration évolutive de données : application à la construction de classifieurs distribués, Thèse de doctorat, Université de Rouen, 1998.

[7] http://www.med.uni-rostock.de/WebServerDaten/IMIB/HTML/english/research/research.html

[8] Koton P., Using experience in learning and problem solving, PhD Thesis, Laboratory of Computer Science MIT, 1988.

[9] Bareiss E. R., PROTOS : A Unified Approach to Concept Representation, Classification and Learning, PhD Thesis, University of Texas, 1988.

[10] Kohonen T., Self-Organizing Maps, Berlin: Springer, 1995.

[11] Cost S., Salzberg S., A weighted nearest neighbor algorithm for learning with symbolic features, Machine learning, 1993.

KES 2002
E. Damiani et al. (Eds.)
IOS Press, 2002

A New Heuristic Genetic-based Algorithm for Constrained MultiCast Tree Construction

Sherin M. Youssef , M. A Ismail, Soheir A. Bssiouny
Computer Science Department, Faculty of Engineering, Alexandria University, Egypt
e-mail: sherin@aast.edu

Abstract. It has been demonstrated that finding a feasible multicast tree with independent additive path constraints is NP-complete [1]. In the field of artificial intelligence, genetic algorithms have emerged as a powerful tool to solve the NP-hard problem. Explored from the fact that deterministic computation of QoS multicast routing is usually time-consuming and most of the currently proposed algorithms either work too *slowly* or *cannot compute delay-constrained multicast with low costs*, we introduce a new Enhanced Genetic-based algorithm that constructs a minimum cost multicast tree that meets bandwidth, end-to-end delay, and inter-destination delay variation constraints, which we call Minimum Cost Delay-DelayVariation-Constrained Multicast Tree (MCDDVCMT). Empirical studies shows that our proposed algorithm performs very well in terms of cost and inter-destination delay variations of the solutions that it generates as compared to some existing algorithms.

1. Introduction

Multicasting [2] refers to the transmission of data to multiple destinations. Multicast routing use trees, called *multicast* routing trees, over the network topology for transmission to minimize resource usage. To support the multimedia applications, multicast routing algorithms are expected to not only generate routes that reach all destinations, but also compute the optimal paths to all destinations and without exceeding the real-time delay constraint imposed by these multimedia applications. In addition, the multicast tree must also guarantee a bound on the variation among the delays along the individual source-destination paths. Practical situations in which the need for bounded variation among the path delays arises includes that of a *teleconference session* and in the *updating of replicated database*. The problem of finding minimum cost multicast routing trees is called the *Steiner Tree problem* and is known to be NP-complete[3]. Near-optimal solutions by heuristic algorithms are feasible way and some heuristics for the delay-constrained multicast routing problem have been presented in Ref.[6,7,8,10,11]. The simulation results given in Ref. [4] have shown, however, that most of the proposed algorithms either work too slowly or cannot compute delay-constrained multicast with low costs. Roukas et al. [4] have proposed a source based multicast algorithm but it does not take the cost of the tree into consideration. Deterministic computation of QoS multicast routing is usually time-consuming. For this reason, methods based on neural networks or genetic computation may be of great help. A proposal using the neural networks to solve such problem has been made in Ref.[8] but there are many coefficients hence the selection of their values is one of the hardest tasks to deal with Hopfield neural networks[9]. Some genetic algorithms have been published [11-14] and also the algorithms mentioned in Ref. [10,12,13,14,15,16] tried to solve the Steiner tree problem but many limitations and weakness are adopted.

The paper is organized as follows: in *section 2*, we formulate the problem, then, in *section 3*, a description of the proposed genetic-based algorithm is adopted. *Section 4* illustrates the experimental results and analysis of the algorithm performance. Finally, *section 5*, summarizes and concludes the paper.

2. Formulation Of The Problem

We try to construct a Bandwidth Delay DelayVariation Constrained Least Cost Steiner tree T rooted at s, that spans the destination nodes di's in the set D, $di \in D$, where $|D|=m$ and m is the number of destination nodes. The tree cost $Cost(T(s,D))$ is the sum of the cost of tree edges given by: $Cost(T(s,D)) = \sum_{e \in T(s,D)} Cost(e)$, *where e: represents a tree edge.* The total path delay from s to any destination di, denoted by $delay(p_T(s,di))$, is evaluated as:
$delay(\ p_T(s,di)) = \sum_{e \in p_T(s,di)} delay(e)$. The bandwidth from s to any destination di, denoted by $bandwidth(p_T(s,di))$, is defined as the minimum of the bandwidth among links along $p_T(s,di)$. The problem can be formulated as follows:

$Min \{\ Cost(T(s,D))\ \}$
$S.t.$
$delay(\ p_T(s,di)\ ) \le DDCF(di) \quad \forall di \in D$
$bandwidth(\ p_T(s,di)) \ge B \qquad \forall di \in D$

$$\left| \sum_{e \in p_T(s,di)} delay(e) - \sum_{e \in p_T(s,dj)} delay(e) \right| \le \ \delta, \quad \forall di,dj \in D, \ i \ne j$$

where $p_T(s,di)$ is the path from s to di on tree T, $\Delta = DDCF(di)$ is the Destination Delay Constraint function that assigns an upper bound to the path delay from the source to each destination di in the tree. B is the bandwidth constraint. We suppose that the bandwidth constraint is the same for all the destinations, while the delay constraint can be different for different destinations. δ is the inter-destination delay variation bound integer value.

3. The Proposed Genetic-Based Algorithm

3.1 The theory of genetic algorithm

The genetic algorithm derives inspiration from the optimization process that exists in nature [13,14,17]. The "survival of fitness" filter is applied to the population, which is constructed by crossover and mutation operators.

3.2 The proposed Genetic-based Algorithm for Constrained Multicast Routing (GCMR)

3.2.1 Coding: In our strategy, we choose the tree structure coding method, in which a chromosome represents a multicast tree, as shown in Fig. 1. By this representation, the coding space is greatly reduced and the coding/decoding operation is omitted. Assuming a network graph of maximum node degree 5, the chromosome structure is represented by a quad-Tree[5]. In the tree, each parent vertex (the representative parent) represents the predecessor of its children. This coding scheme performs better than that of Ref.[14] which uses a $N_n \times N_n$ one-Dim. binary code that leads to a remarkably increase in search space.

3.2.2 Initial population: we first remove the links with bandwidth less than the requirement (this is called a refining process [18]). So, the remaining paths in the refined graph must satisfy the bandwidth constraint. The construction of an initial population of size Np is done by forming a random multicast Steiner trees. A randomized depth-first search algorithm (DFS) was employed for this purpose [19].

3.2.3 Fitness function: The fitness function of a multicast tree T is defined as follows:

$$f(T) = \frac{\alpha}{\sum_{e \in T} cost(e)} \prod_{di \in D} \Phi\big(delay(p_T(s,di)) - DDCF(di)\big) \prod_{\substack{di,dj \in D \\ i \ne j, i \prec j}} \Psi\left(\left| \sum_{e \in p_T(s,di)} delay(e) - \sum_{e \in p_T(s,dj)} delay(e) \right| - \delta\right)$$

$$\Phi(Z) = \begin{cases} 1 & Z \leq 0 \\ r & Z > 0 \end{cases} \qquad \Psi(Z) = \begin{cases} 1 & Z \leq 0 \\ g & Z > 0 \end{cases}$$

Where α is a positive real coefficient, $\Phi(Z)$ is the penalty function associated with the path delay upper-bound constraint, and $\psi(Z)$ is the penalty function associated with the inter-delay variation constraint. The values of r and g determine the degree of penalties associated with the path-delay upper-bound constraint and the inter-delay variation constraint, respectively. Particularly, we have selected r=0.5 and g=0.5 in our experiments.

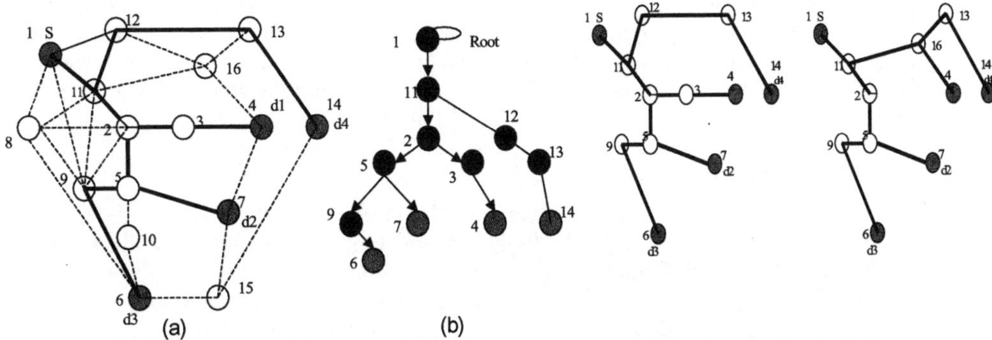

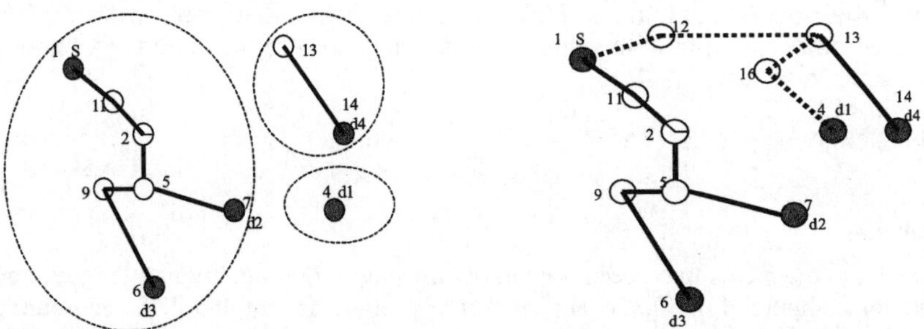

Fig. 1   Fig.2

Fig. 1 (a) represents a Graph of 16 nodes and 4 destination nodes 4,7,6, and 14. Bold face links represent one of the Multi-destination routing tree solutions. (b) The tree-structured chromosome of the multicast tree shown in Fig. 1(a). Fig. 2 represents a two parent multicast trees.

3.2.4 Selection of parents: Our selection pool will be the new resulted offspring multicast trees and their *Np* multicast parents. By the model we first select the optimal individuals and directly copy them to the next generation, and then select the rest by a proportional model. The probability sp_i that an individual tree *Ti* is selected is given by:

$$sp_i = f(T_i) / \Sigma_{(j \in \text{ selection pool})} f(T_j)$$

3.2.5 Crossover Scheme

Fig. 3 (a) The disconnected sub-trees , (b) The resulted connected child tree C after re-connection.

According to the crossover probability *Pc*, two multicast trees are selected as parents to produce an offspring. This process is repeated *Np-1* times, where *Np* is the population size. The crossover operations used in our algorithm are described as follows: *(1)* Select the same links of the parents and copy them to the child directly. *(2)* Connect the separate sub-trees to be a multicast tree: *First*, we randomly select two separate sub-trees among these sub-trees. *Second*, connect them with the least-cost feasible-delay path. If no connecting paths satisfy the delay constraint, we choose the least-delay path, otherwise we choose the least-cost path. ***Third***, a new selection begins again. The selection repeats until a multicast tree is constructed. Clearly, there is no loop in the multicast tree constructed by this connection scheme. The above crossover scheme speeds up the algorithm convergence and delivers the good characteristics (lower cost and delay) to the offspring. Fig. 2 and Fig. 3 explain the crossover steps for the construction of a new offspring tree.

3.2.6 Mutation

The mutation operation is performed according to the probability of mutation *pm*. By our experiments we find that when $pm \in (0.005, 0.05)$ it can speed up the convergence of the algorithm and prevents the stuck of the solution tree chromosome to some sub-graphs structure. The mutation procedure randomly selects a node (called current node) and changes its parent by choosing a new node to be its predecessor. The above mutation operation improves the performance and it is effectively convert the algorithm from a local optimal point into a good state with constrained delay and low cost.

3.2.7 The complete algorithm

```
GCMR (G, s, D)            /* where G is the refined network graph */
{ For (i=1; i<= Np; i++ )   /*create initial population by random Depth-First-Search*/
    A(i) = RandomDFS(G,s,D);

  For (j=1; j<= Ng; j++){              /* Ng is the max. no. of generations */
    For (k=1; k<= Np-1; k++)           /*Select parent trees T_M, T_F*/
    { T_M= MSTSelect(A)
      T_F= MSTSelect(A)
      Select the same links from T_M and T_F
      Put the resulted disconnected sub-trees in a temp pool

      Repeat
      { Randomly select two sub-trees at random from the temp pool
        Choose a feasible-delay min-cost path connecting the two sub trees.
      } Until a multicast tree is connected

      If  (rand( ) < pm) Perform Mutation according to section 3.2.6
    } /* end for k */

    Select the best Nbest (Nbest< Np) individuals, then select the remaining (Np
    -Nbest) next-generation individuals according to the selection probability
    described in section 3.2.4.
  } /* end for j */
} /* end GCMR Algorithm */
```

4. Experiments

The simulation experiments have been implemented using C++ programming language on a PC Pentium computer PII330. To ensure that the simulation of the different routing algorithms were fairly evaluated, random graphs of size n with average degree 4, which would better represent the topologies of common networks, e.g., the NSFNET, were constructed. The DDCF(di) Function for destination di is uniformly distributed in the range of [30-160 ms], the delay of each link is uniformly distributed in [0-50 ms]. The bandwidth requirement is uniformly distributed in [10,50], the bandwidth of each link in [0,100], and the cost of each link in [0,100]. We compared the performance of our proposed algorithm, GCMR, with a distributed Bellman-Ford Shortest Path algorithm (SPT) [19], DDVBM algorithm [20], and Kompella's Distributed Multicast Algorithm [3]. Figure 4(a) shows the fast convergence of our proposed GCMR algorithm compared with the DDVBM algorithm.

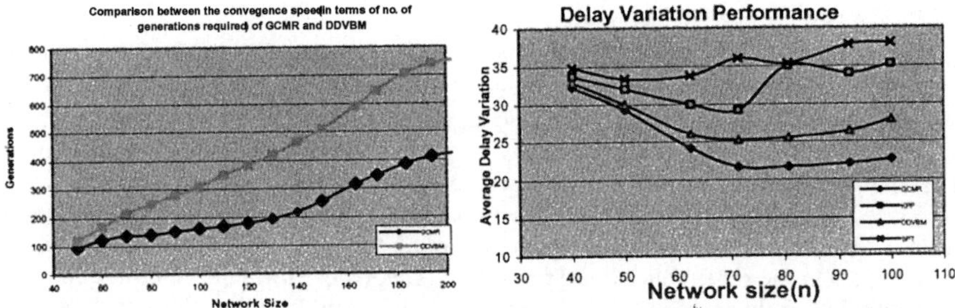

Fig. 4 (a) The Convergence time (in terms of no. of generations) versus Network Sizes, for m=30, Δ=120, and δ = 50, Fig. 4(b) The inter-destination delay variation for different network sizes.

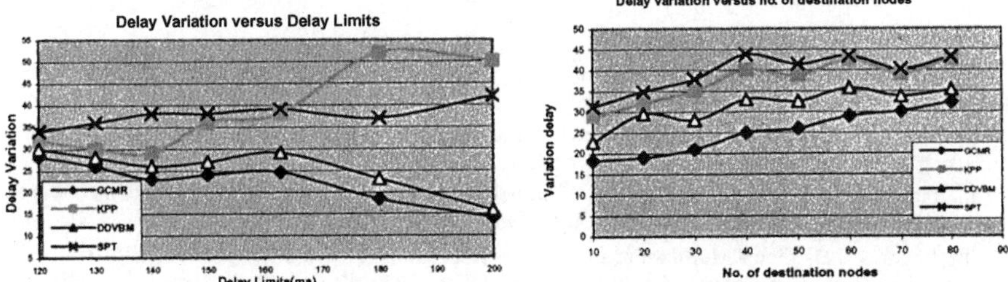

Figure 5(a) The inter-destination delay variation for different delay limits Δ, 5 (b) The inter-destination delay variation for different no. of destination nodes,.

We can notice from figure 4(b) that as the number of nodes (n) increases, new search space area can be explored and so less delay links can be founded. So, more lower delay paths with small inter-delay variations can be explored. Fig 5(a) shows the performance of GCMR algorithm as the Delay limits grows (delay constraints are more relaxed), the delay variation decreases dramatically. Figure 5(b), shows an increase in the delay variations as the no. of destination nodes increases, but the performance of our proposed algorithm overcomes the others in comparison.

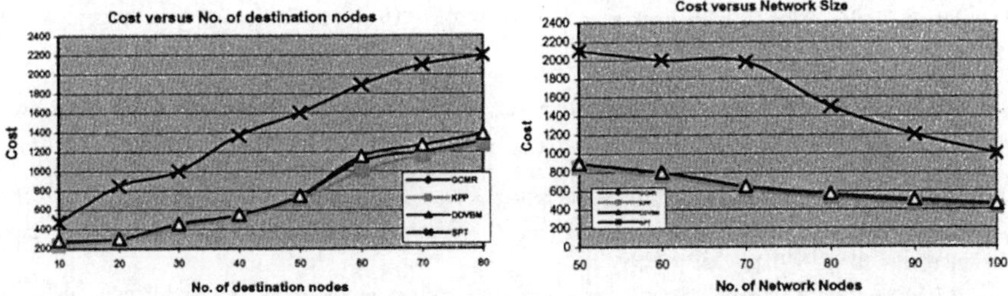

Fig. 6(a) The Cost versus No. of destination nodes (m) for α= 0.2, Δ=160, δ=0, and n=100, and, (b) The Cost versus different Network sizes, for m= 30, α= 0.2, Δ=160,and δ=0.

From Fig. 6(a) we can observe that the cost of the solutions obtained using GCMR are on average slightly higher than those obtained by Kompella's algorithm but are lower than those obtained using SPT algorithm. This phenomenon is expected as GCMR constructs trees that meet the inter-destination delay constraint and at the same time with low cost while the other comparable algorithms does not take these QoS factors into consideration. Fig. 6(b) shows the decrement in the total tree cost as n increases. We can conclude that GCMR, DDVBM, and Kompella's Algorithms produce solutions with comparable cost but they consistently performed better than the SPT's Algorithm.

5. Summary And Conclusions

In this paper, we have introduced a new Genetic-based algorithm to solve the problem of constructing minimum cost multicast trees that guarantee certain QoS bounds. Our empirical study shows that our proposed algorithm performs better than Kompella's algorithm in terms of inter-destination delay variations while at the same time keeping the cost of solutions close to that of Kompella's algorithm. Our empirical study also shows that our proposed algorithm performs better in terms of quick convergence to global optimal solution and avoid stuck to local optimal due to its efficient operators and parallel computations. On the other hand, it has proved its accepted performance for large networks. This in turn will help in reducing the complexity of using distributed algorithms in the construction of distributed multicast trees in large networks [20]. Future work should be focused on applying the proposed algorithm to the multilevel hierarchical routing.

References

[1] Z. Wang, J. Crowcroft, "Quality of Service for Supporting Multimedia Application", IEEE Journal on Selected Areas in Communications, 14(7) (1996), pp. 1228-1234.

[2] M. Parsa, Q. Zhu, J.J. Garcia-Luna-Aceves, "An iterative algorithm for delay-constrained minimum-cost multicasting", IEEE/ACM Transactions on Networking 6(4) (1998), pp. 461-474.

[3] V.P.Kompella, J.C.Pasquale,G.C.Polyzos, "Optimal multicast routing with quality of service constraints", Journal of Network and Systems Management 4(2) (1996), pp. 107-131.

[4] G. N. Rouskas, I.Baldine, "Multicast routing with end-to-end delay and delay variation constraints", IEEE Jounal on Selected Areas in Communications, 15 (3) (1997), pp. 346-356.

[5] Gregory L. Heileman, Data Structures, Algorithms, and Object Oriented programming, MIT Press and MC Graw-Hill, 1996.

[6] R. Sriram, G. Manimaran, S.R. Murthy, "Allgorithms for delay-constrained low-cost multicast tree construction", Computer Communications, 21 (18) (1998), pp.1693-1706.

[7] R. Sriram, G. Manimaran, S.R.Murthy, "A rearrangeable algorithm for the construction of delay-constrained dynamic multicast trees", Proceedings of the Conference on Computer Communications(IEEE Infocom), New York, March, 1999.

[8] P.Chotipat,C.Goutam,S.Norio,"Neural network approach to multicast routing in real-time communiction networks", International Conference on Network Protocols, 1995, IEEE pp. 332-339.

[9] Simon Haykin, Neural Networks – A comprehensive Foundation, 1999.

[10] F.K. Hwang, D.S. Richards, P.Winter, The Steiner Tree Problem, Elsevier Science, Amsterdam, 1992.

[11] Carlos A. Coello, Alan D. Christiansen, H. Aguirre Arturo,"Use of Evolutionary techniques to automate the design of combinational circuits", International Journal of Smart Engineering System Design, 2 (4) (2000), pp.299-314.

[12] H. Esbensen, "Computing near-optimal solutions to the Steiner problem in a graph using a genetic algorithm", Networks 26 (1995), pp. 173-185.

[13] Q. Sun, "A genetic algorithm for delay-constrained minimum-cost multicasting", Technical Report, IBR, TU Braunschweig, Butenweg, 74/75, 38106 Braunschweig, Germany, 1999.

[14] F. Xiang, L. Junzhou, W. Jieyi, G. Guanqun, "QoS routing based on genetic algorithm", Computer Communications, 22 (1999), pp. 1394-1399.

[15] C.P.Ravikumar, R.Bajpai, "Source-based delay-bounded multicasting in multimedia networks", Computer Communications, 21 (1998), pp. 126-132.

[16] Q. Zhang, Y.W. Lenug, "An orthogonal genetic algorithm for multimedi multicast routing", IEEE Transactions on Evolutionary Computation, 3(1) (1999), pp.53-62.

[17] S. Pierre, G. Legault, "A genetic algorithm for designing distributed computer network topologies", IEEE Transactions on Systems Man and Cybernetics – Part B: Cybernetics 28(2) (1998) 249-258.

[18] Wang Zhengying, Shi Bingxin, Zhao Erdun,"Bandwidth-delay-constrained least-cost multicast routing based on heuristic genetic algorithm", Computer Communications, 24 (2001), pp.685-692.

[19] Thomas H. Cormen, Charles E. Leiserson, Ronald L. Rivest, Clifford Stein, Introduction to Algorithms, The MIT Press, McGraw-Hill, 2001.

[20] C.P.Low, Y.J.Lee,"Distributed multicast routing with end-to-end and delay variation constraints", Computer Communications, 23 (2000), pp. 848-862.

KES 2002
E. Damiani et al. (Eds.)
IOS Press, 2002

Boundaries detection based on polygonal approximation by genetic algorithms

Walid Barhoumi and Ezzeddine Zagrouba
Faculté des Sciences de Tunis, Groupe de Recherche Images et Formes de Tunisie,
1060 le Belvédère, Tunis, TUNISIA

Abstract. This paper introduces a genetic algorithm approach to optimise polygonal approximation of regions boundaries in the setting of the lateral stereovision. The segmentation in regions of the original image is obtained by an hierarchical method through adaptive thresholding. Then, polygonal approximation is applied on each region belonging to the given regions map in order to determine the correspondent boundary. The Polygonal approximation is modelled as a combinatory problem which is resolved by employing the genetic algorithms techniques. Finally, the interest of our approach is illustrated through objective and subjective tests showing the robustness of the approach.

1 Introduction

The boundaries extraction is unavoidable step in the case of indoor scenes images since geometric characteristics of objects must be considered in order to realise the ulterior treatments (matching, 3D-reconstruction, objects recognition, $\cdots$). As indoor scenes are formed of manufactured environment with regular-form objects, polygonal approximation is a robust solution for the boundaries extraction which preserves the geometric form and reduces the data requirements.

Thus, the original image is first segmented in regions using our hierarchical algorithm of segmentation by adaptive thresholding [1], which provides better results in the case of indoor scenes images than the classical techniques of segmentation (figure 2.b). Then, the polygonal approximation is modelled as a combinatory problem (section 2.) which is resolved using genetic algorithms techniques (section 3.) in order to estimate the correspondent boundary for each region. Finally, the section 4 deals with results according to some chosen criteria.

2 Polygonal approximation

The polygonal approximation (**PA**) is the representation of a curve ζ by a polygon P which must approximate as well as possible the initial curve in order to preserve all its features and its visual appearance, but also to cover the minimum possible space memory. Given a distance function **d** measuring the error between two polygons, there are two ways to perform the polygonal approximation :

- given a threshold ϕ, one looks for the polygon P including the smallest number of points and not differing of the initial curve ζ more than ϕ [5].

$$(\pi_1) : \begin{cases} \mathbf{PA}(\zeta) = P \ / \ Card(P) = \min_{Q \subset \Lambda} Card(Q) \\ \Lambda = \{Q \subset \zeta \ / \ \mathbf{d}(\zeta, Q) \leq \phi\} \end{cases} \tag{1}$$

- given the number m of angular points that must have the approximated polygon P, one finds the nearest one of the curve ζ [3].

$$(\pi_2) : \begin{cases} \mathbf{PA}(\zeta) = P \ / \ \mathbf{d}(\zeta, P) = \min_{Q \subset \Gamma} \mathbf{d}(\zeta, Q) \\ \Gamma = \{Q \subset \zeta \ / \ Card(Q) = m\} \end{cases} \tag{2}$$

Different qualitative criteria, sometimes inconsistent within themselves, are used in polygonal approximation. In [4], the authors classify these criteria in objective and subjective ones. An essential and objective criterion to evaluate the performances of an approximation consists in being able to ensure that the results provided are "invariants" through a whole of a priori defined transformations (scale factor, rotation, translation and masking). The subjective criteria which seemed to us particularly relevant are : the control of the maximum variation, the preservation of the angles and the preservation of the straight lines.

As far as we are concerned, we have opted for the second method (π_2). In fact, fixing the number of the points does not distort, even could improve in certain cases, the similarity between potentially homologous regions in stereoscopic images. Besides, the second method is invariant by rotation, translation and scale factor.

Thus, for each region a follow-up of pixels located on the region border and characterized by a local maxima of gradient defines the n angular-points forming the initial polygon ζ which represents a first approximation of the region boundary. Then, the polygonal approximation allows to find the polygon P defined by m angular-points forming a subset of ζ. To resolve the combinatory problem (π_2), one looks for the best solution between C_n^m ones. In order to reduce the combinatory complexity, we used the genetic algorithms techniques which are adaptive heuristic search algorithms generally applied to spaces which are too large to be exhaustively explored.

3   Genetic algorithms

A genetic algorithm [6] is an iterative procedure that consists of a constant-size population of individuals, each one represented by a finite string of symbols and encoding a possible solution in a given problem space. The standard genetic algorithm proceeds as follows :

An initial population of individuals is randomly generated. Then, at every evolutionary step the individuals in the current population (*generation*) are evaluated and a new population is selected according to some predefined quality criterion (*fitness function*). New points in the search space are generated by crossover and mutation. Crossover is performed with probability P_{cross} between two selected individuals, called parents, by exchanging parts of their genomes to form new individuals, called offspring. This operator tends to enable the evolutionary process to move toward promising regions of the search space. The mutation operator is introduced to prevent premature convergence to local optima by randomly sampling new points in the search space. It is carried out by flipping bits at random, with some small probability P_{mut}. In the following, we detail the main steps of our approach.

3.1 Data coding and initialisation

Each polygon solution P is represented by a binary chromosome X_P composed of n genes (points), of which the value indicates whether the point is kept in P. Then, we create randomly z $(z \leq C_n^m)$ chromosomes forming the initial population Ω_ζ relatively to ζ.

3.2 Fitness function and selection

The fitness function $\mathbf{F}$ is applied on a chromosome X_P and returns a real value measuring the similarity between the polygon P and ζ [ie $\mathbf{F}(X_P) = (\mathbf{d}(P,\zeta))^{-1}$]. In [2], the authors define the distance $\mathbf{d}$ as the sum of distances between line segment $[S_i, S_{i+1}] \in P$ and the correspondent arc $A_{i,i+1} \in \zeta$ (3), such $S_{m+1} = S_1$ and $\delta\,(p_i, [S_i, S_{i+1}])$ is the orthogonal distance between the point p_i, belonging to the arc $A_{i,i+1}$ and the segment $[S_i, S_{i+1}]$.

$$\mathbf{d}(\zeta, X_P) = \sum_{i=1}^{m} \mathbf{L}_{i,i+1} \,, \quad with \quad \mathbf{L}_{i,i+1} = \sum_{p_i \in A_{i,i+1}} \delta^2\,(p_i, [S_i, S_{i+1}]) \tag{3}$$

Our experimentations showed that the use of this distance eliminates some meaningful angular points specially for low value of m. From there, the shape of a region is better characterized by using the area attribute (4) in order to preserve the visual appearance as shown in figure 1 (by using the equation (3) in (b) and (4) in (c)).

$$\mathbf{L}_{i,i+1} = \sum_{p_i \in A_{i,i+1}} \delta^2\,(p_i, [S_i, S_{i+1}]) + \Delta_{i,i+1} \tag{4}$$

where, $\Delta_{i,i+1}$ is the area of the zone limited by the segment $[S_i, S_{i+1}]$ and the correspondent arc $A_{i,i+1}$.

Then, each polygon P_i $(1 \leq i \leq z)$ in a current generation is selected only if the probability Pr_i (5) is bigger than a number generated randomly. At every selection step, we get a generation of some chromosomes, that we complete randomly by new chromosomes so that all populations have the same number of individuals.

$$\mathbf{Pr_i} = \frac{\mathbf{F}(X_{P_i})}{\sum_{j=1}^{z} \mathbf{F}(X_{P_j})} \tag{5}$$

3.3 Crossover and mutation

First, we choose $z/2$ couples of chromosomes at random. Then, for every couple $(X_P = x_1 \cdots x_n, X_{P'} = x'_1 \cdots x'_n)$, we look for indexes i_1 and i_2 verifying (6). This means that there is the same number of segments in the polygons P and P' to approximate the arc $A_{i1,i2}$.

$$\sum_{j=1}^{i1} x_j = \sum_{j=1}^{i2} x'_j \tag{6}$$

Therefore, it is possible to exchange the genes of X_P and $X_{P'}$ between the positions i_1 and i_2. Finally, every new chromosome which makes better the fitness function replaces the old one in the population.

Figure 1:   comparison of the fitness functions :   (a) initial curve,   (b), (c) final polygons.

For the mutation operator, we choose randomly a point S_l and we look for the two nearest points S_k and S_j ($k \prec l \prec j$). Then, the point S_t minimizing the expression $(L_{k,t} + L_{t,j})$ is searched between S_k and S_j and we permute the binary values of the correspondent genes to S_t and S_l. Finally, the termination criterion is based up on a constant T defining the number of successive iterations where the current fittest polygon is stable.

4   Results and discussion

The obtained region map by the segmentation process is formed of 67 and 3729 segments are used to define the initial boundaries approximation. The results comparison of various methods of polygonal approximation is a delicate problem. Indeed, one seeks to check several contradictory criteria. The numerical criteria that we used to compare the results are :
• the cumulated surface between ζ and P (**CS**) : a low value is desirable which means that final polygon sticks with the curve.
• the mean (**M**) and the variance (**V**) of the orthogonal distance δ between the not kept angular points of ζ and the correspondent segment in P.
• the invariance degree through a rotation r (**RI**r) : it is the number of the kept points in the final polygon P some is the rotation r to be applied to ζ.

The use of the previous numerical criteria is justified by the not optimality of the approximated final solution given by genetic algorithms. The figure 3. illustrates and compares the means and curve's shape of the used functions (**CS, M, V, RI**) by considering the equations (3) (figure 3.a) and (4) (figure 3.b). Indeed, the taking into account of the error on surface introduced into equation (4) improves clearly the robustness of the method. In fact, it allows to deduce an approximation more "invariant" by rotation (**RI**90, **RI**270) and produces a polygon having better localised angular points (**M, V**). Besides, the resulting polygon is more stick (**CS**) to the initial curve and consequently with better visual appearance (figure 2.).

5   Conclusion and perspectives

The method introduced in this paper can be considered as a robust approach for the boundaries detection in indoor scenes. In fact, the convergence of the genetic algorithms is guaranteed and the time of convergence is faster by using the area attribute in the fitness function (about 9 iterations per region). Besides, the optimal or near-optimal solution is obtained for the polygonal approximation problem. Then, we can improve this approach by paralleling the segmentation algorithm and the polygonal approximation algorithm considering the independence of the regions processing.

Figure 2: Contours detection with $z = min(100, C_n^m/4)$, $m = 70\%n$, $Pmut = 0.2$ and $T = 4$: (a) original image, (b) regions map, (c) boundaries extraction (equation (3)) , (d) boundaries extraction (equation(4)).

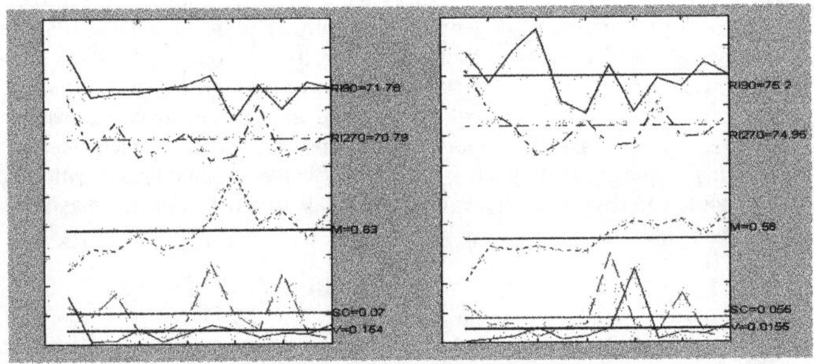

Figure 3: Comparative results by objective and subjective criteria.

References

[1] W. Barhoumi, E. Zagrouba, F. Ghorbel, Hierarchical region segmentation and matching for stereoscopic images, World Multiconference on Systemics, cybernetics and Informatics. Vol. V, USA, (2000) 16–20.

[2] W. Barhoumi, E. Zagrouba, Region's boundaries estimation using genetic algorithms in indoor scenes, 2eme Journees Tunisiennes Electrotechnique et Automatique, Tome I, Sousse, Tunisia, (2002) 144–151.

[3] J. Benois, D. Barba, Image segmentation by region contour cooperation as a basis for efficient coding sheme, SPIE 29, V.18, USA, (1992) 1218-1229.

[4] P. Garnesson, G.Giraudon, l'approximation polygonal : bilans et perspectives, TR1621, INRIA, PASTIS.

[5] Y.N. Sun, S.C. Huang, Genetic algorithms for error-bounded polygonal approximation, Inter Journal of Pattern Recognition and Artificial Intelligence, V.14, (2000) 297-314.

[6] M.D. Vose, The simple Genetic Algorithm : Foundations and theory, The MIT Press, (1999).

KES 2002
E. Damiani et al. (Eds.)
IOS Press, 2002

Skill-based Resource Allocation using Genetic Algorithms and Ontologies

Kushan Nammuni[1], John Levine[2] & John Kingston[2]

[1]*Department of Biomedical Informatics,*
Eastman Institute for Oral Health Care Sciences, University College London, U.K
E-mail: nammuni@hotmail.com
[2]*Artificial Intelligence Applications Institute,*
CISA, Div. Of Informatics, University of Edinburgh, U.K
E-mail: John.Levine, John.Kingston@aiai.ed.ac.uk

Abstract Allocation of tasks to resources is a crucial issue for an organisation. This paper investigates the use of genetic algorithms, combined with an ontology, in producing optimal allocations of tutors to tutorials. The ontology is used to support partial matching; it helps determine which tutors' skills (i.e. resources) are ontologically close to the optimal skill. These are then matched against the skill requirements of the task. The other novel feature of this GA implementation is the steady state duplication elimination algorithm, which (when coupled with 10% best chromosome crossover) produces more solutions with better fitness than a standard steady state GA.

1. Introduction

Project planning is a notoriously difficult task. Typically, a project planner has to cope with all the complexity of combinatorial optimisation problems, along with restricted resources. Where the general problem is to allocate known tasks to limited resources, the problem is effectively a scheduling/timetabling problem [1]. In order to make good allocations in such situations, some method of rating a candidate allocation of resources to tasks, and some method of reasoning about these ratings are needed.

The work described in this paper investigates the use of a genetic algorithm in producing optimal allocations of tasks to resources, using an ontology to provide simple resource ratings. The domain chosen is tutor allocation i.e. the allocation of teaching resources (tutors/demonstrators/teaching assistants) to teaching tasks (tutorials). The tutors have various skills that make them more or less appropriate for different tutorials. For example, a tutor with excellent skill in C++ would be ideal for a C++ tutorial, acceptable for a Java tutorial (since there are similarities between C++ and Java) but of little use for a tutorial on natural language processing. The specific task chosen is the allocation of tutors to tutorials in the School of Artificial Intelligence at the University of Edinburgh, where 40 tutors must service 80 tutorial groups (each tutorial is repeated for different groups of students) that require 14 separate skills. The following entities and relationships were identified for this specific task:

- ***Tutors*** – defined by their knowledge, preferred time and number of tasks, rate per hour and availability.
- *Knowledge (1-5)* – the person's overall knowledge or expertise for a particular skill. Each tutor has a "knowledge score" for each skill of 1-5, corresponding to *aware, familiar, skilled, expert* and *guru*;
- *Rate (integer)* – the hourly rate is the employee's/tutors minimum wages per job. More experienced tutors earn higher rates;
- *Preferred time (morning/afternoon)* – this defines the preferred time of the day that the employee wants his/her tutorial sessions;
- *Preferred number of tasks (1-5)* – a tutor can define a preferred number of tasks/tutorials that he wants in a weekly schedule. The default is 2.
- *Availability (pair of values – day, morning/afternoon)* – specifies those mornings and afternoons each week when a tutor is (un)available.

- ***Module / Tutorials*** – defined by skill requirements, tutorial slots and budget.
- *Skill requirement (0-5)* – defines the skills needed to tutor the particular module;
- *Slots (pair of values: day, morning/afternoon)* – defines the timetable slots available to tutorials;
- *Budget (integer)* – specifies the allocated salary budget per module.

2. Knowledge Representation

Tree Structure	Node Type	Level
◆ Programming	*Parent*	*1*
▪ AI_programming	*Parent*	*2*
• *Lisp*	*Leaf*	*3*
• *Prolog*		
▪ General_programming	*Parent*	*2*
• *C++*	*Leaf*	*3*
• *Java*		
◆ AI_techniques	*Parent*	*1*
▪ Knowledge representation	*Parent*	*2*
• *Fuzzy Logic*	*Leaf*	*3*
• *1ˢᵗ ord. Pred. Calculus*		
▪ Search	*Parent*	*2*
• *Evolutionary Comp.*		
• *Constraint Satis. Prob.*	*Leaf*	*3*
• *Planning*		
▪ Learning	*Parent*	*2*
• *Neural Nets*	*Leaf*	*3*
◆ Cognitive_science	*Parent*	*1*
▪ *Linguistics*		
▪ *Psychology*	*Leaf*	*2*
▪ *Human Computer Interac.*		
◆ Research Techniques	*Parent*	*1*
▪ *Experimental Methodology*	*Leaf*	*2*

Figure 1 - Ontology of skills.

Figure 1 illustrates the ontology of skills that was used in this work. Leaf nodes represent the skills available or required; parent nodes represent the technical area classifications. This ontology is specific to Artificial Intelligence topics that were taught at the University of Edinburgh in 1999/2000.

In order to represent the relationships between leaf nodes, a weight is allocated to each pairing. The weight represents the ability of a tutor to undertake different tasks even if the tutor does not possess the exact job requirements. The weights are allocated on a "family tree" basis: siblings (i.e. two skills with the same parent) are awarded a relative weight of 0.25; first cousins are awarded a relative weight of 0.5. For example, C++ and Java would have a relative weight of 0.25, while Java and Prolog would have a relative weight of 0.5.

Each tutor is assigned a skill level in one or more of these fourteen areas, and required skill levels are also assigned to taught modules. For example, the *Learning from Data* module requires basic knowledge (1) in *Java* and in *Evolutionary Computation* but very good knowledge (4) in *Neural Nets*. The advantages of this ontology-based approach are that it allows the possibility of graded partial matching (which is essential in a resource-constrained task) and that the hierarchical tree representation gives the user a good visualisation of the knowledge. For customisability, the ontology was used to generate a matrix of weights that can be edited by the user, thus allowing scope for preferences or exceptions.

3. Evolutionary Parameters

Algorithm: The main algorithm used is a variant of a standard genetic algorithm with tournament selection and one-point crossover. A novel aspect of this work was the combination of a duplicate elimination algorithm with a steady state GA; for details of other combinations that were considered, see [1]. A standard genetic algorithm works by generating new "chromosomes" (solutions) by mutating or combining existing chromosomes, adding these offspring to the original population, and then sorting the population according to a fitness function and removing the worst chromosomes to keep the population size constant. The steady state duplicate elimination (SSDE) algorithm [2] used on this project uses this method with one addition: at the point at which the worst chromosomes are removed from the population, duplicate chromosomes are also removed.

Encoding: Direct encoding has been chosen to represent the schedule in a chromosome [3, 4]. Each gene represents a possible assignment of a tutor to a tutorial group. The length of the chromosome is equal to the total number of tutorial groups. Tutors can be assigned to multiple tutorial groups, but tutorial groups can have only one tutor.

Selection mechanisms: Tournament [5] and rank based selection mechanisms have been used to select individuals for the genetic operations. A tournament size of $k = 5$ was employed; this is a rank based selection in which chromosomes in the population are sorted according to their fitness values, and then the first k chromosomes are selected for the best chromosome crossover method (see below).

Mutation: The intuition behind the two mutation operators (swap and random) is to introduce some extra variability into the population. A swap mutation [6] swaps two tutor allocations from randomly selected genes in the chromosome; this is the primary method used to introduce variability. Random mutation changes a randomly selected gene value (i.e. an allocated tutor) to another possible value (i.e. another tutor); this method generally has an adverse effect on fitness due to poor 'distribution' (some tutors now being over- or under-worked), but it can introduce 'lost' gene values back into the chromosome, thus helping to prevent premature convergence. A combination of these two methods shows better results than using them individually.

Crossover: The crossover operator is one of the most important genetic functions in evolutionary algorithms. This is a process where chromosomes exchange segments via recombination. We have investigated a standard one point crossover method and a "eugenic" best chromosome crossover method, in which we force a certain percentage of offspring to be generated from the chromosomes with the highest fitness values.

Fitness: The user is allowed to specify the relative importance of three fitness constraints: skill, finance (i.e. staying within budget) and distribution. The fitness function of a schedule is computed by assigning penalty points for the violation of each constraint. The current fitness function awards a low number of penalty points if tutors are overqualified for the skills required for a tutorial; a higher number of penalty points if tutors are have the right skills but are under qualified; and if a tutor has no skill in an area but does have a related skill, then a penalty point value is assigned that is modified by the ontological weight matrix. Penalty points are also assigned for going over the allocated budget; for tutors being allocated to a tutorial group at a non-preferred time; and for each tutor not given any tasks. Finally, a very high number of penalty points are awarded to any job conflicts that arise where a tutor is allocated to two tutorials simultaneously.

4. Results

Based on the School of AI example, our results show that the population size is directly proportional to the fitness value; a larger population size generates better schedules. However, increasing population size also produces a rapid increase in computation time.

The steady state duplicate elimination algorithm has tremendous benefits for this problem. As shown in Figure 2, the standard steady state algorithm produced a lot of duplicates and it prematurely converged into one solution. The SSDE algorithm, however, produced multiple solutions with better fitness values. The drawbacks of the SSDE technique are the overhead involved in carrying out a large number of equality tests whenever an offspring is created, and the reduced probability of reproduction of the best chromosomes due to duplicate elimination. The latter problem is addressed by use of the best chromosome crossover technique; after experiments with 5% and 10% best chromosome crossover, it was determined that forced generation of 10% of offspring from the best chromosomes gave the earliest convergence.

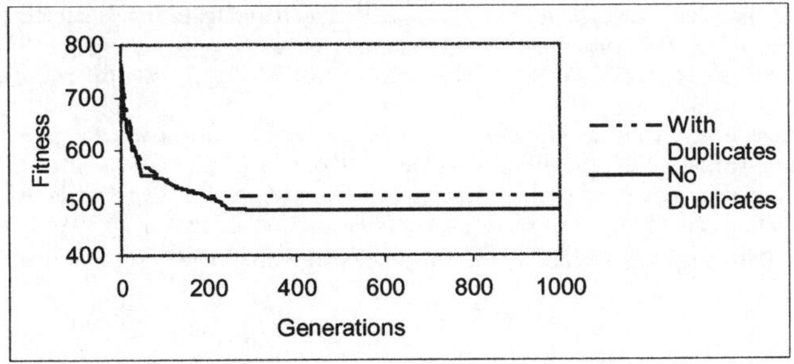

Figure 2 - Standard steady state Vs Steady state without duplicates

In order to compare the quality of the GA produced schedule, we solved a smaller version of the problem by hand. We employed several heuristic methods to allocate tutors to tutorial groups. Initially we allocated the most specialised subjects followed by others. In the assignment of tutors, the most important factor was to match skill requirements. The distribution of the tutors was the next factor considered; all the tutors were assigned to two tutorial groups in the same module. The finance factor was ignored to save time. Even after checking for timetable conflicts, it took 35 minutes to fulfil all the tutorial slots.

Table 1 illustrates the fitness values of the hand generated and GA produced schedules. It shows that individual penalty constraints are lower in GA schedule than in hand implemented schedule. Therefore, it proves that the GA performs much better than the hand produced allocation for the same elapsed time.

	Hand Created	GA Generated
Skill Fitness	**589.0**	**499.5**
Distribution Fitness	**0.0**	**38.0**
Finance Fitness	**24.0**	**22.5**
Pref. Time	**29.3**	**17.3**
No. Tutors Not allocated	**0**	**0**
No. Job Conflicts	14	0
Overall Fitness	**3778.3**	**577.3**
Total Time (Min)	**35**	**35**

Table 1 - Comparison between hand generated and GA produced schedules

5. Discussion

This approach seems promising for the general task of allocating tasks to resources. One possible improvement would be the use of multi-perspective [7] ontologies; for example, considering the similarity between skills based on a "how" criterion (i.e. skill A *is required for* skill B) rather than a "what" criterion (skill A *is in the same category as* skill B). Applying this approach to our domain, Prolog would be required for First Order Predicate Logic, while C++ might be considered a distant relative of Evolutionary Computation. Other ontologies are also possible, and an accurate relative weighting matrix should draw on several ontologies to derive its weights.

References

[1] Hsiao-Lan Fang. *Genetic Algorithms in Timetabling and scheduling.* (Ph.D. thesis) Department of Artificial Intelligence, University of Edinburgh, 1994

[2] Darrell Whitley. *GENITOR: a different genetic algorithm.* In Proceedings of the Rocky Mountain conference on Artificial Intelligence. Denver, Colorado.

[3] M Gröbner, P Wilke. *Optimizing Employee Schedules by a Hybrid Genetic Algorithm.* E.J.W Boers et al. (Eds.) EvoWorkshop 2001, LNCS 2037, pp. 463-472. Springer-Verlag, Berlin, Heidelberg.

[4] C. Di Stefano, A. G. B. Tettamanzi. *An Evolutionary Algorithm for solving the School Timetabling Problem.* E.J.W Boers et al. (Eds.) EvoWorkshop 2001, LNCS 2037, pp. 452-462. Springer-Verlag, Berlin, Heidelberg.

[5] Zbigniew Michalewicz. *Genetic Algorithms + Data Structures = Evolution Programs.* 3rd revised and extended edition. Springer-Verlag, Berlin, Heidelberg, 1996. ISBN 3-540-60676-9.

[6] Hugh M Cartwright. *Getting the Timing Right – The Use of Genetic Algorithms in Scheduling.* Proc. of Adaptive computing and information processing conference. Brunel, London. pp 393-411 (UNISYS 1994)

[7] Kingston J., *Multi-perspective modelling: A Framework for Knowledge Representation.* Forthcoming.

KES 2002
E. Damiani et al. (Eds.)
IOS Press, 2002

Re-usable Knowledge: Development of an Object Oriented Industrial KBS and a Collaborative Domain Ontology

Paul Crowther[1], Gerd Berner[2] and Raymond Williams[2]

[1]*School of Computing and Management Sciences, Sheffield Hallam University, Howard St, Sheffield S1 1WB, UK.*
e-mail: P.Crowther@shu.ac.uk
[2]*School of Computing, University of Tasmania, PO Box 1214 Launceston, Tasmania 7250, Australia.*
e-mail: R.Williams@utas.edu.au

Abstract. Since the early eighties, there has been a gradual shift in the focus of development of knowledge based systems away from the rapid prototyping techniques that had previously prevailed, toward more structured methodologies, including model based reasoning and modeling of knowledge domains. the default standard for the development of these systems has become the CommonKADS methodology. This paper will assess the feasibility of applying the CommonKADS methodology to the knowledge base of an existing legacy system, the subsequent re-useability of knowledge and domain schema that results from the process and the development of collaborative domain ontologies. Currently the CommonKADS methodology embraces reuse by providing standard inferencing primitives and a set of generic task models. The resulting model set is assessed to determine its suitability to form the basis of an ontology for electroplating in the manufacturing industry.

1. Introduction

This study investigates the benefits and feasibility of applying the CommonKADS methodology to an existing knowledge base which was developed using rapid prototyping without a formal methodology for structuring the knowledge base. Knowledge bases of this nature are described by Wielinga and Schreiber [18] as often being idiosyncratic and having little or no generality.

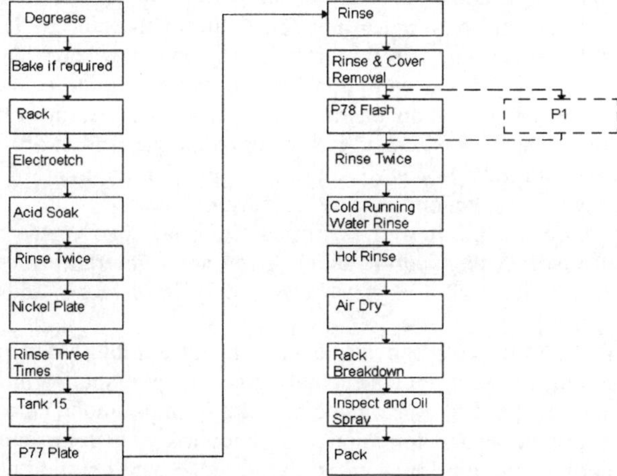

Figure 1. The general arrangement of the ACL Bearing Overlay Plating Section. ACLE Mark 1 was based on the P78 Flash tank process

The problem domain for this study was the Overlay Plating Section (figure 1) of ACL Bearing Company, located at Rocherlea in Northern Tasmania. ACL Bearings is a highly specialised manufacturer which recognises they are susceptible to loss of organisational expert knowledge due to domain experts leaving. A first generation knowledge system, referred to here as ACLE Mark 1, was designed and constructed using rapid prototyping, and was successfully completed to specification but without any facility or structure being implemented for reuse of the domain knowledge. The system incorporated a set of production rules that chain together to form a series of categorised binary trees. At the root of each tree is the general category of problem which is selected by the user of the system upon beginning a consultation.

ACLE Mark 1 operationalises the somewhat outdated approach, typical of rule based systems, whereby the original notion of task comprised both the task specification and the knowledge base. Effectively, the problem solving method is fixed and forms an integral part of the implementation [12]. ACLE Mark 1 could therefore be described as having no function-data decoupling [14]. The system also fails to meet the guidelines set out by Moller [11] with regard to declarative, modular knowledge bases and easy explanations of the knowledge stored within the system. It is in fact doubtful as to whether the original ACLE Mark 1 system can be classified as an expert system, when considering a common definition of expert systems [4].

2. Theoretical Background

KADS was initiated as a "Structured methodology for the development of knowledge based systems"[15]. The limitations of production rules, combined with their inherent non-reusability contributed significantly to the impetus to develop methodologies like KADS. The two central principles that underlie the KADS approach are the introduction of multiple models as a means of coping with the complexity of the knowledge engineering process, and the use of knowledge-level descriptions as an intermediate model between expertise data and system design.

Motta [12] coins the term "knowledge modeling revolution", which refers to the paradigm switch from symbol level (rule based) approaches to knowledge level task centred analysis. This heralded the necessary decoupling of the task specification and the problem solving method.

There has also been the development of methodologies for the re-use of components of knowledge models, which allow for more efficient development of subsequent systems using reusable libraries of domain ontologies. For this to occur there needs to be standardisation of defining complex domains from the full spectrum of human knowledge [2].

Traditionally an ontology is a view of what can possibly exist, concerned with possibility, necessity and contingency, but on an abstract level, rather than dealing with rules and constraints [16]. This definition has been extended to accommodate the field of artificial intelligence, such that an AI ontology is a theory of what entities can exist in the mind of a knowledgable agent [18]. An agent may be a human operator, or a software system.

A knowledge base can be viewed as a model of some part of the world, that allows for reasoning to take place in that world model, given some inference mechanism. An ontology defines the constraints of possible objects expressed in the model in addition to the constraints imposed by the syntax [18].

In order to make statements and ask queries about a subject domain, Farquhar et al [3] explain that an ontology must use a conceptualisation of that domain which names and describes the entities that may exist in that domain and the relationships among them. It therefore provides a vocabulary for representing and communicating knowledge about the domain.

Using ontologies in the development of a knowledge based system allows for a disciplined design to be carried out and facilitates sharing and reuse [5] [17]. In the construction of an ontological model, consideration should be given to adherence to the 5 design criteria, as described by Gruber [7]. The criteria include "clarity, coherence, extendibility, minimal encoding bias and minimal ontological commitment".

3. Methodology

3.1 Knowledge Reuse Strategy

The idea of knowledge reuse is central to the principles of MIKE (Model based and Incremental Knowledge Engineering), a paradigm that recognises the cyclic nature of the knowledge system lifecycle and encompasses the CommonKADS methodology [13]. This is in keeping with Fensel's [4] description of knowledge acquisition as an iterative process that is infinite and approximate. One of the principle goals of the process of the CommonKADS modeling is to provide a catalogue of artefacts that can be reused within the current application, or in others.

3.2 Construction of CommonKADS Models

By carrying out an examination of the categories of knowledge and the decision trees held in the ACLE Mark 1 system, it was possible to identify the individual physical objects. These objects were conceptualised in the CommonKADS model set [14]. Once the complete set of objects in the domain were identified, they were defined to the required level of abstraction by the addition of schema. Schema definition for physical objects in the domain was carried out by examining each internal node in the tree structure of the ACLE Mark 1 system.

The knowledge base was constructed by instantiating each leaf node into a diagnosis (problem defect object) and, from each leaf node, backtracking to the root. Each internal node that is encountered on the path to the root node is included in the set of antecedents that apply to that particular diagnosis node (the root node is also included). The diagnosis stores the question text, the required value and the domain object to which the question refers. In addition to this, each classification of problem or defect was included as a concept [14].

3.3 The Clustering Exercise

The clustering of objects in the Overlay Plating Domain was carried out using a Repertory Grid and Johnson Hierarchical Clustering process [1]. The results were that the P1 flash plate section clustered closely with the P78 flash plate section. When considering the close relationship that these two objects share (they are a mutually exclusive component of the system but occupy the same physical space and interact with the same agents and machinery), this is an expected result. The clustering also shows that no other objects in the domain share as much similarity with the P1 flash plate section. This led to the conclusion that the reuse knowledge candidates for schema and domain knowledge that would be proposed to the domain expert should come exclusively from those that apply to the P78 flash plate section.

Although this was not an unexpected result, the exercise was very important, in that there may have been considerations in choosing the reuse candidates that only the domain expert could make. An example of this is that excluding the P78 Flash Plate, the next most likely objects (to the lay person) that might cluster closely to the P1 Flash Plate would be the other plating baths, the Nickel plating Bath and the P77 plating bath. The results of the formalised clustering exercise show clearly that the differences are much more significant than a casual knowledge of the domain might suggest [13].

3.4 The Reuse Questionnaire

The results from the clustering exercise formed the basis of choosing reuse candidates. All diagnoses that had antecedents or diagnoses that referred to the P78 Flash Plate section were selected and placed in a reuse questionnaire for consideration by the domain expert. The P78

Flash Plate schema was also included for consideration as to its applicability to the P1 Flash Plate Section.

The domain expert was encouraged to consider the reuse candidate diagnosis with regard to potential reuse in the P1 Flash Plate Section, The domain expert was then requested to give a rating of applicability of this diagnosis to the P1 bath (a rating between 1 and 5 where 5 is a perfect match with no changes necessary). Reuse candidates were classified as, full matches, partial matches adapted, partial matches scrapped, outright misses. Hutchinson and Hindley [8] specifically encourage the approach of adapting partial matches for the purpose of reuse of domain knowledge [13].

4. Results and Discussion

4.1 ACLE Mark 2 and CommonKADS

The decision to build a system that embodied the remodeled system and knowledge base, including the reused knowledge as applies to the P1 flash plate bath is supported by Fensel [4], in that it allows evaluation of the resulting knowledge. The ACLE Mark 2 system is effectively a fully instantiated model of a knowledge domain, with sufficient information to perform a reasoning process, with a problem solving strategy that has been decomposed to an atomic level [18]. The ACLE Mark 2 system is also consistent with the 5 design criteria for knowledge models, described by Gruber [7]: clarity, coherence, extendibility, minimal encoding bias and is burdened with minimal ontological commitment.

The structure that has been used in the design and construction of the ACLE Mark 2 system is in fact a "reusable abstract domain-independent problem solving strategy" [10], which is the essence of the CommonKADS methodology.

4.2 Reuse of Knowledge

The results showed that 56% of the reuse candidate diagnoses were directly applicable to P1 process without modification. A further 25% of the reuse candidates could also be applied to P1 with a minor modification such as dropping or adding an antecedent, or modifying the wording of a diagnosis slightly. The final 19% of reuse candidates were totally non-applicable to the P1 plating tank.

Of the total of 32 schema attributes that are used to describe the P1 flash plate section, 56% were derived from the reuse experiment, and could be applied directly without modification. There was 1 schema attribute that required some modification, but was still able to be included in the P1 flash plate schema. In addition to this, there were an additional 3 schema attributes derived from the reuse questionnaires, in the form of a required additional antecedent. This is an extremely significant ratio of reuse, with a total of 68% of the total domain schema for the P1 flash plate section being supplied by the reuse exercise.

4.3 A Formal Ontology for Electroplating

Some shortcomings of the ACLE Mark 2 knowledge base would currently prevent it from being used as the basis for standardised domain ontology, but with the investment of some time by a knowledge engineer and the domain expert, the potential is certainly there. The principle deficiency that should be addressed prior to any ontology being released is the inconsistent 'grain size' of attributes. Some have been decomposed sufficiently, and others have not [14]. There are also the confidentiality issues to be addressed, with respect to the release of information that is proprietary to ACL Bearings.

5. Conclusion

This research has demonstrated that a legacy system, developed using rapid prototyping and built from a foundation of no formal knowledge modeling can still form the basis for a reuse strategy centred on. the CommonKADS methodology. This being so, it follows that legacy systems developed using more rigorous decomposition and modeling techniques, but still employing rapid prototyping, would be more amenable to this technique. The capacity for reuse of knowledge that is provided by CommonKADS, such that new objects that are added to an existing knowledge domain might be modeled and added to the knowledge system with a minimum of effort and expense, has been demonstrated.

The results of applying these techniques to the ACLE Mark 1 system proved very successful. There was also a large number of candidates that were reused with minor changes. This is a very significant result when considering that the documented knowledge base that existed before the reuse exercise represents only 25% of the total knowledge base after the reuse exercise.

These results should be tempered with the knowledge that the P78 and P1 baths are mutually exclusive in the system but occupy the same space and perform very similar functions. This should not discourage the practitioner of knowledge reuse, as many domains in all aspects of human endeavour, particularly manufacturing, embrace the principle of redundancy, whereby a knowledge domain may contain numerous objects of similar characteristics [9] [6].

References

[1] Borgatti, S. P., 1994, *How to Explain Hierarchical Clustering.* http://analytictech.com/networks/hiclus.html.

[2] Doyle, J. and Dean, T., 1997, Strategic Directions in Artificial Intelligence. *AI Magazine*, V18(1), pp. 87-101.

[3] Farquhar, A., Fikes, R. and Rice, J., 1996, *The Ontolingua Server: a Tool for Collaborative Ontology Construction.* Standford Knowledge Systems Laboratory. http://www-ksl-svc.stanford.edu:5915/doc/project-papers.html

[4] Fensel, D., 1995, *The Knowledge Acquisition and Representation Language, KARL*, London: Kluwer Academic Publishers.

[5] Fridman Noy, N. and Hafner, C. D., 1997, The State of the Art in Ontology Design: A Survey and Comparative Review. *AI Magazine* V18(3), pp53 - 74.

[6] Gallagher, S., 1999, *Scalability Through Redundancy*, Information Week, CMP Publications inc.

[7] Gruber, T.R., 1993, *Toward Principles for the Design of Ontologies Used for Knowledge Sharing.* Standford Knowledge Systems Laboratory. Available: http://www-ksl-svc.stanford.edu:5915/doc/project-papers.html

[8] Hutchinson, J. W. and Hindley, P. G., 1988, A Preliminary Study of Large Scale Software Reuse. *Software Engineering Journal*, V3(5), pp. 208 - 212.

[9] Lawley, M., Reveliotis, S., Ferreira, P., 1997, *Correctness and Scalability*, Institute of Industrial Engineers.

[10] Menzies, T., 1995, *Limits to Knowledge Level-B Modeling (and KADS).* http://www.cse.unsw.edu.au/~timm/pub/docs/papersonly.html

[11] Moller, J., 1995, *Operationalisation of KADS models by using Conceptual Graph Modules.* [Online] [2000, June 27]http://nats-www.informatik.uni-hamburg.de/~jum/papers/95/CG-KADS.html

[12] Motta, E., 1997, Trends in Knowledge Modeling: Report on the 7th KEML Workshop. *The Knowledge Engineering Review*, V12(2), pp. 209-21

[13] Pressman, R. S., 1997, *Software Engineering: A practitioner's approach*, The McGraw-Hill Companies, USA.

[14] Schreiber, G., Akkermans, H., Anjewierden, A., de Hoog, R., Shadbolt, N., Van de Velde, W. and Wielinga, B, 2000, *Knowledge Engineering and Management, The CommonKADS Methodology*, Cambridge: MIT Press.

[15] Schreiber, G., Wielinga, B. and Breuker, J., 1993, KADS: *A Principled Approach to Knowledge-Based System Development*, London: Academic Press.

[16] Thro, E., 1991, *The Artificial Intelligence Dictionary*, Microtrend Books, California, USA.

[17] Van der Vet, P. E. and Mars, N. J. I., 1998, Bottom-Up Construction of Ontologies. *IEEE Transactions on Knowledge and Data Engineering*, V10(4), pp. 513 - 526.

[18] Wielinga, B. J. and Schreiber, A. 1993. *Reusable and Sharable Knowledge Bases: A European Perspective.* Proceedings of the International Conference on Building and Sharing of Very Large-Scaled Knowledge Bases '93.

KES 2002
E. Damiani et al. (Eds.)
IOS Press, 2002

Ontology-based Communication Forum

Marian Mach, Peter Macej, Jan Hreno
Technical University, Letna 9, 041 20 Kosice, Slovakia

Abstract. The paper presents a web-based communication forum for exchanging and delivering information. It provides one-way and two-way communication modes, including information publication, contributing to discussions, and opinion polling. In order to organize information available within the system, it uses knowledge modeling and document annotation techniques.

1. Introduction

In many applications information is stored somewhere without a possibility to retrieve and reuse it at a later stage. The aim of the Webocrat system is to empower users with innovative communication mechanism supporting information sharing. This system will support discussion, Internet publishing, browsing and navigation, opinion polling on questions of interest, intelligent retrieval, calculation of summary statistics, and user-friendly access to information in a customized way.

The Webocrat system applies a knowledge-based approach [1]. Information of all kinds produced by various modules is linked to a shared ontology representing an application domain. Such an ontology serves as a means for structuring and organizing available information resulting in improved search capability and contents presentation.

2. Domain Modeling for Document Annotation

To provide a system with intelligence, the system must be able to interpret meaning of information it manipulates with - i.e. some kind of semantic representation is needed. One solution is to annotate documents [2], i.e. an explicit information about the meaning is attached to the documents (either manually or (semi-)automatically). The annotation enables then an intelligent retrieval, giving more relevant results than a pure full-text search. Annotation can be of two basic types: description in natural language and a list of links to a predefined vocabulary. In the former case the full-text search provides more precise results. The latter case assumes existence of some vocabulary of terms or concepts used in the domain of interest.

Using a vocabulary in the form of a list of terms (each term having its own description) has a disadvantage – it does not reflect existing relations among terms. One possible solution is to use an ontology – 'vocabulary' with its own internal structure [3]. Ontology is typically domain specific – it can determine what 'exists' in the domain. Thus, it represents a model of the domain.

Use of an ontology enables to define concepts and relations representing knowledge about a particular document in domain specific terms. In order to express the contents of a document explicitly, it is necessary to create links between the document and relevant parts of a domain model, i.e. links to those elements of the domain model, which are relevant to the contents of the document.

Model elements can be also used for search and retrieval of relevant documents. In case all documents are linked to the same domain model, it is possible to calculate a similarity between documents using the structure of this domain model. Such approach supports also 'soft' techniques, where a search engine can utilize the domain model to find concepts related to those specified by user. The search engine can thus return every document linked to the concepts, which are close enough to the concepts mentioned in the user's query.

3. System Functional Overview

From the functionality point of view, it is possible to break down the system into several parts and/or modules [4]. They can be represented in a layered sandwich-like structure.

The central position of this structure is occupied by a Knowledge Model module (KM). This system component contains one or more ontological domain models providing a conceptual model of a domain. The purpose of this component is to index all information stored within the system in order to describe the content of this information (in terms of domain specific concepts). The central position symbolizes that the knowledge model is the core (heart) of the system – all parts of the system use this module in order to organize information in the system and to access it.

Information stored within the system has the form of documents of different types. Since three main types of documents are expected to be processed by the system, a document space can be divided into three subspaces – publishing space, discussion space, and opinion polling space. These areas contain published documents expected to be read by users, users' contributions to discussions on different topics of interest, and records of users' opinions about different issues, respectively.

Documents stored in these three document subspaces can be inter-connected with hyper-textual links. They can contain links to other documents – to documents stored in the same subspace, to documents located in another subspace, and/or to documents from outside of the system. Thus, documents within the system are organized using net-like structure. Moreover, documents located in these subspaces should contain links to elements of a domain model.

Since each document subspace expects different way of manipulating with documents, three modules are dedicated to them. Web Content Management module (WCM) offers means to manage the publishing space. It enables to prepare documents in order to be published (e.g. to link them to elements of a domain model), to publish them, and to access them after they are published. Discussion space is managed by Discussion Forum module (DF). The module enables users to contribute to discussions they are interested in and/or to read contributions submitted by other users. Opinion Polling Room module (OPR) represents a tool for performing opinion polling on different topics. Users can express their opinions in the form of polling – selecting those alternatives they prefer.

In order to navigate among information stored in the system in an easy and effective way, one more layer has been added to the system. This layer is represented by two modules, each enabling easy access to the stored information in a different way. Citizens' Information Helpdesk module (CIH) is dedicated to search. It represents a search engine based on the indexing of stored documents. Its purpose is to find all those documents which match user's requirements expressed in the form of a query.

The other module performing information retrieval is the Reporter module (REP). This module is dedicated to providing information of two types. The first type represents information in an aggregated form. It enables to define and generate different reports concerning information stored in the system. The other type is focused on providing particular documents – but unlike the CIH it is focused on off-line mode of operation. It monitors content of the document space on behalf of the user and if information the user may be interested in appears in the system, it sends an alert to him/her.

The upper layer of the presented functional structure of the system is represented by a user interface. It integrates functionality of all the modules accessible to a particular user into one coherent portal to the system and provides access to all functions of the system available to a particular user in a uniform way.

In order for the system to be able to provide required functionality in a real setting, several security issues must be solved. This is the aim of the Communication, Security, Authentication and Privacy module (CSAP).

4. Using Domain Model in Webocrat

The main idea behind the whole Webocrat system is to associate information with a part of a domain model – ontology. In this way it is possible to annotate discussions, reports, polling or ordinary WWW pages (all those documents that are published, such as news, announcements, memos and other documents that could be interesting for users). When pieces of information are submitted, they are annotated first, whether manually or semi-automatically. After that they are prepared for intelligent retrieval. When accessing information, user can define his/her query consisting of words for full-text search and of terms (concepts) used in ontology. Using concepts ensures that also hidden meaning will be discovered. Formulation of such query also allows users to define their personal profiles of interest in terms of ontology elements. Then personalized reports and newsletters can be automatically generated and sent to users.

Knowledge about semantics of documents can play more active role during communication as well. Discussions are typical examples in the Webocrat system. A discussion is considered as a thread of documents that are annotated. In order to enable retrieval of discussion contributions according to their content, it is necessary to create links to elements of a domain model when creating a new discussion. These elements will represent topics on which the discussion will be focused. Each contribution, which will be added to this discussion later, will be linked to the same elements from the domain model in an automatic way (contributions inherit links from the discussion they belong to).

In order to enable organizing contributions within the discussion not only according to date and time of submissions or author(s), it is possible to complete the contributions with a set of links. These links can be of two types – links to elements of a domain model and links to other contributions from within the discussion. The former type of links enables to define the content in more detail (not only in the sense that the contribution is about exactly the same issues as the discussion as a whole) – this includes both addition of some more links to the set of links inherited from the discussion definition as well as reduction of this inherited set as well. The latter type of links enables user to determine to which existing contribution(s) he/she responds. In addition, it is possible to enrich a contribution to some discussion with links to documents from inside or outside of the system, e.g. in case when users (submitters) refer in their contributions to those documents.

In order to read particular contributions, it is necessary to access them. User has several possibilities how to complete this task. First of all, he/she can choose from a list of all available discussions. Another alternative way is to use linking of contributions to elements of a domain model in order to create groups of contributions dealing with the same set of issues.

Using links to ontology, system can suggest a discussion on some topic when user reads document on that topic. Or when user contributes to some discussion, system can advise where to find more relevant information. It would be impossible without links to domain model. Even more, when user links his/her contribution to some concepts, overriding linkage of whole discussion (those inherited from the definition of the discussion), system can automatically find more relevant discussion, if such discussion exists, and suggest it. Similarly, if some contributions get more and more different from topics of the original thread, discussion maintainer can be notified to split the discussion. The similarity of contributions is measured using distances of corresponding concepts in the ontology.

5. Coarse System Architecture

Webocrat is being implemented using a client-server architecture. Three types of clients are expected to play a role in the project – Web based clients, mail readers, and standalone applications. Mail clients are specialized on user alerting on events which could be of interest for those users.

Standalone clients represent specialized clients – each of them focuses on a small set of functions provided by the system. Users working with these clients take special positions in maintaining the system and the information stored within it. Examples of such clients are ontology editor for building and maintaining a domain model and annotator for enriching documents by a set of links to elements from the domain model. In addition to that, this type of client can represent an interface to an existing information system – for example some external system can retrieve documents which reside within the Webocrat system.

The most common type of client is a web client. It can be either a standard Web browser or any software which uses the HTTP protocol to communicate with the Webocrat system. Web client is dedicated to main functions offered by the system – accessing information published in the publishing space, sharing ideas and opinions in various discussions, and providing opinions on different issues by participating in some opinion polling.

The server part of the Webocrat system consists of two relatively separate parts splitting the system into two separate servers – Webocrat Information Server and Webocrat Ontology Server. This division of responsibilities for domain models and for documents containing information of interest enables easier maintenance and better utilization of available resources.

6. Related Work

Majority of systems and approaches using ontologies as a platform for semantic description of presented information is focused on web publishing. A rich set of tools for community portals is presented in [5]. Supplied information can be annotated manually using an HTML-A annotation tool or annotation can be automated when one finds regularities in a large number of documents. Querying capabilities are extended using an inference engine enabling to derive new knowledge by a combination of facts and ontologies.

An approach how to build a structured index of a web site using terminology oriented ontology in a semi-automatic way is presented in [6]. The approach is based on employing natural language techniques in connection with the Wordnet thesaurus. Semiautomatic annotation procedure driven by predefined templates based on the typology of events is presented in [7]. It is combined with push technology using user profiles.

OntoWebber [8] uses domain ontology as a reference ontology for mapping source ontologies to it in order to integrate heterogeneous data sources. It focuses not only on the content of a site but the domain ontology together with other ontologies (e.g. for navigation, presentation, content, personalization) enable also to structure, create and generate the site.

The authors are not aware of a tool annotating discussions directly. In [9] each document defining an example problem is complemented with a discussion about the example – but they are annotated as one unit.

Comparing with our approach, the Webocrat system still lacks some features, providing by above mentioned systems dedicated to publishing. On the other hand, our approach tries to provide several different communication channels not only this one.

7. Conclusions

The aim of this paper is to present the Webocrat system. The focus is on using knowledge models to annotate and organize information within the system in order to retrieve this information according to the content.

Currently, the Webocrat system does not exist as a whole – only in the form of several units to be integrated later into one coherent package. At the time of writing this paper, the project is in the phase of testing of some implemented modules (WCM, DF, OPR), implementing some modules (KM, CSAP), and analyzing and designing the other modules (CIH, REP). It is expected to be finished and launched into a routine operation at the end of 2003.

Acknowledgements

This work is done within the Webocracy project, which is supported by European Commission DG INFSO under the IST program, contract no. IST-1999-20364, and within the VEGA project 1/8131/01 "Knowledge Technologies for Information Acquisition and Retrieval" of Scientific Grant Agency of Ministry of Education of the Slovak Republic.

References

[1]   M. Dzbor, J. Paralic and M. Paralic, Knowledge management in a distributed organization. In: Proc. of the 4th IEEE/IFIP International Conference BASYS'2000, Kluwer Academic Publishers, London, 2000, pp. 339-348.

[2]   J. Hreno, Automatic Document Abstract Creation. In: Proc. of the 11th Int. Conference on Information and Intelligent Systems iis2000, Varazdin, 2000.

[3]   B. Chandrasekaran *et al.*, What are Ontologies and Why Do We Need Them, *IEEE Intelligent Systems* 14 (1) (1999) 20-26.

[4]   M. Mach *et al.*, Webocrat System Architecture and Functionality. Webocracy Project Report R2.4, Technical University, Kosice, 2001.

[5]   S. Staab *et al.*, AI for the Web – Ontology-based Communication Web Portals. In: Proc of the 17th National Conference on Artificial Intelligence, AAAI Press, Menlo Park, 2000.

[6]   E. Desmontils and C. Jacquin, Indexing a Web Site with a Terminology Oriented Ontology. In: Proc. of The First Semantic Web Working Symposium, Stanford, 2001, pp. 549–565.

[7]   Y. Kalfoglou *et al.*, myPlanet: an ontology-driven Web-based personalized news service. In: Proc. of IJCAI 2001 workshop on Ontologies and Information Sharing, Seattle, USA, 2001.

[8]   Y. Jin *et al.*, OntoWebber: Model-driven Ontology-Based Web Site Management. In: Proc. of The First Semantic Web Working Symposium, Stanford, 2001, pp. 529–547.

[9]   Z. Zdrahal *et al.*, Sharing engineering design knowledge in a distributed environment, *Journal of Behaviour and Information Technology* 19 (3) (2000) 189–200.

KES 2002
E. Damiani et al. (Eds.)
IOS Press, 2002

Relational Text Mining and Visualization

Boris KOVALERCHUK

Dept. of Computer Science, Central Washington University, Ellensburg, WA 9892, USA
borisk@cwu.edu

Abstract. Discovering hidden patterns in distributed heterogeneous textual databases and unstructured data is a new challenge in data mining. Traditional data mining often assumes that preprocessing is already done -- homogeneous data are available on the needed level. For distributed heterogeneous textual data this is not the case. Complex relations between items/entities (e.g., relations between people in a fraud detection task) should be discovered and generalized. This paper offers a new hierarchical relational clustering method (Φ-method) based on the Φ-equivalence concept. The method permits to process: (i) incomplete relations: (ii) relations without converting them to attributes of individual entities, and (iii) relations presented in distributed heterogeneous databases using XML tags. Clustering produced by the method is invariant, has a clear meaning and natural visualization.

1. Introduction

A new challenge in Data Mining and Visualization (DMV) is developing techniques for discovering hidden patterns in ***distributed heterogeneous textual databases and unstructured data*** [1]. This problem includes two stages:
- *Preprocessing*: Knowledge Representation (KR) and Evidence Extraction (EE) and
- *Actual Discovery*: Link Discovery (LD) and Pattern Learning (PL).

Traditional data mining operates with a much simpler and smaller preprocessing stage. Often it starts with an assumption that preprocessing is already done -- homogeneous (numeric and nominal) data are available. That is data are uniformly organized in a single data table with rows representing cases (items) and columns representing their attributes. Also often, it is assumed that the data table already contains necessary domain knowledge implicitly. For distributed heterogeneous textual data, preprocessing is extremely complex and its result may not be a single homogeneous data table of attributes of individual items/entities. In many cases these data are relations between items/entities. For instance, in fraud detection it can be relations between people with the goal of discovering suspicious relations.

Intensive study on Evidence Extraction and Link Discovery (EELD) is currently underway with DARPA support [1]. This program reflects recommendations of the DARPA-sponsored Workshop on Knowledge Discovery, Data Mining, and Machine Learning [2] at Carnegie Mellon University: "Whereas conventional data mining and knowledge discovery is sometimes described as 'finding a needle in a haystack,' the system discussed at the Workshop must help analysts to 'reassemble needles from pieces that have been deliberately hidden in many haystacks.'" The workshop and EELD program identified parameters of this type of data mining. Below in table 1 we organized them into three categories: (1) general environment, (2) data and prior knowledge, and (3) patterns.

The current research challenge is developing a new class of data mining models and techniques that satisfy parameters described in Table 1.(based on [1]). In [1] such data mining models are called ***Models of Relational Data***. Another similar term for these models is ***Rela-***

tional Data Mining [3]. It was emphasized in [1,3] that these models are very different from traditional "Relational" Models associated with relational databases.

Table 1. Data Mining Parameters

<table>
<tr><td colspan="2">(1) General environment:
Uncircumscribed domain, Multiple simultaneous problems, High costs of failure (e.g., a missed terrorist event).</td></tr>
<tr><td>(2) Data and prior knowledge:
• A fraction of what could be known;
• Vast volume of potentially relevant data items;
• Few actually relevant data items;
• Large volume of explicit and implicit general and domain-specific knowledge;
• Most of data items are examples of normal (unsuspicious) patterns
• Heterogeneous data sources and data items.
• The available data only represent low-level objects and events.</td>
<td>(3) Patterns to be discovered:
• Highly structured (relational), dynamic and not exact;
• Can be deliberately obscured;
• Built on relations that include temporal and spatial relationships;
• Built on the low-level objects, which should be generalized from lower-level recorded data items;</td></tr>
</table>

2. Approach to learn structural (relational) patterns

The following categories and potential approaches for relational pattern learning are identified in [1]:

- *Relational classification*/clustering;
- Probabilistic relational models -- Extend Bayes nets to *richer relational settings*
- Mutual bootstrapping/*co-training;*
- *Stochastic relational learning techniques* (as enhancement of techniques such as Inductive Logic Programming, ILP);
- *Sequence* Discovery and Classification.

This paper focuses on relational classification/clustering. In fact, relational classification tasks have a long history in theoretical considerations, e.g., [4]. New data mining tasks in distributed heterogeneous textual databases and unstructured data foster a renewed and very practical interest in this subject.

The mentioned paper [4] uses model-theoretical knowledge representation [5] to define meaningful classification (clustering) in relational terms using the concept of Φ- equivalence (physical equivalence) [6]. One of the theorems proved in [4] states that every meaningful relational clustering should keep Φ-equivalent objects in the same class. It is proved that if a clustering algorithm assigns two Φ-equivalent objects a and b to different clusters then this algorithm will not be invariant even for renaming objects.

Informally two objects a and b are ***Φ-equivalent*** in a model <A,P> with objects A and predicate $P(x,y)$ if for every x not equal to a and b ($x{\neq}a$ and $x{\neq}b$) $P(x,a)=P(x,b)$, $P(a,x)=P(b,x)$, $P(a,a)=P(b,b)$ and $P(a,b)=P(b,a)$. However, Φ-equivalence *does not require* all properties of the standard equivalence. It is possible that $P(a,b){\neq}P(b,b)$ and $P(b,a){\neq}P(a,a)$. The standard equivalence of a and b requires $P(a,b)=P(b,b)$ and $P(b,a)=P(a,a)$. In the standard equivalence a and b are indistinguishable by predicate P. This difference shows that Φ-equivalence of a and b requires that a and b behave identically in their relation with all other elements of A. But a and b *may not be interchangeable in some of their relations to each other,* that is $P(b,a){\neq}P(a,a)$. In contrast the standard equivalence requires that $P(b,a)=P(a,a)$. For instance, it is possible that $P(a,a)$ and $P(b,b)$ are true, that is "a likes himself ", "b likes himself ", but $P(a,a)$ and $P(b,b)$ are false, that is "a hates b" and "b hates a". For Φ-equivalent objects it may not be possible to say which one we deal with by analyzing their behavior in predicate P. Below we provide an example of Φ-equivalence for a model with a predicate with two arguments. This example permits compare Φ-equivalence approach with *common directed-graph analysis approach* for clustering. Note that graph-based methods are designed to analyze a single predicate with two arguments. In contrast Φ-equivalence in defined for *any number predicates of arguments of predicates.*

3. Example of use of Φ-equivalence

Table 2 presents an extract for textual database with records such as "Mr. Smith hates Mr. Brown", "Ms. Smith likes Mr. Smith and hates every other person that hates Mr. Smith", and so on. Table 2 uses notation a,b,x,y,z,w for persons to store these relations. Table 2 encodes these sentences in a predicate form: $P(s,q)=1$ if s hates q and $P(s,q)=0$ if s likes q. For instance the bold cell can be read that "a hates b".

Table 2

	Entity a	Entity b	Entity x	Entity y	Entity z	Entity w
Entity a	0	**1**	0	0	0	0
Entity b	1	0	0	0	0	0
Entity x	0	0	0	0	0	0
Entity y	1	1	1	0	1	1
Entity z	1	1	1	1	0	1
Entity w	0	0	0	0	0	0

The first picture on Figure 1 presents predicate P in a directed graph form. Picture (2) in Figure 1 shows typical traditional generalization of objects to create larger objects. The most connected nodes are grouped together. It is hard to interpret this grouping.

Use of Φ-equivalence produces better interpretable clustering. This process has several steps.

Step1: Select a first pair of elements from table 2.

Result: Pair a and b is selected.

Step 2. Test pair a and b to be a Φ-equivalent pair, by applying the definition of Φ-equivalent pair.

Result: The pair (a,b) is a Φ-equivalent pair, because

$P(a,a)=P(b,b)=0; P(a,b)=P(b,a)=1;$

$P(a,x)=P(b,x)=0; P(a,y)=P(b,y)=0; P(a,z)=P(b,z)=0; P(a,w)=P(b,w)=0;$

$P(x,a)=P(x,b)=0; P(y,a)=P(y,b)=1; P(z,a)=P(z,b)=1; P(w,a)=P(w,b)=0.$

See informal and formal definitions in this paper.

Step 3. Combine a Φ-equivalent pair into a single node (cluster).

Result: Pair a and b is combined to a single node (see picture (3), Figure1).

Step 4: Interpret the combined node (cluster)

Result: This node/cluster has a clear meaning – a and b hate each other, but like all others and are liked by all others. We can call this node an "internally hating" but "externally loving" but node (HH-LL-node).

Comment: Elements a and b behave identically with respect to all others. This means that we really can combine them into a single node.

Step 5. Select another pair to test for Φ-equivalence.

Result: Pair a and w is selected.

Step 6. Test a and w to be Φ-equivalent.

Result: Test fails. There is an element b such that a and w behave differently with respect to b (a hates b, but w likes b, $P(a,b)=1$, $P(w,b)=0$).

Step 7. Looping steps 6 and 7 thorough all pairs of elements in table 2.

Result: Pair z and y is Φ-equivalent.

$P(z,y)=P(y,z)=1; P(z,w)=P(y,w)=1; P(w,z)=P(w,y)=0; P(z,x)=P(y,x)=1; P(x,z)=P(x,y)=0$

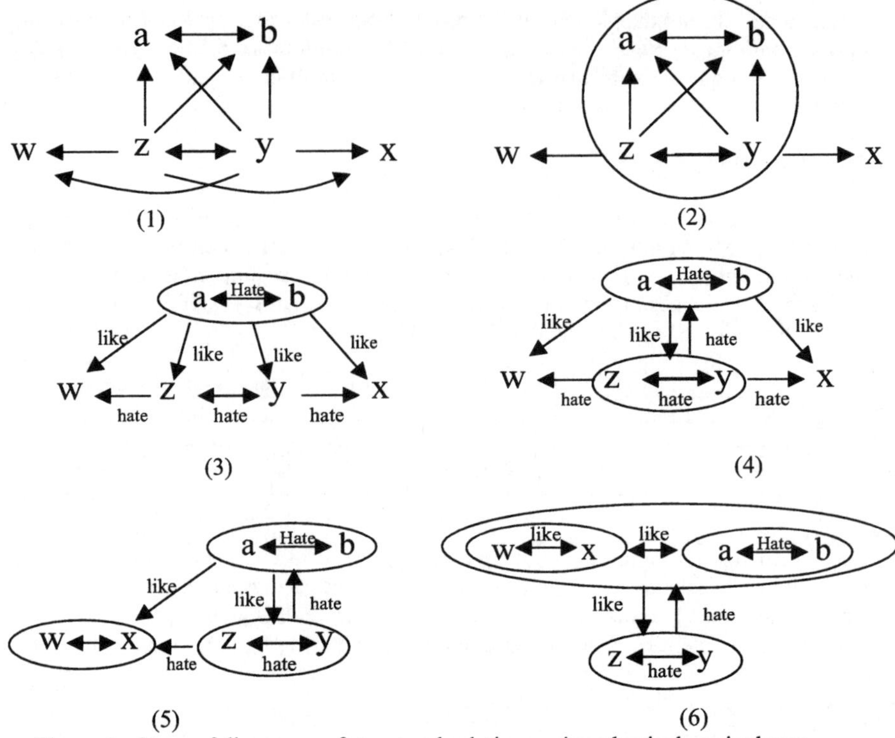

Figure 1. Steps of discovery of structural relations using physical equivalence

Step 8. Combine a Φ-equivalent pair to a single node (cluster).
Result: Pair z and y is combined to a single node (see picture (4), Figure1).
Step 9: Interpret the combined node (cluster)
Result: This node/cluster has a clear meaning – y and z hate each other, both hate all others, and all others like them, e.g., P(a,z)=0, P(b,y)=0. We can call this node an "internally hating" and "externally mixed" node (HH-HL-node).
Comment: Elements y and z behave identically with respect to all others. This means that we really can combine them to a single node.
Step 10. Eliminate redundant links between combined nodes.
Result: Only two links between two combined nodes are shown (see picture (4)) instead of showing four links between (a,y), (a,z), (b,y) and (b,z) (see picture (1)).
Comment: The structure has been simplified in a meaningful way.
Step11. Discover another Φ-equivalent pair.
Result: Pair w and x is discovered.
Step12. Combine a Φ-equivalent pair (w,x) into a single node (cluster).
Result: Pair w and x is combined into a single node (see pictures (5) and (6), Figure1).
Step 13: Interpret the combined node (cluster)
Result: This node/cluster has a clear meaning – both like each other and both like all others. We can call this node an "externally and internally loving " node.
Step 14. Eliminate redundant links between combined nodes.
Result: See (pictures (5) and (6), Figure 1).
Comment: The total structure has been simplified in a significant way. We have two types of nodes with their internal structure: "hating" and "loving". There is one node (z,y) that

hates two other nodes and that two other nodes like all other nodes. Thus, we can say that there are: *Two externally friendly nodes, (a,b) and (w,x); one hostile node (z,y) with mutual internal hatred inside; and friendly node (a,b) has a mutual internal hatred.*

Now we can compare this result with clustering based on finding most connected subgraph (see picture (2), Figure 1). Picture (5) provides meaningful information on data structure. Picture (2) missed discovery that there are two externally friendly nodes and only one externally hostile.

4. Definitions, Generalization, and Related Work

Definition. Elements a and b are $\boldsymbol{\Phi}$*-equivalent* (physically equivalent) if
$$\forall \ <x>_{vi} \ i \in J \ P_i(<x(a,b)>_{vi})=P_i(<x>_{vi}).$$
Here $\{P_i\}$ is a set of predicates, $i \in J$, $<x>_{vi}$ is a sequence of elements of set A of length v_i and $<x(a,b)>_{vi}$ is a modified sequence $<x>_{vi}$ with every occurrence of a is substituted by b and every occurrence of b is substituted by a.

We also introduce a weak Φ-equivalence (*VΦ-equivalence*) that indicates how close a pair to be Φ-equivalent. We count the number of violations $V(a,b)$ of Φ-equivalence for pair (a,b). In general terms, nodes a, b are Φ-equivalent if and only if $V(a,b)=0$. Thus t and s are 3Φ-equivalent if there are three violations, $V(t,s)=3$. We interpret $V\Phi$-equivalence as a kind of **distance** between nodes that can be processed by standard clustering algorithms.

After all Φ-equivalent items are clustered we may want to go to the next level of clustering (generalization) and discover patterns on the next level. For instance, we can join combined nodes (w,x) and (a,b). Both nodes have the same type of relations with node (z,y), that is "like-hate" (see picture (5), Figure 1). Formally, this procedure means that we applied Φ-equivalence to combined nodes and found that nodes (a,b) and (w,x) are Φ-equivalent.

The discovered structure can be written as follows: *one complex externally friendly node, ((a,b), (w,x)); one externally hostile node (z,y) with a mutual hatred internal, and externally friendly node ((a,b), (w,x)) has a mutual internal hatred in (a,b) subnode.*
Note that picture (6) shows only five links, which is less than the number of links on any other pictures in Figure 1. On the hand, picture (6) still captures all information presented in other pictures. Often generalization means loosing some lower-level information. We made the generalization *without missing any lower level information*.

Now we can show the use of Φ*-clustering in distributed data mining* (DDM). Every predicate P(x,y) or even individual statement "Jon likes Mary", Like(Jon,Mary) can be stored separately (in different tables, databases and sites). The fact that data are stored at different locations does not change the method. For instance, let right and left halves of the table 2 are stored in different locations. We can test Φ-equivalence partially in each of these locations and then combine results. If Φ-equivalence of a and b is refuted in one part then there is no need to test it in another part. If a and b appear to be Φ-equivalent in both parts then a and b are Φ-equivalent. This is true for any decomposition of the table 2 for three, four or more parts. This freedom of decomposition may not be the case for other data mining methods. Other methods may require specific decomposition methods. Several decomposition methods have been developed in DDM (e.g., BODHI and JAM systems [7]). These methods also can be applied for Φ-equivalence.

There is also a benefit to use Φ*-clustering for text mining tasks* outlined in [1]. Let us illustrate it by contrasting typical distributed data mining [7] and new text mining tasks [1]. In typical distributed data mining one may want to learn dependency between hepatitis-C (stored in a large DB A) and weather in the US (stored in a large DB B in another location) [7]. In distributed text mining tasks outlined in [1] one may want uncover money laundering scheme or suspicious terrorist activities. Relevant information could be deliberately spread over dozens, hundreds or thousands of "haystacks". There are two large well-organized DBs in the first case, and hundreds of unstructured sources in the second case. Bringing all of these sources to a single location does not help much – information is not structured. Φ-clustering does not requires a rigid structuring of the sources in a single flat file with multi-dimensional records. Note that this is a typical requirement for many data mining methods. An individual 1 Mb text source can contain only a single relevant statement "John hates Brian". Φ-clustering will work with this text without need to bring it to a single location. With proliferation of semantic web it will be more and more common that the statement will be tagged:

<name> John </name> <relation> hates </relation><name>Brian</name>.

In this case converting to the table similar to table 2 is not necessary.

In general the problem of combining data into a single flat file from tables in relational databases is well known in both conventional and distributes data mining. For instance, it is noticed in [7, p.19] that 100 Mb DB results in 2.5 Gb flat file for a fraud detection mining. What is one of the major sources of the problem? The traditional answer is -- data mining methods require a flat file to run. Current works in DDM (BODHI and JAM systems [7]) focus on removing this requirement by redesigning and hybridizing conventional methods to work with a set of *fragments of a flat file*. A recent workshop [10] on multi-relational data mining (MRDM) addresses this issue too. It considers MRDM as the multi-disciplinary field dealing with knowledge discovery from relational databases consisting of multiple tables. "The field aims at integrating results from existing fields such as ILP, KDD, ML and Relational Databases, as well as at producing new techniques" [10].

5. Conclusion

This paper offered a new hierarchical relational clustering method (Φ-method) applicable to discovering hidden patterns in distributed heterogeneous textual databases and unstructured data. The method can process incomplete relations. That is relations between some objects can be unknown. The method does not require converting relations to attributes of individual entities as required by many other methods. Such conversion often is accompanied by explosion of the size of data to be processed. Φ-method generalizes without missing lower level information, clusters produced have a clear meaning and natural graphical visualization as it is shown in Figure 1.

References

[1]   T. Senator, EELD Program, http://www.darpa.mil/ito/research/eeld/EELD_BAA.ppt, 2001

[2]   Workshop on Knowledge Discovery, Data Mining, and Machine Learning (KDD-ML), 1998 http://www.darpa.mil/ito/research/eeld/KDD-ML_Report.doc

[3]   B. Kovalerchuk, Vityaev, E. Data Mining in Finance: Advances in Relational and Hybrid Methods, Kluwer Acad. Publ., Boston, 2000.

[4]   B. Kovalerchuk, Classification structures invariant to renaming of objects, *Computational Systems*, **55** (1973) 90-97, Institute of Mathematics, Novosibirsk (In Russian).

[5]   A.I. Malcev, Algebraic Systems, Springer, 1973.

[6]   J. Czajsner , Equivalence relations determining useful properties, *Studia Logika*, XXIV (1969)

[7]   Advances in Distributed and Parallel Knowledge Discovery, Eds. H. Kargupta, P. Chan, MIT, 2000.

[8]   S. Dzeroski, Inductive Logic Programming and knowledge discovery and data mining, In: Advances in Knowledge Discovery in Databases. Eds. U. Fayad, G. Piatetsky-Shapiro, P.Smyth, R. Uthrusamy, AAAI/MIT press, 1996, pp. 117-152.

[9]   Relational data mining, S. Dzeroski, N. Lavrac, editors, Springer, Berlin, 2001

[10] Workshop Multi-relational Data Mining, MRDM 2001, September 6, 2001, Freiburg, Germany, http://mrdm.dantec.nl/

KES 2002
E. Damiani et al. (Eds.)
IOS Press, 2002

Content-Based Image Retrieval Based On Color Histogram And Discrete Cosine Transform

Golam SORWAR, Manzur MURSHED, and Laurence DOOLEY

Gippsland School of Computing and Info. Tech., Monash University, Churchill Vic 3842, Australia

E-mail: {*Golam.Sorwar,Manzur.Murshed,Laurence.Dooley*}*@infotech.monash.edu.au*

Abstract: In this paper we propose a novel image retrieval method by combining the existing two methods based on color histogram and Discrete Cosine Transform (DCT) coefficients respectively. Experimental results show that the retrieval performance of this combined method is better than the Histogram or the DCT method.

1. Introduction

The rapid development in multimedia and the Internet applications allows us to conveniently access large amount of image and video data in digital formats. However, in order to realize some of the benefits of the digital media, there is a need for developing "intelligent agents" to filter out relevant information from large amount of data, and to facilitate browsing and searching based on content information. But searching, finding desired information, is still a difficult and laborious task. The Traditional keyword annotation approach to accessing image or video information, however, requires manual effort which is time consuming, expensive, inefficient, inadequate, and subjective.

Currently, however, searching for multimedia content is not possible even though text-based search is available. Therefore, the new MPEG project, MPEG-7, is working towards providing solutions for identifying multimedia content [1]. Recently, along with that activity, a number of content-based approach emerged to retrieve image and video efficiently [2][3][4]. There methods have various application areas such as digital library, tourist information, investigation services, medical applications, verification of trade marks or logos, and so on [1][3].

Generally, low level visual features, e.g., shape, color, and texture, have been used as main content features for content-based image retrieval. The use of shape features has largely been limited to specialized retrieval systems, because of its use of domain specific database that have to be set up manually beforehand [5]. The existing approaches utilizing color features mostly use histograms [6][7][8].

In general, neighboring pixels within an image tend to be highly correlated. As such, it is desired to use an invertible transform to concentrate randomness into fewer, decorrelated parameters. The Discrete Cosine Transform (DCT) has been shown to be near optimal for a large class of images in energy concentration and decorrelating. It has been adopted the JPEG and MPEG coding standard [11] [12]. Such the DCT coefficients of an image tend themselves as a new feature, which have the ability to represent the regularity, complexity and some texture features of an image [9], it can be directly applied to image data in the compressed domain. This may be a way to solve the large storage space problem and the computational complexity of the existing methods.

Different visual feature has different advantages in different situations. Hence, if the features are combined appropriately, a more desirable result can be expected. So far color histogram and DCT have been used separately in different systems. In this paper, we propose a novel approach for content-based image retrieval by combining both color histograms and DCT coefficients to improve the performance, compare to considering each of these two features individually.

The remainder of this paper is organized as follows. In Section 2, a general content-based image retrieval system is introduced. Section 3 and 4 discuss color and texture features respectively. Section 5 shows some experiment results.

2. General Image Retrieval System

The general image retrieval system is shown in Fig. 1. The system consists of three main modules: the input module, the query module, and the retrieval module.

In the input module, the feature vector is extracted from each input image. It is then stored in an image database with its input image. On the other hand, when a query image enters the query module, the feature vector of the query image is extracted. In the retrieval module, the extracted feature vector of the query image is compared with the feature vectors of storage images in the image database. As a result of the query, similar images are retrieved according to their similarity with the query image. Finally one or more target images will be obtained from the retrieved images.

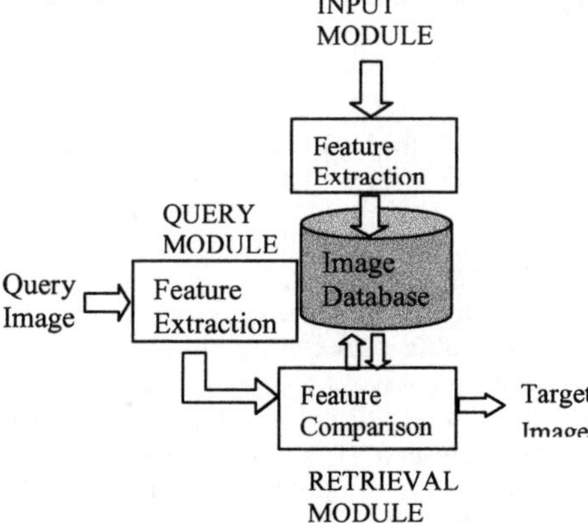

Fig. 1: Block diagram of general image retrieval system.

3. Color Feature Representation

The color feature, in the form of color histograms, is one of the most widely used visual features in the image and video retrieval. This is very effective and widely used method for global image/frame characteristics. Statistically, it denotes the joint probability of the intensities of the three-color channels (R, G, B).

We consider the uniform quantizer for quantizing the RGB color space into N number of bins. N=64 is used in our experiments. Each bin contains the color of similar values and is represented by its centroid $c_i = (r_i, g_i, b_i)$, $i = 0, 1, \ldots, N-1$.

For each pixel in the image, its color is classified into one of the N bins and the number of colors in each bin is counted and normalized in the form of a N-dimensional color feature vector, $P = [P_0, P_1, \ldots, P_{N-1}]^T$ where $\sum_i P_i = 1$.

4.　Texture Feature Extraction

Most existing approaches to texture feature extraction use statistical methods. For the analysis of a texture image, it requires large storage space and a lot of computational time to calculate the matrix of features. For solving the problems, some researchers proposed to use DCT [9] for texture representation.

When a query image is provided, its RGB version is converted into a gray level version for DCT transform. A DCT based transformation is then used for considering spatial localization. Each image is divided into $N{\times}N$ sized subblocks.

The two dimensional DCT can be written in terms of pixel values $f(i, j)$ for i, j=0, 1,..., N-1 and the frequency-domain transform coefficients $F(u, v)$:

$$F(u,v) = \frac{1}{\sqrt{2N}} C(u)C(v) \sum_{i=0}^{N-1}\sum_{j=0}^{N-1} f(i,j) \times \cos\left[\frac{(2i+1)u\pi}{2N}\right] \cdot \cos\left[\frac{(2j+1)v\pi}{2N}\right] \tag{1}$$

for u, v =0, 1, ..., N-1, where

$$c(x) = \begin{cases} \dfrac{1}{\sqrt{2}} & \text{for x=0} \\ \\ 1 & \text{otherwise.} \end{cases}$$

The inverse DCT transform is given by

$$f(i,j) = \sum_{u=0}^{N-1}\sum_{v=0}^{nN-1} c(u)c(v)F(u,v) \times \cos\left[\frac{(2i+1)u\pi}{2N}\right] \cdot \cos\left[\frac{(2j+1)v\pi}{2N}\right] \tag{2}$$

for i, j =0, 1, ..., N-1.

Each sub block contains one DC coefficient and a number of AC coefficients. As the DC coefficient of each sub block represents the average energy of block of an image, we only consider the DC coefficients for reducing the computational cost.

5.　Experimental Results

We consider a database of 1500 images of various different classes: mountains with sky, flowers, fruits, sports and flags etc. These are 256-level color images of size 120×160 pixels each. The subblock size is 8×8 pixels. For the similarity calculation, we use the following Euclidean distance function

$$d(Q,D) = \sqrt{\sum_{i=1}^{n}(D_i - Q_i)^2} \tag{3}$$

where $d(Q,D)$ is the distance between query image Q and the database image D and n is the number of color bins for histogram and number of DCT coefficients for each image.

Let d_{max} = max($d(Q, D)$), then the similarity value between two images for all images in database is given by

$$Similarity(Q,D) = \left(1 - \frac{d(Q,D)}{d_{max}}\right) \times 100 \tag{4}$$

To evaluate the retrieval efficiency of the proposed integrated methods, we use the objective measures, *Recall* and *Precision* [10] as shown in equation (5). Recall is the relevant retrieval rate from all the relevant items in the image database and Precision represents the correct retrieval rate.

$$\text{Recall} = \frac{R_r}{T}, \quad \text{Precision} = \frac{R_r}{T_r} \qquad\qquad (5)$$

where R_r is the number of relevant retrieved items, T is the total number of relevant items in the image database, and T_r is the number of all retrieved items. The retrieval performances for Histogram, DCT, and the Combined methods, in terms of Recall and Precision are shown in Fig. 2.

Form the above Fig. 2, we can make a comparison for the performance of the different methods. According to the graph, it is very clear that only for 10% recall value the precision values for all methods are equal but for other recall values, the precision value for the Combined method is higher than the Histogram or the DCT method. It can also be seen that there is quite fluctuations between the performance of the Histogram and DCT methods. Some cases the DCT method is better and some cases the Histogram method is better; but the performance of the Combined method is always higher than any individual one.

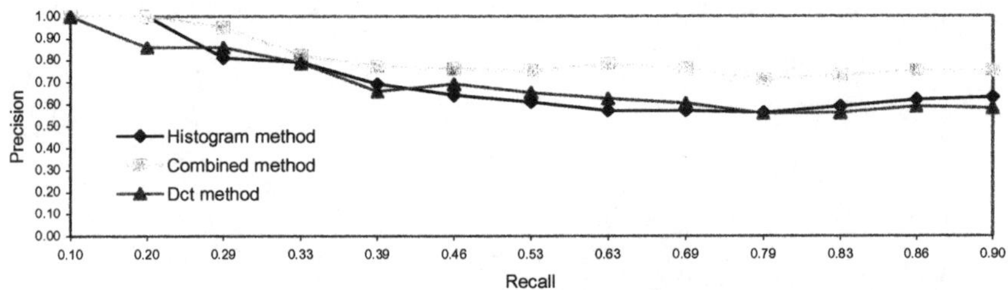

Fig. 2: Recall-Precision curves averaged for 10 set of queries, for Color histogram, DCT coefficient, and Combined method.

Form the above Fig. 2, we can make a comparison for the performance of the different methods. According to the graph, it is very clear that only for 10% recall value the precision values for all methods are equal but for other recall values, the precision value for the Combined method is higher than the Histogram or the DCT method. It can also be seen that there is quite fluctuations between the performance of the Histogram and DCT methods. Some cases the DCT method is better and some cases the Histogram method is better; but the performance of the Combined method is always higher than any individual one.

A sample query result using these methods is shown in Fig. 3., which shows that the Combined method retrieves more similar images to the query image, compared to the Histogram method or the DCT method.

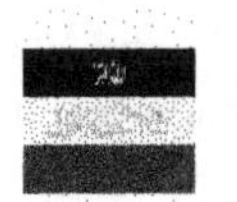

Fig. 3(a): The query image

Fig. 3(b): Combined method result.

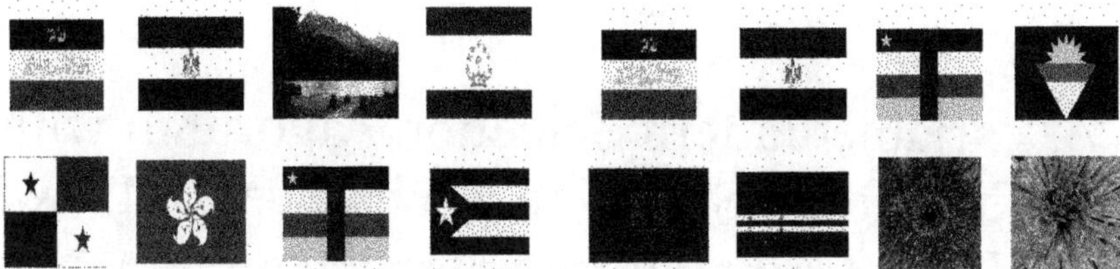

Fig. 3(c): Histogram based method result.   Fig. 3(d): DCT based method result.

6. Conclusions

In this paper we have proposed a novel image retrieval method by combining the existing two methods based on color histogram and DCT coefficients respectively. Experimental results have shown that the retrieval performance of this combined method is better than the Histogram or the DCT method. Testing of this method with huge database is our on going work. Combining other features and considering the non-linear adaptation using neural network can improve the performance of this method.

Referrences

[1] MPEG Video Group, "MPEG-7: Overview(v1.0)," ISO/IEC GTC1/SC29/WG11 N3158, Maui, Hawaii, December 1999.

[2] Flickner M.,. Sawhney H.S., Ashley J., Huang Q., Dom B., Gorkani M., Hafner J., Lee D., Petkovic D., Steele D. and Yanker P., "Query by Image and Video Content: The QBIC System," IEEE Computer 28(9): pp. 23-32, 1995.

[3] Smith J.R. and Chang S.-F., "VisualSEEk: a Fully Automated Content-Based Image Query System," Proceedings, ACM Multimedia '96 Conference, Boston, MA, 1996.

[4] Smith J.R. and Chang S.-F., "Visually searching the web for content," IEEE Multimedia Magazine 4(3), pp. 12-20, 1997.

[5] Bimboo A.D. and Pala P., "Image Indexing using Shape-based Visual Features," Proceedings of ICPR' 96 IEEE, Vol. 3, pp. 251-355,1996.

[6] Jain A.K., Vailaya A. and Wei X., "Query by Video clip," Multimedia Systems 7 (5), pp. 369-384, 1999.

[7] Swain M. and Ballard D., "Color Indexing," international Journal of Computer Vision 7(1), pp.11-32,1991.

[8] Ma W.Y., Deng Y. and. Manjunath B.S, "Tools for texture/color based search of images," SIPE Int. Conf., Human Vision and Electronic Imaging, vol. 3016, pp.496-507, 1997.

[9] Lee S.-M., Bae H._J., and Jung S.-H., "Efficient Content-Based Image Retrieval Methods Using Color and Texture," ETRI Journal, Vol. 20, pp. 272-283, 1998.

[10] Salton G. and Buckley C., "Improving Retrieval Performance by Relevance feedback," Research Report f Cornell University, pp. 1-24, 1998.

[11] Borko F., "Video and Image processing in multimedia Systems," Kluwer Academic publishers, pp 225-249, 1995.

[12] Wallace G.K, "Overview of the JPEG still Image Compression standard," SPIE, Vol. 1244, pp. 220-233, 1990.

KES 2002
E. Damiani et al. (Eds.)
IOS Press, 2002

A Variable Pattern Selection Algorithm with Improved Pattern Selection Technique for Low Bit-Rate Video-Coding Focusing on Moving Objects

Manoranjan Paul, Manzur Murshed, and Laurence Dooley

Gippsland School of Computing and Info. Tech., Monash University, Churchill Vic 3842, Australia
E-mail: {Manoranjan.Paul,Manzur.Murshed,Laurence.Dooley}@infotech.monash.edu.au

Abstract. Research into pattern representation of moving regions in blocked-based motion estimation and compensation in video sequences has focused mainly upon using a fixed number of regular shaped patterns. Recently we presented the *Variable Pattern Selection* (VPS) algorithm, which selects a preset number of best-matched patterns from a pattern codebook of regular shaped patterns. The pattern elimination technique in the selection process of the VPS algorithm however, was to be computationally expensive, especially when the preset number is low. In this paper, the concept is extended to develop the *Extended Variable Pattern Selection* (EVPS) algorithm where the pattern elimination technique is replaced with a fast solution. The complexity analysis confirms that this algorithm can be as much as 8.5 times faster than the VPS algorithm. In order to take advantage of this computational speed-up in eliminating patterns, the pattern codebook size of the EVPS algorithm has also been increased to 32.

Keywords—*Motion estimation and compensation, low bit-rate video coding, moving region detection, pattern matching.*

1. Introduction

Researchers into very low bit-rate digital video coding are often faced with the daunting challenge of meeting two diametrically conflicting requirements—reducing the transmission bit-rate while concomitantly retaining image quality. Computational complexity and system bandwidth limitations of a communication media mean these two factors tend to be inversely proportional. H.26X [6][7][8], MPEG-X [3][4][5] are some of the well-known contemporary standards for video compression. H.26X for example, are widely used in video-telephony and video-conferencing applications, where very low bit-rates to accommodate public switched telephone networks (PSTN) is required

Many of the video coding standards tends to employ block-based techniques because of their implementation simplicity and also because they generally provide good results when the bandwidth requirement is relaxed e.g., in MPEG-1/2. This is however, not the case with low bit-rate block-based video coding such as in H.263. The shape of a moving object is generally arbitrary and may not necessarily be aligned with the hypothetical grid structure created by the fixed-sized, non-overlapping rectangular blocks, termed *macroblock* (MB) in the coding standards. The typical size of a MB being 16×16 pixels, which leads to a large number of blocks, some of which will contain only static background, some will have moving objects and some a combination of the two.

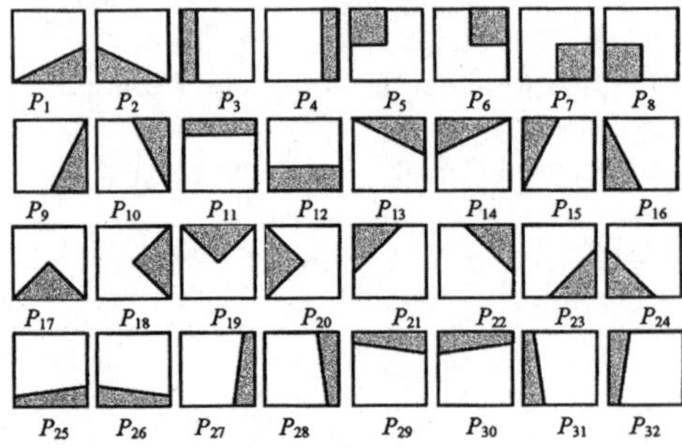

Figure 1: 32 regular shaped 64-pixel patterns, defined in 16×16 blocks, where the shaded region represents 1's and the white region represents 0's.

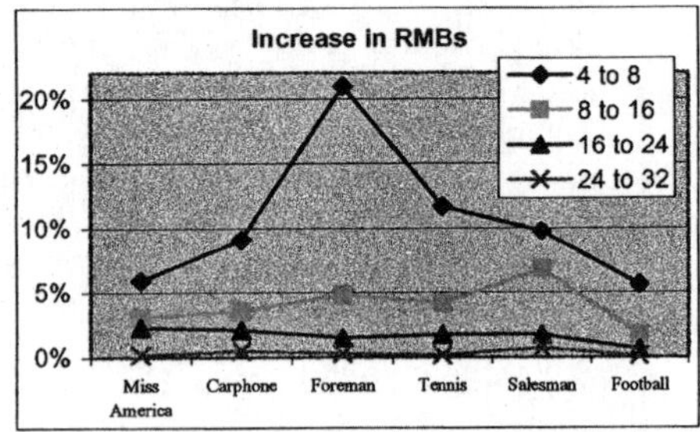

Figure 2: Percentage increase in RMBs when the number of patterns is increased from 4 to 8, 8 to 16, 16 to 24 and 24 to 32.

In [12], *macroblocks* were classified according to the following three mutually exclusive classes:

i) **Static *MB* (SMB)**—Blocks that contain little or no motion;

ii) **Active *MB* (AMB)**—Blocks that contain moving object(s) with little static background;

iii) **Active-*Region MB* (RMB)**—Blocks that contain both static background and some part(s) of moving object(s).

By treating AMB and RMB alike, as is done in H.263/H.263+, leads to coding inefficiencies [1]. In order to improve this efficiency, block size may be reduced only to add additional information to be transmitted due to the increase in the number of blocks [11].

Both [1] and [12] successfully addressed the above issue by segmenting each RMB into two regions using a fixed number of predefined regular patterns with 128-pixels and 64-pixel respectively. Once the segmentation process was complete, motion estimation/compensation was then only performed on moving regions. Each SMB was skipped for transmission (since they did not contain motion and could be copied from the reference frame) and each AMB was treated exactly as defined in H.263 standard, using motion estimation and compensation techniques. The non-coding of SMB patterns and limiting the number of RMBs to a prescribed set of patterns lead to an improved coding efficiency.

Using eight instead of four patterns improved the *peak signal to noise ratio* (PSNR) by up to 0.6 dB [12]. It also meant that better classification of the RMB blocks was achieved, thus contributing to a higher compression ratio, even after compensating for the larger size codebook required for Pattern coding. In [10] we extended this concept by proposing a new *Variable Pattern Selection* (VPS) algorithm, using 24 regular shaped patterns. The VPS algorithm selects a preset number of the best-matched patterns, from a 24-pattern codebook, by eliminating the least frequent pattern per iteration. This strategy is computationally expensive as $24-\lambda$ iterations are required to select λ best-matched patterns. In this paper, an *Extended VPS* (EVPS) algorithm is proposed with improved pattern elimination strategy. In the EVPS algorithm, the best-match pattern selection process is still based upon the matching frequency of each pattern. However, unlike the VPS algorithm, the EVPS algorithm allows to eliminate more than one pattern per iteration under the assumption that λ best-matched patterns lie in the first ρ $(\geq \lambda)$ most frequent patterns such that their cumulative frequency is greater than or equal to 75%. Empirically it has been observed that only $\lceil 16/\lambda \rceil + 1$ iterations are required to select λ best-matched patterns, where $\lambda \in \{4, 8, 16, 24\}$. This improved elimination strategy translates into lowered computational time. It has been proved in Section 2.2 that the EVPS algorithm can be as much as 8.5 times faster than the VPS algorithm..

The pattern frequency values are used to code pattern identifier numbers using a variable length coding technique e.g., Huffman or arithmetic coding. This leads to an improvement in coding efficiency. Similar to the VPS algorithm, the EVPS algorithm naturally exhibits both better compression and PSNR performance compared with using a fixed patterns as presented in [12].

The rest of the paper is organized as follows. In Section 1, low bit-rate video-coding algorithms focusing on moving region using fixed patterns is explained. The EVPS algorithm that uses λ best-matched patterns from a pattern codebook of 32 patterns is developed in Section 2. Section 3 concludes the paper.

1.   Low Bit-Rate Video Coding Using Fixed Patterns

1.1. Moving Region Detection

The basis of this technique is to let the first eight patterns P_1–P_8 in Figure 1 approximate the moving region. Let $C_k(x, y)$ and $R_k(x, y)$, $0 \leq x, y \leq 15$, denote the kth block of the current and the reference frames respectively. The moving region $M_k(x, y)$ in the kth block of the current frame is obtained as follows:

$$M_k(x, y) = T(|\, C_k(x, y) \bullet B - R_k(x, y) \bullet B\,|) \quad (1)$$

where B, a square pattern of size 3×3, is the structuring element of morphological closing operations [2][9], $|v|$ returns the absolute value of v, $T(v)$ returns 1 if $v > 2$ or 0 otherwise, and $0 \leq x, y \leq 15$.

Each block is then classified into SMB, AMB, and RMS according to the following rules. For the k-th block, if $M_k(x, y)$ has less than eight 1's then the block is classified as an SMB; else the block is divided into four sub-blocks and if none of these sub-blocks contain all 0's, the block is classified as AMB. Otherwise, the block is considered as a *candidate* RMB. Each of these candidate RMBs is then matched against all eight prescribed patterns and the best-match pattern is obtained by minimizing the following expression: - $D_{k,n} = \dfrac{1}{256} \sum\limits_{x=0}^{15} \sum\limits_{y=0}^{15} |\, M_k(x, y) - P_n(x, y)\,| \quad (2)$

where $1 \leq n \leq 32$. A candidate RMB is classified as an RMB if $\min(D_{k,n}) < 0.25$; otherwise it is an AMB.

1.2. Motion Estimation and Compensation

Since both each SMB and the static regions of RMBs are considered as having no motion, they can be skipped from coding and transmission as they can be obtained from the reference frame. For each AMB, as well as the moving region of each RMB, motion vector and residual errors are calculated using conventional block-based methods, with the obvious difference in having the shape of the blocks for the moving regions of RMBs as that of the best-match pattern, rather than being square.

1.3. Encoding and Decoding

Having dealt with the coding/decoding of SMB and AMB, the issue of how to process the RMB blocks is now discussed. A motion vector is calculated from only the 64 moving pixels of the best-match pattern. To avoid multiple 8×8 blocks of DCT calculations for only 64 residual error values per RMB, these 64 values are *rearranged* into an 8×8 block. Similarly, an *inverse rearrangement* is performed decoding.

2. Low Bit-Rate Video Coding Using Variable Patterns

As stated in the introduction, the motivation of using more than eight patterns originates from the observation in [12] that using eight patterns instead of four patterns not only improves PSNR, but also classifies more blocks as RMBs, which contributes towards higher coding compression even after compensating for the larger codebook size (one extra bit per RMB) requirement. However, Figure 2 clearly shows for a selection of popular video sequences, the increase in RMB becomes insignificant once the number of patterns is greater than 24.

Another interesting observation is made in Figure 3, which shows that the maximal-frequency pattern sequence is not fixed for all types of video data. For example, the most frequent eight patterns for "Miss America" video sequence is 5, 6, 7, 8, 4, 3, 16, and 1 (in order); whereas the same for "Foreman" video sequence is 6, 1, 14, 8, 7, 12, 5, and 11 (in order). It confirms the general judgment, that the first eight patterns in Figure 1, which were used in [12], do not constitute the optimal set.

2.1. The EVPS Algorithm

The *Extended Variable Pattern Selection* (EVPS) algorithm attempts to select an optimal set of λ best-matched patterns from a 32-pattern codebook presented in Figure 1, where $\lambda \in \{4, 8, 16, 24\}$. However, selecting such an optimal set is not straightforward, as for example, in finding the optimal set of eight patterns, it is not sufficient to simply select the patterns with the highest frequencies as given in Figure 3. All the RMBs that were initially matched against a pattern, outside the selected patterns, should be considered as candidate RMBs to be matched against the selected patterns. Some of these candidate RMBs may not be classified as RMBs and the frequency of the patterns may also be changed. In some cases, this change may lead to a different ordering in the optimal pattern set.

The VPS algorithm [10] selects the optimal set by eliminating the least frequent pattern per iteration. The EVPS algorithm improves this computationally expensive elimination process by allowing more than one pattern to be removed per iteration. This elimination strategy is based on the empirical observation that λ best-matched patterns lie in the first ρ ($\geq \lambda$) most frequent patterns such that their cumulative frequency is greater than or equal to 75%. The full algorithm is formalized in Figure 5. It is interesting to note that as more than one candidate patterns are eliminated in the EVPS algorithm, there always exists a possibility (even if the possibility is negligible) that the λ best-matched patterns selected by the EVPS algorithm will not be identical to the selection made by the VPS algorithm.

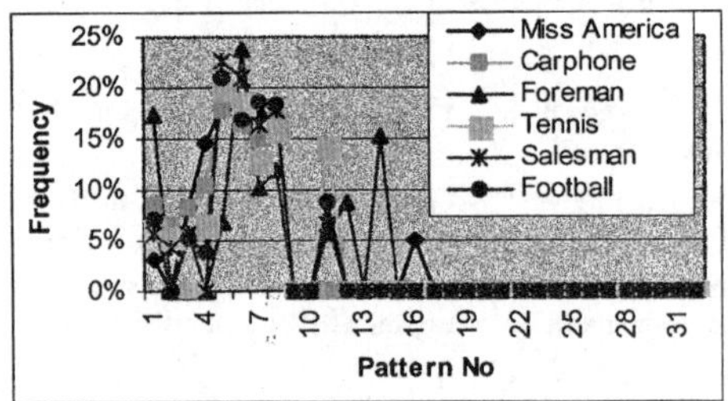
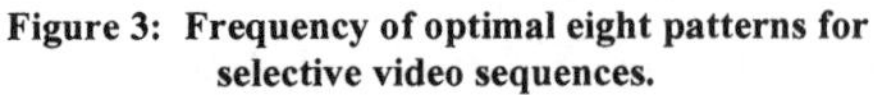

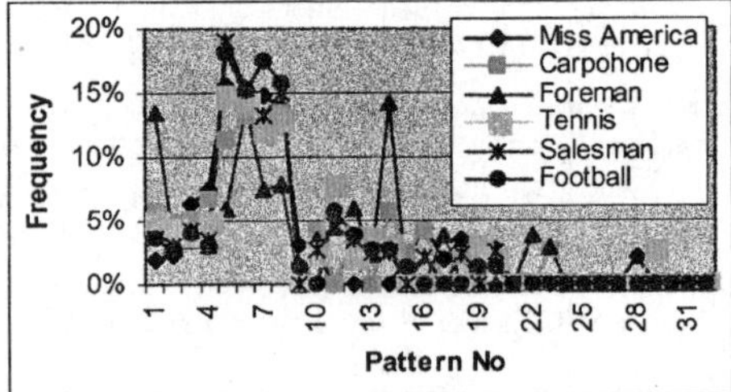

Figure 3: Frequency of optimal eight patterns for selective video sequences.

Figure 4: Frequency of optimal sixteen patterns for selective video sequences.

<table>
<tr><td>

Algorithm EVPS(λ)

Parameter: λ = Size of the optimal pattern set.

Return: Ω = The optimal pattern set.

Step 1: $\Omega = \{P_1, P_2, ..., P_{32}\}$;

Step 2: Calculate the frequency of each pattern in Ω;

Step 3: if $|\Omega| = \lambda$ then exit;

Step 4: Sort Ω to $(P_{i_1}, P_{i_2}, ..., P_{i_{|\Omega|}})$ such that frequency $\mathrm{Freq}(P_{i_j}) \geq \mathrm{Freq}(P_{i_{j+1}})$, for all $j < |\Omega|$;

Step 5: Find the minimum value of $\rho \geq \lambda$ such that
$$\sum_{j=1}^{\rho} \mathrm{Freq}(P_{i_j}) \geq 0.75 ;$$

Step 6: $\Omega = \Omega - \{\forall j > \rho : P_{i_j}\}$;

Step 7: Go to Step 2;

</td><td>

Algorithm VPS(λ)

Parameter: λ = Size of the optimal pattern set.

Return: Ω = The optimal pattern set.

Step 1: $\Omega = \{P_1, P_2, ..., P_{24}\}$;

Step 2: Calculate the frequency of each pattern in Ω;

Step 3: if $|\Omega| = \lambda$ then exit;

Step 4: Find the pattern $P_i \in \Omega$ such that its frequency is the minimum;

Step 5: $\Omega = \Omega - \{P_i\}$;

Step 6: Go to Step 2;

</td></tr>
</table>

Figure 5: The EVPS algorithm.   **Figure 6: The VPS algorithm.**

The approach of selecting suitable patterns based on their relative frequencies has an additional benefit in coding. The frequency information can now be used to code pattern identifier numbers using a variable length coding, such as Huffman or arithmetic coding. Experimental result shows that the optimal eight (sixteen) patterns for six standard test video sequences can be Huffman coded using on average only 2.69 (3.56) bits, instead of using fixed 3 (4) bits.

2.2. Comparative Time Complexity Analysis of the EVPS algorithm

For the VPS algorithm is described in Figure 6. Although the VPS and EVPS algorithms use pattern codebooks of different sizes, for comparison, we assume they both use the same sized codebook of α patterns. Now, the VPS algorithm requires $\alpha - \lambda$ iterations to select λ best-matched patterns, whereas it has been empirically observed on a large number of standard and non-standard test video sequences that the EVPS algorithm requires only $\lceil \alpha/2\lambda \rceil + 1$ iterations to select λ best-matched patterns, where $\lambda \in \{4, 8, 16, 24\}$ and $\alpha = 32$.

Consider that both the VPS and EVPS algorithms are applied on a video sequence with β RMBs in total. Let the pattern codebook at iteration i be denoted by Ω_i. Now consider iteration i. Step 2 of both the algorithms requires $\beta|\Omega_i|$ evaluations of $D_{k,n}$ values while keeping track of the minimum. Therefore, this step can be done in $O(\beta|\Omega_i|)$ time. Step 4 of the VPS algorithm, needs to track down the pattern with minimum frequency from the current pattern set Ω_i and this can be done in $O(|\Omega_i|)$ time. However, Step 4 of the EVPS algorithm requires sorting than can be done in $O(|\Omega_i|\log|\Omega_i|)$ time. Step 5 of the EVPS algorithm requires scanning the sorted pattern set, which can be done in $O(|\Omega_i|)$ time.

Considering the $\alpha - \lambda$ iterations, the VPS algorithm can be completed in the following time:

$$\sum_{i=1}^{\alpha-\lambda+1} O(\beta|\Omega_i|) + \sum_{i=1}^{\alpha-\lambda} O(|\Omega_i|) = O\left(\frac{\beta(\alpha+\lambda)(\alpha-\lambda+1)}{2} \right) \quad \text{where } \beta >> \alpha .$$

Now, consider that there will be at most $\psi = \lceil \alpha/2\lambda \rceil + 1$ iterations by the EVPS algorithm. Obviously, $|\Omega_1| = \alpha$ and $|\Omega_\psi| = \lambda$. Although it has been empirically observed that the value of $|\Omega_i|$ decreases logarithmically, for the sake of simplicity, the decrease is assumed to be linear. Let $|\Omega_i|$ be $\lambda + (\psi - i)\dfrac{\alpha - \lambda}{\psi - 1}$, for all $i = 1, 2, ..., \psi$. Without loosing any generality, this assumption will overestimate the complexity of the EVPS algorithm. Using such high estimate in comparing the EVPS algorithm against the VPS algorithm also guarantees that the strength of the EVPS algorithm is never unfairly overestimated.

Considering the ψ iterations, the EVPS algorithm can then be completed in the following time:

$$\sum_{i=1}^{\psi} O(\beta|\Omega_i|) + \sum_{i=1}^{\psi-1} O(|\Omega_i|\log|\Omega_i|) = O\left(\frac{\psi\beta(\alpha+\lambda)}{2}\right) \qquad \text{where } \beta >> \alpha.$$

It can therefore, be concluded that the EVPS algorithm is at least $\dfrac{\alpha-\lambda+1}{\psi} = \dfrac{\alpha-\lambda+1}{\lceil\alpha/2\lambda\rceil+1}$ times faster than the VPS algorithm. For $\alpha = 32$ and $\lambda = 4, 8, 16$, and 24, the EVPS algorithm is respectively at least 5.8, 8.33, 8.5, and 4.5 times faster than the VPS algorithm.

3. Conclusions

Pattern representation of moving regions in blocked-based video motion estimation and compensation has received considerable attention recently. In general, the pattern representation problem is equivalent to selecting λ best-matched patterns from a pattern codebook of α patterns. [1] used $\alpha = \lambda = 4$ and [12] used $\alpha = \lambda = 8$, i.e., both of these two works used only a fixed number of regular shaped patterns for all video sequences. In [10] we introduced the *Variable Pattern Selection* (VPS) algorithm with $\alpha = 24$ and $\lambda = 16$. The pattern elimination technique in the selection process of the VPS algorithm, however, has been found to be computationally expensive, especially when $\lambda << \alpha$. In this paper, the *Extended Variable Pattern Selection* (EVPS) algorithm has been presented with fast pattern elimination technique where $\alpha = 32$ and $\lambda \in \{4, 8, 16, 24\}$. The complexity analysis has revealed that the EVPS algorithm is at least $\dfrac{\alpha-\lambda+1}{\lceil\alpha/2\lambda\rceil+1}$ times faster than the VPS algorithm. For $\alpha = 32$, the peak performance of the EVPS algorithm is achieved at $\lambda = 16$ when it is at least 8.5 times faster than the VPS algorithm while coding a number of standard test video sequences with better compression ratio as well as *peak signal to noise ratio*.

References

[1]   Fukuhara, T., K. Asai, and T. Murakami, "Very low bit-rate video coding with block partitioning and adaptive selection of two time-differential frame memories," *IEEE Trans. Circuits Syst. Video Technol.*, 7, 212–220, 1997.

[2]   Gonzalez, R.C. and R. E. Woods, *Digital Image Processing*, Addition-Wesley, 1992.

[3]   ISO/IEC 11172, International Standard, 1992.

[4]   ISO/IEC 13818, MPEG-2 International Standard, Video Recommendation ITU-T H.262, 1995.

[5]   ISO/IEC N4030, MPEG-4 International Standard, 2001.

[6]   ITU Telecom. Standardization Sector of ITU, "Video codec for audio-visual services at $p\times$ 64kbits/s," ITU-T Recommendation H.261,1993.

[7]   ITU Telecom. Standardization Sector of ITU, "Video coding for low bitrate communication," Draft ITU-T Recommendation H.263, 1996.

[8]   ITU Telecom. Standardization Sector of ITU, "Video coding for low bitrate communication," Draft ITU-T Recommendation H.263, Version 2, 1998.

[9]   Maragos, P., "Tutorial on advances in morphological image processing and analysis," *Opt. Eng.*, 26(7), 623–632, 1987.

[10]   Paul, M., M. Murshed, and L. Dooley, "A Low Bit-Rate Video-Coding Algorithm Based Upon Variable Pattern Selection," To appear in the Proc. of Int. Conf. on Signal Processing, Beijing, China, August 26–30, 2002.

[11]   Shi, Y.Q. and H. Sun, *Image and Video Compression for Multimedia Engineering Fundamentals, Algorithms, and Standards*, CRC Press, 1999.

[12]   Wong, K.-W., K.-M. Lam, and W.-C. Siu, "An Efficient Low Bit-Rate Video-Coding Algorithm Focusing on Moving regions," *IEEE trans. circuits and systems for video technology*, 11(10), 1128–1134, 2001.

KES 2002
E. Damiani et al. (Eds.)
IOS Press, 2002

1565

Content Based Image Retrieval Using Discrete Wavelet Transform

Xianyu Hong*, S. Belkasim*, and O. Basir**
** Department of Computer Science, Georgia State University, Atlanta, GA., USA, 30303*
Email: sbelkasim@cs.gsu.edu
***Department of Systems Design Engineering, University of Waterloo, Waterloo, Ont., Canada,*
N2L 3G1

Abstract. Image retrieval plays an important role in broad spectrum of applications. Content-based retrieval (CBR) is one of the popular choices in many biomedical and industrial applications. Discrete image transforms have been widely studied and suggested for many image retrieval applications. The Discrete Wavelet Transform (DWT) is one of the most popular transforms recently applied to many image processing applications. The Daubechies wavelet can be used to form the basis for extracting features for retrieving images based on the description of a particular object within the scene. This wavelet is widely used for image compression. In this paper we highlight the common features between compression and retrieval. Several examples are used to test the DWT retrieval system. A comparison between DWT and Discrete Cosine Transform (DCT) is also made. The retrieval system using DWT requires preprocessing and normalization of images, which might slow down the retrieval process. The accuracy of the retrieval using DWT has been significantly improved by incorporating efficient K-Neighbor Nearest Distance (KNND) measure in our system.

1. Introduction

The need to manipulate large amount of multimedia data such as images, video and audio have risen recently due to the high demand to retransmit and reuse this kind of data over several computer networks. As a result, it is more economical to retrieve parts or segments of the contents of such data from a multimedia database. Unfortunately, conventional relational data models based on text and numerical value are not suitable for such requirements[1-4]. Query by example (QBE)[5] on the other hand is intuitive to user, and can represent subtle differences such as shapes, spatial relations, colors, and textures.

In this paper, QBE approach is used to retrieve the most similar image or images in a database for a given unknown image example. The core idea behind the approach is to build the original database of large number of sample images, which will form the base of the database library. The main task of the retrieval system is to compare the unknown example and match it to one or more library samples to find out the most similar one. However, in real world application, especially in real-time application, it is very important to find out the matched image quickly and correctly. Aiming at such goal, we use Discrete Wavelet Transform (DWT) and K Nearest Neighborhood Distance (KNND) to achieve less computation time and high successful retrieval rate. Discrete Wavelet Transform (DWT) has been used in several image retrieval applications[6-7]. Gabor in 1945[8] introduced the basic idea of wavelet theory. A considerable effort has been made to further improve the DWT on several scientific fields. The greatest contribution in DWT development[9,10] has come from signal and image processing.

In this paper, we use Daubechies-4 (DB-4) DWT in image retrieval and compression. DB-4 DWT enables us to decompose an image in both spatial and temporal domains. By DB-4 decomposition, one can compress an image, which in turn results in significant reduction of the feature space of the image. As a result, both the speed of database storing, and image retrieval processes can be greatly improved. The choice of decomposition level is very critical to achieve successful retrieval with relatively less computation time. The K-nearest neighbor distance (KNND) matching procedure is used to improve the retrieval strategy.

The main implementation procedure is shown in section 4 with results, discussions and conclusions in section 5.

2. Discrete Wavelet Transform (DWT)

The fundamental idea behind wavelet is to analyze according to scale and time. It is well known that Fourier Transform can transform a signal from spatial domain to a frequency domain. One big disadvantage of Fourier Transform is that one can only represent a signal with frequency resolution without any time resolution or spatial information.

In wavelet analysis, temporal analysis is performed with a contracted, high-frequency basis function, and frequency analysis is performed with a dilated, low-frequency basis function.

Haar wavelet (daubechies-1) is one of Daubechies wavelets members that is very popular because of its simple interpolation schema. Haar wavelet uses two types of filters. One is a low-pass filter and the other is a high-pass filter. The output of the low-pass filter is obtained by averaging the input, while the output of high-pass filter is obtained from the difference of the input. One can easily conclude that the low-pass filter contains more information than high-pass filter because most of the signal energy is concentrated in low-pass filter. Daubechies-4 wavelet, on the other hand, splits the input signal by using four kinds of filters (LL, LH, HL, HH) with most of the energy concentrating in LL sub band (L stands for low, while H stands for high). It has slightly computational overhead and is more complex than Haar wavelet, but it is capable of including more details than Haar wavelet algorithm.

The Daubechies-4 transform has a wavelet function and a scaling function. The scaling function is given as following.

$$a[i] = h_0 s[2i] + h_1 s[2i+1] + h_2 s[2i+2] + h_3 s[2i+2] \tag{1}$$

where s is discrete signal and four scaling function coefficients are defined as:

$$h_0 = \frac{1+\sqrt{3}}{4}, \ h_1 = \frac{3+\sqrt{3}}{4}, h_2 = \frac{3-\sqrt{3}}{4}, h_3 = \frac{1-\sqrt{3}}{4} \tag{2}$$

In each step of the DB-4 transform, the scaling function is applied to the input signal. If the signal has N values, the scaling function will generate N/2 smoothed values stored in the lower half of the N element input vector.

The wavelet function is given as following.

$$c[i] = g_0 s[2i] + g_1 s[2i+1] + g_2 s[2i+2] + g_3 s[2i+2] \tag{3}$$

where $g_0 = h_3$, $g_1 = -h_2$, $g_2 = h_1$, $g_3 = -h_0$

In each step of the DB-4 transform, the wavelet function is applied to the input signal. If the signal has N values, the wavelet function will generate N/2 difference stored in the upper half of the N element input vector.

When applying DB-4 wavelet to a 2-D image, one can obtain 4 sub bands with most of energy concentrated in LL sub band, which plays a very important role in image compression and retrieval.

3. Reducing Image Feature Space Using Compression

Image compression is closely related to image retrieval. Generally speaking, compression can reduce the feature space of images significantly, and hence increase retrieval speed. However, in compression, one has to reconstruct or decompress the image from the compressed information. Image retrieval, on the other hand, there is no need to decompress the image. Nevertheless, understanding image compression is very critical to image retrieval. The core ideas behind image compression process, is the removal of redundant data the efficient coding of such data.

A typical two-dimensional DWT transform output is illustrated in Fig.1 and Fig.2. One can observe that most of the energy concentrates in LL sub band at each decomposition level.

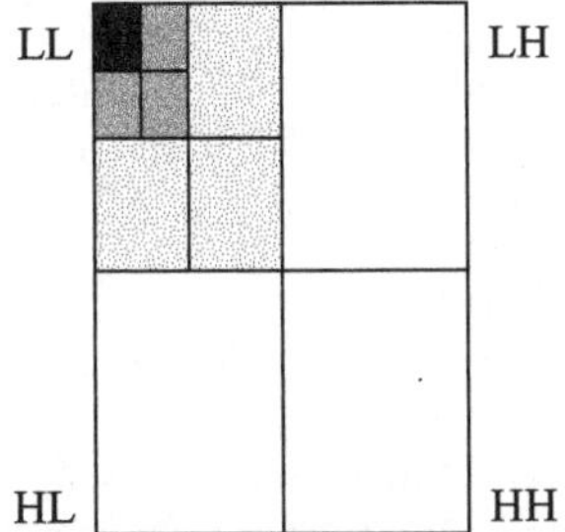

Fig. 1 DWT Components

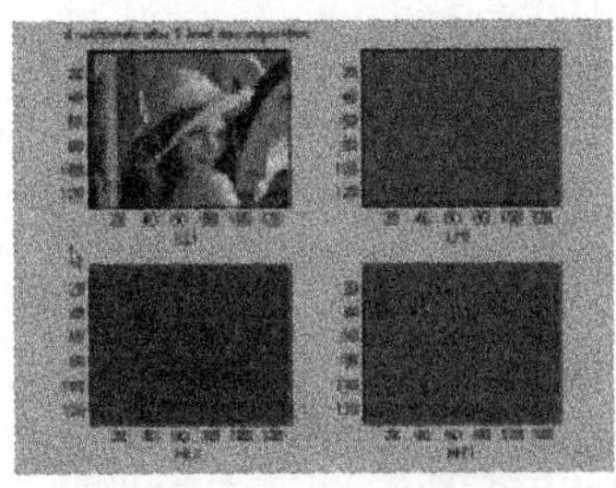
Fig.2 Four sub bands after 2nd level

4. Image Retrieval Using Query-By-Example (QBE)

The core idea behind content-based image retrieval is to compare feature vectors between an unknown image and those in training samples database.

There are two important requirements for image retrieval. One is to store training samples and build the database in an efficient manner. The other is to perform comparison in the shortest possible time. DWT is very suitable for such purposes since it can decompose an image into several parts with most of the energy concentrated in a single sub-band. Comparison between an unknown image and training samples can be performed only on the sub-band that has most of the energy. Therefore, DWT can achieve high retrieval ratio in higher speed.

A database consisting of energy concentration components of several hundreds of sample images is needed to build our database. DB-4 discrete wavelet transform is applied to the training samples to reduce the effective feature space, and keep only the most important components in the database. Once the database is established, a reference or an unknown image can be compared to the database, and the closest match can be retrieved from the database.

The classifier then can search the database and map the unknown sample to its nearest or k-nearest neighbors. The unknown image is in fact the query image (query-by-example).

In the nearest neighbor search problem, the feature vectors of samples that form the database are preprocessed into a multidimensional binary search tree (*k-d* tree) [11, 12, 13, 14], to generate an efficient hierarchical data structure. To simplify the search problem, the Euclidean distance is used in our experimental results. A *k-d* tree is built by recursively bisecting the database using single coordinate position cuts. For a given coordinate, the database is cut at the median of the distribution generated by the projection onto that coordinate. A *2-d* tree decomposition is illustrated in Fig.3, and the corresponding *2-d* tree is shown in Fig.4.

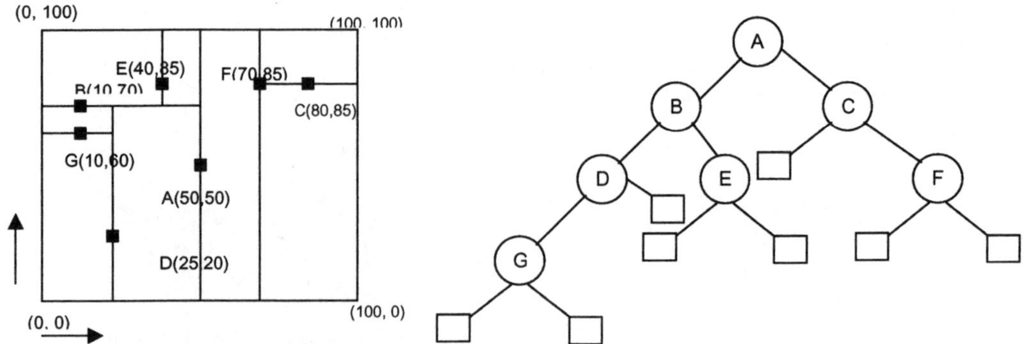

Fig.3 A 2-d tree decomposition   Fig.4 A 2-d tree

The *k-d* tree data structure reduced the complexity of the search from n to $\log(n)$. K-Nearest Neighbor Distance (KNND) method is used in this paper to improve correct retrieval ratio. KNND is based on computing the K-Nearest Neighbor Distance, which is directly related to the minimum Mean Square Error (MSE). In other words, the smaller the MSE, the closer the feature vectors of the two images are. The NND approach for retrieval-by-example, starts with a search for the nearest neighbor of the example image in the *k-d* tree of database samples. Once a matched image is found, the corresponding semantic content is extracted and assigned to the reference image. Such approach is appropriate for a set of sample images with small error bound. The minimum MSE does guarantee the matching of the most similar pair of images. The KNND is used to increase the probability of successful retrieval without increasing the running time dramatically. The KNND is a statistical majority algorithm. For example, by finding the closest 10 samples instead of just one, one can determine which set of training sample is the closest match set. KNND can be implemented by extending the range of nearest neighbor search over the k-d tree.

5. Results and conclusions

To successfully implement and test the proposed retrieval system, we need a relatively large database as well as small feature space.

Using DB-4 wavelet transform, one can decompose an image into 4 sub bands (LL, LH, HL, HH) at each step. Most of the energy is concentrated in the LL band. The HH band has the least energy of the four sub bands. From Fig.1 and Fig.2, one can observe that the LL band is very similar to the original image. However, the other sub bands also have correlation with the original image. This shows that the energy distribution is not random.

If we decompose the LL band further, the energy distribution begins to randomize. The reason is that the decomposition window becomes so small, to contain any significant resolution of the image. A sample of the energy concentration ratio is shown in Fig.5.

One can easily determine that the less the level of decomposition, the smaller the energy concentration. However, there is a tradeoff. Less level of decomposition means relatively larger amount of coefficients needed for computation during retrieval phase. Therefore, one has to compromise somewhere in between. In this paper, since all the images are normalized to 256 x 256.We have also noticed that decomposing an image to the 4th level can achieve both high successful retrieval ratio and smaller running time

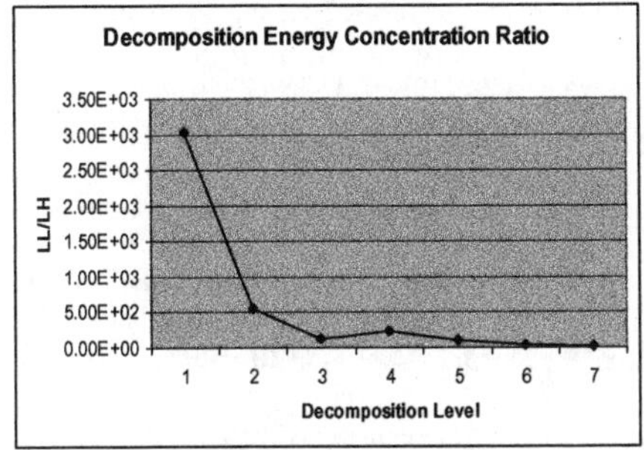

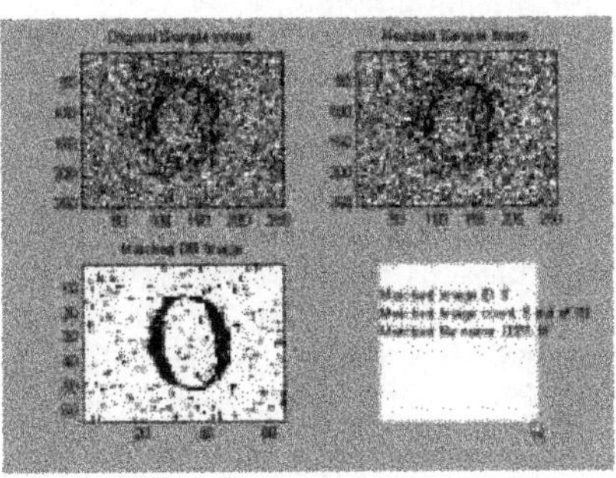

Fig. 5 Energy Concentration Ratio   Fig.6 Numeral 0.

The database of the training samples consists of 1,000 binary images with different sizes ranging from 32x32 to 256x256 pixels. Ten reference images of handwritten characters with different sizes are used as query examples. Since both training samples and reference images have different image sizes, and comparison need to be performed on the same basis, image normalization should be implemented. In this paper, all the information of both training samples and reference images are stored in normalized size of 256 x 256. The DB-4 wavelet decomposition level is 4. An attempt was made to decompose the images to 7th decomposition level. Unfortunately, the successful retrieval ratio turns out to be poor although the computational time is relatively less.

The reference or example images were denoised before being compared to the ones in the database to reduce interference due to noise. The k parameter in the KNND has been set to 10. The successful retrieval ratio is 0.8 or 8 out of the 10 unknown images have been correctly retrieved. A sample of the retrieval results is shown in Fig.6.

Based on our results, we can conclude that, DWT is very effective in reducing the feature space by decomposing an image to an appropriate level. *k-d* tree data structure and the corresponding nearest neighbor search algorithm can achieve less retrieval time compared to naive search. The KNND statistical method has significantly increased the correct retrieval rate.

References

[1]   P.J. Smith, S.J. Shute, and D. Galdes, *"In search of Knowledge-Based Search Tactics"*, 12th Int'l Conf. Research and Development in Information Retrieval, pp. 3-10, 1989

[2]   G.P. Zarri, *"Conceptual Representation for Knowledge Bases and Intelligent Information Retrieval Systems"*, 11th Int'l Symp. Visual Languages, pp. 314-321, Sept. 1997

[3]   M. Atkinson, F. Bancilhon, D. DeWitt, K. Dittrich, D. Maier, and S. Zdonik, *"The Object-Oriented Database System Manifesto"*, First Int'l Conf. Deductive and Object-Oriented Databases, pp. 40-57, 1989

[4]   A. Klinger and A. Pizano, *"Visual Interaction"*, Int'l Conf. Multimedia Information Systems. Pp. 109-120, Elesvier, 1989

[5]   M.M. Zloof, *"QBE/OBE: A language for Office and Business Automation"*, Computer, vol. 14, no. 5 pp. 13-22, 1981

[6] Guojun Lu, Atul Sajjanhar, "Region-based shape representation and similarity measure suitable for content-based image retrieval", Multimedia Systems 7: 165–174 , Multimedia Systems, Springer-Verlag 1999

[7] O. Sukmarg and K.R. Rao, "Fast object detection and segmentation in MPEG compressed domain," TENCON 2000, Kuala Lumpur, Malaysia, Sept. 2000

[8]   Gabor, D. *"Theory of Communication"*, J. Inst. of Electr. Engr., vol 93, pp. 429-457, 1946

[9]   Riol, O., and Vetterli, M., *"Wavelets and Signal Processing"*, IEEE Signal Processing Magazine, vol. 8, No.4, pp. 14-38, Oct. 1991

[10] M.K.Mandal, S.Panchanathan and T.Aboulnsar, Illuminaiton ,"Invariant image indexing using moments and wavelets" , Journal of Electric Imaging , April 1998

[11] J.H. Friedman, F. Baskett, and L.J. Shustek, *"An Algorithm for Finding Nearest Neighbors"*, IEEE Transactions on Computers, Oct. 1975

[12]   J.H. Friedman, J.L. Bentley, and R.A. Finkel, *"An Algorithm for Finding Best Matches in Logarithmic Expected Time"*, ACM Transactions on Mathematical Software, vol. 3, Sept. 1977

[13]   J.L. Bentley and J.H. Friedman, *"Data Structures for Range Searching"*, Computing Surveys, Dec. 1979

[14]   J.L. Bentley, *"Multidimensional Divide-and-Conquer"*, Communications of the ACM, vol. 23, Apr. 1980

Causal Possibility Model Structures

Lawrence J. Mazlack
Computer Science
University of Cincinnati
Cincinnati, Ohio 45221-0030
mazlack@uc.edu

Abstract: Causality occupies a position of centrality in human reasoning. It plays an essential role in commonsense human decision-making. Determining causality has been a tantalizing goal throughout human history. Proper sacrifices to the gods were thought to bring rewards; failure to make the proper observations to led to disaster. Today, data mining holds the promise of extracting unsuspected information from very large databases. The most common methods build association rules. In many ways, the interest in association rules is that they offer the promise (or illusion) of causal, or at least, predictive relationships. However, association rules only calculate a joint occurrence frequency; they do not express a causal relationship. If causal relationships could be discovered, it would be very useful. This paper explores the possible representation of causality drawn from large data sets.

1. Introduction

Causality occupies a position of centrality in commonsense human reasoning. It plays an essential role in human decision-making by providing a basis for choosing an action that may lead to a desired result. Considerable effort has been spent examining causation. Philosophers, mathematicians, computer scientists, cognitive psychologists, psychologists, and others have explored questions of causation beginning at least three thousand years ago.

Determining causality entails recognizing causal relationships; e.g., *A* causes *B* to happen. Whether this can be done at all has been a speculation for thousands of years. At the same time, in our daily lives, we rely on the common sense observation that causality exists. The computational concern is how to recognize and represent common sense causal relationships.

Data mining is an advanced tool for the management of large masses of data. Data mining is a *secondary* analysis. Data mining is a secondary analysis because the data were not collected to answer questions now posed. Secondary analysis of already collected data eliminates the possibility of experimentally varying the data to search for causal relationships.

There are several different data mining products. The most common are *conditional rules* or *association rules*. Of these, the most common are association rules; for example:

<table>
<tr><td>

• *Conditional rule:*

   IF *Age* < 20

     THEN *Income* < *$10,000*

      with *{belief = 0.8}*

</td><td>

• *Association rule:*

   Customers who buy beer and sausage also tend to buy mustard

     with *{confidence = 0.8}*

      {support = 0.15}

</td></tr>
</table>

At first glance, decision structures of this kind seem to imply a causal or cause-effect relationship. For example: *A customer's purchase of both sausage and beer <u>causes</u> the customer to also buy mustard*. In point of fact, when first developed, association rules do not describe causality. Also, the strength of dependency may be very different from a respective association value All that can be said with certainty is that associations describe the strength of co-occurrences. Sometimes, the relationship might be causal; for example, if someone eats salty peanuts and then drinks beer, there is probably a causal relationship. On the other hand, if a dog barks in Kabul and a cat meows in Beijing, there is probably not a causal relationship.

Causal relationships exist in the common sense world. If someone fails to stop at a red light and there is an automobile accident, it can be said that the failure to stop was the accident's cause. Another way to think of causal relationships is counterfactually. For example, if a driver dies in an accident, it might be said that had the accident *not* occurred; they would still be alive.

A common approach to recognizing causal relationships is by manipulating variables through experimentation. While there are known techniques to discover causal relationships among controlled[1] data. How to accomplish causal discovery in purely observational[2] data is not solved. (Observational data is the most likely to be available for data mining analysis.)

Algorithms for discovery in observational data often use correlation and probability independence to find possible causal relationships. If two variables are statistically independent, it can be asserted that they are not causally related. The reverse is not necessarily true.

Real world events are often affected by a large number of potential factors. For example, with plant growth, many factors such as air temperature, chemicals in the soil, types of creatures present, etc., can all affect plant growth. What is unknown is what causal factors will be present in the data; and, how many of the underlying causal relationships can be discouvered among pure observational data. It is possible that some causal factors will not be present in the data.

A largely unexplored aspect of mined association rules is how to determine when one event causes another. Given that A and B are variables and there appears to be a statistical covariability between A and B. Is this covariability a causal relation? More generally, when is a relation a causal relation? Differentiation between covariabilty and causality is difficult.

Some problems with discouvering causality include:

- Defining adequately a causal relation,
- Representing possible causal relations,
- Computing causal strengths,
- Missing attributes that have a causal effect,
- Distinguishing between association values and causal values,
- Infering causes and effects from the representation.

2. Causality

Centuries ago, in their quest to unravel the future, mystics aspired to decipher the cries of birds, the patterns of the stars and the garbled utterances of oracles. Kings and generals would offer precious rewards for the information soothsayers furnished. Today, though predictive methods are different from those of the ancient world, the knowledge that dependency recognition attempts to provide are highly valued. From weather reports to stock market prediction, and from medical prognoses to social forecasting, superior insights about the shape of things to come are prized commodities. [Halpern, 2001]

There are also recently discovered limitations. Godel's theorem, Heisenberg's uncertainly principle, chaos theory and other mathematical insights have shown that knowledge of the world might never be complete because we, as observers, are integral parts of what we observe.

Democritus, the Greek philosopher, once said: "Everything existing in the universe is the fruit of chance and necessity." This seems self evident. Both randomness and causation are in the world. Democritus gave the poppy as an example. Whether the poppy seed lands on fertile soil or on a barren rock is chance. If it takes root, however, it will grow into a poppy, not a geranium or a Siberian Husky. [Lederman, 1993]

2.1 Nature Of Causal Relationships

Some questions about causal relationships that would be desirable to answer are:

- To what degree does a cause b? Is the value for b sensitive to a small change in the value for a?
- Does the relationship always hold in time and in every situation? If it does not hold, can the particular situation when it does hold be discouvered?

$$a \xrightarrow{S_{a,b}} b \qquad b \xrightarrow{S_{b,a}} a$$

Figure 1. Mutual dependency notation.

- Is it possible that there might be mutual dependencies; i.e., $a \to b$ as well as $b \to a$? Is it also possible that they do so with different strengths? They can be described as shown in *Figure 1*. where $S_{i,j}$ represents the strength of the causal relationship from i to j . For example, a could be *short men* and b could be *tall women*. If $S_{a,b}$ meant the romantic attraction that was caused in *short men* by the sight of *tall women*, it might be that $S_{a,b} > S_{b,a}$

It would seem that if there are causal relationships in market basket data, there would often be imbalanced dependencies. For example, if a customer first buys strawberries, there may be a reasonably good chance that she will then buy whipped cream. Conversely, if she first buys whipped cream, the subsequent purchase of strawberries may be less likely.

It is potentially interesting to discouver the absence of a causal relationship. For example, discouvering the lack of a causal relationship in drug treatment's of disease. If some potential cause can be eliminated, then attention can become more focused on other potentials.

---

[1] *Controlled data* means that the actions producing particular values can be reproduced. Typically, this is experimental data.

[2] *Observational data* indicates that the data has been produced and no repeated actions can be taken reproduce further data.

2.2 Types Of Causality

There are at least three ways that things may be said to be related:

•Coincidental: Two things happen to describe the same object and have no determinative relationship between them.

•Functional: There is a generative relationship.

•Casual: One thing causes another thing to happen. There are at least three types of causality:

  •*Chaining:* In this case, there is a temporal chain of events, $A_1, A_2, ..., A_n$, which terminates on A_n. To what degree, if any, does A_i (i=1,.... n-1) cause A_n? A special case of this is a backup mechanism or a *preempted alternative*. Suppose there is a chain of casual dependence, A_1 causing A_2; suppose that if A_1 does not occur, A_2 still occurs, now caused by the alternative cause B_1 (which only occurs if A_1 does not).

  •*Conjunction (Confluence):* In this case, there is a confluence of events, $A_1, ..., A_n$, and a resultant event, B. To what degree, if any, did or does A_i cause B? A special case of this is *redundant causation*. Say that either A_1 or A_2 can cause B; and, both A_1 and A_2 occur simultaneously. What can be said to have caused B?

  •*Network*

Recognizing and defining causality is difficult. The following show some of the difficulties:

•*Example 1: Simultaneous Plant Death:* My rose bushes and my neighbor's rose bushes both die. Did the death of one cause the other to die? (Probably not, although the deaths are associated.)

•*Example 2: Drought:* There has been a drought. My roses and my neighbor's rose bushes both die. Did the drought cause both rose bushes to die? (Most likely.)

•*Example 3: Traffic:* My friend calls me up on the telephone and asks me to drive over and visit her. While driving over, I ignore a stop sign and drive through an intersection. I am hit by another driver. I die. Who caused my death? -- Me? -- The other driver? -- My friend? -- The traffic engineer who designed the intersection? -- Fate?

•*Example 4: Umbrellas:* A store owner doubles her advertising for umbrellas. Her sales increase by 20% What caused the increase? -- Advertising? -- Weather? -- Fashion? -- Chance?

•*Example 5: Poison:* (Chance increase without causation) Fred and Ted both want Jack dead. Fred poisons Jack's soup; and, Ted poisons his coffee; and, each act increases Jack's chance of dying. Jack eats the soup but (feeling rather unwell) leaves the coffee, and dies later. Ted's act raised the chance of Jack's death but was not a cause of it.

2.3 Classical Dependence

Statistical dependence is interesting in this context because it is often confused with causality. Such reasoning is not correct. Two events E_1, E_2 may be statistical dependent because both have a common cause E_0. But this does not mean that E_1 is the cause of E_2.

For example, lack of rain (E_0) may cause my rose bush to die (E_1) as well as that of my neighbor (E_2). This does not mean that the dying of my rose has caused the dying of my neighbor's rose, or conversely. However, the two events E_1, E_2 are statistically dependent.

The general definition of statistical dependence is:

Let A, B be two random variables that can take on values in the domains $\{a_1, a_2, ..., a_i\}$ and $\{b_1, b_2, ..., b_j\}$ respectively. Then a_i is said to be statistically independent of B iff

  prob $(a_i | b_j)$ = prob(a_i) for all b_j and for all a_i.

The formula

  prob$(a_i | b_j)$ = prob(a_i) prob(b_j)

describes the joint probability of a_i AND b_j when A and B are independent random variables. Then follows the law of compound probabilities

  prob(a_i, b_j) = prob$(a_i) \cdot$ prob$(b_j | a_i)$

In the absence of causality, this is a symmetric measure. Namely,

  prob(a_i, b_j) = prob(b_j, a_i)

A causality relationship between two events E_1 and E_2 will always give rise to a certain degree of statistical dependence between them. The converse is not true. A statistical dependence between two events may; but need not, indicate a causality relationship between them. There is a positive correlation if

  prob(a_i, b_j) > prob(a_i) prob(b_j)

However, all this tells us that it is an interesting relationship. It does not tell us if there is a causal relationship.

2.4 Probabilistic Causation

Probabilistic Causation designates a group of philosophical theories that aim to characterize the relationship between cause and effect using the tools of probability theory. A primary motivation is the desire for a theory of causation that does not presuppose physical determinism.

The success of quantum mechanics, and to a lesser extent, other theories employing probability, has brought some to question determinism. Many philosophers were interested in developing a theory of causation that does not presuppose determinism.

One notable feature has been a commitment to indeterminism, or rather, a commitment to the view that an adequate analysis of causation must apply equally to deterministic and indeterministic worlds. Mellor [1995] argues that indeterministic causation is consistent with the connotations of causation. Hausman [1998], on the other hand, defends the view that in indeterministic settings there is, strictly speaking, no indeterministic causation, but rather deterministic causation of probabilities.

Following Suppes [1970] and Lewis [1986], the approach has been to replace the thought that causes are sufficient for, or determine, their effects with the thought that a cause need only raise the probability of its effect. This shift of attention raises the issue of what kind of analysis of probability, if any, is up to the job of underpinning indeterministic causation.

3. Representation

Representation is an open and important question. The representation used constrains the methods that can be used. Several representations have been proposed. The representations allow for the characterization and inference of causal relationships.

3.1 Digraphs

Some authors have suggested that *sometimes* it is possible to recognize causal relations through the use of directed graphs (digraphs) [Pearl , 1991] [Spirtes ,1993]. In a digraph, the vertices correspond to the variables and each directed edge corresponds to a causal influence. An example is shown in *Figure 2*. Pearl (2000) and Sprites (2000) use a form of digraphs called DAGs for representing causal relationships.

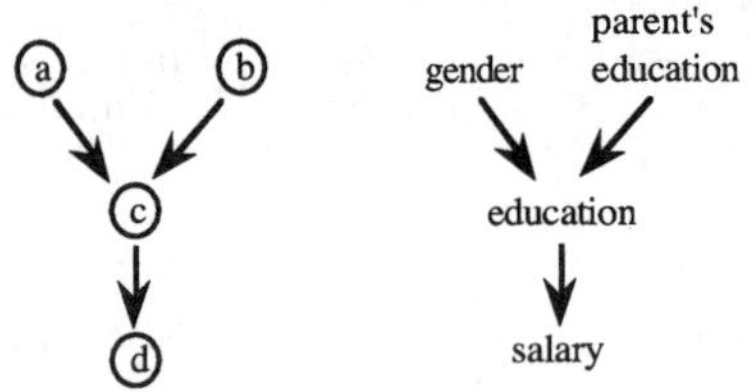

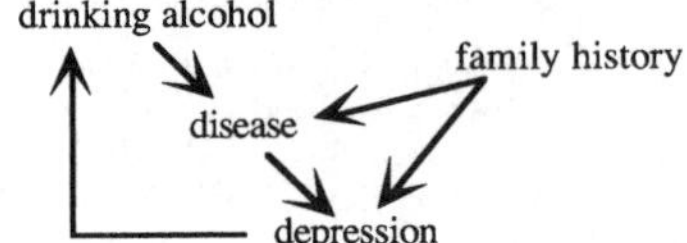

Figure 3. Example of cyclic causal relationships

Figure 2 (a) An example digraph (DAG)
    (b) Example instantiating to (a).

Sometimes, cycles exist. For example, a person's family medical history influences both whether they are depressive and whether they will have some diseases. Drinking alcohol combined with the genetic predisposition to certain diseases influences whether the person has a particular disease, which then influences depression, which in turn may influence the person's drinking habits. *Figure 3* shows the cyclic digraph for this example.

Quantitatively describing the relationships between the nodes can be more complex. One possible general model is an extension of what can be found in Rehder [2002]; shown in Figure 4.

Developing directed acyclic graphs from data is computationally expensive. The amount of work increases geometrically with the number of attributes. For constraint based methods, the reader is directed to Pearl (2000), Sprites (2000), Silverstein (1998), and Cooper (1997). For Bayesian discouvery, the reader is directed to Heckerman (1995) and Geiger (1995).

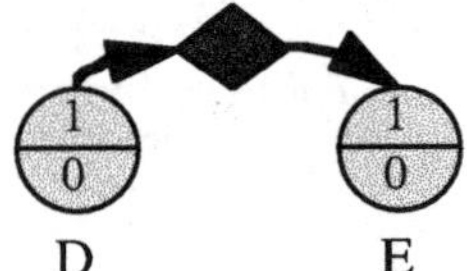

Figure 4. Quantitative relationship between digraph nodes; c = P(D), m = The probability that when D is present, the causal mechanism brings about E, m = The probability that some other (unspecified) causal mechanism brings about E.

3.2 Probability Trees

Probability trees can be used to show causal influences Shafer (1998). A tree is a digraph starting from one vertice, the root. In this case, the vertices represent situations. Each edge represents a particular variable with a corresponding probability.

Time ordering of the variables is represented via the levels in the tree. The higher a variable is in the tree, the earlier it is in time. This can become ambiguous for networked representations; i.e., when a node can have more than two parents and thus two competing paths (and their imbedded time sequences). By evaluating the expectation and probability changes among the situations (or nodes) in the tree, one can decide whether the two variables are causally related.

3.3 First Order Logic

Hobbs (2000) uses first-order logic to represent causal relationships. One difficulty with this approach is that the representation does not allow for any gray areas. For example, if an event occurred when the wind was blowing east, how could a wind blowing east-northeast be accounted for? The causality inferred may be incorrect due to the rigidity of the representation.

Nor can first order logic deal with dependencies that are only *sometimes* true. For example, *sometimes* when the wind blows *hard*, a tree falls. This kind of *sometimes* event description can possibly be statistically described. Alternatively, a qualitative fuzzy measure might be applied.

Another problem is recognizing differing effect strengths. For example, if some events in the causal complex are more strongly tied to the effect? Also, it is not clear how a relationship such as the following would be represented: *a* causes *b* some of the time; *b* causes *a* some of the time; other times there is no causal relationship.

4. Epilogue

Causality occupies a central position in human common sense reasoning. In particular, it plays an essential role in common sense human decision-making by providing a basis for choosing an action that is likely to lead to a desired result. Whether recognizing causality can be done at all has been a speculation for thousands of years. At the same time, in our daily lives, we make the common sense observation that causality exists. Carrying this common sense observation further, the concern is how to computationally recognize a causal relationship. The domain is large data sets.

Data mining holds the promise of extracting unsuspected information from very large databases. Methods have been developed to build association rules. Association rules indicate the strength of association of two or more data attributes. In many ways, the interest in association rules is that they offer the promise (or illusion) of causal, or at least, predictive relationships. However, association rules only calculate a joint occurrence frequency; not causality. There are several open questions across all types of association rules. A fundamental question is determining whether or not an association can lead to causality.

A particular interest is in determining when causality can be said to be stronger or weaker. Either in the case where the causal strength may be different in two independent relationships; or, where in the case where two items each have a causal relationship on the other.

Causality is a central concept in many branches of science and philosophy. In a way, the term "causality" is like "truth" -- a word with many meanings and facets. Some of the definitions are extremely precise. Some of them involve a style of reasoning best be supported by fuzzy logic.

A deep question is when anything can be said to cause anything else. And if it does, what is the nature of the causality? There is a strong motivation to attempt causality discovery in association rules. The research concern is how to best approach the recognition of causality or non-causality in association rules. Or, if there is to recognize causality as long as association rules are the result of secondary analysis?

References

Due to space limitations, the references are not shown. There is a version of this paper on the web that contains the references. It may be found at:
http://www.ececs.uc.edu/~mazlack/professional/IKOMAT.2002.pdf

Author Index

Albolino, S.	1434	Levine, J.	1534
Albus, J.S.	1424	Lo Re, G.	1476
Angiulli, G.	1466	Macej, P.	1544
Arif, I.	1512	Mach, M.	1544
Badalà, A.	1476	Mahjoub, M.A.	1444
Barbera, R.	1476	Martí, L.	1483
Barhoumi, W.	1529	Mazlack, L.J.	1571
Baroglio, C.	1471	Meisel, H.	1439
Barrile, V.	1466	Mesenzani, M.	1434
Basir, O.	1565	Messina, E.R.	1424
Belkasim, S.	1565	Murshed, M.	1555,1560
Ben Dayan Rubin, D.D.	1497	Nammuni, K.	1534
Berner, G.	1539	Navabzadeh Razavi, S.	1488
Bssiouny, S.A.	1523	Omri, M-N.	1444
Cerutti, S.	1497	Palmeri, A.	1476
Compatangelo, E.	1439	Pappalardo, G.S.	1476
Crowther, P.	1539	Paul, M.	1560
De Souza, F.J.	1507	Policriti, A.	1483
Dooley, L.	1555,1560	Pulvirenti, A.	1476
Dvořàk, J.	1449	Riggi, F.	1476
Ennaji, A.	1518	Schael, T.	1434
Evans, J.M.	1424	Schlenoff, C.I.	1424
Fathy, M.	1488	Šeda, M.	1449
Filis, I.V.	1454	Sefion, I.	1518
Gailhardou, M.	1518	Segal, C.	1461
García, L.	1483	Seta, K.	1429
Hong, X.	1565	Shawkat Ali, A.B.M.	1502
Hreno, J.	1544	Sideridis, A.B.	1454
Inbar, G.F.	1497	Sorwar, G.	1555
Ismail, M.A.	1523	The, H.L.	1512
Jermol, M.	1419	Urbančič, T.	1419
Kakusho, O.	1429	Velloso, M.L.	1507
Khotimah, S.N.	1512	Versaci, M.	1466
Kingston, J.	1534	Williams, R.	1539
Kovalerchuk, B.	1549	Yialouris, C.P.	1454
Lavrač, N.	1419	Youssef, S.M.	1523
Lazo, R.	1483	Zagrouba, E.	1529